P9-DMM-834

RIVER, COASTAL AND ESTUARINE MORPHODYNAMICS:
RCEM 2005

BALKEMA – Proceedings and Monographs
in Engineering, Water and Earth Sciences

PROCEEDINGS OF THE 4TH IAHR SYMPOSIUM ON RIVER, COASTAL AND ESTUARINE MORPHODYNAMICS, 4–7 OCTOBER 2005, URBANA, ILLINOIS, USA

River, Coastal and Estuarine Morphodynamics: RCEM 2005

Gary Parker
Department of Civil and Environmental Engineering and Department of Geology, University of Illinois at Urbana-Champaign, Illinois, USA

Marcelo H. García
Department of Civil and Environmental Engineering, University of Illinois at Urbana-Champaign, Illinois, USA

VOLUME 1

LONDON/LEIDEN/NEW YORK/PHILADELPHIA/SINGAPORE

Published by: Taylor & Francis/Balkema
P.O. Box 447, 2300 AK Leiden, The Netherlands
e-mail: Pub.NL@tandf.co.uk
www.balkema.nl, www.tandf.co.uk, www.crcpress.com

ISBN Set (Book + CD-ROM): 0 415 39270 5
ISBN Book: Vol. 1: 0 415 39375 2
Vol. 2: 0 415 39376 0
ISBN CD-ROM: 0 415 39377 9

Printed in Great-Britain

River, Coastal and Estuarine Morphodynamics: RCEM 2005 – Parker & García (eds)
© 2006 Taylor & Francis Group, London, ISBN 0 415 39270 5

Table of Contents

Deltas, estuaries, bays

Fluvial and coastal turbulent flow

Numerical modelling of river morphodynamics

Large-scale morphodynamics

River bends and meandering

River bedforms

Scour and bank erosion

Coastal and shelf bedforms

River, Coastal and Estuarine Morphodynamics: RCEM 2005 – Parker & García (eds)
© 2006 Taylor & Francis Group, London, ISBN 0 415 39270 5

Preface

Consider the following recipe. Add a) a fluid such as water and b) "sediment" such as the disaggregated crustal material of Earth to a "container", also made of sediment. Examples of "containers" include an estuary, a near-shore zone, a river or a continental shelf. Stir vigorously. Voilà; before long the interaction of sediment and fluid creates some beautiful morphologies. These morphologies may be rhythmic or quasi-periodic, such as sediment waves on the continental shelf, beach cusps, river dunes and river meandering. Or they may be expressions of a consistent large-scale shape of the container itself, such as the funnel shape of an estuary, the platform of a continental margin, the upward-concave profile of a river or the channel network of drainage basins. Morphodynamics consists of the study of how fluid-sediment interaction gives rise to such forms.

On January 14, 2005 a momentous event in the history of morphodynamics took place. On that day, the Huygens Probe of the Cassini Spacecraft, launched many years before by the European Space Agency, descended into the smog-shrouded atmosphere of Titan, the largest satellite of Saturn. Until the arrival of the Huygens Probe the surface morphology of Titan was unknown. Within a short amount of time the images transmitted by the Huygens Probe revealed a world that challenges the hypothesis that researchers truly understand the deeper physics underlying morphodynamics.

Figure 1 shows what appears to be a channelized drainage network on the surface of Titan. Figure 2 shows what appear to be clasts of ice that have been well-rounded by fluvial processes. Both images are from the European Space Agency, and were downloaded from http://www.esa.int/SPECIALS/Cassini-Huygens/SEM3SP5DIAE_0.html. The implication is that Titan has "rivers" created by the interaction of a fluid, i.e. liquid methane/ethane and a "sediment," i.e. disaggregated crustal ice. If the morphodynamics of water and sediment on Earth is truly known at a fundamental level, then the same laws should translate in a straightforward way into an understanding of the morphodynamics of liquid methane/ethane and granular ice on Titan. Can morphodynamicists rise to the challenge?

Here on Earth a quantitative knowledge of terrestrial morphologies is obtained by means of field and experimental research, the first of which is not yet possible on Titan. An explanation of the underlying physics is then obtained by means of theoretical and numerical research. The field of morphodynamics arises as a subset of a combination of these four research approaches. More specifically, morphodynamics focuses on the interaction of a sediment-laden fluid flow over an erodible bed and the erodible bed itself. The fluid is water in this volume, but it is air in the case of desert sand dunes, and liquid methane/ethane on Titan.

Quantitative morphodynamics naturally divides into three elements. The first element is the fluid mechanics of the sediment-laden flow over the erodible bed. Depending on the problem, this can be described at a range

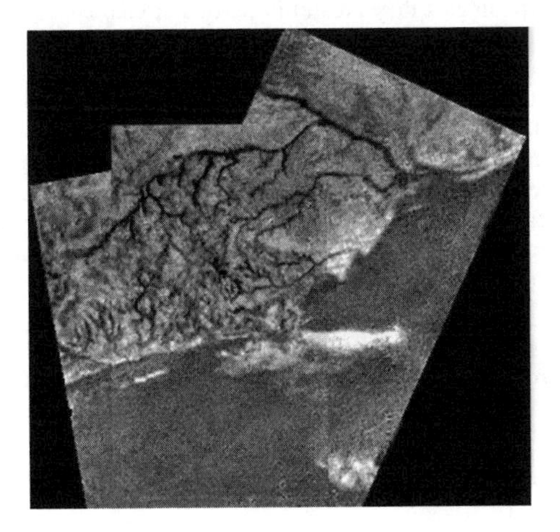

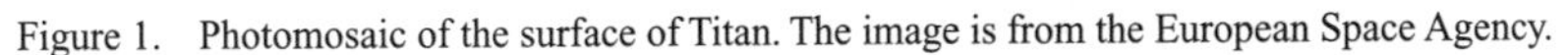

Figure 1. Photomosaic of the surface of Titan. The image is from the European Space Agency.

Figure 2. View of the surface of Titan from the Huygens Probe after landing. The image is from the Eurpoean Space Agency.

of complexities, starting from the simplest, i.e. steady uniform "normal" flow to the most complex, i.e. direct numerical simulations using the Navier-Stokes equations modified for the presence of a solid phase. The second element is the kinematics of the bed, which is described in terms of the many flavors of the Exner equation of sediment continuity, the flavor being chosen appropriately to the problem. The third element includes the various "constitutive" relations by which sediment entrainment from, transport over and deposition onto the bed is related to the fluid parameters that drive the processes. The range of beautiful morphologies that can be described in terms of these three elements is breathtaking. This range is nicely illustrated in this volume.

The present series of symposiums began with the 1st Symposium on River, Coastal and Estuarine Morphodynamics, held in Genoa, Italy in 1999. This first symposium embodied the vision of Giovanni Seminara of the Department of Environmental Engineering, University of Genova. The 2nd Symposium was held in Obihiro, Japan in 2001 under the direction of Syunsuke Ikeda and Yasuyuki Shimizu. The 3rd Symposium was organized by Agustin Sánchez-Arcilla and Allen Bateman in Barcelona, Spain in 2003.

In this 4th Symposium the scope has been broadened to include two new areas: submarine morphodynamics associated with turbidity currents (including deep-sea sediment waves, submarine canyons and meandering on submarine fans) and large-scale morphodynamics (including long profiles of rivers and continental margin morphodynamics). The analogy between turbidity currents (and the morphologies they create) and sediment-laden river flows (and the morphologies they create) in particular challenges our fundamental understanding of morphodynamics. The challenge is identical to that offered by Titan; the basic physics is common to both cases, but the different settings cause analogous morphologies to be expressed in different ways. Fortunately, the relatively recent advent of acoustic technologies for imaging the seafloor and its subsurface has made the submarine world accessible to the researcher. Some of the differences between terrestrial and submarine morphology are subtle (such as channel meandering, which shows similar planforms in both cases) and some are dramatic (such as natural levees, which are much higher in the submarine case).

The editors hope that the reader will find this volume to contain fascinating new developments in the field of morphodynamics. And one day, perhaps at the 8th or 12th Symposium on River, Coastal and Estuarine Morphydynamics, it will be possible to feature research on the morphodynamics of Titan.

Gary Parker
Department of Civil and Environmental Engineering and Department of Geology
University of Illinois

Marcelo H. García
Department of Civil and Environmental Engineering
University of Illinois

River, Coastal and Estuarine Morphodynamics: RCEM 2005 – Parker & García (eds)
© 2006 Taylor & Francis Group, London, ISBN 0 415 39270 5

Acknowledgements

The symposium itself was sponsored by the following organizations: the International Association of Hydraulic Engineering and Research (IAHR), the Hokkaido River Disaster Prevention Research Center (HR-DPRC), the US Department of Navy Office of Naval Research (ONR), the US National Center for Earth-surface Dynamics (NCED) and the University of Illinois Department of Civil and Environmental Engineering. The last four organizations provided funds which helped defray the costs for both the conference and the production of this volume.

The following people at St. Anthony Falls Laboratory, University of Minnesota played an instrumental role in assembling the manuscripts for delivery to the publisher: Alessandro Cantelli, Phairot Chatanantavet, J. Wesley Lauer, Charles Nguyen, Paola Passalaqua, and Miguel Wong. The following individuals at the publishing company, Taylor and Francis/Balkema, are also due a hearty thanks for their patience and assistance: Leon Bijnsdorp, Richard Gundel and Maartje Kuipers.

Both the symposium and the volume represent part of the knowledge transfer activities of the National Center for Earth-surface Dynamics, a Science and Technology Center funded by the US National Science Foundation, and housed at St. Anthony Falls Laboratory, University of Minnesota.

Sediment transport processes

River, Coastal and Estuarine Morphodynamics: RCEM 2005 – Parker & García (eds)
© 2006 Taylor & Francis Group, London, ISBN 0 415 39270 5

Saltating or rolling stones?

Christophe Ancey
Ecole Polytechnique Fédérale de Lausanne, Ecublens, Lausanne, Switzerland

Tobias Böhm & Philippe Frey
Cemagref, Dom. Univ. Saint Martin d'Hères Cedex, France

Magali Jodeau
Cemagref, bis quai Chauveau, Lyon, France

Jean-Luc Reboud
Université Joseph Fourier, LEMD, Grenoble, France

ABSTRACT: A longstanding problem in the study of bed load transport in gravel-bed rivers is related to the physical mechanisms governing the bed resistance and particle motion. Although a number of experimental investigations have been conducted over the last three decades, there seems to be a substantial gap between the field measurements and the predictions of theoretical models, although these models provide a correct description of bed load transport for lab experiments. To elucidate this point, we investigated the motion of coarse spherical glass beads entrained by a shallow turbulent water flow down a steep two-dimensional channel with a mobile bed. This experimental facility is the simplest representation of bed load transport on the lab scale, with the tremendous advantages that boundary conditions are perfectly controlled and a wealth of information can be obtained using imaging techniques. Bed load equilibrium flows were achieved (i.e. neither erosion nor deposition of particles occurred on average, over sufficiently long time intervals). Flows were filmed from the side by a high-speed camera. Using an image processing software made it possible to determine the flow characteristics such as particle trajectories, their state of motion (rest, rolling or saltating motion), and flow depth. In accordance with earlier investigations, we observed that over short time periods, bed load transport appeared as a very intermittent process although the bed load rate was relatively intense. A striking result was that whereas for gentle slopes particles were mainly transported in saltation, the rolling regime played an increasingly important role at steep slopes. These experimental results suggest that to some extent, the mismatch between bed-load formulas and field data may be the consequence of a misinterpretation of the role played by the rolling particles, at least for steep slopes.

1 INTRODUCTION

Despite substantial progress made over the last two decades in the physical understanding of the motion of coarse particles in a turbulent stream, the ability to compute bulk quantities such as the sediment flux in rivers remains poor. For instance, the sediment flow rates measured in gravel-bed rivers differ within one to two orders of magnitude from the bed-load transport equations (Wilcock 2001; Martin 2003; Barry et al. 2004), whereas these equations have been established from flume experiments using regression techniques and are believed to provide a proper evaluation of bed load transport in a well-controlled lab environment. Surprisingly enough, simple power-law models relating the sediment flow rate to the water flow depth can perform better than more sophisticated physically-based models (i.e., models that take into account a number of physical processes such as a threshold for incipient motion, nonlinear interaction with water velocity, etc.) (Barry et al. 2004).

The objective of this paper is to show that a plausible explanation of the discrepancy between field measurements and empirical formulas is that unlike assumptions used in the theoretical support of these equations, the rolling regime can play a much more significant role than it is currently recognized. To be more specific, let us first recall that different forms of transport are produced depending on the particle weight and size relative to the turbulence scale. Light particles are maintained in suspension by turbulence whereas heavy particles roll along the bed or, when ejected from the bed, perform a series of leaps (saltation) (Julien 1994). A pervasive assumption is that for coarse particles, saltation dominates relative to the rolling regime (Bagnold 1973). This assumption is supported by a number of lab observations for various flow conditions. For instance, in our preliminary experimental investigations with the same experimental facility, where we studied the motion of a single particle down a bumpy line and subject to a turbulent water stream, we found that the rolling regime is a marginal mode of transport (Ancey et al. 2002; Ancey et al. 2003). Surprisingly enough, when studying the collective motion of particles in the same flow conditions, the rolling regime is no longer a marginal mode and its contribution to the total bed-load transport increases with slope.

2 EXPERIMENTAL FACILITIES AND TECHNIQUES

2.1 *Channel*

Experiments were carried out in a tilted, narrow, glass-sided channel, 2 m in length and 20 cm in height. Figure 1 shows a sketch of the experimental facility. The channel width W was adjusted to 6.5 mm, which was slightly larger than the particle diameter (6 mm). In this way, the particle motion was approximately two-dimensional and stayed in the focal plane of the camera (see Sec. 2.4 and Sec. 2.5). The channel slope $\tan\theta$ ranged from 7.5% to 15% (i.e., channel representative of steep gravel-bed rivers as encountered in mountain areas). In order to test the influence of the channel slope on bed load, we ran 15 experiments with different inclinations and various flow rates (see Table 1). The channel base consisted of half-cylinders of equal size ($r = 3$ mm), but they were randomly arranged on different levels, from 0 to 5.5 mm, by increments of 0.5 mm. These levels were generated using a sequence of uniformly distributed random numbers. Disorder was essential as it prevented slipping of entire layers of particles on the upper bed surface, which would have induced artificial erosion conditions. Maintaining disorder in monosized spherical particles is difficult, with severe constraints (Bideau and Hansen 1993), here involving mainly how to create disorder in the packing and the bed thickness. For thick beds (typically, whose thickness exceeded 5–6 particle diameters), a regular, crystalline arrangement was observed along the upper part of the bed. This is expected since it is well-known that the disorder range induced by a defect in a crystalline arrangement of monosize spherical particles is a few particle diameters (Bideau and Hansen 1993). Therefore, in order to be able to control the order in the particle arrangement, we built beds whose thicknesses did not exceed 5 particle diameters. The effects of disorder on bed-load transport features have been presented in another paper (Boehm et al. 2004).

Note that this bed thickness condition had severe consequences as regards the collisional interaction of a saltating particle with the bed. Indeed, in his experiments of dune formation, Rioual observed that the value of the effective coefficient of restitution depended on bed thickness (Rioual 2002). This can be physically understood by recalling that the capacity of a particle to retrieve its momentum after the impact depends on the reflection of the elastic waves generated during the collision. For a thick bed, considered as an infinite granular medium, the probability of an

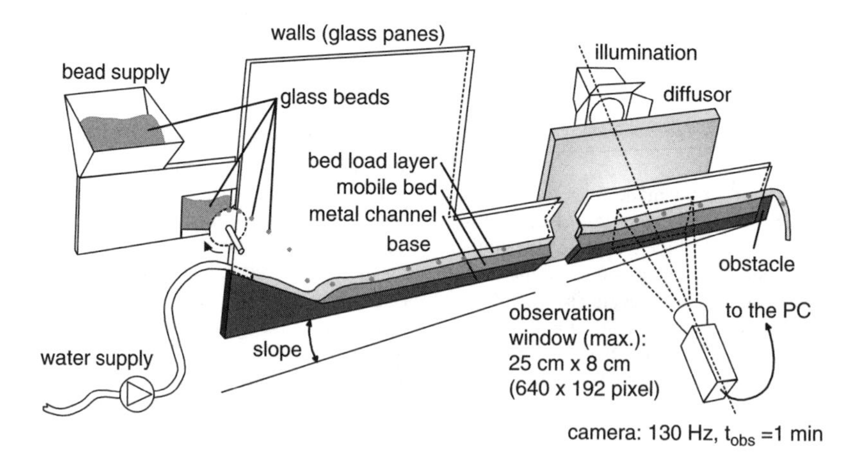

Figure 1. Sketch of the experimental setup.

elastic wave to return to the impacting bead is very low, whereas for a shallow bed, a part of the elastic energy transferred during the collision can be restored to the bead.

An obstacle was set at the channel outlet to enable bed formation and prevent full bed erosion; its height could be adjusted. The procedure used for building the bed is explained in Sec. 2.4.

2.2 *Solid and water supplies*

Colored spherical glass beads with a nominal diameter d of 6 mm and a density ρ_p of 2500 kg/m^3 (provided by Sigmund Lindner GmbH, Germany) were used. They were injected from a reservoir into the channel using a wheel driven by a direct current motor and equipped with 12 hollows on the circumference, as depicted in Figure 1 for the experiments presented here, the injection rate $\dot{n}_0$ ranged from 5 to 22 beads per second, with an uncertainty of less than 5%. This corresponded to a solid discharge per unit width q_s of 9–38 × 10^{-5} m^2/s. The water supply at the channel entrance was controlled by an electromagnetic flow meter provided by Krohne (France). The discharge per unit width q_w ranged from 3 to 10 × 10^{-3} m^2/s. The values for each run are summarized in Table 1.

2.3 *Dimensionless numbers*

The hydraulic conditions can be specified using classic dimensionless numbers. Table 1 reports the time-averaged values of these numbers. In Table 1, h and σ_h denote the time-averaged flow depth and its standard deviations, respectively. Here, to make the flow rate $\dot{n}$ more palpable, we express it in beads/s instead of m^3/s. The flow Reynolds number is defined as $\mathrm{Re} = 4R_h\bar{u}_f/\nu$, where $R_h = Wh/(2h+W)$ denotes hydraulic radius, $\bar{u}_f = q_w/h$ fluid velocity (averaged in the y- and z-directions), ν kinematic viscosity of water, and h water depth. The Froude number $\mathrm{Fr} = \bar{u}_f/\sqrt{gh}$ (where g denotes gravity acceleration) varied significantly over the experiment duration and along the main stream direction. The mean Fr values and the variation scale Δ_{Fr} are reported in Table 1. The variation scale of the Froude number was estimated assuming a constant water discharge and considering the variations in water depth: $\Delta_{\mathrm{Fr}} = |\delta\mathrm{Fr}| = |3\,\mathrm{Fr}\,\sigma_h/2\,h|$. These numbers show that frequent transitions from subcritical to supercritical regimes occurred at steep slopes.

The Shields number is defined as the ratio of the bottom shear stress ($\tau_0 = \rho_p gh\tan\theta$) to the stress equivalent of the buoyant force of a particle lying on the bottom (Julien 1994): $\mathrm{Sh} = \tau_0/(gd(\rho_p - \rho_f))$.

Table 1. Flow characteristics and time-averaged values of dimensionless numbers characterizing bed load and water flow. Varying parameters: Channel inclination $\tan\theta$ and solid discharge $\dot{n}$. The notation E7-6 indicates: $\tan\theta \approx 7\%$ and $\dot{n} \approx 6$ beads/s. Not all experiments made at $\tan\theta = 10\%$ are reported here, see (Boehm et al. 2004; Boehm 2005) for further information.

Experiment	E7-6	E7-8	E7-9	E7-11	E10-16	E12-9	E12-16	E12-21	E15-16	E15-21
$\tan\theta$ (%)	7.5	7.5	7.5	7.5	10	12.5	12.5	12.5	15.0	15.0
$\dot{n}_0$ (beads/s)	5.7	7.8	8.7	10.9	15.4	9.3	15.2	20.0	15.6	21.5
q_w (10^{-3} m^2/s)	10.00	11.54	13.85	26.15	8.19	2.97	3.85	4.46	2.31	2.92
h (mm)	18.9	20.8	24.9	40.8	16.9	7.0	8.2	9.4	4.9	6.7
σ_h(mm)	2.2	2.3	2.5	2.8	2.5	2.2	2.3	2.4	2.0	2.5
u_f (m/s)	0.53	0.55	0.56	0.64	0.48	0.42	0.47	0.48	0.47	0.44
$\dot{n}$ (beads/s)	5.45	7.76	9.20	10.99	15.56	9.52	15.52	19.86	15.45	20.55
$\sigma_{\dot{n}}$ (beads/s)	3.13	3.39	3.72	3.73	3.96	4.28	5.13	5.71	5.18	4.45
Re	5860	6230	6400	7720	5280	3760	4360	4600	3680	3830
Fr	1.26	1.26	1.15	1.02	1.24	2.20	2.09	1.90	3.72	2.63
Δ_{Fr}	0.222	0.209	0.172	0.105	0.276	1.046	0.875	0.716	2.272	1.468
Sh	0.158	0.173	0.207	0.340	0.188	0.098	0.114	0.130	0.082	0.111
Re_p	1050	1150	1350	1990	1000	1120	1140	1060	1720	1270
D	147	147	147	147	147	147	147	147	147	147
C_s (%)	0.95	1.17	1.16	0.73	3.30	5.58	7.02	7.74	11.65	12.23
h/d	3.16	3.47	4.15	6.80	2.82	1.17	1.37	1.56	0.82	1.11
u_r (m/s)	0.078	0.084	0.079	0.078	0.075	0.074	0.075	0.077	0.072	0.079
u_s (m/s)	0.35	0.36	0.33	0.31	0.32	0.24	0.28	0.30	0.18	0.23
κ_{re} (%)	6.9	5.5	6.4	5.1	7.5	12.1	12.1	10.8	10.6	7.3
κ_r (%)	29.0	28.6	31.1	30.7	38.1	68.0	63.8	57.0	84.9	76.2
κ_s (%)	63.9	65.7	62.2	64.1	54.3	19.9	24.1	32.1	4.4	16.4
N_l	849	1183	1484	1981	2242	635	1107	1850	194	1115
l_l	5.39	5.66	5.05	4.74	4.75	2.65	3.33	3.92	1.75	2.58
h_l	0.55	0.54	0.54	0.60	0.48	0.36	0.33	0.37	0.29	0.37

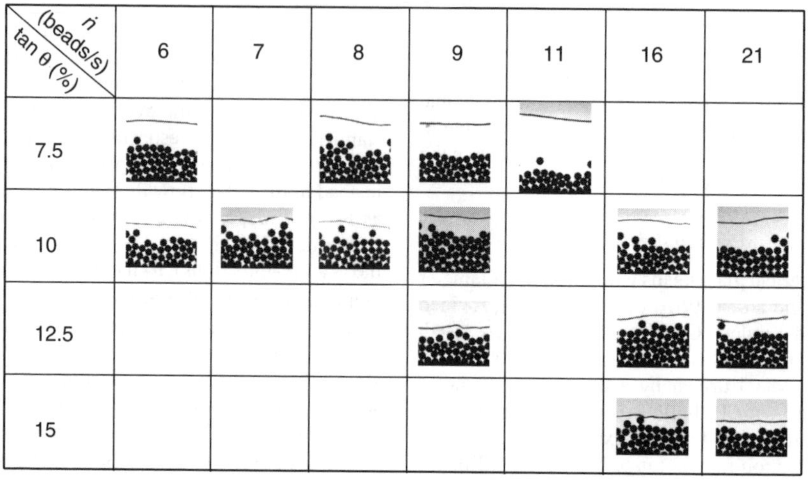

$\dot{n}$ (beads/s) / $\tan\theta$ (%)	6	7	8	9	11	16	21
7.5							
10							
12.5							
15							

Figure 2. Overview of the experiments conducted at various solid discharges $\dot{n}$ and slopes $\tan\theta$. For each experiment, a detail of one filmed image is shown. See Table 1 for the experimental conditions.

For our experiments, the critical Shields number Sh_c corresponding to incipient motion is in the range 0.004–0.005 (Ancey et al. 2002). Here the reduced Shields number (also called the *transport stage*) $T_* = N_{sh}/\mathrm{Sh}_c$ is on the order of 25, which indicates that the flow regime is far above the threshold of motion. The particle Reynolds number can be defined as $\mathrm{Re}_p = |\bar{u}_f - \bar{u}_p|d/\nu$, where $\bar{u}_p$ denotes the mean velocity of a particle in motion. We furthermore introduced the dimensionless particle diameter $D = d\sqrt[3]{(\rho_p/\rho_f - 1)g/\nu^2}$ used in sedimentology to characterize the particle size compared to the turbulence scale (Julien 1994). The solid concentration is defined as the ratio of the solid and the water discharge $C_s = q_s/q_w$. Values reported in Table 1 are low, which indicates that particle flow was dilute. The ratio h/d is low, typically in the range 0.8–7. A few aspects of the free-surface line and the bed configuration are shown in the tabulated Figure 2.

Note that the dimensionless number values differ substantially from the values usually found in the hydraulics literature. The reason is twofold: first we used a short and narrow channel, which led to studying low Reynolds number regimes, whereas in most experiments on bed load transport, one takes care to avoid such regimes; this is the price to pay to have access to the details of particle movements. Since we used coarse particles, particle motion was weakly dependent on the actual value of the flow Reynolds number and turbulence structure. Therefore we think that the small size of the experimental setup is not a handicap. Second we investigated supercritical flows because flow must be energetic enough to carry particles. However, in a supercritical regime, flow depth was low: on the order of the particle size, meaning that particle motion was affected by the water free surface.

2.4 *Experimental procedures*

The preliminary procedure can be split into three major steps. First of all, a particle bed was built along the channel base, which remained stationary on average. To that end, an equilibrium between the water discharge, solid discharge, bed elevation, and channel slope was sought. This equilibrium was reached by using the following procedure:

1. The water discharge q_w was set to a constant value.
2. An obstacle (approximately 20 mm in height) was positioned at the downstream end of the channel. The solid discharge $\dot{n}_0$ at the channel entrance (or the injection rate) was set to a constant value. The solid discharge per unit width q_s was calculated by the relation $q_s = \pi d^3 \dot{n}_0/(6W)$. An initial guess $\hat{q}_s$ for the solid discharge at equilibrium was obtained using an empirical sediment transport equation (Rickenmann 1992). For our case (uniform sediment), this equation can be simplified into the following form

$$\hat{q}_s = 6.27\,(q_w - q_c)\,\tan^2\theta, \tag{1}$$

where $q_c = 0.128\sqrt{g\,d^3}\tan^{-1.12}\theta$ is the critical solid discharge corresponding to incipient motion of particles (here $q_c = 2.46 \times 10^{-3}$ m²/s). The first beads supplied by the feeding system were stopped by the obstacle at the channel outlet and started to form a bed. The bed line rose to the level of the obstacle and beads began to leave the channel.

After approximately 10 minutes, the system arrived at bed load equilibrium, i.e., there was no more bed deposition or erosion over a sufficiently long time interval.

3. In order to make the bed line parallel with the channel base, the water discharge was then adjusted. After several iterations, we arrived at the configuration of a bed that consisted of two to three almost stationary bead layers along the channel, for which the bed line slope matched the channel base inclination. Average equilibrium conditions were sustained over long time periods, basically as long as 30 minutes.

Once bed equilibrium was reached, the particles and the water stream were filmed using a Pulnix partial scan video camera (progressive scan TM-6705AN). The camera was placed perpendicular to the glass panes at a distance of 115 cm from the channel, approximately 80 cm upstream from the channel outlet. It was inclined at the same angle as the channel. Lights were positioned in the backside of the channel. An area of approximately 25 cm in length and 8 cm in height was filmed and later reduced to accelerate image processing.

The camera resolution was 640 × 192 pixels for a frame rate of $f = 129.2$ fps (exposure time: 0.2 ms, 256 gray levels). Each sequence was limited to 8000 images due to limited computer memory; this corresponded to an observation duration of approximately 1 minute.

Each experiment was repeated at least twice in order to spot possible experimental problems and to get an idea of the data scattering.

2.5 *Image processing*

Images were analyzed using the WIMA software, provided by the *Traitement du Signal et Instrumentation* laboratory in Saint-Etienne (France). Positions of the bead mass centers were detected by means of an algorithm combining several image-processing operations. It compared the filmed images with the image of a model bead and calculated the correlation maxima to obtain the bead positions. The water free surface (averaged in the direction perpendicular to the channel walls) was detected using its slim form; missing portions were inter- or extrapolated. Resulting uncertainty on the bead and water line position was less than 1 pixel or 0.4 mm. For more details, the reader can refer to (Boehm et al. 2005; Boehm 2005).

2.6 *Data processing*

Data obtained from the image sequences were analyzed to obtain the particle trajectories. For this purpose, we developed a particle-tracking algorithm, which was integrated into the WIMA software. This algorithm compared the bead positions of two consecutive images to determine the trajectory of each bead step by step. Since the particle movement was nearly two-dimensional and the displacement of a particle between two images was always smaller than a particle diameter, the trajectories (approximately 700 per sequence) could be calculated with no significant error. Problems occurring at the entrance and the exit of the observation zone prompted us to calculate variables such as the solid discharge in a reduced window 580 pixels in length.

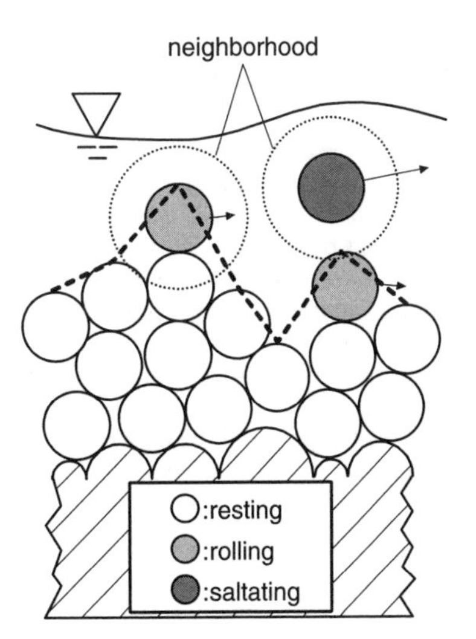

Figure 3. Sketch defining the state of motion and the bed line.

The state of movement of a particle was defined by considering that each bead was always either in a resting, rolling, or saltating regime (see Fig. 3). Such partitioning posed several difficulties from the algorithmic viewpoint. The three states of movement were distinguished as follows:

- The resting beads formed the bed, they were in sustained contact with their neighbors. They were not expected to move or, more precisely, their possible drift velocities (together with fluctuating velocities) were lower than a threshold velocity u_t: $|\mathbf{u_p}| < u_t$, where $\mathbf{u_p}$ denotes the bead velocity averaged over five consecutive frames.
- The beads in the rolling regime were located above the beads at rest: they remained in close contact with the bed and moved at a certain velocity. Our algorithm used two criteria to distinguish rolling beads: $|\mathbf{u_p}| \geq u_t$ (bead in motion) and $d_n/d \leq \varepsilon$ (particles in the vicinity), where d_n is the distance to the next neighbor (averaged over five consecutive frames) and ε is a threshold.
- The beads in saltation leaped above the others. They had no close neighbors except when they

collided with other beads. The algorithm distinguished saltating beads using the criteria $|\mathbf{u_p}| \geq u_t$ and $d_n/d > \varepsilon$.

The values of the threshold parameters u_t and ε were adjusted by trial and error to minimize the differences between the state determined by the algorithm and the state determined by the naked eye. Good agreement was obtained for $u_t = 0.025$ m/s and $\varepsilon = 1.07$.

Since the experiments involved a mobile bed, the water depth was defined as the difference between the free surface and the bed surface elevation. Arbitrarily, we considered that the bed surface profile is the broken line linking the top points of the uppermost resting or rolling beads. Figure 3 depicts such a broken line at a given time.

3 RESULTS

3.1 *Preliminary observations*

It is worth noticing the following points:

- Increasing the water discharge essentially caused an increase in the water depth, whereas the mean fluid velocity u_f slightly increased. Therefore u_f stayed in the narrow range 0.41–0.64 m/s for all the experiments presented here (the maximum was reached at low slopes) due to sidewall friction. The characteristic velocities of the rolling (u_r) and the saltating beads (u_s) varied little and were still closely linked to u_f.
- Various experiments were made to test the influence of the bed configuration on the features of bed-load transport (Boehm et al. 2004). Experiments provided clear evidence that these fluctuations resulted, to a large extent, from the finite size of the observation window, e.g., experiments done with a fixed bed showed that solid discharge inherited stochastic properties exhibited by individual particles. Fluctuations were exacerbated when the bed was mobile, i.e., deposition and entrainment of particles were made possible. Mobile bed experiments were also characterized by a spectacular change in particle transport behavior, notably in the occurrence of rolling and saltating regimes.
- Solid discharge variations with time were intercorrelated to obtain a characteristic time of particle motion. For fixed bed experiments, the characteristic time was nothing but the average time for a particle to travel the length of the observation window. For mobile bed experiments, the characteristic times were found to be larger; it was mainly related to the mean downstream velocity of the saltating phase. A striking result is that, although the probability density function of the solid discharge differed significantly depending on the bed configuration, there was not much difference between their intercorrelation functions. Another notable result is that solid discharge depended a great deal on bed arrangement. For instance, keeping the solid discharge constant, but changing the bed configuration led to water discharge variations (at equilibrium) as wide as 40%.
- Regime transitions (rolling from rest, saltation from rolling) depended a great deal on the bed arrangement: It was shown that (i) the local dynamics (liftoff, settling) was largely dependent on the bed arrangement and (ii) there were space and time correlations in the regime transitions, implying collective changes in particle behavior. This observation is of fundamental importance from the theoretical viewpoint because it emphasizes the importance of taking into account particle arrangement in the study of regime transition. Apart from a few recent experimental investigations (Papanicolaou et al. 2002), there are very few theoretical or experimental works that have been devoted to this topic.
- The Shields number Sh was proportional to the product $h \tan\theta$. In our case, the variation of h outweighed the effect of $\tan\theta$, which explains that Sh was a decreasing function of $\tan\theta$. For experiments E12-9 and E15-16, Sh was below 0.1 and thus very close to the threshold for incipient motion, usually given for coarse particles on a gentle slope ($\mathrm{Sh}_c = 0.047$–0.06) (Buffington and Montgomery 1997).

3.2 *Contribution of the rolling regime to the solid discharge*

We broke down the solid discharge into the contributions due to the saltating and the rolling beads. We introduced κ_s as the ratio of the volume flow rate due to the saltating particles to the total solid discharge.

When working at constant slope (tests were made for $\tan\theta = 10\%$), there was almost no variation in the ratio κ_s with total solid discharge. In contrast, this ratio turned out to be strongly dependent on channel slope and water depth h. This dependence can be shown by plotting κ_s as a function of h/d for the different slopes (see Fig. 4). For $\tan\theta = 7.5\%$, κ_s was higher, but still constant (between 62.2 and 64.1%). We can thus conclude that for mild slopes (7.5 and 10%), κ_s was a function of $\tan\theta$ only. For steeper slopes though (12.5 and 15%), the diagram shows that κ_s was a linear function of h/d independent of $\tan\theta$.

A closer look at the images in Figure 2 helps to understand this relation. For $\tan\theta = 12.5$ and 15%, beads in saltation got in touch with the water free surface quite frequently, which implied that their vertical motion was hindered by the low water depth. This might explain why saltation was impeded here, many particles had to be transported in the rolling or sliding motion, which occupied less space in the vertical

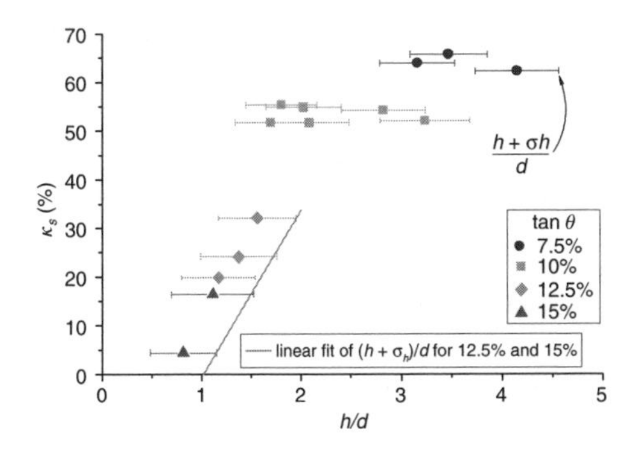

Figure 4. Relative contribution of the saltating particles to the solid discharge κ_s as a function of the ratio h/d, where h and d are the water depth and the particle diameter, respectively. The error bars represent the standard deviation $(h \pm \sigma_h)/d$. Experiment E7-11 has been omitted.

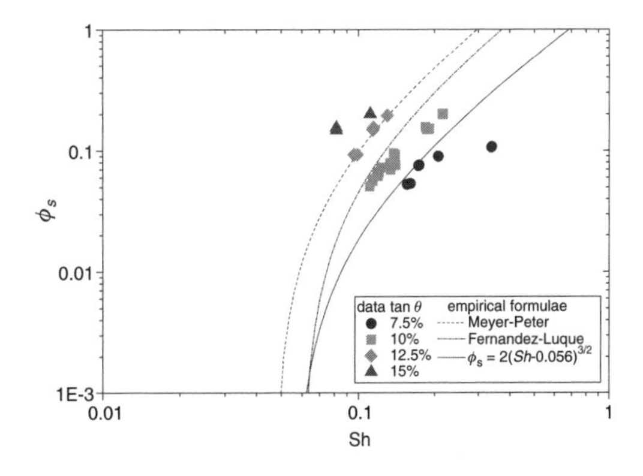

Figure 5. Dimensionless solid discharge ϕ_s as a function of the Shields number Sh.

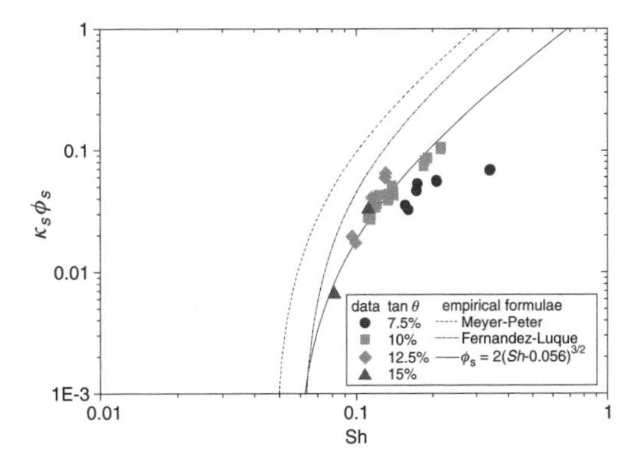

Figure 6. Dimensionless solid discharge due to the saltating particles, $\kappa_s\phi_s$, as a function of the Shields number Sh.

direction. Surprisingly, the data show that saltation occurred even if the time-averaged water depth fell below one particle diameter. This was due to the fluctuations of h, since even in the worst case $h/d = 0.82$ the instantaneous water depth frequently exceeded one particle diameter, which enabled saltation (see the error bars in Fig. 4). To take the effect of water line fluctuations into account, we defined the efficient flow depth as $h + \sigma_h$ and we considered that the experimental trend for slopes in the range 10–15% could be described using the linear fit $(h + \sigma_h)/d$. This experimental trend at higher slopes suggests the existence of a threshold of saltation at $(h + \sigma_h)/d = 1$, as shown in Figure 4 when the fitted straight line crosses the h/d-axis. Note that the saltation threshold is only obtained by the extrapolation of the data presented in Fig. 2, experimentally we were not able to produce a flow where saltation was absent. For a water depth significantly lower than the particle diameter, no stable bed load equilibrium could be achieved.

On the opposite, for the experiments made at milder slopes $\tan\theta = 7.5$ and 10%, the particle leap height was slightly influenced by the water line (see the flow images in Fig. 2). In that case, we would have expected that the ratio κ_s could freely adapt and vary with slope and solid discharge and/or the relative submersion, but it turned out that it depended on the channel slope alone (see Fig. 4).

A widespread way of representing bed load transport rates for different flow conditions is to plot the dimensionless solid discharge ϕ_s as a function of the Shields number Sh. Figure 5 shows our experimental data points compared with two widely used empirical formulas. Note that these empirical formulas were established for slopes much lower than in our case, but this is not a major issue since the physics should remain essentially the same. The solid discharge was made dimensionless by the definition

$$\phi_s = q_s/\sqrt{(\rho_p/\rho_f - 1)gd^3},$$

where q_s is the bed load transport rate per unit width, $q_s = \pi d^3\dot{n}/(6W)$. Both the Meyer-Peter and the Fernandez–Luque formula (Fernandez-Luque and van Beek 1976) make use of a transport threshold, they yield a non-zero ϕ_s if Sh exceeds a critical value, Sh_c.

Our data was quite scattered in this diagram, the points seemed to be affected by slope to large extent. As shown above, the contribution of the saltating particles κ_s depended on the channel slope (at least via the water depth). This led to plot the diagram differently by only reporting the contribution due to the saltating particles, i.e., by plotting $\kappa_s\phi_s$.

In so doing, data scattering is markedly reduced, which allows us to fit a curve to the data. The resulting equation is similar to the Meyer-Peter equation,

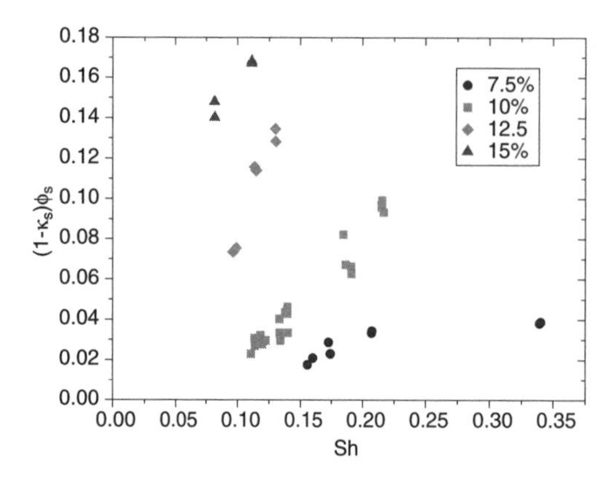

Figure 7. Dimensionless solid discharge due to the rolling particles, $(1-\kappa_s)\phi_s$, as a function of the Shields number Sh.

the only difference lies in the values used for the coefficients:

$$\phi_s = 2\,(\mathrm{Sh} - 0.056)^{3/2}.$$

This result confirms that, at steep slopes, the bed-load transport due to saltation can be reasonably well described using a power law with a threshold. While a different approach is needed for the rolling motion. Figure 7 shows the variations in the dimensionless flow rate due to the rolling particles with the Shields number. This contribution was computed as $(1-\kappa_s)\phi_s$. The effect of slope is obvious in this figure. Not also that the experimental curves that we adjust to the data with the naked eye seem to cross the Sh-axis at the same point, which is $\mathrm{Sh} \approx 0.056$, i.e., the same value as found for incipient motion. Contrary to the saltating regime, we failed to find a scaling that would make it possible to collapse the points onto a single curve.

4 CONCLUSIONS

In this paper, the effect of slope on bed load transport was investigated using an idealized experimental setup. We observed a number of interesting properties:

- Substantial fluctuations of the solid discharge, which can be attributed to the collective motion of particles (Boehm et al. 2004).
- Contrary to the movement of a single particle (for which the saltating regime was predominant), the motion of a set of particles exhibit both rolling and saltation.
- Slope has a significant effect on the total bed-load transport. This dependence is obvious when plotting the solid discharge as a function of the Shields number. However, when the solid flow rate is broken down into a contribution representing the saltating particles and another one related to the rolling motion, we found that slope has little effect on the contribution due to the saltating particles, which makes it possible to derive a single curve linking the solid discharge and Shields number in a way similar to empirical formulas such as Meyer-Peter's equation. In contrast, slope significantly affects the contribution related to the rolling regime.
- At steep slopes, the rolling regime becomes predominant relative to the saltating regime.

The striking result was the predominance of the rolling regime at steep slope. At gentle slopes (in our context, 'gentle' slopes refer to slopes lower than 10%), we observed that the rolling regime, albeit not prevailing, represents 45–30% of the total solid discharge, hence is far from negligible. This contrasts substantially with our earlier observations (Ancey et al. 2002; Ancey et al. 2003) on the movement of a single particle, where it was found that the rolling regime is a marginal mode of transport.

Another remarkable result is that, whereas it is possible to find a scaling law for the solid discharge due to the saltating particles, which is furthermore structurally close to empirical formulas obtained with larger flumes and natural sediment, the contribution of the rolling bead to the total solid discharge reveals a much more complicated dependence on slope and Shields number.

Since the saltation of a single particle has been used as a paradigm to provide theoretical support to empirical bed-load equations (Bagnold 1973; Wiberg and Smith 1985; Wiberg and Smith 1989; Bridge and Bennett 1992; Seminara et al. 2002), we can wonder whether the role of saltation has been overly emphasized. We do not pretend that the experimental results presented here yield sufficiently firm arguments to completely reply to this question, but they provide new insight into the physics of bed-load transport.

ACKNOWLEDGMENTS

This study was supported by the program ECCO/PNRH of INSU. We are grateful to the laboratory TSI UMR 5516 (Christophe Ducottet, Nathalie Bochard, Jacques Jay, and Jean-Paul Schon).

REFERENCES

Ancey, C., F. Bigillon, P. Frey, and R. Ducret (2003). Rolling motion of a single bead in a rapid shallow water stream down a steep channel. *Phys. Rev. E 67*, 011303.

Ancey, C., F. Bigillon, P. Frey, R. Ducret, and J. Lanier (2002). Motion of a single bead in a rapid shallow water stream down an inclined steep channel. *Phys. Rev. E 66*, 036306.

Bagnold, R. (1973). The nature of saltation and of 'bed load' transport in water. *Proc. Roy. Soc. London A 332*, 473–504.

Barry, J., J. Buffington, and J. King (2004). A general power equation for predicting bed load transport rates in gravel bed rivers. *Water Resour. Res. 40*, W10401.

Bideau, D. and A. Hansen (Eds.) (1993). *Disorder and granular media*. Random materials and processes. Amsterdam: North-Holland.

Boehm, T. (2005). *Motion and interaction of a set of particles in a supercritical flow*. Ph.D. thesis, Joseph Fourier University.

Boehm, T., C. Ancey, P. Frey, J.L. Reboud, and C. Duccotet (2004). Fluctuations of the solid discharge of gravity-driven particle flows in a turbulent stream. *Phys. Rev. E 69*, 061307.

Boehm, T., C. Duccotet, C. Ancey, P. Frey, and J.-L. Reboud (2005). Two-dimensional motion of a set of particles in a free surface flow with image processing. *submitted to Experiments in Fluids*.

Bridge, J. and S. Bennett (1992). A model for the entrainment and transport of sediment grains of mixed sizes, shapes, and densities. *Water Resour. Res. 28*, 337–363.

Buffington, J. and D. Montgomery (1997). A systematic analysis of eight decades of incipient motion studies, with special reference to gravel-bedded rivers. *Water Resour. Res. 33*, 1993–2029.

Fernandez Luque, R. and R. van Beek (1976). Erosion and transport of bed-load sediment. *J. Hydraul. Res. 14*, 127–144.

Julien, P.-Y. (1994). *Erosion and Sedimentation*. Cambridge: Cambridge University Press.

Martin, Y. (2003). Evaluation of bed load transport formulae using field evidence from the Vedder River, British Columbia. *Geomorphology 53*, 75–95.

Papanicolaou, A., P. Dipla, N. Evaggelopoulos, and S. Fotopoulos (2002). Stochastic incipient motion criterion for spheres under various bed packing conditions. *J. Hydraul. Eng. 128*, 369–390.

Rickenmann, D. (1992). Hyperconcentrated flow and sediment-transport at steep slopes. *J. Hydraul. Eng. 117*, 1419–1439.

Rioual, F. (2002). *Etude de quelques aspects du transport éolien : processus de saltation et formation des rides*. Ph.D. thesis, University of Rennes I.

Seminara, G., L. Solari, and G. Parker (2002). Bed load at low Shield stress on arbitrarity sloping beds: failure of the Bagnold hypothesis. *Water Resour. Res. 38*, 1249.

Wiberg, P. and J. Smith (1985). A theoretical model for saltating grains in water. *Journal of Geophysical Research (C4) 90*, 7341–7354.

Wiberg, P. and J. Smith (1989). Model for calculating bedload transport of sediment. *J. Hydraul. Eng. 115*, 101–123.

Wilcock, P. (2001). Toward a practical method for estimating sediment-transport rates in gravel bed-rivers. *Earth Surface Processes and Landforms 26*, 1395–1408.

River, Coastal and Estuarine Morphodynamics: RCEM 2005 – Parker & García (eds)
© 2006 Taylor & Francis Group, London, ISBN 0 415 39270 5

Effects of non-hydrostatic pressure distribution on bedload transport

Simona Francalanci
Department of Civil Engineering, University of Florence, Italy

Gary Parker
Departments of Civil and Environmental Engineering and Geology, University of Illinois, Urbana, Illinois, USA

Enio Paris
Department of Civil Engineering, University of Florence, Italy

ABSTRACT: The fact that the hydrostatic pressure distribution is implicit in the definition of the Shields parameter does not seem to be widely recognized. This hypothesis is accurate in the case of uniform and rectilinear flow, but fails where local non-hydrostatic pressure distributions can affect the mobility of sediment and hence the bedload transport rate. The aim of the present work is to investigate the effect of a non-hydrostatic pressure distribution on sediment transport and derive a generalized empirical formula to evaluate the bedload transport rate. Both upward and downward seepage flows through a sediment bed were used to create a non-hydrostatic component of pressure affecting the particles at the bed surface. Experiments were carried out in a sand-recirculating flume with a steady main flow, where we created a groundwater flow in a short reach of the flume. The experimental results showed a clear effect of the groundwater flow, and hence of a non-hydrostatic pressures distribution, on sediment transport.

1 INTRODUCTION

Non-hydrostatic pressures can be quite significant in the case of flow around an obstacle, such as a bridge pier. Experimental work and field observations show evidence of a complex flow field and local scour of the bed around a bridge pier. In this case, the sediment bed mobility is affected by the vertical distribution of the pressure gradient, and an increased or decreased buoyant force can give more or less mobility to the sediment particles.

The main goal of this work is to investigate the effect of a non-hydrostatic pressure distribution on bedload transport. Direct study of the problem of flow around an obstacle can be rather difficult, so a simplification has been used. Groundwater flow induces non-hydrostatic pressure, and hence an upward or downward seepage flow through a sediment bed has been used to create a non-hydrostatic component of pressure affecting the particles at the bed surface.

The effect of an upward seepage flow on incipient sediment motion has been studied by Chen and Chiew (1999). They found that the critical shear velocity is reduced in the presence of upward seepage. The same authors studied a turbulent open-channel flow with upward seepage and found out that the bed shear stress also shows a steady reduction with increasing upward seepage velocity.

In the following both a theoretical approach and experimental results are discussed.

2 THEORETICAL BACKGROUND

2.1 *General expression of the Shields parameter*

The sediment transport discharge is usually quantified by the Shields parameter

$$\tau^* = \frac{\tau_b}{(\rho_s - \rho)gD} \tag{1}$$

where τ_b = boundary shear stress, ρ = water density, ρ_s = sediment density, g = gravitational acceleration and D = particle size. The assumption of hydrostatic pressure is implicit in the above equation. Specifically, the term $-\rho gD$ is due to the Archimedian buoyant force generated by the vertical pressure gradient associated with the hydrostatic hypothesis.

A more general expression for the relation (1) is the following

$$\tau^* = \frac{\tau_b}{\left(\rho_s + \frac{1}{g}\frac{dp}{dz}\right) gD} \quad (2)$$

where p = pressure, z = bed elevation.

Upward or downward groundwater flow in a zone of the flume induces a non-hydrostatic pressure distribution along the vertical. This can be seen as follows. In the case of groundwater flow the vertical pressure gradient can be evaluated from Darcy's law:

$$v_s = -K \frac{\partial h_p}{\partial z} \quad (3)$$

where v_s is the averaged filtration velocity through the porous bed, h_p is the piezometric head $= p/\gamma + z$, γ = specific weight of the fluid and K is the hydraulic conductivity.

In the presence of groundwater flow, then:

$$\frac{dh_p}{dz} = \frac{d}{dz}\left(\frac{p}{\rho g} + z\right) = -\frac{v_s}{K} \quad (4)$$

and hence:

$$\frac{dp}{dz} = \rho g \cdot \left(-1 - \frac{v_s}{K}\right) \quad (5)$$

So, from equation (5), when v_s is positive (upward seepage flow), the term on the right-hand side drops below the hydrostatic value of -1, while when v_s is negative (downward seepage flow), the term on the right side becomes higher than -1. The non-hydrostatic pressure gradient due to seepage induces an increased or decreased buoyant force acting on a bed particle, as shown in Figure 1.

The above equation can be used as an approximate expression for pressure gradient and substituted into

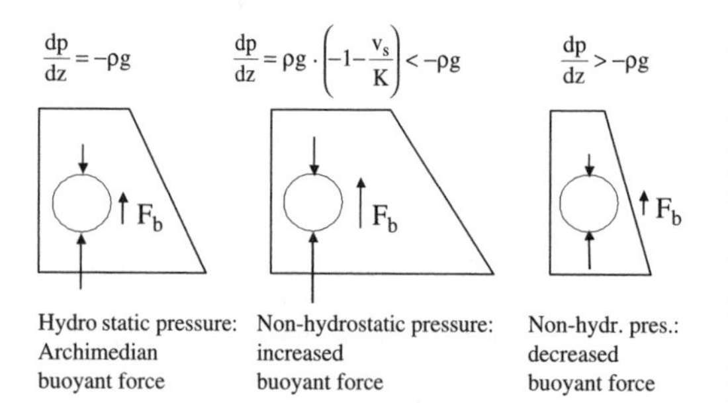

Figure 1. Vertical pressure gradient dp/dz near and buoyant force F_b acting on a bed particle.

(2), yielding the following generalized form for the Shields parameter in the case of vertical seepage:

$$\tau^* = \frac{\tau_b}{\left(\rho_s - \rho\left(1 + \frac{v_s}{K}\right)\right) gD} \quad (6)$$

Hence, when the velocity v_s is positive (upward ground water flow) the particle is effectively lighter and more movable, and the Shields parameter is increased, while when the velocity is negative (downward groundwater flow) the Shield parameter is decreased. The flow condition for the experiments reported here have been set up with a range of values of v_s so as to characterize the effect the Shields parameter.

2.2 *Bed shear stress*

The bed shear stress has a different expression in the case of boundary seepage, because the boundary conditions of the flow are changed. Chen and Chiew (1998a) derived a momentum integral equation in the case of seepage, so that the bed shear stress can be computed using the depth-averaged velocity U, the water depth H, the water surface slope dh/dx and the seepage velocity v_s, according to the following relation:

$$\tau_b = \rho u_*^2 =$$

$$\rho g H \cdot \sin\phi_b - \rho g H \cdot \frac{dH}{dx}\left(\cos\phi_b - \frac{\beta U^2}{gH}\right) - 2\beta\rho U v_s \quad (7)$$

In the above relation u_* = shear velocity, ϕ_b = longitudinal bed slope angle and β = a momentum correction factor.

Hence, there are two combined and opposite effects on the dimensionless Shields parameter in the presence of seepage: the effect of an upward seepage leads to a reduction of the bed shear stress, but also a reduction of the denominator of the Shields parameter due to a non-hydrostatic pressure gradient. The opposite behavior is expected in the case of downward seepage.

3 EXPERIMENTAL SET-UP

3.1 *Open-channel flume*

Experiments were carried out in a glass-sided horizontal flume that was 15 m long, 0.61 m wide and 0.4 m deep. The incoming water flow rate was controlled by a valve and monitored using an electromagnetic flow-meter.

The system was a sediment-recirculating, water-feed flume, with a steady main flow; a short 1.1 m reach of the flume with upward or downward groundwater flow was located 9 m from the upstream end of the flume.

In the sediment-recirculating, water-feed flume the sediment and water were separated at the downstream end. Nearly all the water overflowed from a collecting tank at the end of the flume. The sediment settled to the bottom of the collecting tank, and was recirculated to the upstream end of the flume with a small amount of water as a slurry with a jet-pump system.

In the case of upward seepage, water was delivered by a second jet-pump into a seepage box through a pipeline from the end of the flume. This seepage box was buried under the flume sediment. The groundwater flow discharge to the seepage box was regulated by a valve and measured by an electromagnetic flow meter. In a similar way, in the case of downward seepage the water was sucked from the seepage box by a suction pump. A sketch of the experimental apparatus is shown in Figure 2.

Velocity measurements were performed using a micro-propeller with a diameter of 14 mm. A point-gage was used to measure the bottom elevation of the bed: five points were measured for each transverse cross-section and the average value was assumed as the cross-sectional mean value of the bottom elevation for the purpose of characterizing long profiles.

3.2 *Groundwater flow*

Vertical seepage was induced by means of a seepage box installed within the sediment bed, allowing a seepage flow across the full width of the flume. The seepage zone was 1.1 m long, 0.61 m wide and 0.06 m deep, and was located within the sediment bed. The upper part of the seepage box (Fig. 3) was covered with a perforated sheet, in order to have a uniform groundwater flow. A special fine filter was inserted above the perforated sheet in order to separate the sand layer

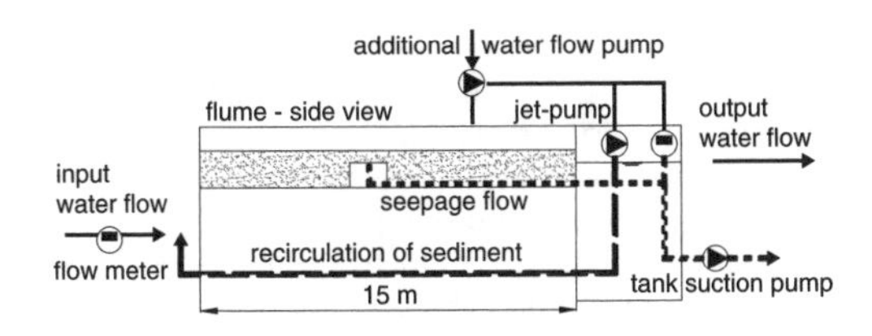

Figure 2. Sketch of the experimental apparatus.

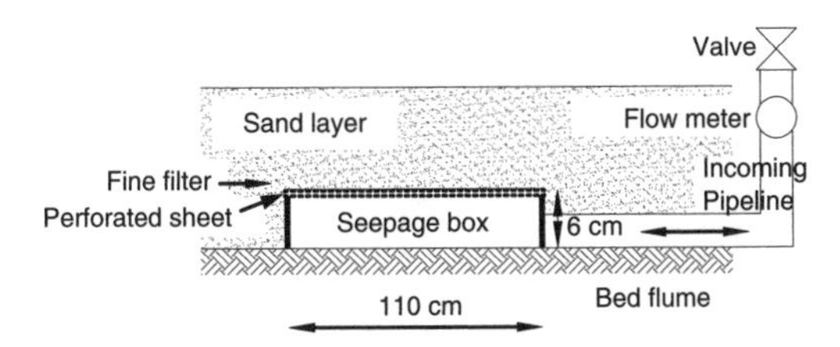

Figure 3. Sketch of the seepage box.

from the water inside the box. The groundwater flow was verified to be uniform at the top of the sand layer above the seepage zone, in order to avoid any effect of non-uniformity during the experiments.

3.3 *Sediment material*

The sediment material was chosen according to the following requirements: uniform size distribution, appropriate hydraulic conductivity of the sand and reduced presence of bedforms.

The hydraulic conductivity K is a parameter related to the mean diameter of the porous material and to its size distribution. Finer sand yields smaller values of hydraulic conductivity, while coarser sand yields bigger values of K and uniform sand has a larger hydraulic conductivity compared to less uniform sand with the same mean diameter. It was necessary to adjust the sediment type in order to obtain optimal values of K for the experiments, and to have a sensible ratio between the seepage velocity and the hydraulic conductivity that could significantly affect the Shields parameter.

Hydraulic conductivity was directly measured with a Darcy tube. The material chosen for the experiments is represented by a uniform sand with mean diameter $D_{50} = 0.9$ mm and $D_{35} = 0.7$ mm. The measured parameter K was found to vary somewhat, but was near 0.25 cm/s.

With this bed material only bedforms with small amplitude were observed in the flume at the equilibrium configuration. The mode of sediment transport was bedload.

3.4 *Equilibrium configuration*

The system described above allows the operator to set the water discharge per unit width (up to the small fraction of water discharge in the recirculation line). The jet-pump usually worked so as to transport out of the tailbox of the flume all but only a small amount of sediment that is stored there. Under equilibrium conditions, then, transport rate of sand in the recirculation line is exactly equal to the bedload transport rate in the flume. Thus the bedload transport rate was measured based on solid discharge measurements in the recirculation line. Sand samples were dried and weighted.

In a sediment-recirculating flume the total amount of sediment is conserved. In addition, a downstream weir controls the downstream elevation of the bed. The combination of these two conditions constrains the bed slope at mobile-bed equilibrium: adding more sediment to the flume increases the equilibrium bed slope.

The mean slope of the flume at the equilibrium configuration was measured for each run from bottom elevation measurements. It was found to be mostly constant around a value of about 0.00384 during the experiments, as shown in Figure 4.

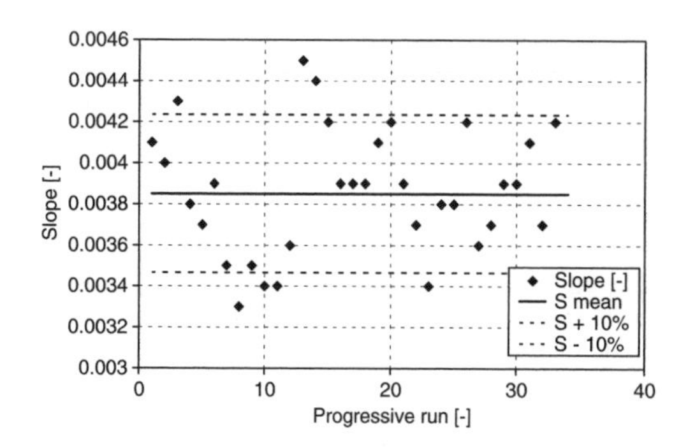

Figure 4. Measured values of slope in progressive runs.

Table 1. Hydraulic conditions for the experiments.

Run code	Q_w (l/s)	$Q_{seepage}$ (l/m)					
1*	20.6	20,	40,	60,	80,	100,	140
2*	16	20,	40,	60,	80,	100,	140
3*	12	20,	40,	60,	80,	100,	140
4**	20.6	30,	60,	90			
5**	16	30,	60,	90			
6**	12	30,	60,	90			

* Upward groundwater flow.
** Downward groundwater flow.

4 EXPERIMENTAL ACTIVITY

Preliminary experiments with the same hydraulic conditions in the flume and different seepage flows allowed a quick evaluation of the effect of seepage flow. Short-term and long-term experiments with more complete data acquisition were then performed. The short-term experiments were focused on the morphodynamic evolution of the bed after a few minutes of constant groundwater flow, while the long-term experiments were focused on the equilibrium steady state of the bed after several hours of constant groundwater flow.

Different sets of experiments were carried out in order to study both the effects of upward and of downward groundwater flow on sediment transport. In Table 1 the hydraulic conditions of the experiments are summarized in terms of the inflow discharge Q_w and seepage discharge $Q_{seepage}$.

4.1 *Experimental procedure*

First the flume was allowed to develop a mobile-bed equilibrium configuration in the absence of seepage, after which measurements of water flow, solid discharge and water surface depth were performed. The system was stopped to measure the bottom topography along the flume, and additional cross-sections were measured upstream, across and downstream of the seepage reach, to allow a better description of local phenomena.

Immediately after this a short-term experiment with seepage was performed: the system was started again with the previous value of water flow Q_w, the pump for the groundwater flow was then turned on and the seepage discharge was gradually adjusted to the predetermined value. The short-term experiment usually lasted about 10 minutes. At the end the bottom topography was measured.

After this, a long-term experiment was performed: the groundwater flow was run for several hours in order for the mobile bed to achieve an equilibrium steady state in the presence of seepage. At the end of the long-term run, measurements of flow velocity profiles along the flume were performed: five velocity profiles were measured upstream of, at the beginning, in the middle, at the end of and downstream of the seepage reach. Water surface depth was measured as well.

Finally, the system was stopped and the bottom topography of the entire flume was measured.

5 BEDFORMS AND FRICTION COEFFICIENT

At the equilibrium condition bedforms of small amplitude were observed in the bottom topography. The drag force acting on the sand bed can be decomposed into *skin friction* and *form drag*. The former is generated by the viscous shear stress acting tangentially to the bed. The latter is generated by the normal stress (mostly pressure) acting on the bed. Only the skin friction is thought to directly contribute to sediment transport. Hence, when bedforms such as dunes are present the form drag associated with flow separation behind dunes is subtracted out in performing bedload calculations, as it does not contribute directly to the bedload transport of bed grains.

The friction coefficient was estimated according to the empirical formulation proposed by White, Paris and Bettess (1980). The total shear velocity was estimated with the following relationship:

$$u_* = \sqrt{g \cdot R \cdot S} \tag{8}$$

where S = longitudinal slope of the bed, R = hydraulic radius, evaluated with the Vanoni (1975) correction for sidewall effects:

$$R = \begin{cases} \dfrac{B}{4} \text{ for } H > \dfrac{B}{2} \\ \dfrac{1}{B}\left[H^2 + H(B - 2H)\right] = \dfrac{BH\text{-} H^2}{B} \text{ for } H < \dfrac{B}{2} \end{cases} \tag{9}$$

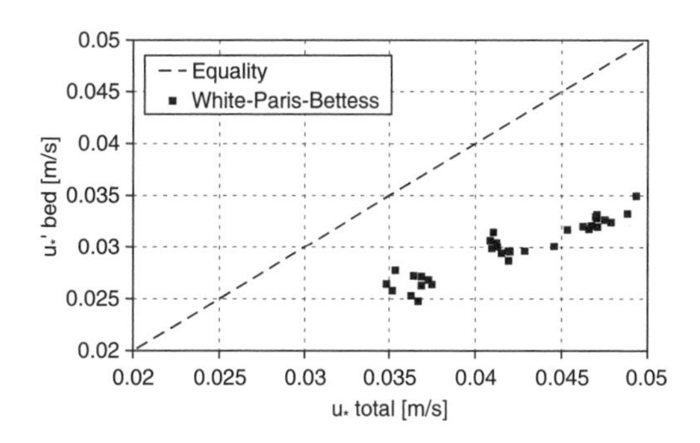

Figure 5. Reduction of the shear velocity due to bedforms.

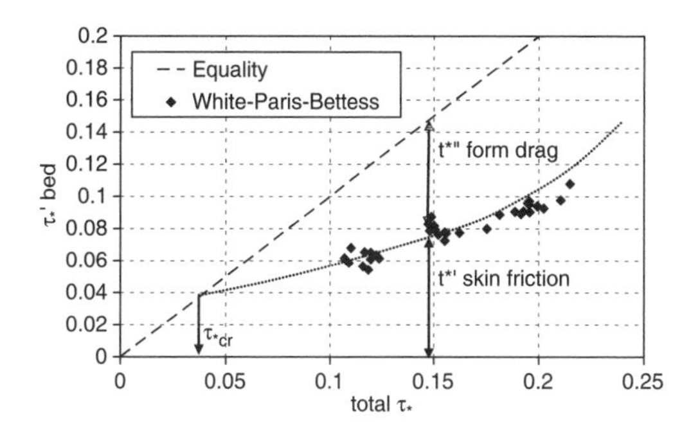

Figure 6. Decomposition of the total dimensionless shear stress into a component due to skin friction and a component due to form drag. The data are for the experiments reported here.

where B = width of the flume. The shear velocity associated with skin friction is evaluated as follows:

$$u_*' = \frac{U_{measured}}{C'} = \frac{U_{measured}}{5.65\log\left(10.6\frac{H}{D_{35}}\right)} \tag{10}$$

The values of u_* and u_*' are compared in Figure 5.

The shear velocity can be used to decompose the total dimensionless Shields number into a contribution τ_*' due to bed roughness, and hence to grains motion, and a contribution due to form resistance τ_*''.

The values of the effective Shields parameter τ_*', to be used in any sediment transport relationship may be compared to the total value of the Shields parameter τ_* to determine form drag. This decomposition is illustrated in Figure 6.

The values of the effective shear stress due to skin friction seem to collapse to the value of the total shear stress, when the total shear stress reduces to the critical value of about 0.039 for the onset of motion, predicted by the criterion of Ackers and White. Also a reduced contribution of the form drag can be noted for smaller values of the total shear stress, in good agreement with the reduced amplitude of bedforms experimentally observed for the lower flow conditions.

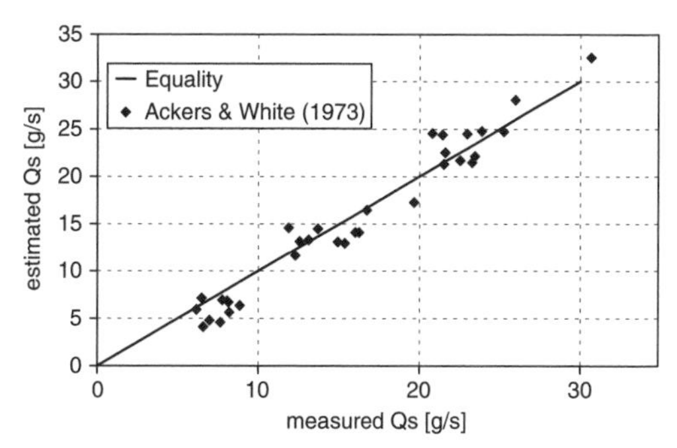

Figure 7. Estimated values of solid discharge from Ackers and White relationship (1973) versus measured values.

6 RELATIONSHIP BETWEEN HYDROSTATIC PRESSURE AND SEDIMENT TRANSPORT

Following the above approach for friction coefficient, the Ackers and White (1973) formulation for bed-load transport was used to estimate solid discharge and compare predicted with measured values. Solid transport measurements were collected at the equilibrium condition with no seepage flow and thus with a hydrostatic pressure distribution.

A very good agreement between the Ackers & White relationship and the experimental data is shown in Figure 7. It should be noted that the sediment transport relation of Ackers and White contains a driving term F_{gr} which is proportional to $(\tau_*)^{1/2}$, and thus might be expected to be sensitive to the effect of seepage in the way described above.

7 RESULTS

7.1 *Reduction of the total shear stress*

The total shear stress in the case of upward seepage flow was estimated with the Chen and Chiew (1998a) relationship, i.e. equation (7). In doing so the water surface slope and the depth-averaged velocity were computed numerically, as the relationship seems to be sensitive to small variations in experimental data. All other input data were directly measured. In Figure 8 the ratio between the shear stress in the presence of seepage τ_b and without seepage τ_o is plotted against the relative seepage velocity v_s made dimensionless with the upstream averaged velocity.

Although not shown in Figure 9, the opposite behavior was observed in the case of downward seepage flow.

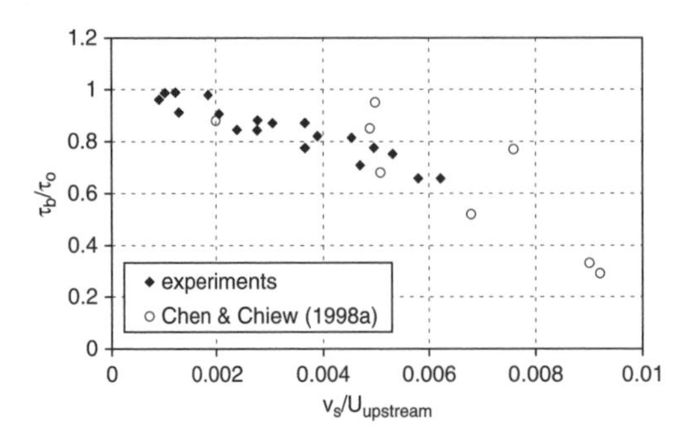

Figure 8. Reduction of the shear stress due to upward seepage.

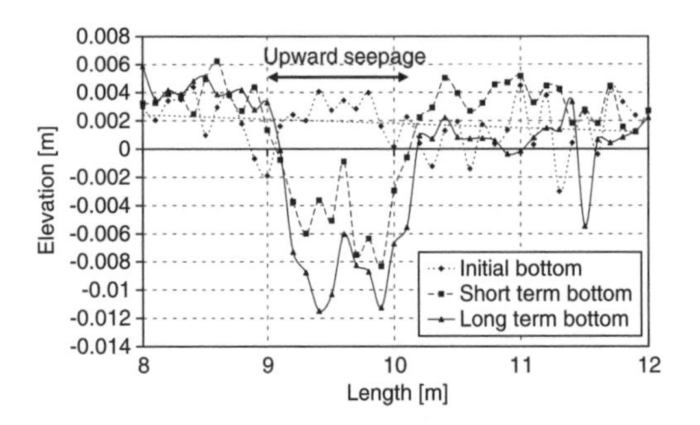

Figure 9. Local scour bed due to upward seepage (Run 1-2, $Q_w = 20.6$ l/s, $Q_{seepage} = 40$ l/m). The effect of longitudinal slope has been subtracted.

7.2 *Effect of an upward seepage flow on bed dynamics*

The bottom elevation of the bed was measured after each short and long-term experiment in order to evaluate the effect of upward seepage flow on bed dynamics. The local bed profile in the seepage zone shows a pattern of scour that increases with time, as seen in Figure 9.

The equilibrium long-term profile (Fig. 10) again shows erosion in the seepage area, as well as a pattern of general deposition upstream due to the backwater effect on the flow profile. A reduced erosion can also be observed downstream of the seepage zone.

In order to summarize the results of the experiments done to evaluate the effect of upward seepage flow on bed dynamics, the maximum values of scour depth are plotted versus the discharge ratio between the seepage flow and the main flow in the flume, i.e. $Q_{seepage}/Q_w$ in Figures 11 and 12. The maximum scour depth is evaluated as the maximum difference between the initial and the final bed, made dimensionless with the water depth. The maximum scour depth was observed to increase with the discharge ratio, and thus with higher groundwater flow.

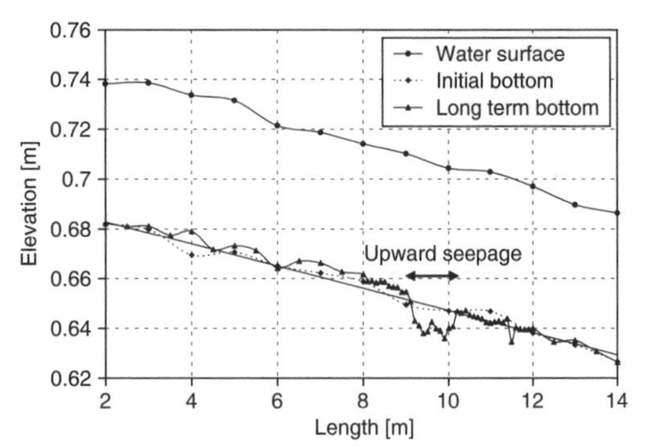

Figure 10. Equilibrium configuration due to upward seepage after a long-term experiment (Run 1-2, $Q_w = 20.6$ l/s, $Q_{seepage} = 40$ l/m).

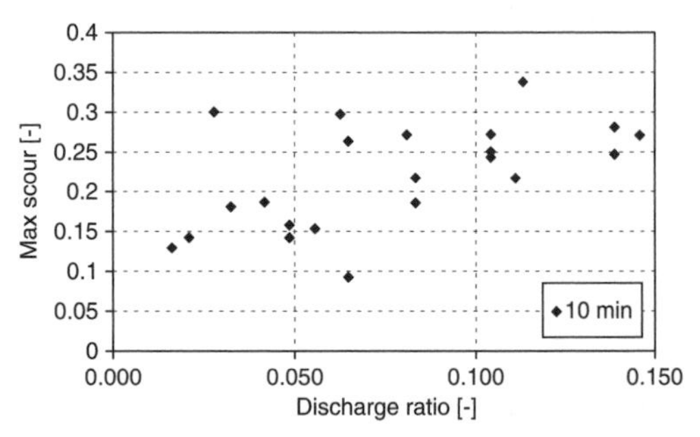

Figure 11. Maximum dimensionless scour depth plotted against discharge ratio $Q_{seepage}/Q_w$ for the short-term experiments.

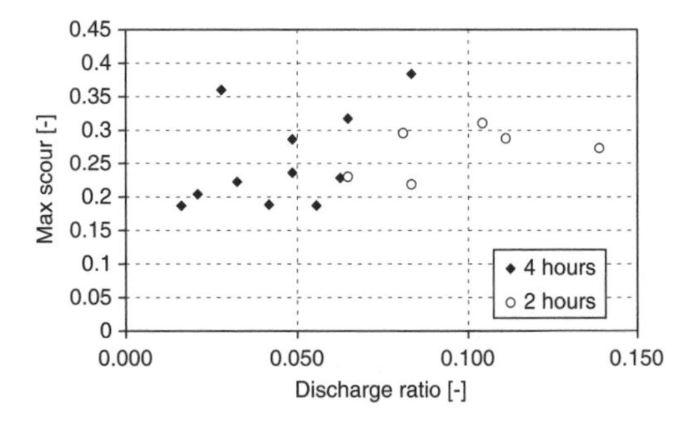

Figure 12. Maximum dimensionless scour depth plotted against discharge ratio $Q_{seepage}/Q_w$ for long-term experiments.

Some of the long-term experiments were performed for a duration of 2 hours, instead of 4 hours, to guarantee uniform flow conditions. This is because at sufficiently large values of the discharge ratio $Q_{seepage}/Q_w$ the bed would eventually fluidize in the form of migrating, upwelling boils, creating highly non-uniform conditions. The condition of a completely fluidized bed was not addressed in this research.

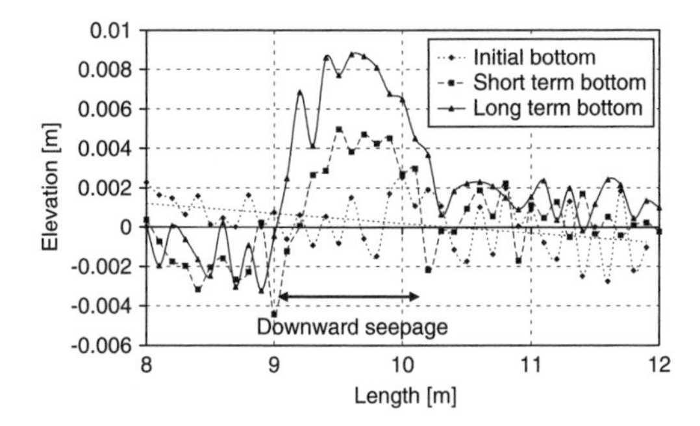

Figure 13. Local deposit bed due to downward seepage (Run 5-2, $Q_w = 16$ l/s, $Q_{seepage} = 60$ l/m).

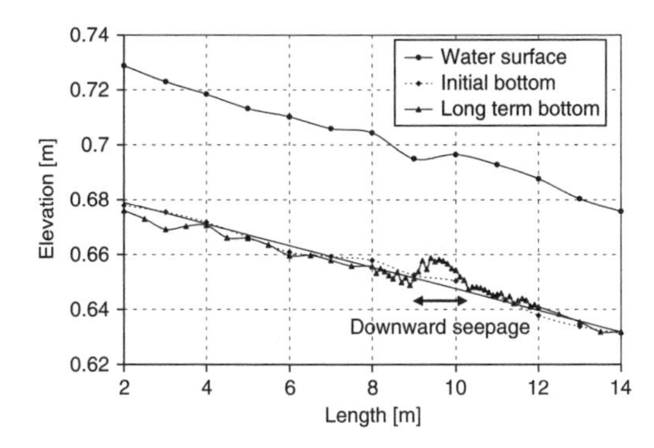

Figure 14. Equilibrium configuration due to downward seepage after a long-term experiment (Run 5-2, $Q_w = 16$ l/s $Q_{seepage} = 60$ l/m).

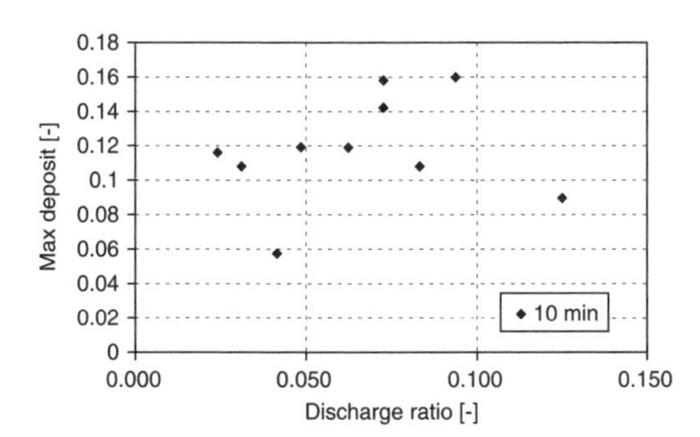

Figure 15. Maximum dimensionless deposit plotted against discharge ratio for the short-term experiments.

7.3 *Effect of downward seepage flow on bed dynamics*

The opposite effect on bed dynamics was observed for the case of downward seepage. The local bed profiles show net fill (deposition) in the area of downward suction (Fig. 13), while the long-term profiles show

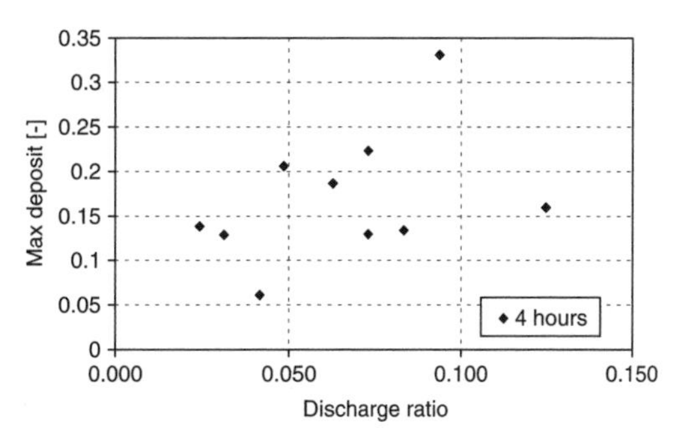

Figure 16. Maximum dimensionless deposit plotted against discharge ratio for the long-term experiments.

an increased deposit in the suction area that affects also the zone farther downstream, as well as inducing general net erosion farther upstream (Fig. 14). As before, the results are summarized in terms of maximum deposit versus discharge ratio. The maximum deposit is evaluated as the maximum difference between the final and the initial elevation of the bed, and made dimensionless with the water depth.

8 THEORETICAL MODEL

A one-dimensional morphodynamic model that considers the momentum equation, the continuity equation and the Exner equation has been implemented in order to study the effect of seepage on bed morphodynamics. The equations, written in integral form, are the following:

$$\frac{\partial H}{\partial t} + \frac{\partial UH}{\partial x} = v_s \tag{11}$$

$$\frac{\partial UH}{\partial t} + \frac{\partial U^2 H}{\partial x} = -gH\frac{\partial H}{\partial x} + gHS - \frac{\tau_b}{\rho} \tag{12}$$

$$(1-\lambda_p)\frac{\partial \eta}{\partial t} = -\frac{\partial q_t}{\partial x} \tag{13}$$

The integral continuity equation shows the presence of the seepage velocity v_s as boundary velocity at the bottom. The total shear stress and the effective shear stress for grains motion have been estimated according to equations (8) and (10). Note that the above formulation yields the relation for boundary shear stress of Chen and Chiew (1998a) for the case of steady flow.

The solid discharge q_t was evaluated with the Ackers and White (1973) relationship, based on the hydrostatic pressure assumption that agrees well with experimental data for which the hydrostatic assumption can be expected to apply.

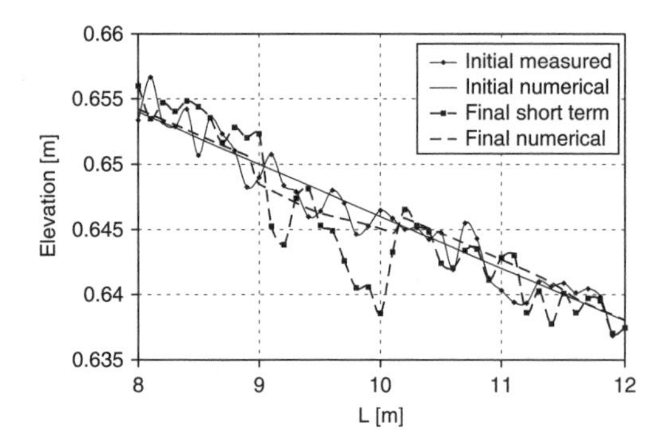

Figure 17. Local profiles of measured and numerically simulated bed elevation (Run 1-1, $Q_w = 20.6$ l/s, $Q_{seepage} = 20$ l/m).

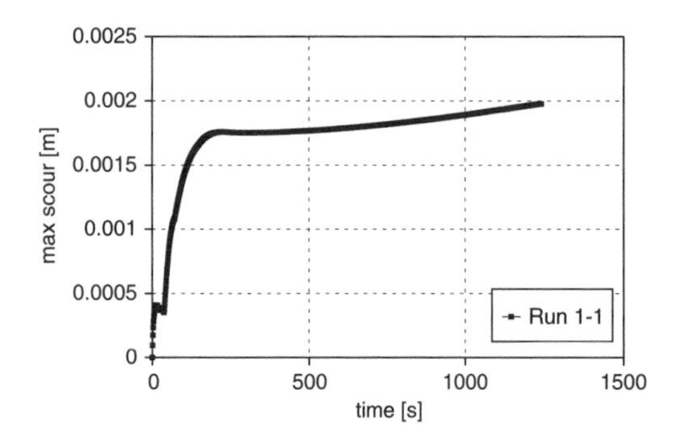

Figure 18. Predicted maximum scour depth versus time (Run 1-1, $Q_w = 20.6$ l/s, $Q_{seepage} = 20$ l/m).

9 DISCUSSION

The system of equations was solved using the quasi-steady approximation and a decoupling between flow dynamics and morphodynamics. Two input parameters for the model have been set as follows: porosity $\lambda_p = 0.3$, $K = 0.0025$ m/s.

As a preliminary result from the numerical experiments, a measured short-term bed profile was compared to the computed one (Fig. 17).

As from Figure 17, both the experiments and the model show net scour in the seepage zone.

The measured scour in the seepage area is observed to be higher then the computed one. It can be deduced that a standard approach to the evaluation of sediment transport that is known to be accurate to describe hydrostatic conditions may not be sufficient to warrant extension to the case of non-hydrostatic pressure distributions. This result is, however, preliminary.

An example of the time evolution of the maximum scour depth predicted by the numerical model is given in Figure 18. After a rapid increase at the beginning, the maximum scour increases more slowly as the experiment proceeds.

These results suggest that an appropriately modified sediment transport relationship may be needed in order to extend the range of validity to non-hydrostatic conditions.

10 CONCLUSIONS

Experimental results of a study of sediment transport over a bed with vertical groundwater flow show a clear effect of seepage, and hence of a non-hydrostatic pressure distribution, on sediment transport. In the case of upward seepage net scour was observed in the zone of seepage; the opposite effect was observed in the case of downward seepage. This result is consistent with the expected effect of seepage on the denominator of the Shields parameter.

A theoretical and numerical model has been implemented to analyze bed morphodynamics in the presence of seepage. Preliminary results from the numerical model show that a general formulation for bedload transport that has been validated in the case of a hydrostatic pressure distribution is able to capture the qualitative aspects of the effect of seepage on bed morphodynamics. The same formulation appears, however, to significantly underestimate the effect of seepage.

A future goal will be to develop a modified sediment transport relationship that does allow extension at least to the simple non-hydrostatic pressure distribution created by vertical seepage flow.

ACKNOWLEDGEMENTS

The experimental activity was carried out at St. Anthony Falls Laboratory, University of Minnesota Minneapolis, USA. The research was funded by the National Center for Earth-surface Dynamics, a Science and Technology Center of the US National Science Foundation. Special thanks is given to Benjamin Erickson for his help with the experimental set-up.

REFERENCES

Ackers, P. & White, W., 1973, Sediment transport in open channel: Ackers and White update. *Journal of Hydraulic Engineering*, 99(HY4), 1–37.

Cheng, N.H. & Chiew, Y.M., 1998a, Turbulent open-channel flow with upward seepage, *Journal of Hydraulic Research*, 36(3), 415–431.

Cheng, N.H. & Chiew, Y.M., 1998b, Modified logarithmic law for velocity distribution subjected to upward seepage, *J. Hydr. Eng.*, 124(12), 1235–1241.

Cheng, N.H. & Chiew, Y.M., 1999, Incipient sediment motion with seepage, *Journal of Hydraulic Research*, 37(5), 665–681.
Chiew, Y.M. & Parker G., 1994, Incipient sediment motion on non-horizontal slopes, *Journal of Hydraulic Research*, 32(5), 649–660.
Vanoni, V.A. & Brooks, N.H. (1957). Laboratory studies of the roughness and suspended load of streams, *Rep. E68*, Sedimentation Lab., California Institute of Technology, Pasadena, Calif.
White, W.R., Paris, E. & Bettess, R., 1980, The frictional characteristics of alluvial streams: a new approach, *Proc. of Inst. of Civil Eng.*, Part 2, Vol. 69, Sept.

River, Coastal and Estuarine Morphodynamics: RCEM 2005 – Parker & García (eds)
 ISBN 0 415 39270 5

Suspended sediment load variation in a sub-arctic watershed in Interior Alaska

H.A. Toniolo & P. Kodial
Civil and Environmental Engineering Department, University of Alaska, Fairbanks, AK, USA

ABSTRACT: A study on suspended sediment load in small streams located in Interior Alaska started last year. The research was conducted in the Caribou-Poker Creeks Research Watershed, which is located 160 km south of the Artic Circle. Permafrost is discontinuous in the basin. The research focused on streams located in sub-basins underlain by different percentages of permafrost. Two sub-basins with similar areas (5.2–5.7 km^2) were considered in the study. Permafrost distributions were 4% (sub-basin C2) and 53% (sub-basin C3). Suspended sediment concentration oscillated from 40 to 70 mg/l with a couple of peaks above 100 mg/l. Maximum and minimum discharges in the streams were 55 and 28 l/s (sub-basin C2); 129 and 4 l/s (sub-basin C3). In general, suspended sediment concentration values in the permafrost-rich area were higher than the values obtained in the sub-basin with lower percentage of permafrost. However, computed sediment load were higher in C2.

1 INTRODUCTION

Established in 1969 to better understand the sub-arctic hydrologic regimes, the Caribou Poker Creeks Research Watershed (CPCRW) is located 50 km north-east of Fairbanks, Alaska. Its coordinates are 65° 10' N and 147°30' W. The CPCRW basin is a northeast-southwest trending oval about 16 km long and 8 km wide (Fig. 1). The watershed has been divided in several sub-basins according to physical and hydrologic characteristics (Lotspeich & Slaughter 1981). Permafrost, perennially frozen ground, distribution is discontinuous in the watershed. No other research area in US presents similar characteristics.

The tributary basin of Caribou Creek was arbitrarily designated with a "C". The research focuses on two sub-basins in the Caribou Creek area, namely C2 and C3. The C2 sub-basin contains the least amount of area underlain by permafrost (4%), in the entire watershed. The watershed trends to the south, with well-drained slopes and permafrost-underlain treeless muskeg in valley bottom. The C3 sub-basin has the greater percentage of area with permafrost (53%). The watershed trends to the northeast, with black spruce/feather moss slopes and permafrost-underlain treeless muskeg in valley bottom. Sub-basins areas are 5.2 km^2 and 5.7 km^2, C2 and C3 respectively. Due to the difference in permafrost percentage, several hydrologic studies were conducted on these two watersheds simultaneously (Slaughter & Long 1974, Haugen et al. 1982, Hilgert & Slaughter 1983 and 1987, Yoshikawa & Hinzman 1999, among many others).

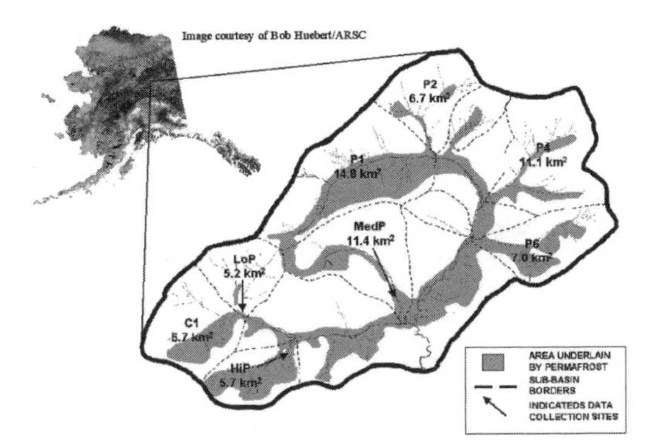

Figure 1. Caribou-Poker Creeks Research Watershed. Location and permafrost distribution (from Haugen et al. 1982). HiP and LoP represent high (C3) and low (C2) permafrost percentage.

As indicated by previous studies (see Permafrost Task Force Report 2003), permafrost plays an important role in the hydrology of sub-arctic watersheds. Ice-rich conditions at the permafrost table do not allow significant surface percolation, a condition which results in decreased response time to precipitation events, limited subsurface storage, and low base flows between precipitation inputs compared to permafrost-free areas. As a consequence, the resultant stream flow in permafrost-underlain terrain is much more responsive to precipitation than is permafrost-free terrain (Bolton et al. 2000). Sediment load is conditioned by stream flows and sediment availability. Therefore, permafrost also influences sediment transport in rivers. This article presents initial results on suspended sediment load in two streams located in areas with different percentages of permafrost.

2 METHODOLOGY

This ongoing research involves not only field work but also laboratory work. Field work consists of deploying instruments in the streams (i.e.: autosamplers, pressure transducers, and dataloggers); frequently velocity measurements and sample collection. Laboratory work mainly involves the calculation of suspended sediment concentration.

Autosamplers ISCO, model 3700, were deployed during summer of 2004 in the streams. Water samplers were collected every 6 hours. Four consecutive samples were placed in a single bottle. Thus, an integrated daily sample was obtained from each bottle in the autosamplers. Samples from the samplers were collected every two weeks. A plastic hose (¾ inches interior diameter) was used to take the water from the stream. The hose was located in the middle of the stream at approximately 10 cm above the stream bed.

Water velocity was measured in several verticals along the channel width. Spacing between verticals was set equal to 5 cm. A portable flowmeter Flo-mate made by Marsh-McBirney, INC was used to measure flow velocity. Stream depths were below bankfull depth in all velocity measurements taken during 2004. Discharge was estimated from the available data (i.e.: velocity and area defined between consecutive verticals). A water depth – water discharge relationship for each section was developed at the end of the summer season. Figure 2 shows the data points and fitted linear function obtained for the study site located in C2. The graph indicates that discharges for water depths smaller than bankfull depth can be adequately estimated by the equation.

Pressure transducers connected to dataloggers CR10X made by Campbell Scientific were deployed in each stream location. Near-continuous water depth data (every 15 minutes) provided by the transducers along with the fitted equation for the elevation-discharge relationship obtained from velocity measurements in the field were used to estimate the daily water discharge variation in the studied streams. Figure 3 shows the discharge variation in sub-basin C2 and C3.

The collected water samples were analyzed in laboratory at the Water and Environmental Research Center (WERC), University of Alaska Fairbanks (UAF) to determine the suspended sediment concentration. Suspended sediment concentration ranged from 37 to 129 mg/l and from 45 to 80 mg/l for C2 and C3 respectively. Figure 4 shows the calculated concentrations for

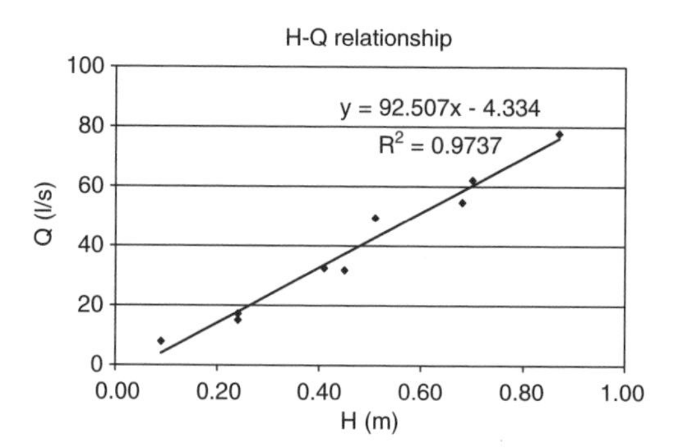

Figure 2. Water depth – water discharge relation obtained from velocity measurements in C2.

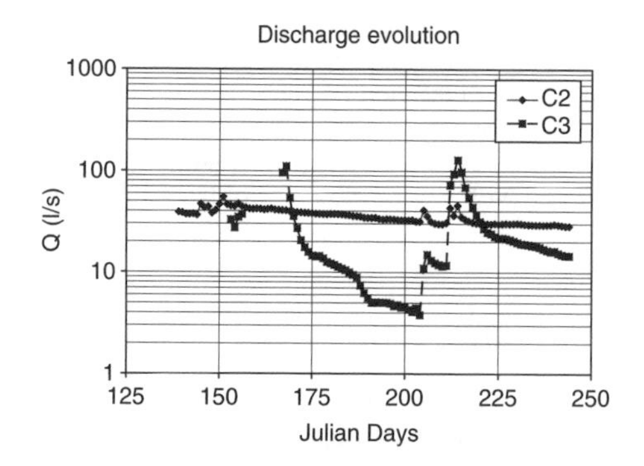

Figure 3. Flow discharge evolution during 2004 in streams located in both sub-basins: C2 and C3.

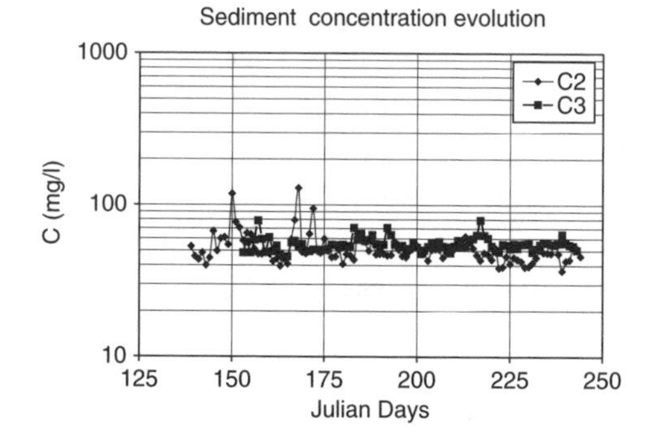

Figure 4. Sediment concentration variation during 2004 in streams located in sub-basins C2 and C3.

the 2004 summer season. Suspended sediment load was estimated through the combination of water discharge and sediment concentration. This approach is in some way limited because the sediment concentration is considered equally distributed in the cross-sectional area. However, due to the geometric channel dimensions, it is estimated that the results are representatives

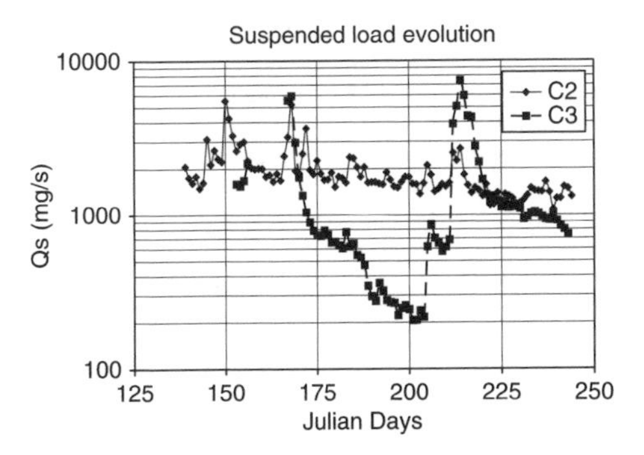

Figure 5. Suspended sediment load variation during 2004 in streams located in sub-basins C2 and C3.

of the sediment transport in both creeks. Figure 5 illustrates the sediment load in both streams.

3 RESULTS

Instruments were deployed in the field on May 18, 2004 and retrieved on August 31, 2004. In spite of early efforts in the deployment of the instruments, no data were continuously collected in C3 before the second part of June. Continuous data for both streams were available after that initial troublesome period.

Estimated discharge variation shows that the stream located in C2 is less responsive to rainfalls than the stream placed in C3. This is originated in the difference of underlain permafrost in both sub-basins. Areas with high permafrost (C3) present low groundwater storage due to the location of the impermeable boundary formed by the permafrost close to the surface. Thus, baseflows are very small in C3. On the contrary, areas with low percentage of permafrost (C2) have high groundwater storage and consequently, sustained baseflows. The dry conditions recorded during last summer allowed a clear visualization of these facts. Flow data agree with previous published results in the same area (Bolton et al. 2000, 2004). Maximum and minimum discharges in the streams were 55 l/s and 28 l/s (sub-basin C2); 129 l/s and 4 l/s (sub-basin C3). Two peaks were well defined in the case of C3 sub-basin. Both of them were originated by two rainfall events in the watershed recorded by an existent weather station.

Suspended sediment concentration evolution indicated an increasing trend in the concentration calculated for C3 as the summer progressed. It may be an indication that more sediment was available to be entrained by the flows as the thawing of top soil layer (active layer) increases. Values for C2 were extremely fluctuant during the first part of the warm season. In general, the level of fluctuation in C2 was higher than in C3. As time passed, concentration in the stream decreased.

Sediment load in C2 was consistently above load estimated for C3. These results could indicate higher sediment production in areas with low percentage of permafrost. Nevertheless, there were two peaks in C3. They occurred on June 15 and on August 1, respectively. The peaks were driven by the increase in discharge in the streams. Rainfalls were recorded in the weather station.

4 CONCLUSIONS

Preliminary results of an ongoing research on suspended sediment load in a sub-artic research watershed were presented in this article. The study was carried in two sub-basins of the Caribou Creek. The study watersheds have similar areas and different percentage of permafrost.

Discharge, suspended sediment concentration, and suspended sediment load were calculated for the studied streams. Discharge variation in C2 and C3 clearly shows the effect of permafrost in the stream flows. Daily suspended sediment concentration values oscillated between 40 mg/l and 70 mg/l in both streams. A couple of peaks above 100 mg/l were calculated in C2. Apparently, sediment load in C3 is dominated by rainfall events (i.e.: high water discharge peaks). It seems that water discharge and sediment concentration are equally important for suspended sediment load in C2. Available data tend to indicate that changes in sediment concentrations are dominant in early spring/summer and discharges are leading the modifications in late summer. These considerations deserve more investigation in a later stage.

Future work involves the continuation of both field and laboratory work during 2005. New data (2004 and 2005) will be compared with available sediment load data obtained in late '70s.

ACKNOWLEDGMENTS

Authors are thankful to Professor Jeremy Jones (Biology Department, University of Alaska Fairbanks) who provided the autosamplers used in the field. USGS through a NIWR grant provided funds to support Mr. Kodial.

REFERENCES

Bolton, W., Hinzman, L. & Yoshikawa, K. 2000. Stream flow studies in a watershed underlain by discontinuous permafrost. *Proceedings AWRA Water Resources in Extreme Environments*, Anchorage Alaska: 31–36.

Bolton, W., Hinzman, L. & Yoshikawa, K. 2004. Water balance dynamics of three small catchments in a Sub-Arctic boreal forest. In D. Kane and D. Yang (eds), *Northern*

Research Basins Water Balance, IAHS Series of Proceedings and Reports. No 290: 213–223.
Haugen, R. K., Slaughter, C. W., Howe, K. E. & Dingman, S. 1982. Hydrology and climatology of the Caribou-Poker Creek Research Watershed, Alaska. *CRREL Report 82-26.* US Army Cold Regions Research and Engineering Laboratory, Hanover: 42 p.
Hilgert, J. W. & Slaughter, C. W. 1983. Water quality and streamflow in the Caribou-Poker Creeks Research Watershed, central Alaska, 1978. *Res. Note PNW-405.* Portland, OR: U.S. Department of Agriculture, Forest Service, Pacific Northwest Forest and Range Experiment Station: 36 p.
Hilgert, J. W. & Slaughter, C. W. 1987. Water quality and streamflow in the Caribou-Poker Creeks Research Watershed, central Alaska, 1979. *Res. Note PNW-RN-463.* Portland, OR: U.S. Department of Agriculture, Forest Service, Pacific Northwest Research Station: 34 p.
Lotspeich, F. B. & Slaughter, C. W. 1981. Preliminary results of a study on the structure and functioning of a taiga research watershed. *AERS Working Paper 30 (CERL 014).* College, AK: U.S. Environmental Protection Agency, Arctic Environment Research Station: 41 p.
Slaughter, C. & Long, L. 1974. Upland climatic parameters on sub-arctic slopes, central Alaska. *Proceedings 24th Alaska science conference:* 276–280.
U.S. Arctic Research Commission Permafrost Task Force. 2003. Climate Change, Permafrost, and Impacts on Civil Infrastructure. *Special Report 01-03*, U.S. Arctic Research Commission. Arlington: Virginia.
Yoshikawa, K. & Hinzman, L. D. 1999. The installation of long-term ground temperature monitoring and hydrological well sites in the Caribou-Poker Creeks Research Watershed. *Water and Environmental Research Center technical Report* 75: 30 p.

River, Coastal and Estuarine Morphodynamics: RCEM 2005 – Parker & García (eds)
© 2006 Taylor & Francis Group, London, ISBN 0 415 39270 5

Estimation of suspended sand concentrations with an ADP, Colastiné river – Argentina

R.N. Szupiany & M.L. Amsler
Consejo Nacional de Investigaciones Científicas y Técnicas (CONICET), Argentina
Facultad de Ingeniería y Ciencias Hídricas, UNL, Argentina

J.J. Fedele
Saint Anthony Falls Laboratory, University of Minnesota, Minneapolis, USA

ABSTRACT: The standard procedures used for sampling suspended sands in large streams require the vessel to be anchored at several verticals during a certain period of time. These methods are usually time-consuming, rather expensive, and have a limited spatial resolution. Acoustic Doppler technology has been recognized as a potential tool that can give quantitative information on suspended sediment concentrations through the analysis of the intensity of the backscattering strength. In this paper the backscatter intensity of an ADP and measured suspended sand concentrations obtained using a classic depth-integrating sampler are correlated with fairly good results when compared with the theoretical straight line slope suggested by the equipment manufacturer. The size of the suspended particles was a relevant factor that affected the consistency of results since the backscatter intensity is strongly influenced by the coarser particles carried in suspension (sands). Further recommendations are given to appraise the results extent.

1 INTRODUCTION

Spatial and temporal variations of bed suspended sediments in braided rivers have been shown to have important morphologic consequences such as node and bar formation and stability (Bristow & Best 1993). Therefore, reliable field measurements of suspended sediment are essential for better understanding the relations between flow structures, suspended particles and channel geometry.

The standard procedures used for sampling suspended sediment in large streams like the Paraná River, in Argentina (mean discharge of 19500 $m^3 s^{-1}$, widths between 600–2500 m and usual depths between 5 m up to 25 m) require point or depth-integrating samplers and the vessel to be anchored at several verticals during a certain period of time. These methods are reliable for obtaining depth-averaged concentrations but are usually time-consuming, rather expensive, and have a limited spatial resolution. Furthermore, in large waterways like the Paraná River, maintaining the vessel at a fixed position to ensure accurate measurements may be difficult and sometimes dangerous.

Acoustic Doppler technology used to measure accurate water discharge values at large streams in recent years, has been also recognized as a potential tool that can give quantitative information on suspended sediment concentrations through the analysis of the intensity of the backscatter strength. Two approaches were followed to solve the problem. One of them is based on the equations of the acoustic theory which describes the sound propagation in water (Thorne et al. 1994, Holdaway et al. 1999, Thorne & Hanes 2002). The other focuses in the sonar equation to relate changes in the backscatter intensity with particles concentrations variations measured in field with standard equipment or optical instruments (Creed et al. 2001, Deines 1999, Filizola & Guyot 2004, Kostaschuk et al. 2005, Poerbandono & Mayerle 2002, Nortek 2001, Sontek 1996, Sontek 1997, Thevenot & Kraus 1993).

Although these authors have reported encouraging results, the use of acoustic devices to accurately measure suspended sediment concentrations, still requires further testing concerning the influence on the backscatter intensity of the quality and non-uniform sizes of suspended particles. In principle, it is well known that in sand-bed rivers the suspended load is composed by the washload (the finer fraction, normally silts and clays) and the suspended bed material (sands). It is also widely accepted the central role played by the coarser fractions in the morphodynamic processes operating within the stream channel.

In this paper we present a correlation between the backscatter intensities of an ADP of 1000 kHz and measured suspended sand concentrations obtained using a depth-integrating sampler. The simplified sonar equation was used to correct the data for the effects of geometric spreading and absorption. Measurements were performed in the Colastiné River, an important secondary channel of the Paraná fluvial system in its middle reach. The characteristic acoustic wave length of the ADP we used (~0.47 mm) when compared with the usual sizes of the suspended sediment particles in the Colastiné River, would explain largely the fairly good resulting regression.

2 ESTIMATION OF SUSPENDED SEDIMENT CONCENTRATIONS WITH AN ADP – BASIC PRINCIPLES

The acoustic Doppler profiler (ADP) used in the experiments described herein, operates based on the same principles as all the instruments of this type commercially available and has similar restrictions. Thus only some features of it, considered of interest to appraise properly the extent of the results, are given next.

The equipment is manufactured by Sontek. It has a frequency of 1000 kHz (12 pings per second) and 3 divergent monostatic transducer (beams) oriented 25° respected the vertical and spaced 120° each other (Sontek 2000, Fig. 1). The cell size and the averaging interval or ensemble may usually be varied between 0.5 m to 1 m and 5 s to 30 s, respectively.

As the pings transmitted by the traducers of a given ADP move across the core of an alluvial stream, any suspended particle (of sediments, air bubbles or particulate organic matter) will scatter a certain part of the sound energy. Thus the backscatter intensity will be related with the number of particles (i.e. with the concentrations) present in the water column. Based on this principle the suspended concentrations could be measured with an accurate spatial (depending on the selected cell size) and timed resolution.

The backscatter intensity will be a function of the equipment characteristics (frequency, transmitted power, measured volume range, receive sensitivity) and water conditions (concentration and size of sediment particles, quantity of organic matter, dissolved solids and air bubbles) (Sontek 1997). Therefore, for a given instrument and assuming a constant sediment type and size distribution, absence of air bubbles or particulate organic matter, the signal strength will be directly proportional to sediment concentration.

The backscatter intensity may be computed by the simplified sonar equation (Sontek 1997):

$$EL = SL + 10 * \log_{10}(PL) - 20 * \log_{10}(R) - 2 * \alpha * R + Sv + RS$$

where EL is the signal intensity as measured by the instrument; PL, SL and RS are parameters whose values depend solely of the equipment characteristics; R ($= d/\cos\theta$, d: water depth, $\theta = 25°$) is the transducer-measured volume distance; $\alpha (\approx 0.26\,\text{dB m}^{-1}$ for an

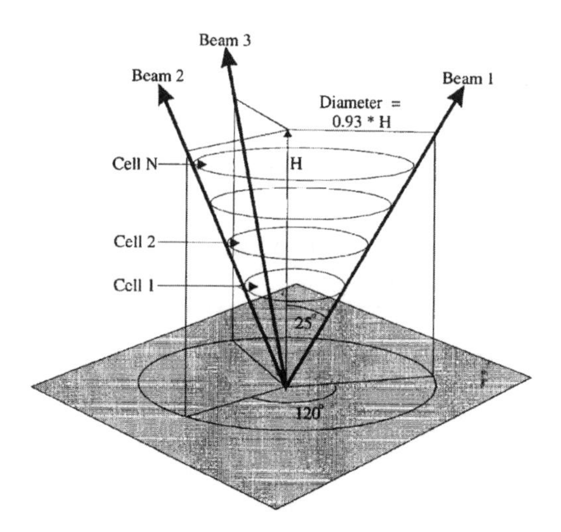

Figure 1. Beam geometry of an ADP (Sontek 2000).

ADP of 1000 kHz and negligible water salinity) is the sound absorption coefficient and, Sv, is the volume scattering strength (dB). In the Sontek's ADP the intensity is measured in internal logarithmic units named "count" (1 count = 0.43 dB).

This equation suggests that comparisons between different types of equipments would not be possible since each one requires its own calibration process to set the values of their specific parameters (SL, etc.). Thus the best way to quantify the suspended sediment would be a relationship obtained directly between concentrations measured in situ with standard methods, and the corresponding backscatter intensities. These intensities should be corrected (see the equation) due to the decay effects of the geometric spreading ($-20 \log_{10} R$) and absorption ($-2 \times \alpha \times R$). The procedure also requires negligible quantities of organic matter and a specific cell size since it affects the length of the acoustic pulse, PL (a longer pulse results in larger signal strengths).

Finally, there exists a backscatter sensitivity depending on the suspended particle size for a given concentration. The simplified relationship as given by Sontek (1997) is shown in Figure 2. Here, k is the wave number ($= 2\pi/\lambda$, λ = acoustic wave length) and a is the particle radius. It is seen that the maximum sensitivity stands when λ is similar to the particle perimeter ($k \times a \approx 1$). As sensitivity is intimately related with λ, instruments with different frequencies will have different sensitivities. In the case of the Sontek's ADP of 1000 kHz, it will have the largest sensitivity for particle radius of 235 μm. Particles with a $<\sim 12$ μm would not be detected ($k \times a < \sim 0.05$; Sontek 1997).

Under these conditions and for a given cell size, the relationship between the backscatter intensities versus $10 \log(C)$, will be a straight line with a slope

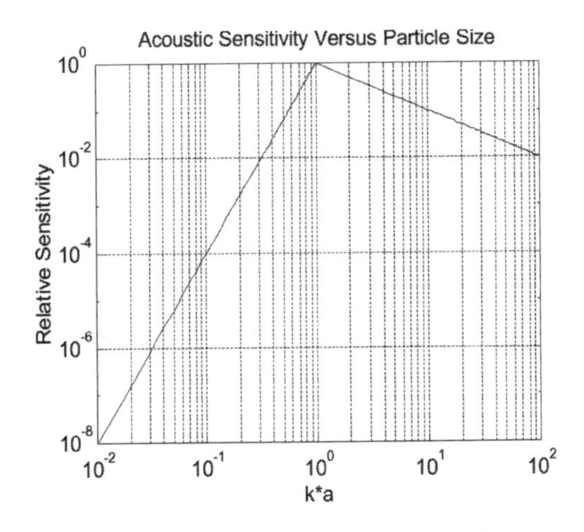

Figure 2. Sensitivity and particle size (Sontek 1997).

near 1 (Sontek 1997). Concentration values larger than ~500 mg l^{-1} would increase the absorption of the backscatter intensity introducing an upper limit for the relationship validity.

3 DESCRIPTION OF FIELD EXPERIMENTS

The field experiments were performed on September 10, 2004, at a cross section of the Colastiné River, an important secondary sand bed channel of the Paraná River floodplain between Santa Fe and Entre Ríos provinces, Argentina (Fig. 3). The river discharge was 1460 $m^3 s^{-1}$. The study cross section was 350 m width with depths up to 8 m. The suspended sediment is composed largely by washload (65–95 % depending on the annual period) and quartzose sands. The bottom sand had a mean diameter of 320 μm and a $\sigma_g \approx 1.15$ (Ramonell et al. 2000). The washload has 6% of particles between 16–31 μm, 60.8% between 8–16 μm and 33.2% between 4–8 μm (Drago & Amsler 1988, Mangini et al. 2003).

Depth-integrated samples were obtained at 8 verticals with a depth – integrating sampler. Simultaneously fixed-vessel measurements during 7 minutes were made with the ADP described above. A cell size of 0.75 m and an averaging interval of 10 sec. were adopted. A DGPS coupled with a Raytheon echosounder enabled to position the vessel at each vertical.

These verticals were located across the section according to the velocity distribution trying to get as a wide sand concentration range as possible for the measured river stage (Fig. 4).

Concentrations values of suspended sand (C_s), washload (C_w), particle organic matter (C_{mo}) and dissolved solids (C_{ds}) were obtained by means of standard methods of analysis (Table 1).

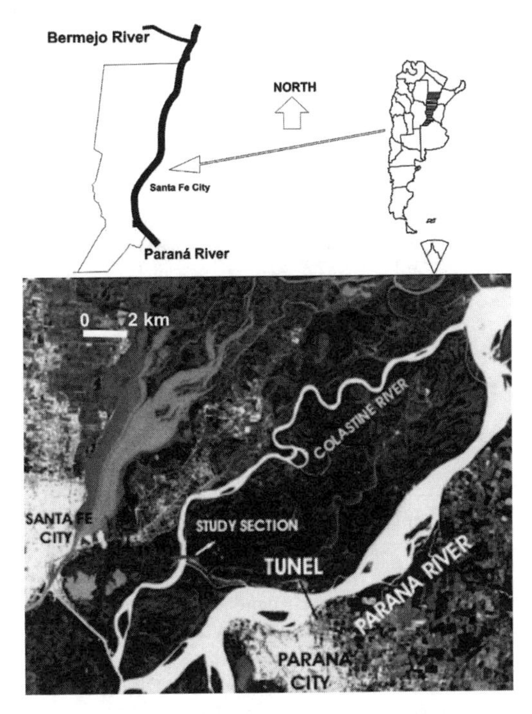

Figure 3. Study site.

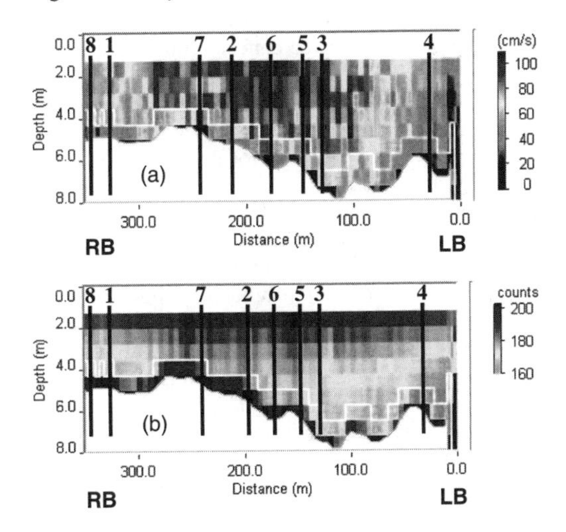

Figure 4. Cross section (Colastiné River). (a) Horizontal velocity field. (b) Backscatter intensities.

4 RESULTS

The horizontal velocity and the backscatter intensity fields together with the measuring verticals, are presented in Figures 4a, b. Although the backscatter intensities are not corrected due to the decay effects, it is clearly seen how their maximum values

Table 1. Concentration values of suspended material. Colastiné River.

Vertical	Depth m	Cs mg l^{-1}	Cw mg l^{-1}	Cmo mg l^{-1}	Cds mg l^{-1}
1	5.0	9.5	–	–	–
	5.0	8.6	–	–	–
	5.0	9.8	109.9	0.00	72.0
2	6.1	31.1	–	–	–
	6.1	32.3	–	0.41	65.2
3	7.5	31.9	–	–	–
4	7.2	20.5	140.9	0.88	62.8
5	7.3	29.0	–	–	–
6	6.4	35.7	132.6	–	–
7	4.8	20.1	123.7	–	–
8	4.7	7.0	91.4	–	–

are correlated with the high velocities flow regions, i.e. where the suspended bed material is transported at highest rates (between ~100 m and ~280 m from left bank). The largest suspended sand concentrations were measured in the verticals located across that band of the section (Table 1).

The relationship between sand concentrations and backscatter intensities is presented in Figure 5a. The backscatter intensities of each cell for a given vertical were averaged in order to correlate with the corresponding depth-integrated concentration measured at that site.

In Figure 5b available data of suspended sand concentrations measured at 4 sections of the Bermejo River (Table 2; see Fig. 3 for location) were included together with those of the Colastiné experiments. The Bermejo data were obtained with the same equipments already described on April 20, 2004 (FICH 2004). Most of the wash load transported by the Paraná River in its middle and lower reaches is supplied by the Bermejo River (Drago & Amsler 1988, Alarcon et al. 2003). At the moment of the measurements it had a discharge of 800 $m^3 s^{-1}$. The bed sands have a mean diameter of 120 μm with $\sigma_g = 1.7$.

Note that the Bermejo data arrange properly around a relationship very similar to that found for the Colastiné River but with a considerably larger scatter. Reasons to explain this increased dispersion could be:

- The backscatter intensities were recorded from a single moving-vessel transect at each cross section with the boat passing near the concentration measuring verticals.
- The large concentrations of the Bermejo River (>500 mg l^{-1}) which increase the decay effects due to absorption making difficult the backscatter correction.
- A reduced sensitivity consequence of the Bermejo River sand sizes, nearly four times smaller than the optimum particle radius.

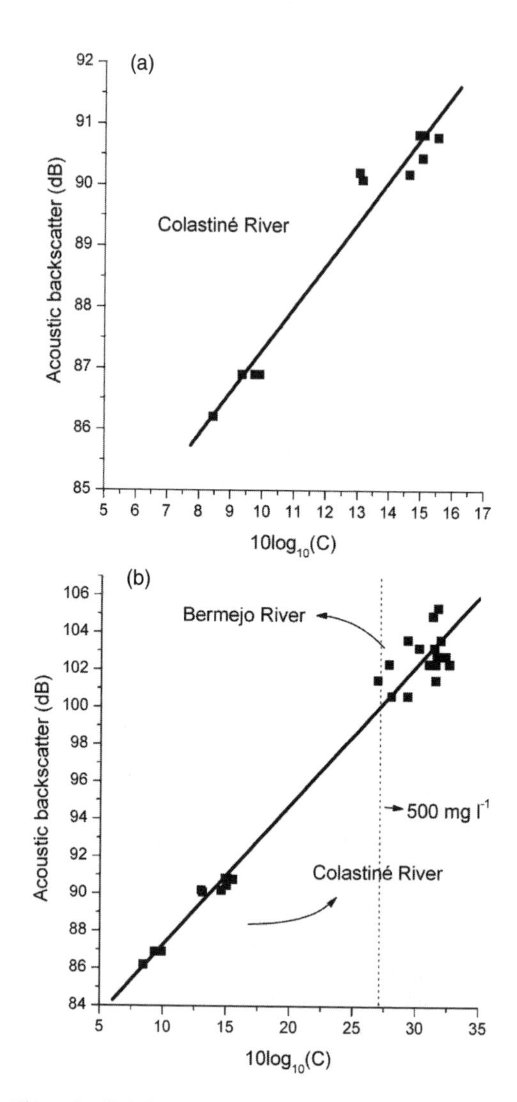

Figure 5. Relationship between corrected ADP backscatter and measured suspended sand concentrations. (a) Colastiné River (regression slope: 0.7; $R^2 = 0.96$), (b) Colastiné and Bermejo Rivers (regression slope: 0.75; $R^2 = 0.98$).

5 CONCLUDING REMARKS

The water characteristics of the Colastiné River (low particle organic matter and dissolved solids concentrations) together with the quantity and texture of its suspended sediments, help to explain the fairly good results obtained when the backscatter intensities of the Sontek's ADP are related with measured suspended sand concentrations. Similar conditions are usual in many secondary channels of the Paraná River floodplain and in the main channel itself.

Table 2. Concentration values of suspended material. Bermejo River.

Section	Vertical	Depth m	Cs mg l^{-1}	Cw mg l^{-1}
1	1	4.0	1484.7	10135.9
	2	3.2	1403.0	10323.3
	3	2.6	1698.9	9892.5
	4	2.4	1834.5	8205.4
2	1	3.4	1375.6	10334.7
	2	3.1	1352.2	10571.9
	3	2.8	1546.8	10734.7
	4	2.6	1404.6	10238.0
3	1	3.9	1438.0	9091.3
	2	3.1	1252.1	9682.7
	3	2.5	856.9	9120.6
	4	2.3	487.0	8469.5
4	1	3.8	857.9	9442.9
	2	3.0	1051.3	9487.5
	3	2.4	607.6	9465.8
	4	1.9	631.6	9470.1

The regression curve slope (0.7) is 30% lower than the theoretical suggested by Sontek (1997) a fact that, in principle, may be attributable to the representative sizes of suspended sands transported by the Colastiné River (in the order of 300 μm), smaller than the optimum diameter (470 μm) to obtain the maximum sensitivity to the backscatter intensity.

The inclusion of the Bermejo River data with still smaller sand particles diameters, concentrations larger than the limit suggested by Sontek (1997) and did not measure and treat with procedures as detailed as those performed in the Colastiné experiments, arranged essentially following the Colastiné relationship though with a higher dispersion.

A general encouraging conclusion would arise since, in spite of differences in the suspended sand particles, sizes and concentrations values, a consistent relationship with prediction purposes could be adjusted complementing properly the equipment performance limits with careful field measurements. The last should include tests comparing backscatter intensities obtained from fixed- and moving-vessel measurements together with the averaging intervals of these intensities best correlated with the mean suspended sand concentrations. For the moment each model of ADP requires a calibration procedure of this type.

Extreme care should be taken in the field measurements and subsequent laboratory analysis to fit the relationship of Figure 5 due to the logarithmic x-coordinate. Note that rather small differences in the backscatter intensity may imply large variations in suspended sand concentrations.

Future field measurements are planned in the Paraná River itself, in order to increase the number of data points and to check the consistency of the relationships of Figure 5.

REFERENCES

Alarcón J., Szupiany N., Montagnini D, Gaudin H., Prendes H. & Amsler M. 2003. Evaluación del transporte de sedimentos en el tramo medio del Río Paraná. *Primer Simposio Regional sobre Hidráulica de Ríos*. Buenos Aires, Argentina.

Bristow C. S. & Best J. L. 1993. *Braided Rivers: Perspectives and Problems*. In Braided Rivers, Best and Bristow (eds), Geological Society Special Publication, 75, 1–12.

Creed E. L, Pence A. M. & Rankin K. L. 2001. Inter-Comparison of Turbidity and Sediment Concentration Measurement from an ADP, an ABS-3, and a LISST. *Oceans 2001 MTS/IEEE Conference Proceedings*, Honolulu, HI. Vol. 3, pp. 1750–1754.

Drago E. C. & Amsler M. L. 1988. Suspended sediment at a cross section of the Middle Paraná River: concentration, granulometry and influence of the main tributaries. Sediment Budgets. *Proc. of the Porto Alegre Symposium*, December, IAHS Publ. No. 174.

Deines K. L. 1999. Backscatter estimation using broadband acoustic Doppler Current Profilers. *IEEE Conference*. San Diego. California.

FICH – Facultad de Ingeniería y Ciencias Hídricas. 2004. Mediciones hidrológicas y sedimentológicas en un tramo del río Bermejo en la zona de la ciudad de El Colorado provincia de Formosa. *Direccion Provincial de Vialidad*. Provincia de Formosa.

Filizola N. & Guyot J. L. 2004. The use of Doppler technology for suspended sediment discharge determination in the River Amazon. *Hydrological Sciences – Journal-des Sciences Hydrologiques*, 46(1).

Holdaway G. P., Thorne P. D., Flatt D., Jones S. E. & Prandle D. 1999. Comparison between ADCP and transmissometer measurement of suspended sediment concentration. *Contineltal Shelf Research*, 19, 421–441.

Kostaschuk R., Best J., Villard P., Peakall J. & Franklin M. 2005. Measurement flow velocity and sediment transport with an acoustic Doppler current profiler. *Geomorphology*, in press.

Mangini S. P., Prendes H. H., Amsler M. L. & Huespe J. 2003. Importancia de la floculación en la sedimentación de la carga de lavado en ambientes del río Paraná, Argentina. *Ingeniería Hidráulica en México*, Vol. XVIII, núm. 3, II Época.

Nortek 2001. *Monitoring Sediment Concentration with acoustic backscattering Instrument*. Nortek Technical Note, N° 003, Author: Atle Lohrmann.

Poerbandono & Mayerle R. 2002. Preliminary Result on the Estimation of Suspended Sediment Concentration from Acoustical Profilers. ADCP. *Anwender Workshop*, Warnemunde, Januar.

Ramonell C. G., Amsler M. L. & Toniolo H. 2000. *Geomorfología del cauce principal*. In: El Río Paraná en su tramo medio. Contribución al conocimiento y prácticas ingenieriles en un gran río de llanura. Centro de Publicaciones, U.N.L., Santa Fe, Argentina.Tomo 1, Cap. 4: pp. 173–233.

Sontek 1996. ADP Versatility in San Felipe, Mexico Deployment In: www.sontek.com.
Sontek 1997. Sontek Doppler Current Meters – Using Signal Strength Data to Monitor Suspended Sediment Concentration. Sontek Application Note.
Sontek 2000. ADP Acoustic Doppler Profile. Technical Documentation.
Thevenot M. M. & Kraus N. C. 1993. Comparison of acoustical and optical measurement of suspended material in the Chesapeak Estuary. *J. Marine Env. Eng*, Vol. 1, Gordon and Breach Science Publishers, pp. 65–79.
Thorne P. D., Hardcastle P. J., Flatt D. & Humphery 1994. On the use of Acoustics for measuring shallow water suspended sediment processes. *IEEE Journal of Oceanic Engineering*. Vol. 19, N° 1.
Thorne P. D. & Hanes D. M. 2002. A review of acoustic measurement of small-scale sediment processes. *Contineltal Shelf Research*, 22, 603–632.

River, Coastal and Estuarine Morphodynamics: RCEM 2005 – Parker & García (eds)
 ISBN 0 415 39270 5

Paradoxical discussions on sediment transport formulas

S. Egashira & T. Itoh
Department of Civil and Environmental Systems Engineering, Ritsumeikan University, Shiga, Kusatsu, Japan

ABSTRACT: One of classic study topics is chosen for discussions. It is well known that bed load formulas which are proposed using Bagnold's idea and associated method are characterized by $\tau_*^{3/2}$. Structure of such formulas are analyzed by separating sediment volume and average velocity from bed-load layer. Results suggest that the sediment volume of bed-load layer increases linearly with bed shear stress and the average velocity is determined using the velocity profile over rigid beds, which yields a specific form of $\tau_*^{3/2}$. Such treatment seems unreasonable from a view point of dynamics of continuum flow field. To overcome such kind of treatment, the authors introduced a method proposed by Egashira et al. (1997) into present topics, and proposed a very different formula for bed-load estimation.

1 INTRODUCTION

One of classic research topics is chosen for discussions. A large number of bed load formulas have been proposed since Du Boys' equation (1879) was published, and parts of them such as Einstein's and Bagnold's equations play an important role on the progress of associated study field (Bagnold 1957, Einstein 1950).

Bagnold's formula may have stimulated many researchers to study mechanics of sediment transportation and yields several refined formulas such as Ashida & Michiue's formula (1972). The sediment transport equations developed by using Bagnold's idea predict bed-load rate that is proportional to $\tau_*^{3/2}$ in the region where bed shear stress, τ_*, is large enough for the critical shear stress of bed-load initiation.

In formulating the bed-load equations, generally they employ the dynamic principles of a continuum media composed of granular material as well as of mass point system. Most studies show that the thickness of bed-load layer or sediment volume is proportional to bed shear stress and its average velocity is proportional to shear velocity, which yields a form of $\tau_*^{3/2}$ in bed-load rate.

Present study describes how they derive a form of $\tau_*^{3/2}$ and, in addition, show that a different form, $\tau_*^{5/2}$, is obtained if we employ the same dynamic principle in formulating both of the bed-load layer thickness and its average velocity.

2 MEANING OF BED-LOAD FORMULA CHARACTERIZED BY $\tau_*^{3/2}$

2.1 *Governing equations*

A uni-directional, uniform and sediment laden flow over erodible bed is shown in Figure 1. Sediment particles can be dispersed in whole flow area when the bed slope is over some critical value. If sediment particle size is coarse so that sediment cannot be suspended due to turbulent diffusion, the upper clear water flow zone will separate from water-sediment mixed flow with the decrease of bed slope, and a bed-load layer will be formed clearly. Finally, a pure water flow is formed over a rigid bed if bed slope decreases so that all of sediment particles cannot be transported.

Referring to the axis of Figure 1, x- and z-components of momentum conservation equations are described for the mean flow, respectively.

$$\rho_m g \sin\theta + \frac{\partial \tau}{\partial z} = 0 \tag{1}$$

$$\rho_m g \cos\theta + \frac{\partial p}{\partial z} = 0 \tag{2}$$

in which ρ_m = mass density of water-sediment mixture; g = acceleration due to gravity; θ = bed slope; p = hydrostatic pressure; and τ = shear stress. In

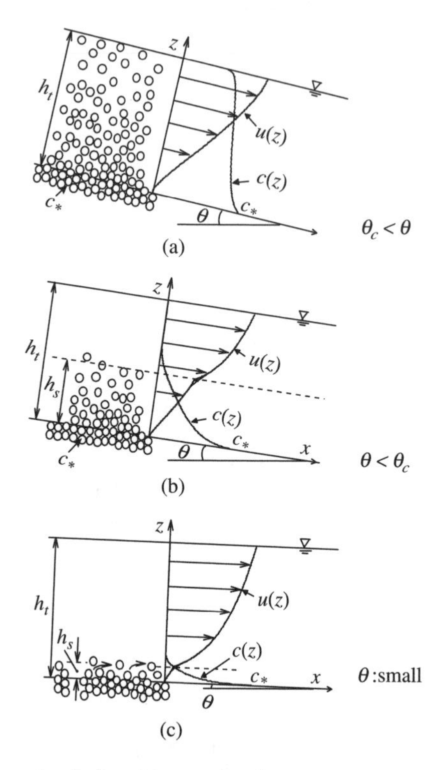

Figure 1. Sediment transport mode.

addition, supposing that the mixture is incompressible, energy conservation equation reduces to

$$\rho_m g \sin\theta u + \frac{\partial \tau u}{\partial z} - \Phi = 0 \tag{3}$$

in which u = local velocity and Φ = energy dissipation rate described by

$$\Phi = \tau \frac{\partial u}{\partial z} \tag{4}$$

The mass density of water-sediment mixture is given by

$$\rho_m = \sigma c + (1-c)\rho \tag{5}$$

in which σ = mass density of sediment particles; ρ = mass density of water; and c = sediment concentration by volume.

2.2 *Formulation of bed-load formula with $\tau_*^{3/2}$*

There are several ways in defining bed-load rate, q_b. From a view of continuum flow field, it is defined as

$$q_b = \int_0^{h_s} cu\,dz \cong \bar{c}_s h_s \bar{u} \tag{6}$$

in which the quantities with over-bar are spatially averaged, h_s = thickness of bed-load layer and $\bar{c}_s h_s$ = sediment volume of bed-load layer in unit area, including saltations.

According to the papers (Bagnold 1957, 1966, 1973), $\bar{c}_s h_s$ can be described by

$$\bar{c}_s h_s = \frac{a\tau_b}{(\sigma-\rho)g}\frac{1}{\mu} \tag{7}$$

and Eq. (7) is transformed into

$$\frac{h_s}{d} = \frac{a}{\bar{c}_s \mu}\tau_* \tag{8}$$

in which d = particle size; τ_b = bed shear stress; τ_* = non-dimensional bed shear stress; μ = dynamic friction coefficient that is $\tan\alpha$ in the original paper; and a = correction factor that is unity in a stage of active sediment transportation.

Ashida and Michiue (1972) modified Bagnold's idea and obtained the relation describing $\bar{c}_s h_s$. Their relation can be transformed into

$$\frac{h_s}{d} = \frac{1}{\bar{c}_s}\frac{1}{\mu}(\tau_* - \tau_{*c}) \tag{9}$$

in which τ_{*c} = non-dimensional critical bed shear stress of initiation of motion.

Egashira and Ashida (1992) introduced constitutive relations of τ and p for debris flow into Eqs. (1) and (2), and obtained

$$\frac{h_s}{d} = \frac{1}{\bar{c}_s(\sigma/\rho - 1)\{\tan\phi/(1+\alpha) - \tan\theta\}}\tau_* \tag{10}$$

in which α = ratio of dynamic and static pressure defined by p_d/p_s. The proposed constitutive relations were slightly modified by Egahira, Miyamto and Itoh (1997), which is described in Chapter 3. Referring to the relation, $p_d/(p_s + p_d)$, illustrated in Chapter 3, the pressure ratio, p_d/p_s, reduce to zero at the bed surface. Thus,

$$\alpha = 0 \tag{11}$$

at the bed surface.

Sediment concentration of bed-load layer must be c_* at the bed surface and 0 at the upper boundary of bed-load layer. Correspondingly, the averaged value, $\bar{c}_s$, can be estimated at around

$$\bar{c}_s \cong 0.5 c_* \tag{12}$$

In Figure 2, the thickness predicted by Eq. (9) with Eq. (12) as well as by Eq. (10) with Eqs. (11) and (12) are compared with flume data. Results show that

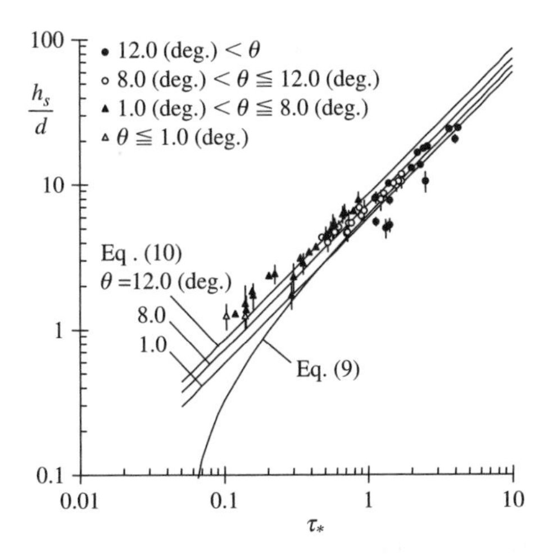

Figure 2. Relationship between non-dimensional bed shear stress and thickness of bed-load layer.

flume data increase linearly with bed shear stress and are predicted well in terms of both equations.

From a theoretical view, the condition illustrated in Eq. (12) may seem to be troublesome. Exact solution (Egahira & Ashida 1992) for thickness of bed-load layer suggests that Eq. (10) yields adequate results in wide area of bed slope, although bed-load layer spreads to free surface at the bed slope smaller than the exact solution.

It can be concluded from the discussions illustrated as above that the idea proposed by Bagnold gives the same results as those predicted by the method proposed by Egashira et al. (1992, 1997) for the thickness of bed-load layer.

A bed-load will be formulated reasonably if either the velocity profile or the averaged velocity of bed-load layer is given to Eq. (6). The averaged velocity, $\bar{u}_s$, has been studied in relation to the velocity profile of pure water flow over 'a rigid bed', i.e. logarithmic velocity law. Bagnold (1973) specified $\bar{u}_s$, referring to the velocity at the height of saltating grains. Ashida et al. (1972) employed an equation of motion for mass point system and logarithmic velocity profile over the rigid bed to evaluate $\bar{u}_s$ of bed-load layer, which is shown in Figure 3.

These studies provided that $\bar{u}_s$ should be proportional to shear velocity, $\bar{u}_s \sim u_*$. It is natural that combining two results on $\bar{c}_s h_s$ and $\bar{u}_s$ leads to the bed-load formula characterized by $\tau_*^{3/2}$; i.e.

$$q_{b*} \sim \tau_*^{3/2} \tag{13}$$

in which q_{b*} = non-dimensional bed-load rate defined as $q_{b*} = q_b/\sqrt{(\sigma-\rho)gd^3}$. The notation such as '$\sim$' means 'proportional to'.

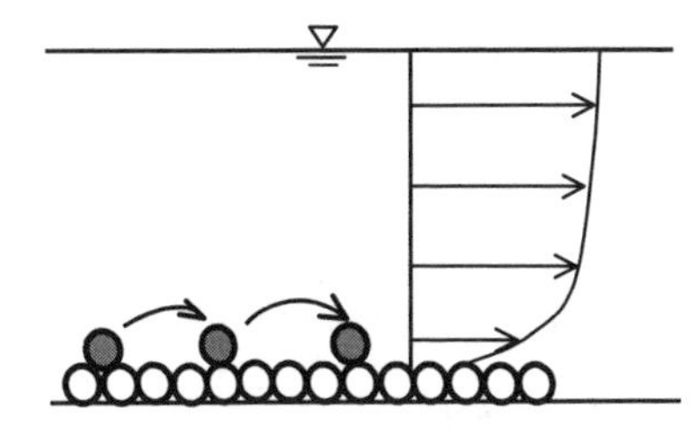

Figure 3. Logarithmic velocity profile over the rigid bed.

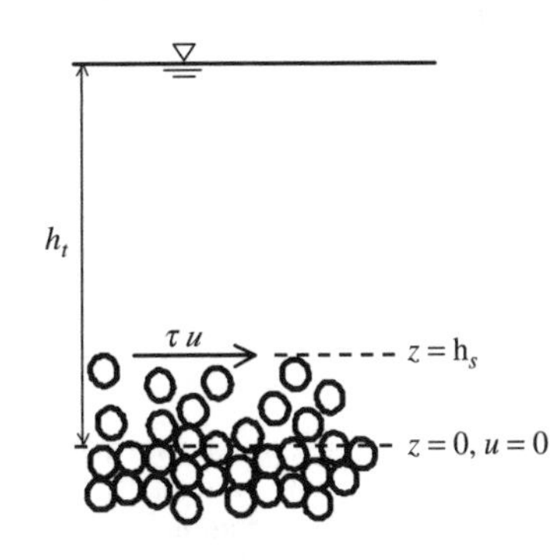

Figure 4. Control volume and control surface in bed-load layer.

In addition, it is widely known that studies in terms of 'stream power' proposed several bed-load equations, and all formulas reduce to a type of Eq. (13). This must be natural, because the stream power is defined as the product of bed shear stress and flow velocity. If these researches were to be conducted using a clearly defined control volume with a clearly defined bed surface, which is shown in Figure 4, several different forms would be proposed.

2.3 *Possible forms of the bed-load equation*

Substitution of Eqs. (4) and (5) into energy conservation equation, Eq. (3), yields

$$g\sin\theta\{(\sigma-\rho)c+\rho\}u + \frac{\partial \tau}{\partial z}u = 0 \tag{14}$$

Even if the bed-load layer is thin, Eq. (14) can be integrated from bed surface, $z = 0$, to the upper boundary, $z = h_s$, referring to Figure 4. Thus,

$$(\sigma-\rho)g\sin\theta\, q_b + \rho g\sin\theta\int_0^{h_s} u\,dz + \int_0^{h_s}\frac{\partial \tau}{\partial z}u\,dz = 0 \tag{15}$$

in which q_b = the bed-load rate defined as $q_b = \int_0^{h_s} c\,u\,dz$ as shown in Eq. (6).

Equation (15) can be deduced to

$$q_b \sim \bar{u}\, h_s \quad (16)$$

in which $\bar{u}$ = average velocity of bed-load layer.

In order to assure that $\bar{u}$ is not proportional to shear velocity, u_*, let us assume a linear velocity profile;

$$u = \beta z \quad (0 \le z \le h_s) \quad (17)$$

Correspondingly,

$$\bar{u} = \int_0^{h_s} \beta z\, dz \Big/ h_s \sim h_s \quad (18)$$

In this case, Eq. (16) suggests q_b is proportional to h_s^2. Taking the relation between h_s and τ_* into account, we will obtain

$$q_b \sim u_*^4 \quad \text{or} \quad q_{b*} \sim \tau_*^2 \quad (19)$$

This result shows that a bed-load formula different from the type of $\tau_*^{3/2}$ will be deduced if the flow structure is investigated in relation to the dynamics describing the bed-load layer.

3 BED-LOAD FORMULA DERIVED FROM CONSTITUTIVE RELATIONS

Several constitutive relations have been proposed for the mixed flow of water and coarse grains. Egashira and Ashida (1992) proposed a bed-load formula which was formulated using their constitutive relations and Eqs. (1) and (2). Egashira, Miyamoto and Itoh (1997) modified Egashira et al.'s constitutive relations to solve differences between the flow over erodible bed and rigid bed. The modified constitutive relations (Egashira et al. 1997) are as follows, for a simple flow shown in Figure 1.

$$\tau = \tau_y + \tau_d + \tau_f \quad (20)$$

$$p = p_s + p_d + p_w \quad (21)$$

in which τ_y = yield stress due to particle-particle contacts; τ_d = shear stress due to particle-particle collisions; τ_f = shear stress due to shearing of pore water such as Reynolds stress; p_w = pore water pressure, p_s = static pressure due to particle-particle contacts; and p_d = dynamic pressure due to particle-particle collisions. These quantities were formulated as follows.

$$\tau_y = p_s \tan\phi_s \quad (22)$$

$$\tau_d = k_d\, \sigma \left(1 - e^2\right) c^{1/3} d^2 \left|\frac{\partial u}{\partial z}\right| \frac{\partial u}{\partial z} \quad (23)$$

$$\tau_f = k_f\, \rho \frac{(1-c)^{5/3}}{c^{2/3}} d^2 \left|\frac{\partial u}{\partial z}\right| \frac{\partial u}{\partial z} \quad (24)$$

$$p_w = \rho g \cos\theta (h_t - z) \quad (25)$$

$$p_d = k_d\, \sigma e^2 c^{1/3} d^2 \left(\frac{\partial u}{\partial z}\right)^2 \quad (26)$$

$$\frac{p_s}{p_s + p_d} = \left(\frac{c}{c_*}\right)^{1/n}, \quad (n = 5) \quad (27)$$

in which ϕ_s = internal friction angle of sediment particles; e = restitution coefficient; k_d = 0.0828; k_f = 0.16; and h_t = total flow depth (see Figure 1).

Integrating Eqs. (1) and (2) from an arbitrary level (z) to the upper boundary ($z = h_s$) and using Eqs. (20) and (21), we obtain

$$\tau_y(z) + \tau_d(z) + \tau_f(z) = \rho g (h_t - z) \sin\theta + \rho g \sin\theta \int_z^{h_s} \{(\sigma/\rho - 1)c + 1\} dz \quad (28)$$

$$p_s(z) + p_d(z) = \rho g \cos\theta \int_z^{h_s} (\sigma/\rho - 1) c\, dz \quad (29)$$

in which the following relations are employed;

$$\tau_y(h_s) = 0$$

$$\tau_d(h_s) = 0$$

$$\tau_f(h_s) = \rho g (h_t - h_s) \sin\theta$$

$$p_s(h_s) = 0$$

$$p_d(h_s) = 0$$

$$p_w(h_s) = \rho g (h_t - h_s) \cos\theta$$

In order to obtain exact solutions for $u(z)$ and $c(z)$, Eqs. (28) and (29) have to be solved numerically with Eqs. (22) to (27) in terms of iteration method (Egashira et al. 1997). However, the approximated solutions can be deduced, supposing $c(z) \cong \bar{c}_s$. Introducing $\bar{c}_s$ instead of c in Eq. (28) and (29), these equations are reduces to

$$\tau_y(z) + \tau_d(z) + \tau_f(z) = \rho g (h_t - h_s) \sin\theta + \rho g \sin\theta \{(\sigma/\rho - 1)\bar{c}_s + 1\}(h_s - z) \quad (30)$$

$$p_s(z)+p_d(z)=\rho g\cos\theta(\sigma/\rho-1)\bar{c}_s(h_s-z) \quad (31)$$

Substituting $z=0$ into Eqs. (30) and (31), these reduce to

$$\tau_y(0)=\rho g h_t\left\{1+(\sigma/\rho-1)\bar{c}_s\frac{h_s}{h_t}\right\}\sin\theta \quad (32)$$

$$p_s(0)=\rho g h_s(\sigma/\rho-1)\bar{c}_s\cos\theta \quad (33)$$

The relation of h_s/h_t is obtained, taking $\tau_y(0)=p_s(0)\tan\phi_s$ into account.

$$\frac{h_s}{h_t}=\frac{\tan\theta}{(\sigma/\rho-1)\bar{c}_s}\frac{1}{\tan\phi_s-\tan\theta} \quad (34)$$

Equation (10) which is shown in Chapter 2 was derived by Eq. (34).

In addition, local and average velocities of the bed-load layer are solved using Eqs. (30) and (31) with Eqs. (22) to (27). The average velocity is

$$\frac{\bar{u}}{u_*}=\frac{4}{15}\frac{K_1K_2}{\sqrt{f_d+f_f}}\tau_* \quad (35)$$

Using the results of h_s and $\bar{u}$, and the definition of bed-load transport rate described by Eq. (6), a bed-load formula is given as

$$q_{b*}=\frac{4}{15}\frac{K_1^2K_2}{\sqrt{f_d+f_f}}\tau_*^{5/2} \quad (36)$$

in which

$$K_1=\frac{1}{\cos\theta}\frac{1}{\tan\phi_s-\tan\theta}$$

$$K_2=\frac{1}{\bar{c}_s}\left[1-\frac{h_s}{h_t}\right]^{1/2}$$

$$f_d=k_d(1-e^2)(\sigma/\rho)\bar{c}_s^{1/3}$$

$$f_f=k_f(1-\bar{c}_s)^{5/3}\bar{c}_s^{-2/3}$$

$$\bar{u}=\frac{1}{h_s}\int_0^{h_s}u\,dz$$

$$u_*=\sqrt{gh_t\sin\theta}$$

$$\tau_*=\frac{u_*^2}{(\sigma/\rho-1)gd}$$

In Figure 5, a curve predicted by Eq. (36) is shown using the parameters; $\theta=8$ degrees, $\phi_s=34$ degrees, $e=0.85$ and $\bar{c}_s=c_*/2=0.26$. These values were specified in the analyses of debris flow (Egashira et al. 1997). Moreover, several empirical formulas (Brown 1950, Shinohara et al. 1959, Graf et al. 1987, Meyer-Peter & Müller 1948) are shown. The formulas proposed by Shinohara et al. and Graf et al. explain flume data (Itoh 2000) almost successfully.

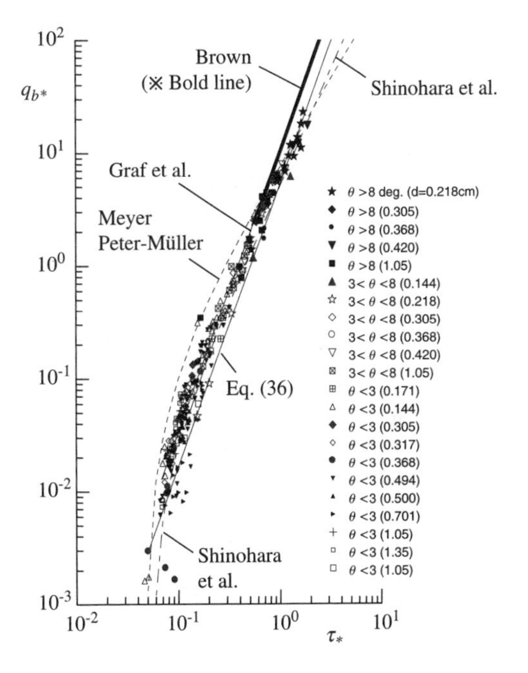

Figure 5. Bed-load formulas.

In comparison with the results obtained from the formulas of Shinohara et al. and Graf et al., the predicted line with Eq. (36) locates near the lower boundary of flume data. However, it must be important to point out that no parameter fitting was made in deriving Eq. (36).

4 CONCLUSIONS

There are many bed-load formulas. Some of them are used to estimate bed-load transport rate and associated river changes in practical problems as well as in sediment researches. Therefore, bed-load formula may seem to be developed to some level, and as a result, such kind of study has been reducing. However, this does not mean that the mechanics of sediment transportation has been developed and well understood. It is considered that theoretical treatment reaches to some kinds of stiff wall due to complex sediment phenomena. In contrast, researches based on simple dynamics such as mass point system and DNS become active in accordance with the development of computation method. In these researches, kinematic conditions which should be determined automatically by the field's dynamics are still used, although some interesting phenomena such as sediment sorting and flow pattern around a sediment particle are produced numerically. Therefore, long years may need to extend the results to practical problems.

Present study is aimed to stimulate bed-load study from a view point of continuum mechanics. Bagnold's type bed-load formulas are chosen for discussion, firstly. It is pointed out that the velocity of bed-load layer is formulated independently from the mechanism describing the thickness of the bed-load layer and is specified as a linear function of shear velocity. In addition, it is suggested that if the velocity of bed-load layer in relation to the same dynamic frame as well as to the corresponding control volume of bed-load layer, a bed-load formula different from the type of $\tau_*^{3/2}$ will be deduced.

Corresponding to this suggestion, it is shown that a bed-load formula characterized by $\tau_*^{5/2}$ is derived based on momentum conservation equations and constitutive relations for water and coarse grain mixture. The proposed formula can predict flume data to some level without performing corrections of parameters and coefficients although it has some problems to be discussed from a practical view as well as from a view of dynamics.

ACKNOWLEDGEMENT

Parts of this study were supported by Grant-in-Aid for Scientific Research (B) (Representative: Shinji EGASHIRA) from the Japan Society for the Promotion of Science.

REFERENCES

Ashida, K. & Michiue, M. 1972. Study on hydraulic resistance and bed-load transport rate in alluvial streams. *Proc. JSCE*. No. 206: 59–69 (in Japanese).

Bagnold, R. A. 1957. The flow of cohesionless grains in fluids, *Philosophical Trans*., Royal Soc. of London. Vol. 249: 235–297.

Bagnold, R. A. 1966. An approach to the sediment transport problem from general physics. *USGS Prof. Paper*. 422-J: 1–37.

Bagnold, R. A. 1973. The nature of saltation and of 'bed-load' transport in water. *Proc. Royal Soc. of London*. A. 332: 473–504.

Brown, C. B. 1950. *Engineering Hydraulics*. edited by H. Rouse. John Wiley & Sons. Inc. New York: N. Y.

Du Boys, M. P. 1879. Le Rohne et les rivieres a lit affouillable. *Anales des Ponte et Chausses*, Series 5. 18: 141–195.

Egashira, S. & Ashida, K. 1992. Unified view of the mechanics of debris flow and bed-load. *Advances of micromechanics of granular materials*: 391–400. Amsterdam: Elsevier.

Egashira, S., Miyamoto, K. & Itoh, T. 1977. Constitutive equations of debris flow and their applicability, *Proc. 1st Int. Conf. on Debris-Flow Hazards Mitigation*, New York: ASCE: 340–349.

Einstein, H. A. 1950. The bed-load function for sediment transportation in open channel flows. *Tech. Bull*. No. 1026. USDA. Soil Conservation Service: 1–70.

Graf, W. H. & Suszka, L. 1987. Sediment transport in steep channel. *Jour. Hydroscience and Hydr. Eng*., JSCE. 5(1): 11–26.

Itoh, T. 2000. *Constitutive equations of debris flow and their applicabilities*. Ph. D.-Thesis of Ritsumeikan University (in Japanese).

Meyer-Peter, E. & Müller, R. 1948. Formulas for bed-load transport. *Proc. 2nd IAHR Meeting*, Stockholm: 39–64.

Shinohara, K. & Tsubaki, T. 1959. On the characteristics of sand waves formed upon the beds of open channels and rivers. *Reports of Res. Insti. Appli. Mech*. Kyushu Univ. Vol. VII. No. 25: 15–45.

River, Coastal and Estuarine Morphodynamics: RCEM 2005 – Parker & García (eds)
© 2006 Taylor & Francis Group, London, ISBN 0 415 39270 5

A particle tracking technique to study gravitational effects on bedload transport

S. Francalanci & L. Solari
Department of Civil Engineering, University of Florence (Italy)

ABSTRACT: In the present study we investigate the bedload transport mechanism due to the action of water flow when transversal and longitudinal slope of the bed are not negligible and in the case of low values of the applied Shield stress. Average properties of saltating particles are studied by means of an image based technique here developed. Results show intensity and direction of average particle velocity as a function of the applied Shield stress. In order to test the validity of the present technique, measurements have been compared with field particle motion obtained through laser Doppler anemometer.

Finally results are compared with theoretical findings from the model of bedload transport on an arbitrarily sloping bed proposed by Parker et al. (2003).

1 INTRODUCTION

Mechanics of sediment transport is a fundamental issue in morphodynamics. When a shear flow acts on a cohesionless granular bed, sediment particles are intermittently and randomly entrained by the flow provided some convenient measure of the hydrodynamic stress acting on the bed surface exceeds a conventional threshold value. Once the sediment particles are entrained by the stream they can move mainly as bedload or suspended load.

In the case of gravel bed streams, bedload appears to be the dominant mode of sediment transport. The mechanics of bedload is in this case governed by the interaction of the particles with the bed. When the latter is characterized by relatively high slopes, such as near the banks and along the bar fronts, the particles dynamics is crucially affected by the downslope force component of gravity. In the case of steep slopes various authors have proposed equations to evaluate sediment transport similar to the ones derived for gentle slopes (Smart, 1984; Damgaard et al., 1987) corrected through the introduction of an empirical function depending on the local bed slope.

Deterministic descriptions of bedload transport (e.g. Ashida & Michiue, 1972, Engelund & Fredsoe, 1976, Wiberg & Smith, 1985, Sekine & Kikkawa, 1992, Nino & Garcia, 1994) usually are developed through two main steps.

In the first step an estimate of the average speed of saltating particles is provided employing either saltation models, studying the trajectories of saltating grains over beds usually slightly inclined on the horizontal direction, or more simply, considering an equivalent uniform translatory motion of particles in which the effects of particle rebound are taken into account by means of a bulk dynamic coefficient of Coulomb friction. The latter bulk formulations, though they provide a more schematic view of the particle dynamics, are generally sufficient to give an estimate of the mean particle velocity in reasonable agreement with the experimental observations also in the case of steep slopes of the bed.

The second step is devoted to the evaluation of the average areal concentration of saltating particles, i.e. the average volume of saltating particles per unit area; such quantity is usually evaluated employing Bagnold hypothesis. Seminara et al. (2002) have theoretically shown by means of a three dimensional generalization of Bagnold assumption, that this assumption cannot be valid. Based on the experimental findings of Fernandez Luque & Van Beek (1976), an alternative model for evaluating bedload transport on arbitrarily sloping beds has been elaborated by Parker G., Seminara G. & Solari L. (2003). In the latter model a 3-D description of bedload transport on an arbitrarily sloping bed is proposed based on an entrainment approach according to which bedload transport occurs only when shear stress at the bed is above the critical value, on the contrary when the bed shear stress reduces to the critical value there is no motion.

Theoretical results of Parker et al. (2003) model about particle velocity in the case of an arbitrarily sloping bed allow to quantify the average velocity intensity changes and the average direction of particle motion. The latter appears to deviate from the direction of the average bottom stress; in particular the deviation of particle motion from the direction of the applied stress decreases as the Shield stress increases.

Such features are fundamental in the evaluation of the transverse gravitational component of sediment transport since the estimate of the latter quantity is a crucial point in any morphodynamic model that considers bottom evolution and bedform dynamics. Usual linearized approaches of bedload transport as a function of local slope can lead to great understimations (Parker et al., 2003).

In the present work the effect of gravity on bedload transport due to water flow has been investigated by means of an experimental activity. PTV techniques have been employed to study average properties of the motion of sediment particles on an arbitrarily sloping fixed bed, in terms of both intensity and direction of average particle velocity. Experimental observations regarding the deviation angle between the average direction of particle velocity and the direction of bottom stress are presented. Such results provide a novel contribution to overcome the lack of experimental data in the literature on this specifical issue.

Results obtained from PTV techniques regarding the intensity of particle velocity have been compared with measurements carried out with laser techniques (Francalanci et al., 2004) in order to test the capabilities of the proposed image based technique.

Finally, the experimental observations have been compared with the theoretical findings of Parker et al. (2003) model.

2 EXPERIMENTAL SET-UP

The experiments were carried out in a tilting, recirculating flume, 10 m long, 0.365 m wide and 0.45 m deep, equipped for clear water flows up to 28 l/s and slope adjustment up to 5°. In order to obtain higher bed inclinations, a 2 m long 0.23 m wide plane has been built inside the flume with inclination adjustable up to 15°. Due to entrance effects the length of measuring reach is about 1 m.

Water flow was supplied at one end of the flume from a centrifugal pump and the water discharge was monitored by an electromagnetic flow-meter.

The sediment particles were fed at a constant discharge by means of a sediment feeder.

The particles employed satisfied the following requirements in all the range of investigated bed slopes: mobility at low Shields stress; ratio water depth to particle size sufficiently high to allow particle saltation, even in the case of high slope; relatively high friction angle thus allowing particles disposition according to a given transversal inclination.

The selected material was represented by steel particles with a density of 7850 kg/m^3, shaped like disks

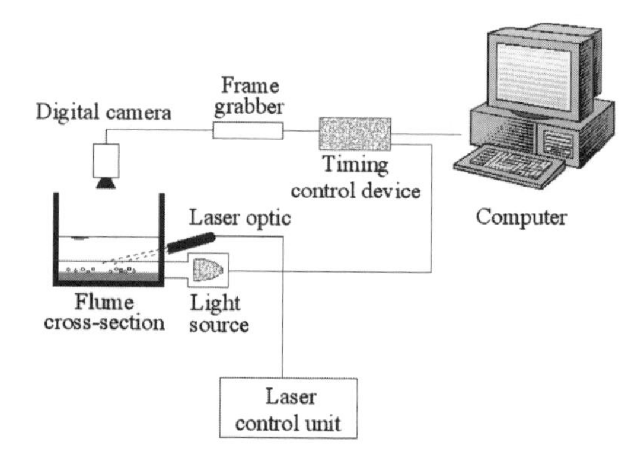

Figure 1. Sketch of the experimental setup.

with almost uniform size of 3 mm of diameter and of 0.6 mm of thickness.

Particle velocity field was studied through PTV (particle tracking velocimetry), in order to measure the intensity and direction of the averaged velocity vectors.

A schematic representation of the experimental setup is shown in Figure 1. In particular, the computer controlled the digital camera and the illumination system, for the acquisition of particle images, through a timing control device; video images of bedload particles saltating over the bed were taken from the upper side of the flume and acquired by the frame grabber.

In addition to the PTV system, a laser Doppler system was also employed to measure flow velocity profiles.

2.1 *PTV system*

The measurements of fluid velocity using digital image processing can be classified into two categories, PIV (particle image velocimetry) and PTV (particle tracking velocimetry), depending on how the displacement vectors are extracted from particle images. In both techniques, the flow is seeded with small tracer particles and then illuminated by a series of laser pulses at a known time interval. The scattered light from the seeding particles in the illuminated plane is recorded as a particle image by a device such as a CCD.

The PTV tracks individual particles images in consecutive image frames and evaluates the directionally resolved displacement vector for each matched particle pair.

In the present experiments, measurements of particle motion were obtained through a PTV technique, which was based on the following components: seeding, illumination, image recording and image processing.

The seeding was based on particle tracers lightened from pulsating light and visible from digital camera: a collimated light source, located on a side of the flume, illuminated the bedload layer, not interfering with free water surface. In the experiments the steel sediment particles acted as tracers and their velocity was then measured. In fact, steel material was highly reflecting and so visible to the camera, while the fixed bed was black dyed to observe the particles moving on it.

The illumination system, with stroboscopic pulsating light, was able to produce couples of lamp light to acquire couples of images. The frequencies range of couple images was of 0.5–5 Hz with a resolution of 0.1 Hz, while the interframing time between two images of a single couple was in the range 1–10 ms with a resolution of 0.1 ms.

The digital camera for the image recording was characterized by high resolution (max resolution 1360 × 1024 pixel), that allowed to take images in which single particle consisted of a bulk of pixel, and a short interframing rate, that allowed to capture two images in which particle tracers covered a relative small distance into the bed plane. The interrogation area had a longitudinal length of about 16 cm and a width of 22 cm. Both digital camera and system illumination were controlled and synchronized by a computer. The CCD (Charge Coupled Device) plane was oriented parallel to the sloping bed layer. The particle velocity measured field was the one projected on a plane parallel to the bed.

3 EXPERIMENTAL ACTIVITY

The experimental activity was focused on the investigation of the bedload transport mechanism when transversal and longitudinal slopes are not negligible and low values of the applied Shields stress occurs.

The experiments that were carried out consist of several sets of data: (i) preliminary set of experiments on a plane, fixed, artificially roughened bed, with no transversal slope; the channel was fed at a constant sediment rate; (ii) fixed bed experiments with laterally tilted bottom; the longitudinal inclination α was 5°, 10°, 15°, while the transversal inclination φ was 5°, 10°, 15°, 20°, combined together. For all these combinations of sloping bed and for different values of the flow discharge Q, the following measurements were collected: bedload transport rate at equilibrium condition along the cross-section, particle velocity profiles and flow velocity profiles with laser Doppler. For some of the experimental flow conditions, experiments with PTV technique were performed, in order to compare independent sets of data regarding particle velocity and to obtain an estimation of the velocity deviation angle. The flow discharge and bed slope inclinations for the images technique experiments are reported in Table 1.

Table 1. Flow experimental conditions for PTV.

α [°]	φ [°]	Q [l/s]					
5	5	2.68	3.1	4.06	4.98	5.98	7.49
5	10	2.68	4.07	5.5	7.02		
5	15	3.67	5.05	6.1	7.5		
10	5	1.4	2.1	3.0	4.5		
10	10	1.5	2.35	3.11	4.5		
10	15	2.53	3.53	4.5			
15	5	1.4	2.1	3.0			
15	10	1.4	2.1	3.0			

4 IMAGES ANALYSIS

4.1 *Tracking analysis*

The tracking process is divided into two steps to analyze a single couple of images. The first step is to locate all objects of interest in each frame, determine each object's position (center of mass) and store that information; the second step is to determine which of these objects can be unambiguously identified in successive frames, and finally assign unique numerical labels to each distinguishable object.

The used approach requires that the distance an object move (from frame to frame) be smaller than the typical distance between objects and that there are few particles in motion (about less than 100). In this way particle centroids position can be easily identified in each frame and the ambiguousness related to the tracked direction is reduced.

Some criterions need to be set for the analysis. First, all the objects that have a smaller area than diameter size in equivalent pixels are eliminated: in this way the noise of the background is avoided, as in general the particles are much bigger than the diameter due to light reflection.

Second, the maximum distance an object can move from frame to frame needs to be set. In that area only one object is unambiguously tracked, that is why the approach requires few particles in motion.

An example of a single couple of images is shown in Figure 2a-b. Numbered labels indicate particles displacements in each frame.

The result of analysis process is a list of particles locations for each couple of images. In Figure 3 an example of particle trajectory is shown, particles locations in each frame are connected by a line. The increasing direction of the bed shear stress in the transversal cross-section is due to the transversal slope of the bed. It is shown that particles deviate from the direction of bed shear stress, aligned with the flow direction, due to the lateral component of gravity.

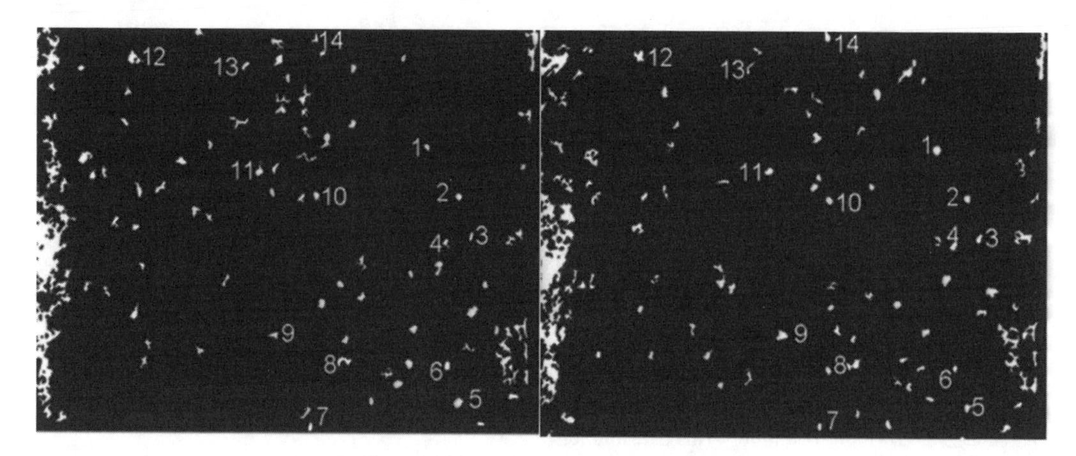

Figure 2a-b. A couple of analysed images ($\alpha = 5°$, $\varphi = 10°$, Q = 7.02 l/s). Labels indicate a sample of tracked objects.

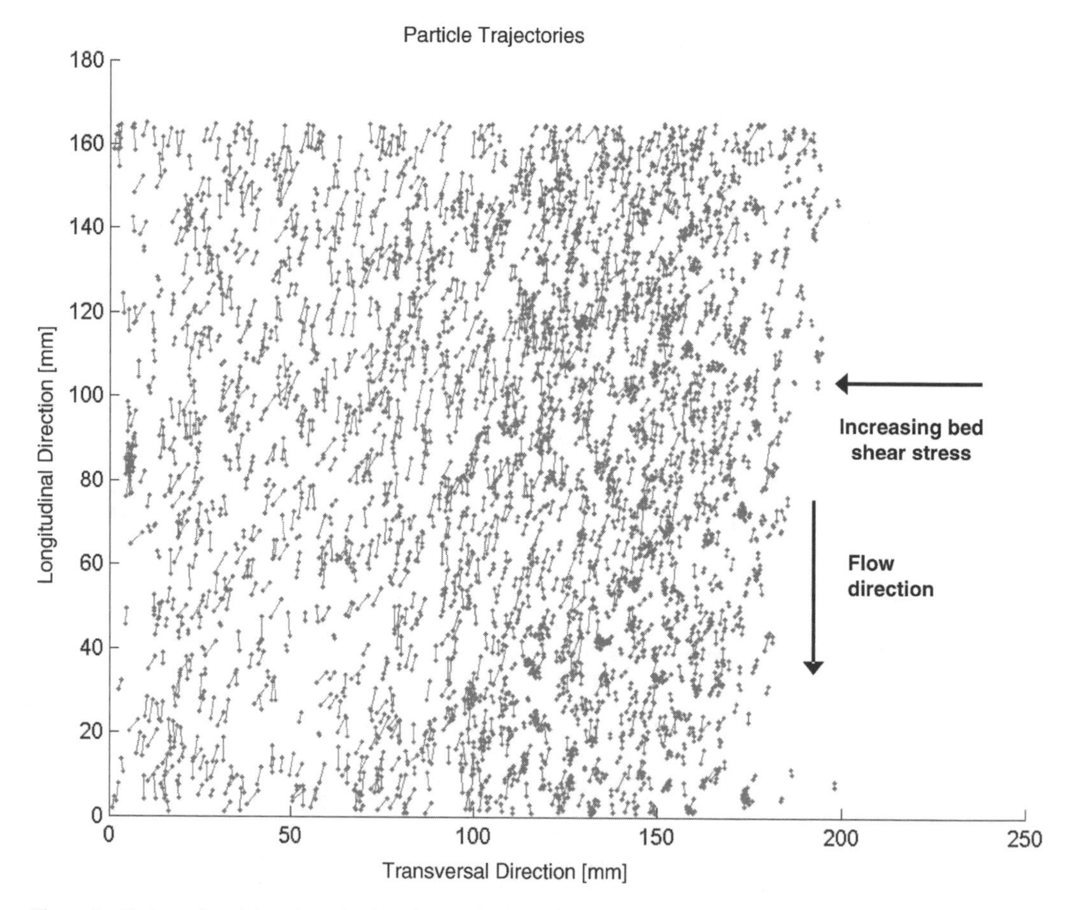

Figure 3. Pattern of particle trajectories ($\alpha = 5°$, $\varphi = 5°$, Q = 7.02 l/s).

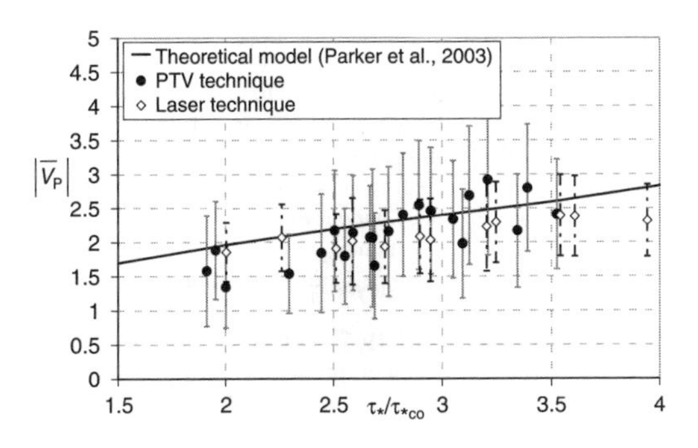

Figure 4. Dimensionless particle velocity ($\alpha = 5°$, $\varphi = 5°$).

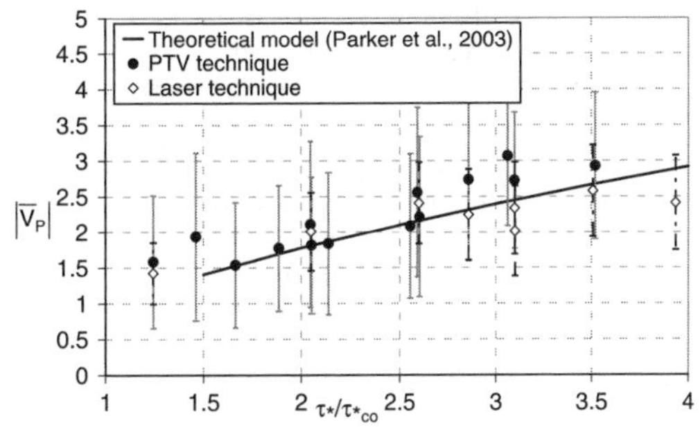

Figure 5. Dimensionless particle velocity ($\alpha = 5°$, $\varphi = 10°$).

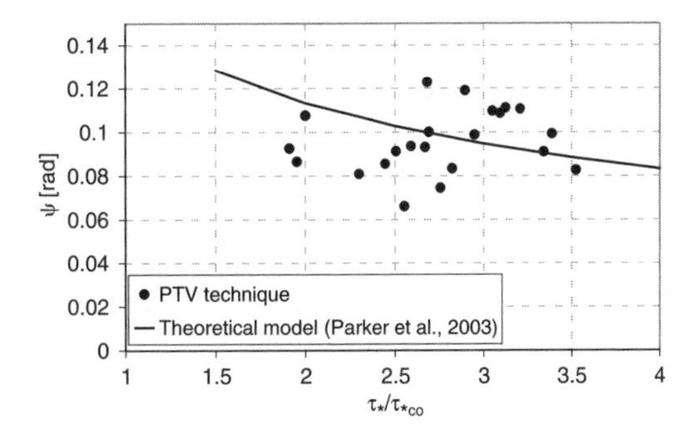

Figure 6. Angle of particle deviation ($\alpha = 5°$, $\varphi = 5°$).

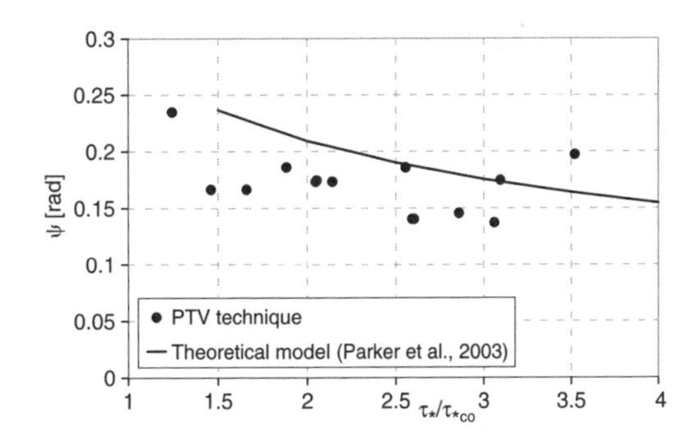

Figure 7. Angle of particle deviation ($\alpha = 5°$, $\varphi = 10°$).

4.2 *Post-processing*

The post-processing of data evaluated the averaged particles velocity field in terms of intensity and direction: the velocity vectors have been averaged according to their location on the cross-section, and related to the corresponding values of the shear stress.

The first step is to remove outliers from data in order to improve the quality of data; several criterions have been used to post-process data: particles move according to the flow direction, particles cannot move against gravity even if they can be carried by the flow for small distances, particles that do not move or oscillate around their positions are removed from the averaging.

The second step is the evaluation of particle velocity from the tracked locations and the interframing time between images, and the averaging of velocity vectors according to their position.

5 RESULTS

The experimental data are presented in Figures 4–7, where dimensionless particle velocity and deviation angles are plotted versus the shear stress ratio, for the experimental conditions $\alpha = 5°$, $\varphi = 5°$ and $\alpha = 5°$, $\varphi = 10°$.

The shear stresses are estimated from the measured flow velocity profiles with laser Doppler anemometer, in the hypothesis of logarithmic velocity profile near the bedload layer. All results are plotted as a function of the ratio between the applied Shields stress τ_* and the critical Shields stress τ_{*co} for the onset of sediment motion in the case of vanishing bed slopes, experimentally estimated to be about 0.03.

Dimensionless particle velocity is seen to increase with the applied Shields stress and with higher values of the transversal inclination of the bed for the same value of α. Particle velocity intensity from PTV technique and from laser technique are compared as independent sets of data: the trend of the data is shown to be the same; also the root mean squares of the data are of the same order of magnitude for both techniques. This latter finding suggests the capability of the present image technique to adequately describe the phenomenon. Finally, the experimental data showed a good agreement with the theoretical model of Parker et al. (2003).

The particle velocity deviation angle is expected to slightly decrease with higher values of the applied Shields stress, and to increase with higher values of the transversal inclination. Experimental results from

PTV technique are presented for two sets of data in order to show the latter dependencies.

Experimental data cannot describe very well the dependency of deviation angle with the Shields stress because of data scattering and of high values of the root mean squares. This latter results could be improved with further progresses in post-processing analysis. However the mean value of the particle velocity deviation angle is shown to be of the same order of the one predicted by the theoretical model. Also experimental data show a clear dependency of deviation angle with the transversal inclination, in good agreement with the theoretical model. This results is a fundamental issue for the estimation of the transversal sediment transport discharge.

6 CONCLUSIONS

The present contribution is devote to better understand the mechanism of the bedload transport when transversal and longitudinal slope are not negligible and in the case of low values of the applied Shields stress.

The experiments were focused on bedload transport rate and particle velocity measurements, and a particle tracking technique was applied for the observation of the particle velocity field. The employed PTV technique allowed the estimation of particle velocity in terms of intensity and direction.

Experimental data seem to confirm the theoretical model proposed by Parker et al. (2003): velocity data acquired with images and laser techniques show a good agreement; the measured deviation angle is of the same order of magnitude than the theoretical one.

Such findings allow to estimate the gravitational effects on bedload transport and the transversal sediment transport discharge, that is a crucial point in any morphodynamic model.

ACKNOWLEDGEMENTS

This work has been developed within the framework of the National Project cofunded by the Italian Ministry of University and of the Scientific and Technological Research and by the University of Florence (COFIN 2001) "Morphodynamics of fluvial networks". The Authors are grateful to Prof E. Paris, University of Florence (Italy) for his comments and suggestions.

REFERENCES

Ashida, J.E. & Michiue, M. 1972. Saltation and suspension trajectories of solid grains in a water stream. *Philos. Trans. R. Soc. London*. Ser. A 284: 225–254.

Bagnold, R.A. 1956. The flow of cohesionless grains in fluids. *Philos. Trans. R. Soc. London*. Ser. A 249: 235–297.

Bagnold, R.A. 1973. The nature of saltation and "bed-load" transport in water. *Proc. R. Soc. London*. Ser. A 332: 473–504.

Bridge, J.S. & Bennett, S.J. 1992. A model for the entrainment and transport of sediment grains of mixed sizes, shapes and densities. *Water Resour. Res.* 28(2): 337–363.

Damgaard, J.S. & Whitehouse, R.J.S. & Soulsby, R.L. 1997. Bed-load sediment transport on steep longitudinal slopes. *J. Hydraul. Eng.* 123(12): 1130–1138.

Engelund, F. & Fredsoe, J. 1976. A sediment transport model for straight alluvial channels. *Nord. Hydrol.* 7: 293–306.

Fernandez Luque, R. & van Beek, R. 1976. Erosion and transport of bedload sediment. *J. Hydraul. Res.* 14(2): 127–144.

Francalanci, S. Paris, E. & Solari, L. Bed load transport on arbitrarily sloping bed at low Shields stress: preliminary experimental observations. *Proceedings River Flow 2004*, Taylor & Francis Group, London, 713–719.

Kovacs, A. & Parker, G. 1994. A new vectorial bedload formulation and its application to the time evolution of straight river channels. *J. Fluid Mech.* 267: 153–183.

Nino, Y. & Garcia, M. 1994. Gravel saltation, 1, Experiments. *Water Resour. Res.* 30(6): 1907–1914.

Nino, Y. & Garcia, M. 1994. Gravel saltation, 2, Modeling. *Water Resour. Res.* 30(6): 1915–1924.

Parker, G. & Seminara, G. & Solari, L. 2003. Bedload at low Shields stress on arbitrarily sloping beds: alternative entrainment formulation. *Water Resour. Res.* 39(7), 1183, doi:10.1029/2001WR001253.

Seminara, G. & Solari, L. & Parker, G. 2002. Bed load at low Shields stress on arbitrarily sloping beds: Failure of the Bagnold hypothesis. *Water Resour. Res.* 38(11), 1249, doi: 10.1029/2001WR000681.

Sekine, M. & Kikkaua, M. 1992. Mechanics of saltating grains. II. J. Hydraul. Eng. 118(4): 536–558.

Sekine, M. & Parker, G. 1992. Bedload transport on transverse slopes. *J. Hydraul. Eng.* 118(4): 513–535.

Smart, G.M. 1984. Sediment transport formula for steep channels. *J. Hydraul. Eng.* 110(3): 267–276.

Wiberg, P.L. & Smith, J.D. 1985. A theoretical model for saltating grains in water. *J. Geophys. Res.* 90(C4): 7341–7354.

Wiberg, P.L. & Smith, J.D. 1989. Model for calculating bedload transport of sediment. *J. Hydraul. Eng.* 115(1): 101–123.

River, Coastal and Estuarine Morphodynamics: RCEM 2005 – Parker & García (eds)
© 2006 Taylor & Francis Group, London, ISBN 0 415 39270 5

Methodological considerations for studying self-weight fluidization in a sedimentation column

A.N. Papanicolaou
The University of Iowa, IIHR-Hydroscience and Engineering, Iowa City, Iowa, USA

A.R. Maxwell
Pacific Northwest National Laboratory, Sequim

ABSTRACT: The objective of this research is to develop a laboratory methodology that facilitates an improved understanding of self-weight fluidization in a batch column. Observations of the fluidization pipes and of the mudline interface are made by means of a digital camera, naked eye and gamma radiation source through repeated runs for a pure kaolinite-water mixture under different initial concentrations and initial suspension heights. Analysis of the experimental results show the onset of self-weight fluidization occurs at an earlier period, during the first falling rate than the second falling rate which has been reported by others. Despite the formation and upward propagation of fluidization pipes no inflection points are observed anywhere atop the mudline interface for the kaolinite mixture. Numerical simulations made via a 1-D sedimentation model that accounts for the upward propagation of fluid further validate the observation of the authors that for the kaolinite mixture, self-weight fluidization cannot be always detected by observing the mudline interface.

1 INTRODUCTION

Fluidization is a phenomenon that is linked to the sedimentation process (Allen, 1982), and is also known as channelling or fingering flow; it may occur during the collapse of a sediment layer under its own weight, as fluid is squeezed out from the interstices, or may be due to external forcing of some type (as in a fluidized bed). Fluidization may generate distinctive structures, grain size sorting, and density segregation effects within the sediment layer, as observed in various laboratory and field studies (Druitt, 1995; Best, 1989). A ubiquitous feature of such fluidization is a vertical fluid channel, known also as a "vent," or "elutriation pipe," and a schematic is presented in Figure 1.

Up to the present day, the characteristics of the self-weight fluidization process have primarily been investigated in the laboratory by performing batch sedimentation tests under controlled conditions (e.g. known or prescribed soil properties, pH, and temperature).

In general, researchers have pointed to the mudline interface variation with time (batch curve) and solids flux vs. volume fraction of solids (flux curve) as indicators of the occurrence of self-weight fluidization (e.g. Vesilind and Jones, 1990; Fitch, 1993). According to Been (1980); Vesilind and Jones (1993); Fitch (1993) and Holdich and Butt (1995b), during

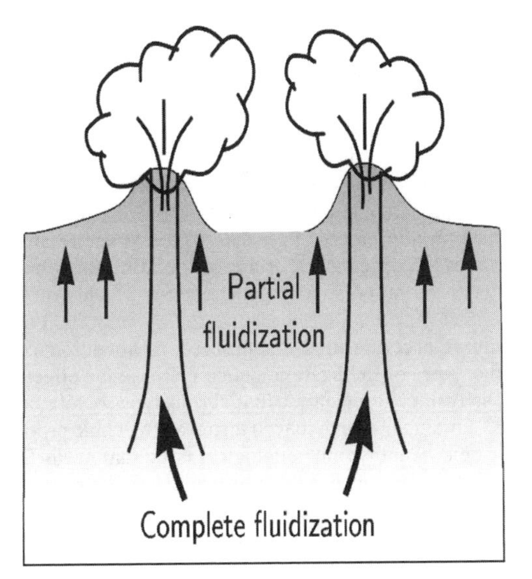

Figure 1. Schematic of vertical fluidization pipes, after Mount (1993).

batch sedimentation tests of different slurries, short, upward-moving channels were observed. With the meeting of the ascending zone of channels and the descending mudline interface, a concomitant increase in the interfacial (mudline) settling velocity was observed, and therefore an increase in the flux of the settling material (Vesilind and Jones, 1993). Other researchers (e.g. Gaudin et al., 1959; Cole, 1968), noticed no inflection point at the mudline interface despite recording the eruption of small flocs. Recently, Maxwell et al. (2003) performed a series of preliminary batch settling tests for a kaolinite sediment mud and found that during self-weight fluidization, no inflection point was recorded in the mudline interface curve. Maxwell et al. (2003) noted that formation of fluidization occurred in the early stages of a test, and not within the second falling rate region as other researchers have reported. Volcanoes, however, were not visible until the supernate had cleared; this generally took place after the second falling region. Lastly, Holdich and Butt (1995b) noted that the top of the fluidized region coincided with a solids concentration characteristic and indicated that stirring during settling had an effect on channel formation and ultimately on the settling rate of suspensions of calcite; Holdich and Butt (1995b) observed an inflection point in the mudline interface curve only for the stirred test. Fitch (1993) noted self-weight fluidization (channelling) in sedimentation tests and attributed changes in the batch and flux curves to the presence of channels; however, he used a glass rod to stir the sediment, or mixed it by repeated inversions of the sediment column, both of which are relatively gentle means of stirring.

The above literature review suggests that while the majority of research completed in the area of self-weight fluidization agrees with the notion that batch sedimentation tests may be the best approach in examining fluidization in a controlled environment, the results are inconsistent across various studies. There are several reasons for these inconsistencies. The implication for mixing methods is that a high-speed mixing process may disrupt the structure and remove the dependence of the gel point of a suspension on its initial solids concentration, which would improve repeatability. Furthermore, it is important to ensure that certain mixing approaches do not introduce other types of fluidization that could mask the effects of self-weight fluidization (e.g. certain types of mixing can entrain air, which may in turn lead to bubble propagation). Another methodological issue that needs to be further examined is the utility of the mudline interface curve to determine the occurrence of self-weight fluidization. The fact that during self-weight fluidization, in some cases, no inflection point was recorded in the mudline interface curve, it raises questions about the efficacy of the mudline interface to detect such fluidization. The use of other technologies needs to be considered to detect the occurrence of fluidization.

The overarching objective of this research is to develop a laboratory methodology that facilitates an improved understanding of self-weight fluidization in a batch column. The first objective of this research is to select first the mixing technique that could facilitate repeatable observations for self-weight fluidization. Second to examine the efficacy of "traditional" means, such as the observation of the formation of infection points atop the mudline interface, to detect the occurrence of fluidization.

2 METHODOLOGICAL PLAN

The methodological plan involved the following steps: 1) Selection of the material: A pure kaolinite clay was employed in the controlled experiments of this study. The available literature on the sedimentation behavior of kaolinite provides a reference point for comparisons with the tests performed in the present study; 2) Selection of the mixing technique: Two methods of mixing were used, namely, pneumatic mixing by compressed air and rotary mixing by high-speed propeller to elucidate the role of mixing on the formation of channels in a kaolinite mixture; 3) Development of an experimental protocol: Qualitative tests were performed for the pure kaolinite-water mixture at initial concentrations of 10, 25, 50, and 75 g/L using visual observations, in order to develop an experimental protocol and gain familiarity with the self-weight fluidization process; 4) Performance of detailed fluidization test for a mixture with initial concentration 125 g/L. These runs are used to examine the efficacy of the mudline interface to detect the occurrence of fluidization and to perform observations of the fluidization pipes and of the mudline interface by means of a digital camera, naked eye and gamma radiation source through repeated runs; 5) Verification of the experimental findings: The experimental component of the research is complemented with numerical modeling to verify the experimental results. An existing batch sedimentation model developed by Diplas and Papanicolaou (1997) and later enhanced by Papanicolaou and Diplas (1999) is employed in order to simulate the sedimentation process for tests 75 and 125 g/L by reproducing the mudline interface and sediment-suspension interface over time (the Papanicolaou and Diplas model is later refereed to as P-D model).

1) The sediment used in this study is prepared from dry kaolin powder (Hydrite PX, Georgia Kaolin Company) with typical median particle size of 0.68 m and specific gravity of 2.58, according to the manufacturer. A specific gravity test placed the actual value at 2.63, and a value of 2.6 was used in calculations. Kaolinite is a 1:1 clay mineral consisting of layered tetrahedral (typically silica) and octahedral (typically alumina) sheets, bonded by hydrogen bonding between hydroxyl and oxygen ions of adjacent layers, as well as van der Waals attraction. Due to these bonds, kaolinite is a stable clay

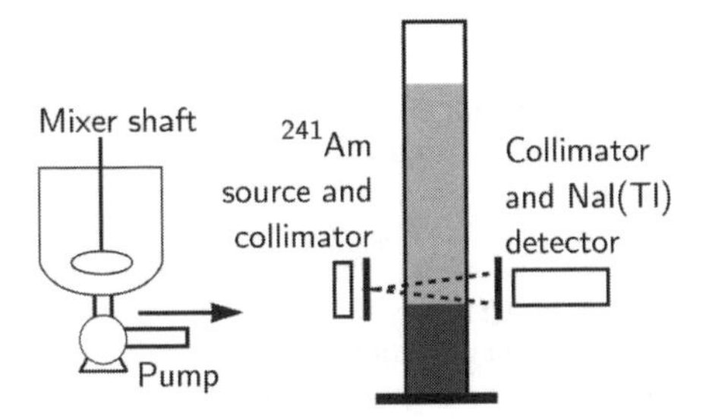

Figure 2. Schematic of instrumentation setup (not to scale).

mineral (van Olphen, 1977), with low cation exchange capacity (CEC) and surface area (McBride, 1994). As kaolinite layers tend to not separate except in "extremely polar solvents," (McBride, 1994), it should give well-defined, repeatable results for settling and consolidation. Kaolinite has also been widely used in laboratory studies (e.g. Michaels and Bolger, 1962; Cole, 1968; Dell and Kaynar, 1968; McConnachie, 1974; Austin and Challis, 1998), and thus a range of literature for qualitative comparison exists. The tests are performed in graduated sedimentation columns constructed of cast acrylic tubing, 140 mm ID by 152 mm OD. According to Cole (1968), diameter (wall) effects are minimal for this size and larger diameter columns. The tallest column used here is 2 m and the shortest is 0.25 m in height. Tap water of near neutral pH (7.6) is used.

2) Two methods of mixing are used: pneumatic and a high-speed propeller. In the former case, compressed air is utilized by connecting an air hose to section of perforated PVC pipe, which is then used to vigorously stir and bubble the kaolin-water suspension directly in the settling column. Unfortunately, it is very difficult to produce repeatable results with this pneumatic mixing method and therefore all tests reported here have been conducted with the high-speed mixer. The mixer uses a 1725RPM motor to drive a 95 mm propeller; due to the size and speed of this system, it is necessary to mix the kaolin-water suspension in a separate container, in order to avoid breaking the batch acrylic cylinder (Figure 2). A small rotary vane pump is then used to transfer the suspension from the mixer to the settling column, a process which took few seconds to complete. In order to determine if weathering of the kaolinite particles occurred during the mixing process, particle size analyses are performed (using a Mastersizer, Malvern Instruments Ltd., Worcestershire UK) on samples mixed for 5, 10, 15, and 20 minutes; no significant difference in particle size was observed by varying the mixing duration.

3) Qualitative preliminary sedimentation tests are performed at initial concentrations of 10, 25, 50, and 75 g/L using visual observations, in order to gain familiarity with the procedure and approximate the time of occurrence of self-weight fluidization. It is observed from the side of the clear acrylic cylinder that fluidization is present in all cases with the formation and propagation of channels being prominent at the early part of the runs. No infections points are recorded atop the mudline interface despite the complete upward propagation of channels in some cases. Quantitative measurements are necessary to substantiate the above findings.

4) A mixture with concentration, 125 g/L is examined in greater detail with the aid of advanced instrumentation, namely, a digital camera with special features and a gamma system for different initial mixture heights. Specifically, a Sony DFW-X700 digital camera equipped with an optical zoom macro lens is used to capture images of the sedimentation process, in order to complement the gamma system readings with respect to the mudline height variation with time. This camera outputs square-pixel images at 1024 × 768 resolution, directly to a computer hard drive via an IEEE-1394 (FireWire) interface. Fluidization activity is captured at the highest possible frame rate (15 frames per second), while overall mudline settling is captured at much longer intervals (10 minutes to 1 hour, depending on the region of the mudline and concentration of the test). The camera is controlled by intervalometer software which captured each image with a time stamp at specified intervals. Frame sequences are analyzed using the FlowJ plugin for ImageJ developed by Abràmoff et al. (2000). Quantitative measurements of the visible features such as the upward-propagating fluid velocity within the channels is carried out with NIH ImageJ, a Java-based image processor available from http://rsb.info.nih.gov/ij/.

Visual measurements (naked eye, camera) of the variation of the mudline interface with time are complemented with an automated gamma radiation scanning system. The system consists of a 550 mCi gamma source (60 keV) used to provide the radiation beam, and a Harshaw 6S2/2-X NaI(Tl) detector with integrated photomultiplier for detection of the radiation. The signal from the detector is amplified and then passed through a single-channel analyzer (Harshaw NC-22) operated in windowed mode to filter out noise. Collimation is a key element of radiation scanning (Been, 1982), and in this case is provided by a lead plate 9.5 mm thick with a 6.35 mm circular hole for the source beam, and a 0.889 by 36.8 mm slit for the detector, machined in a 31.8 mm deep block of lead. These are shown on opposite sides of a sedimentation column in Figure 2. The gamma system results minimize the subjective error which is inherent in visual observation of the mudline interface and the region where inflection points in the mudline interface may occur. Vertical motion of the source and detector is provided by a step motor, and measurements are taken at discrete points. A QuickBASIC program controlled the stepper motor and gamma counter; the photon count was recorded at each specified height, over a two-second interval (the interval was determined via a χ^2 test according to Knoll (1979)) beginning at a specified time. The time between traversals was set at a time interval of ~10–20 minutes, depending on the total traverse time, and a given test could last up to three weeks. By examining profiles and comparing with visual observation, the spatial accuracy of the system is ±1 mm at a 5 mm scan interval, with respect to determining the location of the mudline interface.

The system is calibrated before each experiment by mixing suspensions of known solids content and recording the gamma count rate; these data were then fit to an equation of the form

$$\phi_s = K_1 e^{-K_2 I} - K_3 \quad (1)$$

where ϕ_s is the volume fraction of solids and $K_{1,2}$ are constants which describe the attenuation characteristics of the material. Error of 3–5% of volume fraction of solids was typical, and was highest at very low volume fractions, due to the statistical nature of radiation interactions.

A Sony DFW-X700 digital camera equipped with an optical zoom macro lens was used to capture images of the sedimentation process, in order to complement the gamma system readings with respect to the mudline height variation with time. This camera outputs square-pixel images at 1024 × 768 resolution, directly to a computer hard drive via an IEEE-1394 (FireWire) interface. Fluidization activity was captured at the highest possible frame rate (15 frames per second), while overall mudline settling was captured at much longer intervals (10 minutes to 1 hour, depending on the region of the mudline and concentration of the test). The camera was controlled by intervalometer software which captured each image with a time stamp at specified intervals. The images were renamed, sorted, and numbered sequentially with a custom computer program written for this purpose, in order to enable import of the images into QuickTime as a continuous movie. Adjustment of image histograms was performed (http://quicktime.apple.com/) to view the image sequence using TIFFany3 on Mac OS X, which allows flexible batch processing of images. Frame sequences were analyzed using the FlowJ plugin for ImageJ developed by Abràmoff et al. (2000).

3 EXPERIMENTAL PROCEDURE

Prior to testing with the gamma system, it was necessary to calibrate the equipment with known concentrations of kaolinite in water. For this purpose, mixtures are prepared at 0, 25, 50, 100, 150, 200, and 400 g/L, in an acrylic cylinder of identical diameter and wall dimensions as the test cylinder. This cylinder is placed in the appropriate location in the traversing system, and three replicate gamma counts of 2 s duration are taken for each mixture. An exponential curve fit to concentration vs. count rate resulted in an equation similar in form to Equation 1, which could be used for post-processing the raw count data from the gamma system. After calibrating the gamma equipment, the settling column is carefully cleaned, and steps are taken to ensure that all equipment was ready to start as soon as the clay-water suspension was prepared. For a specified initial concentration, the appropriate volume of water was measured, using a settling column as a graduated cylinder. Likewise, the appropriate mass of kaolin powder is weighed, and placed in the mixer.

In each of these tests, dry kaolin powder was mixed with tap water for a minimum of five minutes, either by compressed air or a rotary device. Samples mixed by means of compressed air were mixed in the sedimentation column, while samples mixed with the rotary mixer were mixed in an external vessel for 5 minutes and then pumped into the sedimentation column already in place for measurements (Figure 2**Error! Reference source not found).**

4 RESULTS

Table 1 summarizes only the results of the experiment which is performed at initial concentrations of 125 g/L-500 mm. In Table 1, "ϕ_{s0}" refers to initial volume fraction, "C_0" refers to initial concentration, "H_0" refers to initial height, "$\bar{u}_{s0}$" is the average initial settling velocity determined from the gradient of the linear increment of the mudline interface, "T_p" is the time spent in the initial (linear) rate period, "Mixing" denotes the type of mixing, and "Measure" indicates whether purely visual measurement or the gamma system was used.

Visual observations made by the digital camera suggest the occurrence of self-weight fluidization. Fluidization pipes are recorded during the early stages

Table 1. Summary of experimental results for the tests discussed in this article.

ϕ_{s0}	C_0 (g/L)	H_0 (mm)	$\bar{u}_{s0}$ (mm/s)	T_p (sec.)	Mixing
0.05	125	500	0.011	16500	Rotary

of the test (a similar condition was recorded for the 75 g/L runs) while remnant volcano-like structures are observed at the end of the test atop the sediment layer at the fluidization pipe's exit into the overlying fluid. A photograph of a remnant volcano, taken when the mudline interface is at 210 mm is shown in Figure 3, a clear example of the remnant volcanoes from the self-weight fluidization process. Few remnants of air bubbles (pockets) were also visible on the mud surface, in spite of degassing with the vacuum system prior to the test.

Figure 4 shows the isoconcentration lines detected by the gamma system. The total duration of the test is approximately twelve days.

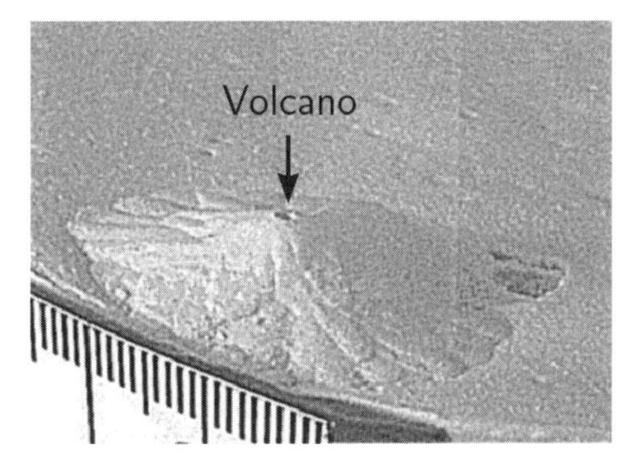

Figure 3. Photograph of a remnant volcano, in pure kaolinite. Initial concentration of 125 g/L, initial height of 500 mm; scale markings visible in lower-left corner are 1 mm apart.

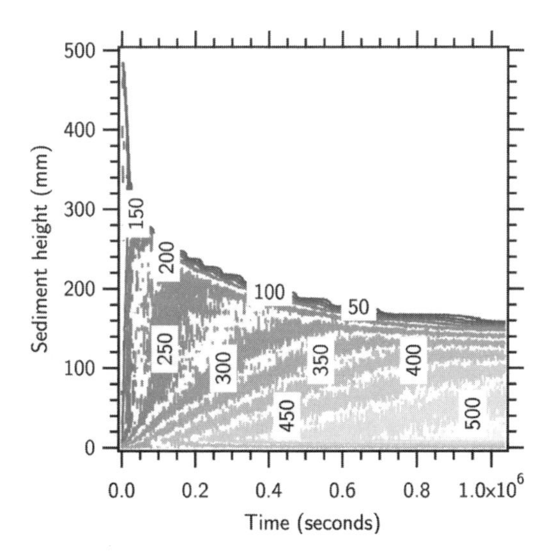

Figure 4. Isoconcentration lines for 125 g/L-500 mm test, from gamma system.

In Figure 4, height is plotted on the vertical axis, as a function of time (horizontal axis), and the contour parameter is concentration in grams per liter. The isoconcentration lines are plotted using Igor Pro, by interpolating between density values in the height-time plane. Self-weight fluidization is observed in this test, once again with no systematic local variations in the batch curve which could be attributed to fluidization.

Knowing that self-weight fluidization is present, in the tests (reported in Table 1), it is desired to compare the results of the experiments where fluidization is pronounced with a numerical simulation.

The initial input parameters to the model include the initial mixture volume fraction of solids "ϕ_{s0}", the initial mixture height, "H_0", and a linear distribution for the effective stress (see Diplas and Papanicolaou 1997). The values of the compressibility parameters (defined in section 2) and fluid properties used in the model are summarized in Table 2.

Figure 5 illustrates the experimental mudline and L-curves, plotted with the results from the numerical test. Overall, experimental and numerical results for the mudline interface compare well with each other. The underprediction shown in Figure 5 is primarily

Table 2. Parameters used in numerical simulation cases.

Parameter	75g/L-1500 mm	125g/L-500 mm
$1/\alpha$	200	150
ϵ_{s0}	0.062	0.07
β	0.323	0.323
δ	1.145	1.145
K_0	$8.5 \infty 10^{-13}$	$6 \infty 10^{-13}$
ρ_s	2630	2630
μ	0.001	0.001

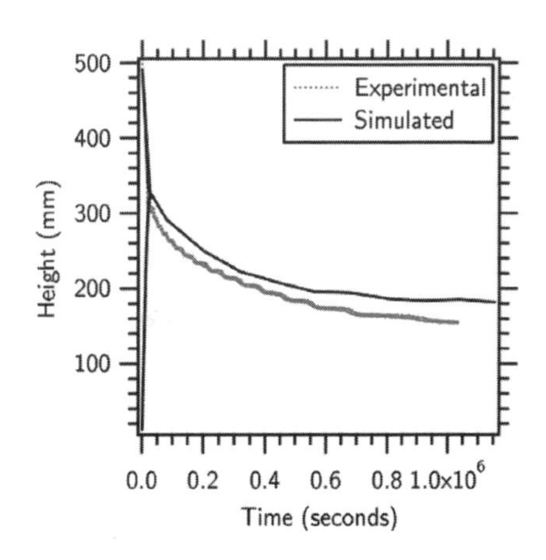

Figure 5. Numerical simulation compared with experimental results for the 125 g/L-500 mm case.

attributed to our inability to identify the exact location of the sediment–suspension interface from the experiment (L-curve) (Tiller, personal communication). In any case fluidization is primarily observed in the early stages of the sedimentation tests (first falling rate region) therefore, the underprediction in the settlement rate for the 125 g/L case can not be attributed to fluidization.

In addition, one can notice that although the effects of self-weight fluidization are not directly considered in the numerical model, in an overall sense, the present model may be able to quantify the sedimentation behavior of a fluid mud, if the necessary boundary conditions are known. The present simulations clearly show that the model predictions seem to be unaffected by the pronounced presence of self-weight fluidization especially in the early stages of these tests. The results of the numerical model experiments further strengthen the observation of the authors that for the materials in question (kaolinite and water), self-weight fluidization cannot be detected by "traditional" means of observing the batch sedimentation curve or even the more sensitive flux curve, as the models bulk property predictions agree with the experimentally measured properties.

5 CONCLUSIONS

The present study examined the results of laboratory experiments in batch sedimentation of water-kaolinite mixture. The overarching objective of this research is to develop a blueprint methodology for batch sedimentation tests that allows qualitative and quantitative identification of self-weight fluidization by removing the effects of mixing on the evolution of fluidization. Other questions that are addressed here are the efficacy of conventional tools such as a batch curve to describe the onset of self-weight fluidization. The role of the initial conditions (concentration, slurry height) in affecting the temporal and spatial characteristics of fluidization is also assessed. The most important findings of this laboratory/numerical investigation can be summarized as follows:

1. Batch sedimentation tests constitute a reliable laboratory procedure for studying self-weight fluidization if and only if the effects of mixing on masking the process are minimal.
2. For the range of the initial concentrations examined here (10 g/L to 125 g/L) it is found that fluidization is present and at the early stages of the tests. The effects of fluidization pipes on the mudline interface are not detectable. Despite the presence of ubiquitous fluidization pipes propagating upwards during the rotary mixing tests, no inflection points are recorded in the mudline. As a result, no increase in the localized settling velocity was recorded for the kaolinite cases reported herein. This finding is against the finding reported by Vesilind and Jones (1990) and Holdich and Butt (1995a) who suggested that fluidization channels increase the settling velocity of a calcium carbonate slurry with initial concentrations of 400 g/L. Perhaps this is attributable to the fact that the above investigators used calcite slurries, rather than the kaolinite used in the present study.
3. The results of the numerical model experiments further validate the observation of the authors that for the materials in question (kaolinite and water), self-weight fluidization cannot be detected by "traditional" means of observing the batch sedimentation curve or even the more sensitive flux curve, as the models bulk property predictions agree with the experimentally measured properties. Hence, a more sophisticated examination of microstructure and small-scale processes will be necessary to determine the effects of fluidization. This includes the use of a X-Ray Computer Tomographer that can facilitate the 3-D representation of the structure.

ACKNOWLEDGMENTS

This study was partially funded by the Office of Naval Research, under Award Number N00014-02-1-0043.

REFERENCES

Abràmoff, M. D., Niessen, W. J., and Viergever, M. A. (2000). Objective quantification of the motion of soft tissues in the orbit. *IEEE Trans. Med. Imag.*, 19(10):986–995.

Allen, J. R. L. (1982). *Sedimentary structures, their character and physical basis*. Elsevier Scientific Pub. Co., New York, NY.

Allison, M. A., Nittrouer, C. A., and Kineke, G. C. (1995). Seasonal sediment storage on mudflats adjacent to the Amazon River. *Marine Geology*, 125:303–328.

Austin, J. C. and Challis, R. E. (1998). The effects of flocculation on the propagation of ultrasound in dilute kaolin slurries. *Journal of Colloid and Interface Science*, 146–157.

Been, K. (1980). *Stress Strain Behaviour of a Cohesive Soil Deposited Under Water*. PhD thesis, University of Oxford.

Been, K. (1982). Nondestructive soil bulk density measurements by X-ray attenuation. *Geotechnical Testing Journal*, 4(4):169–176.

Best, J. L. (1989). Fluidization pipes in volcaniclastic mass flows, Volcan Hudson, southern Chile. *Terra Nova*, 1: 203–208.

Blunt, M. J. (2001). Flow in porous media – pore-network models and multiphase flow. *Current Opinion in Colloid Interface Science*, 6(6):197–207.

Channell, G. M., Miller, K. T., and Zukoski, C. F. (2000). Effects of microstructure on the compressive yield stress. *AIChE Journal*, 46(1):72–78.

Cole, R. F. (1968). *Experimental Evaluation of the Kynch Theory*. PhD thesis, University of North Carolina, Chapel Hill.
Dell, C. C. and Kaynar, M. B. (1968). Channelling in flocculated suspensions. *Filtration & Separation*, pages 323–327.
Den Haan, E. J. (1992). Formulation of virgin compression of soils. *Géotechnique*, 42(3):465–483.
Dionne, J.-C. (1973). Monroes: a type of so-called mud volcanoes in tidal flats. *Journal of Sedimentary Petrology*, 43(3):848–856.
Diplas, P. and Papanicolaou, A. N. (1997). Batch analysis of slurries in zone settling regime. *Journal of Environmental Engineering*, 123(7):659–667.
Druitt, T. H. (1995). Settling behaviour of concentrated dispersions and some volcanological applications. *Journal of volcanology and geothermal research*, 65:27–39.
Fitch, B. (1993). Thickening theories – an analysis. *AIChE Journal*, 39(1):27–36.
Font, R. (1991). Analysis of the batch sedimentation test. *Chemical Engineering Science*, 46(10):2473–2482.
Gaudin, A. M., Fuerstenau, M. C., and Mitchell, S. R. (1959). Effect of pulp depth and initial pulp density in batch thickening. *Mining Engineering*, 613–616.
Holdich, R. G. and Butt, G. (1995a). Compression and channelling in gravity sedimenting systems. *Minerals Engineering*, 9(1):115–131.
Holdich, R. G. and Butt, G. (1995b). An experimental study of channelling and solid concentration during the batch sedimentation of calcite suspensions. *Trans. IChemE, Part A*, 73:833–841.
Jackson, R. (2000). *The Dynamics of Fluidized Particles*. Cambridge University Press.
Jepsen, R., McNeil, J., and Lick, W. (2000). Effects of gas generation on the density and erosion of sediments from the Grand River. *J. Great Lakes Res.*, 26(2): 209–219.
Jones, G. N. (1986). *Channelling During Gravity Thickening*. PhD thesis, Duke University, Durham, North Carolina.
Knoll, G. F. (1979). *Radiation detection and measurement*. John Wiley & Sons, Inc., New York.
Kuehl, S. A., Nittrouer, C. A., and DeMaster, D. J. (1988). Microfabric study of fine-grained sediments: observations from the Amazon subaqueous delta. *Journal of Sedimentary Petrology*, 58(1):12–23.
Kynch, G. J. (1952). A theory of sedimentation. *Trans. Faraday Soc.*, 48(166):166–176.
Lee, S.-C. and Mehta, A. J. (1997). Problems in characterizing dynamics of mud shore profiles. *Journal of Hydraulic Engineering*, 123(4):351–161.
Liu, J.-C. and Znidarčić, D. (1991). Modeling one-dimensional compression characteristics of soils. *Journal of Geotechnical Engineering*, 117(1):162–169.
Maxwell, A. R., Papanicolaou, A. N., and Hou, Z. (2003). Microventing in underconsolidated sediments. In *Proc. ASCE 16th Engineering Mechanics Conference*, Seattle, WA.
McBride, M. B. (1994). *Environmental Chemistry of Soils*. Oxford University Press, Inc., New York.
McConnachie, I. (1974). Fabric changes in consolidated kaolin. *Géotechnique*, 24(2):207–222.
Mehta, A. J. (1989). On estuarine cohesive sediment suspension behavior. *Journal of Geophysical Research*, 94(C10):14303–14314.
Michaels, A. S. and Bolger, J. C. (1962). Settling rates and sediment volumes of flocculated kaolin suspensions. *I&EC Fundamentals*, 1(1):24–33.
Mount, J. F. (1993). Formation of fluidization pipes during liquefaction: examples from the uratanna formation (lower cambrian), south australia. *Sedimentology*, 40:1027–1037.
Papanicolaou, A. N. and Diplas, P. (1999). Numerical solution of a non-linear model for self-weight solids settlement. *Applied Mathematical Modeling*, 23:345–362.
Richardson, J. F. and Zaki, W. N. (1954). Sedimentation and fluidisation: Part I. *Trans. Instn Chem. Engrs*, 32: 35–53.
Roche, O., Druitt, T. H., and Cas, R. A. F. (2001). Experimental aqueous fluidization of ignimbrite. *Journal of Volcanology and Geothermal Research*, 112:267–280.
Schultheiss, P. J. (1990). In-situ pore-pressure measurements for a detailed geotechnical assesssment of marine sediments: state of the art. In Demars, K. R. and Chaney, R. C., editors, *Geotechnical Engineering of Ocean Waste Disposal, ASTM STP 1087*. American Society for Testing and Materials, Philadelphia.
Tiller, F. M. (1981). Revision of Kynch sedimentation theory. *AIChE Journal*, 27(5):823–829.
Tiller, F. M. and Khatib, Z. (1984). The theory of sediment volumes of compressible, particulate structures. *Journal of Colloid and Interface Science*, 100(1):55–67.
Tory, E. M. (1961). *Batch and Continuous Thickening of Slurries*. PhD thesis, Purdue University, Lafayette, Indiana.
van Olphen, H. (1977). *Clay Colloid Chemistry*. John Wiley & Sons Inc., 2nd edition.
Vesilind, P. A. and Jones, G. N. (1990). A reexamination of the batch-thickening curve. *Research Journal of the Water Pollution Control Federation*, 62:887–893.
Vesilind, P. A. and Jones, G. N. (1993). Channelling in batch thickening. *Wat. Sci. Tech.*, 28(1):59–65.
Yamamoto, T., Konig, H. L., and Hijum, E. V. (1978). On the response of a poro-elastic bed to water waves. *Journal of Fluid Mechanics*, 87:193–206.
Yun, J. W., Orange, D. L., and Field, M. E. (1999). Subsurface gas offshore of northern California and its link to submarine geomorphology. *Marine Geology*, 154: 357–368.

Mountain rivers

River, Coastal and Estuarine Morphodynamics: RCEM 2005 – Parker & García (eds)
© 2006 Taylor & Francis Group, London, ISBN 0 415 39270 5

The hydraulic scaling of step-pool systems

P.A. Carling
School of Geography, University of Southampton, Southampton, UK

W. Tych & K. Richardson
Lancaster Environment Centre, Lancaster University, Lancaster, UK

ABSTRACT: The hydraulic control of the down channel spacing of step-pool pairs within bedrock channels has been considered using advanced statistical analysis of two spatial data series from separate bedrock channels in Thailand. Bedload and suspended sediment loads are unmeasured, but the load is evidently small with only very small quantities of pebbles and sand-sized sediments stored in pools during periods of low flow. Importantly, coarse gravel, such as cobbles and large blocks, is rare or absent and consequently few steps, if any, are characterized by groups of boulders. Rather, in each case, pools are deeply eroded into homogeneous bedrock whilst steps consist of less eroded bedrock. Both systems carry substantial, high-energy 'clear-water' flows each monsoon season during which the small sediment load abrades the channel bed. The presence of neither significant structural control nor a boulder bedload, ensures that the morphology of step-pool pairs is conditioned by hydraulic action alone rather than by the presence of joints, faults, bedding planes and boulder deposition as steps. A strong periodic signal in the spacing of steps in both systems is taken as evidence for a controlling inherent flow instability within step-pool systems that conditions the response of the bed, including the height of the steps. This is in contrast to several prior studies of alluvial step-pool systems wherein emphasis is placed on the size of the mobile boulders in scaling alluvial step heights. It is proposed that the optimum spacing is such that the energy generated at an upstream step is effectively dissipated before the next downstream step and an ensemble of step-pools sustains an equilibrium reach-scale energy slope. Effective dissipitation is yet to be quantified but any treatment will constitute a joint consideration of the threshold of clast ejection from the pools and the threshold of bedrock erosion in the pools.

1 INTRODUCTION

Step-pools are common in mountain river systems and act to dissipate energy (Heede, 1981; Chin, 1989). Consequently it is important to understand their hydraulic function and there is a burgeoning literature that considers the scaling of step-pool systems in steep gravel-bedded alluvial channels. Predominately the literature determines that the geometry of the step-pool system can be described as a function of the step spacings and step heights, with the latter determined by the size of the largest clasts forming the steps (see Curran and Wilcock, 2005). The free variable is the step spacing, which is measured as the distance between the crests of the successive clast accumulations that constitute the steps. High flow events may destroy some or all of the clast accumulations, which are re-established in part during the recession limb of the flood or during subsequent competent flows (e.g. Chin, 1998). Thus step spacings might adjusted through time. In some systems, step spacing is relatively uniform (Chin, 2002), whilst in other systems the spacing is more irregular (Zimmermann and Church, 2001; Curran and Wilcock, 2005).

Local variability in the locus of steps might be controlled by singular peculiarities of the channel, for example by bed roughness impeding clast transport, by woody debris, or by channel width or slope variations. However, although local bed roughness and the character of mobile sediment might exert an influence (Curran and Wilcock, 2005), it is posited here that it is the inherent hydraulic character of the flows that primarily determines the step-pool geometry. This conclusion is derived from the limited published observations that well-formed step-pool systems develop in bedrock (e.g. Bowman, 1977; Wohl and Grodek, 1994; Duckson and Duckson, 1995) wherein a mobile sediment load largely may be absent. The conclusion is used herein to generate an hypothesis to be tested.

1.1 *An hypothesis*

Hypothesis: Step-pool systems of a defined geometry develop in streams running across homogeneous bedrock wherein the sediment load is negligible.

Figure 1. Examples of step-pool systems. a: homogeneous soil. b: bedrock. Note absence of clast accumations on steps in each example.

A defined geometry (i.e. spacing) would imply a dominant hydraulic control on step-pool spacings in bedrock systems and potentially within alluvial systems as well. To test the hypothesis, steep streams running across rock or soil were examined in a number of geographical localities worldwide. Several different rock-types were examined but all were selected to exhibit a high degree of homogeneity in-as-much as discontinuties such as bedding planes, joints, fractures, faults and dikes were largely absent. Fissured rock-types were not examined because the fissuring may induce variations in bed roughness, channel width and slope and influence the local flow patterns. In all cases selected there was little evidence of a significant suspended load or bedload. In each case, there was a strong, visible apparent periodicity in the spacing of pools eroded into the bedrock. In all examples studied, steps did not consist of accumulated boulders but were constituted of less eroded bedrock or soil (e.g. Figure 1). Thus qualitative evidence is widely available within the fluvial environment to support the hypothesis.

2 FIELD SURVEY

To test the hypothesis quantitatively, the long profiles of two bedrock streams in Thailand were surveyed using a Trimble™ 5600 Total Station. The river channels selected are Huai Nang Rong developed within rhyolitic agglomerate and Than Rattana developed in flow-banded andesite. In the latter case although flow banding is evident no obvious geological control was exerted on the channel form. Huai Nang Rong is located in Tambon Hin Tang, Amphoe Muang district with a channel width of around 50 m. Than Rattana is located in Khao Yai National Park in Tambon Noen Hom District, with a channel width of around 15 m.

Both streams have a high discharge during the wet season, but are ungaged. In each system, a single straight reach was identified visually as consisting of a series of well-defined step-pool pairs. In the Huai Nang Rong a *c.* 588 m reach was surveyed to define the along-channel topographic variation over about 29 step-pool pairs; this obtained 206 survey points with an average spacing of 2.85 m. These data included key reference points such as step crests and the base of pools. In Than Rattana, a *c.* 420 m reach was surveyed to define the topography of about 15 step-pool pairs; this obtained 192 survey points with an average spacing of 2.18 m. Figure 2 shows the unprocessed field data for Huai Nang Rong. To obtain equally-spaced data series required for statistical analysis, the series were resampled at 1m intervals using cubic spline interpolation, giving good results as shown for Than Rattana (Figure 3). Additionally, where there were wider gaps between measurements, missing values flags were set, so as

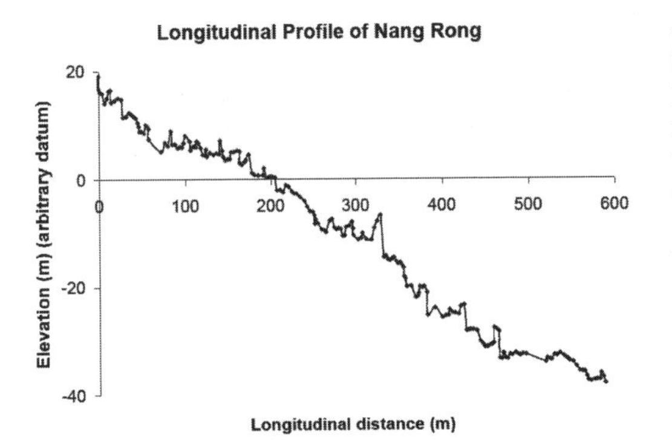

Figure 2. Example of long profile field survey of about 30 step-pools for Huai Nang Rong.

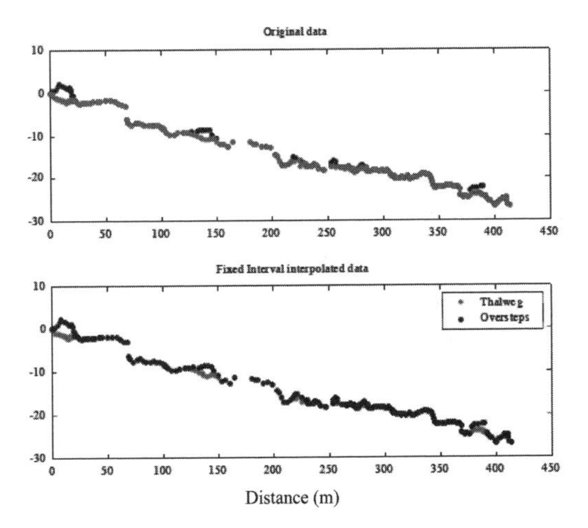

Figure 3. Resampling of field data at 1m intervals for Than Rattana.

to not prejudice the spatial-varying spectrum results (described below). Finally each series was detrended to remove the overall down channel gradient. In the case of Than Ratanna some steps were breached by low-flow inner channels. The survey data accounted for these breaches by including both the thalweg and the profile of the step either side of the inner channel (Figure 3). A statistical method (DBM) described below then was used to search for preferred periodicity in the topographic data series and the periodicity results were compared to visual estimates of step-spacings that might be obtained from Figures 2 and 3.

3 DATA-BASED MECHANISTIC MODEL

Chin (2002) applied time series (Fourier analysis) techniques to determine periodicities in step-pool profiles. Unfortunately, Fourier analysis averages through all space and any spatial-variance distribution implied by the fluctuating nature of the topography is lost. Data-based Mechanistic (DBM) modeling is an alternative methodology that gives stochastic descriptions which are appropriate to the often limited or partial data available as linear spatial data series, such as surveyed topographic transects. The DBM approach uses advanced methods of statistical identification and estimation, a particular form of Generalized Sensitivity Analysis based on Monte Carlo Simulation, and Dominant Mode Analysis; the latter involving a new statistical approach to combined model linearization and order reduction (Young, 1999). The data-based approach utilizes objective statistical analysis of the output (topography) from which inferences may be drawn in respect of an unknown input (flow structure). The methodological tools that underpin the DBM modeling can be unified in terms of an Unobserved Components (UC). In the UC model, the observed spatial-series y_t is assumed to be composed of an additive or multiplicative combination of different components, all of which have a physical interpretation and defined statistical characteristics but which cannot be observed directly. A typical, and fairly general, UC model for a scalar spatial series is the following:

$$y_t = T_t + C_t + S_t + f(u_t) + N_t + e_t$$

where y_t: the observed spatial series; T_t: a trend component; C_t: a damped cyclical or AutoRegressive-Moving Average (ARMA) component; S_t: a low frequency component; $f(u_t)$: captures the influence of a exogenous variables vector u_t, if necessary including nonlinear dynamic relationships N_t: a stochastic perturbation model (coloured noise); e_t: a normally distributed Gaussian sequence with zero mean value and variance σ^2.

The Dynamic Harmonic Regression (DHR) model used here is a special example of an UC model and is a recursive interpolation, extrapolation and smoothing algorithm developed for non-stationary time-series (Young et al., 1999) and equally applicable to spatial-series (e.g. Carling and Orr, 2000). The DHR model identifies three components in the series, i.e.,

$$y(k) = t(k) + s(k) + e(k)$$

as they are described by Young et al. (1988), where: $y(k)$: the observation at sample k (as in $y(k) = y(k\Delta t), k = 1, \ldots, T$); $t(k)$: the smooth trend (usually of the IRW type); $s(k)$: a low frequency component of Dynamic Harmonic Regression type:

$$s(k) = \sum_{j=1}^{Ns} \{a_j(k)\cos(\omega_j k) + b_j(k)\sin(\omega_j k)\}$$

where a_j and b_j: the Spatial-Variable-Parameters (TVP's) of the model; Ns: the number of low frequency components; ω_j: the set of frequencies chosen by reference to the spectral properties of the spatial-series.

This component is a linear combination of harmonics, similar to the harmonic regression used to interpret the spectrum. The main difference is in the fact that the coefficients may now vary in space. Finally, $e(k)$: is a zero mean, serially uncorrelated white noise component.

4 RESULTS

In each system, the fitted DBM statistical model well-described the resampled data series (Figure 4). In each system, using an initial spectral estimation (constant parameters AR spectrum with AR order estimated using Akaike criterion), a strongly defined preferred component of periodicity was identified with subsidiary periodicities (Figures 5 and 6), that in one system (Than Rattana) were phase-shifted in the downstream portion of the channel (Figure 7).

The spatiall varying spectral analysis demonstrates that for each system there is a prefered and regular spacing of topographic high points. For Huai Nang Rung this is around 23 m (Figure 8). In the case of Than Rattana it is around 50 m in the upper reach but this is shifted suddenly to around 25 m in the lower portion of the system (Figure 7). No explanation readily can be provided for this change in preferred spacings based on reach-scale bed slope which is constant (Figure 3) or channel width. However it is know that step spacings are inversely related to channel slope, becoming closer for steeper gradient streams (Grant et al, 1990; D'Agostino and Lenzi, 1997; Chin, 1999). In the case of Than Rattana the downstream limit of the surveyed reach is determined by a vertical waterfall some 60 m in hieght. It is possible that the water surface slope steepens close to the falls (*c.f.* Gardner, 1983) and this locally steepened energy gradient may translate to a bed adjustment extending over some 100 m upstream within a 'draw-down' reach. In contrast, Chin

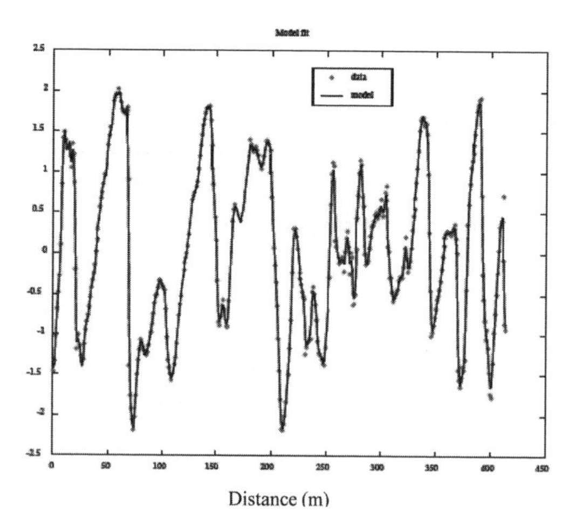

Figure 4. Example of the model fit to the resampled data for Than Rattana.

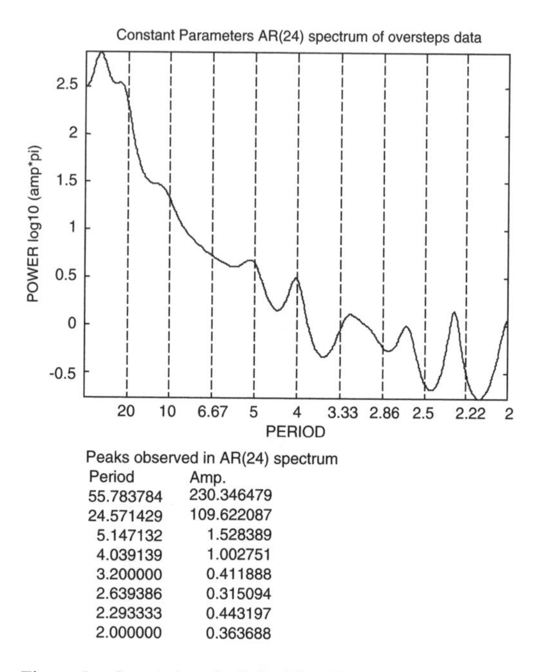

Period	Amp.
55.783784	230.346479
24.571429	109.622087
5.147132	1.528389
4.039139	1.002751
3.200000	0.411888
2.639386	0.315094
2.293333	0.443197
2.000000	0.363688

Figure 5. Spectral analysis for Than Rattana.

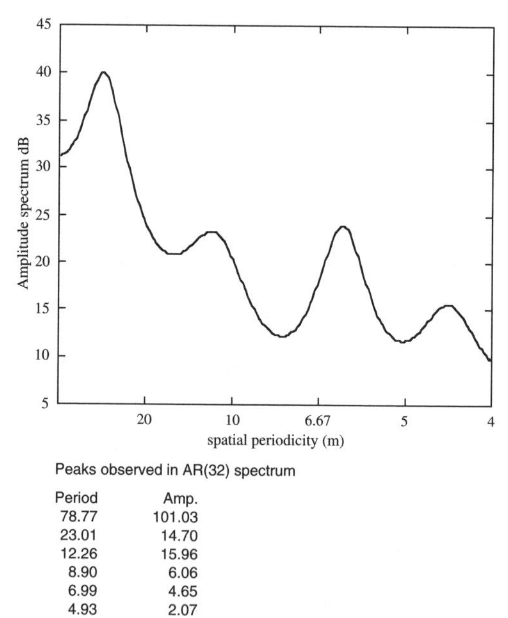

Period	Amp.
78.77	101.03
23.01	14.70
12.26	15.96
8.90	6.06
6.99	4.65
4.93	2.07

Figure 6. Spectral analysis for Huai Nang Rong.

(2002) using conventional spectral analysis applied to several alluvial channels noted that a high variance in step-pool spacings tended to mask underlying central tendencies in periodicities. The difference in the results presented here and those of Chin may represent differences in the alluvial systems studied by Chin (2002) and the bedrock systems reported here. However the greater sensitivity of DBM modelling to spatial trends in step-pool spacings in contrast to conventional spectral methods should be noted.

Wohl and Ikeda (1997) reported that they, and others, had only observed defined step-pool spacings in natural bedrock channels with pronounced heterogeneities caused by structural controls such as bedding and joints. This observation is contrary to the results presented herein which demonstrate that structural controls or deposition of boulder steps are not required to define strongly periodic step-pool spacings. Rather, as in laboratory studies on homogeneous substrata (Wohl and Ikeda, 1997) the periodic bed structures observed in the field are interpreted to reflect the presence of large-scale periodic flow structures. Specifically in the case of Huai Nang Rong step spacing developed in homogeneous bedrock is strongly defined throughout the surveyed reach; being around 23 m. However, it should be noted that the strength of this periodic signal varies with distance with four strong (power) peaks with values of 20 or more defining the "23 m" step spacing and one strong peak indicative of a greater spacing at about 450 m down channel. Weak signals at lower periodities (*ca.* 4+, 6 and 12 are regarded as harmonics of the main periodicity (Figure 8). The DBM models provide a sensitive means of isolating any prefered periodicity within the topography including shifts in dominant periodicity down-channel. In this respect Figures 7 and 8 may be interpreted in the same manner. Both demonstrate subtle spatial variation in spectral signatures with an overprint of noise. In neither case is there any evidence of a control on the spectral signature exerted by gradient or by channel width, for example. Consequently, the DBM model could be applied profitably to longer spatial step-pool topographic series, wherein covariation in potential external controls (slope; width) could also be examined within the DBM framework.

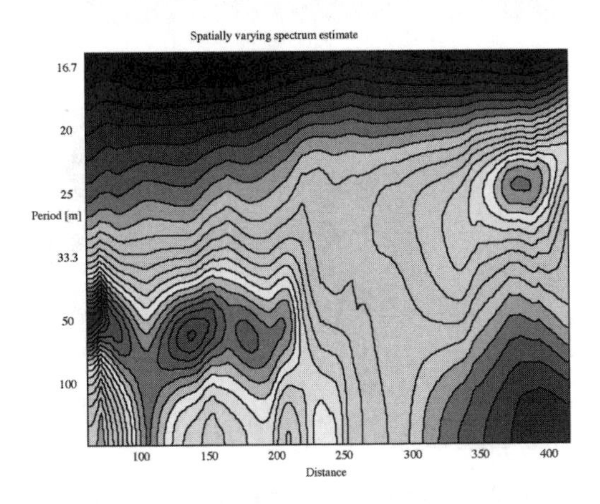

Figure 7. Dynamic AutoRegressive spatially-varying 2D visualization spectrum shows evolution of the peaks visible in the standard spectrum as steps develop along the river Than Rattana (see Figure 4 for values of power amplitudes).

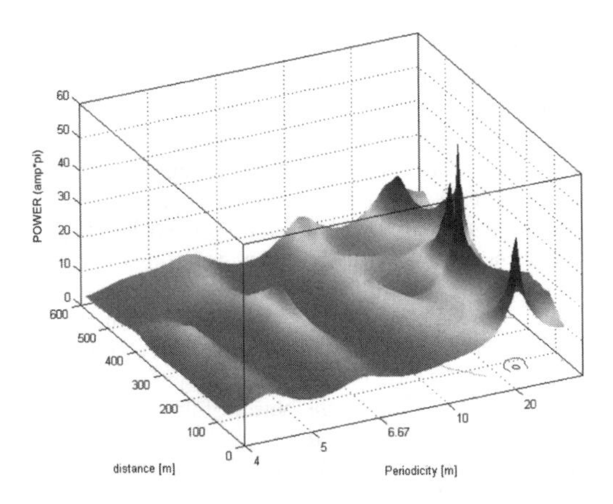

Figure 8. Dynamic AutoRegressive spatially-varying 3D visualization spectrum shows evolution of the peaks visible in the standard spectrum as steps develop along the river Huai Nang Rong.

Within Than Rattana *some* bedrock steps were breached (Figure 3), in-as-much as well-defined steep-walled low-flow narrow (*c.* 1 m maximum width) central channels cut through the steps (the later up to 20 m wide). However, many steps showed no evidence of central channel development. It is postulated that the central channels are cut by sustained low and moderate flows through dissolution of the rock matrix and through direct abrasion of the constituent grains by a small bedload. Evidently these features might be enhanced during high flows when coarse bedload will follow the thalweg. However, their absence on some steps and their narrow and 'delicate' steep-walled forms seem to militate against their evolution primarily by high flow power during high flow events (*cf* Wohl, 1993) and in the experience of the authors, inner channels in steps always possess knickpoints at their heads, implying that inner channel formation occurs through knickpoint retreat. Further it is known that during high flows within plunge pools, energy and flow dissipates radially outwards towards the distal ends of pools (Minor et al., 2002) (see Figure 10b) and so concentrated flow during high flows is unlikely to be the explanation for breached-step morphology. Thus this delicate breached morphology is taken as strong evidence that high flows constrain step-pool spacing through abrasion and dissolution of the deep pools when steps and pools are completely inundated.

5 DISCUSSION

The qualitative and quantitative results taken together indicate that coarse bedload accumulation is not required to define steps in bedrock channels and by corrolary this component also, in principle, need not be a prime requirement to determine the geometry of alluvial step-pools. Rather the periodicity of the bedrock step-pools reflects self-similar geometries of the ensemble of individual step-pool units. This repeated morphological structure is a response to an inherent hydraulic periodicity that determines the spacing of zones of high fluid stressing (high erosion) and low stressing (less erosion) such that fluid forcing largely dictates the locus of less-eroded bedrock steps in bedrock systems and possibly step deposition in alluvial systems. In each of several units, energy is dissipated in a similar manner downstream of each step. A conceptual framework is provided below.

The initial condition for step-pool development is envisaged as a uniform v-shaped valley incised within bedrock. The steepness of the slope ensures that unsteady transcritical flow is common for a range of discharges including those which are competent to erode bedrock through entraining and transporting a bedload over the bedrock surface. For transcritical flows, standing waves will develop with a defined periodicity and spatial variation in energy expenditure at the bed surface (see Richardson and Carling, 2005: there Fig. 108). Such a model for incipient step-pool sequences in alluvial systems has been proposed prior where deposition of coarse grains as nascent steps is anticipated to occur beneath the standing waves (e.g. Whittaker and Jaeggi, 1982; Allen, 1983; Chartrand and Whiting, 2000) with flushing of finer material from the intervening pools on falling stages (Curran and Wilcock, 2005). Similarly, it is proposed here that the intensity and frequency of bedrock erosion will be spatially in accordance with the spatial development of diverging-converging flow and standing waves (*vis* Ashton and Kennedy, 1972; Tinkler, 1993; Tinkler and Wohl, 1998), being at a maximum beneath the wave troughs and at a minimum beneath wave crest. The spacing of the standing waves initially will be dependent on water depth (Allen, 1983), but because bedrock step-pools are abrasional forms (or rarely dissolutional) it follows that only flows competent to erode the bedrock are formative. Thus a threshold flow should be isolated below which there is no erosion and no development of topography. Above this threshold, as the topography develops pronounced amplitude, the process should become self-perpetuating as the evolving topography reinforces the pattern of the incident flow structure, reducing the wave-spacing dependence on water depth. Thus in this model incipient step-pools are envisaged as erosional equivalents of antidunes in loose sediments. For the well developed step-pools, and on steep bedrock surfaces, two flow models are envisaged that promote further bedrock erosion. The first model is characterised by flow, passing over a step riser that is inclined, as an inclined jet (Figure 9a) termed a chute (Richardson and Carling, 2005). In this case the flow at the lip of the step does not separate from the inclined riser of the step but enters the downstream pool at an acute angle (Figure 10a). The degree of interaction of the plunging jet and the bed will vary depending on the power of the incident discharge but for competent discharges erosion will be at a maximum in the pool base and decline downstream.

Figure 9. (a) Example of deep pools separated by steps functioning as inclined overfalls, Japan. (b) Example of deep pools separated by steps functioning as free overfalls, Australia.

The second model considers a vertical or under-cut step from which the flow descends as a near-vertical free overfall (waterfall) into the downstream pool (Figures 9b and 10b). The free overfall jet will interact with the bed of the pool to varying degrees depending on

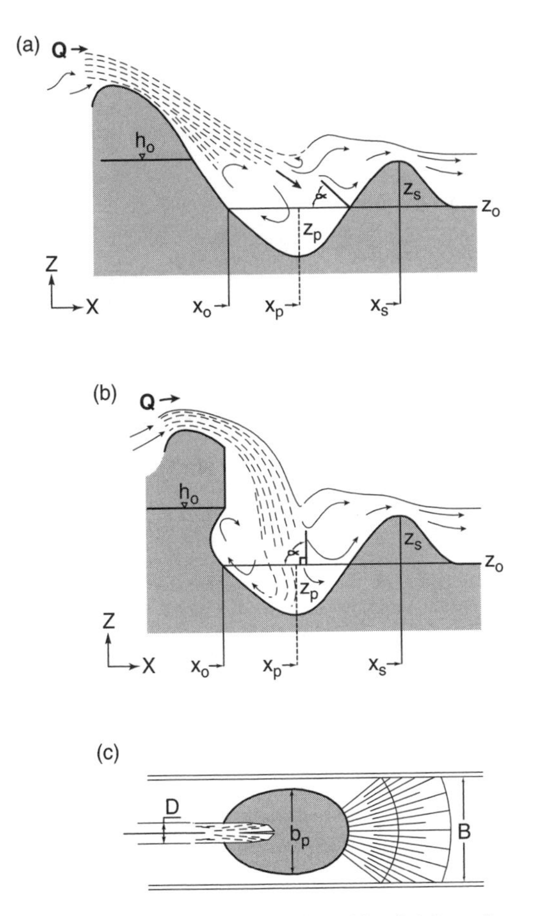

Figure 10. (a) Free overfall, (b) Chute, (c) radial dispersion of jet in pool.

the power of the vertically plunging flow (see Carling and Grodek, 1994).

The erosion of bedrock is a complex phenomenon addressed recently by the collected works edited by Scheiss and Bollaert (2002). Often in factured, fissile and jointed rocks pressure propagating into cracks is responsible for rock flake and block entrainment into the flow. Whipple et al. (2000) contend that plucking is the rate-limiting style of erosion in rocks that are well-jointed at the sub-meter scale, whereas abrasion processes dominate in massive, unjointed rocks or those with wide joint spacings relative to the magnitude of mean shear stress. Here the consideration is of largely homogeneous bedrock and the case to be considered is erosion of small grains or crystals of the constituent rock base. Three mechanisms of rock erosion are then envisaged. For soluble rock, dissolution will be most effective in highly turbulent flow and will reduce in less turbulent flow. However, in either case, for both soluble and insoluable rocks, abrasion might be the main cause of erosion and would reach a maximum within the base of the pools beneath the jet core impact as bedload and suspended load is circulated. Coarse bedload will be churned around until reduced in size and ejected downstream over the step. However pressure fluctuations can also dislodge individual grains from the rock matrix by a process termed evorsion whereby fluid stressing occurs as the maximum pressures at the plunge pool bottom are transferred through the micro-fractures and suture lines around individual grains to beneath the grains. These maximum pressures combine with the minimum pressures at the plunge pool base to create net uplift pressure. The ultimate bedrock erosion threshold is reached when the net pressure difference on any single grain is not able to eject it from the base of the pool (Schleiss, 2002).

Thus two critical thresholds of pool basal-erosion can be defined. The first relates to a coarse but minor fraction of the total load which as in other bedrock systems (Howard, 1980, 1987) can come to exert the principal control on bedrock erosion. Coarse gravel deposited in the pool base has to be entrained before erosion of the underlying bedrock by abrasion and evorsion can occur at high stage (vis, Howard and Kerby, 1983). Usually this will also entail ejection of gravel from the pool. In such a case, bedrock incision depends upon sediment transport capacity and is therefore transport-limited. Excess sediment transport capacity, relative to the sum of sediment stored within the pool and any sediment supply from upstream, results in exposed bedrock in the base of the pool at which time bedrock incision depends upon the capacity of the sediment-laden flow to detach bedrock: a detachment-limited condition (Howard, 1980; Whipple and Tucker, 1999). This latter condition pertained in most of the bedrock pools examined in this study. Further, the power of the flow in the pool and measures of the turbulence intensity, will all reduce in a downstream direction as the energy of the plunging jet dissipates downstream (Schleiss, 2002). Consequently, clasts entrained or detached from the pool base have to be ejected along an adverse slope and, in many cases, the larger ones will not be ejected but will be returned to the pool-base to be further abraded until they are small enough to be ejected. By these joint means the frequency and intensity of the bed erosion process reduces downstream of the point of maximum pool depth and the rockbed elevation rises to constitute the downstream step. The full quantitative model cannot be developed here but will be published in due course and a summary is provided here. The model is based upon the supposition that the reach-scale bed topography is adjusted to maintain critical flow with a constant minimized reach-scale specific energy (*vis* Henderson, 1966) at the formative discharge. The latter is defined as that discharge that is both competent to flush clasts from the base of the pools and to erode the bedrock. Thus step-pool sequences asymptotically approach an equilibrium state where erosion rates are spatially uniform. Erosion is initially greatest in the base of the pools, but as pool depths increase, the bed is protected by the depth of the water and the armoring of the bed by sediment such that the erosion rate reduces there until it is no greater than the erosion rate on the steps, at which point equilibrium is reached and the profile relief is maintained.

6 CONCLUSIONS

Step-pools with defined periodic spacings occur in bedrock channels wherein the structure of the bedrock is not a major control on step spacings. Consequently, the periodic topography reflects the imposed hydraulic structure. The initial flow condition is envisaged as a plane bed beneath transcritical undular flow. Periodic pressure fluctuations induce erosion of the bedrock and result in periodic variation in the bed elevation. Step spacings are determined through the mechanism of energy dissipation downstream of each step. It is proposed that the optimum spacing is such that the energy generated at an upstream step is effectively dissipated before the next downstream step and an ensemble of step-pools sustains an equilibrium reach-scale energy slope. Effective dissipitation is yet to be quantified but any treatment will constitute a joint consideration of the threshold of clast ejection from the intervening pool and the threshold of bedrock erosion in the pool base. Such a model is commensurate with the observation that steps are more closely spaced on steeper reach-scale slopes, such that the reach energy dissipation rate is optimized by the development of periodic prefered step spacings.

ACKNOWLEDGEMENTS

This project was funded in part through grants from the British Council. Hiroshi Ikeda (Tsukuba University, Japan) and Kosit Lorsirirat (Royal Irrigation Department, Thailand) are thanked for hospitality and discussions in the field.

REFERENCES

Allen, J.R.L. 1983. A simplified cascade model for transverse stone-ribs in gravelly streams. *Proc. R.Soc. London A* 385: 253–266.

Ashton, G.D. and Kennedy, J.F. 1972. Ripples on the underside of river ice covers. *A.S.C.E. Journal of the Hydraulics Division* 98: 1603–1624.

Bowman, D.A. 1977. Stepped-bed morphology in arid gravelly channels. *Bulletin, Geological Society of America* 88: 291–298.

Carling, P.A. and Grodek, T. 1994. Indirect estimation of ungauged peak discharges in a bedrock channel with reference to design discharge selection. *Hydrological Processes* 8: 497–511.

Carling, P.A. and Orr, H. 2000. Morphology of pool-riffle sequences in the River Severn, UK. *Earth Surface Processes & Landforms* 25: 369–384.

Chartrand, S.M. and Whiting, P.J. 2000. Alluval articheture in head-water streams with special emphasis on step-pool topography. *Earth Surface Processes and Landforms* 25: 583–600.

Chin, A. 1989. Step-pools in stream channels. *Progress in Physical Geography* 13: 391–408.

Chin, A. 1998. On the stability of step-pool mountain streams. *Journal of Geology* 106: 59–69.

Chin, A. 1999. The morphological structure of step-pools in mountain streams. *Geomorphology* 27: 191–204.

Chin, A. 2002. The periodic nature of step-pool mountain streams. *American Journal of Science* 302: 144–167.

Curran, J.C. and Wilcock, P.R. 2005. Characteristic dimensions of the step-pool bed configuration: An experimental study. *Water Resources Research* 41: W02030, doi:10.1029/2004WR003568.

D'Agostino, V. and Lenzi, M.A. 1997. Oringine e dinamica della morfologia a gradinata (step-pool) nei torrenti alpini ad elevata pendenza, *Riv. Semestr. Dell'Assoc. For. Del Trentino*: 7–39.

Duckson, D.W. and Duckson, D.W. 1995. Morphology of bedrock step pool system. *Water Resources Bull.* 31: 43–51.

Gardner, T.W. 1983. Experimental-study of knickpoint and longitudinal profile evolution in cohesive, homogeneous material. *Bulletin, Geological Society of America* 94: 664–672.

Grant, G.E., Swanosn, F.J. and Wolman, M.G. 1990. Pattern and origin of stepped-bed morphology in high-gradient streams, Western Cascades, Oregon. *Bulletin, Geological Society of America* 102: 340352.

Heede, B.H. 1981. Dynamics of selected mountain streams in the western United States of America. *Zeitschrift für Geomorphologie* 25: 17–32.

Henderson, F.M. 1966. *Open Channel Flow*. New York: Macmillan.

Howard, A.D. 1980. Thresholds in river regimes. In D.R. Coates & J.D. Vitek (eds), *Thresholds in Geomorphology*: 227–258. Boston: Allen and Unwin.

Howard, A.D. 1987. Modelling fluvial systems: rock-, gravel-, and sand-bed channels. In: K.S. Richards (ed), *River Channels: Environment and Process*: 69–94. Oxford: Blackwell. Oxford.

Howard, A.D. and Kerby, G. 1983. Channel changes in badlands. *Bulletin, Geological Society of America* 94: 739–752.

Minor, H.-E., Hager, W.H. and Canepa, S. 2002. Does an aerated water jet reduce plunge pool scour? In A. J. Schleiss & E. Bollaert (eds), *Rock Scour Due to Falling High-Velocity Jets*: 117–124. Lisse: Balkema.

Richardson, K. and Carling, P.A. 2005. A typology of sculpted forms in open bedrock channels. *Geological Society of America Special Paper* 392.

Schleiss, A.J. 2002. Scour evaluation in time and space – the challenge of dam designers, In A.J. Schleiss & E. Bollaert (eds), *Rock Scour Due to Falling High-Velocity Jets*: 3–22. Lisse: Balkema.

Schleiss, A.J. and Bollaert, E. 2002. *Rock Scour Due to Falling High-Velocity Jets*. Lisse: Balkema.

Tinkler, K.J. 1993. Fluvially sculpted rock bedforms in 20 Mile Creek, Niagara Peninsula, Ontario. *Canadian Journal of Earth Sciences* 30: 945–953.

Tinkler, K.J. and Wohl, E.E. 1998. A primer on bedrock channels. In K.J. Tinkler & E.E. Wohl (eds), *Rivers Over Rock: Fluvial Process in Bedrock Channels*. Geophysical Monograph 107: 1–18. Washington: American Geophysical Union.

Whipple, K.X., Anderson, R. and Hancock, G.S. 2000. River incision into bedrock: mechanisms and relative efficacy of plucking, abrasion, and cavitation. *Bulletin, Geological Society of America* 112: 490–503.

Whipple, K.X. and Tucker, G.E. 1999. Dynamics of the stream-power river incision model: implications for the height limits of mountain ranges, landscape response timescales, and research needs. *Journal of Geophysical Research* 104: 17661–17674.

Wittaker, J.G. and Jaeggi, M.N.R. 1982. Origin of step-pool systems in mountain streams, *A.S.C.E. J. Hydraulics Division* 108: 758–773.

Wohl, E.E. 1993. Bedrock channel incision along Piccaninny Creek, Australia. *Journal of Geology* 101: 749–761.

Wohl, E.E. and Grodek, T. 1994. Channel bed-steps along Nahal Yael, Negev Desert, Isreal. *Geomorphology* 9: 117–126.

Wohl, E.E. and Ikeda, 1997. Experimental simulation of channel incision into a cohesive substrate at varying gradients. *Geology* 25: 295–298.

Young, P. 1999. Data-based mechanistic modelling generalised sensitivity and dominant mode analysis. *Computer Physics Communications* 117:113–129.

Young, P.C., Pedregal, D.J. and Tych, W. 1999. Dynamic harmonic regression. *Journal of Forecasting* 18: 369–394.

Zimmermann, A. and Church, M. 2001. Channel morphology, gradient profiles and bed stresses during flood in a step-pool channel, *Geomorphology* 40: 311–327.

River, Coastal and Estuarine Morphodynamics: RCEM 2005 – Parker & García (eds)
© 2006 Taylor & Francis Group, London, ISBN 0 415 39270 5

A novel gravel entrainment investigation

G.M. Smart
National Institute of Water & Atmospheric Research, Christchurch, New Zealand

ABSTRACT: Gravel entrainment is instigated by lift and drag pressures. *In situ* pressure measurements were made with a differential pressure sensor installed in a gravel-bed river with one port buried 70 mm into the bed and the other flush with the bed surface. The instrument recorded uplift (and downdraft) pressures being advected by the overlying flow and static pressures within the bed. At the same time as pressure measurements were made, artificial tracking stones containing motion sensors and radio transmitters were monitored. Measurements taken over 6½ hours during the passage of a moderate flood showed that the fluctuating vertical pressure differential in the bed surface layer reached a level sufficient to lift bed particles at the same time as first particle movement was detected. Conclusions are: (a) particle entrainment can result from advecting pressure fluctuations, without lift and drag forces generated as a result of a particle's shape; (b) dilating pressures may occur in the bed surface layer during strong particle movement.

1 INTRODUCTION

1.1 *Background*

Rivers, coasts and estuaries are shaped by interactions between wind, water, tectonics and geology. The wind and water entrain, transport and deposit sediments in accordance with basic laws of physics. Provided these physical processes are understood, models can be built to predict and manage future evolution of the land-water interface and to study conditions that existed in the past. This paper reports an investigation to help develop a better understanding of the physical processes governing sediment entrainment in gravel-bed rivers.

1.2 *Conventional entrainment models*

Common criteria for calculating entrainment of bed particles are based on Shields (1936) results which relate sediment movement to a dimensionless bed shear stress $\theta = \tau_o/(\gamma(s-1)d)$ where τ_o is the bed shear stress, γ the specific gravity of the fluid and s the relative density of bed particles with size d. Measurements of steady, uniform flows indicate that gravel bed movement commences when θ reaches a critical value, $\theta_{cr} \approx 0.05$ (e.g. Meyer-Peter & Mueller, 1948). More recent research, however, indicates that shear stress alone is a poor predictor of sediment movement in all but the simplest flows (Nelson et al. 1999) and there is no universally applicable value for θ_{cr} (Buffington & Montgomery, 1997).

1.3 *Entrainment by turbulence*

In 1943 Kalinske proposed an entrainment model incorporating fluctuations in shear velocity. Einstein and El-Samni (1949) used a stochastic approach where turbulent stresses cause particle entrainment. Naden (1987) suggested that using the instantaneous vertical velocity instead of streamwise velocity would improve calculation of particle entrainment in gravel-bed rivers. Papanicolaou et al (2002) proposed a model using a joint (horizontal and vertical) local velocity distribution. Wu and Chou (2003) used the probability distribution of local approach velocity to calculate moments acting on spherical particles to give rolling and lifting probabilities.

In these approaches, the turbulent, local velocity is considered to be a crucial parameter. However, on the experimental side, flume measurements of Nelson et al (2001) show that instantaneous lift forces on a gravel particle do not correlate well with local instantaneous velocity (horizontal or vertical) or with the drag on the particle.

1.4 *Differential pressure approach*

From a mechanistic point of view, in a turbulent flow on an alluvial bed, the dominant lift and form drag forces are caused by pressure differences across individual bed particles. The relation between local velocities and pressures can be extremely complex, especially where the bed is permeable and where it comprises a range of different shapes, sizes and particle orientations. This study therefore avoids velocity and approaches the problem by directly measuring turbulent fluid pressures within a natural river bed. It investigates the way in which instantaneous local fluctuations in pressure may affect or determine particle motion.

2 EXPERIMENTAL MEASUREMENTS

2.1 *Description of the novel measurements*

Local pressure measurements and particle tracking experiments were carried out in a gravel-bed river. A differential pressure sensor was buried in the surface layer of the river bed with one port 70 mm below the bed surface and the other flush with the crests of bed-surface particles (Fig. 1). Contrary to previous investigations which have monitored forces or pressures on a particular particle, the upper port was situated in the middle of a flat, 50 mm diameter horizontal plate to avoid any local particle-induced pressure variations. The instrumentation recorded the uplift (and downdraft) pressures being advected by the overlying flow and the static pressures within the bed. At the same time as the pressure measurements were made, tracking stones beside the buried instrument were monitored. These were four artificial gravel particles, containing battery-powered radio transmitters and motion sensors. Measurements were taken over a 6½ hour period whilst a moderate flood passed the measurement site.

2.2 *Location of measurements*

Measurements were made on 14–15 May 1999 in the Waimakariri River, New Zealand at "Cross Bank" (Fig. 2), where the channel is highly braided and the bed slope is around 0.3%. Bed material is greywacke with $d_{50} = 26$ mm, $d_{90} = 75$ mm. The differential pressure measurement device and particle tracking stones were placed near the outside of a curved braid at the foot of a 300 mm high eroding bank (Fig. 3).

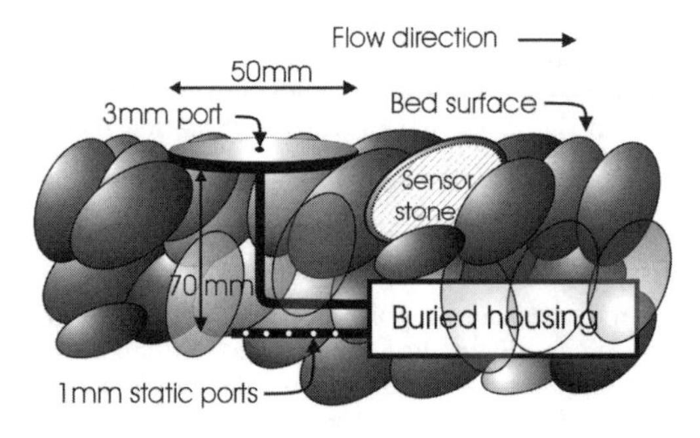

Figure 1. Placement of the pressure detection apparatus in the river bed. Stones were packed under the flat plate to represent the natural bed.

2.3 *Instrumentation*

Pressures were gauged with Motorola MPX series transducers with ±0.25% linearity and a response time of 1 millisecond. Transducers were located in a housing (Fig. 1) buried in the river bed (Fig. 3) with a shielded cable leading to an onshore PC datalogger.

The size and placement of pressure ports can affect the recorded pressures. A small port is necessary to sense the finer structure of turbulence but is also more sensitive to sheltering or blocking effects from nearby stones. The 3 mm port is protected from such effects by the surrounding plate and senses pressures that include dynamic effects from passing eddies. The sub-bed sensor detects background, in-bed pressures. To prevent proximity effects and blocking it comprised 14 ports each 1 mm in diameter and spaced over 24 mm along a 5 mm o.d. tube. A differential pressure transducer

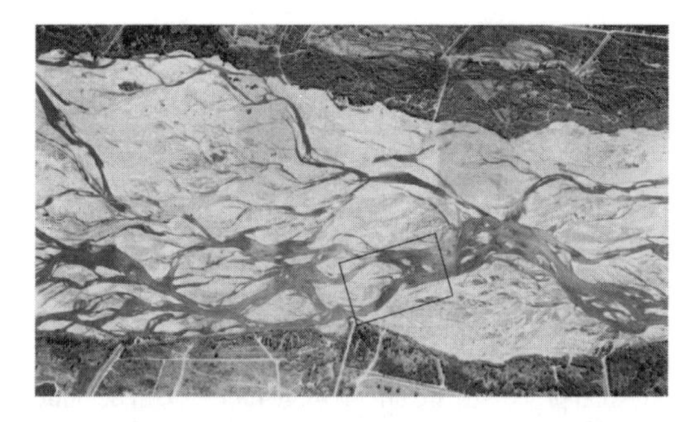

Figure 2. Waimakariri reach in which measurements were made. Photo taken on 19 March 1999 with gravel fairway ~1000 m wide, $Q = 65\ m^3/s$, flow direction from left to right. The rectangular area shown is enlarged in Figure 3.

Figure 3. Approximate site of the buried pressure sensor. At the time of the measurements the low bank at the arrowhead moved towards the arrow and became more concave than shown in this photo. Scale: the arrow is ~60 m long.

measured the instantaneous difference in pressure between the 3 mm port on the bed surface and the 1mm static ports 70 mm beneath the river bed surface. An absolute pressure transducer also measured the static head at the 1mm ports within the bed. Measurements were recorded at 112 Hz over 6½ hours.

Technical details of the stone tracking sensors are given by Habersack (2001). The density of the tracking stones was 2600 kg/m^3. Two stones were 62 mm in diameter and spherical. Two were shaped to be representative of particles forming the surface armour layer of the Waimakariri River and had an intermediate diameter of 52 mm. Motion of the tracking stones was continuously monitored while the flood passed.

Water surface slope was calculated as the gradient between water surface elevations measured upstream and downstream of the pressure detection instrument. The water surface measurement sites were located 30 m apart.

3 MEASUREMENT RESULTS

3.1 *Flood hydrograph*

A downstream gauging station indicated that river flow increased from 48 m^3/s to around 160 m^3/s and started to fall again during the monitoring period. River stage at the sites of water surface elevation measurements is shown in Figure 4 along with the mean sub-bed pressure recorded at the 1 mm static head ports. The sub-bed pressures indicate flow depth at the buried instrument if pressures within the top 70 mm of the bed are hydrostatic. Non-hydrostatic conditions could arise if the bed surface is not permeable or if vibration and shear from bulk bed movement causes changes in interstitial pore pressures. In Figure 4 some anomalous behavior in the sub-bed pressure is indicated between 22:40 hrs on 14 May and 01:00 on 15 May.

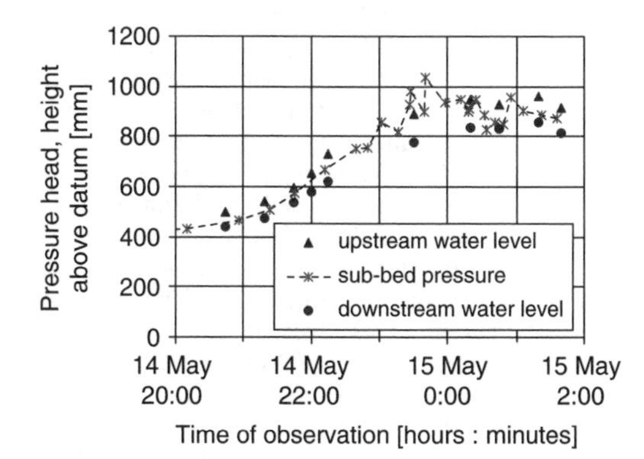

Figure 4. Stage hydrograph and average sub-bed pressure head from 8 pm on 14 May to 2 am on 15 May showing the rising flood wave.

3.2 *Bed movement*

The artificial sensor stones indicated movement (jiggling or displacement) of bed material between 22:40 on the 14th May and 00:14 am on the 15th May and again between 1:26 am to 1:36 am on 15 May. These periods are shown above the time axis of Figure 5. Movement was most intense just before midnight. The second period of movement would have indicated conditions downstream of the pressure sensor because the tracking stones were displaced during the first period of motion.

3.3 *Slope and shear stress*

The water surface slope, S, indicated by surface elevations measured upstream and downstream of the pressure sensors is shown in Figure 5. By assuming $\tau_o = \gamma HS$ (where water depth H above the instrument is interpolated from the surface elevation measurements) an estimate of Shields entrainment parameter θ may be calculated as described in section 1.2. These estimates of θ between 8 pm on 14 May and 2 am on 15 May are shown on Figure 5.

Shields criteria appears to correctly predict the onset of bed movement at 22:40 when $\theta = .05$ for median sized bed material, however, following the hydrograph peak, bed motion ceased at around 00:15 am with $\theta > 0.07$.

3.4 *Pressure fluctuations*

Pressures at the bed surface can be calculated by adding the recorded differential pressure to the recorded sub-bed pressure. Fluctuating pressure heads, calculated in this way, are shown in Figure 6 for a 7 second period prior to the arrival of the flood wave.

Pressure fluctuations arising from advecting eddies are evident in the trace of bed pressure head shown on Figure 6. Similar but less pronounced fluctuations are evident in the sub-bed pressure trace.

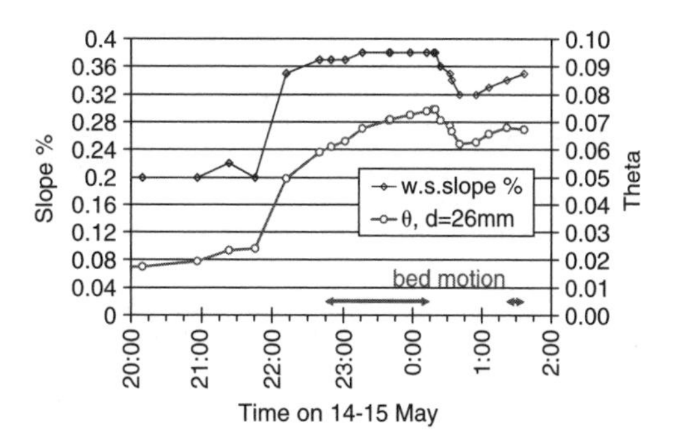

Figure 5. Water surface slope and Shields entrainment parameter. Arrows indicate when bed motion occurred.

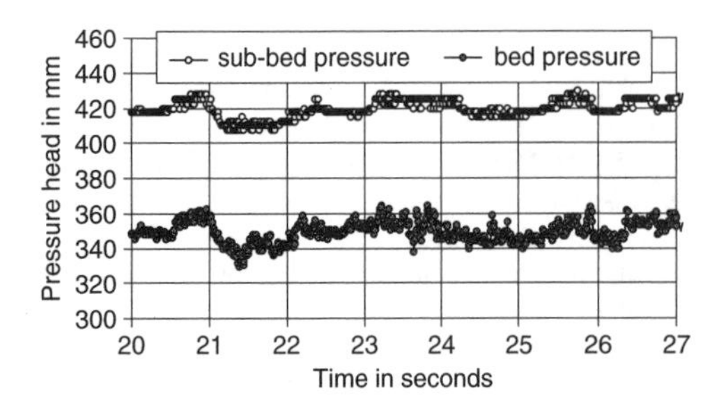

Figure 6. Pressure head at bed surface and 70 mm below bed surface from 19:22:20 (prior to flood wave). Sub-bed background pressures are a damped version of the bed surface fluctuations.

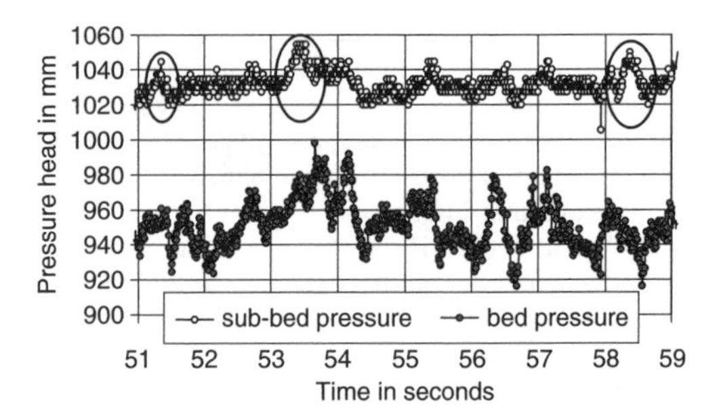

Figure 8. Pressure head at bed surface and 70 mm below bed surface from 23:40:51 (near peak bed motion). Pulses occur in the sub-bed pressure which do not come from the bed surface.

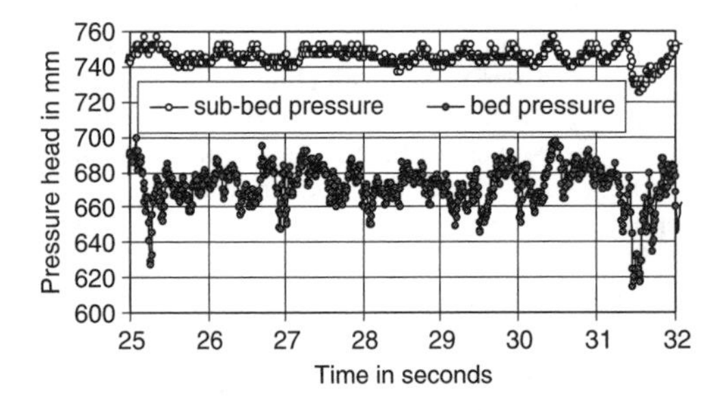

Figure 7. Pressure head at the bed surface and 70 mm below the bed surface from 22:40:25 (first bed motion).

When averaged over several seconds or more, the mean pressure head difference between the bed and sub-bed pressures is 70 mm indicating that the time-averaged pressure distribution is hydrostatic within the stationary bed surface layer and the net uplift force is zero.

Figure 7 shows bed and sub-bed pressure heads over 7 a second period on the rising flood at the time first bed motion was detected. Pressure fluctuations at the bed surface have larger amplitude than pre-flood fluctuations (Fig. 6). Large, abrupt fluctuations at the bed surface are not detected at the sub-bed sensor (e.g. at 25.2 s and 27 s on Fig. 7). The water depth above the instrument plate was ~679 mm, – similar to the depth calculated from interpolating the upstream and downstream water levels (Fig. 4).

Figure 8 shows bed and sub-bed pressure heads over an 8 second period around the time of the flood peak when the bed was very active. Figure 8 pressure fluctuations are larger than on the rising flood (Fig. 7). Most bed surface fluctuations with periods > ~0.5 s are reflected in the sub-bed pressures. Pressure surges are now evident in the sub-bed pressure that have no counterpart in the bed surface pressures. Examples of such "bed generated" pressure surges are evident in the sub-bed pressure trace at 51.35 s, 53.4 s and 58.3 s on Figure 8.

4 DISCUSSION

4.1 *Pressure fluctuations*

Uplift pressures are indicated in Figures 6–8 by the difference between the sub-bed pressure and the bed surface pressure less the 70 mm hydrostatic pressure difference. During the period shown in Figure 7 (the time that sediment jiggling was first detected by the motion sensors in the tracking stones) the maximum uplift head was 52.5 mm (about 0.25 seconds from the start of Fig. 5). This head is equivalent to a pressure of 515 Pa. The immersed weight per unit area of spherical particles diameter d is given by 2/3 $(\rho_s - \rho_w)$g d. Thus with $\rho_s = 2600\,\text{kg/m}^3$ an uplift of 515 Pa could overcome the weight force on spherical particles having diameters less than 49 mm. This is similar to the median diameter of the bed surface layer (~50 mm) and larger than the overall median grain size of bed gravel (26 mm). Thus, measured uplift pressures became high enough to lift particles that would be found on the bed surface, at the same time that the first jiggling motion of the tracking particles was detected.

A question remains as to whether the uplift pressures detected by a 3 mm port, are large enough to lift bed particles that are typically 26–50 mm in diameter. The duration of the spikes shown in Figure 7 are ~0.02 s (left side of Fig. 7) and ~0.08 s (right side of Fig. 7). Taking a frozen turbulence assumption that pressure surges advect at mean flow velocity (~2.2 m/s) indicates that a .02 s pressure "structure" would have a size of 44 mm. The .08 s uplift pulse would have a size of 176 mm. This indicates that the pressure fluctuations cover an area sufficient to lift bed particles.

As bed surface pressures were measured at the centre of a flat plate, only advected pressure fluctuations were being detected. The occurrence of advected pressures creating forces sufficient to lift bed particles at the same time as particle movement was first detected, is strong evidence that eddy induced pressures can be responsible for sediment entrainment.

4.2 *Pressure abnormalities*

As previously indicated in Figure 4 and Figure 8, there are some unusual features that appear to be associated with bed movement. Figure 4 shows that if hydrostatic conditions are assumed, when the bed is very active, 1 minute averaged water level above the pressure sensor becomes greater than the water level further upstream at the surface elevation site.

Possible explanations for the apparent depth increase are that pressure within the bed is not hydrostatic or that the flow resistance increases at the location of active bedload movement.

Figure 8 shows uplift pressure pulses equivalent to 20–30 mm head are generated within the bed during the period near peak bed motion. Such "deep seated" pressure surges are unlikely to arise from overlying flow – particle interaction but could be caused by inter-particle interactions within the mobile bed.

5 CONCLUSIONS

Uplift pressures, measured in the top 70 mm of a gravel bed river, became strong enough to lift particles found on the river bed at the same time that first motion of bed particles occurred. This gives strong evidence that particle entrainment can result from advected pressure fluctuations, without lift and drag forces generated by particle shape. While this phenomenon raises questions over some conventional velocity-based lift and drag force theories for entrainment, it explains how particles apparently shielded from the flow, can be suddenly ejected from the bed (Smart, 1984).

During periods when strong bed motion was detected, pressure surges occurred in the sub-bed that had no counterpart in pressures at the bed surface and it is concluded that they were generated within the moving bed. At the same time, according to upstream and downstream water levels, the sub-bed pressures deviated from expected hydrostatic pressures (Fig. 4). It is postulated that dilationary particle pressures, similar to those occurring in earthquakes, avalanches and lahars, were being detected in the active river bed.

ACKNOWLEDGEMENTS

This research was partly funded by contracts CO1X10014 and CO1X0218 from the Foundation for Research, Science and Technology (New Zealand). A complementary report on these experiments is being prepared for the IAHR, J. Hydraulic Research.

REFERENCES

Buffington, J.M. and Montgomery, D.R. 1997. A systematic analysis of eight decades of incipient motion studies, with special reference to gravel-bedded rivers. *Water Resources Research*, 33(8), 1993–2029.

Einstein, H.A. and El-Samni E.A. 1949. "Hydrodynamic forces on a rough wall", Rev. mod. Phys., 21, 520–524.

Habersack, H.M. 2001. Radio-tracking gravel particles in a large braided river in New Zealand: a field test of the stochastic theory of bed load transport proposed by Einstein, *J. Hydrological Processes*, Vol. 15/3, 377–391.

Kalinske, A. 1943. "The role of turbulence in river hydraulics." Bull. Univ. Iowa Studies Eng., 27, 266–279.

Meyer-Peter, E. and Muller, R. 1948. "Formula for bed load transport", Proc., 2nd, Intern. Assco. Hyd. Res., Vol. 6.

Naden, P. 1987. "An erosion criterion for gravel-bed rivers", Earth Surf. Processes Landforms, 12, 83–93.

Nelson, J.M., Schmeeckle, M. Shreve, R.L. and McLean, S. 1999. Sediment entrainment and transport in complex flows. *IAHR Symposium on River, Coastal & Estuarine Morphodynamics*. University of Genova. Vol. 1, 3–4.

Nelson, J.M., Schmeeckle, M.W. and Shreve, R.L. 2001. Turbulence and particle entrainment. In "Gravel Bed Rivers 5", Mosley M.P. (ed.) N Z Hydrological Soc., Wellington New Zealand, 221–240.

Papanicolaou, A.N., Diplas, P., Evanaggelopoulos, N. and Fotopoulos, S. 2002. Stochastic incipient motion criterion for spheres under various packing conditions. J. Hyd. Eng. 128(4), 369–379.

Shields, A. 1936. Anwendung der Aenlichkeitsmechanik und der Turbulezforschung auf die Geschiebbewegung, *Mitteilungen der Preussischen Versuchsanstal fur Wasserbau und Schiffbau*, Berlin.

Smart, G.M. 1984. "Sediment transport formula for steep channels", Journal of Hydraulics, ASCE 110(3).

Wu, F.C. and Chou, Y.J. 2003. "Rolling and lifting Probabilities for Sediment Entrainment", J. Hyd. Eng. 129(2), 110–119.

River, Coastal and Estuarine Morphodynamics: RCEM 2005 – Parker & García (eds)
© 2006 Taylor & Francis Group, London, ISBN 0 415 39270 5

Quantifying particle organization in boulder bed streams

Katherine Clancy
University of Wisconsin – Stevens Point, Stevens Point, WI

Karen Prestegaard
University of Maryland, College Park, MD

ABSTRACT: Methods are needed to characterize and quantify the effect of large particles in streams, where the particles protrude well above the flow. In boulder bed streams, the upper tail of the particle size distribution includes the largest particles and is very heterogeneous; therefore it is important that any particle metric include a consideration of both particle size and heterogeneity. In this paper we present a technique to characterize these large particles that is linked to the particle size distribution and includes a consideration of both particle size and the particle size distribution. In aggregate the particles in the upper tail of the particle size distribution can occupy a significant portion of the stream width. We defined lower limit of the upper tail of the particle distribution at the 84th percentile. To incorporate both the effects of particle size and particle heterogeneity, we used the sum of the upper tail of the particle size distribution, which we term, *topsum*. We find that topsum is a useful variable when normalized by stream width. In addition, topsum provides the framework to characterize bed particle organization such as clustering.

1 INTRODUCTION

A considerable percentage of natural streams in the United States are boulder-bed channels. They are admired for their beauty, stability, and diverse aquatic habitat. Boulder-bed streams are unique systems where an individual particle is large enough to constrict the stream width by over 10–30 percent (Whittaker, 1992), change the flow direction, and create considerable flow resistance. By contrast, the flow resistance of bed forms such as dunes is distinct from grain flow resistance in sand bed channels. As particle size increases in the channel bed, these two types of flow resistance become less distinct. In boulder-bed reaches where the particles are not clearly organized into step-pool morphology, there few options to characterize and quantify the boulder run morphology. Most of these methods use a form of the Wolman pebble count, which is used to calculate particle size statistics in gravel bed stream. The dual purpose that boulders serve in streams highlights the importance of the particle size distribution.

In brief, the Wolman pebble count is method to sample the particle size distribution of the stream bed particles. This is performed by measuring the intermediate axis of approximately 100 particles, which are chosen at random along the stream cross section. While there have been efforts to improve upon the accuracy of the sampling technique (Wolcott, J. & Church, M., 1991), most field workers and researcher continue with little or no change to method described by Wolman in 1955.

From this particle sample, the particle statistics for the stream reach are determined. Particle statistics calculated from the particle size distribution are used to quantify grain and particle resistance. The D_{84}, which represents the 84th particle size percentile in the grain size distribution, is one of the most frequently used grain statistics used is, because it is substituted for the roughness height. The roughness height is the distance above the bed where the vertical velocity profile approaches zero. Particle size on the bed affects the height of the boundary layer. This height is referred to as the roughness height. Grain resistance is often expressed as a ratio of particle diameter to depth of flow. This ratio also is called relative roughness.

It was chosen because field workers had observed that particles larger than the median play the most important role in flow resistance. The choice of particle size was validated by subsequent research that expanded the data set and found agreement with D_{84} as the choice for bed roughness (Limerinos, 1970, and Marchand et al, 1984). Later Wiberg & Smith (1991) developed a probabilistic model that yielded D_{84} as a best-fit sediment vertical length scale for gravel bed streams.

2 STATEMENT OF THE PROBLEM

2.1 *Particle size roughness height*

Representing D_{84} as the average roughness height solves the problem in gravel bed streams where the particles are well submerged. In streams where the particles of size D_{84} and larger have a very low relative submergence or protrude above the flow, the current conception of roughness height is unsuitable. In channels with cobbles and boulders, or large scale roughness features, previous methods to quantify flow resistance have included a measurement of the frequency of the "large" particles or packing density. The main problem with this method is that the size of the sampling grid will affect the packing density depending upon how the particles are distributed throughout the streambed. Fluvial processes may sort the bed such that large particles may be more prevalent in one portion of the stream (Clancy, 2003); therefore, sample size must be determined methodically.

2.2 *Particle heterogeneity*

In addition, there is the heterogeneity of the particle size distribution to consider. The particle sizes in boulder-bed channels range from sand size (1 mm) to particles over 1 meter in diameter (Bathurst, 1978, and Clancy, 2003). This range in particle heterogeneity is significant because particle sorting or heterogeneity affects the forces acting on bed particles, particularly particle lift during transport initiation (Wiberg & Smith, 1987). Calculating particle resistance and transport in boulder-bed streams requires detailed information about the particle sizes and sorting. The upper and lower tails of the particle size distribution represent the extremes in the bed particle sample; however, for boulder bed streams, the upper tail is of greater importance as these particles represent the largest particles within the channel.

Finally, the particles in the upper tail of a boulder and cobble bed stream can be quite large and in aggregate occupy a significant portion of the channel width (Clancy, 2003). Current methods do not quantify the width constriction from the large particles or address the spatial organization of the boulders. For example, three uniformly spaced particles across the channel width have a very different affect to three particles clustered in the channel center.

It is our hypothesis the characterization of boulder and cobble bed channels requires a metric that incorporates particle heterogeneity, particle size, a systematic method to determine sampling size, and can lead to the develop of other methods to characterize the spatial organization of bed particle. For sampling size, it seems appropriate to return to the geomorphology of the stream to determine the sampling area rather than use a standard or arbitrary sample size, especially if one considers that the particle samples are taken traditionally over a cross section. For the particle size, one idea points toward developing a horizontal length scale. Like the vertical length scale, it would be dependent upon the particle distribution. Unlike the vertical length scale, a horizontal length scale would consist of a sum of large particles rather a single particle statistic. The sum of the particle diameters in upper tail of the particle size distribution incorporates both a consideration of particle size and particle heterogeneity.

3 TOPSUM DEFINITION

In developing a horizontal length scale, we chose the lower limit of the upper tail at the 84%. In boulder-bed streams, the particles in the upper tail (≥84%) represent particles that rarely move and may have been deposited by non-fluvial processes. In boulder and cobble bed channels, the distribution above the 84% becomes more heterogeneous compared to the lower percentiles. For these reasons, the sum of the large particles greater than 84% emerges as a good choice to represent the constriction in width or horizontal roughness length. The sum of the upper tail does not have a name, so in this document it is referred to as the *topsum* and is defined in equations 1 and 2 below. Equation 1 is for data that has not been divided into bins. Equation 2 is for data that is collected in bins.

$$Topsum_{rank} = \sum_{84}^{\infty} y_i \qquad (1)$$

Where y is the random variable (particle diameter size in this case) summed from the rank of 84 or greater. The bin topsum is defined as follows:

$$Topsum_{bin} = \sum_{84}^{\infty} y_m n_i \qquad (2)$$

Where y is the random variable and mid-point of the bin size and n is the count in the bin. The bin data is the sum of the product of the mid-point of the bin and the count in the bin for all bins in the 84% or greater. If the diameter of the entire particle size distribution sample is summed, the upper tail usually accounts for an average 30–60% of the sample. Depending upon the heterogeneity and particle size this percentage can be higher.

4 TOPSUM: EXPERIMENTAL DATA

4.1 *Experimental data generation*

To compare and differentiate topsum from other statistics such as standard deviation, D_{50}, D_{84}, and D_{95} we simulated particle size distribution by using a random

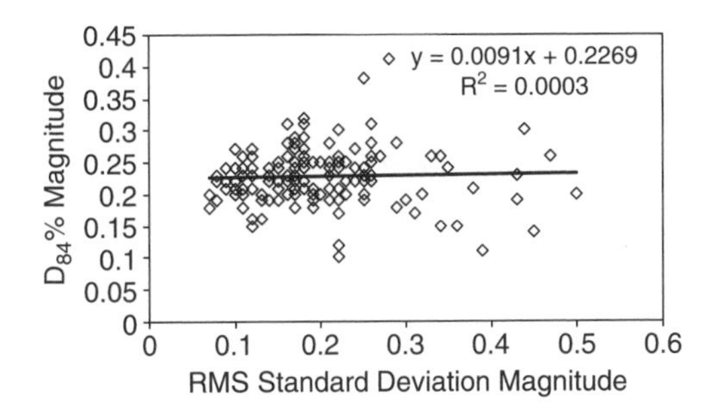

Figure 1. Experimental data: D_{84} versus standard deviation.

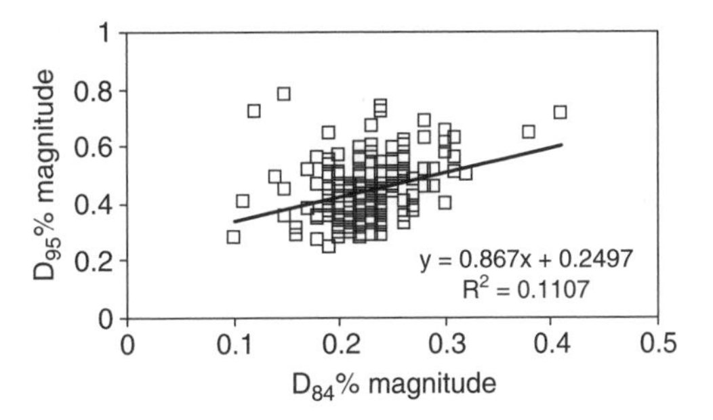

Figure 2. Experimental data: the 95th percentile versus the 84th percentile to.

number generator to generated 200 lognormally distributed data sets with 100 points. The magnitude of the generated data is meant to simulate cobbles and boulders measured in meters. We set the standard deviation range of 0.1–0.5 and a mean range of 0.8–0.21. From these data sets we calculated, the root mean square standard deviation, D_{50}, D_{84}, D_{95} and topsum. We calculated the particle statistic sizes by using a rank method rather than binning the data. To determine the standard deviation for the field data we used the root mean square standard deviation, which is shown to be a less bias and consistent way to calculate standard deviations for rank particle distribution data (Clancy, 2003).

We chose a lognormal distribution to approximate the particle size distribution in stream beds. By setting the standard deviation and mean range, we are able to systematically compare a different standard deviation and mean combinations.

4.2 *Experimental data analysis and results*

We analyze the statistics calculated from the experimental data to determine the range and limits of topsum's relation to other well known statistical parameters. In Figure 1, we compared the 84 percentile particle with the distribution's standard deviation. From Figure 1, it is clear that the two statistics are poorly correlated; therefore, D_{84} is not a good choice to represent the particle heterogeneity in boulder bed streams.

We compare the 84th percentile to the 95th percentile in Figure 2. This comparison examines how the lower limit of the upper tail compares to a higher magnitude statistic (95th percentile) in the upper tail. From Figure 2, the 84th percentile is poorly correlated to the 95th percentile. What this implies is that there is a wide range of values for the upper tail, which is controlled by the heterogeneity of the particle distribution. If we consider an extreme case from the results in Figure 2 where the 95th percentile is nearly 8 times the size of the 84th percentile, this causes a significant loss of information if the roughness height or other important hydraulic values are represented only by the

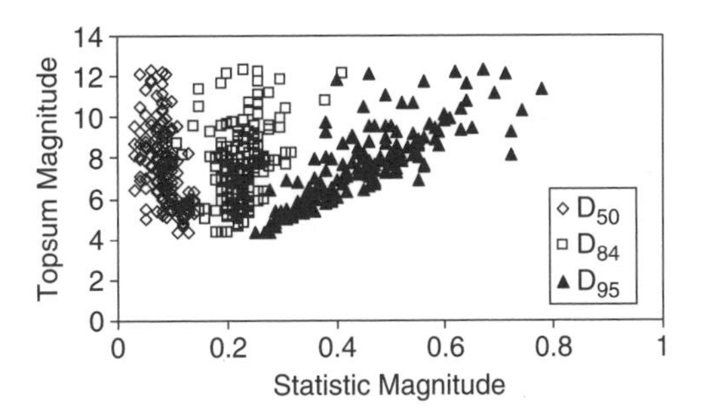

Figure 3. Experimental data: topsum versus particle statistics.

D_{84}. In boulder bed channels, these magnitudes would be similar to a D_{84} particle of size 0.10 m and the D_{95} particle of size 0.8 m. Clearly the importance of the D_{84} particle is diminished in this particle distribution.

In Figure 3, the topsum of each distribution is compared to the 50th, 84th, and 95th percentile of the distribution. For the 50th percentile, there is no relationship with topsum. This is also the case for the 84th percentile, which is of interest as the 84th percentile is the lower boundary for topsum.

The 95th percentile is also poorly correlated with topsum, but the relationship between the two is significantly stronger. This is to be expected as the 95th percentile is close to the upper limit of the topsum, so its magnitude has a greater effect on the summation. Even though there is a stronger correlation to the 95th percentile compared to the other statistics, topsum is additionally influenced by the standard deviation.

In Figure 4, the topsum is compared to the distribution's standard deviation. To determine the standard deviation for the data we used the root mean square standard deviation, which is shown to be a less bias and consistent way to calculate standard deviations for rank particle distribution data (Clancy, 2003). Topsum and the standard deviation show a high correlation. In general this result is expected as it simulates boulder

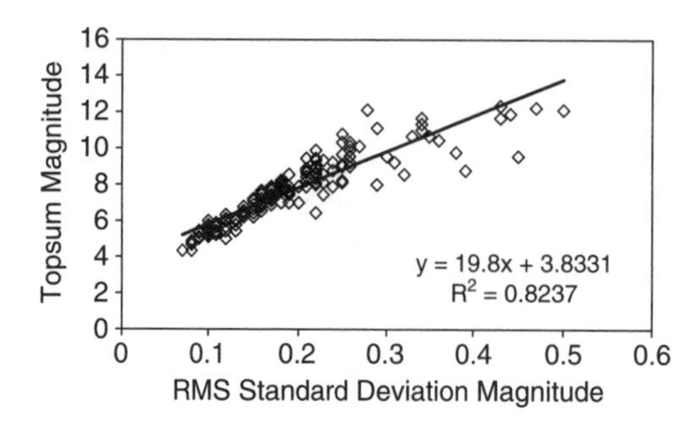

Figure 4. Experimental data: topsum versus the root mean square standard deviation.

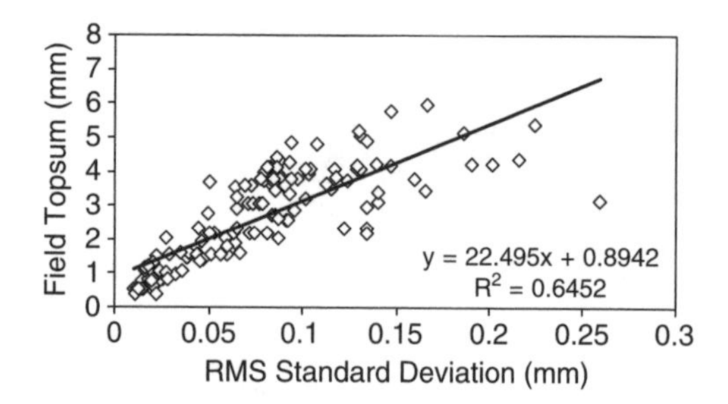

Figure 5. Field data: topsum versus root mean square standard deviation.

bed channels. Usually in boulder bed channels, particle heterogeneity will increase when an anomalously large particle is in the channel. In field observation, particle size increase is associated with the standard deviation of the particle distribution also increasing (Dusterhoff, 1997, Stoner, 1999, Clancy, 2003). In the following section these data will be compared to field measurements.

5 TOPSUM: FIELD DATA

5.1 *Field data methods*

One data set was collected from gravel and boulder-bed streams in the Maryland Piedmont and Coastal Plains, Montana, Northern California, Washington State, and Northern Virginia from 28 different streams for a total of 153 particle size distributions from 153 different cross sections. For each stream reach, particle size data were collected from a minimum of three cross sections; however, many streams were sampled along 6 or more cross sections. For eight-five of the particle counts the particles were measured randomly without recording position along the cross section using the traditional method prescribed by Wolman (1955). For the remaining 68 cross sections, particle position along the cross section was recorded. To prevent duplicate sampling and to assure the bed particles are uniformly represented, the particles were sampled at 2 cm intervals and 25 cm intervals. Particle position along the cross section was also recorded. For all cross sections, the sample size range is between 100–210 particles with an average of 128 particles sampled per cross section.

One hundred and three of the cross sections were sampled using the Wolman particle method. Sixty-eight of the particle distributions were sampled using the traditional binning method of data collection. The bin method consists of recording the number of occurrences of a particle within a size or bin range. Rather than record the actual particle size, the number of occurrences of a bin size is recorded. This method is used to reduce the bias of the sampling technique; however, for sampling streams with large particles a rank method may be preferable. For the rank method the actual particle sizes are recorded. The particles are ranked according to size. If there are more or less than 100 particles, the rank is normalized. The normalized rank is used to determine the particle size statistics; thus, a particle with a rank of 50 is the D_{50}.

5.2 *Field data and experimental data comparison*

As in the analysis of the experimental data, we used the root mean square of the particle distribution. The correlation between the standard deviation and the topsum shows a correlation but it is lower in the field data than in the experimental data as seen in Figure 5. This is expected due to some of the artificial constraints on the experimental data set. First, the experimental data were exactly 100 data points. The same size data set results in the number of particles in the upper tail are always the same. Second, the all experimental data were calculated using a rank method for particle size statistics rather than binning the data; whereas many of the field particle distributions are from data collected using the traditional binning method. In these cases, equation 1 rather than 2 was used to calculate the topsum. The difference between the methods yields an average of a twenty percent in the value of topsum (Clancy, 2003). Finally, the range of the standard deviation of the experimental data is much greater than that of the field data, which affects the correlation.

6 TOPSUM: JAMMING CRITERIA

By itself, topsum combines particle size and particle heterogeneity. In order to assessing the impact of the large particles on the hydraulics, it is necessary understand how much of the width is affected by the large particles. The large particles affect on the channel width is illustrated in Figure 6, which is a line drawing based on a photograph of a boulder bed stream.

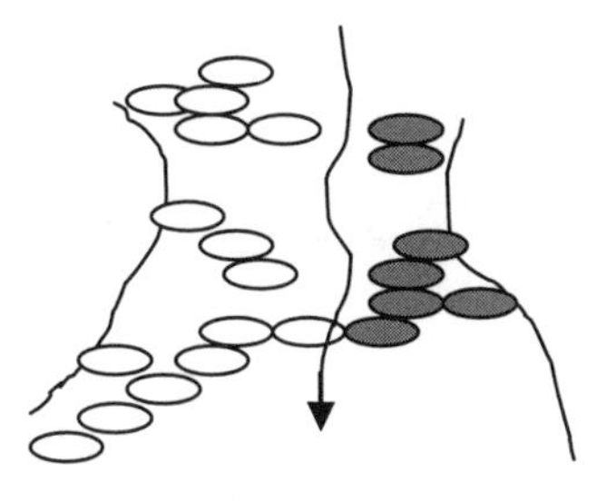

Figure 6. Large particles occupancy of channel.

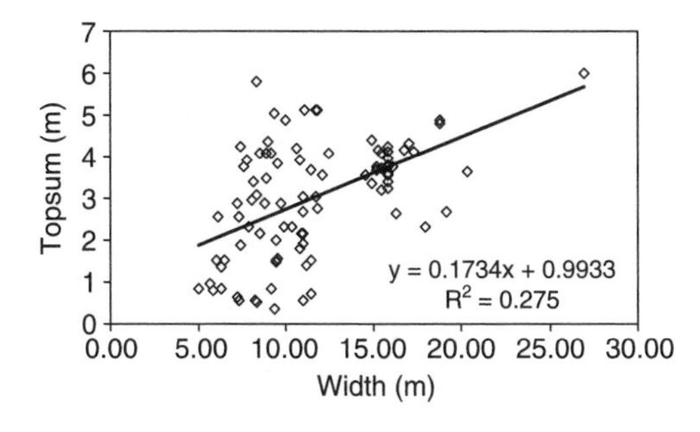

Figure 7. Field data: topsum versus channel width.

For topsum to become a horizontal length scale, it must be compared to the channel width. Figure 7 is a graph of topsum versus width, which shows no correlation between the two variables.

The two variables are expected to have some relationship; however, wide channels can have small particles or large particles, which affect the magnitude of topsum. In addition, a wider range of data, collected at uniform intervals may indicate a higher correlation.

In boulder bed channels, the large particles can be so large that they skew the upper tail; therefore, the larger the particles in the stream, the larger the standard deviation. Comparing the width and topsum to the standard deviation quantifies the percentage of the width constriction as well as the heterogeneity of the particle distributions. These three variables together can characterize boulder and cobble bed channels.

We examined the standard deviation of the particle size distribution to both the channel width normalize by the topsum as well as topsum normalized by channel width. The reason for the first method, width normalized by topsum, is that it more clearly gives the limits of topsum's usefulness as a variable. The second method, topsum normalized by width is this gives the percentage of the width constriction and is expected method to examine topsum.

In Figure 8, the width is normalized by topsum and is compared to the standard deviation. As the width to topsum ratio becomes smaller, the channel is jammed with more or large particles. For the width to topsum

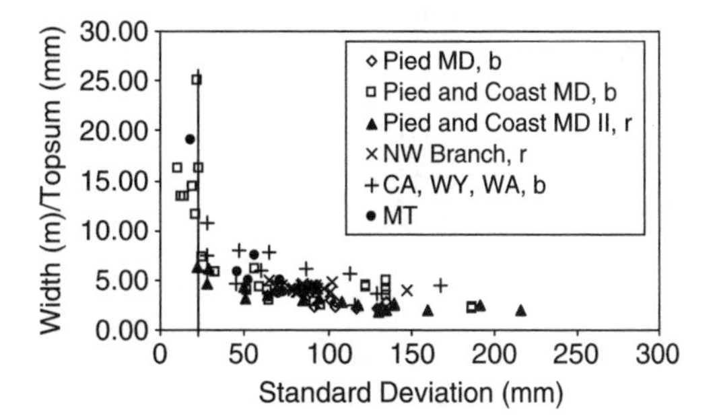

Figure 8. Field data: width/topsum versus standard deviation.

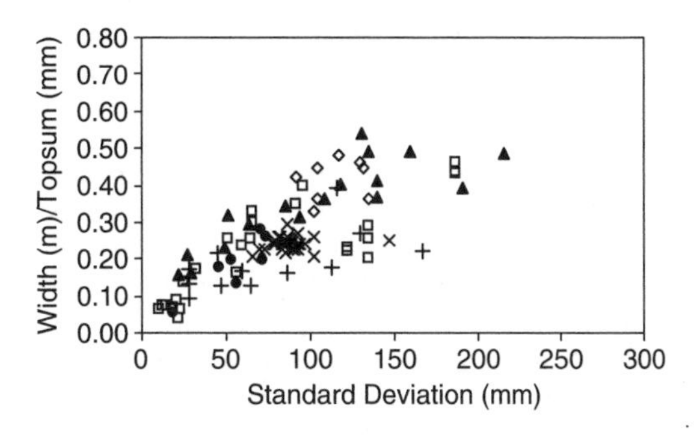

Figure 9. Field data: topsum/width versus standard deviation.

ratio of 5–25, the standard deviation is the approximately 20–25 mm for this entire range. As the large particles dominate the channel, the general trend is that the standard deviation increases. What this suggests is that streams with a very low standard deviation are not well represented by topsum. In general, the streams on Figure 8 with low heterogeneity are also streams with small particles.

In Figure 9, topsum is normalized by the width and is compared to the standard deviation. The topsum to width ratio represents the percentage of the width taken up by the upper tail of the particle size distribution. In general, this region represents the percentage of the width taken up by large particles.

The overall trend in both Figures 8 and 9 is that the standard deviation increases as the channel is jammed with more or larger particles. The relationship is somewhat hampered by the difference in the bin and rank method of calculating particle size statistics. The other inherent problem in the data is sampling technique. Nearly 60% of the data are from pebble counts that randomly sample any particle on the bed. When the topsum method was conceived, we minimized field errors by sampling the bed uniformly. This avoided particle resampling and gave a uniform sample of the entire cross section.

7 PARTICLE SPATIAL DISTRIBUTION: CLUSTERS AND CHAINS

7.1 *Particle chain definition*

While the topsum/width ratio gives some indication of the amount of large particles jamming the stream's width, it still does not yield much information about how these particles are arranged along the cross section. In general, the morphology of boulder-bed channels is often paired with step-pool resistance features. Although boulder steps have been observed on slopes as low as 1%, many boulder-bed reaches do not have a step-pool morphology or even well defined riffle and pool sequences (Judd and Peterson, 1969). Without recognizable morphological features, the emphasis is placed on the individual bed particle's organization.

The topsum relates a horizontal length scale to the particle distribution's size and sorting, therefore a starting point to examine the organization of clustering particles begins with those particles in the topsum. We examined the particle organization along the cross stream direction. We defined clustering or particle chains along the cross section in the following manner:

1) Particles were in the 84% of the particle size distribution, and therefore part of topsum;
2) Two or more particles closely spaced together; and
3) Closely spaced together is defined as having a spacing between the particles equal to the D_{84} or smaller. We present the initial phase of the particle chain study and its relationship to particle clusters. A thorough analysis will be presented in a forthcoming paper.

7.2 *Particle chain field methods*

To collect information about particle organization, we revised the Wolman pebble count. In addition to measuring the intermediate bed particle diameter, and recording the particle diameter size (not frequency of occurrence in a bin size), we also recorded particle position along the cross section. To ensure adequate and uniform sampling we sampled every two centimeters. For streams with particles larger than 2-cm, we simply noted the extent of the particles occupancy on the bed. Data were collected at seven sites along the Maryland coastal Plain and Piedmont region. The D_{84} range for the data set is 5–25 cm with stream width range of 6 to 17 meters. At each site a minimum of three cross sections were sampled. We selected sites that exhibited little or no bedforms along straight portions of the stream.

7.3 *Particle chains and topsum*

The total particle chain size is the aggregate of the particles in topsum that meet the definition outlined in the

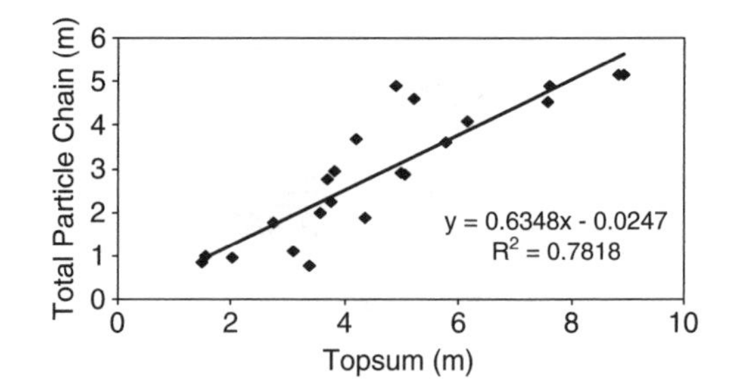

Figure 10. Field data:topsum versus particle chains.

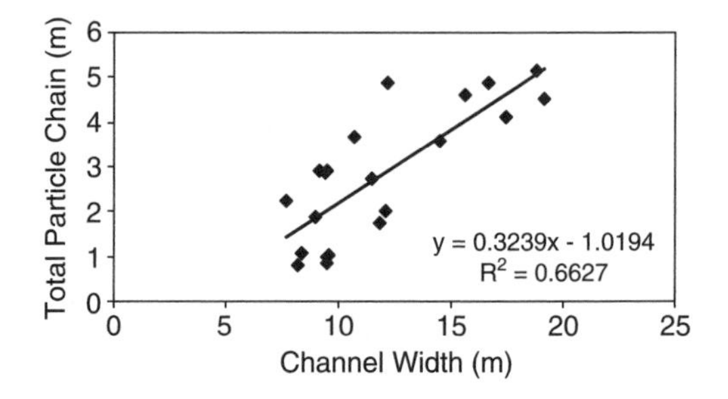

Figure 11. Field data: total particle chain versus channel width.

section above. As only particles in the topsum are analyzed for particle organization (i.e. particle chains), it is expected that topsum and the total particle chain size be highly correlation. This relationship is featured in Figure 10, which does show a high correlation between the topsum and particle chains. The relationship between particle chains and topsum; however, is not trivial, as it indicates that particles are not uniformly place along the bed cross section, but rather there are regions of high flow resistance and low flow resistance.

The relationship between total particle chain and width is a little unexpected. Figure 11 shows the total particle chain versus width with a relatively high correlation. If we reconsider Figure 7, topsum versus width, which has a significantly lower correlation, then we may consider this result surprising. What this may indicate wider channels have more room for particle chains to organize. A fuller analysis of these results and other data will be presented and discussed in a forthcoming paper.

8 DISCUSSION AND CONCLUSIONS

Topsum is a useful variable for streams with large scale roughness features. It combines channel geomorphology and the bed particle size distribution. We recommend that the following modifications to the Wolman pebble count for calculating topsum: 1) Particle diameter be uniformly sampled across the bed at a distance between 2–10 cm; 2) Particle size should be recorded rather than frequency of bin size; 3) Particle size distribution should be comprised of approximately 100 particle or more (no less than 85); 4) A rank method should be used to calculate the particle size statistics; and 5) Topsum should be a sum of the particle diameters and not the bin sizes.

What has not been developed for topsum is a full analysis of the particle statistic lower limit. In other words, there may be streams where the D_{50} and larger particles protrude above the flow. We have found that the limiting cases for topsum are extremely narrow channels (<5 m) filled with large particles where the particle size distribution is relatively homogenous. When sampling particles in these cases it may be difficult to obtain approximately 100 samples per cross section. An example of this situation is a drainage ditch filled with rip rap. Further examination is required for the limits of topsum in streams with large particles and a particle size distribution with low heterogeneity.

Another use of topsum is to examine particle organization. Generally, particle organization such as clustering is examined in the longitudinal or stream wise direction (Dal, 1968). Given our analysis of these boulder channels, we choose to examine the cross stream direction. The results presented in this paper suggest that particle clustering may be controlled by channel width. In addition, the data also suggest that particle organization is not random and should be considered within the overall evaluation of flow resistance. Using the methods that we have outlined above, particle organization such as particle chanin can be examined with the traditional particle size statistics such as D_{84}. A more detailed analysis of particle chains will be presented in a forthcoming paper.

ACKNOWLEDGMENTS

I would like to thank E-an Zen, Glenn Moglen and Kaye Brubaker for their help, insights, and suggestions.

REFERENCES

Bathurst, J.C. 1978. Flow resistance of large-scale roughness. *Journal of Hydraulic Engineering.* ASCE, 104(12): 1587–1603.

Bathurst, J.C., Li, R.M. and Simons, D.B. 1981. Resistance equation for large scale roughness. *Journal of Hydraulic Engineering.* ASCE, 107(12): 1593–1613.

Clancy, K.F. 2003. *The spatial distribution of large particles in boulder-bed streams and consequences on sediment transport and hydraulics.* Ph.D. dissertation, University of Maryland, College Park, MD.

Dal, C. 1968. Pebble Clusters: Their origin and utilization in the study of paleocurrents. *Sedimentary Geology*, 2, 1991–226.

Dusterhoff, S.R. 1996. Effect of bed particle size distribution on bed and hydraulic roughness in a natural stream environment. Undergraduate Thesis, University of Maryland.

Judd, H.E. and Peterson, D.F. 1969. Hydraulics of large bed element channels. *Report PRWG* 17-6, Utah Water Resources Lab., Utah State Univ., Logan.

Limerinos, J.T. 1970. Determination of Manning's coefficient for measured bed roughness in natural channels. *Water Supply Paper I1898-B*, Washington D.C.

Marchand, J.P., Jarrett, R.D. and Jones, L.L. 1984. Velocity profile, water surface slope, and bed-material size for selected streams in Colorado. *U.S. Geological Survey Open File Report*, 84–733.

Stoner, E.C. 1999. Flow resistance in gravel bed channels. Master's Thesis, University of Maryland as College Park.

Whittaker, J.G. 1992. Sediment transport in step-pools. In Thorne, C.R., Bathurst, J.C., and Hey, R.D. (eds), *Sediment transport in gravel-bed rivers*. Chichester: Wiley, 545–570.

Wiberg, P.L. and Smith J.D. 1987. Calculations of the critical shear stress for motion of uniform and heterogeneous sediments. *Water Resources Research*, 23, 1471–1480.

Wiberg, P.L. and Smith J.D. 1991. Velocity distribution and bed roughness in high gradient streams. *Water Resources Research*, 27, 825–838.

Wolcott, J. and Church, M. 1991. Strategies for sampling spatially heterogeneous phenomena: the example of river gravels. *Journal of Sedimentary Petrology*, 61(4): 534–543.

Wolman, M.G. 1955. A method of sampling coarse river-bed material. *Transactions of American Geophysical Union*, 35(6): 951–956.

River, Coastal and Estuarine Morphodynamics: RCEM 2005 – Parker & García (eds)
© 2006 Taylor & Francis Group, London, ISBN 0 415 39270 5

Sediment patches, sediment supply, and channel morphology

W.E. Dietrich, P.A. Nelson, E. Yager, J.G. Venditti & M.P. Lamb
Department of Earth and Planetary Science, University of California, Berkeley, Berkeley, CA, USA

L. Collins
Watershed Science, Berkeley, CA, USA

ABSTRACT: Bed surface particle size patchiness may play a central role in bedload and morphologic response to changes in sediment supply in gravel-bed rivers. Here we test a 1-D model (from Parker ebook) of bedload transport, surface grain size, and channel profile with two previously published flume studies that documented bed surface response, and specifically patch development, to reduced sediment supply. The model over predicts slope changes and under predicts average bed surface grain size changes because it does not account for patch dynamics. Field studies reported here using painted rocks as tracers show that fine patches and coarse patches may initiate transport at the same stage, but that much greater transport occurs in the finer patches. A theory for patch development should include grain interactions (similar size grains stopping each other, fine ones mobilizing coarse particles), effects of boundary shear stress divergence, and sorting due to cross-stream sloping bed surfaces.

1 INTRODUCTION

Most gravel-bed rivers, and many gravel-bedded flumes, display distinct patchiness in the surface sorting of particles. Figure 1 shows a detailed map of Wildcat Creek, near Berkeley, California, in which the organized heterogeneity of the bed surface is quantitatively displayed. This is a small (~7 m bankfull width), modestly steep (1.5% slope) channel that receives a very high sediment load (~8000 tonnes/km^2/yr in recent years; San Francisco Estuary Institute 2001). Much of the sediment is from deep-seated landslides that periodically push massive amounts of debris ranging from boulders to clay into the channel. It is difficult to look at this map and not wonder how the river arranges unsorted debris into many distinct patches that range from boulder fields to sand patches. This map was made in 1987 following a period of particularly active landsliding and several years of high flows. Repeat visits to this reach since then have revealed two very different responses. In 1988, despite large amounts of bed surface mobilization, the distribution of patches remained relatively stable (except for where wood fell into the channel and redirected the flow). Yet over a longer time period, significant changes in the bed took place. Some of the high-flow coarse patches became buried with sand and subsequently vegetated. Occasional wood jams formed and caused local scour downstream and backwater-induced deposition upstream. Some of the boulders moved downstream. Now 17 years later, although the channel has changed, the assembly of patches, although distributed differently, by and large remains. This implies that some set of processes segregates a heterogeneous mixture of sediment into patches of distinct grain sizes. Some patches retain their location for decades while others come and go, but the range and number of classes persist. What leads to this patchiness, what sets the size distribution for a given patch, and how many patches should arise in a given reach of river? Presently there is limited theory and observation to guide an answer to these questions.

These questions address what we see so distinctly when walking along the channel, but they also raise other questions. All bedload transport models require a grain size distribution, be it of the surface or the subsurface (e.g. Parker & Klingeman 1982, Parker 1990, Wilcock & Crowe 2003). What grain size distribution should one use when there are distinct patches of sediment? One approach is to map the areal extent of each of the patch types and then determine an area-weighted grain size distribution. Some field methods guidelines specifically call for statistical procedures to sample through this variation to get a representative size distribution (see Bunte & Abt (2001) for an excellent review of gravel-bed sampling methods). Although this procedure is straightforward, no local region of the bed surface consists of this size distribution, which makes it nearly useless. Instead, bedload transport rates and sizes vary locally due to the interaction of local flow fluctuations with the spatially heterogeneous bed surface texture. The sum of these

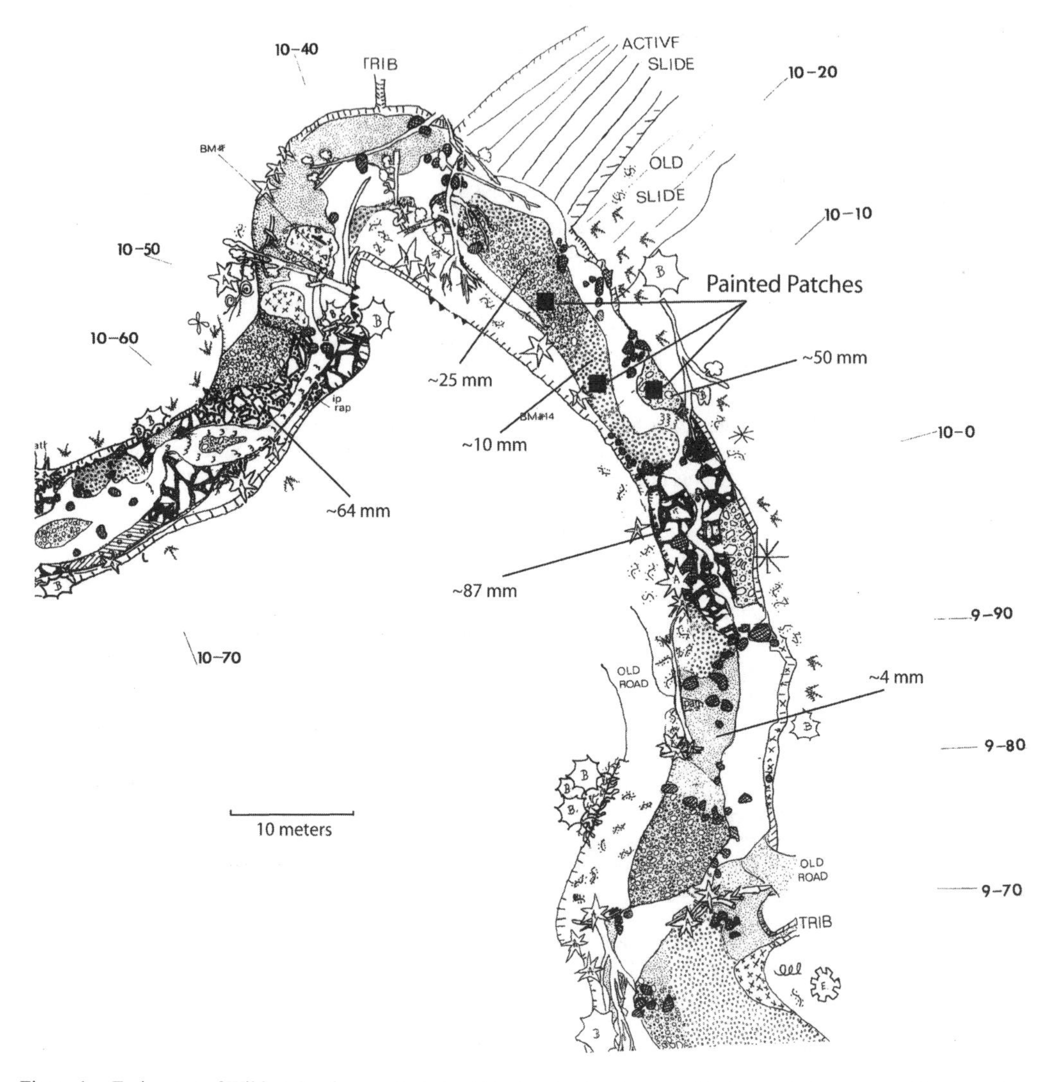

Figure 1. Facies map of Wildcat Creek, near Berkeley, CA, showing locations used in the painted rock study. Numbers along border are distances in meters. Tree locations along the bank are indicated by various symbols. Large woody debris is shown roughly to scale. Average median grain sizes are indicated for each facies type, and area with no pattern was covered by water. Flow is from left to right.

fluxes across a channel yields the actual total bedload discharge.

Several studies of bedload transport and bed surface textures have suggested that the stage-dependent size distribution of bedload in gravel-bedded rivers may reflect the mobilization of progressively coarser patches with increasing discharge (e.g. Lisle & Madej 1992, Lisle 1995, Garcia et al. 1999). The widely used bedload transport equation of Parker (1990) captures the observed stage-dependent mobility through a non-linear dependency of bedload flux on a hiding function and boundary shear stress. Some of this stage-dependent behavior may arise more from patch dynamics than from one-dimensional grain size adjustments, as portrayed in the model. How well does the model perform when patchiness is an important part of the channel dynamics? The interaction between random fields of boundary shear stress and sediment patches will increase transport of fine material and selective transport of coarse sediment (Paola & Seal 1995). Hence, it is not sufficient to simply characterize the size distribution of the patches without estimating the appropriate shear stresses on the patches. In this regard, the coupled modeling and field study of Lisle et al. (2000) is particularly instructive. They found that the bankfull stage boundary shear stress field and the low-flow mapped grain size distribution were uncorrelated. Areas of finer sediment and local high boundary shear stress would then be expected to carry a disproportionate amount of the bedload because of the nonlinear flux dependency. Furthermore, a recent analytical model incorporating statistical variation in shear stress and bed patchiness (Ferguson 2003) suggests that cross-stream variance in hydraulic and critical shear stress for incipient motion can produce bedload fluxes substantially greater than those predicted in a one-dimensional, width-averaged calculation.

Experimental and field studies of gravel-bed channels have shown that the primary response to changes in sediment supply may be through the extent and size distribution of patches (Dietrich et al. 1989, Kinerson 1990, Lisle & Madej 1992, Lisle et al. 1993, Lisle et al. 2000). Furthermore, in steep channels, boulders create a relatively immobile framework across which finer gravel passes. The boulders create large spatial deviations in flow that lead to local finer sediment deposition or scour. These streams are often supply-limited, which will strongly influence the extent of patch deposition. Successful prediction of bedload transport in such channels requires accounting for the area and size distribution of dynamic patches and the influence of boulders on the flow (Yager et al. 2004, 2005).

Here we examine the patch issue in three ways: 1) a model comparison between predicted and observed bed surface response to sediment supply, 2) a description of field observations on patch surface dynamics, and 3) a discussion of the controls on patch occurrence. The model and field studies highlight the importance of patch occurrence and dynamics.

2 SURFACE RESPONSE TO REDUCED BEDLOAD SUPPLY: THE ROLE OF PATCHES

In the late 1980's and early 1990's a series of papers were written based on experiments performed in the same flume in Japan that explored how a channel responds to reduced bedload supply (Dietrich et al. 1989, Iseya et al. 1989, Kirchner et al. 1990, Lisle et al. 1991, Lisle et al. 1993). The flume was straight, 0.3 m wide and 7.5 m long, and in all experiments it was given a constant water discharge, but a stepwise decrease in sediment feed rate. Details about the experimental observations and methods can be found in the published papers. The initial experiments (Dietrich et al. 1989, Kirchner et al. 1990) attempted to force a 1-dimensional response by running the flume at a low width-to-depth ratio ($w/d \approx 3$). Although alternate bars did not form, at high feed rates well-defined, spatially-sorted bedload sheets (Whiting et al. 1988) developed and, with decreasing supply, lateral zones of coarser inactive bed emerged. With this decreasing

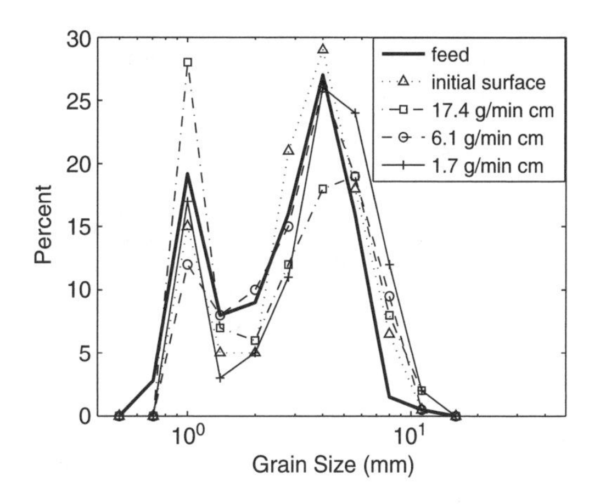

Figure 2. Grain size distributions of the feed and bed surface in the Dietrich et al. (1989) experiments.

sediment supply the area of active transport also narrowed and remained a relatively fine-textured surface.

Figure 2 shows the size distribution function of the feed, the initial bed surface (no sediment feed, and a short period of discharge sufficient to shift loose particles but cause no significant transport), and the bed surface at the high, medium and low feed rates. The feed and resulting bed surfaces were biomodal, which contributed to the strong development of bedload sheets and, as subsequently proposed by Paola & Seal (1995), the overall patchiness of the bed surface. It is important to understand how the bed surface size distribution reported in Figure 2 (and previously in Dietrich et al. 1989) was generated. The bed surface tended to sort into three to four distinct patches ("congested" (coarse), "transitional", "smooth" (fine) and "inactive" (coarse zones during the lower feed rates) see Iseya & Ikeda (1987) for discussion of terms). For each run, the entire bed surface was mapped into each of these categories (by eye) and then point counts were done on each patch. The size difference between the patches could be large. For example, for the highest feed case (17.4 g/min-cm) the median grain size of the smooth patch was 2.7 mm, the transitional was 3.4 to 4.3 mm and the congested was 4.7 mm. The congested areas tended to be more unimodal compared to the generally bimodal smooth and transitional areas. The single grain size distributions reported in Figure 2, however, are the area-weighted average grain size distributions of the entire bed.

The median grain size of the spatially-averaged surface progressively increased with diminishing load from 3.7 mm (at 17.4 g/min-cm feed rate), to 4.3 mm (at 6.1 gm/min-cm), and then to 4.9 mm (at 1.7 gm/min-cm). This coarsening resulted from increased areas of congested and inactive patches. Except for the highest feed rate, the transitional bed surface predominated, leading to a modal value close to 4 mm (Fig. 2). Coarsening with reduced supply results from a great increase in the occurrence of the coarser fractions on the surface (Fig. 2).

Dietrich et al. (1989) suggested that the occurrence of the coarse surface layer in streams can be quantitatively linked to sediment supply. They reasoned that the primary response to a supply deficit is scour and hence surface coarsening, rather than channel degradation and slope reduction. The experiments suggest that this coarsening may be accomplished through expansion of coarse patches and narrowing of the finer-textured active bedload zone. They did note, however, that slope changed from 0.0052 (high feed rate) to 0.0046 (intermediate) and finally to 0.0035 (lowest). The water surface slope, bed surface slope, and bedload flux out of the flume reached approximate steady state in less than 1.5 hours, but considerable temporal fluctuations in these properties persisted throughout the experiments (e.g. 5 minute bedload transport samples gave approximately 5 to 33 g/min-cm for the 17.4 g/min-cm case, 0.7 to 12 g/min-cm for the 6.1 g/min-cm case and 0.3 to 7 gm/min-cm for the 1.7 gm/min-cm case). Some of this variation is due to inevitable small variations in feed rates and texture of the feed, which through grain interactions were sometimes amplified in the flume.

Lisle et al. (1993) used the same flume, but with a shallow flow such that the width-to-depth ratio was about 23. Stationary bars formed during their experiments. The grain size of the feed was nearly unimodal (Fig. 3), and finer than the narrow w/d case of Dietrich et al. (1989). Lisle et al. describe how coarse grain interactions controlled bar development, and, with decreasing sediment supply, the main flow incised, leaving the bars as inactive terraces. The bed surface was mapped into three distinct patch types based on percent gravel (>50%, 5–50%, and <5%) and a single, area-weighted grain size distribution was reported. Feed rates decreased from 16.3 g/min-cm, to 5.2 g/min-cm, then to 1.6 g/min-cm, which were comparable to the Dietrich et al. (1989) experiments. Median grain size systematically coarsened from ~2.3 mm at the high feed rate, to 3.0 mm for the moderate and 3.8 mm for the lowest rate. The slope declined slightly from 0.031 to 0.028. Lisle et al. emphasized that nearly all the adjustment to reduced load occurred through bed surface coarsening (as compared to slope change), in a manner consistent with the simple theory proposed by Dietrich et al. (1989). The bed surface initially coarsened through expansion of the intermediate size class (5–50% gravel) and ultimately through expansion of the coarse size class. As in the small width-to-depth case, the zone of active transport narrowed considerably and was the area of finest bed surface. Additionally, there was considerable temporal variation in bedload flux throughout the experiment, with a tendency toward less fluctuation with time.

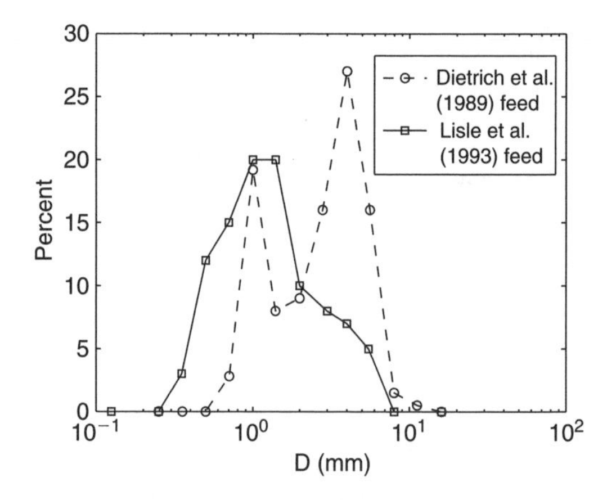

Figure 3. Grain size distributions of the Dietrich et al. (1989) and Lisle et al. (1993) experiments.

These two flume studies show that patch dynamics strongly influence the channel response to changes in sediment supply. There is some field support for these findings. Kinerson (1990) surveyed six channels with varying amounts of coarse sediment supply. He noted that the ratio of the median grain size of the surface relative to that of the subsurface was greater in channels with lower sediment supply, although patchiness caused large variations in this ratio. Following the proposal of Dietrich et al. (1989), he reasoned that the spatial extent of finer, unarmored patches may correlate with sediment supply.

Recently G. Parker has made available on his web page an ebook (G. Parker, 2005, *1D Sediment Transport Morphodynamics with Applications to Rivers and Turbidity Currents*, hereinafter referred to as Parker ebook) and included in that ebook are useful Excel spreadsheets for performing morphodynamic calculations. One, entitled "RTebookAgDegNormGravMixPW.xls", is specifically set up to model the profile and grain size response of a river to altered bedload supply. The bedload equation used is either that of Parker (1990) or the recent Wilcock & Crowe (2003) modification that specifically accounts for the presence of sand. Here we explore how well the model predicts the results in the Japan flume studies. This exploration reveals further the importance of patches.

The ebook model was used to simulate the small w/d experiments of Dietrich et al. (1989) and the large w/d experiments of Lisle et al. (1993). We modeled the initial high feed rate and then subsequent runs in which the feed rate was reduced. In each case the predicted surface grain size distribution (at the furthest upstream computational node) and average bed slope of the previous run were used as the initial conditions in the next run. Both experiments were modeled such

Table 1. Comparison of the Dietrich et al. (1989) flume observations with numerical model results. Δ denotes relative change, calculated as the change in the variable's magnitude between the two runs divided by the variable's magnitude in the earlier run, and qbT/qbTf is the ratio of the bedload transport rate at the downstream end of the flume to the sediment feed rate.

Supply (g/min-cm)	17.4		6.1		1.7
Observations					
D_{50} (mm)	3.7		4.3		4.9
Slope	0.0052		0.0046		0.0035
ΔD_{50}		0.16		0.14	
Δ Slope		−0.12		−0.24	
Model–experimental duration					
D_{50} (mm)	4.55		4.83		5.13
Slope	0.0054		0.0039		0.0029
ΔD_{50}		0.06		0.06	
Δ Slope		−0.28		−0.26	
qbT/qbTf	0.99		0.99		1.07

Table 2. Comparison of the Lisle et al. (1993) flume observations with numerical model results. Symbols and abbreviations are the same as in Table 1.

Supply (g/min-cm)	16.3		5.2		1.6
Observations					
D_{50} (mm)	2.3		3.0		3.8
Slope	0.031		0.027		0.028
ΔD_{50}		0.30		0.27	
Δ Slope		−0.13		0.04	
Model–experimental duration					
D_{50}(mm)	1.97		2.6		3.32
Slope	0.0439		0.0436		0.0406
ΔD_{50}		0.32		0.28	
Δ Slope		−0.01		−0.07	
qbT/qbTf	0.35		1.02		3.07
Model–steady state					
D_{50} (mm)	1.99		2.6		3.43
Slope	0.0632		0.0443		0.0342
ΔD_{50}		0.31		0.32	
Δ Slope		−0.30		−0.23	

that the model duration equaled the experiment duration. For each run in the small w/d case, the model approximately reached steady state (the median grain size varied down the flume by only about 10% and the sediment flux out of the downstream end was within 10% of the feed in (see Table 1)). The large w/d case, however, was far from steady state even at the end of each experimental run: the sediment flux out of the flume ranged from 35% to over 300% of the feed, depending on which stage of the experiment was being modeled (see Table 2). We therefore ran the model again until steady state conditions were met for each feed rate (typically orders of magnitude longer than the actual flume run durations). Table 3 provides the relevant input values used in each model run.

The results of the small w/d flume experiments and model predictions are shown in Figures 2 and 4 and summarized in Table 1. The model predicts a higher initial median grain size and less coarsening with decreasing feed rate than what was observed experimentally. The differences between predicted and observed grain sizes for the two reduced feed cases, however, are small, and perhaps within the range of error in flume measurement. The model predictions

Table 3. Input parameters for numerical model runs. All runs were performed using the Wilcock & Crowe (2003) bedload transport relation. Variables are defined as: qbTf: gravel input rate; qw: water discharge/width; I: intermittency; Sfbl: initial bed slope; L: reach length; dt: time step; n_a: factor by which surface D_{90} is multiplied to obtain active layer thickness; Mtoprint: number of steps until a printout of results is made; Mprint: number of printouts after initial one. See Parker (ebook) for further details.

Variable	Units	Dietrich et al.			Lisle et al. – experiment duration			Lisle et al. – steady state		
qbTf	$m^2/s \times 10^{-6}$	10.9	3.82	1.07	10.2	3.29	1.03	10.2	3.29	1.03
qw	m^2/s	0.06	0.06	0.06	0.00194	0.00194	0.00194	0.00194	0.00194	0.00194
I		1.0	1.0	1.0	1.0	1.0	1.0	1.0	1.0	1.0
Sfbl		0.0046	0.0054	0.0039	0.0310	0.0439	0.0436	0.0310	0.0632	0.0443
L	m	7.5	7.5	7.5	7.5	7.5	7.5	7.5	7.5	7.5
dt	days $\times 10^{-5}$	2.6	2.5	1.4	1.0	1.0	1.0	1.0	1.0	1.0
n_a		1	1	1	1	1	1	1	1	1
Mtoprint		2000	3000	3000	4167	3516	2735	40,000	100,000	50,000
Mprint		6	6	6	7	8	8	7	8	8
Calc time	hr	7.5	10.8	6.0	7.0	6.75	5.25	67.2	192.0	96.0

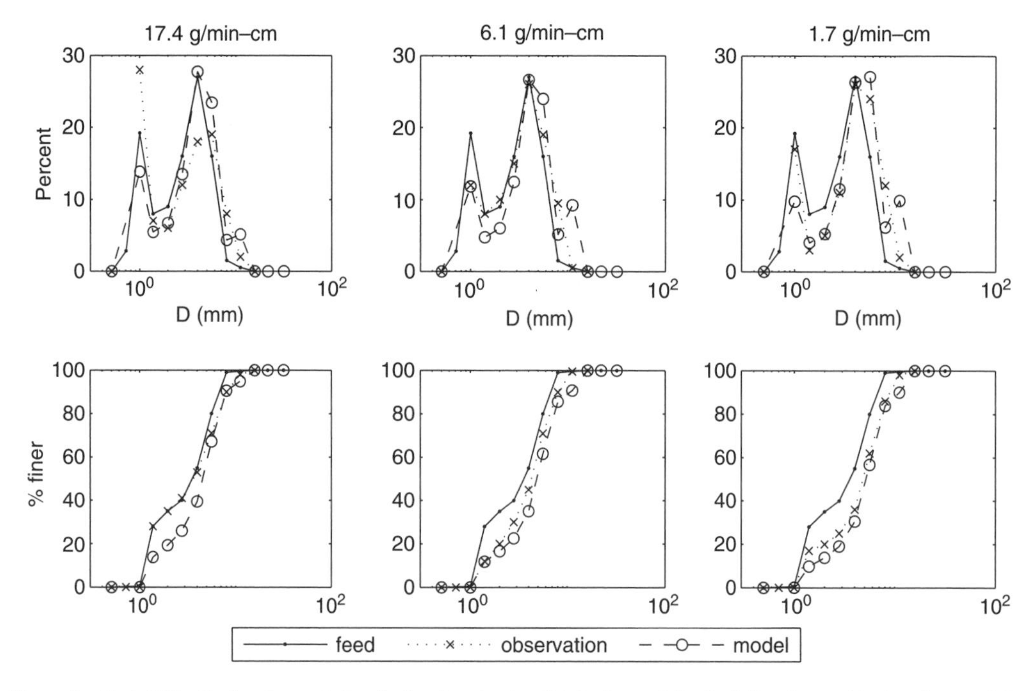

Figure 4. Probability (top) and cumulative (bottom) grain size distributions of the feed, the observed bed surface, and the bed surface predicted by the 1-D model, for the Dietrich et al. (1989) experiments.

match the shape of the surface grain size distributions reasonably well (Fig. 4). The slope measured during the experiment's initially high sediment supply is matched by the model, but both reduced-feed rate slopes are predicted to be much less than those observed. The model also predicts a larger slope reduction than what was observed.

For the Lisle et al. experiments (high w/d), Table 2 shows the values of slope and median grain size for the observed case, the numerical model run for the same duration as the experiments, and the steady state model results. Although the absolute values differ between the flume and model results for the experiment duration, the relative amount of change in slope and median grain size with decreased sediment supply is quite similar. In this case, however, the numerical model predicts that the flume would be far from steady state conditions. If the model is run to steady state, the median

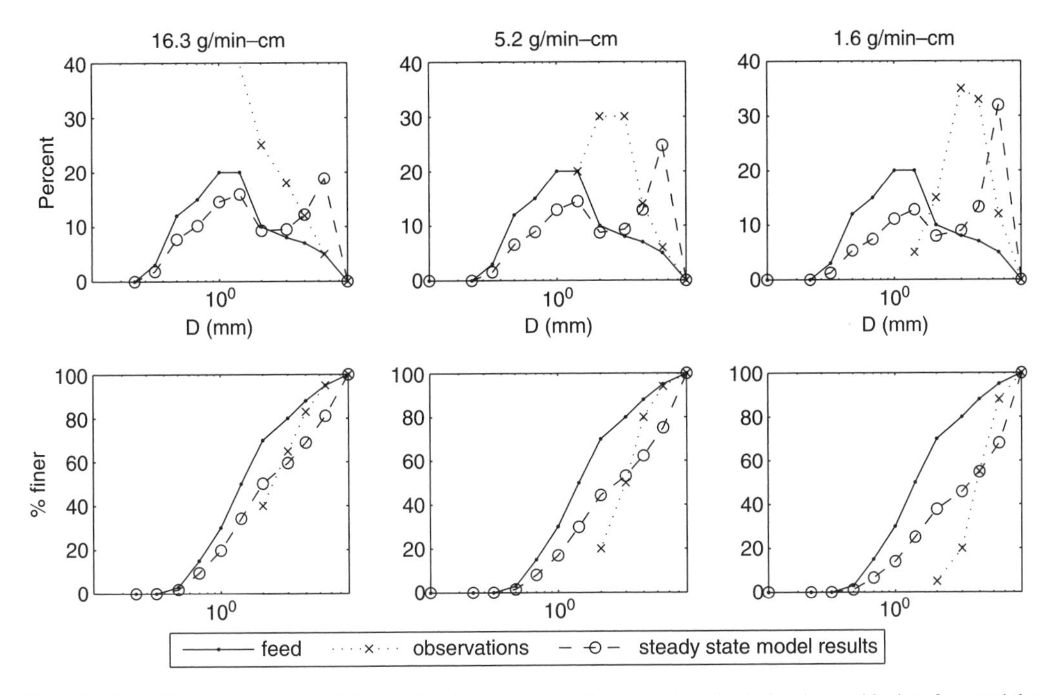

Figure 5. Probability (top) and cumulative (bottom) grain size distributions of the feed, the observed bed surface, and the steady state bed surface predicted by the 1-D model, for the Lisle et al. (1993) experiments.

grain size differs little from the experimental observations, but the slope differs from them considerably. The initial slope is predicted to be 0.063 instead of the observed 0.031 and the final slope at the lowest feed rate is predicted to be 0.034, compared to 0.028. The numerical calculations, then, contrast with the Lisle et al. findings in that steady state took much longer to develop and the reduction in sediment feed led to the slope reducing to half its initial value, whereas only minor slope changes took place in the flume.

The surface distributions measured by Lisle et al. were unimodal; however, the steady state numerical model predicts a bimodal surface size distribution (Fig. 5). As a result, the median grain size of the surface is predicted rather well, but the model tends to over predict the relative amount of fine and coarse material and under predict the amount of intermediate material close to the median grain size. The Wilcock & Crowe (2003) expression used in the Parker ebook model is based on an empirical fit to data from flume experiments in which the sediment size distributions were bimodal and trimodal (Wilcock et al. 2001). This may explain why the application of this expression to other experimental data may predict biomodality even when it does not occur.

In summary, for the small w/d case the model is able to achieve steady state conditions on a timescale equivalent to that of the actual experiments, but for the large w/d case the model requires much more time to reach steady state than what was observed in the flume. In both cases, the model shows a tendency to over predict the slope change in response to a reduction in sediment supply. The model is fairly accurate in predicting changes in the median surface grain size, but it is far more successful at predicting the shape of the surface grain size distribution in the small w/d case than it is for the large w/d case. It is unclear to what degree the model's relative success in predicting the surface grain size distribution depends on the use of a unimodal or bimodal feed or the presence or absence of two-dimensional sediment transport dynamics. Lisle et al. also noted that bar development (something not treated explicitly in the Parker model) created additional drag and influenced the boundary shear stress available for sediment transport. Given that there was no calibration in application of the numerical model, overall it performed reasonably well in capturing the general magnitude of the response of grain size and slope change.

In light of this comparison, we propose that patch size adjustment in response to reduced sediment load caused the amount of scour and slope change to be less than what would otherwise occur; this effect was greatest in the high width-to-depth case. As both Dietrich et al. (1989) and Lisle et al. (1993) noted, the response to reduced load is the establishment of inactive coarse patches and a narrow zone of finer texture bedload transport. The development of this narrow zone may be the primary reason the experimental flumes reached steady state sooner than did the numerical model calculations, since the narrow zone allowed the adjusted bedload transport rate to propagate more quickly downstream.

The maps in Lisle et al. (1993) suggest that in the moderate to low feed cases, as sediment travels downstream it travels through successive fine and coarse patches. Hence, patches are exchanging sediment but remaining texturally distinct. We explore that possibility in field study in the following section.

3 SEDIMENT DYNAMICS IN PATCHES: A FIELD STUDY

The map in Figure 1 raises several questions about patch dynamics. As the stage rises, do the patches experience the onset of transport at the same discharge? It seems likely that fine patches would be more mobile than coarse ones, as Lisle & Madej (1992) have suggested and Garcia et al. (1999) have inferred. If so, does sediment released from the fine patch cross into the coarser patches and influence the mobility of particles there? Given that finer patches are of finite extent, do they tend to scour and disappear if modest flow events persist? Here we describe some simple observations based on painted rock measurements for Wildcat Creek, CA (the site in Figure 1).

We painted 1 m^2 areas in three patches (Fig. 1) at the end of the dry season of 1987. We also painted large numbers on the boulder field at the downstream end of the site. This was done following a stormy previous year in which large quantities of sediment had entered the channel via deep-seated landsliding. Each of the painted patches had moved in the previous season. Each patch was point counted to determine its size distribution. Over the next two wet seasons these patches were recounted two to seven times (Fig. 6). These two years were relatively dry, with peak flow reaching only 0.37 and 0.93 of the 1.5-year recurrent discharge of about 5.7 m^3/s. In the first year, movement took place in the two finer patches, but not in the coarsest, and no significant change took place in the patch grain size distributions. The two finer patches were then repainted in the fall of 1988 and subsequent point counts kept track of both the painted and unpainted particles in the patch. All the painted particles were transported from the finest patch (Sta. 1010 m) during the first significant runoff event. Subsequent point counts show that the patch coarsened and then fined some relative to the initial values, but generally maintained the same size distribution. Grain size counting was done as much as possible by measuring the particles in place, rather than by picking them up, in order to minimize disturbance from measurement.

The intermediate patch (Sta. 1018 m) progressively lost painted particles through the winter runoff. The

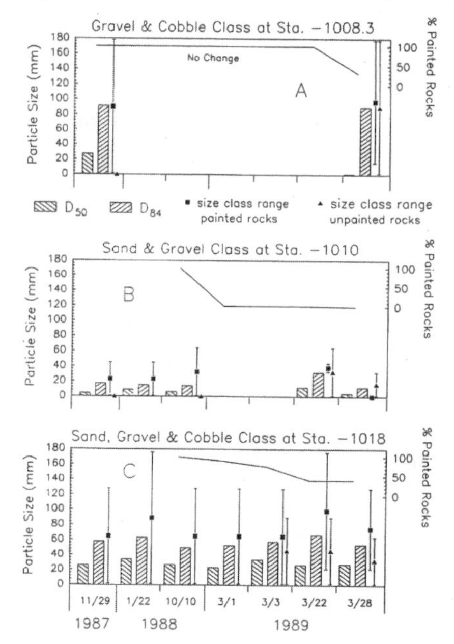

Figure 6. Data from the painted rock study performed at the three patches illustrated in Figure 1. Station numbers increase in the upstream direction, and are shown along the border of the channel in Figure 1 (e.g. 10-10 = 1010 m).

unpainted rocks observed in this patch were both exposed rocks from the bed below and replacement rocks from upstream. These unpainted rocks had fewer large rocks than the original bed and the size distribution of the remaining unpainted rocks was generally unchanged. Overall, the patch lost most of its original particles but changed little in grain size.

The coarsest patch (Sta. 1008.3 m) experienced no transport. This patch lay at a somewhat higher location along the channel bank and in the spring of 1989, a high flow brought sand onto the patch, burying much of it. This led to the patch being colonized by vegetation, and it has remained buried since then.

These observations demonstrate that patch dynamics are size-dependent, with finer patches experiencing frequent movement and coarser ones less so. Patch size distribution can remain about the same despite complete replacement of all particles, and patches can persist in the same location for several years despite transmitting large quantities of sediment.

We painted two patches at a second site 5 km upstream (Fig. 7) where the channel is narrower. In this reach the patches span the width of the bed and are locally arranged in series due to sequential downstream changes in width and slope. We painted one fine patch of sand and gravel and a downstream patch of sand, gravel and cobble (Fig. 8). As we found at

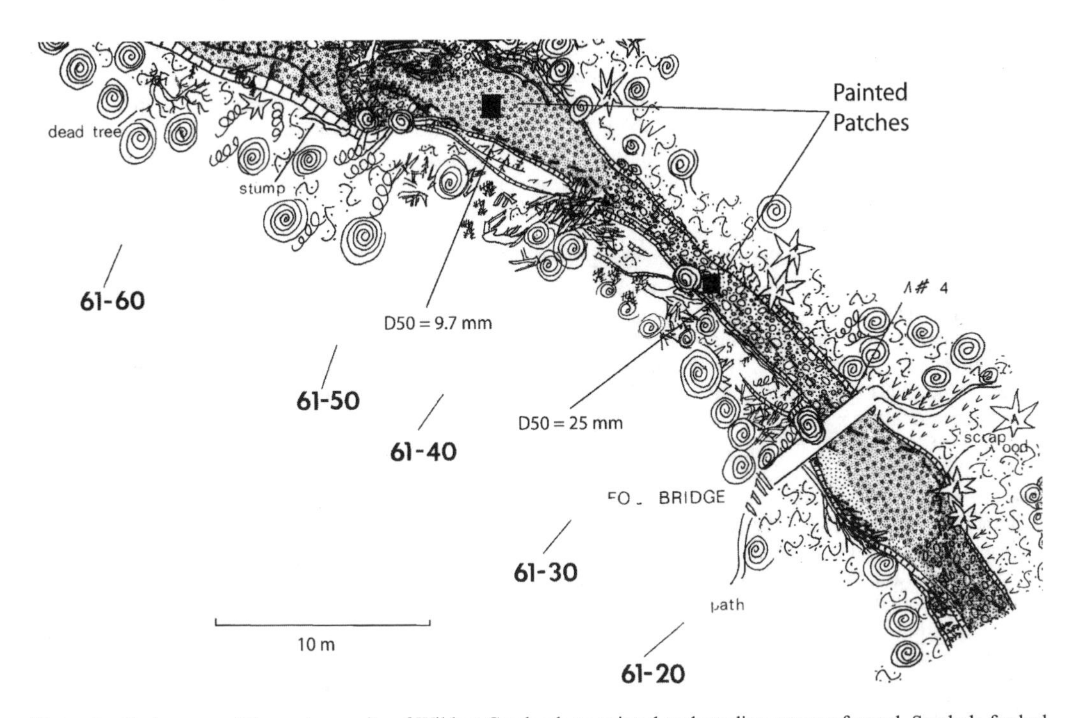

Figure 7. Facies map of the upstream site of Wildcat Creek where painted rock studies were performed. Symbols for bed texture are the same as in Figure 1. Spiral forms along banks are the location of trees. A footbridge crosses the channel near 61–20. Flow is from left to right.

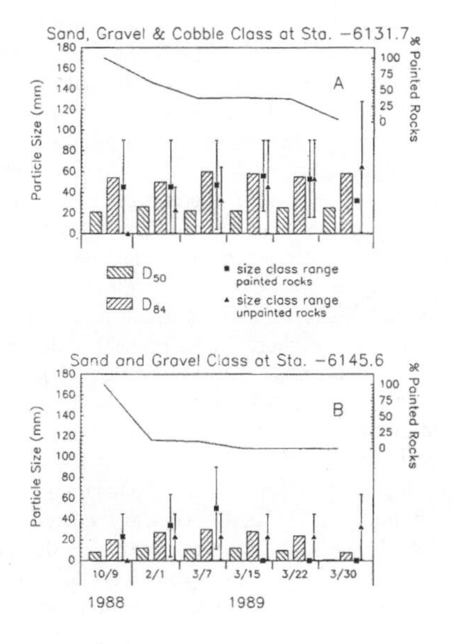

Figure 8. Data from the painted rock study performed at the two patches illustrated in Figure 7. Station numbers increase in the upstream direction, and are shown in Figure 7.

the downstream site (Fig. 1), the finer patch (Sta. 6145.6 m) experienced complete replacement with the first flow event but remained texturally the same. The downstream displacement of these particles was traced. Most of the particles released during the first runoff event from the upstream fine patch crossed the painted coarser patch downstream, although some of the finer particles came to rest in this coarser patch. In contrast, the coarser patch (Sta. 6131.7 m) progressively lost painted rocks, with finer painted particles experiencing net loss first (note the range of sizes reported in the plots). The unpainted rocks were finer at first, but then coarsened, eventually exceeding the range of the original surface. All sizes in the coarse patch moved, and similar sizes arrived from upstream having crossed the fine patch located upstream. These observations indicate that the alternative fine-coarse sequence remained stable while actively exchanging with all size ranges passing downstream. Fine particles crossed the coarse patch without fining it and coarse particles crossed the fine patch without coarsening it.

4 DISCUSSION

Flume and field studies reported and cited above show that grain size surface patches are common, and multiple patch sizes will develop even in unimodal sediment. Observations in straight flumes suggest that patch development affects bed surface dynamics and favors the topographic shoaling or emergence of coarse-textured patches. Experiments in which sediment supply is reduced favor the stagnation and emergence of these coarse patches and the concentration of finer bedload transport into the fine patches. Field studies suggest that patchiness may be correlated with sediment supply (Kinerson 1990). Tracer studies reported here show that coarse and fine particles may begin transport at similar stages, but that finer patches are significantly more mobile. This supports results from other field studies (e.g. Lisle & Madej 1992, Garcia et al. 1999). The tracer studies also show that the patches exchange particles with the through going bedload, yet can remain unchanged. Particles from fine patches cross coarse patches and vice versa. The finer patches are fully mobile while the coarse patches probably vary from partial to selective transport with increasing local boundary shear stress.

Spatial structure to bed surface sorting is typically strongly developed in river meanders with bed surfaces coarsening from the outside to the inside bank on the upstream end of the bed, and fining inward on the downstream end of the bend (e.g. Dietrich & Smith 1984, Bridge 1992, Hoey & Bluck 1999). In sand-bedded meandering rivers, the bed surface can be fully mobile, and yet this spatial structure of the grain sorting is temporally constant. Dietrich & Smith (1984) concluded that topographically-induced stress divergences, which would tend to create scour or deposition (depending on sign), were balanced by cross-stream fluxes of sediment. Across the point bar slope into the pool, particles roll outward against inward secondary circulation, causing large particles to roll into zones of high boundary shear stress and fine particles to be carried into zones of low boundary shear stress. Sand-bedded rivers are typically not described as having a patchy bed because the grain sorting tends to be more continuous, but the grain size across the channel can vary by a factor of 6 (Dietrich & Smith 1984). In gravel-bedded rivers, topographically-induced boundary shear stress divergences may be primarily compensated by grain size adjustments, and the bedload transport field may be uncorrelated with the boundary shear stress field (Dietrich & Whiting 1989, Lisle et al. 2000). We could not find any field studies of bedload transport through meandering gravel-bedded rivers that document controls on grain size adjustments.

These studies suggest that there should be three components of a complete patch theory: grain interactions, stress-divergence response, and lateral sorting effects. Whiting et al. (1988) and Dietrich et al. (1989) argued that a "catch and mobilize" process gives rise to mobile, downstream spatially-sorted bedload sheets. Seminara et al. (1996) used a stability analysis to show that the formation of bedload sheets may be highly dependent upon the deviation from equal mobility of different grain sizes in the surface layer, and that their growth is strictly associated with grain sorting. In a heterogeneous mixture of sediment, coarse particles will tend to "catch" other coarse particles in their wake and cause them to stop moving. Fine particles also trapped by the coarse particles will mobilize the coarse particles, perhaps through smoothing or altering the exposure or friction angle of the coarse particles. Once mobile, the coarse particles will run across the smoothed areas until stopped again by other coarse particles. The Wilcock & Crowe (2003) bedload transport model explicitly includes sand mobilization of coarse particles. As Paola & Seal (1995) reasoned, for a given mixture of sediment there must be some upper size class that will not be mobilized by the finer fraction and also there should be some lower size class that either goes into suspension or is too small and numerous to be slowed by the wakes of the coarser particles. These observations suggest that patchiness is an intrinsic tendency in sufficiently heterogeneous sediment. Although patchy bed surfaces form under steady uniform flow in straight narrow channels, unsteady hydrographs, which can cause transient preferential movement of patches, probably strongly reinforce patchiness (e.g. Lisle & Madej 1992, Parker et al. 2003).

The response to local boundary shear stress divergence in channels actively transporting sediment is through some combination of grain size adjustment, topographic adjustment, and cross-stream compensating sediment transport. It is not clear if current models would predict a purely grain size response. The Parker ebook discussion on armoring nicely illustrates how both slope and bed surface grain size are expected to covary for a constant bedload size distribution in response to varying sediment supply. Comparison with flume experiments here illustrates that the model tends to favor topographic change rather than grain size adjustment. It may be that the empirical structure of the underlying nonlinear dependency of bedload flux on the hiding function term does not permit a predominantly grain size response. Furthermore, the use of the Wilcock-Crowe bedload function predicts bed surface bimodality where sand is present. This was not consistent with the Lisle et al. observations. The empirical nature of the Wilcock-Crowe expression makes its application to sorting problems less certain.

Lateral sorting effects arise when there is cross-stream topography, where a cross-stream slope and commonly associated secondary currents create an ideal condition for grain sorting as described above. Well-developed theory exists for this process (Sekine & Parker 1992, Kovacs & Parker 1994). For channels with well-developed bars and pools or scour holes and lobate deposits, inclusion of this lateral sorting effect may be of first-order importance.

5 CONCLUSIONS

Patch dynamics appears to be a primary response to altered sediment supply. The Parker ebook spreadsheet comparison with the two flume studies shows that the model can predict the median grain size reasonably well, but it may incorrectly predict the surface grain size distribution and it tends to greatly over predict the steady state channel slope at low feed rates. The model also predicts a much longer time to steady state than observed in the flume. Lateral coarsening and narrowing of the finer textured zone of active bedload transport leads to less topographic change and more rapid response to altered sediment supply than predicted from the model. Application of the Wilcock & Crowe (2003) bedload expression in the Parker model consistently predicts a bimodal bed surface size distribution, even when the sediment source is unimodal. The unimodal sediment feed can nonetheless lead to strong patch development, especially in high width-to-depth channels in which bars emerge, interact with the flow, and strongly influence particle sorting.

Field studies using painted rocks illustrate that particles from different size patches must cross each other, even while the individual size distribution of the patches remains roughly constant. Grain interactions, transport and grain size adjustment associated with boundary shear stress divergence fields, and sorting processes on cross-channel slopes need to be coupled for a complete patch theory.

ACKNOWLEDGEMENTS

This work was supported in part by the California Water Resources Center and the NSF National Center for Earth Surface Dynamics.

REFERENCES

Bridge, J.S. 1992. A revised model for water flow, sediment transport, bed topography, and grain size sorting in natural river bends. *Water Resources Research* 28(4): 999–1013.

Bunte, K. & Abt, S.R. 2001. Sampling surface and subsurface particle-size distributions in wadable gravel- and cobble-bed streams for analyses in sediment transport, hydraulics and streambed monitoring, *US Forest Service, General Techn. Report RMRS-GTR-74*, 450 pp.

Dietrich, W.E., Kirchner, J.W., Ikeda, H. & Iseya, F. 1989. Sediment supply and the development of the coarse surface layer in gravel-bedded rivers. *Nature* 340: 215–217.

Dietrich, W.E. & Smith, J.D. 1984. Bed load transport in a river meander. *Water Resources Research* 20: 1355–1380.

Dietrich, W.E. & Whiting, P.J. 1989. Boundary shear stress and sediment transport in river meanders of sand and gravel. In S. Ikeda & G. Parker (eds), *River Meandering, American Geophysical Union Water Resources Monograph 12*. Washington D.C.: American Geophysical Union.

Ferguson, R.I. 2003. The missing dimension: effects of lateral variation on 1-D calculations of fluvial bedload transport. *Geomorphology* 56: 1–14.

Garcia, C., Laronne, J.B. & Sala, M. 1999. Variable source areas of bedload in a gravel-bed stream. *Journal of Sedimentary Research* 69(1): 27–31.

Hoey, T.B. & Bluck, B.J. 1999. Identifying the controls over downstream fining of river gravels. *Journal of Sedimentary Research* 69(1): 40–50.

Iseya, F. & Ikeda, H. 1987. Pulsations in bedload transport rates induced by a longitudinal sediment sorting: a flume study using sand and gravel mixtures, *Geografiska Annaller* 69(1): 15–27.

Iseya, F., Ikeda H. & Lisle, T.E. 1989. Fill-top and fill-strath terraces in a flume with decreasing sediment supply of sand-gravel mixtures. *Transactions Japanese Geomorphological Union* 10(4): 323–342.

Kinerson, D. 1990. Surface response to sediment supply, M.S. thesis, Univ. of Calif., Berkeley.

Kirchner, J.W., Dietrich, W.E., Iseya, F. & Ikeda, H. 1990. The variability of critical shear stress, friction angle, and grain protrusion in water-worked sediments. *Sedimentology* 37: 647–672.

Kovacs, A. & Parker, G. 1994. A new vectorial bedload formulation and its application to the time evolution of straight rivers. *Journal of Fluid Mechanics* 267: 153–183.

Lisle, T.E., Ikeda, H. & Iseya, F. 1991. Formation of stationary alternate bars in a steep channel with mixed-size sediment: a flume experiment. *Earth Surface Processes and Landforms* 16(5): 463–469.

Lisle, T.E., Iseya, F. & Ikeda, H. 1993. Response of a channel with alternate bars to a decrease in supply of mixed-size bed load: a flume experiment. *Water Resources Research* 29(11): 3623–3629.

Lisle, T.E. & Madej, M.A. 1992. Spatial variation in armouring in a channel with high sediment supply. In P. Billi, R.D. Hey, C.R. Thorne, & P. Tacconi (eds), *Dynamics of Gravel-Bed Rivers*. New York: John Wiley.

Lisle, T.E., Nelson, J.M., Pitlick, J., Madej, M.A. & Barkett, B.L. 2000. Variability of bed mobility in natural, gravel-bed channels and adjustments to sediment load at local and reach scales. *Water Resources Research* 36(12): 3743–3755.

Paola, C. & Seal, R. 1995. Grain size patchiness as a cause of selective deposition and downstream fining. *Water Resources Research* 31(5): 1395–1407.

Parker, G. & Klingeman, P.C. 1982. On why gravel bed streams are paved. *Water Resources Research* 18(5): 1409–1423.

Parker, G. 1990. Surface-based bedload transport relation for gravel rivers. *Journal of Hydraulic Research* 28(4): 417–436.

Parker, G., Toro-Escobar, C.M., Ramey, M. & Beck, S. 2003. Effect of floodwater extraction on mountain stream morphology. *Journal of Hydraulic Engineering* 125(11): 885–895.

San Francisco Estuary Institute. 2001. Wildcat Creek: a scientific study of physical processes and land use effects, report, 85 pp.

Sekine, M & Parker, G. 1992. Bedload transport on transverse slopes. *Journal of Hydraulic Engineering* 118(4): 513–535.

Seminara, G., Colombini, M. & Parker, G. 1996. Nearly pure sorting waves and formation of bedload sheets. *Journal of Fluid Mechanics* 312: 253–278.

Whiting, P.J., Dietrich, W.E., Leopold, L.B., Drake, T.G. & Shreve, R.L. 1988. Bedload sheets in heterogeneous sediment. *Geology* 16: 105–108.

Wilcock, P.R. & Crowe, J.C. 2003. A surface-based transport model for sand and gravel. *Journal of Hydraulic Engineering* 29: 120–128.

Wilcock, P.R., Kenworthy, S.T. & Crowe, J.C. 2001. Experimental study of the transport of mixed sand and gravel. *Water Resources Research* 37(12): 3349–3358.

Yager, E., McCardell, B.W., Dietrich, W.E. & Kirchner, J.W. 2005. Measurements of flow and transport in a steep, rough stream. *Geophysical Research Abstracts* 7(05836).

Yager, E., Schmeekle, M., Dietrich, W.E. & Kirchner, J.W. 2004. The effect of large roughness elements on local flow and bedload transport. *Eos Transactions AGU* 85(47), Fall Meeting Supplement, Abstract H41G-05.

River, Coastal and Estuarine Morphodynamics: RCEM 2005 – Parker & García (eds)

River incision due to gravel mining: a case study

J.P. Martín-Vide & C. Ferrer-Boix
Technical University of Catalunya, Jordi Girona, Barcelona, Spain

ABSTRACT: Historical information was used to study river degradation in a case study in NE Spain. The accumulation of good topographical information since the 1940s for the Gállego river (a tributary of the Ebro river, which drains about 4,000 km^2 of the southern slopes of the Pyrenees) enables a comparison of longitudinal bed profiles. Historical research has also revealed the volume of gravel mined, which amounts to ≈1 million m^3 compared to the volume of alluvium lost due to incision in the same period, which is ≈2 million m^3. This unbalance is explained by a simple model based on a bedload equation and an algorithm to determine whether effective bedload transport is controlled by transport capacity or by the supply of sediment. It follows from the analysis that the incision process has changed the magnitude of the shear stresses on the bottom. This is because, as the river becomes degraded, higher discharges fit into the main channel before spilling.

1 INTRODUCTION

The Gállego river drains about 4,000 km^2 of the southern slopes of the Pyrenees. It empties into the Ebro river, the largest river in Spain, very close to Zaragoza, which has 800,000 inhabitants (Fig. 1). The city is expected to grow and reach the shores of the river in the near future. This was the motivation behind a major study covering flooding risks and river morphodynamics as its main topics. The data, discussion and conclusions regarding the morphodynamics of this gravel-bed river are dealt with in this paper.

The Gállego river is regulated by three dams: La Peña (built in 1913, capacity 25 hm^3), Ardisa (1932, 5.3 hm^3) and Búbal close to its source (1971, 64 hm^3). A gauging station with good data from the period 1913–1987 is located at the toe of the Ardisa dam. The 20 km reach close to the Ebro river is the area we are interested in (Fig. 1), where major changes in the plan and longitudinal profile have occurred in recent decades. There is a small weir (Urdán) in this reach, 11 km upstream from the river mouth. The discharge records at the Ardisa gauging station (both annual flows and floods) are appropriate for the reach we are interested in since the rest of the drainage basin is small and mostly dry. Analysis of mean daily flow rates over the 76 years gives a discharge $Q_1 = 374$ m^3/s representative of the 76 highest data, a discharge $Q_2 = 241$ m^3/s representative of the second highest 76 data and likewise, $Q_3 = 207$ m^3/s and $Q_4 = 177$ m^3/s. A peak discharge of 1765 m^3/s corresponds to a return period of 500 years and 1210 m^3/s to 50 years. In recent decades one flood reached a peak of 1260 m^3/s.

Figure 1. Sketch of the situation and the stretch under study.

2 DESCRIPTION OF INCISION

Severe incision has taken place in recent decades along the reach we are interested in. Figure 2 is a photograph taken in 2004 of the river channel looking downstream at around 9 km upstream from the mouth (coordinate x = 9 km). The left bank stands at the original floodplain elevation, whereas the right bank has been eroded. Note the coarse gravel deposits and banks everywhere. A maximum degradation of more than

Figure 2. Photograph of the river channel showing its incision.

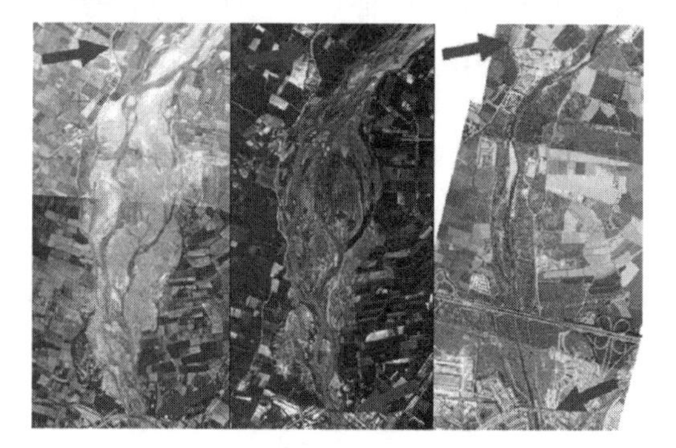

Figure 3. Changes in the Gállego river plan: 1946 (left), 1957 (centre) and 2004 (right).

5 m has been recorded in this area. The vegetation along the banks and on the floodplains has almost disappeared. The incision has been accompanied by a dramatic change in plan patterns, from a very wide, shifting, highly braided channel from the start of the 20th century until the 1960s to a single, mostly straight channel in 2004. Figure 3 shows this plan change.

Close scrutiny of cartografic maps from different periods has enabled longitudinal profiles along the channel thalweg to be compared. Figure 4 is a summary of this information, which is from the middle of the century, the early 1970s and the years 1992 and 2004. The figure draws our attention to certain facts. Firstly, there was no significant incision prior to the early 1970s; the bed profile of that "pristine" braided river was slightly convex. Secondly, since then there has been severe incision downstream from the Urdán weir. The weir seems to act as a fixed bed boundary condition for the upstream reach. The degradation is not uniformly distributed: a sort of nickpoint with a bed lowering of more than 5 m is located around x = 7.5 km in 1992, which is somewhat smoothed in the 2004 profile. In the 1992–2004 period, the channel degraded very close to the weir and aggraded around the nickpoint. At present, the slope is becoming milder close to the weir and steeper around the nickpoint and downstream from it, although maximum incision is still more than 5 m.

3 GRAVEL MINING

Relevant information has been inserted into Figure 4 to explain what happened in the river. The archives of the Water Administration provided reliable information on the exact places, dates and amounts of gravel mined from the channel or from the functional floodplains. The amount of gravel mined from 1963 onwards is plotted at four different reaches along the river profile in Figure 4. The arrows point to the centre of every reach where mining has been computed. Two assumptions are worth noting. Firstly, the entire volume of gravel was extracted during the same year the permit was issued and secondly, the amount was equal to the amount that was allowed to be extracted. There is a clear relationship between gravel mining and channel incision, as shown in Figure 4. No other causes can explain such an incision.

An interesting fact that can be gleaned from Figure 4 is that degradation proceeds upstream from the centre of major gravel mining pits or zones. This suggests an upstream propagation of the bed disturbance. For example, the severe degradation close to the weir must be caused by the gravel mined close to the nickpoint but not by that mined along the reach itself, which is minor. We have also inserted into Figure 4 a history of floods in the Gállego river since 1959. This shows the main events that might have activated gravel transport and channel dynamics, although the high annual flows (Q_1, Q_2, etc.) are not negligible with respect to gravel transport.

Table 1 summarizes the amount of degradation in two time periods (the minus sign indicates lowering). The figures in the table have been obtained by averaging the bed lowering along each reach. The first period (of 26 years) includes all the known permits for gravel mining while there are no records of gravel mining for the second period (of 17 years). An interesting point is that degradation did not stop although gravel mining operations did. This suggests a sort of "inertia" in the behavior of the bed profile. Similarly, it must be borne in mind that degradation had not started in the early 1970s despite the mining and floods took place in the latter half of the 1960s (Fig. 4).

Table 2 shows the gravel sizes at different x-positions in the channel. Sub-surface samples weighing more than 200–300 kg each and surface samples of 100 particles each (using the Wolman method) were taken from the active channel in 2004. The surface results were not corrected by any conversion to

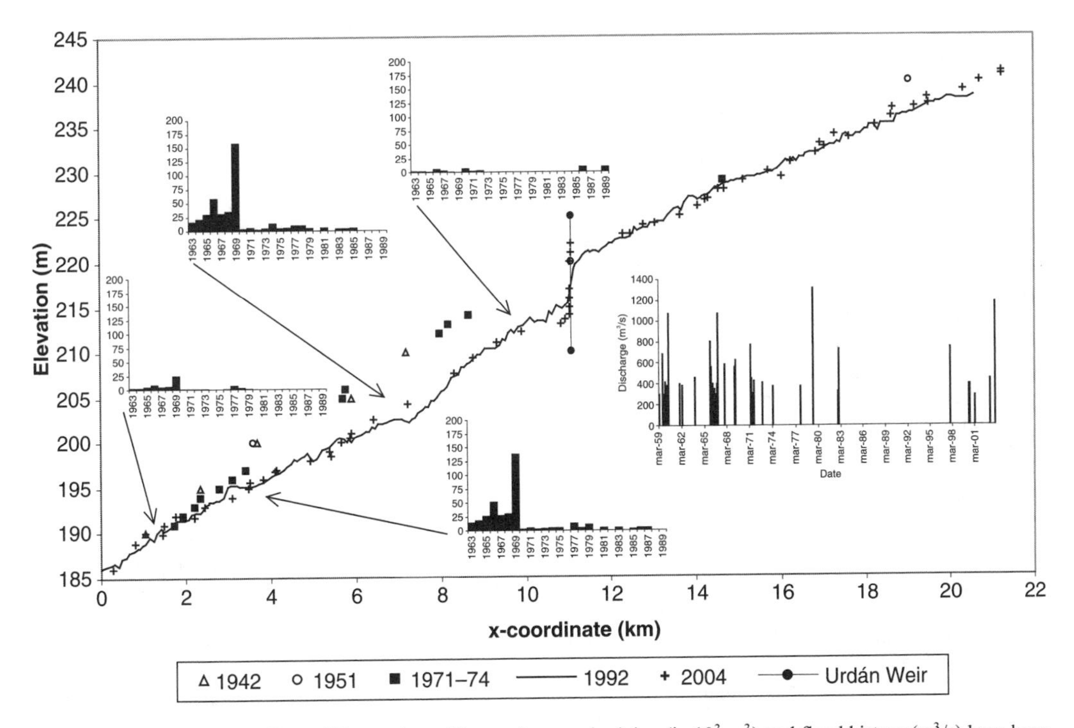

Figure 4. Longitudinal profile at different times. Figures for gravel mining (in 10^3 m^3) and flood history (m^3/s) have been inserted.

Table 1. Mean values of incision at different times.

Distance km	1962–1987 incision m	1988–2004 incision m
0–1.83	−0.99	–
1.83–3.50	−2.00	–
3.50–5.80	−3.64	–
5.80–8.30	−4.86	−0.40
8.30–11.09	−3.36	−1.47

Table 2. Grain sizes along the river.

x-coord km	D_m mm	σ_g mm	D_m* mm	σ_g* –
1.48	19.7	1.84	58.3	1.53
5.38	14.1	2.00	102.6	2.11
7.27	10.6	2.61	104.9	1.77
9.30	11.9	2.38	145.2	1.38
12.99	20.1	2.64	95.5	1.41
17.64	25.5	2.16	89.9	1.92

*Surface samples.

volumetric sampling. The mean diameter, D_m, was computed using the ψ parameter. Table 2 shows that the bed is armoured – note the large difference in surface and sub-surface D_m. This difference is greater downstream from the weir than upstream of it, which might be related to the process of degradation that would have promoted the armouring of the bed. The mean diameters did not show very significant fining (or coarsening) along the distance between the x-coordinate. The standard deviations of the grain sizes are smaller for the armoured surface layer than for the sub-surface material, as expected.

4 GRAVEL TRANSPORT MODELLING

A simple mass balance or budget model for sediment transport has been applied to this case study. The 20 km reach we are interested in has been divided into 8 sub-reaches (the five that are farthest downstream are shown in Table 1). The model is based on a bedload equation and an algorithm to determine whether the effective bedload transport is controlled by the transport capacity or by the supply of sediment from stream, as shown in Martín-Vide (2002). The mass balance compares the volume coming into and going out of the sub-reach. Four different bedload equations were tested: Einstein-Brown (E-B), Meyer-Peter and Müller (MPM), Smart-Jaeggi (S-J) taken from Rickenmann (1991) and Smart (1984), and finally Parker from Parker (1990). The latter is capable of computing the partial transport of each size fraction. However, all the equations have been applied by size fractions, because the balance is sometimes controlled by the transport capacity and sometimes by the sediment supply. The supply of sediment is computed by the same equation, which is applied to the cross-sectional and grain size data upstream. Shear stresses caused by the flow are computed by averaging the results of the steady gradually varied Hec-Ras model, using the cross-sectional information from 2004. The model has been run for every flow of the annual distribution (Q_1, Q_2, etc.), although the sediment transport might be zero most of time for medium and low flows, depending on the equations. The model has been also run for actual flood hydrographs with a time step of 1 h. Both results are summed up.

A main general result of the gravel mining analysis shown in Figure 4 is now given for the 11 km reach between the Urdán weir and the mouth at the Ebro river. In the period 1962–1987, the volume of gravel extracted was 965,100 m^3. From 1988 to 2004, it was zero. The volumes of alluvium lost in the same periods and along the same reach (obtained by geometrical calculations -as in Table 1 above for incision) were 1,818,200 m^3 from 1962 to 1987 and 346,000 m^3 from 1988 to 2004. These figures are compared or combined with the model results using the following sediment budget:

$$(\text{Volume in}) - (\text{Volume out}) - (\text{Volume mined}) = (\text{Volume change})$$

The first two quantities are those computed by the model. The last two are the gross data given above. The volume change is negative (volume of alluvium lost). Note that the volume of alluvium lost does not have to exactly balance the volume mined, but the unbalance must be explained by the difference in transport in and out. If the volume mined was around 1 million m^3 from 1962 to 1987, why was its effect in terms of lost alluvium around 2 million m^3? Can this difference of 1 million m^3 be explained by the gravel transport equations? Table 3 presents the balance results. If the balance is verified, the figure in the table is zero. Note that the MPM equation can explain most of this one million m^3 difference between volume mined and volume lost, because the volume coming in and going out does reduce this difference to only 0.2 million m^3. The Parker and S-J equations do not explain the difference at all, whereas E-B would predict an even greater unbalance.

It follows from this analysis that gravel mining may change channel dynamics. The mining itself directly accounts for half of the incision, but might be indirectly responsible for the other half. As incision continues, greater flow can fit into the channel and less flows are forced to spill onto the floodplain. Therefore, the magnitude of the shear stresses on the bottom, averaged over space and time, are increased. This is thought to be the reason for this phenomenon.

Table 3. Sediment budget using the four bed-load equations.

Period	E-B budget m^3	MPM budget m^3	S-J budget m^3	Parker budget m^3
1962–1987	−1,010,000	203,000	776,000	831,000
1988–2004	−1,130,000	181,000	321,000	331,000

5 CONCLUSIONS

The case study of the Gállego river shows incision amounting to more than 5 m between the early 1970s and 1992. The mining of its gravel (15–20 mm) is the only plausible cause of this severe degradation. There is some sort of "inertia" between cause and effect, as incision has not ceased yet in spite of mining operations having stopped. Degradation seems to propagate upstream. High annual flows and floods are thought to be the link between cause and effect because they activate gravel transport and river dynamics. A gravel budget along the 11 km reach of interest and real data spanning 26 years suggest ranking the Meyer-Peter and Müller equation first from among the equations tested. The loss of alluvial material due to incision (2 million m^3 in 26 years) is twice the gravel mined. This is explained by the difference between the materials transported in and out. Gravel mining "directly" causes incision. This incision, in turn, can cause higher average shear stresses on the bed under the same flows and, therefore, may aggravate the process of degradation.

REFERENCES

Martín-Vide, J.P. 2002. *River Engineering* (in Spanish). Barcelona: UPC.

Parker, G. 1990. Surface-based bedload transport relation for gravel rivers. *Journal of Hydraulic Research*, 28, (4), 417–436.

Rickenmann, D. 1991. Hyperconcentrated Flow and Sediment Transport at Steep Slopes. *Journal of Hydraulic Engineering ASCE*, 117, (11), 1419–1439.

Smart, G.M. 1984. Sediment Transport Formula for Steep Channels. *Journal of Hydraulic Engineering ASCE*, 110, (3), 267–276.

River, Coastal and Estuarine Morphodynamics: RCEM 2005 – Parker & García (eds)
© 2006 Taylor & Francis Group, London, ISBN 0 415 39270 5

Do perennial and ephemeral streams represent parts of the same continuum?

P. Diplas
Dept. of Civil and Environmental Engineering, Virginia Polytechnic Institute and State University, Blacksburg, USA

J. Almedeij
Dept. of Civil Engineering, Kuwait University, Safat, Kuwait

C.L. Dancey
Dept. of Mechanical Engineering, Virginia Polytechnic Institute and State University, Blacksburg, USA

ABSTRACT: Numerous observations from perennial gravel-bed streams indicate the persistence of a surface bed layer, called armor layer that is coarser than the subsurface material. Several theories have been proposed in an effort to explain the existence of this layer and its possible evolution under different flow and sediment transport conditions. Recent observations though from some ephemeral gravel streams in Australia and Israel do not corroborate the existence of an armor layer. In some cases they even attest to a reverse layering of the bed material, with finer sediment overlying coarser material. Another difference that has been identified between the two stream types is regarding the efficiency in transporting bedload, which is higher in ephemeral streams compared to their perennial counterparts. It is suggested here that the behavior of the two stream types, in spite of the apparent differences, when considered in terms of dimensionless parameters, represents parts of the same river continuum.

1 INTRODUCTION

A distinctive feature of perennial gravel-bed streams is the stratification of their bed material in terms of grain size, with a thin coarse layer overlying finer sediment (Fig. 1a). This feature has been reproduced in several flume studies that properly account for the wide range of particle sizes typically encountered in natural gravel streams (e.g. Proffitt 1980, Parker et al. 1982, Diplas & Parker 1992). Several researchers, based on field measurements, have reported values of the surface to subsurface ratio of the corresponding median sizes as high as 15 (e.g. Mueller et al., 2005).

Until recently, it was believed that the armor layer characterized all gravel streams. However, field data from some ephemeral gravel-bed streams indicate that bed armoring is not a universal feature (e.g. Laronne et al. 1994). Actually, in these cases a layer of finer sediment positioned on top of coarser material has been reported (Fig. 1b). Ephemeral streams are found in arid and semiarid environments and remain dry except for brief periods of time, when high intensity and short duration thunderstorms often generate excessive surface runoff. Occasionally, stream flows of uncommon strength can be generated as a result of such events.

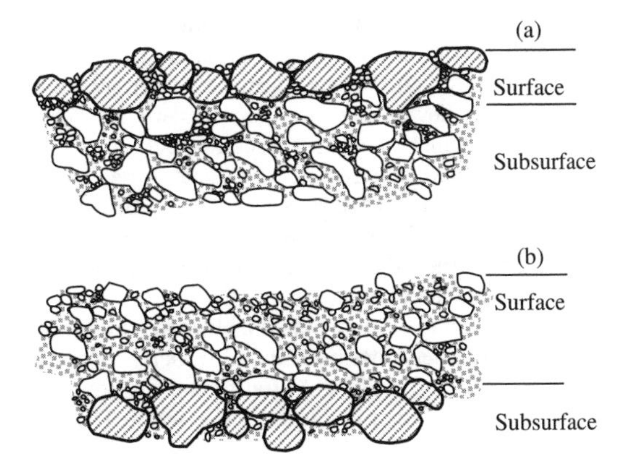

Figure 1. Schematic of the bed structure typically observed in perennial streams (a) and in some ephemeral streams (b).

Another difference between ephemeral and perennial gravel streams that has been reported in the literature is with regard to the higher bedload transport efficiency exhibited by the former type (Laronne & Reid 1993). Given the interdependence between streambed structure and bedload transport rate, it is reasonable to assume that the two differences mentioned here should be examined together.

Several researchers have emphasized the features distinguishing ephemeral from perennial streams and have concluded that they exhibit different behavior (e.g. Reid & Laronne 1995, Powell et al. 1998). This study points out the inability of the selective transport reasoning (e.g. Dietrich et al. 1989) to explain the formation of inverse layering in some ephemeral streams and proposes instead a mechanical process of particle size segregation. It is suggested here that when both the make-up of the bed material and bedload transport are examined in terms of dimensionless parameters, the apparent distinct trends mentioned here can be viewed as parts of the same river continuum.

2 BED MATERIAL STRATIFICATION

Vertical winnowing and selective transport of the finer grains have been proposed as the two mechanisms responsible for the development of the armor layer observed in perennial gravel streams (e.g. Parker & Klingeman, 1982). The thickness of the armor layer, based on field and flume measurements, is estimated to be around D_{90} (e.g. Petrie & Diplas 2000), where D_{90} is the armor layer grain size that is coarser than 90% of the available material by weight (Fig. 1a). It has been advocated that selective transport is a response to the imbalance between sediment supply and sediment transport capacity of a stream reach (Dietrich et al. 1987).

The coarsest state of the armor layer is reached under starved sediment supply conditions, when the bed material surface will reach, eventually, a threshold of motion condition (static armor). Stream sections immediately downstream of dams experience such conditions, which identify the configuration at the lower limit of shear stresses. The selective sediment transport mechanism becomes less pronounced as the boundary stress increases and coarser particles get entrained by the flow, resulting in finer composition of the armor layer. In the upper limit, at very high shear stress values when, presumably, equal mobility prevails, the finest possible state of the bed surface is attained, when no grain size stratification is present, $(D_{50}/D_{s50}) = 1$ (Dietrich et al. 1987); where D_{s50} is the subsurface material median size. The trend of increasing degree of bedload coarseness with applied shear stress is shown in Figure 2, where D_{50}/D_{l50} and D_{s50}/D_{l50} are plotted against φ_{50}, a normalized Shields stress parameter, for data obtained by Milhous (1973) from Oak Creek, a perennial gravel stream in Oregon; where D_{l50} is the median size of the material transported as bedload.

Most, if not all the data reported in the literature from field and flume studies dealing with perennial gravel streams, and they do represent an impressive

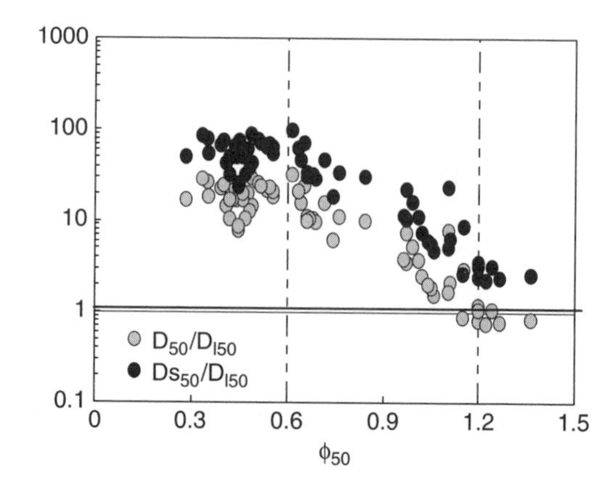

Figure 2. Variation of bedload median size with respect to increasing values of normalized shear stress measured in Oak Creek.

effort from a large number of sites, point to the common observation that even under bankful flow conditions the boundary shear stresses experienced by these streams modestly exceed the corresponding critical stress values (e.g. Emmett & Wolman, 2001; Mueller et al. 2005). Therefore perennial stream data cannot be used to test the selective transport model for very high Shields stresses.

Observations from some ephemeral gravel streams reported during the last decade indicate a bed material configuration with reverse particle size segregation. For example, in Nahal Yatir, a stream located in Israel which drains a watershed of $19\,km^2$, it was found that $(D_{50}/D_{s50}) = 0.6$ (Reid et al. 1995). The concept of selective transport, endorsed by many (e.g. Dietrich et al. 1987), cannot be employed in this case to explain the existence of a finer surface layer. Such an approach, based on mass balance arguments, will require that for very high shear stresses the coarser particles become more mobile than the finer ones, which does not seem reasonable and has never been reported in the literature. It is therefore suggested here that for stresses that substantially exceed the corresponding bedload threshold conditions, while at the same time remain below the suspension threshold values, conditions resembling a temporarily fluidized bed might prevail (Diplas & Almedeij, in review). Such conditions would lead to a segregated sediment deposit, with the finer grains on top and the coarser particles sinking to the bottom of the entrained layer of bed material. This phenomenon is encountered in several chemical engineering applications. While in the latter case fluid drag is known to be the mechanism responsible for fluidizing the bed, the lift force has been recently proposed as the mechanism of accomplishing similar results in river flows operating under very high boundary stresses (Diplas & Almedeij, in review). It is worth

Table 1. Characteristics of two gravel-bed streams.

Characteristics	Oak Creek	Nahal Yatir
Stream type	Perennial	Ephemeral
D_{50} (mm)	54	6
D_{s50} (mm)	20	10
Slope S	[a]0.0083–0.0108	[b]0.007–0.0101
Depth d (m)	0.11–0.45	0.10–0.47
τ_0 ($N\ m^{-2}$)	8.93–43.22	8.67–36.92
Shields stress τ^*	0.00877–0.0424	0.092–0.39
Bedload q_B ($kg\ m^{-1}\ s^{-1}$)	1.26×10^{-7}–0.115	0.20–7.05

[a] Channel bed slope.
[b] Water surface slope.

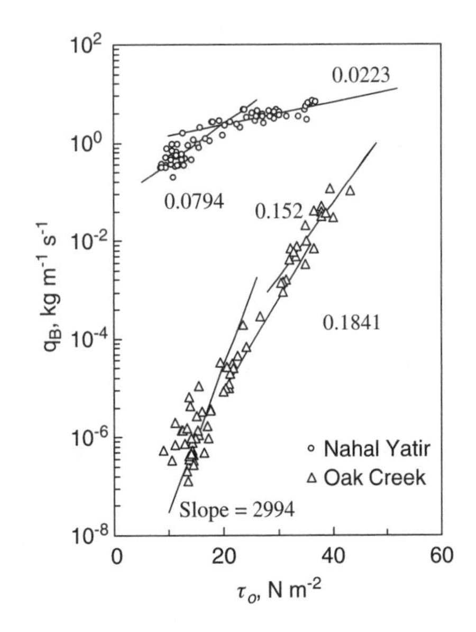

Figure 3. Bedload transport versus boundary shear stress for Oak Creek and Nahal Yatir data.

pointing out that such a phenomenon is more likely to occur in the case of gravel and coarser particles where the difference between bed and suspension load threshold values is substantial (e.g. Sklar & Dietrich 2004); to wit, for 6 mm material the threshold Shields stress value for suspension is about 1.30, compared to a value around 0.03 that is widely assumed as the corresponding bed load threshold value. So it can be surmised that the traditional models based on bed load transport break down beyond a certain shear stress value and a different mode of transport, fluidization, might become important before suspension takes over.

A casual examination of the dimensional shear stress values of Oak Creek and Nahal Yatir indicates that the two streams operate in a similar range of shear stress values, τ_o, (Table 1). However, in dimensionless terms, τ^*, it is evident that the flows in Nahal Yatir generate Shields stress values about an order of magnitude higher than those in Oak Creek, which in turn, result in much higher bedload transport rates per unit channel width, q_B (Table 1) for the ephemeral stream. It is therefore proposed here that the differences in bed material stratification between perennial and ephemeral streams are due to the significant differences in dimensionless shear stress values experienced by the two stream types. If such high Shields stresses were to occur in perennial streams, it is speculated that a reverse grain size stratification would be encountered there as well.

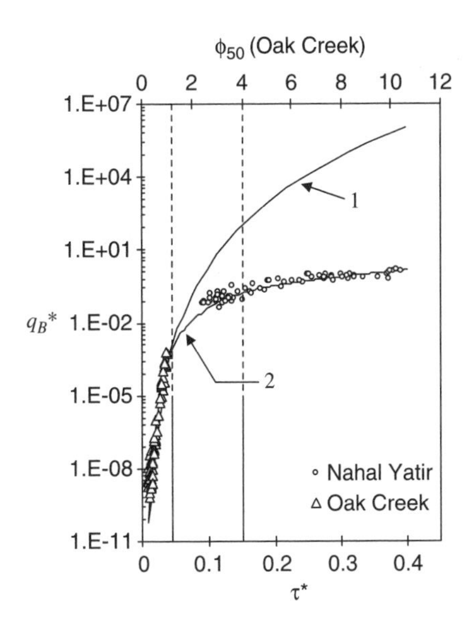

Figure 4. The bedload data shown in Figure 3 replotted in terms of dimensionless parameters.

3 BEDLOAD TRANSPORT

The presence of an armor layer in most gravel-bed streams complicates the response of bedload transport with increasing boundary stress in terms of both its composition and its volume. Several researchers have advocated that the change in the coarseness of the material transported as bedload (Figure 2) is accompanied by a change in the texture of the surface layer.

When bedload data from Oak Creek and Nahal Yatir are plotted in dimensional terms in Figure 3, the picture we get is confusing. Apparently the ephemeral stream is far more efficient in transporting sediment for the same value of shear stress, suggesting distinct behavior for each stream case. Replotting the data in terms of dimensionless parameters (dimensionless bedload transport rate per unit channel width, q_B^* versus Shields stress) in Figure 4, it appears that the trends exhibited by the two streams can be reconciled and possibly the bedload variation with Shields stress could be expressed through a single-valued function (Diplas & Almedeij, in review).

4 CONCLUSIONS

Data from a representative perennial gravel-bed stream are compared against that of an ephemeral stream operating under shear stresses well in excess of critical values. A casual way of pursuing such a comparison suggests certain dissimilarities in the behavior of these two stream types. Two of these issues, streambed structure and bedload transport, were highlighted in this study. A more careful analysis of the data though, through the use of appropriate dimensionless parameters, indicates that previously considered inconsistencies are removed and furthermore reveals that the two streams exhibit consistent behavior when accounting for the significantly different ranges of Shields stress values they operate under. It is therefore concluded that the trends shown by the two stream types belong to the same overall stream behavior continuum for gravel-bed streams.

REFERENCES

Dietrich, W.E., Kirchner, J.W., Ikeda, H. & Iseya, F. 1989. Sediment supply and the development of the coarse surface layer in gravel-bedded rivers. *Nature*, 340: 215–217.

Diplas, P. & Parker, G. 1992. Deposition and removal of fines in gravel-bed streams. In P. Billi, R.D. Hey, C.R. Thorne & P. Tacconi, *Dynamics of Gravel-Bed Rivers*: 313–329. John Wiley, New York.

Emmett, W.W. & Wolman, M.G. 2001. Effective discharge and gravel-bed rivers. *Earth Surf. Process and Landforms*, 26: 1369–1380.

Laronne, J.B. & Reid, I. 1993. Very high rates of bedload sediment transport by ephemeral desert rivers. *Nature*, 366: 148–150.

Milhous, R.T. 1973. Sediment transport in a gravel- bottomed stream. Unpublished Ph.D. thesis, Oregon State Univ., Corvallis.

Mueller, E.R., Pitlick, J. & Nelson, J.M. 2005. Variation in the reference Shields stress for bed load transport in gravel-bed streams and rivers. *Water Resour. Res.* 41, 4006, doi:10.1029/2004WR003692.

Parker, G. & Klingeman, P.C. 1982. On why gravel bed streams are paved. *Water Resour. Res.* 18(5): 1409–1423.

Parker, G., Dhamotharan, S. & Stefan, H. 1982. Model experiments on mobile, paved gravel bed streams. *Water Resour. Res.* 18(5): 1395–1408.

Petrie, J. & Diplas, P. 2000. Statistical approach to sediment sampling accuracy. *Water Resour. Res.* 36(2): 597–605.

Powell, D.M., Reid, I. & Laronne, J.B. 2001. Evolution of bed load grain size distribution with increasing flow strength and the effect of flow duration on the caliber of bed load sediment yield in ephemeral gravel bed rivers, *Water Resour. Res.* 37(5): 1463–1474.

Proffitt, G.T. 1980. Selective transport and armouring of non-uniform alluvial sediments, *Report no. 80/22*, 203 pp., Department of Civil Engineering, University of Canterbury, New Zealand.

Reid, I. & Laronne, J.B. 1995. Bed load sediment transport in an ephemeral stream and a comparison with seasonal and perennial counterparts, *Water Resour. Res.* 31(3): 773–781.

Reid, I., Laronne, J.B. & Powell, D.M. 1995. The Nahal Yatir bedload database: sediment dynamics in a gravel-bed ephemeral stream, *Earth Surf. Processes Landforms*, 20: 845–857, 1995.

Sklar, L.S. & Dietrich, W.E. 2004. A mechanistic model for river incision into bedrock by saltating bed load. *Water Resour. Res.* 40(6), p 21.

River, Coastal and Estuarine Morphodynamics: RCEM 2005 – Parker & García (eds)
© 2006 Taylor & Francis Group, London, ISBN 0 415 39270 5

Modeling the bedrock river evolution of western Kaua'i, Hawai'i, by a physically-based incision model based on abrasion

P. Chatanantavet
St. Anthony Falls Lab, Dept. of Civil Engineering, University of Minnesota, Minneapolis, MN, USA

G. Parker
Dept. of Civil & Environmental Engineering & Dept. of Geology, University of Illinois, Urbana, IL, USA

ABSTRACT: Channel incision into bedrock by abrasion plays an important role in orogenesis by redistributing material within drainage basins. It is often the dominant erosional mechanism in bedrock streams. The purpose of this paper is to qualitatively apply a physically-based incision model of abrasion to bedrock streams in western Kaua'i. These streams have various long profiles despite their homogeneous lithology throughout. We begin by analytically and numerically developing a physically-based bedrock incision model by abrasion (wear) driven by collision that includes a cover factor for alluvial deposits, the capacity bedload transport rate of effective tools, sediment delivery from hillslopes, and hydrology. The numerical results in the present study are analyzed together with Digital Elevation Model data of stream longitudinal profiles and field observations due to DeYoung (2000) and Seidl et al. (1994). The model qualitatively simulates the main features of the bedrock rivers in western Kaua'i, most of which have profiles with a convex shape, a straight shape, or with knickpoints. The results support an analytical result that knickpoints found in bedrock rivers may be autogenic in addition to being driven by base level fall and lithologic changes. This supports the concept that bedrock incision by knickpoint migration might not be separate from the normal abrasion process. Moreover, the results show that the upstream distance from a channel head to its divide is an important factor affecting the shape of the long profile of a bedrock river. The distance to channel head of the streams flowing into the Mana Plain region of Kaua'i was reduced as Waimea Canyon migrated upstream and captured their headwaters. This reduction in channel head region, combined with a relative sea level fall of 10 m over the last 5000 years, may have played a role in the formation of the Mana Plain itself. In the Napali region just north of the Mana Plain, on the other hand, where there is no limit on the upstream distance from the channel head to the divide, all streams show distinct long profiles with knickpoints which may have developed both autogenically and due to base level fall.

1 INTRODUCTION

Surprisingly little is known about what processes or factors control the long profile shapes of bedrock rivers, which can be concave, convex, straight, or with knickpoints. In addition, even though the importance of river incision into bedrock in driving landscape evolution is now well known, relatively few studies supported by field data exist to predict knickpoint genesis in river long profiles.

The starting point of much modeling research is based on a search for a profile for which incision is in steady-state balance with uplifting. Some modelers have interpreted their results as realistic when the simulated profiles were found to be concave upward, a common feature in nature (e.g. Slingerland & Snow, 1988). A thorough examination of bedrock river profiles, however, reveals that a concave region often co-exists with steep or convex parts, also known as knickpoints. Some bedrock river profiles are even convex throughout the reach (Seidl et al., 1994).

Knickpoints are often found in bedrock rivers at locations unrelated to lithology, tributary convergence, or uplift (Woodford, 1951). Knickpoint migration remains a poorly understood process in terms of controlling factors despite the fact that the incision rate associated with knickpoints is relatively usually high, such as in the region of Kaua'i analyzed here (DeYoung, 2000).

The western part of the Hawaiian island of Kaua'i has various geomorphologic features that have intrigued several researchers on bedrock rivers. Seidl et al. (1994) were the first to thoroughly investigate bedrock streams in this region by field surveys, topographic maps, and a stream power model. They suggested that the processes controlling Hawaiian channel downcutting may be composed of two separate processes; (1) abrasion of the channel bed by transported particles and (2) step-wise lowering caused by knickpoint migration. Accordingly, they came to the conclusion that a theory for knickpoint evolution is needed in conjunction with their stream power model to generate incision across a knickpoint. Their work has provided useful information and understanding of streams in Kaua'i. This notwithstanding, it is suggested here that their stream power model may be too simple to explain some important features of bedrock rivers. Subsequently Seidl et al. (1997) also used the cosmogenic isotope dating method to determine surface exposure ages in one of the bedrock streams in western Kaua'i, Kaulaula stream, to test their hypothesis that knickpoints at this site can propagate upstream rather than simply stabilize at a point of lithological change. They found that the measured exposure ages can neither confirm nor reject the knickpoint migration hypothesis. However, the boulder ages downstream of knickpoints were found to be consistent with a wave of erosion sweeping upvalley. They also suggested that some large boulders may remain in the valley downstream of the knickpoint for hundreds of thousands of years.

DeYoung (2000) was perhaps the second researcher to investigate the western Kaua'i site in detail. She developed a numerical model using both a linear stream power formulation and a linear undercapacity formulation by Chase (1992) to test the hypothesis that the geomorphology of western Kaua'i resulted partly from the processes of coastal retreat and/or catastrophic landslides. From her modeling results, she concluded that the knickpoints at this site should have resulted from catastrophic landslides rather than the process of constant coastal retreat. She was, however, convinced by the evidence of Moore et al. (1989) that catastrophic landslides only occurred near the streams in the north of the island rather than the west. DeYoung also found that the undercapacity model could describe incision both downstream and upstream of the knickpoints, so that the additional erosion laws proposed by Seidl et al. (1994) are not required to represent incision at this site. Both the stream power model and the linear undercapacity model that she used in her treatment are still rather simple, even though the latter constitutes an attempt to include the role of sediment flux in bedrock incision models. The undercapacity model that she used assumes that the capacity transport rate is only linearly proportional to water discharge and channel slope. More importantly, the undercapacity model allows the incision rate only to drop with increasing sediment transport rate in a channel (cover effect). As Sklar & Dietrich (1998, 2004) have shown, however, there is also a second effect (tool effect) operative at lower transport rates, according to which a higher sediment transport rate in a channel leads to a higher incision rate. In addition, DeYoung (2000) mentions that both stream power and undercapacity models have problems maintaining the steepness of knickpoints.

Sklar & Dietrich (1998) and Sklar & Dietrich (2004) were among the first researchers to develop a detailed mechanistic model for river incision into bedrock by abrasion and consider the importance of both the tool effect and the cover effect. They did not, however, apply their results to Hawaiian bedrock-streams.

In spite of their valuable contributions, Seidl et al. (1994), DeYoung (2000), and Sklar & Dietrich (2004) did not demonstrate that knickpoints can be autogenic in addition to being driven by base level fall. Moreover, Seidl et al. (1994) and DeYoung (2000) overlooked the importance of the upstream boundary condition on the evolution of these bedrock channels. Here we hypothesize that the shortening of the upstream distance from a channel head to its divide may lead to dramatically different morphologies. The hypothesis is supported by numerical modeling.

2 FIELD SITE: WESTERN KAUA'I, HAWAI'I

The bedrock rivers in western Kaua'i provide an ideal site for this study since they have various river long profiles, including convex profiles, straight profiles, and profiles with various shapes of knickpoints. The site is also ideal because a) it has a homogeneous basaltic lithology throughout, b) the dominated incision process appears to be wear by coarse-grained particles moving over a bedrock surface, c) it has never been glaciated (Porter, 1979), d) the preserved paleosurface allows measurement of incision depth and e) tectonic processes are secondary (Seidl et al., 1994; DeYoung, 2000). The Napali coast is located on the northwestern part of the island. The Mana plain is a sediment depositional apron forming a broad flat surface along the southwestern part of the island (Figs. 1a and 1b), just south of the Napali coast. Unlike the rivers in the Napali region, most rivers adjacent to the Mana plain region have either slightly convex or straight shapes without a knickpoint (Seidl et al., 1994; DeYoung, 2000). At the upstream end of all streams in the Mana plain region and a few streams in the Napali region, the distances from the water head to the original divide were drastically shortened by the Waimea canyon as it elongated (Fig. 1a). Waimea canyon is about 1100 m deep and 16 km long. It was formed not only by the steady process of erosion, but also by a catastrophic collapse of the volcano creating Kaua'i (source: NASA website). The work by Moore et al. (1989) manifests these catastrophic landslides in the southern part offshore of Kaua'i in the form of submarine mass wasting deposits by using the GLORIA side-scan sonar system.

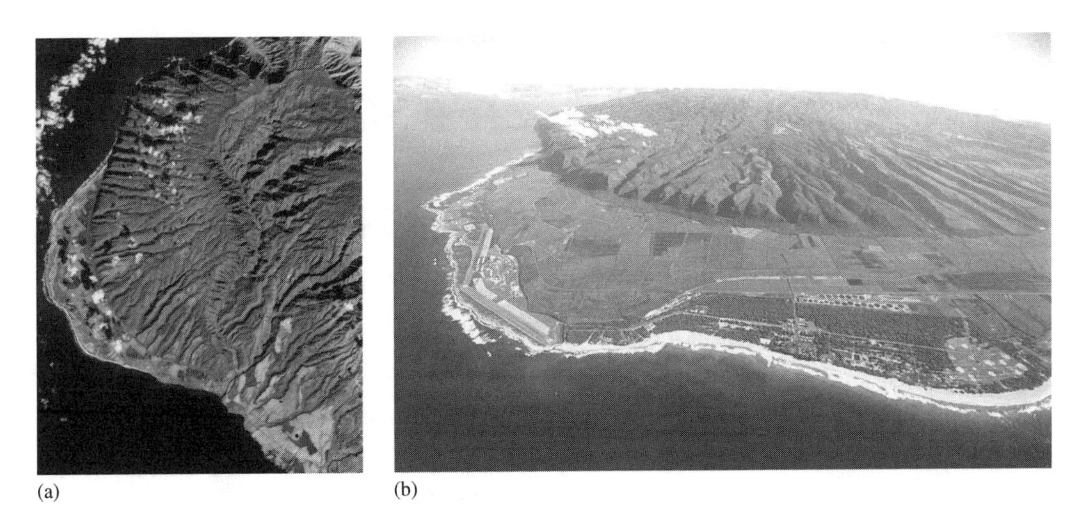

Figure 1. (a) Plan view of the study area: a series of bedrock rivers with Mana Plain on the left and Waimea Canyon on the right. From NASA website. (b) Areal photo of bedrock rivers in western Kaua'i and the Mana Plain. From the PMRF website.

Some researchers (e.g. Baker, 1990; Kochel, 1988) have argued that Hawaiian valleys may have formed mainly by groundwater sapping or subsurface-erosional systems associated with U-shape valleys, linear river long profiles, steep (theater-like) valley heads, steep walls that meet the valley floor at nearly right angles, non-dendritic drainage patterns, and short tributaries. Seidl et al. (1994) found no field evidence for seepage processes on the western side of Kaua'i. Our analysis is based on that conclusion: seepage is neglected. A perusal of NASA satellite data and aerial photos suggest that many valleys on O'ahu island may have been formed by groundwater sapping. The same data, however, suggest that many sites in other Hawaiian islands including western Kaua'i seem to have been formed by fluvial or overland flow processes, and are characterized by V-shaped valleys, tapered valley heads, and a dendritic drainage pattern (Figs. 1a and 1b).

The bedrock rivers in western Kaua'i are incised into gently dipping, 5.2 Ma basalt flows (Figs. 1a and 1b) (Clague & Dalrymple, 1987). Long-term erosion rates can be estimated along the river long profiles based on the 5.2 Ma age of the lava flows and the preservation of ridges above the streams marking the initial lava flow surfaces (Seidl et al., 1994). Annual precipitation in this region of western Kaua'i ranges from about 500 mm near the coastline to 1000 mm near Waimea Canyon and the divide (Macdonald et al., 1960).

Considering the western Kaua'i site in Figs. 1a and 1b, an immediate question comes to rise: why there is a plain or depositional area (i.e. the Mana Plain) in the figure? Why does it not extend farther to the north? In fact, of all eight Hawaiian Islands, this is the single extensive plain of its type. Looking carefully, it is seen that the north end of the plain matches the end of the migrating headcut of the Waimea canyon, which shortens the upstream boundary of the bedrock streams along its valley edge. We suggest here that the shortening of the upstream distance from a channel head to its divide may have led to the formation of the unique Mana Plain, a result that follows from our numerical model presented below.

3 MODEL DEVELOPMENT

3.1 *Abrasion process driven by collision*

The model of incision presented here derives from the original work of Sklar & Dietrich (1998, 2004). Wear or abrasion is the process by which bedrock is ground to sand or silt. The stones that do the wear are assumed to have a characteristic size D_w. Let $q(x)$ denote the volume transport rate of sediment in the stream per unit width during the storm events coming from the adjacent hillslope. Let the fraction of this load that consists of particles coarse enough to do the wear be α. The volume transport rate per unit width q_w of sediment coarse enough to wear the bedrock is then given as

$$q_w = \alpha q \tag{1}$$

For simplicity, α might be set equal to the fraction of the load that is gravel or coarser, yet capable of moving during the floods of interest. Consider the case of saltating bedload particles. Let E_{saltw} denote the volume rate at which saltating wear particles bounce off the bed per unit bed area and L_{saltw} denotes the characteristic saltation length of wear particles. It follows from simple continuity that

$$q_w = E_{saltw} L_{saltw} \tag{2}$$

The mean number of bed strikes by wear particles per unit bed area per unit time is equal to E_{saltw}/V_w, where V_w denotes the volume of a wear particle. It is assumed that with each collision a fraction r of the particle volume is ground off the bed and a commensurate, but not necessary equal amount is ground off the wear particle. The rate of bed incision v_{Iw} due to wear is then given as (number of strikes per unit bed area) × (volume removed per unit strike), or

$$v_{Iw} = \frac{E_{saltw}}{V_w} r V_w \tag{3}$$

Reducing (3) with (2), it is found that

$$v_{Iw} = \beta_w q_w \quad , \quad \beta_w = \frac{r}{L_{saltw}} \tag{4a,b}$$

Here the parameter β_w has the dimensions [1/L]. It is approximated as a specified constant here. It has exactly the same status as the abrasion coefficients used to study downstream fining by abrasion in rivers (e.g. Parker, 1991). L_{saltw} depends on flow conditions and r depends on rock type and perhaps the strength of the collision.

3.2 *Cover factor of alluvial deposits*

Eq. (4a) is valid only to the extent that all wear particles collide with exposed bedrock. If wear particles partially cover the bed, the wear rate should be commensurately reduced. This effect can be quantified in terms of the ratio q_w/q_{wc}, where q_{wc} denotes the capacity transport rate of wear particles. Let p_o denote the areal fraction of surface bedrock that is not covered with wear particles. In general p_o can be expected to approach unity as $q_w/q_{wc} \to 0$, and approach zero as $q_w/q_{wc} \to 1$. Various authors, (e.g. Sklar & Dietrich, 1998), have proposed the form

$$p_o = \left(1 - \frac{q_w}{q_{wc}}\right)^{n_o} \tag{5}$$

Eq. (4a), the relation for incision due to abrasion, is then modified to

$$v_{Iw} = \beta_w q_w p_o \quad \text{or} \quad v_{Iw} = \beta_w \alpha q \left(1 - \frac{\alpha q}{q_{wc}}\right)^{n_o} \tag{6a,b}$$

Note that (6a,b) include both the tool effect (incision increases with increasing q) and the cover effect (incision decreases with increasing q). Note also that v_{Iw} drops to zero when αq becomes equal to q_{wc}, downstream of which a completely alluvial gravel-bed stream is found. Therefore, the above formulation can describe the end point of the incisional zone as well as the incision rate. From preliminary experimental results conducted by the authors, the exponent n_o seems to be proportional to bedrock bumpiness (or deviation of the bed about a mean elevation) and inversely proportional to channel slope.

3.3 *Capacity bedload transport rate of effective tools, boundary shear stress, and hydrology*

The parameter q_{wc} can be quantified in terms of standard bedload transport relations. A generalized relation of the form of Meyer-Peter & Müller (1948), for example, takes the form

$$q_{wc} = \sqrt{RgD_w} D_w \gamma_T \left(\frac{\tau_b}{\rho R g D_w} - \tau_c^*\right)^{n_T} \tag{7}$$

where g denotes the acceleration of gravity, ρ denotes water density, τ_b denotes bed shear stress, $R = (\rho_s/\rho) - 1$ where ρ_s denotes sediment density, τ_c^* denotes a dimensionless critical Shields number, γ_T is a dimensionless constant and n_T is a dimensionless exponent. For example, in the implementation of Fernandez Luque and van Beek (1976), $\gamma_T = 5.7$, $n_T = 1.5$ and τ_c^* is between 0.03 and 0.045. The standard formulation for boundary shear stress places it proportional to the square of the flow velocity $U = q_f/H$ where q_f denotes the flow discharge per unit width and H denotes flow depth. More precisely,

$$\tau_b = \rho C_f \frac{q_f^2}{H^2} \tag{8}$$

where C_f is a friction coefficient for a completely alluvial bed, which is here assumed to be constant for simplicity. In the case of the characteristically steep slopes of bedrock streams the normal flow approximation, according to which the downstream pull of gravity just balances the resistive force at the bed, should apply, so that momentum balance ultimately reduces with Eq. (8) to the form

$$\tau_b = \rho C_f^{1/3} g^{2/3} q_f^{2/3} S^{2/3} \tag{9}$$

where S is the slope of the bedrock river. Now let i denote the precipitation rate, $B_c(x)$ denote channel width, and $A(x)$ denote drainage area. Assuming no storage of water in the basin, the balance for water flow is

$$q_f B_c = iA \tag{10}$$

Between (7), (9), and (10), the capacity bedload transport rate of effective tools for wear is given as

$$q_{wc} = \sqrt{RgD_w} D_w q_{wc}^*$$

$$q_{wc}^* = \gamma_T \left[\frac{C_f^{1/3} i^{2/3} \chi^{2/3}}{Rg^{1/3} D_w} S^{2/3} - \tau_c^* \right]^{n_T} \quad (11a,b)$$

where $\chi = A/B_c$ serves as a surrogate for distance x in this model.

3.4 *Sediment transport in bedrock rivers*

A routing model is necessary to determine q, and thus q_w. It is assumed here that the local fluvial incision pulls down the adjacent hillslope at the same rate as the channel incises. The equation of sediment conservation on a bedrock reach can then be written as

$$\frac{d}{dx}(qB_c) = q_h \quad (12)$$

where q_h denotes the volume of sediment per stream length per unit time entering the channel from the hillslopes (either directly or through the intermediary of tributaries). Note that a more sophisticated model might later include the development of regolith on hillslopes. Several models can be postulated for q_h depending on hillslope dynamics. Here, it is assumed that the watershed consists of easily-weathered rocks, so that bed lowering by channel incision results in hillslope lowering at the same rate. In this case, we have

$$q_h = v_I \frac{dA}{dx} \quad (13)$$

as illustrated in Fig. 2. Note that in (13), v_I is the total incision rate and not just that due to wear. (Here, however, only incision due to wear is modeled.) Moreover, this equation serves only as a simple example that must later be generalized to include e.g. hillslope diffusion, hillslope relaxation due to landslides driven by earthquakes or saturation in the absence of uplift.

Between (12) and (13) it is found that

$$\frac{d}{dx}(qB_c) = v_I \frac{dA}{dx} \quad (14)$$

In the case of steady-state incision in response to spatially uniform (piston-style) uplift, (14) becomes

$$q = v_I \chi \quad (15)$$

To obtain an approximate treatment of the case of deviation from steady-state incision in response to piston-style uplift, we start from an empirical relation

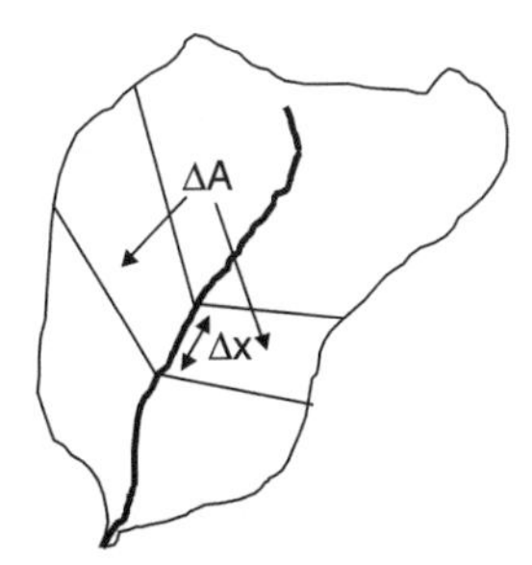

Figure 2. Illustration of a drainage basin with hillslope lowering after channel incision.

between channel width and the drainage area upstream (e.g. Montgomery & Gran, 2001), defined as

$$B_c = \alpha_b A^{n_b} \quad (16a)$$

or equivalently

$$B_c = \tilde{\alpha}_b \chi^{m_b}, \quad \tilde{\alpha}_b = \alpha_b^{1/(1-n_b)}, \quad m_b = \frac{n_b}{1-n_b} \quad (16b)$$

where

$$\chi = A/B_c \quad (16c)$$

Generally, $\alpha_b \sim 0.002$ to 0.088 and $n_b \sim 0.3$ to 0.5 where channel width is in meters and drainage area is in square meters (Montgomery & Gran, 2001; Whipple, 2004). Drainage area A, can be written as a function of down-channel distance x using Hack's law (1957) below;

$$A = K_h x^{n_h} \quad (17)$$

where according to Whipple & Tucker (1999), when channel length is in meters and drainage area is in square meters, $K_h \sim 6.7$ and $n_h \sim 1.7$, based on the work of Hack (1957). Seidl et al. (1994) analyzed the western Kaua'i site and found that K_h ranges widely from 1 to 100 and $n_h \sim 1.0$ to 2.0 where channel length is in km and drainage area is in km^2.

Between (14) and (16b), one finds

$$q = \frac{m_b + 1}{\chi^{m_b}} \int_0^{\chi} v_I \chi'^{m_b} d\chi' \quad (18)$$

3.5 *Exner equation*

If a river is assumed to be morphologically active only intermittently (during floods of interest), the Exner equation of sediment continuity for bedrock rivers becomes

$$(1-\lambda_p)\frac{\partial \eta}{\partial t} = \upsilon - I v_I \quad (19)$$

where υ is uplift rate, λ_p is porosity of bedrock (~ 0), t is time, η is elevation of the river bed, and I is the intermittency of large flood events (fraction of time the river is experiencing a large flood). Uplift is not really continuous, but it is treated as such here. In the case of a more general hydrology, the formulation can be generalized to

$$(1-\lambda_p)\frac{\partial \eta}{\partial t} = \upsilon - \sum_{k=1}^{N} I_k v_{I,k} \tag{20}$$

in which I_k is fraction of time the flood flow is in the kth flow range.

3.6 *Assembling all submodels*

Here a wear-dominated bedrock river is considered, and so only wear is assumed to cause incision. From (6b) and (18) one finds that

$$v_{Iw} \sim v_I = \left[\beta_w \alpha \left(\frac{m_b+1}{\chi^{m_b}} \int_0^{\chi} v_I \chi'^{m_b} d\chi'\right)\right] * \left[1 - \frac{\alpha\left(\frac{m_b+1}{\chi^{m_b}} \int_0^{\chi} v_I \chi'^{m_b} d\chi'\right)}{q_{wc}}\right]^{n_o} \tag{21}$$

In order to solve this equation, it is useful to introduce a new auxiliary variable

$$W = \int_0^{\chi} v_I \chi'^{m_b} d\chi' \tag{22}$$

from which,

$$v_I = \frac{1}{\chi^{m_b}} \frac{dW}{d\chi} \tag{23}$$

From (21), one then obtains the following first-order ordinary differential equation:

$$\frac{dW}{d\chi} = \left[\beta_w \alpha (m_b+1) W\right] \left[1 - \frac{\alpha\left(\frac{m_b+1}{\chi^{m_b}} W\right)}{q_{wc}}\right]^{n_o} \tag{24}$$

This equation can be solved numerically, i.e. by the Runge-Kutta method upon the specification of a single boundary condition.

3.7 *Upstream boundary condition*

Applying (14), at the headwater of the bedrock river or point of stream inception, it is found that

$$q_b B_{c,b} = v_{I,b} A_b \quad \text{or} \quad q_b = v_{I,b} \chi_b \tag{25a,b}$$

where the subscript b denotes the beginning point of a stream. Substituting the above relations back into (6b), it is found that

$$v_{I,b} = \frac{q_{wc,b}}{\alpha \chi_b} \left[1 - \left(\frac{1}{\beta_w \alpha \chi_b}\right)^{1/n_0}\right] \tag{26}$$

From (18) and (22), the sediment transport rate q can be represented in terms of the variable W, and likewise the headwater value q_b can be related to the headwater value W_b as

$$q = \frac{m_b+1}{\chi^{m_b}} W \quad \text{and} \quad q_b = \frac{m_b+1}{\chi_b^{m_b}} W_b \tag{27a,b}$$

From (25b), (26), and (27b), one finds the upstream boundary condition for (24) to be

$$W_b = \frac{q_{wc,b} \chi_b^{m_b}}{\alpha (m_b+1)} \left[1 - \left(\frac{1}{\beta_w \alpha \chi_b}\right)^{1/n_0}\right] \tag{28}$$

3.8 *Numerical model*

The sediment transport rate and incision rate along the reach can be calculated using (11a, b), (24), (28), (27a), and (6b). Then by using the Exner equation (19), the bed elevation at one time step later can be determined along the reach. The new bed slope at any spatial point can be then calculated. The time evolution of the long profile of the river can then be computed by cycling (11a, b), (24), (28), (27a), and (6b) for as many time steps as desired. In the case of continuous uplift at a constant rate, the long profile must eventually evolve to an equilibrium (steady) state where incision rate and uplift rate are balanced everywhere. It is shown here, however, that transient but nevertheless long-lived autogenic knickpoints may develop during this process.

4 SAMPLE MODEL RESULTS

The model results in this paper focus primarily on the qualitative trends of bedrock river evolution. These qualitative results help answer questions such as a) what happens if the distance from the channel head to the divide is shortened? (Figs. 3 and 4) and b) how does a river profile evolve after experiencing a sudden base level fall? (Fig. 5).

The results below (Figs. 3 to 5) are based on numerical modeling using the following input parameters: uplift rate = 0 mm/yr (for a volcanic island), initial river bed slope = 0.10 (very steep reach), effective rainfall rate = 15 mm/hr, flood

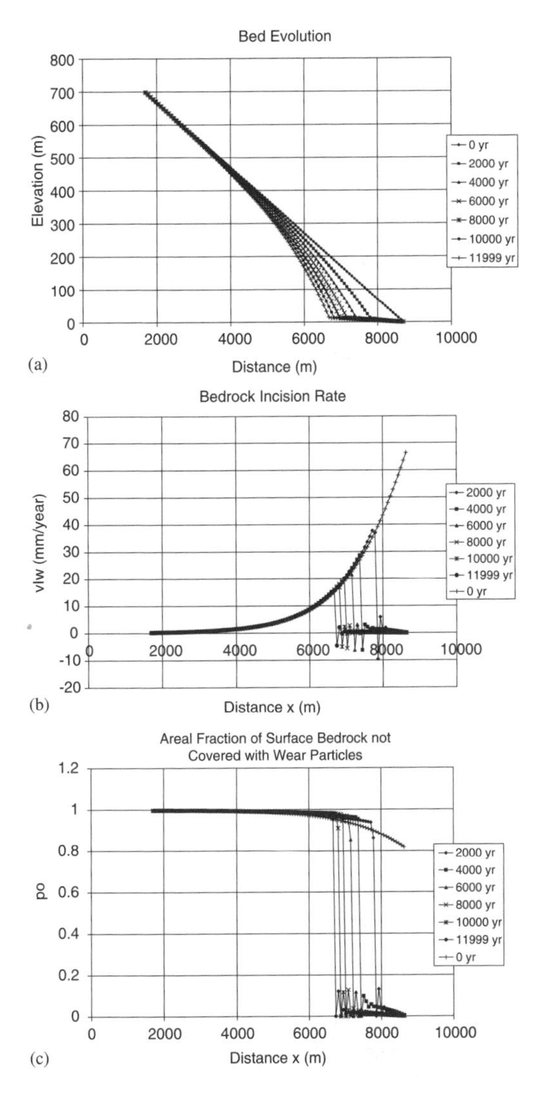

Figure 3. Numerical results for long profiles of bed elevation, mean annual incision rate and fraction of the bed not covered by alluvium, showing the autogenic formation of a flat reach at the downstream end. The profiles resemble rivers flowing to Mana Plain.

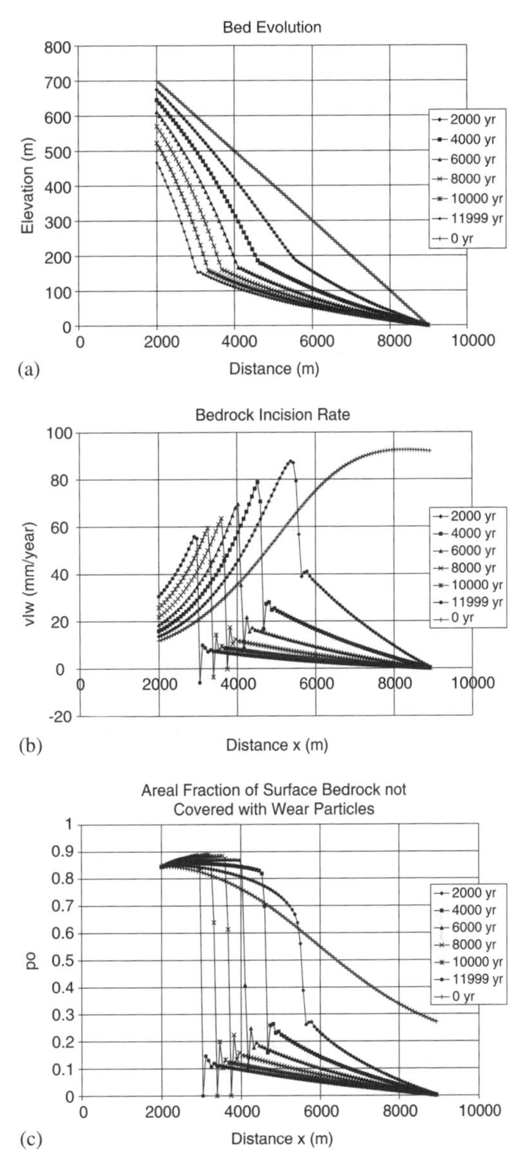

Figure 4. Numerical results for long profiles of bed elevation, mean annual incision rate and fraction of the bed not covered by alluvium, showing the evolution of an autogenic knickpoint resembling Kauhao channel #13 of DeYoung (2000). The differences between Figure 3 and Figure 4 result from a longer distance from the channel head to the divide in the case of Figure 4.

intermittency = 0.001 (infrequent floods of sufficient strength to drive incision), wear coefficient = $6.5 \times 10^{-5}\,m^{-1}$, effective size of particles that do the wear = 50 mm, fraction of load consisting of sizes that do the wear = 0.05, and bed friction coefficient = 0.01. The total river length ranges from 7 to 10 km. The total time of calculation is 12000 years for Figs. 3 and 4. The only difference in input between the numerical runs of Figs. 3 and 4 is that the upstream distance from the channel head to the divide for the case shown in Fig. 3 is 300 m shorter than the one for Fig. 4. This is intended to represent the effect of Waimea Canyon in shortening the streams flowing onto the Mana Plain (Fig. 1). The results in Fig. 3 document the autogenic formation of a flat, low-slope reach (plain) at the downstream end. The amalgamation of a low-slope zone from a dozen or so bedrock streams in the area, combined with

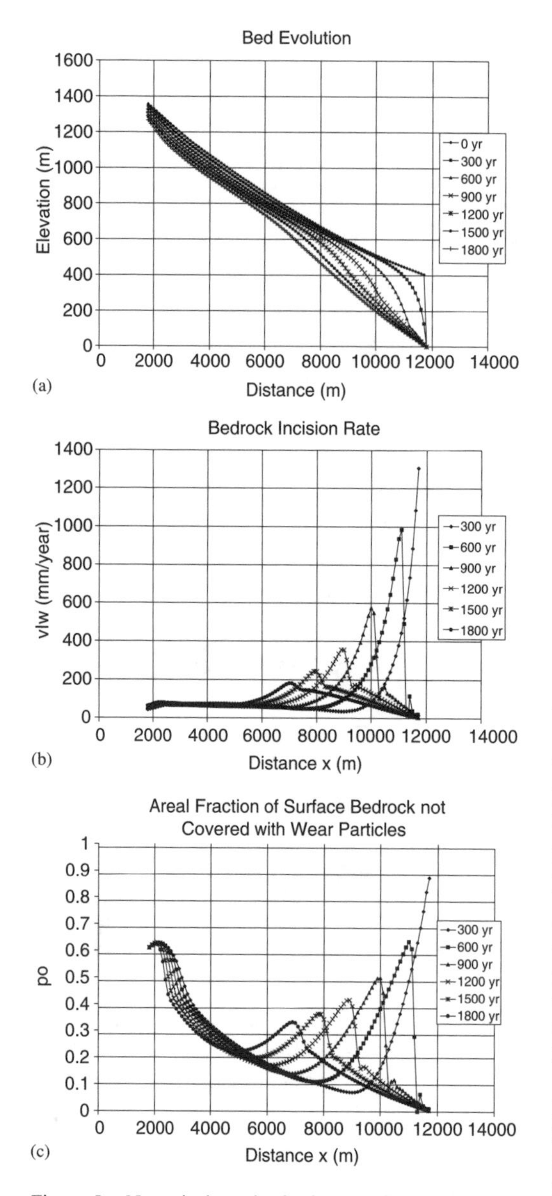

Figure 5. Numerical results for long profiles of bed elevation, mean annual incision rate and fraction of the bed not covered by alluvium, with a knickpoint resulting from sudden base level fall. The profiles resemble rivers #15, 16, and 17 of DeYoung (2000). Note the shorter time scale for morphological change.

coastal retreat and submergence/re-emergence associated with minor relative sea level change may thus result in a plain like the Mana plain today. The plots of Fig. 3 include long profiles of a) elevation η, b) mean annual incision rate v_{Iw} and c) fraction of the bed p_o not covered with alluvium.

With a longer distance from the channel head to the divide, the profile results show the autogenic knickpoint shown in Fig. 4. The zone downstream of the knickpoint is much steeper than that in Fig. 3. The long profile resembles a bedrock river in Napali region, i.e. Kauhao #13 of DeYoung (2000). The results in Fig. 5 have the same input parameters as those used for Figs. 3 & 4 except that the river experiences a 400-meter sudden base level fall generated by e.g. a catastrophic landslide at the initial year. The total time of calculation is only 1800 years for Fig. 5, by which time drastic morphological changes are evident. The profile results resemble the rivers #15, 16, and 17 of DeYoung (2000), which are in the Napali area.

5 COMPARISONS WITH FIELD DATA, ANALYSIS, AND DISCUSSION

Only streams from the south up to Makaha, i.e. #10 of DeYoung (2000) in the Napali region, have been considered in this study since the streams farther north are so steep (i.e. more than 20%) that the dominant incision process may be debris flows rather than fluvial processes (W. Dietrich, pers. comm.). All of the bedrock river long profiles in this section are from DeYoung (2000), who used the Digital Elevation Model (DEM) technique to obtain a series of long profiles in this area. In all figures, the lower solid line represents the river bed profile and the upper dotted and solid lines represent the adjacent valley ridges. Figs. 6 and 7 show satellite images from the NASA website (http://zulu.ssc.nasa.gov/mrsid/). The bedrock channels are numbered after DeYoung (2000), together with a name for some streams. Fig. 6 shows the streams in the Napali region (#10, 13, 15, 16, and 17) and the transitional area from Napali to the Mana Plain region (streams #19 and 20). Fig. 7 shows the streams in the Mana Plain region (#21 to 35).

Fig. 8 shows the representative of streams in the Mana Plain region (#21 to 38) (plot from DeYoung, 2000), of which all of them have either straight or slightly convex profiles. This feature combined with the presence of the Mana Plain at the downstream end is consistent with the results of Fig. 3. After experiencing a combination of coastal retreat and sequences of submergence/reemergence due to minor relative sea level changes, alluviation onto and amalgamation of flat reaches downstream may have resulted in the Mana Plain of today. Coastal retreat here is likely a consequence of shoreline erosion due to waves acting on the edge of a surface of low slope. A relative 10-meter sea level fall over the last 5000 years (Atlas of Hawai'i, 1998) and/or uplift over 2 Ma due to the formation of Oahu and Moloka'i (Atlas of Hawai'i, 1998; Grigg & Jones, 1997) may have also abetted the emergence of the Mana Plain. Subsequent to the

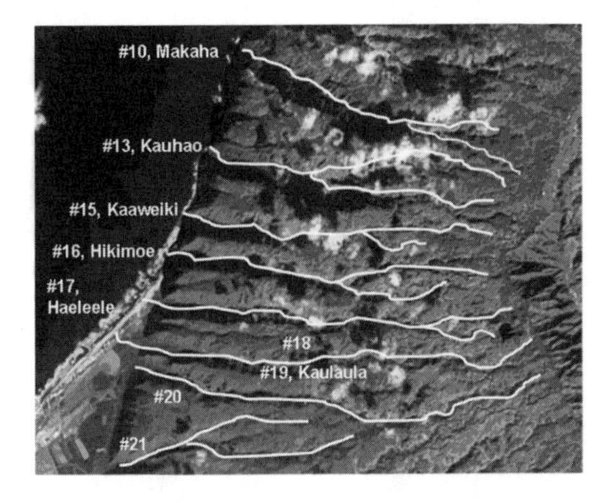

Figure 6. A series of bedrock rivers from Napali Coast in the north to the northern edge of the Mana Plain in the south. From NASA website. Numbered after DeYoung (2000).

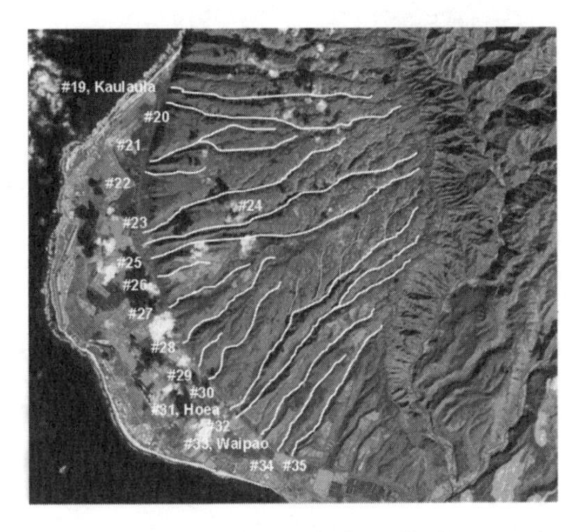

Figure 7. A series of bedrock rivers flowing onto the Mana Plain. From NASA website. Numbered after DeYoung (2000).

formation of a plain in this region, catastrophic landslides were no longer able to cause sudden base level falls as in the Napali region, because the Mana Plain itself is not much higher than sea level. In addition, the upstream migration of the Mana Plain by the process described in Fig. 3 would have shortened the upstream bedrock reach of any stream flowing onto the Mana Plain, thus suppressing the formation of a local reach of upward convexity just upstream of the plain. As a result, (probably combined with the expansion of the width of Waimea Canyon through time) the profiles of the streams presently flowing into the Mana Plain have rather straight profiles (Fig. 8), as compared to those in the transition region between the Mana Plain and the Napali Region (Fig. 9), which would have formed later.

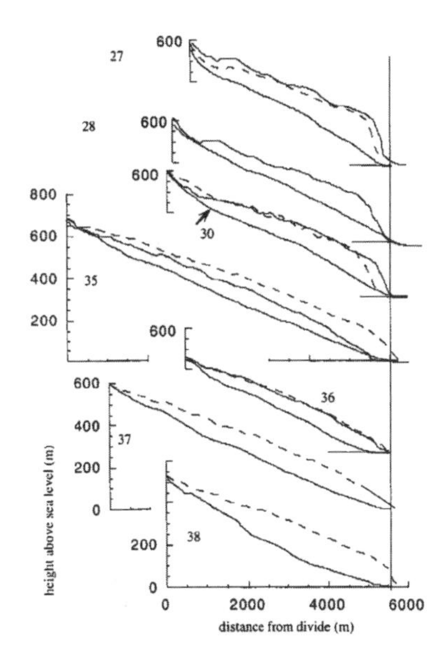

Figure 8. River profiles in the Mana Plain region. The Mana Plain itself is excluded. Plot from DeYoung (2000).

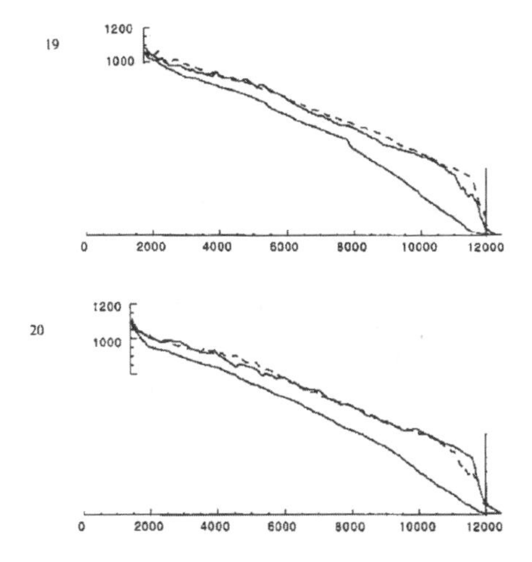

Figure 9. River profiles with convex shapes at the transition between the Mana Plain region and the Napali region. Plot from DeYoung (2000).

Fig. 9 shows streams #19 and 20 (transition to the Napali region; plot from DeYoung, 2000), which have strongly convex profiles. They may have developed this convexity as indicated in Fig. 3, and the development may have been later than most streams farther south, since the headcut of Waimea Canyon likely propagated through time. This may explain why the downstream plain in this transitional area is shorter than the one to the south.

Fig. 10 shows channel profiles for four very short secondary channels that do not head up near the divide

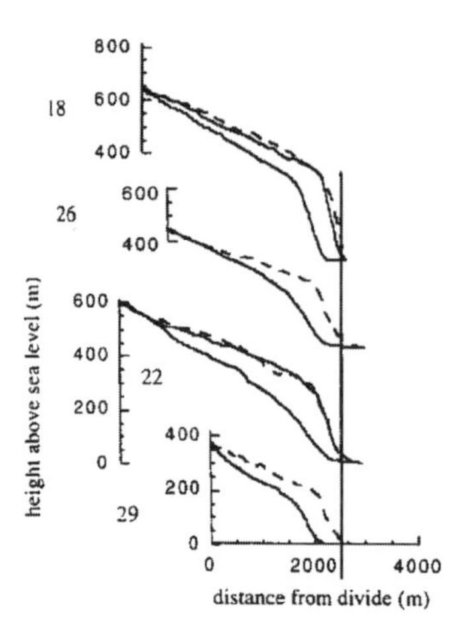

Figure 10. Secondary channel profiles with convex shapes. The distance from the channel head to the divide is very limited (see Fig. 7). Note that all of these streams have flat regions downstream. Plot from DeYoung (2000).

with Waimea Canyon (plot from DeYoung, 2000). These profiles are all upward convex. Their locations are shown in Figs. 6 and 7. Note that all of them have a flat reach at the downstream end. Presumably the distance from channel head to divide is limited for these streams and hence, their behavior resembles the larger streams flowing onto the Mana Plain.

The similarity between results in Fig. 3 and the profiles of rivers flowing onto the Mana Plain in Figs. 8, 9, and 10 becomes more convincing if we look at the eastern part of the outlet of Waimea Canyon in the south (Fig. 1a). In that area, the bedrock streams show dendritic networks, but the tributaries show a morphology that differs from those to the west. Even though the shoreline at the connection of the western and the eastern parts of the outlet of Waimea Canyon appears to be continuous, the Mana plain disappears abruptly just at the outlet of the canyon. It is likely that since there is no flat reach downstream of the eastern part of the Waimea Canyon, shoreline erosion cannot easily erode the coast, resulting in preservation of a steep paleosurface.

Fig. 11 shows river profiles #15, 16, & 17 in the Napali region, with knickpoints likely resulting from sudden base level fall (plot from DeYoung, 2000). Fig. 12 shows experimental profiles of a channel in cohesive sediment resulting from sudden base level fall (Begin et al., 1981). Note how the field river profiles in Fig. 11 and the experimental results in the Fig. 12 are similar to our model results for the case of sudden base level fall in Fig. 5. Begin et al. (1981) offer conclusions similar to those from our numerical results (Fig. 5), i.e. a) the main impact of erosion is felt in the early stages after initiation of sudden base level fall, and primarily near the downstream end of the

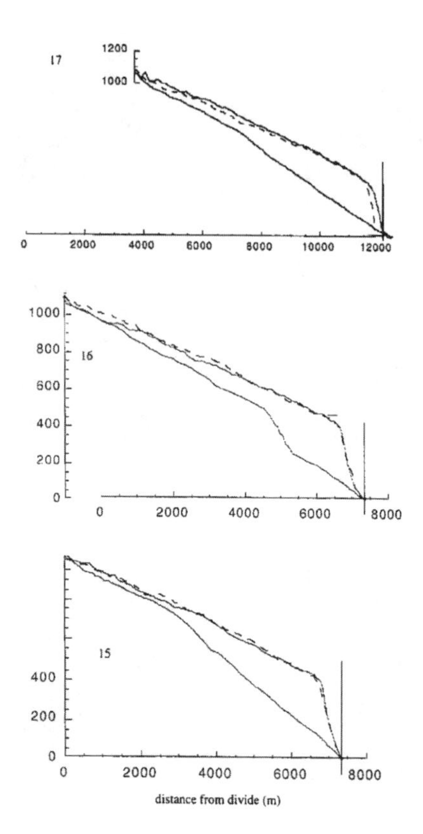

Figure 11. River profiles in the Napali region with knickpoints likely resulting from base level fall. Plot from DeYoung (2000).

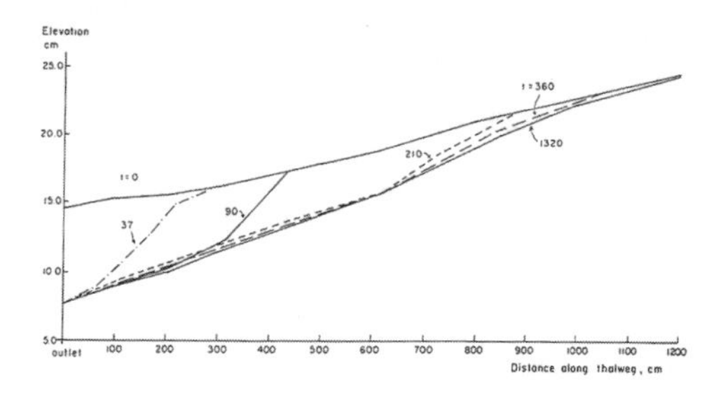

Figure 12. Development of long profiles in an experimental channel in cohesive sediment in response to base level lowering; from Begin et al. (1981).

river, b) the rate of degradation or incision at any point along the channel reaches a peak and then gradually decreases with time, and c) the peak rate of incision is attenuated with distance from the outlet.

Considering both Figs. 10 and 11, and especially the profile of stream #16 of Fig. 11, it appears that base level fall should have been around 250 m or more in order to drive knickpoint formation. Since the maximum sea level change due to glacio-eustatic

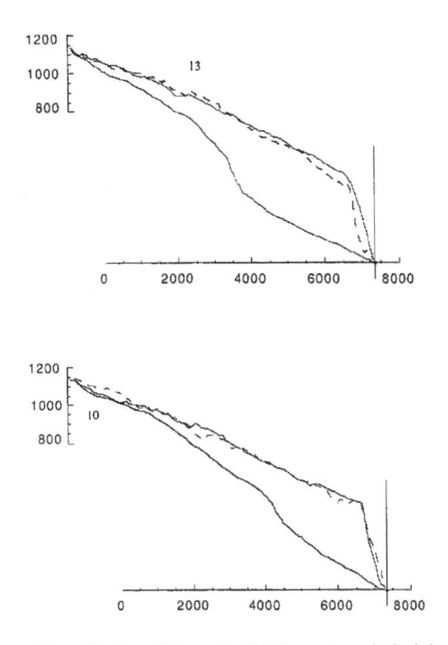

Figure 13. Kauhao (#13) with likely autogenic knickpoint and Makaha (#10) with likely initiation of an autogenic knickpoint in the lower part and the remnant of base level fall in the upper part. Plot from DeYoung (2000).

fluctuations in the last 900,000 years is not more than 150 m (Atlas of Hawai'i, 1998), it seems certain that knickpoints resulting from base level fall in western Kaua'i must have been generated by another process in this region (e.g. catastrophic landslides) rather than by glacio-eustatic fluctuations as claimed by Seidl et al. (1994). The work by Moore et al. (1989) indicates that catastrophic landslides occurred in the northern and southern parts of the island. The results here suggest that they should have occurred in the western part as well. Such catastrophic landsliding would have begun early, when Kaua'i was a small submarine seamount. It may have peaked near the end of subaerial shield building and continued long after dormancy, occurring approximately every 0.1–0.2 million years (Moore et al., 1989).

Fig. 13 shows stream #13, with a knickpoint resembling the autogenic one in Fig. 4, and stream #10, which appears to show the initiation of an autogenic knickpoint in the lower part and the remnant of base level fall in the upper part (plot from DeYoung, 2000). Note again that streams #10 and 13 have always had longer distances from the channel head to their divide than ones to the south (Fig. 6). This may cause both streams to form autogenic knickpoints as shown by the numerical model (Fig. 4).

Considering Figs. 4 and 5, we can see that in the zone upstream of the knickpoints both the incision rate and the fraction of bedrock exposure are much higher than the zone downstream. This is consistent with the field observation by DeYoung (2000) and Seidl et al. (1994) that considerable portions of the channel lengths below the knickpoints are mantled with large boulders. It should be noted from Figs. 4 and 5 also that even though the incision rates in later years are minimal, especially downstream of the knickpoints, incision rates toward the downstream end are very high in the initial period (i.e. year 0). This is consistent also with the relative incision depth of the streams in Figs. 11 and 13 relative to the paleosurfaces.

How can an autogenic knickpoint form? The process can be briefly explained as follows. From the Exner equation (19), taking the second derivative in x and assuming a constant uplift rate results in

$$\frac{\partial}{\partial t}\left(\frac{\partial^2 \eta}{\partial x^2}\right) = -I\frac{\partial^2 v_I}{\partial x^2} \quad \text{or} \quad \frac{\partial}{\partial t}\left(\frac{\partial S}{\partial x}\right) = I\left(\frac{\partial^2 v_I}{\partial x^2}\right) \tag{29a,b}$$

Now consider the plot of incision rate in Fig. 4b at year zero. Note that the shape of the curve of the incision rate v_{Iw} changes from concave-upward upstream to convex-upward downstream at a point near 5000 m. Thus the term $\partial^2 v_I/\partial x^2$ changes from positive to negative near this point. Considering (29a), a stream with such a shape of the incision curve must gradually form an autogenic knockpoint such that the term $\partial^2\eta/\partial x^2$ has the opposite sign of $\partial^2 v_I/\partial x^2$. This results in an elevation curve that changes from upward convex in the upstream reach to upward concave in the downstream reach. The point of the shock condition (knickpoint) then automatically migrates upstream (Fig. 4). Note that the size of an autogenic knickpoint can vary depending on the input parameters.

It should be noted that this study examines bedrock rivers in Kaua'i only in a qualitative sense. More field work is required for further quantification. Some parameters (e.g. wear coefficient characterizing rock strength, β_w and the fraction of load doing the wear, α) could be at least partially quantified in the laboratory.

These very preliminary results suggest that the time scale of morphological changes in this site may be in thousands of years to hundred thousands of years rather than million years as suggested by Seidl et al. (1994). Especially, in the case of 400-meter base level fall as a result of a catastrophic landslide, the downstream end of the river would tend to reach sea level first, which would result in an incision rate at the downstream end that might be as high as ~one meter per year near the knickpoint (Fig. 5b). However, the streams may be less active than indicated by rainfall data, so that morphological changes may take much longer than the calculation time estimated here. For instance, during her field survey, DeYoung (2000) found that Waipao stream (#33) was completely dry, and local people claimed that they have not seen water in the southwest streams even during southwest-origin storms. Nonetheless, during glacial times, the

Hawaiian Islands were generally known to experience a wetter and cooler environment (e.g. Porter, 1979). In addition, boulder armouring of the channels may have greatly slowed down the incisional process compared to that modelled here.

6 CONCLUSION

The numerical model presented here indicates that knickpoint generation may be an intrinsic autogenic response to the mechanics of bedrock incision, in addition to being an allogenic response to e.g. base level fall. Here both types of knickpoints develop from the same model of incision due to wear. The model is applied in a qualitative way to bedrock streams in the Hawaiian island of Kaua'i, USA. The comparison is encouraging. In particular, the results suggest that the Mana plain could have formed at least partly due to autogenic processes associated with the rivers flowing into it.

ACKNOWLEDGEMENTS

This research was funded by the National Center for Earth-surface Dynamics (NCED), based at the St. Anthony Falls Lab at the University of Minnesota, USA. We thank R. Slingerland, L. Sklar, and W. Dietrich for the discussion regarding the formulation of the wear model using in the analysis. We give special thanks to N. DeYoung for her permission to use the figures of Kaua'i river long profiles from her master thesis; these were extremely helpful for this study.

REFERENCES

Atlas of Hawai'i, Department of Geography, University of Hawai'i at Hilo. 1998. Juvik, S.P. & Juvik, J.O., editors, University of Hawai'i Press, Honolulu.

Baker, V.R. 1990. Spring sapping and valley network development, with case studies Kochel, R.C., Baker, V.R., Laity, J.E. & Howard, A.D. in Higgins, C.G. & Coates, D.R., eds. Groundwater geomorphology: the role of subsurface water in earth-surface processes and landforms. Geological Society of America Spec. Paper 252, 235–265.

Begin, Z.B., Meyer, D.F., & Schumm, A.S. 1981. Development of longitudinal profiles of alluvial channels in response to base-level lowering. Earth Surface Processes and Landforms, 6, 49–68.

Chase, C. 1992. Fluvial landsculpting and the fractal dimension of topography. Geomorphology, 5, 39–51.

Clague, D.A. & Dalrymple, G.B. 1987. The Hawaiian-Emperor volcanic chain part I geologic evolution. In Decker, R.W., Wright, T.L. & Stauffer, P.H. (editors) Volcanism in Hawai'i. US Geol. Survey Prof. Paper, 1350, 5–54.

DeYoung, N.V. 2000. Modeling the geomorphic evolution of western Kaua'i, Hawai'i; a study of surface processes in a basaltic terrain. Master Thesis, Dalhousie University, Nova Scotia, Canada.

Grigg, R. & Jones, A. 1997. Uplift caused by lithospheric flexure in the Hawaiian Archipelago as revealed by elevated coral deposits. Marine geology. 141, 11–25.

Hack, J.T. 1957. Studies of longitudinal stream profiles in Virginia and Maryland. U.S. Geol. Soc. Prof., Paper 294-B, 45–97.

Kochel, R.C. 1988. Role of groundwater sapping in the development of large valley networks on Hawaii, In Howard, A.D., Kochel, R.C. & Holt, H., eds. Sapping features of the Colorado Plateau: a comparative planetary geology field guide. NASA Spec. Pub. SP-491, 100–101.

Macdonald, G.A., Davis, D.A. & Cox, D.C. 1960. Geology and groundwater resources of the island of Kauai, Hawaii. Hawaii Division of Hydrography Bulletin, 13, 212 p.

Montgomery, D.R. & Gran, K.B. 2001. Downstream variations in the width of bedrock channels. Water Resources Research, 37, 6, 1841–1846.

Moore, J.G. 1987. Subsidence of the Hawaiian ridge. U.S. Geological Survey Professional Paper, 1350, 85–100.

Moore, J.G., Clague, D.A., Holcomb, R.T., Lipman, P.W., Normark, W.R. & Torresan, M.E. 1989. Prodigious submarine landslides on the Hawaiian ridge. Journal of Geophysical Research, 94 (B12), 17465–17484.

Parker, G. 1991. Selective sorting and abrasion of river gravel I: Theory. Jour. of Hydraulic Eng. 117, 2, 131–149.

Porter, S. 1979. Hawaiian glacial ages. Quaternary Research, 12, 161–187.

Seidl, M.A., Dietrich, W.E. & Kirchner, J.W. 1994. Longitudinal profile development into bedrock: an analysis of Hawaiian channels. Journal of Geology, 102, 457–474.

Seidl, M.A., Finkel, R.C., Caffee, M.W., Hudson, B.G. & Dietrich, W.E. 1997. Cosmogenic isotope analyses applied to river longitudinal profile evolution: problems and interpretations. Earth Surf. Proc. & Landforms. 22, 195–209.

Sklar, L.S. & Dietrich, W.E. 1998. River longitudinal profiles and bedrock incision models: Stream power and the influence of sediment supply, in River over rock: fluvial processes in bedrock channels. Geophysical Monograph Series, 107, edited by Tinkler, K. and Wohl, E.E., 237–260, AGU, Washington D.C.

Sklar, L.S. & Dietrich, W.E. 2004. A mechanistic model for river incision into bedrock by saltating bed load. Water Resources Research, 40, W06301, 21 p.

Slingerland, R.L. & Snow, R.S. 1988. Stability analysis of a rejuvenated fluvial system: Zeitschrift fur Geomorphologie, N. F. Supplement Band 67, 93–102.

Whipple, K.X. & Tucker, G.E. 1999. Dynamics of the stream-power river incision model: Implications for height limits of mountain ranges, landscape response timescales, and research needs. Jour. of Geophy. Res., 104, B8, 17661–17674.

Whipple, K.X. 2004. Bedrock rivers and the geomorphology of active orogens. Annual Reviews Earth Planet Sciences, 32, 151–185.

Woodford, A.O. 1951. Stream gradients and Monterey Sea Valley. Geological Society of America Bulletin, 62, 799–852.

River, Coastal and Estuarine Morphodynamics: RCEM 2005 – Parker & García (eds)
© 2006 Taylor & Francis Group, London, ISBN 0 415 39270 5

Effect of debris flow discharge on equilibrium bed slope

T. Itoh & S. Egashira
Department of Civil and Environmental Systems Engineering, Ritsumeikan University, Shiga, Kusatsu, Japan

ABSTRACT: Results obtained from both of flume data and theories suggest that equilibrium bed slope in flow over an erodible bed are determined uniquely by supplying sediment discharge rate when the movements of sediment particles are laminar and thus no suspended transportation take place. This means that the static friction force is dominant in debris flow and that sediment concentration is determined by shear stress balance on the bed surface as seen in our previous studies. On the other hand, if part of sediment particles in debris flow body is transported in suspension, sediment concentration will be larger and the equilibrium bed slope will decrease. These facts are supported by Egashira et al.'s flume data and others' experimental data.

The present study discusses experimentally an influence of flow discharge on an equilibrium bed slope and flow structure, based on experimental data and emphasizes that the equilibrium bed slope decreases with increasing of flow discharge if part of debris flow body is turbulent.

1 INTRODUCTION

In case that the sediment concentration is small and that the flow condition is turbulent in the flow with fine sediment, the flows are well known as 'suspended flows'. There are a lot of classic studies (e.g. Rouse 1936, Lane et al. 1939, Vanoni 1944, Einstein 1950) for suspended flows, and flow structures such as local velocity and sediment concentration are somewhat clarified using logarithmic velocity profile and sediment concentration profile of Rouse's type. On the other hand, 'mud flow' or 'hyper-concentrated flow' is named for quite fine sediment-water mixture flow, and its treatment is similar to the one for suspended flow (e.g. Chen 1988, O'Brrien 1988, Arai et al. 1986a, Arai et al. 1986b, Winterwerp et al. 1990).

Arai et al. (1986a & 1986b) proposed flow model in terms of shear stress due to momentum transportation of Plandtl-type and shear stress due to sediment collisions of Bagnold-type, and analyzed flow resistance and Karman constant for flume data of sediment-water mixture flows with uniform sediment whose diameter is 0.099 mm to 0.99 mm, respectively. Winterwerp et al. (1986) discussed relationships between sediment concentration and velocity gradient in terms of flume data of sediment-water mixture flow with uniform sediment whose diameter is 0.12 mm to 0.225 mm, respectively. In those studies, flume tests are conducted for the flow over rigid bed, and logarithmic velocity distribution is assumed for analyzing flume data. According to those studies, it is reported that the mixing length decreases with increase of sediment concentration, and its length increases again when sediment concentration exceeds a certain critical value, as discussed in flow with suspended sediment.

There are some researches for fine sediment and water mixture flow in the flow over erodible bed (e.g. Sumer et al. 1996, Hashimoto et al. 1992, Takahashi 1992, Egashira et al. 1992). Sumer et al. (1996) reported the experimental results in pipe and open channel flows using sediment in 0.13 mm and artificial material in 0.6, 2.6 and 3.0 mm, respectively, and reported the possibility of formulation of two layers in the flow. The hyper-concentrated layer (we call the layer 'lower layer') is formed near the bed and the layer of the flow with suspended sediment (we call the layer 'upper layer') is formed near free surface. It is proposed that velocity profile is power-law type and sediment concentration distribution is linear in the lower layer and that the velocity profile is logarithmic type and the sediment concentration profile of Rouse's type is formed in the upper layer. Hashimoto et al. (1992 & 1997) reported experimental results using non-dimensional parameter defined by their constitutive equations of debris flow in the hyper-concentrated flow over rigid bed. Two layers are formed such that the inertia force is dominant in the upper layer and the dynamic shear stress due to sediment particles collisions are dominant in the lower layer. It is proposed that the position of interface of both layers is experimentally determined with the non-dimensional parameter. Takahashi (1992 & 2000) proposed that the equation of logarithmic type's flow resistance is applicable for the flow when flux sediment concentration exceeds about 0.3 and relative depth, h/d, exceeds about 30. It is reported the two layers, consisted of the layer in which sediment particles collisions are dominant in lower layer and of the layer in which Reynolds stress due to turbulent mixing is dominant in upper layer, are formed in the flow. The equations for velocity and sediment concentration profiles are proposed for each layer. Egashira et al. (1992 & 1994) proposed the following results using flume data in the open channel flow over erodible bed such that sediment size is 0.017, 0.030 and 0.066 cm, respectively and that the sediment concentration has a range from 0 to 0.3. The turbulent diffusion coefficient decreases with increase of sediment concentration. The flow resistance increases in case that the sediment concentration exceeds around 0.1, and the equilibrium bed slope decreases in comparison with the value obtained in the flow condition that sediment particle movements are laminar.

Egashira, Miyamoto and Itoh (1997a,b) have already suggested that the wide flow regime from debris flow to flow with bed loads can be explained using their constitutive relationship of debris flow for the coarse sediment-water mixture, and that the movements of sediment particles are laminar motion. Comprehensively discussing previous studies, those experimental results show that there are two flow characteristics such as laminar and turbulent flow as shown in Figure 1 in the fine sediment-water mixture flow over erodible bed. In the figure, h_t is the flow depth, h_f is the thickness of upper layer, h_l is the thickness of lower layer, θ is the bed inclination, τ is the shear stress, u is the local mean velocity, c is the sediment concentration by volume, c_* is the sediment concentration of stationary sediment layer by volume, τ_y is the yield stress. However, it seems that their flow characteristics are not clarified theoretically and experimentally.

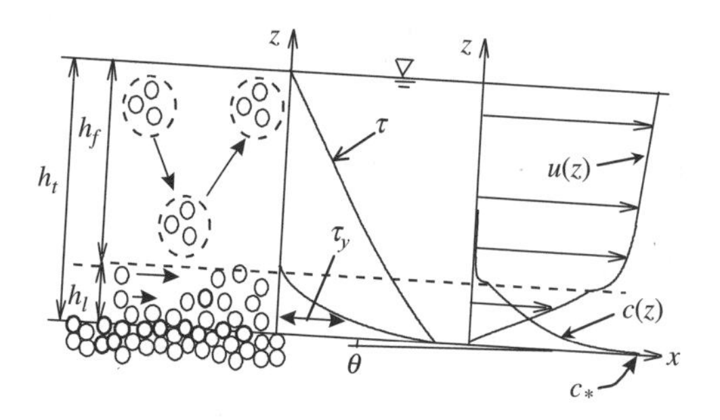

Figure 1. Schematics of flow model.

In present study, effect of flow discharge on the hyper-concentrated flow is discussed using flume data such as velocity distributions and equilibrium bed slope in the flow over erodible bed, taking into account that the increase of sediment size is essentially equivalent to the increase of flow discharge. In addition, non-dimensional parameter is defined with Egashira et al.'s constitutive equations of debris flows in order to analyze flow characteristics. The interface between turbulent layer (upper layer) and laminar layer (lower layer) is evaluated experimentally with the non-dimensional parameter, focused on the relationship between equilibrium bed slope and flux sediment concentration.

2 FLUME TESTS

Figure 2 shows the experimental flume, which is 5.0 cm wide, 20.0 cm high and 12.0 m long. A 10.0 cm high sand stopper is set at the downstream of the flume, and at the upstream, a mixer is set in order to make the sediment-water mixture. The sediment particles steadily supplied using the sand hopper are mixed with steadily supplied water in the mixer, and the steady flow condition is produced. The debris flow produced in the upstream reach forms an equilibrium bed, and it runs long reach over the erodible bed in order to obtain uniform flows longitudinally. The uniform sediment used in the experiment is $d_{50} = 0.0292$ cm in diameter, $\sigma/\rho = 2.65$ in specific weight, $\phi_s = 38.3°$ in interparticle friction angle, $c_* = 0.537$.

Though it seems important to make the sediment size to be small in this study, it is quite difficult to analyze the movements of sediment particles. As

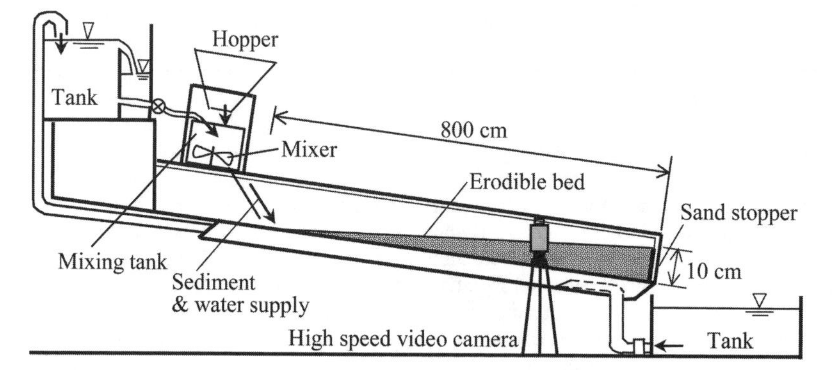

Figure 2. Schematics of experimental flume.

mentioned above, in present study, the 4 kinds of mixture discharge by unit width, q_m, is set such as 10, 30, 50 and 100 cm^2/s respectively, considering that the decrease of sediment particle size is equivalent to the increase of flow discharge.

In flume tests, the length of erodible bed reach should satisfy the condition of $h_t/L \ll w_0/u_\tau$. In which L = length of erodible bed. In present tests, it ranges 5.0 to 7.0 m, which can neglect influence of flume length on suspended transportation.

The discharge of sediment-water mixture is measured at the downstream end of the flume. The flow depth, velocity distribution and equilibrium bed slope at the section where is 65 cm upstream from downstream end are measured. Especially, the velocity distribution is analyzed with the image obtained by high-speed video camera shown in Figure 2, and the elevation of free surface is measured by point gauge.

Table 1 shows parts of the experimental condition. In the table, d = diameter of sediment particles, θ_e = equilibrium bed slope, q_m = sediment-water discharge rate in unit width; h_t = flow depth; c_f = flux sediment concentration defined as $c_f = \int_0^{h_t} c\,u\,dz/h_t$; u_τ = shear velocity defined as $u_\tau = \sqrt{gh_t \sin\theta_e}$; τ_* = non-dimensional bed shear stress defined as $u_\tau^2/\{(\sigma/\rho - 1)gd\}$; w_0 = settling velocity of sediment particle in the clear water; R_{e*} = shear velocity Reynolds number defined as $R_{e*} = u_\tau d/\nu$; F_r = Froude number; R_e = Reynolds number defined as $R_e = \bar{u}h_t/\nu$; $\bar{u}$ = depth-averaged velocity; and T = temperature of water. As seen in Table 1, it seems that sediment suspension takes place actively for every case because the ratio, w_0/u_τ, is smaller than unity ($w_0/u_\tau < 1$) referring to knowledge of suspended flows.

3 ANALYSES OF EXPERIMENTAL DATA AND DISCUSSIONS

3.1 *Equilibrium bed slope, flux sediment concentration and velocity distributions*

Figures 3(a)–(d) show the velocity distributions in case that equilibrium bed slope is about 10 degrees. In those figures, the calculated lines for velocity distributions are illustrated, which are called 'exact solutions'. The exact solutions of these figures are obtained by numerically solving for the constitutive equations of debris flows and the momentum conservation equations (Egashira et al. 1992, 1997a,b). Notice that the constitutive relations proposed by Egashira et al. were presented for coarse sediment-water mixture flow and that sediment particles behave as laminar motion.

Figure 4 shows the relationships between equilibrium bed slope and flux sediment concentration. In this figure, flume data in cases of q_m = 10, 30, 50, 100 cm^2/s are plotted respectively, and the exact solution is drawn. In the case of q_m = 10 cm^2/s, the flume data for velocity and flux sediment fit very well with exact solutions, and this shows that behaviors of sediment particles are laminar. On the other hand, in the cases of $q_m > 10\ cm^2/s$, there are some discrepancies between experimental data and exact solutions as shown in Figures 3–4. These discrepancies become larger with increase of flow discharge, and the fluctuations of local mean velocity near free surface become larger with increase of flow discharge as shown in Figure 3. Thus, it seems that turbulent diffusion is dominant due to increasing of flow discharge near the free surface. In additions, the velocity distribution near the bed in each case tends to be a debris flow's velocity profile in the flow over erodible bed, and is convex near bed surface.

Those results show that the flow with two layers is formed as reported by previous researches. Moreover, it is found that the convex velocity profile near the bed surface shows existence of static yield stress due to sediment contacts as proposed by Egashira et al. (1992, 1997a, b). In other words, shear stress due to sediment particle's contacts is dominant in lower laminar layer and shear stress due to turbulence is dominated in the turbulent upper layer. It seems that the sediment transportation by suspended flow is active with increase of flow discharge.

In Figure 4, the flume data are not plotted for bed slope smaller than about 9 degrees and for flow discharge larger than 10 cm^2/s in present studies, because sand waves such as anti-dunes and chutes & pools take place. Parts of the omitted data should be plotted in region of 'flat bed' and the others should belong to the region of 'anti-dunes' according to the results of bed forms (e.g. Fukuoka et al. 1982, Hydraulic Equations Handbook 1999).

Table 1. Parts of experimental condition.

Run	d(cm)	θ_e (deg)	q_m (cm^2/s)	h_t (cm)	h_t/d	c_f	u_τ (cm/s)	τ_*	w_0 (cm/s)	w_0/u_τ	R_{e*}	F_r	R_e	T (°C)
003	0.0292	9.72	9.71	0.336	11.5	0.117	7.46	1.20	3.67	0.492	16.7	1.59	746	11.0
113	0.0292	9.75	29.8	0.539	18.4	0.176	9.45	1.93	3.67	0.388	19.0	2.41	2049	7.2
001	0.0292	9.49	49.8	0.870	29.8	0.189	11.9	2.98	3.67	0.310	39.4	1.96	5668	24.0
017	0.0292	9.83	102	1.19	40.9	0.257	14.1	4.31	3.67	0.260	27.5	2.50	6812	6.2

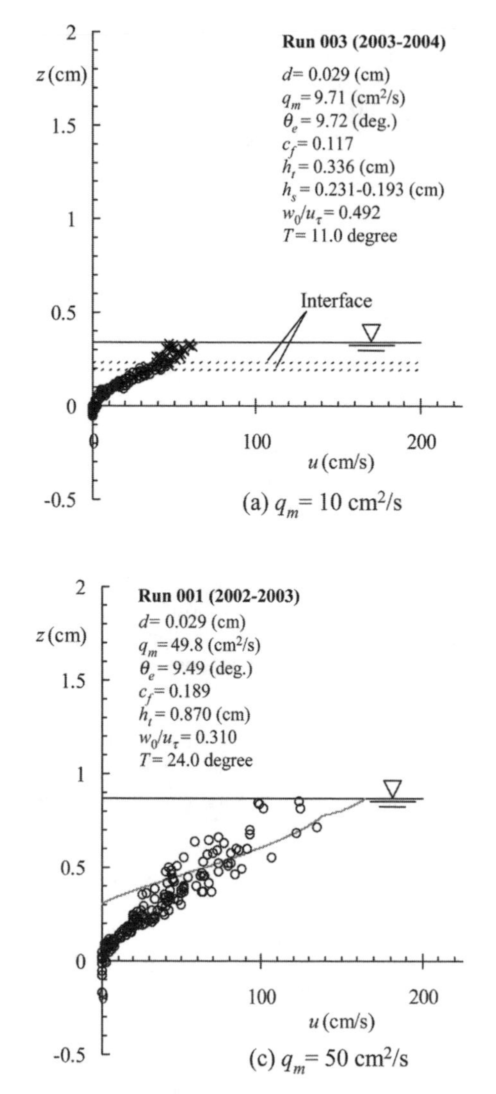

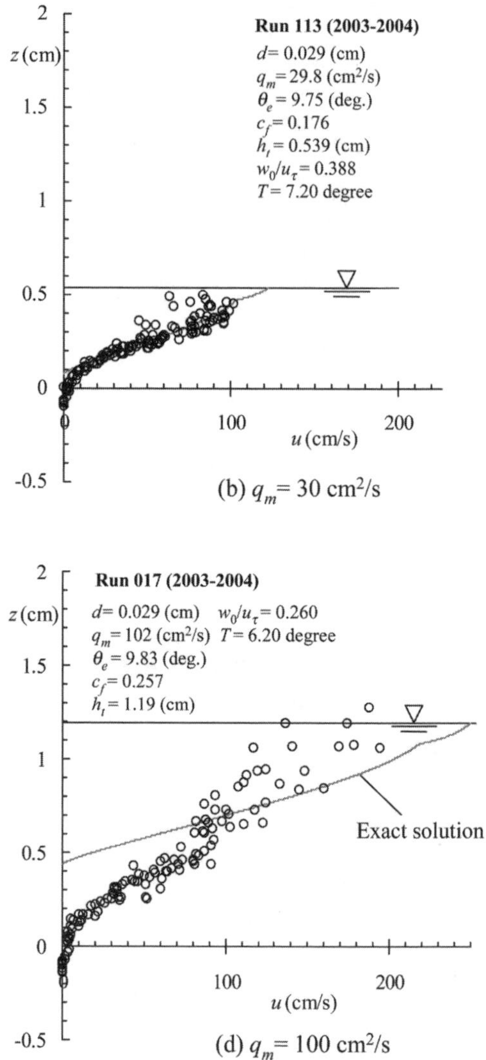

Figure 3. Velocity distributions.

3.2 *Two layer model and equilibrium bed slope*

According to Figures 3–4, sediment particles behave as laminar motion in the case of $q_m = 10\,\text{cm}^2/\text{s}$ and a part of the flow body can be turbulent with increase of flow discharge. It is suggested that the two layers such as laminar and turbulent flows are formed, and that the increase of sediment transportation as shown in Figure 4 is caused by turbulence in the turbulent upper layer.

Egashira et al. proposed the two-layers model as shown in Figure 1 (Egashira et al. 1994). In the turbulent upper layer, turbulent diffusion is dominant and shear stress due to particle-particle contacts is dominant in the lower layer, and those flow characteristics are taken into account in the model. The momentum conservation equation in the uniform flow is described as follows, referring to Figure 1.

Upper layer ($h_l \leq z \leq h_t$):

$$\tau(z) = \int_z^{h_t} \rho_m g \sin\theta dz \tag{1}$$

$$p_w(z) = \int_z^{h_t} \rho_m g \cos\theta dz \tag{2}$$

in which $\rho_m =$ mass density of sediment and water mixture ($\rho_m = (\sigma - \rho)c + \rho$); $\sigma =$ mass density of sediment; $\rho =$ mass density of clear water; $c =$ volumetric sediment concentration; $g =$ acceleration due to gravity; $\theta =$ bed inclination; and $p_w =$ hydro-static pressure when the turbulent diffusion is active in the flow.

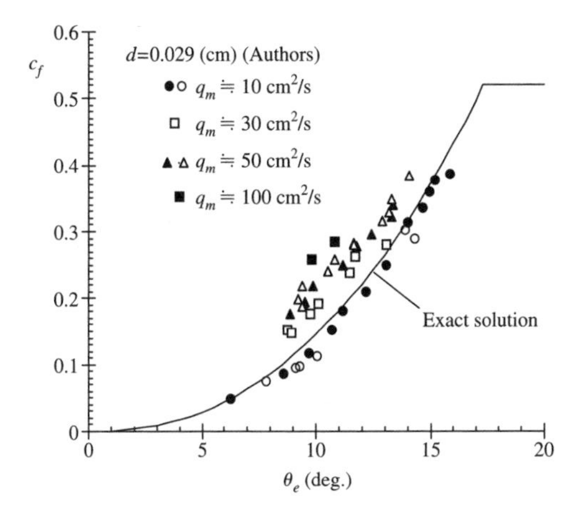

Figure 4. Relationship between equilibrium bed slope and flux sediment concentration.

For the lower layer, substituting authors' constitutive relationship into momentum conservation equation yields the following equations.
Lower layer ($0 \leq z \leq h_l$):

$$p_s(z)\tan\phi_s + \tau_f(z) + \tau_d(z) = \int_z^{h_t} \rho_m g \sin\theta dz \tag{3}$$

$$p_s(z) + p_d(z) + p_w(z)$$
$$= \int_z^{h_l} \rho_m g \cos\theta dz + p_w(h_l) \tag{4}$$
$$p_w(z) = p_w(h_l) + \int_z^{h_l} \rho g \cos\theta dz$$

in which τ_y = yield stress ($= p_s \tan\phi_s$); ϕ_s = interparticle friction angle; τ_d = dynamic shear stress due to particle collisions; τ_f = shear stress in the interstitial water; p_s = static pressure; p_d = pressure due to inelastic collision; and p_w = hydro-static pressure. τ_d and τ_f mean particle size Reynolds stress with pore scales and particle size scale.

Taking into account the shear stress balance of the static yield stress and external shear stress on the theoretical bed surface: i.e. substituting $z = 0$ and $\partial u/\partial z = 0$ into Eqs. (3)–(4) yields the following relationship between equilibrium bed slope, θ_e, and the relative laminar layer thickness, h_l/h_t.

$$\tan\theta_e = \frac{(\sigma/\rho - 1)\bar{c}\tan\phi_s}{(\sigma/\rho - 1)\bar{c} + 1}\frac{h_l}{h_t} \tag{5}$$

in which h_l = thickness of laminar layer of sediment moving layer near bed surface. In case that sediment particles behave as laminar motion, equilibrium bed slope, θ_{e0}, is shown as follows (Egashira et al. 1997a, b).

$$\tan\theta_{e0} = \frac{(\sigma/\rho - 1)\bar{c}\tan\phi_s}{(\sigma/\rho - 1)\bar{c} + 1} \tag{6}$$

in which $\bar{c}$ = depth-averaged sediment concentration. Substituting Eq. (6) into Eq. (5) in the same sediment concentration yields the following equation concerning to the relative equilibrium bed slope.

$$\frac{\tan\theta_e}{\tan\theta_{e0}} = \frac{h_l}{h_t} \tag{7}$$

It is desired that the relationship in the right hand side of Eq. (7) is deduced generally. In this study, the relationship between the relative equilibrium bed slope and flow discharge in terms of flume data.

3.3 *Definition of non-dimensional parameter in terms of constitutive equations*

The influence of flow discharge on the relative equilibrium bed slope: i.e. the relative laminar layer thickness, h_l/h_t, is analyzed with flume data.

First of all, let us define the non-dimensional parameter using constitutive equations proposed by Egashira et al. (1997 a,b). Equations for shear stress are shown as follows.

$$\tau = p_s \tan\phi_s + \rho(f_d + f_f)d^2\left|\frac{\partial u}{\partial z}\right|\frac{\partial u}{\partial z} \tag{8}$$
$$f_d = k_d(\sigma/\rho)(1 - e^2)c^{1/3}, \quad f_f = k_f(1 - c)^{5/3}c^{-2/3}$$

in which d = sediment particle size; u = local mean velocity; e = restitution coefficient of sediment particles; $k_f = 0.16$; and $k_d = 0.0828$. The dynamic shear stress with sediment particle scale is the second term in right hand side of Eq. (8). For simplicity, we call the dynamic shear stress, τ_D, particle size Reynolds stress including collisional shear stress.

$$\tau_D = \rho(f_d + f_f)d^2\left|\frac{\partial u}{\partial z}\right|\frac{\partial u}{\partial z} \tag{9}$$

Reynolds stress with turbulent upper layer scale, τ_T, is shown using the momentum mixing length, and is as follows.

$$\tau_T = \rho_m l^2\left|\frac{\partial u}{\partial z}\right|\frac{\partial u}{\partial z}, \quad l \sim h_t \tag{10}$$

in which the notation of '$\sim$' means 'is proportional to'. The following non-dimension parameter, R_D, is defined using Eqs. (9)–(10).

$$R_D = \frac{\rho_m h_t^2|\partial u/\partial z|(\partial u/\partial z)}{\rho(f_d + f_f)d^2|\partial u/\partial z|(\partial u/\partial z)} \tag{11}$$

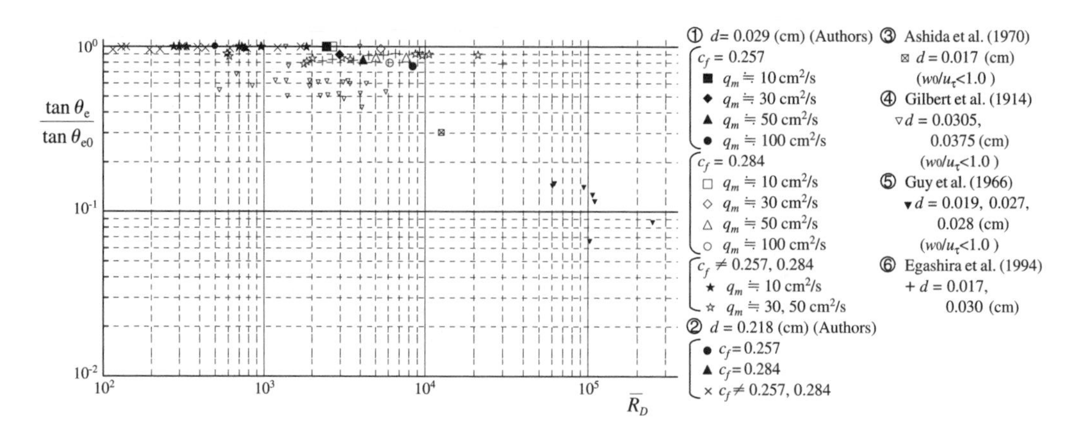

Figure 5. Relationship between non-dimensional parameter and the relative equilibrium bed slope, $\tan\theta_e/\tan\theta_{e0}$.

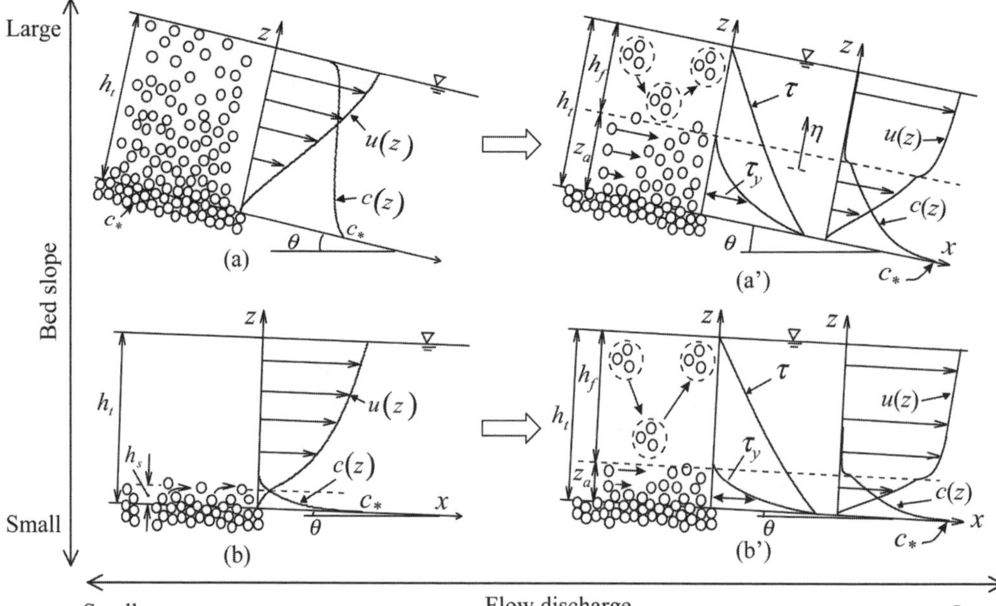

Figure 6. Flow regime of sediment–water mixture.

The following non-dimensional parameter, $\bar{R}_D$, is redefined using the depth-averaged quantities such as $c \sim \bar{c}$ and $\partial u/\partial z \sim \bar{u}/h_t$ in Eq. (10).

$$\bar{R}_D = \frac{\bar{\rho}_m/\rho}{f_f(\bar{c}) + f_d(\bar{c})}\left(\frac{h_t}{d}\right)^2 \tag{12}$$

in which $\bar{c}$ = depth-averaged sediment concentration; $\bar{u}$ = depth-averaged velocity; and $\bar{\rho}_m$ = depth-averaged mass density of sediment-water mixture ($\bar{\rho}_m = (\sigma - \rho)\,\bar{c} + \rho$).

The non-dimensional parameter as defined in Eq. (11)–(12) means the ratio of particle size Reynolds stress to Reynolds stress with turbulent upper layer scale. For example, the particle size Reynolds stress, τ_D, is dominant in the flow when $\bar{R}_D$ takes small value. The non-dimensional parameter, $\bar{R}_D$, is the same parameter to Reynolds number defined in the discussions of debris flow similarity (Miyamoto et al. 2003).

Secondly, the relative equilibrium bed slope, $\tan\theta_e/\tan\theta_{e0}$, is calculated as follows in terms of flume data as shown in Figure 4. The relative equilibrium bed slope in case of $q_m > 10$ cm^2/s to one in case of $q_m = 10$ cm^2/s (laminar motion of sediment particles) in the same flux sediment concentration in the figure.

Figure 5 shows the relationships between non-dimensional parameter, $\bar{R}_D$, and the relative equilibrium bed slope. In the figure, the flume data in cases of $c_f = 0.257$ and 0.284 are shown. In addition, several kinds of flume data are shown; data except the condition of $c_f = 0.257$ and 0.284 in present flume tests, data of coarse sediment ($d = 0.218$ cm) whose motion is laminar (Itoh 2000), data collected by Egashira et al. (1994) and data of suspended flow over flat beds (Gilbert 1914, Guy et al. 1966, Ashida et al. 1970). The value of equilibrium bed slope, θ_{e0}, obtained by computations of exact solution is used in case that there are no data for θ_{e0} in data of the suspended flows. Furthermore, the flux sediment concentration, c_f, is used instead of the depth-averaged sediment concentration, $\bar{c}$, if there are no sediment concentration data directly measured.

It is shown that the value of relative equilibrium bed slope, $\tan\theta_e / \tan\theta_{e0}$, takes less than unity in case that the value of $\bar{R}_D$ exceeds about 4000 to 5000 in spite of scattered data in Figure 5. Furthermore, it seems that the value of the relative equilibrium bed slope monotonically decreases with increase of $\bar{R}_D$. Incidentally, when $\bar{R}_D$ is over 4000 to 5000, the turbulent upper layer is formed in the laminar debris flow body and the value of h_l/h_t takes less than unity.

In addition, it seems that flow regime of sediment–water mixture is classified as schematically shown in Figure 6, based on previous Egahira et al.'s studies and present discussions. For examples, sediment particles behave as laminar motion as shown in Figures 6(a)–(b). However, when flow discharge increases and $\bar{R}_D$ becomes larger, parts of sediment particles can be transported in the turbulent upper layer as shown in Figure 6(a').

4 CONCLUSIONS

In present study, the influence of flow discharge on equilibrium bed slope is experimentally discussed using flume data in the flow over erodible bed. The results are summarized as follows.

(1) The flume data of flux sediment concentration are not plotted for bed slope smaller than 9 degrees and for flow discharge larger than 10 cm^2/s in present studies, because sand waves such as antidunes and chutes & pools take place.

(2) In the case of $q_m = 10$ cm^2/s, the behaviors of sediment particles are laminar motion using exact solutions and flume data. On the other hand, in the cases of $q_m > 10$ cm^2/s, there are some discrepancies between experimental data and exact solutions as shown in Figures 3–4. These discrepancies increase with flow discharge, and the fluctuations of local mean velocity near free surface become larger with increase of flow discharge. Those results show that the turbulent upper layer is formed with increase of flow discharge, and that shear stress due to sediment particle's contacts is dominant in laminar lower layer and Reynolds stress of sediment–water mixture flow is dominant in turbulent upper layer. Moreover, flume data suggest that the convex velocity profile near the bed surface shows existence of static yield stress due to sediment contacts. The increase of flux sediment concentration with increase of flow discharge is caused by turbulence in the turbulent upper layer.

(3) The relative equilibrium bed slope, $\tan\theta_e / \tan\theta_{e0}$, means the relative laminar layer thickness, h_l/h_t, as shown in Eq. (7). In order to evaluate h_l/h_t, non-dimensional parameter, $\bar{R}_D$, as shown in Eqs. (11)–(12) is proposed using constitutive equations proposed by Egashira et al. The non-dimensional parameter, $\bar{R}_D$, means the ratio of particle size Reynolds stress to Reynolds stress with upper layer scale. The value of the relative equilibrium bed slope, $\tan\theta_e / \tan\theta_{e0}$, takes unity in case that $\bar{R}_D$ is smaller than 4000 to 5000, and takes less than unity in case that $\bar{R}_D$ exceeds 4000 to 5000. In addition, the relative equilibrium bed slope monotonically decreases with increase of $\bar{R}_D$.

The collection of flume data is needed in order to evaluate the flow structure precisely. The shear stress structure will be examined by using non-dimensional parameter such as Richardson number using directly measured sediment concentration profiles and the knowledge concerning to suspended sediment transportation.

ACKNOWLEDGEMENT

Authors should be thankful for Mr. Atsuo UTSUMI, Mr. Shogo KURODA, Mr. Yusuke ETOH and Mr. Keisuke YOSHIDA who were graduate school students of Ritsumeikan University to help the flume test and data analyses.

Parts of this study were supported by Grant-in-Aid for Scientific Research of Young Scientists (B) (Representative: Takahiro ITOH) from the Japanese Ministry of Education, Culture, Sports, Science and Technology.

REFERENCES

Arai, M. & Takahashi, T. 1986. The karman constant of the flow laden with high sediment. *Third Int. Sym. on River Sedimentation*. The University of Mississippi: 824–833.

Arai, M. & Takahashi, T. 1986. The mechanics of mud flow. *Proc. JSCE* (Proceedings of JSCE). No. 375: 69–77 (in Japanese).

Ashida, K. & Michiue, M. 1970. Study on the suspended sediment (1) – Concentration of the Suspended Sediment near the Bed Surface. *Ann., D.P.R.I.*, Kyoto Univ. 13B-2: 233–242 (in Japanese).

Chen, Cheng-Lung. 1988. Generalized viscoplastic modeling of debris flow. *Jour. Hydr. Eng.* ASCE. 114. No. 3: 237–258.

Egashira, S., Ashida, K., Tanonaka, S. & Satoh, T. 1992. Studies on mud flow – Stress Structure. *Ann., D.P.R.I. (Annuals Disaster Prevention Research Institute)*, Kyoto University. 35B-2: 79–88 (in Japanese).

Egashira, S. & Ashida, K. 1992. Unified view of the mechanics of debris flow and bed-load. *Advances of micromechanics of granular materials*: 391–400. Amsterdam: Elsevier.

Egashira, S., Satoh, T. & Chishiro, K. 1994. Effect of particle size on the flow structure of sand-water mixture. *Ann., D.P.R.I.*, Kyoto University. 37B-2: 359–369 (in Japanese).

Egashira, S., Miyamoto, K. & Itoh, T. 1997a. Bed-load rate in view of two phase flow dynamics. *Proc. Hydraulic Eng.* JSCE. 41: 789–794 (in Japanese).

Egashira, S., Miyamoto, K. & Itoh, T. 1977b. Constitutive equations of debris flow and their applicability. *Proc. 1st Int. Conf. on Debris-Flow Hazards Mitigation, San Francisco, California, 7–9 August*, New York. ASCE: 340–349.

Einstein, H. A. 1950. Soil Conservation Service, *Tech. Bull.* 1025: 1–71.

Fukuoka, S., Okutsu. K. & Yamasaka, M. 1982. Dynamic and kinematic features of sand waves in upper regime. *Proc. JSCE.* No. 323: 77–89 (in Japanese).

Gilbert, G. K. 1914. The transportation of debris by running water, *USGS Prof. Paper*, 86.

Guy, H. P., Simons, D. B. & Richardson, E. V. 1966. Summary of Alluvial channel data from flume experiments, 1956–61, *Professional Paper 462-I*, U.S.G.S.

Hirano, M., Hashimoto, H., Fukutomi, A., Taguma, K. & Pallu, Muh. Saleh. 1992. Nondimensional parameters governing hyper-concentrated flow in a open channel. *Proc. Hydraulic Eng. (Proceedings of Hydraulic Engineering)*. JSCE. 36: 221–226 (in Japanese).

Hashimoto, H. & Hirano, M. 1997. A flow model of hyper-concentrated sand-water mixtures. *Proc. 1st Int. Conf. on Debris-Flow Hazards Mitigation, San Francisco, California, 7–9 August*, New York. ASCE: 464–473.

Itoh, T. 2000. *Constitutive equations of debris flow and their applicabilities*. Ph. D.-Thesis of Ritsumeikan University (in Japanese).

Japan Society of Civil Engineers (JSCE) 1999. *Hydraulic Equations Handbook*. 183–184. JSCE. Tokyo.

Lane, E. W. & Kalinske, A. A. 1939. Engineering calculation of suspended sediment. *Trans., A.G.U., Reports and papers, Hydrology*: 603–607.

Miyamoto, K. & Itoh, T. 2003. Numerical simulation and similarity of debris flow. *JSECE*. Vol. 55. No. 6: 40–51 (in Japanese).

O'Brien, J. S. & Julien, P. Y. 1988. Laboratory analysis of mudflow properties. *Jour. Hydr. Eng.*, ASCE. Vol. 114. No. 8: 877–887.

Rouse, H. 1936. Modern conceptions of the mechanics of fluid turbulence. *Trans. ASCE*. No. 1965: 463–543.

Sumer, B. M., Kozakiewicz, A., Fredsøe, J. & Deigaard, R. 1996. Velocity and concentration profiles in sheet-flow layer of movable bed. *Jour. Hydr. Eng.*, ASCE. 122. No. 10: 549–558.

Takahashi, T. 1992. Mechanics of debris and mud flow. *Proc. of sediment movement symp., Kyoto, 22 May*: 39–55 (in Japanese).

Takahashi, T. 2000. Initiation and flow of various types of debris-flow. *Proc. 2nd Int. Conf. on Debris-Flow Hazards Mitigation, Taipei, Taiwan, 16–18 August*, Rotterdam. Balkema: 15–25.

Vanoni, V. A. 1944. Transportation of suspended sediment by water. *Trans., ASCE*. 2267: 67–133.

Winterwerp, J. C., de Groot, M. B., Mastbergen, D. R. & Verwoert, H. 1990. Hyperconcentrated sand-water mixture flows over flat bed. *Jour. Hydr. Eng.*, ASCE, Vol. 116, No. 1: 36–54.

River, Coastal and Estuarine Morphodynamics: RCEM 2005 – Parker & García (eds)
© 2006 Taylor & Francis Group, London, ISBN 0 415 39270 5

Toward prediction of the spatial distribution of clusters in a watershed via geomorphic and hydraulic methods

Kyle B. Strom & Athanasios N. Papanicolaou
Dept. Civil and Envirn. Engr., IIHR-Hydroscience & Engineering, University of Iowa, IA USA

ABSTRACT: Small-scale bedforms, or microforms such as particle clusters, have been shown to affect sediment transport and stream flow characteristics when present in laboratory experiments, and have been documented in various mountain rivers throughout the world. In addition to accessing the impacts on mean and turbulent flow, sediment transport, and aquatic habitat of clusters, there is need for a basic understanding of the hydraulic, sedimentary, and geomorphic properties prompting existence of clustered topography. With such knowledge we might be able to predict the onset of clusters in stream reaches and therefore watersheds. The present research aims at addressing this shortcoming through a field study on a mountain stream. Traditional geomorphic characterization of clusters and the site at which they were located was conducted. Five main types of clusters with differing visual and quantitative characteristics were identified; namely the comet, heap, line, pebble, and ring shaped clusters. Additionally, hydraulic characterization of a cluster at the study site was conducted using a 3D acoustic Doppler velocimeter to identify particular fluid signatures left by clusters. The investigation is then integrated with existing data in the literature to present a combined dataset of river slope and sediment d_{84} where clusters have been recorded. Results show that clusters have been found in a wide range of slopes and particles sizes ($0.004 \leq S \leq 0.06$ and in $50 \leq d_{84} \leq 500$ mm), but are most likely to occur in cobble size sediment at slopes typical of riffle-pool to plane bed morphologies. We hypothesize that although cluster microform dynamics are an outcome of local sediment and fluid conditions, these local dynamics are in turn a function of the larger-scale watershed dynamics; as watershed and climate properties govern topography, sediment input, and hydrology of a local stream reach. Implications of these results and hypothesis suggest that given the proper input from remote sensing satellite data, we will be able to predict, in an average sense, where cluster bedforms will develop in a stream corridor/watershed.

1 INTRODUCTION

Particle groupings or clusters of gravel to cobble size sediment in gravel-bed rivers have been observed in both field and laboratory settings as prevalent features of the surface layer in some gravel-bed streams (Teisseyre 1977; Brayshaw 1984; Reid et al. 1992; de Jong 1995; Church et al. 1998; Papanicolaou et al. 2003; Strom et al. 2004). These bedforms are termed "microforms" since their local size is on the order of the grain size of the bed material. We hereafter referred to these morphologic bed features as clusters or cluster microforms.

Cluster microforms have been shown to play a role in bedload transport rates, downstream fining, and flow field characteristics (Brayshaw et al. 1983; Hassan and Reid 1990; Clifford et al. 1992; Strom et al. 2004; Strom et al. 2005). However, there remains a void of detailed field characterization of such bedforms in the literature. Additionally, with the lack of characterization of clusters and the streams in which they are found, we have no means by which to predict when and where cluster microforms will occur in a stream. Other gravel-bed river bedforms such as pool-riffle and step-pool sequences have received extensive characterization in relation to cluster microforms and this has aided in our understanding and prediction of such features.

The two objectives of the study are: 1) to characterize cluster microforms in greater detail; and 2) to determine where and under what conditions cluster topography will develop in a stream reach. Addressing the first objective will help to better define cluster topography and some of the associated effects they might have on sediment transport and the frictional characteristics of the river. This more detailed characterization will then help in obtaining the overarching second objective of determining when and where clusters from. This second objective has far reaching implications for stream modeling, assessment, and management. For instance, if we can conclude the location and conditions that cause clustered topography in relation to a mountain stream morphology continuum, such as the one presented by Montgomery and Buffington (1997), we might be able to predict clustered topography in numerical models and account for their effects on sediment transport and resistance.

2 METHODOLOGY

There are two ways to characterize a bedform. The first way is geomorphically, i.e. what do the bed forms look like, and what are their geometric and sedimentary characteristics? This traditional approach is typically carried out on a reach scale and placed within the context of a stream continuum type characterization. The second way to characterize a bed form is from a fluid mechanics angle. It has been shown that various configurations of bed roughness elements can leave distinctive patterns in mean and turbulent quantities (Krogstad and Antonia 1999; Papanicolaou et al. 2001). Therefore, it might be possible to characterize a bedform by its particular imprint signature in the flow. This would then allow for identification of bedform type simply from velocity measurements.

In this study we attempt to characterize cluster topography by both geomorphic and hydraulic characterization methods. Information gained through this characterization will then be used to try to and understand where clusters might form in a stream.

To meet our objectives, a field study was conducted on the American River in the Cascade mountains of Washington State. The paper will first describe the study site. This will be followed by geomorphologic and hydraulic characterization of clusters and concluded with a discussion on the location of clustered topography in a stream reach.

3 SITE DESCRIPTION

3.1 *Watershed and reach scale*

The American River is a mountain stream located on the eastern slopes of the Cascade mountains of Washington State in the Yakima drainage basin. The headwaters of the American begin at 1600 m (5250 ft) near Chinook Pass at the crest of the range, and flow down to its confluence with the Naches River. The site is located approximately 17 river km downstream of its source at an elevation of 1012 m (3320 ft), and approximately 15 km upstream of USGS gaging station 12488500. Mean annual discharge at the USGS gaging station is 6.7 m^3/s with an average annual peak discharge of 45.4 m^3/s. Flow at the site is approximately equal to the flow at the gaging station as there are no major tributaries to the American between the study site and the gaging station. Runoff is snowmelt dominated with multiple peaks occurring between October and April due to rain and snow melt events with steady high flows from April to July due to bulk snowmelt.

The study site is located in a riffle-pool reach (Montgomery and Buffington 1997) with average slope, $S = 0.004$, and is comprised mainly of gravel to cobble size sediment. Channel slope was determined by fitting a regression line to survey data over 1 morphologic unit, i.e. top of riffle to top of riffle. Cluster microforms were documented in two regions, the main study bar (MB) and a side channel (SC) through which water flows at higher discharges (Fig. 1). The slope of the main bar and side channel are $S_{MB} = 0.002$ and $S_{SC} = 0.014$ respectfully.

3.2 *Bar scale*

Grain size distributions of the bar and side channel surface sediment were measured via Wolman (1954) pebble counts, and an automated grain size extraction (AGSE) method similar to Butler et al. (2001) that determines sediment size distributions from digital images of the bed. Particles were binned in 0.5ϕ increments according to particle intermediate b-axis length, where $\phi = -\log_2 d$ and d is particle diameter. The AGSE results were weighted with a d_i^2 weighting to transform the areal, frequency-by-number grain size distribution to an equivalent Wolman (1954) pebble count distribution (Kellerhals and Bray 1971; Strom and Papanicolaou 2005).

Local downstream fining occurred on the bar, with coarser material depositing at the head of the bar. The main bar and the side channel have similar grain size distributions, ranging from coarse gravel to small cobbles; for the main bar $d_{16} = 15$, $d_{50} = 45$, and $d_{84} = 80$ mm, and $\sigma_g = \sqrt{d_{84}/d_{16}} = 2.18$. The main bar

Figure 1. Aerial photograph of the American River study site. LWD stands for large wood debris deposit. The black lines trace the current course of the river.

and the side channel both have an average particle shape factor $SF = c/\sqrt{ab} = 0.49$, where a is the long, b is the intermediate, c is the short axis of a particle, and the shape factor of a sphere is $SF = 1$. Following the Zingg (1935) classification of sediment particle shape as a function of b/a and c/b, 42% of the particles are disc shaped, 21% are rods, 19% are blades, and 18% are spheres. Figure 2 and Table 1 present summaries of the grain size data for the American River study site.

4 GEOMORPHIC CHARACTERIZATION

4.1 *Cluster shape*

Discrete cluster topography was differentiated from connected, net-like reticulate microtopography (de Jong 1995; Church et al. 1998). Reticulate structures were also identified at the sites, and tended to occur in similar sediment and reach scale morphology as discrete cluster topography. The differing condition between discrete clusters and reticulate structures seemed to be higher local slopes, e.g. the perimeter of the bar that dropped of more steeply into the main channel. According to Papanicolaou et al. (2003), all microforms with different geometric shapes were considered as cluster microforms. Within discrete cluster topography, individual clusters were identified as one of the following five types: pebble, heap, comet, line, or ring clusters (Fig. 3).

Pebble clusters are microforms that align with the Brayshaw (1984) definitions where the presence of a large stable clast particle gives the cluster its stability. Other medium size particles then deposit in an imbricated fashion in the stoss, and fine particles deposit in the sheltered wake region of the obstacle (Fig. 4(a)). A line cluster is similar to the pebble cluster but is comprised of uniform size sediment in single profile. No clearly defined wake or stoss region is present in line clusters (Fig. 4(b)). Comet clusters are broader than both pebble and line clusters, and consist of a stable head from which stable tails grow downstream on either side. Wake regions of comet clusters are usually loosely packed (Fig. 4(c)). In plan view, ring clusters look like a "ring" of stable connected particles (Fig. 4(d)). While the hedge of a ring cluster contains larger more stable particles, the center usually consist of finer material (Kozlowski and Ergenzinger 1999).

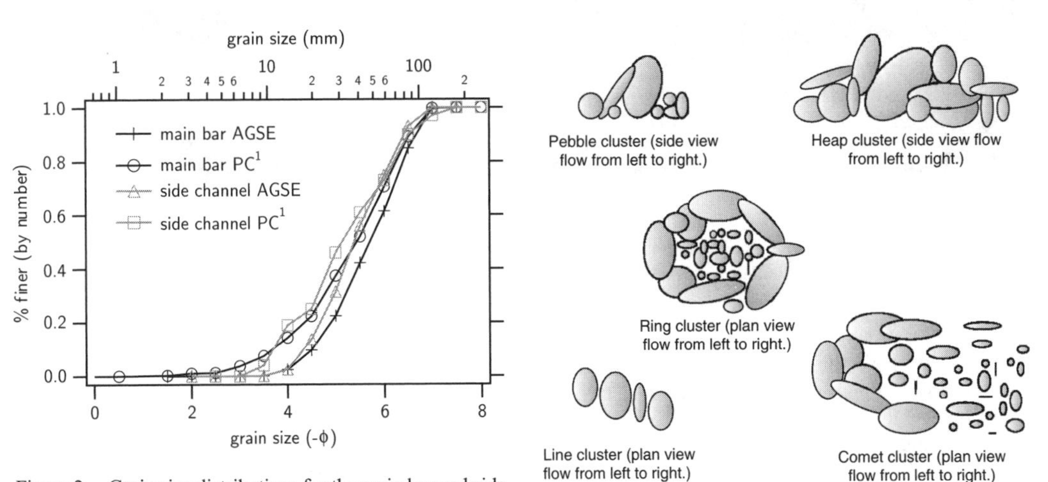

Figure 2. Grain size distributions for the main bar and side channel on the American River using AGSE and Wolman (1954) pebble counts (PC[1]).

Figure 3. Schematic of the five types of identified clusters.

Table 1. Grain size distribution statistics for the American River (AM). PC[1] = pebble counts measured with a ruler, and PC[2] = pebble counts measured with a gravelometer.

River	Location	Method	d_{16} (mm)	d_{50} (mm)	d_{84} (mm)	σ_g	*SF*
AM	Main bar	AGSE	25	50	90	1.82	–
AM	Main bar	PC[1]	15	45	80	2.18	0.49
AM	Side channel	AGSE	25	55	90	1.94	–
AM	Side channel	PC[1]	20	40	80	2.07	0.49
AM	Side channel	PC[2]	25	45	80	1.73	–

Figure 4. Examples of clusters on the American River. Scale in image is approximately 0.3 m. (a) pebble cluster (flow is from left to right), (b) line cluster (flow is from right to left), (c) comet cluster (flow is from left to right), and (d) ring/comet cluster (flow is from right to left).

Heap clusters have the largest areal footprint of all discrete cluster topography types observed. They are typically several grain diameters in width and can have multiple grains stacked atop each other in the direction normal to the bed.

4.2 *Cluster geometric properties*

At the site, clusters were visually identified, marked and digitally imaged. The average height, h_c, length, l_c, width w_c, and b-axis of the largest particle in the cluster, d_o, were measured. In addition, based on the definition described earlier, each cluster was tagged as either a pebble, comet, line, heap, or ring type cluster. Figure 5 shows the allotment of each cluster type.

On the American River the most dominate cluster forms are the heap and pebble clusters; followed by comet shaped and line clusters. With the most rare cluster form being ring shaped clusters.

Distributions of measured geometric properties of all clusters are plotted in Fig. 6. Cluster properties from the American River main bar and side channel sites were very similar and have therefore been combined to represent overall American River cluster properties. Note: in all graphics containing box plots, the center line represents the mean of the distribution, and the upper and lower bounds mark off the 75th and 25th percentiles.

Figure 7 depicts cluster geometric properties broken down by cluster type. In these figures the distribution of cluster height, length, and width are sorted by cluster type to reveal different properties of individual cluster types.

4.3 *Discussion: cluster properties compared to other studies*

In general, cluster height is slightly $>d_{84}$ of the bar sediment, but is smaller than the b-axis of the largest

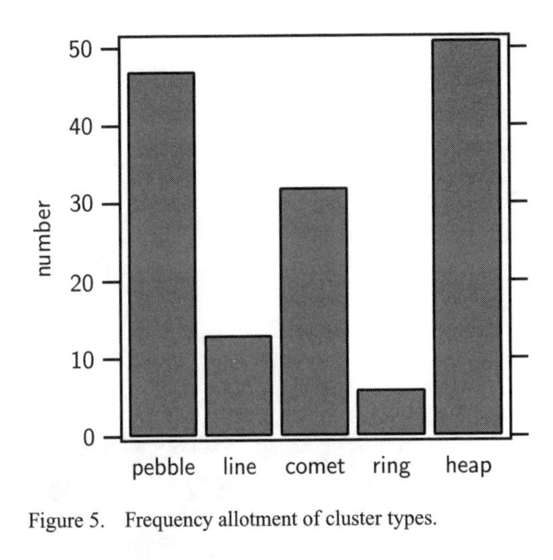

Figure 5. Frequency allotment of cluster types.

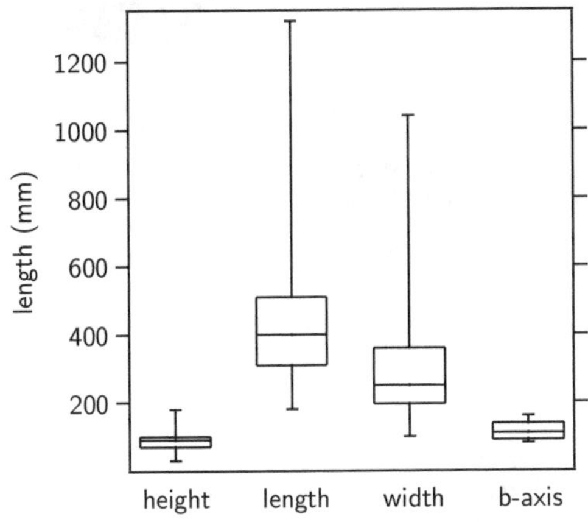

Figure 6. Overall average cluster geometric properties.

particles in the cluster, d_o (Table 2). Brayshaw (1984) also notes that the size of the largest particle in a cluster is quite large, typically being $>d_{95}$. It is reasonable to assume that c-axis of the largest particle in a cluster will govern cluster height, as the c-axis of a particle is typically the axis projected normal to the boundary (Komar and Li 1986) (Fig. 4(c) & 4(b)). However, due to imbrication this is not always the case. On the American River, a few clusters were observed with the a-axis of the largest particle oriented perpendicular to the bed (Fig. 4(a)). Heap type clusters can also accumulate particles in the vertical so that their height is greater than the c-axis of the largest particle in the cluster. Based on the heap cluster's ability to be comprised of grains stacked in the vertical, one would suppose that it would tend to be the tallest cluster form. This however is not the case (Fig. 7(a)). Heap cluster heights might be less because they tends to be made up of a larger accumulation of smaller particles rather than by

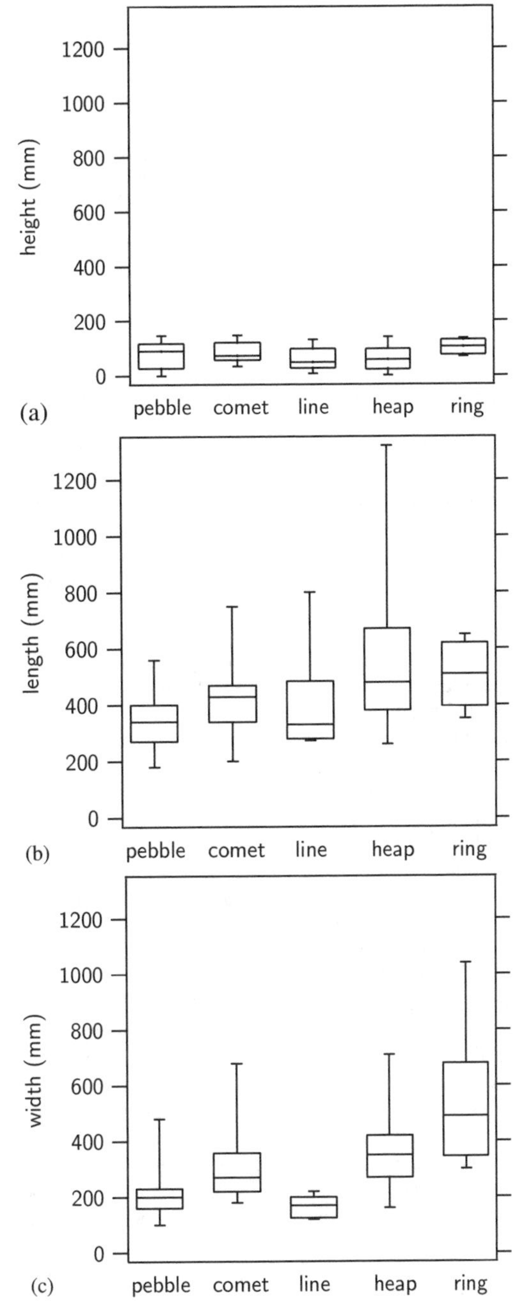

Figure 7. American River cluster geometric properties broken down by cluster type for (a) height, (b) length, and (c) width.

a few large dominate particles as is the case with the pebble and comet clusters.

With the exception of the ring cluster ($l_c/w_c = 0.94$), clusters are longer than they are wide. Average l_c/w_c

Table 2. Ratios of geometric cluster properties where l_c is cluster length, h_c is cluster height, w_c is cluster width, d_o is the b-axis of the largest particle in the cluster, and d_{84} is the sediment size of the surface material for which 84% of the sediment is finer than.

River	Type	l_c/h_c	l_c/w_c	w_c/h_c	d_o/h_c	d_o/w_c	d_o/d_{84}	h_c/d_{84}
AM	avg. of all clusters	4.94	1.53	3.22	1.28	0.40	1.40	1.10
AM	Pebble	4.28	1.64	2.61	1.46	0.56	1.40	0.98
AM	Comet	5.03	1.40	3.59	1.20	0.33	1.28	1.04
AM	Line	6.46	2.39	2.70	2.62	0.97	1.95	0.73
AM	Heap	8.94	1.55	5.76	2.10	0.36	1.59	0.73
AM	Ring	4.92	0.94	5.24	–	–	–	1.28

ratios for the American Rivers is $l_c/w_c = 1.53$. If broken down by cluster type, l_c/w_c ranges from 1.40–2.39. Brayshaw (1984) had similar findings to these; reporting cluster lengths ≈2 times their width. Specifically, the l_c/w_c ratios of Brayshaw (1984) varied from 1.43–2.08 over the three rivers examined in the study ($0.24 \leq SF \leq 0.61$). In the study by Buffin-Bélanger and Roy (1998), cluster length to height ratio was recorded as $l_c/h_c = 5$, which closely matches the overall l_c/h_c ratio for the American River. This comparison shows that in general, the clusters examined in this study align with those previously reported. As such, the information from this study adds data that can be compared and combined with some of the previous studies regarding clusters.

Based on our definition of cluster types, it is speculated that if more particles are added to the width of a pebble or line cluster it might become a comet or heap type cluster. Following this logic, comet, heap, and ring type clusters might be pebble or line clusters that evolved in shape due to further deposition of sediment. Or conversely, a pebble cluster might be the evolutionary outcome of a comet or heap cluster under high fluid shearing action.

A basis for this argument comes from Papanicolaou et al. (2003) where they noted a change in cluster shape as a function of incoming sediment and bed shear stress (Fig. 8). They hypothesized that individual clusters underwent a change in shape over the course of their existence. In a controlled laboratory setting using uniform size particles they noted that cluster shape underwent a change as bed shear stress increased from a 2-particle cluster to a comet cluster to a triangle cluster to a rhomboid cluster (Papanicolaou et al. 2003). There are some similarities in cluster shape between the observed forms of Papanicolaou et al. (2003) and the current field study (e.g. 2-particle cluster = line, comet = comet, and rhomboid = pebble). This suggests the possibility that clusters shape might represent stages of a cluster evolutionary cycle and previous flow and sediment conditions.

Whether or not clusters in the field undergo the same evolutionary process as has been observed in the lab is speculative. What is certain is that varying cluster shapes will have differing effects on the flow field and hence sediment transport processes, overall resistance, and hydraulic characterization. This has implications for correctly analytically or numerically modeling flow over a clustered bed. Drag force will vary with cluster type. One can conceptualize cluster bedforms as automobiles with different shapes and hence different drag coefficients. The accumulative effect of a cluster bed therefore is dependent on the shape of the clusters in the cluster patch. The accumulative resistance of the bed to fluid motion is then the summed effect of the obstacles or clusters on the near-bed flow structure. If we wish to accurately predict river roughness due to changing in the bed, then we must accurately predict not only the development of cluster, but also cluster shape and spacing as this will effect the structure of the turbulent flow and hence drag properties.

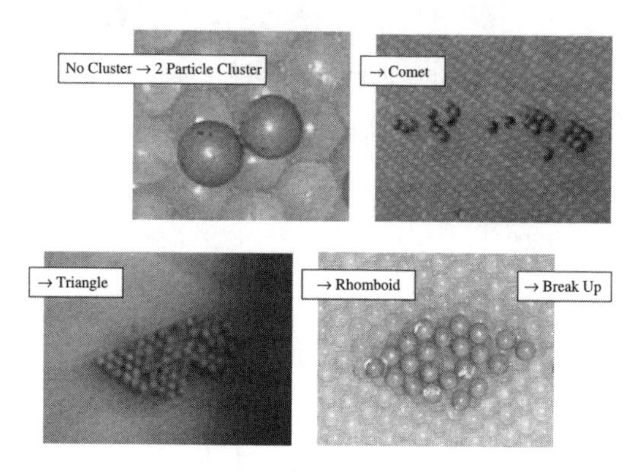

Figure 8. Laboratory study of cluster evolutionary cycle for uniform size sediment clusters with increasing bed shear stress (after Papanicolaou et al. 2003).

5 HYDRAULIC CHARACTERIZATION

The goal of the hydraulic characterization is to determine specific effects, or signatures, that clustered

Figure 9. ADV transverse. Flow is from right to left.

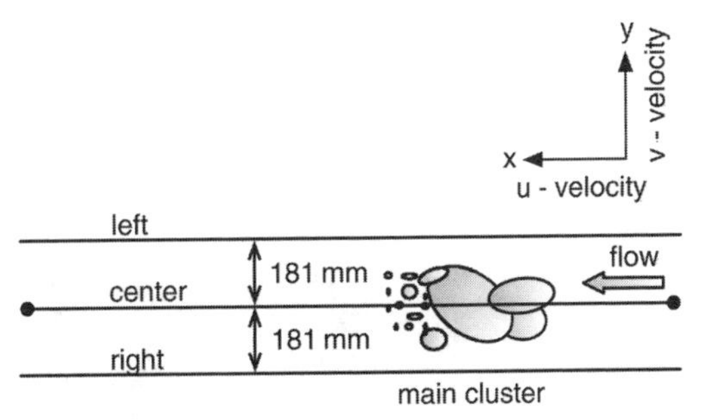

Figure 10. Plan view of ADV longitudinal measurement transects.

topography have on the flow. These signatures might be patterns in mean averaged flow, or in certain turbulent scales of motion. Knowledge of a particular cluster induced signature not only has implications for sediment transport and resistance to flow, but might also allow us to "map" the bed features through their velocity signature when they are covered by water.

5.1 *Velocity measurements*

Velocity measurements were carried out around a representative cluster on the American River. A pebble cluster in a cluster patch was chosen as the representative cluster for the measurements because this shape was found to be very prominent on the American River (Fig. 5) and has often been cited in the literature as the primary cluster form.

3D velocity measurements were collected around the cluster using a laboratory 10 MHz SonTek acoustic Doppler velocimeter (ADV) sampling at 25 Hz with a flexible cable mount probe. A ridged, portable 3D traverse to hold the ADV was built on site, placed over the cluster of interest, and leveled (Fig. 9). Three transects of data were taken around the main cluster, namely a center transect and two transects down the flanks of the cluster (Fig. 10). Velocity profiles were made at 11 cross section across the three transects. Note: the ADV probe had to be rotated several times during the data acquisition to obtain the measurements due to conditions of shallow flow and larger bed protrusions.

The resulting measurements were time series of longitudinal, vertical, and transverse velocity $U = U(x, y, z, t)$, $W = W(x, y, z, t)$ and $V = V(x, y, z, t)$ respectfully. These series where then ensemble averaged to produce $u(x, y, z)$, $w(x, y, z)$, $v(x, y, z)$, and second order moments, $\overline{u'u'} = \overline{u'u'}(x, y, z)$, $\overline{w'w'} = \overline{w'w'}(x, y, z)$, $\overline{v'v'} = \overline{v'v'}(x, y, z)$, $\overline{u'w'} = \overline{u'w'}(x, y, z)$, $\overline{u'v'} = \overline{u'v'}(x, y, z)$, and $\overline{w'v'} = \overline{w'v'}(x, y, z)$.

5.2 *Velocity signature results*

At the measuring site, the average bulk velocity approaching the cluster is $U_{bulk} \approx 0.5$ m/s, and the average flow depth, h, is $h = 155$ mm. This flow depth and cluster height of $h_c = 75$ mm results in a relative submergence of $h/h_c = 2$. $h/h_c = 2$ implies that at, these flow conditions, cluster generated effects on the flow should be felt at the free surface (Shamloo et al. 2001).

A sample of three profiles taken along the centerline transect of the cluster (Fig. 10) are presented in Fig. 11. The first profile, taken at cross section 1 (xs 1), can be thought of as the velocity distribution unaffected by the cluster. At this location there is retardation in u close to the bed, comparatively to a standard logarithmically distributed velocity profile. This is due to momentum extraction caused by the variable rough bottom protruding into the flow and generating separated flow pockets at the grain scale level. Peak turbulent stress, $-\overline{u'w'}$, occurs at a level slightly above the bed due to the rough boundary induced shear layer. Furthermore, the velocity distribution does not reach a free stream state where $\partial u/\partial y$ is very small at some location up away from the bed, indicating that the flow is depth limited (Nowell and Church 1979). This implies that interactions at the bed/fluid interface are likely to propagate and significantly effect the flow throughout the entire water column. Cross section 4 (xs 4) depicts the flow directly atop the cluster obstacle. Here flow is accelerated and turbulent stress is damped by the acceleration in the mean flow. In the cluster wake, cross section 6 (xs 6), mean flow and turbulent stress are both reduced in the near bed region compared to the unaffected state. Here the peak stress level occurs at the top of the cluster in the shear layer at the interface of the dead zone behind the cluster and the accelerated flow coming over the top of the cluster.

Preliminary results presented here and in previously conducted studies show that clusters do systematically leave their imprint on the flow by significantly altering flow characteristics compared to a planar rough gravel-bed.

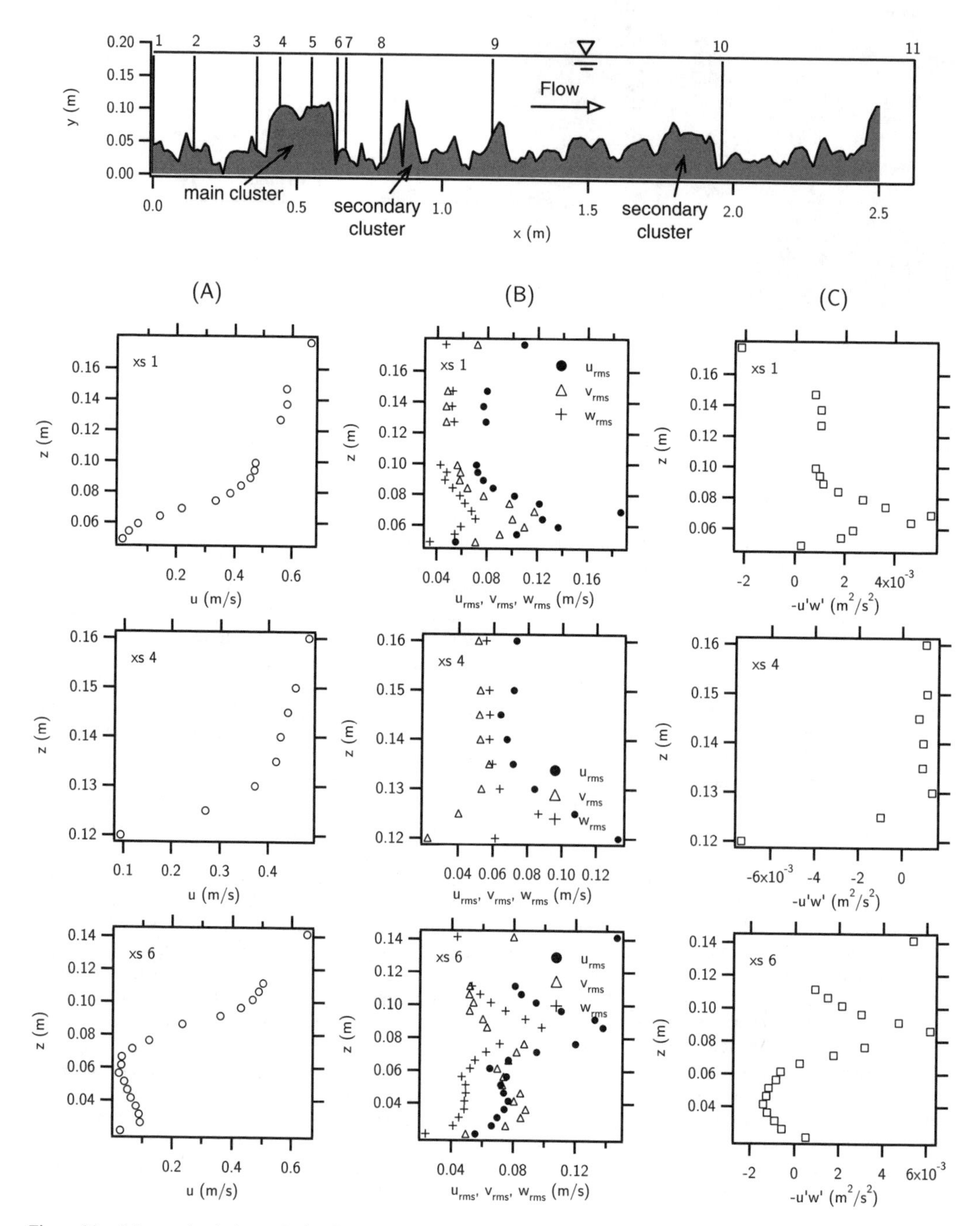

Figure 11. Mean and turbulent velocity distribution around the main cluster. Column (A) shows the mean streamwise velocity u; column (B) the rms turbulent intensities $u_{rms} = \sqrt{\overline{u'^2}}$, $v_{rms} = \sqrt{\overline{v'^2}}$, $w_{rms} = \sqrt{\overline{w'^2}}$; and column (C) the Reynolds stress $-\overline{u'w'}$.

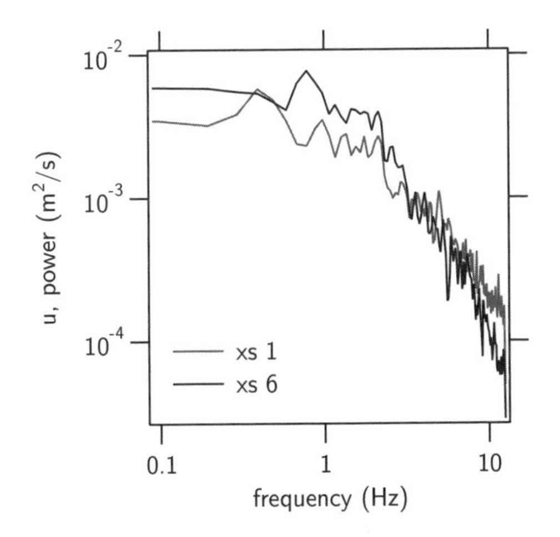

Figure 12. *U* power spectral density at cross sections 1 and 6. Both points are located in the vertical at the peak in turbulent stress (Fig. 11).

In a similar study of flow over a cluster with $h/h_c = 2$, Buffin-Bélanger and Roy (1998) proposed that flow around a cluster could be classified by 6 flow regions. Namely an acceleration region in the cluster stoss, a recirculation region in the near cluster wake, a shedding zone extending from the top of the recirculation region to the free surface, a reattachment point, an area of upwelling downstream of the reattachment point, and return to an unobstructed state. The majority of these flow regions were also observed here in the field and in a complementary laboratory/numerical study conducted on the flow around a self occurring cluster in the laboratory (Strom et al. 2005). The high relative cluster submergence of Strom et al. (2005), i.e. $h/h_c = 11$, resulted in effects due to the presence of a cluster not being felt on the free surface. However, the $h/h_c = 2$ relative cluster submergence in this study and that of Buffin-Bélanger and Roy (1998) did produce effects from the cluster that propagated throughout the depth of flow.

Of interest in determining the signature of cluster induced flow would be the scale of motion produced by incoming flow and that flowing in the lee of a cluster. Preliminary results from power spectral density analysis of the shear layer at the peak in turbulent stress at cross section 1 and 6 suggest that the cluster induced shear layer produces smaller dominate scales and slightly alters the decay in scales slope comparatively to a planar gravel-bed. Power spectral density peaks in the shear layer in the lee of the cluster occur at approximately 1 Hz, whereas the peak from the shear layer at cross section 1 is at a frequency of 0.3 Hz.

While the preliminary data presented here on hydraulic signature of clusters is inconclusive as to a definite particular pattern in turbulent scales of motion or mean flow regions, it does point to the possibility of deriving such a signature once a greater number of measurements have been added to the database. Of interest would be characterization of cluster effects on the free surface via particle image velocimetry (PIV). Such a characterization might lead to the mapping of geomorphic bed features from patterns and turbulent scales extracted from the free surface by a digital camera.

6 CLUSTER GEOMORPHIC SETTING

In this study we have attempted to carefully document the geomorphic conditions of clusters and the river reach in which they were found. In doing so, we hope to add valuable information to a database who's end would lead to prediction of clusters in a reach based on known input parameters.

One hindrance in the evaluation of data on clusters present in the literature is the variety of reported cluster types and documented variables. The first question that needs to be asked in a comparison and compilation of the data from the literature is, "are the cluster features reported similar from study to study?" That is, are clusters defined and recorded in a similar fashion from study to study? Needing to answer a question such as this highlights the need for a more standardized definition and method for identification of clusters in the field and laboratory. Furthermore, similar variables need to be measured in each study in order to combined datasets and obtain larger amounts of information from which to draw conclusions about such things as where clusters might be located in a watershed.

All known published data concerning clusters, where clusters are features similar to those presented here in the geomorphic characterization, were combined in an attempt to discover the governing variables for cluster formation. The most common, and debatably the most important, variables measured from study to study, where clusters were documented, are channel slope, S, and the d_{84} of the surface material. Figure 13 presents the slope and d_{84} of streams where clusters were found for this study and all studies on clusters published in the literature.

Figure 13 shows that clusters have been found in a wide range of slopes and bed material sizes, ranging in slope from 0.004–0.06 m/m and in d_{84} from 50–500 mm. Figure 13 is gridded with the slope based morphology demarcations of Montgomery and Buffington (1997) and the Wentworth grain size groupings to aid in classification of the geomorphic conditions where clusters might form. This figure shows that the majority of studies have document clusters in cobbles size material within the slope range typical of riffle-pool to plane bed morphologies. However, significant spread in the data hints that a classification based on slope and d_{84} alone is not enough to predict

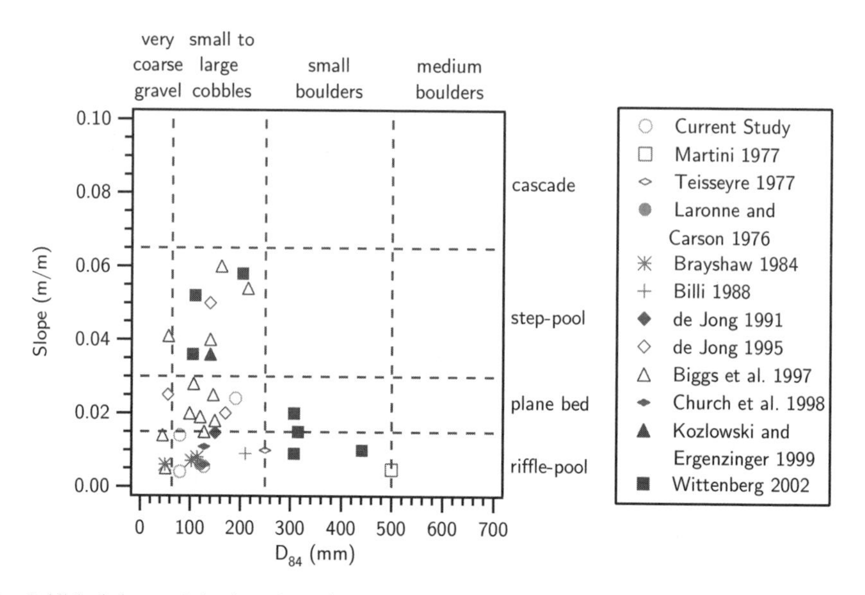

Figure 13. Published slope and d_{84} data where clusters have been recorded. Reach morphology demarcations are based on those of Montgomery and Buffington (1997).

the occurrence of clusters, assuming that all clusters recorded are similar features and not outcomes of different processes. The later comment is difficult to asses as each investigators can define what is and is not a cluster slightly differently. This points to a need for a more precise definition of clusters and methodological procedures for identification and classification in the field. Additionally, there is a need to examine additional parameters that might offer additional control to the formation of clusters.

7 CONCLUSIONS

The study has attempted to rigorously characterize clusters and the streams in which they are found through a field study on the American River. This was done following the more traditional geomorphological approach, and supplemented with a hydraulic characterization of the signature left in the flow by clusters. An overarching goal for the outcome of the study is to understand and predict the spatial location of cluster topography in a stream and to be able to quickly map the occurrence of clusters in a watershed based on a small number of measured or calculated variables.

Five main types of discrete cluster microforms were observed, those being the pebble, line, comet, heap, and ring type clusters. Although deciphering cluster shape is subjective, examining cluster geometric properties reveals that there are physical differences among cluster types. These physical differences might be related to a point at which a cluster is in it's evolutionary cycle. In addition, individual cluster types and their associated shapes are also likely to effect the way in which a particular cluster type will effect sediment transport and the near bed flow field.

From an assessment of the current study and previously reported data in the literature, it has been purposed that cluster topography is likely to develop from the larger size fractions of cobble size sediment in riffle-pool to plane bed morphologies. Further work needs to be conducted to determine a more accurate and precise means of predicting the occurrence of clusters. It is here hypothesized that since local controlling variables such as sediment availability, slope, and flow conditions are outcomes of larger watershed variables such as climate, topography, and geology, prediction of clusters might be improved by considering these larger scale variable.

If a particular cluster signature can be identified from the flow field, it might be possible to map the morphologic bed features from surface measurements of velocity. This would allow simultaneous mapping of velocity and bed morphologies from a single instrument, such as a PIV, and might lend itself to quick characterization of the bed and flow structure from measurements at the surface.

REFERENCES

Biggs, B. J. F., M. J. Duncan, S. N. Francoeur, and W. D. Meyer (1997). Physical characterisation of

microform bed cluster refugia in 12 headwater streams, New Zealand. *New Zealand journal of marine and freshwater research 31*, 413–422.
Billi, P. (1988). A note on cluster bedform behavior in a gravel-bed river. *Catena 15*, 473–481.
Brayshaw, A. C. (1984). Characteristics and origin of cluster bedforms in coarse-grained alluvial channels. In E. H. Koster and R. J. Steel (Eds.), *Sedimentology of Gravels and Conglomerates*, Number 10, pp. 77–85. Canadian Society of Petroleum Geologist.
Brayshaw, A. C., L. E. Frostick, and I. Reid (1983). The hydrodynamics of particle clusters and sediment entrainment in coarse alluvial channels. *Sedimentology 30*, 137–143.
Buffin-Bélanger, T. and A. G. Roy (1998). Effects of a pebble cluster on the turbulent structure of a depth-limited flow in a gravel-bed river. *Geomorphology 25*, 249–267.
Butler, J. B., S. N. Lane, and J. H. Chandler (2001). Automated extraction of grain-size data from gravel surfaces using digital image processing. *Journal of Hydraulic Research 39*(5), 519–529.
Church, M., M. A. Hassan, and J. F. Wolcott (1998). Stabilizing self-organized structures in gravel-bed stream channels: field and experimental observations. *Water Resources Research 34*(11), 3169–3179.
Clifford, N. J., K. S. Richards, and A. Robert (1992). The influence of microform bed roughness elements on flow and sediment transport in gravel bed rivers: comment on a paper by Marwan A. Hassan and Ian Reid. *Earth Surface Processes and Landforms 17*, 529–534.
de Jong, C. (1991). A reappraisal of the significance of obstacle clasts in cluster bedform dispersal. *Earth Surface Processes and Landforms 16*, 737–744.
de Jong, C. (1995). Temporal and spatial interactions between river bed roughness, geometry, bedload transport and flow hydraulics in mountain streams – examples from Squaw Creek, Montana (USA) and Schmiedlaine/Lainbach (Upper Germany). *Berliner Geographische Abhandlungen 59*, 229.
Hassan, M. A. and I. Reid (1990). The influence of microform bed roughness elements on flow and sediment transport in gravel bed rivers. *Earth Surface Processes and Landforms 15*, 739–750.
Kellerhals, R. and D. I. Bray (1971). Sampling procedures for coarse fluvial sediments. *Journal of the Hydraulics Division, Proceedings of the American Society of Civil Engineers 97*(HY 8), 1165–1180.
Komar, P. D. and Z. Li (1986). Pivoting analysis of the selective entrainment of sediments by shape and size with application to gravel threshold. *Sedimentology 33*, 425–436.
Kozlowski, B. and P. Ergenzinger (1999). Ring structures – a specific new cluster type in steep mountain torrents. *XXVII IAHR Congress Proceedings Graz*, 410.
Krogstad, P. A. and R. A. Antonia (1999). Surface roughness effects in turbulent boundary layers. *Experiments in Fluids 27*, 450–460.
Laronne, J. B. and M. A. Carson (1976). Interrelationships between bed morphology and bed material transport for a small, gravel-bed channel. *Sedimentology 23*, 67–85.
Martini, I. P. (1977). Gravelly flood deposits of Irvine Creek, Ontario, Canada. *Sedimentology 24*, 603–622.
Montgomery, D. R. and J. M. Buffington (1997). Channel-reach morphology in mountain drainage basins. *Geological Society of America Bulletin 109*(5), 596–611.
Nowell, A. R. M. and M. Church (1979). Turbulent flow in a depth-limited boundary layer. *Journal of Geophysical Research 84*(C8), 4816–4824.
Papanicolaou, A. N., P. Diplas, C. L. Dancey, and M. Balakrishnan (2001). Surface roughness effects in near-bed turbulence: Implications to sediment entrainment. *Journal of Engineering Mechanics 127*(3), 211–218.
Papanicolaou, A. N., K. Strom, A. Schuyler, and N. Talebbeydokhti (2003). The role of sediment specific gravity and availability on cluster evolution. *Earth Surface Processes and Landforms 28*, 69–86.
Reid, I., L. E. Frostick, and A. C. Brayshaw (1992). Microform roughness elements and the selective entrainment and entrapment of particles in gravel-bed rivers. In P. Billi, R. D. Hey, C. R. Thorne, and P. Tacconi (Eds.), *Dynamics of Gravel-bed Rivers*, pp. 253–275. John Wiley & Sons Ltd.
Shamloo, H., N. Rajaratnam, and C. Katopodis (2001). Hydraulics of simple habitat structures. *Journal of Hydraulic Research 39*(4), 351–366.
Strom, K. and A. N. Papanicolaou (2005). Case study: Bed stability around the east caisson of the Tacoma Narrows Bridge. *Journal of Hydraulic Engineering 131*(2), 75–84.
Strom, K., A. N. Papanicolaou, N. Evangelopoulos, and M. Odeh (2004). Microforms in gravel bed rivers: Formation, disintegration, and effects on bedload transport. *Journal of Hydraulic Engineering 130*(6), 554–567.
Strom, K. B., A. N. Papanicolaou, and G. S. Constantinescu (2005). Flow heterogeneity over a 3d cluster microform: a laboratory and numerical investigation. *in review Journal of Hydraulic Engineering*.
Teisseyre, A. K. (1977). Pebble clusters as a directional structure in fluvial gravels: modern and ancient examples. *Geologia Sudetica 12*(2), 79–89.
Wittenberg, L. (2002). Structural patterns in coarse gravel river beds: Typology, survey and assessment of the roles of grain size and river regime. *Geografiska Annaler 84 A*(1), 25–37.
Wolman, M. G. (1954). A method of sampling coarse riverbed material. *Transactions American Geophysical Union 35*(6), 951–956.
Zingg, T. (1935). Beitrage zur schotteranalyse. *Schweiz. Min. u. Pet. Mitt. 15*, 39–140.

River, Coastal and Estuarine Morphodynamics: RCEM 2005 – Parker & García (eds)
© 2006 Taylor & Francis Group, London, ISBN 0 415 39270 5

Flume experiments with tracer stones under bedload transport

M. Wong
University of Minnesota, Department of Civil Engineering, St. Anthony Falls Laboratory, Minneapolis, Minnesota, USA

G. Parker
University of Illinois at Urbana, Departments of Civil and Environmental Engineering and Geology, Urbana, Illinois, USA

ABSTRACT: A channel-averaged deterministic approach is commonly used to study bedload transport in gravel-bed streams. Here an alternative formulation based on tracking the vertical and streamwise displacement of individual bed particles, and mass continuity of the sediment in the bed and bedload expressed in probabilistic terms is presented. Flume physical modeling of bedload transport with tracer stones is used for the development of the new theory. All experiments presented here were conducted under normal flow equilibrium transport conditions. The results obtained have been used to derive predictors for the probability density functions of bed elevation fluctuations, particle entrainment into bedload transport, and streamwise displacement of particles from entrainment to deposition. The formulation links tracer conservation with conservation of gravel as a whole, and in addition establishes a connection between vertical and streamwise dispersion of tracers with the overall bedload transport.

1 INTRODUCTION

A common way to estimate the bedload transport rate in a mountain stream is via empirically derived relations, which are based on the mean characteristics of the driving force (i.e. of the flowing water) and the corresponding resistance properties of the bed material. One limitation of these channel-averaged models is that they do not contain the mechanics necessary to describe the displacement patterns of individual particles, hence they lack the option of explicitly linking changes in the composition and surface configuration of the bed deposit with the overall response of the river in terms of its channel morphometry. As such, the standard approach is somewhat deficient in formulating a process-oriented coupling of the governing equations for bed sediment continuity, bedload transport and flow momentum (Blom 2003, Cao & Carling 2002).

These channel-averaged models are based on mass conservation equations for the bed deposit and the bedload that are expressed in deterministic terms. However, it has been long and amply recognized that bedload transport, as well as particle entrainment and deposition, are intrinsically stochastic processes (Einstein 1950, Paintal 1971), mainly because of the turbulent characteristics of the flow velocity near the bed and the different degrees of exposure and support conditions of individual bed particles (Roy et al. 1996, Schmeeckle & Nelson 2003, Shvidchenko & Pender 2000).

A probabilistic formulation that connects the displacement history of individual bed particles with channel hydraulic parameters and bed resistance properties may also relate the main statistics of the vertical and streamwise dispersion of these particles to the bedload transport rate in the stream. Consequently, this formulation may provide an alternative way to evaluate the evolution of the morphology of a mountain stream. With this purpose, following the theory proposed by Parker et al. (2000), predictors for the probability density functions of bed elevation fluctuations (or effective thickness of the active layer), particle entrainment and deposition, and particle step length (or particle virtual velocity) need to be developed.

Tracer stones have been extensively used in the past two decades on field studies of bedload transport (Ferguson & Hoey 2002, Hassan & Church 2000, Wilcock 1997), most of them aiming to derive the aforementioned predictor functions. The general goal of these studies has been to find an alternative but reliable method to estimate the bedload transport rate, and the interpretation of the results has been centered around the effects of macro-scale bedforms and channel morphology as well as of the spatial and temporal scales of analysis on the measured dispersion of the tracers. This paper deals with a simpler case implemented in a laboratory-controlled setting, so a first understanding of the key concepts explaining the entrainment, transport and deposition of individual bed particles can be attempted without need to refer to the influence of, for instance, the bed sediment size distribution, the presence of alternate bars or the existence of pool-riffle sequences.

A basic review of the theoretical framework used here is introduced in the following section, focusing on the estimate of bedload transport rate rather than on the accounting for bed sediment continuity. Results of experiments with tracer stones for lower-regime plane-bed normal flow and equilibrium bedload transport conditions of uniform gravel material in a straight flume of constant width are presented subsequently. Preliminary predictor functions for bed elevation fluctuations, particle entrainment and particle virtual velocity are then proposed, culminating in a discussion of how motion patterns of individual particles can be related to the overall sediment transport response of the system.

2 BASIC THEORETICAL FRAMEWORK

One of the formulae most widely used in laboratory and field investigations as well as in numerical simulations of bedload transport is the empirical relation proposed by Meyer-Peter & Müller (1948). It allows estimation of the bedload transport rate in an open-channel, as a function of the bed shear stress applied by the flowing water. A power law relation of this type is used here in the following general form:

$$q^* = \alpha_b\left(\tau^* - \tau^*_{cr}\right)^{n_b} \tag{1a}$$

with,

$$q^* = \frac{q_b}{\sqrt{RgD_{50}}\,D_{50}} \tag{1b}$$

and

$$\tau^* = \frac{\tau_b}{\rho RgD_{50}} \tag{1c}$$

In the above relations q^* = dimensionless volume bedload transport rate per unit channel width (Einstein number); q_b = volume bedload transport rate per unit channel width; R = submerged specific gravity of bed sediment; g = acceleration of gravity; D_{50} = median particle size of bed sediment; τ^* = dimensionless boundary shear stress (Shields number); τ_b = boundary shear stress, after applying the Vanoni-Brooks sidewall correction (Vanoni 1975); ρ = density of water; α_b and n_b are fitting parameters; and τ^*_{cr} = critical Shields number.

Two other forms prove useful for estimating the bedload transport rate for this one-dimensional approximation of normal (uniform and steady) flow equilibrium transport of bed sediment under plane-bed conditions. The first one makes use of the entrainment formulation of sediment continuity as follows:

$$q^* = \alpha_1\hat{E}_b\hat{L}_s \tag{2a}$$

with,

$$\hat{E}_b = \frac{E_b}{\sqrt{RgD_{50}}} \tag{2b}$$

and

$$\hat{L}_s = \frac{L_s}{D_{50}} \tag{2c}$$

In the above relations E_b = volume rate of entrainment per unit bed area; L_s = particle step length, i.e. the particle travel distance from entrainment to deposition; the 'hatted' terms are the corresponding dimensionless parameters; and α_1 is a fitting coefficient.

The second one assumes that the bedload transport rate is the product of a spatially and temporally constant thickness of the bed deposit that contributes to the transport of sediment (similar to but not the same as the concept of the active layer from Hirano 1971), and of a virtual velocity of the bedload particles which accounts for both moving and resting times. The expression is the following:

$$q^* = \alpha_2\hat{L}_a\hat{U}_{mr} \tag{3a}$$

with,

$$\hat{L}_a = \frac{L_a}{D_{50}} \tag{3b}$$

and

$$\hat{U}_{rm} = \frac{U_{rm}}{\sqrt{RgD_{50}}} \tag{3c}$$

In the above relations L_a = standard deviation of bed elevation fluctuations; U_{rm} = particle virtual velocity, i.e. the average velocity of a stone including the time it spends at rest either on the bed surface or buried within the bed, in addition to the time it spends moving as bedload; the 'hatted' terms are the corresponding dimensionless parameters; and α_2 is a fitting coefficient, which also includes the effect of the porosity of the bed deposit.

The experiments with tracers to be presented in the next section allow the derivation of preliminary predictors for the 'hatted' terms representing E_b and U_{rm}. A description of the procedure used to process the experimental data for deriving these two parameters follows in the next paragraphs.

Given that the analysis herein is for cases of equilibrium bedload transport under uniform and steady flow conditions, the entrainment formulation of mass continuity can be applied to any region of the bed deposit. If such region corresponds to the one occupied by tracer stones, then:

$$z_{tr}\frac{df_{tr}}{dt} = D_{tr} - E_{tr} \tag{4a}$$

with,

$$D_{tr} = 0 \tag{4b}$$

and

$$E_{tr} = E_b f_{tr} \tag{4c}$$

So,

$$z_{tr}\frac{df_{tr}}{dt} = -E_b f_{tr} \tag{4d}$$

In the above relations z_{tr} = thickness of the region occupied by tracers; f_{tr} = fraction of tracers in control volume at time t; D_{tr} = volume deposition rate of tracers per unit bed area; and E_{tr} = volume entrainment rate of tracers per unit bed area. The implicit assumption in this mass balance is that z_{tr} is large enough to cover the depth of the bed deposit effectively subject to entrainment and deposition. Moreover, if exported tracers stones are not replaced by identical tracer stones in their seeding positions, D_{tr} vanishes. Integrating Equation (4d) from time t_1 to t_2, the expression for E_b becomes:

$$E_b = -\frac{z_{tr}}{(t_2 - t_1)}\ln\left[\frac{f_{tr}(t_2)}{f_{tr}(t_1)}\right] \tag{4e}$$

Again making use of the fact that the experiments are for equilibrium bedload transport under uniform and steady flow conditions, it is reasonable to consider that the virtual velocity of the tracer stones is independent of the duration of the test. Then:

$$U_{rm} = \frac{L_{tr}(t_2) - L_{tr}(t_1)}{(t_2 - t_1)} \tag{5}$$

where L_{tr} = total mean travel distance of tracers for time (run duration) t; and $t_2 > t_1$. An issue to highlight here with respect to the computation of U_{rm}, and also thus for E_b, is that the above comparison between travel distances avoids working with $t_1 = 0$, i.e. the beginning of the experiment. The reason for this is to reduce the influence of the initial surge of water when the experiment starts.

The standard deviation of bed elevation fluctuations is here computed from sonar measurements carried out for each experimental equilibrium state. It will be assumed, as demonstrated in the next sections, that measurements at different streamwise locations can be aggregated into one single dataset for each experimental equilibrium state; stationarity in the second statistical moment of the bed elevation can be reasonably assumed for plane-bed equilibrium conditions.

3 EXPERIMENTAL LAYOUT AND PROCEDURES

Physical modeling of bedload transport in mountain streams with tracer stones was conducted in one flume facility located at St. Anthony Falls Laboratory (SAFL), University of Minnesota. The general arrangement of the experimental facility included: (i) 0.5-m wide, 22.5-m long flume; (ii) regulated water supply, for a maximum discharge of 0.120 m^3/s; (iii) two point gauges, to measure longitudinal bed profiles; (iv) pressure tubes every 0.5 m in the streamwise direction, to measure water surface levels; (v) controlled sediment supply, for a maximum feed rate of 0.150 kg/s; (vi) automated recirculation of the sediment collected at the downstream end of the flume to a buffer box installed above the sediment feeder, so the system operates as sediment-feed not sediment-recirculating; and (vii) six sonar transducer probes at fixed streamwise positions, to simultaneously measure bed elevation fluctuations at high temporal resolution.

The sediment used in all experiments was uniform gravel, with geometric mean size $D_g = 7.2$ mm, geometric standard deviation $\sigma_g = 1.2$, median particle size $D_{50} = 7.1$ mm, particle size for which 90% of the sediment is finer $D_{90} = 9.6$ mm, and specific gravity $G_s = 2.55$ ($R = 1.55$). Tracer stones were used to study the vertical and streamwise dispersion of individual bed particles. These tracers were marked by painting gravel particles with spray paint, indelible colored pens or a combination of both. Sixteen groups of tracers were formed, each of a different color. About 200 tracers of each group were used on each experiment, for a total of approximately 3200 tracers.

The sequence followed in all the experimental runs reported in this paper included four main steps. The first one involved running the experiment a time span sufficiently long to reach lower-regime plane-bed normal flow equilibrium transport conditions. For this purpose the water discharge and sediment feed rate were set at constant values. It took between 30 and

Table 1. Location of ultrasonic transducer probes.

Probe No.	Streamwise position m*
1	6.75-R**
2	6.75-L**
3	7.75
4	8.75
5	12.75
6	16.75

* Distance from entrance weir at upstream end of flume measuring region.
** R = 0.08 m to the right (looking downstream) of the center of the channel; L = 0.08 m to the left; the others at the center.

Table 2. Location of tracer stones.

Spot	Streamwise position m*	Vertical location (color)***
1	6.75-R**	Orange, blue, yellow-purple, yellow normal
2	6.75-L**	Yellow, pink-purple, red, yellow-blue
3	7.75	Green, yellow-orange, red-black, yellow-red
4	8.75	Pink, green-purple, orange-black, yellow-brown

* Distance from entrance weir at upstream end of flume measuring region.
** R = 0.08 m to the right (looking downstream) of the center of the channel; L = 0.08 m to the left; the others at the center.
*** Colors are listed from first to fourth (top to bottom) layers.

50 hours of run to reach equilibrium conditions, which were denoted by streamwise water surface and bed slopes that were constant in time and space. During this preparation time and as complementary measures to validate the equilibrium, water levels were frequently recorded, the top elevation of the gravel at the buffer box of the sediment feeder was continuously monitored, and three long profiles of the bed were obtained after every half-day of run. The latter measurements also served to check that no bedforms developed. Once equilibrium had been reached, simultaneous measurements of bed elevation fluctuations were recorded for the six streamwise positions indicated in Table 1.

All ultrasonic transducer probes were of videoscan immersion type, with a nominal element size of 1 in (25.4 mm). The first four probes worked at a frequency of 0.5 MHz, while the other two worked at 1.0 MHz. The values of the basic parameters characterizing the 0.5 MHz-probes and 1.0 MHz-probes, respectively, are: (i) the near field distance is 55 mm and 109 mm; (ii) the beam angle is 6.9° and 3.4°; and (iii) the standard deviation from measurements in standing water and a fixed surface bottom is 0.16 mm and 0.04 mm. The distance between the bottom of the probes and the mean surface bed elevation was not constant neither for all probes nor for all experimental runs; the main setting concern was to avoid air bubbles being entrained below the ultrasonic transducer probes. Minimum, average and maximum beam spread (footprint) values were: (i) 4.6 mm, 7.2 mm and 10.1 mm for the 0.5 MHz-probes; and (ii) 2.4 mm, 3.8 mm and 4.9 mm for the 1.0 MHz-probes. All sonar measurements were carried out at a frequency of once every 3 seconds, and produced two independent sets of 1-hr duration time series for each of the streamwise positions and experimental equilibrium states.

The second step consisted of excavating the bed deposit at four different spots and carefully replacing the extracted sediment with evenly placed layers of tracers that were identical to the non-tracer sediment except for color. Four layers of tracers were installed at each spot, each containing stones of a different color. The first layer was exposed on the bed surface and parallel to the surface of the ambient sediment. The second, third and fourth layers were buried in order below it. The thickness of each layer so installed was between D_{50} and D_{90} of the sediment. Thus the color of sediment in each layer became a proxy for initial depth of burial (or initial vertical location) and streamwise position in the bed (see Table 2). Each excavation spot had a surface area of 0.10 × 0.10 m, and a total depth of approximately 3.0–3.5 cm, i.e. about $4D_{50}$ to $4D_{90}$.

The third step demanded re-running the experiment under the same combination of water discharge and sediment feed rate for which equilibrium transport had been achieved at the end of the first step, but now with the tracers seeded in the bed deposit and a run duration set in advance. In order to avoid an 'artificial' over-entrainment of the tracers, the flume tank was initially filled with water at a very slow pace. Water discharge was then gradually increased up to the experimental design value, and the sediment feeder was turned on. During the run, tracers displaced all the way out of the system were picked out by hand at the buffer box above the sediment feeder, so that no particle made more than one trip through the flume. Seven different run durations were completed for each equilibrium state, as indicated in Table 3. A preparation run without tracers, following a similar procedure to the one described in the first step, was performed in between each of the runs with tracers.

The fourth step comprised the identification of the final location of the tracers after completion of each experimental run. Overall four groups of final locations were considered: (HD1) tracers did not move from their seeding position; (HD2) tracers were entrained into bedload transport and were displaced out of the system; (HD3) tracers were entrained into

Table 3. Duration of experimental runs with tracers for each equilibrium state.

Type	Duration min.	Information recorded
Long	120	Vertical entrainment and streamwise displacement
Long	60	Vertical entrainment and streamwise displacement
Long	30	Vertical entrainment and streamwise displacement
Long	15	Vertical entrainment and streamwise displacement
Short	2	Vertical entrainment
Short	1	Vertical entrainment
Short	Variable	Vertical entrainment and streamwise displacement

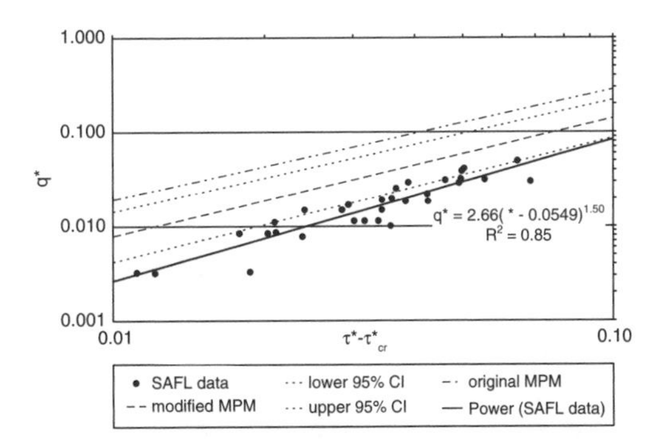

Figure 1. Bedload transport relation derived from experiments at SAFL. MPM = Meyer-Peter & Müller (1948); CI = confidence interval.

bedload transport, were displaced to a flume location upstream of the sediment trap and were found at the bed surface; and (HD4) tracers were entrained into bedload transport, were displaced to a flume location upstream of the sediment trap and were found buried in the bed deposit. The identification included counting the number of tracers of every color that corresponded to each of the four classes listed above, and recording the streamwise position for groups (HD3) and (HD4). Tracer recovery always exceeded 99% of the amount seeded, with only less than 0.25% tracers lost for most of the experimental runs.

4 EXPERIMENTAL RESULTS AND PREDICTOR FUNCTIONS

Experiments for ten different equilibrium states have been completed; all seven runs of Table 3 were performed for each state. Channel-averaged parameters for these tests corresponded to constant input values of water discharge ranging between 0.034 and 0.102 m^3/s, and sediment feed rate between 0.026 and 0.150 kg/s. Associated equilibrium states resulted in streamwise bed slopes ranging between 0.0081 and 0.0152, with Shields numbers between 0.0815 and 0.1370.

These results, in combination with an additional set of twenty-one experiments conducted in a different flume facility (of the same width but of shorter length than that of the flume used for the runs with tracers reported here), allowed determining the best fit for Equation (1a) as follows:

$$q^* = 2.66(\tau^* - 0.0549)^{1.50} \tag{6}$$

As seen in Figure 1, Equation (6) predicts a transport rate that is lower than that of Meyer-Peter & Müller (1948), and also slightly smaller than the lower 95%-confidence interval results of Wong (2003). The main difference in this case is that the critical Shields number in Equation (6) is greater than that considered in both references cited above; however, it is very close to the value of 0.0550 computed using the relation proposed by Yalin & Scheuerlein (1988).

The results of the experiments with tracers for each of the ten equilibrium states, differentiated in terms of the initial vertical position of the tracers seeded for a sample of two of the seven tests listed in Table 3, are presented in Figure 2.

It is evident from these graphs that the fraction of tracers moved per layer increases with excess shear stress, as well as with the run duration. Moreover, the effective depth of tracers subject to entrainment increases with both excess shear stress and run duration. Note, for instance, that when the dimensionless sidewall-corrected shear stress is twice the critical Shields number, the fraction of tracers moved can be as large as 0.59 and 0.24 for the third and fourth (bottom) layers, respectively. Although not worked out in this paper, these results open the interesting possibility of deriving depth-specific entrainment functions.

Figure 3 shows the fraction of tracers moved (summed over all four layers at the four excavation sites) as a function of: (i) excess Shields number (horizontal axis); and (ii) run duration (legend). This figure includes all ten experimental conditions and six out of the seven run durations (Table 3).

These results have been used to derive a power law relation between the dimensionless entrainment rate (from Equations (4e) and (2b)) and excess Shields number. A total of fifteen different estimates of the volume entrainment rate per unit bed area, E_b have been obtained using the allowable different combinations of t_1 ($t_1 > 0$) and t_2 ($t_2 > t_1$). Figure 4 presents the results of the average and maximum of these estimates for

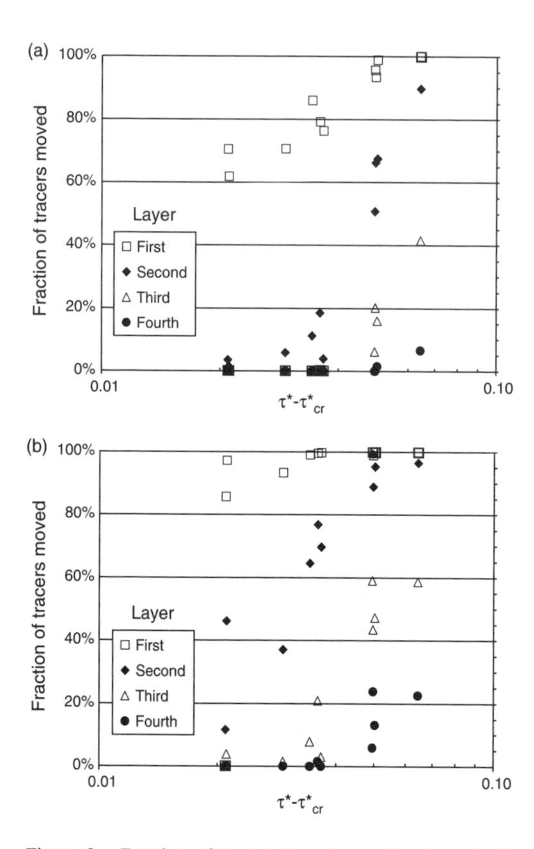

Figure 2. Fraction of tracers moved as a function of: (i) excess Shields number (horizontal axis), and (ii) initial depth of burial (see legend: the first layer was flush with the bed surface and the fourth layer was buried the deepest). Figure 2(a) shows results for 15-min duration flows; Figure 2(b) shows results for 120-min duration flows.

each experimentally measured value of excess Shields number.

A power law predictor based on the average of the estimated values for the dimensionless E_b resulted in:

$$\hat{E}_b = 0.06\left(\tau^* - 0.0549\right)^{1.97} \tag{7}$$

The maximum measured values of the dimensionless E_b provide estimates that are about half-an-order of magnitude larger than the ones obtained with Equation (7), but show the same slope of increase as a function of excess dimensionless shear stress. The average estimate was selected for deriving a predictor function for the dimensionless E_b, as it is less biased to potential over- or under-entrainment rates in runs of specific duration. In addition, it compares well with the relation proposed by Fernandez-Luque & van Beek (1976), which is also included in Figure 4.

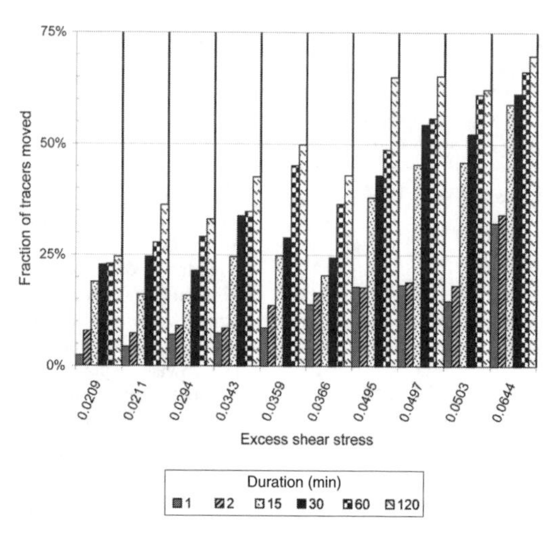

Figure 3. Fraction of tracers moved as a function of excess shear stress and run duration.

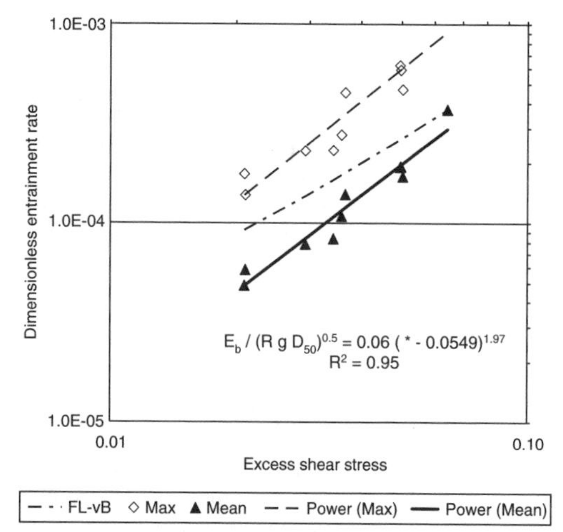

Figure 4. Measured and estimated values of dimensionless volume entrainment rate as a function of dimensionless excess shear stress. FL-vB = Fernandez-Luque & van Beek (1976); Max = maximum value estimated for dimensionless E_b; Mean = average value estimated for dimensionless E_b.

One important thing to clarify at this point is that the dimensionless E_b already accounts for any mobility factor corresponding to the way Wilcock (1997) defines partial transport. In other words, the dimensionless E_b already quantifies the fraction of the bed area per unit time that is in transport. Moreover, the method employed for estimating it precludes the potential mistake of confusing actual particle entrainment with net entrainment (i.e. entrainment minus deposition).

An even more promising finding derived from the formulation of the predictor for the dimensionless E_b

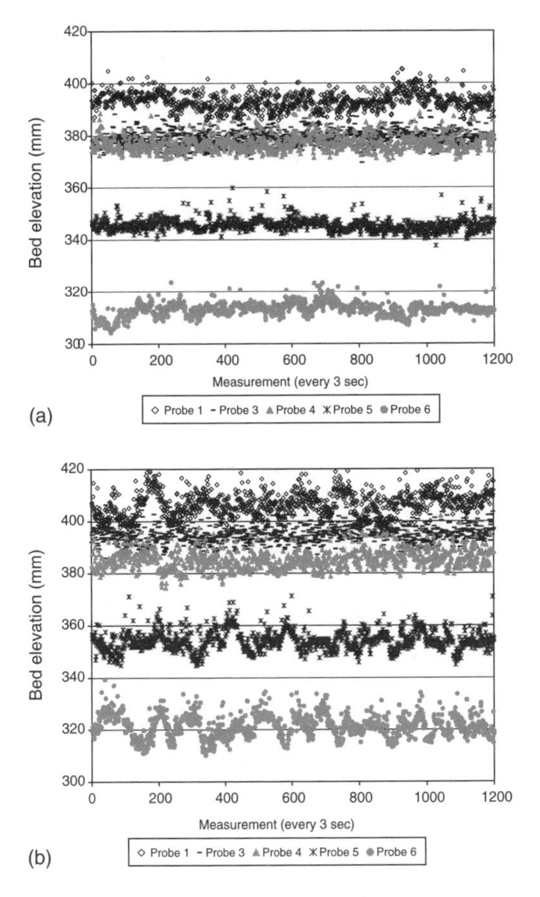

Figure 5. Time series of bed elevation measurements with ultrasonic transducer probes. (a) $\tau^* - \tau^*_{cr} = 0.0294$; (b) $\tau^* - \tau^*_{cr} = 0.0495$. Probe numbers correspond to the streamwise locations listed in Table 1.

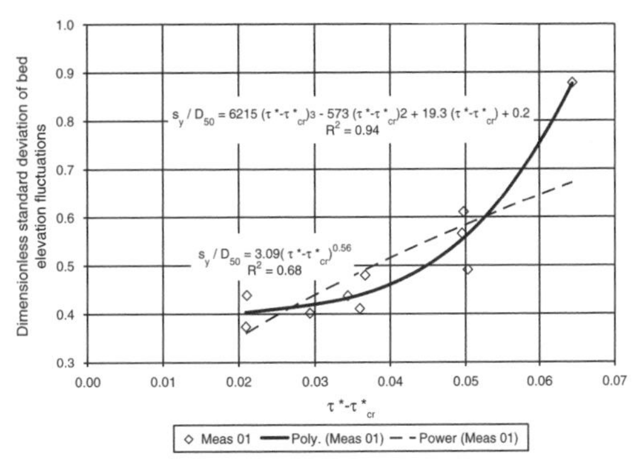

Figure 6. Measured and estimated values of dimensionless L_a as a function of excess dimensionless shear stress.

is given by the combination of Equations (1a), (2a), (6) and (7), which leads to the relation:

$$\hat{L}_s \cong 44.33\left(\tau^* - 0.0549\right)^{-0.47} \tag{8}$$

Equation (8) is contrary to what has been considered as true since the work by Einstein (1950); the step length L_s does not only depend on particle size diameter but also on excess shear stress. And what appears to be counterintuitive with respect to the latter relation is that the dimensionless L_s decreases as the excess Shields number increases; the dimensionless L_s varies between 160 and 270 for the range of experimental equilibrium states presented in this paper.

This result however, is not wrong. Indeed, it has an important physical meaning: not only the entrainment rate increases but also the deposition rate does when the bedload transport rate increases, in order to keep the equilibrium configuration. A more extended explanation supporting this finding follows after the derivation of the predictors for bed elevation fluctuations and particle virtual velocity. It can be said here that an increase in bedload transport rate is mainly due to an increase in the entrainment rate, or for that matter, as a result of a greater fraction of the bed surface area being in transport per unit time.

Two sample datasets of time series of bed elevation obtained from the sonar measurements are presented in Figure 5. Time series of bed elevation are plotted for Probes 1, 3, 4, 5 and 6 (Probe 2 was excluded due to possible errors). The positions of these probes are given in Table 1; the probe numbers are ordered such that a higher number corresponds to a location farther downstream.

The sample dataset presented in Figure 5(a) is for a dimensionless shear stress that is about 1.5 times the critical Shields number; whereas the sample dataset in Figure 5(b) is for a dimensionless shear stress approximately two times the critical value of 0.0549. It is evident in Figure 5 that the intensity of the bed elevation fluctuations increases with excess dimensionless shear stress.

Figure 6 shows the dimensionless standard deviation of bed elevation fluctuations (dimensionless L_a, which is defined by Equation (3b)) as a function of excess Shields number. Although dimensionless L_a increases with increasing excess dimensionless shear stress, the relation evidently does not follow a power law. Thus the data were fitted using a third-order polynomial, as shown in Figure 6.

Insofar as this predictor is still very preliminary because an improved data filtering is currently underway, a power law in terms of excess shear stress will be proposed here as a loose approximation:

$$\hat{L}_a = 3.09\left(\tau^* - 0.0549\right)^{0.56} \tag{9}$$

This 'forced' fitting as a power law allows a simple derivation of a predictor for the particle virtual velocity

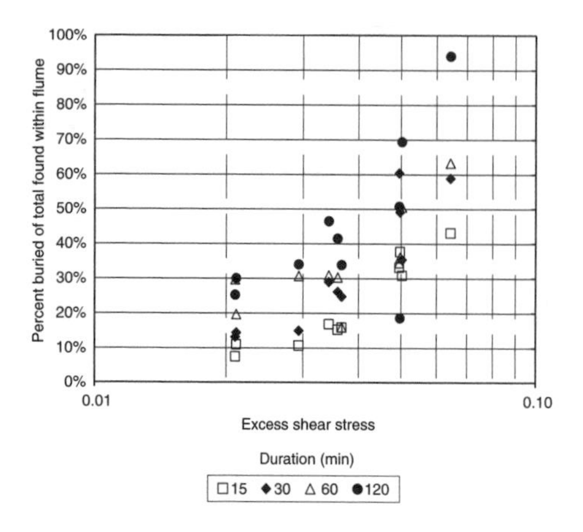

Figure 7. Measured percent of tracer stones displaced that were found buried in the bed deposit (HD3) with respect to total number of tracer stones displaced that were found within the flume limits (HD3 + HD4).

from the combination of Equations (1a), (3a), (6) and (9). The result is:

$$\hat{U}_{rm} \cong 0.86(\tau^* - 0.0549)^{0.94} \tag{10}$$

Equation (10) shows that particle virtual velocity increases with excess Shields number. More importantly, the combination of Equations 1a, 3a, 9 and 10 show that an increase in bedload transport rate, q_b is translated into an increase in particle virtual velocity, U_{rm} at a faster rate than the increase in the thickness of the bed deposit that actively participates in the transport of sediment, L_a. This is likely because as the bedload transport rate increases, the time a particle remains in a resting position, either at the bed surface or buried in the bed deposit, becomes shorter. This is partly explained by the increase in the intensity of the bed elevation fluctuations, which in turn is accompanied by an increase in both particle entrainment and deposition rates (which equal each other for the equilibrium flows discussed here).

The interpretation of the experimental results given in the paragraph above is in agreement with previous discussion about the entrainment rate and step length. Because the chance that a moving particle will be captured and deposited into a resting position increases with transport rate (see Figure 7), it is reasonable to assume that the step length becomes smaller when the excess dimensionless shear stress gets larger, as indicated by Equation (8).

5 CONCLUSIONS

The results of seventy experimental runs with tracer stones for conditions of lower-regime plane-bed normal flow and bedload equilibrium transport have been presented in this paper. These experiments with tracers, in combination with ultrasonic measurements of bed elevation at a high temporal resolution, have allowed the derivation of preliminary predictor functions for bed elevation fluctuations, active layer thickness, particle entrainment rates, particle step lengths and particle virtual velocities. All of these functions except step length show a positive correlation (power laws) with excess dimensionless shear stress.

ACKNOWLEDGMENTS

This work was supported by the National Science Foundation via Agreement Number EAR-0207274. Additional support was derived from the STC program of the National Science Foundation via the National Center for Earth-surface Dynamics under Agreement Number EAR-0120914. This paper represents a contribution of the research of the National Center for Earth-surface Dynamics in the area of channel dynamics. Special thanks go to Andrew Fyten and Danielle Trice, undergraduate students from University of Minnesota, who actively participated in the design and execution of the runs with tracers.

REFERENCES

Blom, A. 2003. *A vertical sorting model for rivers with non-uniform sediment and dunes*. Twente: University of Twente (Ph.D. thesis).

Cao, Z. & Carling, P. 2002. Mathematical modeling of alluvial rivers: reality and myth. Part I: General review. *Water and Maritime Engineering* 154(3), 207–219.

Einstein, H.A. 1950. *The bed-load function for sediment transportation in open channel flows*. Washington, D.C.: U.S. Department of Agriculture, Soil Conservation Service (Technical Bulletin No. 1026).

Ferguson, R.I. & Hoey, T.B. 2002. Long-term slowdown of river tracer pebbles: Generic models and implications for interpreting short-term tracer studies. *Water Resources Research* 38(8): 10.1029/2001WR000637.

Fernandez Luque, R. & van Beek, R. 1976. Erosion and transport of bed-load sediment. *Journal of Hydraulic Research* 2, 127–144.

Hassan, M.A. & Church, M. 2000. Experiments on surface structure and partial sediment transport on a gravel bed. *Water Resources Research* 36(7), 1885–1895.

Hirano, M. 1971. River bed degradation with armouring. *Transactions Japan Society of Civil Engineering* 195, 55–65.

Meyer-Peter, E. & Müller, R. 1948. Formulas for bed-load transport. In International Association of Hydraulic Engineering and Research (eds), *Proceedings of the 2nd. Meeting IAHR*: 39–64. Stockholm: IAHR.

Paintal, A.S. 1971. A stochastic model of bed load transport. *Journal of Hydraulic Research* 9(4), 527–554.

Parker, G., Paola, C. & Leclair, S. 2000. Probabilistic Exner sediment continuity equation for mixtures with no active layer. *Journal of Hydraulic Engineering* 126(11), 818–826.

Roy, A.G., Buffin-Bélanger, T. & Deland, S. 1996. Scales of turbulent coherent flow structures in a gravel-bed river. In P.J. Ashworth, S.J. Bennett, J.L. Best & S.J. McLelland (eds), *Coherent flow structures in open channels*: 147–164. New York: John Wiley & Sons Ltd.

Schmeeckle, M.W. & Nelson, J.M. 2003. Direct numerical simulation of bedload transport using a local, dynamic boundary condition. *Sedimentology* 50(2), 279–301.

Shvidchenko, A.B. & Pender, G. 2000. Initial motion of streambeds composed of coarse uniform sediments. *Water and Maritime Engineering* 142(4), 217–227.

Vanoni, V.A. 1975. *Sedimentation Engineering*. New York: ASCE (ASCE Manuals and Reports on Engineering Practice No. 54).

Wilcock, P.R. 1997. Entrainment, displacement and transport of tracer gravels. *Earth Surface Processes and Landforms* 22: 1125–1138.

Wong, M. 2003. Does the bedload equation of Meyer-Peter and Müller fit its own data? In International Association of Hydraulic Engineering and Research (eds), *Proceedings of the XXX Congress IAHR – JFK Student Paper Competition*: 73–80. Thessaloniki: IAHR.

Yalin, M.S. & Scheuerlein, H. 1988. *Friction factors in alluvial rivers*. Munich: Institut für Wasserbau und Wassermengenwirtschaft und Versuchsanstalt für Wasserbau Oskar v. Miller in Obernach (59).

Deltas, estuaries, bays

River, Coastal and Estuarine Morphodynamics: RCEM 2005 – Parker & García (eds)

The morphodynamics of deltas and their distributary channels

James P.M. Syvitski
Environmental Computation and Imaging Facility, INSTAAR, University of Colorado at Boulder, Boulder CO, USA

ABSTRACT: A consistent database was established to characterize delta morphology, and includes: (1) location, basin morphology, fluvial and sediment discharge, delta morphology, ocean energy, and shelf depth reached by the sub-aqueous delta. Fifty-five deltas were selected to cover the global parameter range, constrained by seasonal satellite image availability. Delta area is best predicted from average discharge and total sediment load feeding the delta. The gradient of a delta plain, measured from the delta apex to the coast along the main channel, is best predicted with a ratio of sediment supply to sediment retention, sediment concentration, and mean water discharge. The number of distributary channels is best predicted by maximum monthly discharge and delta gradient. Widths of distributary channels form a lognormal distribution, with total width a function of water and sediment discharge; mean the tidal range and the maximum monthly-wave height additionally influence channel width.

1 INTRODUCTION

The architecture of deltaic systems is controlled by the interaction between boundary conditions and forcing factors (Coleman and Wright, 1975; Orton and Reading 1993; Overeem et al., in press): (1) sediment supply of bedload and suspended load: reflecting drainage characteristics, water discharge and sediment yield; (2) accommodation space: reflecting sea-level fluctuations, offshore bathymetry, tectonics, subsidence, compaction, and isostasy; (3) coastal energy: waves and tides, longshore and cross-shelf transport; and (4) density differences between effluent and receiving waters critical in defining the dynamics of plumes. A plethora of classification schemes have tried to sort out the morphology of deltas through an understanding of transport dynamics, i.e. through morphodynamics. A major problem associated with these schemes has been the difficulty in surveying a global spectrum of deltas in a consistent way. Often delta morphology has been rendered as cartoons of surface patterns, with a generalized discussion of environmental factors described in a qualitative manner (Fig. 1). Some of the newer classifications add complexity by adding new dimensions for grain size (Orton and Reading, 1993), or changes in sea level (Dalrymple et al., 1992; Postma, 1995), or depth of accommodating space (Postma, 1990). Researchers now recognize that forcing factors vary in time and space, and with these changes will be a delta's placement on the classic wave-tide-river dominated triangle (Fig. 1): e.g. a delta may begin its life as a tidal-dominated system but later evolve to a wave-dominated system as it progrades onto the open, continental shelf. Various distributary outlets of a delta may also be differentially influenced by river discharge, wave action and tides (Restrepo et al., 2002; Bhattacharya and Giosan, 2003).

To address the cartoon-like approach to deltaic morphodynamic shown in Figure 1, Syvitski et al. (in press) began to establish a consistent database of key environmental factors associated with representative deltas. The database included the following factors: (1) drainage area (A) feeding the delta; (2) maximum relief (R) of the drainage basin; (3) water discharge (Q_{av}) across the delta; (4) sediment yield (Y) within

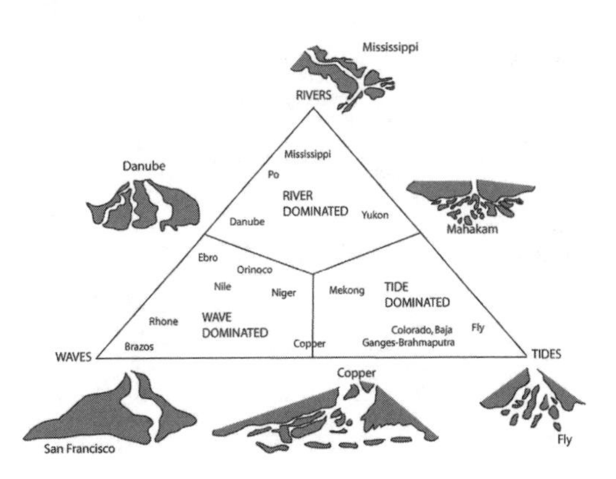

Figure 1. Typical cartoon used in textbooks to explain the morphodynamics of the earth's various deltas (after Galloway, 1975).

the drainage basin; (5) length (L) of the main stem of the river; (6) gradient of the delta plain (D_{grd}); (7) number of distributary channels (C_N); (8) spring ($K_1 + M_2$) tidal range (T_i); (9) maximum monthly wave height (W_a); and (10) miscellaneous factors such as permafrost (presence/absence) or bedrock control (significant presence in the vicinity of the delta). With this study the database has been refined further, along with new data added: (11) basin climate (temperature (T), precipitation (P), runoff coefficient (Q_{av}/PA), (12) monthly maximum fluvial discharge (Q_{mx}) onto the delta along with its variability, (13) sediment discharge to the delta (suspended sediment concentration (C_s) and load (Q_s), average (Q_b) and maximum monthly (Q_{bmx}) bedload), (14) delta area (A_D), (15) width of distributary channels (C_w), (16) bank-full width (R_w) of river entering the delta, and (17) shelf depth (D_{sh}) reached by the sub-aqueous delta. Fifty-five deltas were selected to cover the parameter range: e.g. discharge (9 m^3/s to 2×10^5 m^3/s), sediment yield (5 $T/km^2/yr$ to 4.2×10^3 $T/km^2/yr$), wave heights (0 m to 4 m), tidal range (0 m to 6 m), polar to tropical, and covering the major oceans and coastal seas.

2 METHODS

Seasonal satellite images of each of the 55 deltas were processed at INSTAAR (e.g. Fig. 2) from primary LANDSAT7 data (15 to 30 m pixel resolution or mpr) obtained from the Global Land Cover Facility of the University of Maryland, and MODIS data (250 mpr) obtained from GFSC/NASA.

For resolving features on smaller deltas, aerial photographs (1 mpr), IKONOS imagery (1 to 4 mpr), and SPOT imagery (10 mpr) were examined. More than ten temporal images from space of each deltaic system were examined to identify bankfull conditions prior to flooding. The number and size of distributary channels with connections to the fluvial river were obtained, along with other geomorphic information. Channel information on the Nile, Ebro, Colorado (CA/Mexico), and Colorado (TX) were obtained from published maps prior to reservoir construction.

Long-term monthly discharge data was obtained from the WMO Global Runoff Data Center (GRDC) (http://grdc.bafg.de/servlet/is/857/), or from the Composite GRDC-Water Balance Model (Syvitski et al., 2005), or from various government archives, or from HydroTrend runs (Syvitski and Kettner, in review) to determine the pre-dam discharge of the small Apennine rivers. Care was taken to obtain information pertaining to pre-dam discharge values so that maximum monthly discharge values (Q_{mx}) would be meaningful, as would be the discharge variability (Q_{mx}/Q_{av}) for each river (Table 1). Other morphodynamic databases (e.g. Oxford Global Sediment Flux Database) have not always made that distinction (see Fig. 3).

Suspended sediment concentration and load data came from various long-term national surveys (Syvitski et al., 2005). Where possible, early values (1960's or earlier) were used as these were considered less influenced by human activities (for discussion see Syvitski et al., 2005).

Bedload Q_b (kg/s) is calculated using a modified Bagnold (1966) equation:

$$Q_b = \frac{\rho_s}{\rho_s - \rho} \frac{\rho g Q_{av}{}^{\beta} D_{grd} e_b}{\tan \phi} \quad \text{when } u \geq u_{cr} \qquad (1)$$

where ρ_s and ρ = density of sediment and water (kg/m^3), respectively; g = acceleration due to gravity (m/s^2); D_{grd} = delta gradient (m/m), measured from the apex of the delta to the coast along the

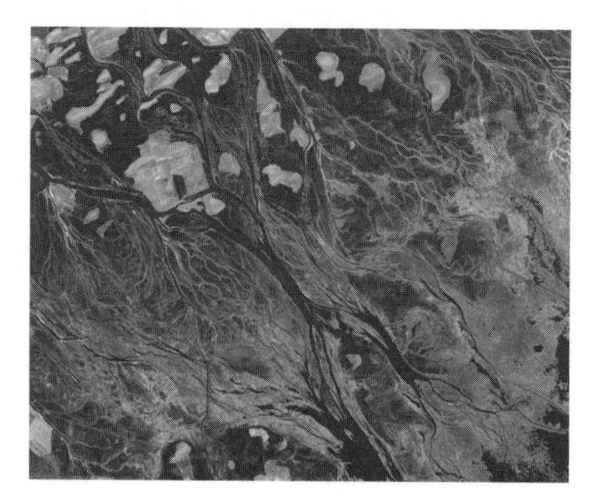

Figure 2. Volga Delta with no tides and low waves leads to feathering of distributary channels: $C_N = 100$, $TC_w{:}R_w = 1.7$, $Q_{mx}/Q_{av} = 2.7$.

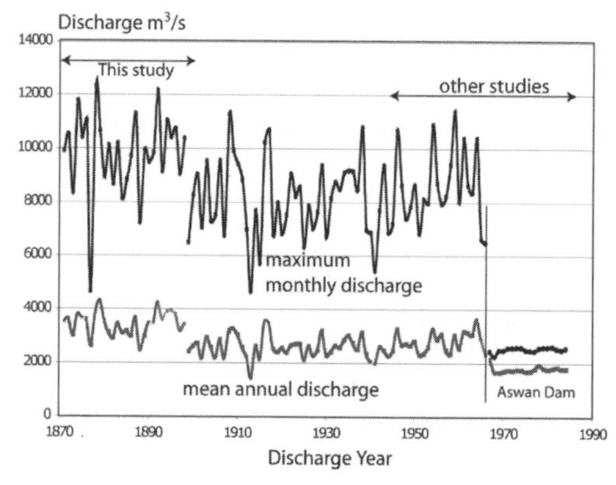

Figure 3. Mean annual and maximum monthly discharge for the Nile River. This study uses early pre-dam values.

Table 1. Parameters (11 of 40) determined for 50 of 55 global deltas, sorted by increasing Q_{mx} a proxy for bankfull discharge. P_r is a proxy of river power ($Q_{mx}D_{grd}$); P_m is a proxy of marine power ($W_a^2 + T_i^2$). Q_{s+b}:P_m a proxy for the ratio of sediment supply to marine dispersal. For the Colorado (TX) and Nile, the first channel number (CN) is for the period after human interference.

River	Q_{av} $10^3 m^3/s$	Q_{mx} m^3/s	Q_b %	Q_{s+b} T/s	D_{grd} m/km	P_r	C_N –	W_a m	T_i m	TC_w: R_w	Q_{s+b}:P_m	P_m: P_r
Var (Fr)	53	250	18.2	0.06	0.81	0.20	1	2.0	0.5	1.0	0.01	21
Arno (Ita)	103	250	3.9	0.07	0.11	0.03	1	2.0	0.7	1.0	0.02	163
Waipaoa (NZ)	41	275	13.8	0.34	4.66	1.28	1	3.0	1.2	1.0	0.03	8.1
Colorado (TX) Pre-dam	142	326	13.6	0.07	0.27	0.09	1–8	1.5	0.7	1.0	0.03	31
Ceyan (Tur)	222	420	4.0	0.18	0.13	0.06	3	1.5	0.7	4.5	0.07	49
Brazos (TX) Pre-dam	227	472	3.1	0.69	0.38	0.18	2	1.5	0.7	3.3	0.25	15
Squamish (Can)	250	550	11.8	0.06	0.12	0.07	2	1.0	3.0	2.4	0.01	148
Orange (SA) Pre-dam	289	560	2.4	2.92	1.00	0.56	1	2.5	1.5	1.0	0.34	15
Homathko (Can)	253	700	9.2	0.15	0.22	0.15	2	1.0	3.6	4.0	0.01	91
Eel (CA)	235	750	9.4	0.62	1.00	0.75	1	3.5	3.0	1.0	0.03	28
Klinaklini (Can)	330	800	14.6	0.19	0.33	0.27	5	1.0	4.0	4.7	0.01	64
Klamath (OR)	490	980	17.1	0.09	0.13	0.13	1	3.5	3.0	1.0	0.00	167
Limpopo (Moz)	169	1425	2.1	1.07	0.54	0.77	1	2.5	4.0	1.0	0.05	29
Chao Phrya (Thai)	963	1800	1.7	0.35	0.03	0.05	3	1.5	2.0	14.0	0.06	139
Vistula (Pol)	1050	1910	6.7	0.08	0.02	0.04	3	2.0	1.0	3.4	0.02	119
Rhone (Fr) Pre-dam	1700	2050	2.2	2.03	0.11	0.22	2	2.0	0.5	1.7	0.48	20
Colorado CA 1904–23	694	2182	1.8	3.87	0.40	0.87	7	0.5	5.0	36.2	0.15	29
Ebro (Spain) 1913–62	1400	2475	11.0	0.65	0.21	0.51	4	1.5	0.2	1.3	0.28	4.5
Po (Ita) 1933–39	1904	2846	3.5	0.56	0.04	0.12	7	1.5	0.7	2.0	0.21	23
Huanghe (PRC) 1921–60	1880	2860	0.0	34.9	0.01	0.03	12	1.5	0.8	3.3	12.1	92
Fly (PNG)	2510	3500	5.1	2.34	0.03	0.09	5	1.5	4.0	36.7	0.03	37
Yana (Rus)	906	3500	5.6	0.10	0.19	0.68	14	1.5	1.0	12.6	0.13	27
Tigris-Euphrates (Iraq)	1500	4000	4.6	1.76	0.22	0.88	5	1.0	3.0	8.2	0.18	11
Copper (AL)	1240	5440	7.7	2.40	0.60	3.26	10	3.0	3.4	7.7	0.12	6.3
Indigirka (Rus)	1710	5570	12.5	0.51	0.15	0.84	28	1.0	1.0	11.8	0.25	2.4
Krishna (Ind) 1901–60	1784	6501	9.2	2.23	0.47	3.03	7	2.0	3.0	2.8	0.17	4.3
Mahanadi (Ind)	1970	6900	2.5	1.95	0.10	0.69	9	2.0	3.5	3.2	0.12	24
Hong Ho (Red) (Viet)	3800	7000	4.4	4.31	0.20	1.40	10	1.5	3.5	4.9	0.30	10
Fraser (Can)	3560	7100	8.3	0.69	0.07	0.46	8	1.5	4.5	5.8	0.03	49
Danube (Rom) 1931–55	6420	8900	4.4	2.22	0.06	0.55	9	1.5	0.0	3.7	0.99	4.1
Nile (Egy) 1871–98	3484	9858	1.5	5.01	0.09	0.87	4–15	1.5	0.4	1.9	2.08	2.8
Magdalena (Col)	7530	10000	3.9	4.75	0.10	1.00	7	1.5	1.5	1.7	1.06	4.5
Kolyma (Rus) 1942–65	2840	10100	12.7	0.46	0.08	0.85	10	1.5	1.0	8.6	0.14	3.8
Godavari (Ind)	2650	11810	3.1	5.56	0.27	3.14	11	2.0	3.0	12.0	0.43	4.1
Niger (Nig)	6130	12000	7.7	1.37	0.07	0.84	15	1.4	3.0	8.5	0.13	13
Pechora (Rus)	3370	13810	5.8	0.21	0.01	0.20	23	2.0	3.0	3.3	0.02	66
Pearl Xijiang (PRC)	9510	16000	6.2	2.33	0.06	0.98	15	1.5	5.0	26.4	0.09	28
Parana (Arg)	13600	18000	7.5	2.71	0.06	1.09	20	0.5	4.0	6.1	0.17	15
Indus (Pak) 1941–62	3171	18100	0.7	12.8	0.11	1.99	6	3.5	5.0	19.3	0.33	3.1
Yukon (AL)	6120	18100	8.0	2.07	0.12	2.17	43	2.0	1.5	5.8	0.34	17
MacKenzie (Can)	9750	21295	2.9	4.12	0.05	1.06	23	1.2	0.2	4.5	2.78	1.4
Volga (Ukr) Pre-dam	8200	22400	6.0	0.64	0.02	0.43	100	1.0	0.0	1.7	0.64	2.3
Irrawaddy (Bur)	13558	25000	3.0	8.50	0.08	1.93	16	1.5	4.2	23.2	0.43	10
Mississippi (LA)<1960	18400	30482	0.5	20.6	0.02	0.70	71	0.5	0.4	10.0	50.2	0.6
Mekong (Viet) Pre-dam	14770	31000	1.4	5.14	0.02	0.62	9	1.5	2.6	17.0	0.57	14
Yangtze (PRC) Pre-dam	28278	47500	2.5	15.6	0.06	2.66	3	1.5	4.5	15.0	0.69	8.5
Orinoco (Ven)	34500	58800	3.0	5.66	0.02	1.18	27	1.3	1.9	15.8	1.09	4.4
Ganges/Brahm. Pre-dam	31000	72000	1.9	35.6	0.09	6.48	20	1.0	3.6	28.5	2.55	2.2
Lena (Rus)	16240	74360	7.0	0.41	0.01	0.53	115	1.0	1.0	17.0	0.20	3.8
Amazon (Bra)	200000	212000	1.6	38.6	0.01	2.65	7	2.0	6.0	8.9	0.97	15

main channel; e_b = bedload efficiency (dimensionless), β = bedload rating term (dimensionless and set to 1 for this study); $\updownarrow$ = limiting angle of repose of sediment grains lying on the river bed; and u_{cr} = critical velocity (m/s) below which no bedload transport occurs. Deltas in our database are relatively fine-grained and thus Bagnold's approach is appropriate. To calculate the maximum monthly bedload we substitute Q_{mx} for Q_{av}.

Total sediment load (Q_{s+b}) is simply the observed suspended load plus the calculated bedload. The percent bedload to total sediment load averaged 6.5% for our 55 rivers, with a range of 0.5% to 17.1%. The sediment volume delivered to the delta (V_s) in a thousand years (km^3/kyr) is a conversion of the total sediment load to volume assuming a sediment porosity (φ) of 50%.

3 TYPE DELTAS

Deltas influenced by low marine energy, i.e. no to low tides and wave action (Mississippi, Atchafalaya, Volga), are truly a unique subset of deltas, displaying growth by crevasse splays leading to channel splitting and or channel feathering (Figs. 1, 2). With channel splitting, the total channel width to river width is large, and with feathering TC_w:R_w is near unity. Both the Volga and Mississippi deltas have very low ratios of marine power to river power (P_m:P_r, Table 1). These deltas are in the upper end of the spectrum of sediment supply to marine power (Q_{s+b}:P_m, Table 1), with the Mississippi an order-of-magnitude larger than any other river except Huange He.

By definition, the supply-to-dispersal ratio assigns smaller deltas to the dispersal end of the spectrum, partly because sediment supply scales with river size, but partly because ocean energy ranges more narrowly. Even the Waipaoa, having one the world's highest sediment yields, falls on the dispersal end of global deltas.

Other "type" deltas include the polar deltas powered by the river (e.g. Lena, Fig. 5A). Such rivers have some of the highest Q_{mx}/Q_{av} ratios, comparable to desert rivers. Many of their distributary channels are active for less than one month. Their delta surface is rich with abandoned overflow channels and thermokarst lakes. Tropical deltas experience intense convective rainfall and establish runoff channels that influence the pathway of their distributary channels (e.g. Orinoco, Fig. 5B, and Niger, Fig. 5D). Where wave energy is high, barriers and beaches develop (e.g. Copper, Fig. 5C, Niger, Fig. 5D, Danube, Fig. 5E). Where tidal energy is high, funnel channels develop with widened-mouths kept open by the tidal currents (e.g. Niger, Fig. 5D, Fly, Fig. 5F): the larger the tidal energy, the larger the channel mouths in relation to the river channel width.

Figure 4. Low tide and wave Atchafalaya (top) and Mississippi (bottom) deltas, Q_{s+b}:P_m = 50, TC_w:R_w = 10, grow by levee maintenance of crevasse splays.

4 PREDICTIVE RELATIONSHIPS

To test our understanding of the morphodynamics of deltas we examine patterns in controlling factors: bedload, suspended load, and their ratio; bankfull discharge (average maximum monthly discharge), average discharge, and their ratio; marine power, river power and their ratio; wave energy; tidal energy; and the ratio of sediment supply to marine power. Most of these factors, along with other boundary conditions (climate, accommodation space), range across orders-of-magnitude. With so many degrees of freedom, no two deltas in our database experience near similar controlling factors; deltas that are somewhat close are the Ebro and the Po, and the Danube and the Nile (Table 1).

In addition, geomorphologic response is often conditioned by the sequence of events, not just the number and magnitude of events.

Examining our environmental database, area of a delta A_D (km^2) is best predicted from average discharge Q_{av} (m^3/s) of its feeder river, and the total sediment load Q_{s+b} (kg/s) with:

$$A_D = 0.02 Q_{av}^{1.05}\, Q_{s+b}^{0.44} \tag{2}$$

Thus a delta's size scales with its delivery system (Fig. 6). Most of the world's deltas are coeval,

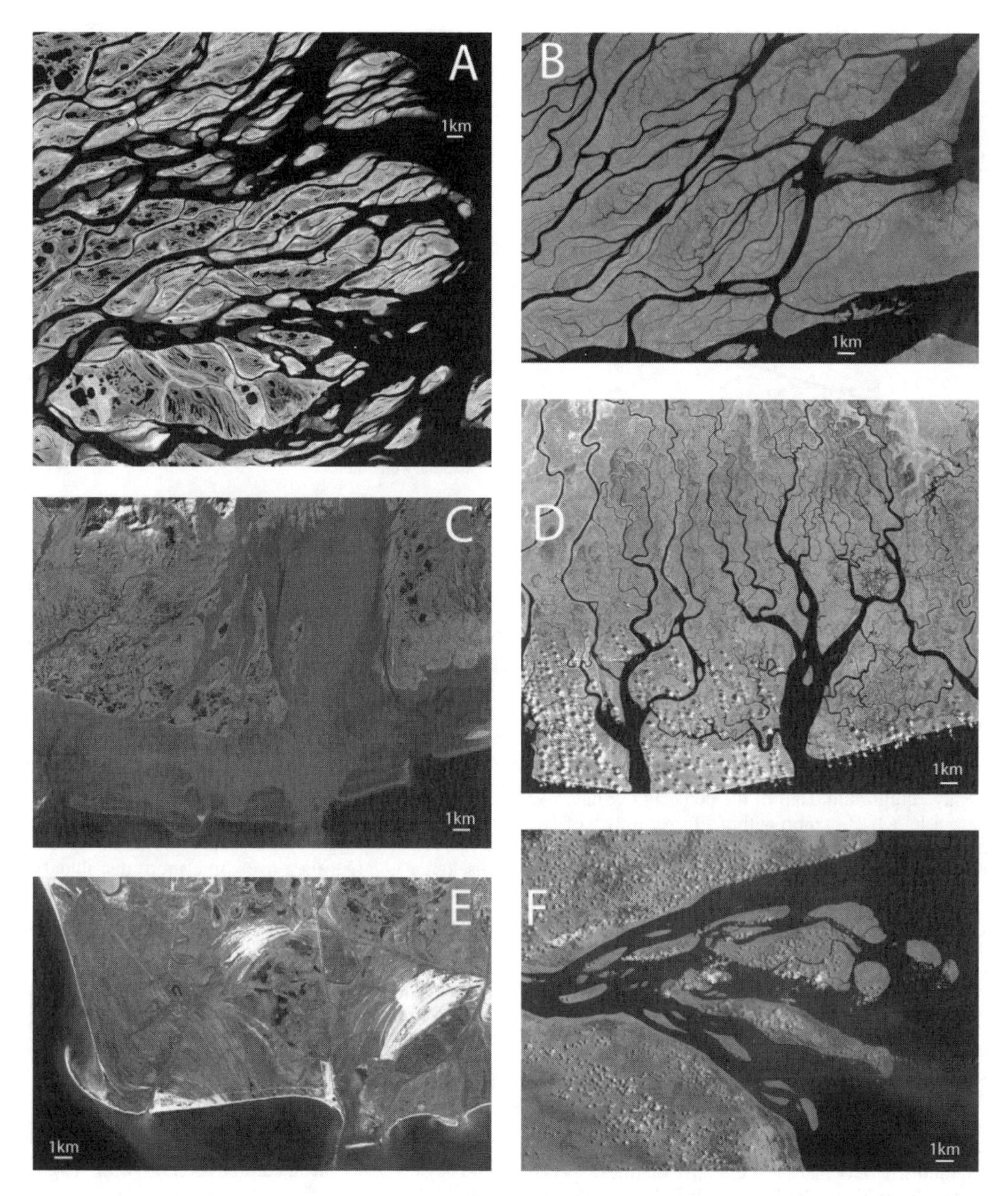

Figure 5. A) Polar Lena delta, Q_{s+b}:$P_m = 0.2$; P_m:$P_r = 3.8$. B) Tropical Orinoco delta, Q_{s+b}:$P_m = 1.1$; P_m:$P_r = 4.4$. C) Polar Copper delta, Q_{s+b}:$P_m = 0.1$; P_m:$P_r = 6.3$. D) Tropical Niger delta, Q_{s+b}:$P_m = 0.1$; P_m:$P_r = 13$. E) Temperate Danube delta, Q_{s+b}:$P_m = 1.0$; P_m:$P_r = 4.1$. F) Tropical Fly delta, Q_{s+b}:$P_m = 0.1$; P_m:$P_r = 27$.

having formed when sea level stabilized ≈ 6 kyr before present (Amorosi and Milli, 2001).

The gradient of a delta plain, is best predicted (Fig. 6) with

$$D_{grd} = 0.001(V_s/(A_D D_{Sh}))^{0.25} C_s^{0.28} Q_{av}^{0.23} \quad (3)$$

where V_s = sediment volume delivered to the delta (km^3/kyr); D_{sh} = water depth (m) reached by the submerged delta determined from shelf profiles and provides a measure of accommodation space; and C_s = discharge weighted suspended sediment concentration (kg/m^3) used as a proxy of topset aggradation.

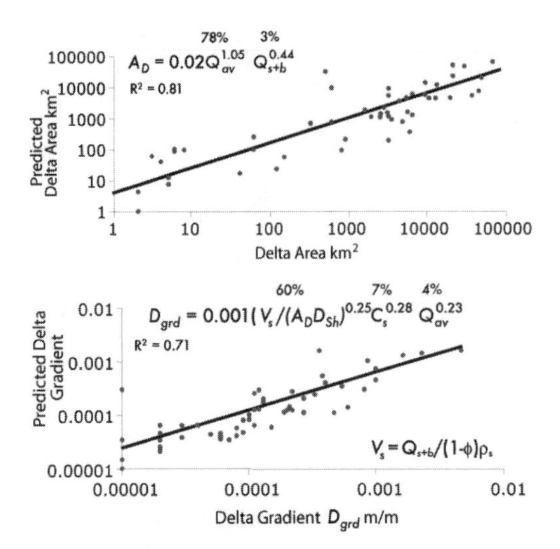

Figure 6. A river's average discharge accounts for 78% of the variance in predicting a delta's area; a river's sediment load accounts for another 3% of the database variance. Bottom: A delta's gradient is largely (60%) determined from a ratio of sediment supply to retention.

$V_s/(A_D D_{Sh})$ is the ratio of sediment supply to sediment retention. If we take the sediment supply over 6 kyr, equivalent to the time when sea level stabilized and world deltas starting rapidly prograding, then those deltas with this ratio near unity are in some quasi-steady equilibrium. Half of the deltas in our database have a $V_s/(A_D D_{Sh})$ ratio within a factor of two of 1.0, and would be considered in balance. Nine of the deltas have a $V_s/(A_D D_{Sh})$ ratio <0.45. These rivers systems may have had a larger supply of sediment in the earlier part of this six thousand year history (possibly the Russian rivers: Lena, Yana, Kolyma, Volga), or have a very effective marine dispersal system (e.g. Amazon, Niger, Chao Phrya, Mekong, Fraser). In contrast 17 of our 55 deltas have a $V_s/(A_D D_{Sh})$ ratio >2.8. These rivers systems may have had a smaller supply of sediment in the earlier part of the six thousand year history (possibly before the strong influence of human agriculture and land clearing: e.g. Ebro, Huanghe, Var, Indus, Arno, Waipaoa), or experienced a drier conditions (Eel, Klamuth), or once had a less effective marine dispersal system (possibly the Red, Colorado (TX), Magdelena, Colorado (CA), Brazos).

Syvitski et al. (in press) determined that a (power-law) relationship ($R^2 = 0.47$) exists between the number of distributary channels covering a delta and the feeder river's length. Their hypothesis was that a river's length scales with the intra-annual variability in discharge (Morehead et al. 2003), with the exception of polar and desert rivers. They suggested that discharge variability could play a role in channel levee breakthroughs and thus help determine the number

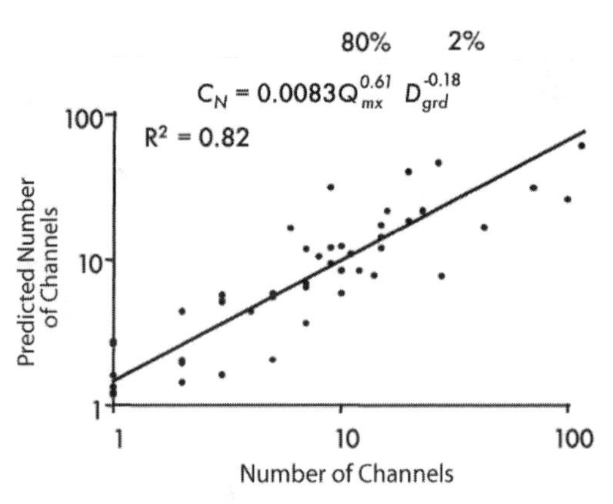

Figure 7. The maximum monthly discharge accounts for 80% of the data variance in predicting the number of distributary channels.

of active channels. They did not have the data to test this hypothesis. Syvitski et al. (in press) also found an inverse power-law relationship ($R^2 = 0.56$) between the number of distributary channels and a delta's gradient. To test their hypotheses discharge variability was obtained for each river (Table 1) and included in the analysis. The number of distributary channels is now better predicted ($R^2 = 0.82$) using maximum monthly discharge Q_{mx} (m^3/s) and inversely by delta gradient (Fig. 7), such that:

$$C_N = 0.0083 Q_{mx}^{0.61} D_{grd}^{-0.18} \quad (4)$$

Thus, if a delta's gradient is small and its maximum discharge is large, then over time more distributary channels will evolve.

The distributary channels form a lognormal distribution in terms of their channel widths (Fig. 8). The total width of these channels (TC_w in m) is a function of bankfull discharge and inversely with the total sediment load (Fig. 9):

$$TC_W = Q_{mx}^{0.76} Q_{s+b}^{-0.22} \quad (5)$$

The average channel width (MC_w in m) is additionally influenced positively by the tidal range (T_i in m) and negatively by the maximum monthly wave height (W_a in m):

$$MC_W = 32.1 Q_{av}^{0.1} Q_{s+b}^{-0.21} e^{0.37 T_i} e^{-0.15 W_a} \quad (6)$$

Tidal currents work to keep the channel mouths enlarged, while littoral transport works to close in river mouths.

To date, our database on deltas (Table 1) contains an estimate of the grain size (D_{mm}) of topset deposits for just 25 of the 55 deltas. Based on this reduced dataset a simple power-law relationship is found between a

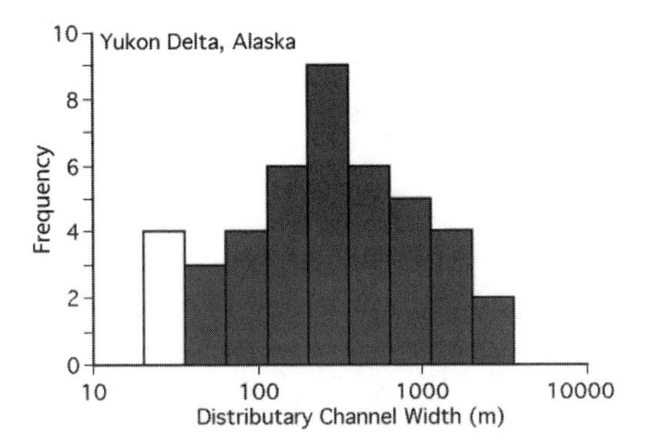

Figure 8. Typical lognormal distribution of channel widths, measured at the coast, for the Yukon delta, Alaska.

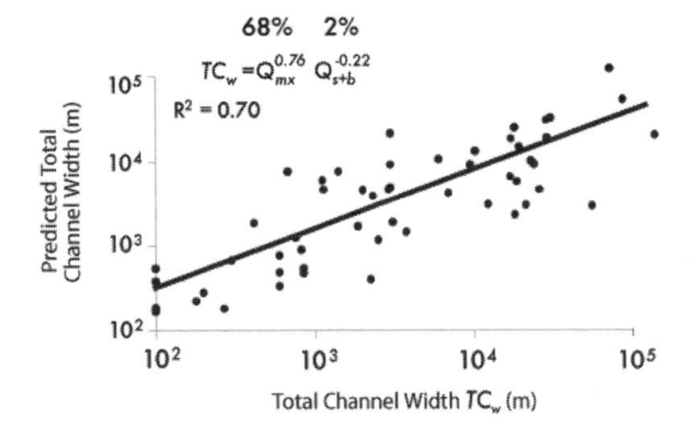

Figure 9. The total width of a delta's distributary channels, measured at the coast, is largely (68%) a function of the maximum-monthly discharge.

river's length (L) and the size of its bed-material load (Fig. 10):

$$D_{mm} = 106\ L^{-0.9} \tag{7}$$

Downstream fining is a well-established relationship in river hydrology (van Niekerk et al., 1992; Vogel et al., 1992), and apparently even with the action of tides and waves that could winnow away the fines, the relationship still holds for deltaic systems. Long rivers thus feed muddy fine-grained deltas, such as the Amazon, Mekong, and Orinoco.

5 SUMMARY

This study provides predictive relationships for morphodynamic features of deltas, such as the number and size of distributary channels, a delta's surface area, and a delta's gradient. Other relationships such as the grain size of the topset deposit are considered preliminary. These statistical relationships are suitable for hypothesis testing, or to constrain or verify numerical models used to simulate the evolution of coastal systems.

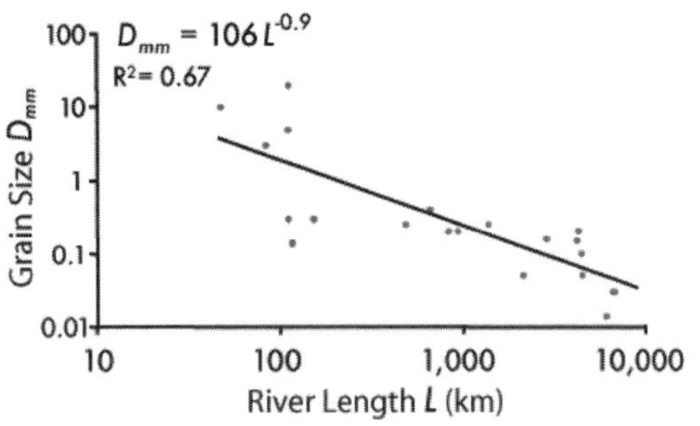

Figure 10. A preliminary relationship based on 25 deltas, between average grain size of a delta's topset deposit and the length of its feeder river.

A large number of environmental factors were investigated, but only a few were found to be important or defining, and include a river's length, average and maximum discharge, total sediment load, suspended sediment concentration, accommodation space, tidal range and maximum monthly discharge. The study also suggests that a simple ternary diagram (Fig. 1) is inadequate to capture the true range in delta morphology. The approach suffers from a scaling issue: all small deltas except those enclosed by bays or inlets, cannot compete with the typical power of marine waves and their longshore currents, or tidal currents. Another scaling issue is that there are too few large deltas to show the full dimension of possibilities. Nevertheless, as the quality of modern environmental databases improves, so will our understanding of deltaic morphodynamics. Future investigations will need to address these patterns in terms of theory, and verified through the judicious application of numerical formulation. The time sequencing of floods and storms should then be more easily tested.

ACKNOWLEDGEMENTS

This study was supported by the U.S. Office of Naval Research through the office of Dr. Tom Drake, and by NASA (IDS/NNG04GH75G). The author also thanks the INSTAAR Delta Force for their many discussions on this study.

REFERENCES

Amorosi, A. & Milli, S. 2001. Late Quaternary depositional architecture of Po and Tevere river deltas (Italy) and worldwide comparison with coeval deltaic successions. *Sedimentary Geology* 144: 357–375.

Bagnold, R.A. 1966. An approach to the sediment transport problem from general physics. *U.S. Geol. Survey Prof. Paper* 422-I, 37.

Bhattacharya, J.P. & Giosan, L. 2003. Wave-influenced deltas: geomorphological implications for facies reconstruction. *Sedimentology* 50: 187–210.

Coleman, J.M. & Wright, L.D. 1975. Modern River Deltas; variability of processes and sand bodies. In M.L. Broussard (ed), *Deltas, models for exploration*. Houston Geological Society pp. 99–149.

Dalrymple, R.W., Zaitlin, B.A. & Boyd, R. 1992. Estuarine facies models: conceptual basis and stratigraphic implications. *Journal of Sedimentary Petrology* 62: 1130–1146.

Galloway, W.E. 1975. Process framework for describing the morphologic and stratigraphic evolution of deltaic depositional systems. In M.L. Broussard (ed), *Deltas, models for exploration*. Houston Geological Society, pp. 87–98.

Morehead, M.D., Syvitski, J.P.M., Hutton, E.W.H. & Peckham, S.D. 2003. Modeling the inter-annual and intra-annual variability in the flux of sediment in ungauged river basins. *Global and Planetary Change* 39: 95–110.

Orton, G.J. & Reading, H.G. 1993. Variability of deltaic processes in terms of sediment supply, with particular emphasis on grain size. *Sedimentology* 40: 475–512.

Overeem, I., Syvitski, J.P.M. & Hutton, E.W.H. in press. Three-dimensional numerical modeling of deltas. *Sedimentary Geology*.

Postma, G. 1990. Depositional architecture and facies of river and fan deltas: a synthesis. In A. Colella and D.B. Prior (eds), *Coarse-Grained Deltas* Intl. Association of Sedimentologists Spec. Publication 10: 13-28.

Postma, G. 1995. Sea-level-related architectural trends in coarse-grained delta complexes. *Sedimentary Geology* 98: 3–12.

Restrepo, J.D., Kjerfve, B., Correqa, I.D. & Gonzalez, J. 2002. Morphodynamics of a high discharge tropical delta, San Juan River, Pacific coast of Colombia. *Marine Geology* 192: 355–381.

Syvitski, J.P.M., Vörösmarty, C., Kettner, A.J. & Green, P. 2005. Impact of humans on the flux of terrestrial sediment to the global coastal ocean. *Science* 308: 376–380.

Syvitski, J.P.M., Kettner, A.J., Correggiari, A. & Nelson, B.W. in press. Distributary channels and their impact on sediment dispersal. *Marine Geology*.

van Niekerk, A., Vogel, K.R., Slingerland, R.L. & Bridge, J.S.1992. Routing of heterogeneous sediments over movable bed: Model development. *Journal of Hydraulic Engineering ASCE* 118(2): 246–262.

Vogel, K.R., van Niekerk, A., Slingerland, R.L. & Bridge, J.S. 1992. Routing of heterogeneous size-density sediments over movable stream bed: Model verification and testing. *Journal of Hydraulic Engineering ASCE* 118(2): 263–279.

River, Coastal and Estuarine Morphodynamics: RCEM 2005 – Parker & García (eds)
© 2006 Taylor & Francis Group, London, ISBN 0 415 39270 5

Morphological evolutions of a macrotidal bay under natural conditions and anthropogenic modifications

F. Cayocca
DYNECO/PHYSED, IFREMER, Plouzané, France

ABSTRACT: Mont Saint Michel Bay exhibits one of the world's greatest tidal amplitudes (up to 14 m). The 30 km wide bay is characterized by a 4 km wide tidal flat in its central area (up to 11 km in the east). This setting makes it a privileged environment for shellfish farming. However, the natural trend of infill combined with hindering of the flow due to farming installations have altered shellfish production. This study aims at understanding general patterns of fine sediment transport in the natural environment. Changes in these patterns due to anthropogenic activities are also investigated. A schematized representation of the hydrodynamic conditions and of the sediment environment (waves are neglected, a single fraction of fine sediments is considered) allows a numerical model to reproduce the infill of the upper tidal flats as well as the changes in sedimentation patterns due to farming structures.

1 INTRODUCTION

Mont Saint Michel Bay lies on the French coast of the English Channel in a macrotidal environment (one of the world's largest tidal ranges, up to 14 m during spring tides). The Bay is one of the largest farming grounds for oysters and mussels on the French side of the English Channel. Since shellfish production has significantly decreased in the past 10 years, assessing the trophic capacity of the Bay has become an ecological priority while, at the same time, the restructuring of shellfish farming was required to counteract the unmanageable mud deposits encountered in oyster beds. These ecological as well as economic concerns are related to turbidity levels (for plankton growth as well as shellfish ecophysiology) and sediment transfers in the Bay (for wild benthic habitat concerns and mud deposits induced by the farming structures). As part of a multidisciplinary project, a major hydrodynamic and sedimentological study was initiated in 2001 to describe the local conditions and explain the physical processes driving sediment transfers.

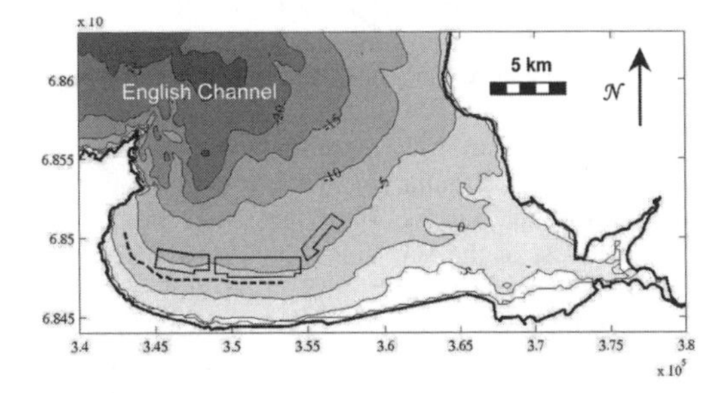

Figure 1. General setting of Mont Saint Michel Bay. Water depths are relative to the mean sea level. Location of the fish traps (dashed line) and contours of the mussel farms (solid polygons) can be seen.

2 CONTEXT AND SETTING

Freshwater input to the Mont Saint Michel Bay arrives principally from its southwest corner where 3 small rivers of fairly limited discharge (total annual average discharge on the order of $25\,m^3 \cdot s^{-1}$) compared to the volume of the bay (tidal prism of $2300\,Mm^3$ for a mean tide, whereas low tide volume is $2000\,Mm^3$, see Figure 1). The Bay is characterized by a 4 km wide tidal flat in its central area (up to 11 km in the east) with a 1/550 slope. Maximum tidal current velocities vary from $40\,cm \cdot s^{-1}$ on the tidal flats to $1\,m.s^{-1}$ at the northern limit of the Bay, with local maxima encountered in the channels and by the cape to the north-west of the domain. The largest waves come from propagation in the Atlantic Ocean, with offshore significant wave heights between 80 cm and 2 m occurring 50% of the time.

The Bay has undergone continuous infill in the most recent centuries, as illustrated by the seaward progression of lowlands now used for agriculture. This feature results from the flood dominance of the tide, and illustrates what Le Hir *et al.* (2000) called intrinsic asymmetry (*i.e.* deposition on the upper flat during high water slack, followed by partial consolidation which does not allow erosion to occur during the ebb) and large-scale asymmetry (non-linear distortion of the tide during propagation). The present rate of infill has been estimated to be 0.2 cm/year on average (Migniot, 1998). Waves enter the domain from the north-west, and their refraction around the north-western cape are responsible for a longshore gradient of bottom shear stress (shear stresses increase eastwards). This process explains the variable sediment coverage of the Bay, from mud to the West to mixed sediments in the centre and fine to coarse sand to the East (Bonnot-Courtois *et al.*, 2002) .

Several man-made structures were introduced in the environment (Fig. 2), above the level of lowest low tide (Bahé, 2003):

- Large fish traps were used between the XIth century and the 1960s. These traps are V-shaped structures made of 2 meter-tall woven branch fences open to the outgoing tide, each branch of the V being 250 to 300 m long. They were meant to catch fish during the ebb, and direct them to cages placed downstream of the tapered end. Even though they are not used anymore, they have altered the water circulation and sediment deposits for several centuries. There is indeed a sharp frontier between very muddy sediments above the fences and muddy sands lower down.
- Oyster tables are light metallic wire structures sitting on the ground, on which lie coarse nets enclosing the oysters. They have been used to the north-west of the Bay since the 1960s. They hinder considerably water circulation and sediment erosion, and are responsible for drastic morphological changes. Deposits under the oyster tables and between rows of tables range from 60 cm to 1 m.
- Mussel farms were installed in the central part of the bay in the 1960s. They consist of rows of 2.4 meter-tall posts around which lines of mussel spawn are rolled spirally. The posts eventually grow to 60 cm in diameter. They lay in the centre of the bay in a fine sand environment and experience very local temporary mud deposits at their bases.

Increased sedimentation around farming structures is prejudicial to shellfish growth. Farming structure transfer is often found as to be the only short term solution when shellfish production has decreased to an unacceptable level. However, up till now, no numerical tool was capable of predicting long term morphodynamic evolutions of such environments, where the sediment behaviour is further modified by the presence of bio-deposits.

This work is a first step towards understanding processes responsible for the natural and man-made related infill of the Bay by means of numerical modelling. The main simplifications applied here lie in the disregard of wave action, the consideration of a single mud fraction and no accounting of consolidation.

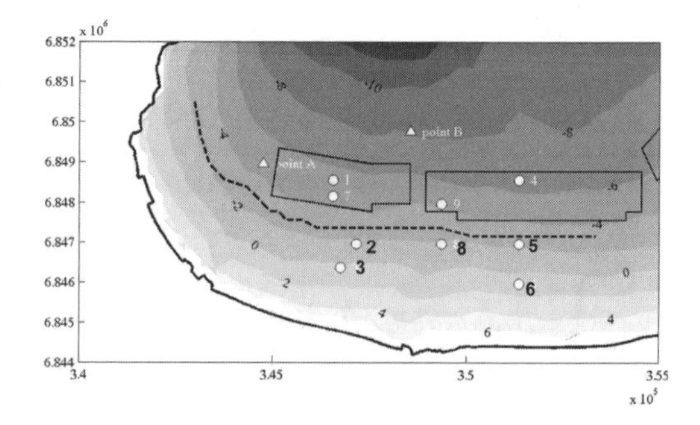

Figure 2. Zoom on the mussel farms (solid polygons) and fish traps (dotted line) locations with the position of the control points mentioned in the text.

3 NUMERICAL MODEL

A finite difference numerical model solving the shallow water equations on an irregular Cartesian grid was used to compute vertically averaged tidal currents in the area. The fairly small fresh water inputs as well as intense vertical mixing due to strong tidal currents justify this bidimensional approach. A SAMPLE device (multi-parameter sampling station, Jestin *et al.*, 1994) was deployed at various locations on the tidal flat in 2003 and 2004. Currents were measured 30 cm above the sea floor with an electromagnetic current meter, turbidity was recorded from 2 OBS sensors at 10 cm and 30 cm above the sea floor, water height was measured from a pressure sensor. In order to compare vertically integrated computed velocities with local measurements, a logarithmic velocity profile was assumed. An equivalent vertically integrated "measured" velocity was therefore used for comparisons with model outputs. The roughness length was computed from the ripple characteristics ($z_0 = 0.002$ m). Figure 3 shows the "integrated" measured and computed velocity components and water heights for a mean tide in the central part of the Bay. Tidal amplitude is well reproduced although the model predicts a late flood. This phase shift is systematic in the whole domain and reflects a similar shift in the boundary conditions (tidal harmonics computed from the French Office of Naval Research, Le Roy and Simon, 2003). Recorded velocities are only valid when the current meter is immerged, *i.e.* when the water height exceeds 30 cm. The measured eastward component of the velocity is directed to the west during the whole cycle, which is not the case for the model. This discrepancy was interpreted to be due to the wind blowing eastward that day, which could explain the absence

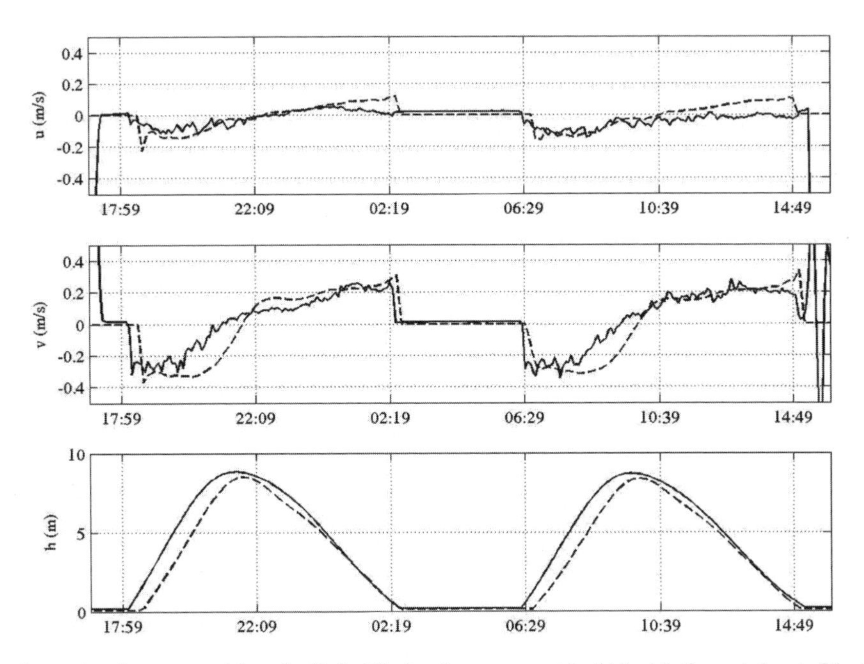

Figure 3. Comparison between model results (dashed line) and measurement (solid line) in the central part of the Bay; from top to bottom: East-West velocity (positive eastwards), North-South velocity (positive northwards) and water height.

of reversal for fairly small eastward components of the velocity ($<10\,\mathrm{cm \cdot s^{-1}}$). The maximum magnitude of the northward component is reproduced very well, with the time shift corresponding to the late tidal phase. The modelled velocity rapidly reaches a maximum during the flood, and remains constant at its maximum value for over 2 hours. This could be due to an underestimation of the friction for larger water heights (the friction was computed with a uniform Strickler coefficient of $45\,\mathrm{m^{1/3}s^{-1}}$, and no particular tuning was done). When maximum velocities exceed an erosion threshold, we can therefore expect to overestimate the time during which erosion occurs, and overestimate the characteristic distances of transport in suspension.

The transport of cohesive sediments is computed by solving an advection-diffusion equation in the water column, where deposition and erosion fluxes constitute the sediment source and sink terms (Le Hir *et al.*, 2001). Hindered settling and viscosity increase with concentration are taken into account.

3.1 *Hydrodynamic effects of mussel farming*

Very few studies have investigated the effects of mussel farms on the flow. Recently, SEAMER (2000) used a numerical model with a 10 cm mesh to compute the hydrodynamic effects of 2.4 m tall posts of 40 cm in diameter representing average size mussel posts ("empty" posts are 20 cm in diameter, they can be as large as 70 cm in diameter when covered with mature mussels). For a single post, the current velocity was shown to increase on both sides of the post as long as it was not submerged, while the velocity decreased upstream of the post as well as in a downstream wake. The wake of an individual post where the velocity is decreased by 80% was a few meters long depending on the ambient velocity and the water height. A row of such posts aligned with the current direction induced a 10% velocity decrease over 1–2 m on both sides of the row. When the current direction was not aligned with the row, the width of the wake increased significantly and the overall velocity could decrease by up to 30%. However a row of posts covered by one meter of water induced very local accelerations and decelerations of the flow, but had no influence on the overall current velocity (which was probably due to the fact that velocities were vertically integrated). Unfortunately, these simulations were only carried out for fairly restrictive conditions (current velocity of $1\,\mathrm{m \cdot s^{-1}}$, one single diameter, 3 water heights), were not validated and do not allow for a general parameterisation of the effects of the posts on the flow.

SOGREAH (1986) carried out physical experiments in order to quantify the overall slowing down of the flow expressed through a variable Strickler coefficient. This coefficient was shown to vary from 18 to 32 depending on the distance between posts and between

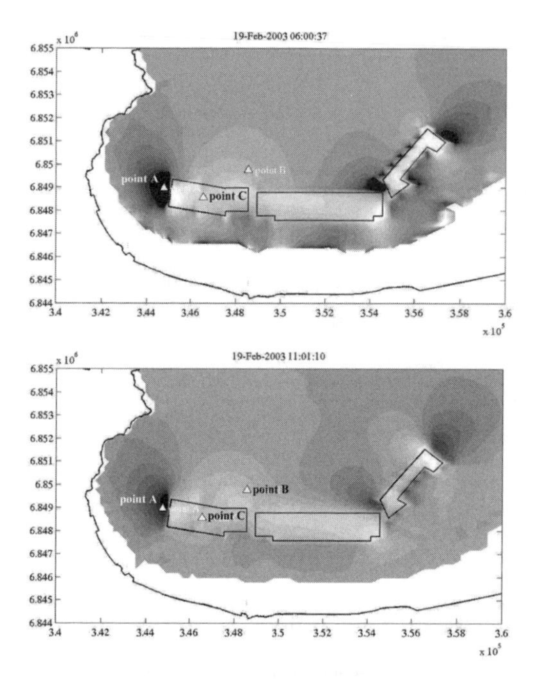

Figure 4. Contours of velocity differences between a situation with no farms and a situation with farms during the flood (top) and the ebb (bottom). Darker shades represent a velocity increase, lighter shade a velocity decrease. See Figure 5 for quantitative values.

rows, the size of the posts, the incidence of the current compared to the posts and the reference Strickler coefficient of a domain without mussel farms. The orientation of the tidal ellipses in the Mont Saint Michel Bay follows bathymetric gradients, whereas rows of mussel posts are oriented north-south in the whole western and central farming areas, and north-west to south-east in the eastern farming area. This disposition was chosen to minimize impacts on the tidal propagation, however the angle between the rows and the tidal current varies throughout the Bay. As a first approach, we considered a constant and uniform Strickler coefficient of $25\,m^{1/3}s^{-1}$ in the farming areas, which is representative of an average post dimension.

Figure 4 shows the maximum effect of the farms on current velocities during the flood and ebb, of a spring tide. The maximum effect is observed within the farms (deceleration), and on the eastern and western ends of the farms (acceleration of the flow that goes around the areas of increased friction); these effects are felt several kilometres away from the farm. Figure 5 gives a quantitative estimate of the velocity changes for a mean tide. Similar patterns are actually observed for spring tides, i.e. acceleration of up to 30% on the edge of the farms (point A), deceleration of 5% up to 2 km north of the farms (point B) and deceleration of 10% within the farms (point C). These differences are likely to substantially modify deposition and erosion patterns.

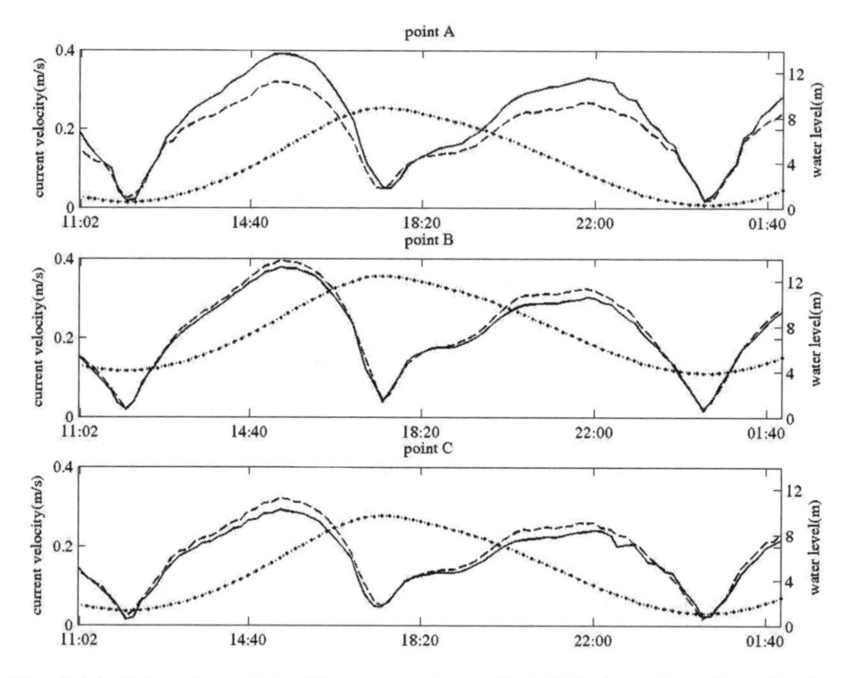

Figure 5. Water height (dots) and magnitude of the current velocity with (solid line) or without (dashed line) mussel farms. Top: point A; center: point B, bottom: point C.

3.2 *Hydrodynamic effects of fish traps*

The effect of fish traps was also investigated. Their setup had to be greatly simplified for 2 reasons:

- the trap sides are oblique compared to the model grid cells (which are rectangles aligned on cardinal directions) in most of the domain.
- the spacing between 2 consecutive traps is smaller than the smallest grid cells (which are 200 m × 200 m in the south west part of the domain).

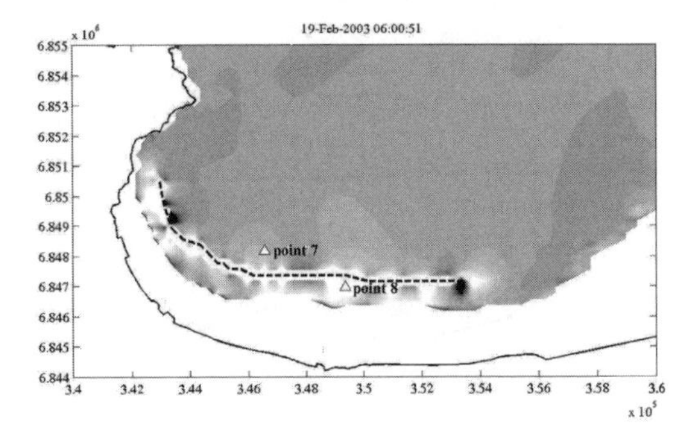

Figure 6. Contours of velocity differences between a situation with no fish traps and a situation with traps during the flood of a spring tide. Darker shades represent a velocity increase, lighter shades a velocity decrease. See Figure 7 for quantitative values.

The fish traps were therefore represented by a 2 m tall submersible dyke following the edges of the grid cells as close as possible to their actual position. The dyke is continuous except for 2 openings in the western part.

Figure 6 shows the effect of the modelled fish traps on the current magnitude during the flood. Velocities are mostly reduced in the immediate vicinity of the dyke. They are significantly increased where the dyke is discontinuous (dark spot towards the west end of the dyke). Since vertical velocities are averaged, velocities are not significantly perturbed as soon as the dyke is submerged by 2 to 4 meters depending on the tidal range. Figure 7 shows a quantitative estimate of this velocity perturbation 800 m north and 500 m south of the dyke. It reaches a maximum of 10% during one to two hours.

The modelled effect of the fish traps on the currents is over-estimated because of the absence of a corridor between traps and because the model dykes are water-tight whereas the actual woven wooden fences are not. However, the vertical integration of the velocities artificially decreases the impact very close to the dyke, where actual ebb velocities near the bottom are likely to be significantly reduced at all times of the ebb (let us recall that the V-shaped fish traps are open to the outgoing tide).

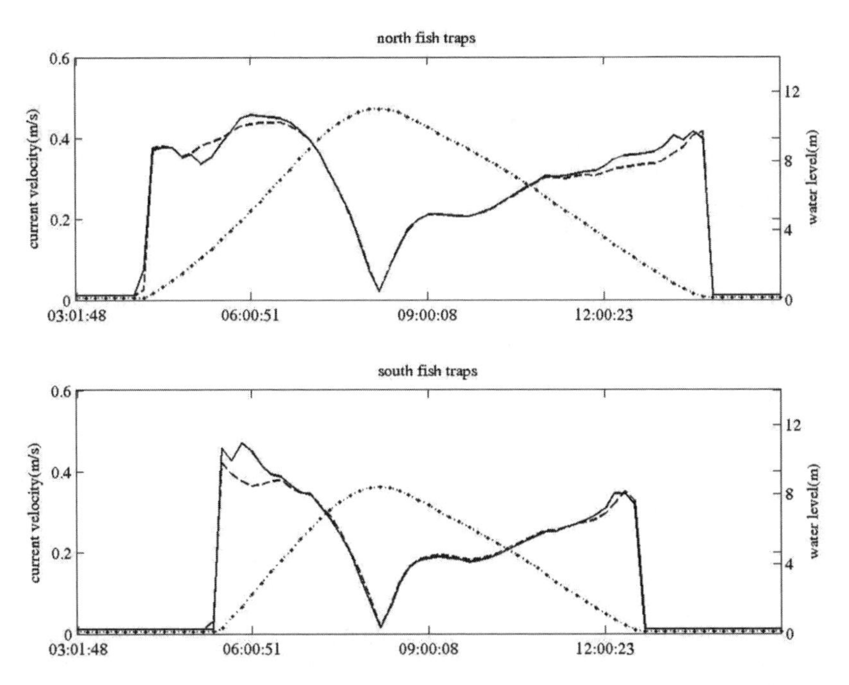

Figure 7. Water height (dots), reference velocity (solid line) and velocity with fish traps (dashed line) in points 7 (north of the traps) and 8 (south of the traps).

4 SEDIMENT TRANSFERS

The sediment coverage in the Bay exhibits strong contrasts: sandy bottom in the eastern side where maximum wave heights and current velocities are encountered, muddy bottom in the western side. Similarly, sediments generally coarsen offshore. While longshore gradients are mainly due to waves, cross-shore gradients are likely due to the tide dynamics. We aimed at studying this process by considering an initial uniform fine sediment coverage of 10 cm throughout the domain. The model is then run for one year, driven by tides only (recorded succession of the tidal conditions of 2003). While bottom friction is computed through a Strickler friction coefficient for hydrodynamic calculations, the actual bottom roughness is used to compute the bottom shear stress that drives sediment erosion. This roughness is assumed to be uniform and constant throughout the runs. The initial sediment concentration is taken to be $900\,\mathrm{g}\cdot\mathrm{l}^{-1}$, which corresponds to a typical concentration measured in the field. Erosion is parameterised according to Partheniades'law:

$$E=E_0\,(\tau-\tau_{cr})/\tau_{cr},$$

where τ is the bottom shear stress and the erosion threshold τ_{cr} depends on the sediment concentration in the sediment surface layer c_s, and is parameterised according to

$$\tau_{cr}=\alpha_1\, c_s^{\ \alpha 2}$$

$\alpha_1=2\cdot 10^{-7}\,\mathrm{m}^2\cdot\mathrm{s}^{-2}$ and $\alpha_2=2.2$ are chosen to fit experimental data from Migniot (1998). The concentration of fresh deposits is prescribed to be $400\,\mathrm{g}\cdot\mathrm{l}^{-1}$. Consolidation is not accounted for, therefore the surface concentration of deposited sediments remains constant (and so does the critical erosion shear stress). Deposition fluxes are computed according to

$$D_{flux}=w_s.c_w.\,(\tau_{crdep}-\tau)/\tau_{crdep} \quad \text{if } \tau<\tau_{crdep}$$
$$D_{flux}=0 \quad \text{if } \tau>\tau_{crdep},$$

where w_s is the settling velocity, c_w is the sediment concentration in the water column and τ_{crdep} is a critical deposition shear stress above which no deposition occurs. Since erosion and deposition occur simultaneously in nature, high values of τ_{crdep} can be prescribed, in which case sediment is always allowed to be deposited, should it be eroded at the next computational step (this cannot be done when compaction is taken into account, since sediment deposited at a previous time step will have a higher critical erosion shear stress at the next time step and may not be able to be eroded anymore).

No morphodynamic coupling is taken into account here: the sediment compartment is treated independently from the water column. It acts as a source and sink of sediments, and erosion or deposition do not affect the bathymetry.

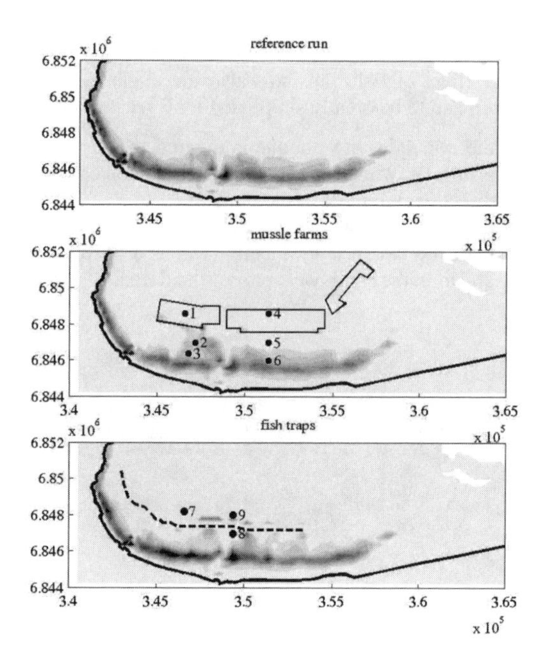

Figure 8. Sediment thickness after one year of simulation. Darker shades indicate thicker deposits. The same scale is used for all plots. See Figure 6 for quantitative estimations. Top: reference simulation; Center: with mussel farms; Bottom: with fish traps.

Results are shown in Figure 8: after one year, the sediment has been eroded from most of the domain. Deposits are restricted to areas higher than 3 m below mean sea level, where mud deposits are observed in the field. The main effect of the mussel farms is to increase deposition between them and the shore. The reduction of velocities offshore is not significant enough to prevent initial erosion nor to allow subsequent deposition.

The local slowing down of the flow in the vicinity of the dyke for the test on the fish traps results in local deposits on both sides of the trap. This occurs during periods of the flood and the ebb when the dyke emerges. Deposition seems otherwise to be restricted to a narrower area than without fish traps.

Figure 9a shows the evolution of the deposits with time for points 1 to 3 (results for points 4 to 6 are shown in Figure 9b). We have to bear in mind that the total amount of sediments in the computational domain is large, considering that the presence of the initial uniform layer of 10 cm is not in equilibrium with the hydrodynamic conditions of the area. Since the sediment is eventually eroded from most of the domain, the amount that is left in shallow water where the presence of these sediments is compatible with the

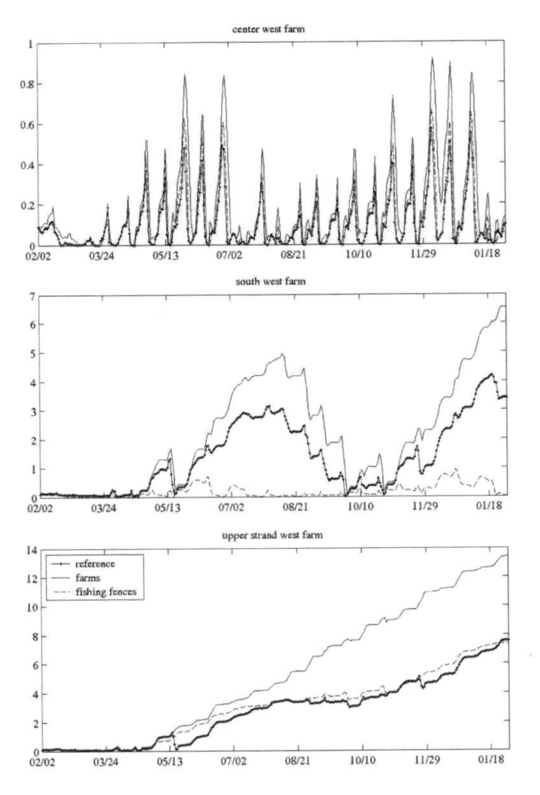

Figure 9a. Evolution of the deposits with time from February 2003 to February 2004 in the middle of the farms (top), south of the farms (centre) and higher up on the tidal flat (bottom). Reference points 1, 2 and 3. Thick dotted line: reference run; continuous line: with mussel farms, dashed line: with fish traps.

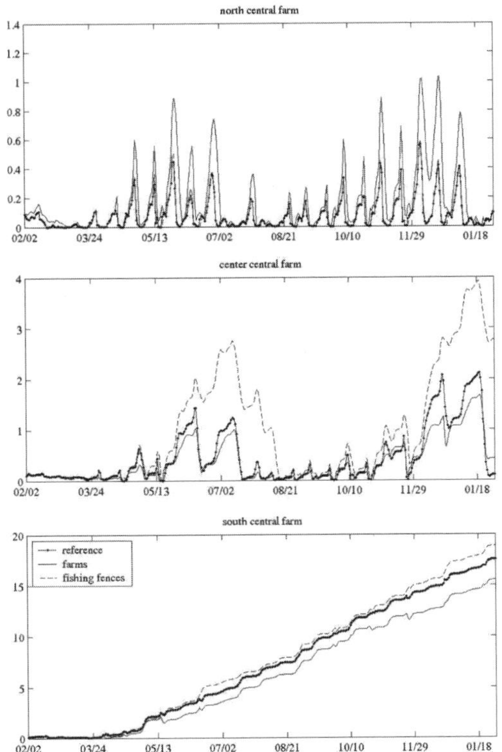

Figure 9b. Evolution of the deposits with time in the middle of the farms (top), south of the farms (center) and higher up on the tidal flat (bottom). Reference points 4, 5 and 6. Dotted line: reference run; continuous line: with mussel farms, dashed line: with fish traps.

hydrodynamic forcing is not realistic. Since no compaction is taken into account, the whole thickness of the deposits is considered as "fresh deposits", which allows massive erosion at each tide, and therefore very high concentrations in the water column. Results can therefore only be interpreted qualitatively.

As a general pattern, spring tides erode large amounts of sediments that migrate to the water column where the concentration increases. During neap tides, bottom shear stresses decrease and deposition can occur. This pattern is clearly visible for point 1 or 4 (centre of the mussel farms) for all runs (Fig. 9). Deposits at this location are thin compared to areas closer to the shore, and the deceleration of velocities observed due to mussel farms or fish traps increases deposition during neap tides. The situation is different on the highest part of the tidal flat (*e.g.* points 3 or 6). There, neap tides are barely energetic enough to put any sediment in suspension (Fig. 10). This results in alternate periods of accretion (during spring tides) and stability (during neap tides) for deposit thickness. In the long run, Figure 9 therefore shows a permanent accretion trend on the higher tidal flat throughout the simulation: sediments are progressively transported to the higher part of the flat. This trend is more acute when mussel farms are present, particularly south of the western farms (Point 3). Fish traps do not seem to have a very significant effect that high on the tidal flat.

Points 2 and 5 reflect an intermediary situation where bottom shear stresses are close to the critical erosion shear stress. Erosion and deposition patterns at this location are very sensitive to the maximum amplitude of the fortnightly spring tide. In 2003, the maximal spring tide amplitude exceeded 13 m in Cancale (small harbour north-west of the Bay) for every month from mid-February to mid-April. From mid-April to mid-August, the tidal amplitude was always under 12 m, whereas it exceeded 13 m again every other fortnightly cycle after mid-August. Since tidal velocities, and therefore bottom shear stresses, are positively correlated with tidal amplitude, we infer that continuous deposition occurs at point 2 (the same pattern is observed at point 5) as long as the maximum tidal range of a fortnightly cycle does not exceed a certain

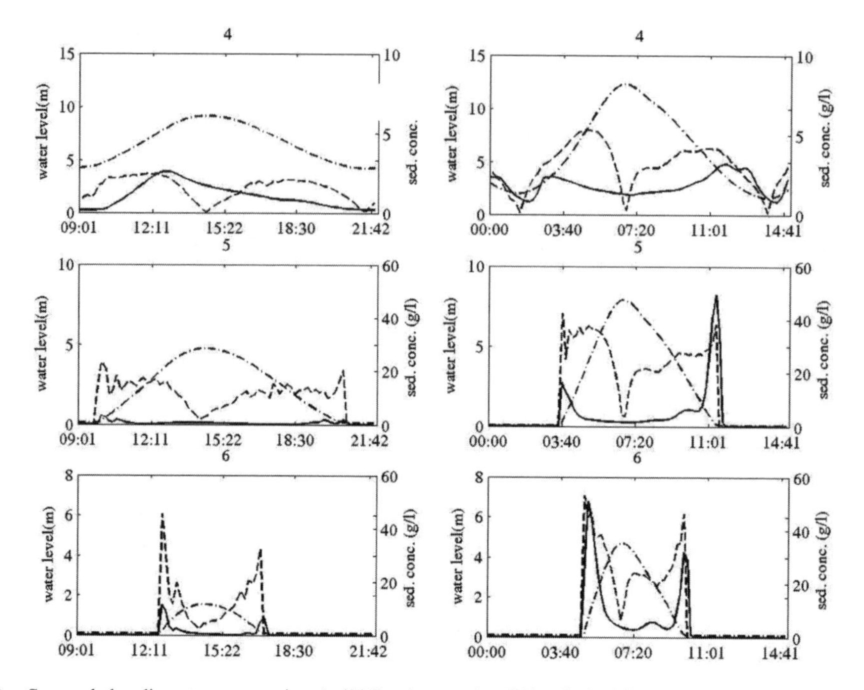

Figure 10. Suspended sediment concentration (solid lines) water level (dot-dashed lines) and 20*velocity magnitude (left scale in $m \cdot s^{-1}$, dashed lines) during neap tide (left) and spring tide (right). From top to bottom: points 4, 5 and 6.

threshold. The presence of mussel farms magnifies this pattern. On the other hand, when fish traps are taken into account, deposition never occurs at point 2 (which was already visible on Figure 8). However, this pattern occurs at point 4 where velocities are reduced by the fish traps compared to the reference run.

5 DISCUSSION, PERSPECTIVES AND CONCLUSION

From the hypothetical initial condition of a uniform mud coverage of Mont Saint Michel Bay, we reproduced by means of numerical modelling the role of the tide in filling in the upper part of the tidal flats while not allowing deposition of mud below 3 m below mean sea level (which is approximately the location of the fish traps). The presence of mussel farms was shown to encourage very limited sedimentation within the farms themselves, where tidal currents are energetic enough to erode any mud that settles during neap tides. However, the presence of the farms increases the width of the tidal flat south of the western farms. The very schematised representation of the fish traps did not lead to satisfying results.

Results obtained for natural conditions or while investigating the influence of mussel farms are encouraging. However, they need to be interpreted while keeping in mind the approximations that were made for the numerical modelling:

1. Waves were not taken into account. Wave measurements as well as preliminary computations of wave propagation show that wave-induced bottom shear stress in Mont Saint Michel Bay can be as high as 10 times the maximum tide-induced shear stress. This reason in itself could explain why the infill rate in nature is not nearly as high as the simulated deposition might suggest.
2. Only the muddy fraction was taken into account. As a response to a longshore and cross-shore variability of wave-induced bottom shear stress, the sediment coverage in the bay is very dependant on the location. This environment typically requires the use of a mixed grain size model, which is an option that we did not include for these runs. The parameterisation of erosion thresholds and erosion fluxes of natural sediments in the Bay is under investigation (Le Hir *et al.*, 2005).
3. Compaction was not taken into account. The very "fluid" density of fresh deposits that was used for these simulations therefore induces unrealistic concentrations in the water column.
4. The aforementioned simplifications made it superfluous to include morphodynamic feedback in the simulations. Updating the bathymetry naturally dampens sharp trends like the "continuous"

deposition simulated here, either by actually making the coastline prograde by infill, or by increasing wave effects in very shallow water, in which case the tidal flat is eroded.

The numerical model used for these applications was designed to take into account all these processes (Waeles *et al.*, 2005): mixed sediments, waves and compaction effects will be included in future works. Our first results nonetheless confirm the main role of the tide in building up tidal flats and show the significant influence of the mussel farms in increasing sediment deposition on parts of the flats.

REFERENCES

Bahé, S. 2003, Conchyliculture et dynamique morpho-sédimentaire en Baie du Mont Saint-Michel: mise en place d'une base de données géographiques, *Master's thesis*, EPHE, Laboratoire de Géomorphologie et Environnement Littoral, 161 pp.

Bonnot-Courtois, C., Caline, B., L'Homer, A., Le Vot, M. 2002, *Baie du Mont Saint-Michel et Estuaire de la Rance: Environnements sédimentaires, aménagements et évolution récente*. Editions TotalFinaElf. Mémoire no 26, 256 pp.

Ehrhold, A. 1999, Dynamique de comblement d'un basin sédimentaire soumis à un régime mégatidal: exemple de la Baie du Mont Saint Michel, *Thesis*, Université de Caen, France, 294 pp.

Jestin, H., Le Hir, P., Bassoullet, P. 1994, The 'SAMPLE system', a new concept of benthic station, *Oceans 94 Proceedings*, Brest, 13–16 September 1994, Vol. III, pp. 278–283.

Le Hir, P., Cann, P., Waeles, B., Bassoullet, P. 2005, Erodability of natural sediments: towards an erosion law for sand/mud mixtures from laboratory and field erosion tests (submitted).

Le Hir,P., Roberts, W., Cazaillet, O., Christie, M., Bassoullet, P., Bacher, C. 2000. Characterization of intertidal flat hydrodynamics. *Continental Shelf Research*, 1433–1459.

Le Roy, R., Simon, B. 2003. Réalisation et validation d'un modèle de marée en Manche et dans le Golfe de Gascogne; Application à la realisation d'un nouveau programme de réduction des sondages bathymétriques, Rapport d'études SHOM 002/03.

Migniot, C. 1998. Rétablissement du caractère maritime du Mont Saint Michel, Synthèse des connaissances hydro-sédimentaires de la Baie, Direction Départementale de l'Equipement de la Manche, Report, 111 pp.

SEAMER, 2000. Etude d'impact de la restructuration conchylicole en Baie du Mont Saint Michel, Etude courantologique et sédimentologique, Rapport SRC Bretagne Nord, 41 pp.

SOGREAH, 1986. Amélioration de la mytiliculture dans la Baie de l'Aiguillon, Rapport LCHF.

Waeles, B., Le Hir, P., Lesueur, P., Delsinne, N. 2005. Modelling sand/mud transport and morphodynamics in the Seine river mouth (France): an attempt by using a process-based approach (accepted for publication).

River, Coastal and Estuarine Morphodynamics: RCEM 2005 – Parker & García (eds)

Long term evolution of self-formed estuarine channels

Ilaria Todeschini, Marco Toffolon & Marco Tubino
Department of Civil and Environmental Engineering, University of Trento, Italy

ABSTRACT: The aim of this paper is to investigate the long term configuration attained by a tidal channel, namely its longitudinal bed profile and shape. Channel evolution is simulated through a one-dimensional numerical model, where the role of intertidal areas is neglected. The channel has a rectangular cross section and it is closed at one end. The major novelty of the work is that the width of the channel is allowed to change in time: in the present model lateral erosion is computed as a function of bed shear stress, provided it exceeds a threshold value within the cross section. The system, forced by a prescribed free surface oscillation at the seaward boundary, reaches an equilibrium profile of both bottom elevation and channel bank. The equilibrium geometry of the channel is determined for different initial and boundary conditions. Furthermore, the spatial variations of the cross section at equilibrium are analyzed in order to identify those conditions under which the commonly used exponential law of channel width variation is reproduced. Finally, a comparison is pursued between model predictions and bathymetries of real estuaries.

1 INTRODUCTION

The definition of long-term equilibrium conditions of tidal channels is an issue of great importance for many aspects related to the management of tidal environments. In spite of its relevance, the problem still awaits for a complete investigation.

In previous works (Schuttelaars and de Swart 2000; Lanzoni and Seminara 2002; Todeschini et al. 2003) the attention has been mainly focused on the case of well-mixed, tide-dominated estuaries, characterized by a given funnel shape, negligible river discharge and vanishing bottom slope (Perillo 1995; Savenije and Veling 2005). Furthermore, the influence of tidal flats has been ignored. The above works suggest that, when a reflective boundary is assigned at the landward end, the morphological evolution of a tidal channel, starting from a horizontal bed profile, is characterized by the formation of a sediment wave that migrates slowly landward until it leads to the emergence of a beach. This condition generally inhibits the further development of the channel and determines an asymptotic intrinsic length of the estuary. The resulting bed profile is characterized by an increasing bottom elevation in the landward direction, as confirmed also by the experimental observations of Bolla Pittaluga et al. (2001). It is important to note that channel convergence directly controls the above equilibrium length; in fact, it strongly affects both the hydrodynamics (Friedrichs and Aubrey 1994) and the consequent morphodynamical evolution (Lanzoni and Seminara 2002). The interest for the study of such class of estuaries is motivated by the fact that, in nature, many estuaries display a typical funnel shape. The Thames and the Bristol Channel, whose plan view can be seen in Figure 1, are examples of this estuarine category.

Understanding the reason why tidal channels are convergent and defining the conditions under which the exponential law for width variation, which is so often observed in nature, is reproduced are the main objectives of the present work. We then remove the main assumption on which previous models are based, namely that of fixed banks, and let the channel bank vary with time. We note that various attempts to reproduce the equilibrium cross section of tidal channels have been pursued in previous works (e.g. Gabet 1998, Fagherazzi and Furbish 2001). However, the above analyses mainly focus on the local scale, hence the full coupling with the morphological evolution of the channel is not taken into account. On the other hand, several contributions already exist in the case of rivers. For instance, a numerical model of widening and bed deformation has been proposed by Darby and Thorne (1996) who considered both planar and rotational failures and calculated channel widening by coupling bank stability with flow and sediment transport algorithms.

In our work a strongly simplified approach is adopted, whereby only the effects related to flow and sediment transport processes within the tidal channel

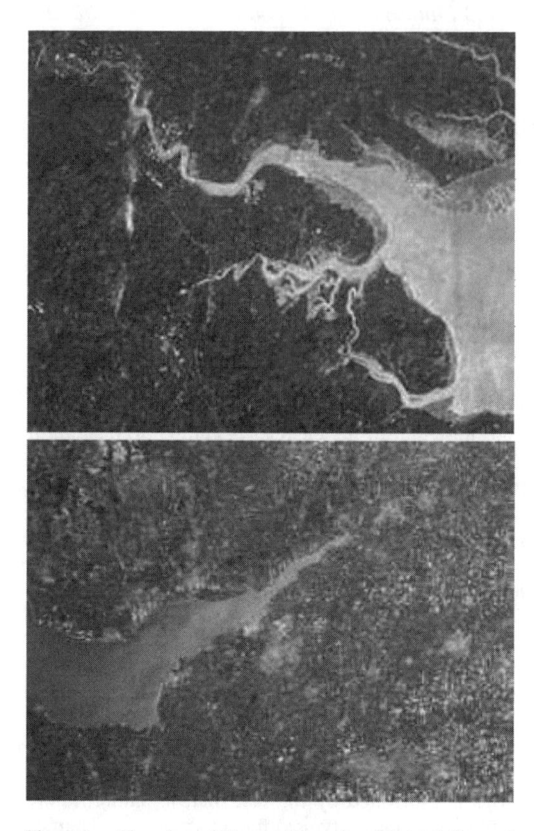

Figure 1. Plan view of Thames (above) and Bristol Channel (below).

are retained and further ingredients, like the control exerted by tidal flats or the direct effect of sea currents in the outer part of the estuary, are discarded. Furthermore, a one-dimensional model is used to investigate the long-term evolution of the channels. The lateral erosion is taken into account and computed as a function of bed shear stress, provided it exceeds a threshold value within the cross section. In the paper we test the effect of different initial and boundary conditions on the widening process of the channel, along with the role of the incoming river discharge.

2 FORMULATION OF THE PROBLEM

We consider a tidal channel of length L_e^* and rectangular cross section (hereafter the asterisk denotes dimensional quantities and the subscript e indicates tidally averaged values at the mouth of the estuary at the beginning of the simulation). The basic notation is reported in Figure 2. We adopt a one-dimensional model to study the propagation of the tidal wave along

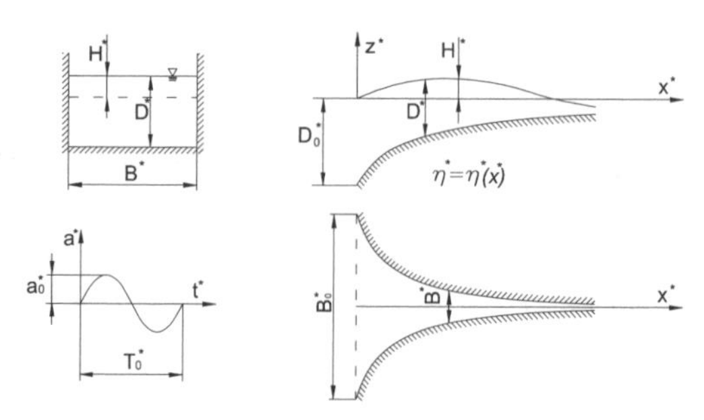

Figure 2. Sketch of the estuary and basic notation.

the channel. The momentum and continuity equations read:

$$\frac{\partial Q^*}{\partial t^*} + \frac{\partial}{\partial x^*}\left(\frac{Q^{*2}}{\Omega^*}\right) + g^*\Omega^*\frac{\partial H^*}{\partial x^*} + g^*\Omega^* j = 0\,, \qquad (1)$$

$$\frac{\partial Q^*}{\partial x^*} + \frac{\partial \Omega^*}{\partial t^*} = 0\,, \qquad (2)$$

where t^* is time, Q^* the water discharge, Ω^* the area of the cross section, H^* the free surface elevation and g^* is gravity; furthermore

$$j = \frac{U^*|U^*|}{C_h^2 g^* D^*} \qquad (3)$$

is the frictional term, where C_h is the dimensionless Chézy coefficient, D^* is the water depth and U^* is the cross-sectionally averaged velocity. The morphodynamic evolution of the channel is governed by the sediment continuity equation, which reads:

$$(1-p)B^*\frac{\partial \eta^*}{\partial t^*} + \frac{\partial (B^* q_s^*)}{\partial x^*} = \frac{\partial B^*}{\partial t^*}D^*\,, \qquad (4)$$

where p is sediment porosity, B^* is the channel width, q_s^* is the sediment flux per unit width and $\eta^* = H^* - D^*$ is bottom elevation. The term on the right hand side represents the amount of sediments eroded from the banks, which can finally settle at the bottom of the channel.

In the results presented herein the dimensionless sediment flux is evaluated at any time using the relationship proposed by Engelund and Hansen (1967)

$$\phi = 0.05\, C_h^2 \left(\frac{U^{*2}}{C_h^2\, g^* \Delta\, d_s^*}\right)^{5/2}\,. \qquad (5)$$

where d_s^* is the characteristic particle diameter and $\Delta \simeq 1.65$ the relative density of sandy sediments with respect to water. In order to evaluate the tidally averaged value of the sediment flux at the mouth, ϕ_0, we assume a sinusoidal velocity $U^* = U_0^* \sin(2\pi t^*/T_0^*)$;

using such expression and averaging equation (5) we obtain a suitable estimate of the scale of sediment flux, per unit width, in the following form:

$$\phi_0 = 0.05 \frac{16}{15\pi} C_h^2 \left(\frac{U_0^{*2}}{C_h^2 g^* \Delta d_s^*} \right)^{5/2} . \tag{6}$$

The problem is then set in dimensionless form, using the length of the estuary L_e^*, the tidal period T_0^*, the width B_0^* and the depth D_0^* at the mouth as representative scales; moreover, a scale U_0^* is assumed for the velocity. Hence we define

$$t = \frac{t^*}{T_0^*}, \quad x = \frac{x^*}{L_e^*}, \quad B = \frac{B^*}{B_0^*},$$

$$(H, \eta, D) = \frac{(H^*, \eta^*, D*^*)}{D_0^*}, \quad U = \frac{U^*}{U_0^*},$$

$$Q = \frac{Q^*}{B_0^* D_0^* U_0^*}, \quad \Omega = \frac{\Omega^*}{B_0^* D_0^*}, \tag{7}$$

and we assume C_h as constant.

We note that the erosion/deposition process typically occurs on a time scale which is much slower than the tidal period T_0^*. As a consequence, the morphodynamic problem can be decoupled from the hydrodynamics and the sediment continuity equation (4) can be scaled using a morphological time scale for bed evolution T_b^*. Recalling that

$$q_{s0}^* = \sqrt{g^* \Delta d_s^{*3}}\, \phi_0 \tag{8}$$

is the scale of the total sediment flux per unit width, where ϕ_0 depends on the empirical relationship used to evaluate the sediment flux (6), the above time scale reads

$$T_b^* = \frac{(1-p)\, D_0^* L_e^*}{q_{s0}^*} . \tag{9}$$

We then set further dimensionless variables in the following form:

$$\tau = \frac{t^*}{T_b^*}, \quad q_s = \frac{q_s^*}{q_{s0}^*} . \tag{10}$$

Using the above scalings, equations (1), (2) and (4) can be written as:

$$\frac{\partial Q}{\partial t} + \frac{1}{S_t} \frac{\partial}{\partial x} \left(\frac{Q^2}{\Omega} \right) + \frac{1}{F_r^2 S_t} \Omega \frac{\partial H}{\partial x} + \frac{\delta}{F_r^2 S_t} \Omega j = 0, \tag{11}$$

$$\frac{\partial H}{\partial t} + \frac{1}{S_t B} \frac{\partial Q}{\partial x} = 0, \tag{12}$$

$$\frac{\partial \eta}{\partial \tau} + \frac{\partial q_s}{\partial x} + \frac{q_s}{B} \frac{\partial B}{\partial x} = \frac{D}{B} \frac{\partial B}{\partial \tau} . \tag{13}$$

We note that two dimensionless parameters arise, namely the Strouhal number $S_t = L_e^*/(U_0^* T_0^*)$ and the Froude number $F_r = U_0^*/\sqrt{g^* D_0^*}$.

The differential system (11)–(12), written in semi-conservative form in terms of the variables Q and H in order to enhance the conservation of mass and momentum, is discretized through finite differences and solved numerically using the explicit MacCormack method. The numerical scheme is second order accurate both in space and in time provided the usual Courant-Friedrich-Levi stability condition is respected. A suitable artificial viscosity is introduced through a TVD filter (Total Variation Diminishing) in order to remove the spurious oscillations around discontinuities, which may arise since the tidal wave tends to break during its propagation due to friction and convergence.

The sediment continuity equation (13) is discretized through finite differences and solved numerically (at every time step of the hydrodynamic part) using a first-order upwind method, with the same time step imposed by the CFL condition for the hydrodynamic problem.

In the numerical simulations we assume a horizontal bed profile as initial condition. At the seaward boundary we impose that the free surface level is only determined by the large scale component of the tide. Hence, we characterize the sea level oscillation with the semi-diurnal M_2 tide, whose dimensional amplitude is a_0^*, and set:

$$H(t)|_{x=0} = \varepsilon \sin(2\pi t), \tag{14}$$

where $\varepsilon = a_0^*/D_0^*$. Moreover we assume that the sediment flux entering through the mouth exactly balances the transport capacity.

At the landward boundary we examine two different conditions: in the first case we assume a negligible river discharge, that corresponds to a reflecting barrier condition, and therefore to negligible sediment flux; in the second case we consider a non vanishing river discharge and we set the incoming sediment flux at the equilibrium value corresponding to the transport capacity of the incoming flow.

It is worth noting that when the channel banks are assumed to be fixed (i.e. the right hand side of equation (13) vanishes), the governing problem reduces to that tackled in previous works (Lanzoni and Seminara 2002; Todeschini et al. 2003). In this case the bottom profile reaches a dynamical equilibrium, which is characterized by negligible residual values of the sediment transport all along the estuary; furthermore, its

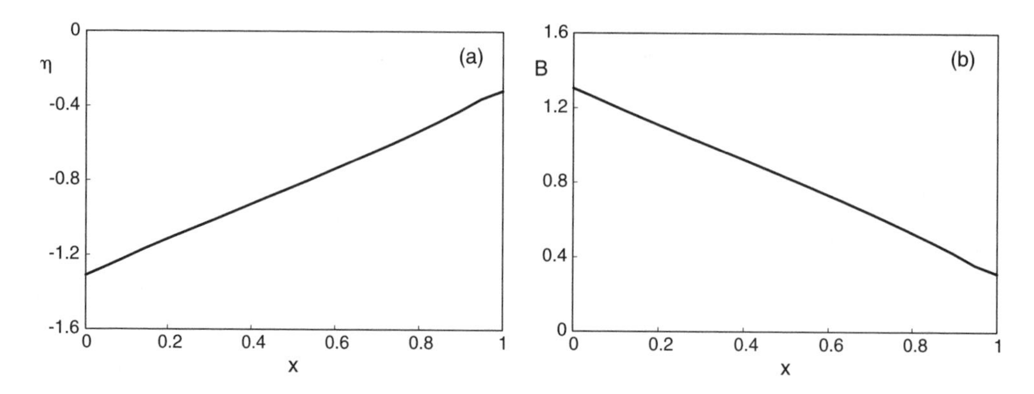

Figure 3. Bank erosion evaluated assuming β as constant using equation (15): the long-term configuration of (a) the bottom profile $\eta(x)$ and (b) the banks profile $B(x)$ [$L_e^* = 20$ km, $D_0^* = 5$ m, $a_0^* = 1$ m, $B_0^* = 100$ m, $C_h = 20$, $d_s^* = 10^{-4}$ m, $\gamma^* = 10^{-8} s^{-1}$, $\beta = 20$].

configuration mainly depends on the values of external parameters, among which the degree of convergence and the amplitude of tidal forcing play an important role.

3 A MODEL FOR WIDTH CHANGE

The aim of this work is to investigate the long term morphological evolution of a tidal channel whose width can vary in time. Hence, we need to supplement the governing problem with a bank erosion law suitable for a tidal context, whose definition is not obvious in a one-dimensional framework. Differently from the bed-erosion process, which can be treated in a satisfactory, albeit simplified, way within a one-dimensional framework, width changes are inherently influenced by the hydrodynamic behavior close to the banks, which can differ significantly from the cross-sectionally averaged flow.

A first possibility is to abridge the transverse behavior into the concept of asymptotic equilibrium shape of the cross-section, assuming that the width-to-depth ratio of the channel may attain a given value β. Consequently, the bank erosion can be modelled in the following form

$$\frac{\partial B^*}{\partial t^*} = \gamma^* \left(B_{eq}^* - B^* \right) , \tag{15}$$

where the equilibrium width

$$B_{eq}^* = \beta D_{av}^* \tag{16}$$

is related to the tidally averaged depth D_{av}^*. We note that the parameter γ^* represents the inverse of the time scale of width evolution. In fact, if we fix B_{eq}^*, the solution of equation (15) can be written as

$$B^*(t^*) = B_{eq}^* + (B^*|_{t^*=t_0^*} - B_{eq}^*) \exp[-\gamma^*(t^* - t_0^*)] \,. \tag{17}$$

As a first approximation we can assume β to be constant along the entire channel, though this is quite a crude simplification (Lawrence et al. 2004; D'Alpaos et al. 2005). In Figure 3 an example of the equilibrium bottom and bank profiles, evaluated according to the above assumption, is shown: the bottom profile is linearly increasing landward and consequently, through equation (16), the tidally averaged depth and the bank profile exhibit the same shape. A more realistic solution would require a further morphological model able to estimate the width-to-depth ratio β in equation (16) on the basis of hydrodynamic and geometrical conditions. The degree of uncertainty affecting the above estimate along with the constraint imposed by the application of equation (16) suggest the opportunity to search for a more physically-based erosion law.

In the case of rivers several existing models can be found in the literature, whereby width adjustment is essentially related to the excess bed shear stress with respect to some threshold value. Following Darby and Thorne (1996), we assume that channel widening may occur in a given section provided the bottom shear stress $\sigma^* = \rho\,(U^*/C_h)^2$ exceeds a threshold value σ_{cr}^*:

$$\frac{\partial B^*}{\partial t^*} = k^* \left(\frac{\sigma^*}{\sigma_{cr}^*} - 1 \right) \quad \text{for } \sigma^* > \sigma_{cr}^* \,. \tag{18}$$

If we write (18) in dimensionless form we obtain:

$$\frac{\partial B}{\partial \tau} = k \left(\frac{U^2}{U_{cr}^2} - 1 \right) \quad \text{for } |U| > U_{cr} \,, \tag{19}$$

in which $k = k^* T_b^* / B_0^*$. It is worth noting that equations (18)–(19) represent pure erosional laws. In our simulations a value of $0.1\,N/m^2$ of the critical bed shear stress σ_{cr}^* is adopted.

The definition of a suitable range of values for the lateral erosion rate k^* poses an important and non trivial problem, since only few contributions can be found

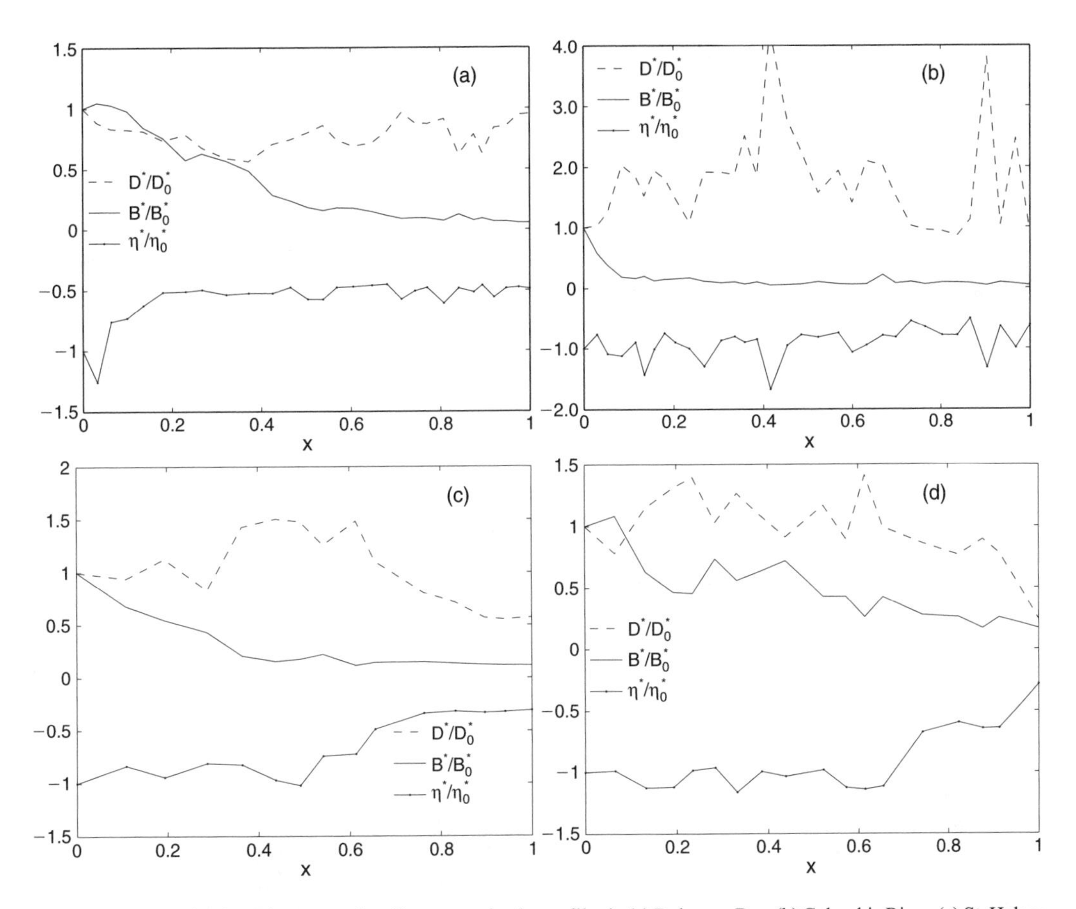

Figure 4. Bottom, banks and cross-sectionally average depths profiles in (a) Delaware Bay, (b) Columbia River, (c) St. Helena Sound and (d) St. Andrew Sound.

in the literature dealing with bank erosion processes in tidal channels. According to the estimates reported by Darby and Thorne (1996) in the case of rivers, we can infer that a value for k^* of the order of 10^{-8} m/s is suitable for the present case. Another estimate for k^* can be deduced from Gabet (1998), who investigated the process of bank erosion of a saltmarsh creek, reporting a lateral migration rate of the order of 10^{-11} m/s. An intermediate value $k^* = 10^{-10}$ m/s is adopted in our numerical simulations where equation (19) is solved using an explicit scheme.

4 RESULTS

We now investigate under which conditions the simple one-dimensional model formulated in the preceding section is able reproduce the main distinctive features of tide-dominated estuaries. In order to examine their morphology more closely, we have utilized the bathymetric Digital Elevation Model (DEM, at one arc second resolution) of some estuaries in North America, provided by the U.S. National Ocean Service.

In Figure 4 we plot the longitudinal profiles of the channel width B^*, cross-sectionally averaged bottom elevation η^* and depth $D^* = \Omega^*/B^*$ of four different estuaries (Delaware, Columbia, St. Helena Sound, St. Andrew Sound). The variables are scaled with their values at the mouth of the estuary (B_0^*, η_0^* and D_0^*); furthermore, the reported values refer to tidally averaged conditions. As it can be seen, all the selected estuaries have a funnel shaped geometry. Moreover, the bed profile typically displays a relatively mild landward slope and in some cases the average bed slope is almost vanishing. This behavior seems to contradict both numerical and experimental findings obtained with fixed convergent geometries (Todeschini et al. 2003; Bolla Pittaluga et al. 2001) where the bottom slope is invariably found to attain relatively large values. We may note that the fixed banks approach is essentially based on the premise that the time scale of bed development is much faster than that of bank, such

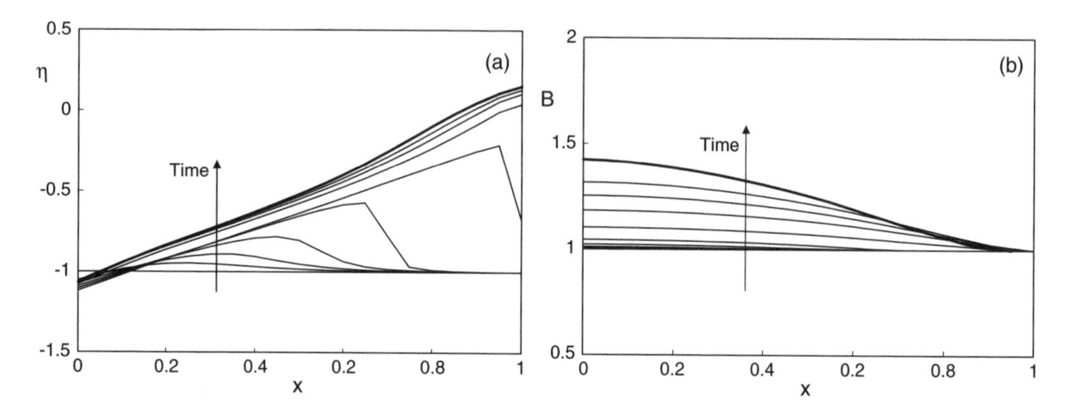

Figure 5. The long-term evolution of: (a) the bottom profile $\eta(x)$; (b) the banks profile $B(x)$ [$L_e^* = 20$ km, $D_0^* = 5$ m, $B_0^* = 50$ m, $a_0^* = 1$ m, $C_h = 20$, $k^* = 10^{-10}$ m/s, $d_s^* = 10^{-4}$ m, $U_0^* = 0.7$ m/s, $T_b^* = 157$ years].

that bed adjustment is driven by the channel geometry. The above discrepancy between prediction and observations could be related to the effect of river discharge which has been typically discarded in the above works. Furthermore, it may also indicate that the mutual interaction between bed and bank development in tidal channels cannot be completely neglected as in the case of fixed banks simulation, at least when the time scales of the two processes are not so far apart.

To investigate the long-term evolution of the channel cross-section we now use our one-dimensional numerical model relaxing the hypothesis of fixed banks. At first we need to define a threshold value for the bottom variation within a certain period of time, below which we can assume that a quasi-equilibrium configuration has been achieved. In fact, within a one-dimensional approach it is not obvious to reproduce the conditions which determine the stability of the banks. If the channel width B varies in time according to (19), there is no mechanism ensuring that an equilibrium configuration can be reached, because the erosion law substantially depends on the peak values of the velocity and not on the tidally averaged values, as in the case of bottom elevation η through (13). Thus, while the bottom configuration becomes relatively stable after a period of time comparable with the morphological time scale, the width continues to increase because the velocities are only weakly influenced by the channel width.

An example of this evolution can be seen in Figure 5 and in Figure 6, where the differences in the adaptation process of bottom elevation and width are quite evident: the bottom evolution is concentrated in a smaller period of time, which is required to let the sediment wave to reach the last section of the channel; on the contrary channel widening proceeds with a velocity that is roughly constant in time. Despite of this fact, when the bottom attains a quasi-equilibrium configuration we assume that the bank profile has achieved a sort of characteristic configuration, which determines the planimetric shape.

In principle we can expect the equilibrium profiles to be mainly dependent on the external forcing. In fact the tidal amplitude a_0^* has a strong influence on the solution, since it controls the scale of velocity U_0^* in the estuary (Toffolon 2002); furthermore, the propagation of the tidal wave along the channel determines its hydrodynamical and morphological behavior. Given the same geometrical conditions, a larger tidal amplitude corresponds to a larger rate of widening, because it involves larger velocities in (19). Hence, we do expect the solution to be only weakly dependent on the initial depth at the mouth D_0^*, for a given tidal forcing. This is confirmed by the results reported in Figure 7, where we plot the equilibrium profiles corresponding to four different values of the initial depth for the same dimensional tidal amplitude a_0^*. The resulting profiles are quite similar.

A special attention has to be paid in the choice of the constant k^* which controls the lateral erosion rate of the banks. As pointed out before, it is very difficult to obtain a reliable estimate of this parameter; on the other hand its value deeply influences the solution. Imposing different values of the parameter k^*, we obtain different equilibrium bottom profiles and consequently different bank configurations (Figure 8). This implies that k^* plays a non negligible role, not only in the transient phase of evolution but also in the equilibrium profiles, though the bottom and the banks evolve on different time scales. The degree of widening influences the bottom profile until it has reached an equilibrium configuration, while it is no longer relevant when the residual sediment transport that governs the bottom evolution becomes negligible.

It is worth noting that in all the examples reported above the resulting bank profiles $B(x)$ are characterized by a concave shape (i.e. with a decreasing rate of widening seaward) and by a significant bottom

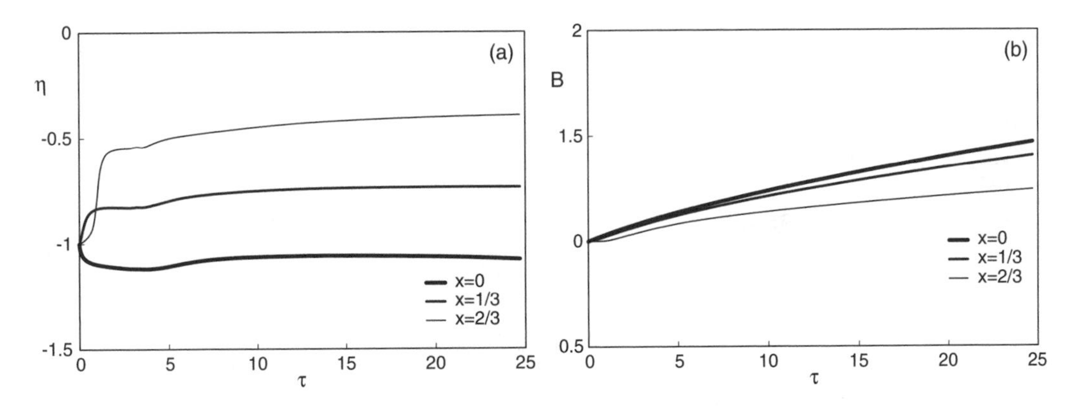

Figure 6. The changing with time of: (a) the bottom elevation $\eta(\tau)$; (b) the width $B(\tau)$, at $x=0, x=1/3$ and $x=2/3$ [$L_e^*=20$ km, $D_0^*=5$ m, $B_0^*=50$ m, $a_0^*=1$ m, $C_h=20$, $k^*=10^{-10}$ m/s, $d_s^*=10^{-4}$ m, $U_0^*=0.7$ m/s, $T_b^*=157$ years].

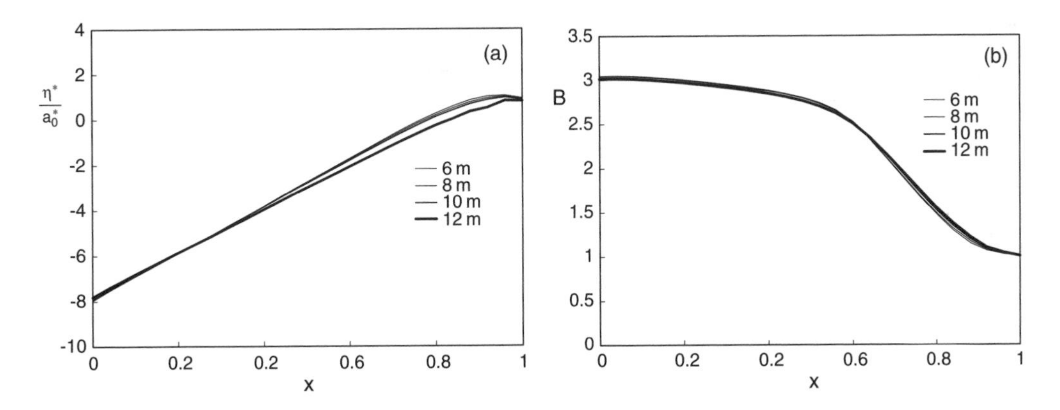

Figure 7. The long-term configuration of: (a) the bottom elevation scaled with the tidal amplitude η^*/a_0^*; (b) the banks profile $B(x)$, for different values of initial depth at the mouth, D_0^* [$L_e^*=50$ km, $a_0^*=1$ m, $B_0^*=50$ m, $C_h=20$, $L_e^*=50$ km, $k^*=10^{-10}$ m/s, $d_s^*=10^{-4}$ m].

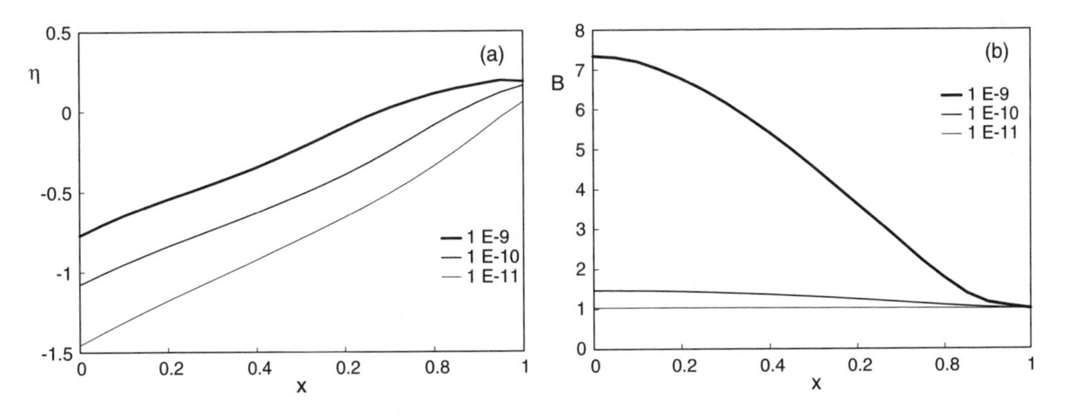

Figure 8. The long-term configuration of: (a) the bottom profile $\eta(x)$; (b) the banks profile $B(x)$, for different values of the constant used in the lateral erosion law k^* $(1\cdot 10^{-11}, 1\cdot 10^{-10}, 1\cdot 10^{-9}\ m/s)$ [$L_e^*=20$ km, $D_0^*=5$ m, $a_0^*=1$ m, $B_0^*=50$ m, $C_h=20$, $d_s^*=10^{-4}$ m].

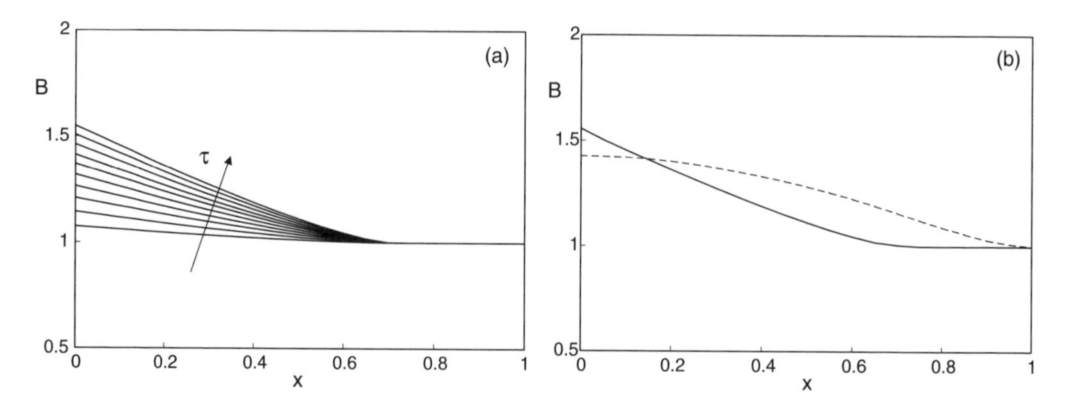

Figure 9. Banks profile $B(x)$: (a) evolution obtained with fixed horizontal bed at τ varying from 2 to 20; (b) comparison between the case of fixed bed (continuous line) and mobile bed (dashed line) at $\tau = 20$ [$L_e^* = 20$ km, $D_0^* = 5$ m, $a_0^* = 1$ m, $B_0^* = 50$ m, $C_h = 20$, $d_s^* = 10^{-4}$ m, $k^* = 10^{-10}$ m/s, $U_0^* = 0.7$ m/s, $T_b^* = 157$ years].

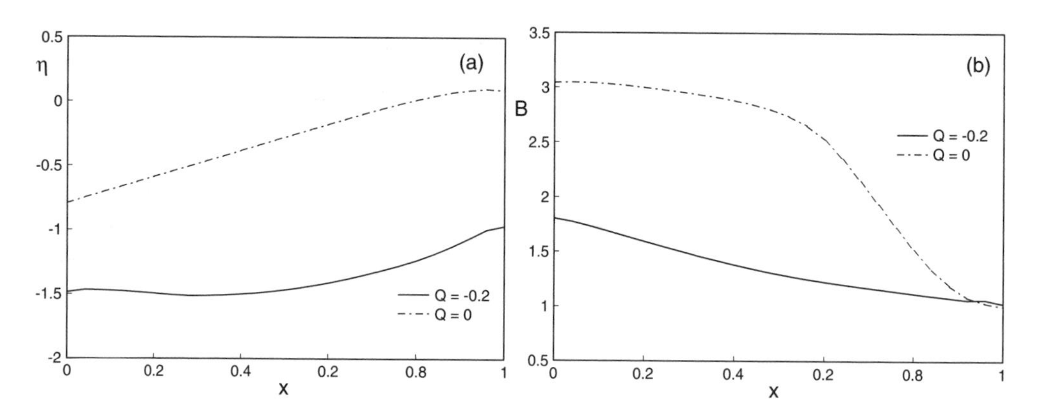

Figure 10. Comparison between the long-term configurations of: (a) the bottom profile $\eta(x)$ and (b) the banks profile $B(x)$, for a vanishing river discharge (dash-dot lines) and a non-negligible value of the river discharge at the landward boundary, $Q|_{x=1} = -0.2$ (continuous lines) [$L_e^* = 50$ km, $D_0^* = 10$ m, $a_0^* = 1$ m, $B_0^* = 50$ m, $C_h = 20$, $d_s^* = 10^{-4}$ m, $k^* = 10^{-10}$ m/s, $U_0^* = 1$ m/s].

slope; on the other hand observed bank profiles in tide dominated estuaries, like those presented in Figure 4, display a convex shape (i.e. with an increasing rate of widening seaward) while the bottom slope keeps relatively small. In order to understand the role of the bottom profile on the bank shape we have tested our model imposing a fixed horizontal bed. The resulting bank profile $B(x)$ is plotted in Figure 9a for different values of τ. It appears that in this case the tendency toward the establishment of an exponential law for the bank profile can be reproduced. The difference with the case of mobile bed is even more evident if we compare the banks profile obtained in this case with the one reported in Figure 5, both plotted at time $\tau = 20$ (Figure 9b).

There are several other factors which can affect the morphological aspect of tidal channels. Here we focus our attention on the role of the river discharge, which has been neglected in previous simulations but can be easily included in a one dimensional model. When we impose a non-vanishing freshwater discharge at the landward boundary, the formation of a beach within the estuary is inhibited and the resulting bottom profile can be significantly different. In Figure 10 we compare the equilibrium profiles in the case of a closed end (the example reported in Figure 8 for $k^* = 10^{-10}$ m/s) and in the case of a non-negligible value of the river discharge ($Q^* = 100$ m^3/s). In the latter case the equilibrium bottom profile is characterized by a smaller slope and a larger depth because the seaward-directed discharge contrasts with the landward-directed sediment transport due to the tidal wave. However, the most important difference with the previously illustrated simulations is that the presence of a discharge, imposed at the landward end, determines a banks profile $B(x)$ with a convex shape, which resembles the exponential profile of real estuaries. Although these simulations are not enough to draw general conclusions on this issue, anyway we can infer that the river discharge influences the solution in a significant way. In particular, its presence induces a morphological evolution that makes the planimetric shape of the channels more similar to that typical of real estuaries, even if the ratio between the river supply and the tidal prism at the mouth remains relatively limited, thus remaining in the tide-dominated class of estuaries.

5 CONCLUSIONS

In nature tide-dominated estuaries are characterized by a funnel-shape, negligible river discharge and small bottom slopes. With this work we investigate the long-term evolution of an estuarine channel, whose width is allowed to vary with time, through the use of a one-dimensional numerical model. The choice of the erosion law, however, is not a trivial question. We have explored two ways to model this phenomenon. Firstly, we have considered the idea of the existence of an equilibrium cross-section described by the width-to-depth ratio β as in (15)–(16). Though this can be regarded as a phenomenological description, it hides the problem of determining the values of β along the estuary; on the other hand, considering a constant value of this parameter gives the unrealistic results that the width is tightly related to the depth.

As a second conceptual scheme, we adopted the physically-based law (18), which relates the velocity of bank erosion to the excess of the bottom shear stress with respect to a threshold value, as it is common in the formulation of the bed erosion due to sediment transport. However, besides the problem to determine the correct value of the parameter k^* that controls the intensity of erosion, the adopted relationship hinders the possibility to reach a stable bank configuration because a purely erosive law vanishes only when the bottom shear stress reaches its critical value. We could have avoided this problem considering a process of bank reconstruction, which usually occurs on a large time scale. Instead of using an unrealistic instantaneously varying depositional law, this process could be modelled with a fixed rate of reconstruction due to vegetation activity, for instance. At the moment, we have chosen not to introduce a new parameter whose evaluation is totally uncertain. On the other hand, through the simple law (18) we can examine the influence of the width on the velocity. In fact, differently from the case of the bed evolution, where the bottom profile has a strong influence on the values of the velocity developing within the channel, the width seems to control the flow field mainly through the influence of the channel convergence rather than through the local value of the width in a given section. For instance, when we assume an exponential variation of the width and fixed banks, the solution is independent of the local value of the width (Todeschini et al. 2003). In this way, even large variations of channel width do not change the values of velocity significantly.

Despite of the uncertainties in the bank erosion law, an interesting result is shown by the present analysis. The typical funnel shape of most estuaries seems to be related to the occurrence of a small longitudinal slope of the bottom, which can results from the effect of an incoming discharge at the head of the estuary. However, we expect that other features, like bi- or three-dimensional circulations within the channel, the formation of large scale bed forms or the exchange with the outer sea, which have not been considered in this analysis, could play a role in the long-term planimetric evolution of estuaries; they certainly should be the subject of further research.

ACKNOWLEDGMENTS

The present work has been funded by the University of Trento and the Italian Ministry of Education, University and Research (MIUR) under the National program 'Idrodinamica e morfodinamica di ambienti a marea' (Cofin 2002).

REFERENCES

Bolla Pittaluga, M., N. Tambroni, C. Zucca, L. Solari and G. Seminara (2001). Long term morphodynamic equilibrium of tidal channels: preliminary laboratory observations. In *Proceedings of 2nd RCEM Symposium*, Obihiro (Japan), pp. 423–432.

D'Alpaos, A., S. Lanzoni, M. Marani, S. Fagherazzi and A. Rinaldo (2005). Tidal network ontogeny: Channel initiation and early development. *Journal of Geophysical Research 110*(F2), F02001 10.1029/2004JF000182.

Darby, S. E. and C. R. Thorne (1996). Numerical simulation of widening and bed deformation of straight sand-bed rivers. i: Model development. *Journal of Hydraulic Engineering 122*(4), 184–193.

Engelund, F. and E. Hansen (1967). *A monograph on sediment transport in alluvial streams.* Danish Technical Press.

Fagherazzi, S. and D. J. Furbish (2001). On the shape and widening of salt-marsh creeks. *J. Geophys. Res. 106*(C1), 991–1003.

Friedrichs, C. and D. Aubrey (1994). Tidal propagation in strongly convergent channels. *J. Geophys. Res. 99*, 3321–3336.

Gabet, E. J. (1998). Lateral migration and bank erosion in a saltmarsh tidal channel in san francisco bay, California. *Estuaries 21*(4B), 745–753.

Lanzoni, S. and G. Seminara (2002). Long term evolution and morphodynamic equilibrium of tidal channels. *J. Geophys. Res. 107*, 1–13.

Lawrence, D. S. L., J. R. L. Allen, and G. M. Havelock (2004). Salt marsh morphodynamics: an investigation of tidal flows and marsh channel equilibrium. *Journal of Coastal Research 20*(1), 301–316.

Perillo, G. M. E. (1995). *Geomorphology and sedimentology of estuaries.* Elsevier.
Savenije, H. H. G. and E. J. M. Veling (2005). Relation between tidal damping and wave celerity in estuaries. *J. Geophys. Res. 110*(C4), C04007 10.1029/2004 JC002278.
Schuttelaars, H. M. and H. E. de Swart (2000). Multiple morphodynamic equilibria in tidal embayments. *J. Geophys. Res. 105*, 24105–24118.
Todeschini, I., M. Toffolon, G. Vignoli and M. Tubino (2003). Bottom equilibrium profiles in convergent estuaries. In *Proceedings of 3nd RCEM Symposium*, Barcelona (Spain), pp. 710–722.
Toffolon, M. (2002). *Hydrodynamics and morphodynamics of tidal channels*. Ph. D. thesis, University of Padova, Italy.

River, Coastal and Estuarine Morphodynamics: RCEM 2005 – Parker & García (eds)
© 2006 Taylor & Francis Group, London, ISBN 0 415 39270 5

Band width analysis morphological predictions Haringvliet Estuary

G. Dam & A.J. Bliek
Svašek Hydraulics, Rotterdam, The Netherlands

A.W. Bruens
National Institute for Marine and Coastal Management, The Hague, The Netherlands

ABSTRACT: Process based morphological models are used as a policy tool to predict the morphological changes due to (human) impacts. Due to computational time of these models sensitivity of the model results are often not considered. This paper presents a case study in the Haringvliet Estuary (The Netherlands) and shows that bandwidth in morphological predictions cannot be neglected. The questions to be answered in estuarine morphology should not only concentrate on the bathymetry after a period of (say) 10 years, but also on the accuracy ranges of the prediction.

1 INTRODUCTION

1.1 *General*

In recent years, different process based morphological models have been developed to predict seabed changes in tidal estuaries. Especially the models that dynamically predict seabed changes over periods of years are very attractive to be used as a tool in the prediction of the impact of human activities on the long term estuarine development. However, this type of models asks for heavy computer capacity to obtain the required spatial resolution in the area of concern. Budget and time frame constraints are the main reasons that the sensitivity of the forecast for model uncertainties and/or sediment properties are often not considered. This paper presents a case study in the Haringvliet Estuary (the Netherlands) that emphases on the uncertainties of the forecast. To determine the bandwidth of the morphological predictions was one of the key items of the scope.

1.2 *Haringvliet case study*

The Haringvliet case study is carried out with the morphodynamic model FINEL2D. The seabed changes over the last 30 years (after closure of part of the estuary and other human activities) were used to calibrate and validate the model. After calibration and validation the model was applied to forecast the long term development of the estuarine bathymetry for different development schemes, like re-opening of the inner part of the estuary and a further extension of the Rotterdam Port area.

The forecast model was not only used with the set of (calibrated and validated) model parameters and boundary conditions, but a range of calculations was done with varying input data. In this way also an accuracy range for the morphodynamic forecast was obtained. Accuracy ranges in terms of future seabed bathymetry have been determined for the following types of uncertainties:

- Uncertainties related to physical parameters like hydrodynamic factors and sediment properties;
- Uncertainties related to modelling principles and assumptions;
- Uncertainties related to climatological factors (like the occurrence of heavy storms or river floods).

2 THE HARINGVLIET ESTUARY IN PAST AND FUTURE

2.1 *General*

The Haringvliet estuary is located in the South West part of the Netherlands, see figure 1. In 1970 the major (inner) part of the estuary has been closed off from the sea to protect the hinterland against flooding. As a part of the closure dam, a fresh water discharge sluice was built to discharge the fresh water of the rivers Rhine and Meuse to the sea. During low tide at sea the sluices in the dam are opened and large quantities of river water are discharged into the sea. Due to the closure of the inner estuary by the dam the tidal volume of the remaining part has decreased by about 70%. This has caused a large accumulation of

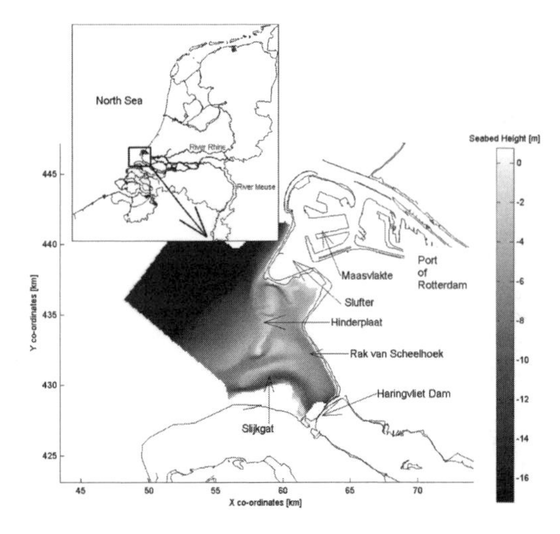

Figure 1. The location of the Haringvliet Estuary (Bathymetry of the year 2000).

sand and silt in the area. See the figures 2, 3 and 4 where the seabed of the years 1970, 1986 and 2000 are shown. The main channels in the area are called 'Slijkgat' and 'Rak van Scheelhoek'. The Slijkgat is the largest channel nowadays through which most of the tidal and rivervolume is transported. In the Rak van Scheelhoek a depth decrease of approximately 5 m has occurred since 1970. Most of the sediment that settled in this channel is silt.

On the sea side, the shallow sand bar (called 'Hinderplaat') has moved eastwards and has grown above mean sea level due to the relative dominance of the wave action since the tidal currents have dropped significantly after the closure.

Also the river discharge plays a role in the morphology of the area. Due to the regular discharge of fresh water into the sea vertical salinity gradients occur, resulting in density currents and vertical exchange of sediments seaward of the Haringvliet sluices.

Just north of the estuary the Port of Rotterdam is located. The port has reclaimed large areas from the sea since 1973. The development started in 1973 by reclaiming the first seaward extension called Maasvlakte and the closure of a secondary channel in the north of the estuary. In 1986 another area is reclaimed called the Slufter.

Table 1 shows the observed areas (in hectare) of different depth ranges for the years 1970, 1986 and 2000. All levels are expressed in NAP, which is approximately Mean Sea Level.

The sedimentation of the channels can best be seen in the -10 m/-3 m depth range, where a decrease of 3000 has occurred. Partly because of the growth of the Hinderplaat, the shallow zone and the intertidal area show a large increase since 1970.

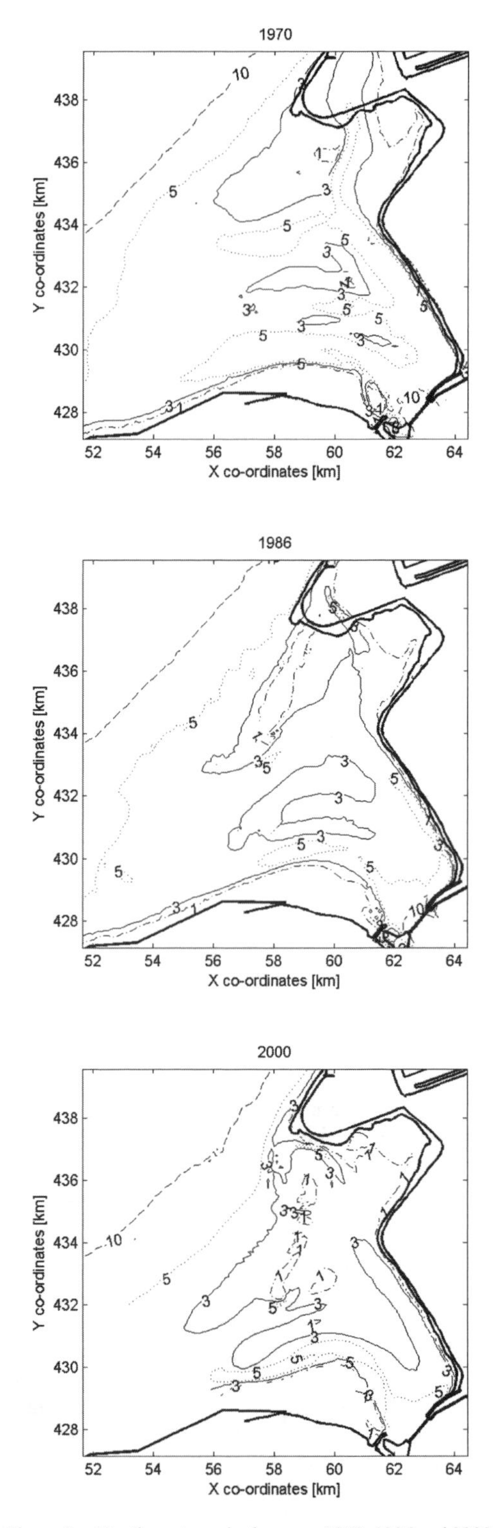

Figure 2. Depth contours in the years 1970, 1986 and 2000.

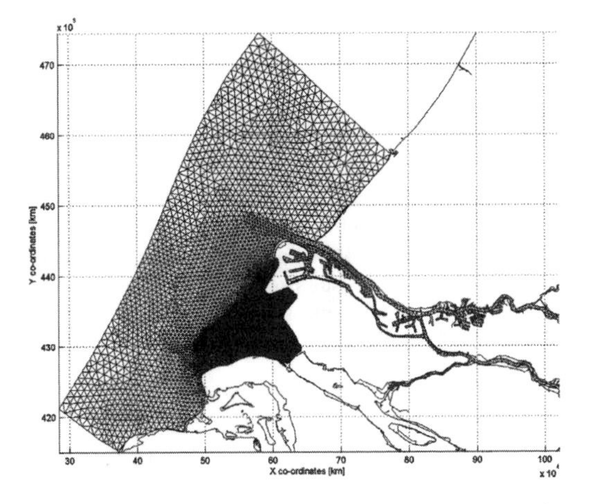

Figure 3. Overall grid of the FINEL2D model.

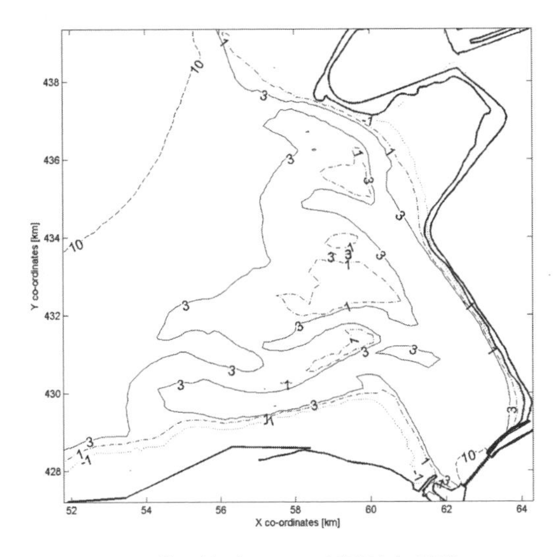

Figure 4. Predicted bathymetry of SET 1 in 2010.

Table 1. Observed depth areas in 1970, 1986 and 2000.

		Meas.		
Area	year: unit:	1970 (ha)	1986 (ha)	2000 (ha)
<−10 m		514	582	911
−10 m/−3 m		10788	9965	7302
−3 m/−1 m		2131	2008	2871
−1 m/+1 m		547	853	1070
+1 m/+2 m		380	412	347
>+2 m		194	308	641
Total		14554	13275	13142

The reduction of the total area in course of time is mainly caused by the reclamation works for the Port of Rotterdam.

The total sedimentation of the Haringvliet area from 1970 till 2000 is approximately 100 million m^3.

2.2 *Future development schemes*

Future development schemes for the estuary refer to:

- partial re-opening the Haringvliet closure dam (MER Beheer Haringvlietsluizen, 1998). This should bring back the tidal influence in the inner part of the estuary as a part of an ecological restoration project;
- further extension of the Port of Rotterdam by 1000–1500 ha (gross area), which is planned west of the existing Maasvlakte and Slufter (called 'Maasvlakte2').

Each of these schemes may have a significant impact on the development of the estuary. An expansion of the port reclamation could lead to a decrease of the wave action in the estuary, while the partial re-opening of the dam could stop the sedimentation trend and enhance the tidal motion again.

A good knowledge of the morphological impact of the different (combinations of) schemes is crucial for the authorities when applying for permits to implement the schemes. Estuarine environments and wetlands have been given international status as protected zones and sea reserves. In the Netherlands among others the Bird and Habitat Directive (issued by the European Union), are applicable. These directives give strict guidelines for any development scheme in wetlands. Mitigation and compensation of adverse impacts are key words in EIA and permit applications. Especially any expected loss of intertidal areas should be compensated. Creating new intertidal zones with the same habitats and biodiversity is an option then.

In other words: the knowledge of the morphological impact of the development schemes is directly related to the extent of compensation and mitigation that is required. Accuracy of morphological predictions and bandwidth of intertidal area forecasts are no longer items of scientific interest only, but have to do directly with obtaining development permits or not, and with the costs involved in compensation and mitigation works.

3 THE MORPHODYNAMIC MODEL FINEL2D

3.1 *General*

The morphological analysis of the Haringvliet estuary is carried out with the hydrodynamic and morphological model FINEL2D.

Under the FINEL2D umbrella a series of modules exists for different applications. In the Haringvliet case the following modules were applied:

- Hydrodynamic modules, focusing on tidal levels and currents, river discharge, wave action and wave induced currents;
- Sediment transport modules for the calculation of bed and suspended load of non-cohesive and cohesive sediments and for the interaction between cohesive and non-cohesive sediments at the seabed;
- Morphodynamic modules for online adaptation of the seabed level in the model grid points to the calculated sedimentation and erosion rates and for online restarting of the dynamic loop.

Each of the modules is discussed in more detail in the next sections.

3.2 *Hydrodynamic modules*

3.2.1 *FINEL2D*

FINEL2D is a 2DH numerical model, based on finite elements. This model is developed by Svašek Hydraulics. FINEL2D makes use of unstructured triangular grids. The advantages of such a unstructured triangular mesh in comparison to a finite difference grid are obvious: the major computational effort takes place in the area of interest, no nesting techniques are required and a triangular mesh can describe complicated coastlines very good.

3.2.2 *Wave model SWAN*

Waves can play an important part in morphological changes. For this reason the wave model SWAN is integrated in the model suite. See for a further description of SWAN Booij et al. (1999).

The wave model SWAN is used in the model in 3 ways:

- The spatial distribution of wave forces, is input for the hydrodynamic module to include the effect of wave driven currents in the total flow pattern. Because of the varying water level during the tide, the wave calculations are done for several water levels during the tidal cycle, for intermediate water levels the wave fields are interpolated.
- The spatial distribution of the orbital motion at the seabed (amplitude and direction), to be added to the current shear stresses (tide, wind or wave driven) to obtain the total shear stress as input for the sediment transport module. Here again, the results of the wave field calculations at varying water levels are used and interpolated if required during the calculation process.
- The cross shore distribution of the wave field in the coastal zone as input for the calculation of sediment transport according to Bailard (1981) and Nipius (1998). This option is applied in this study as an extra calibration parameter in the dynamic coastal section with breaking waves.

3.3 *Sediment transport modules*

3.3.1 *Sand module*

The transport formula for non-cohesive sediments (sand) in the Haringvliet study is the Bijker formula. We refer to Bijker (1967) for details. The formula is applied to make the results comparable with previous studies in the same estuary. Other transport formulae can easily be incorporated in the sand transport module.

3.3.2 *Silt module*

The basis of the silt module is the well known formula of Krone (1962) for deposition and of Partheniadis for erosion (1962, 1965). The module takes into account the availability of silt at the seabed.

3.3.3 *Sand-silt module*

The separate sand and silt modules as discussed above have the disadvantage that the transport processes of both sediments are treated completely independent of each other. In this approach the availability of sediment at the seabed is the only parameter that can prevent unrealistic erosion of one of both sediment types (for instance excessive erosion of silt at locations where no silt is present). To refine this approach a module is developed that takes into account the interaction between cohesive and non-cohesive sediments. This is the so-called sand-silt module.

The sand-silt module in the Haringvliet study is based on Van Ledden (2003). The main elements of the 'Haringvliet' sand-silt module are:

- The formulae of Bijker for sand and Krone-Partheniadis for silt are applied (similar to the separate modules).
- The module administrates the history of the mix of sand and silt in the seabed on the basis of field data and/or morphodynamic calculation results in a number of layers.
- In course of time the sand-silt ratio of succeeding bottom layers is adapted related to physical processes like bioturbation (so even without seabed exposure the composition of the bottom layers is not constant).
- The erosion characteristics of the top layer at the seabed (being the layer that is directly exposed to the water forces) depend on the sand/silt ratio of the top layer at the moment of exposure.
- This dependency is based on the approach of Van Ledden (2003) that triggers on cohesive or non-cohesive behavior of the top layer only. A silt content below 30% declares the seabed to be non-cohesive and erosion is generally easily. In return, if the silt content at the seabed exceeds the 30% level,

the seabed characteristics are defined as cohesive, which implies that the erosion of this layer is much harder. The availability of sand for erosion and transport is determined by the silt component.

This sand-silt approach has many entrances for refinement. But, in the Haringvliet case, this relatively straightforward approach has proven to be a suitable instrument to contribute to the assessment of reliability and bandwidth of morphological forecasts.

3.4 *Morphodynamic modules*

3.4.1 *General*

In the morphodynamic modules of the model no additional calculations of physical processes are done. The results of the hydrodynamic and sediment transport modules are used to come to a statistically representative mix of conditions and processes. Also the sequence of execution of the different calculations, the method to cope with daily conditions and extremes and with the co-incidence of different events are organised by the morphodynamic modules. Finally, this part of the model arranges for the gradual adaptation of the seabed level and composition in the model to the changing conditions and calculated sediment movements.

3.4.2 *Representative mix of conditions*

In an estuarine environment like the Haringvliet estuary often a series of statistical independent physical processes and phenomena are present. These phenomena are:

- The tidal motion; this phenomenon is well predictable by using the harmonic components of the astronomical tide. In this way (for instance) spring-neap cycles can be defined which are basically forecastable over a period of decades.
- The wave climate; the character of the wave climate is stochastic. On the long run a statistical distribution of wave conditions can be defined, but the occurrence of specific conditions (like a storm from a certain direction and with certain intensity) may coincide with any phase of the tide in the spring-neap cycle.
- The river discharge; this phenomenon is also stochastic. The occurrence is statistically independent from both the tide and the waves. The time scale of changes in river discharge is in the order of several days, compared to one or two days for the waves and to 12.5 or 25 hours for the well-defined tidal cycle.

The easiest way to deal with all of these conditions is to take a sufficiently long period and to run the model in real time mode. The historical joint occurrence of tide, wind, wave and river discharge will pass the computer then without further mixing problems. There is one constraint in this approach, and it is a showstopper. Its name is computer time. Depending on the statistical distribution of extreme waves and river discharges, the minimum calculation period should be in the order of several years to achieve a representative mix of conditions. This makes the real time mode as a standard solution to be not realistic.

A complete morphological calculation consists of several blocks. In each block a constant river discharge, wind and wave condition is applied. Each block calculates a certain hydrodynamic timeframe of 1 or more tides. A morphological acceleration factor (N) is applied to the calculated bottom change each timestep the morphological module is called. So for example when the hydrodynamic part of the model has calculated one timestep (dt) the morphological part has calculated N*dt timesteps. The reason for using an acceleration factor is computational time.

The choice for the number of tides which are calculated, the constant river discharge, wave conditions and wind is based on the long term observations. When combining all different parameters, for example northerly directed waves and high river discharge and southerly directed waves and low river discharge, the complete mix of these statistical variables are treated and should in theory approach realistic statistical conditions. All blocks are calculated after each other.

In this study two different approaches for the combination are followed, called SET 1 and SET 2.

3.5 *Methodology SET 1*

Because of limitations of computational time the input of morphological calculations are often very schematised. The modeller often uses a morphological tide, schematised discharge and meteorological conditions. This approach is called 'SET 1' in this paper.

The assumptions used for SET 1 are derived from previous studies, which are carried out in this area (Roelvink et al., 1998; Steijn et al., 2001). These assumptions include:

- A morphological tide of 12 hr and 25 minutes. In this study the same morphological tide is used as the previous studies;
- Three wave conditions; Northern directed wave conditions; Southern directed wave conditions; Conditions without waves. All three wave conditions are applied with a constant wind of that direction;
- Constant river discharge.

These assumptions are used to calculate the calibration period (1970–1986), the validation period (1986–2000) and the prediction period (2000–2010).

3.6 *Methodology SET 2*

On the other hand when taking into account a complete neap-spring cycle, varying discharge and varying meteorological conditions should give better results. This approach is called 'SET 2'. Besides these boundary conditions the calculations are carried out with the sand-silt module instead of the separate sand and silt module.

The basic principle of SET 2 is to take into account as many relevant aspects as possible in comparison to SET 1 to get a maximum band width between these two sets.

The difference between SET 1 and SET 2 are:

- The basic difference between these two sets is that SET 2 uses a complete neap-spring tidal cycle instead of a schematised morphological tide.
- SET 2 takes into account the influence of variable meteorological conditions instead of averaged meteorological conditions (SET 1). Storms are calculated separately in SET 2, while SET 1 is averaging the storms in the wave conditions. The wave conditions in the past decades show that the averaged number of storms in one year is 2.
- SET 2 takes into account the influence of variable river discharge, while SET 1 calculates an average river discharge. The observed river discharge is used as the boundary condition. For the prediction in the future the average discharge of the last 3 decades is taken, since the discharge of the future is unknown.
- SET 2 uses a sand-silt interaction model, while SET 1 uses a different model for sand and silt.

Since SET 2 uses different way of modelling than SET 1 the calibration and validation has to be carried out again with this model. The calibrated model is then used to carry out a prediction. The difference between SET 1 and SET 2 gives a band width due to different boundary conditions and physical processes.

3.7 *Methodology SET 2 meteo*

Future meteorological conditions are not known. This causes a natural band width of morphological predictions. The prediction of SET 2 uses an averaged number of storms each year and an averaged river discharge. Therefore three more predictions are carried out:

1. SET 2 extreme: This prediction uses the settings of SET 2, but with 3 storms each year, instead of 2 storms;
2. SET 2 mild: This prediction uses the settings of SET 2, but with 1 storms each year, instead of 2;
3. SET 2 river: This prediction uses the settings of SET 2, but with a higher river discharge and 0 storms.

The difference between these predictions gives a band width in meteorological conditions.

4 BUILDING, CALIBRATION AND VALIDATION OF THE MORPHODYNAMIC MODEL

4.1 *Building of the grids*

The sea boundaries of the computational grid were chosen such that the results of existing models could be used as boundary condition for this model. The overall grid is shown in figure 3. At the sea boundaries the grid is coarse. Near the area of interest the grid becomes finer, until a maximum resolution of approximately 100 m is reached.

Please note that 3 different grids are build for this study: the first for the calibration phase (1970–1986) in which the Maasvlakte is being build, the second for the validation phase (1986–2000) in which the Slufter is completed and the third for the future layout of Maasvlakte 2 (2000–2010).

4.2 *Calibration of the watermovement*

The first and important step in the calibration of a morphological model is the calibration of the watermovement. In this case the model is calibrated on observed water levels during a spring tide and neap tide and observed discharges of channels in the Haringvliet during a neap tide. The boundary conditions of the calibration periods of the watermovement are obtained from a hydrodynamic model of the complete coast of the Netherlands.

The shape and magnitude of the calculated waterlevels match the observed waterlevels well. The waterlevels are usually calculated within an accuracy of 10 cm. The observed discharges in the Haringvliet channels correspond very good to the FINEL2D discharges. The quality of the calculated waterlevels and discharges are good enough to begin the calibration of the morphology.

4.3 *Calibration SET 1 & SET 2 (1970–1986)*

Starting point of the calibration of SET 1 and SET 2 is the geometry of 1970. This observed bathymetry is used as input. The calculated bathymetry in 1986 is used for calibration against the observed bathymetry. Assumed is that no silt is present at the initial sea bed.

For more information about the calibration we refer to Dam (2004).

The overall morphology could be reproduced well for SET 1. Most of the relevant morphological phenomena could be reproduced by the model such as the sedimentation of the channels, the eastward movement and growing of the Hinderplaat. The total sedimentation in the period 1970–1986 is 75 Mm^3 in the area. SET 1 calculates a sedimentation of 67 Mm^3.

The calibration of SET 2 was performed on the first 2 years because of a lack of time. After 2 years the morphological changes could be reproduced well, however when calculating the complete 16 years of the calibration period the calibration effort was less successful as SET 1. Too much sediment settles in the estuary. A further calibration effort can substantially improve the results, because of the overall processes. SET 2 calculates a sedimentation of 158 Mm^3, instead of the observed 75 Mm^3. Because of the overestimation of the sediment volume the inter tidal area is overestimated about 5 times.

4.4 *Validation SET 1 & SET 2 (1986–2000)*

The validation period was chosen for the period 1986–2000 and is used to verify if the calibration still applies for this period. In this period the sedimentation of the estuary still going on. SET 1 is calculation, a sedimentation of 40 Mm^3, while the observation shows a measurement of 43 Mm^3, so the overall sedimentation is calculated good. When looking at different patterns like the morphologic change of the Hinderplaat the validation of SET 1 shows less good results than the calibration of SET 1. This also applies for the results of SET 2. A total sedimentation of 122 Mm^3 is calculated using SET 2.

5 PREDICTION OF GETEMD GETIJ (2000–2010)

5.1 *Bandwidth SET 1 & SET 2*

Both SET 1 and SET 2 are used to calculate a ten year prediction including the partly opening of the Haringvliet dam (Getemd Getij) and a Maasvlakte 2 variant. When the sluices of the dam are opened a lot more water is transported through the channels each tide. It is therefore expected that erosion of the channels might occur.

It is known that the sedimentation in the last 30 year of the Rak van Scheelhoek is silt. This silt layer of approximately 5 m thickness is used as input in SET 1 and SET 2. The major difference for this channel between SET 1 and SET 2 is the use of the sand-silt interaction module, which is used in SET 2, while SET 1 uses a 'normal' silt model. Because the silt percentage of this channel is very high the sand–silt interaction module assumes cohesive behaviour of the channel. The other major channel called 'Slijkgat' is a sandy channel.

The calculated bathymetry in 2010 for SET 1 is shown in figure 4. The initial bathymetry of 2000 can be seen in figure 2. The most important change is that the silt layer in the Rak van Scheelhoek has completely eroded. Since the Rak van Scheelhoek has eroded the northern part of the area shows a strong morphologic change.

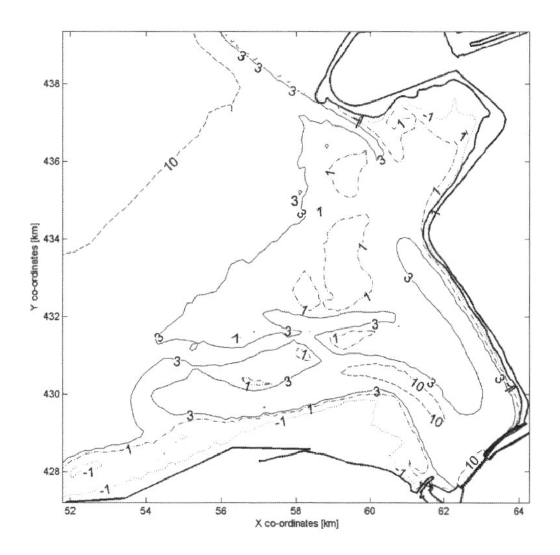

Figure 5. Predicted bathymetry of SET 2 in 2010.

The prediction of SET 2 shows a complete different behaviour as can be seen in figure 5. Instead of the Rak van Scheelhoek the Slijkgat has eroded. This is explained by the fact that cohesive behaviour of the silt layer is assumed in the Rak van Scheelhoek, which makes it harder to erode. The Slijkgat becomes the main channel in the area.

The difference between SET 1 and SET 2 shows a transition in channel development; In SET 1 the Rak van Scheelhoek is eroded, in SET 2 the Slijkgat is eroded.

To calculate the influence of the sand-silt module (which takes into account the cohesive behaviour of the Rak van Scheelhoek) alone versus the rest of the differences between SET 1 and SET 2, like the boundary conditions (neap-spring cycle, storms etcetera) a separate run was carried out. The results showed that the sand-silt module was responsible for the major difference between SET 1 and SET 2. The other differences are of a second order.

The erosion parameters of the silt layer in the Rak van Scheelhoek are not known, since the channel has shown a sedimentating trend since 1970 and therefore calibration of the erosion parameters are almost impossible. The erosion of silt is high in SET 1, while the erosion is slow in SET 2. In this way the possible outcomes are covered.

A possible solution for the calibration of the silt erosion constants lies in the periods with high discharge. A high discharge might give the same morphologic response as the opening of the sluices, since a lot of water is transported through the sluices in both cases. In the 1990's a severe high water occurred in the Dutch rivers. The difference in bathymetry before and after the high water period showed an erosion in the Slijkgat, while the Rak van Scheelhoek remained

Table 2. Comparison area of SET 1 and SET 2 prediction period (2000–2010) for Getemd Getij.

		Meas.	SET 1	SET 2	Diff.*	
Area	year: unit:	2000 (ha)	2010 (ha)	2010 (ha)	2010 (ha)	2010 (%)
<−10 m		911	689	1008	+319	46
−10 m/−3 m		7302	7244	6710	−534	−7
−3 m/−1 m		2871	2429	2940	+511	21
−1 m/+1 m		1170	924	1494	+570	+62
+1 m/+2 m		347	1315	449	−866	−66
>+2 m		641	641	643	+2	0

* Difference is defined as SET 2 − SET 1 in the year 2010.

Table 3. Comparison area of prediction SET 2 meteo (2000–2010) for Getemd Getij.

				Diff.*		
		Meas.	SET2	extreme	mild	river
Area	year: unit:	2000 (ha)	2010 (ha)	2010 (%)	2010 (%)	2010 (%)
<−10 m		911	1008	2	0	6
−10 m/−3 m		7302	6710	−1	1	2
−3 m/−1 m		2871	2940	0	0	−5
−1 m/+1 m		1170	1494	−1	−2	−1
+1 m/+2 m		347	449	11	−3	−8
>+2 m		641	643	0	0	0

* Difference is defined as this run minus SET 2.

stable. This gives an indication that the erosion parameters of SET 2 are more realistic, although this cannot be said for certain. A high level of uncertainty remains. It is clear that future research for this area should concentrate on the silt parameters of the Rak van Scheelhoek.

Table 2 shows the differences of the depth areas of the two sets.

5.2 *SET 2 meteo*

Future meteorological forcing are unknown. This causes a natural band width in the morphological predictions. Three calculations were carried out using different meteorological forcings:

1. SET 2 extreme: As SET 2, but instead of 2 storms each year, 3 storms are forced.
2. SET 2 mild: As SET 2, but instead of 2 storms each year, 1 storm is forced.
3. SET 2 river: As SET 2, but no storms and a higher river discharge is forced.

The results in depth areas are shown in table 3. The columns shows a difference in % in relation to the normal SET 2 run. The difference is not high, a maximum difference of 11% can be seen. This is not high in comparison to the differences between SET 1 and SET 2.

The cumulative sediment volume changes of all the runs of the prediction 2000–2010 are presented in table 4.

Table 4. Cumulative sediment volume changes prediction period (2000–2010) in Mm^3.

	Volume change
SET 1	35
SET 2	5
SET 2 extreme	7
SET 2 mild	4
SET 2 river	0

The difference between SET 1 and SET 2 is large. The difference is mainly caused by the large erosion of the Rak van Scheelhoek and a large sedimentation in the rest of the area. SET 2 shows a much calmer development in volume, mainly because the erosion of the Rak van Scheelhoek is slower. The runs using different meteorological forcings show that storms are transporting the sediment inside the system, resulting in a positive sedimentation. A higher river discharge on the other hand results in less sedimentation. A higher river discharge is transporting the sediment out of the estuary. The differences after ten years are in the order of a few million m^3.

6 DISCUSSION

6.1 *Bed roughness intertidal areas*

The water movement is the basis of process based morphological models. The water movement is usually calibrated on observed water levels and some current/discharge observations in the main channels. The main parameter which is used to calibrate hydraulic models is the bed roughness. The global water movement is calibrated in this way. The intertidal areas are usually not taken into account when calibrating the hydraulic model, since no data is available and these areas are not important for the global water movement. It was found that the velocities of the model in intertidal areas are very sensitive of the hydraulic bed roughness. Since no data is available to calibrate the model in the intertidal areas the same roughness as found in the channel is applied to this areas. Since sediment transport formulas often use a higher order power (3–5) of the current a small error in the velocity has big consequences for the resulting morphology. In SET 1 and

Table 5. Comparison* area between SET 1 and SET 2.

Area		Calibration	Validation	Prediction
	year:	1986	2000	2010
	unit:	(%)	(%)	(%)
<−10 m		−53	−42	46
−10 m/−3 m		−18	−20	−7
−3 m/−1 m		45	51	21
−1 m/+1 m		86	73	62
+1m/+2 m		−2	4	−66
>+2 m		6	4	0

* Difference is defined as SET 2 − SET 1.

SET 2 a constant bed roughness is applied, since this could not be calibrated, but a few sensitivity runs with a hydraulic bed roughness which was varied within realistic ranges showed that this could dominate the complete morphologic solution of the intertidal areas.

It is a paradox that these models are developed and calibrated for global hydrodynamic results (modeller), while the area of interest is shifting more and more to the intertidal zone (manager).

6.2 *Calibration of SET 1 and SET 2*

Table 5 contains the difference in depth area between SET 1 and SET 2 for the calibration/validation and prediction.

The calibration and validation shows more or less the same differences between SET 1 and SET 2, while the prediction is completely different. It can be concluded that the calibration in another regime like the opening of the sluices cannot be applied by definition.

The results also show that although the calibration of SET 2 is not good, the results of the prediction are in this case still valuable, since the solution is dominated by another effect, which could not be calibrated, namely the cohesive erosion behaviour of the Rak van Scheelhoek.

7 CONCLUSION

Two sets of model assumptions/boundary conditions and processes have been calibrated, validated and used for a prediction. The results show a large band width in the prediction of the development of the Haringvliet estuary for Getemd Getij situation. This is mainly caused by taking into account the cohesive erosion behaviour of the silt in the Rak van Scheelhoek. This is almost impossible to calibrate since the estuary is a sedimentating since 1970.

The calibration and validation effort of the model from 1970 till 2000, in which the system is sedimentating, is no longer valid when predicting the morphology of the opening of the sluices (Getemd Getij situation), in which the system is eroding. It is better to also calibrate on a period with a high river discharge, which has probably the same morphological result as the Getemd Getij situation.

Meteorological effects are of minor importance for the band width than the difference between SET 1 and SET 2.

The results clearly show the necessity to take band width into account when predicting morphological developments. The management of the area needs to deal with this band width and use it properly. At the same time the analysis of the results gives a good insight in the path for model development.

REFERENCES

Bailard, J.A., 1981. An Energetics Total Load Sediment Transport Model for Plane Sloping Beaches. Journal of Geophysical Res., Vol. 86 (C11).

Bijker, E.W., 1971. Longshore Transport Computations. Journal of the Waterways, Harbours and Coastal Engineering Division, Vol. 97, No. WW4.

Booij, N., Ris R.C., Holthuijsen, L.H., 1999. A third-generation wave model for coastal regions, Part I, Model description and validation, J. Geoph. Research C4, 104, 7649–7666.

Dam, G., 2004. Bandbreedte morfologische voorspellingen Haringvlietmonding (in Dutch). Svasek Hydraulics. gd/04459/1306.

Glaister, P., 1993. Flux difference splitting for open-channel flows, Int. J. Num. Meth. Fluids, 16, 629–654.

Harten, A., Lax, P.D., Leer, van, B., 1983. On upstream differencing and Godunov-type schemes for hyperbolic conservation laws, SIAM review, Vol. 25, No. 1, January 1983.

Krone, R.B., 1962. Flume studies of the transport of sediment in estuarial shoaling process. University of California.

Ledden, M., van., 2003. Sand-mud segregation in estuaries and tidal basins. PhD-Thesis. Delft University of Technology.

Nipius, K.G., 1998. Dwarstransportmodellering m.b.v. Bailard toegepast op de Voordelta Grevelingen-monding (in Dutch). Msc. Thesis, Delft University of Technology.

MER Beheer Haringvlietsluizen, 1998. Over de grenzen van zoet en zout, hoofdrapport (in Dutch), ISBN 903694802.

Partheniades, E.A., 1962. Study of erosion and deposition of cohesive soils in salt water. PhD Thesis, University of California.

Partheniades, E.A., 1965. Erosion and deposition of cohesive soils, J. of Hydr. Div. Vol 91., No. HY 1.

Roelvink, J.A. et al. 1998. Kleinschalig Morfologisch Onderzoek MV2 (in Dutch). ref. MCM1766/Z2428/arh. WL|Delft Hydraulics.

Steijn, et al. 2001. Bandbreedte morfologische effectvoorspelling – MV2: een onderzoek ten behoeve van natuurtypering (in Dutch). ref. A792/Z3127. WL|Delft Hydraulics & Alkyon.

Fluvial and coastal turbulent flow

River, Coastal and Estuarine Morphodynamics: RCEM 2005 – Parker & García (eds)
© 2006 Taylor & Francis Group, London, ISBN 0 415 39270 5

Direct numerical simulation of an oscillatory boundary layer close to a rough wall

F. Fornarelli & G. Vittori
Dept. of Environmental Engineering, University of Genova, Genova, Italy

ABSTRACT: In the present contribution, the boundary layer generated close to a rough wall by an oscillatory, uniform pressure gradient is studied. The flow simulates the boundary layer generated at the sea bottom by a monochromatic propagating wave. The problem is tackled by numerical means and detailed information on flow dynamics is obtained. In particular, the evolution of vorticity is considered and the coherent vortex structures which are formed within the boundary layer are identified. The force exerted by the fluid on the bed and on the roughness elements is computed along with its pressure and viscous components.

1 INTRODUCTION

The flow in the boundary layer at the bottom of sea waves is known to influence both hydrodynamic aspects of wave propagation (wave damping, generation of steady streaming, etc) and sediment transport. Therefore, the coastal engineering community has devoted a lot of attention to the study of the flow structure in this boundary layer at high values of the Reynolds number as those encountered in practical applications.

The boundary layer at the bottom of a propagating sea wave of small amplitude can be modeled as that generated by the harmonic oscillations of a fluid close to a wall. The geometry of the wall can be plane or can show undulations named "ripples", depending on hydrodynamic and sediment parameters. In the present contribution attention is focused on a plane rough bed.

On the theoretical side, detailed information is available only for oscillatory boundary layer close to a plane smooth wall. In particular, the flow field was determined analytically in the laminar regime (Stokes, 1855) and a large number of works has been devoted to the investigation of the transition to turbulence both by analytical means (Hall, 1978, Blondeaux & Seminara, 1979, Blondeaux & Vittori 1994) and by means of direct numerical simulations (Akhavan et al., 1991, Vittori & Verzicco 1998, Costamagna et al., 2003).

Direct numerical simulations have also allowed to disclose peculiar aspects of the flow field over a smooth wall and in particular of the transition to turbulence and of turbulence characteristics for moderate values of the Reynolds number. In particular, it has been shown that the critical value of the Reynolds number for transition to turbulence is related to the amplitude of wall imperfections and that the unsteady nature of the flow influences the time development of turbulent quantities (Vittori & Verzicco, 1998). Indeed equilibrium conditions are not attained as the accelerating parts of the cycle are characterized by turbulence production while dissipation of turbulence is dominant during the decelerating parts of the cycle.

An analysis of the formation and development of coherent structures in a Stokes boundary layer has shown many similarities with that taking place in steady boundary layers (Costamagna et al., 2003). At the end of the accelerating part of the cycle low-speed streaks appear close to the wall and their subsequent dynamics creates a sequence of streamwise vortices of alternating circulation which contribute to the vertical momentum flux.

The information available on the boundary layer close to a planar rough wall is based only on experimental investigations. Investigations of the flow field close to natural rough walls (Sleath, 1987) showed that turbulence intensities fluctuate significantly during the cycle even though the variation with distance from the bottom is qualitatively similar to that observed in steady flows. Moreover turbulent intensities as well as Reynolds stresses are significantly increased with respect to the corresponding smooth wall flow even though the effect tends to disappear at increasing distances from the wall (Jensen et al., 1989). The interpretation of peculiar aspects of the experimental results suggested the presence of jets of fluid associated with vortex formation and ejection over the roughness elements (Sleath, 1987).

Experiments carried out using beds with a regular roughness made with spheres glued to an oscillating plate (Keiller & Sleath, 1976) showed that close to the rough wall the velocity profile shows two relative maxima during each half cycle. One maximum is in phase with the maximum of the velocity of the plate while the second maximum is observed close to flow reversal. The second peak, which is related to the presence of a bottom roughness of large size, rises from zero at the bed to a maximum at a distance of about one-eighth of the roughness size and falls off again. It was suggested that the observed secondary peaks correspond to the strong vertical velocities produced by the incipient vortex formation around the spheres on the bed. This hypothesis is apparently confirmed by measurements and visualizations carried out over a wall where the roughness was made with square roughness elements (Krstic & Fernando, 2001). The visualizations showed that the boundary layer is fed with dipole-like vortex structures generated by flow separation at the roughness elements.

In the present contribution direct numerical simulations of the oscillatory flow close to a rough wall are carried out with the aim of providing more insight on vorticity dynamics over a rough wall and on its effect on averaged flow quantities.

The rough wall is obtained by arranging semi-spheres on a smooth wall. The characteristic of the semi-spheres as well as flow parameters are chosen in such a way to have accurate results with an affordable computational effort. Moreover the chosen geometry resembles that used by Keiller & Sleath (1976) so that the code can be validated by comparing the obtain results with the experimental data.

2 THE PROBLEM AND THE NUMERICAL APPROACH

The sea bottom is approximated as a solid wall covered with semi-spheres aligned in rows to form a hexagonal pattern (see figure 1). The geometrical characteristics of the pattern are given in the figure and have been chosen in an attempt to reproduce the bed configuration used in the experiment n.41 of Keiller & Sleath (1976). This choice allows to compare the numerical results with the experimental measurements and, hence, to validate the numerical code.

The bed bounds an incompressible, homogeneous viscous fluid, the density and the kinematic viscosity of which are denoted by ρ and ν respectively. The fluid, far from the wall, oscillates harmonically. Introducing a Cartesian coordinate system (x_1, x_2, x_3) with the (x_1, x_2)-plane coincident with the plane bed (the x_1-axis being in the direction of the fluid velocity oscillations far from the wall and the x_3-axis pointing upward), the fluid velocity far from the wall turns out to be:

$$(u_1, u_2, u_3) = (-U_0 \cos(\omega t), 0, 0) \tag{1}$$

where U_0 and ω are the amplitude and the angular frequency of the fluid velocity oscillations, respectively.

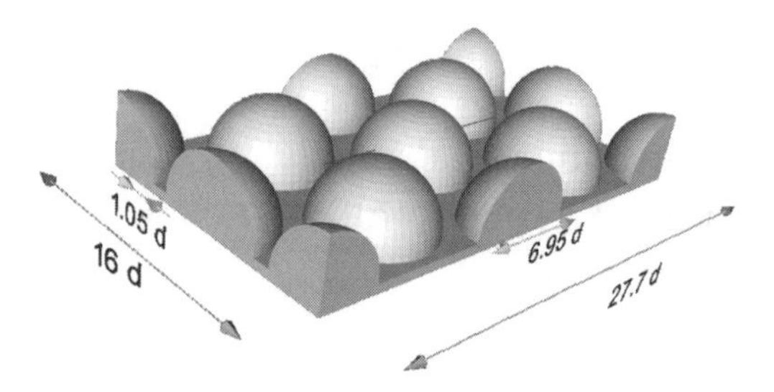

Figure 1. Sketch of the bottom configuration ($d = \delta$).

The temporal development and spatial distribution of the velocity and pressure fields follow from continuity and Navier-Stokes equations along with the no-slip condition at the wall and at the surface of the roughness elements. The numerical approach solves the problem in primitive variables using standard centered second-order finite difference approximations for the spatial derivatives. The time advancement of the Navier-Stokes equations employs the fractional-step method extensively described by Kim & Moin (1985) and Rai & Moin (1991): a non-solenoidal intermediate velocity field is computed by means of a third-order Runge-Kutta scheme to discretize convective terms together with a Crank-Nicholson scheme for the diffusive terms. The implicit treatment of the viscous terms would require for the inversion of large sparse matrices which are reduced to three tridiagonal matrices by a factorization procedure with an error of order $(\Delta t)^3$ (Beam & Warming, 1976).

The equations are integrated on a numerical box the length, width and height of which are $27.7\,\delta$, $16\,\delta$, $30\,\delta$ respectively, where the quantity δ, chosen as length scale, is the order of magnitude of the thickness of the oscillatory boundary layer over a plane wall ($\delta^2 = 2\nu/\omega$). The number of grid-points is 161, 161 and 81 in the stream-wise, cross-stream and spanwise directions, respectively.

Along the x_1 and x_2 directions, periodic boundary conditions are enforced. The use of periodic boundary conditions in the homogeneous directions x_1 and x_2 is justified if the computational box is large enough to include the largest eddies of the flow. In the present geometry, the size of the largest vortex structures is either that of the vortices shed by the roughness elements, which can be supposed to scale with their size, or that of the coherent vortices generated by the intrinsic instability of the oscillatory boundary layer. The work by Costamagna et al. (2003) shows that in Stokes boundary layers over flat walls the largest flow structures which appear when transition to turbulence takes place are elongated in the streamwise direction and characterized by a length of the order of $10\,\delta$. Hence it appears that the box size is large enough to capture the main features of the flow.

The pressure field is obtained from Poisson equation which follows from continuity equation and Navier-Stokes equation and is readily solved by taking advantage of the imposed periodicity in the stream-wise and span-wise directions. The forcing of the no-slip condition at the "rough" wall is achieved by using boundary body forces that allow the imposition of the boundary condition on a given surface not coincident with the computational grid. Therefore, the governing equations can be discretized and solved on a regular mesh, thus retaining the advantages and the efficiency of the standard solution procedures (Fadlun et al., 2000). The solution is advanced in time with a variable time step but keeping the Courant number constant and equal to 0.7.

As already pointed out, the geometrical and flow parameters are chosen in order to reproduce experiment n. 41 of Keiller & Sleath (1976), even if Keiller & Sleath (1976) used arrays of spheres glued to the bottom of their experimental facility. Hence, the diameter of the semi-spheres is set equal to 6.95 δ and the Reynolds number $Re = (U_0)^2/(\nu\,\omega)$ is set equal to 4560. Then, further runs are made for larger values of Re.

Since, as expected, large variations on the computed quantities are observed from cycle to cycle because of turbulence presence, a phase averaging procedure is introduced to compute the mean quantities. The phase averaged value $\overline{f}$ of a generic quantity f is defined as follows:

$$\overline{f}(x_1, x_2, x_3, t) = \frac{1}{N_p} \sum_{n=1}^{N_p} f\left(x_1, x_2, x_3, t + \frac{2\pi n}{\omega} \right) \qquad (2)$$

where N_p indicates the number of simulated cycles. Strictly, the value of N_p should tend to infinity. Unfortunately, the high computational costs do not allow a large number of cycles to be simulated. Therefore, where appropriate, in order to make more evident the trend of the mean quantities, it has been decided to "filter" the phase averaged values with a low pass filter, thus removing high frequency fluctuations.

3 RESULTS

In this preliminary investigation, the roughness size is kept fixed and equal to that used in the experiment n. 41 by Keiller & Sleath (1976) and two values of the Reynolds number are considered, namely $Re = 4560$ and $Re = 2 \times 10^4$. Looking at the results obtained by Sleath (1988) and in particular at figure 13 of Sleath (1988), it can be seen that the Reynolds numbers presently considered are close to the transition limit between the laminar and turbulent regimes. Hence, particularly for the lower values of the Reynolds number, the flow is expected to be dominated by the vortex structures shed by the roughness elements, even though the intrinsic instability of the flow should affect the phenomenon and give rise to turbulent eddies.

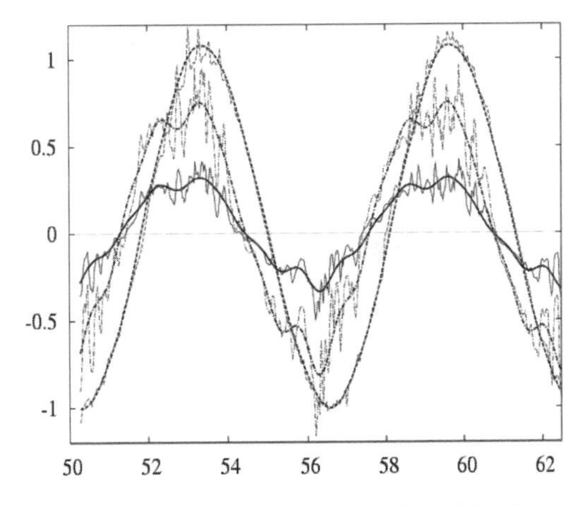

Figure 2. Actual and phase-averaged values of the dimensionless stream-wise velocity u_1/U_0 plotted versus ωt at 0.02 δ from the crest (solid line), 0.27 δ from the crest (dash-dotted line), 15 δ from the crest (dashed line). $Re = 2 \times 10^4$.

A first investigation of the velocity field can be made by looking at the outputs of numerical velocity probes located just above the crest of a roughness element at different distances from it. Figure 2 shows the actual streamwise velocity component along with its phase-averaged and filtered values versus time at different distances from the crest for $Re = 2 \times 10^4$. Comparing the actual velocity signals with the averaged values, it can be appreciated that the largest turbulent fluctuations are observed close to the phases of maximum velocity. Moreover, the strength of the turbulent fluctuations decays moving far from the wall and, at $x_3 = 15\,\delta$, the turbulent fluctuations are much smaller than those at 0.02 δ. A similar finding characterizes the measurements of Sleath (1987).

To compare our results with the measurements of Keiller & Sleath (1976), we consider the quantity V which is the projection of velocity in a vertical plane and compute its modulus |V|. Figure 3a–b shows the time development of |V| at different distances from a crest measured by Keiller & Sleath (1976) at $x_2 = 8\,\delta$, together with the numerical results. The quantity $U_0 \cos(\omega t)$ is added to the numerical results shown in figure 3b since in Keiller & Sleath (1976) the bed was oscillating in a still fluid.

In figure 3 it can be seen that the velocity shows two maxima during each half oscillating cycle. The first maximum is related to the maximum of the velocity of the plate while the second one is related to the presence of the roughness and increases in strength as the distance from the crest is increased up to $x_2 = 0.79\,\delta$ and then decreases moving far from the bed. For experiment n. 41, Keiller & Sleath (1976) measured a

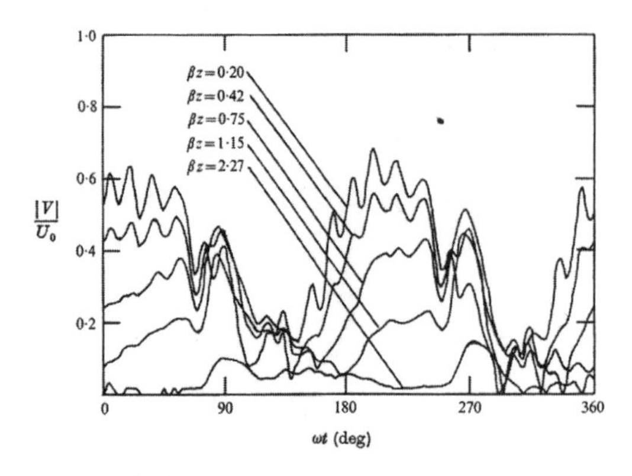

Figure 3a. $|V|/U_0$ at different distances from a crest on the vertical plane $x_2 = 8\,\delta$. Experimental measurements (adapted from Keiller & Sleath, 1976) $\beta z = x_3/\delta$. Re = 4560.

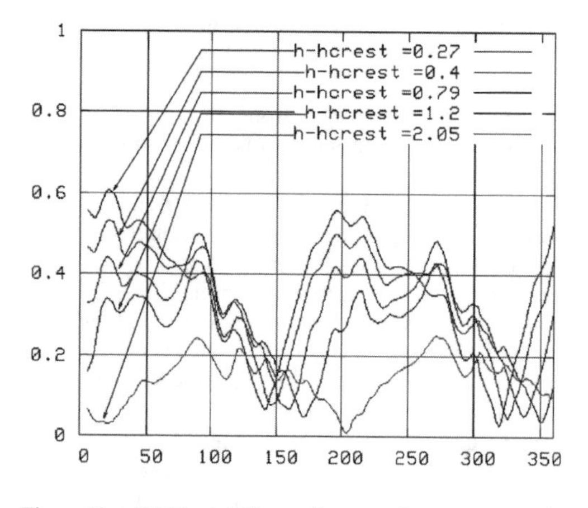

Figure 3b. $|V|/U_0$ at different distances from a crest on the vertical plane $x_2 = 8\,\delta$. Numerical results. Re = 4560.

maximum secondary peak of amplitude equal to 0.49 U_0 at a distance 0.70 δ from the crests and characterized by a relative phase of 88 degrees. The present results show a maximum equal to 0.54 U_0 which is attained at a distance from the semi-sphere crests equal to 0.79 δ and at a relative phase equal to 86 degrees. Figure 4a shows the computed maximum values of $|V|$ above a crest for different values of x_3, together with the measurements of Keiller & Sleath (1976). The computed and measured relative phases of the maxima are shown in figure 4b.

Figures 3 and 4 show that the numerical code provides accurate results even if small discrepancies are present between actual results and the experimental data by Keiller & Sleath (1976). However, it should be kept in mind that in the present investigation the bottom as well as the measuring points are kept fixed and the fluid moves to and fro while in the experiments of Keiller & Sleath (1976) the measuring probes

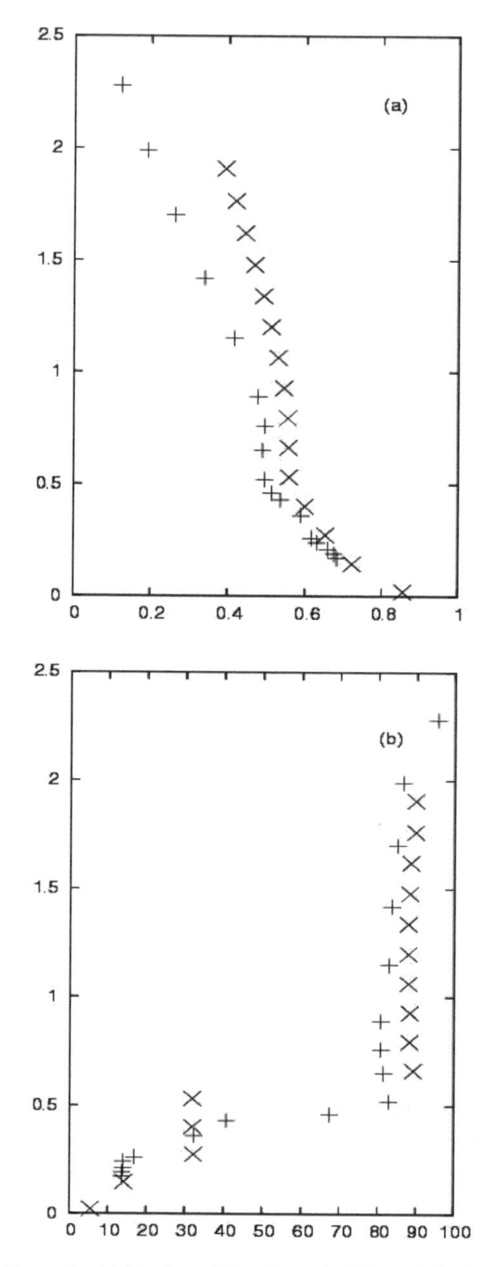

Figure 4. (a) Maxima of $|V + U_0 \cos(\omega t)|/U_0$ and (b) their phase (degrees) plotted versus the dimensionless (x_3/δ) distance from the crests of the roughness elements. (× computed values, + measured values of Keiller & Sleath (1976).) Re = 4560.

were fixed and the bed was oscillating. Finally, the bed used by Keiller & Sleath (1976) was composed by spheres glued on a plane wall, hence the overall geometry used in the numerical simulation is similar to the experimental one but the details are different.

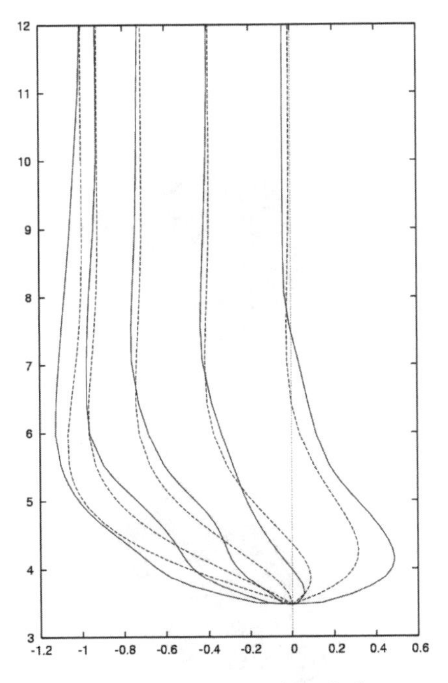

Figure 5a. Dimensionless streamwise velocity component (u_1/U_0) above the roughness element plotted versus x_3/δ, during the decelerating phase. Continuous lines = present results; dashed lines = velocity profiles over a plane wall, Re = 4560. From the left to the right: $\omega t = 69.1$, $\omega t = 69.5$, $\omega t = 69.9$, $\omega t = 70.3$, $\omega t = 70.7$.

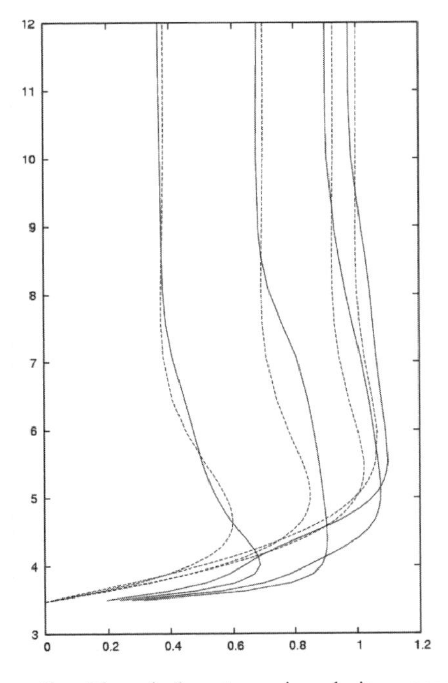

Figure 5b. Dimensionless streamwise velocity component (u_1/U_0) above the roughness element plotted versus x_3/δ, during the accelerating phase. Continuous lines = present results; dashed lines = velocity profiles over a plane wall. Acceleration phase. Re = 4560. From the left to the right: $\omega t = 71.1$, $\omega t = 71.5$, $\omega t = 71.9$, $\omega t = 72.2$.

Once the reliability of the numerical code is verified, the main advantage of using numerical simulations to investigate the phenomenon is the possible access to velocity and pressure fields in the three-dimensional space and time. Hence, the numerical results make it possible to evaluate quantities such as vorticity, strain rate, stress tensor components, ... which would be impossible to measure accurately in laboratory experiments.

The effect of the bed roughness on the velocity field can be seen in Figure 5a–b where the vertical profile of the streamwise velocity component, averaged over different crests, is shown at different phases during the 11th oscillation cycle together with the corresponding profiles obtained assuming a flat wall.

It can be seen that the effect of the roughness is more pronounced close to the wall and during the accelerating parts of the cycle when it affects distances from the top of the half-spheres larger than 9 δ. In particular, the roughness appears to induce larger velocities close to the bed. Figure 6 shows that the vertical velocity component is one order of magnitude smaller than the horizontal one and always positive close to the roughness crests while it attains also negative values for large distances from the wall.

An analysis of the vorticity field shows that the second relative maximum of velocity, previously described and displayed in figure 3, is the result of the vortex structures shed by the crests of the roughness elements which create a jet of fluid, close to flow reversal, and induce a sudden increase of both the vertical and the stream-wise velocity components. The existence of jets of fluid created by the roughness elements at flow reversal was first shown by flow visualizations and by the measurements of Sleath (1987), who discussed also the flux of momentum, induced by these jets, from one fluid layer to another.

In figure 7, the modulus of the vorticity $|\mathbf{\Omega}|$ is shown for Re = 4560 at different phases of the 8th-cycle. The largest contributions to the vorticity field appears to be related to the shear layers shed from the top of the half-spheres.

Indeed, when the external flow accelerates, vorticity is generated at the top of the half-spheres, (see figure 7a) is convected streamwise, interacts with the adjacent crests and increases in intensity (see figure 7b,c). In the gaps between the rough elements, significant values of vorticity can be observed, which are generated by the high shear induced by flow acceleration between adjacent roughness element. The

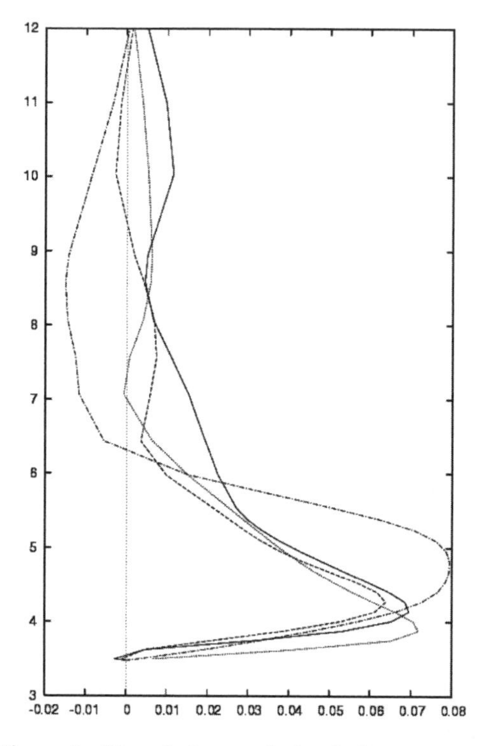

Figure 6. Dimensionless vertical velocity component (u_3/U_0) above the roughness crests plotted versus x_3/δ. —— $\omega t = 69.1$, - - - $\omega t = 69.9$, - - - - $\omega t = 70.7$, ... $\omega t = 71.5$.

modulus of vorticity attains the largest values during the accelerating part of the cycle and then decreases in intensity and elongates in the streamwise direction (see figure 7d).

A similar scenario of the vorticity time development is observed for Re equal to 2×10^4. However, for $Re = 2 \times 10^4$, a much stronger interaction of the vortex structures released from one crest with the adjacent roughness elements and a much stronger diffusion in the vertical direction are observed. Moreover, the vortex structures shed by the roughness elements at flow reversal are smaller, incoherent and randomly distributed in space.

More insight into the dynamics of the coherent vortex structures is obtained by considering the regions characterized by two negative values of the eigenvalues (λ_1, λ_2, λ_3 with $\lambda_1 > \lambda_2 > \lambda_3$) of the tensor made by summing the squares of the symmetric and antisymmetric parts of the velocity gradient. As suggested by Jeong & Hussain (1995), these regions capture the pressure minimum in a plane orthogonal to the axis of a vortex structure and correctly detect the axis of the vortex.

The inspection of the regions where λ_2 assumes negative values shows that during the last parts of the accelerating phases of the external flow, when

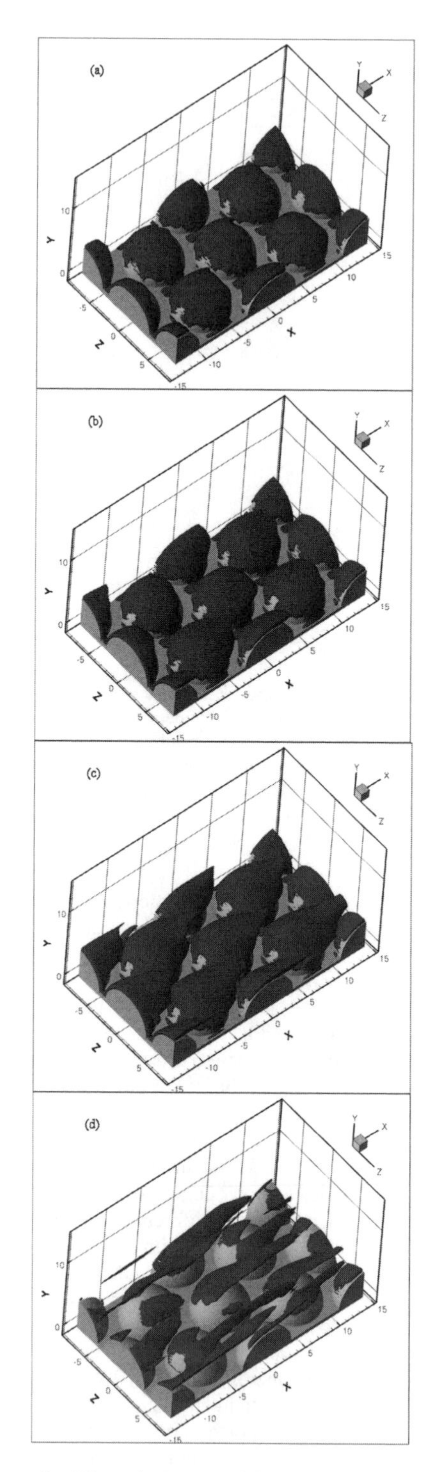

Figure 7. Dimensionless modulus $|\mathbf{\Omega}|$ of the vorticity at a) $\omega t = 52.2$, b) $\omega t = 52.4$, c) $\omega t = 52.8$, d) $\omega t = 54.0$. Iso-surfaces characterized by $|\mathbf{\Omega}| = 0.7\ U_0/\delta$. $Re = 4560$.

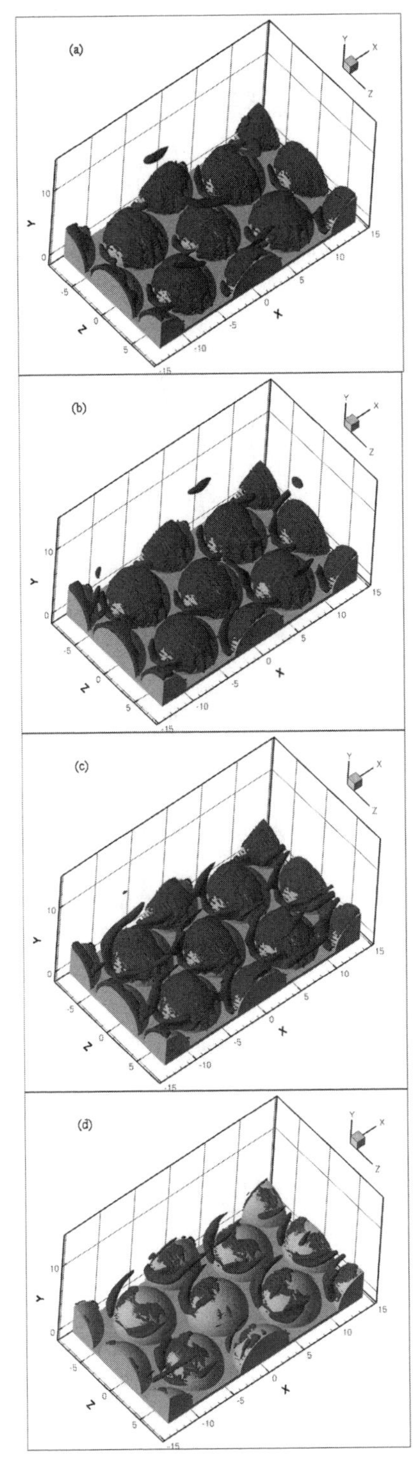

Figure 8. Iso-surfaces of λ_2 ($\lambda_2 = -0.02$) at a) $\omega t = 52.2$, b) $\omega t = 52.4$, c) $\omega t = 52.8$, d) $\omega t = 54.0$. Re = 4560.

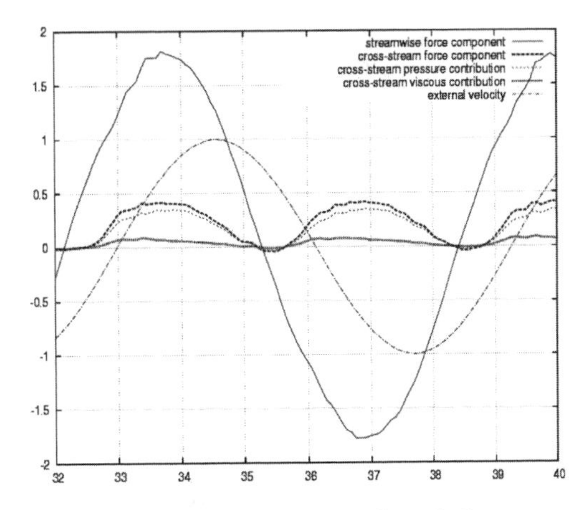

Figure 9. Time development of the dimensionless streamwise ($F_1/(\rho U_0^2 \delta^2)$) and cross-stream ($F_3/(\rho U_0^2 \delta^2)$) components of the force exerted on the generic roughness element together with the pressure contribution and the viscous contribution to the cross-stream force component and the external velocity. Re = 4560.

vorticity reaches its minimum negative value, horse-shoe vortices are formed close to the upstream surface of the roughness elements (see figure 8a). As the flow accelerates, the two branches of the horse-shoe vortices tend to embrace the half-spheres and to elongate in the stream-wise direction (see figure 8b). Before the external velocity reaches its maximum value, the horse-shoe vortices lift from the bed (see figure (8c,d). During the decelerating phases the vortex structures released from the bed break up into smaller structures and dissipate. In the following half-cycle a vortex is generated at the other side of the roughness elements and undergoes a similar dynamics.

The horse-shoe vortices are found to influence the force exerted by the fluid on the semi-spheres at the bottom. Indeed, looking at figure 9, where the stream-wise and vertical components of the force acting on a single half-sphere are plotted as a function of time, it can be appreciated that the maximum values are reached just before the external flow reaches its maximum, approximately when the horse-shoe vortices reach their maximum intensity and wrap around the half-spheres. The spanwise component of the force turns out to be negligible throughout the whole cycle. It is worth to mention that the vertical component of the drag force in all the computational box always vanishes. Hence while the roughness element (the semi-spheres) tend to be lifted from the bed, the fluid exerts a compression force in the gaps between the roughness elements.

The tendency of the fluid to lift the particles from the bed is stronger during the accelerating phases of the cycle while during the decelerating parts of the cycle, the vertical force drops even if it remains positive.

Considering the viscous and pressure contributions to the component of the force in the x_3 direction, it can be seen (figure 9) that the viscous contribution is much smaller than that due to pressure. Moreover the viscous contribution reaches the maximum close to flow reversal while the pressure part as well as the two components of the force reach their maximum during the last part of the accelerating part of the cycle.

4 CONCLUSIONS

In the present paper, the results of direct numerical simulations of the oscillatory flow above a rough wall are presented. The geometry of the wall consists of half-spheres attached to a plane wall in an hexagonal pattern in order to model the geometry used by Keiller & Sleath (1976). An analysis of the flow field has allowed to detect the coherent vortex structures which form close to each half-sphere as the flow far from the bed accelerates. The horse-shoe vortices created close to the upstream part of the half-spheres, together with the pressure minimum induced by the shear layer shed from the crests of the roughness elements, induce a force which tends to lift the half-spheres from the wall while in the gaps between the roughness elements the fluid exerts a force directed toward the bottom.

This preliminary investigation will be continued in the future by investigating flow dynamics for larger values of Reynolds number and by taking into account different geometries of the rough wall.

REFERENCES

Akhavan R., Kamm R.D. and Shapiro A.H. 1991. An investigation of transition to turbulence in bounded oscillatory Stokes flows. Part 2. Numerical simulations. *J. Fluid Mech.* 225, 423–444.

Beam R.M. and Warming R.F. 1976. An implicit finite-difference algorithm for hyperbolic system in conservation-law form. *J. Comput. Phys.* 22, 87.

Blondeaux P. and Seminara G. 1978. Transizione incipiente al fondo di un'onda di gravità. *Rendiconti Accad. Naz. Lincei* 67, 407–417.

Blondeaux P. and Vittori G. 1994. Wall imperfections as a triggering mechanism for Stokes-layer transition. *J. Fluid Mech.* 264, 107–135.

Costamagna P., Vittori G. and Blondeaux P. 2003. Coherent structures in oscillatory boundary layers. *J. Fluid Mech.* 474, 1–33

Fadlun E.A., Verzicco R., Orlandi P. and Mohd-Yusof J. 2000. Combined immersed-boundary finite-difference methods for three-dimensional complex flow simulations. *J. Comp. Phys.* 161, 35–60.

Hall P. 1978. The linear stability of flat Stokes layers. *Proc. R. Soc. Lond.* A259, 151–166.

Jensen B.L., Sumer B.M. and Fredsoe J. 1989. Turbulent oscillatory boundary layers at high Reynolds numbers. *J. Fluid Mech.* 206, 265–297.

Jeong J. and Hussain F. 1995. On the identification of a vortex. *J. Fluid Mech.* 285, 67–94

Keiller D.C. and Sleath J.F.A. 1976. Velocity measurements close to a rough plate oscillating in its own plane *J. Fluid Mech* 73, 673–691.

Kim J. and Moin P. 1985. Application of a fractional-step method to incompressible Navier-Stokes equations. *J. Comput. Phys.* 59, 308.

Krstic R.V. and Fernando H.J.S. 2001. The nature of rough-wall oscillatory boundary layers. *J. Hydraulic Res.* 39, 655–666.

Rai M.M. and Moin P. 1991. Direct simulations of turbulent flow using finite-difference schemes. *J. Comput. Phys.* 96, 15.

Sleath J.F.A. 1987. Turbulent oscillatory flow over rough beds. *J. Fluid Mech.*, 182, 369–409.

Sleath J.F.A. 1988. Transition in Oscillatory Flow over Rough Beds. *J. Waterway., Harbors and Coastal Eng. Div., ASCE*, 114, 18–33.

Stokes G.G. 1855. On the effects of internal friction of fluids on the motion of pendulums. *Trans. Camb. Phyl. Soc.* 9.

Vittori G. and Verzicco R. 1998. Direct simulation of transition in an oscillatory boundary layer. *J. Fluid Mech.* 371, 207–232.

River, Coastal and Estuarine Morphodynamics: RCEM 2005 – Parker & García (eds)
© 2006 Taylor & Francis Group, London, ISBN 0 415 39270 5

Density currents in the Chicago River, Illinois

Carlos M. García & Claudia Manríquez
Graduate Research Assistants, Ven Te Chow Hydrosystems Laboratory, Dept. of Civil Engineering, University of Illinois at Urbana-Champaign

Kevin Oberg
U.S. Geological Survey, Office of Surface Water, Urbana, IL

Marcelo H. García
Chester and Helen Siess Professor, Ven Te Chow Hydrosystems Laboratory, Dept. of Civil Engineering, University of Illinois at Urbana-Champaign

ABSTRACT: Bi-directional flow observed in the main branch of the Chicago River is due, in most cases, to density currents generated by density differences between the water in the North Branch Chicago River and the Chicago River. An upward-looking 600-KHz acoustic Doppler current profiler was installed by the U.S. Geological Survey (USGS) in the center line of the Chicago River at Columbus Drive at Chicago, IL, to characterize these flow conditions. Bi-directional flow was observed eight times in January 2004 at Columbus Drive. Three bi-directional flow events, with temperature stratification of approximately 4°C, were also observed on the North Branch Chicago River. Analysis of these data indicates that the plunging point of the density current moves upstream or downstream on the North Branch Chicago River, depending on the density difference. Complementary water-quality and meterologic data from January 2004 help confirm the mechanism causing the formation of the density currents.

1 INTRODUCTION

The City of Chicago, Illinois (IL) and many of its suburbs lie within the glacial Lake Chicago Plain. The Lake Chicago Plain encompasses the Chicago, Des Plaines, and Calumet Rivers. Early explorers discovered and used the Chicago Portage, an area within Mud Lake that was only 4.6 meters (m) above the level of Lake Michigan and near the watershed divide between the Mississippi River and the Great Lakes basins. Because of the low relief, the area was poorly drained. The level of Lake Michigan in the late 1800s was only 0.61 m below the river banks, making subsurface drainage ineffective (Juhl, 2005).

Flow from the North Branch Chicago River (NB) and the South Branch Chicago River (SB) joined just north of present-day Lake Street (Fig. 1) and flowed eastward into Lake Michigan. Sewage discharged into the Chicago River caused serious health hazards during the late 1800s as this sewage affected the drinking water supply from Lake Michigan. In 1900, a canal dug by the Sanitary District of Chicago linking the Chicago River to the Des Plaines River (Mississippi River basin) was completed and reversed the flow in the Chicago River. This canal, the Chicago Sanitary and Ship Canal (CSSC) is 45 kilometers (km) from the SB to Lockport, IL. The CSSC carries waste waters away from the city and Lake Michigan.

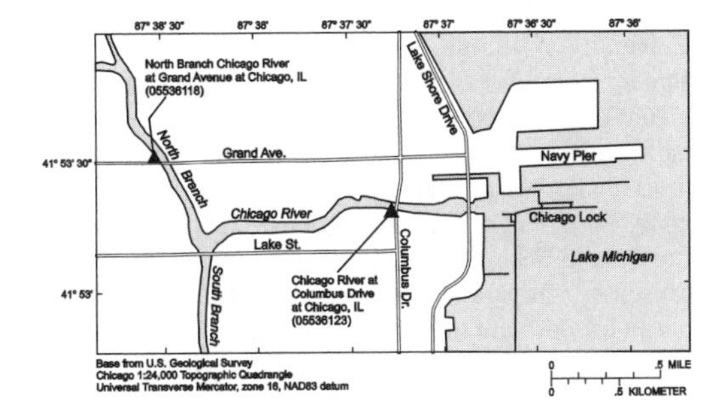

Figure 1. The Chicago River and the North and South Branches of the Chicago River at Chicago, IL.

Today (2005), the Chicago River (CR) flows west from Lake Michigan, through downtown Chicago, and joins flow coming from the NB where it enters the SB/CSSC. Flow in the CR is controlled by the Lockport Powerhouse and Controlling Works (near Joliet, IL) and by the Chicago River Controlling Works (CRCW) and the Chicago Lock. During summer, water from Lake Michigan flows into the CR through sluice gates in the CRCW and, because of lockages, through the Chicago Lock at CRCW. Flow of water from Lake Michigan into the CR during the summer months, called discretionary diversion; is used to preserve or improve the water quality in the CR and CSSC. During winter, flow from Lake Michigan into the CR is small and typically results from leakage through the gates and sea walls at CRCW and some lockages. Other contributions to the CR discharge include water from direct precipitation and discharges of water used for cooling purposes from neighboring buildings. The NB carries runoff from the watershed up-stream and treated municipal sewage effluent released by a water-treatment plant located 16 km upstream from the confluence of the branches. Most or all of this effluent is transported down the SB into the CSSC and then to the Des Plaines and Illinois Rivers.

Discharge measurements made by the U.S. Geological Survey (USGS) beginning in 1998, indicated bi-directional flow in the CR. Although the duration of this bi-directional flow was not known, it indicated the possibility that water from the NB might be flowing into the CR and perhaps even into Lake Michigan. The possibility of flow from the NB entering the CR meant that water quality in the CR, and hence Lake Michigan, might be impaired.

The Metropolitan Water Reclamation District of Greater Chicago (MWRDGC) manages the flow and quality of the CR. Because of the possible effects that bi-directional flow may have on water quality, the MWRDGC contacted researchers at the Ven Te Chow Hydrosystems Laboratory (VTCHL) at the University of Illinois at Urbana-Champaign to investigate the possible causes of these flows. The staff of the VTCHL suggested that the bi-directional flows could indicate the presence of density currents in the CR. It was hypothesized that these density currents developed because of differences in density between waters from the NB and CR. Density differences might be caused by temperature differences, or by the presence of salt or sediment in suspension or some combination thereof. Density currents are well known for having the capacity to transport contaminants, dissolved substances, and suspended particles for long distances (Garcia, 1994).

The hypothesis of density currents in the CR was supported initially by the results from a three-dimensional hydrodynamic simulation conducted by Bombardelli and Garcia (2001). Field information is presented herein to support this hypothesis through the analysis of a unique set of water-velocity measurements (vertical profiles of three dimensional water-velocity components) collected continuously by the USGS near Columbus Drive on the CR (Fig. 1) during January 2004. In addition, hydrological, water-quality, and meteorological data collected by the USGS and MWRDGC are used as complementary information to both characterize the flow conditions at the NB and to evaluate the boundary conditions in the CR.

In this paper, we first present a description of the available data. This presentation is followed by an analysis for the flow conditions observed during the entire month of January 2004. Density current events are identified based on the water-velocity records, and the main characteristics of each identified event are described (for example, duration, flow conditions in the NB, and aver-age air temperature). Finally, one of the observed density current events in January 2004 is analyzed in detail, and includes an estimation of the force driving the underflow and the description of the time evolution of both the vertical velocity profiles and the bed shear stress. Causes for the observed flows are established based on analysis of the boundary conditions.

2 DATA DESCRIPTION

Most of the data analyzed for this study are velocity, backscatter, and temperature data obtained from a 600-KHz acoustic Doppler current profiler (ADCP) manufactured by RD Instruments (Fig. 2). The instrument was installed in an upward-looking configuration on the bottom of the CR at Columbus Drive (CR_CD), in the center of the channel, approximately 0.8 km downstream from the Chicago River Lock. The CR is 55 m wide at this location. The water depth at CR_CD is held at a nearly constant value of 7.0 m in the center of the channel throughout the year. The center of the ADCP transducers were located about 0.30 m above the stream-bed. The ADCP was connected by means

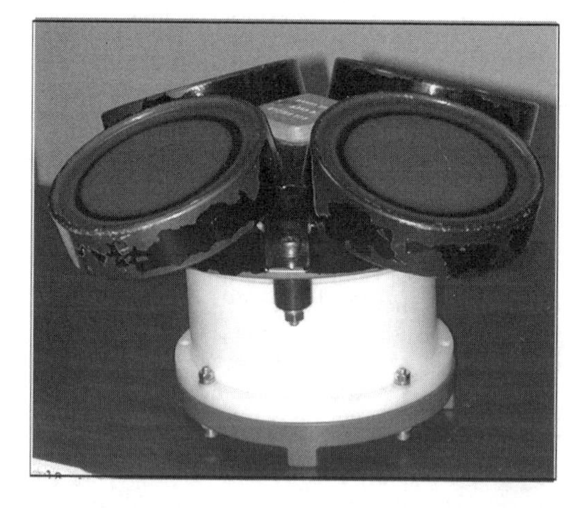

Figure 2. A 600-KHz acoustic Doppler current profiler installed at Chicago River at Columbus Drive at Chicago, IL.

of an underwater cable to a computer in the USGS streamflow gaging station located on the south side of the CR_CD. Data measured by the ADCP were transferred from the computer to the USGS office in Urbana, IL, by way of a dedicated high-speed Internet connection.

The ADCP provided continuous three-dimensional velocity profiles at a sampling frequency of 0.2 Hz (a complete water-velocity profile is recorded every 5 seconds (s)). The 600-KHz ADCP was configured to collect velocity profiles using a pulse-coherent technique known as mode 5 in RD Instruments profilers. Depth-cell size for the velocity measurements was 0.1 m and the blanking distance was set to 0.25 m. With this depth-cell size and blanking distance, the deepest velocity measurement was made in a depth cell centered approximately 0.65 m above the stream-bed. Therefore, no velocity data were available for analysis in the first 0.65 m above the stream-bed. Velocity measurements were also not possible near to the water surface because of side-lobe interference (Simpson, 2001) and decorrelation near the surface. This unmeasured region extended 1.3 m below the free surface for most of the profiles analyzed. For the normal water depth of approximately 7.0 m, valid water-velocity measurements in each profile were obtained for nearly 72 percent of the total depth.

A temperature sensor mounted near the ADCP transducers (0.3 m above the streambed) measured the temperature of the water close to the transducer at the same sampling frequency as the velocity data (0.2 Hz). These temperatures measurements are primarily used by the ADCP to compute the speed of sound at the transducer face. The measurements from this sensor are also used herein to characterize the water temperature of the underflow caused by the density current. The technical specifications provided by the manufacturer state that the temperature sensor operates in a temperature range from −5° Celsius (C) to 45°C, with a precision of ±0.4°C and a resolution of 0.01°C (RD Instruments, 2001). In addition to water velocity and temperature, the ADCP also can be used to measure backscatter. Sound emitted from the ADCP is scattered by suspended sediments and other material in the water. The ADCP measures the intensity of echoes returned to the instrument, called received signal strength indicator (RSSI), using an arbitrary scale of 0 to 255. RSSI profiles for each beam may be converted to backscatter, as measured in decibels, by accounting for propagation losses and other characteristics of the ADCP. Backscatter can be used as a surrogate of the vertical distribution of the suspended sediment in the flow (Gartner, 2004).

Complementary hydrological and meteorological data are analyzed herein to characterize the flow conditions at the NB and to evaluate the boundary conditions presented in the CR system during a density current event. The discharge and water temperature at the NB are measured by the USGS at a streamgaging station located on right upstream side of Grand Avenue bridge in Chicago (Lat 41°53′30″, Long 87°38′30″), about 1,000 m upstream from the confluence with the main stem of the CR. The streamflow gaging station, NB at Grand Avenue at Chicago, IL (NB_GA), has a SonTek/YSI Argonaut-SL (Side-Looking) current meter that is located at about half of the depth (1.55 m above the streambed) on the right bank. This side-looking sensor is used to quantify the discharge by the index-velocity method. In addition, a string of six water-temperature sensors are used to measure water temperatures at various elevations along the right bank at Grand Avenue. The water-temperature probes are located at 0.6 m increments in the vertical with sensor 6 (T6) being the lowest in the water column (located at 0.15 m above the streambed at the wall) and sensor 1 (T1) being the highest. Water-temperature time series for each sensor are available at a sampling frequency of 5 minutes.

Daily effluent discharges to the NB from North Side Water Reclamation treatment Plant (NS WRP) are used here to validate the USGS discharge data collected at NB_GA. The plant is located 16 km upstream from the NB_GA streamflow gaging station. A preliminary study (Manriquez, 2005) showed that more than 75 percent of the observed discharge measured at NB_GA is from the NS WRP effluent discharge. The plant has a design capacity of 1,260 million liters per day or a maximum daily discharge equal to 14.6 cubic meters per second (m^3/s) (Manriquez, 2005).

Wind speed and direction used in the analysis were recorded at the Chicago meteorological station that is operated as part of the real-time meteorological observation network operated by the Great Lakes Environmental Research Laboratory (GLERL). This station is located approximately 5 km offshore from the City of Chicago. The anemometer is located at 25.9 m above station elevation. Meteorological measurements are made every 5 seconds at the station and then averaged together and recorded every 5 minutes. Meteorological data recorded by the Chicago O'Hare International Airport meteorological station (ORD), located 24 km from the measurement site, are used to evaluate spatial variability in the aforementioned properties. The ORD data, which include wind speed and direction, air temperature, snow depth, are available in 3-hour intervals.

Results of preliminary studies (Bombardelli and Garcia, 2001, Manriquez, 2005) indicated that density current events in the CR are more likely to occur during the winter season. For that reason, the analysis reported herein focuses on flow conditions observed in January 2004.

Table 1. Characteristics of the density current events at the Chicago River at Columbus Drive at Chicago, IL.

No.	Start time	Dur. [hs]	Water NB_GA	Temp [°C] CR_CD	Mean daily discharge [m^3/s] NB_GA	Mean air temp [°C]
1	1/1 01:50	21.1	8.5	5	9.74	3.3
2	1/2 15:19	5.0	9.3	7	9.51	14.3
3	1/4 03:41	24.0	10.1	5	9.23	−1.2
4	1/7 13:37	64.7	6.0**	2	8.88***	−5.1
5	1/13 12:18	21.8	7.9	4	10.52	−1.8
6	1/15 11:23	56.6	6.9	3	9.84	−2.8
7	1/19 12:45	55.4	6.1**	2	8.97***	−7.5
8	1/23 23:21	228.3*	4.6**	3	8.81***	−10.8

* Estimated 192.6 hours of the total duration included in January 2004.
** Stratified conditions in NB during the event at CR_CD.
*** Discharge estimated using discharge from NS WRP plant.

3 DATA ANALYSIS

3.1 *Analysis for January 2004*

Water-velocity profiles recorded every 5 seconds during January 2004 at CR_CD were analyzed to identify density current events. Based on the presence of bi-directional flow, a total of eight density currents events were identified in January 2004. Characteristics of these events are shown in Table 1 and include: starting time (in military time); duration of the bi-directional flow conditions at CR_CD; water temperature before the density current event occurred at both the sensor nearest to the streambed at NB_GA and at the ADCP transducer at CR_CD; the mean daily discharge at NB_GA; and the mean air temperature during the event. The ADCP water-velocity data were not available during some periods during event 8; however, water-temperature data indicate that the bi-directional flow persisted for more than 9 days, ending on February 2, 2005, at 11:38.

A contour plot of velocity in the east-west direction from profiles recorded by the ADCP during 5 days (starting January 5, 2004, at 00:09 and ending on January 5, 2004, at 23:58) is shown in Figure 3. A bi-directional flow is detected in this plot starting January 7, 2004, at 13:37 (ensemble 3524002). During this event, the lowest part of the flow is moving east (to the lake) whereas the upper part is moving west (to the confluence). The bi-directional flow conditions persisted at CR_CD for more than 64 hours.

The time series of water temperatures recorded for the near-bed (T6) and the near-surface (T1) temperature sensors at the NB_GA streamgaging station are shown in Figure 4. Three temperature stratification events were observed at NB_GA in January 2004, occurring simultaneously with the density current events 4, 7, and 8. The maximum observed differences in water temperature between the deep and the shallow sensors at NB_GA were 3.8°C, 3.7°C, and 3.0°C for density current events 4, 7, and 8, respectively. For each event, the upper part of the water column was coldest. Meteorological data also indicate that the 3 days with the lowest mean daily air temperature for January 2004 occurred at the beginning of each of these three events.

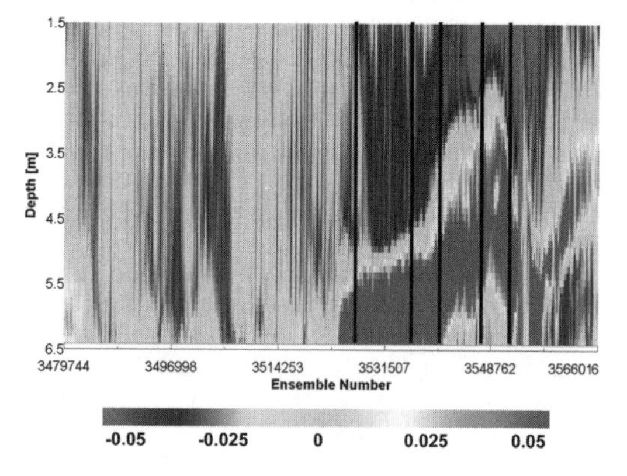

Figure 3. Time series of east velocity [m/s] profiles measured at Chicago River at Columbus Drive at Chicago, IL, starting January 5, 2004, at 00:09 (ensemble 3479744). Dark lines define different stages in the density current event starting January 7, 2004, at 13:37.

During the periods when thermal stratification was observed at NB_GA, it appears that there were also bi-directional flows at this location as well. The index velocities measured at this site by the Argonaut SL confirm this flow condition. Negative velocities (in other words water was flowing upstream at the elevation measured by the Argonaut SL) were measured during the periods with thermal stratification conditions at NB_GA (Fig. 5). Negative velocities were measured in the upper layer of a bi-directional flow.

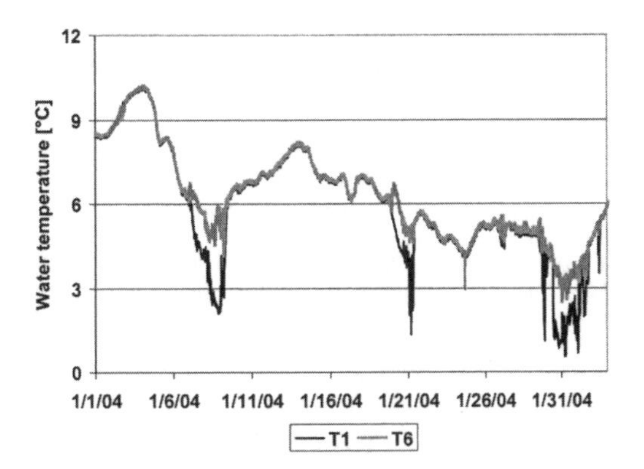

Figure 4. Near-surface (T1) and near-streambed (T6) time series of water temperature recorded at North Branch Chicago River at Grand Ave, Chicago, IL.

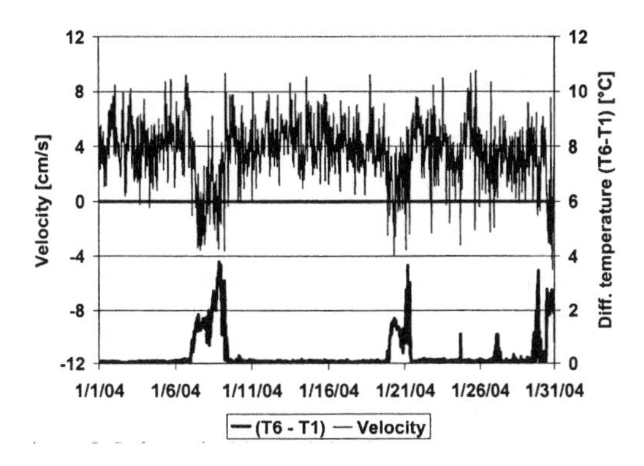

Figure 5. Index velocities and the difference between near-bed (T6) and near-surface (T1) water temperatures measured at the North Branch Chicago River at Grand Avenue at Chicago, IL.

If the discharges at NB_GA are computed using only the existing index-velocity rating developed by the USGS, negative net discharge will be computed for these time periods. This result, however, does not agree with the records of the effluent discharge from the NS WRP located 16 km above this section. For these events, the discharges computed using the measured index velocities are not valid and the values for discharge at NB_GA reported in table 1 correspond to the daily mean effluent discharge at NS WRP.

The time series of wind-velocity data from the GLERL station were analyzed, and no relation was found between the occurrence of the bi-directional flow at either CR_CD or NB_GA and the presence of wind blowing in the upstream or downstream direction in each branch. Some wind effects, however, were observed on the shape of the vertical profiles of water velocity recorded at CD_CR during a density current event. These effects are discussed later in this paper, for the density current event starting on January 7, 2004.

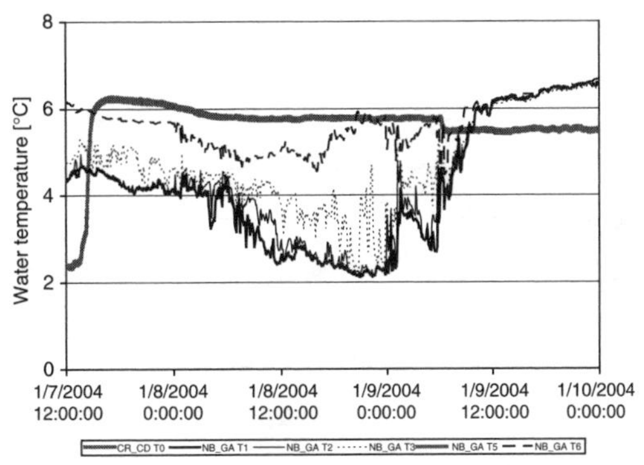

Figure 6. Water temperature measured at North Branch Chicago River at Grand Avenue (NB_GA) and the Chicago River at Columbus Drive (CR_CD), Chicago, IL. For NB_GA, sensor T6 is located nearest to the streambed and T1 is located nearest to the water surface for the six-temperature string.

3.2 *Analysis of density current event starting January 7, 2004*

Density current event 4 (see table 1 for details) starting on January 7, 2004, at 13:37 is described in detail in this section. As previously mentioned, stratified conditions were present at NB_GA during this event. At the time when the event was observed, extremely low air temperatures were observed in Chicago (<−5°C) (see table 1). The time series of water temperature recorded by five different sensors (T6 is nearest to the streambed and T1 is nearest to the water surface) located at NB_GA, along the ADCP water-temperature sensor (T0) located at CR_CD are shown in Figure 6.

The average air temperature recorded at ORD meteorological station during this event was −5.1°C and the minimum was −9.4°C. As the bi-directional flow starts (January 7, 2004, at 13:37), the water temperature recorded by the ADCP at the centerline at CR_CD increases from 2.4°C to 6.2°C, approximately the same as the water temperature observed by the water-temperature sensor located nearest to the streambed at NB_GA at the same time. This result seems to indicate that underflow water coming from NB is present as an underflow at CR_CD.

Stages in the development of the density current within the event were delimited based on periods with homogenous vertical velocity-profile patterns for velocities in the east direction (see Fig. 3). Stage 1 is completely unsteady and includes the time during which the velocity profile changes from a well-mixed condition to a developed density current profile (stage 2). For stage 2, the scatter of the instantaneous velocity profile recorded by the ADCP when compared with the mean computed profile in the stage is shown in Figure 7. The mean velocity profiles observed in each stage are shown in Figure 8. Stage 6 also presented

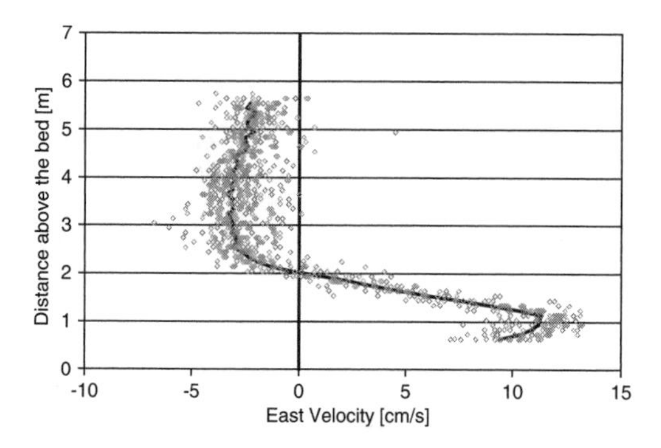

Figure 7. Instantaneous and mean velocity profiles in the easterly direction measured at the Chicago River at Columbus Drive,Chicago, IL, during stage 2 of the January 7, 2004, event.

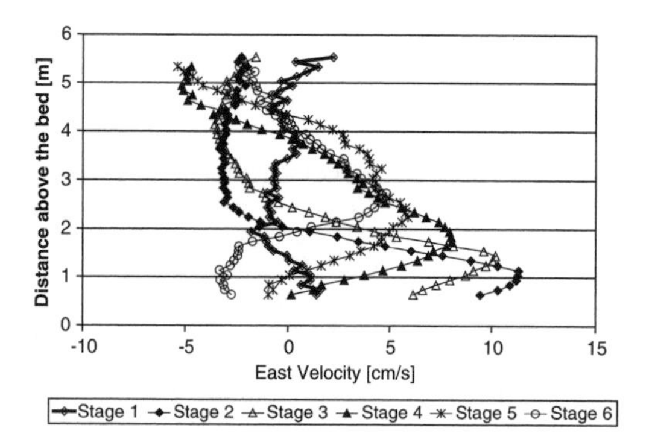

Figure 8. Mean velocity profiles in the easterly direction measured at the Chicago River at Columbus Drive, Chicago, IL, during different stages of the density current event number 4.

a highly unsteady behavior; the mean profile for this stage, included in Figure 8 and characterized in Table 2, represents the flow conditions observed on January 9, 2004, between 16:18 and 19:05.

The main flow characteristics observed at CR_CD for each stage of the density current event starting January 7, 2004 at 13:37 are included in table 2. The flow characteristics included in this table are: starting time (in military time) of each stage, duration, discharge per unit width in the underflow q_u, depth of the underflow H, the layer-averaged velocity in the underflow U, an estimate of the relative difference in density driving the flow $R = (\rho_u - \rho_0)/\rho_0$ (ρ_0 is the ambient water density) and the shear velocity, u_s. Average values for stage 1 and 6 are not included in Table 2 because of the strong unsteadiness observed in those stages.

The layer-averaged velocity, U, is defined from the observed east velocity (*Veast*) profiles via a set of moments (Garcia, 1994) through the ratio of the

Table 2. Flow characteristics for each stage of the density current event observed at Chicago River at Columbus Drive at Chicago, IL, starting January 7, 2004.

Stage no.	Start time	Dur. [hs]	q_u [m²/s]	H [m]	U [m/s]	R %	u_s [m/s]
1	1/7 13:37	4.17	–	–	–	–	–
2	1/7 17:47	11.8	0.13	2.0	0.09	0.044	0.022
3	1/8 05:35	7.00	0.14	2.5	0.08	0.023	0.023
4	1/8 12:35	8.97	0.14	3.9	0.06	0.008	0.034
5	1/8 21:33	4.87	0.12	3.3*	0.04	–	–
6	1/9 02:25	27.9	–	–	–	–	–

*H is computed for this stage as the portion of the flow depth with positive east velocities.

integrals indicated in equations (1) and (2) below assuming that $\beta = 1$.

$$UH = \int_0^H Veast \cdot dy \tag{1}$$

$$\beta \cdot U^2 H = \int_0^H (Veast)^2 \cdot dy \tag{2}$$

The parameter R is computed using the observed average profile for each stage based in the following relation:

$$R = \frac{\Delta\rho}{\rho} = \frac{\rho_u - \rho_0}{\rho_0} = \frac{U^2}{gH} \tag{3}$$

The shear velocity u_s is computed for each stage based on the assumption that the lower portion of the underflow velocity profile presented a logarithmic shape. The number of point pairs (velocity, distance above the streambed) used in the log-law fitting were 3, 8, and 12 for stages 2, 3, and 4, respectively.

Wind data at GLERL station were analyzed to evaluate the boundary conditions during the different stages in event 4. Time series of the average wind speed and the wind direction (0° = from north, 90° = from east) recorded during the analyzed density current event are shown in Figure 9. At stage 1 and 2, the wind blew from west to east at speeds ranging between 5 and 10 meters per second (m/s). The wind changed direction with very slow speed in stage 3 and then blew from east to west most of the time in stages 4, 5, and 6 (average wind speed around 10 m/s). Wind data of ORD meteorological station also were analyzed and included in Figure 9 to represent the spatial variability of the data and to validate the assumption that the wind conditions at CD_CR are acceptably well represented by GLERL data (recorded around 5 km away from the measurement site). The same pattern is observed in Figure 9 for the time evolution of the wind data for both locations.

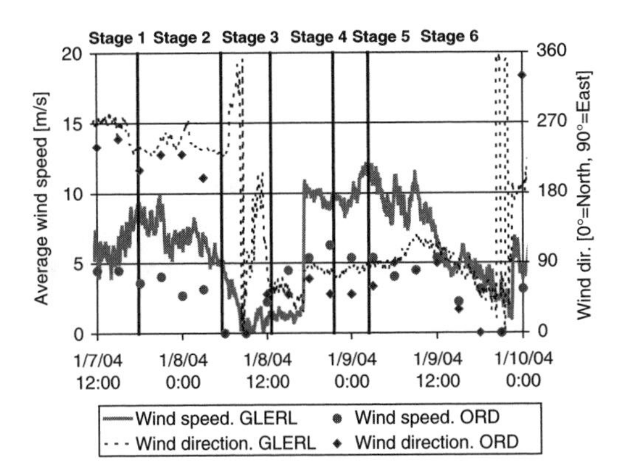

Figure 9. Time series of average wind speed and direction recorded at GLREL and ORD meteorological station.

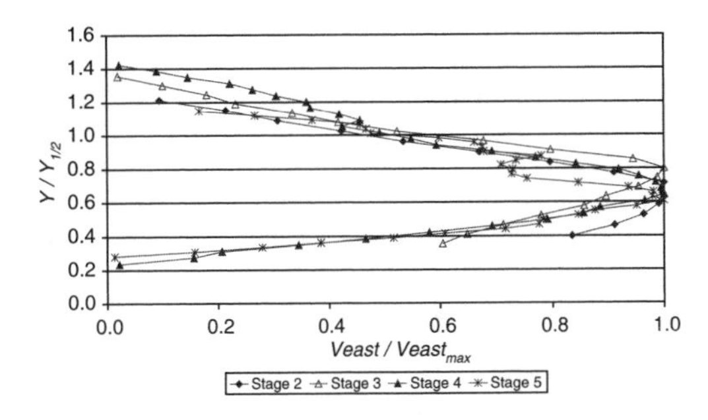

Figure 10. Dimensionless water-velocity profiles recorded at Chicago River at Columbus Drive at Chicago, IL, during different stages of event number 4.

Underflow velocity profiles recorded during stages 2 to 5 of the analyzed density current event were made dimensionless and plotted in Figure 10 based on the parameters commonly used in turbulent wall jet studies. This approach has been used in other density current studies (Kneller and Buckee, 2000, Gray and others, 2005). East velocities (*Veast*) were made dimensionless by dividing them by the maximum observed value ($Veast_{max}$) whereas the distance above the stream-bed (Y) was divided by the distance above the streambed corresponding to the point where the underflow velocity value is equal to half of the maximum observed value ($Y_{1/2}$).

For stages 2 to 5, the dimensionless profiles all had approximately the same shape. This result indicates the density current is the process driving the underflow for stages 2 to 5. For example, a wind-driven flow would result in different dimensionless profiles than those obtained from these data.

The average acoustic backscatter levels observed for the four ADCP beams in each stage are plotted in Figure 11. These profiles can be used as a surrogate of the vertical distribution of the suspended sediment in

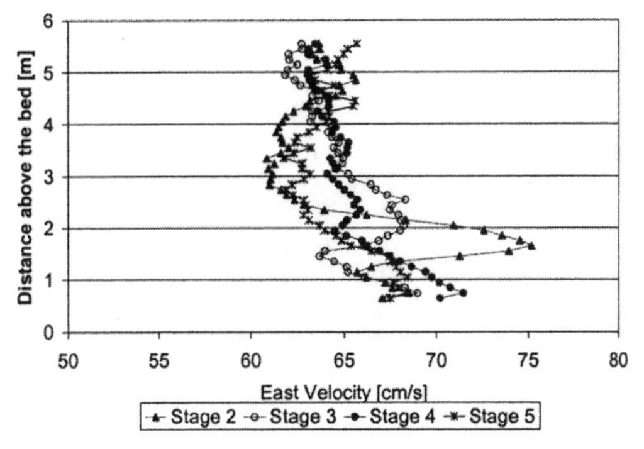

Figure 11. Vertical profiles of acoustic backscatter reported by ADCP at Chicago River at Columbus Drive at Chicago, IL, for different stages of the density current event, number 4. The values included in the plots are averaged values reported for the four ADCP beams.

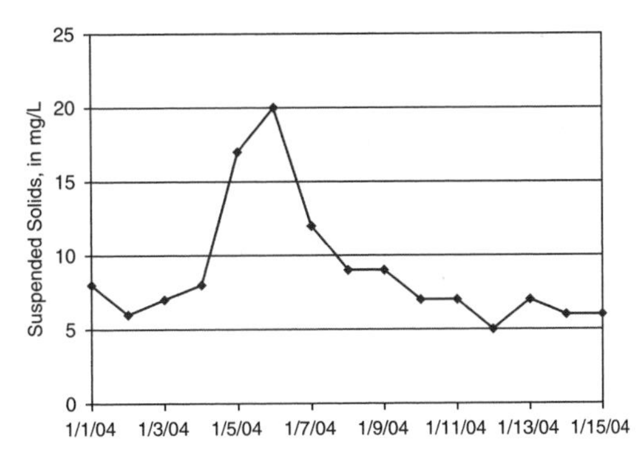

Figure 12. Time series of suspended-solid concentrations in effluent of the North Side Water Reclamation Plant on the North Branch Chicago River, Chicago, IL. (data provided by MWRDGC).

the flow (Gartner, 2004). During the density current event, the profiles indicate that the greatest acoustic backscatter occurs in the underflow, which in turn indicates that the highest levels of suspended sediment are in this region (0 to 2.5 m above the streambed for stage 2). Inconsistency, however, is observed in some of the deepest bins for stages 2 and 3 that could be generated by the technique used to compute backscatter for observations in the near field.

The presence of suspended sediment in the underflow during the density current event agrees with an increase in the suspended sediment in the NB according to daily water-quality measurements of the treated effluent from the NS WRP plant at the NB (Catherine O'Conner, MWRDGC, written comm., 2005). The SS values [in milligrams per liter (mg/L)] for the first 2 weeks of January 2004 are shown in Figure 12. The water-quality data indicate elevated suspended-solids concentration (SS) values (above background levels)

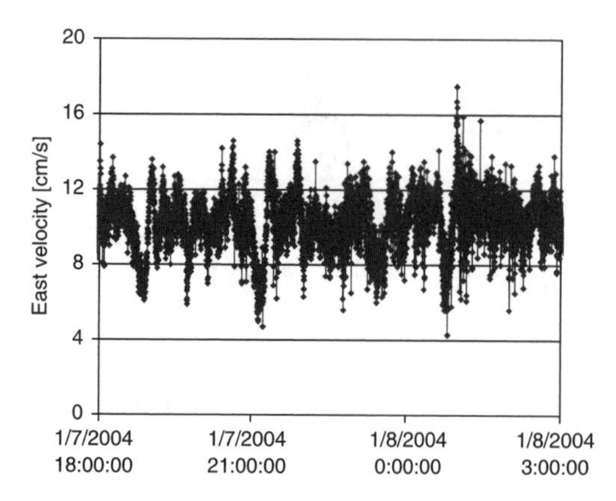

Figure 13. Time series of east water velocity recorded at Chicago River at Columbus Drive at Chicago, IL, at bin 1 (0.64 m above the streambed).

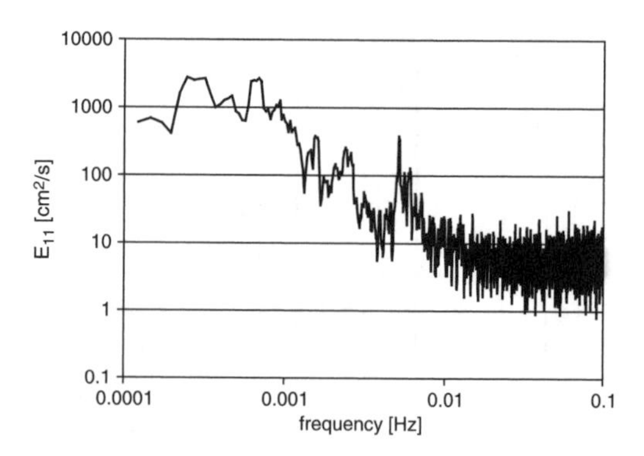

Figure 14. Power spectrum in the frequency domain of the time series of east water velocity recorded at Chicago River at Columbus Drive at Chicago, IL, represented in figure 13.

during the period from January 5, 2004, and January 10, 2004.

Finally, the time series of east water-velocity values recorded by the ADCP at different depths were analyzed to describe the different random process (for example, turbulence, internal waves, driven-force pulsing, and others) presented in the flow at the time that each signal was recorded. The time series of east water velocity recorded by the ADCP at the deepest bin (0.64 m from the streambed) at CR_CD during stage 2 of event 4 (starting January 7, 2004) is shown in Figure 13. A steady signal is observed for this stage with an average east velocity value of 9.4 centimeters per second (cm/s).

Garcia and others (2005) presented some tools to evaluate the capabilities of acoustic sensors to sample the flow turbulence under different flow conditions. One of these tools, a dimensionless power spectrum, is used here. A minimum dimensionless sampling frequency $F = fL/U_c = 1$ is required to get a minimum description of the flow turbulence (Garcia and others, 2005, Fig. 7). The parameters f, L, *and* U_c represent the instrument sampling frequency, the length scale of the energy containing eddies in the flow, and the flow-convective velocity, respectively. For stage 2 of event 4, a value of $F = 4$ (higher than the minimum required value) is computed by using $f = 0.2$ Hz and assuming L is of the order of $H = 2$ m and U_c is of the order of the observed east velocity, or 0.1 m/s. Higher values of F (larger H and slower velocities) were obtained for other stages of the analyzed event.

A value of F higher than the minimum required indicates that some turbulence characterization would be possible from the recorded signal. The dimensionless plot, however, presented by Garcia and others (2005) is based on the assumption that the signal noise is much smaller than the turbulence energy for all the analyzed frequencies. High values of Doppler noise, which are included in the recorded signal, reduce the value of maximum useful frequency (Garcia and others, 2005). The power spectrum computed from the velocity signal plotted in Figure 13 is shown in Figure 14. The Doppler noise energy level is detected as a flat plateau (white noise characteristics) at frequencies higher than $f_n = 0.01$ Hz. A value of $F = 0.2 < 1$ is obtained using the value of $2 f_n$ instead of f in the computation of dimensionless frequency. This result indicates that no description of the turbulence can be made because the turbulence process occurs at smaller time scales where energy is dominated by the Doppler noise.

Although no description of the flow turbulence can be performed on the basis of the water-velocity signal recorded by the ADCP, other periodic random components in the flow with frequency smaller than 0.01 Hz (length scales longer than the turbulence structures) can be described using the recorded series.

The random periodic processes presented at CR_CD during stage 2 of event 4 can be detected based on the computed autocorrelation functions plotted in Figure 15 for the recorded east velocity signals at six different distances above the streambed. Two main random periodic processes can be detected for all the flow depths with periods around 180 and 4,800 s. In addition, the signal recorded close to the density current interface at stage 2 (2.54 m and 3.54 m from the bottom) present a new random periodic component with a period of the order of 1,200 s.

The vertical profile of standard deviation of the time series of east velocity signal for stage 2 of event 4 (January 7, 2004), is shown in Figure 16. A vertical profile of standard deviation generated only by the turbulence process during a density current event should show the highest values of turbulent energy near the bottom and in the interface between the under and overflow. A uniform vertical profile, however, is observed because no turbulence description is achieved by the

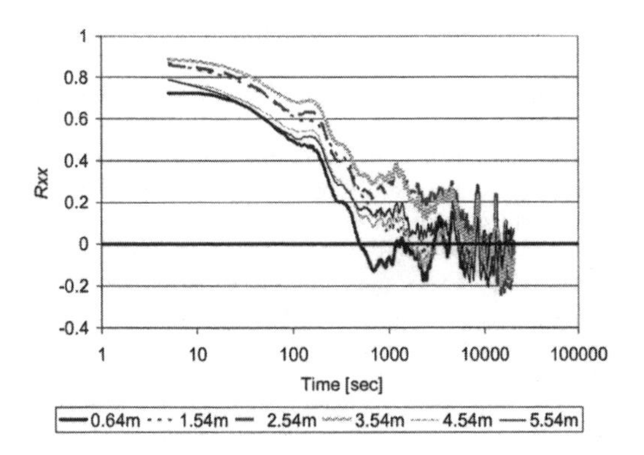

Figure 15. Autocorrelation function of the east velocity time series recorded Chicago River at Columbus Drive at Chicago, IL, at different distance above the streambed for stage 2 of event 4.

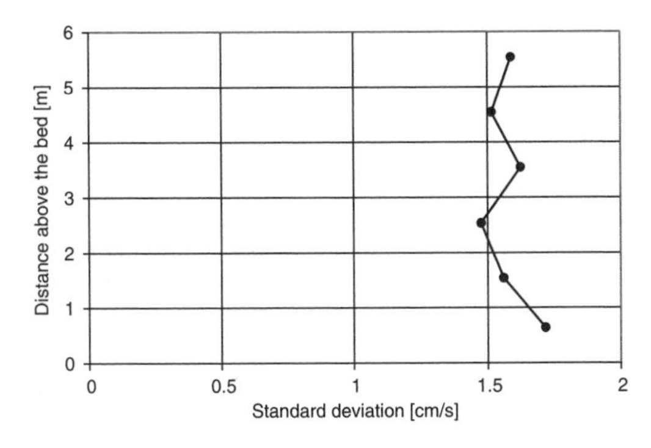

Figure 16. Vertical profile of standard deviation of the time series of east velocity signal recorded at Chicago River at Columbus Drive at Chicago, IL, for stage 2 of event 4 (January 7, 2004).

instrument and because almost the same random periodic components were observed throughout the flow depth.

4 DISCUSSION

A total of eight density currents events were detected at CR_CD based on ADCP velocity-profile measurements. Bi-directional flow conditions were observed for a total of 441 hours in January 2004 (59 percent of the time). These conditions indicate that density current events occur frequently at CR during winter periods. Events, as short as 5 hours and as long as 9 days, were observed in the period with a wide variability in the main characteristics of each event (see table 1). The water temperature recorded near the streambed at the centerline of the CR_CD cross section increased at the time when the density current event started from the ambient fluid temperature at CR_CD

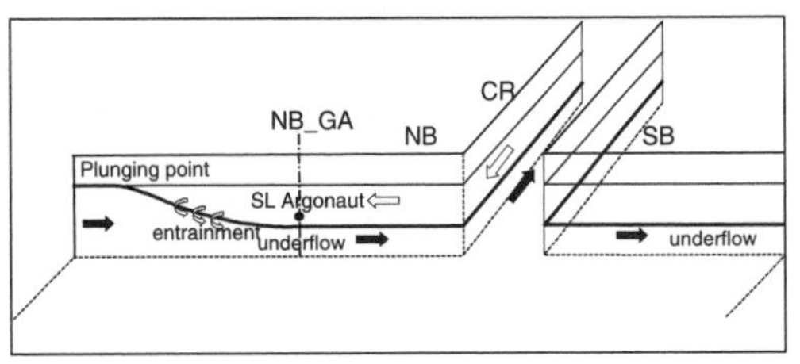

Figure 17. Schematic view of flow conditions observed during event 4 at junction of the Chicago River, North Branch Chicago River, and the South Branch Chicago River, Chicago, IL.

to the temperature recorded by the sensor nearest to the streambed at NB_GA at the same time. This result indicates that the underflow coming from NB is also detected as an underflow at CD_CR.

Three bi-directional flow events with temperature stratification conditions were also observed at NB_GA (see figs. 4 and 5). Each time that a bi-directional flow event was observed at NB_GA, a density current event was also observed at CR_CD. Flow conditions observed at NB and CR during event 4 are illustrated in Figure 17.

Other density currents events were observed at CR_CD when no temperature stratification was observed at NB_GA. For such events, this result may be explained if the plunging point for the density current occurred downstream of NB_GA.

The location of the plunging point depends on the characteristic of each density current event (relative difference in density) and the NB flow discharge. Manriquez (2005) analyzed this behavior through laboratory experiments performed on a distorted physical model (horizontal scale 1:250; vertical scale 1:20) of the CR system. Manriquez (2005) found that for the same discharge at NB, increasing the density of the underflow caused the plunging point to migrate upstream in the NB, sometimes upstream of NB_GA. Therefore, for the weakest density current event, bi-directional flow would not likely be observed at NB_GA. The presence of the bi-directional flow at NB_GA would require changes to the methodology for computing discharge at this station.

The driving force for the observed density currents events could be caused by different factors, including temperature differences, salinity differences, and differences in suspended-sediment concentration. Water temperatures at NB_GA are always higher than at CR_CD in the winter period according to historical data for both branches (James J. Duncker, USGS, written comm., 2005; see table 1). Flow in the NB at NB_GA consists mostly of effluent discharge from the NS WRP plant. The average daily temperature of the effluent during January 2004 was 11.8°C with a standard deviation of 1.1°C. The only time that warmer

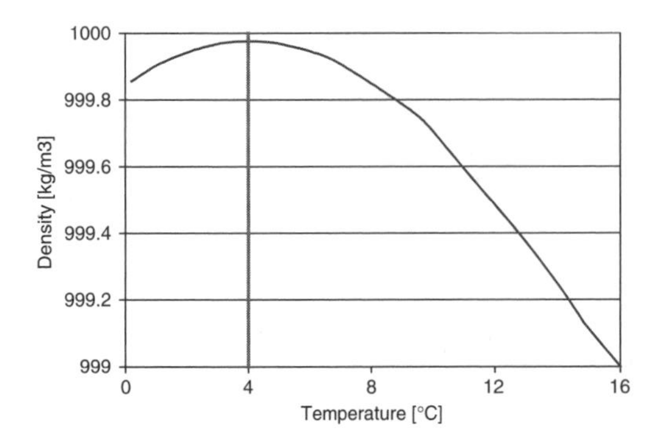

Figure 18. Water density as a function of the water temperature (suspended-solid concentration or salinity effects are not included in this analysis).

water results in denser water (without the presence of high suspended-sediment concentrations or salinity) is when the temperature ranges between 0°C and 4°C (see Fig. 18). The maximum possible relative difference in density R (assuming temperatures of 4°C and 0°C for the NB and CR, respectively), however, is 0.012 percent smaller than the values of R estimated for some of the events. This result indicates that other factors are causing the difference in density.

An increase in suspended-solid concentration of the effluent of the NS WRP similar to the increase observed for the density current event 4 started at January 7, 2004, (see Fig. 12) was observed for events 7 and 8 (see table 1). This increase was also detected in the acoustic backscatter profiles (see Fig. 11 for profiles of event 4).

Salinity in the effluent discharged to the river could increase as a result of salt applications to roads and subsequent inflow from streets or other areas to the treatment plant. Based on data from the ORD meteorological station, 0.1 m of snow depth was observed on January 5, 2004 and the snow stayed on the ground (decreasing in depth) until January 14, 2004. It seems likely that, because of this snowfall, salt was applied to the roadways. Water-quality data for the NS WRP indicate elevated chloride concentrations (above normal levels) for at least two of the eight observed density currents events in January 2004 (events 4 and 8).

Event 4 is one of the strongest density current events observed during January 2004. The maximum underflow velocity recorded at CR_CD during this event was 0.12 m/s. The shape of the bi-directional, vertical velocity profile changes during the event. Six stages were defined during the event in order to define periods with quasi-steady behavior that were longer than 4.8 hours.

Different processes were investigated as a possible cause of the unsteady behavior in the vertical profiles of event 4 (for example, changes in time series of difference in density, wind effects, and backwater effects from locks). A quasi-uniform underflow discharge per unit width observed in stages 2 to 5 of event 4 (see table 2) would indicate that the unsteady behavior does not result from changes in the difference in density-driven flow. It is possible that the steadiness could result because of backwater effects from the locks, which would result in the depth (thickness) of the underflow being less at the beginning and greater towards the end for the entire event. The January 15, 2004, event (number 6) indicated an inverse time evolution (the depth of the under-flow was greater at the beginning and less towards the end). An analysis of event 6, similar to the one performed for event 4, showed the same pattern in the relation between the depth of the underflow and the wind speed and direction (see Fig. 9): wind blowing in the downstream or up-stream direction would generate underflows that have a greater or smaller thickness, respectively. Thus, wind speed and direction provided the best explanation to the time evolution of the east velocity profiles.

Underflow velocity profiles recorded during stages 2 to 5 of the analyzed density current event were made dimensionless in Figure 10 using the parameters commonly used in turbulent wall jet studies (Gray and others, 2005, Kneller and Buckee, 2000). Although the average east velocity profiles for the different stages within event 4 had different shapes, a good collapse was obtained when these event profiles were made dimensionless using the above-mentioned approach.

No turbulence parameters could be estimated based on the measured east velocity time signal because of the noise energy level of the signal; however, main features of the other random periodic processes presented in the flow field could be represented for stage 2 of event 4. Two scales were observed for all flow depths. The first scale, of the order of 180 s, could be generated by a local effect and the second scale (around 4,800 s or 1.33 hours) could be generated by a large-scale process. Random periods of the order of 1,200 s (20 minutes) were observed in the velocity signals recorded near the interface. These signals could represent the period of the interfacial waves.

5 SUMMARY AND CONCLUSIONS

This paper presents evidence for the occurrence of density current events in the Chicago River system. A unique set of water-velocity profiles recorded at the Chicago River at Columbus Drive provide the primary basis for the analyses presented here.

Analysis of the complementary meteorological and water-quality data collected in January 2004 indicate that the presence of bi-directional flow is not restricted only to the main stem of the Chicago River, but also could travel upstream on the North Branch depending on the density difference of the generated underflow. The greater the density difference, the farther upstream on the North Branch the plunging point is observed. This result is an important consideration for projects designed to minimize the occurrence of the density current events because design alternatives need to consider all parts of the Chicago River system, including both the North Branch and South Branch.

The data obtained from velocity and temperature measurements indicate that bi-directional flow is a common feature at Chicago River at Columbus Drive (CR_CD) and at North Branch Chicago River at Grand Avenue (NB_GA) during January 2004. The characteristics of the observed events indicate that there is no single cause for the development of density currents. For one event analyzed, the difference in temperature, suspended sediment, and salinity contributed to the driving force of the underflow. Analysis of meteorological information measured simultaneously with the water-velocity data indicated that vertical profiles of the underflow at CR_CD are strongly affected by wind speed and direction. The continuous record of water-velocity data recorded by the ADCP allowed detection of three different random, periodic processes.

On the basis of the analyzed data, the following suggestions are offered for consideration regarding future monitoring and streamflow computation.

- An upward-looking ADCP should be installed in the river at the USGS streamflow gaging station, NB_GA.
- A string of water-temperature sensors should be installed at the USGS streamflow gaging station, CR_CD.
- Consideration should be given to the installation of salinity and other water-quality probes at NB_GA and CR_CD.
- The methods used to compute streamflow at the streamflow gaging station, NB_GA, should be altered to account for the presence of bi-directional flows.

The use of trade, product, or firm names in this paper is for descriptive purposes only and does not imply endorsement by the U.S. Government.

ACKNOWLEDGEMENTS

The research summarized in this paper was supported by both the U.S. Geological Survey (Office of Water Surface) and the Metropolitan Water Reclamation District of Greater Chicago. Comments from Dr. Jim Best provided valuable insights for the data analysis.

REFERENCES

Bombardelli, F.A., & García, M.H. 2001. Simulation of density currents in urban environments. Application to the Chicago River, Illinois. 3rd. Int. Symposium on Environmental Hydraulics, Tempe, Arizona, USA.

García, M.H. 1994. Depositional turbidity current laden with poorly sorted sediment. Journal of Hydraulic Engineering 120 (11): 1240–1262.

García, C.M., Cantero, M., Niño, Y., & Garcia, M.H. 2005. Turbulence Measurements with Acoustic Doppler Velocimeters. Approved to be published in the Journal of Hydraulic Engineering.

Gartner, J.W. 2004. Estimating suspended solids concentration from backscatter intensity measured by acoustic Doppler current profiler in San Francisco Bay, California. Marine Geology 211: 169–187.

Gray, T.E., Alexander, J., & Leeder, M.R. 2005. Quantifying velocity and turbulence structure in depositing sustained turbidity currents across breaks in slope. Sedimentology 1–22.

Juhl, A. 2005, A History of Flood Control & Drainage in Northeastern Illinois: accessed June 27, 2005, at http://dnr.state.il.us/owr/OWR_chicago.htm#owrnortheast

Kneller, B., & Buckee, C. 2000. The structure and fluid mechanics of turbidity currents: a review of some recent studies and their geological implications. Sedimentology 47: 62–94.

Manriquez, C. 2005. "Hydraulic model study of Chicago River density currents". Master's thesis, University of Illi-nois, 2005.

RD Instruments, Inc., 2001, Workhorse installation guide: San Diego, Calif., RD Instruments P/N 957-6152-00 (January 2001), 64 p.

Simpson, M.R. 2001, Discharge measurements using a Broad-Band Acoustic Doppler Current Profiler: U.S. Geological Survey Open-file Report 01-01, 123 p.: accessed June 28, 2005, at http://water.usgs.gov/pubs/of/ofr0101/

River, Coastal and Estuarine Morphodynamics: RCEM 2005 – Parker & García (eds)
© 2006 Taylor & Francis Group, London, ISBN 0 415 39270 5

Cross-section periodicity of turbulent gravel-bed river flows

M.J. Franca & U. Lemmin
Laboratory of Environmental Hydraulics (LHE), École Polytechnique Fédérale de Lausanne, Switzerland

ABSTRACT: A 3D wide gravel-bed river flow is investigated using ADVP data from the Swiss river, Chamberonne. The flow aspect ratio was $W/h = 23$, and the relative roughness $D_{84}/h = 0.324$. The flow is mainly 3D near the most important bed forms and near the surface. Bed forms present a wave-like periodic pattern across the section indicating the possible existence of streamwise ridges. All hydraulic characteristics of the flow, such as bed shear, velocity deviations and turbulence production respond to the periodic riverbed morphology. We detected a permanent flow structure in the upper layer of the flow ($z/h > 0.80$), hereinafter called Surface Layer Organized Motion (SLOM), also conditioned by the bed morphology: it consists of periodically distributed low and high velocity regions across the section associated with a mass-compensation secondary motion. This SLOM seems be an inviscid flow response to the bed forms confined to the layer $z/h > 0.80$ due to a large-scale roughness wake-effect.

1 INTRODUCTION

Wide rivers are generally considered two-dimensional since the bank influence is negligible. However, in the presence of macro bed forms such as boulders and gravel stripes, 2D approaches may no longer be valid and three-dimensionality has to be considered. Recent studies provide some insight into different features of river flow dynamics: mean velocities, turbulence intensities, shear stresses, bed shear, friction velocity, roughness parameters and velocity spectra (Babaeyan-Koopaei et al. 2002); bed-load transport across the river section (Powell et al. 1999); vertical distribution of turbulent energy dissipation and characteristic turbulent scales (Nikora & Smart 1997); turbulence features between weakly mobile and fixed gravel bed cases (Nikora & Goring 2000); coherent structures in rivers and their influence on transport and mixing (Buffin-Bélanger et al. (2000); Hurther et al. (2002)); turbulence in the wake of obstructions (Buffin-Bélanger & Roy 1998); flow structures present in gravel bed rivers (Roy et al. 2004).

Two different regions in the water column have been found to be affected by 3D flow: an inner region, the roughness layer (Nikora & Smart 1997), where wake effects from macro-roughness and skirting effects around the gravel boulders produce deviations from the logarithmic profile (Robert et al. 1992, Nelson et al. 1993, Baiamonte et al. 1995, Buffin-Bélanger & Roy 1998); and an outer region ($z/h > 0.80$) where an organized cellular movement resulting from a flow response to the bed forms is present. Secondary motion occupying the whole water depth may be associated with a response to the riverbed roughness (Studerus 1982, Nezu & Nakagawa 1993). The hydraulic response to the morphology of the riverbed may be confined to the outer layer of the flow due to a general wake produced by the gravel macro-roughness (Baiamonte & Ferro 1997). This response relates to the occurrence of the so-called d-shaped velocity profiles (Ferro & Baiamonte 1994, Yang et al. 2004). Despite the three-dimensionality of the flow in the lower layers, Nikora & Smart (1997), Smart (1999) and Franca & Lemmin (2004) showed that occasionally the log-law may still be used locally to describe velocity profiles in gravel-bed river flows.

This study is based on field measurements made with the deployable 3D Acoustic Doppler Velocity Profiler (ADVP) in the Swiss river Chamberonne. Several aspects of the flow are assessed: cross-section velocity distribution; influence of bed forms on flow resistance and on velocity distribution; influence of bed morphology on the turbulent flow structure (scales and production).

2 FIELD MEASUREMENTS

3D ADVP field measurements were taken in the Swiss gravel-bed river Chamberonne. They were made on a single day during a low water period, under stationary flow conditions (confirmed by the gauging of the Swiss Hydrological and Geological Services). The cross-section was at 385 m before the river mouth. The main river characteristics at the time of the measurements are given in Table 1.

S is the mean river slope; Q, the discharge; h, the water depth (corresponding to the deepest profile); W, the river width; Re, the Reynolds number; Fr, the Froude number; D_{50}, D_{50} and D_{84}, the bottom grain size for which 40%, 50% and 84% of the grain diameters are smaller, respectively. The riverbed material was sampled according to the Wolman (1954) technique. The water depth varied between 0.15 and 0.29 cm. Considering the average water depth, the aspect ratio (W/h) is 23, corresponding to shallow water or wide channel. The river slope was determined with topographic data from the Gesrau system (SESA – Canton of Vaud). The riverbed is composed of coarse gravel and randomly spaced macro-roughness producing a rough turbulent flow with no sediment transport during the measurements. The profiles were taken across a section starting from the right riverbank ($y/W = 0$). We assumed symmetrical hydraulic conditions and only measured across half of the river width, taking 24 velocity profiles with a horizontal spacing of 5 to 12.5 cm. Data were acquired during five minutes at each point.

3 INSTRUMENTATION

The 3D Acoustic Doppler Velocity Profiler (ADVP) measures quasi-instantaneous velocity profiles. A detailed description of the ADVP working principle is given in the literature (Rolland & Lemmin 1997). In the present study we used the multistatic configuration of the ADVP: four transducers receive the backscattered signals emitted from one emitter, providing one redundancy in the transformation of the captured Doppler frequency into the three Cartesian velocity components. This redundancy is used for signal noise elimination and for data quality control (Hurther & Lemmin 2000). A Pulse Repetition Frequency (PRF) of 2000 Hz and a Number of Pulse Pairs (NPP) equal to 64 was used resulting in an unfiltered velocity sampling frequency of 31.25 Hz. Even though a low degree of dealiasing was expected in the data, a rectifying algorithm based on the signal time history developed by the authors was applied. A portable metallic structure was used to displace the instrument across the section. The same vertical distance from the water surface was maintained during the measurements, and the vibration of the ADVP was minimized by the mounting.

Table 1. River Chamberonne flow characteristics.

S (%)	Q (m^3/s)	h (m)	W	Re ($\times 10^4$)	Fr (–)	D_{40} (mm)	D_{50}	D_{84}
0.26	0.55	0.29	5.75	4.5–12.3	0.24–0.44	41	49	81

4 BED FORMS

The bed shape is found to be periodic across the section, with a wavelength of about twice the water depth ($\lambda_{b,y} \approx 2h$). This suggests the existence of streamwise sediment stripes which may have formed during higher water periods when the aspect ratio is lower and Prandtl secondary currents of the second kind occur. Flow characteristics during different flood events (provided by SESA – Canton of Vaud) for a cross-section located about 10 m upstream from the measurements are summarized in Table 2.

$\overline{U}$ is the section averaged velocity, and q the unit discharge. The Schoklitsch relation is most appropriate to described bed load conditions in these rivers (Graf & Altinakar 1998), based on the slope S and D_{40} of the bed sediments. Assuming uniform conditions, the critical discharge for incipient sediment motion according to Schoklitsch is given by:

$$q_{cr} = 0.26(s-1)^{5/3} \frac{D_{40}^{3/2}}{S^{7/6}} = 4.8 \ m^2/s \quad (1)$$

s is the relative density of the sediment particles, here considered equal to 2.6. Given the uncertainty in the empirical formula for the calculation of critical bed load conditions and using the values of unit discharge in Table 2, gravel movement over the riverbed may already occur for low return periods. Second type Prandtl secondary motion (Nezu & Nakagawa 1993) may take place due to the decrease in aspect ratio. Periodic streamwise sediment stripes with a wavelength equivalent to the water depth would appear across the section, generated by lateral sediment transport (Tsujimoto 1989).

The periodic bed forms detected during the measurements (Fig. 1) might be signatures of streamwise bed stripes, produced during high water periods when

Table 2. Flood event characteristics.

Return period (years)	Q (m^3/s)	h (m)	$\overline{U}$ (m/s)	W/h (–)	q (m^2/s)
2	18.00	0.55	4.52	13.2	2.48
3	20.60	0.59	4.78	12.4	2.82
5	23.35	0.63	4.99	11.8	3.14
10	26.81	0.68	5.24	11.1	3.56
30	30.21	0.72	5.47	10.7	3.93
100	35.00	0.78	5.75	10.0	4.48
300	39.15	0.83	5.97	9.5	4.95

lateral transport conditions by cellular secondary currents are fulfilled. High water events will hence condition the flow characteristics during low water periods.

5 MEAN VELOCITY FIELD

The streamwise velocity distribution in the cross-section is represented by contour lines in Figure 2. In particular near the most important bed forms and in the surface layer ($z/h > 0.80$), strong transversal gradients indicate that the structure of the mean flow is 3-D, making any 2-D approach questionable. The same conclusions were reached in a similar study made in another Swiss river, the Venoge (Franca & Lemmin 2004).

The section-averaged value of the vertical velocity represents 16.5% and the spanwise component 29.1% of the streamwise velocity, respectively. The high spanwise component is mainly orientated towards the right riverbank which corresponds to the outer bank of the river curve and may be explained by the presence of a smooth river bend upstream from the measuring site. No circular cells from Prandtl's secondary flow of the first kind were observed. This might be due to macro-roughness elements in the riverbed which reorient the flow canceling the vorticity generated by the river curvature.

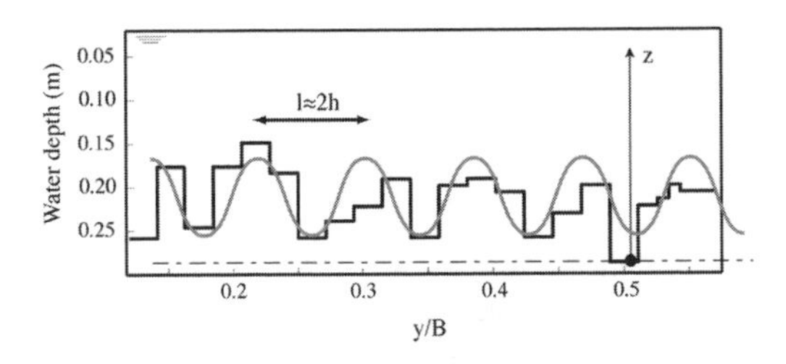

Figure 1. Periodic cross-section variation of the riverbed with hypothetical associated sinusoidal function (gray line) corresponding to an ideal model of the cross-section bed morphology.

In Figure 2, the most interesting observation is the periodical occurrence of low velocity regions across the section in the surface layer. In wide channel flows high velocity regions were previously observed near the surface above rougher bed regions (Studerus 1982, Franca & Lemmin 2004). If bed forms mainly determine the roughness, these regions locally compensate the additional bed resistance produced by shallower profiles or those with higher roughness. An increased velocity in the upper layer will compensate for the reduction of the available flow depth. In the present case, however, higher velocity regions are less well developed near the surface ($z/h > 0.80$) compared to the regions of low velocity (Fig. 2). In the following, we call the high and low velocity regions in the upper layer of the flow high velocity cells (CH) and low velocity cells (CL).

Nine CL can be identified which are uniformly and periodically distributed across the section, with a density of about 4 per meter. Due to lateral momentum compensation, weak CH exist adjacent to the CL. Although the CL seem to be distributed independent of the local bed forms, a stronger retarding effect is observed where the water depth is greater and profiles are less rough, corresponding to local minima of k^+ ($k^+ = ku^*/\nu$ is the dimensionless roughness; k is the Nikuradse equivalent roughness, u^* the friction velocity and ν the cinematic viscosity). The periodicity of the CL distribution has a wavelength equal to the water depth ($\lambda_{CL,y} \approx h$). If, however, we consider only the more intense cells this becomes about two times the water depth ($\lambda_{CL+,y} \approx 2\,h$). These two complementary periodicities may occur due to continuity.

CL correspond to a deviation in the velocity profile in the upper layer due to different hydraulic conditions (the so-called velocity dip phenomenon, Yang

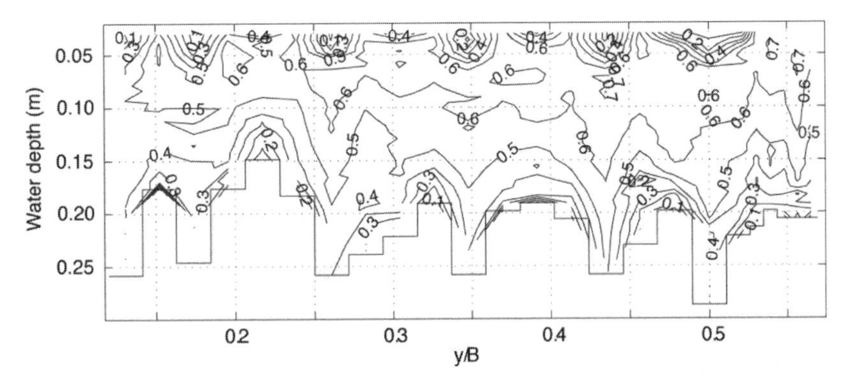

Figure 2. Mean longitudinal velocity field (m/s). The horizontal extent of the measured section is 3.25 m (y/B = 0.565). The spanwise and vertical length scales are different.

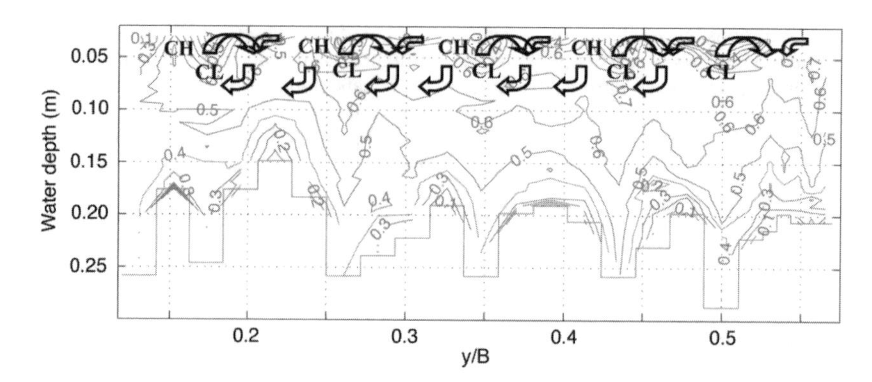

Figure 3. Interpretation of the SLOM (Surface Layer Organized Motion, z > 0.80 h) considering all three components of the mean velocity.

et al. (2004)). Velocity profiles with a reduction of the velocity near the surface are known as d-shaped. For mass conservation reasons, the organized co-existence of CH and CL has to be compensated for by secondary motion in the flow (Ferro & Baiamonte 1994).

Eliminating the mean cross-section trend of the time mean vertical and spanwise velocity components (spatial averaging, which corresponds to subtracting the cross-section convective velocity), it is possible to identify a permanent cellular organized structure in the surface layer flow (here called SLOM – Surface Layer Organized Motion), connected to the CL and CH regions suggesting interaction between the three velocity components. An interpretation of this three-dimensional flow is given in Figure 3.

The mechanism involved in the SLOM creates a circulation layer at z/h > 0.80 and is described as follows:

- Most of the CL-profiles coincide with an inversion of the vertical mass flux (mainly in the deeper profiles).
- Vertical downward jets coincide with weak CH.
- Lateral mass transfer in the direction of the CH occurs above the CL; this contributes to the CH and to the associated downward jets.
- In the spanwise direction, SLOM cells scale with the water depth (density of 4 cells per meter).
- The lateral mass transfer induces a rotating movement. Clearly symmetrical rotating cells might be found with more accurate measurements under well-controlled laboratory conditions.

As expected in such a rough flow, a boundary layer is formed near the bed around the forms with at least a 2D geometry (Fig. 2). In the roughness sublayer (Nikora & Smart 1997) the flow is highly three-dimensional and deviations from the log-law may occur. These deviations result from a so-called generalized macro-roughness wake and correspond to s-type profiles (Marchand et al. 1984, Bathurst 1988, Ferro & Baiamonte 1994, Katul et al. 2002, Carravetta & Della Morte 2004 and Franca 2005). Nevertheless, the logarithmic profile still describes reasonably well the velocity distribution in about 65% of the measured profiles up to $z/h < 0.40\bar{u}$. An application of the log-law to the whole cross-section section is difficult due to the high variability of the local parameters. In particular, the local drag conditions vary greatly with the local bed configuration.

6 CROSS-SECTION NIKURADSE ROUGHNESS DISTRIBUTION

The local log-law parameterization (determination of local Nikuradse roughness k and friction velocity u*) might be used to quantitatively map the roughness across the river section. The logarithmic mean velocity distribution is defined as (Schlichting 1968):

$$\frac{\bar{u}(z)}{u^*} = \frac{1}{\kappa}\ln\left(\frac{z}{k}\right) + A \tag{2}$$

$\bar{u}$ is the time averaged velocity, κ the Von Karman constant (taken as $\kappa = 0.41$), and A is taken as 8.5 for turbulent rough flows (Monin & Yanglom 1971). The friction velocity and Nikuradse roughness were calculated by regressing the velocity against the logarithm of the water depth. Although the application of the log-law and the assumption of the parameter $A = 8.5$ in this flow case may not be without reservation, it may be used to locally approximate the velocity distribution in order to provide an estimate of the respective flow resistance. The majority of the measured profiles corresponds to completely rough bed cases ($k^+ > 70$, Nezu & Nakagawa 1993). The mean k obtained corresponds to D_{84}. Figure 4 shows a plot of the cross-section distribution of the

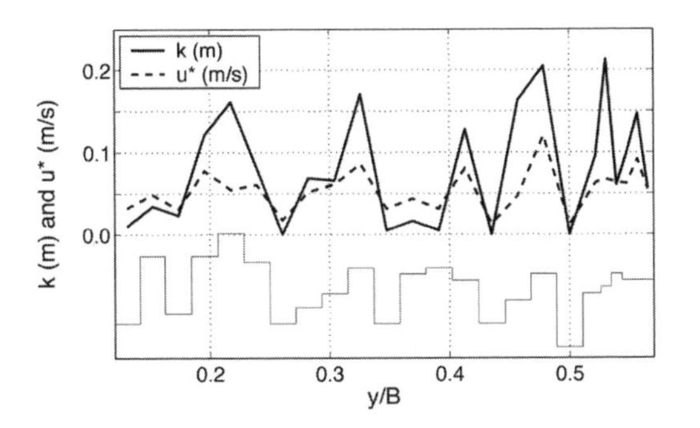

Figure 4. Nikuradse roughness and friction velocity cross-section distribution.

Nikuradse roughness and friction velocity. The riverbed roughness cross-section evolution is found to also have a periodicity, in phase with the bed forms ($\lambda_{k,y} \approx 2$ h). This confirms two observations: the flow resistance has a strong form dependence and the cross periodicity of the riverbed is intrinsically related to the flow resistance.

7 FLOW RESPONSE TO THE BOTTOM FORM

A strong dependence of the flow on riverbed morphology was established above. We define a dimensionless variable (the velocity dip, dU) that allows to identify d-type profiles and to evaluate the magnitude of CL, as follows:

$$dU(y) = \frac{\bar{u}_{max}(y)}{u(y, z = 0.90h)} \qquad (3)$$

The level at $z/h = 0.90$ is considered in the middle of the SLOM layer. In Figure 5 we plot the observed velocity dips against the respective local Froude number, defined as:

$$Fr(y) = \frac{U(y)}{\sqrt{gh(y)}} \qquad (4)$$

U is the depth-averaged velocity for each profile across the section and g the gravity acceleration. There is clear tendency of a more intense velocity dip for lower local Froude numbers. A limit of around $Fr \approx 0.35$ may be established for the beginning of the occurrence of a significant dip. The local characteristics of the mean flow in the surface layer are thus dependent on the flow regime.

The origin of the d-shaped profiles or CL in this river is not yet clear. Possible reasons for their occurrence are: the presence of stationary breaking surface waves, such as boulder produced waves (ship- and lee-waves) and stationary hydraulic jumps or the inviscid

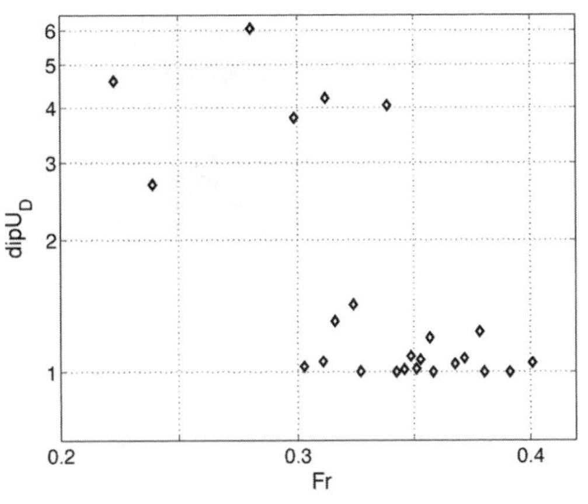

Figure 5. Inviscid response to the bottom shape: relation between the velocity dip (dU ratio) and the Froude number (Fr).

response of the flow to the bed forms (potential flow assumption – barotropic flow). Given the local Froude number dependence, the latter one seems to be the more likely origin.

8 TURBULENCE SCALES

Since the ADVP measurements have sufficient resolution to resolve turbulence scales, a cross section power spectral analysis of the energy distribution has been carried out in order to determine whether periodic bed forms affect the intensity of turbulence. For the calculation of the power spectral density the data was segmented into blocks of 30 s duration. We used the Welch method with 50% overlapping. The results (Fig. 6) are presented for two levels in the water column: ≈2 cm from the water surface and ≈2 cm from the riverbed. This figure allows to determine the variation of the production plateaus in the spectral domain across the section.

The level of energy in the low frequency production range also varies periodically across the section corresponding to the bed form distribution ($\lambda_{Su',y} \approx 2$ h). Depending on the distance to the water surface, two distinct patterns exist: near the surface, the lower production energy levels correspond to the deeper profiles, whereas near the riverbed the opposite occurs. This indicates that bed forms also has a pronounced effect on the generation and the redistribution of turbulence. Close to the riverbed, ADVP observations correspond to locally produced turbulence. It is evident that the redistribution of turbulent energy within the water column is strongly affected by the 3D structure of the flow field. Near the surface, the low energy regions coincide with the larger CL.

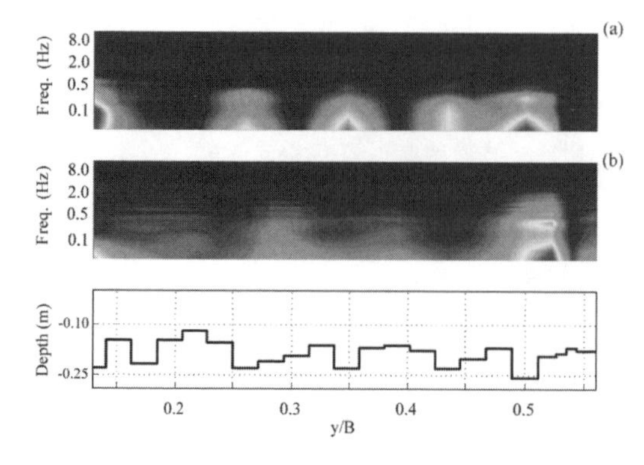

Figure 6. Cross-section variation of the power spectral density ($m^2/s^2/Hz$) of the instantaneous streamwise velocity, as a function of the occurrence frequency, calculated for two levels; (a) ≈2 cm from the water surface and (b) ≈2 cm from the bottom. Lighter areas indicate higher energy (the plots do not have the same color scale).

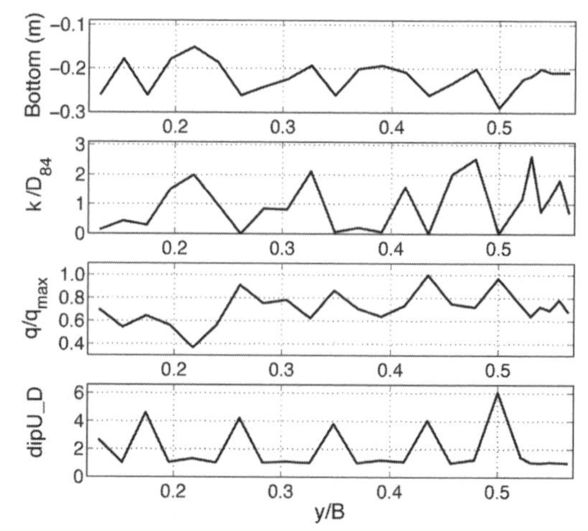

Figure 7. Periodicity of the flow characteristics across the investigated river cross-section.

9 DISCUSSION

The riverbed morphology presents a periodic character across the section, with a wavelength of two times the water depth ($\lambda_{b,y} \approx 2h$; Fig. 1). Given the flood prediction conditions, the bed forms seem to be signatures of streamwise coarse material stripes which may already be produced during high water events at low return periods.

The measurements show a case of a wide river where two-dimensional open-channel models are no longer valid, due to macro-roughness caused by riverbed forms. A logarithmic form of the velocity profile is occasionally observed near the riverbed. However, a general parameterization of the logarithmic law to the whole cross section becomes impossible, because the studied flow is essentially three-dimensional near the most important bed forms and in the outer layer of the flow (Fig. 2).

The analysis of the three components of the time-averaged velocity indicates a strong dependence of the hydraulic characteristics on the extreme values of the bed morphology. The analysis of the roughness parameters based on the application of the log-law confirmed the roughness dependence on the bed forms (Fig. 4): $\lambda_{k,y} \approx 2h$ and $\lambda_{u^*,y} \approx 2h$. Thus, the flow resistance mainly resulting from bottom drag, presents the same periodic character of the bed forms. Given the relation between friction velocity and shear stress distribution $u^* = u'w'|_{bed}$, we may conclude that turbulence production across the section is also related to the periodic bed forms.

Figure 7 compares the cross-section variation of several hydraulic characteristics with the bottom variation: dimensionless Nikuradse equivalent roughness (k/D_{84}); local unitary discharge (q/q_{max}), which is related to the lateral momentum distribution; and velocity dip as defined above by expression (3). Since the roughness (flow resistance) is in phase with the bottom height variation, the lateral momentum distribution is obviously out of phase with these two, with an expected increasing tendency when moving to the river axis. As indicated above, the bottom shape conditions the occurrence of the d-shaped profiles (associated to the CL cells). The smaller CL regions are visible in Fig. 7 (bottom), situated between the larger CL regions. This reinforces the hypothesis of the origin of the SLOM as an inviscid response of the flow to the riverbed shape.

Based on the observed local Froude number dependence (Fig. 5), we defined the SLOM as an upper layer flow response to bed forms. A low velocity (CL) region in the upper layer of the flow is found to be associated with the periodically spaced deeper profiles ($\lambda_{CL+,y} \approx 2h$). Adjacent to these, discreet high-velocity jets (CH) are formed due to a compensatory secondary motion. For continuity reasons, weaker CL occur between two larger ones, reducing the CL periodicity to half the wavelength ($\lambda_{CL,y} \approx h$). As illustrated in Figure 3, the SLOM is thus constituted by alternating large CL/CH and small CL/CH. The turbulent mechanism responsible for the link between the bed forms and the upper layers of the flow is still unknown. Further laboratory work on the turbulence structure of the flow should be done to characterize this interaction.

Based on the observations, in these kinds of macro-rough flows, a general wake is formed by the continuous and superimposed effect of the bed shape/gravel boulders, similar to the wake-interference flow (Baiamonte & Ferro (1997)). An intermediate shielded region of the flow is formed between the roughness layer as defined by Nikora & Smart (1997) and the SLOM layer, z/h > 0.80, similar to the blending layer concept used in atmospheric flows analysis (Wieringa 1976).

10 CONCLUSIONS

From the present field measurements in the river Chamberonne we observed the following main features of the interaction between the bottom profile and the flow characteristics:

- The riverbed cross-section observation highlighted equally spaced roughness elements in the bottom profile with a wavelength of $\lambda_{b,y} \approx 2h$. This indicates the possible existence of streamwise ridges composed of coarse material.
- The structure of the mean flow is essentially 3D near the most important bed forms and in the outer layer of the flow.
- Although the log-law may represent to some extent the velocity distribution in this river, 2D cross-section concepts are not valid and are to be used with caution in shallow gravel-bed flows.
- All river hydraulic characteristics are strongly conditioned by the bed forms, in phase with its periodic cross-section variation, and consequently also the bed-load capacity and transversal momentum distribution.
- We identified a permanent three-dimensional flow structure in the upper layer of the flow (z/h > 0.80), the Surface Layer Organized Motion (SLOM), which is strongly conditioned by the bed forms. This motion field consists of periodically distributed low (CL) and high (CH) velocity regions associated with a mass compensation secondary motion. This structured outer layer motion seems to be due to an inviscid response to bed form variations since it is dependent on the local Froude number. A general wake effect might exist due to large-scale roughness, confining this flow response to the layer z/h = 0.80.
- The transversal distribution of the turbulence structures is also influenced by the bed morphology. Production intensity changes across the section and changes with water depth depending on the periodic bed shape.

High water events may condition the flow characteristics during low water periods. Bed form changes such as streamwise gravel-stripes which were produced during a high water period, determine the complex 3D flow patterns during shallow water periods.

Mixing is widely present within this flow given the strong shear zones created by the mean flow variability across the river section. Turbulence production is influenced by the existence of shear zones in addition to those resulting from riverbed drag, thus affecting mixing and transport processes. The SLOM mechanism may have several effects: the creation of a stagnation zone in the upper layers of the flow ($z/\delta > 0.80$) where mass, heat or momentum inputs may be concentrated; promotion of gas transfer through the free surface. Additional studies shall be made to investigate the nature of the permanent structures existing in the layer z/h > 0.80 to determine how they might possibly enhance gas transfer between the free surface and the inner flow areas.

The existence of cellular secondary currents in shallow water flows still is an open subject in geophysical fluid mechanics. Thus, 2D concepts are usually considered. We have shown that there are limits to that approach. The bottom roughness height threshold for the onset of 3D flow effects and its influence on mixing and transport processes are not yet well established. Only through further detailed 3D profile measurements as were carried out in this project, progress will be made in understanding and a correct interpretation of the hydraulic processes in such flows.

ACKNOWLEDGEMENTS

The authors acknowledge the financial support of the Swiss National Science Foundation (2000-063818) and the FCT (BD 6727/2001).

REFERENCES

Babaeyan-Koopaei, K., Ervine, D.A., Carling, P.A. & Cao, Z. 2002. Velocity and turbulence measurements for two overbank flow events in river Severn. Journal of Hydraulic Engineering 128(10): 891–900.

Baiamonte, G. & Ferro, V. 1997. The influence of roughness geometry and Shields parameter on flow resistance in gravel-bed channels. Earth Surface Processes and Landforms 22: 759–772.

Baiamonte, G., Giordano, G. & Ferro, V. 1995. Advances on velocity profile and flow resistance law in gravel bed rivers. Excerpta 9: 41–89.

Bathurst, J.C. 1988. Velocity profile in high-gradient, boulder-bed channels. Proceedings of the International Conference on Fluvial Hydraulics, Budapest. 30 May–3 June 1988.

Buffin-Bélanger, T. & Roy, A.G. 1998. Effects of a pebble cluster on the turbulent structure of a depth-limited flow in a gravel-bed river. Geomorphology 25: 249–267.

Buffin-Bélanger, T., Roy, A.G. & Kirbride, A.D. 2000. On large-scale flow structures in a gravel-bed river. Geomorphology 32: 417–435.

Carravetta, A. & Della Morte, R. 2004. Response of velocity to a sudden change of the bed roughness in sub critical open channel flow. Proceedings of the Riverflow 2004, Naples, 23–25 June 2004. Balkema.

Ferro, V. & Baiamonte, G. 1994. Flow velocity profiles in gravel-bed rivers. Journal of Hydraulic Engineering 120(1): 60–80.

Graf, W.H. & Altinakar, M.S. 1998. Fluvial Hydraulics. West Sussex: John Wiley and Sons.
Franca, M.J. & Lemmin, U. 2004. A field study of extremely rough, three-dimensional river flow. Proceedings of the 4th IAHR International Symposium on Environmental Hydraulics, Hong-Kong, 15–18 December 2004. Balkema.
Franca, M.J. 2005. Flow dynamics over gravel riverbed. Accepted for the JFK International Student Paper Competition, XXXI IAHR Congress, Seoul.
Hurther, D., Lemmin, U. & Blanckaert, K. 2002. A field study of transport and mixing in a river using an acoustic Doppler velocity profiler. Proceedings of the Riverflow 2002, Louvain-la-Neuve, 4–6 September 2002. Balkema.
Katul, G., Wiberg, P., Albertson, J. & Hornberger, G. 2002. A mixing layer theory for flow resistance in shallow streams. Water Resources Research 38(11): 1250.
Marchand, J.P., Jarret, R.D. & Jones, L.L. 1984. Velocity profile, surface slope, and bed material size for selected streams in Colorado. U.S. Geol. Surv. Open File Rep., 84–733.
Monin, A.S. & Yaglom, A.M. 1971. Statistical Fluid Mechanics: Mechanics of Turbulence – Vol. 1. Camridge (MA): MIT Press.
Nelson, J.M., McLean, S.R. & Wolfe, S.R. 1993. Mean flow and turbulence fields over two-dimensional bed forms. Water Resources. Research 29(12): 3935–3953.
Nezu, I. & Nakagawa, H. 1993 Turbulence in open-channel flows: IAHR monograph. Rotterdam: Balkema.
Nikora, V. & Goring, D. 2000. Flow turbulence over fixed and weakly mobile gravel beds. Journal of Hydraulic Engineering 126(9): 679–690.
Nikora, V. & Smart, G.M. 1997. Turbulence characteristics of New Zealand gravel-bed rivers. Journal of Hydraulic Engineering 123(9): 764–773.
Powell, D.M., Reid, I. & Laronne, J.B. 1999. Hydraulic interpretation of cross-stream variations in bed load transport. Journal of Hydraulic Engineering 125(12): 1243–1252.
Robert, A., Roy, A.G. & De Serres, B. 1992. Changes in velocity profiles at roughness transitions in coarse grained channels. Sedimentology 39: 725–735.
Rolland, T. & Lemmin, U. 1997. A two-component acoustic velocity profiler for use in turbulent open-channel flow, Journal of Hydraulic Research 35(4): 545–561.
Roy, A.G., Buffin-Belanger, T., Lamarre, H. & Kirkbride, A.D. 2004. Size, shape and dynamics of large-scale turbulent flow structures in a gravel-bed river. Journal of Fluid Mechanics 500: 1–27.
Schlichting, H. 1968. Boundary Layer theory. New York: McGraw-Hill.
Smart, G.M. 1999. Turbulent velocity profiles and boundary shear in gravel bed rivers. Journal of Hydraulic Engineering 125(2): 106–116.
Studerus, X. 1982. Sekundärströmungen im offenen Gerinne über rauhen Längsstreifen: Ph.D. thesis. Institut für Hydromechanik und Wasserwirstschaft – ETH, Zurich (Switzerland).
Tsujimoto, T. 1989. Longitudinal stripes of alternate lateral sorting due to cellular secondary currents. Proceedings of the XXIII IAHR Congress, Ottawa, 21–25 August 1989.
Wieringa, J. 1976. An objective exposure correction method for average wind speeds measured at a shelter location. Quarterly Journal of the Royal Meteorological. Society 102(431): 241–253.
Wolman, M.G. 1954. A method of sampling coarse riverbed material. Transactions American Geophysical Union 35(6): 951–956.
Yang, S.Q., Tan, S.K. & Lim, S.Y. 2004. Velocity distribution and Dip-phenomenon in smooth uniform open channel flows. Journal of Hydraulic Engineering 130(12): 1179–1186.

River, Coastal and Estuarine Morphodynamics: RCEM 2005 – Parker & García (eds)
© 2006 Taylor & Francis Group, London, ISBN 0 415 39270 5

Spatially-averaged turbulent flow over square-rib roughness

S.E. Coleman
The University of Auckland, Auckland, New Zealand

V.I. Nikora
National Institute of Water and Atmospheric Research, Christchurch, New Zealand

S.R. McLean
The University of California at Santa Barbara, California, United States of America

E. Schlicke
The University of Auckland, Auckland, New Zealand

ABSTRACT: In order to investigate the influence of changing bed morphology on flow structure (e.g. velocity and momentum-flux profiles), a series of 11 tests of fully-rough turbulent subcritical flow over two-dimensional transverse repeated square ribs was undertaken in which the ribs were varied (for a single flowrate) to give uniform rib spacing-to-height ratios of $\lambda/h = 1$–16. The double- (time- and space-) averaged velocity profile is found to be (quasi) logarithmic above roughness tops, changing below roughness tops with increasing rib spacing from exponential ($\lambda/h < 10$) to linear ($\lambda/h \geq 10$) to logarithmic ($\lambda/h \gg 10$). The experimentally-determined momentum balance, including effects of secondary currents, flow acceleration, and Reynolds and form-induced stresses, is found to agree well with theoretical expectations. It shown that the knowledge of the effects of form-induced stresses, secondary currents, and flow non-uniformity can be particularly important for describing and modeling flows over roughness elements.

1 INTRODUCTION

Fluvial flow transporting sediment causes the river bed to deform, which alters the flow through changed boundary conditions, in turn influencing sediment transport, bed formation, etc. Understanding and prediction of each of the components of this continuous feedback loop needs to incorporate recognition of the mechanics of the feedback. In regard to bed morphology, for example, questions arise as to how bedforms grow in response to a flow (e.g. what are the key controlling flow structures), what these bedforms do to the flow (how do they alter the key flow structures), and how these process-response mechanisms develop to give an equilibrium flow-bed system. In the following, one aspect of this feedback loop, namely the influence of changing bed morphology on flow structure (e.g. velocity and momentum-flux profiles), is studied for fully-rough turbulent subcritical flow over two-dimensional transverse repeated square-rib roughness.

2 DAM FRAMEWORK AND VELOCITY PROFILE PARAMETERISATION

Although the vertical distributions of velocity and momentum flux over rough beds have been extensively studied, confusion and debate continue regarding even the most basic flow-structure definitions and properties. For example, how does velocity vary below roughness tops; what is the fluid stress and how does this vary with depth; how does total shear stress vary with depth; how should total bed shear stress be determined and where does it act on the bed; how do the boundary drag and skin friction arise in the momentum balance; and how do these latter forces act with depth to balance the gravity-induced flow momentum. As will be shown below, the double- (time- and space-) averaging methodology (DAM) provides a framework within which these questions can be properly addressed and a consistent picture of flow-boundary interaction can be obtained.

The starting point for considerations is the double- (time- and space-) averaged continuity and momentum conservation equations (Nikora et al. 2005a,b). For steady water flow over fixed boundaries these equations reduce to

$$\frac{\partial \phi \langle \bar{u}_i \rangle}{\partial x_i} = 0 \tag{1}$$

$$\langle \bar{u}_j \rangle \frac{\partial \langle \bar{u}_i \rangle}{\partial x_j} = g_i - \frac{1}{\phi \rho} \frac{\partial \phi \langle \bar{p} \rangle}{\partial x_i} - \frac{1}{\phi} \frac{\partial \phi \langle \overline{u_i' u_j'} \rangle}{\partial x_j}$$
$$- \frac{1}{\phi} \frac{\partial \phi \langle \tilde{u}_i \tilde{u}_j \rangle}{\partial x_j} + \frac{1}{V_f \rho} \iint_{S_{int}} \bar{p} n_i dS - \frac{1}{V_f} \iint_{S_{int}} \left(\nu \frac{\partial \bar{u}_i}{\partial x_j} \right) n_j dS \tag{2}$$

For some applications, it is convenient to integrate the double-averaged equations in the vertical direction, integration from any level z to the water surface z_{ws} giving

$$gS_{ws} \int_z^{z_{ws}} \phi dz = -\int_z^{z_c} \frac{1}{V_o} \iint_{S\,int} \left[\frac{\bar{p}}{\rho} n_i - \nu \frac{\partial \bar{u}_i}{\partial x_j} n_j \right] dSdz$$
$$+ \int_z^{z_{ws}} \left[\phi \langle \bar{u}_j \rangle \frac{\partial \langle \bar{u}_i \rangle}{\partial x_j} + \frac{\partial \phi \langle \overline{u_i' u_j'} \rangle}{\partial x_j} + \frac{\partial \phi \langle \tilde{u}_i \tilde{u}_j \rangle}{\partial x_j} \right] dz \tag{3}$$

where the gravity and fluid pressure gradient terms have been combined for the present tests of uniformly accelerating flow, and S_{ws} is the mean water-surface slope. In (1) to (3), u_i is the ith velocity component (in direction x_i); the straight overbar and angle brackets denote the time and spatial averaging of flow variables, respectively; the prime denotes turbulent fluctuations; the wavy overbar denotes the spatial fluctuations in the time-averaged flow variable, e.g. $u_i = \bar{u}_i + u_i' = \langle \bar{u}_i \rangle + \tilde{u}_i + u_i'$; $\phi = V_f / V_o$ is the roughness geometry function ($1 \geq \phi \geq 0$); V_f is the volume of fluid within the averaging volume V_o; S_{int} is the roughness-fluid interfacial surface within the averaging volume; g is gravity; p is pressure; ρ and ν are fluid density and kinematic viscosity; n_i is the ith component of the unit vector normal to the surface element dS and directed into the fluid; z_c is rib crest elevation; and $\rho \langle \overline{u_i' u_j'} \rangle$ and $\rho \langle \tilde{u}_i \tilde{u}_j \rangle$ are a spatially-averaged Reynolds stress and a form-induced stress, respectively. Viscous stresses have been ignored as minor except at the roughness-fluid interface where they determine the skin friction. The right-handed coordinate system is implied herein, i.e., the x-axis (u-velocity component) is oriented along the main flow parallel to the averaged bed, the y-axis (v-velocity component) is oriented to the left bank, and the z-axis (w-velocity component) is pointing towards the water surface. Here the spatial averaging is carried out over a slab of small thickness dz. The horizontal extent of the thin-slab averaging volume V_o is typically chosen to be much larger than the roughness or flow geometry scales, but smaller than the larger geometric features, such as channel curvature and widening or narrowing. When these averages do not change along the flow, the flow can be considered uniform. Owing to measurements having only been made over a single roughness wavelength for the present tests, the averaging volume is limited to this single wavelength, and all averaged quantities are assumed not to vary in the flow direction for these tests. Key advantages of the double-averaged equations of (1)–(3) over conventional Reynolds-averaged Navier-Stokes equations include explicit accounting for both the form and viscous drag on the rough bed, and also for (form-induced) momentum fluxes due to roughness-induced near-bed flow heterogeneity.

Equation (3), used herein to investigate the above questions regarding fluid and shear stresses and momentum balance, can be seen to describe the flux of gravity-induced momentum via fluid stresses (e.g. Reynolds and form-induced stresses), secondary currents and flow nonuniformity to the boundary where the momentum is removed through form drag and skin friction (for a fixed boundary).

Along-channel velocities are empirically found to reasonably approximate a logarithmic behaviour throughout most of the depth of open-channel flows. This is despite theoretical considerations requiring a deep flow for existence of a genuine overlap logarithmic layer (e.g. Raupach et al. 1991). In line with previous studies, the distribution of the double- (time- and space-) averaged along-channel velocity $\langle \bar{u} \rangle$ in a free-surface channel flow over roughness elements is described herein as (Coleman et al. 2005a)

$$\frac{\langle \bar{u} \rangle}{u_*} = \frac{1}{\kappa} \ln \left(\frac{z-d}{h} \right) + f \left(\frac{h}{\lambda_v}, \frac{h}{H_m}, \frac{L_i}{h} \right) \tag{4}$$

where shear velocity $u_* = (\tau/\rho)^{0.5}$; τ is bed shear stress; von Karman's $\kappa = 0.4$–0.42; the origin of the z axis is taken to be the cavity base for the present rib roughness; d is the elevation of the zero-velocity plane for the logarithmic profile (e.g. Nikora et al. 2002); h is roughness height; viscous length scale $\lambda_\nu = \nu/u_*$; H_m is mean flow depth; h/H_m is roughness submergence; and L_i are roughness plan lengths (e.g. crest length, crest width, and crest spacing λ). Equation (4) is used herein in assessing velocity profiles (particularly below roughness tops) for varying roughness spacing, where for the present transverse ribs, L_i/h becomes λ/h, and h/λ_ν can be neglected.

3 EXPERIMENTS

For a single combination of flowrate (0.28 m^3/s), water depth above rib trough ($H = 0.165$ m), and square-rib profile (crest length and height $h = 15$ mm), a series of 11 experiments (Table 1) was undertaken in which the rib spacing λ was varied to give $\lambda/h = 1$–16 (with

Table 1. Experimental parameters.

λ/h	H_m (m)	U (m/s)	u_* (m/s)	f_T
1	0.150	0.285	0.0147	0.021
2	0.158	0.271	0.0211	0.049
3	0.160	0.267	0.0244	0.067
4	0.161	0.265	0.0273	0.085
5	0.162	0.264	0.0300	0.104
6	0.163	0.263	0.0326	0.123
7	0.163	0.262	0.0328	0.125
8	0.163	0.262	0.0331	0.128
10	0.164	0.261	0.0331	0.128
12	0.164	0.261	0.0317	0.118
16	0.164	0.260	0.0311	0.114

$h/H = 0.09$, and $\lambda/h = 1$ being a solid bed). These spacings were designed to span d-type (skimming flow) to k-type (interactive flow) roughness (Perry et al. 1969, Jiménez 2004), with a maximum resistance to flow expected within the range of λ/h tested. Each of the experiments, of individual h/λ, involved centerline measurements of 3D velocity vectors (using a Sontek MicroADV), water surface profiles (using an upward-looking ultrasonic transducer), and bed pressures (using a Validyne differential pressure transducer connected to tappings on a rib surface) and forces. Velocity measurements in planes parallel to the centerline were also undertaken in order to assess secondary current effects. The tests were carried out in a non-recirculating glass-sided horizontal-bed flume measuring 0.45 m × 0.455 m (wide) × 20 m. The test section of hardwood ribs was 7.1 m long, with the focus rib (of pressure and velocity measurements) located 5.9 m from the upstream end of the test section. Velocities were measured at 50 Hz for 90s at each measurement point, at 2–20 mm vertical spacings (focusing near the rib), and 2–40 mm horizontal spacings (increasing with rib spacing) from the downstream edge of the focus rib to the same edge of the immediately upstream rib. Pressure measurements were made on the vertical focus-rib surfaces at 5, 7.5 and 10 mm from the rib top. Table 1 summarises hydraulic parameters for the tests, where U is depth-averaged velocity; shear velocity was determined from the hydraulic radius m and the measured free-surface slope S_{ws}, $u_* = (gmS_{ws})^{0.5}$; and friction factor $f_T = 8(u_*/U)^2$. All tests were fully rough turbulent flows of $h/\lambda_\nu > 70$ and Reynolds number $R = 4mU/\nu = 99-103 \times 10^3$, with centerline Froude number $F = U/(gH_m)^{0.5} = 0.21$–$0.23$. The f_T values of Table 1 indicate that maximum drag due to wall roughness occurs at $h/\lambda \approx 0.125$ for two-dimensional transverse ribs (see also Leonardi et al. 2003; Jiménez 2004).

For each test, Reynolds and form-induced stresses and secondary current effects were determined from the measured velocities, skin friction was estimated

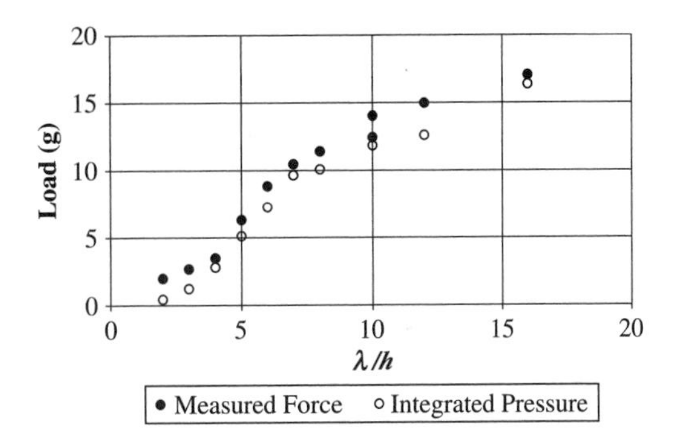

Figure 1. Measured rib forces.

assuming turbulent boundary-layer development over the rib surface, and form drag was assessed from integration of the measured pressures. Based on the pressure measurements (Coleman et al. 2005a), and the indications of numerical simulations (Ikeda and Durbin 2002; Leonardi et al. 2003), constant pressure distributions were assumed over the roughness height for each test. The resulting form drag trends and values were confirmed by subsequent direct measurements of forces on the ribs (Figure 1). These force measurements were carried out (Coleman et al. 2005a) by suspending the focus rib on a knife edge via thin profile arms, with micrometer adjustment of knife-edge elevation to ensure minimal distance of the rib to the underlying bed while the rib remained free to move. Balance arms were connected to the knife edge to aid system stability, with one of the balance arms resting, via an extension of adjustable length to ensure minimal load for no flow, on high-precision weigh scales. Subtracting load measurement prior to flow, plus the load for the rib support arms alone in the flow, from the load measured during the flow, the force on the rib was calculated knowing the respective lever arms to the weigh scales and the rib via the knife edge.

4 VELOCITY PROFILE

As expected from the form of (4), the vertical distribution of double-averaged velocity for flow over ribs displays a logarithmic behaviour (Figure 2) above roughness tops (roughness offset being a function of rib spacing). Within the interfacial layer between roughness crests and troughs, Nikora et al. (2004) propose that the double-averaged velocity profile can take several forms dependent on flow conditions and roughness geometry, namely: constant, exponential, linear, or a combination of these. Raupach et al. (1991) found an exponential profile for within-canopy air flows of close roughness spacing. Coleman et al. (2005b) found a slightly exponential to linear profile below roughness tops for water flow over closely-spaced ($h/\lambda \approx 0.1$) two-dimensional fluvial sand waves, with a linear profile occurring for fluvial sand waves of more-natural wider spacing ($h/\lambda \approx 0.05$). The present results reinforce these trends, an exponential profile occurring (with some flow reversal mid-rib) for the closely-spaced ribs of $h/\lambda > 0.1$, and a linear profile occurring for the widely-spaced ribs of $h/\lambda \leq 0.1$, with the profile transitioning between the two for decreasing h/λ (Figure 3). For open-channel flow over two-dimensional roughness elements, the double-averaged velocity profile below roughness tops can be concluded to change with increasing element spacing from exponential ($h/\lambda > 0.1$) to linear ($h/\lambda \leq 0.1$) to logarithmic (with increasing roughness element isolation, $h/\lambda \ll 0.1$). Further discussion of the double-averaged velocity profile for flow over ribs, including definition of the zero-velocity elevation d and the additive function $f(h/\lambda_\upsilon, h/H_m, \lambda/h)$ for the logarithmic portion of the profile, is given in Coleman et al. (2005a). Further discussion of the double-averaged velocity profile for flow over fluvial dunes, including predictive expressions for the logarithmic and linear portions of the profile, above and below roughness tops respectively, is given in Coleman et al. (2005b).

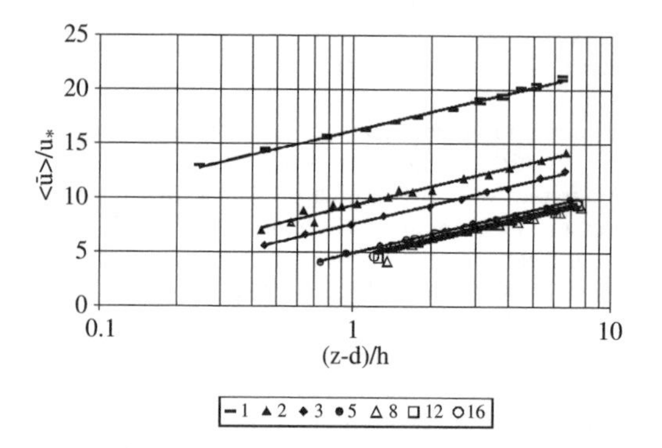

Figure 2. Logarithmic representation of above-rib velocity profiles (legend values are λ/h).

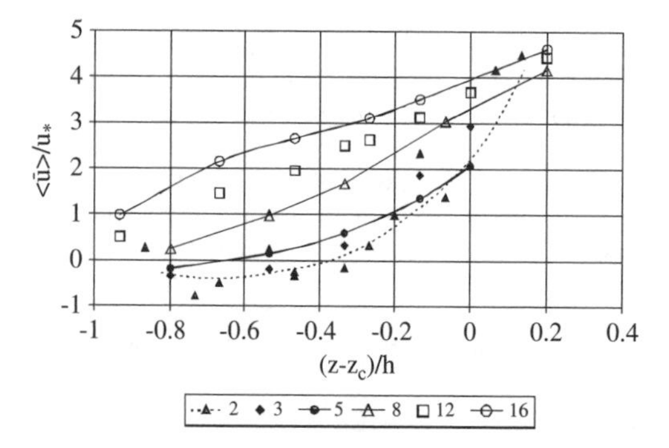

Figure 3. Velocity profiles below rib tops (legend values are λ/h).

Plotting fields of $(\tilde{u}, \tilde{w})$ highlights 'coherent' structures in the mean flow, namely organised patterns in mean-flow disturbances introduced by the ribs. For the present tests (Figure 4), the flow structure then consists of a pair of counter-rotating vortices (of opposite orientations for d-type roughness of $h/\lambda > 0.2$ and k-type roughness of $h/\lambda < 0.2$) set up by the ribs, superimposed on an overall double-average velocity profile of interfacial-layer (below roughness tops, Figure 3) and quasi-logarithmic form-induced-layer (above-roughness tops, Figure 2) sections. Flow over dunes (Coleman et al. 2005b) similarly consists of a pair of k-type (Figure 4) counter-rotating vortices superimposed on an overall average velocity profile of linear and quasi-logarithmic sections.

5 MOMENTUM FLUX

In keeping with the form of (3), stresses and momentum-flux components are normalised herein by $\rho u_{*H}^2 = \rho g H_m S_{ws}$. The variations of spatially-averaged Reynolds stress $\rho\langle\overline{u'w'}\rangle$ with elevation are given in Figure 5 for the present tests. For $\lambda/h = 1$ (solid bed), Reynolds stress is zero at the rib top. For $\lambda/h = 2$, Reynolds stress approximates zero near the rib top, increasing to a peak and then decreasing to zero with decreasing elevation owing to within-cavity circulation. With increasing rib spacing (λ/h), the below-roughness-top stress distribution transitions to a linear decay (for $\lambda/h \geq 5$) from the rib top to the cavity base. For $\lambda/h \leq 5$, Reynolds stress above the rib varies

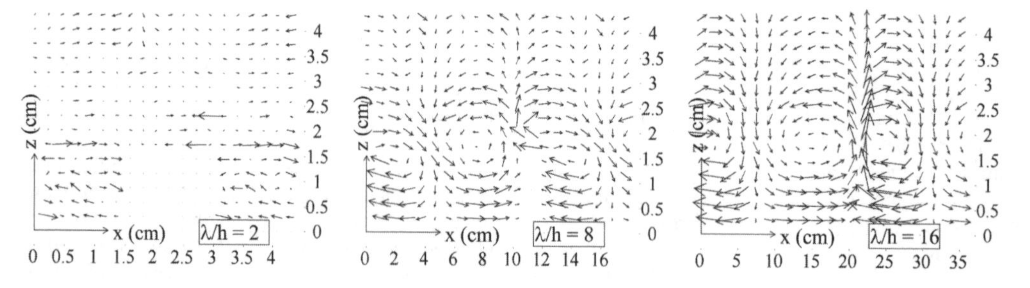

Figure 4. Fields of $(\tilde{u}, \tilde{w})$, each image covering 1.5λ (x) by $3h$ (z) (the legend value indicates relative velocity vector (cm/s) magnitude).

approximately linearly with depth and increases with rib spacing. For $\lambda/h > 5$, above-rib Reynolds stress variation with depth remains approximately linear, with a surge in Reynolds stress (larger for increased rib spacing) apparent as the rib top is approached from above. In line with flow-resistance indications from f_T values (Table 1) and the velocity profiles (Figure 2), this surge and the linear profile would be expected to reach a maximum for $\lambda/h \approx 8$, decreasing thereafter with increasing rib spacing (Figure 5). For their modeling of transverse square ribs of $\lambda/h = 10$, similar results, including the surge in Reynolds stress in the vicinity of the rib top, are given by Ikeda and Durbin (2002).

Variations of normalised spatially-averaged form-induced stress with elevation are presented in Figure 6. For $\lambda/h \leq 5$, form-induced stresses are essentially zero above the rib, and increase to a positive peak below the rib, this peak decreasing for increasing rib spacing. For $\lambda/h \geq 6$, form-induced stresses are essentially zero for $(z-z_c)/h \geq 3$, below which a negative peak is found just above the rib top, with a positive peak arising about half a rib height below the rib top, the magnitudes

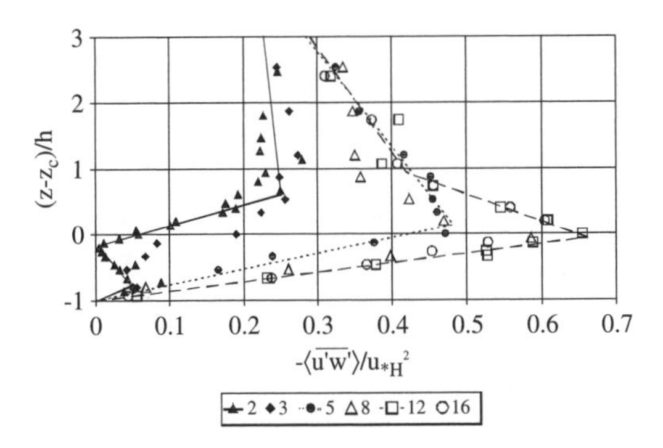

Figure 5. Normalised Reynolds stress (legend values are λ/h).

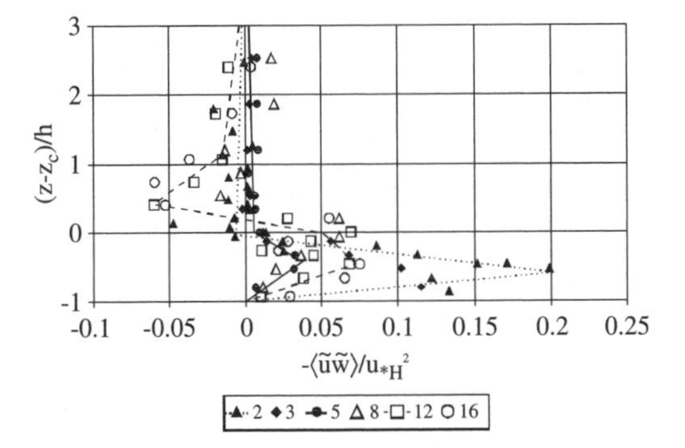

Figure 6. Normalised form-induced stress (legend values are λ/h).

of these peaks increasing with rib spacing. From Figures 4 and 5, it can be seen that form-induced stresses complement variations in Reynolds stresses, being of similar magnitudes to and acting to balance variations in these latter stresses, particularly above roughness crests. Knowledge of form-induced stress is thereby indicated to be potentially significant in describing or modeling flow over rough beds.

For the rib spacing of $\lambda/h = 6$, Figure 7 summarises the variations with elevation (referenced to the cavity base level z_t) of momentum flux components and momentum balance. Primary Reynolds and form-induced stresses are presented, along with form drag and skin friction effects. Secondary current terms are collected together (namely the integral from z to z_{ws} of $[\phi\langle\bar{w}\rangle\partial\langle\bar{u}\rangle/\partial z + \phi\langle\bar{v}\rangle\partial\langle\bar{u}\rangle/\partial y + \partial(\phi\langle\overline{u'v'}\rangle)/\partial y + \partial(\phi\langle\tilde{u}\tilde{v}\rangle)/\partial y]$), with nonuniform flow terms unable to be determined for the limited measuring windows adopted in the present tests (refer above). Also shown is the roughness geometry function $\phi(z)$, and the total momentum efflux through flow and boundary effects ("Total") is compared with influx through gravity ("Gravity").

The gravity-induced momentum flux of Figure 7 is predicted by (3) to be a straight line from zero at the mean water surface to the roughness crest, this line then changing slope to give bed shear stress $\rho u_{*H}{}^2 = \rho g H_m S_{ws}$ at the roughness trough (Nikora et al. 2001, 2005b). The equivalent measured momentum-flux-component sum of Figure 7 is of a concave-upwards shape consistent with the accelerating-flow nature of this test. Within limitations of experimental methods (e.g. ADV sampling-volume and probe-alignment uncertainties, rib alignment and spacing uncertainties) and measurement accuracies (e.g. ADV accuracies and measurement of very small pressures), the sum of momentum-flux components approximates the theoretical gravity-induced expectations, particularly allowing for unaccounted-for nonuniform flow effects.

From (3) and Figure 7, secondary currents and flow nonuniformity act to reduce the momentum

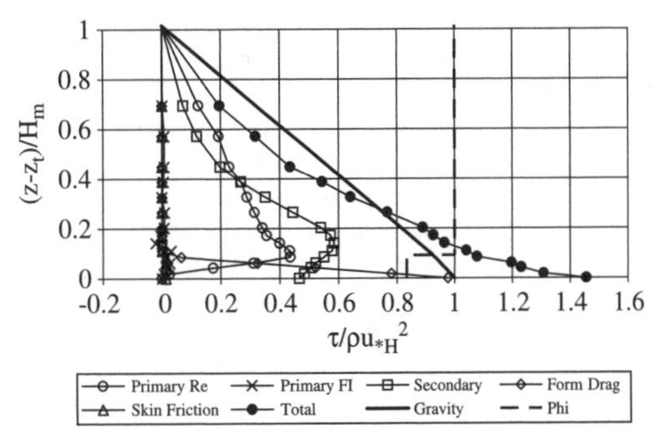

Figure 7. Momentum flux components and balance for the rib spacing of $\lambda/h = 6$.

flux occurring through Reynolds and form-induced stresses for a flow. Secondary flows for the present tests arose principally out of a combination of a slight lateral slope in the flume floor, and a flow aspect (width-to-depth) ratio of 3.0. Without measurement of the resulting crossflows, or of bed pressures, the influence (Figure 7) of the associated secondary currents for this test, not expected for the high relative bed roughness, would not have been detected. From Figure 7, secondary current effects on momentum transfer and conservation considerations can potentially be significant, with conventional extrapolation of measured Reynolds stresses to the mean bed level giving less than half of the true bed shear stress for this test.

The relative magnitudes, trends and role of form-induced stresses, and the agreement between the sum of momentum-flux components and theoretical gravity-induced expectations, is reinforced for fluvial-dune bed roughness in Coleman et al. (2005b).

6 CONCLUSIONS

The framework provided by the double- (time- and space-) averaging methodology (Nikora et al. 2001, 2004) description of momentum conservation is found to be central to describing flows over heterogeneous roughness.

For flow over the present ribs (and also over dunes), the flow structure consists of a pair of counter-rotating vortices (of opposite orientations for d-type roughness of $h/\lambda > 0.2$ and k-type roughness of $h/\lambda < 0.2$) superimposed on an overall double-average velocity profile of interfacial-layer (below roughness tops) and quasi-logarithmic form-induced-layer (above-roughness tops) sections. The interfacial layer (below roughness tops) velocity profile changes with increasing element spacing from exponential ($h/\lambda > 0.1$) to linear ($h/\lambda \leq 0.1$) to logarithmic (with increasing roughness element isolation, $h/\lambda \ll 0.1$).

In terms of momentum fluxes, the sum of momentum-flux components approximates theoretical gravity-induced expectations. In this regard, secondary currents and flow nonuniformity act to reduce the momentum flux occurring through Reynolds and form-induced stresses for a flow, secondary current effects on momentum transfer and conservation considerations being significant for the present tests, with conventional extrapolation of measured Reynolds stresses to the mean bed level giving less than half of the true bed shear stress. In addition, form-induced stresses complement variations in Reynolds stresses, being of similar magnitudes to and acting to balance variations in these latter stresses, particularly above roughness crests. Knowledge of the effects of form-induced stresses, secondary currents, and flow nonuniformity can be particularly important for describing and modeling flows over roughness elements.

ACKNOWLEDGMENTS

This research was partly funded by the Marsden Fund (UOA220) administered by the Royal Society of New Zealand. The writers are grateful to G. Kirby, J. Luo, R. Lau, L. Lou, and S. Blackbourn for assistance with the experimental work and analyses. The writers are further grateful to Dr I. McEwan, Dr D. Pokrajac, Dr L. Campbell, Dr G. Constantinescu and Prof. V. C. Patel for useful discussions.

REFERENCES

Coleman, S. E., Nikora, V. I., Mclean, S. R., Clunie, T. M., Schlicke, E., & Melville, B. W. 2005b. Equilibrium hydrodynamics concept for developing dunes. Submitted for review to *Journal of Fluid Mechanics*.

Coleman, S. E., Nikora, V. I., Mclean, S. R., & Schlicke, E. 2005a. Spatially-averaged turbulent flow over square ribs. Submitted for review to *Journal of Engineering Mechanics, ASCE*.

Ikeda, T., & Durbin, P. A. 2002. Direct simulations of a rough-wall channel flow. *Report No. TF-81,* Department of Mechanical Engineering, Stanford University, Stanford, California, USA, 142pp.

Jiménez, J. 2004. Turbulent flows over rough walls. *Ann. Rev. Fluid Mech*., 36, 173–196.

Leonardi, S., Orlandi, P., Smalley, R. J., Djenidi, L., & Antonia, R. A. 2003. Direct numerical simulations of turbulent channel flow with transverse square bars on one wall. *J. Fluid Mech*., 491, 229–238.

Nikora, V., Goring, D., McEwan, I., & Griffiths, G. 2001. Spatially-averaged open-channel flow over a rough bed. *J. Hyd. Eng., ASCE*, 127(2), 123–133.

Nikora, V., Koll, K., McEwan, I., McLean, S., & Dittrich, A. 2004. Velocity distribution in the roughness layer of rough-bed flows. *J. Hyd. Eng., ASCE*, 130(10), 1036–1042.

Nikora, V., Koll, K., McLean, S., Dittrich, A., & Aberle, J. 2002. Zero-plane displacement for rough-bed open-channel flows. *Proc., International Conference on Fluvial Hydraulics, River Flow 2002, Louvain-la-Neuve, Belgium, September 4–6*, 83–92.

Nikora, V. I., McEwan, I., McLean, S. R., Coleman, S. E., Pokrajac, D., & Walters, R. 2005a. Double averaging concept for rough-bed open-channel and overland flows: theoretical background. Submitted for review to *J. Hyd. Eng., ASCE*.

Nikora, V. I., McLean, S. R., Coleman, S. E., Pokrajac, D., McEwan, I., Campbell, L., Aberle, J., Clunie, T. M., & Koll, K. 2005b. Double averaging concept for rough-bed open-channel and overland flows: applications. Submitted for review to *J. Hyd. Eng., ASCE*.

Perry, A. E., Schofield, W. H., & Joubert, P. N. 1969. Rough wall turbulent boundary layers. *J. Fluid Mech*., 37(2), 383–413.

Raupach, M. R., Antonia, R. A., & Rajagopalan, S. 1991. Rough-wall turbulent boundary layers. *Appl. Mech. Rev., ASME*, 44(1), 1–25.

River, Coastal and Estuarine Morphodynamics: RCEM 2005 – Parker & García (eds)
© 2006 Taylor & Francis Group, London, ISBN 0 415 39270 5

A new integrated, hydro-mechanical model applied to flexible vegetation in riverbeds

D. Velasco, A. Bateman & V. DeMedina
Hydraulic, Maritime and Environmental Dept. Universidad Politecnica de Catalunya (UPC). Barcelona, Spain

ABSTRACT: The present paper suggests a simple, new numerical scheme to calculate vertical velocity profile, turbulent shear stresses distribution and canopy deflection in flow through vegetated channels. The scheme is derived from the simplified, steady Reynolds equation (momentum balance) and its integration along the vertical direction. This study includes an experimental work in order to compare and calibrate the numerical model to flume data. A set of 14 runs was performed in a laboratory flume, where 9 runs included natural grass (cultivated barley) and 5 runs included plastic plants (PVC). Turbulent diffusion coefficient, mixing length, and a resistance equation (drag coefficient Cd vs. Reynolds Number) are the input parameters for the numerical model and, as along as they have physical meaning, they were calibrated. The advantage of this scheme is a low computation time as well as a good estimation in plant deflection and velocity and stresses profiles.

1 INTRODUCTION

The study of fluvial hydrosystems requires to focus our attention not only into the different, isolated elements which are involved in the process (water, sediment, vegetation and fauna), but to deepen our knowledge into their common relationships. In this sense, the aim of this group is to study the effect of flexible vegetation in the flow. Presence of flexible canopy (grass, bushes and reeds) in the riverbed and banks changes the flow resistance and, in consequence, river hydrodynamics. Many authors (Kouwen, 1969; Nepf, 1999) have experimented with plastic and natural canopy in their laboratories to obtain some empirical relationships. The complexity of physical models studies and the extension and popularization of numerical schemes in hydraulics has been the starting point for the development of applications focused on flow through vegetation. Three different lines of numerical research have been followed. First is the application of well-known 3D turbulent schemes to special geometric boundaries which simulate obstruction due to vegetation. Lopez and García (2001) and Fischer-Antze (2001) adjusted the 2 equations κ–ε closure model to flow through vertical, rigid cylinders and simulations are quite adjusted. Cui And Neary (2002) compare numerical results of RANS (Reynolds Averaged Navier-Stokes) and LES (Large Eddy Simulation) in order to reproduce particular anisotropy generated in flow trough rigid, submerged cylinders. Finally, Choi and Kang (2004) use an RSM (Reynolds Stress Modeling) to improve mean and turbulent field results. High accuracy numerical schemes will be developed in the future, but the elevated computational cost is still a disadvantage for their practical apply to large scale hydraulic problems. A second group of researchers (Klopstra, 1997, Carollo, 2002) have developed analytical expressions to adjust velocity profiles $U(z)$ in very specific conditions, but input parameters involved in the problem are hard to estimate. The third line of research is very interesting in the sense that incorporates the flexural properties of vegetation and calculates the reduction of relative depth h/k due to stem bending. Kutija & Hong (1996) have developed this line of research in the basis of very simple turbulent models applied to 1-D Reynolds momentum equation. Deflected plant height k is calculated by a function that reproduces the elastic behaviour of a vertical beam or cantilever which is loaded by drag forces. Unfortunely, no data was available for the verification of the model with flexible vegetation, and only data from Tsujimoto's test with rigid cylinders were used in the calibration.

The present paper tries to improve this third line of research. The author develops a new integrated momentum balance scheme to calculate the velocity and turbulent stress profile including a mechanical subroutine of stem deflection. Results are then verified and calibrated with own experimental data, which have been measured in a flume covered with flexible vegetation.

2 GOVERNING EQUATIONS

The numerical model is based on the integrated equations of conservation of momentum, eq.(1) to (4), which are obtained from the vertical integration of Reynolds x-equation. At this point, it is necessary to recognize and separate 4 different regions or zones in the flow, so resulting equations are wrote as eq.(1) to (4):

$$0=\underbrace{\rho.g.So.(h-z)+\tau_{xz}(z)}_{\tau_G(z)} \qquad \text{ZONE 1. external region}, \quad h>z>k$$

$$0=\rho.g.So.(h-z)+\tau_{xz}(z)+\tau_{Cd}(z) \qquad \text{ZONE 2. internal region}, \quad k\geq z>p$$

$$0=\rho.g.So.(h-z)+\tau_{Cd}(z) \qquad \text{ZONE 3. shearless region}, \quad p\geq z>\delta_o$$

$$0=\rho.g.So.(h-z)-\nu.\frac{\partial U(z)}{\partial z}+\tau_{Cd}(z) \qquad \text{ZONE 4. laminar region}, \quad \delta_o\geq z$$

(1),(2),(3),(4)

Zone 1 is called de "external region", and corresponds to heights z from h to k, above canopy, and, in consequence, without drag forces. Integral eq.(1) shows how the gravity shear term is balanced uniquely to turbulent shear stress τ_{xz}, and maximum turbulent stress is located, then, at $z=k$. Below $z=k$ appears Zone 2, that is called "internal region" and it is characterized by the presence of canopy. The vertical integration of drag forces f_{cd} from the top of canopy k to height z is the amount of shear stress that is absorbed by the plant (τ_{cd}). This vegetative stress is defined in eq.(5), where C_d = drag coefficient, B = plant width and a = stem spacing. From eq.(2) its deduced that gravity shear stress is balanced both by vegetative stress and turbulent stress. Hence, turbulent stress τ_{xz} is been reduced at $z<k$, and , if τ_{cd} equals to gravity shear τ_G, turbulent stress τ_{xz} disappears (see Fig. 1). This means that the absorbing momentum capability of canopy is higher than hydraulic gravity shear. Height $z=p$ is called penetration depth (Nepf, 1999), and delimitates Zone 3, called "shearless region" because turbulent stress τ_{xz} is negligible (eq.(3)). In this region there is no vertical momentum exchange, but turbulence is completely developed; the flow is governed by drag forces which balance gravity shear completely.

$$\tau_{Cd}(z)=\int_z^k f_{cd}(z)dz=\int_z^k \frac{1}{2}.\rho.C_d(z).\frac{B(z)}{a^2}.U^2(z)..dz \qquad (5)$$

The vertical domain of computation is discretisated uniformly with a Δz step, as presented in Fig. 2, and all the variables are calculated in these nodes. Water depth (h), undeflected plant height (h') and deflected plant height (k) are denoted as ih, ihp and ik indexes, respectively. Finite difference approximations (forward differences) are used in the explicit method when computing vertical velocity U profiles, and first order integral methods (Simpson's rule) is applied in the integral operations. Due to the explicit nature of the scheme, the number of nodes must be considerable.

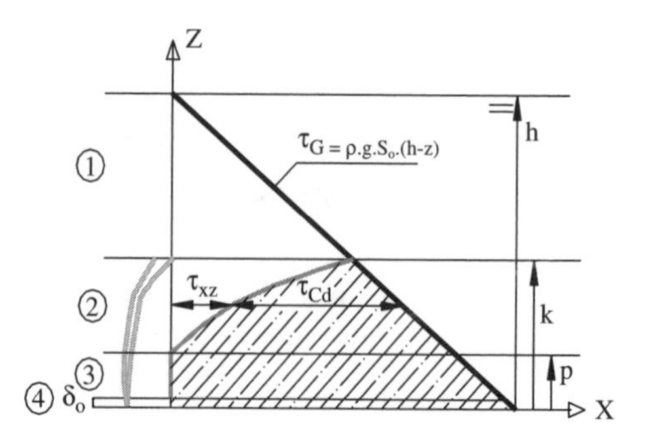

Figure 1. Shear balance scheme for high vegetative capability for absorbing momentum.

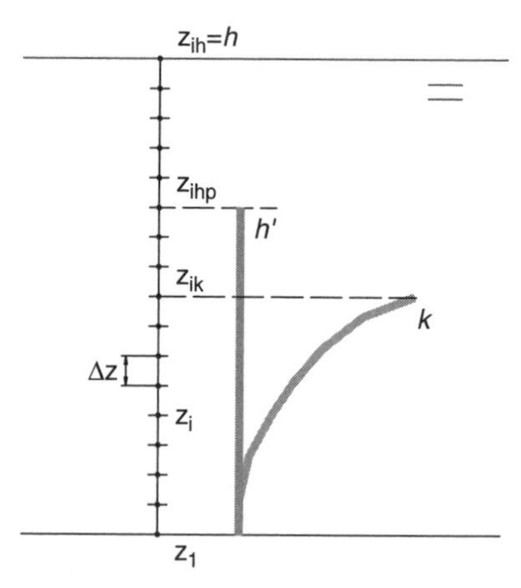

Figure 2. Scheme of vertical discretisation.

Two different boundary conditions must be set in the computational domain: 1) $\tau_{xz}=0$ at water surface ($z=h$), and 2) $\partial\tau_{xz}/\partial z=0$ at the shearless region ($0<z<p$). The application of former boundary condition to the Reynolds equation in x direction leads to an expression which can be managed to estimate explicitly velocity U. So, at the very fists node of calculation, and for all the shearless zone ($z_i<p$), velocity U_o is determined by eq.(6) as:

$$U_o=\left(\frac{2.g.So.a^2}{C_{d,o}.B_o}\right)^{1/2} \qquad (6)$$

Where drag coefficient $C_{d,o}$ and plant width B_o correspond to node 1. This boundary condition in velocity is critical because the general method is based in the hypothesis that the shearless region (Zone 3 of Fig. 1) is developed.

In order to relate Reynolds turbulent stresses to velocity gradients, the mixing-length turbulent theory of Von-Karman has been implemented. This one equation turbulence closure reflects momentum diffusion in a very simple way and could be considered of low accuracy method, but, as a first approximation to the problem of turbulence, goodness of results are quite good, according to the aim of this work. Mixing length equation of turbulence is expressed as eq.(7):

$$\tau_{xz} = \rho . l^2 . \left| \frac{\partial U}{\partial z} \right| . \frac{\partial U}{\partial z} \qquad (7)$$

where ρ = water density and l = mixing length. Some theories have been found in the literature about vertical profiles of mixing length l(z) with and without vegetation, and, finally, eq.(8) has been adopted to estimate the variation of mixing length l in depth. The concept of integral scale L could be understood in analogy to the popular mixing length l of Von-Kárman turbulence closure equation because both concepts reflect a characteristic eddy scale, and so, the mechanism of diffusion should not be so different. In this sense, vertical distribution of Integral scale L was analysed from experimental profiles and the conclusion was that , in spite of dispersion of estimated data, a linear tendency above $z = p$ was guessed. The proposed distribution of mixing length l is described in eq.(8).

$$\begin{bmatrix} l(z) = l_o + \kappa'.(z-p) & \text{for } z > p \\ l(z) = l_o & \text{for } z \leq p \end{bmatrix} \qquad (8)$$

where l_o is the mixing length at shearless zone, and κ' represents a diffusion coefficient. Numerical parameters l_o (bed mixing length) and κ' (diffusion coefficient) are unknown and must be calibrated, but the advantage is that both of them have a very particular physical meaning.

Furthermore, the intrinsic effects of secondary currents and turbulent anisotropy of flow over vegetated covers are introduced in the model by a first, simple approximation. Experimental data show the effective reduction of energy slope ($S_{o\,max\,shear}$), so, it is proposed the use of a secondary current factor α_{sc}, defined as $\alpha_{sc} = S_{o,max\,shear}/S_o$, where S_o = channel slope (in uniform flow conditions). The value of α_{sc} is always less than one, and , as a turbulent parameter, it must be also calibrated.

Finally, the drag stress term τ_{cd} in the governing equations (2) and (3) is estimated through eq.(5), and the numerical integral is expressed as eq.(9):

$$\tau_{Cd}(z_i) = \frac{\rho}{2.a^2} . \int_{z_i}^{z_{ik}} C_{d.j} . B_j . U_j^2 . dz \qquad (9)$$

where $C_{d,j}$, B_j and U_j are drag coefficient, plant width and velocity, respectively, at node j.

Special attention is required for the estimation of drag coefficients $C_{d,j}$ because it is intimately linked to the nature of resistance to flow over vegetated covers.

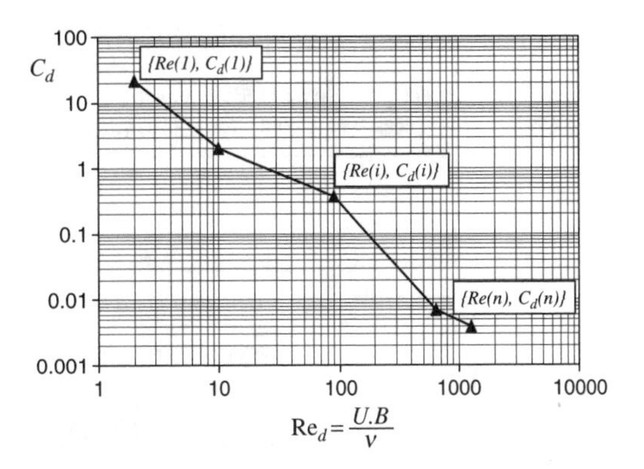

Figure 3. Estimation of $C_d(\text{Re}_d)$ law as a linear approximation between a series of discrete points {Re_d(i), C_d(i)} in a log-log graph.

The value of C_d is very sensible to changes in the flow turbulent regime, which is governed by the obstacle or stem Reynolds number $\text{Re}_d(= U \cdot B/\nu)$, where ν = water cinematic viscosity. So, a resistance function of the type $C_d(\text{Re}_d)$ is introduced in the model to take into account this important factor, and Re_d is calculated at every computational node i as eq.(10):

$$\text{Re}_{d,i} = \frac{U_i . B_i}{\nu} \qquad (10)$$

Unfortunately, classical resistance laws $C_d(\text{Re}_d)$ of 2-D rigid bodies that has been widely studied in the past are not suitable to the heterogeneous shape of stems and leaves and new relations must be investigated. In order to further fitting-optimization procedure, the estimation of $C_d(\text{Re}_d)$ law has been introduced as a linear approximation between a series of discrete points {Re_d (i), C_d (i)}, for i = 1,n, in a log-log graph. Fig. 3 shows the resistance law proposed in function of a series of n = 5 points.

The numerical estimation of drag coefficient C_d at node i is based on previous calculation of $\text{Re}_{d,i}$ by eq.(10) and the application of the linear, discrete function $C_{d,i} = \Psi_n(\text{Re}_{d,i})$. Finally, this $C_{d\,i}$ value is incorporated to drag in eq.(9).

3 DESCRIPTION OF NUMERICAL PROCEDURE

3.1 *General iterative scheme*

The main idea of this numerical model is to create an iterative succession of flow well-balanced solutions that can generate the final stabilisation or convergence of the deflected plant height k. The computational flow diagram is presented in Fig. 4. At every iteration i, the deflected plant height k^i is calculated and, as a consequence of the geometric change, a new flow solution

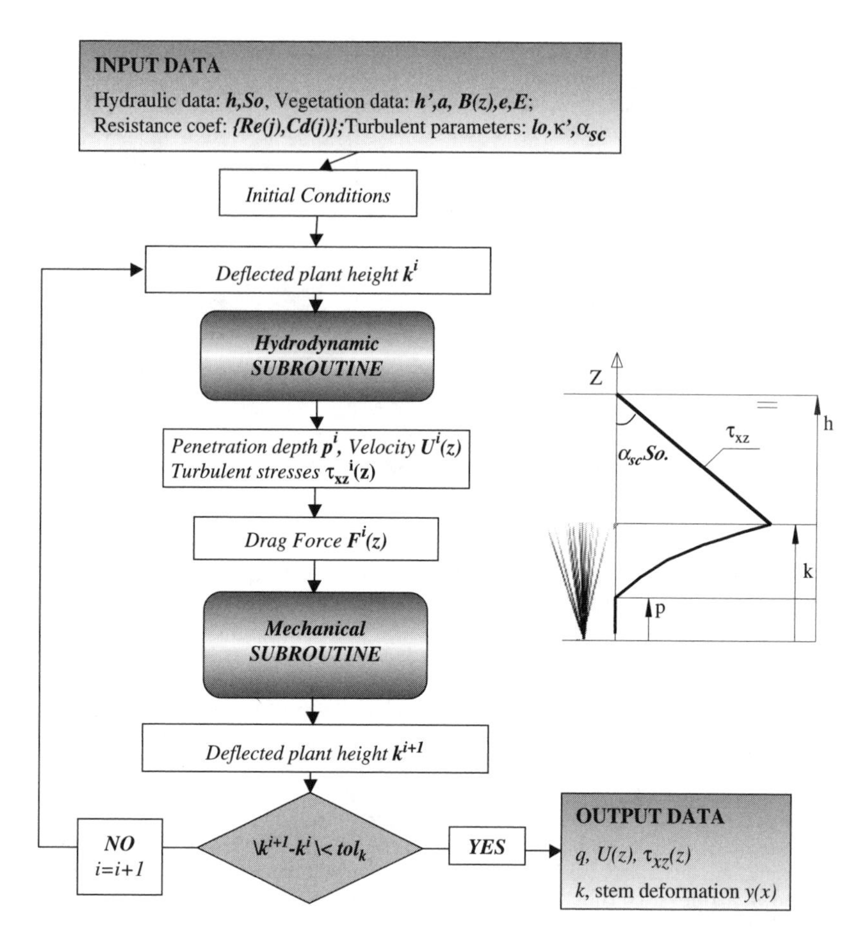

Figure 4. General flow diagram.

$U^i(z)$, $\tau^i_{xz}(z)$ is obtained. At this point, deformation of plant is re-calculated using former flow conditions, and a new deflected plant height k^{i+1} is incorporated at iteration $i+1$. Computation is finished when a convergence in $\{k^i\}$ is set $(\backslash k^{i+1} - k^i \backslash < tol_k)$; then, the flow solution is considered dynamically and mechanically balanced.

Input data is composed by hydraulic variables: water depth h, energy slope S_o), vegetation geomechanic properties (undeflected plant height h', plant spacing a, vertical distribution of plant width $B(z)$, thickness of stems e, stiffness modulus E, the Resistance law $\{\mathrm{Re}(j), C_d(j)\}$ and three turbulent parameters (bed mixing-length l_o, diffusion coefficient κ', secondary currents factor α_{sc}). Results of the model (Output data) are unit discharge q, velocity $U(z)$, Reynolds stresses $\tau_{xz}(z)$, deflected plant height k and stem deformation $y(x)$.

The model can estimate the unit discharge q given an initial water depth h and an energy slope S_o, that is equivalent to solve the resistance to flow equation for any kind of vegetation. Friction factors (Darcy-Weisbach f or Manning's n) can be easily deduced from hydraulic results.

The general scheme is very simple in its conception, but the most important and critical part is the creation of adequate modules to calculate velocity profiles and stem deformation. The hydrodynamic subroutine calculates flow profiles $\{U^i(z), \tau_{xz}(z)\}$ for a particular plant height k condition and the mechanical subroutine estimates the deflected plant height k as a function of drag forces $F(z)$. Next, it is presented a brief explanation of the numerical methods used in both subroutines.

3.2 *Description of the hydrodynamic subroutine*

The numerical strategy used in the estimation of well-balanced velocity profile is based on a predictor-corrector technique in Reynolds stresses τ_{xz} and a minimization function. Fig. 5 shows a flow diagram and a schematic picture where predicted τ_{xz} and

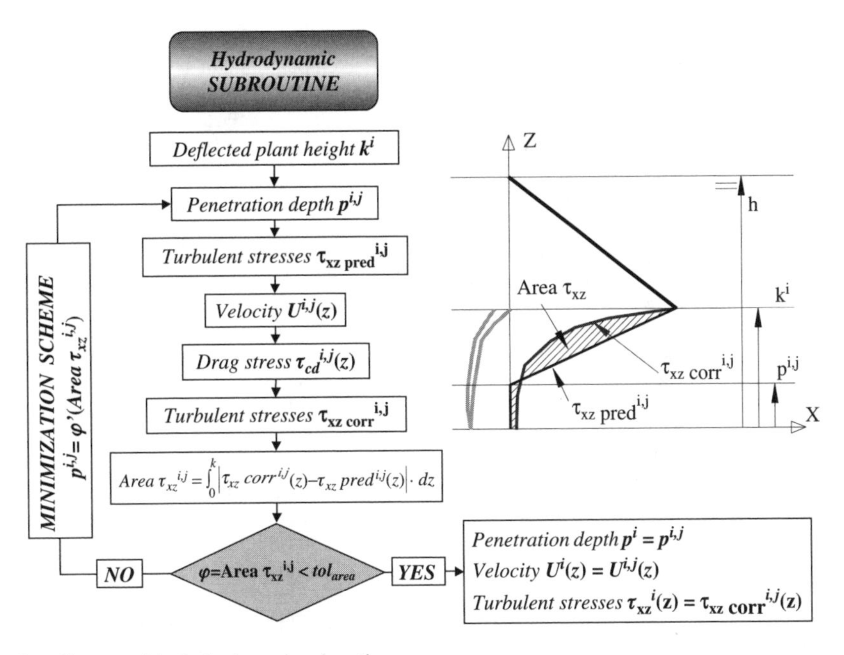

Figure 5. Flow diagram of the hydrodynamic subroutine.

corrected τ_{xz} profiles are shown. On the basis of constant deflected plant height k^i (input of this subroutine), a penetration depth $p^{i,j}$ is proposed and a linear distribution of turbulent stresses $\tau^{i,j}_{xz,pred}(z)$ is then predicted (as seen in Fig. 5). Eq.(6) and (7) are then used to calculate the velocity profile $U^{i,j}(z)$ and eq.(9) to estimate drag stresses $\tau^{i,j}_{cd}(z)$. Then, the corrected $\tau^{i,j}_{xz,corr}(z)$ profile is balanced from eq. (1), (2) and (3). Next step is to calculate the area existing between both turbulent stresses profiles- predicted and corrected, as eq. (11).

The value of $Area\ \tau^{i,j}_{xz}$ is then related to the penetration depth $p^{i,j}$, and a positive function is defined as $\varphi(p) = Area\ \tau_{xz}(p)$. Under this scheme, the minimization of function φ implicates a well-balanced solution in $\tau_{xz,corr}(z)$ and $U(z)$ profiles. An iterative, first order minimization method is applied to $\varphi(p)$ and the solution (penetration depth p^i, velocity $U^i(z)$ and stresses $\tau^i_{xz}(z)$) is sent to main program. The accuracy of the solution depends basically of the definition of function $\varphi(p)$ (existence and uniqueness of absolute minimum); the analysis of this function shows a good behaviour and a sufficient efficiency of the search method.

$$Area\,\tau_{xz}{}^{i,j} = \int_0^k \left|\tau_{xz,corr}{}^{i,j}(z) - \tau_{xz,pred}{}^{i,j}(z)\right| .dz \tag{11}$$

3.3 *Description of the mechanical subroutine*

This second subroutine consists in a numeral code which reproduces load-deformation processes in a stem. Every canopy element is modelled as a vertical, isolated beam or cantilever, in order to simplify the structural problem. Forces on the stem are imported from the hydrodynamic subroutine and they are applied to every calculation node to obtain an approximation $y(x)$ of stem deformation. Classical, analytical expressions of deformation on beams are not valid here, because they are derived only for small deformation conditions. Natural canopy bend in river floods easily, as well as brushes and aquatic species in normal flows, and the deformation of the top of canopy is often the same order than the deflected plant height ($y(h') \cong k$). Large deformations are required in the simulation to estimate adequately the vertical deflected plant height k.

A discretisation of the equation of elasticity in beams is applied in the stem (adopted x as the vertical, longitudinal direction of the stem) and an explicit forward finite differences scheme is computed to solve transversal deformation profile y(x), as eq.(12):

$$\frac{\partial^2 y}{\partial x^2} = -\frac{Mf}{E.I} \tag{12}$$

where, y = transversal deformation, Mf = Moment (N·m), E = Stiffness Modulus (N/m^2) and I = Cross Area Inertial Modulus (m^4). The large deformation hypothesis implicates the conservation of total stem length h', and then, vertical distribution of load (drag forces) must be redistributed to accomplish a well-balanced equilibrium. Under this consideration, a general iterative scheme is established in $\{y_i(x), k_i\}$ to converge to the deflected plant height k. Deformation profile $y_i(x)$ is calculated from eq.(12) and the total stem length is measured along the deformed stem to define the deflected plant height k_i. New nodes are defined according to k_i, a new distribution of forces F is used and, then, $y_{i+1}(x)$ and k_{i+1} are calculated. Figure 6 presents a scheme of the iterative procedure. The convergence into the solution is fast enough (less than 15 iterations, in general). An important concept to take into account in large deformation calculation is the additional bending due to Moments of Second Order in the beam. New moments Mf' appear in the beam as a consequence of non orthogonality between loads F and longitudinal direction of the deformed beam (s). This effect is introduced as follows: (1) First Order Moments Mf are used, and a deformed profile $y(x)$ is obtained; (2) secondary moments Mf' are calculated from derivates $y'(x)$ and vertical distances, and (3) a new calculation of the beam is done for total moments $Mf+Mf'$. The final result is a decrease in deflected plant height k from first order calculations, and an increase in the curvature of bending (observed in real elastic materials). Fig. 7 shows the importance of including secondary moments to reproduce more accurately the bending of elastic stems.

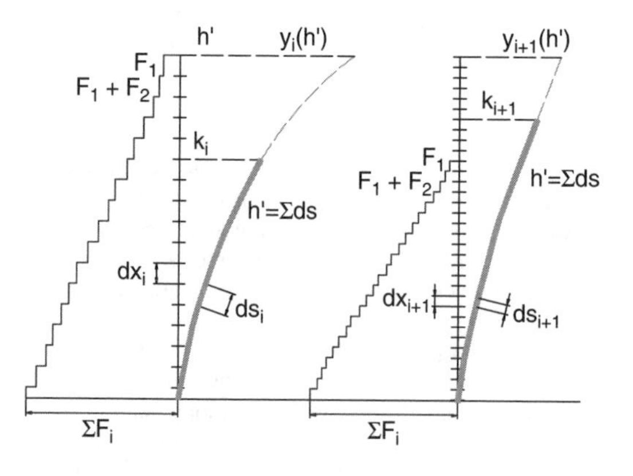

Figure 6. Numerical discretisation and iteration process for the mechanical subroutine (large-deformations).

4 MODEL VERIFICATION

In the present study, the model calibration and verification has been possible in an accurate, controlled way due to a previous experimental work. Own data has become of vital importance and though tests are limited in number (more variety of plant density and shapes should be desired), detailed measurements were obtained.

In order to calibrate the mechanical subroutine, load-deformation tests were done in the laboratory in very controlled conditions. PVC cylinders of well-known geo-mechanical were first tested to check that large deformation hypothesis and numerical algorithm was acceptable. Geo-mechanical properties of canopy (both, PVC palms and natural barley stems), which consist in length, average width and thickness, stiffness modulus E and inertial modulus I were directly measured or estimated. They were introduced as an input data in the mechanical model. Results are quite good for all load steps, and though curvatures of the bending stem are hard to reproduce (heterogeneity of real stem properties) the discrepancies are assumed and the model is finally accepted. Limitations are present for larger deformations ($k/h' < 0.6$) and the adjustment between measured and calculated deformation profile presents lowest accuracy (Fig. 8), and the error on k is 12%.

Numerical results of the integrated model must be compared to experimental data in order to verify the

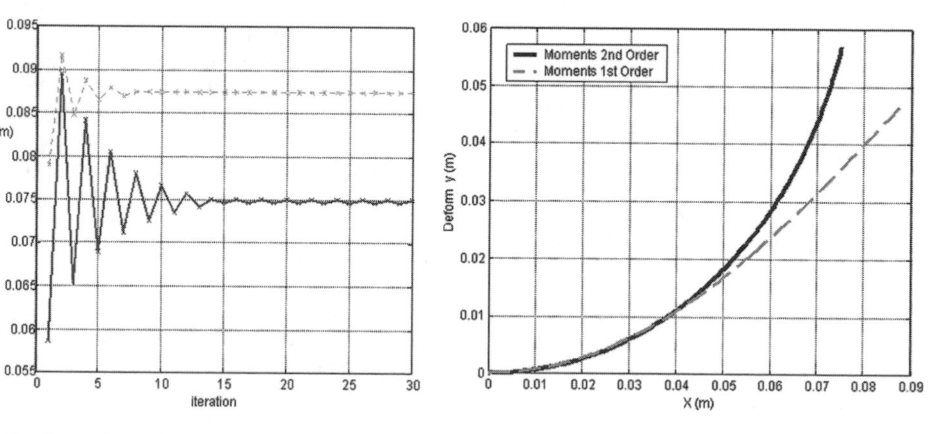

Figure 7. Comparison of computational stem deformation with secondary moments calculation (solid line) and without (dashed line).

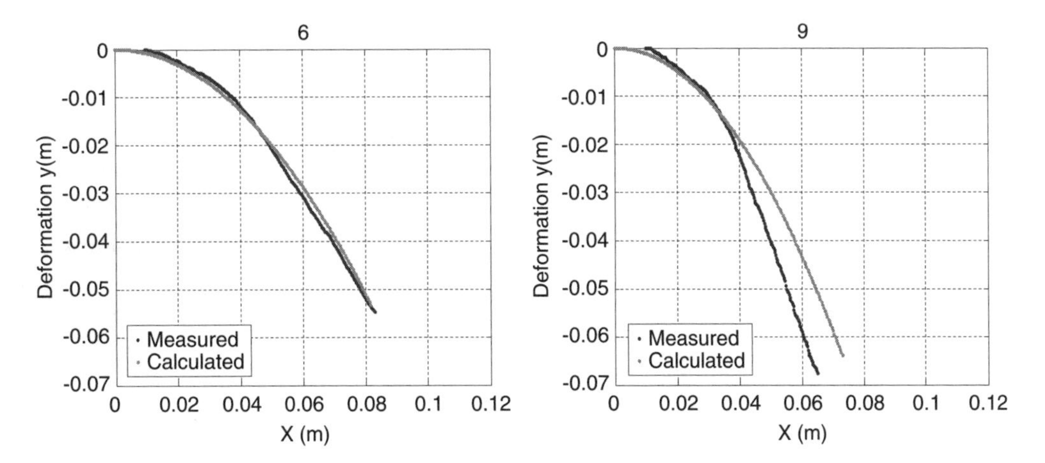

Figure 8. Comparison between measured and calculated deformation of a barley stem (length $h' = 0.125$ m) for increasing loads.

accuracy of the method. Two examples are presented in Figure 9. First, it is adjusted the experimental data obtained by Tsujimoto (1990) in flow through rigid cylinders. Run A11 consist in vertical cylinders of height 4.59 cm and 1.5 mm diameter, in a 2 cm spacing staggered pattern. Second run in Fig. 9 corresponds to data from own tests (barley cover). Flexible barley stems are 18 cm tall, spacing between stems is $a = 6.7$ mm, and submergence ratio $h/k = 1.83$. Vertical distribution of plant width $B(z)$ was defined as a constant for cylinders, and variable for barley (calculated from a frontal image). Drag coefficients $C_d(z)$ were set as a unique value for cylinders ($C_d = 1.5$), but the experimental values of $C_d(z)$ obtained from the barley test were directly introduced in the numerical model.

Input computational parameters like mixing length l_o, momentum diffusion constant, κ' and the secondary currents factor α_{sc} are adjusted for every run; for Tsujimoto's Run they are $l_o = 2.5$ mm, $\kappa' = 0.17$ and $\alpha_{sc=}1.0$. For barley Run TN-h' = 0.18 m, parameters are $l_o = 10$ mm, $\kappa' = 0.08$ and $\alpha_{sc=}0.78$. Fig. 9 shows a very good agreement between measured and calculated velocity profiles and curvatures of velocity profile are quite well reproduced.

An important work was done to calibrate the general model. A particular set of general computational parameters must be found to fit all different tests. The list of unknown parameters was composed by Drag Coefficients points, $\{Re(j), Cd(j)\}, j = 1,5$, mixing length l_o, momentum diffusion constant, κ' and the secondary currents factor α_{sc}. A Multi-parametric optimization technique was applied to 14 runs with vegetation (T3 and TN runs) where canopy was deflected and totally submerged. Measured data used in the optimization routine was unit discharge q_{mea}

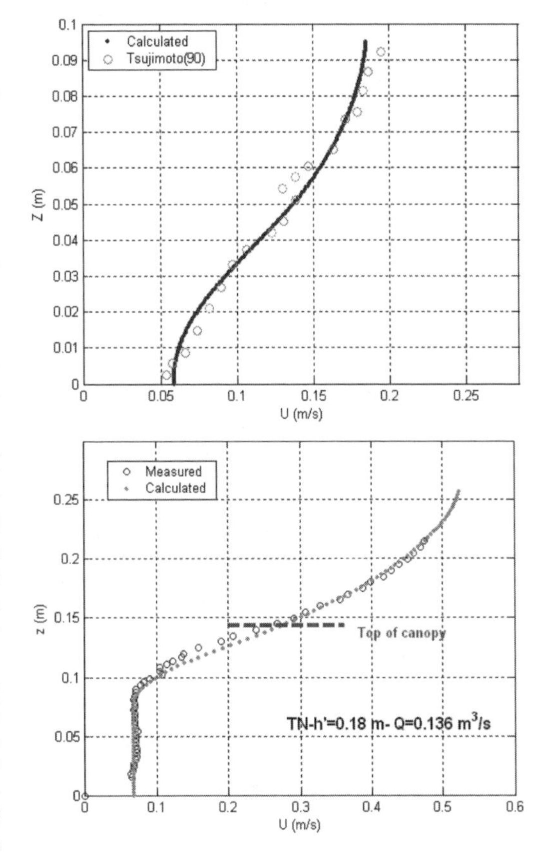

Figure 9. Comparison between measured and calculated velocity profiles. (Up) A11 Run for vertical rigid cylinders from Tsujimoto (1990) experiments. (Down) flexible barley Run (TN-h' = 0.18 m) from our own experiments.

Table 1. Adjusted parameters of drag coefficients law and turbulent parameters.

Resistance law		Turbulent parameters		
Re	C_d	l_o (m)	κ'	α_{sc}
2	22.5	0.01	0.04	0.54
10	3.14			
90	0.37			
632	0.009			
1275	0.004			

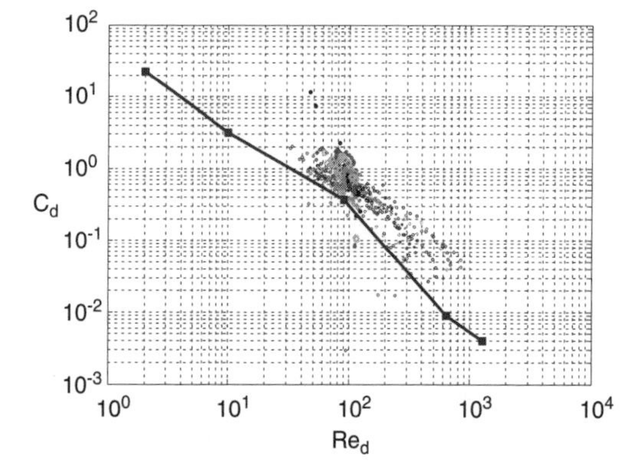

Figure 10. Optimally adjusted resistance law (C_d vs. Re_d) compared to experimental data (circles).

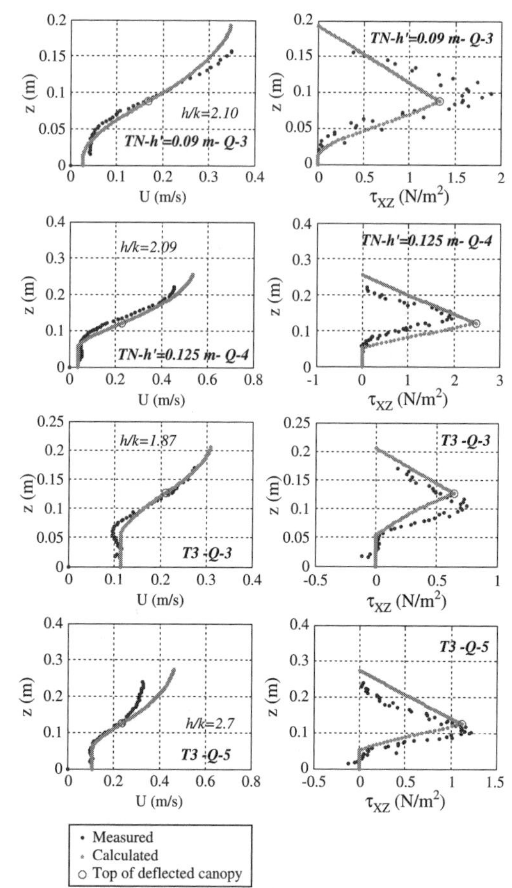

Figure 11. Comparison between calculated and measured U(z) and $\tau_{xz}(z)$ profiles.

(m^3/m/s) and velocity profiles $U_{mea}(z)$. Some discrepancies were obtained between the measured unit discharge q_{mea} and the vertical integral of $U_{mea}(z)$, due to 3D flow characteristics, and both measures were integrated in the optimization function. This function Φ was defined as a quadratic function of difference between output calculated data and measured data. A first order, intelligent-conjugate-gradients technique was computed to find the minimum of Φ. Optimal, adjusted parameters are presented in Table 1.

These parameters are coherent respect to physical measures, but some comment must be done about the resistance law. The values of adjusted $Re(j)$ and, $Cd(j)$ are shifted respect to measured data, as can be observed in Fig. 10. Adjusted drag coefficients C_d are lower than measured C_d and the reason for the shifting is probably a numerical diffusion in the model.

5 NUMERICAL RESULTS

Next, the results of the model calibration are commented. In Fig. 11 it is shown 4 different runs and experimental data is compared to computed data. Velocity $U(z)$ and Reynolds turbulent stress $\tau_{xz}(z)$ profiles are presented and, in general terms, both velocities and shear are well-adjusted. A very good agreement is observed for the penetration depth p, and shearless zone is well defined in velocity and shear. Velocity U_o in this zone is calculated as a uniform profile and the agreement with measured U_o is good enough. Then, it is possible to conclude that the proposed boundary condition of velocity (eq.(6)) is efficient. At Zone 2 (internal zone $p < z < k$), calculated stress τ_{xz} shows some deviation respect to measured data, but vegetative absorption of momentum seems to be accomplished. Top of canopy at the figure (circle) corresponds to the peak of calculated stress ($\tau_{xz,max}$) and approximately agrees with the peak of measured τ_{xz}. In the external zone ($z > k$) the slope on calculated τ_{xz} does not fit data, but the effects of secondary currents are simulated and it is believed to constitute an important goal of the numerical model and the calibration work. The calculated velocity $U(z)$ reproduces quite well the gradients in the internal zone,

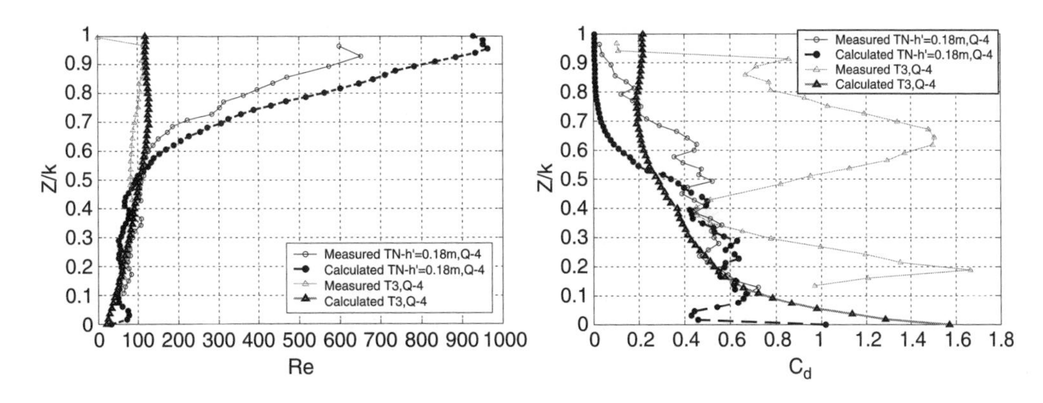

Figure 12. Measured and calculated Re and C_d vertical profiles.

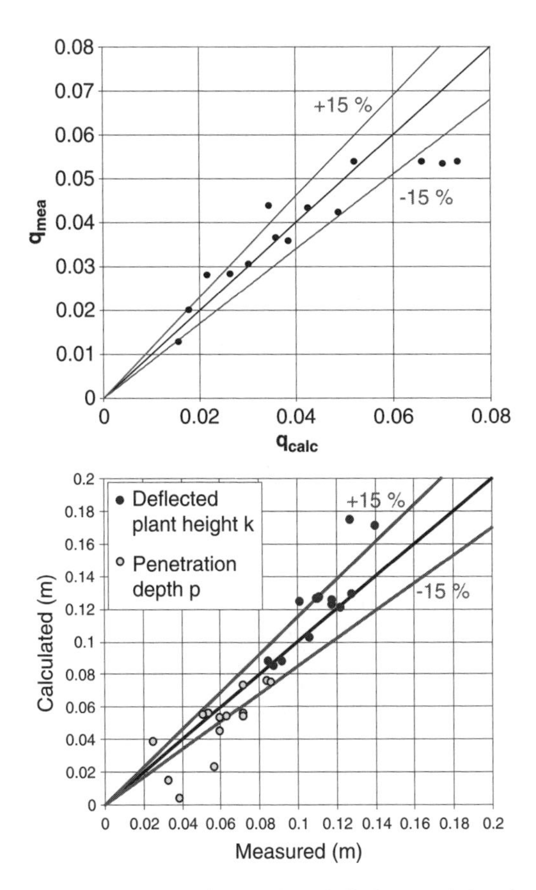

Figure 13. Numerical results in unit discharge q (up) and deflected plant height k and penetration depth p (down).

but some discrepancy is observed close to the free surface.

It is shown in Fig. 12 calculated Re and C_d vertical profiles, and they are compared to those measured experimentally. There is a good agreement in obstacle Reynolds number (as can be seen, the plant shape is controlling Re distribution) both for natural and artificial plants. But as commented before, numerical drag coefficients are lower than experimental C_d, and this effect is much important in the artificial plants (T3 runs), as can be observed in the figure. Lower drag coefficients should be traduced into a reduction in drag stresses, (see eq.(9)), but, in opposition, balance of momentum is finally well adjusted. This discrepancy should be revised in the future through the verification of the adjusted parameters to new tests with diverse configuration of vegetation (the important influence of spacing a) and submergence ratio.

Next step in the analysis of results is the mass conservation. Water discharge is an output of the numerical model and it is compared to the experimental discharge in Fig. 13. The comparison between calculated unit discharge q_{calc} (m^2/s) and measured q_{mea} for all 14 tests shows certain dispersion. The average error of estimation is 14.8 %, but the experimental error between q_{mea} and the vertical integral of experimental $U_{mea}(z)$ was set it 10%. The conclusion is that the accuracy in discharge q data is not much greater than typical errors in laboratory with vegetation covers, and uncertainty is almost limited. Results about deflected plant height k and penetration depth p are presented in Fig. 13, too. Some dispersion of data is detected, but in general, estimations are into an error range of 15%. We must insist in the necessity of more experimentation to check the adjustment of calculated deflected plant to reality (especially in large deformation conditions, $h'/k > 0.6$).

6 LIMITATIONS OF THE MODEL

The integrated model of vegetation is thought as a very useful tool to reproduce vegetative covers, both in emergent and submerged plant conditions. This model can be easily incorporated to any 1D or 2D general hydraulic code to estimate friction factors in riverbed and floodplains, due to the low computational cost. On the other hand, some limitations of the numerical code must be commented. Governing equations are only defined for the existence of a shearless zone, so, vegetation must be capable of absorbing total gravity shear τ_G. So, this model is recommended for high-middle plants density (high A_o/a ratio). The model should be modified for flooding scenarios where canopy is totally prone, frictions factors are reduced and an effective shear stress is acting on the riverbed. Second, the presence of real convective currents in the flow trough canopy is introduced in the model (α_{sc} factor) but it is hard to evaluate for different flow conditions, still. Finally, a more intense experimental campaign is also needed to verify the proposed drag coefficient law C_d (Re_d) for new plant shapes and configurations.

7 SUMMARY AND CONCLUSIONS

This paper presents a combination of experimental and numerical work to provide a deeper knowledge on resistance to flow through flexible vegetation. The objective of the paper is the computation of a numerical model to simulate vertical velocity profiles and plant deflection. The scheme is derived from the simplified, steady Reynolds equation, which is integrated in vertical direction, above and along the deflected plant height k. A mechanical module has been computed to estimate elastic large deformation in stems, and, mixing length theory is used as a closure turbulent equation for the hydrodynamic subroutine. An iterative scheme of convergence for plant deformation k as a function of balanced drag forces is proposed. Finally, the well-balanced vertical profile of velocity $U(z)$ is calculated by a simple, explicit finite differences scheme.

In the experimental chapter, two different experimental flumes were used and both natural (barley) and artificial (PVC strips) vegetative covers were fixed on the channel bed. Vertical profiles of velocity U and turbulent shear stress τ_{xz} (by ADV measurement techniques) provide valuable experimental data for every flow and vegetative configuration. A set of 14 experimental runs were used to verify and calibrate the numerical model. Turbulent diffusion coefficients, mixing length, and, a resistance equation (drag coefficient C_d vs. Re_d) are input parameters for the model and, as along as they have physical meaning, they were calibrated. A multi-parametric minimization algorithm was applied to adjust computed profiles to experimental data and parameters were finally estimated. In theory these computational parameters would be universal, independent of the vegetation shape or configuration. In practice, the verification of the uniqueness of this set of general parameters should be done in the future with different species of vegetation.

The advantage of this integrated model is a low computation cost as well as a reasonable estimation in plant deflection and velocity and stresses. Obviously, the accuracy of vertical profiles is lower than in any 3D-turbulent scheme, but errors in the approximation are quite limited and would be perfectly assumed in a general floodplain calculation. Then, this quick model is very suitable and fast enough to be included in a general 1D or 2D hydraulic model as a resistance function for vegetation, to develop in the future.

ACKNOWLEDGMENTS

This work was completed under the "UPC Per la Recerca" Grant (Universitat Politècnica de Catalunya), supported by the REN2000-1013/HID Project. Special thanks to Prof. J.M.Redondo, who was very helpful in the initial phase of the experimental work.

NOTATION

The following symbols are used in this paper:

a	= spacing between plants.
A_o	= Frontal Area of plant.
b	= channel width.
B	= average plant width.
C_d	= drag coefficient.
E	= Stiffness Modulus.
h	= water depth.
h'	= undeflected plant height.
I	= Second order geometrical momentum.
k	= deflected plant height.
l	= mixing length.
L	= Integral scale.
M	= density of vegetation (plants/m^2).
Mf	= Moment.
p	= penetration depth.
q	= unit discharge ($= Q/b$).
Q	= discharge.
Re	= Reynolds number ($= U.h/\nu$).
Re_d	= obstacle Reynolds number ($= U.B/\nu$).
R_h	= Hydraulic radius.
S_f	= Energy slope.
$S_{o\,max\,shear}$	= Experimental energy slope obtained from τ_{xz}, vertical profile.
S_o	= Geometric slope.
U, V, W	= longitudinal, transversal and vertical mean velocity, respectively.
u', v', w'	= longitudinal, transversal and vertical velocity fluctuation, respectively.
u^*	= shear velocity.

α_{sc} = secondary current factor.
γ = water specific weight.
κ = Von Karman universal diffusion constant.
κ' = diffusion coefficient.
ν = kinetic viscosity.
ρ = water density.
$\tau_{xz}, \tau_{xy}, \tau_{yz}$ = Reynolds Stresses.
τ_{cd} = vegetative drag stress.
τ_* = bed shear stress.

REFERENCES

Carollo, F., Ferro, V. 2002. "Flow velocity Measurements in Vegetated Channels" *ASCE Journal of the Hydraulics Engineering*, Vol. 128, 664–673.

Cui, J., Neary, V.S. 2002. Large eddy simulation (LES) of fully developed flow through vegetation. *Hydroinformatics 2002: Proceedings of the 5th International Conference on Hydroinformatics*, Cardiff, UK.

Chen, C. 1976. Flow resistance in broad shallow grassed channels. *ASCE Journal of the Hydraulics Division*, Vol. 102, No HY3, 307–322.

Choi, S., Kang, H. 2004. Reynolds stress modeling of vegetated open-channel flows. *Journal of Hydraulic Research,* Vol 42, No. 1, 3–11.

Dunn, C., López, F. 1996. Mean flow and turbulence in a laboratory channel with simulated vegetation. *Civil Engineering Studies, Hydraulic Engineering Series*, No. 51. Urbana-Champaign, Ilinois University.

Finnigan, J. 2000. Turbulence in plant canopies. *Annu. Rev. Fluid Mech.*, Vol. 32, 519–571.

Fischer-Antze, T., Stoesser T., Bates, P. 2001. 3D numerical modelling of open-channel flow with submerged vegetation.*Journal of Hydraulic Research*, Vol 39, No. 3. 303–310.

Ikeda, S., Kanazawa, M. 1996. Three- dimensional organized vortices above flexible water plants. *ASCE Journal of Hydraulic Engineering*, Vol. 122, No 11, 634–640.

Klopstra, D., Barneveld, H.J. 1997. Analytical model for hydraulic roughness of submerged vegetation. *Proceedings of the 27th Congress of the IAHR*, San Francisco (USA).

Kouwen, N., Unny, T.E, Hill, H.M. 1969. Flow retardance in vegetated channels. *ASCE Journal of Irrigation and Drainage Division*, Vol. 95, No IR2, 329–342.

Kouwen, N., Unny. T.E. 1973. Flexible roughness in open channels. *ASCE Journal of the Hydraulics Division*, Vol. 99, No HY5, 713–727.

Kouwen, N. Li, R. T.E. 1980. Biomechanics of vegetative channel linings. *ASCE Journal of the Hydraulics Division*, Vol. 106, No HY6, 1085–1103.

Kouwen, N. 1992. Flow resistance in vegetated waterways. *ASCE Journal of Irrigation and Drainage Engineering*, No 5, 733–743.

Kutija, V., Minh Hong, H.T. 1996. A numerical model for assessing the additional resistance to flow introduced by flexible vegetation. *Journal of Hydraulic Research*, Vol 34, No. 1, 99–114.

López, F., García, M.H. 2001. Mean Flow and turbulence structure of open-channel flow trough non-emergent vegetation. *ASCE Journal of the Hydraulics Division*, Vol. 127, No HY5, 392–402.

Naot, D., Nezu, I., Nakagawa, H., 1996. Hydrodynamic behaviour of partly vegetated open channels. *ASCE Journal of Hydraulic Engineering*, Vol. 122, No 11, 625–633.

Nepf, H.M. 1999. Drag, turbulence and diffusion in flow through emergent vegetation. *Water Resources Res*. 35(2).

Nepf, H.M., Vivoni, E. 2000. Flow structure in depth-limited, vegetated flow. *Journal of Geophysical Research*, vol.105.

Nezu, I., Nakagawa, H. 1993. Turbulence in open channel flows. *Monograph series IAHR*. Ed. Balkema.

Nezu, I., Onitsuka, K. 2001. Turbulent structures in partly vegetated open-channel flows with LDA and PIV measurements. *Journal of Hydraulic Research*, Vol 39, No. 3. 629–642.

Petryk, S., Bosmajian, G. 1975. Analysis of flow through vegetation. *ASCE Journal of the Hydraulics Division*, Vol. 101, 871–884.

Raupach, M.R., Thom, A.S. 1981. Turbulence in and above plant canopies. *Annual Reviews Fluid Mechanics*. 97–129.

Rodi, W. Turbulence Modeling and their application in Hydraulics. *Monograph, IAHR*, Delft, The Netherlands.

Rouse, H. 1946. Elementary Mechanics of fluids. *Dover Publications*.

Tsujimoto, T., Kitamura, T. 1990. Velocity profile of flow in vegetated-bed channels. *KHL Progressive Report*, Hydraulic Laboratory, Kanazama University, Japan.

Velasco, D., Bateman A. 2003. An open channel flow experimental and theorical study of resistance and turbulent characterization over flexible vegetated linnings. *Flow, Turbulence and Combustion*, No 70. FtaC Special Issue: Fluxes and Structures in Fluids. Kluwer Publications, 69–88.

Wu, F.C., Shen, H.W., 1999. Variation of roughness coefficients for unsubmerged and submerged vegetation. *ASCE Journal of Hydraulic Engineering*, Vol. 125, No 9, 934–942.

River, Coastal and Estuarine Morphodynamics: RCEM 2005 – Parker & García (eds)
 ISBN 0 415 39270 5

Morphodynamics and hydraulics of vegetated river reaches: a case study on the Müggelspree in Germany

A. Sukhodolov & T. Sukhodolova
Institute of Freshwater Ecology and Inland Fisheries, Berlin, Germany

ABSTRACT: Submersible plants are abundant in fluvial systems and cause seasonal perturbation of flow patterns and flow transport functions. This study examines dynamics of vegetative cover in a lowland river reach during the vegetative period, subsequent hydraulic alterations of flow and sediment transport resulting in specific changes of riverbed morphology. The results of the study indicate that macro-forms of the riverbed morphology, particularly alternate bars, represent an important factor controlling formation of vegetative cover on the river reach. Reduced depths, sorted sediments and specific flow patterns over alternate bars provide favorable conditions for initiation of plants growths. Contrary, the enhanced depths and active bedload transport in the pools adjacent to the bars are viewed as limiting factors. When the vegetative cover is maximally developed it substantially modifies the riverbed morphology by substituting oblique transverse dunes with longitudinally drop-shaped forms. Morphodynamic changes of the riverbed increase bulk hydraulic roughness of the river channel, decrease water surface slope and persisted for the period of three-four months after vanishing of the vegetative cover. The study provides substantial amount of quantitative information that can form a factual basis for further development of coupled biological and hydraulic models predicting grows of submersible plants, their spatial distributions, effects on flow and riverbed morphology.

1 INTRODUCTION

Fluvial systems, and particularly lowland rivers, provide habitats for many species of freshwater macrophytes – a valuable ecosystem's component and significant factor for sedimentary processes due to mutual interactions of plants with the flow and riverbed deposits (Carpenter & Lodge, 1986; Cushing & Allan, 2001). Until recently freshwater macrophytes and their interactions with flow were studied predominantly separately: biologists were mostly concerned with the population dynamics of aquatic plants (Sculthorpe, 1967; Hynes, 1970; Spencer & Whitehand, 1993) while engineers were focused on the problem of flow resistance (Kouwen & Unny, 1973; Kouwen, 1990; Stephan & Gutknecht, 2002). During the last decade a tendency for multidisciplinary approach to the problem was clearly evidenced (Sand-Jensen & Pedersen, 1999; Nepf, 1999) and some other important aspect, for instance, interactions of flow and vegetation over moveable beds (Sand-Jensen, 1998; Tsujimoto, 1999) received more attention.

Sand-Jensen (1998) reported development of certain bedforms in mono-specific plant patches in lowland Danish streams. These bedforms were the result of sequential erosion and deposition perturbations induced by flow in response to flow's interaction with plants in the patches. Tsujimoto (1999) elaborated two-dimensional depth-averaged model for flow and sediment transport over vegetative patches and successfully compared model predictions to the laboratory experiments. Qualitatively these two studies exhibit noticeable agreement outlining significance of morphodynamic changes caused by vegetation and naturally inspire a question on how significant are vegetation-induced morphological changes in the river channel in respect to the other bedforms (ripples, dunes, bars), and what are the consequences of morphological changes for bulk characteristics of the flow. Although Tsujimoto (1999) has proposed a general scheme for flow-vegetation-sediment transport-river morphology interrelating river system, there is practically no study addressing those questions quantitatively for a river reach hitherto.

Development of modern methods in freshwater biology and experimental hydrodynamics and their success in river flow applications (Sukhodolov et al., 1998; Sukhodolova et al., 2004; Engelhardt et al., 2004) provide technical and methodological grounds for advanced field research. This study was designed to address, to a certain extant, the following problems: (1) to explore morphological structures and their seasonal dynamics in a river reach of a lowland river with submersible vegetation; (2) to estimate quantitatively the effect of vegetation-induced morphodynamic processes on the balk parameters of the flow; (3) to investigate spatio-temporal dynamics of the vegetative cover and its relationships to the morphology, sedimentology and hydraulics of the river channel and flow.

2 FIELD MEASUREMENTS

2.1 *River reach*

Field measurements for this study were carried out in a river reach of the Müggelspree River near the village Freienbrink, 10 km east of Berlin, Germany. In the study reach the channel of the river is straight with stone armored banks and with a bed covered by sands (mean diameter about 1 mm). The channel has a trapezoidal shape that measures 20 m by the bottom and is sided the banks sloping as 1:5. Hydraulic regime of the flow is regulated by the weir located approximately 20 km upstream the experimental reach. Maximum discharges of 15–20 m^3/s are normally supplied during the winter season while in summer the water supply is reduced to only 3.5–5 m^3/s. This regime supports mean velocities of flow in the range from 20 to 60 cm/s with mean depths ranging from 0.8 to 1.8 m. *Sagittaria Sagittifolia* (Figure 1) is a dominating macrophyte species inhabiting the study river reach (Figure 2). Submersible form of this plant has elongated leaves that grow up to 2 meters long and reproduce by vegetative propagation. Measurements were performed in the period of 2000-2004 and some results of those measurement studies have been already published (Sukhodolov et al., 1998; Fischer et al., 2003; Sukhodolova et al., 2004).

Figure 1. *Sagittaria Sagittifolia.*

Figure 2. A vegetative patch in the Müggelspree river.

2.2 *Measurement program*

Measurement program consisted of riverbed and sedimentary deposits surveys, measurements of hydraulic characteristics, and surveys of vegetative cover. Riverbed surveys consisted of cross- and longitudinal profiles. River depth was recorded with a spatial resolution of 1m for cross-sectional profiles and 5–10 cm for longitudinal profiles. Water depth was measured with portable echosounding system. Plane coordinates for the profiles were obtained with a total station EltaR55. Bottom sediments were sampled with a hand sampler at 24 locations uniformly distributed over the river reach, the coordinates of which were determined with a total station. Water levels were measured at three gauging stations equipped each with 4 piles. The vertical elevations of the piles were leveled with the total station enabling determination of water surface slope with an accuracy of 5–20% depending on the water level on the reach. Velocities were recorded with the acoustic Doppler velocimeters (ADV) manufactured by Sontek, Inc. The ADV probe was mounted on a special tripod enabling accurate vertical alignment of the device and preventing flow-induced vibration. Vegetative cover was mapped by observations in 6 permanent cross-sections uniformly distributed over the river reach and in one cross-section macrophytes were harvested for determination of biomass and plants morphology by divers.

Riverbed survey was performed on the reach of approximately 150 m long on 15 cross-sections uniformly distributed along the reach. Seven longitudinal profiles were measured to determine characteristics of the sand dunes. A spatial survey over a patch of vegetation was performed on fine spatial grid of 25 × 25 cm. Sediment samples were processed by standard sieving technique. Observations of the water levels and water surface slope were performed over the period May 2004–April 2005 approximately biweekly. Velocities were measured at 10 vertical profiles uniformly spaced across the river width. Eight velocity profiles included 5-point measurements uniformly distributed over the local river depth and two profiles comprised

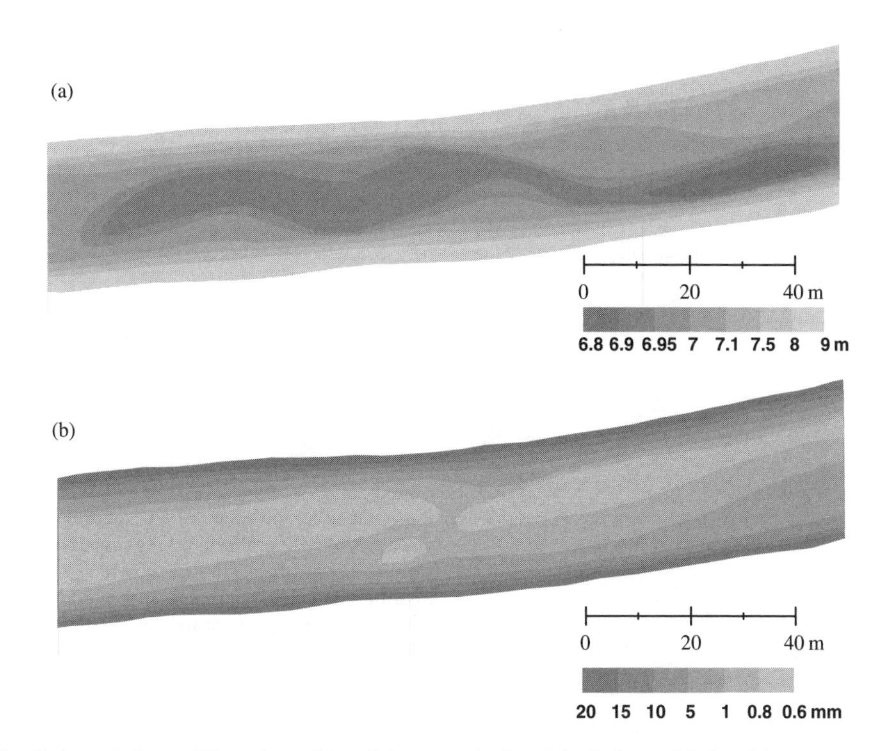

Figure 3. Bed morphology, arbitrary datum (a), and characteristic size of riverbed material, $d_{50\%}$ (b).

of 12 points each. The time of velocity recording was 3 minutes. Plants were harvested in a cross-section in the corridor of 1.5 m wide. The corridor was progressively dislocated upstream ensuring that a previous sampling would have no effect on the successive. Harvested plants were manually measured to obtain the maximum length of the plant, the maximum leaf width, the diameter of the steam near the roots, and the number of leaves. Wet weight of the plants was determined in field on portable scales, and later plants were dried for 24 hours at 100°C and weighted on laboratory scales.

3 RESULTS AND ANALYSIS

3.1 *Riverbed morphology*

A detailed map of river bed elevations was obtained from the data of riverbed survey, Figure 3a. The map displays the riverbed geometry typical for channels with alternate bars. The lengths of chess-board like located waves range from 1.5 to 3 flow widths and their heights range from 20 to 40 cm (Figure 3a). The alternate bars on this reach are relatively small and stable features as compared to a frequently reported in literature. This difference is due to a scarce supply of sediment trapped by the reservoirs located upstream.

The data of riverbed material sampling together with the data of near banks longitudinal profiling of banks riprap were generalized in the map of grain size distribution, Figure 3b. This map displays distribution of characteristic ($d_{50\%}$) sizes of sediments over the river reach. At the banks covered with a riprap the characteristic size was determined as a standard deviation of riprap protrusions into flow obtained from longitudinal profiles measured near banks. The data displayed in Figure 3 were collected in spring before the vegetative period.

Detailed longitudinal profiling was performed with a high spatial resolution (10 cm) to determine smaller scale morphological structures – sand dunes. Longitudinal profiling was carried out in spring at higher water levels that enabled active bedload transport. An example of the typical longitudinal profile is shown in Figure 4. An array of 6 parallel longitudinal profiles was used to map large-scale sand dunes in the horizontal plane, Figure 5. Large-scale sand dunes were $10 \div 12$ meter long, $15 \div 20$ cm high oblique (20–30° with normal to flow direction) features fully developed and mobile in spring before the vegetative period and were determined to decline during the vegetative period. The large-scale dunes were superimposed with small-scale sand waves and ripples (Figure 4). The sand waves were $2 \div 5$ meter long, $7 \div 10$ cm high,

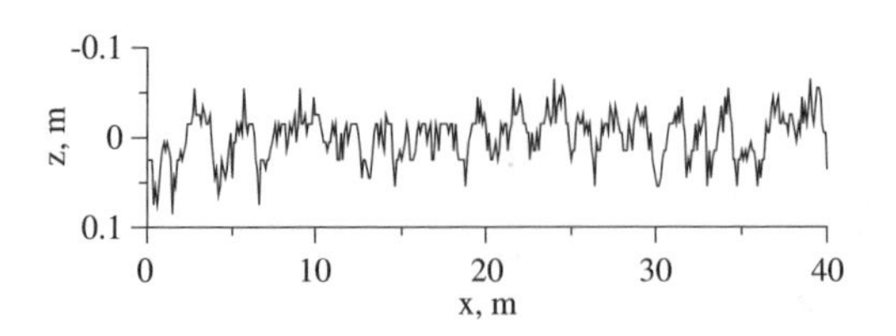

Figure 4. An example of a longitudinal profile of the riverbed, May 2002.

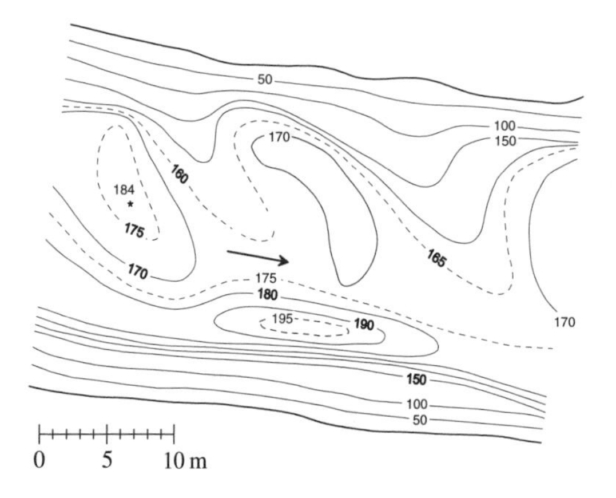

Figure 5. Oblique sand dunes at the riverbed, April 2001.

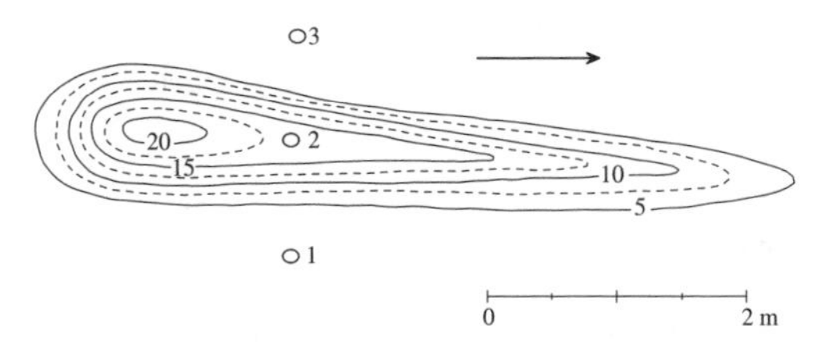

Figure 6. Bedform developed near a vegetative path (circles mark positions for riverbed material sampling, Figure 7).

approximately two dimensional and parallel to the flow direction. These bedforms were observed only in spring at water discharges of 10 m^3/s, and they were absent during the vegetative period.

A spatial survey of bed elevation in patches of vegetation was performed on a fine spatial grid determined with a total station. Data of these surveys were used to plot small scale maps displaying specific bedforms shaped as a water drop, Figure 6. The bedform shown in Figure 6 was developed near isolated small patch of vegetation that included 3 ÷ 4 plants growing closely to each other. The patches of longer longitudinal extant (10 ÷ 15 m) formed long (up to 20 ÷ 25 m) longitudinal ridges of 20 ÷ 25 cm in height. Such vegetation-induced bedforms were observed to develop over the sand wave of 10 ÷ 12 m long remaining from previous

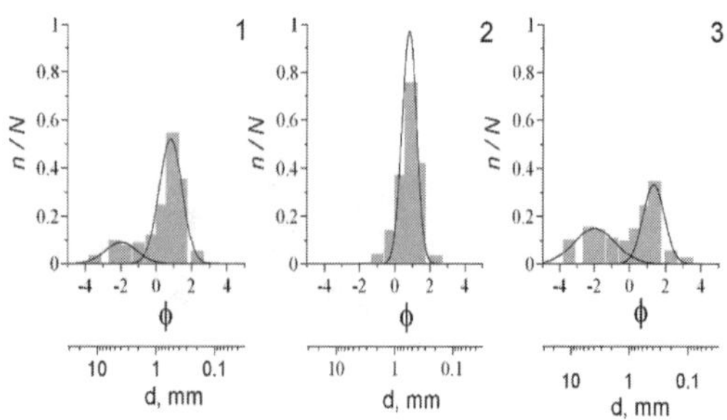

Figure 7. Distribution of grain size in the samples of riverbed material collected across the bedform in a vegetative patch (positions are indicated in Figure 6).

high-water periods. Gradually during the vegetative period those remnants of sand waves were vanishing and at the end of vegetative period (in October) were completely replaced by vegetation-induced bedforms like one shown in Figure 6 and by much larger longitudinal ridges. Samples of the riverbed material taken across a vegetative-induced bedform, Figures 6 and 7, indicate that such forms are formed by selective deposition of fine sediments, $d_{50\%} \approx 0.8$ mm due to shadowing effect of plants. Deflecting of flow over plants steams promotes scouring on the sides of the bedform where coarser ($d_{50\%} \approx 4$ mm) riverbed material predominates, Figure 7.

3.2 *Dynamics of the vegetative cover*

Dynamics of the vegetative cover can be explored with the successive maps of area occupied by plants and characteristics of plants (lengths of stems, number of leaves, stem diameter, biomass related to the unit area of the riverbed). To create such maps the data from regular measurements in the permanent cross-sections distributed uniformly along the experimental reach were used, Figure 8. Seasonal evolution of individual plants forming the vegetative patches is presented in Figure 9. Patches of vegetation initially appear as narrow (2 ÷ 3 m in width), elongated (length up to a 100 m) strips that widen and merge during the vegetative period. Noteworthy, the plants forming patches get their maximum characteristics quite fast – practically during the first three weeks and change non-significantly during the rest of vegetative period, Figure 9. At the same time the plants biomass grows quite monotonously and reaches its maximum at the end of summer. Some deviations are due to the spatial non-uniformity of the vegetative cover in the area of plants harvesting. Although the most characteristics of individual plants remain practically constant during the vegetation period, the diameter of plants near the root increases 1.5 ÷ 2 times. This characteristic indicates rigidity of the plants and therefore controls

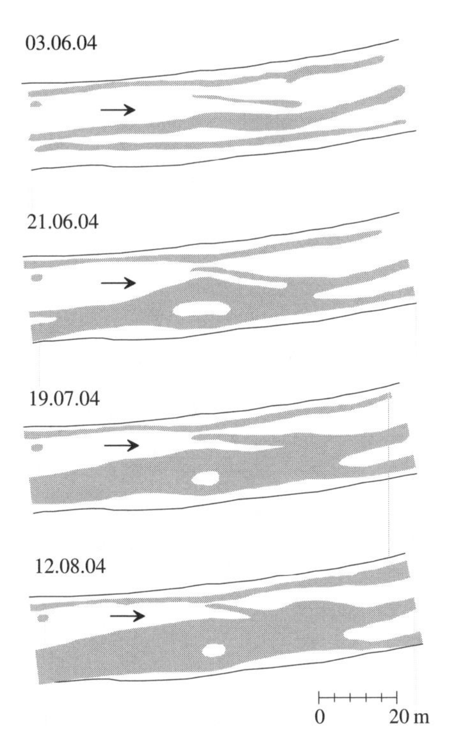

Figure 8. Grows of the area covered by plants during the vegetative period.

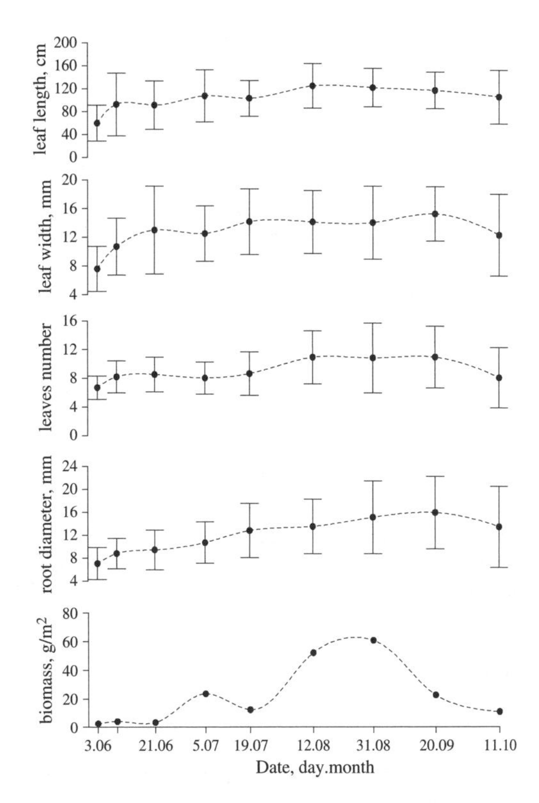

Figure 9. Characteristics of plants and their changes during the vegetative period.

bending angle – one of the most important parameters modeling vegetation-flow interactions.

3.3 *Flow patterns and hydraulic characteristics*

Velocity measurements taken in the cross sections on the study reach at various flow conditions were processed to obtain spatial distributions allowing comparison of flow patterns on the river reach with and without submersible vegetation. An example of such patterns is presented in Figure 10. Figure 10a displays mean velocity pattern measured in the vegetation-free channel. The pattern indicates a remarkable decrease of velocity at the distance of 1/3 from the right bank. This decrease is produced by the secondary circulation driven by turbulence anisotropy (Sukhodolov et al., 1998). Noteworthy, the zone of reduced velocities was colonized by submersible plants while the part of the channel located closer to the bank remain free of vegetation during the whole season, Figure 10b.

Integrally the effect of vegetation on flow characteristics and riverbed morphology appeared to be pronounced in the seasonal changes of the free surface slope (S), Figure 11. The data on Figure 11 are represented in dimensionless form where S_0 is the maximum slope corresponding to the bank full discharge,

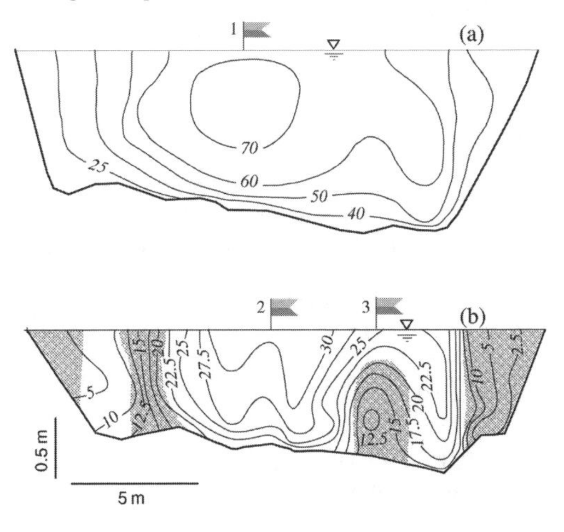

Figure 10. Downstream velocity isovels (cm/s) in the vegetation-free cross-section in May (a), and in the cross-section with vegetation (shaded areas) in July (b).

Δ is an averaged height of alternate bars and h is the mean water depth on the reach. Thus, the dimensionless slope S/S_0 is represented as a function of the relative roughness of the channel Δ/h assuming that

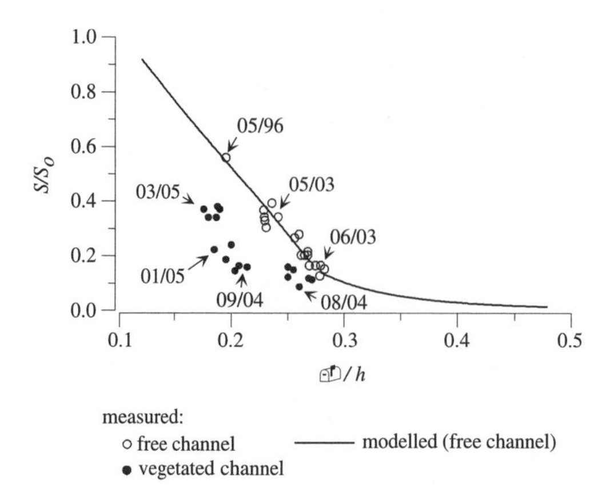

Figure 11. Seasonal changes of the free surface slope.

roughness on the reach scale is mainly controlled by alternate bars. The height of alternate bars was taken constant for this plot and equals 0.25 m. Figure 11 summarizes the data of different measurements and compares the measured data to the results of numerical computations. Measurements marked with open circles (May–June 2003) were completed in a channel which was purposely cleaned from plants. For this open channel flow without vegetation a series of numerical computations was performed with a steady-state model accounting for spatial non-uniformity of roughness characteristics. The model was calibrated for conditions of a vegetation free channel and assisted interpolation of the channel roughness function, Figure 11.

3.4 *Vertical structure of turbulence*

Distributions of turbulence primary shear stress over flow depth $-\overline{u'w'} = f(z/h)$ measured at verticals 1 and 2 (Figure 10) are shown in Figure 12a. The values of stress for the profile 1 were scaled with the local shear velocity $U_* = 3.82$ cm/s; the value determined fitting the linear regression as (Nezu & Nakagawa, 1993)

$$-\overline{u'w'} = U_*^2(1 - z/h), \quad (1)$$

where u', w' = velocity fluctuations in streamwise and vertical direction respectively, U_* = shear velocity, z = distance from the bed, and h = flow depth on the vertical. The data for vertical 2 were scaled with shear velocity 1.38 cm/s. The data measured in the flow without vegetation (profile 1) agrees with equation (1) at distances $z/h > 0.2$ and respectively the upper boundary of the roughness sublayer is located at distance 29 cm. In the vegetation-free path between macrophyte strips the location of roughness sublayer was 48 cm, respectively (Figure 12). Noteworthy that roughness

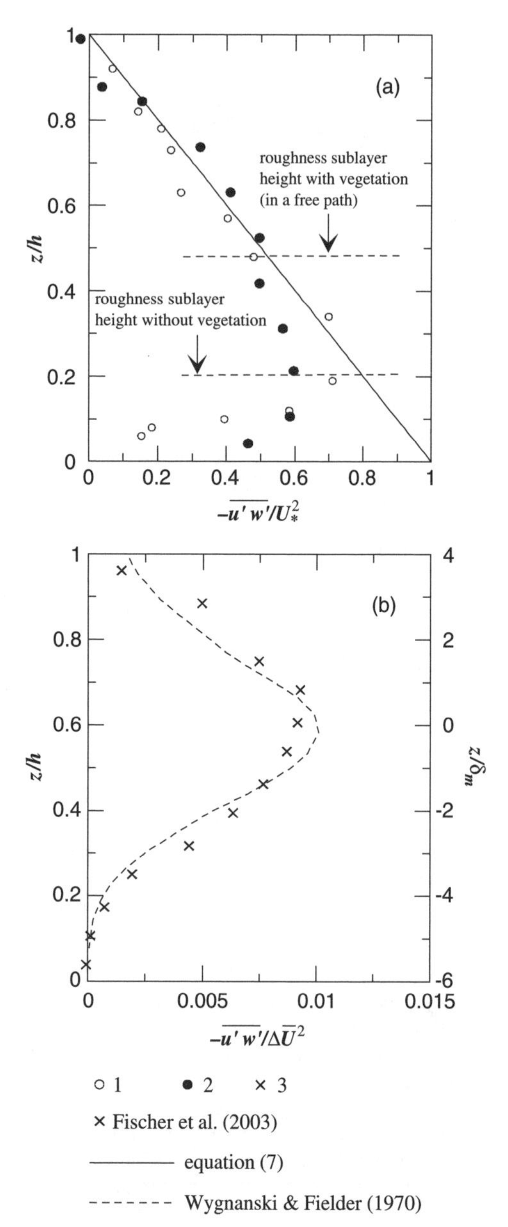

Figure 12. Distributions of turbulence shear stresses over the flow depth in vegetation-free channel and a vegetation-free path (a), and inside the stand of macrophytes (b).

height in vegetation-free channel agrees with the characteristic vertical scale of alternate bars and during the vegetative period its vale substantially increases.

Stresses measured in the macrophyte stand (profile 3, Figure 10b) were scaled with the mean velocity difference across the mixing layer $\Delta\overline{U} = 30$ cm/s to fit the similarity curve by Wygnanski & Fielder (1970).

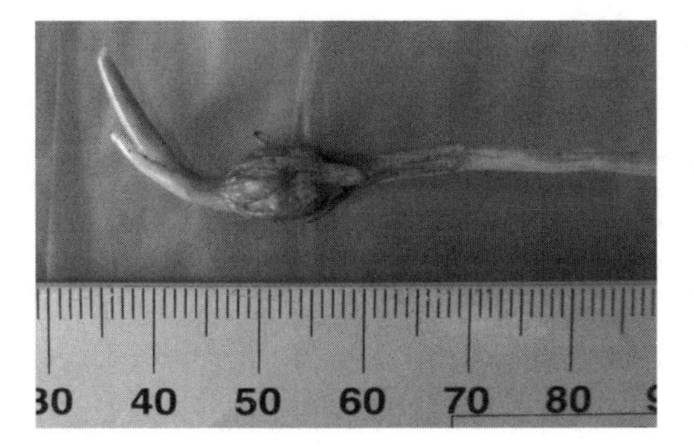

Figure 13. A tuber of *Sagittaria Sagittifolia*.

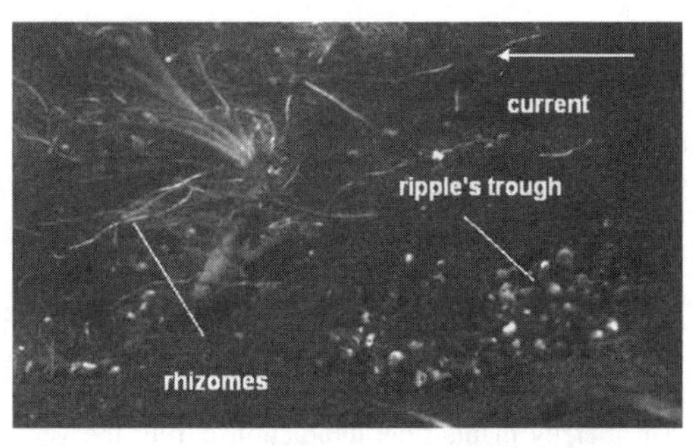

Figure 14. Underwater picture of a plant at the margin of a macrophyte strip.

Momentum thickness of the mixing layer δ_m equaled 11 cm, and its average center position was located at $z/h = 0.6$.

4 DISCUSSION

The results of field observations presented in this paper indicate some important details about such biophysical systems as the reach of a fluvial system subjected to the seasonal evolution of submersible plants. Comparison of riverbed morphology (Figure 3a), distributions of size for riverbed material (Figure 3b), and maps showing seasonal evolution of the vegetative cover (Figure 8) indicates that submersible form of *Sagittaria Sagittifolia* is quite sensitive to the characteristics of the riverbed substrate, specifically at the beginning of the vegetative period. The plants occupied the area of riverbed where mean grain size ranged from 0.8 to 5 mm. This selectivity can be explained by the physiological properties of the plants. In lowland rivers *Sagittaria Sagittifolia* reproduces by vegetative propagation and has rhizomes about 3 mm in diameter with tubers of $10 \div 15$ mm (Figure 13). Plants can find a stable substrate for rooting in case when a grain size is comparable in diameter with their rhizomes. Substrates with smaller grain size ($0.6 \div 0.8$ mm) are less stable and were observed in vegetation-free areas of channel (Figure 3b and Figure 8) to move in form of ripples ($20 \div 30$ cm long and $1.5 \div 3$ cm high) during the vegetative period. Rhizomes of plants subjected to the flow area with finer substrate, active bedload transport, and relatively stronger current bend downward (Figure 14), and therefore lateral colonization of macrophytes became limited.

Morphological structures of investigated river reach are represented by alternate bars ($30 \div 50$ m long, $20 \div 40$ cm high), oblique sand dunes ($10 \div 12$ m long, $15 \div 20$ cm high), nearly two-dimensional sand waves ($2 \div 5$ m long, $7 \div 10$ cm high), and by vegetation-induced forms of $20 \div 25$ m long, $4 \div 6$ m wide and $20 \div 25$ cm high. Observed alternate bars were quite stable and their contours change insignificantly during the year. Oblique dunes were appearing in the spring after passage of high water discharges and remained without movement till midsummer when they were gradually distorted by other bedforms, particularly by vegetation-induced forms. 2-D sand waves were relatively short time living structures determined only at high discharges (>10 m^3/s). Ripples were observed on the riverbed at areas covered with fine sands ($0.6 \div 0.8$ mm) during the vegetative period and at relatively low water discharges (less than 5 m^3/s). Thus, for investigated river reach, vegetation-induced morphological structures attained significant sizes and appeared to be comparable in their extant with the alternate bars. Moreover, distribution of plants patches (Figure 8) coincided with the elevated parts of alternate bars (Figure 3a) and plants were absent or their number was significantly reduced in the pools adjacent to bars. Apart of substrate peculiarities, the elevated parts of the bars naturally enable larger portion of sunlight for plants colonizing bars rather than pools.

Unlike sand dunes and waves which are created by longitudinal sequences of erosion-deposition, the vegetation-induced morphological structures are mainly formed by lateral erosion-deposition sequences. Therefore secondary circulations driven by turbulence anisotropy in the river flow can have especially important role for formation of vegetation patches and associated morphological structures. Unfortunately such secondary circulations are investigated quite scarcely (Nezu and Nakagawa, 1993; Sukhodolov et al., 1998) and there is no data demonstrating the effect of vegetation on the lateral structure of flow. Though this paper presents some illustrations of the secondary currents effect (Figure 10), further research is necessary to provide quantitative information on the lateral components of the turbulence stresses tensor and local transport capabilities governing development of morphological structures.

Aggravation of alternate bars because of deposition of riverbed material beneath the vegetative canopies causes an increase in the height of the roughness sublayer even in the non-vegetated areas of the flow (Figure 12) and consequent reduction of the free surface slopes (Figure 11). Decreased values of slope during the winter season most probably indicate that depositions of the riverbed material formed during the vegetative period disappear/transform gradually. In early spring period with increasing discharges vegetation-induced structures almost vanish and the channel morphology attains the vegetation-free state. These observations indicate that influence of macrophytes on flow hydraulics in lowland rivers is more than merrily plants-flow interaction during the vegetative period and due to complex morphodynamic processes the effect extends almost over the whole year cycle.

Spatial heterogeneity of vegetative covers in natural streams provides possibilities for the development of non-uniform distributions of turbulence characteristics and as a consequence promotes conditions for secondary flow circulation shown to govern morphodynamical processes. Unfortunately due to the serious difficulties in determining of secondary currents in natural conditions they were not examined in the frame of present study and represent the topic of the ongoing research.

5 CONCLUSIONS

Spatial patterns of vegetative cover formed by submersible freshwater macrophyte *Sagittaria Sagittifolia* in the initial stage of development depend on the riverbed morphology, morphodynamic processes and associated distributions of riverbed material.

Complex plant-flow interactions lead to significant modifications of morphodynamic processes and hydraulic characteristics of flow. In the study reach the interactions resulted in the development of specific morphological structures of the size comparable with dimensions of alternate bars – the larger-scale structures on the reach.

Superposition of alternate bars and vegetation-induced bedforms increases the thickness of the roughness sublayer up to two times and significantly reduces slopes of the free surface.

Morphological structures induced by the plants-flow interactions persist quite a long period after vegetative period and impose considerable effect on flow hydraulics. Therefore, the influence of macrophytes on flow hydraulics in lowland rivers is more than merrily plants-flow interaction during the vegetative period and due to complex morphodynamic processes the effect extends almost over the whole year cycle.

ACKNOWLEDGEMENTS

Mischael Thiele, Heinz Bungartz, Christof Engelhardt, Helmut Fischer, Nicholas Nikolaevich, and Werner Sauer are thanked for their assistance with field surveys. We also wish to express our appreciation to Jan Köhler for helpful discussions, friendly encouragement and financial support in the frame of EU program BUFER. The research was supported in parts by Deutsche Forschungsgemeinschaft (BU 1442/1-1) and by Collaborative Programme of NATO (ESP.NR.EV 981608).

REFERENCES

Carpenter, S.R., & Lodge, D.M. 1986. Effects of submersed macrophytes on ecosystem processes. *Aquatic Botany* 26: 341–370.

Cushing, C.E., & Allan, D. 2001. *Streams. Their Ecology and Life.* San Diego: Academic Press.

Engelhardt, C., Krüger, A., Sukhodolov, A.N., & Nicklisch, A. 2004. A study of phytoplankton spatial distributions, flow structure and characteristics of mixing in a river reach with groynes. *Journal of Plankton Research* 26: 1351–1366.

Fischer, H., Sukhodolov, A., Wilczek, S., & Engelhardt, C. 2003. Effects of flow dynamics and sediment movement on microbial activity in a lowland river. *River Research and Applications* 19: 473–482.

Hynes, H.B.N. 1970. *Ecology of Running Waters.* Liverpool: Liverpool University Press.

Kouwen, N., & Unny, T.E. 1973. Flexible roughness in open channels. *Journal of the Hydraul. Div.* 99: 713–728.

Kouwen, N. 1990. Modern approach to design of grassed channels. *Journal of Irrigation and Drainage Engineering* 118: 733–743.

Nepf, H.M. 1999. Drag, turbulence, and diffusion in flow through emergent vegetation. *Water Resources Research* 35: 479–489.

Nezu, I., & Nakagawa, H. 1993. *Turbulence in Open Channel Flows.* Rotterdam: Balkema.

Sand-Jensen, K. 1998. Influence of submerged macrophytes on sediment composition and near-bed flow in lowland streams. *Freshwater Biology* 39: 663–679.

Sand-Jensen, K., & Pedersen, O. 1999. Velocity gradients and turbulence around macrophyte stands in streams. *Freshwater Biology* 42: 315–328.

Sculthorpe, C.D. 1967. *The biology of aquatic vascular plants.* London: Edward Arnold Ltd.

Spencer, D.F., & Whitehand, L.C. 1993. Experimental design and analysis in field studies of aquatic vegetation. *Lake and Reserv. Manage.* 7: 165–174.

Stephan, U., & Gutknecht, D. 2002. Hydraulic resistance of submerged flexible vegetation. *Journal of Hydrology* 269: 27–43.

Sukhodolov, A.N., Thiele, M., & Bungartz, H. 1998. Turbulence structure in a river reach with sand bed. *Water Resorces Research* 34: 1317–1334.

Sukhodolova, T.A., Sukhodolov, A., & Engelhardt, C. 2004. A study of turbulent flow structure in a partly vegetated river reach. *Proceedings of the second international conference on fluvial hydraulics River Flow 2004* Vol. 1: 469–478.

Tsujimoto, T. 1999. Fluvial processes in streams with vegetation. *Journal of Hydraulic Research* 37: 789–803.

Wygnanski, I., & Fielder, H.E. 1970. The two dimensional mixing region. *Journal Fluid Mech.* 41: 327–361.

Numerical modelling of river morphodynamics

River, Coastal and Estuarine Morphodynamics: RCEM 2005 – Parker & García (eds)
© 2006 Taylor & Francis Group, London, ISBN 0 415 39270 5

The morphodynamics of super- and transcritical flow

Y. Zech
Université catholique de Louvain, Dept Civil & Environmental Engineering, Louvain-la-Neuve, Belgium

S. Soares-Frazão
Université catholique de Louvain, Dept Civil & Environmental Engineering and Fonds National de la Recherche Scientifique, Belgium

B. Spinewine, M. Bellal & C. Savary
Université catholique de Louvain, Dept Civil & Environmental Engineering, Louvain-la-Neuve, Belgium

ABSTRACT: The fluvial regime prevails in most of the alluvial rivers, during most of the time. However, drastic alterations in river morphology happen during floods: the velocities increase, with possible appearance of supercritical regime. Intense erosion and deposition may disrupt the river feature forming a sequence of breaks in bottom slope. The common Saint-Venant–Exner approach may fail in reproducing such intense transport and also in properly handling the discontinuities in regime. The influence of the hydraulic regime on the solid discharge is analyzed and a two-layer model is proposed, able to cope with intense transport or severe transients. Some results are presented as well in uniform conditions as in dam-break modeling. The model is also promising in modeling the generation of antidunes. Moreover the two-layer approach helps to correctly pose the boundary conditions in morphodynamics, above all if transcritical flow is expected.

1 INTRODUCTION

Fluvial hydraulics generally focuses on subcritical flow. Indeed, this regime, which is often identified as the *fluvial* regime, prevails in most of the alluvial rivers, during most of the time.

However, the most drastic alterations in river morphology happen during floods and are thus associated with sudden changes in discharge and water level. Moreover, intense erosion and deposition may disrupt the river features forming a sequence of breaks in bottom slope. Regime discontinuities may appear, such as hydraulic jumps and control sections.

In order to predict conveniently the river morphological evolution, it is thus required to model supercritical flow as well as subcritical. Also transitions from one regime to the other have to be captured. The present paper aims to present some specific difficulties in modeling super- and transcritical flow on mobile bed, and to propose a two-layer approach, addressing some of those problems.

Most of the available models describing the geomorphic evolution of movable-bed rivers in presence of transient flow rely on the Saint-Venant equations for the hydrodynamic flow routing (see for example: Cunge et al. 1980):

$$\frac{\partial z_b}{\partial t}+\frac{\partial h}{\partial t}+\frac{\partial q}{\partial x}=0 \tag{1}$$

$$\frac{\partial q}{\partial t}+\frac{\partial}{\partial x}\left(\frac{q^2}{h}\right)+g\,h\frac{\partial h}{\partial x}+g\,h\frac{\partial z_b}{\partial x}+g\,h\,S_f=0 \tag{2}$$

where z_b is the bottom level, h is the water depth, q the discharge per unit width and S_f the friction slope (for sake of simplicity the developments are made for a unit-width channel).

The rate of sediment movement is described by the Exner equation:

$$(1-\varepsilon_0)\frac{\partial z_b}{\partial t}+\frac{\partial q_s}{\partial x}=0 \tag{3}$$

where q_s is the solid discharge by unit width and ε_0 the porosity.

The three main variables of the system (1–3) are h, q and z_b. Two closure equations are thus needed to determine S_f and q_s.

Knowing that the flow regime is characterized by the Froude number, the question to be answered is now: how will a sub- or supercritical regime affect the system solutions and thus the modeling of river morphology?

The increase of the mean velocity in supercritical regime induces three consequences that cannot be disregarded in modeling:

1. The solid transport becomes more intense.
2. In supercritical flow, some flow instabilities may appear; the bed surface enters in phase with the water profile, possibly producing specific bed forms known as antidunes.
3. The change in sign of the characteristics associated to the equations may complicate the accurate expression of boundary conditions.

According to this, the paper is organized in the following way. Section 2 focuses on the influence of the hydraulic regime on the solid discharge q_s, and the limitations of common approaches in high-velocity flows. Section 3 describes a two-layer model able to cope with intense transport or severe transients. Then, section 4 concentrates on the flow resistance and the associated friction slope S_f, attempting to take into account the influence of the bed forms and their modeling. Finally, section 5 points out the difficulty to accurately pose the boundary conditions in morphodynamics, above all if transcritical flow is expected.

2 REGIME AND SOLID TRANSPORT

2.1 *Regime and grain mobilization*

Bed-load transport models generally rely on the interaction between hydrodynamic forces (typically proportional to the square of the mean velocity u^2 and of the characteristic grain diameter d^2) and grain resistance (proportional to d^3). The rate between acting and resisting forces is the mobility number (or Shields number) τ_{*b}:

$$\tau_{*b} = \frac{\tau_b}{(\gamma_s - \gamma)d} = \frac{\rho u_*^2}{(\gamma_s - \gamma)d} \tag{4}$$

where γ_s and γ are the volumetric weight of grain and water, respectively, τ_b and u_* are the bottom shear stress and shear velocity.

The dimensionless particle parameter d_* represents the grain diameter:

$$d_* = d\left(\frac{g(s-1)}{\nu^2}\right)^{1/3} \tag{5}$$

where $s = \rho_s/\rho$ is the grain specific density, ρ_s and ρ being the volumetric mass of grain and water, respectively, and ν is the kinematic viscosity.

In this model of equilibrium between acting and resisting forces on the grain, the Froude number does not play any role and the common formulations of the bed-load transport ignore its influence. In a Shields-van Rijn diagram (van Rijn 1984) where the mobility number τ_{*b} is represented as a function of the particle parameter d_* (Fig. 1), the change in regime from subcritical (Froude number $\mathbf{Fr} < 1$) to supercritical ($\mathbf{Fr} > 1$) does not induce any critical change in the grain mobility.

2.2 *Regime and solid transport*

The same observation may be done in Figure 2, representing the dimensionless solid discharge

$$q_{s*} = \frac{q_s}{\sqrt{g(s-1)d^3}} \tag{6}$$

as a function of the Froude number, for some popular transport formulae in uniform flow conditions.

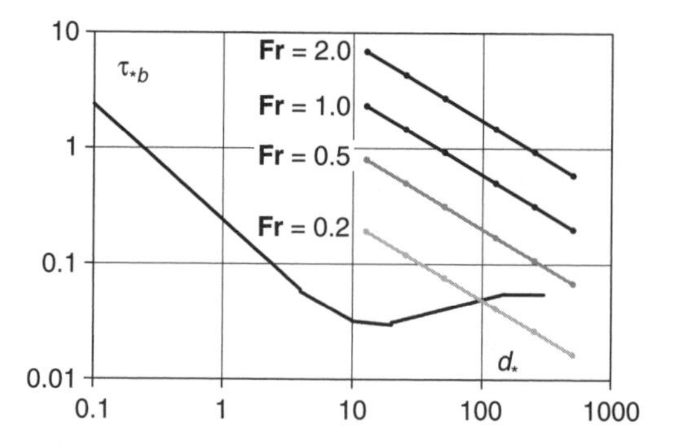

Figure 1. Shields-van Rijn diagram – $d = 0.5$, 1, 2, 5, 10 mm; $q = 10\,m^3/s/m$. The curve corresponds to the movement initiation, according to van Rijn (1984).

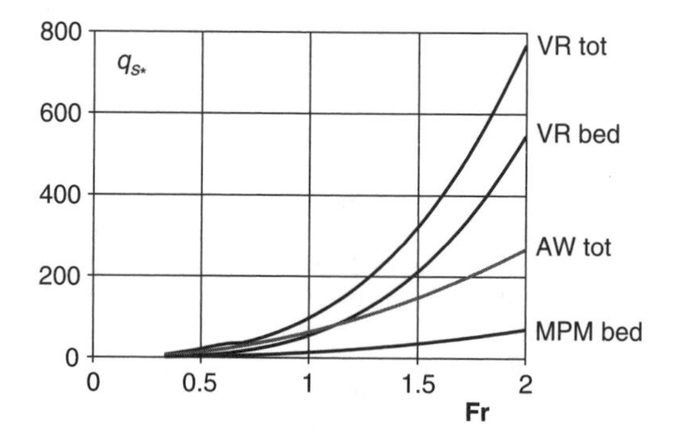

Figure 2. Dimensionless solid discharge in uniform flow conditions – $d = 1$ mm; $q = 10\,m^3/s/m$; VR = van Rijn formula (bed load and total load), AW = Ackers-White formula (total load), MPM = Meyer-Peter–Müller formula (bed load).

Nothing special occurs at the critical stage, except the fact that the solid transport seriously increases in the supercritical domain. Also the divergence between various formulations is still greater for large Froude numbers.

Supercritical flow implies large velocities inducing intense solid transport. This can turn in debris flow if the interaction between grains (frictions and collisions) exceeds the interaction between individual grains and liquid flow (drag and submerged weight). Even if such a debris flow does not form, the transport becomes intense and the grain mobilization involves more complex mechanism.

For example, in the case of a sudden increase in bed slope, the upstream subcritical conditions lead to a limited transport while the downstream supercritical reach is intensely mobilized. The discontinuity in solid discharge results in a rather fast regression of the nick-point. That means that, at a given section, the amount of mobilized sediment suddenly increases. In severe transients, such as dam-break waves, this phenomenon is further amplified.

The initiation of the movement of such an amount of granular material implies momentum exchanges between water and sediments, which can be taken into consideration only by a two-layer model. Such a model is presented in the next section.

3 TWO-LAYER MODEL FOR INTENSE TRANSPORT AND SEVERE TRANSIENTS

3.1 *The two-layer model*

The two-layer model presented here, based on the pioneering work by Capart (2000) is representative of flow conditions in which the sediments are moving in a dense layer, with negligible suspension in the upper layer. The flow is thus represented by the following layers: (a) the upper layer consisting in clear water of depth h_w, (b) the moving sediment layer h_s and (c) the fixed-bed layer, having the bed level z_b as upper limit (Fig. 3).

In the original model, the concentration of sediment was assumed to be constant ($C_s = C_b$) and the upper

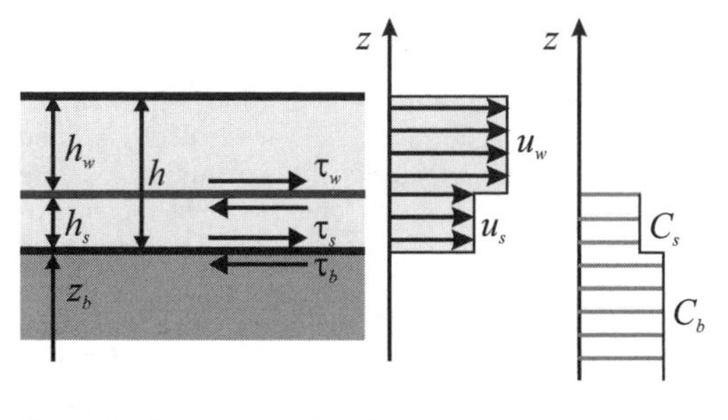

Figure 3. Two-layer model scheme.

part of the water/sediment mixture (h_s) was assumed to be in movement at the same uniform velocity as the clear-water layer ($u_s = u_w = u$).

Some improvements were brought to the model. New degrees of freedom were given to the velocities: $u_s \neq u_w$ (Capart & Young 2002) and to the concentrations: $C_s \neq C_b$ (Spinewine 2003, Spinewine & Zech 2002a) between the three layers. The distinction between the concentration of the bottom and the moving layer ($C_s \neq C_b$) accounts for the expansion required by the grain mobilization.

Defining the erosion rate (negative in the case of deposition) as

$$\frac{\partial z_b}{\partial t} = -e_b \tag{7a}$$

it is possible to express the continuity of both the water and the water/sediment mixture layers:

$$\frac{\partial h_w}{\partial t} + \frac{\partial}{\partial x}(h_w u_w) = -e_b \frac{C_b - C_s}{C_s} \tag{7b}$$

$$\frac{\partial h_s}{\partial t} + \frac{\partial}{\partial x}(h_s u_s) = e_b \frac{C_b}{C_s} \tag{7c}$$

Using a shallow-water approach, with the assumption that the pressure distribution is hydrostatic, the momentum conservation of those moving layers writes:

$$\frac{\partial (h_w u_w)}{\partial t} + \frac{\partial}{\partial x}\left(h_w u_w^2 + \frac{g h_w^2}{2}\right) + g h_w \frac{\partial}{\partial x}(z_b + h_s) = -\frac{\tau_w}{\rho_w} - e_b \frac{C_b - C_s}{C_s} u_1 \tag{7d}$$

$$\frac{\partial (h_s u_s)}{\partial t} + \frac{\partial}{\partial x}\left(h_s u_s^2 + \frac{g h_s^2}{2}\right) + g h_s \left(\frac{\partial z_b}{\partial x} + \frac{\rho_w}{\rho'_s}\frac{\partial h_w}{\partial x}\right) = \frac{\tau_w}{\rho'_s} - \frac{\tau_b}{\rho'_s} + \frac{\rho_w}{\rho'_s}\frac{C_b - C_s}{C_s} e_b u_1 \tag{7e}$$

where $\rho'_s = \rho_w(1 - C_s) + \rho_s C_s$ is the volumetric mass of the moving mixture layer h_s.

The asymmetrical exchange of momentum in case of erosion or deposition is taken into consideration in the following way: $u_1 = u_w$ if $e_b > 0$ while $u_1 = u_s$ if $e_b < 0$. For details about the above equations see Spinewine (2005).

In this model the five dependent variables are h_w, h_s, z_b, $(h_w u_w)$ and $(h_s u_s)$. Closure equations are required for e_b, τ_w, τ_s and τ_b. The erosion rate e_b (negative in the case of deposition) results from the inequality between the shear stresses τ_s and τ_b on both faces of the bed interface:

$$e_b = \frac{1}{\rho'_b |u_s|}(\tau_s - \tau_b) \tag{8}$$

where $\rho'_b = \rho_w(1 - C_b) + \rho_s C_b$ is the volumetric mass of the bottom layer at rest. The shear stresses τ_w and τ_s are evaluated from the turbulent friction:

$$\tau_w = \rho_w C_{fw} (u_w - u_s)^2 \qquad (9)$$

$$\tau_s = \rho'_s C_{fs} u_s^2 \qquad (10)$$

while τ_b depends on the critical shear stress c, and the internal friction angle ϕ:

$$\tau_b = c + \sigma' \tan\phi = c + (\rho'_s - \rho_w) g h_s \tan\phi \qquad (11)$$

where σ' is the effective pressure.

3.2 *Application to uniform flow*

In uniform steady flow conditions, (7d) and (7e) simplify since the partial derivatives $\partial/\partial t = 0$ and $\partial/\partial x = 0$, except $\partial z_b/\partial x = -S_0$:

$$- g h_w \frac{\partial z_b}{\partial x} = g h_w S_0 = \frac{\tau_w}{\rho'_w} \qquad (12)$$

$$- g h_s \frac{\partial z_b}{\partial x} = g h_s S_0 = \frac{\tau_b}{\rho'_s} - \frac{\tau_w}{\rho'_s} \qquad (13)$$

The solid transport is in equilibrium, which means that neither erosion nor deposition occurs; thus (8) reduces to the condition

$$\tau_b = \tau_s \qquad (14)$$

If we now define the unit liquid and solid discharges

$$q_w = u_w h_w \qquad (15)$$

$$q_s = u_s h_s \qquad (16)$$

equations (9–16) form a set of 8 uniform-flow equations linking 12 parameters: S_0, q_w, q_s, h_w, h_s, u_w, u_s, τ_w, τ_s, τ_b, C_{fw} and C_{fs}. In uniform-flow condition, if four of them are known, the others may be deduced.

It is thus possible to calibrate the two friction coefficients, C_{fw} and C_{fs}, either from measurement or from estimation of only four parameters.

For example, for a given discharge q_w in a channel of slope S_0 with mobile bed whose characteristic diameter is d, the Manning roughness coefficient n may be evaluated from the Strickler formula

$$K = \frac{1}{n} = \frac{26}{d^{1/6}} \qquad (17)$$

and the water depth h_w may be deduced from the Manning uniform-flow relation:

$$q_w = \frac{1}{n} h_w^{5/3} S_0^{1/2} \qquad (18)$$

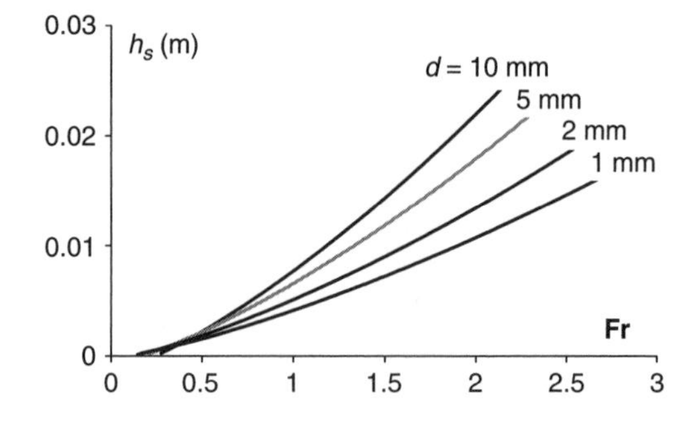

Figure 4. Uniform flow ($q_w = 10\,\text{m}^3/\text{s/m}$): relationship between sediment-layer depth and Froude number (two-layer model).

If we now compute the solid discharge q_s from any empirical formula, we obtain a set of four parameters (the water depth h_w, the discharges q_w and q_s and the slope S_0) required to calibrate C_{fw} and C_{fs}.

Once those coefficients are calibrated, one may investigate the behavior of the two-layer model in other flow conditions by considering various combinations of q_w and S_0. For each combination, it is possible to compute q_s, h_w, h_s, u_w, u_s, τ_w, τ_s and τ_b.

To analyze the two-layer model behavior, an example was computed with the following data: a unit water discharge $q_w = 10\,\text{m}^3/\text{s/m}$ in a channel, whose bed is constituted of uniform grains of diameter d varying from 1 to 10 mm; the density $\rho_s/\rho_w = 2.65$, and the internal friction angle $\phi = 30°$.

For each diameter, the calibration of C_{fw} and C_{fs} was done for a bottom slope S_0 corresponding to the value of the Froude number $\mathbf{Fr} = 1$. The corresponding transport q_s and critical shear stress c are evaluated from Meyer-Peter–Müller formula.

Then, for each diameter a range of bottom slopes is explored. Using the set of equations (9–16), it is possible to compute the various parameters as a function of the Froude number

$$\mathbf{Fr} = \frac{u_w}{\sqrt{g h_w}} \qquad (19)$$

From Figures 4 and 5, it clearly appears that as well the sediment-layer depth h_s as its velocity u_s increase with increasing Froude number (thus with increasing bottom slope S_0, decreasing water depth h_w, and increasing water velocity u_w). The resulting increase of solid discharge q_s is due for a part to the increase of sediment velocity u_s but mainly to the development in depth of the sediment layer.

Computing the solid discharge from (16) yields a result very close to the Meyer-Peter–Müller formula (Fig. 6). The latter formula was used to calibrate the two-layer model for $\mathbf{Fr} = 1$. Then the way to compute

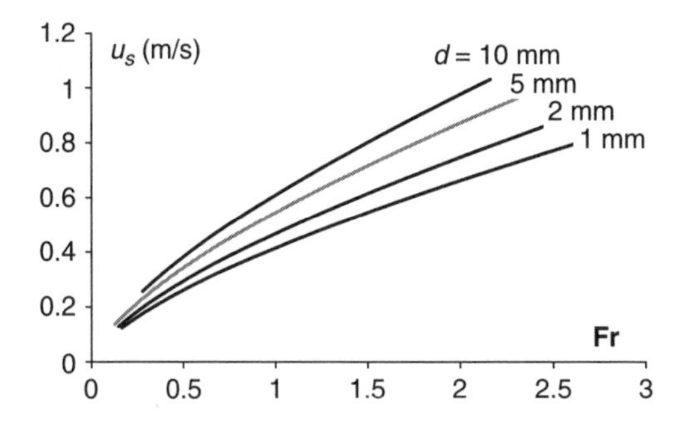

Figure 5. Uniform flow ($q_w = 10\,\text{m}^3/\text{s/m}$): relationship between sediment-layer velocity and Froude number (two-layer model).

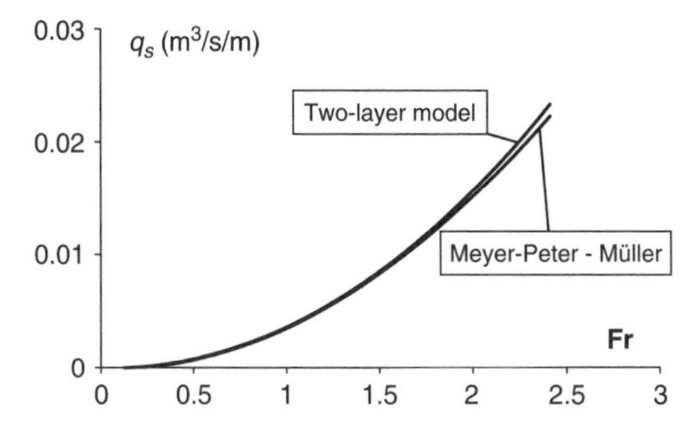

Figure 6. Uniform flow: solid discharge: comparison between two-layer model and Meyer-Peter–Müller formula ($d = 5\,\text{mm}$, $c = 3.8\,\text{N/m}^2$, $q_w = 10\,\text{m}^3/\text{s/m}$, $C_{fw} = 0.00245$, $C_{fs} = 0.0735$).

q_s for other Froude numbers was completely independent for the two curves in Figure 6. That means that the two-layer model and its closure equations (9–11) are at least consistent with the predictions of Meyer-Peter and Müller. This consistency is confirmed for other diameters ($d = 1$, 2 and 10 mm) and other unit discharges ($q_w = 1\,\text{m}^3/\text{s/m}$), at least if the calibration conditions are not too close to the threshold $\tau_{b,crit}$.

3.3 *Application to dam-break wave*

The two-layer model is also able to reproduce the sediment response to a severe transient flow, for instance a dam-break induced wave.

This kind of wave is typically transcritical, the control section being theoretically located at the initial dam section. The part of the wave downstream from this section is thus supercritical with a so intense solid transport that, during the first instants, the whole depth may be constituted of a water/sediment mixture, very similar to a debris flow.

As an example, the model was compared to experiments carried out at the *Université catholique de Louvain* (Spinewine and Zech, 2002b). The set-up is sketched in Figure 7 with the following characteristic dimensions: a water layer of depth $h_{w,0} = 0.10\,\text{m}$ in the reservoir, and a fully saturated bed of thickness $h_{b,0} = 0.05\,\text{m}$. The bed material consisted of cylindrical PVC pellets with an equivalent diameter of 3.5 mm and a density of 1.54, deposited with a bulk concentration of about 60%.

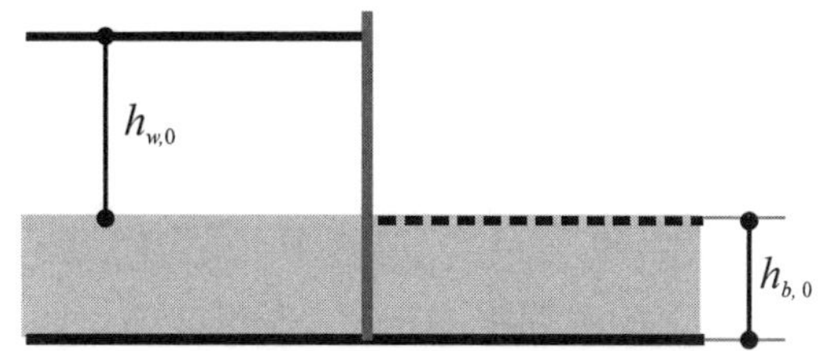

Figure 7. Scheme of the dam-break experiment (UCL).

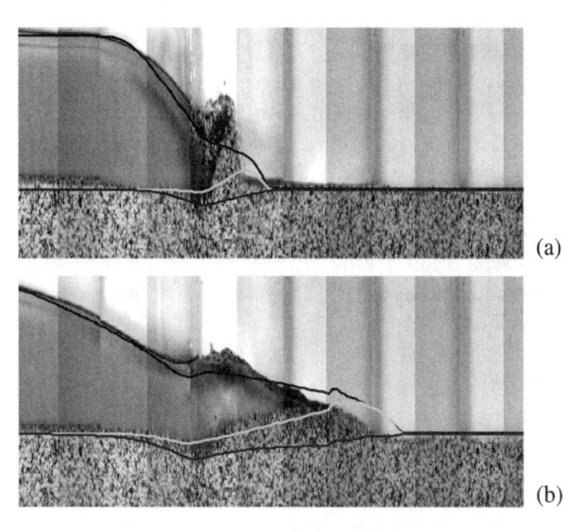

Figure 8. Comparison between experiments and numerical results at times (a) $t = 0.2$ s and (b) $t = 0.6$ s.

Figure 8 presents a comparison between experimental observation and the model presented above. The first picture (Fig. 8a: time $t = 0.2$ s) clearly evidences the limitation of the shallow-water model for the earlier stage of the dam-break: some features linked to the vertical movements are missed, like the splash effect on water and sediment. The erosion depth is slightly underestimated, partly due to the movement of the gate.

All those phenomena induce energy dissipation that is not accounted for in the model, what explains that the modeled front has some advance compared to the actual one.

Looking at the second picture (Fig. 8b: time $t = 0.6$ s), it appears that some characters of the movement are really well modeled, such as the jump at the water surface, the scouring at the dam location, and the moving-layer thickness. The modeled front is yet ahead but this advance is the same as at the former time, which implies that the front celerity is correctly estimated.

4 SUPERCRITICAL FLOW AND ANTIDUNES

Even though the solid transport is not directly influenced by the flow regime in the sense that the mechanism is the same for super- as for subcritical flow, appearance of bed forms may alter this first conclusion.

Indeed, bed forms induce additional resistance to the flow, which distracts a part of the energy available for sediment mobilization. This may be modeled by a fictitious partition of the bottom shear stress τ_b in a part τ_b' relative to the grain resistance and a part τ_b'', which corresponds to the form resistance. Various approaches may be used, introducing a fictitious separation of either the hydraulic radius R_b (e.g. Einstein and Barbarossa, 1952) or the friction slope S_f (Engelund, 1966):

$$\tau_b = \gamma R_b S_f = \tau_b' + \tau_b'' = \gamma (R_b' + R_b'') S_f = \gamma R_b (S_f' + S_f'') \tag{20}$$

or considering bed forms as large-scale grains with equivalent roughness (Brownlie, 1983).

Appropriate mobility numbers are associated to the grain and the form shear stresses, respectively:

$$\tau_{*b}' = \frac{\tau_b'}{(\gamma_s - \gamma) d} = \frac{\gamma R_b' S_f}{(\gamma_s - \gamma) d} = \frac{\gamma R_b S_f'}{(\gamma_s - \gamma) d} \tag{21a}$$

$$\tau_{*b}'' = \frac{\tau_b''}{(\gamma_s - \gamma) d} = \frac{\gamma R_b'' S_f}{(\gamma_s - \gamma) d} = \frac{\gamma R_b S_f''}{(\gamma_s - \gamma) d} \tag{21b}$$

The velocity distribution along a vertical is governed by the grain roughness, in such a way that the depth-averaged velocity U writes (see e.g. Chang, 1992):

$$\frac{U}{\sqrt{g R_b' S_f}} = A + \frac{1}{\kappa} \ln\left(\frac{R_b'}{k_s}\right) \tag{22}$$

where A is a constant ($A = 6.0$ for Engelund and Hansen 1967), κ the von Kármán constant and k_s is the Nikuradse's sand roughness, whose value is linked to a characteristic grain diameter (for Engelund and Hansen, $k_s = 2\,d_{65}$).

4.1 *Empirical approach*

The separation between grain and bed-form shear stresses generally relies on statistical inference from laboratory or field data. For example such data, fitted by Engelund and Hansen (1967) and extended by Brownlie (1983), give the following relationship for the various flow regimes (Fig. 9):

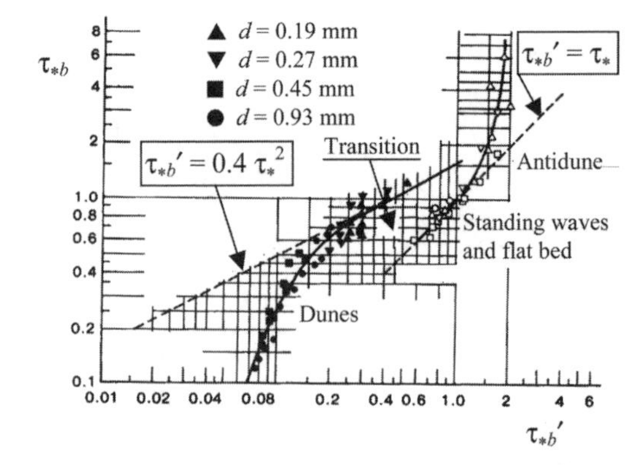

Figure 9. Engelund's relationships between dimensionless grain shear stress and total shear stress (after Chang 1992).

– lower regime: dunes ($\tau_{*b}' < 0.55$):

$$\tau_{*b}' = 0.4\,\tau_{*b}^2 + 0.06 \tag{23a}$$

– transition: flat bed ($0.55 < \tau_{*b}' < 1$):

$$\tau_{*b}' = \tau_{*b} \tag{23b}$$

– upper regime: antidunes ($\tau_{*b}' > 1$):

$$\tau_{*b}' = \left(0.702\,\tau_{*b}^{-1.8} + 0.298\right)^{0.556} \tag{23c}$$

It is interesting to notice that, in (23), τ_{*b}' serves as a regime indicator, instead of the common Froude number. Following Athaullah (1968), the common idea to link dunes to subcritical flow and antidunes to supercritical flow is only valid if the depth ratio R_b/d is in the average range 200 to 400.

4.2 *Energy approach*

A less empirical approach is based on the division of the friction slope $S_f = S_f' + S_f''$. The bed form friction S_f'' is due primarily to expansion losses due to changes in water depth. The expansion head loss $\Delta H''$ may be estimated from the formula

$$\Delta H'' = \zeta \frac{|U_1 - U_2|^2}{2g} \tag{24}$$

where ζ is the loss coefficient, U_1 the mean velocity above the crest and U_2 the mean velocity over the trough.

In upper flow regime, the water surface and the antidunes are in phase in such a way that the changes in mean velocity are less developed than in lower regime.

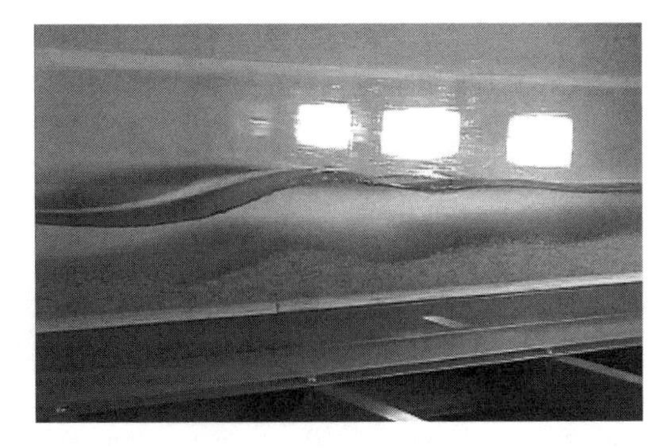

Figure 10. Antidunes at the toe of a dam-break wave.

Thus the grain resistance is predominant in upper regime, compared to the form resistance. Passing from sub- to supercritical flow generally corresponds to a decrease of the apparent roughness. However, if the Froude number still rises, the velocity also increases and the determination of both grain and bed-form head losses is required for accurate predictions.

Using an expression like (24) requires the knowledge of both water and bed profiles, which is generally not available from solving St-Venant–Exner relations (1–3).

For dunes, some approaches exist to model bed form shaping and migration, starting for example from an initial sinusoidal shape (see e.g. Yalin & Ferreira da Silva 2001).

Such models are less developed for antidunes, mainly due to the fact that antidunes generally feature two-dimensional characteristics that cannot be captured by a one-dimensional approach. The above-mentioned two-layer model, extended to a two-dimensional frame, seems a promising way to cope with the problem.

4.3 *Antidune modeling*

While dunes and bars are generated by turbulent bursts and associated large eddies (Yalin & Ferreira da Silva 2001), the mechanism of antidune generation seems different. The increase in the velocity causes the water surface to become unstable, since, for supercritical flows, a slight change in energy may result in significant changes in depth and velocity. Antidunes are the replication of this instability (Fig. 10).

The transport mechanism in such situations becomes rather complex. This transport is relatively intense and continuous. The particles are entrained in the downstream direction, while the sand wave itself may travel in the upstream direction. As the water velocity is large, the sediments are unable to reach the same velocity: the clear water and the sediments thus present distinct velocities. If two-dimensional features appear, also the velocity direction may be distinct between the water and the sediment layers. A two-layer model is thus well adapted to reproduce this behavior.

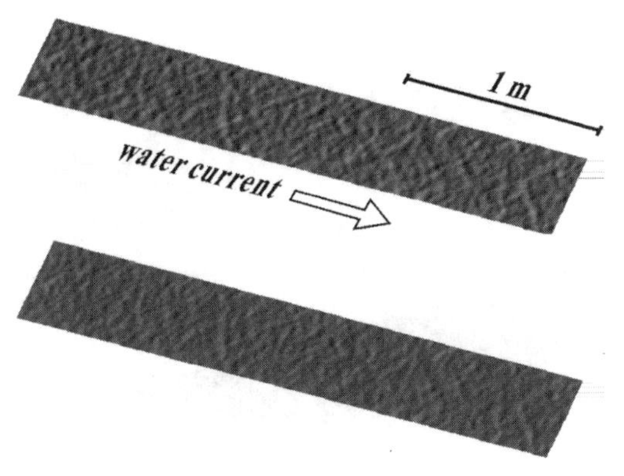

Figure 11. Water and bed surface at time $t = 1$ s (Capart & Young 2002).

4.3.1 *Two-dimensional two-layer model*

This model was developed by Capart and Young (2002). In their approach, the difference in concentration between the moving layer and the sediments at rest is neglected ($C_s = C_b$).

Simple two-dimensional extension of relations (7b) and (7c) yields mass conservation for both moving layers. For the momentum conservation, extension of relations (7d) and (7e) has to account for cross terms in momentum exchanges. For instance the water-layer momentum conservation in the x-direction reads

$$\frac{\partial(h_w u_w)}{\partial t} + \frac{\partial}{\partial x}\left(h_w u_w^2 + \frac{g h_w^2}{2}\right) + \frac{\partial(h_w u_w v_w)}{\partial y} + g h_w \frac{\partial z_s}{\partial x} = -\frac{\tau_{wx}}{\rho_w} \quad (25)$$

The closure relations are similar to equations (8–11), with the particularity that the water velocity u_w and the sediment velocity u_s can have distinct directions, which implies that the water friction τ_w could have a distinct direction than the sediment friction τ_s. More details can be found in Capart and Young (2002).

4.3.2 *Antidune generation modeling*

This two-dimensional model is able to generate antidunes from a flat bed situation. However, since antidunes are generated by instabilities, the initial plane bed has to be perturbed by small uncorrelated random variations (Fig. 11).

The small-scale variations of sand bed- and free surface grow simultaneously, progressively evolving

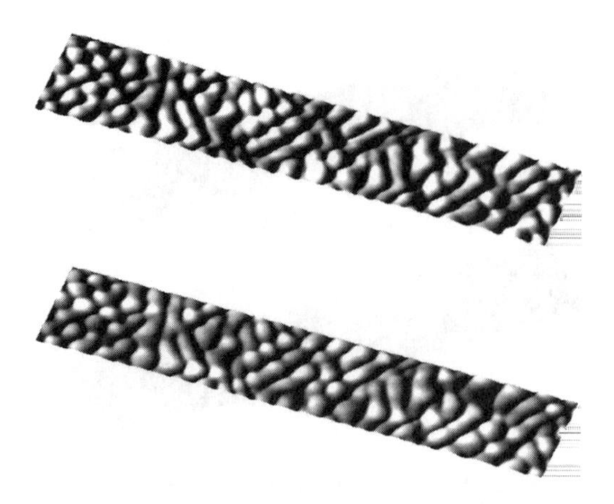

Figure 12. Water and bed surface at time $t = 20$ s (Capart & Young 2002).

to waves with stable amplitude and length (Fig. 12). The antidunes clearly appear, in phase with the water surface. The forms slowly migrate in the upstream direction, according to the generally accepted description of antidunes. The two-dimensional feature could also be observed in the velocities: both the magnitude and the direction of velocities in water and sediment are distinct (Capart & Young 2002).

5 SUPER- AND TRANSCRITICAL FLOW AND BOUNDARY CONDITIONS

5.1 *Common approach*

Vreugdenhil and de Vries (1967) proposed the first eigenvalue analysis of the Saint-Venant–Exner equations (see also Jansen et al. 1979). For this analysis, the variation in time of the bottom level $\partial z_b/\partial t$ is neglected compared to the water-depth variation $\partial h/\partial t$, in such a way that (1) reduces to

$$\frac{\partial h}{\partial t} + u\frac{\partial h}{\partial x} + h\frac{\partial u}{\partial x} = 0 \tag{26}$$

Decomposing the momentum equation (2), and subtracting (26) multiplied by u yields

$$\frac{\partial u}{\partial t} + u\frac{\partial u}{\partial x} + g\frac{\partial h}{\partial x} + g\frac{\partial z_b}{\partial x} + g S_f = 0 \tag{27}$$

With the assumption that the solid discharge q_s depends on the velocity and the water depth, (3) reads:

$$(1-\varepsilon_0)\frac{\partial z_b}{\partial t} + \frac{\partial q_s}{\partial u}\frac{\partial u}{\partial x} + \frac{\partial q_s}{\partial h}\frac{\partial h}{\partial x} = 0 \tag{28}$$

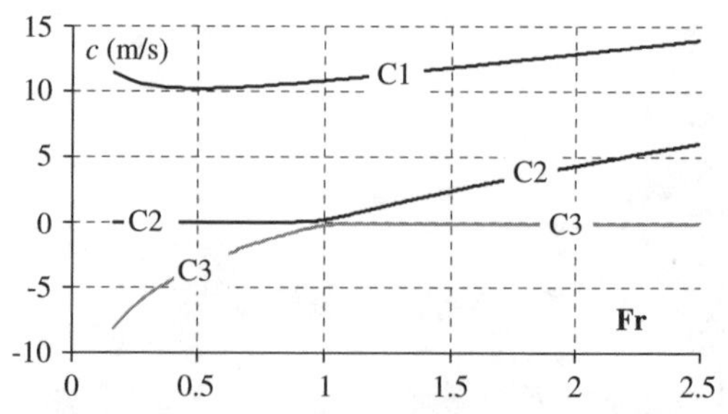

Figure 13. Characteristic celerities according to the Vreugdenhil & de Vries (1967) analysis ($q = 16\,\mathrm{m^3/s/m}$, $n = 0.0277$, $d = 0.0274$ m, $\rho_s/\rho_w = 2.65$, $\varepsilon_0 = 0$).

The system (26–28) counts three main variables: u, h and z_b. Completing the system by the decomposition of du, dh and dz_b in partial derivatives yields

$$\begin{pmatrix} 1 & u & 0 & g & 0 & g \\ 0 & h & 1 & u & 0 & 0 \\ 0 & \frac{\partial q_s}{\partial u} & 0 & \frac{\partial q_s}{\partial h} & 1-\varepsilon_0 & 0 \\ dt & dx & 0 & 0 & 0 & 0 \\ 0 & 0 & dt & dx & 0 & 0 \\ 0 & 0 & 0 & 0 & dt & dx \end{pmatrix} \times \begin{pmatrix} \frac{\partial u}{\partial t} \\ \frac{\partial u}{\partial x} \\ \frac{\partial h}{\partial t} \\ \frac{\partial h}{\partial x} \\ \frac{\partial z_b}{\partial t} \\ \frac{\partial z_b}{\partial x} \end{pmatrix} = \begin{pmatrix} -gS_f \\ 0 \\ 0 \\ du \\ dh \\ dz_b \end{pmatrix} \tag{29}$$

The condition determining the characteristic velocities $c = dx/dt$ is that the determinant of the set of equations (29) be zero, which yields

$$c^3 + \beta c^2 + \gamma c + \delta = 0 \tag{30}$$

with three roots: c_1, c_2 and c_3, being the celerities of three characteristics.

To give an idea of those celerities, the following example was analyzed, inspired by Rahuel (1988): a discharge $q = 16\,\mathrm{m^3/s/m}$, presenting a uniform flow in a channel with mobile bed (roughness $n = 0.0277$, grain diameter $d = 0.0274$ m, $\rho_s/\rho_w = 2.65$). The solid discharge q_s is computed from the Meyer-Peter–Müller formula.

The three celerities are given in Figure 13 as a function of the Froude number. The result is clear: two characteristics (C1 and C2) are always positive and the third one (C3) always negative.

$$\mathbf{Fr} < 1 \Rightarrow \begin{cases} c_1 > 0 \\ c_2 > 0 \quad (c_2 \approx 0) \\ c_3 < 0 \end{cases} \tag{31a}$$

$$\mathbf{Fr} > 1 \Rightarrow \begin{cases} c_1 > 0 \\ c_2 > 0 \\ c_3 < 0 \quad (c_3 \approx 0) \end{cases} \tag{31b}$$

If the solid discharge q_s does not exist (fixed bed, or mobile bed below the transport threshold), the characteristic C1 is unaffected, while the part of C3 corresponding to subcritical flow (**Fr** < 1) merges with the part of C2 corresponding to supercritical flow (**Fr** > 1), the other part of both characteristics vanishing. The two vanishing parts correspond to very low celerities, which confirms that the celerity of sediment-related information is significantly less than the hydrodynamic information.

The common result in pure hydrodynamics is that, for subcritical flow, one boundary conditions has to be imposed upstream (typically the liquid discharge q) and the other one downstream (typically a water level z_w), while, for supercritical flow, both conditions are imposed at the upstream channel end.

According to the analysis by Vreugdenhil and de Vries, one *sediment* condition has now to be added: upstream for subcritical and downstream for supercritical flow.

What could be a sediment boundary condition? The concerned variable is the bottom level z_b, but, in practice, this parameter is often difficult to impose. In practical cases, the solid discharge q_s is often to be imposed, but this variable is not a main unknown of the set of equations (26–28). The solid discharge q_s and the friction slope S_f are predicted as closure equations. The solid discharge may thus be imposed externally, but that means that the propagation of such an information will be driven by the arrangement of the computation cells rather than by the physics. Two examples will illustrate the problem.

5.1.1 *Steep-sloped aggradation or degradation*

This is a typical example of supercritical flow. Consider a steep-sloped channel of finite length, with a steady water discharge and a given initial bed profile (Capart et al. 1998).

If a stable supply of sediment is imposed upstream, overloading or underloading respectively results in channel aggradation or degradation. Tending to equalize sediment supply and transport capacity, the system will evolve towards a state of morphological equilibrium characterized ultimately by a constant slope (Fig. 14).

In such a configuration, the boundary conditions clearly match the theoretical situations. Upstream two hydrodynamic boundary conditions prevail: the liquid discharge q and a relationship $q = f(h)$ characterizing a control section ($h = h_c$) at the entrance. Downstream, the *sediment* boundary condition is the bottom level z_b at the end edge of the channel where sediments are supposed to fall down.

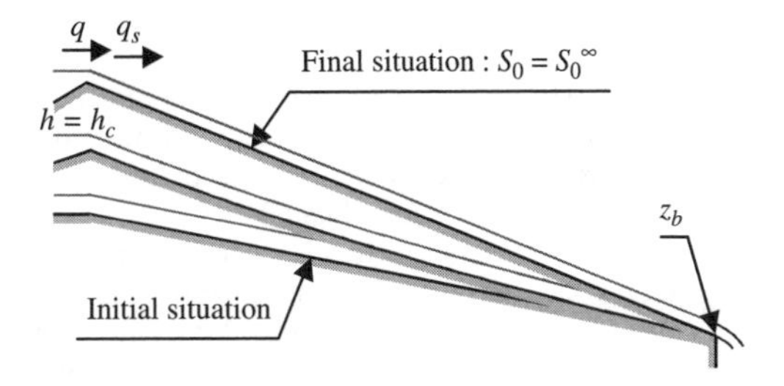

Figure 14. Steep-sloped aggradation (Capart et al. 1998).

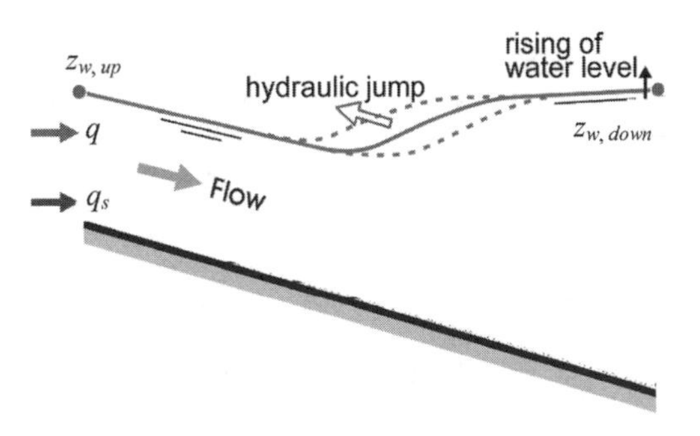

Figure 15. Initiation of a hydraulic jump over a mobile bed.

The solid discharge q_s (assumed constant in this case) is imposed upstream as closure of the equation set. That means that at the entrance of the upstream computational cell, the solid discharge is imposed, while, at the exit of the same cell, the solid discharge is assumed to be the equilibrium one. In general cases, it may result in exaggerated change in bottom profile in this upstream cell.

This is the case in the present example at the beginning of the aggradation process, when the actual slope S_0 is far from the final equilibrium slope S_0^∞. This may be addressed by using a spatial and/or time lag expressing that the solid discharge progressively, and not instantaneously, adapts to the transport capacity (Phillips & Sutherland 1989). But this kind of solution is empirical and no complete agreement exists about the best-suited approach.

5.1.2 *Hydraulic jump over mobile bed in the Saint-Venant–Exner approach*

Consider a supercritical flow progressively submerged by a downstream water rise forcing a jump to appear (Fig. 15).

This problem is common in pure hydrodynamics with fixed bed: the jump is the transition between the upstream supercritical flow, controlled by two upstream boundary conditions: the upstream discharge q and the upstream water level $z_{w,up}$, and the downstream subcritical flow controlled by the downstream boundary condition: the downstream water level $z_{w,down}$. The jump thus resolves the contradiction between otherwise incompatible boundary conditions.

Now considering the movement of sediments, the situation becomes more complicated.

For the upstream supercritical reach the sediment boundary condition would be a downstream one ($c_3 < 0$), while for the downstream subcritical reach this sediment condition would be an upstream one ($c_2 > 0$). The morphodynamics of the system is thus depending on an internal condition at the section of the jump, acting as a kind of sediment control section.

Of course such a morphological control section does not really exist and the problem is thus ill-posed, in what concerns the boundary conditions.

Moreover, the physics of the phenomenon does not match the above considerations. In fact, what is really imposed is the upstream solid discharge q_s (for instance the discharge corresponding to the transport capacity of the upstream supercritical flow).

Once again, the two-layer model is able to address some of the above inconsistency, thanks to its flexibility.

5.2 *The two-layer model approach*

The relations (17), including the definition of the erosion rate (17a), constitute a five-equation system, which can be written in a vector formulation:

$$\frac{\partial \mathbf{U}}{\partial t}+\frac{\partial \mathbf{F}(\mathbf{U})}{\partial x}+\mathbf{H}(\mathbf{U})\frac{\partial \mathbf{U}}{\partial x}=\mathbf{S}(\mathbf{U}) \qquad (32a)$$

with

$$\mathbf{U}=\begin{pmatrix} z_b \\ h_w \\ h_s \\ h_w u_w \\ h_s u_s \end{pmatrix}, \quad \mathbf{F}(\mathbf{U})=\begin{pmatrix} 0 \\ h_w u_w \\ h_s u_s \\ h_w u_w^2+\frac{1}{2}g h_w^2 \\ h_s u_s^2+\frac{1}{2}g h_s^2 \end{pmatrix} \qquad (32b\text{-}c)$$

$$\mathbf{H}(\mathbf{U})=\begin{pmatrix} 0 & 0 & 0 & 0 & 0 \\ 0 & 0 & 0 & 0 & 0 \\ 0 & 0 & 0 & 0 & 0 \\ g h_w & 0 & g h_w & 0 & 0 \\ g h_w & \frac{\rho_w}{\rho_s'} g h_s & 0 & 0 & 0 \end{pmatrix} \qquad (32d)$$

For the eigenvalue analysis the source terms are not considered. Neglecting this term, the system (32) can be written

$$\begin{aligned}\frac{\partial \mathbf{U}}{\partial t}+\frac{\partial \mathbf{F}(\mathbf{U})}{\partial x}+\mathbf{H}(\mathbf{U})\frac{\partial \mathbf{U}}{\partial x} &=\frac{\partial \mathbf{U}}{\partial t}+\mathbf{A}(\mathbf{U})\frac{\partial \mathbf{U}}{\partial x}+\mathbf{H}(\mathbf{U})\frac{\partial \mathbf{U}}{\partial x} \\ &=\frac{\partial \mathbf{U}}{\partial t}+\mathbf{A}'(\mathbf{U})\frac{\partial \mathbf{U}}{\partial x}=0\end{aligned} \qquad (33)$$

where $\mathbf{A}(\mathbf{U}) = \partial\mathbf{F}(\mathbf{U})/\partial\mathbf{U}$ is the Jacobian matrix associated to the system, and $\mathbf{A}'(\mathbf{U}) = \mathbf{A}(\mathbf{U}) + \mathbf{H}(\mathbf{U})$ is a kind of pseudo-Jacobian matrix:

$$\mathbf{A}'(\mathbf{U})=\begin{pmatrix} 0 & 0 & 0 & 0 & 0 \\ 0 & 0 & 0 & 1 & 0 \\ 0 & 0 & 0 & 0 & 1 \\ g h_w & -u_w^2+g h_w & g h_w & 2u_w & 0 \\ g h_w & \frac{\rho_w}{\rho_s'} g h_s & -u_s^2+g h_s & 0 & 2u_s \end{pmatrix} \qquad (34)$$

whose eigenvalues are the celerities of the five associated characteristics.

The eigenstructure of $\mathbf{A}'(\mathbf{U})$ has no straightforward analytical solutions, and it is impossible to express the compatibility equations linked to the characteristics. The eigenvalues of the pseudo-Jacobian matrix $\mathbf{A}'$ are thus computed numerically.

From such numerical computation it was observed (Savary & Zech submitted) that (a) one of the celerities is zero; (b) the two extreme characteristics have a real celerity; and (c) the two intermediate eigenvalues are also real in most of the fluvial hydraulics problems, when the lower layer is relatively thin in comparison to the upper layer ($h_s << h_w$). Some examples may illustrate those considerations and their consequences in term of boundary conditions.

5.2.1 *Uniform flow*

Two cases of uniform flow are simulated over a granular bed characterized by a density $\rho_s = 2650\,\text{kg/m}^3$ and a concentration $C_s = 0.60$, in such a way that the sediment layer density $\rho_s' = 1990\,\text{kg/m}^3$. In both cases, the two velocities are imposed: $u_w = 1$ m/s and $u_s = 0.6$ m/s, and the two depths: h_w, varying from 0.01 to 2.55 m, and $h_s = 0.01$ m in the first case (thin-layer case), and 0.1 m in the second case (thick-layer case).

Since four parameters are imposed, the set of equations (9–16) yields the eight other ones: S_0, q_w, q_s, τ_w, τ_s, τ_b, C_{fw} and C_{fs}.

From numerical eigenanalysis of $\mathbf{A}'$ (34), it is possible to compute the eigenvalues. Figures 16 and 17 give their values as a function of the water Froude number, defined as

$$\mathbf{Fr}_w = \frac{u_w}{\sqrt{g\,h_w}} \tag{35}$$

similar to (19).

In the thin-layer case (Fig. 16), the smallest eigenvalue becomes negative for water Froude numbers lower than 1.1, showing that the Froude number only is no more sufficient to define the regime.

It can be seen that for lower water Froude numbers the five eigenvalues are real and distinct, while for higher values, two of them become conjugate complex (same real part, opposite imaginary part).

In the thick-layer case (Fig. 17), one eigenvalue is always negative and, for water Froude numbers lower than 0.5, a second one becomes negative.

Those results are rather distinct of the common approach. For instance in supercritical flow all the characteristics may be positive in contrast with the common analysis issuing two positive characteristics and one negative. More details are given in Savary & Zech (submitted).

Moreover the five-equation approach offers new degrees of freedom to conveniently pose the boundary conditions, as exemplified in the next section.

5.2.2 *Hydraulic jump over mobile bed in the two-layer model approach*

Consider a supercritical flow as initial situation (Fig. 18). A unit discharge $q = 0.024\,\text{m}^3/\text{s}$ is flowing on an initial bottom slope $S_0 = 3\%$ composed of uniform sediments with grain diameter $d = 1.65$ mm.

All the non-zero eigenvalues are positive. Four boundary conditions are thus to be imposed upstream: the liquid discharge q, the upstream water level z_w (corresponding to the critical depth in this particular case), the solid discharge q_s, the sediment-layer thickness h_s (computed from the uniform flow in equilibrium situation, see Section 3.2). A fifth condition, corresponding to the zero eigenvalue, could be the bottom level $z_s = z_b + h_s$, which fixes all the levels of the problem. It can be demonstrated (Savary & Zech, submitted) that the compatibility equation corresponding to this zero eigenvalue reduces to the form dz_b/dt, showing that this boundary condition may be considered apart of the other ones.

Suppose that this initial situation is perturbed by the rapid raise of a weir at the downstream end of the flume, imposing a subcritical flow in the downstream reach. A hydraulic jump results between the supercritical (five upstream conditions) and the subcritical (one downstream condition) reach (Fig. 19).

The so-generated hydraulic jump propagates in the upstream direction, altering the sediment transport capacity. In the supercritical part of the flow, the

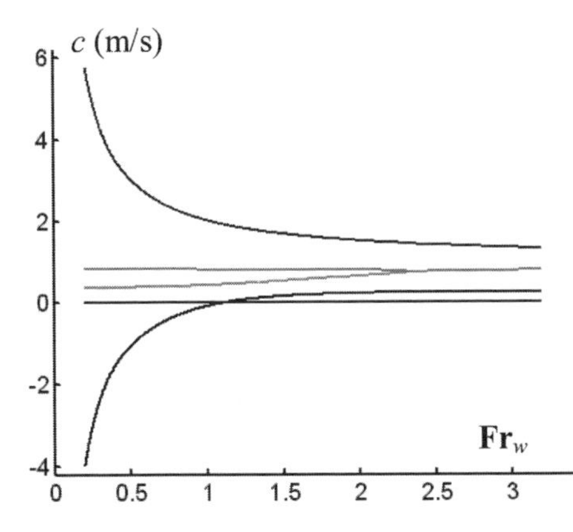

Figure 16. Eigenvalues (real part) of the pseudo-Jacobian matrix as a function of the water Froude number $\mathbf{Fr}_w$. Thin-layer case: $u_w = 1$ m/s, $u_s = 0.6$ m/s, $h_s = 0.01$ m and h_w varies from 0.01 to 2.55 m.

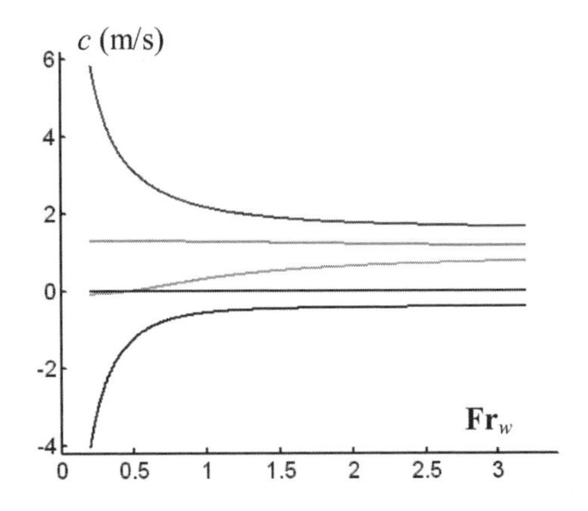

Figure 17. Eigenvalue of the pseudo-Jacobian matrix as a function of the water Froude number $\mathbf{Fr}_w$. Thick-layer case: $u_w = 1$ m/s, $u_s = 0.6$ m/s, $h_s = 0.1$ m and h_w varies from 0.01 to 2.55 m.

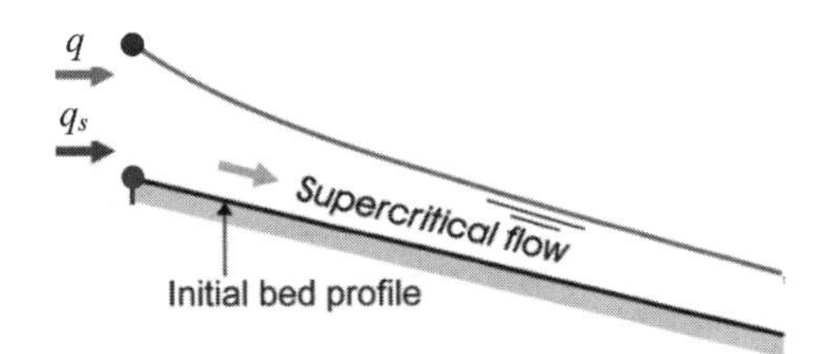

Figure 18. Hydraulic jump: initial situation.

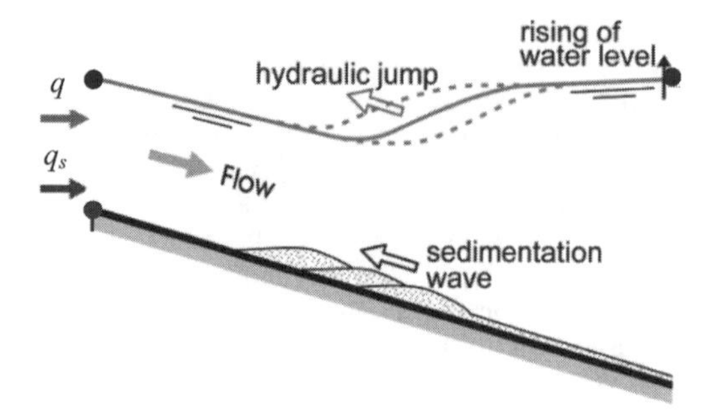

Figure 19. Initiation of hydraulic jump.

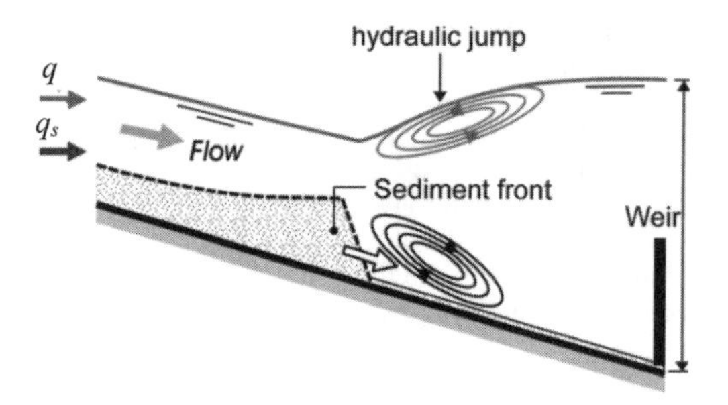

Figure 20. Progression of the sediment front.

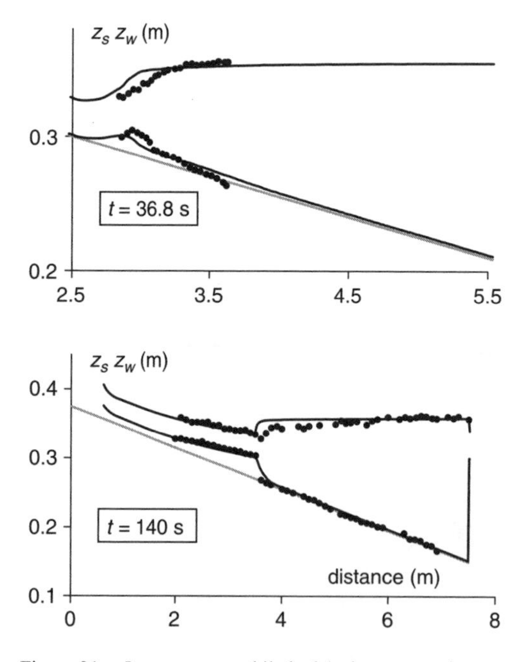

Figure 21. Jump over a mobile bed (points = experiments, solid line = two-layer model, with $C_{fw} = 0.1$ and $C_{fs} = 0.02$).

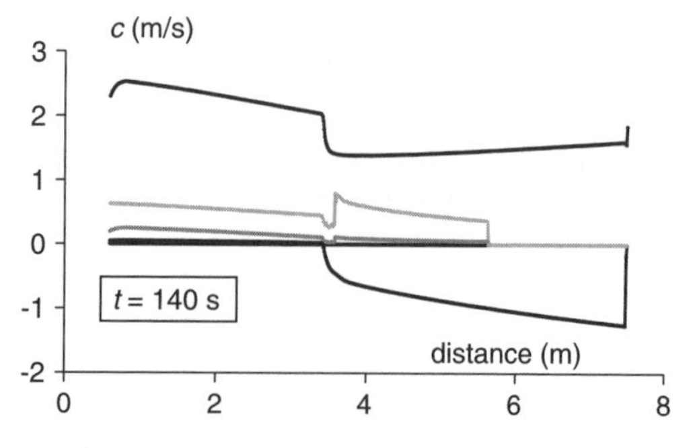

Figure 22. Eigenvalues along the channel.

sediments remain in motion, but not anymore in the subcritical part, implying that sediments depose at the transition, under the surge.

Once the jump location stabilizes, sediments accumulate at its upstream end, forming a sediment front progressively moving downwards (Fig. 20).

The comparison (Fig. 21) between experimental (Bellal et al. 2003) and computed data is rather good, above all taking into consideration that the whole process is modeled, including the appearance of the jump and of the resulting sediment bump ($t = 36.8$ s).

The plot of the eigenvalues along the channel (Fig. 22) shows the change in regime, passing from supercritical zone (four positive eigenvalues) to the zone just following the jump (one eigenvalue has become negative) and then to the subcritical flow (the two characteristics associated to solid transport vanish, as the transport itself vanishes). The downstream end is a control section over the weir, leading the negative characteristic to be zero.

6 CONCLUSIONS

Interference between supercritical flow and morphodynamics is generally less studied than the subcritical case, even though changes in regime often accompany major floods. In principle, this change in regime does not affect the transport mechanism. However the dramatic increase of transport intensity, the appearance of antidunes, and the difficulty to pose correctly the boundary conditions in transcritical flow, justify investigating the supercritical regime.

An approach based on a two-layer model is proposed. It seems to be able to address most of those challenges.

REFERENCES

Athaullah, M. 1968. *Prediction of bed forms in erodible channels.* Ph.D. thesis, Colorado State University, Fort Collins, Colorado, USA.

Brownlie, W.R. 1983. Flow depth in sand-bed channels. *J. Hydraulic Engineering ASCE*; 109(7): 959–990.
Bellal, M., Spinewine, B., Savary, C. & Zech Y. 2003. Morphological evolution of steep-sloped river beds in the presence of a hydraulic jump: Experimental study. *Proceedings of XXX IAHR Congress*, C-II: 133–140.
Capart, H. 2000. *Dam-break induced geomorphic flows*. Ph.D. thesis, Université catholique de Louvain, Louvain-la-Neuve, Belgium.
Capart, H., Bellal, M., Boxus, L., De Roover, C. & Zech, Y. 1998. Approach to morphological equilibrium for steep-sloped river beds. *Proceedings of the seventh International Symposium on River Sedimentation, Hong Kong*: 231–237.
Capart, H. & Young, D.L. 2002. Two-layer shallow water computations of torrential geomorphic flows. *Proceedings River Flow 2002*, Louvain-la-Neuve, Belgium: 1003–1012.
Cunge, J.A., Holly, F.M. Jr. & Verwey, A. 1980. *Practical aspects of computational river hydraulics*. London, UK: Pitman.
Chang, H.H. 1992. *Fluvial Processes in River Engineering*. Malabar, Florida: Krieger Publishing Company.
Einstein, H.A. & Barbarossa, N. 1952. River Channel Roughness. *Trans. ASCE* 117: 1121–1146.
Engelund, F. 1966. Hydraulic resistance of alluvial streams. *J. Hydraulic Division ASCE*. 92(HY2): 315–326.
Engelund, F. & Hansen E. 1967. *A monograph on sediment transport in alluvial streams*. Copenhagen, Denmark: Teknisk Verlag.
Fraccarollo, L. & Capart, H. 2002. Riemann wave description of erosional dam-break flows. *Journal of Fluid Mechanics* 461: 183–228.
Jansen, P.Ph., Van Bendegom, L., Van Den Berg, J., de Vries, M. & Zanen, A. 1979. *Principles of River Engineering. The non-tidal alluvial river*. London, UK: Pitman.
Phillips, B.C. & Sutherland, A.J. 1989. Spatial and lag effects in bed load sediment transport. *Journal of Hydraulic Research* 27: 275–292.
Rahuel, J.L. 1988. *Modélisation de l'évolution du lit des rivières alluvionnaires à granulométrie étendue* (in French), Institut National Polytechnique de Grenoble, Grenoble, France.
Savary, C. & Zech, Y. (submitted). Boundary conditions in a two-layer geomorphical model. Application to a hydraulic jump over a mobile bed. Submitted to *J. Hydraulic Research*.
Spinewine, B. 2003 *Ecoulements transitoires sévères d'un milieu diphasique liquide/granulaire (eau/sédiments), en tenant compte du couplage rhéologique entre les deux phases et des composantes verticales du mouvement* (in French). M.Sc. thesis. Université catholique de Louvain, Louvain-la-Neuve, Belgium.
Spinewine, B. 2005. Effects of granular dilatancy on dam-break induced sheet-flow. In Ph.D. thesis. Université catholique de Louvain, Louvain-la-Neuve, Belgium.
Spinewine, B. & Zech, Y. 2002a. Geomorphic dam-break floods: near-field and far-field perspectives. *Proceedings 1st IMPACT Project Workshop*, Wallingford, UK, (CD-ROM).
Spinewine, B. & Zech, Y. 2002b. Dam-break waves over movable beds: a 'flat bed' test case. *Proceedings 1st IMPACT Project Workshop*, Wallingford, UK (CD-ROM).
van Rijn, L.C. 1984. Sediment transport: bed-load transport. *J. Hydraulic Engineering ASCE*, 110(10): 1431–1456.
Vreugdenhil, C.B. & de Vries, M. 1967. *Computations of non-steady bed-load transport by a pseudo viscosity method*. Delft Hydr. Lab. Publ. 45, Delft, The Netherlands.
Yalin, M.S. & Ferreira da Silva, A.M. 2001. *Fluvial processes*. Delft, the Netherlands: IAHR monograph.

River, Coastal and Estuarine Morphodynamics: RCEM 2005 – Parker & García (eds)

Two-dimensional morphological simulation in transcritical flow

J.A. Vasquez & R.G. Millar
Department of Civil Engineering, University of British Columbia, Canada

P.M. Steffler
Department of Civil and Environmental Engineering, University of Alberta, Canada

ABSTRACT: A two-dimensional Finite Element hydrodynamic model with capabilities for transcritical flow has been coupled with a sediment transport model to simulate bed elevation changes in alluvial beds. The model has been tested for a knickpoint migration experiment that involved a short and very steep reach of 10% bed slope. The initial Froude was Fr = 3.5 downstream from the knickpoint with a hydraulic jump formed close to the toe of the oversteepened reach. The results computed by the model agreed with the experimental data. The new upstream location of the knickpoint was correctly predicted and the computed bed profile seemed to follow the average trend of the measured profile. The model seems as a promising morphodynamic tool for alluvial streams subject to transcritical flow.

1 INTRODUCTION

Although several river morphology models are currently available, most of them are intended for applications in lowland alluvial rivers with small gradients (Papanicolaou, in press). Such models usually assume subcritical flow conditions and therefore cannot be applied in cases where supercritical flow or hydraulic jumps are present.

A transcritical flow condition, where flow regime changes between subcritical and supercritical, usually involves the presence of a sharp shock wave or hydraulic jump, as sketched in Figure 1.

Some practical examples of transcritical flow over alluvial beds are: sediment deposition upstream of dams (Brusnelli et al. 2001, Bellal et al. 2003); dam break (Fraccarollo & Capart 2002, Spinewine & Zech 2002); dam removal (Cantelli et al. 2004, Wong et al. 2004, Cui & Wilcox, in press, Cui et al. a, b, in press); dyke breach (Spinewine et al. 2004); and knickpoint migration (Brush & Wolman 1960, Bhallamudi & Chaudhry 1991).

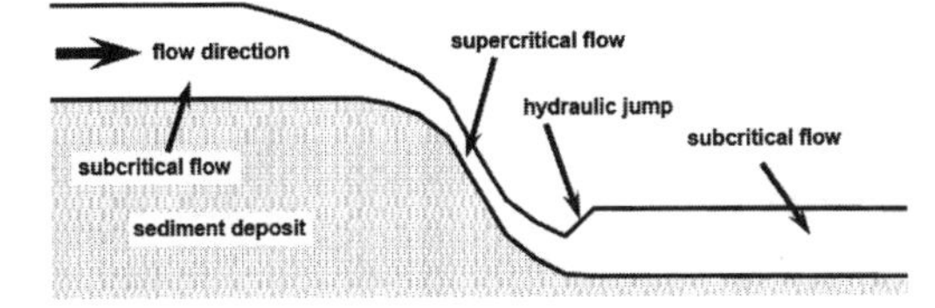

Figure 1. Sketch of transcritical flow after dam removal (Cui & Wilcock, in press).

Knickpoints are points along longitudinal profiles of streams where the slope increases abruptly from mild to steep (Brush & Wolman 1960). Knickpoints may appear after meander cut-off (Brush & Wolman 1960) or dam removal (Fig. 1). Their presence can lead to transcritical flow conditions and intense scour and deposition.

Morphodynamic models with transcritical flow capabilities are uncommon, and normally limited to one-dimensional (1D) flow (e.g. Papanicolaou, in press, Brusnelli et al. 2001, Cui et al. a, b, in press).

The main objective of this paper is to assess the capabilities of a new two-dimensional (2D) depth-averaged river morphology model for simulating scour and deposition in transcritical flow generated by a knickpoint. Comparisons with the experimental data of Brush & Wolman (1960) suggested that model is a promising tool for alluvial streams subject to transcritical flow.

2 THE NUMERICAL MODEL

2.1 *River2D hydrodynamic model*

River2D is a two-dimensional (2D) depth-averaged hydrodynamic model intended for use on natural streams and rivers and has special features for supercritical/subcritical flow transitions, ice covers, and variable wetted area. For the spatial discretization it uses a flexible Finite Element unstructured mesh composed of triangular elements. River2D is based on the 2D vertically averaged Saint Venant equations expressed in conservative form; which form a system of three equations representing the conservation of water mass and the two components of the momentum vector (Steffler and Blackburn, 2002). It has been developed at the University of Alberta and it is freely available at www.River2D.ca.

The Finite Element Method used in River2D's hydrodynamic model is based on the Streamline Upwind Petrov-Galerkin (SUPG) weighted residual formulation (Hicks & Steffler 1992, Ghanem et al. 1995). In this technique, upstream biased test functions are used to ensure solution stability under the full range of flow conditions, including subcritical, supercritical, and transcritical flow. A fully conservative discretization is implemented which ensures that no fluid mass is lost or gained over the modeled domain. This also allows implementation of boundary conditions as natural flow or forced conditions. The equations solved by the River2D are (Steffler and Blackburn 2002):

The water continuity equation:

$$\frac{\partial h}{\partial t}+\frac{\partial q_x}{\partial x}+\frac{\partial q_y}{\partial y}=0 \tag{1}$$

The vertically averaged momentum equation in the x-direction:

$$\frac{\partial q_x}{\partial t}+\frac{\partial (uq_x)}{\partial x}+\frac{\partial (vq_x)}{\partial y}+\frac{g}{2}\frac{\partial h^2}{\partial x}=$$

$$gh\left(S_{ox}-S_{fx}\right)+\frac{1}{\rho}\left(\frac{\partial (h\tau_{xx})}{\partial x}+\frac{\partial (h\tau_{xy})}{\partial y}\right) \tag{2}$$

The vertically averaged momentum equation in the y-direction:

$$\frac{\partial q_y}{\partial t}+\frac{\partial (uq_y)}{\partial x}+\frac{\partial (vq_y)}{\partial y}+\frac{g}{2}\frac{\partial h^2}{\partial y}=$$

$$gh\left(S_{oy}-S_{fy}\right)+\frac{1}{\rho}\left(\frac{\partial (h\tau_{yx})}{\partial x}+\frac{\partial (h\tau_{yy})}{\partial y}\right) \tag{3}$$

where h = water depth; (u, v) = depth-averaged velocities in the (x,y) directions; $q_x = uh$ = flow discharge in x-direction per unit width; $q_y = vh$ = flow discharge in the y-direction per unit width; (S_{ox}, S_{oy}) = bed slopes in the (x,y) directions; S_{fx} and S_{fy} = corresponding friction slopes; τ_{xx}, τ_{xy}, τ_{yx}, τ_{yy} = components of the horizontal turbulent stress tensor; ρ = water density; and g = gravitation acceleration.

The basic assumptions in equations (1) through (3) are (Steffler and Blackburn 2002):

- The pressure distribution is hydrostatic, which limits the accuracy in areas of steep slopes and rapid changes in bed slopes.
- The horizontal velocities are constant over the depth. Information on secondary flows and circulations is not available.
- Coriolis and wind forces are assumed negligible.

The friction slope terms depend on the bed shear stresses which are assumed to depend on the magnitude and direction of the vertically averaged velocities. For example, in the x-direction:

$$S_{fx}=\frac{\tau_{bx}}{\rho gh}=\frac{\sqrt{u^2+v^2}}{ghC_*^2}u \tag{4}$$

τ_{bx} is the bed shear stress in the x direction and C_* is the dimensionless Chezy coefficient, which is related to the effective roughness height k_s through

$$C_*=5.75\log\left(12\frac{h}{k_s}\right) \quad ; \frac{h}{k_s}\geq\frac{e^2}{12} \tag{5a}$$

$$C_*=2.5+\frac{30}{e^2}\left(\frac{h}{k_s}\right) \quad ; \frac{h}{k_s}<\frac{e^2}{12} \tag{5b}$$

$e = 2.7182$; C_* is related to Chezy's C coefficient through

$$C_*=\frac{C}{\sqrt{g}} \tag{6}$$

The vertically-averaged turbulent shear stresses are modeled with a Boussinesq type eddy viscosity. For example:

$$\tau_{xy}=\upsilon_t\left(\frac{\partial u}{\partial y}+\frac{\partial v}{\partial x}\right) \tag{7}$$

where υ_t = eddy viscosity coefficient assumed by default as

$$\upsilon_t=0.5\frac{h\sqrt{u^2+v^2}}{C_*} \tag{8}$$

Unlike other popular 2D hydrodynamic models based of Finite Elements (e.g. RMA2, FESWMS), River2D does not require artificial eddy viscosity for convergence. In fact, River2D is rather insensitive to the value of eddy viscosity.

2.2 *River2D-Morphology*

River2D was extended by including a solver for the bed load sediment continuity (Exner) equation

$$(1-\lambda)\frac{\partial z_b}{\partial t}+\frac{\partial q_{sx}}{\partial x}+\frac{\partial q_{sy}}{\partial y}=0 \tag{9}$$

where z_b = bed elevation; λ = porosity of the bed material; t = time; and (q_{sx}, q_{sy}) = components of the volumetric rate of bedload transport per unit width q_s (ignoring the effects of secondary flow and bed slope):

$$q_{sx} = \frac{u}{\sqrt{u^2 + v^2}} q_s \tag{10a}$$

$$q_{sy} = \frac{v}{\sqrt{u^2 + v^2}} q_s \tag{10b}$$

Equation (9) is discretized using a conventional Galerkin Finite Element Method (GFEM). The time advance is performed using a Runge-Kutta second order scheme. Since GFEM lacks upwinding properties (equivalent to a central difference scheme) it is known to be adequate for diffusive problems, but not for advection-dominated problems (Hirsch 1987, Chung 2002, Donea & Huerta 2003).

This extended version is known as River2D-Morphology and has been successfully applied to simulate aggradation, degradation and bend scour in alluvial channels (Vasquez et al. 2005a, b). However, the model should not be applied to problems where advection is important, such as migrating bedforms or sediment waves.

3 EXPERIMENTAL TEST CASE

3.1 *Experimental data*

Brush & Wolman (1960) carried out a laboratory study of the behavior of knickpoints in noncohesive sediment. The experiments were conducted at the University of Maryland in a flume 15.85 m (52 ft) long and 1.22 m (4 ft) wide. Before starting each experimental run a trapezoidal channel 0.21 m (0.7 ft) wide and 0.03 m (0.1 ft) deep with rounded corners was molded in noncohesive sand. A short over-steepened reach with a length of 0.30 m (1 ft) and a fall of 0.03 m (0.1 ft), was located 10.8 m from the flume entrance. This oversteepened reach immediately downstream of the knickpoint had a 10% slope which was significantly steeper than the adjacent reaches. A total of 5 runs were performed with different slopes in the upstream and downstream reaches (between 0.125 and 0.88%). In every run, the slope of both water surface and bed below the knickpoint decreased with time, causing the position of the knickpoint to migrate upstream. As the knickpoint moved upstream the channel directly above it narrowed. At the lower end of the steep reach, sediment eroded from above deposited as a dune, which moved downstream causing the channel to locally widen.

Run 1, which has more detailed information, was selected for the numerical simulation. For Run 1 the

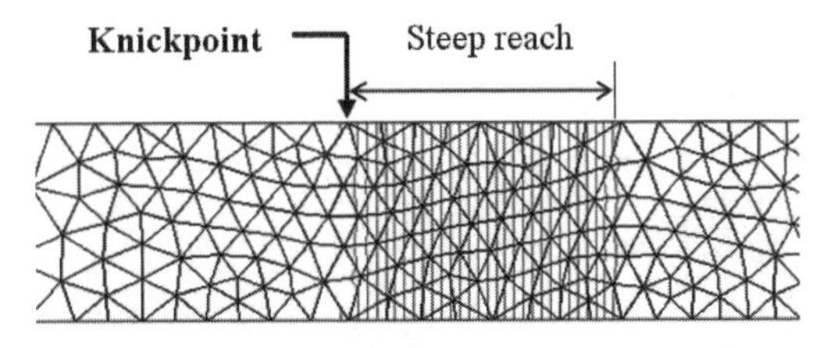

Figure 2. Resolution of Finite Element mesh around knickpoint and steep 10% slope reach (parallel gray lines are bed contours every 0.001 m).

longitudinal bed slope changed at a knickpoint from 0.125% to 10% and back to 0.125% again (Fig. 3). The channel was molded in sand 0.67 mm in diameter with initial water depth h_o = 0.0137 m and constant flow rate Q = 0.59 L/s. During Run 1 the width of the channel increased downstream of the knickpoint by bank erosion to about 0.25 m; while it got deeper and narrower upstream.

At the beginning of Run 1, no sediment was moving in the reaches above and below the steep reach. The erosion and deposition resulted solely from the presence of the knickpoint. Sediment eroded above the knickpoint was deposited below. The head and toe of the steep reach moved upstream and downstream respectively. Sediment transport rate was not measured.

3.2 *Numerical model*

In the numerical model the channel was assumed as rectangular with a fixed width of 0.21 m. An irregular mesh was used; the average size of the elements was about 0.20 m away from the steep reach and about 0.04 m around the knickpoint (Fig. 2). The mesh was refined close to the knickpoint to capture the strong flow gradients and the hydraulic jump. The mesh had 869 triangular elements and 563 nodes. The roughness height was set to k_s = 0.3 mm and porosity to λ = 0.4.

The boundary conditions for the hydrodynamic model were constant discharge in the upstream inflow section, and constant water surface elevation in the downstream outflow section.

4 RESULTS

Figure 3 shows the initial water surface profile computed before the beginning of the morphological simulation. Away from the steep reach, the flow remained subcritical with a Froude around 0.5; which corresponds to a velocity of about 0.2 m/s and a depth of 0.014 m, in agreement with observed values (Table 3 in Brush & Wolman 1960). As the flow approaches the knickpoint, it accelerates and becomes supercritical. The maximum velocity of about 0.7 m/s is reached

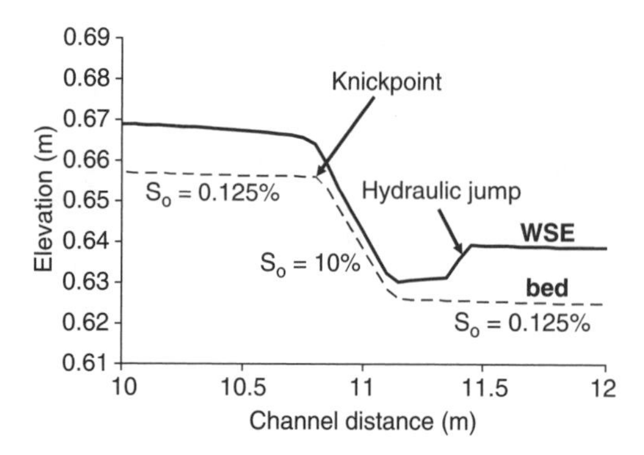

Figure 3. Initial profiles of bed elevation and water surface elevation (WSE) around knickpoint and steep reach ($t = 0$).

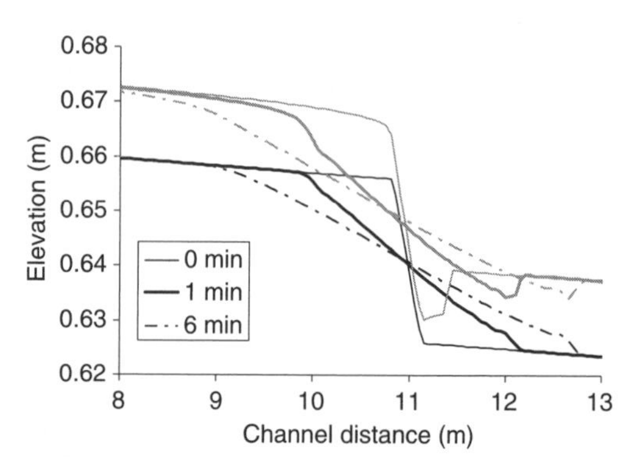

Figure 4. Computed profiles of bed (black lines) and water surface (gray lines) at the beginning of the simulation: $t = 0$, 1 and 6 min.

over the steep reach, the Froude number being as high as 3.5 with a water depth of only 0.004 m.

Downstream from the toe of the steep reach, the flow returns to subcritical through a hydraulic jump. The sharp front of the jump was captured by the model, as shown in Figure 3, without any noticeable spurious oscillation.

Since most sediment transport equations are derived for subcritical conditions, there is considerable uncertainty in the sediment transport equation for this experiment. Some preliminary runs were made testing 3 different sediment transport equations: Engelund-Hansen, Van Rijn and the empirical equation used by Bhallamudi & Chaudhry (1991) for this particular experiment:

$$q_s = (S_f)^{1.7}\left(\sqrt{u^2+v^2}\right)^{4.2} \tag{11}$$

None of the equations provided a perfect fit with the measured data; but Engelund and Hansen and Van Rijn equations notoriously underpredicted the upstream migration of the knickpoint (results not shown). Therefore, equation (11) was adopted.

A very small time step was used $\Delta t = 0.01$ s to prevent large bed changes during the initial period of the simulation. The results showed that both the computed bed and water surface profiles were rather smooth, without any noticeable numerical oscillation.

Figure 4 shows the computed bed and water surface profiles at the early stages of the simulation. It can be noticed that the most dramatic changes happen during the first minute. The profiles seem to pivot around the middle point of the steep reach. The intensity of the hydraulic jump decreases rapidly. The toe of the steep reach and the hydraulic jump moved rapidly downstream; while the knickpoint migrated upstream.

A comparison between the computed profiles and the bed profiles measured by Brush & Wolman (1960) at $t = 160$ min is shown in Figure 5. By this time, the intensity of the hydraulic jump has notoriously

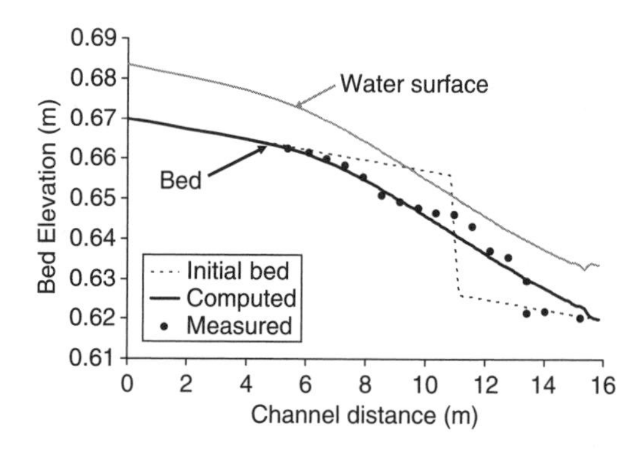

Figure 5. Comparison between computed and measured bed profiles at $t = 160$ min.

decreased since the maximum value of the Froude number was only slightly larger than 1.0. The steep bed slope has reduced to only 0.44% from its original value of 10%. The computed bed profile was smoother than the observed profile and tended to follow the average trend of the measurements. The new upstream location of the knickpoint agreed well with the observed value.

Bhallamudi & Chaudhry (1991) also simulated this experiment using a 1D model. However, they did not observe a hydraulic jump in their numerical simulations (Bhallamudi, pers. comm.); likely because the space discretization used $\Delta x = 0.30$ m was too large to capture the sharp front of the jump, and also because their downstream water level was higher ($h_o = 0.03$ m).

5 DISCUSSION

The morphodynamic simulation of the knickpoint migration experiment is a challenging problem because of the steep 10% bed slope that gives rise to a hydraulic jump and transcritical flow conditions.

1D morphological model with transcritical flow capabilities exist (e.g. Busnelli et al. 2001, Fraccarollo & Capart 2002, Papanicolaou et al., in press, Cui et al. a,b, in press); but they are still rare or inexistent in 2D or 3D models (Spinewine & Zech 2002). The shock-capturing techniques incorporated in River2D, added to the flexibility of the Finite Element method to increase mesh resolution around areas with strong gradients (Fig. 2), made possible to simulate the sharp hydraulic jump front (Fig. 3), without any numerical oscillation. This test case would be intractable for most 2D commercial models currently available (e.g. RMA2, MIKE 21C, FESWMS).

Busnelli et al. (2001) argue that hydraulic jumps tend to rapidly disappear in alluvial beds under constant water discharge. The experiments reported by Bellal et al. (2003) show that amplitude of the hydraulic jump is drastically reduced by a movable sediment front. At least qualitatively, the model seems to show a rapid reduction in the intensity of the jump as bed evolves (Fig. 4). In the final profiles shown in Figure 5, the jump is barely noticeable.

The agreement between measured and computed bed profiles is reasonable (Fig. 5), considering the large uncertainty in the sediment transport equation and the fact that both bank erosion and deposition have been completely ignored by assuming a fixed width. Brush & Wolman (1960) reported that channel became wider downstream of the knickpoint and narrower upstream; the maxima changes in width were about ±20% of the original width. Recently, Cantelli et al. (2004) and Wong et al. (2004) have demonstrated the importance of width changes in the morphological evolution of steep channels after dam removal; which can be considered as a knickpoint-like problem.

A very interesting and potential application of River2D-Morphology is the simulation of sediment transport after dam removal, which is a topic of high current interest (Cantelli et al. 2004, Wong et al. 2004, Cui & Wilcox, in press, Cui et al. a,b, in press). The hydraulic conditions after dam removal shown in Figure 1 (from Cui & Wilcox, in press) are practically identical to those depicted in Figure 3, with the over-steepened reach having a slope close the angle of repose of sediment. The computed bed profiles shown in Figure 4 also have some similarities with the initial bed profiles after dam removal computed by Cui & Wilcox (in press) for the Marmot Dam.

6 SUMMARY AND CONCLUSIONS

The recently developed River2D-Morphology model (Vasquez et al. 2005a, b) has been applied to the morphodynamic simulation of a knickpoint migration experiment in transcritical flow with successful results.

The main features of the model are:

- It uses a two-dimensional unstructured mesh based on the Finite Element method.
- It incorporates shock capturing techniques to simulate the sharp fronts of hydraulic jumps.
- It can dynamically update the bed elevation as scour and deposition progresses, in both subcritical and supercritical regimes.

Currently, the main limitations of the sediment transport model are:

- Suspended sediment is ignored.
- Sediment size is assumed uniform.
- Sediment waves (advection-dominated problems) can not be properly simulated.

However, the present limitations listed above only reflect the early stage of development of this model, which is probably one of the first Finite Element depth-averaged models with capabilities for transcritical flow morphodynamics.

The potential of River2D-Morphology to simulate alluvial rivers in both lowland and mountain reaches looks promising.

REFERENCES

Bellal, M. Spinewine, B., Savary, C. & Zech, Y. 2003. Morphological evolution of steep-sloped river beds in the presence of a hydraulic jump: experimental study. *Proceedings 30th IAHR Congress, Thessaloniki, Greece, 24–29 August 2003*. (C): 133–140.

Bhallamudi, S.M. & Chaudhry, M.H. 1991. Numerical modeling of aggradation and degradation in alluvial channels. *Journal of Hydraulic Engineering* 117: 1145–1164.

Brush, L.M. & Wolman, M.G. 1960. Knickpoint behaviour in noncohesive material: A laboratory study. *Bulletin of the Geological Society of America* 71: 59–73.

Busnelli, M.M., Stelling, G.S. & Larcher, M. 2001. Numerical morphological modeling of open-check dams. *Journal of Hydraulic Engineering* 127(2): 105–114.

Cantelli, A., Paola C. & Parker, G. 2004. Experiments on upstream-migrating erosional narrowing and widening of an incisional channel caused by dam removal. *Water Resources Research*, 40(3).doi:10.1029/2003WR002940, 2004.

Chung, T.J. 2002. *Computational Fluid Dynamics*, Cambridge University Press.

Cui, Y. & Wilcox, A., in press. Development and application of numerical modeling of sediment transport associated with dam removal. In M. Garcia (ed.), *Sedimentation Engineering*, ASCE Manual 54, Volume II.

Cui, Y., Parker, G., Braudrick, C., Dietrich, W.E. & Cluer, B. In press a. Dam Removal Express Assessment Models (DREAM). Part 1: Model development and validation. *Journal of Hydraulic Research*.

Cui, Y., Braudrick, C., Dietrich, W.E., Cluer, B. & Parker, G. In press b. Dam Removal Express Assessment Models (DREAM). Part 2: Sensitivity tests/sample runs. *Journal of Hydraulic Research*.

Donea, J. & Huerta, A. 2003. *Finite Element Methods for Flow Problems,* John Wiley and Sons, England.
Fraccarollo, L. & Capart, H. 2002. Riemann wave description of erosional dam-break flows. *Journal of Fluid Mechanics,* 461: 183–228.
Ghanem, A., Steffler, P., Hicks, F. & Katopodis, C. 1995. Two-dimensional modeling of flow in aquatic habitats. *Water Resources Engineering Report No. 95-S1.* University of Alberta, Canada.
Hicks, F.E. & Steffler, P.M. 1992. Characteristic dissipative Galerkin scheme for open-channel flow. *Journal of Hydraulic Engineering* 118: 337–352.
Hirsch, C. 1987. *Numerical computation of internal and external flows. Volume 1: Fundamentals of numerical discretization*, John Wiley and Sons, Brussels, Belgium.
Papanicolaou, A.N., Bdour, A. & Wicklein, E., in press. One-dimensional hydrodynamic/sediment transport model applicable to steep mountain rivers. *Journal of Hydraulic Research.*
Spinewine, B., Delobbe, A., Elslander, L. & Zech, Y. 2004. Experimental investigation of the breach processes in sand dikes. In Greco, Carravetta & Della Morte (eds), *River Flow 2004* (2): 983–991. Rotterdam:Balkema.
Spinewine, B. & Zech, Y. 2002. Geomorphic dam-break floods: near-field and far-field perspectives. 1st Impact Workshop, HR Wallingford.
Steffler, P. & Blackburn, J. 2002. *River2D: Two-dimensional depth-averaged model of river hydrodynamics and fish habitats*. University of Alberta, Edmonton, Canada.
Vasquez, J.A., Millar, R.G. & Steffler, P.M. 2005a. River2D Morphology, Part I: straight alluvial channels. *17th Canadian Hydrotechnical Conference.* Edmonton: CSCE.
Vasquez, J.A., Steffler, P.M. & Millar, R.G. 2005b. River2D Morphology, Part II: curved alluvial channels. *17th Canadian Hydrotechnical Conference.* Edmonton: CSCE.
Wong, M., Cantelli, A., Paola, C. & Parker, G. 2004. Erosional narrowing after dam removal: theory and numerical model. *Proceedings, ASCE World Water and Environmental Resources 2004 Congress, Salt Lake City, June 27–July* (1): 10.

River, Coastal and Estuarine Morphodynamics: RCEM 2005 – Parker & García (eds)
© 2006 Taylor & Francis Group, London, ISBN 0 415 39270 5

Validation of a computational model to predict suspended and bed load sediment transport and equilibrium bed morphology in open channels

Jie Zeng, George Constantinescu & Larry Weber
Civil & Environmental Engineering Department, IIHR-Hydroscience and Engineering, The University of Iowa, M. Stanley Hydraulics Laboratory, Iowa City, IA

ABSTRACT: Prediction of flow and sediment transport in open channels and rivers, in particular the equilibrium scour bathymetry, is of critical importance in many river engineering applications due to the possible effects on the river eco-system, structural integrity of hydraulic structures, navigability, etc. This paper focuses on the validation of the suspended sediment transport module and its coupling with the non-equilibrium bed load module that was described in a previous paper (Zeng et al., 2005). The non-hydrostatic fully three-dimensional (3D) model solves the incompressible, Reynolds-Averaged Navier-Stokes (RANS) equations in generalized curvilinear coordinates and integrates the equations up to the wall such that the use of wall-functions and the approximations associated with them are avoided. One of the novelties of the revised form of the model is the implementation of the Spalart-Allmaras model in a near-wall version that can account for bed roughness effects (Spalart 2000) besides the original k-ω model. The solver uses movable grids in the vertical direction to account for changes in the free surface elevation where the proper kinematic and dynamic conditions are imposed and in the bathymetry due to erosion/deposition at the bed as the code converges toward steady state. A non-equilibrium bed load sediment transport model similar to the one used by Wu et al. (2000) is used with the additional introduction of down-slope gravitational force effects. The suspended sediment is modeled using an advection diffusion equation with a settling velocity term. Validation of the suspended sediment module is accomplished through simulation of two test cases in straight channels, one with net entrainment and one with net deposition (zero entrainment) at the bed. The model is then used to predict the suspended and bed loads as well as the equilibrium bed bathymetry through an 180° open channel bend for which detailed experimental data are available (Odgaard & Bergs, 1988). Overall, the model predictions of the sediment transport and equilibrium bathymetry are found to be accurate.

1 INTRODUCTION

Successful simulation of flow and sediment transport in open channels (including those of variable section or meandering in the streamwise direction for which the effect of secondary motions and flow separation/reversal can be very important) and river reaches requires the use of complex numerical models that can accurately capture not only the mean flow but also the effect of presence of bed roughness of different scales including vegetation, the suspended and bed load sediment components, the deformation of the free surface and the associated morphological changes.

As in the present work we are interested mainly by predictions of flow and sediment in channels of complex geometry in which strong secondary currents exist we are going to briefly review only previous simulation efforts conducted using 3D codes. One of the first models in which the sediment transport was calculated in 3D was proposed by van Rijn (1986). However, for the hydrodynamics he used a 2D depth-averaged model. A further step toward a fully 3D model was the 3D layer-integrated model of Lin & Falconer (1996) that used Cartesian meshes. It was used to predict tidal flows and sediment transport in estuaries. Gessler et al. (1999) were one of the first to use a 3D hydrostatic solver (CH3D) and couple it with a very complex mobile-bed module which can account for aggradations and scour, bed-material sorting, and the movement of non-uniform sediment mixtures as bed load and suspended load, respectively. More recently, Olsen (2003) used a fully 3D non-hydrostatic model to predict the meandering of alluvial channels using a finite volume time-accurate solver that can employ unstructured grids. Models for both the suspended load and the bed load were incorporated into the code. The standard k-ε model with wall functions was used. Typical mesh sizes used in the applications of the model were around $2 \cdot 10^4$ cells. Wu et al. (2000) implemented a fairly complex non-equilibrium bed load model along with an advection-diffusion based suspended sediment model into a fully non-hydrostatic 3D finite volume code. A particularity of the code was that the water surface deformation was obtained from a 2D Poisson equation for the surface height obtained from the depth-averaged 2D momentum equations. Again, the standard k-ε model with wall functions was used and wall roughness effects were accounted by modifying the wall-function formulas used for smooth walls based on correlations for rough walls deduced from experiment. The main validation test case was the prediction of the open channel bend flow of Odgaard & Bergs (1988) at equilibrium scour conditions. The simulation was run on a mesh with 36,000 elements. Same code was used by Fang & Rodi (2003) to predict the flow and suspended sediment transport for the reservoir generated by the dam of the Three Gorges Project on the Yangtze River. The grid size was around 200,000 elements.

The approach that is advocated in this paper is the use of a fully 3D non-hydrostatic viscous flow solver (Constantinescu & Squires, 2004) and turbulence models that: 1) avoid the use of the widely used wall functions approach which is known to introduce significant errors in flows in separated flows or flows in which large adverse pressure gradients are present and 2) have the capability to account for roughness effects via the boundary conditions for the transported quantities in the model. The present paper describes only the RANS part of the model that is used to predict the equilibrium steady state flow including the equilibrium scoured bathymetry (the proposed model uses deformable meshes in which the points are redistributed in the vertical direction during convergence to steady state), however we intend to extend the model to be used within a Detached Eddy Simulation (DES) formulation, in which case a fully time accurate version has to be developed. DES is a hybrid method in which the base RANS model is modified in the regions away from the solid boundaries such that it acts as a Large Eddy Simulation model allowing the eddy viscosity to scale with the square of the local grid spacing as in classical LES. As the most popular formulations of DES use the Spalart-Allmaras (SA) and the k-ω models, and the original solver (Constantinescu & Squires, 2004) already has DES capabilities, we choose to concentrate on the use of these two models in our RANS simulations which can also predict bed roughness effects. The critical parts of the sediment transport module, meaning the ones that introduce most of the empiricism and uncertainties, are the choosing of the bed load model, the way in which the suspended sediment transport model is coupled with the bed load model and the bed entrainment model. The bed load model is very similar to the one used by Wu et al. (2000), however a correction is introduced to account for down-slope gravitational force effects (Sekine & Parker, 1992). To our knowledge this is the first time this coupling is done in the context of using near-wall models in a fully 3D code, in which the first point off the wall is situated well within the thickness of the bed load layer. As the advection-diffusion equation for the suspended sediment is solved only in the domain over the bed load layer, and the thickness of the bed load layer is variable, the correct handling of the boundary conditions and estimation of variables at the interface between the suspended and bed load layers is very important for the robustness and accuracy of the model. At the free surface the usual kinematic and dynamic boundary conditions are imposed to calculate the free surface deformation.

2 HYDRODYNAMIC MODEL AND SEDIMENT TRANSPORT MODEL

The 3D incompressible RANS and turbulence/sediment transport equations are first transformed in generalized curvilinear coordinates, but the primitive variables for which the momentum equations are solved are still the Cartesian velocity components, V_i. The continuity and momentum equations are:

$$J\frac{\partial}{\partial \xi^j}\left(\frac{V^j}{J}\right)=0 \tag{1}$$

$$\frac{\partial Q}{\partial t}+A_j\frac{\partial Q}{\partial \xi^j}-J\frac{\partial E_{vj}}{\partial \xi^j}+H=0 \tag{2}$$

where x_i are the Cartesian coordinates and ξ^i are the curvilinear coordinates.

In eqn. (2), $Q=(V_1, V_2, V_3)^T$ is the Cartesian velocity vector; J is the Jacobian of the geometric transformation $J=\partial(\xi^1,\xi^2,\xi^3)/\partial(x_1,x_2,x_3)$, $V^i=V_j\xi^j_{x_i}$ are the contravariant velocity components, $A_j=\mathrm{diag}(V^1,V^2,V^3)$ and $E_{vj}=(E^1_{vj},E^2_{vj},E^3_{vj})$ are the vectors containing the viscous and turbulence fluxes and, finally, H is the pressure gradient vector

$$H=\left(\xi^k_{x1}\frac{\partial \psi}{\partial \xi^k},\xi^k_{x2}\frac{\partial \psi}{\partial \xi^k},\xi^k_{x3}\frac{\partial \psi}{\partial \xi^k}\right)^T \tag{3}$$

The contravariant metric tensor of the transformation is $g^{ij}=\xi^i_{x_k}\xi^j_{x_k}$ (summation over k). The governing equations including the ones solved part of the turbulence and suspended sediment transport models are non-dimensionalized using a length scale (the inlet water depth) and a velocity scale (bulk velocity in the inlet section) such that formally the Reynolds number ($\mathrm{Re}=UL/\nu$) replaces the molecular viscosity ν, and the Froude number ($\mathrm{Fr}=U/(gL)^{0.5}$) is used to scale the free surface deformations. The modified pressure ψ is defined as $\psi = p + 2/3k$ and the effective piezometric pressure P is defined as $P = p/\rho + z/\mathrm{Fr}^2 - 2/3k$, where p is the pressure, z is the free surface elevation and k is turbulence kinetic energy. As we shall see later, the Froude number enters the solution through the free-surface boundary condition assuming that the deformable free surface model is turned on (for more details on this see Zeng et al., 2005). At the outflow all variables are extrapolated from the interior. The pressure is extrapolated from the interior of the domain on all boundaries except the free surface.

The eddy viscosity is provided by the SA model in the test cases presented in the present paper, however the k-ω model is also available (Zeng et al., 2005). In this model only one transport equation is solved for the modified viscosity $\tilde{\nu}$:

$$\frac{\partial \tilde{\nu}}{\partial t} = c_{b1}\tilde{S}\tilde{\nu} + \frac{1}{\sigma}[\nabla\cdot((\nu+\tilde{\nu})\nabla\tilde{\nu}) + c_{b2}(\nabla\tilde{\nu})^2] - c_{w1}f_w[\frac{\tilde{\nu}}{d}]^2 \quad (4)$$

where $\tilde{S} \equiv S + (\tilde{\nu}/\kappa^2 d^2)f_{v2}$ with S the magnitude of the vorticity and $f_{v2} = 1 - \tilde{\nu}/(1/Re + \tilde{\nu}f_{v1})$.

The eddy viscosity ν_t is obtained from

$$\nu_t = \tilde{\nu} f_{v1} \quad (5)$$

where $f_{v1} = \chi^3/(\chi^3 + C_{v1}^3)$, $\chi = \tilde{\nu}/\nu + 0.5\frac{k_s}{d}$,

$$f_w = g[\frac{1+C_{w3}^6}{g^6 + C_{w3}^6}]^{\frac{1}{6}},\ g = r + C_{w2}(r^6 - r),$$

$$r \equiv \frac{\tilde{\nu}}{\tilde{S}\kappa^2 d^2} \quad \text{and} \quad f_{t2} = c_{t3}\exp\left(-c_{t4}\chi^2\right).$$

To account for roughness effects the distance to the wall is redefined (see also Spalart, 2000) as:

$$d = d_{\min} + 0.03k_s \quad (6)$$

where $d_{\min}$ is the distance to the nearest wall and k_s is the equivalent roughness height.

The model constants in the above equations are:

$$C_{b1} = 0.135,\ C_{b2} = 0.622,\ \sigma = 0.67,\ \kappa = 0.41,$$
$$C_{v1} = 7.1,\ C_{t3} = 1.1,\ C_{t4} = 2.0,\ C_{w2} = 0.3,\ C_{w3} = 2.0,$$
$$C_{w1} = C_{b1}/\kappa^2 + (1 + C_{b2})/\sigma.$$

All velocity components and $\tilde{\nu}$ are set equal to zero at the (side) walls which are simulated as smooth surfaces. At the bed, that is treated as a rough surface, the value of $\tilde{\nu}$ is estimated (see also Spalart, 2000) by solving the following equation:

$$\partial\tilde{\nu}/\partial n = \tilde{\nu}/d \quad (7)$$

This makes the modified viscosity and the eddy viscosity to be formally non-zero on rough walls. The equivalent bed roughness k_s may be estimated using a formula proposed by van Rijn (1984) in the case in which small bed forms are present:

$$k_s = 3d_{90} + 1.1\Delta(1 - e^{-25\psi}) \quad (8)$$

where the first term is the sand grain roughness contribution and the second term represents the bed form contribution. In eqn. (8), $\psi = \Delta/\lambda$ where Δ and λ are the height and length of the sand waves, respectively. Following van Rijn (1984) and Wu et al. (2000) the length λ is assumed to be $\lambda = 7.3\,h$ and the parameter ψ is estimated from:

$$\psi = \Delta/\lambda = 0.015(d_{50}/h)^{0.3}\left(1 - e^{-0.5T}\right)(25 - T) \quad (9)$$

in which h is water depth, d_{50} and d_{90} are the median and 90% diameters of the bed material and T is the non-dimensional excess shear stress given by van Rijn (1987). Its expression is given below in eqn. (17).

A fractional step method is used to solve the RANS equations. All terms in the pressure Poisson equation are discretized with second-order accurate central differences. The momentum and turbulence transport equations are discretized using second order accurate upwind differences for the convective terms, whereas all other operators are calculated using second-order central discretizations. All terms are treated implicitly, including the source terms in the transport equations for turbulence quantities ($\tilde{\nu}$). To accelerate the convergence of the resulting system of equations, local time-stepping techniques are used. An approximate factorization technique is used to simplify the inversion of the discrete form (left hand side) of the momentum, pressure-Poisson and turbulence transport equations. The equations are solved implicitly using the alternate-direction-implicit (ADI) method.

The kinematic and dynamic boundary conditions are applied on the free surface to determine the free surface deformation. The implementations of these boundary conditions along with a validation test case for the deformable free surface module were described in details in Zeng et al. (2005).

An advection-diffusion scalar transport equation with an additional settling-velocity source term is solved to determine the local sediment concentration C and the suspended sediment load. The equation is not solved up to the wall as are the momentum equations but rather up to the interface with the bed load layer

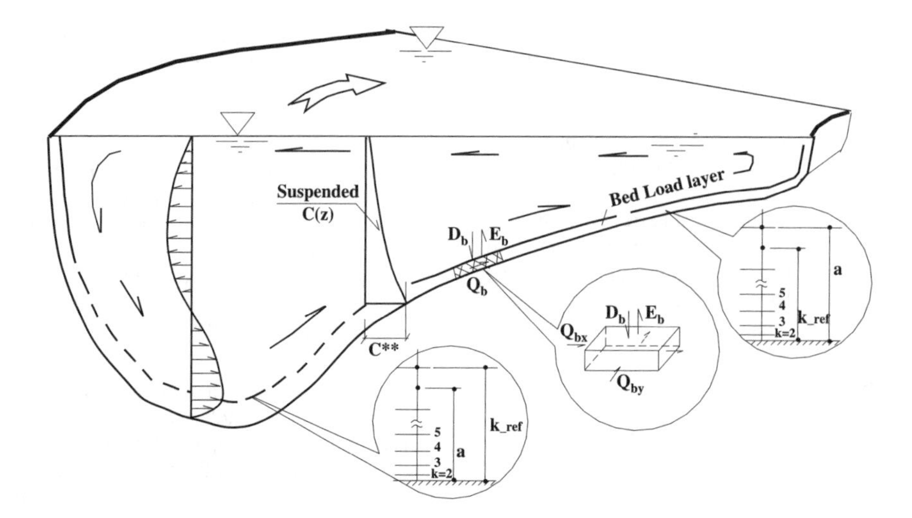

Figure 1. Sketch showing the flow, sediment transport process, the suspended and bed load layers.

($\delta(x, y)$) which typically corresponds to a k = constant surface in the computational grid (see also sketch in Fig. 1). In curvilinear coordinated the equation is:

$$\frac{\partial C}{\partial t}+V^j\frac{\partial C}{\partial \xi^j}-\omega_s\delta_{3i}\frac{\partial C}{\partial \xi^j}\frac{\partial \xi^j}{\partial x_i}-J\frac{\partial}{\partial \xi^j}\left(\frac{g^{ij}}{J}\left(\frac{\nu+\nu_t}{\sigma_c}\right)\frac{\partial C}{\partial \xi^i}\right)=0 \tag{10}$$

where w_s is the setting velocity of the sediment, δ_{3j} is the Kronecker delta symbol with $j=3$ indicating the vertical direction and σ_c is the Schmidt number. Its role is to relate the sediment (turbulent) diffusivity to its (eddy) viscosity.

At the free surface, the total vertical flux of suspended sediment is set to zero, giving:

$$\frac{\nu+\nu_t}{\sigma_c}\frac{\partial C}{\partial \xi^j}\frac{\partial \xi^j}{\partial x_3}+w_sC=0 \tag{11}$$

At the interface between the suspended sediment and the bed load layers situated not far but not necessarily at the reference level $z=a$, following van Rijn (1987), and Wu et al. (2000), the entrainment rate is assumed to be equal to the one under equilibrium conditions. The total vertical sediment flux at the interface is equal to the net sediment transport (deposition minus entrainment) across the interface.

$$\frac{\nu_t}{\sigma_c}\frac{\partial C}{\partial \xi^j}\frac{\partial \xi^j}{\partial x_3}+w_sC=D_b-E_b=w_sC-w_sC_{b**} \tag{12}$$

or

$$\frac{\nu_t}{\sigma_c}\frac{\partial C}{\partial \xi^j}\frac{\partial \xi^j}{\partial x_3}+w_sC_{b**}=0 \tag{13}$$

where C_{b**} is the equilibrium concentration at $z=\delta$, $D_b=w_sC_b$ is the deposition rate and $E_b=w_sC_{b**}$ is the entrainment rate at the top of the bed layer. The value of C_{b**} can be calculated assuming Rouse formula to be valid between the reference level where $C=C_{b*}$ ($z=a$) and the interface between the bed and suspended load layers ($z=\delta(x, y)$). There is no unique way to choose the reference level a. In some cases it is taken equal to the bed roughness determined from eqn. (8), in other cases it is taken as a percentage of the total depth (see also Lin & Falconer, 1996). At the inlet section generally a concentration profile needs to be prescribed, while at the sidewalls and outlet a zero gradient boundary condition is imposed.

Similar to the method used to solve the discretized form of the turbulence transport equation for $\tilde{\nu}$, eqn. (10) is first discretized in delta form, the left hand side is then factorized and solved for ΔC using ADI. On the right hand side the convective terms are discretized using the second order upwind scheme while the diffusion terms are discretized using second order central differences scheme. The settling velocity term is treated as a source term. The time derivative $\partial C/\partial t$ is used to advance the equation in pseudo-time during the iteration process and local time stepping techniques are used to accelerate the convergence.

The empirical formulas proposed by van Rijn (1987) are used in the present study to estimate the equilibrium concentration at the reference level (C_{b*}) the equilibrium bed load transport rate (Q_{b*}) and the non-equilibrium adaptation length (L_s).

$$C_{b*}=0.015\frac{d_{50}T^{1.5}}{aD_*^{0.5}} \tag{14}$$

$$Q_{b*}=0.0053(Rg)^{0.5}\left(d_{50}^{1.5}T^{2.1}\right)/\left(D_*^{0.3}\right) \tag{15}$$

$$L_s = 3d_{50}D_*^{0.6}T^{0.9} \quad (16)$$

The non-dimensional excess bed shear stress T is defined as:

$$T = [(u'_*)^2 - (u_{*cr})^2]/(u_{*cr})^2 \quad (17)$$

In the above equations $R = (\rho_s/\rho) - 1$, ρ_s is the sediment density and ρ is the water density. The particle-size diameter is $D_* = d_{50}[Rg/\nu^2]^{1/3}$, $u_{*_{cr}}$ is the critical bed-shear velocity for sediment motion given by Shields diagram. In eqn. (17) u'_* can be estimated as $u'_* = g^{0.5}\hat{U}/C'$, where $\hat{U}$ is the local depth-averaged streamwise velocity. The sand grain Chezy coefficient is calculated as $C' = 18 \cdot \log(12R_b/3d_{90})$, where R_b is the hydraulic radius. A more exact and general way to estimate the shear stress for cases in which the version of the SA model that can account for roughness effects is used is to simply equate u'_* to u_* where $u_* = \sqrt{\tau_*/\rho}$ and the bed shear stress is calculated using its definition $\tau_* = \rho(\nu + \nu_t)(\Delta V/\Delta z_1)$ where ΔV is the velocity magnitude at the first point off the wall (normal component is very small and velocity on the wall is zero) and, as mentioned, the eddy viscosity is non-zero on rough boundaries.

For the bed-load transport and the calculation of the bed changes due to global deposition/erosion from the bed, we are using an approach similar to Wu et al. (2000). However, there are some differences in the way we account for the suspended sediment contribution to the modification of the bed level. We start with the same mass balance equation for the sediment within the bed load layer (van Rijn, 1987) in which we neglected the storage term:

$$(1-p')\frac{\partial z_b}{\partial t} - D_b + E_b + J_b\left(\frac{\partial}{\partial \xi}\left(\frac{Q_{b\xi}}{J_b}\right) + \frac{\partial}{\partial \eta}\left(\frac{Q_{b\eta}}{J_b}\right)\right) = 0 \quad (18)$$

Then we assume that the relation between the bed change due to deposition from the bed load layer is of the form $((1 - p')\partial z_{b\_bed\_load}/\partial t = (Q_b - Q_{b*})/L_s)$ such that the total change in bed elevation is:

$$(1-p')\frac{\partial z_b}{\partial t} = D_b - E_b + \frac{Q_b - Q_{b*}}{L_s} \quad (19)$$

The equation for the bed load rate Q_b is derived from eqn. (18) in which we made the substitution corresponding to the model assumed in eqn. (19).

$$\frac{(Q_b - Q_{b*})}{L_s} + J_b\left(\frac{\partial}{\partial \xi}\left(\frac{Q_{b\xi}}{J_b}\right) + \frac{\partial}{\partial \eta}\left(\frac{Q_{b\eta}}{J_b}\right)\right) = 0 \quad (20)$$

If T in equation (17) becomes negative at a certain location, then a very small value for L_s is used in eqn. (20) such that in fact the solution is $Q_b = Q_{b*}$. One should make the important observation that the equation (20) which is solved to determine the bed load is the same for both the case when the suspended sediment transport module is active and for the case when the suspended bed load is neglected. However, in the first case the total bed elevation change will be partly due to the net entrainment/deposition of suspended sediment at the top of the bed load layer. If only the bed load transport is considered, D_b and E_b are set to be zero in eqn. (19). In the above equations p' is the porosity, J_b is the Jacobian of the geometric transformation in the bed load layer, $(J_b = \partial(\xi,\eta)/\partial(x,y))$, z_b is the bed level above a datum, $Q_{b\xi}$ and $Q_{b\eta}$ are the components of the bed-load transport in the two directions. In a first approximation one can use $Q_{b\xi} = \alpha_{b\xi}Q_b$ and $Q_{b\eta} = \alpha_{b\eta}Q_b$ where $\alpha_{b\xi}$ and $\alpha_{b\eta}$ are direction cosines in the streamwise ξ and transversal η directions of the bed shear stress. However, we found that more accurate results can be obtained by accounting for bed slope effects on the bed load transport. This is because when the bed becomes sloped the gravitational force on the particles will resist the shear force to further carry the particles to the upper part of the slope. In this study, the relationship of Delft Hydraulics (Sekine & Parker, 1992) is used:

$$Q_{b\xi}/Q_{b\eta} = \left(\sin\theta_b - \beta\frac{\partial z_b}{\partial l_\xi}\right)\Big/\left(\cos\theta_b - \beta\frac{\partial z_b}{\partial l_\eta}\right) \quad (21)$$

where $\theta_b = \tan^{-1}(\tau_{b\eta}/\tau_{b\xi})$ is the angle between the bed shear stress vector and the streamwise direction, $\beta = \beta^*(\tau_c^*/\tau_L^*)^m$, τ_L^* is the longitudinal Shield stress $\tau_L^* = \tau_*\cos\theta_b/\rho Rgd_{50}$, $\tau_c^* = \rho u_{*cr}^2$, m is a coefficient equal to unity in Delft's formula, $\beta^* = \alpha_G/\tau_c^*$ and $\partial z_b/\partial l_\xi$, $\partial z_b/\partial l_\eta$ are the streamwise and transverse bed slopes. Finally, α_G is a coefficient typically in the range of 0.45–0.7. In our simulations with the SA model we used $\alpha_G = 0.6$. Equation (20) in which (21) was used to express $Q_{b\xi} = f(Q_b)$ and $Q_{b\eta} = f(Q_b)$ can be discretized in delta form and then solved in 2D for ΔQ_b^0 using ADI. Finally, an implicit residual smoothing operator is applied to obtain ΔQ_b and thus the bed load discharge at the next iteration. In solving this equation, the bed load transport rate at the inlet is the only variable needed to be specified. A zero gradient condition is used at the outlet.

Once z_b is calculated, the grid points between the bed level and the free surface are redistributed vertically based on the positions of the water surface and bed level elevations using a hyperbolic stretching function.

3 VALIDATION

3.1 *Flow in a straight channel with net deposition of suspended load*

Wang & Ribberink (1986) reported concentration measurements in a straight channel with an initially rigid bottom region followed by the test region in which the bottom was perforated such that no entrainment from the bed was possible ($E_b = 0$). A sketch of the experiment is shown in Fig. 2(a). So, over the test section there is a net deposition of suspended load. The mean velocity was $U = 0.56$ m/s and the water depth $L = 0.215$ m. Correspondingly, the Reynolds and Froude numbers were 12,040 and 0.386, respectively. The characteristic diameters of the sediment material were $d_{50} = 0.095$ mm, and $d_{90} = 0.105$ mm. In the present numerical simulation, the fall velocity was taken as $w_s = 0.0065$ m/s and the bottom roughness height $k_s = 0.0025$ m ($k_s^+ = 87.1, k_s/d_{90} \sim 25$) following van Rijn (1986) and Wu et al. (2000). The mesh size was 251 × 40 in the streamwise and vertical direction, respectively with the first point off the wall situated at approximately $\Delta z^+ = 1$. First, a fully developed SA solution in a channel of identical section, at same Reynolds number was calculated. The calculated bed shear velocity was 0.0348 m/s, which matches well with the value determined from measurements (0.033 m/s). Then the velocity and modified viscosity profiles were used to specify the inlet conditions in the inflow outflow simulation corresponding to the experiment. At the bottom boundary, a no-slip condition was specified for the velocity along with the value of the bottom roughness, k_s, needed to account for wall roughness effects in the SA model. The value of the Schmidt number was $\sigma_c = 1.0$. Using the zero entrainment condition ($E_b = 0$), eqn. (13) reduces to implementing a zero gradient boundary condition for the concentration at the bottom boundary $((\nu + \nu_t)/\sigma_c) \cdot \partial C/\partial z = 0)$. The measured concentration profile at the most upstream station (x/h = 0.0 in Fig. 2(b)) was used as the inflow boundary for the suspended sediment.

The calculation results at various downstream stations (x/h = 4.65 to x/h = 72.1) are compared with the measured concentration profiles in Fig. 2(b).' The overall agreement is good. The concentration values are somewhat overestimated for 10 < x/h < 60, however the same phenomenon was observed in the simulation of Wu et al. (2000).

3.2 *Flow in a straight channel with net entrainment of suspended load*

A simple experiment in a straight flume in which clear water was convected over a sediment bed was conducted by van Rijn (1981). A net sediment entrainment from the loose bed into suspension was observed at all test sections. The concentration profiles at various locations downstream of the start of the sediment bed were recorded after the fully transport sediment capacity was reached. The experiment configuration is sketched in Fig. 3(a). In the experiment, the water depth was $L = 0.25$ m and the mean velocity was $U = 0.67$ m/s. The Reynolds and Froude numbers were 167,500 and 0.427 respectively. The bed material consisted of sand with $d_{50} = 0.23$ mm and $d_{90} = 0.32$ mm. The representative particle size for determining the fall velocity was chosen equal to 0.2 mm. The corresponding fall velocity was 0.022 m/s.

In our numerical simulation a mesh with 250 × 40 points in the streamwise and vertical directions was used. The first point off the bottom surface was situated at approximately $\Delta z^+ = 1$. Similarly with the previous test case, a fully developed SA solution in a channel of identical section, at same Reynolds number was first calculated. Then the velocity and modified viscosity profiles were used to specify the inlet conditions in the inflow-outflow simulation corresponding to the experiment. At the top surface a symmetry boundary condition was used (normal velocity component was set equal to zero and shear stress component in directions parallel to the surface was set equal to zero). At the bottom boundary, the value of the bottom roughness was taken equal to the one used by van Rijn (1986) $k_s = 0.01$ m ($k_s{}^+ = 517, k_s/d_{90} \sim 31$). The value of the Schmidt number was $\sigma_c = 1.0$. Based also on a parametric study, the same height of the reference level as the one employed by van Rijn (1986) was used ($a = 0.0025$ m), which corresponds to 1% of the channel depth ($a/d_{90} \sim 8$). About 14 points were distributed between the bed level and the reference level. The general boundary condition given by eqn. (13) was imposed at the interface between the bed and suspended sediment layers. At the inlet section (x/h = 0.0) the concentration C was set equal to zero.

The calculation results at various downstream stations (x/h = 4.0 to x/h = 40.0) are compared with the measured concentration profiles in Fig. 3(b). As clearly observed in these plots, the present model successfully captures the concentration distribution at all downstream stations.

3.3 *Flow in a curved U-shaped channel with mobile bed and deformable free surface*

Test 1 experiment performed by Odgaard & Bergs (1988) in an U-shaped open channel (Fig. 4) is simulated. The channel width was 2.44 m and its length was 80 m. The cross section of the channel was trapezoidal with vertical side walls. The channel bottom was covered with sand with $d_{50} = 0.35$ mm and $d_{90} = 0.53$ mm. The porosity p' value was 0.65. The average water depth at inlet was $L = 0.15$ m, the inflow mean velocity was $U = 0.43$ m/s (Fr = 0.371, Re = 65,500) and

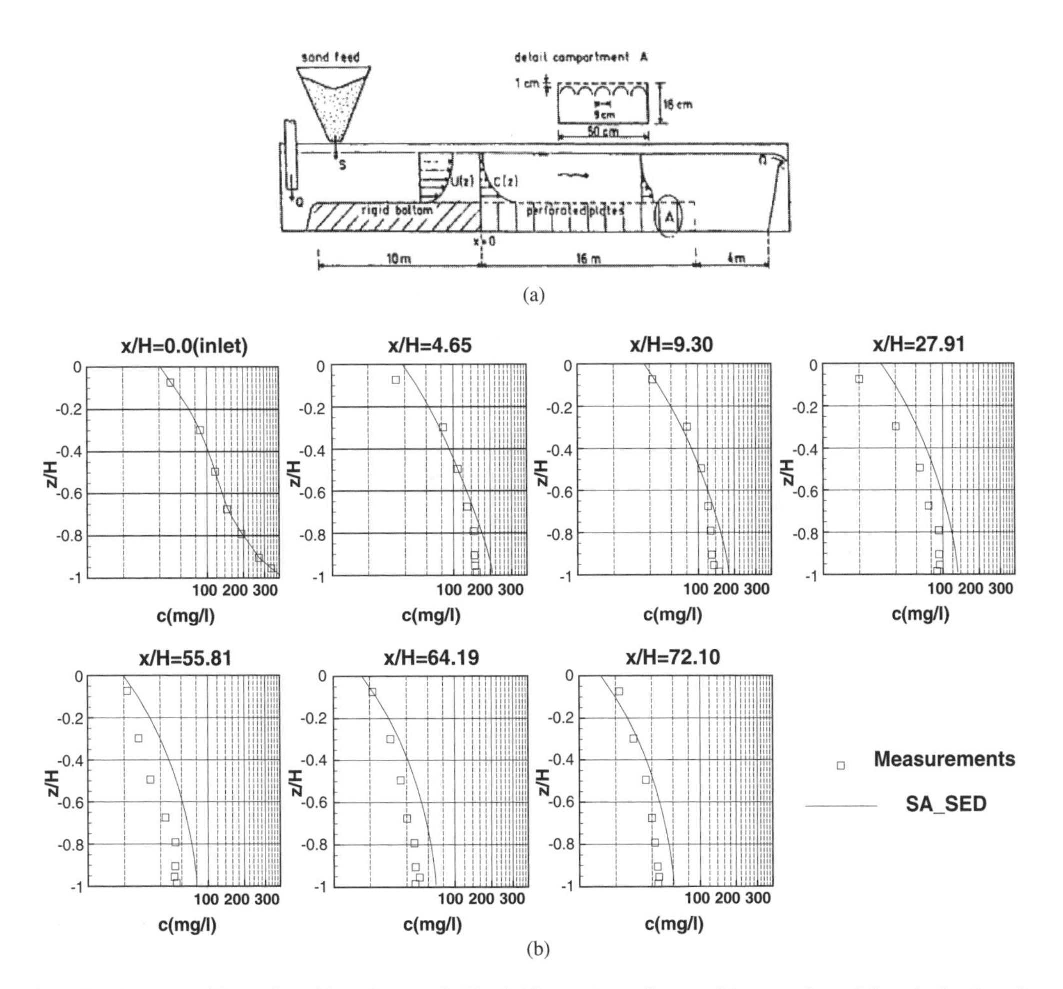

Figure 2. Test case with net deposition of suspended load. (a) experimental setup; (b) comparison of the calculated results with measurements.

the streamwise slope of water surface was 0.00116. The experiment was run until an equilibrium steady state for the bed bathymetry and the flow was reached. The bed and water surface elevations and depth averaged velocities at several sections were measured for equilibrium conditions. The present simulations were performed on a grid with 360,000 $(101 \cdot 101 \cdot 35)$ mesh points, with the first point off the lateral walls (smooth surface) and bed situated at $\Delta z/h \sim 0.00038$ $(\Delta z^+ \sim 1)$. The simulations started with a flat sand bed. Using formulas (8) and (9), the bed roughness was estimated to be $k_s = 0.02$ m which is well into the fully rough regime $(k_s^+ = 920)$. A fully developed SA solution in a channel of identical section, at same Reynolds number was first calculated. The velocity and modified viscosity distributions were used to prescribe the corresponding profiles at the inlet section of the U-shaped channel. The sediment moved in the experiment mainly as bed load. The measured inflow rate of 3.7 g/(cm · min) was specified at the inflow section in the simulations.

Two simulations were performed. In the first one, only the bed load transport was considered. This is justified as in the experiment the bed load was identified as the main mechanism responsible for sediment transport in the experiment. Thus the suspended sediment module was turned off and D_b and E_b were assumed to be zero in eqn. (19). In the second simulation, the suspended load and bed load modules were turned on. In our simulation the value of a was equal to 0.00375 m $(a/d_{90} \sim 7.5)$ corresponding to 2.5% of the water depth at the inlet section. With this choice, matching a to a k = constant plane in the inlet section resulted in a value of k = 17 (k = 1 at the bed and k = 34 at the free surface) which defined the interface between the two layers in the whole domain. A value of the Schmidt number of $\sigma_c = 0.85$ was used. At the inlet section (x/h = 0.0) the concentration C was set equal to zero

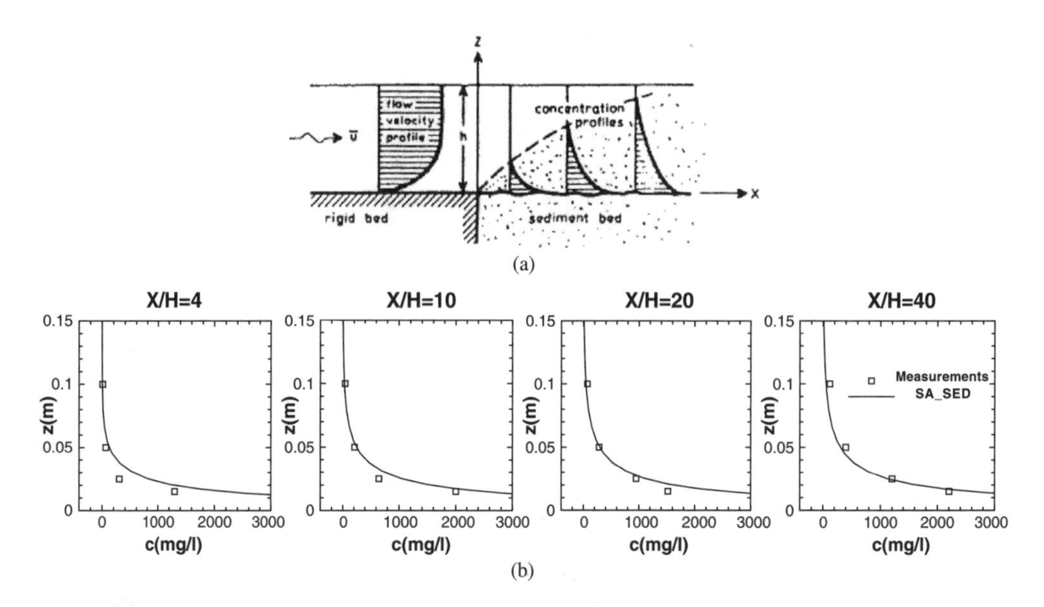

Figure 3. Test care with net entrainment of sucspended load. (a) experimental setup; (b) comparison of the calculated results with measurements.

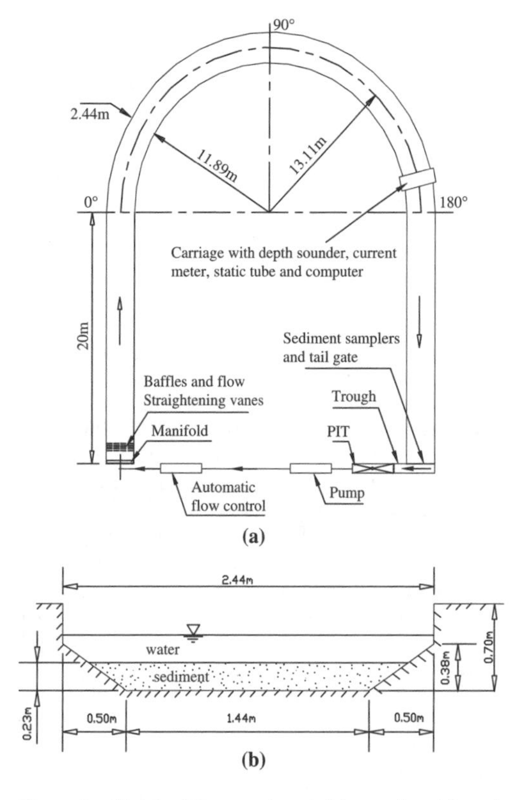

Figure 4. Sketch of the experiment. (a) experiment layout; (b) cross-section of the flume.

(no experimental measurements were available). On the lateral walls and on the free surface a zero flux boundary condition was used.

Results of two simulations performed by Wu et al. (2000) that used the standard k-ε model with wall functions and accounted for both bed load and total load, respectively, are included to get a better idea about the performance of the present model.

Figure 5 shows the variation of the water depth along the streamwise direction (θ) at three positions situated respectively at 3b/8 (near outer bank), 0 and $-3b/8$ (near inner bank) from the centerline. Due also to the scattering of the experimental data, it is fairly difficult to draw a clear conclusion on the relative performance of the present SA_SED simulations over the k-ε simulations as well as on the effects of taking into account the suspended load sediment transport. For instance for $0° < \theta < 80°$ the SA_SED simulation which accounts for the suspended component appears to be very close to the experimental data at the 3b/8 position. It also appears to be the only one to capture the presence of a second hump in the downstream region at $\theta \sim 160°$ albeit little bit early compared to the measurements where the hump is situated at $\theta \sim 175°$. None of the simulations capture the relative minimum observed for the bed elevation in the measurements at $\theta \sim 60°$. With the exception of the k-ε simulation without the suspended load component that appears to predict a too faster decay of the bed elevation variation for $150° < \theta < 210°$ (3b/8) the agreement between the other models and the experimental data is similar for the 3b/8 line. On the inner side of the channel where deposition takes place,

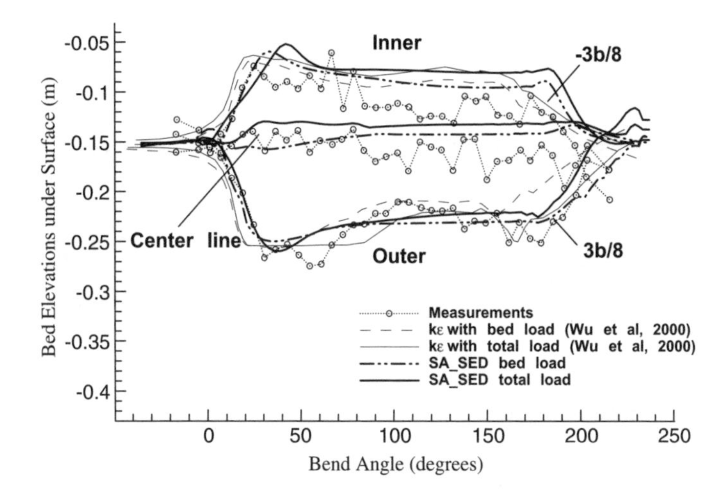

Figure 5. Comparison of longitudinal water depth at three positions in the 180° channel bend. The upper lines are at -3b/8. The middle lines are at center line. The lower lines are at 3b/8 where b is the bottion width of the channel.

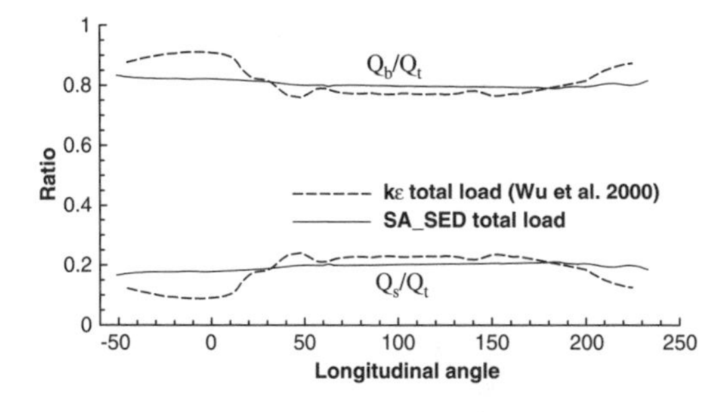

Figure 6. Calculated proportions of bed load and suspended load in different channel sections along longitudinal direction (Q_s is suspended load; Q_b is bed load, Q_t is total load).

all models appear to overpredict the deposition for $80° < \theta < 160°$, with the SA_SED simulation which accounts only for the bed load transport giving the overall best agreement. The adding of the suspended load component appears in both the SA_SED and the k-ε simulations to increase the deposition and to produce a slightly poorer agreement with the experimental data.

Figure 6 shows the variation in the streamwise direction of the ratios between the bed load and the total load and between the suspended load and the total load, respectively. It is observed that both simulations predict the suspended load contribution to be approximately 20% of the total load, which confirms the fact that in the present flow the bed load dominates. The results of Wu et al. (2000) show a somewhat larger dependence of this ratio with the longitudinal angle (streamwise position) than the SA_SED results.

In Figure 7 the simulated equilibrium channel cross sections at five representative locations ($0° < \theta < 180°$) are compared with the measurements and the k-ε simulation (only the one in which both the suspended and the bed load components were considered). The water level is situated at the zero mark in all cross sections. The presence of the mean streamwise slope was accounted for when the experimental data was plotted. The free surface deformation effects on the overall shape of the channel cross sections at different streamwise positions were found to be relatively small compared to those due to erosion/deposition. Away from the upstream part which maintains the original trapezoidal shape (e.g., $\theta = 0°$), the bed changes via erosion to a triangular shape. As expected, the largest depths are observed near the outer bend. Near the inner bank deposition takes place and a short sand bar is present. In the experiment (see 3b/8 line in Fig. 5) the maximum scour depth ($d_{max} \sim 0.27$ m) inside the bend is reached near the outer edge over a relatively small distance ($\theta \sim 45° - 55°$) from the bend entrance after which the maximum scour depth in the cross sections starts decaying smoothly with θ to $d_{max} = 0.22$ m before increasing again to $d_{max} = 0.25$ m at $\theta \sim 160°$. Both SA_SED simulations and the one using the k-ε model appear to satisfactorily capture not only qualitatively but also, in many respects, quantitatively the variations of the water depth and bed elevation inside the bend. However, some differences can be observed. With the exception of the fact that the SA_SED simulation that accounts for the suspended sediment predicts very accurately the maximum depth and its position in the $\theta = 45°$ section, the SA_SED simulation that takes into consideration only the bed load appears to produce slightly better agreement with the measurements in the sections represented in Figure 7. The largest error present in the k-ε simulation results are observed near the inner bank in the $\theta = 0°$ section where the model

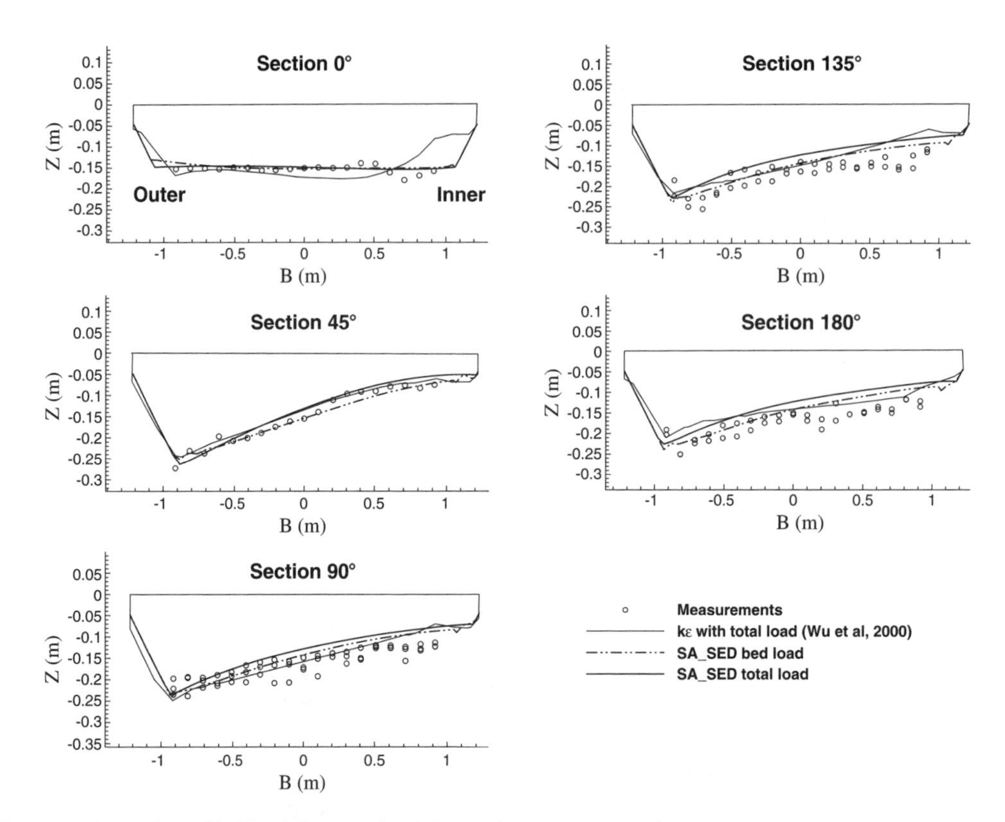

Figure 7. Comparison of bed levels between simulations and measurement results.

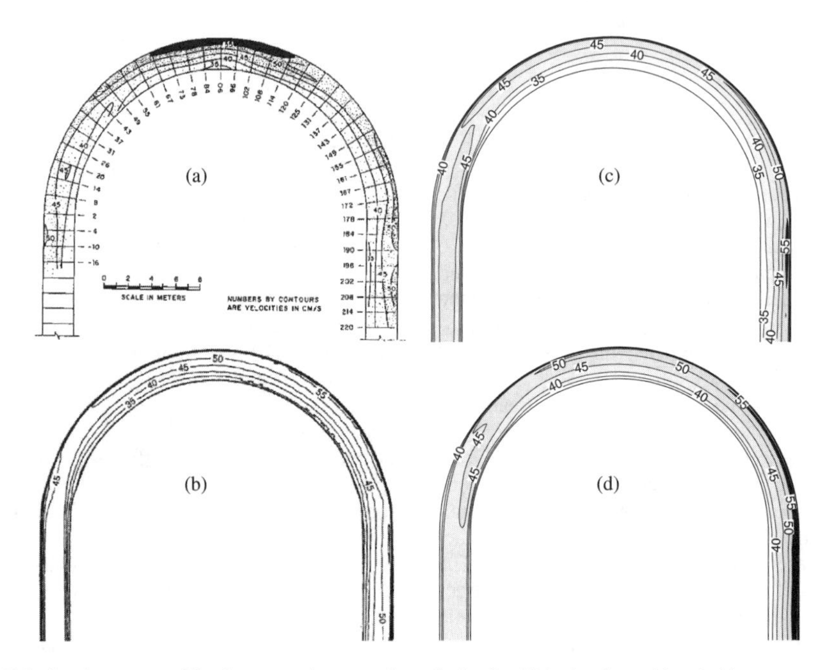

Figure 8. Calculated contours of depth-averaged streamwise velocity (cm/s) in the channel bend; (a) measurements; (b) k-ε with total load model (Wu et al. 2000); (c) SA_SED bed load model; (d) SA_SED total load model.

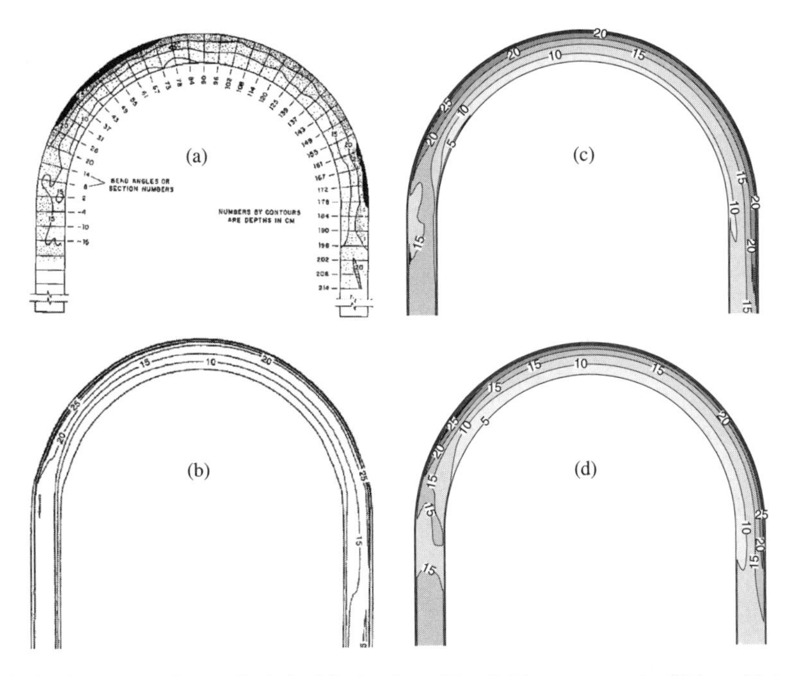

Figure 9. Calculated contours of water depth (cm) in the channel bend; (a) measurements; (b) k-ε with total load model (Wu et al. 2000); (c) SA_SED bed load model; (d) SA_SED total load model.

predicts the formation of a sandbar way too early compared to experiments and to the SA_SED simulations. At the $\theta = 45°$ section, the k-ε and the SA_SED simulations, both accounting for suspended sediment load, are quite close, while at the $\theta = 90°$ section, the k-ε predictions appear to be slightly closer to the measured bathymetry. For the other two downstream sections no clear conclusion can be reached on the accuracy of the k-ε simulation vs. the one of the SA_SED simulations.

Figure 8 shows the measured and simulated contours of the streamwise depth-averaged velocity in the channel bend. As expected, larger velocities are observed near the outer bank where the water depth is also larger, and smaller ones near the inner bank. Though the comparisons in Figure 7 appeared to slightly favor the SA_SED simulation which accounts only for the bed load transport, analysis of Figure 8 appears to favor the SA_SED simulation which also accounts for the suspended sediment component. The other interesting observation is that the k-ε model (total load) predictions of the depth averaged streamwise velocity are quite close to the ones given by the SA_SED (total load) simulation. The SA_SED (bed load only) simulation underpredicts the depth averaged streamwise velocity especially near the outer bank for $\theta > 60°$. Analysis of Figure 9 which shows the water depth contours in the channel bend confirms the trends observed in Figure 8. In particular, the SA_SED (total load) is able to capture the two areas of high erosion observed in the experiment situated very close to the outer bank around $160° < \theta < 180°$ and $0° < \theta < 180°$, respectively. Overall, a fairly good agreement has been observed between the measurements and the modeling results.

4 CONCLUSION

A fully 3D non-hydrostatic model developed by Zeng et al. (2005) to predict the flow, free surface deformation, bed load transport and bed morphology in open channels was extended to account for the suspended sediment load by solving an advection-diffusion scalar transport equation with a settling velocity term for the suspended sediment concentration and by incorporating proper boundary conditions at the interface between the bed load layer and the suspended sediment layer. Additionally a new version of the Spalart-Almaras (SA) RANS model capable of accounting for roughness effects without using wall functions was implemented and tested.

Two simulation cases in straight open channels, one with net entrainment and the other with net deposition (zero entrainment) at the bed surface were carried to test the suspended sediment module. The agreement with the experiment was found to be satisfactory. Finally, the flow, sediment transport and bathymetry at equilibrium conditions in an U-shaped open channel were calculated. Two simulations were performed. In the first one only the bed load transport was considered, as this component was observed to be the main mode of sediment transport in the experiment, while in the second one both the bed load and the suspended load transport were considered. Overall, the performance of the newly developed model in predicting this complex test case was quite satisfactory, with both simulations predicting relatively close results.

For the present test cases in which massive separation was not present, the simulations using the newly proposed model showed only a slight improvement over simulations using 3D RANS models with wall functions. However, it is expected that when applied to flows characterized by massive separation (e.g., scour around hydraulic structures like bridge piers/abutments or spur dikes) the use of near wall RANS models will give more accurate hydrodynamics predictions and, ultimately, sediment predictions compared to models that use wall functions as the underlying assumptions associated with the wall function approach are not satisfied in these flows.

ACKNOWLEDGEMENTS

The authors would like to thank Professor A. Jacob Odgaard from IIHR, The University of Iowa for his advice on different aspects related to this project and to Dr. Weiming Wu from NCCHE, The University of Mississippi for discussing many detailed aspects of his model and allowing us to use his results for comparison. The authors would also like to thank the National Center for High Performance Computing in Taiwan for providing the computational resources needed to perform some of the simulations.

REFERENCES

Constantinescu, G. & Squires, K. 2004, Numerical investigation of flow over sphere in the subcritical and supercritical regimes, *Physics of Fluids*, 16(5), 1446–1465.

Fang, H. & Rodi, W. 2003, 3D calculations of flow and suspended sediment transport in the neighborhood of the dam for the Three Gorges Project Reservoir in the Yangtze River, *J. Hydraulic Research*, 41(4), 379–394.

Gessler, D, Hall, B., Spasojevic, M., Holly, F., Pourtaheri, H, Raphelt, N. 1999 Application of a 3D mobile bed, hydrodynamic model, *J. Hydraulic Engineering*, ASCE, 125, 737–749.

Lin B. L. & Falconer, R. 1996, Numerical modeling of three-dimensional suspended sediment for estuarine and coastal waters, *J. Hydraulic Research*, 34(1), 435–455.

Odgaard, A.J. & Bergs, M. 1988, Flow processes in a curved alluvial channel, *Water Resources Research*, 24(1), 45–56.

Olsen, N.R. 2003, 3D CFD modeling of self-forming meandering channel, *J. Hydraulic Engineering*, ASCE, 129(5), 366–372.

Sekine, M. & Parker, G. 1992, Bed load transport on transverse slope, *J. Hydraulic Engineering*, ASCE, 118(4), 513–535.

Spalart. P.R. 2000, Trends in Turbulence Treatments, *AIAA Paper 2000–2306*, Fluids 2000, Denver.

van Rijn, L. C. 1981, Entrainment of fine sediment particles; development of concentration profiles in a steady, uniform flow without initial sediment load. *Rep. No M1531, Part II*, Delft Hydraulic Lab., Delft, The Netherlands.

van Rijn, L. C. 1984, Sediment transport, Part III: Bed forms and alluvial roughness, *J. Hydraulic Engineering*, ASCE, 110(12), 1733–1754.

van Rijn, L. C. 1986, Mathematical modeling of suspended sediment in nonuniform flow transport, *J. Hydraulic Engineering*, ASCE, 112(6), 443–455.

van Rijn, L. C. 1987, Mathematical modeling of morphological processes in the case of suspended sediment transport, *Delft Hydraulics Communication* No. 382.

Wang, Z. B. & Ribberink, J. S. 1986, The validation of a depth-integrated model for suspended sediment transport. *J. Hydraulic Research*, 24(1), 53–66.

Wu, W., Rodi, W. & Wenka, T. 2000, 3D numerical modeling of flow and sediment transport in open channels, *J. Hydraulic Engineering*, ASCE, 126(1), 4–15.

Zeng, J., Constantinescu, G. & Weber, L. 2005, A fully 3D non-hydrostatic model for prediction of flow, sediment transport and bed morphology in open channels, *XXXIst International Association Hydraulic Research Congress*, Seoul, Korea, September 2005.

River, Coastal and Estuarine Morphodynamics: RCEM 2005 – Parker & García (eds)
© 2006 Taylor & Francis Group, London, ISBN 0 415 39270 5

Numerical modeling of potential erosion of the lower Sheboygan River, Wisconsin during extreme flood flows using CH3D-SED

Qimiao Lu & Robert B. Nairn
Oakville, Ontario, Canada

Jim Selegean
U.S. Army Corps of Engineers, Detroit, USA

ABSTRACT: The lower Sheboygan River is a tributary to west coast of Lake Michigan in Wisconsin. Contaminated sediments were found to be sequestered well below the riverbed level. This paper will present an analysis of the morphodynamic response of the lower river to predicted 100-year scour events. The investigation included surveys, physical modeling and numerical modeling. The field surveys included measurements of river flow velocities, water levels and suspended sediment load. In addition, relatively undisturbed duplicate core samples were extracted from the areas of expected erosion for erodibility testing. Blind tests of erodibility for the extracted cores were completed in unidirectional flow flumes at two separate laboratories. The results of the erodibility testing were used to define the critical shear stress and erosion characteristics for the mixed sediment type (silty sand) for input to the numerical model. The US Army Corps of Engineers CH3D-SED model was setup, tested, and applied to determine the scour and sedimentation patterns expected for extreme events. After model setup and testing was completed, several snapshots of bathymetry were provided for validation testing of the morphodynamic response of the model. A key aspect of the investigation was the prediction of maximum bed scour for a 100-year flow event.

1 INTRODUCTION

This paper describes an investigation of sediment transport processes for the Lower Sheboygan River and the Inner Harbor of Sheboygan in Wisconsin. The primary objective of the investigation was to evaluate the mobility of riverbed sediments and potential erosion during extreme flood events on the river.

The Sheboygan River and Harbor site includes the lower 22.5 km (14 miles) of the river from Sheboygan Falls Dam downstream to, and including the Inner Harbor. This segment of the river flows through Sheboygan Falls, Kohler, and Sheboygan before entering Lake Michigan. The Sheboygan River runs from west to east through east central Wisconsin, emptying into Lake Michigan.

The USEPA divided the river into three sections during the Remedial Investigation (RI) based on physical characteristics such as average depth, width, and level of polychlorinated biphenyl (PCB) sediment contamination. The Lower River of three EPA sections extends 5 km (3 miles) from the C&NW railroad bridge to the Pennsylvania Avenue bridge in downtown Sheboygan. The Inner Harbor includes Sheboygan River from Pennsylvania Avenue Bridge to the river's outlet at the Outer Harbor (see Figure 1). The Outer Harbor is defined as the area formed by the two breakwaters.

The record of decision (ROD) for this site was signed on May 12, 2000. The ROD called for the removal of PCB-contaminated sediment in the Upper River, Middle River and Lower River and Inner Harbor components of the site. With regards to the Inner Harbor, the ROD called for the removal of the top two feet of sediment between the Pennsylvania Avenue and 8th Street bridges. Additional sediment will be removed where hot spots exist at the 2-foot (60 cm) depth or annual bathymetric surveys have indicated areas susceptible to scour. All excavated areas of the Inner Harbor will be backfilled with clean sediment.

While the 20-year history of bathymetric surveys showed the effects of a number of relatively extreme flood events in the harbor, there remain questions surrounding the impacts that a 100-year flood event may have on the Inner Harbor sediment.

The ultimate purpose of this study is to develop a sediment transport model for the Inner Harbor of the Sheboygan River and Harbor Superfund site. The study area begins at the mouth of Sheboygan River and extends upstream for 5 km (3 miles). Numerical model simulations were completed to evaluate the scour

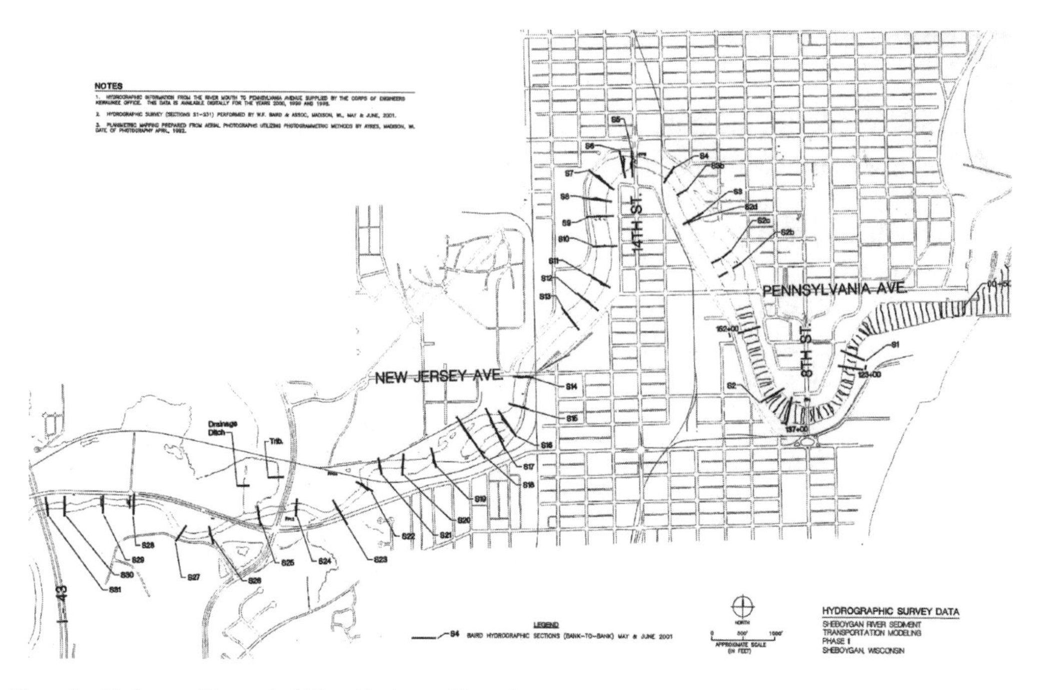

Figure 1. Sheboygan River and additional hydrographic sections.

potential of these sediments during a 100-year flood event.

2 DATA COLLECTIONS, FIELD SURVEYS, AND ERODIBILITY TESTS

The objective of this study is to evaluate the stability of the Sheboygan River bed sediment by using a numerical model. The reliability of modeling results is greatly dependent on the correction of the data used for model setup. In order to obtain the most representative data for model setup, the existing data was collected from a variety of sources and data gaps were identified. Additional field surveys were carried out to fill the data gaps. The surveys included hydrographic survey, topographic survey, hydrological survey, and suspended and bed sediment survey. The lab tests for erodibility of the river bed sediment were performed and will be described below.

2.1 *Data collections and data gaps*

The pre-existing data were collected at USGS gages, NOAA stage gages, and from the other sources. The collected data includes hydrographic and topographic survey data, flow survey data, lake level records, and sediment information.

Hydrographic information is essential for 3D modeling setup. Hydrographic information from the river mouth to Pennsylvania Avenue is available through USACE Kewaunee Office. This information includes cross sections located on 100 ft intervals with bank-to-bank coverage for years 1979 through 1997. Hydrographic information upstream of Pennsylvania Avenue to the USGS gaging station located at I43 was available from a previous HEC-2 model developed by the USACE. However, the cross-section information extracted from the HEC-2 model was too coarse to set up the model. Additional hydrographic surveys were required in that area.

Lake level analysis was completed to evaluate water level fluctuations in the Inner Harbor area. A pressure gage used to record water surface elevations was installed opposite the US Coast Guard station in the Sheboygan River mouth for the period from May through June 2001. The recorded water surface elevation data were verified by comparing with two NOAA data sets from Milwaukee and Kewaunee Harbors.

Sediment information including total suspended sediment load, particle size distribution for both suspended sediment and bed sediment, and bed soil density was required to set up boundary conditions of river bed and upstream conditions of incoming sediment. These data was collected from the USGS gaging stations, the WDNR collections, and Sheboygan River Sediment Sampling Logs. However, these data were insufficient to set up temporally varied incoming sediment and the erodibility of bed sediment.

2.2 *Field surveys*

Hydrographic and land-based survey was performed to supplement the bathymetry upstream of Pennsylvania Avenue. The thirty-seven additional cross sections along the river channel were measured in May/June 2001 (see Figure 1).

Particle size distributions of suspended sediment were measured utilizing both a horizontal sampler and a continuous profile sampler at the USGS I43 gaging station (#04086000) and at 14th Street. Eighteen samples were collected at these locations for the three flow events. Suspended sediment samples were collected midstream at 0.2, 0.6 and 0.8 times the total depth. Suspended sediment concentration levels necessitated combining all three-depth samples to yield grain size percentages. Tests were obtained for sediment concentration and particle size information for percent sand (% > 0.062 mm) and percent silt and clay (% < 0.062 mm). Currents were measured at the same cross sections and dates as the concentration profile characteristics.

It was recommended that the sediment sampling sites be selected in consideration of the river hydrodynamic condition and PCB distribution in the sediment as reported (BBL, 1998). Sediment samples were collected to be representative of areas with a range of high to low erosion potentials.

Four cores were collected from cross sections S2b (LT Island), 152+00, 137+00, and 123+00 (see Figure 1). All cores were taken at the center of the channel immediately adjacent to each other. Two cores were taken at each location using Shelby tubes; one for flume testing and one as a backup if the original was disturbed in the lab or during transport, or for an additional lab test. The other two cores were clear PVC tube samples for performing geotechnical tests. One set of the PVC tube samples were analysed using three representative slices per core.

The geotechnical laboratory analysis of the cores found that the clay content ranged from 0 to 16% and was on average 9%. Organic content ranged from 4 to 23% and was on average 9%. The median grain sizes for the sediment samples ranged from 0.027 to 0.28 mm (medium silt to fine to medium sand) with an average D50 of 0.060 mm (coarse silt).

2.3 *Erodibility tests*

This section provides the findings of the laboratory experiments completed at the University of Wisconsin (UW) and Sandia Laboratory in New Mexico to determine the erosion characteristics of relatively undisturbed core samples extracted from the river bed of the Lower Sheboygan River and Inner Harbor.

There were two primary results from the laboratory experiments that were required to parameterize the erosion process in the numerical modeling: (1) the

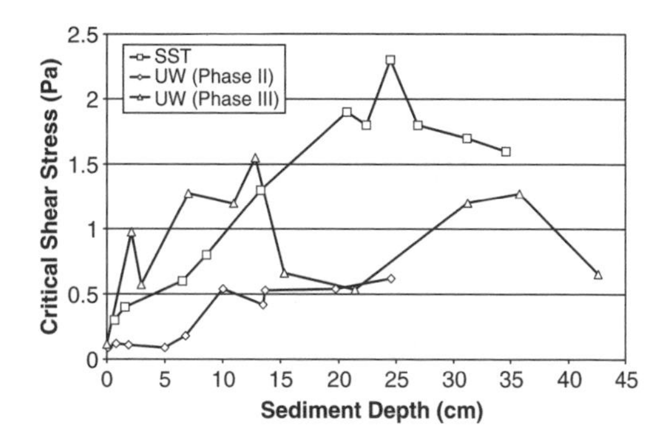

Figure 2. Critical shear stress measured at STA 137+00.

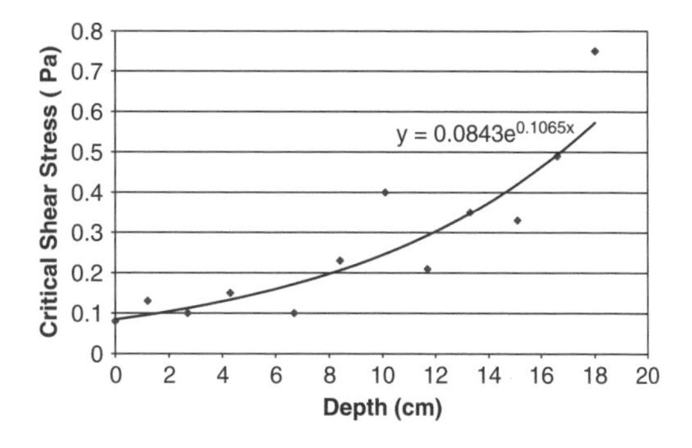

Figure 3. Variation in critical shear stress with depth.

critical shear stress; and (2) the erosion rate – excess shear stress relationship, the erodibility.

2.3.1 *Critical shear stress*

A key finding of the laboratory experiments was the increase in critical shear stress with depth. This kind of increase in critical shear stress with depth has been observed by others and attributed to consolidation and the associated increase in bulk density (Parchure and Mehta, 1985). Figure 2 shows the measured critical stress with depth. The critical shear stress for the active (surface) layer compare well to the Shields' curve that was derived from a large data set of cohesionless sediment experimental results. The influence of consolidation with depth is clearly evident. This information provided the basis for the parameterization of an increase in critical shear stress with depth and different sediment/soil types.

The variation in critical shear stress with depth was plotted for one of the three cores with continuous data through the depth are shown in Figures 3. It is evident that there is an exponential increase with depth (a similar rate for two of the cores and different for the third). These changes were reviewed to determine how the rate of increase was influenced by measured grain size distribution (i.e. in addition to the influence of depth or bulk density). Based on the results of these

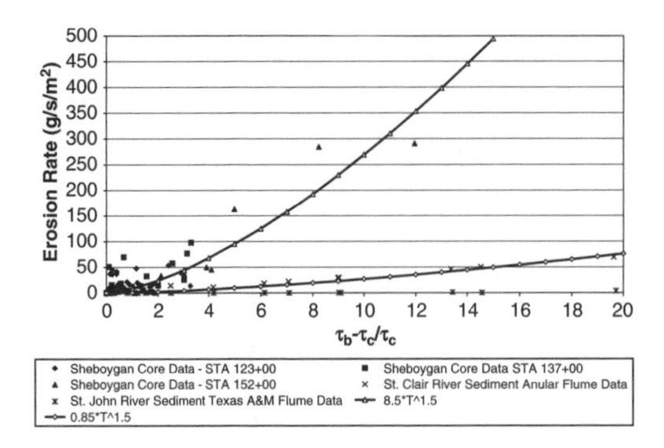

Figure 4. Measured erosion rate.

analyses, the critical shear stress was determined for seven representative soil types and four depth layers which will be described below.

2.3.2 *Erodibility*

The erosion rate – excess shear stress relationship (i.e. erodibility) was determined on a fine vertical resolution for the four cores tested. The results of the UW experiments in a conventional form that relates the rate of erosion in $g/s/m^2$ to the dimensionless excess shear stress are shown in Figure 4. Also plotted in this figure are some results from annular flume tests of similar sediments (i.e. mixed sand, silt and clay) for the other sites and two possible applications of the van Rijn (1989) formulation for erosion of coarse silt/fine sand sediment.

The van Rijn formulation requires the definition of a reference height above the bed to determine a reference concentration that significantly effects the predicted rate of resuspension of sediment from the bed. Van Rijn (1989) indicates that the reference concentration height should not be less than 1% of the water depth. For the Sheboygan River and the Inner Harbor the water depths range from 0.5 to 5 m. Therefore, the reference concentration should be not less than 5 mm anywhere in the model domain. Typically the reference concentration height is equated to half the bedform roughness height or ks (the effective grain roughness equal to $2D_{90}$).

For the minimum recommended value of reference height "a" (i.e. 1% of the depth), the van Rijn equation for erosion reduces to (see Figure 4):

$$E = 0.85 \cdot T^{1.5} \tag{1}$$

This formulation is shown in Figure 4 to match the St. Clair River and Swan Creek Lake data relatively well. However, it significantly under-predicts the Sheboygan River data determined by the laboratory test. To match the Sheboygan River results from UW the following version of the van Rijn equation is required (see Figure 4):

$$E = 8.5 \cdot T^{1.5} \tag{2}$$

For the numerical model it is necessary to describe the reference concentration height at all grid calculation locations within the model domain. In summary, the UW data imply much higher erosion rates than determined from other experiments with similar mixed sand/silt/clay freshwater river sediments and from the van Rijn formulation that has been developed from a very large data set for erosion of fine sands.

3 NUMERICAL MODELING

CH3D-SED has been selected for the modeling evaluations. CH3D-SED merges the 2D mobile bed modeling techniques developed by the Iowa Institute of Hydraulic Research (IIHR) and the CH3D three-dimensional hydrodynamic simulation code of the US Army Corps of Engineers – Waterways Experiment Station now referred to as the Coastal and Hydraulics Laboratory (Gessler et al, 1999 and Chapman et al, 1996). It is a finite difference model with curvilinear grid transformation in the horizontal plane and standard σ-grids in the vertical direction. It is well suited to the investigation of erosion and sedimentation in rivers and estuaries. In addition, the model has the ability to simulate freshwater and salt-water interface, thermal diffusion, contaminant transport, and wind stress. The sediment transport module of CH3D-SED in a strict sense is applicable only to sand-bed rivers. Van Rijn's formula (van Rijn, 1986) is used to calculate sediment entrainment from the riverbed. Bed consolidation, bed armoring and bed load are considered in the model. It is a public domain model but with restricted distribution. The model applies an older numerical technique and this may cause some numerical problems such as inaccuracy and instability (Singh & Ghosh, 2000). Some modifications were implemented in this application to address these and other issues.

3.1 *Model domain*

The domain of this sediment transport model extends from the river mouth to New Jersey Avenue, 3.8 km (2.4 miles) in total. The curvilinear grids were generated and consist of 110 cells longitudinally and 10 cells laterally. The minimum grid size is about 4 m laterally and 23 m longitudinally.

A σ-grid system in which the grid thickness is varied with total water depth was applied by CH3D-SED in the vertical direction. There are five Sigma layers for all cells. A minimum water depth of 30 cm was imposed to ensure model stability under all conditions.

Table 1. Definition of representative soil types.

Soil type	Gravel	Sand	Silt	Clay	
Avg. grain size (mm)	4.760 (%)	0.250 (%)	0.062 (%)	0.002 (%)	D50 (mm)
1	8	86	4	2	0.159
2	4	73	20	3	0.132
3	2	59	33	6	0.096
4	1	47	44	9	0.058
5	0	37	49	14	0.046
6	0	26	60	14	0.038
7	0	20	68	12	0.036

The bathymetric data was determined from the HEC-6 model data developed by BBL (1998), the USACE survey data from 2001, and supplemented by the Baird survey of 2001. This information was used to interpolate water depths at the center of each cell. All data was adjusted to Low Water Datum (LWD) of Lake Michigan. Since the model functions with negative water depth (i.e. bed elevation above LWD) as dry land, the model datum is set as 5 meters above LWD. All depth and stage (water level elevation) data were converted to the model datum.

3.2 *Bed sediment*

The information of bed sediment in terms of grain size distribution, dry density (or porosity), critical shear stress, and erodibility are required for the morphodynamic modeling. The spatial variations of these parameters in sediment depth are also required. The grain size distribution was interpolated within the model domain from the field surveys to obtain an initial grain size identification (ID) at each grid cell. The variation in sediment grain size with depth was defined based on the four core samples taken as part of this investigation together with the trends from the older HEC-6 data.

Based on the results of the grab and core sampling, four grain size classes were used to define seven representative soil types (see Table 1). The bed throughout the model domain and in depth was classified according to these seven soil types. Figure 5 shows an inverse behavior between silt-clay and gravel-sand grain sizes that suggest a general trend of decreasing grain size from upstream to downstream, as expected.

The bulk densities measured by the geotechnical firm CGC Inc. in the four sediment cores taken in the field survey were used to define the spatially varied bulk density (actually porosity) in the model.

The depth-varied critical shear stresses determined from the results of the UW lab tests described in above section were applied. In order for the model to calculate the erosion flux, the concentration reference height, described by van Rijn (1984) were specified throughout the domain. This parameter was calculated by using the measured shear stress and erosion rate from the UW flume test. The sediment thickness that can be eroded in the model was assumed for all practical purposes to be unlimited.

Figure 5. Spatial distribution of grain size (Red – Yellow – Green represents soil type from 1 to 7).

3.3 *Boundary conditions*

Discharge data from USGS Gage 0408600 at Sheboygan on the Sheboygan River was obtained. The data set consisted of a 48-year daily time series record and annual maximum peak flows. The annual maximum peak flow data set was used to complete a flood frequency analysis. Table 2 lists the different return periods and their associated discharge. As observed, the maximum recorded peak flow at the USGS gage station (1998) for the period of record approximately corresponds to a 25-year flood event.

Several storm events were selected from the Sheboygan River gage discharge time series (with daily average values) to construct a 100-year event hourly hydrograph (duration and peak). The 100-year event was defined by scaling up the 1966 event by a uniform scaling factor across the full time series of this event. Although this approach mixes the annual peak value data set with the daily record data set it is the only feasible approach given the lack of long-term hourly data records for the gage.

A rating curve from previous HEC-6 model, which is described by Equation 3, was imposed as the upstream boundary for the time series model test. When event conditions are simulated, a clear water assumption was made. It was decided that this would be more appropriate and conservative (in the sense that erosion predicted for the event simulated with clear water would be over-estimated if anything, see below for more explanation) considering the lack of upstream sediment load data for extreme flow conditions.

Table 2. Estimated discharge for different extreme event return periods.

Return period (years)	Estimated discharge	
	cfs	m^3/s
2	2900	82
5	4800	140
10	6100	170
25	7800	220
50	9000	250
100	10200	290
200	11400	320
500	13000	370

$$TSS = 0.00022Q^{2.079} \tag{3}$$

The flow velocities in the lower part of the river and the Inner Harbor will have a strong dependence on the lake levels. Four representative conditions were selected for model simulations: (1) a high lake level of 1.7 m above Low Water Datum (LWD); an average mean lake level of 1 m above LWD; a low lake level of 0.3 m below LWD; and an extreme low condition that combines a constant setdown of 15 cm with the representative low lake level for a level of 0.45 m below LWD.

3.4 *Incoming sediment load for extreme events*

Measured suspended sediment concentrations provide an indication of the degree of resuspension and settling throughout the entire length of the river. For the numerical model domain extending from just downstream of the New Jersey Avenue bridge to the boundary between the Inner and Outer Harbors, the suspended sediment concentrations are the result of two key processes: (1) upstream loading represented as a boundary condition that relates incoming suspended sediment concentration (and grain size distribution) to discharge; and (2) the balance between resuspension and settling within the model domain.

The upstream loading is treated as a boundary condition and often in sediment transport modeling on rivers is a key factor influencing sediment load downstream. For most rivers an understanding of the sediment loading at any point in the river is poorly defined and only available at the low flow condition. It usually consists of constructing a rating curve that relates sediment load (and in some rare cases grain size distribution for the sediment load) to discharge based on a very limited number of measurements. Such a relationship exists for the USGS gage station in Sheboygan. However, the extrapolation with the rating curve developed by using data at low flow condition may incorrectly estimate the sediment load in an extreme flow event. In fact, the sediment loading at different locations in the river is related to many more factors in the watershed (condition of agricultural lands at the given time of year and extent of bank erosion among many other factors) and discharge is usually a poor descriptor overall. Therefore, in the model runs completed for the 100-year event a conservative assumption was made whereby the sediment loading at the upstream boundary was assumed to be zero – a clear water assumption. As will be presented below, this did not appear to have a significant influence on the results (in other words the suspended sediment concentration quickly adjusted to equilibrium within the upstream part of the model domain owing to the high flow velocities).

3.5 *Model modifications*

To achieve the primary objectives of the study it was necessary to make three main modifications to the CH3D-SED model.

CH3D-SED uses constant porosity, dimensionless critical stress and concentration reference height all of which are key parameters for the parameterization of erosion processes. The model was modified in this study to allow the input of these three variables in a spatially varied manner. Different values for each of the three parameters can now be set for each cell and for each sediment bed layer. To be compatible with original version, negative or zero values set for these parameters sets the parameter to the default or calculated value in CH3D-SED. This change was necessary to incorporate the erodibility characteristics determined from the laboratory experiments.

There is no calculation for cohesive sediment concentration in CH3D-SED. Lick's function (see Lick et al., 1995) was added to calculate the erosion rate for the clay class of sediment. The erosion rate in Lick's formulae is given as

$$E = \frac{a_0}{T_d^n}\left(\frac{\tau_b - \tau_{cr}}{\tau_{cr}}\right)^m \tag{4}$$

in g/cm^2, where τ_b is bed shear stress ($dynes/cm^2$, note $1\ dyne/cm^2 = 0.1\ Pa$), τ_{cr} is critical shear stress for erosion, T_d is time after deposition in days, a_0 is approximately 0.008, n and m are approximately 2. Because the concept of reference concentration is applied in CH3D-SED, the erosion rate calculated from above formulae should be converted to reference concentration, as

$$C_a = \frac{E}{3600\rho_s\omega_s} \tag{5}$$

The above equation is derived under the assumption that the entrainment process takes about one hour at a constant shear stress, a condition imposed in many laboratory experiments. The constant settling velocity for cohesive settling is assumed to be 0.01 cm/s that corresponds to an average floc settling velocity in fresh water.

In order to use the Baird in-house software Spatial Data Analyzer (SDA) to interpret the model results (and produce plots and animations), the CH3D-SED model output format was modified.

3.6 *Model calibration and verification*

Calibration is the process of modifying model parameters in order to achieve satisfactory comparisons between measurements and predictions for the various key parameters being simulated including: velocities, suspended load and erosion/deposition. Verification model runs are completed for two purposes: (1) to test the calibrated model under different conditions; and (2) to test a model that has not required calibration. Therefore, in this section the results from the CH3D-SED model are compared to measurements in order to provide some level of site-specific verification of the model (or calibration if necessary).

3.6.1 *Flow velocity calibration*

It has been noted that the primary focus of this investigation is to estimate erosion during a 100-year flood condition on the river. Therefore, ideally model verification should be performed with velocities measured in an extreme flow event. Unfortunately, it is very difficult to be in the right place at the right time to record flow velocities for extreme events.

The model was run for the 1.5 month field data collection period and the simulated and measured velocities are compared in Table 3. Generally, CH3D-SED does well at matching the measured flow velocities, there is a tendency for slight under-prediction. The peak flow speed of 0.35 m/s measured for the average flow event of June 13, 2001 compares to the predicted peak flow speed at this location of 2.3 m/s for a 100-year event with a low lake level. This comparison underscores the fact that the largest flow event recorded during the field program was in no way an extreme event. Nevertheless, and importantly, CH3D-SED correctly matches the distribution of flow across the river cross-section. A common feature in the 100-year event simulations is a pattern of higher velocities on the inside of bends on the upstream side. The model reproduces this feature well for the June 13, 2001 event.

In summary, the most important finding of the model verification tests was the fact that CH3D-SED reproduced well the flow distribution across the river coming into a bend. In addition, the predicted flow speeds matched the measured values quite well. No model calibration of velocities was necessary. However, the measured flow speeds were much less

Table 3. Comparison of measured and predicted velocities.

Event		Right bank (80%)	Center of river (50%)	Left bank (20%)
		Velocity (m/s)		
May 24th, 2001	Measured	0.09	0.30	0.24
	Predicted	0.08	0.12	0.17
June 11th, 2001	Measured	0.04	0.27	0.14
	Predicted	0.14	0.14	0.10
June 13th, 2001	Measured	0.27	0.49	0.29
	Predicted	0.32	0.35	0.30

than what is expected during a 100-year event. Nevertheless, the hydrodynamic component of CH3D-SED model is well tested in many other applications and there is a degree of confidence that predictions will be accurate even for flows during extreme events.

3.6.2 *Bed change verification*

The most direct measure of the ability of a sediment transport model to predict erosion during an extreme event is the change in riverbed elevation at a given location. Clearly, the relatively low flow events experienced during the monitoring program did not lead to any measurable bed change, especially for bed erosion. Previous modeling efforts such as the HEC-6 model application (BBL, 1998) relied on comparison of long term model predictions of deposition to dredging records for the lower part of the river. Baird & Associates (1999) found that the success of the model to correctly predict the observed deposition rates was largely due to calibration of the upstream loading.

In a review of the available data we were unable to identify periods of rapid bed change that would provide an unambiguous data set for verification of the bed change predicted by the model at the time we calibrated the model. After finished the project, EPA provided the bed change map in the Inner Harbor from 1997 to 1998, as shown in Figure 6. In this figure, the colors from yellow to dark red represents bed scours from 1 ft to 7 ft while the colors from light green to dark blue represents the bed deposition from 1 ft to 7 ft. Therefore, the direct comparison for bed change had not been carried out at the time we finished the study.

Note that there was a 25-year return flow (about 220 m^3/s) from 1997 to 1998. The bed change shown in Figure 6 was the results of integrating erosion during this large flow event and deposition during low flow condition. Therefore, the erosion during that 25-year return flow event should be larger than that shown in Figure 6, in both surface area and erosion depth. The locations with transparency (black color) where the bed change is less than ± 0.5 ft were likely eroded in

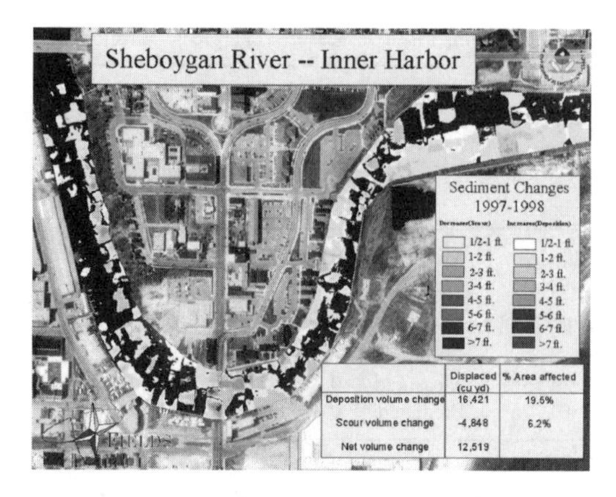

Figure 6. Measured bed change from 1997 to 1998 (including a 25-year return flow).

Figure 7. Modeled bed change for a 100-year return flow event.

the 25-year return flow event. By comparing the measured bed change (see Figure 6) with the bed change modeled for a 100-year return flow (see Figure 7), the locations of erosion and deposition predicted by the model agrees well with the bed change measured from 1997 to 1998. Note that the colors from green to blues in Figure 7 represents the erosion and the colors from yellow to red represent the deposition.

Therefore, we are confident that the predicted bed change is accurate within the range of simulation results presented in Section 4 and that the laboratory experiment results provided the necessary information for calibration of the resuspension algorithm in the model.

4 MODEL SIMULATION RESULTS

This section presents a description of the results for the various CH3D-SED simulations of sediment

Figure 8. Modeled bed change for a 100-year return flow and low lake level using Equation 2.

transport and associated erosion/deposition on the lower Sheboygan River and Inner Harbor.

4.1 *Model tests*

Three model runs were carried out for erodibility sensitivity tests. The first model test consisted of the simulation of the 100-year event for low lake level (downstream boundary) conditions using the erodibility algorithm determined from the UW laboratory experiments (i.e. Equation 2). This parameterization resulted in unusually high erosion rates – much higher than those determined for similar sediments from two other rivers using different laboratory facilities. As it turned out the application of this algorithm resulted in the development of instability in the model upstream of Pennsylvania Avenue. This instability took the form of unrealistically large scour in this upper reach of the river bed from upstream of Pennsylvania Ave and unrealistically large deposition in the lower reach (see Figure 8). Ultimately the model was crashed before completion of simulations. Referring to Figure 5, this upper reach of the river bed from upstream of Pennsylvania Ave to New Jersey Ave consists of sand, gravel and cobbles. It is inappropriate to apply Equation 2 developed from the erodibility of mixed sand/silt/clay sediment cores extracted from the reaches downstream of Pennsylvania Ave into entire model domain.

The second test was completed for the 100-year event using the more conventional Van Rijn (1986) formulation for erodibility as represented by Equation 1. The test featured a low lake level and was identical in all respects to the first group of tests with the exception that the Equation 2 algorithm was replaced by Equation 1. The results are presented in Figure 9. Looking at Figure 9 (and comparing it to Figure 8) it is evident that the bed erosion depth is slightly less than the corresponding tests with Equation 2 and much less deposition than that predicted with the model run that utilizes the Equation 2 parameterization. This may

Figure 9. Modeled bed change for a 100-year return flow and low lake level using Equation 1.

Figure 10. Flow vectors for a 100-year flow and low lake level.

be the result of less sediment load generated from the upstream by using Equation 1.

The third test consists of applying Equation 1 throughout the model domain (i.e. the bed was not armored upstream of Pennsylvania Ave) for a low lake level. The results indicate that armoring the upstream section of the river bed in the model does not have a significant influence on the erosion in the downstream. This provides additional confidence that the armoring assumption required for model stability will not have influenced the results significantly.

4.2 *Model runs*

Based on the model tests, it is inappropriate to apply Equation 2 which was developed from the erodibility of mixed sand/silt/clay sediment cores to upstream reach. Therefore, it was decided to consider this upper reach of the river bed as armored and applied Equation 1 for erodibility to avoid the instability issues when the model was run using Equation 2 for erodibility. The assumption that this would not have a significant influence on the results was tested in the model simulations with Equation 1. It is noted that all of the major hot spots are located downstream of Pennsylvania Ave in the area known as the Inner Harbor.

Four model runs that consist of a 100-year flow event and four different lake levels as mentioned above were carried out for final model runs. The results for one of four runs (low lake level) are presented in Figure 7. All simulations were reviewed and there was very little difference between the maximum instantaneous erosion and the erosion at the end of the simulation.

4.2.1 *Flow patterns*

A general description of the velocity patterns for all the 100-year tests is presented here with particular reference to the low water simulation (refer to Figure 10). In general the trends in flow speed along the channel and across the channel were somewhat similar for the four different lake level simulations. The variations in flow speed with lake level were relatively significant and are discussed at the end of this paragraph.

Downstream of Pennsylvania Ave. the channel narrows and deepens. These two factors combine to result in relatively high flows (2 to 2.5 m/s) that are relatively uniform across the channel (i.e. there is not a distinct parabolic distribution).

The river widens again as it approaches the bend at the 8th St bridge and the flow speeds range from 1.5 m/s on the right bank increasing to 2 m/s at the left bank. The flow speeds are highest on the inside of the bend where the depths are shallow over an apparent point bar feature. On the downstream side of the bend at the 8th St bridge the channel narrows again and the flow speeds are fairly uniform across the channel at 1.5 m/s. The flows are locally higher (approximately 2 m/s) over the shallow point bar along the left bank.

As the river approaches the wider section of channel at the lower end of the Inner Harbor the flow distribution across the channel takes on a more parabolic form with flow speeds in the range of 0.5 to 1.7 m/s. The flow speeds are higher near the right bank. As the river channel expands rapidly into the downstream section of the Inner Harbor the flow speeds are reduced to 0.2 to 1.2 m/s from bank to the middle of the channel taking a pronounced parabolic form. At the downstream end of the model domain the channel is constricted between two jetties and the velocities range from 1.2 to 1.7 m/s and are relatively uniform across this deep section of the channel.

Flow speeds on the river decreased for simulations with higher lake levels at the downstream boundary of the model (particularly for the lower part of the river). As an example, the mid-channel flow speeds with the high lake level test were 1.6 to 2 m/s from New Jersey Ave down to Boat Island (with the exception of the reach upstream of the bend before the 14th St bridge where speeds were 1.3 to 1.5 m/s), in the range of 1.2 to 1.4 m/s down to the wider part of the inner channel where the speeds dropped to 0.8 m/s. In general, the flow speeds for the high lake level condition are 50 to 75% of the speeds for the low lake level simulation.

4.2.2 *Bed change*

The accumulated bed change (i.e. the total erosion/deposition at the end of the model simulation is presented in Figure 7 for one of the four runs (low lake level). As noted above, the river bed above Boat Island was armored in these initial simulations to avoid model instability. The pattern of erosion and deposition in the Inner Harbor area is similar for each of the four runs. There are four primary areas of bed change.

A long scour hole develops along the left bank upstream of the bend at 8th St. This hole is quite narrow for the high lake level test and almost half the channel width for the low lake level test plus setdown. The average maximum depth of scour in this channel ranges from 0.6 to 1.5 m from the high lake level test to the low lake level plus setdown test.

A relatively wide point bar feature develops starting at the middle of the 8th St bend and extending downstream mostly in the left half of the channel for the high lake level test, and a relatively narrow point bar for the low lake level plus setdown, continuing all the way along the left side of the bend to the wider part of the Inner Harbor. For the high lake level test the thickest deposition is in the range of 0.8 m compared to 1.0 m for the low lake plus setdown simulation.

This second long wide scour hole extends from downstream of the 8th St bend all the way to the area where the channel widens at the downstream end of the Inner Harbor. The average maximum depth of this scour hole ranges from 1.6 m for the high lake level test to 2.3 m for the low lake level plus setdown test.

Finally, there is a broad depositional area in the wider downstream end of the Inner Harbor with the thickest part of the deposit emanating from the right hand bank at the beginning of this wider area (i.e. where the second scour hole ends). The deposition in this wider area of the channel ranges from 0.1 to 0.2 m for the high lake level test to 0.5 to 0.8 m for the low lake level plus setdown test.

4.2.3 *Model limitations*

The primary model limitation related to the instability of the model during the 100-year event simulations with Equation 2 representing the erodibility. These instabilities were not related to the hydrodynamic component of the model but instead to the bed change (both erosion and deposition) that became much too large and eventually caused the model to crash before it reached the end of the simulation. Nevertheless, this limitation was partially overcome by armoring the riverbed upstream of Boat Island. It was shown in comparative tests with the Equation 1 simulation that this approach would not have a significant influence on the predicted magnitude and distribution of erosion and deposition downstream of Pennsylvania Ave where the primary PCBs hot spots are located.

Figure 11. Bed shear stress for a 100-year flow and low lake level.

The range of shear stresses that were predicted in the numerical model exceeded the range of shear stresses tested in the UW laboratory facility (see Figure 11). Therefore, the erodibility characteristics were extrapolated for high bed shear stress that was found in the upstream of the 8th St. It is our opinion that this extrapolation was completed in a conservative manner (i.e. if anything resulting in an over-estimate scour depths).

Another key limitation relates to the manner in which CH3D-SED determines shear stress at the bed. The depth-averaged velocity is used in this calculation. It is much more correct to determine the shear stress from the gradient in velocity near the bed. The approach used in CH3D-SED will tend to underestimate reductions in shear stress where scour holes develop and therefore the erosion of these scour holes may be somewhat over-estimated.

The bed load term in the model has an important influence on the development of the scour holes in the Inner Harbor. This term from Van Rijn (1989) does not include a gravity or slope term. Therefore, it may overestimate the quantity of sediment that is removed in an upslope direction on the downstream side of the scour holes, thereby resulting in an over-estimate in the total scour depth.

Calculations of bed change from gradients of sediment transport are made based on an explicit solution scheme that only considers the sediment transport in the immediately adjacent grid cells. This approach can result in stability problems for the sediment transport component of the model as was experienced for some of the model simulations completed in this investigation. This approach also precludes or makes it very difficult to simulate non-equilibrium sediment transport (i.e. conditions of overloading or underloading).

Another limitation of the model is the lack of functionality with respect to wetting and drying at the perimeter of the flow (i.e. to simulate inundation).

However, the absence of this functionality did not significantly influence the results in these simulations.

In general the limitations encountered did not significantly impact the ability of the model to complete the necessary simulations to define with a degree of confidence the extent of potential river bed scour in the Inner Harbor area during extreme events.

5 CONCLUSIONS

An understanding of the erosion or scour potential of river bed sediment in the Inner Harbor (Pennsylvania Ave to the mouth) reach of the Sheboygan River has been developed in this study. This section summarizes the key findings and limitations.

A review of the existing data identified some key gaps in information that would need to be addressed to develop an understanding of scour potential in the Inner Harbor. At the top of the list were the erodibility characteristics of the sediment at and below the river bed surface in the Inner Harbor.

The field data collection phase provided the necessary information on bed characteristics. The sediment cores and grab samples provided information on the grain size distribution (and bulk density for the cores) of the sediment. A limitation of the field data collection was the length of the cores extracted for testing – generally less than 30 cm. It was important to define the characteristics below this level in the model.

The key findings of the erodibility tests were that the critical shear stress increased significantly with depth below the river bed surface and that erosion rates could potential be very high for the mixed sand/silt/clay soils. For the shear stresses tested the erosion rates were higher than those measured for samples from two other rivers with similar sediment types.

Numerical model simulations were completed for a range of different scenarios combining a 100-year flow event with four lake levels. A 100-year flood hydrograph was constructed based on a review of historic hydrographs for extreme events on the Sheboygan River together with a flood frequency analysis. The greatest scour was experienced with the low lake level and low lake level plus setdown conditions. The lake level has significant impact on the bed change.

Two key areas of river bed scour developed: (1) upstream of the 8th St bridge varying in width depending on the model run but mostly confined to the left side of the channel; and (2) downstream of the 8th St bridge mostly confined to the right side of the channel. Both scour holes (each about 300 m long) are the result of the pattern of flow conditions through the bend at the 8th St bridge and the fact that the sediment in this area is relatively fine (average median grain size of coarse silt). The predicted average maximum scour depth in these two scour holes was approximately 2 to 2.5 m for the low lake level tests.

REFERENCES

Baird & Associates (2001). *Sediment Transport Modeling Sheboygan River, Wisconsin. Phase 1: Data Collection and Model Selection*. Technical Report.

Baird & Associates (1999). *Sediment Trap Assessment On Sheboygan River And Harbor, Wisconsin*, Technical Report.

BBL (Blasland, Bouck & Lee, Inc.) (1998). *Feasibility Study Report*, Technical Report.

Briaud, J.L., Ting, F.C.K., Chen, H.C., Cao, Y., Han, S.W. and Kwak, K.W. (2001). Erosion function apparatus for scour rate predictions, *Journal of Geotechnical and Geoenvironmental Engineering*, 127(2), pp. 105–113.

Chapman, R.S., Johnson, B.H. and Vemulakonda, S.R. (1996). *User's Guide for the Sigma Stretched Version of CH3D-WES*. US Army Corps of Engineers Technical Report HL-96-21. 25 p.

Gessler, D, Hall, B., Spasojevic, M., Holly, F., Pourtaheri, H. and Raphelt, N. (1999). Application of 3D Mobile Bed, Hydrodynamic Model, *J. Hydr. Engrg*., ASCE, Vol. 125, No. 7, pp. 737–749.

Parchure, T.M. and Mehta, A.J. (1985). "Erosion of soft cohesive sediment deposits." *J. Hydr. Engrg*., ASCE, 111(10), pp. 1308–1326.

Singh, B.C. and Ghosh, L.K. (2000). Discussion on "Application of 3D Mobile Bed, Hydrodynamic Model" by Gessler etc., *J. Hydr. Engrg*, ASCE, November, pp. 858–860.

Van Rijn, L. C. (1984). Sediment transport, part II: Suspended load transport. *J. Hydr. Engrg*., ASCE, Vol. 110, No. 11 pp. 1613–1639.

Van Rijn, L.C. (1986). Mathematical Modeling of suspended sediment in nonuniform flows, *J. Hydr. Engrg*., ASCE, 117, 433–455.

Van Rijn, L.C. (1990). *Handbook Sediment Transport by Currents and Waves*. Delft Hydraulics.

River, Coastal and Estuarine Morphodynamics: RCEM 2005 – Parker & García (eds)
© 2006 Taylor & Francis Group, London, ISBN 0 415 39270 5

1D morphodynamic model for natural rivers

M. Catella, E. Paris & L. Solari
Department of Civil Engineering, Florence University, Florence, Italy

ABSTRACT: A one-dimensional numerical model to study river morphodynamics is presented. The St. Venant-Exner equations, written in a conservative form, are solved by employing a finite volume method. The numerical scheme is extensively tested under steady and unsteady flow conditions by reproducing various channel geometries providing a great versatility, stability and robustness. The river model is applied to evaluate riverbed evolution involving steep and non-prismatic channels, and including subcritical, supercritical and transcritical flows with and without shocks.

1 INTRODUCTION

Mobile-bed numerical models solve the flow governing equations along with the sediment continuity equation by means of three different procedures: coupled, semicoupled and decoupled (Kassem & Chaudhry, 1998, Cao et al., 2002, Lyn & Altinakar 2002). Despite of decoupled models have been subject to a great number of criticisms, their performance on the modeling of river morphodynamics is not yet fully understood (Cao et al., 2002). The present analysis shows that decoupled models are not necessarily unstable and they could give results comparable with theoretical behaviors and experimental data.

In a decoupled model, at a given time step the continuity and momentum equations are firstly solved by assuming a fixed bed, and then the sediment continuity equation is solved by using the hydraulic variables previously obtained.

The St. Venant-Exner equations are modeled by assuming the riverbed to be composed by an uniform particle size and in the case of dominant bed-load sediment transport. Besides suspended loads or bed armoring processes are avoided.

The water-sediment mixture continuity equation is assumed to be almost identical to that for a fixed-bed by considering the bed morphological evolution time scale of a lower order of magnitude than that of the flow changes with adequately low sediment concentration. The water-sediment mixture momentum equation over a mobile bed is reduced to that for a fixed-bed clear-water flow.

In the sediment continuity equation the rate of sediment exchange between bed and flow in the water column is neglected.

Several benchmark problems reproducing various open channel geometries under steady and unsteady sediment transport are simulated to evaluate the riverbed aggradation and degradation processes. Since the extraordinarily conceptual and algorithm simplicity, the proposed model provides a great versatility, stability and robustness.

2 GOVERNING EQUATIONS AND CLOSURES

Consider a system of orthogonal Cartesian axis (x,y,z) with unit vectors $(\bar{i},\bar{j},\bar{k})$. The vertical z-axis is positive in the uprising direction and the horizontal x-axis, corresponding to the projection of the thalweg line on the plan $z=0$, is positive downstream. Considering weak values of bottom slope, channel cross sections are assumed to be orthogonal to the x-axis. As usual in the one-dimensional scheme the flow is straighten in the plan.

The control volume V at the j_{th} cell is defined by the channel reach included between two vertical cross sections j and $j+1$ orthogonal to the x-axis. The control surface S is then compounded respectively by the upstream cross section S_j, the downstream cross section S_{j+1}, the wetted surface S_l, delimited by S_j and S_{j+1}, and the water free surface S_{pl}.

The momentum equation applied to the control volume V and projected on the horizontal x-axis, reads:

$$\rho \frac{\partial}{\partial t} \int_{x_j}^{x_{j+1}} Q dx + S_{tj+1} - S_{tj} + \int_{S_l} (p\bar{n} - \bar{\tau}) \cdot \bar{i} dS_l = 0 \quad (1)$$

where $\rho=$ water density; $Q=$ liquid discharge; $p=$ pressure; $\bar{n}=$ unit vector outward normal to S; $\bar{\tau}=$

shear stress (viscous + turbulent); and $\overline{S}_{tj}$ is the momentum function defined as follows:

$$\overline{S}_{tj} = \int_{S_j} (p + \rho v^2)\bar{i}\, dS \qquad (2)$$

Particular attention is posed to the treatment of the source terms of the Equation (1) related from one hand to the hydrostatic pressure acting on the wetted surface S_l and on the other hand to the shear stress. Following Catella & Solari (2005), Equation 1 reads:

$$\frac{\partial}{\partial t}\int_{x_j}^{x_{j+1}} Q dx + \left[\beta \frac{Q|Q|}{A} + gI_1\right]_{x_j}^{x_{j+1}} + $$
$$- g\left[A_{j+1}(h_m - z_{gj+1}) - A_j(h_m - z_{gj})\right] + \qquad (3)$$
$$+ \frac{\Delta x_j}{2}\left[B_{j+1}\left(\frac{Q|Q|}{(CA)^2}\right)_{j+1} + B_j\left(\frac{Q|Q|}{(CA)^2}\right)_j\right] = 0$$

where β = Boussinesq velocity distribution coefficient; A = wetted cross sectional area; g = gravitational acceleration; I_1 = first moment of the wetted cross section with respect to the free surface; h_m = average elevation of the water surface within the j_{th} cell; z_g = center mass elevation of the wetted cross section; Δx_j = distance between the vertical cross sections j and $j+1$; B = wetted perimeter; and C = dimensionless Chézy coefficient.

The integral form of the continuity equation under the assumption of no lateral inflow or outflow is:

$$\frac{\partial}{\partial t}\int_{x_j}^{x_{j+1}} A dx + \left[Q\right]_{x_j}^{x_{j+1}} = 0 \qquad (4)$$

Finally, the integral form of the sediment continuity equation under the assumption of no lateral sediment supply or shut-off may be written as:

$$\frac{\partial}{\partial t}\int_{x_j}^{x_{j+1}} z_b dx + \frac{1}{(1-\lambda)b_{mj}}\left[Q_s\right]_{x_j}^{x_{j+1}} = 0 \qquad (5)$$

where z_b = elevation of bed channel; λ = porosity of bed material; b_m = average channel width of the water surface within the j_{th} cell; and Q_s = solid discharge.

To close the governing Equations (3)–(5) the sediment discharge must be specified. In the present study the equilibrium transport hypothesis is applied, i.e. the solid load is able to adapt instantaneously to mild spatial or temporal variations of flow. The sediment transport capacity is evaluated by applying an arbitrary bed-load discharge formula.

In particular, the Meyer-Peter and Müller formula (1948) and Parker formula (1990) are employed herein.

Introducing the usual discretization:

$$\frac{\partial}{\partial t}\int_{x_j}^{x_{j+1}} G dx \cong \frac{G_m^{n+1} - G_m^n}{\Delta t^n}\Delta x_j \qquad (6)$$

(where G is an arbitrary variable, G_m its average value within the j_{th} cell and n the time step index) we obtain a 3(N-1) equations system function of 3(N-1) variables A_m, Q_m and Δz_{bm} average values of the wetted cross sectional area A, the liquid discharge Q and the bed elevation variation Δz_b in each N-1 cells defined by a N grid points approximation of the entire reach length. The N grid points are generally provided by topographic surveys at not equidistant discrete locations x_j.

The main characteristic of the present method is the employment of the average value of A, Q and Δz_b within each cell as conservative variables of the governing equations independently from the adopted spatial grid. Thus the developed method appears to be particular suitable for natural river modeling, where wetted cross sectional areas, due to the irregularities of topography, can be very different at the extreme vertical interfaces of each cell.

The assignment of the 3N variables A, Q and Δz_b to the N grid points from the 3(N-1) average values within each cell follows a criterion based on the Froude number of the average flow of the upstream and downstream cells between the interface to assign.

3 NUMERICAL SCHEME

The solution of the present equations system is conducted by following a decoupled procedure.

Firstly, by assuming a fixed-bed, the solution of water continuity and momentum equations is conducted by applying an explicit two steps predictor-corrector conservative scheme based on a finite volume method (Catella & Solari, 2005).

The proposed technique, composed by a sequence of two sub-steps in which the spatial derivatives are taken in the same directions, reads:

$$\mathbf{U}_j^p = \mathbf{U}_j^n - \frac{\Delta t^n}{\Delta x_j}[\mathbf{F}_{j+1}^n - \mathbf{F}_j^n] + \Delta t^n \mathbf{S}(\mathbf{U}^n, x) \qquad (7)$$

$$\mathbf{U}_j^c = \mathbf{U}_j^n - \frac{\Delta t^n}{\Delta x_j}[\mathbf{F}_{j+1}^p - \mathbf{F}_j^p] + \Delta t^n \mathbf{S}(\mathbf{U}^p, x) \qquad (8)$$

where the superscripts p and c indicate the predictor and the corrector steps, while **U**, **F** and **S** are the vectors of conserved variables, fluxes and source terms, respectively.

The solution at the new time level $n+1$ is then evaluated as an average between the two sub-steps:

$$\mathbf{U}_j^{n+1} = 0.5 \cdot \left(\mathbf{U}_j^p + \mathbf{U}_j^c\right) \tag{9}$$

In the present model Q is always assigned by allocating Q_m to the downstream intercell starting from the inflow boundary condition. To assign A, a criterion based on the Froude number of the average flow of the upstream and downstream cells across the interface to assign is applied in addition to an appropriate boundary condition at the upstream or downstream edge of the channel. In particular, if the average flows in the j_{th} and $j+1_{th}$ cells are both subcritical, A at the $j+1_{th}$ intercell takes the value of A_{mj+1}, in the opposite case, both supercritical, the value of A_{mj}. In the case of transcritical flows, if the average flows in the j_{th} and $j+1_{th}$ cells are subcritical and supercritical respectively, A at the $j+1_{th}$ intercell takes the critical value, in the opposite case, supercritical and subcritical respectively, a criterion based on the momentum function of the flows in the upstream and downstream cells across the interface to assign is applied.

Secondly, the sediment continuity equation is solved by an explicit conservative scheme based on a finite volume method and by employing the flow variables previously computed:

$$\Delta z_{bm\,j}^{n+1} = \Delta z_{bm\,j}^{n} - \frac{\Delta t^n}{\Delta x_j} \frac{1}{(1-\lambda) b_m} [Q_{s\,j+1}^{n+1} - Q_{s\,j}^{p}] \tag{10}$$

To assign Δz_b, a criterion based on the Froude number of the average flow of the up- and down-stream cells across the interface to assign is applied again in addition to a proper boundary condition.

In particular, if the average flows in the j_{th} and $j+1_{th}$ cells are both subcritical, Δz_b at the $j+1_{th}$ intercell takes the value of Δz_{bmj}, in the opposite case, both supercritical, the value of Δz_{bmj+1}. According to De Vries (1965), under the assumption of continuous solutions of the hyperbolic model, bed level disturbances propagate downstream or upstream if the flow is subcritical (Froude number <0.8) or supercritical (Froude number >1.2) respectively.

In the case of transcritical flows, if the average flows in the j_{th} and $j+1_{th}$ cells are subcritical and supercritical respectively, Δz_b at the $j+1_{th}$ intercell takes the average value of Δz_{bmj} and Δz_{bmj+1}, in the opposite case two different scenarios can be depicted: the $j+1_{th}$ intercell can be supercritical or subcritical as results from the criterion based on the momentum function. In the first case Δz_b at the $j+1_{th}$ and $j+2_{th}$ intercells take the value of $\Delta z_{bmj+1}/2$, in the second case Δz_b at the j_{th} and $j+1_{th}$ intercells take the value of $\Delta z_{bmj}/2$. At discontinuity points for transcritical flow regimes, according to Sieben (1999), the bed level disturbances propagate both down- and up-stream by inducing a more or less symmetrical deformation of the bed level.

For the stability of the present explicit scheme, the Courant-Friederich-Lewy number N_{CFL} must satisfy the following condition:

$$N_{CFL} = \Delta t \cdot \frac{\max_i\left[\left|\frac{Q}{A}\right|_i\right] + \max_i\left[\sqrt{\frac{g \cdot A}{\alpha \cdot b}}\Big|_i\right]}{\min_i\left[\Delta x_i\right]} < 1 \quad 1 \le i \le N \tag{11}$$

where $\alpha =$ Coriolis velocity distribution coefficient.

4 BOUNDARY CONDITIONS

Irrespective of the value of the Froude number, it is easy to prove that two celerities are always positive (downstream direction) and one is always negative (upstream direction). Therefore, two boundary conditions must be imposed at the upstream edge in relation to the two characteristics entering the domain, while one boundary condition needs to be specified at the downstream edge.

Boundary conditions can be expressed in term of liquid discharge Q, water depth y, and bed level z_b or a combination of the former variables.

In subcritical flow, the liquid discharge, the bed level are specified as upstream boundary conditions, while the free water surface elevation is imposed at the downstream boundary.

In supercritical flow, the liquid discharge and the water depth are specified as upstream boundaries, and the bed level as downstream boundary.

5 APPLICATIONS

To verify the applicability of the decoupled model and the goodness of the assignment criterions to predict river morphodynamics, some hypothetical benchmark tests are conducted.

The riverbed evolution due to channel width and bottom topography variations is investigated separately involving different flow regimes.

In all the following numerical tests the initial water level profile corresponds to the equilibrium steady state obtained under fixed-bed condition. The inflow liquid discharge is set equal to $500\,\text{m}^3/\text{s}$ and the sediment transport rate is evaluated with the Parker formula (1990). The inflow sediment discharge is set equal to the sediment rate evaluated in the first cross

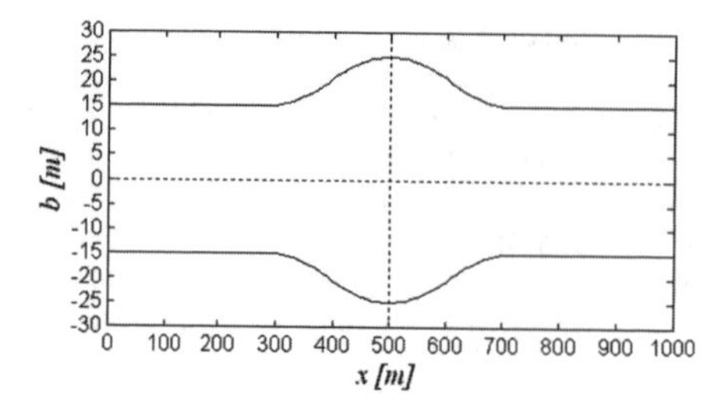

Figure 1. Plan of the channel.

section. The median diameter of bed material D_{50} is 0.01 m and the sediment porosity 0.3.

In the computations the rectangular channel 1000 m long is outlined by 100 cells ($\Delta x = 10$ m) and N_{CFL} is set equal to 0.9. Note that no changes must be introduced in the numerical scheme to extend the present model to natural irregular geometries.

In mild channel the bottom slope is 0.001 and the flow regime is subcritical, while in steep channel the bed slope is 0.01 and the flow regime is supercritical.

5.1 *Channel width enlargement*

Consider a channel characterized by a mild variation of the channel width (Fig. 1), ranging between 30 and 50 m in a reach 440 m long located across the channel midpoint.

It can be shown (Seminara, 1997), that regardless of mild or steep bottom slope, water depth variations must attenuate channel width variation effects. Thus, due to the channel width enlargement, at the bed equilibrium the water depth inside the widen reach must decrease, while the bed level must increase.

In the mild slope test case (Fig. 2), the channel geometry ensures a fluvial regime throughout the channel and consequently the expansion causes an increase in the flow depth. At the beginning of the bed evolution, inside the enlarged area, the decelerated flow triggers a deposition phenomenon at the upstream end, while the accelerated flow induces an erosion phenomenon at the downstream end. Both aggradation and degradation increase their amplitude during their slow propagation in the downstream direction. At the end, the deposit gains the entire widen reach and still moving downstream it reestablishes the initial bed profile.

In the steep slope test case (Fig. 3), the channel expansion causes a decrease in the flow depth and at the downstream edge, since the flow changes from supercritical to subcritical state, a hydraulic jump occurs. At the beginning of the bed evolution, inside the enlarged area, the accelerated flow causes an erosion phenomenon at the upstream end, while a deposition process occurs at the edge of the hydraulic jump, which starts to migrate downstream reducing its intensity. The hydraulic jump disappears in few minutes and a supercritical flow is established along the entire channel. Both degradation and aggradation increase their amplitude during their migration in the upstream direction. At the end, the deposit reaches the entire enlarged reach and moving upstream it reestablishes the initial bed profile.

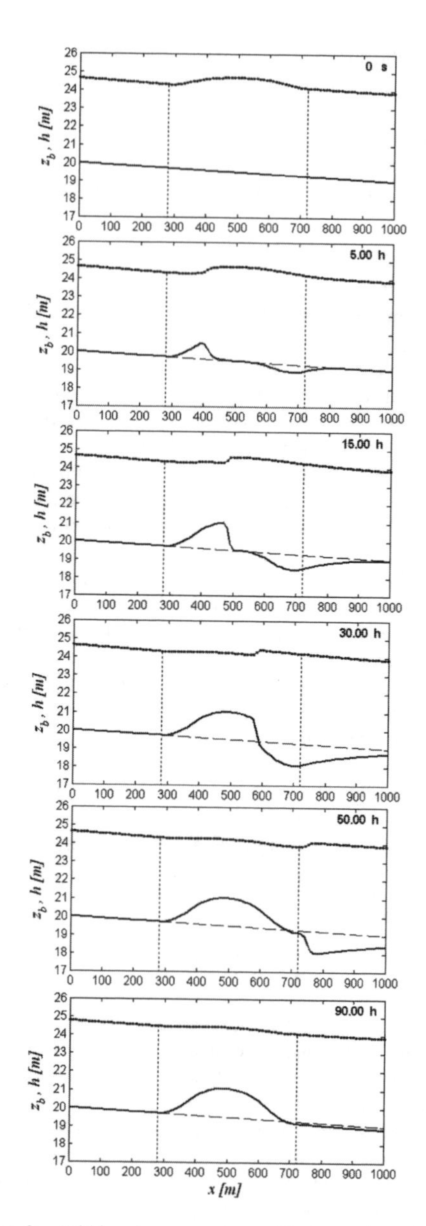

Figure 2. Width enlargement in a mild channel: evolution of bed level (solid line) and water surface level (line plus square). Dot lines delimitate the enlargement; dashed line indicates the initial bed profile.

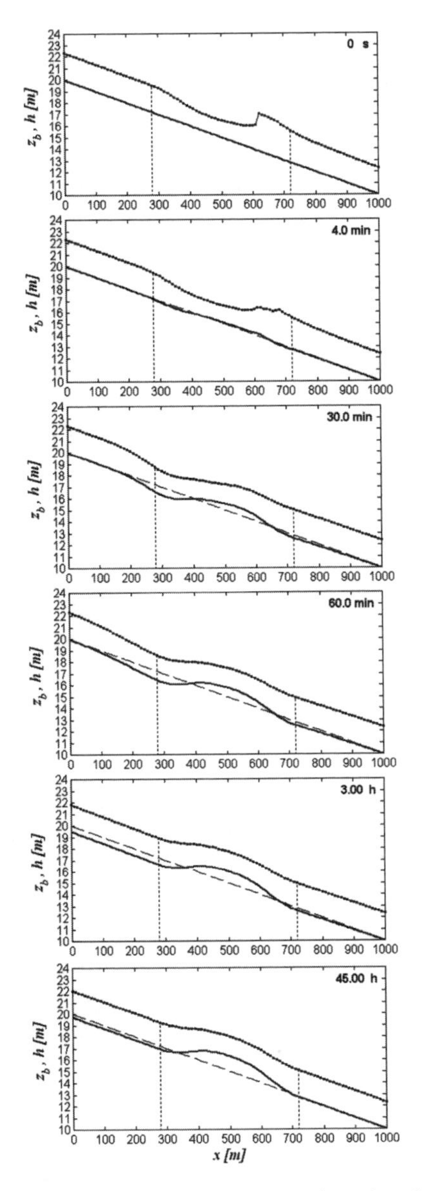

Figure 3. Width enlargement in a steep channel: evolution of bed level (solid line) and water surface level (line plus square). Dot lines delimitate the enlargement; dashed line indicates the initial bed profile.

5.2 *Channel width narrowing*

Consider a channel characterized by a mild variation of the channel width (Fig. 4), ranging between 50 and 30 m in a reach 440 m long located across the channel midpoint.

As reported in the previous test case, it can be shown (Seminara, 1997), that regardless of mild or steep bottom slope, water depth variations must attenuate channel width variation effects. In the present case due to the channel width narrowing, at the bed equilibrium the water depth inside the restrict reach must increase, while the bed level must decrease.

In the mild slope test case (Fig. 5), the channel geometry ensures the formation of the critical depth inside the constricted region. Downstream, the flow, which is subcritical, changes locally into supercritical and a hydraulic jump occurs. Upstream a continuous backwater profile is formed. At the beginning of the bed evolution, more and more sediments come to settle at the edge of the hydraulic jump, and the deposit slowly takes the form of a bore (Bellal et al., 2003). At the same time, under the accelerated flow in the upstream region of the constricted reach, the scour of the bottom channel begins. The hydraulic jump, together with the bore, starts to migrate downstream. Moreover, due to degradation and aggradation processes, the hydraulic jump reduces its intensity until the flow transition is totally washed-out. Thus, since the flow is subcritical along the whole channel, the bed perturbation starts to move only downstream. Besides, the scour hole increases its amplitude by gaining the entire restricted reach.

In the steep slope test case (Fig. 6), the channel constriction induces at the upstream edge a hydraulic jump. At the beginning of the bed evolution, a deposition phenomenon occurs at the edge of the hydraulic jump, while an erosion process begins at the downstream edge of the restricted area. Also in this case, the hydraulic jump disappears in few minutes and a supercritical flow is established along the entire channel. At the end the scour hole achieves the entire restricted reach and upstream the initial bed profile is reestablished.

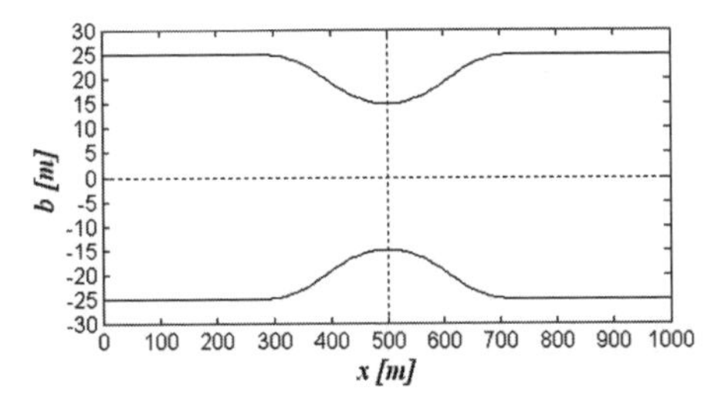

Figure 4. Plan of the channel.

5.3 *Bump*

Consider a rectangular channel 50 m wide characterized by an abrupt rise of the bed elevation located across the channel midpoint. In the mild channel the bump is 60 m long and 1 m height, while in the steep channel the bump is 120 m long and 1.56 m height.

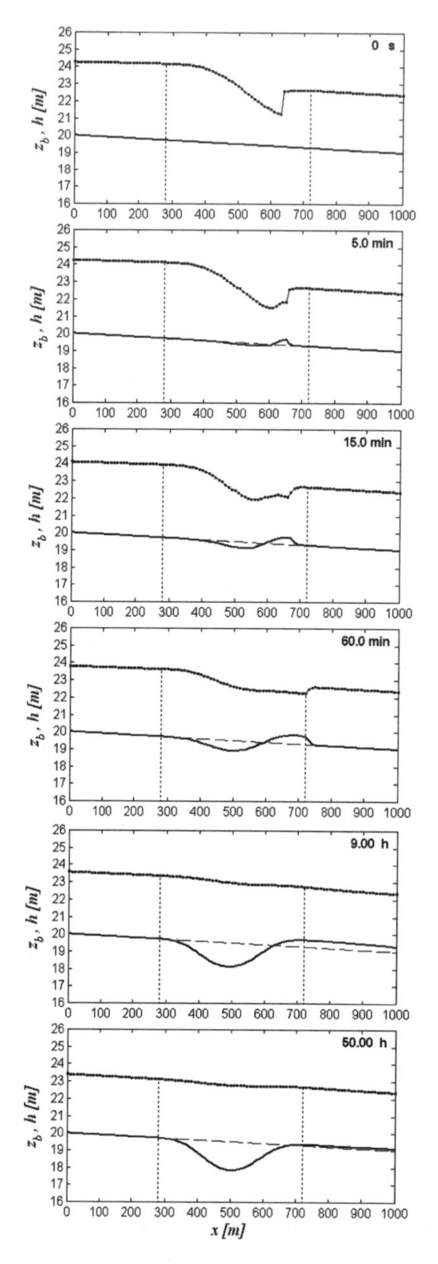

Figure 5. Width narrowing in a mild channel: evolution of bed level (solid line) and water surface level (line plus square). Dot lines delimitate the enlargement; dashed line indicates the initial bed profile.

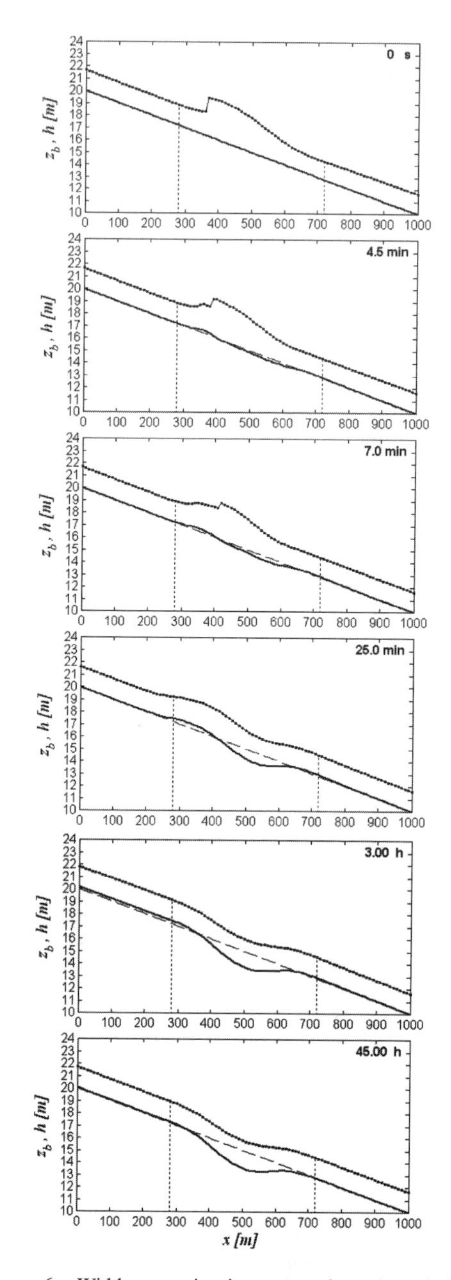

Figure 6. Width narrowing in a steep channel: evolution of bed level (solid line) and water surface level (line plus square). Dot lines delimitate the enlargement; dashed line indicates the initial bed profile.

In the mild slope test case (Fig. 7), the channel geometry ensures the formation of the critical depth on the bump and consequently a backwater profile upstream and a hydraulic jump downstream. At the beginning of the bed evolution, the top of the bump starts to be eroded and a deposition phenomenon develops just downstream. Erosion and deposition processes evolve delimited upstream by the subcritical flow and downstream by the supercritical flow, until the hydraulic jump by reducing its intensity vanished. Thus the perturbation of the bed level migrates downstream decreasing its height and increasing

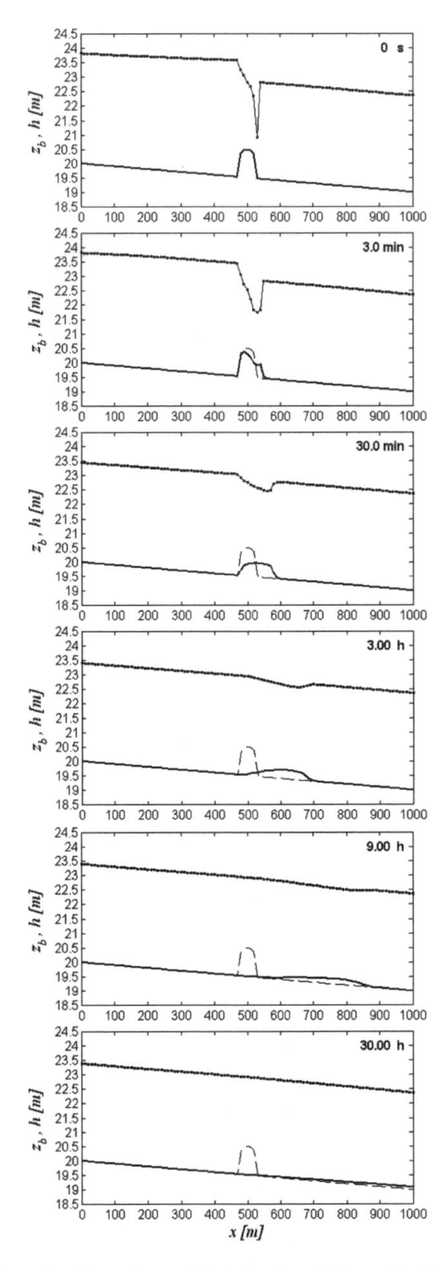

Figure 7. Bump in a mild channel: evolution of bed level (solid line) and water surface level (line plus square). Dashed line indicates the initial bed profile.

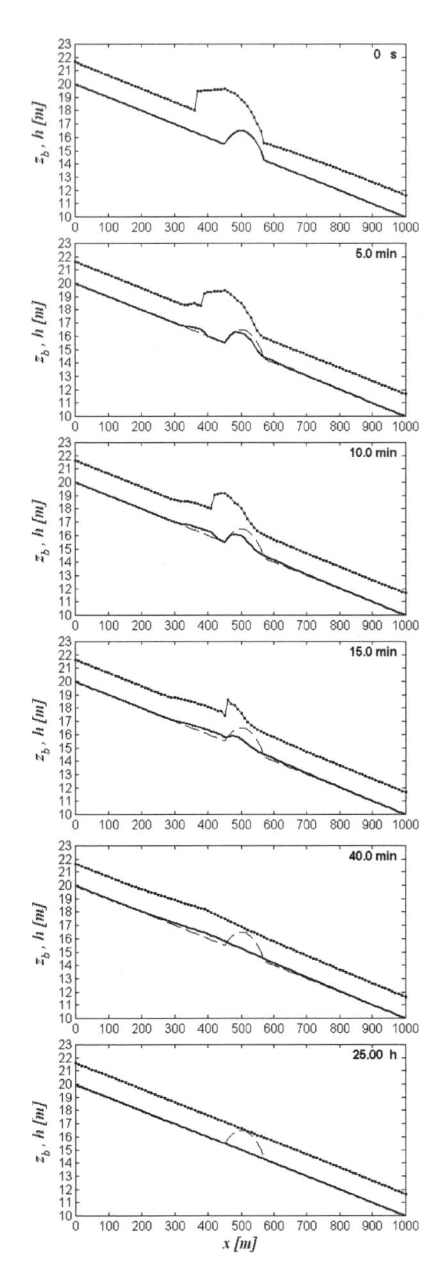

Figure 8. Bump in a steep channel: evolution of bed level (solid line) and water surface level (line plus square). Dashed line indicates the initial the bed profile.

its amplitude by reestablishing the undisturbed bed profile.

In the steep slope test case (Fig. 8), the channel geometry ensures the formation of a decelerated flow and a transition to a subcritical regime immediately upstream the bump. The change from supercritical to subcritical flow upstream is obtained by means of a hydraulic jump. At the beginning of the bed evolution, a deposit is formed at the edge of the hydraulic jump and it propagates downstream, while the top of the bump is eroded. Both these phenomena bring to the formation of a single deposit and to the hydraulic jump vanishing. When the flow regime starts to be supercritical throughout the channel, the bed level perturbation propagates upstream decreasing its height and increasing its amplitude by reestablishing the undisturbed bed profile.

6 APPLICATION TO OPEN-CHECK DAM

The numerical scheme is here applied to simulate the morphological evolution of the deposit that occurs upstream an open-check dam interested by a flood wave.

The present model is verified with the experimental data described by Armanini & Larcher (2001). The experiments were carried out in a rectangular flume 12 m long, 0.50 m wide (working height 0.45 m), with a bottom slope of 0.05. The tests were done with an uniform particle size, whose median diameter is 0.003 m. The sediments were transported as bed load. The open-check dam had a thickness of 0.018 m and it was located at $x = 10$ m.

Tests were performed reproducing a linear flood hydrograph: the inflow liquid discharge was 0.001 m^3/s at t = 0 s (initial value) and 0.00668 m^3/s at t = 46 min (peak value). The inflow solid hydrograph is built by means of the liquid hydrograph considering the peak solid discharge equal to 0.171 kg/s.

In the present analysis we refer to TEST C (Busnelli et al., 2001), in which the slit opening was equal to 0.20 m. Experimental data of bed level and free water surface are available at t = 23 min and t = 46 min.

In the present test, the sediment rate is evaluated by means of the Meyer-Peter and Müller formula (1948), in which the k constant is taken equal to 5.2 instead of 8 (original value). In the computations the channel is outlined by 384 cells, N_{CFL} is set equal to 0.7 and both the flood wave and the bed evolution do not start until a steady state for the initial liquid discharge is gained.

At the first instants a hydraulic jump develops upstream the open-check dam and it migrates in the upstream direction. At the same time a sediment bore occurs at the edge of the hydraulic jump. Then the deposition front starts to move forward the dam, while an aggradation process arise also in the upstream direction inducing a progressive adaptation of the bottom slope to the new equilibrium.

The proposed numerical model shows a quite good agreement between computed and measured bed level and water depths as reported in Figures 9–10.

7 APPLICATION TO DAM REMOVAL

Finally the numerical scheme is applied to reproduce the morphological evolution of a reservoir deposit induced by a sudden removal of a dam. The model is tested by comparing the numerical results with the data collected by Cantelli et al. (2004) at St. Anthony Falls Laboratory, University of Minnesota, Minneapolis.

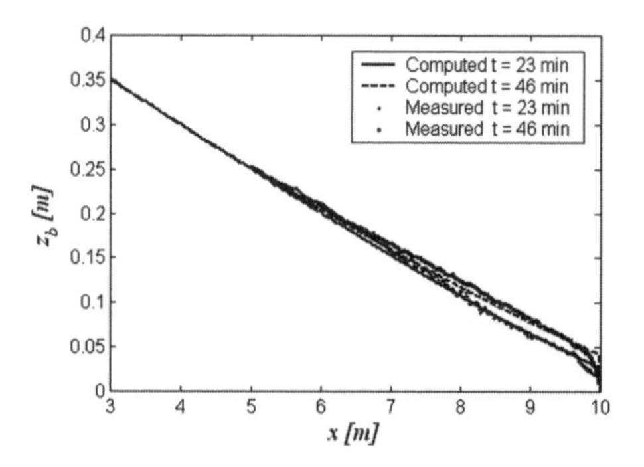

Figure 9. Comparison between computed (line) and measured (marker) bed levels evaluated during the rising part of the flood hydrograph.

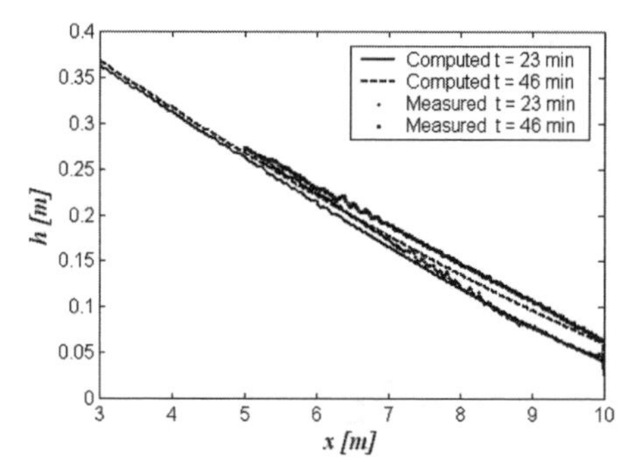

Figure 10. Comparison between computed (line) and measured (marker) water surface levels evaluated during the rising part of the flood hydrograph.

The experiments were carried out in a rectangular flume 14 m long, 0.61 m wide (working height 0.48 m), with a bottom slope of 0.018. Sediment had a median diameter equal to 0.0008 m and a specific gravity of 2.67. The dam was located at $x = 9$ m. At the end of the sedimentation process the deposit reached the dam and it was characterized by a slope of 0.004. In the present analysis we refer to RUN 6: the constant inflow liquid discharge is set equal to 0.0003 m^3/s, while the initial channel width is imposed equal to 0.275 m.

The bed-load discharge is estimated by means of the Meyer-Peter and Müller formula (1948), which is calibrated to allow a sediment rate of 0.002 kg/s with a liquid discharge of 0.0003 m^3/s and with a bed slope of 0.018. Thus, the k constant is taken equal to 4.7 and the threshold value of Shields number equal to 0.024 instead of 0.047. In the computations the channel is outlined by 140 cells and N_{CFL} is set equal to 0.3.

According to Sieben (1999), we observe in Figure 11 the contribution of an upstream-propagating

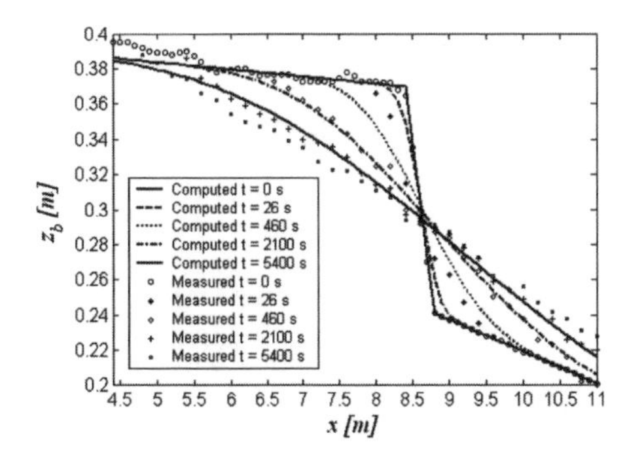

Figure 11. Evolution of the bed levels after dam removal. Comparison between computed and measured bed profiles evaluated at different temporal instants.

expansion wave and the contribution of a downstream-propagating shock wave of the bed level. The evolution mechanism of the bed level is reasonable catched on the overall simulation time, even if the discrepancies, specifically in the early stages of the erosion processes, since the channel narrowing phenomenon occurring at the downstream edge of the deposit, can not be ignored. Therefore, it may suppose that these discrepancies between numerical results and experimental data, are due to limitations of the 1D analysis rather than to limitations of the numerical scheme.

Future improvements of the numerical scheme may be focused on the modeling of channel incision and consequently narrowing involving bank erosion process.

8 CONCLUSION

A decoupled one-dimensional numerical model to study river morphodynamics is presented. The analysis reported here shows that decoupled models are not necessarily unstable and they give results comparable with theoretical behaviors and experimental data.

The St. Venant-Exner equations are solved by employing a finite volume method.

The governing equation system is solved in the following unknowns: wetted cross sectional area and liquid discharge of the average flow in each cell, and average bed elevation variation within each cell. The assignment of these unknowns along the river inter-cells is based on a Froude number criterion of the upstream and downstream average cell flow across the node point.

The main characteristic of the present method is the employment, as conservative variables of the governing equations, of average quantities within each cell independently from the adopted spatial grid.

The scheme is successfully applied to various steady and unsteady benchmark problems including subcritical, supercritical and transcritical flows modeling aggradation and degradation processes induced by the variability of channel width and bed elevation.

Since the extraordinarily conceptual and algorithm simplicity, the proposed model provides a great versatility, stability and robustness moreover vanishing transcritical sections.

ACKNOWLEDGMENTS

The authors wish to acknowledge Prof. Luigi Montefusco for his help and his valuable comments and suggestions.

REFERENCES

Armanini, A. & Larcher, M. 2001. Rational Criterion for Designing Opening of Slit-Check Dam. *Journal of Hydraulic Engineering* 127(2): 94–104.

Bellal, M., Spinewine, B. Savary, C. & Zech, Y. 2003. *Proc., XXX Int. Congress, Int. Association for Hydraulic Research, Tessaloniki, Greece.*

Busnelli, M.M., Stelling, G.S. & Larcher, M. 2001. Numerical morphological modeling of open-check dams. *Journal of Hydraulic Engineering* 127(2): 105–114.

Cantelli, A., Paola, C. & Parker, G. 2004. Experiments on upstream-migrating erosional narrowing and widening of an incisional channel caused by dam removal. *Water Resources Research* 40, W03304, doi:10.1029/2003WR-002940.

Cao, Z., Day, R. & Egashira, S. 2002. Coupled and decoupled numerical modeling of flow and morphological evolution in alluvial rivers. *Journal of Hydraulic Engineering* 128(3): 306–321.

Catella, M. & Solari, L. 2005. Conservative scheme for numerical modeling of flow in natural geometry. *Proc., XXXI Int. Congress, Int. Association for Hydraulic Research, Seoul, Korea.*

Kassem, A.A. & Chaudhry, M.H. 1998. Comparison of coupled and semicoupled numerical models for alluvial channels. *Journal of Hydraulic Engineering* 124(8): 794–802.

Lyn, D.A. & Altinakar, M. 2002. St. Venant-Exner equations for near-critical and transcritical flows. *Journal of Hydraulic Engineering* 128(6): 579–587.

Meyer-Peter, E. & Müller, R. 1948. Formulas for bed load transport. *Proc., 2nd Int. Congress, Int. Association for Hydraulic Research, Stockholm, Sweden.*

Parker, G. 1990. Surface-based bedload transport relation for gravel rivers. *Journal of Hydraulic Engineering* 28(4): 417–436.

Seminara, G. 1997. Equilibrio morfodinamico, stabilità ed evoluzione dei corsi d'acqua. *Proc., Nuovi sviluppi applicativi dell'idraulica dei corsi d'acqua, Bressanone, Italy* (in Italian).

Sieben, J. 1999. A theoretical analysis of discontinuous flow with mobile bed. *Journal of Hydraulic Research* 37(2): 199–212.

Vries, M. de 1965. Considerations about non-steady bed-load transport in open channels. *Proc., 11th Int. Congress, Int. Association for Hydraulic Research, Leningrad, URSS.*

River response to floods

River, Coastal and Estuarine Morphodynamics: RCEM 2005 – Parker & García (eds)
© 2006 Taylor & Francis Group, London, ISBN 0 415 39270 5

An experimental investigation of the channel adjustment process due to the passage of floods

C. Ershadi
Former research student, School of Civil Engineering and Geosciences, University of Newcastle upon Tyne, Newcastle upon Tyne, U.K.

E.M. Valentine
Professor of Civil Engineering, School of Engineering and Logistics, Charles Darwin University, Darwin, Australia

ABSTRACT: A series of laboratory experiments have been carried out to investigate the channel forming process under unsteady flow conditions, when channel boundaries are allowed to develop freely. To simulate unsteady flows, a stepped triangular unit hydrograph was used. The peak flow rates were set such that the flood plains were not mobile. The base flow rates were also set such that the main channel dimensions were stable. The change in width of the main channel, the water surface elevation, and sediment transport rate, as channel adjustment criteria, were recorded during the experiment. The results show that the width of the main channel and the channel side slope increased, and the depth decreased in response to the passage of the flood. The sediment concentration was usually larger on the rising than on the falling limb for the equivalent discharge. When channel conditions changed from bankfull to overbank, the sediment concentration decreased. A significant observation was that the stage was higher on the falling limb than on the rising limb for the equivalent discharge. As the discharge increased, two-dimensional dunes developed and during the recession they were suppressed or their migration ceased as sediment transport rates reduced. The results of this work may also improve our understanding of river channel response under flood flows.

1 INTRODUCTION

Although the quantities of water and sediment in natural rivers vary through space and time, most studies have addressed the response of a channel to steady flow conditions. Engineers wish to employ stable channel theories which use steady flow to represent the effects of the natural unsteady flow conditions in the river. Relationships for steady flow conditions are then used to predict the channel form. The main difficulty that arises in this hypothesis is to determine which steady flow represents the integrated effects on channel morphology of the natural variation in water and sediment discharges.

Previous laboratory work on unsteady flows is limited. For example, Tominaga et al (1995) studied the hydraulic response of a rectangular channel to flash floods with fixed boundaries in the laboratories. Kabir (1993), Graf & Song (1995) and Bestawy (1997) examined sediment transport and velocity distributions in flash floods in laboratory conditions with mobile beds.

2 PROGRAMME AND PROCEDURE

Experiments were carried out in a flume 22 m long, 2.5 m wide and 0.6 m deep, filled with uniform sand of 0.94 mm median size. The flume has been used over a period of 15 years to investigate channel forming and adjustment processes under steady inbank and overbank flow conditions (see Haidera & Valentine (1999) and Valentine & Shakir. (1992)).

The experiments with unsteady flow conditions were carried out with a slope of 0.002 m/m and the same initial conditions (straight bankfull channel with a discharge of 6 l/s). The channel was then subjected to unsteady flow. This was simulated by a triangular hydrograph with increasing flow rates to peak, followed by decreasing flow rates in a finite series of steady steps. The peak flow rate of less than 25 l/s was selected such that the floodplains were kept below the threshold of motion. The base flow rate of 2 l/s was selected such that the main channel boundary shear stress was below the threshold of motion for this inbank condition. The rates of widening, the water

Figure 1. Views of a typical experiment: (a) initial bankfull condition, (b) overbank condition during passing flood, (c) bankfull condition in recession, and (d) the final shape showing some developed bedforms.

Table 1. The resulting channel characteristics after the passage of the flood.

Exp.	Q_p l/s	$S \times 10^3$ m/m	B_b mm	z m/m	h mm	B_t mm	T (°C)
1	20.0	0.20	509	4.33	26.0	734	20.0
2	20.0	0.21	511	4.52	26.8	753	21.5
3	20.0	0.20	529	4.18	24.9	737	22.5
4	20.0	0.22	512	3.52	27.7	707	22.0
5	20.0	0.20	533	4.08	24.9	736	21.0
6	20.0	0.19	558	4.30	25.2	775	22.5
7	25.0	0.23	522	6.11	24.8	825	22.0
8	20.0	0.20	544	3.06	28.0	715	17.0
9	15.0	0.21	519	3.22	27.2	694	15.0
10	10.0	0.22	448	3.18	29.6	637	15.0
11	6.0	0.23	416	2.20	32.6	559	17.0
12	20.0	0.22	598	2.71	25.9	738	20.5
13	20.0	0.22	509	4.65	27.1	761	21.3
14	20.0	0.20	622	3.40	24.2	787	17.0
15	15.0	0.21	553	3.42	25.2	725	15.5
16	10.0	0.20	489	3.00	27.7	655	17.0
17	20.0	0.20	523	3.87	26.9	731	16.5
18	20.0	0.21	533	3.64	27.1	730	17.0

surface elevation and the sediment concentration were measured at the end of each time step before changing the flow rate. For more detail see Valentine & Ershadi (2003). Figure 1 shows a typical test and Table 1 shows the final hydraulic parameters of all tests, where Q_p is the peak flow rate of the hydrograph, S is the bed slope of the channel, z is the bank slope of the main channel, B_b is the bottom width of the main channel, B_t is main

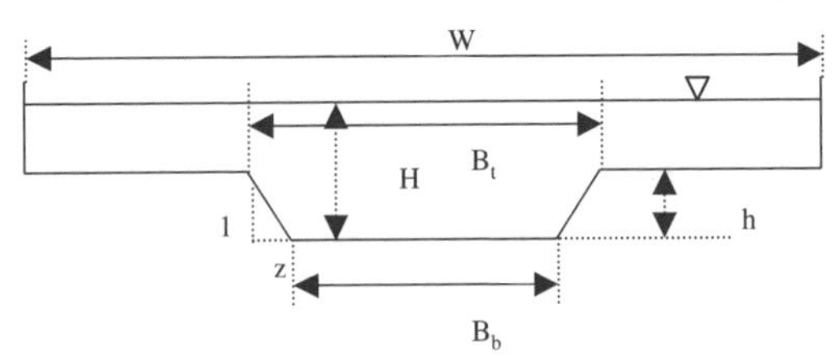

Figure 2. Compound channel cross-section with definition of variables.

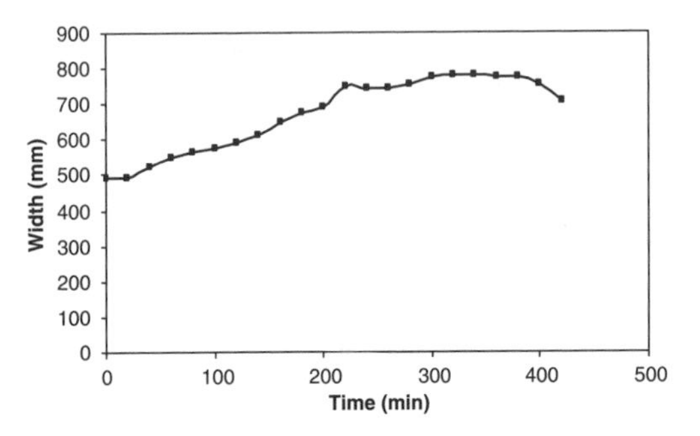

Figure 3. Variation in the width of the channel over time for a typical experiment.

channel top width, h is depth of main channel, and T is temperature (see Fig. 2).

3 EXPERIMENTAL RESULTS AND DISCUSSION

The channel development was considered in terms of change in channel configuration. According to the measurements and observations, carried out during the test, the channel development was discussed in the terms of change in the main channel width and depth.

3.1 *Flood effect on main channel width*

The hydrograph was initiated with a steady flow rate as base flow. The discharge was increased gradually. It was observed that the main channel widened rapidly when the channel became overbank. The width of the channel was defined as the water surface width of the main channel in inbank or bankfull conditions and the width of the main channel in overbank conditions. Figure 3 shows the variation in widths of the main channel at one cross-section at the central test reach during the passage of the flood in a typical experiment.

It was observed that initially the width of the channel gradually increased with the change in the water elevation. Then the width of the main channel rapidly increased due to the overbank conditions. After this significant adjustment in the width of the main channel, the main channel gradually continued to widen after passage of the peak flow rate until the flow rate fell to 0.75 Q_p (peak flow rate). When the water surface returned to the main channel in the falling limb, the width of the main channel did not change.

It was observed that the channel became successively wider (see Fig. 3) and no reduction of the main channel width was observed. This observation is consistent with Pitlick et al (2004) and Gupta & Fox (1974). Pitlick et al (2004) conducted a series of experiments in a 16-meter flume, which was filled with sand and pea-gravel, and in which water was circulated. These experiments were in the range of 10.0 l/s as a bankfull discharge to 23.0 l/s as a peak flow rate. They stated that the channel widened from 400 mm initial width to roughly 800 mm by the end of the experiments. They suggested that deposition of sediment suspended in low velocity areas can quickly narrow the channel or build-up the height of the floodplain, thus establishing different channel geometry.

Gupta & Fox (1974) studied four major floods that occurred in the Patuxent River basin in the Maryland piedmont during the summers of 1971–1972. They stated that the effect of the floods on the channel banks was remarkable. They observed strong lateral scouring and associated collapse of the channel banks with temporary increases in channel width. They observed that bank collapse under the effect of a flood is related to eddies which form during high flows, especially at pre-existing elliptical scour holes in both banks. The existence of these eddies causes lateral shear development between fast streamlines in midstream and slower ones closer to the bank. As slower velocity streamlines tend to turn onto the banks creating eddies, the banks erode and thus give rise to elliptical scour marks (see Fig. 4). Scoured material is removed, and generally a weak toe slope develops between the channel bed and the foot of the elliptically scoured steep bank. On the falling stage, coarse materials were deposited on these silt-clay toe slopes in lenses. Finally, a narrow strip of deposited material formed a bar at the foot of the collapsed steep bank. In the experiments reported here, this type of bank collapse was observed (see Fig. 5).

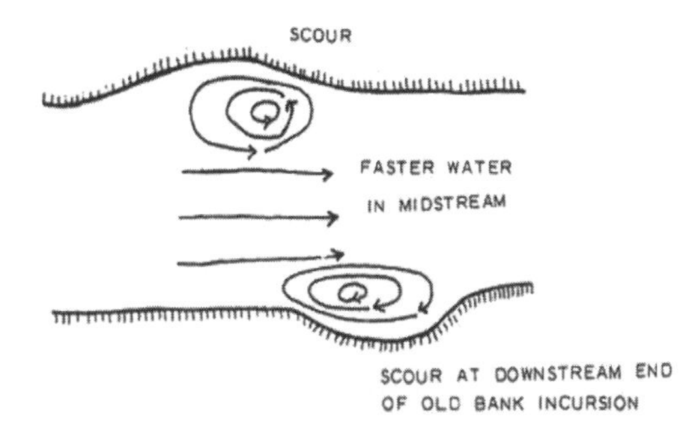

Figure 4. Sketch of bank scouring (after Gupta and Fox, 1974).

Figure 5. The bank scouring due to eddies formation in bank.

3.2 *Flood effect on main channel depth and planform*

During the rising limb of the hydrograph bedforms appeared and developed; during the falling limb, the bedforms were suppressed or their migration ceased. Due to the main channel depth changing over time, it was very difficult to determine the change in the main channel depth. However, after passage of the flood, the trace of bedforms remained and the main channel depth varied along the channel. These changes in bedform affected the planform of the channel. Figure 6 shows the variation in width and depth of the main channel. The depth was measured at the centre of the channel.

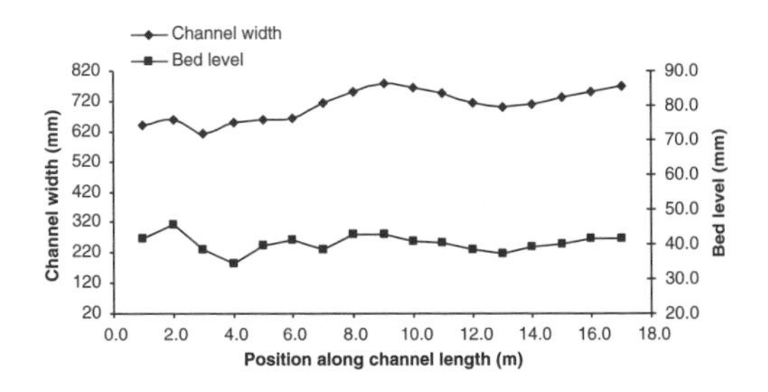

Figure 6. The change of bed profile and channel width following of passage the Flood for a typical experiment.

Yalin (in Best & Ashworth 1999) hypothesised that any fluid moving along an inclined boundary will

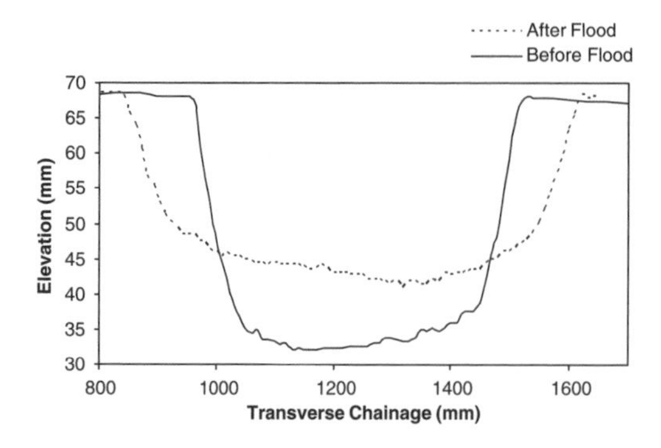

Figure 7. Averaged cross-section for experiment 7, before and after passing flood.

develop alternating zones of fast, accelerating, and slow, decelerating flow, with subsequent sediment erosion and deposition creating a sequence of topographic lows (pools) and highs (riffles). These changes in the bed are accompanied with changes in the width.

Gupta & Fox (1974) observed that in the field, the river was competent to move boulders, which are normally carried as bedload at high flow. In addition to the increase in competence, sorting of bed materials took place in association with the pool and riffle sequence. Immediately following a flood, riffles of sand and finer materials were exposed. These sand and finer materials in turn were found in the pool. There was no change in the location of the pools and riffles or bars despite the fact that the floods often swept out the channel during peak stages. They found that very few changes occurred in the low-flow pattern of rivers, with none of the rivers changing course. They also stated that when banks collapse, scoured material is removed, and generally a weak toe slope develops between the channel bed and the foot of scoured steep bank. On the falling stage, coarse materials are deposited on these silt-clay toe slopes in lenses. Finally, a narrow strip of deposited material forms a bar at the foot of the collapsed steep bank. The same process was observed in these experiments with developing bars in the toe. The experiments reported here support the observations of Yalin (Best & Ashworth 1999) and Gupta & Fox (1974).

In addition, it was observed that the average channel depth decreased after the flood passed (see Fig. 7). Therefore, the final cross-section tended to be wider and shallower than the initial channel. Although the channel dimensions changed after passing the flood, there is no significant change in the final cross-sectional area of the main channel in comparison to the initial channel. The stable cross-sectional area implies no net loss of sand in the recirculating system.

The results indicate that the cross-sectional area of the channel (A) was approximately constant and the wetted perimeter increased. Increases in the wetted perimeter are due to the channel widening. Therefore,

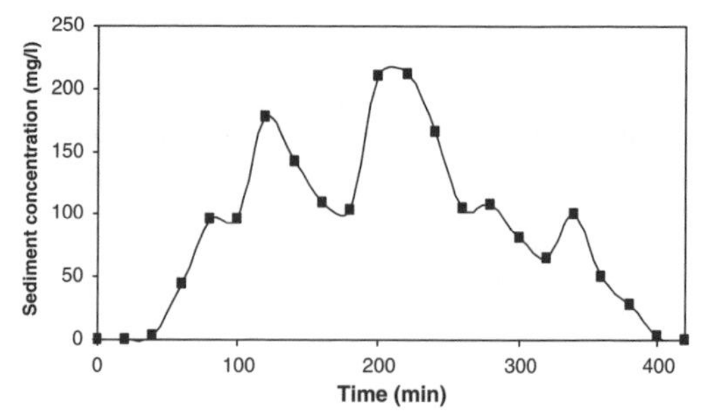

Figure 8. Variation in sediment concentration over time for test 8.

the hydraulic radius of the main channel reduces. Due to the reduction in the hydraulic radius of the channel, the channel capacity reduces.

3.3 *Flood effect on sediment concentration*

Sediment transport was measured during the experiments at the end of each time step. The experiment was started by introducing the base flow rate which was at threshold conditions. Figure 8 shows the variation in sediment transport during the passage of the flood.

It was observed that initially the threshold of motion for the developed channel was about 3.8 l/s, when the channel was inbank, velocity was 0.32 m/s and water depth was 24.4 mm. The sediment transport rate increased due to the change in flow rate affecting the shear stress. When channel conditions changed from bankfull to overbank flow condition, the sediment concentration decreased. These observations agree with Ackers's hypothesis (1992). He stated that when channel flow is overbank, the interaction effect between the main channel and the floodplain flows influences the sediment transport capacity in the main channel. He explained that there is a significant change in the sediment transport function when flow goes out of bank, with the main channel becoming less effective than it would be in the absence of floodplains or channel berms. In other words, when channel flow goes out of bank the charge of sediment reduces because of interference effects from the floodplain. The sediment concentration would be rising again as discharge increases.

After passing the peak flow rate, the sediment transport reduced in the falling limb, due to reduction in flow rate. The sediment concentration on the falling limb was smaller than for the same flow rate on the rising limb (see Fig. 9). These changes are the result of changes in the channel shape such as widening. In addition, when the water returned to the main channel in the falling limb, there was no discontinuity in the sediment concentration, as was observed in rising limb. This may be because the alluvial channel

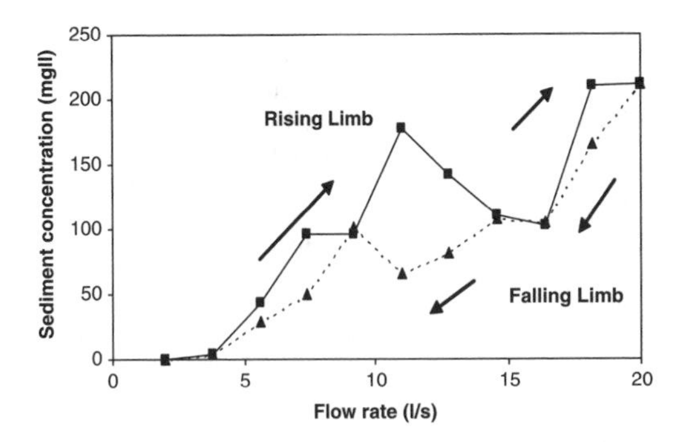

Figure 9. Change in sediment concentration with flow rate for test 8.

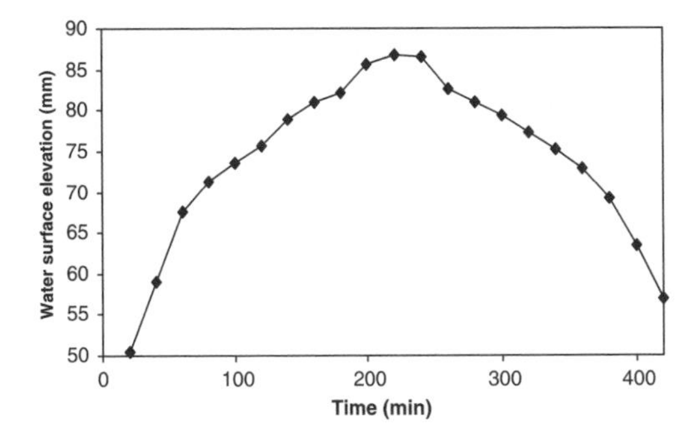

Figure 10. Development in water level surface over time for test 8.

in overbank conditions will respond to reducing the interaction effect between the main channel and the floodplain.

The fluctuation of sediment transport in this zone can be explained by the migration of the dunes and scouring banks, but it generally increased.

This observation agrees with De Sutter et al (2001). They stated that according to the result of artificial flood experiments, the rate of sediment transport is higher in the rising limb of a hydrograph than in the falling limb for the same flow rate, therefore a clockwise hysteresis loop is induced. They stated that the clockwise loop is also produced by the unsteadiness of the flow. They hypothesized that anti- or counter-clockwise hysteresis occurs less frequently, but is often prominent when sediment originates from a distance source or when the valley slopes form the most important source.

3.4 *Water elevation*

The water elevation was measured while running the experiment, at the end of each time step. Figure 10 shows the variation in water level surface during the passage the flood.

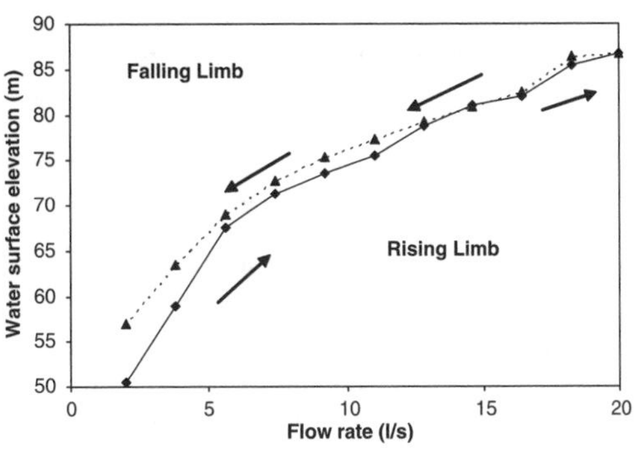

Figure 11. Rating curve for test 8.

The results show that the water surface elevation of the rising limb was generally lower than the falling limb for the same flow rate (see Fig. 11). This agrees with Griffiths & Sutherland's (1977) observations. They stated that the change of bedform such as dunes and different bedform behaviour between rising limb and falling limb have an effect on the stage discharge relations by the production of an asymmetrical loop of the rating curve especially around peak flow rate.

This behaviour was reflected in the bankfull discharge by reduction of bankfull discharge in the falling limb, as explained.

4 CONCLUSIONS

For this limited series of laboratory experiments the following conclusions have been drawn:

1. A stable alluvial bankfull channel, when subjected to an unsteady flow, tends to a new stable channel section. Further work will be required to determine the significance of morphological response times.
2. In response to the passage of the flood, the width of the main channel increased, the depth decreased and the channel side slope increased.
3. When the channel is inbank, there is no significant change in the width of the main channel, but the width of the main channel rapidly increases in flows above bankfull. After passage of the peak flow rate, the main channel gradually widens and this continued until the flow rate fell to 0.75 of the peak flow rate.
4. The sediment concentration is usually larger on the rising than on the falling limb for the equivalent discharge. This is due to the channel adjustment.
5. When channel conditions change from bankfull to overbank on the rising limb, the sediment concentration decreases. After passage of the peak flow rate, the sediment transport reduces on the falling limb.

6. The stage is higher on the falling than on the rising limb. This is due to the channel adjustment.
7. Following the passage of a flood, the hydraulic capacity of the main channel reduced when compared to the initial channel conditions. It is important to determine if this effect will be a potential outcome of adjustment to more frequent, higher flood levels which may be a feature of climate change.

The results of this work may help to improve our understanding of the response of a river to flood flows and to develop rational regime predictions for prototype unsteady conditions.

REFERENCES

Ackers, P. 1992. Hydraulic Design of Two-Stage Channels. *Proceedings of the Institution of Civil Engineering, Water. Maritime and Energy* 96: 247–257.

Best, J. & Ashworth, P. 1999. EPSRC Project Investigation of Flow and Sediment Dynamics of the Pool-Riffle Unit *in http://earth.leeds.ac.uk/research/dynamics/index.htm.*

Bestawy, A. 1997. Bed Load Transport and Bed Forms in Steady and Unsteady Flow. *PhD Thesis. Catholic University Leuven*. Heverlee. Belgium.

De Sutter, R., Verhoeven, R. & Krein, A. 2001. Simulation of Sediment Transport during Flood Events: Laboratory Work and Field Experiments. *Hydrological Science-Journal -des Sciences Hydrologiques* 46(4): 599–610.

Graf, W.H. & Song, T. 1995. Bed-Shear Stress in Non-Uniform and Unsteady Open-Channel Flows. *Journal of Hydraulic Research* 33(5): 699–704.

Griffiths, G.A. & Sutherland, A.J. 1977. Bedload Transport by Translation Waves. *Journal of the Hydraulics Division, Proceedings of the American Society of Civil Engineers* 103(HY11): 1279–1291.

Gupta, A. & Fox, H. 1974. Effects of High-Magnitude Floods on Channel Form: A Case Study in Maryland Piedmont. *Journal of Water Resources Research* 10(3): 499–509.

Haidera, M.A. & Valentine, E.M. 1999. Behaviour of Alluvial Channel with Overbank Flow. *IAHR Symposium on River, Coastal and Estuarine Morphodynamics*. Genova, Italy.

Kabir, M.R. 1993. Bed Load Transport in Unsteady Flow. *PhD Thesis. Catholic University Leuven.* Heverlee. Belgium.

Pitlick, J., Pizzuto, J. & Marr J. 2004. Width adjustment in alluvial channels. *in http://www.colorado.edu/geography/geomorph/nsf_safl.html*

Tominaga, A., Liu, J., Nagao, M. & Nezu, I. 1995. Hydraulic Characteristics of Unsteady Flow in Open Channels with Flood Plains. *The 26th IAHR Congress, HYDRA 2000.* Thomas Telford, London.

Valentine, E.M. & Ershadi, C. 2003. A Laboratory Study of Alluvial Channels With Unsteady Flow. *30th IAHR Congress, Technical session of Theme C.* Thessaloniki, Greece.

Valentine, E.M. & Shakir, A.S. 1992. River Channel Planform: An Appraisal of the Rational Approach. *The 8th, congress of the Asia and Pacific, Division of IAHR.* Poona, India.

River, Coastal and Estuarine Morphodynamics: RCEM 2005 – Parker & García (eds)
© 2006 Taylor & Francis Group, London, ISBN 0 415 39270 5

Flood simulation for the Appetsu River in Hokkaido, Japan, during Typhoon Etau of 2003

Yoshihiko Hasegawa & Yasuyuki Shimizu
Graduate School of Engineering, Hokkaido University, Japan

ABSTRACT: Typhoon Etau arrived in Hokkaido, Japan, on 9 August 2003. Because of incomplete embankment protection, there was flooding, bank erosion, and transport of sediment and driftwood in great quantity, particularly on the Appetsu River in the Hidaka district. This study attempts to reproduce flood and deposition on the floodplain by numerical calculation. Steady flow calculation is conducted using the peak discharge of this flood event. Unsteady calculation is simulated using the calculated hydrograph. Sediment transport is calculated using the transport equation for suspended load. Discharge, bed load, and bed height change in the flood region are calculated using a generalized coordinate system. The characteristics of the flood during Typhoon Etau on the Appetsu River are clarified.

1 INTRODUCTION

Typhoon Etau arrived in Hokkaido, Japan, on 9 August 2003. Rainfall of 277 mm was recorded at the Hidaka observation point in the Hidaka district. River improvement on the Appetsu River had not been completed, and flooding, bank erosion, and sand and driftwood transport were remarkable there. Figure 1 shows a hydrograph of the Appetsu River from 8–10 August 2003 (Wongsa et al.). Figure 2 shows the location of the Appetsu River and the calculation area, which is 4 km from the river mouth. This study reproduces flood flow and sediment deposition on the floodplain by numerical calculation, to examine their characteristics during this flood event. Steady-flow calculation was conducted using the peak discharge. Flow bifurcation and convergence was found on the floodplain, which suggests that multiple-row bars formed. To reproduce sediment erosion and deposition on the floodplain, unsteady-flow calculation was conducted using a hydrograph. Suspended and bed load sediment transport were calculated, and floodplain deformation was calculated for the period covered by the flood hydrograph. Sediment was supplied from the upstream end of the calculation region. All the calculations were made using a generalized coordinate system. Flood flow configuration and bed deformation were accurately reproduced. As a result, the characteristics of flood flow and the geometric formation of floodplain morphology could be studied.

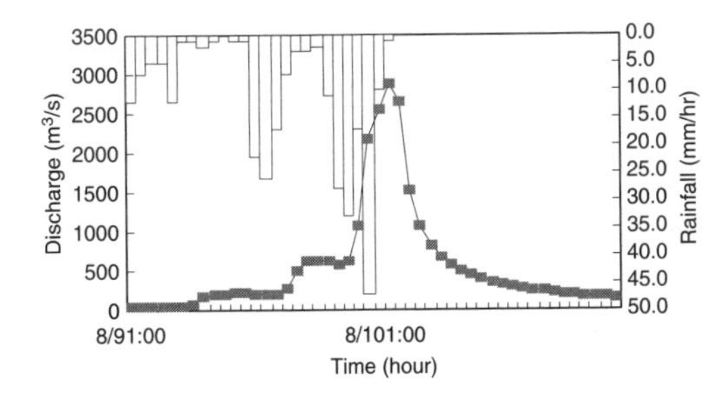

Figure 1. The hydrograph of the Appetsu River.

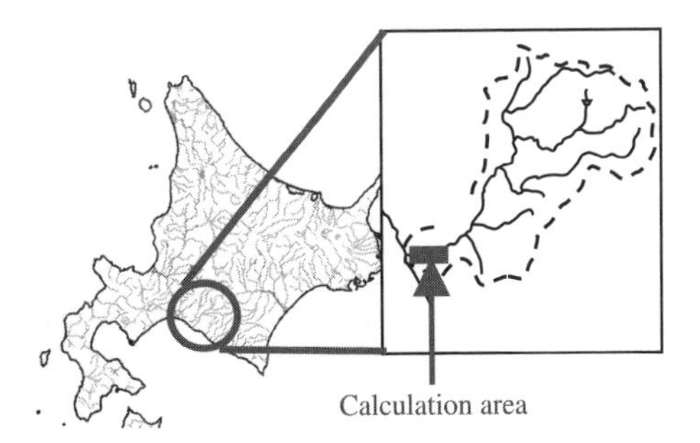

Figure 2. The location of the Appetsu River and the calculation area.

2 INVESTIGATION

Flood damage investigation was made on 11 August 2003. Figure 3 is an aerial photo of the Appetsu River 1 km from the mouth. Arrows show the velocity vectors during the flood. Figures 4 and 5 are photos of points P and Q in Figure 3. Large amounts of driftwood were produced by the flood flow (Figure 4), and houses were destroyed (Figure 5).

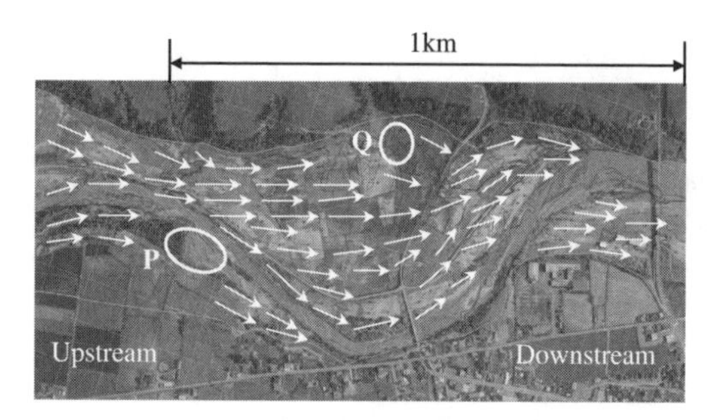

Figure 3. An aerial photo of the Appetsu River 1 km from the mouth.

Figure 4. Photo of points P in Figure 3.

Figure 5. Photo of points Q in Figure 3.

3 CALCULATION OF FLOOD FLOW

The two-dimensional, unsteady-flow field is calculated using the continuity equations. The momentum equations are separated into three phases: advection, friction and pressure, and diffusion. In the advection phase, a scheme with high-order accuracy known as the CIP method is employed. In the friction and pressure phase, equations are coupled with the continuity equation and solved iteratively. In the diffusion phase, a central difference scheme is employed. Equations are transformed into a general coordinate system.

The flow is calculated for the low-water channel and the flooded region. Figures 6 and 7 show the calculation area, which is 4 km from the mouth of the Appetsu River. The flooded region is identified by comparing Figures 6 and 7. The white line in Figure 6 indicates the flooded region. Computational grids divide the flood region into 28 grids in the flow direction and 14 in the transverse direction. Figure 8 shows the computational grids. The blue line in Figure 8 indicates the low-water channel. The riverbed height was obtained from field survey. Figure 9 is the contour map of grand elevation. The uniform flow depth and velocity calculated by Manning's formula are given as the initial condition. The flow velocity at the upstream end and the water level at the downstream end are kept constant during the calculation as the boundary condition. Manning roughness coefficient is set as 0.045 in the high-water channel and 0.035 in the low-water channel.

Since the hydraulic characteristics are shown at the peak discharge, the discharge of 2884 (m^3/s) is given as the constant discharge. The discharge of 2884 m^3/s is shown for the hydrograph in Figure 1. The water level at the downstream end is 1.46 m which is the observed water level. This water level is kept constant during the calculation.

Discharge at the upstream end is varied in the unsteady-flow calculation. The changes in the discharge (the range is from 73 m^3/s on 9 August to 156 m^3/s on 10 August) are shown.

From the start of the calculation to the 1500th sec, the discharge is set at 73 m^3/s. From that point in time, discharge was increased to match the discharge from the hydrograph, since the calculation reaches equilibrium at the 1500th second. The water level at the downstream end was changed from the initial condition to the observed water level (1.46 m) together with the increase in discharge.

Figure 10 shows the water surface elevation along the center line of the low-water channel in the steady condition. Figure 10 shows also the observed water level. Figure 11 shows the flow velocity vector in the steady condition. Figure 12 shows contours of flow velocity plotted using particles.

Particles transported at the same velocity are put into the flow and tracked in order to visualize the flow

Figure 6. Calculation area, which is 4 km from the mouth of the Appetsu River (take a picture before the flood).

Figure 7. Calculation area, which is 4 km from the mouth of the Appetsu River (take a picture after the flood).

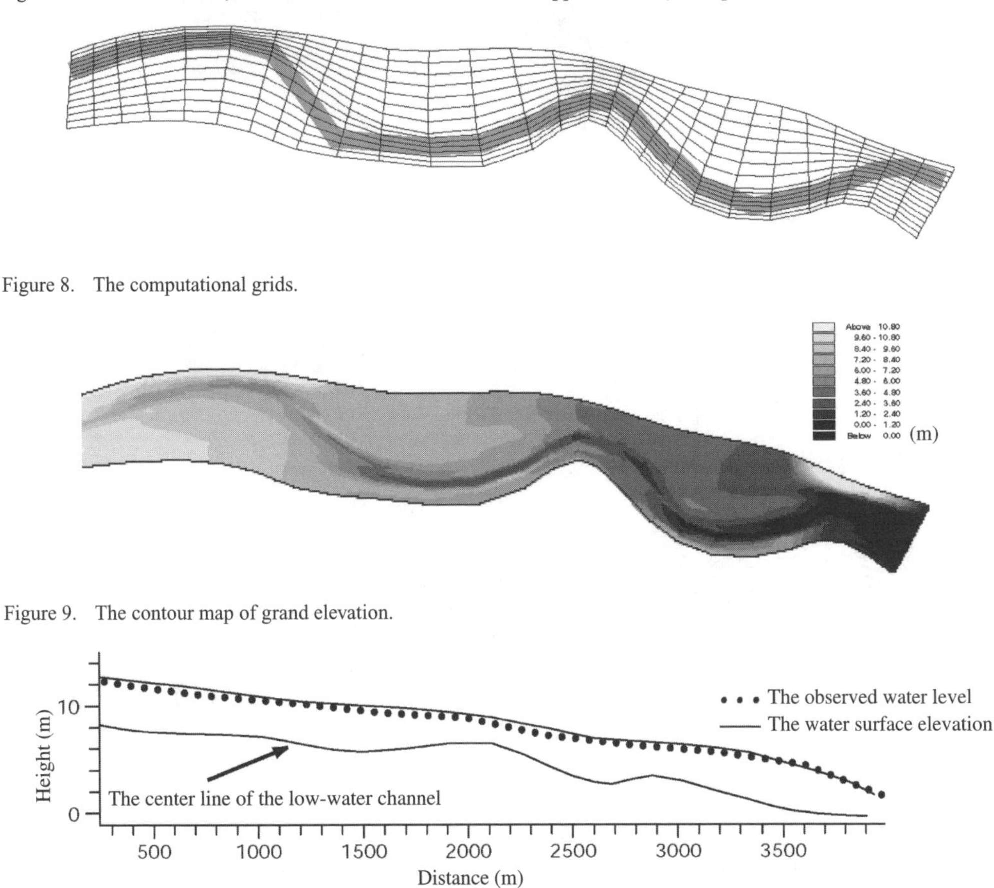

Figure 8. The computational grids.

Figure 9. The contour map of grand elevation.

Figure 10. The water surface elevation and the observed water level along the center line of the low-water channel in the steady condition.

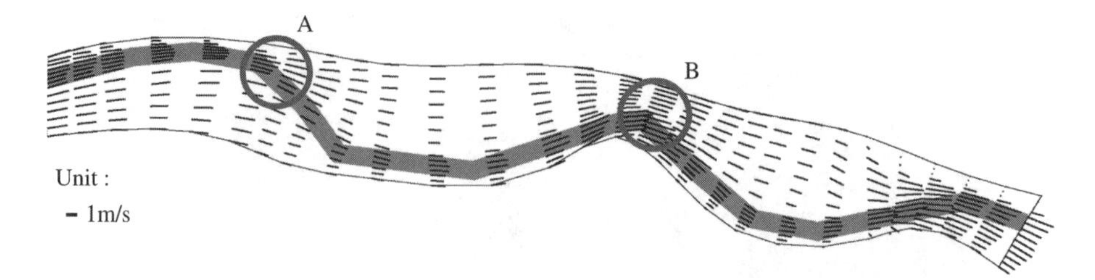

Figure 11. The flow velocity vector in the steady condition.

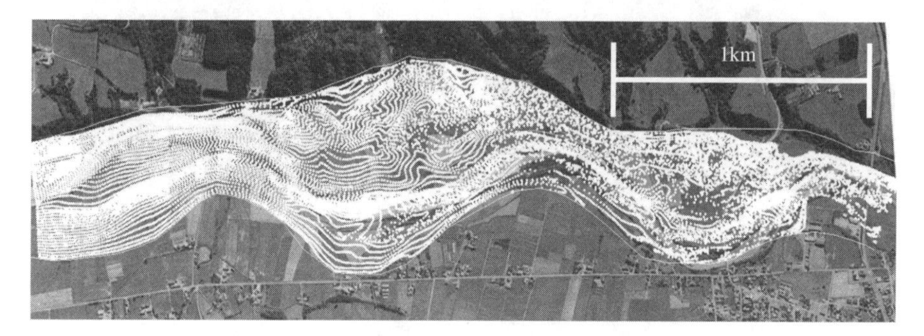

Figure 12. Contours of flow velocity plotted using particles.

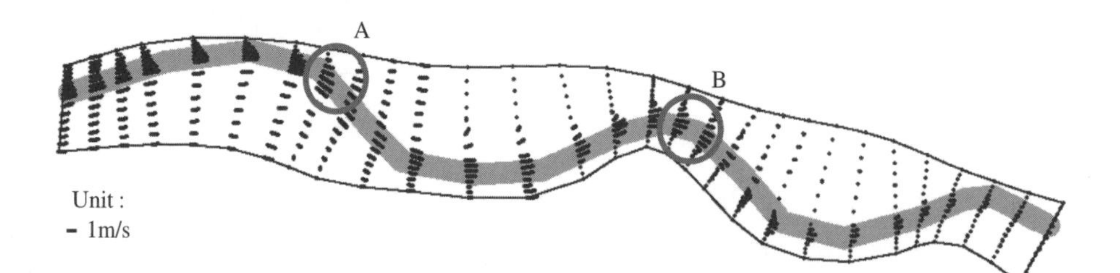

Figure 13. The velocity vector of 1810 (m^3/s) on 11 p.m. Aug 9th.

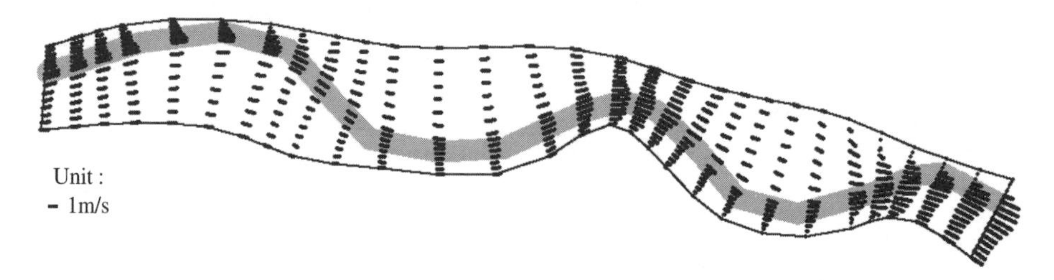

Figure 14. The velocity vector of 2884 (m^3/s) on 2 a.m. Aug 10th.

characteristics. Since the calculation reaches equilibrium at the 500th second, calculated values are time-averaged from 500th sec to 1500th sec. Figures 13 to 15 show the velocity vector in the unsteady condition. Figure 13 shows the results for 1810 m^3/s at 23:00 on 9 August. Figure 14 shows the results for 2884 m^3/s at 02:00 on 10 August. Figure 15 shows the results for 156 m^3/s at 12 noon on 9 August.

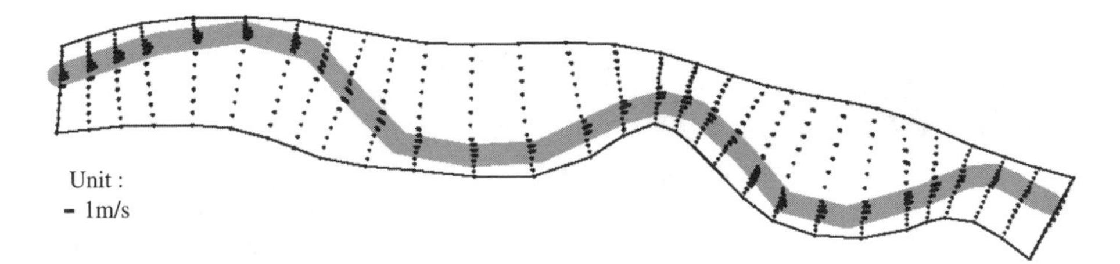

Figure 15. The velocity vector of 156 (m^3/s) on 12 p.m. Aug 10th.

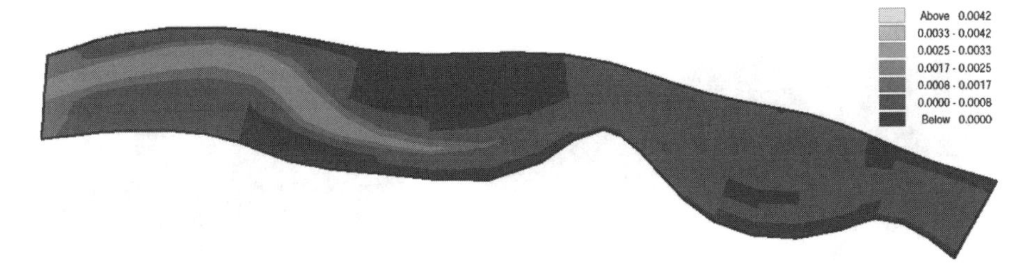

Figure 16. The distribution of the suspended load of 1810 (m^3/s) on 11 p.m. Aug 9th.

4 CALCULATION OF THE DEFORMATION ON THE FLOOD PLANE

Suspended load sediment transport and deformation of the floodplain are calculated for the period covered by the flood hydrograph. Sediment is supplied from the upstream end as the equilibrium condition. The low-water channel is also at equilibrium. The floodplain deformation is calculated in the unsteady condition, and sediment is supplied from the upstream end from the point at which the discharge increases. Sediment transport is calculated using a transport equation for suspended load. There is no suspended load in the initial condition on the floodplain, but from the point at which discharge increases, sediment begins to be suspended and deposited. The grain size is 1 mm. It is clear that the grain size of suspended sediment during flooding is less than 0.1 mm. But most of the sediment with grain size <0.1 ran off from the mouth of the river as wash load, so the grain size of the calculation is 1 mm, which is main grain size in the calculation area.

Sediment transport was calculated using the transport equation for suspended load. The equation, shown below, is transformed into general coordinate system.

$$\frac{\partial}{\partial t}\left(\frac{\bar{c}h}{J}\right)+\frac{\partial}{\partial \xi}\left(\frac{u^{\xi}\bar{c}h}{J}\right)+\frac{\partial}{\partial \eta}\left(\frac{u^{\eta}\bar{c}h}{J}\right)=\left(\frac{q_{su}}{J}-\frac{w_f c_b}{J}\right)+D_c \qquad (1)$$

where $\bar{c}$ is the distribution of suspended load at average water depth, q_{su} is the amount of suspended bed load, w_f is the rate of sedimentation, c_b is the distribution of suspended load near the bed and D_c is the diffusion phase. q_{su} is calculated using an equation from Kishi and Itakura. w_f is calculated using the equation of Rubey.

$$\frac{\partial}{\partial t}\left(\frac{z_b}{J}\right)+\frac{1}{1-\lambda}\left(\frac{q_{su}}{J}-\frac{w_f c_b}{J}\right)=0 \qquad (2)$$

where Z_b is bed elevation and λ is void ratio.

Figure 16 to 18 show the distribution of suspended load at each discharge. Figure 19 shows the change in bed elevation.

5 DISCUSSION AND CONCLUSION

The water overflows the left bank of the high-water channel at point A, and flows to the mouth of the river (Figure 12). The water also overflows the left bank of the high-water channel at point B, and flows to the mouth of the river. The channel floods at point A and point B when the discharge is 1810 m^3/s (Figure 13). The Appetsu River channel meanders under normal conditions, and the wavelength is about 1 km. According to the simulation, flow bifurcation and convergence occurred on the floodplain when the discharge increased rapidly, which suggests that multiple-row bars formed. Figure 20 shows the area

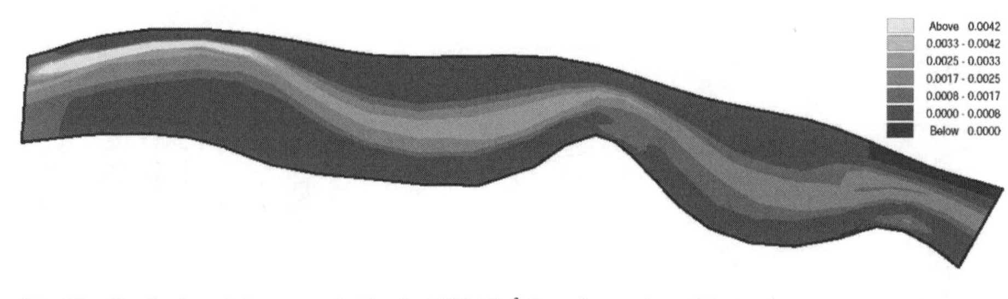

Figure 17. The distribution of the suspended load of 2884 (m^3/s) on 2 a.m. Aug 10th.

Figure 18. The distribution of the suspended load of 156 (m^3/s) on 12 p.m. Aug 10th.

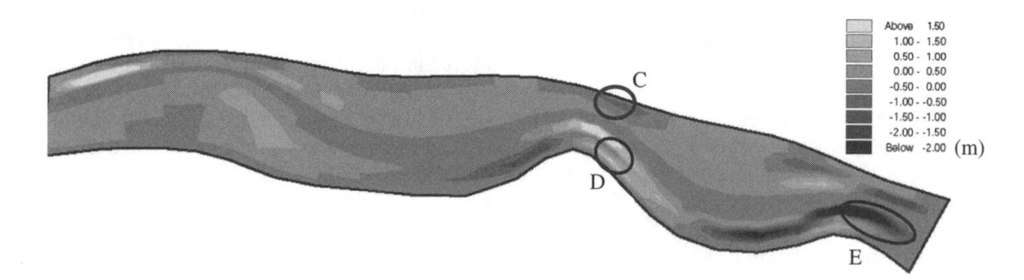

Figure 19. The change in bed elevation.

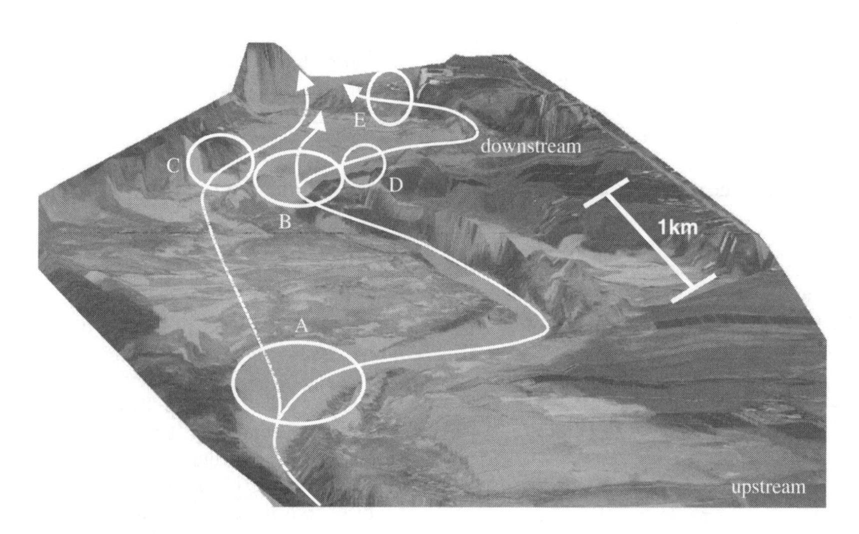

Figure 20. The area for which calculations were made.

Figure 21. Photo of points C taken in the investigation.

Figure 22. Photo of points D taken in the investigation.

for which calculations were made. It was made by integrating photos of that area with bed elevation data. The bank is eroded at points A and B, and floodwaters exit from those points. It is found that flow indicated by the white arrow in Figure 20 forms, as is the case for the calculation result. Figures 21 and 22 are photos of points C and D taken in the investigation. The erosion at point C and the deposition at point D agree with the calculated results. The erosion at point E agrees with the calculated results. The elevation change in the high-water channel agrees closely with the calculated results.

This model accurately reproduces the flood and bed elevation change of the flood region. Calculation is possible even when there is an embankment or groyne. This will enable more rational flood control.

REFERENCES

Sanit Wongsa, Yasuyuki Shimizu, Yasuhiro Murakami (2004): Runoff and sediment transport characteristics in a river system in the Hidaka region of Hokkaido during Typhoon Etau of 2003. Annual Proceedings of Hydraulic Engineering, JSCE, Vol. 48(2), 1099–1104.

Yu'ichiro Ito, Yasuyuki Shimizu (200): A study on the numerical calculation, replicating the model experiment of the Ishikari River. Annual Journal of Hydraulic Engineering, JSCE, Vol. 47, 661–666.

Report of the damage in Hokkaido from Typhoon Etau in 2003 JSCE.

River, Coastal and Estuarine Morphodynamics: RCEM 2005 – Parker & García (eds)
© 2006 Taylor & Francis Group, London, ISBN 0 415 39270 5

The effects of a catastrophic flood event on the morphodynamics of an Austrian river

C. Hauer & H. Habersack
Inst. of Water Management, Hydrology and Hydraulic Engineering, Dept. of Water-Atmosphere-Environment
BOKU University of Natural Resources Applied Live Sciences, Vienna, Austria

ABSTRACT: The aim of this paper is the analysis of the morphodynamic processes which happened during the catastrophic flood in 2002 at the Austrian River Kamp. Discharges with a recurrence interval between 500 and 2000 years caused widening and channel shifting of the river bed. The interesting point was that erosion processes were found in spatially restricted areas of inundation zones. Instead of side erosion in curved sections these morphodynamic processes were frequently found in straight parts of the river. For analysing the reason of these dislocations hydrodynamic – numerical models were used and combined with field measurements. The calculated hydraulic conditions during the flood event and the numerical results for bottom shear stress, bankfull discharge and a comparison of pre- and post flood situation allowed a cause–effect study. The analysis showed that the morphodynamic effects of the flood in 2002 were influenced by the special geomorphologic characteristics of the river Kamp and additionally in some cases by anthropogenic influences. Contraction and expansion of the channel geometry combined with scouring and aggradation during the flood event and the high spatial variability of bankfull discharge were responsible for these channel shifting and dislocations in inundation zones.

1 INTRODUCTION

There are a lot of theories which morphodynamic processes happen to river morphology during a flood event (IKEDA & PARKER, 1989; HUANG & WAN, 1997; JIN et al., 1997). Most of them are generalized studies for braided or meandering rivers. At the river Kamp a specific situation occurred during the catastrophic flood in August 2002 (DVWG, 2002; HABERSACK & MOSER, 2002; ARGE ELBE, 2003). The special meteorological conditions for Lower Austria and the South of Czech Republic was responsible for the extreme flood events in summer 2002 (HOLZMANN, 2002). In the investigated area a precipitation of 150 to 370 mm was documented in the first 12 days of August 2002, being 3 to 4 times higher than the average monthly precipitation in the catchment area of the River Kamp. These massive rain falls caused a flood at the described River Kamp with two distinctive peaks. The first flood wave occurred at the 8th of August. Within 24 hours the discharge increased from $100\,m^3s^{-1}$ up to $804\,m^3s^{-1}$ (gauging station Stiefern, flood with recurrence interval of 100 years $= 490\,m^3s^{-1}$). At the 14th of August the second wave occurred. The maximum discharge was about $500\,m^3s^{-1}$ at that time. After the flood hit the valley the river bed showed widening, several dislocations, channel shifts and overbank scours at special restricted areas. At the beginning it was difficult to classify the morphodynamic processes of these areas with erosion features because of partly non visible boundary conditions. Therefore the main question of this paper was:

Is it possible to find a physically based approach to characterize the flood induced morphodynamic features of the river Kamp?

2 STUDY REACH

The study reach is situated in the northern part of Lower Austria. The origin of the River Kamp is near the village Karlstift (920 m.a.sl). Its 160 km long course discharges into the River Danube (182 m.a.sl) at Altenwörth. The total catchment area is defined with 1.753 km^2 (Fig. 1). In this part of Lower Austria the basic geological features were mainly formed by processes of the "variszidischen" land forms 300 mill. years ago. But the important landform processes for the development of the river system in this area were occurring millions of years later in the tertiary (Miozän and Pliozän; 23.8–1.8 mill. years before present). The whole area of the "Bohemian mass" was at that time several hundred meters below the heights from present time (GRILL, 1957, STEININGER & ROETZEL, 1991) and a prehistoric sea was found till the peaks of the bohemian mass. Since that time over a 2.5 mill. year long period the valley of the river Kamp was formed

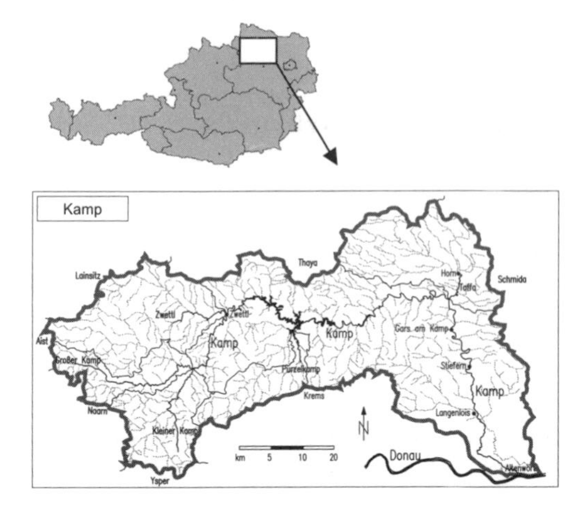

Figure 1. Study reach.

(STEININGER, 1999) with its nowadays character. The course was determined by the breaking tectonic forms in this area. These geological requirements were also responsible for the V – shaped forms of the tributaries to the river Kamp. Caused by different bed rock material the river course can be divided further into two parts of different geological boundary conditions for morphodynamic processes. The first river stretch (until the city of Zwettl) (Fig. 1) is built up by granite and has a good resistance against erosion.

The following 2nd part of the river course, between Zwettl and the city of Langenlois, consists of waxier bedrock material. Crystalline schist, like "Paragneis" with good erosion features avail a degradation of the river bed in the tectonic breaking forms over the last 1.8 million years. An uplift of the whole bohemian mass during the upper Miozän amplified these processes. During these land surface processes the river valley was formed to a canyon with meandering character defined as "valley meanders" (Fig. 1). In the recent past (1950–1957) three large power plants were build. The reservoirs of Ottenstein, Dobra and Thurnberg influenced the sediment transport and the hydrological regime drastically. The whole part of the River Kamp from river station 73 + 500 km to the River Danube can be described as a residual flow zone with no sediment input from upstream and reduced input from tributaries since that time. The release of water from the bottom of the power plants into the River Kamp with 4 degrees temperature by the three power plants causes a reduction of maximum water temperature from 25 degrees to 16 degrees in July at the gauging station Stiefern (STOISS, 1992). Therefore the fish region changed from epipotamal to meta – hyporhithral (WIESBAUER, 1999). The hydrological characteristics of the catchment area are depicted in Table 1.

Table 1. Hydrological characteristics (Hydrologisches Jahrbuch, 1999).

Gauging station	NQ* $[m^3s^{-1}]$	MQ** $[m^3s^{-1}]$	HQ*** $[m^3s^{-1}]$	NQ:HQ
Neustift	0.04	0.87	18.5	1:462
Zwettl	0.21	5.88	140	1:677
Rosenburg	1.36	7.93	153	1:113
Stiefern	1.84	7.48	162	1:88

* NQ = low flow, ** MQ = mean flow, *** HQ = high flow.

3 METHODS

The erosion features of the river Kamp were investigated considering of the river scaling concept by HABERSACK (2000). It was found that these local forms of erosion which happened during the flood in 2002 could only be analysed and defined correctly if the superior aspects of river dynamics were determined also. Therefore an Airborn – Laserscan with a field resolution of 1 m * 1 m was used to investigate the 73 + 500 km long study reach. Combined with terrestrial surveys of the river bed a high quality digital terrain model (DTM) was constructed. Based on this DTM model areas with erosion features, sediment tributaries and anthropogenic influences on the river Kamp could be mapped in GIS and CAD programs. It was possible to investigate the superior aspects of the morphodynamic processes of the river Kamp at the sectional scale. For analysis which physical parameters were responsible for the special, documented morphological changes in August 2002, 1 – dimensional and 2 – dimensional hydrodynamic – numerical models were used. The one dimensional hydraulic approach was applied for analysing the physical conditions in a steady state calculation for longitudinal analysis on the sectional scale. The scientific interest was focused on distinguishing the main channel processes from the hydraulic conditions during the overbank flow, which were responsible for the development of secondary channels. To analyze the physical parameters before the flood 2002 the secondary channels in the model were artificially filled. The calculation of the bottom shear stress in the one dimensional model is based on following formula:

$$\tau = \gamma \cdot \overline{R} \cdot \overline{S_f} \qquad (1)$$

where γ = unit weight of water, $\overline{R}$ = average hydraulic radius, $\overline{S_f}$ = Slope of energy grade line (friction slope).

Combined with the calculation for the critical shear stress after MEYER-PETER & MÜLLER (1948) for certain grain sizes the effects of erosion and aggradation could be analysed.

$$\tau_{cr} = 0.047 \cdot (\rho_F - \rho_W) \cdot g \cdot d_m \qquad (2)$$

Table 2. Dislocation and channel shifts in August 2002.

	Station [km]	Length [m]	Max. width [m]	Anthr. infl.	Form of erosion
Steinegg 1	63 + 100	157	19	no	secondary channel
Steinegg 2	61 + 500	241	26	bridge	s. ch./chute cut off
Altenburg 1	57 + 750	217	44	no	chute cut off
Altenburg 2	56 + 650	233	56	no	chute cut off
Umlauf	53 + 850	433	118	no	secondary channel
Rosenburg 1	52 + 200	740	55	no	secondary channel
Rosenburg 2	50 + 500	81	30	no	chute cut off
Stallegg	48 + 050	155	46	bridge	s. ch./chute cut off
Buchberg 1	42 + 500	441	34	bridge	secondary channel
Buchberg 2	42 + 050	236	32	bridge	s. ch./chute cut off
Buchberg 3	41 + 300	204	33	dam	secondary channel
Tobelbach	39 + 600	351	33	dam	secondary channel
Altenhof	34 + 650	302	34	dam	secondary channel
Stiefern	31 + 350	177	37	bridge	s. ch./chute cut off
Schönberg	29 + 350	96	36	bridge	secondary channel

* anthr. Infl. = anthropogenic influences, ** s. ch. = secondary channel.

where ρ_F = density of sediment (2665 kg m^{-3}), ρ_W = density of water (1000 kg m^{-3}), d_m = mean diameter of sediment (m), g = acceleration due to gravity (9.81 m s^{-1}).

For detailed analysis of the hydraulic situations in curved sections a 2D model was implemented in the modelling process. The model is a two dimensional, depth averaged, unsteady, turbulent flow model. The model was developed by NUIJC (1999) and calculates the hydraulic conditions on a linear grid by a finite volume approach. The convective flow is based on the Upwind – scheme by PIRONNEAU (1989) and the discretisation of time is done by an explicit Runge Kutta method in second order. The shear stress was calculated for each node of the grid with following formula (NUIJC, 2004):

$$\tau = \rho \cdot g \cdot h \cdot S_f \tag{3}$$

$$S_f = \frac{v^2}{k_{str}{}^2 \cdot h^{\frac{4}{3}}} \tag{4}$$

where ρ = density of water, v = velocity at node (ms^{-1}), k_{str} = roughness number by Strickler (1/Mannings n), h = water depth (m), g = acceleration due to gravity (9.81 ms^{-1}).

The simulations were performed with the parabolic eddy viscosity model and a constant turbulent viscosity coefficient of 0.6.

4 RESULTS

Based on an Airborne Laserscan the dislocations and channel shifts of the river Kamp were mapped and surveyed at the local scale. The results of this analysis is shown in Table 2. The names of areas with erosion features were given according to nearby villages. The table is listed depending on river stationing containing length of the erosion areas and their maximum width. Furthermore the areas were described where anthropogenic influences to the river hydraulics occurred.

It was found that areas with major erosion features can be divided into two different responsible processes. One of them is the increase of the energy slope at the inner bend of curved sections during flooding. Such processes are often found at meandering rivers (HABERSACK et al., 2000) and were proofed by 2d – hydraulic modelling (Fig. 2).

Figure 2. Chute cut off at Altenburg 2 (a) Photo (b) 2D hydrodynamic model (463 m^3/s).

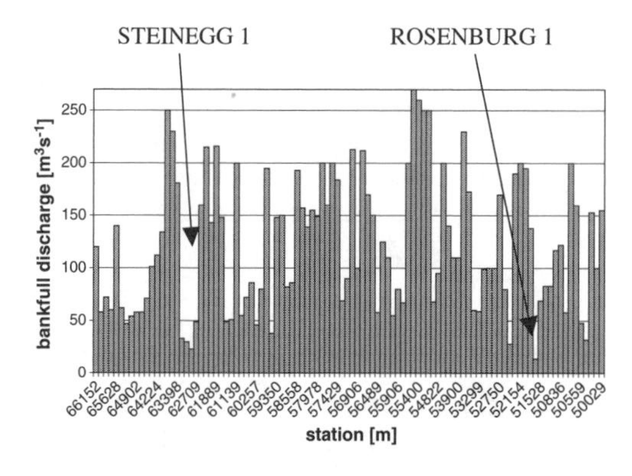

Figure 3. Natural variability of bankfull discharge.

The other one is the high variability of bankfull discharge and the consequential physical effects of sudden change in overbank flow. In Figure 3 the natural high variability of bankfull discharge between station 73 + 500 and 49 + 500 is shown. In the graph STEINEGG 1 and ROSENBURG 1 are shown as example.

These hydraulic effects of high variability of bankfull discharge cause partly high and sudden changes in

Table 3. Results of the simulations of the flood event 2002 at STEINEGG 1 with the geometric data before the flood.

	Station [m]	$Shear_{chan.}$ [Nm^{-2}]	$Shear_{LOB}$ [Nm^{-2}]	$Q_{chan.}$ [%]	Q_{LOB} [%]
C.s. 39	1640.97	124.59	5.08	94.45	–
C.s. 37	1524.14	122.78	–	85.56	–
C.s. 35	1442.05	215.07	52.61	77.65	0.43
C.s. 33	1389.99	200.44	113.01	71.87	6.77
C.s. 31	1328.37	150.57	100.46	69.65	13.97
C.s. 29	1267.59	92.82	53.15	77.94	5.70
C.s. 27	1206.65	64.04	38.62	87.63	4.74
C.s. 25	1148.62	61.32	29.69	87.17	1.67
C.s. 23	1083.67	168.95	120.52	50.88	33.10
C.s. 22	1061.76	338.99	199.09	58.34	30.05
C.s. 21	1025.18	227.30	157.42	51.78	30.62
C.s. 20	989.58	193.20	129.93	79.76	8.17

C.s. = Cross sections; LOB = Left overbank; chan. = channel.

overbank flow that can be further effectuated by artificial structures. Such areas were documented near bridges or in the surrounding of dams (Tab. 2). It was found that the development of secondary channels during the flood in August 2002 was similar in natural and anthropogenically influenced reaches, but different processes of sediment transport and deposition forming these erosion features could be defined. This cognitions are based on the analysis of hydrodynamic simulations at the local scale and will be explained for four examples. For the non influenced reach the dislocations STEINEGG 1 (63 + 100) and ROSENBURG 1 (52 + 200) were selected (Tab. 2, Fig. 3). As anthropogenically influenced the development of secondary channels at TOBELBACH (39 + 600) and ALTENHOF (34 + 650) are presented (Tab. 2). In Table 3 the results of the 1 – dimensional hydraulic simulations for the dislocation STEINEGG for 700 m^3s^{-1} (estimated discharge flood 2002) are depicted.

4.1 *Steinegg*

The reach where contraction of discharge occurred in a curved section is found from river station 39 to river station 35. It is possible to see the lowering of the overbank flow in this section during the simulated flood of August 2002 (Tab. 3, C.s. 39 – C.s. 25). The results of the hydrodynamic modelling for the channel and the left overbank flow are presented in form of the percentage of discharge and the arising bottom shear stresses in Table 3. The left overbank values were chosen since only in the left overbank area the development of a side channel occurred. The results in Table 3 originate from the reconstructed geometry before the flood 2002. The developed side channel was filled artificially in the model.

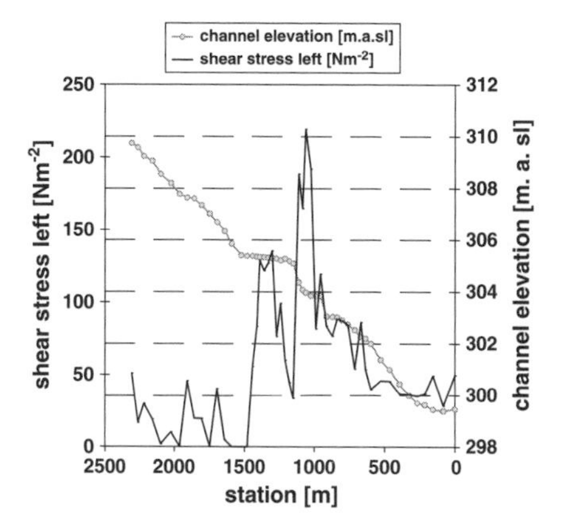

Figure 4. Results of hydrodynamic modelling with geometric data including the secondary channel after the flood.

Figure 5. Picture of STEINEGG 1.

The increase of left overbank flow downstream from C.s. 37 to C.s. 21 of 30.62% caused aggradation of the transported bed load material with an average diameter of 0.15 m. That can be seen in the decrease of the bottom shear stress between Cross section 33 and Cross section 25 in the main channel. These results are also illustrated in Figure 4 where the river bed elevation is pictured in combination with the bottom shear stress during the left overbank flow. The results from Figure 4 refer however to the condition after the flood 2002 and allow a comparison of the effects of the river-morphologic changes on the bottom shear stress in the left overbank area. The result was that it came, effected by the dislocation, to an increase of the bottom shear stress of around 20 Nm^{-2} (Fig. 4, Tab. 3).

In Figure 5 a photograph of the area STEINEGG 1 is pictured. The flow direction is from left to right.

It is obvious to see that bed material was transported and deposited (Fig. 4, Fig. 5). Downstream of the deposited material the energy slope increases, especially for the left overbank flow, therefore the bottom shear stress rises from 29.69 Nm^{-2} at Cross section 25 to 199.09 Nm^{-2} at Cross section 22 (Tab. 3). The consequence of these dynamic processes of high increase of overbank flow combined with aggradation of bed load material was the forming of a side channel with a depth of 1.46 m and a length of 157 m (Tab. 2).

Table 4. Results of the simulations of the flood event 2002 at ROSENBURG 1 with the geometric data before the flood.

	Station [m]	$Shear_{chan.}$ [Nm^{-2}]	$Shear_{LOB}$ [Nm^{-2}]	$Q_{chan.}$ [%]	Q_{LOB} [%]
C.s. 41	1740.42	44.51	19.79	89.02	10.98
C.s. 39	1654.31	65.40	17.53	88.72	11.28
C.s. 37	1540.32	156.90	9.87	99.93	0.07
C.s. 36	1498.16	106.56	9.43	99.43	0.57
C.s. 34	1404.03	107.42	34.02	84.18	15.53
C.s. 32	1308.01	72.64	28.83	90.07	9.78
C.s. 30	1209.02	52.13	20.51	88.77	10.78
C.s. 28	1104.12	60.51	34.08	74.19	23.12
C.s. 26	994.83	66.70	42.46	59.83	39.45
C.s. 24	885.09	60.39	34.82	69.02	30.81

C.s. = Cross sections; LOB = Left overbank; chan. = channel.

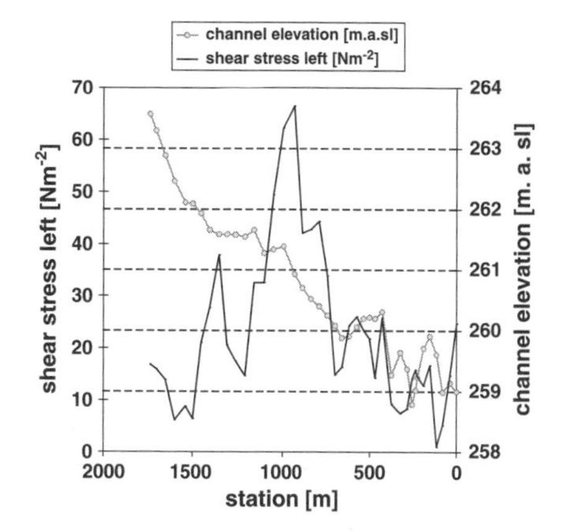

Figure 6. Results of hydrodynamic modelling with geometric data including the secondary channel after the flood.

4.2 *Rosenburg 1*

The results of the hydrodynamic numerical analysis for ROSENBURG 1 are shown in Table 4.

In this section the flood discharge is carried mainly within the bankfull area between Cross sections 41–36 (Tab. 4). The steeper bottom downward gradient of the river bed within this range (Fig. 6) caused an additional acceleration of the discharge. In the following

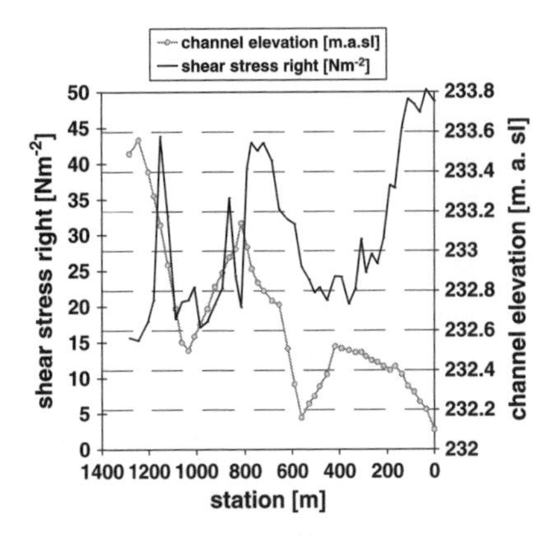

Figure 7. Results of hydrodynamic modelling with geometric data including the secondary channel after the flood.

stretched section of the modelling area an increase of the overbank flow occurred again which was responsible for the emergence of the secondary channel in this section. The results of Figure 6 refer however to the condition after the flood 2002. The secondary channel caused an increase of 24 Nm^{-2} for bottom shear stress at overbank flow in this modelling stretch (Fig. 6, Tab. 4).

The areas with erosion features STEINEGG 1 and ROSENBURG 1 were found in morphologically not influenced sections and their occurring is based on the naturally high variability of bankfull discharge in combination with aggradation during the flood event. In opposition to that the development of a side channel downstream river station 49 + 500 were influenced by artificial structures.

4.3 *Tobelbach*

As anthropogenically influenced the secondary channels of TOBELBACH (39 + 600) and ALTENHOF (34 + 650) are presented (Tab. 5, Tab. 6). what is obvious at these areas is, that in terms of contraction of the flow the river bed was eroded (Fig. 7, Tab. 5), (Tab. 6, Fig. 10). parts of the eroded material were deposited at the beginning of the following cross section enlargement. The results of the hydrodynamic numerical analysis for TOBELBACH are shown in Table 5.

At the dislocation TOBELBACH the maximum percentage of discharge for right overbank flow during the flood increased from 2.5% at C.s. 35 to 29.32% at C.s. 30 (Tab. 6). Additionally the results at Figure 8 are representing the hydraulic condition after the flood 2002. The secondary channel caused an increase of 14 Nm^{-2} in this part (Figure 7, Table 5).

The increase of the maximum overbank flow and the river bed aggradation downstream of the contraction till its maximum at C.s. 30 continue parallel. That fact emphasises the hydraulic processes of sudden change in overbank flow. In Figure 9 a photograph of the area TOBELBACH is shown. The flow direction is from north to south. It is possible to see the stretch of contraction effectuated by the railway dam and the area with the developed secondary channel, formed as chute cut off, downstream.

Table 5. Results of the simulations of the flood event 2002 at TOBELBACH with the geometric data before the flood.

	Station [m]	$Shear_{chan.}$ $[Nm^{-2}]$	$Shear_{ROB}$ $[Nm^{-2}]$	$Q_{chan.}$ [%]	Q_{ROB} [%]
C.s. 37	984.75	64.06	16.63	90.83	3.93
C.s. 36	954.91	66.36	17.40	92.11	3.52
C.s. 35	923.04	77.48	19.27	90.68	2.50
C.s. 34	892.91	86.36	21.42	77.36	5.14
C.s. 33	863.54	107.07	30.12	74.86	8.19
C.s. 32	837.80	87.35	21.49	78.71	7.72
C.s. 31	814.94	50.60	17.44	60.34	13.34
C.s. 30	790.63	80.57	35.03	58.15	29.32

C.s. = Cross sections; ROB = Right overbank; chan. = channel.

Table 6. Results of the simulations of the flood event 2002 at ALTENHOF with the geometric data before the flood.

	Station [m]	$Shear_{chan.}$ $[Nm^{-2}]$	$Shear_{ROB}$ $[Nm^{-2}]$	$Q_{chan.}$ [%]	Q_{ROB} [%]
C.s. 25	585.12	105.92	38.64	77.30	6.61
C.s. 24	564.83	107.22	43.46	77.39	6.65
C.s. 23	536.58	97.64	40.71	73.55	7.64
C.s. 22	518.42	83.62	39.70	76.43	6.67
C.s. 21	494.66	77.39	32.92	77.14	5.31
C.s. 20	471.85	77.87	22.91	80.39	2.65
C.s. 19	450.60	89.31	47.26	67.30	17.16
C.s. 18	426.82	85.63	53.45	59.66	27.80
C.s. 17	405.07	84.34	64.31	57.49	33.20
C.s. 16	380.49	95.76	69.56	52.38	39.94
C.s. 15	354.55	57.02	48.86	39.90	41.64

C.s. = Cross sections; ROB = Right overbank; chan. = channel.

4.4 *Altenhof*

The degradation of the river bed and the following effect of the expansion of flow at ALTENHOF are depicted in Table 6.

In Figure 9 the results are representing the hydraulic condition after the flood 2002. The secondary channel caused an increase of 6 Nm^{-2} in this part for right overbank flow (Fig. 9, Tab. 6).

Figure 8. Picture of TOBELBACH.

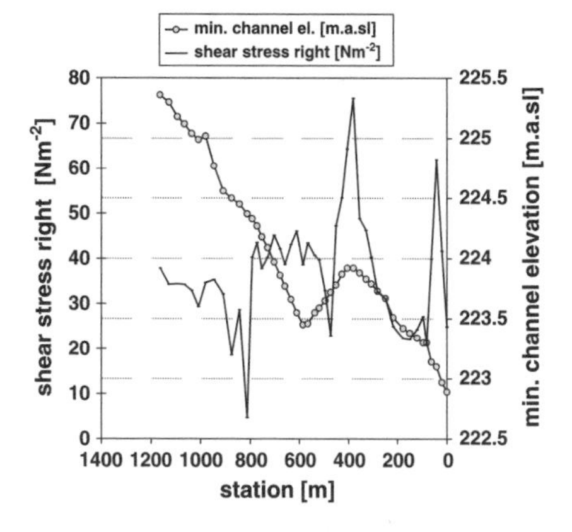

Figure 9. Results of hydrodynamic modelling with geometric data including the secondary channel after the flood.

In Figure 10 a photograph of the area ALTENHOF is pictured. The flow direction is from north to south. On the right side, in flow direction, the secondary channel is visible.

Both areas with erosion features ALTENHOF and TOBELBACH were influenced by the dams of the "Kamptal" railway. The anthropogenic influence in combination with a naturally, geomorphologically developed valley caused these effects of change in overbank flow.

5 DISCUSSION

Based on the results of Table 2 combined with the GIS/CAD analysis and the results of the hydrodynamic modelling the following classification of the areas with erosion features could be made (Fig. 11).

Figure 10. Picture of ALTENHOF.

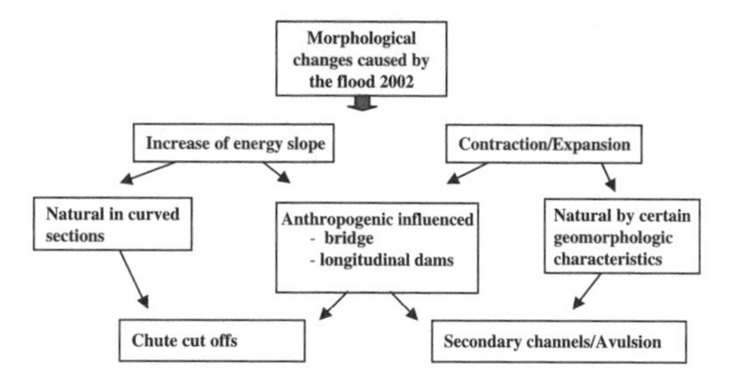

Figure 11. Classification of erosion features at the river Kamp.

In comparison to other studies which describe the effects of flood processes to river morphology (IKEDA & PARKER, 1989; HUANG & WAN, 1997; Jin et al., 1997) the results of this paper show that specific geomorphologic conditions caused specific morphodynamic processes at the River Kamp during the flood 2002. It was found that the high variability of bankfull discharge and the resulting hydraulic effects of sudden change in overbank flow were responsible for the development of secondary channels. Similar effects could be documented at bridges (Table 2). In literature bankfull discharge is associated with a momentary maximum flow which, on the average, has a recurrence interval of 1.5 years as determined using a flood frequency analysis (DUNNE & LEOPOLD, 1978). Further exists, however, a range of return periods for bankfull discharge from 1 to 25 years (WILLIAMS, 1978; NASH, 1994). At the river Kamp

Table 7. Longitudinal profile of hydrologic data (REGIONAL GOVERNMENT OF LOWER AUSTRIA, 2003).

	Catchment area					
Gauging station	[km^2]	HQ_{500} [m^{3}s^{-1}]	HQ_{100} [m^{3}s^{-1}]	HQ_{30} [m^{3}s^{-1}]	HQ_{10} [m^{3}s^{-1}]	HQ_1 [m^{3}s^{-1}]
Rosenburg	1150.2	550	380	250	150	30
Hirschbach	1443.8	644	474	310	210	64
Tobelbach	1466.5	651	481	315	215	67

* HQ_x = flood with recurrence interval of x years.

a high natural variability of bankfull discharge occurs at short distances and that's specific for this river. Based on Table 7 the following classification of recurrence intervals for bankfull discharge could be found at the investigated areas STEINEGG 1, ROSENBURG 1, TOBELBACH and ALTENHOF.

At Steinegg 1 the bankfull discharge decreases on a distance of 420 m (Tab. 7) from a recurrence interval of 30 years to 1 year, at dislocation Rosenburg 1, the second non influenced section, the bankfull discharge decreased from HQ_{30} to a HQ_{2-3} at a length of 325 m (Tab. 7). The anthropogenically influenced section of TOBELBACH shows a decline from HQ_{10} to a discharge of $32\,m^3s^{-1}$ being the half of HQ_1 at a distance of 47 m (Tab. 7). ALTENHOF as the last presented section features a reduction from HQ_{10} to $14\,m^3s^{-1}$, which can be described as the double of mean discharge (gauging station Stiefern) at a distance of 109 m. These effects led to the development of the characteristic erosion forms during the flood 2002.

6 CONCLUSIONS

By mapping the dislocations and hydrodynamic modelling, the special characteristics of the River Kamp with respect to morphodynamic changes could be found. It was documented that areas with major erosion features can be divided into two different releasing factors. One of them is the increase of energy slope at the inner bend of curved sections during flooding. The other one and that's specific for the River Kamp is the high variability of bankfull discharge and the consequential physical effects of sudden change in overbank flow. Therefore the development of side channels in straight parts of the river was possible during the flood event in 2002.

ACKNOWLEDGEMENTS

The authors want to thank the Regional Government of Lower Austria for financing the monitoring program of the River Kamp.

REFERENCES

ARGE Elbe, 2003 "Hochwasser August 2002, Einfluss auf die Gewässergüte" Abschlussbericht, März 2003

Dunne, T., & Leopold, L.B., 1978 "Water in Environmental Planning". W.H. Freeman and Co., San Francisco, CA: 818 pp

DVWG., 2002 "Dokumentation und Handlungsempfehlungen" – Studie des DVWG

Grill, R., 1957 Die stratigraphische Stellung des Hollenburg – Karlstettner Konglomerats (Niederösterreich). – Verh. Geol. Bundesanstalt., 1957: 113 – 120. Wien

Habersack, H.M., 2000. "The River Scaling Concept – RSC: a basis for ecological assessments", Hydrobiologia 422/423: p49–60, Kluwer Academic Publishers, The Netherlands

Habersack, H. M., 1999 "Interactions between natural morphological structures and hydraulics" Proceeding on the 3rd International Symposium on Ecohydraulic, Salt Lake City

Habersack, H., & Moser. 2002 "Ereignisdokumentation Hochwasser August 2002" Zenar, Universität für Bodenkultur

Hauer, R., 1952 Die Flusssysteme des n.- ö. Waldviertels, Ein Beitrag zu ihrer Entwicklungsgeschichte; Verlag der Stadtgemeinde, Kulturreferat; Gmünd 1952

Holzmann, H., 2002. AP- Hydrologi; in Habersack H. & Moser, 2002 "Ereignisdokumentation Hochwasser August 2002" Zenar, Universität für Bodenkultur

Huang, J. & Wan, Z., 1997 '2-D Numerical Model for flood forecast in rivers with heavily sediment laden flow'; International Journal of Sediment Research, Vol. 12, No. 3, Dec. 1997

Ikeda, S., & Parker, G., 1989 "River Meandering" AGU Monograph

Jin D., Chen H., Zhang O., 1997 "An experimental study on bedmaking and catastrophic processes in island braided channels" International Journal of Sediment Research, Vol. 12, No. 3, Dec. 1997

Meyer – Peter, E., & Müller, P., 1948 Formulas for bed – load transport International Association of Hydraulic Research, 2nd Meeting, Stockholm

Nash, D.B., 1994 Effective sediment-transporting discharge magnitude-frequency analysis. Journal of Geology, Vol. 102: 79–95

Regional Government of Lower Austria, 2003 longitudinal hydrologic characteristic of the River Kamp unpublished

Rosgen, D., 1996 Applied river morphology Colorline, Lakewood, Colorado

Steininger, F., "Erdgeschichte des Waldviertels" Schriftreihe des Waldviertler Heimatbundes Nr. 38

Steininger, F., & Roetzel, R., 1991 "Die tertiären Molassesedimente am Ostrand der Böhmischen Masse.- In: Roetzel & Nagel (Hg.): Exkursionen im Tertiär Österreichs. Molassezone – Waschbergzone – Kornneuburger Becken – Wiener Becken – Eisenstädter Becken. – Österr. Paläont. Ges.: 59 – 141 (Schindler), Wien

Stoiss, C., 1992 "…" diploma thesis, University of Agricultural Science

Wiesbauer, H., 1999 "GBK Unterer Kamp: Teil 1 – Naturräumliche Grundlagen" Im Auftrag des Amtes der Niederösterreichischen Landesregierung, Abteilung Wasserbau und des BMLF, Sektion IV

Wiliams, G.P., 1978 Bankfull discharge of rivers. Water Resources Research. Vol. 14, No. 6: 1141–1153 Amsterdam, 713 p

River, Coastal and Estuarine Morphodynamics: RCEM 2005 – Parker & García (eds)

Changes of the material transport depending on the flood scale

Y. Yoshikawa & Y. Watanabe
Civil Engineering Research Institute of Hokkaido, Sapporo, Japan

ABSTRACT: Nutrients transported in rivers are important in terms of river ecosystems. Understanding of Nutrients behavior in the river environment is important in the maintenance and management of the rivers. Transport of Nutrients and other materials increases during floods. To better understand the influence of flooding on the river environment, this study aims to clarify the relationships between suspended solids (SS) and Nutrients, both of whose behaviors in the river water are not clear, and to estimate the amount of Nutrients deposited in the high water channel as a result of flood. The study area was the Saru River basin, and the study was conducted in three steps. First, toward understanding the differences in material transport in four floods (two rainfall floods and two snowmelt floods) observation data were examined. Second, equations were derived from the observation data in order to estimate the amounts of Nutrients deposited on the flood plain as a result of flooding events. Third, calculation for unsteady flow in a compound cross-section was performed to clarify how the amounts of Nutrients deposited on the flood plain differ by the flood scale. The following were found: (1) The volume of material supplied from the channel relates to the discharge, and we could hypothesize that when the discharge reaches a certain threshold, the material being transported reaches a saturation concentration. (2) From the estimation equation that we derived for Nutrients, we found a close correlation between the amount of silt deposition and the amount of water quality components in particulate form. (3) The calculations revealed that the influence of floods of different scale on the river environment varies by location.

1 INTRODUCTION

Material transport in rivers increases under flood conditions. Transported materials include suspended solids (SS) and Nutrients. These materials originate at the upper reaches and in tributaries, irrigation facilities behind sluices, and the river channel. SS and Nutrients repeatedly tract up and settle on the riverbed and flood plain as they make their way downstream. This study examines the relationship between SS and Nutrients and estimates the amount of Nutrients deposited as the result of a flood. The study area was the Saru River basin, and the study involved three steps. In the first step, we closely examined observation data, to better understand the difference in material transport in four floods: two rainfall floods of different scales (one large scale, and one very large scale) and two snowmelt floods (one before the very-large-scale flood, and the other after that flood). In the second step, equations for estimating Nutrients deposited on the flood plain based on measurements of SS in rivers were derived. In the third step, we clarified how the amount of Nutrients deposited on the flood plain differs by the scale of flood by calculating unsteady flow in a compound cross section.

2 STUDIED RIVER BASIN FLOODS

2.1 *River basin*

The study area is the Saru River, which is in the westernmost part of Hidaka sub-prefecture, Hokkaido. The river has a catchment area of 1,350 km^2, a trunk river channel length of 104 km and a bed slope ranging from 1/500 to 1/800, making it one of the steepest rivers in Hokkaido. The river passes through an alluvial plain. We have been conducting observations of water discharge and water sampling for SS [1][2] since 2001 at the points marked with circles in Figure 1, and we chose a 14.2-km-long river section between the Shin-Biratori Ohashi Bridge (KP16.0) and the Sarugawa Bridge (KP2.8) as the study area. Water sampling points include one at Nukibetsu near the Nukibetsu gauging station, and another at Horokeshi near the confluence of the Saru and Taushinai rivers. These are upstream from the studied section. Water sampling points for tributaries were one at about KP6.5 on the Shirau River (a relatively large tributary), one at about KP7.5 on the Biratori Dai-ni River (a small tributary), and one at about KP5.5 on the Fukuman River. Water sampling for the sluice was done at Tomikawa D Sluice (about KP3.5). Moreover, an important observation point name was changed in this thesis. Observation point name was changed in this manner. It is Sarugawa Bridge was Site1, Shin-Biratori Ohashi Bridge was Site2, Horokeshi Site3, Nukibetsu was Site4.

2.2 *Floods*

We studied four floods: two rainfall floods and two snowmelt floods. The first rainfall flood was a large-scale flood in September 2001, and the second was a very-large-scale flood in August 2003. The first snowmelt flood was in May 2003, before the August 2003 rainfall flood, and the second was in May 2004, after the August 2003 rainfall flood. Table 1 shows the peak discharges observed at the Site1 and on the Shirau River during the four floods. The peak discharge for the 2003 rainfall flood (Table 1) was calculated (on the hour) using the *H-Q* formula, because observation at the peak hour could not be made for safety reasons. These discharge values were revised for accuracy and are the official values. From Table 1, we see that the peak discharge at the Site1 during the 2003 rainfall flood is about 2.2 times greater than that of the 2001 rainfall flood. The discharge of the May 2004 snowmelt flood is about 1.6 times that of the May 2003 snowmelt flood. The tributaries had peak discharges of about the same scale for the four flood events. To show the secular variations in discharge, the annual maximum discharges from 1969 to 2004 at the point near the Site1 (Tomikawa gauging station) are shown in Figure 2. The 2001 rainfall flood was the third largest flood in the past 35 years, and the 2003 rainfall flood was the largest on record. It exceeded the design flood. The observation periods for the four floods are as follows: 04:00 on September 11 to 15:00 on September 13, 2001 Total: 60 hours, 00:00 on April 30 to 08:00 on May 1, 2003 Total: 32 hours, 10:00 on August 9 to 09:00 on August 11, 2003 Total: 48 hours, 20:00 on May 3 to 09:00 on May 5, 2004 Total: 37 hours.

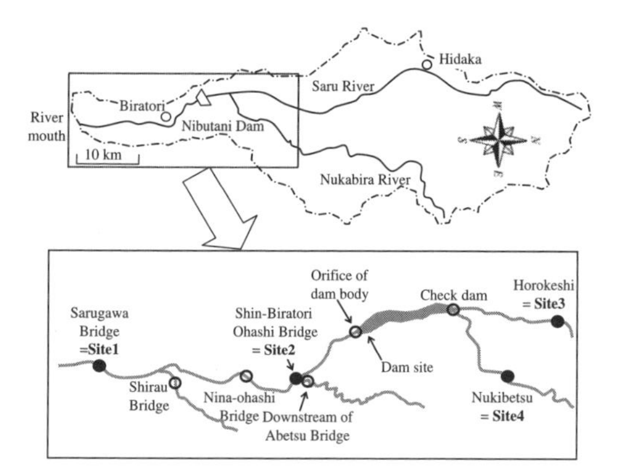

Figure 1. Saru river basin and observation locations.

Table 1. Peak discharge of studied flood (m^3/s).

	Flood type	Site1 KP2.8	Shirau River KP6.5
2001/09	Rainfall	2,383.6	27.7
2003/05	Snowmelt	373.0	0.5
2003/08	Rainfall	5,177.5	27.8
2004/05	Snowmelt	606.0	1.3

3 DIFFERENCE IN MATERIAL TRANSPORT

We used three methods to better understand the differences in material transport between the four floods. First, we investigated the relationships between factors of material transport. Then we investigate the volume of material passing each point. Finally, we investigated the material behavior at each observation point.

3.1 *Relationship between factors of material transport*

The key factors related to material transport at a time of flooding are discharge (Q), suspended sediment (SS) concentration, and Nutrients concentration (TN is a

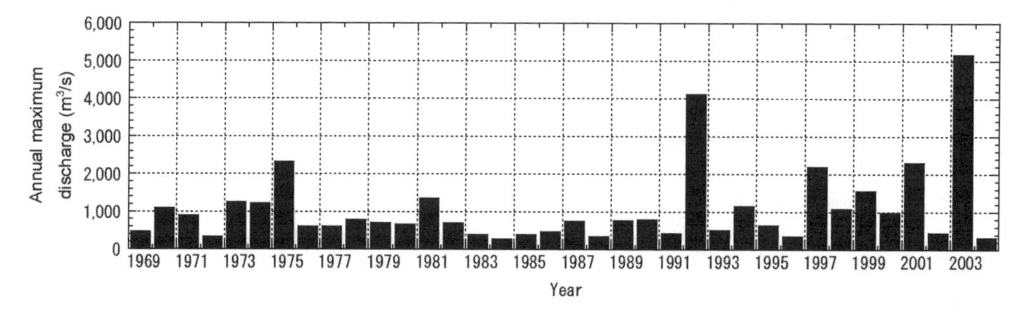

Figure 2. Annual maximum discharge from 1969 to 2004.

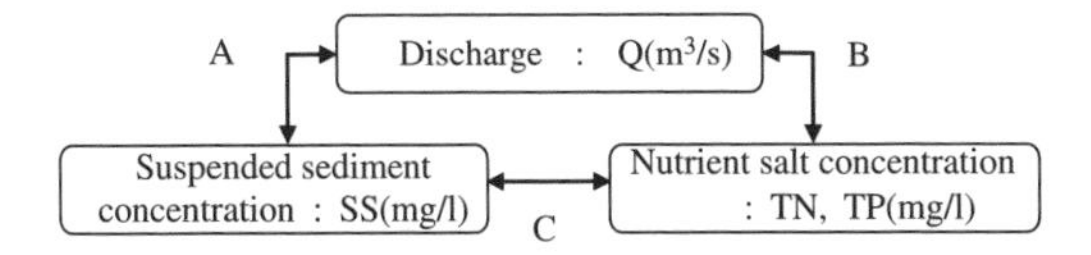

Figure 3. Correlation of the key factors of material transport in river water.

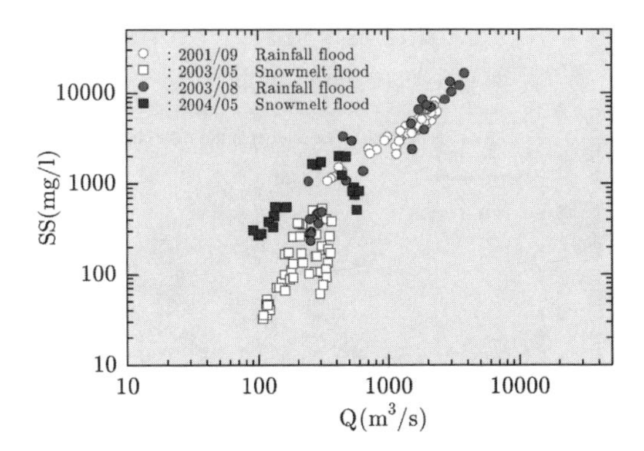

Figure 4. Relationship A during each flood.

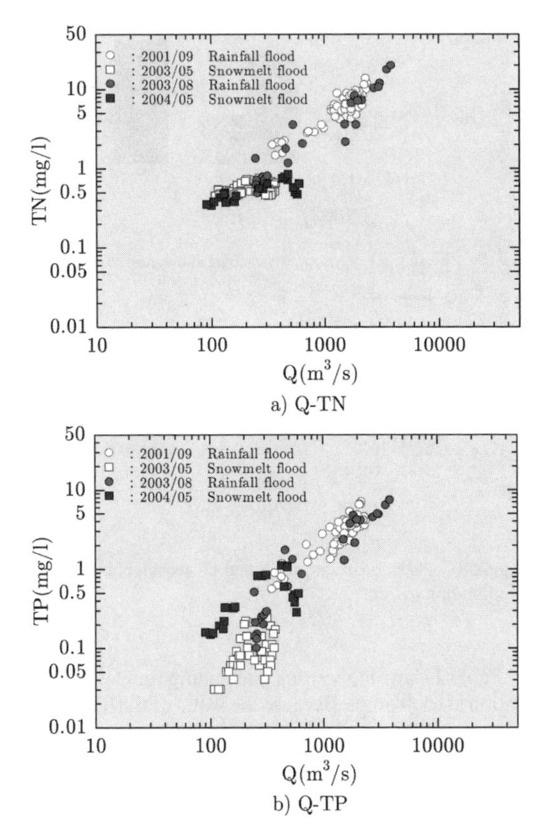

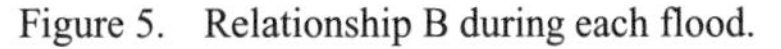

Figure 5. Relationship B during each flood.

clipped form of total nitrogen and TP is a clipped form of total phosphorus) (Figure 3). We investigated the following relationships between factors: Q and SS (relationship A), Q and TN, and Q and TP (relationship B), and SS and TN, and SS and TP (relationship C). Figure 4, Figure 5, and Figure 6 show relationships A, B and C. Figure 4 shows that the ratios of SS to discharge are higher for the May 2004 snowmelt flood (i.e., the flood after the 2003 rainfall flood) than those of snowmelt flood before that rainfall flood, and that the ratios of SS to discharge of the May 2004 snowmelt flood (i.e., the flood after the 2003 rainfall flood) are close to those of the two rainfall floods. A similar relationship is observed between TP and discharge (Figure 5b)). This suggests that the concentration of material transported during the two snowmelt floods differed. The investigated relationships did not differ greatly between the two rainfall floods.

3.2 *Passing volume of material*

To clarify how the material is transported in the entire river basin, the volumes of SS, TN and TP passing

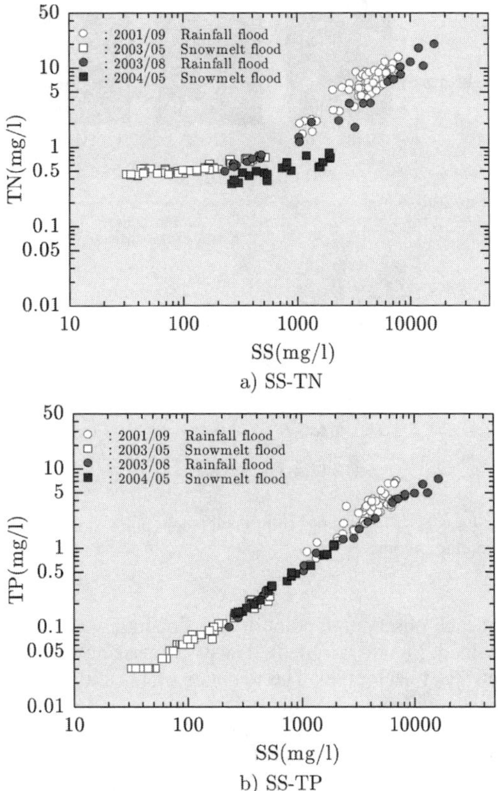

Figure 6. Relationship C during each flood.

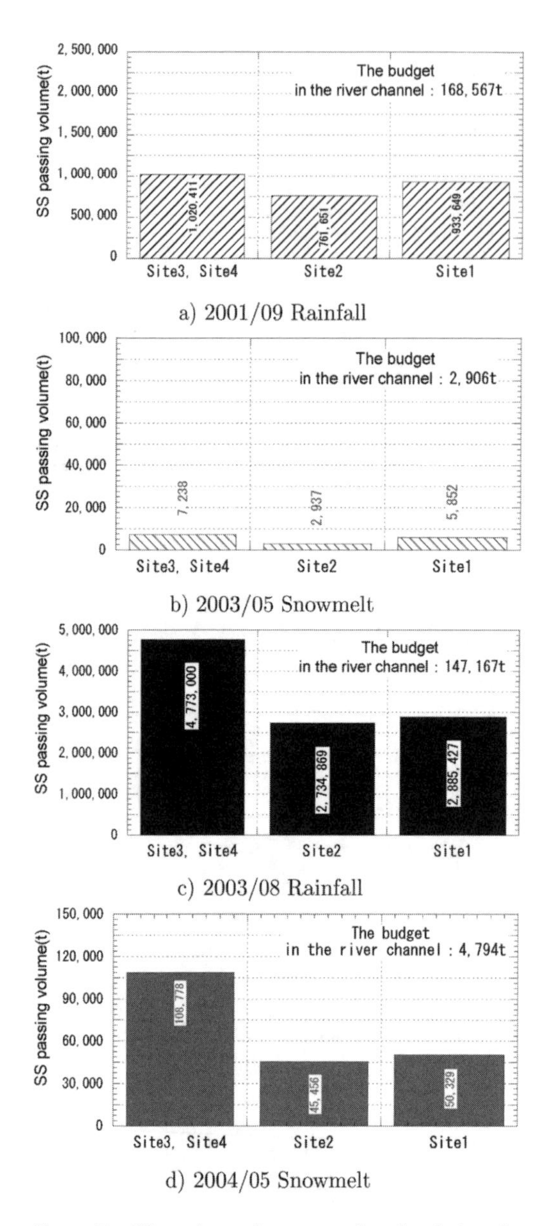

Figure 7. SS passing volume at each point during the flooding periods.

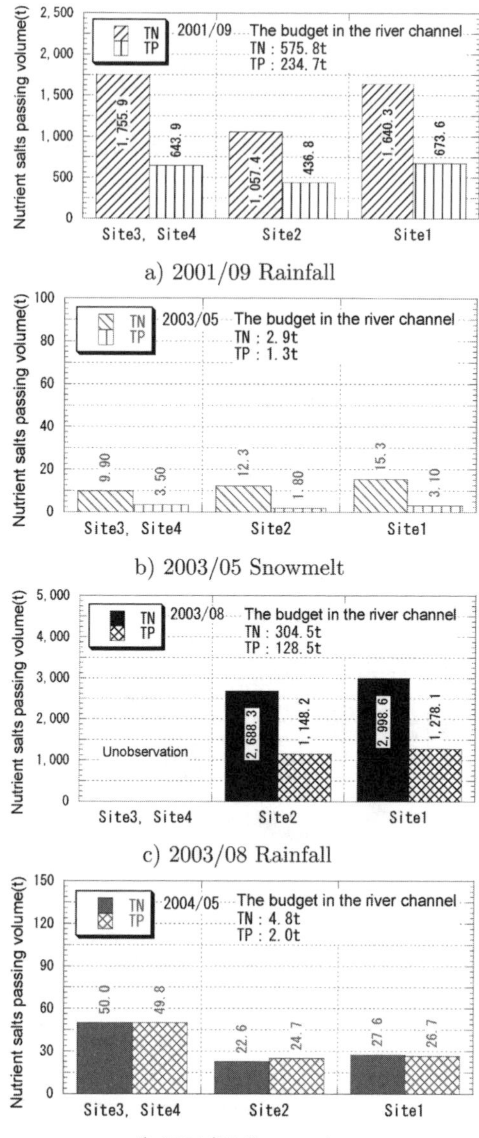

Figure 8. Passing volume of TN and TP at each point during the flooding periods.

at each observation point during flooding were compained. Figure 7a), b), c), d) lists the passing volume of SS at each point. The volumes of TN and TP are shown in Figure 8a), b), c), d).

To compare it easily, the vertical axis of each figure is made for the following. The spindle of Figure 7 was changed based on the discharge. It is a magnification of about flowing quantity, c) is twice a), and d) is 1.5 times b). Figure 7 takes 1,000 times Figure 8.

Table 2 show the starting and ending time of observation at each point. Because the value of the tributary stream and the sluiceway is small, the value is not put on Figures 7 and 8.

The budget in the river channel of Figures 7 and 8 is the values obtained by subtracting the passing volume of SS, TN and TP at observation points in the section from the Site1 (KP2.8) to the Site2 (KP16.0),

Table 2. Time of observation at each point during the flooding periods.

Flood	Observation	Site3 Site4	Site2 KP16.0	Biratoridai-ni River KP7.5	Shirau River KP6.5	Fukuman River KP5.5	Tomikawa D Sluice KP3.5	Site1 KP2.8
2001/09	start	09/11 11:00	09/11 04:00	–	09/11 05:00	–	–	09/11 04:00
	end	09/03 15:00	09/13 15:00	–	09/13 16:00	–	–	09/13 15:00
2003/05	start	04/30 00:00			04/30 00:00			
	end	05/01 05:00			05/01 08:00			
2003/08	start	08/09 10:00			08/09 10:00			
	end	08/11 09:00			08/11 09:00			
2003/05	start	05/03 21:00		05/03 20:00			–	05/03 20:00
	end	05/04 21:00		05/05 09:00			–	05/05 09:00

including the tributaries and sluice observation points, from the passing volume of SS, TN and TP at the Site1. The budget in the channel for SS, TN and TP exhibits positive values, which shows that these materials were supplied from the channel at the time of flooding. The values obtained at Site4 and Site3 are combined and used as a reference, because these observation points are upstream of the studied section. Nibutani Dam is located between the Shin-Biratori Bridge and the Nukibetsu and Horokeshi sites. To avoid the influence of the dam, the examinations of passing volume (SS, TN, TP) were conducted using the data at the Sarugawa Bridge and the Shin-Biratori Bridge which exists immediately downstream of the dam.

3.2.1 *Passing volume of SS*

As we see from Figure 7a), c), the passing volume of SS from the upper reaches during the 2003 rainfall flood is about 4.7 times that of the 2001 rainfall flood. As described above, whereas the peak discharge of the former flood at the Site1 is about 2.2 times that of the latter flood, the SS budget in the channel is similar: The value for the former is about 0.9 times that of the latter. This suggests that SS budget in the channel was not influenced by changes of the passing volume of SS from the upper reaches nor by changes in the discharge.

As we see from Figure 7b), d), the SS passing volume for the 2004 snowmelt flood is about 15.0 times that for the 2003 snowmelt flood, while the peak discharge of the former at the Site1 is about 1.6 times that of the latter. The SS budget in the channel of the former is about 1.6 times that of the latter, which is similar to the rate of increase in discharge. We consider that the difference in discharge had a greater effect on SS budget in the channel than the difference in passing volume of SS flowing from upstream.

It is found, as seen above, that the amount of SS supplied in the channel (SS budget in the channel) relates to discharge and that when the discharge exceeds a certain threshold, the SS in the water reaches a saturated concentration at which the SS supplied by the channel reaches its limit.

3.2.2 *Passing volume of TN and TP*

In Figure 8a), c), the budgets of TN and TP in the channel during the 2003 rainfall flood are smaller than those of the 2001 rainfall flood (are about 0.5 times those), which is a larger-scale flood.

In Figure 7b), d), the discharge of the 2004 snowmelt flood was greater than that of the 2003 snowmelt flood, and the budgets of TN and TP in the channel for the 2004 snowmelt flood were about 1.7 and about 1.5 times those for the 2003 snowmelt flood.

From this, it seems that, as in the case of the SS passing volume, the amount of TN and TP supplied from the channel (TN and TP budgets in the channel) relates to the discharge, and that when the discharge exceeds a certain threshold, TN and TP reach saturated concentration. Therefore, it was assumed that the amount of TN and TP supplied in the channel has an upper threshold.

3.3 *Behavior of SS*

To find how materials vary by discharge, we examined the temporal changes in SS at each point. In Figure 9a), b), the temporal changes in discharge (vertical axis) and SS (horizontal axis) are shown. These were observed at Site2 (at the upstream end of the studied section) and at Site1 (the lower reaches of the studied section) during the four floods. Because the discharge was small during the snowmelt flood, the graphs for the two snowmelt floods are shown in larger scale than those for the two floods. The line that indicates time-series behavior of SS in each flood in Figure 9 is explained having two directions: gclockwiseh and gcounterclockwiseh (see the table at the top of the figure). SS at the Site2 and Site1 observation points showed similar behavior in each flood in time series. The interval of observation at each point was as follows: 2001 rainfall flood: about one hour. 2003 rainfall flood: about two hours, with no observation at the near peak stage. 2003 snowmelt flood: about one hour. 2004 snowmelt flood: about three hours.

The time series of SS during the two rainfall floods is expressed by the counterclockwise lines, in

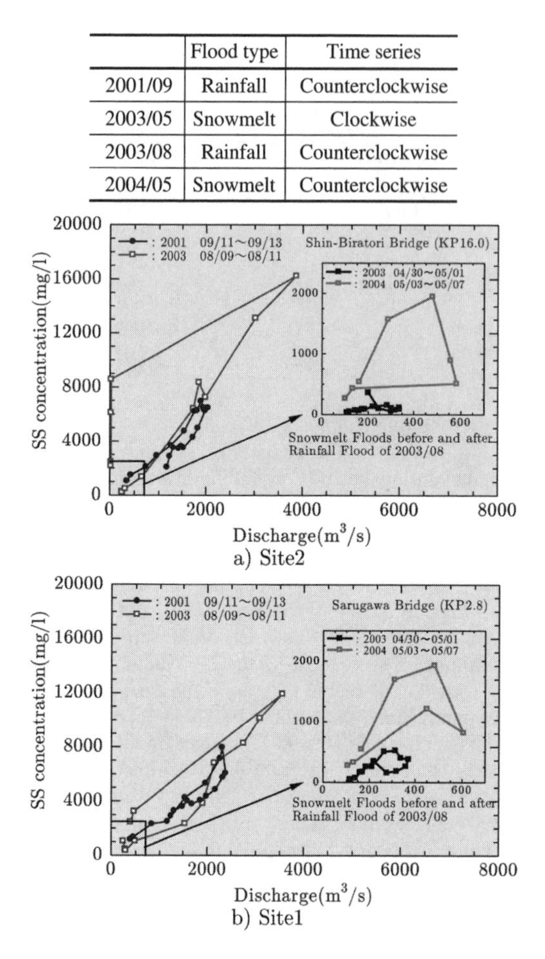

	Flood type	Time series
2001/09	Rainfall	Counterclockwise
2003/05	Snowmelt	Clockwise
2003/08	Rainfall	Counterclockwise
2004/05	Snowmelt	Counterclockwise

Figure 9. Behavior of SS during flooding.

Figure 9a), b). For the snowmelt floods, the temporal behavior of SS during the snowmelt flood before the 2003 rainfall flood is plotted as a clockwise line, but that for the snowmelt flood after the 2003 rainfall flood is plotted as a counterclockwise line. From this result and Figure 7b), d), it can be assumed that between the two snowmelt floods there were differences in the passing volume of SS supplied from upstream of the observation points and in the SS budget in the channel.

4 DERIVING AN ESTIMATION EQUATION FOR NUTRIENTS

To estimate the amount of Nutrients deposited on the flood plain during a flood, an estimation equation for the deposition of nutrient slats was derived based on the analysis of water quality and grain size of SS in river water sampled during the floods of September 2001 and August 2003.

Table 3. Classification of SS and Nutrients.

		Grain size mm
Clay		0.000–0.005
Silt		0.005–0.075
Fine sand		0.075–0.250
Medium sand		0.250–0.850
Coarse sand		0.850–2.000
	Nitrogen : N	Phosphorus : P
Total	TN	TP
Particulate organic matter	PON	POP
Particulate inorganic matter	–	PPO4-P
Dissolved organic matter	DON	DOP
Dissolved inorganic matter	DIN	DPO4-P

4.1 *Analysis of flowing water*

The data necessary for derivation of the estimation equation includes those on SS, sediment grain-size distribution, detailed data on water quality components, and bottom materials from the flood plain after the flood. "Detailed water quality analysis is analysis" that enables the classification of material into particulate, dissolved, organic and inorganic matters.

Analyses of SS and grain size were performed for the 2001 rainfall flood and 2003 rainfall flood. Detailed water quality analysis was made only in September 2001. Although bottom materials from the flood plain were sampled a month after this flood, the data were insufficient and analysis was not made. For the 2003 rainfall flood, only the ignition loss was observed. It is difficult to extract the Nutrients alone by the ignition loss. Since it has been reported [3] that the proportion and quality of Nutrients adhered to SS in river water during the flood and in the bottom material deposited on the flood plain are equal, we assumed that the Nutrients in SS in the river water and those in the sediment on the flood plain were identical in quality and quantity. Furthermore, we used the data on SS and Nutrients from the Mu River that was used in the above-mentioned report, and compared them with the observation data for the Saru River.

On the basis of analytical results of SS and Nutrients in the 2001 rainfall flood, the amount of Nutrients entering the flood plain during the flooding was estimated.

4.2 *SS and nutrients*

Table 3 shows the classification of SS and Nutrients we used in the analysis. We calculated the amount of TN, DTN, NH4-N, NO2-N, NO3-N, TP, DTP, DPO4-P and IP (PO4-P) for each category of SS, using addition and subtraction. Inorganic nitrogen (a Nutrients) was ignored in this analysis because most of the inorganic nitrogen in river water is in dissolved form.

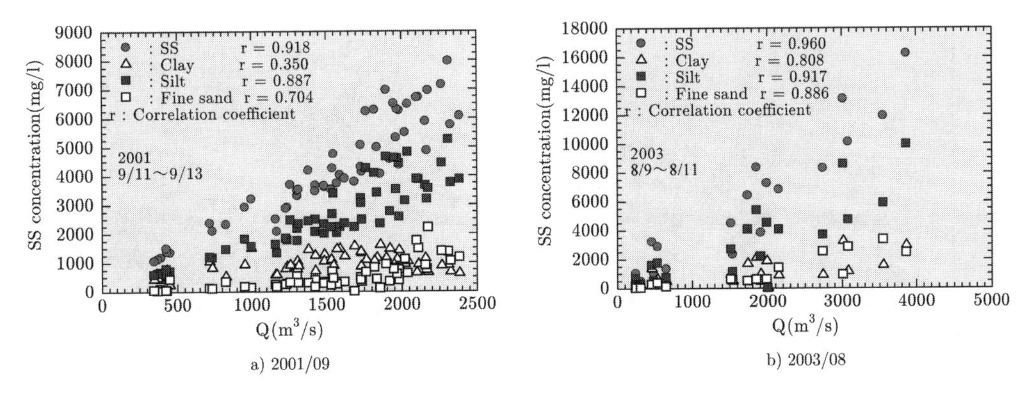

Figure 10. Relationship between discharge and each SS in the 2001 and 2003 rainfall floods.

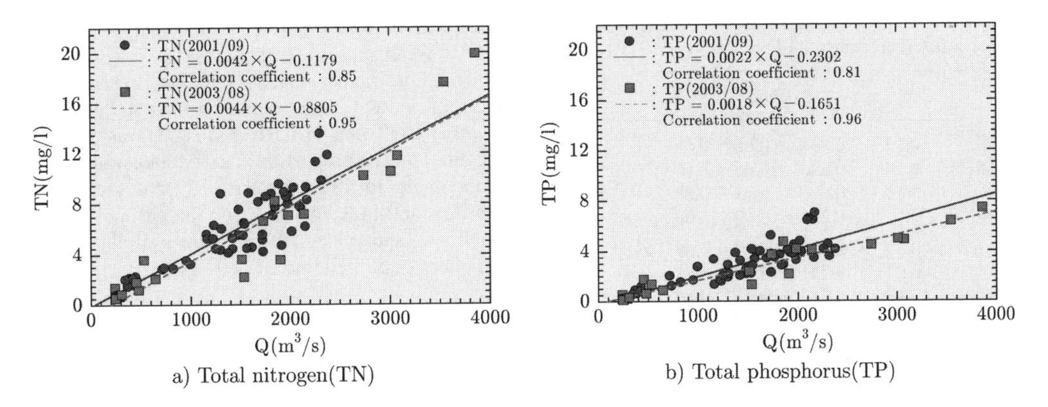

Figure 11. Relationship between discharge and total nitrogen and total phosphorus during the 2001 and 2003 rainfall floods.

4.2.1 *SS*

The behavior of each type of SS with respect to the discharge during the two floods is given in Figure 10a), b). There is a positive correlation between discharge and concentration for each type of SS. The total amount of SS is broken down as follows: 2001 rainfall flood: clay (26), silt (61), fine sand (11), medium sand (2), coarse sand (0). 2003 rainfall flood: clay (24), silt (58), fine sand (14), medium sand (4), coarse sand (0). As seen above, the ratios of each type of SS for the two floods are nearly equal.

4.2.2 *Nutrients*

The behavior of TN with respect to discharge during the 2001 and 2003 rainfall floods is shown in Figure 11a), and that of TP during the same floods is shown in Figure 11b). It is seen that the concentrations of TN and TP tend to increase with the discharge. We expressed this behavior in a simple regression equation. It can be said that the behaviors of TN and TP are independent of flood scale, at least in the rainfall floods studied.

Figure 12a) shows the behavior of each type of nitrogen with respect to the discharge during the 2001 rainfall flood. Figure 12b) shows the behavior of each type of phosphorus. The particulate components show a positive correlation with the discharge, increasing in concentration as discharge increases. However, the dissolved components show no correlation with the discharge, showing little changes in concentration even when the discharge greatly increases. Nitrogen composition is as follows: PON (88), DON (3) and DIN (6). Phosphorus composition is POP (37), PP04-P (60) and DOP (2) and DPO4-P (1). From this, we understand that 88 of the nitrogen and 97 of the phosphorus are particulate components during floods.

4.3 *Correlation coefficient*

It can be hypothesized that there is a positive correlation between the discharge and concentration of each type of Nutrients and SS. The correlation coefficients were determined for each pair of Nutrients and SS concentration. The results are shown in Table 4. The table

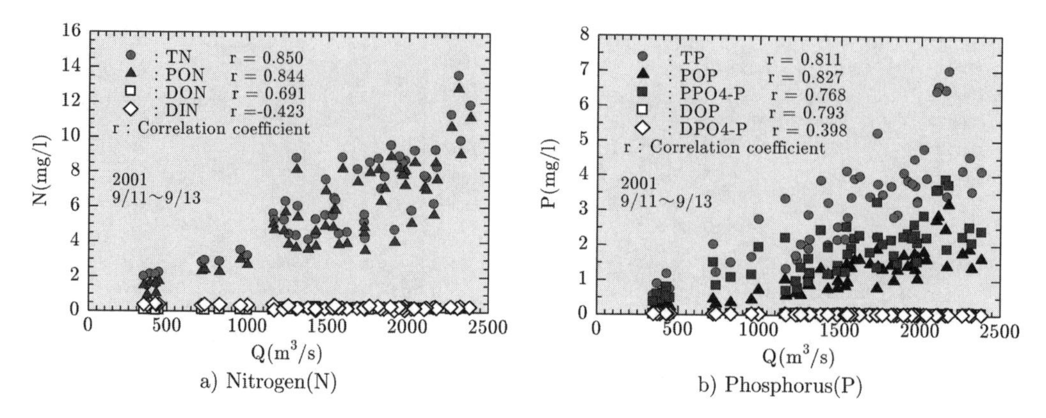

Figure 12. Relationship between discharge and each Nutrients during the 2001 snowmelt flood.

Table 4. Correlation coefficients for Nutrients and suspended solids (September 2001).

	Clay	Silt	Fine*	Medium*	Total SS
TN	0.19	0.83	0.65	0.65	0.83
PON	0.19	0.82	0.65	0.65	0.83
DON	0.34	0.51	0.33	0.22	0.52
DIN	−0.37	−0.25	−0.09	0.03	−0.26
TP	0.28	0.73	0.79	0.69	0.83
POP	0.27	0.72	0.82	0.71	0.82
PPO4-P	0.29	0.71	0.75	0.66	0.80
DOP	0.38	0.66	0.47	0.43	0.69
DPO4-P	0.16	0.38	0.27	0.31	0.39

* Sand.

distinguished between total SS and each component SS. From Table 4, we see that silt correlates closely with each Nutrients and that the particulate components (PON, POP and PPO4-P) also correlate well with each type of SS. It seems that most of the particulate components of Nutrients are transported by adhering to silt particles in SS.

4.4 *Mu River and Saru River*

It has been reported in the study on the Mu River (Catchment area is 1,270 km^2 and trunk river channel length is 135 km) that Nutrients adhered to SS on the flood water are in equal proportion to those in sediment on the flood plain, sampled after flooding, and are of identical composition. Based on this report [3] and using the relationship between the SS and the Nutrients in the floodwater, we derived an estimation equation for Nutrients, and estimated the amount deposited on the flood plain. By comparing the data from the Mu and the Saru rivers, we examined the relationship between the amounts of SS and the Nutrients in the particulate form in the river water during the flood, and that between the amounts of SS in the river water and the sediment on the flood plain.

4.4.1 *SS in river water and Nutrients (particulate water quality components)*

The relationship between SS in the flowing river water and Nutrients during the two snowmelt floods (Mu River in 1998, and Saru River in 2003) was compared (Figure 13). The proportion of particulate water quality components to SS in the Saru River is greater than that for the Mu River; however, the particulate water quality component and the SS show a similar increase (gradient of plotted line).

4.4.2 *SS in river water and sediment on the flood plain*

The relationship between SS and Nutrients is roughly the same in the rainfall flood as in the snowmelt flood. From this, we can infer that the ratio of Nutrients to SS in the flood water of the Saru River is roughly equal to the ratio of Nutrients to SS in sediment on the flood plain of the Mu River. This is because in the Mu River, the proportion of Nutrients adhered to SS in the river water during a flood is roughly equal to those adhered to sediment on the flood plain. Based on the data on the SS and Nutrients in the river water of 2001 rainfall flood on the Saru River and those on the sediment on the flood plain of the Mu River sampled after the 1997 rainfall flood, the values for particulate TN and TP were made dimensionless using the SS values and were categorized by the average gain size (Figure 14). We found that the data on the Saru River overlaps the data plotted for the Mu River, which were obtained by using a simple linear regression equation. This indicates that, as is the case in the Mu River, the Nutrients adhered to the SS in the flood water of the Saru River are roughly the same as those adhered to the sediment on the flood plain after the flood in that river.

4.5 *Estimation equation for Nutrients*

α and β were obtained by deriving a simple regression equation expressing the relationship between the

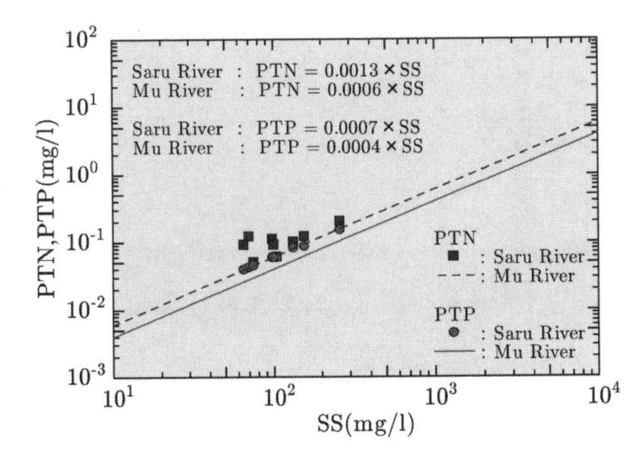

Figure 13. Relationship between SS and particulate water quality components.

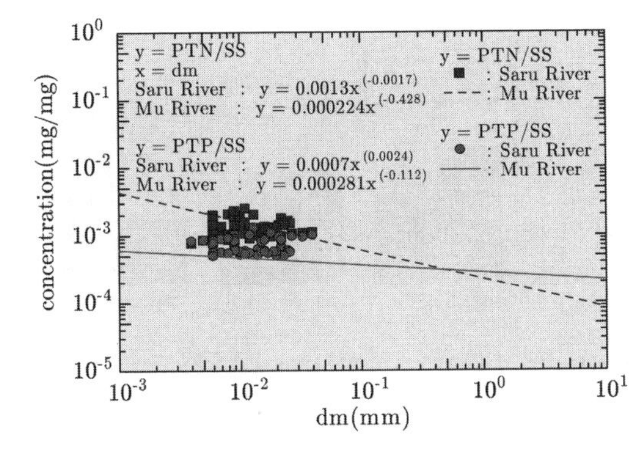

Figure 14. Average grain size of SS vs. concentration of particulate components.

concentration of each Nutrients, L_m[mg/l], and each SS_j[mg/l] (Equation (1)). As SS are present in water in bivalent form [4], two simple regression equations are necessary to be derived; however, for simplicity, one simple regression equation below was used:

$$L_m = \alpha_{mj} SS_j^{\beta_{mj}} \quad (1)$$

α,β: the multiplier and the index with respect to each type of Nutrients and SS, m: type of Nutrients, j: type of SS.

To quantitatively estimate the Nutrients deposited on the flood plain, Equation (2) is used. Equation (2) was derived by substituting the amount of each Nutrients W_m[t] for L_m in Equation (1), and the amount of each sediment W_j[t] for SS_j. To determine amount of each sediment W_j, Equation (3) was used, where the volume of each type of SS is V_j[m^3], the soil particle density is p_s[t/m^3] and the void content is λ. Table 5 shows the values of r^2 and α, β in equations (1) and (2).

$$W_m = \alpha_{mj} W_j^{\beta_{mj}} \quad (2)$$

$$W_j = p_s (1 - \lambda) V_j \quad (3)$$

Table 5. Coefficients in estimation equation for Nutrients (2001/09).

	Clay			Silt			Fine sand			Medium sand			Total SS		
	α	β	r^2	α	β	r^2	α	β	r^2	α	β	r^2	α	β	r^2
TN	0.14593	0.53475	0.15	0.01101	0.80364	0.78	1.41030	0.23761	0.33	4.40227	0.11352	0.33	0.00317	0.90645	0.77
PON	0.06844	0.62602	0.16	0.00394	0.91872	0.78	1.00733	0.27237	0.33	3.73659	0.12716	0.32	0.00095	1.03535	0.78
DON	0.00221	0.63605	0.21	0.00108	0.65062	0.50	0.06247	0.17072	0.17	0.15111	0.04951	0.06	0.00041	0.72917	0.49
DIN	2.21113	−0.32661	0.09	1.10815	−0.19779	0.08	0.29507	−0.03613	0.01	0.23391	0.01169	0.01	1.34874	−0.20971	0.07
TP	0.01076	0.81112	0.27	0.00263	0.89465	0.76	0.53815	0.27865	0.35	2.07427	0.12610	0.32	0.00043	1.06131	0.83
POP	0.00105	1.00364	0.28	0.00019	1.10117	0.77	0.13982	0.33657	0.35	0.71725	0.14967	0.30	0.00002	1.30300	0.84
PPO4-P	0.01121	0.72962	0.25	0.00294	0.81379	0.71	0.36381	0.25748	0.34	1.26319	0.11749	0.31	0.00056	0.96571	0.78
DOP	0.00078	0.60133	0.32	0.00058	0.56539	0.64	0.02146	0.13405	0.17	0.04116	0.05966	0.15	0.00024	0.63971	0.64
DPO4-P	0.00965	0.16315	0.10	0.00714	0.18223	0.29	0.02146	0.05387	0.12	0.02759	0.02911	0.16	0.00529	0.20775	0.30

5 ESTIMATING THE AMOUNT OF NUTRIENTS DEPOSITED ON THE FLOOD PLAIN

Calculation was performed to clarify how the amount of Nutrients deposited on the flood plain differed between the two floods of different scales (2001 and 2003 rainfall floods on the Saru River). Changes in the amount of SS on the flood plain as the result of flooding were calculated, and an estimation equation for Nutrients was obtained from the results of this calculation.

5.1 *Calculation method*

We performed calculations for unsteady flow with a compound cross section. The calculations covered a duration of 60 hours that included the observation period, and the section was between KP0.4 and KP21.2.

5.1.1 *Difference in estimated amounts of deposited Nutrients*

Figure 15 shows the variation in sediment [t] determined by adding up the amount of SS scoured up and that deposited on the flood plain section with a longitudinal length of 200 m. Figure 15 shows that in the upper part of the 200-m section, the SS deposition is greater during the 2001 rainfall flood than during the other flood, and that in the lower part of that section the SS deposition is greater during the 2003 rainfall flood than during the other flood. This seems to be due to the difference in the hydraulic quantities. At the upstream end of the calculated section, the 2001 rainfall flood recorded less discharge and less appreciable peak discharge than the 2003 rainfall flood, which probably facilitated the SS deposition on the flood plain. At the end of the section, there was a 30 cm difference in peak water level between the floods. The 2003 rainfall flood had a higher water level than the other flood, and this allowed more water to flow into the flood plain. It is likely that this made it easy for SS to be deposited before and after the peak, particularly after the peak. It is, therefore, possible to hypothesize from the calculation that the amount of SS deposited on the flood plain is greatly influenced by the difference in flood scale, particularly by the difference in hydraulic quantities.

We then chose a long section between Site2 (KP16.0) and Site1 (KP2.8) to minimize the effect of discharge at the upstream end and water levels at the downstream end. We arbitrarily divided the section into three subsections and calculated the SS deposition for each subsection. The calculation results for the two floods (2001 and 2003 rainfall floods) are given in Figure 16. Figure 16 shows that the SS deposition on the flood plain varies between the three subsections

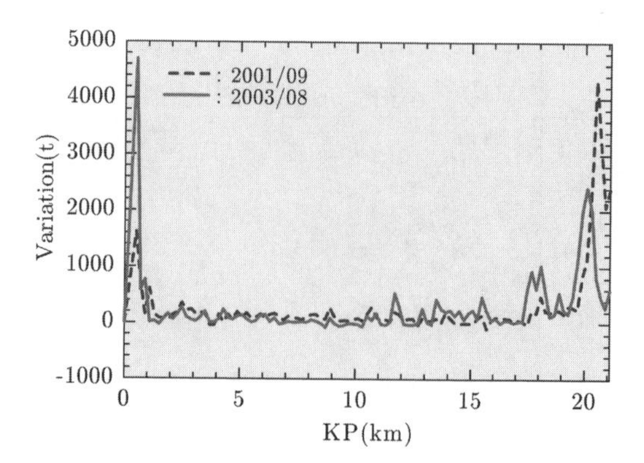

Figure 15. Variation of SS deposition in flood plain in flood (calculated).

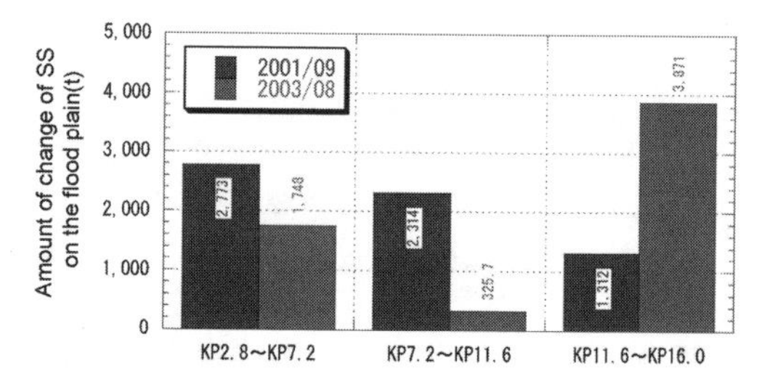

Figure 16. Difference in the amount of SS deposition in flood plain in flood.

and between the two floods. The same can be said for Nutrients (TN and TP), as there is a positive correlation between the deposited SS and Nutrients. However, the total amounts of Nutrients deposited on the flood plain for the entire section during the 2001 and 2003 rainfall floods (obtained through calculation using Equation (2) and the SS values) are 13.3 tons and 12.0 tons for TN, for the respective years, and 3.8 tons and 3.7 tons for TP, for the respective years. For the section as a whole, it can be said there was no considerable difference in Nutrients deposition on the flood plain.

6 CONCLUSION

6.1 *Difference in material transport*

Comparison of material transport during the four floods was performed through three approaches. As a result, it was hypothesized that the amount of material supplied from the river channel is related to the discharge, when the discharge exceeds a certain level, however, the material transported in the river water reaches the saturated concentration. Therefore, the concentration of materials in flowing water has an upper threshold.

6.2 *Deriving an estimation equation for Nutrients*

To derive the estimation equation for the deposition of Nutrients, the SS and Nutrients were classified based on the observation result from the 2001 rainfall flood. The calculation showed that there is a high correlation between silt and particulate water quality components. In addition, the comparison of the data from a study on the Mu River and from observation on the Saru River revealed that the relationship between the SS in flowing water and Nutrients was roughly equal for the two rivers.

6.3 *Estimated Nutrients deposition on the flood plain*

Estimation calculation using the derived equations for the two floods (2001 and 2003 rainfall floods on the Saru River) clarified that the amount of Nutrients deposited on the flood plain of the entire river section studied do not differ between the two floods. However, the amount varied by subsection. This suggests that, depending on the scale of a flood, the influence of the flood on the river environment varies locally.

REFERENCES

[1] Yasuharu Watanabe, Takehiro Ogawa: Characteristics of SS Transport during the Saru River Flood of Summer 2001, Proceedings, Hokkaido Chapter, JSCE, 2002.

[2] Takehiro Ogawa, Yasuharu Watanabe: Observation of August 2003 Flood in Saru River Basin, Proceedings of Hydroscience and Hydraulic Engineering, JSCE, Vol. 48, pp. 955–960, 2004.

[3] Yasuharu Watanabe, Ryuichi Shinme, Daisaku Saito, Takeshi Tamagawa: Study on Mass Transport in Mu River Snowmelt Flood of 1998, Proceedings of Hydroscience and Hydraulic Engineering, JSCE, Vol. 43, pp. 587–592, 1999.

[4] Committee on Hydraulics, JSCE: Report by Investigation Team on Heavy Rainfall Disaster in Hokkaido from Typhoon Etau (Typhoon No. 15) of 2003, pp. 129–141, 2004.

River, Coastal and Estuarine Morphodynamics: RCEM 2005 – Parker & García (eds)
© 2006 Taylor & Francis Group, London, ISBN 0 415 39270 5

Flood control of the Vara River (North-western Italy)

A. Stocchino, A. Siviglia & M. Colombini
Dipartimento di Ingegneria Ambientale, Università di Genova, Via Montallegro, Genova, Italy

ABSTRACT: Aspects related to the sediment transport through the river system and sediment management are still rarely addressed in the design of flood control and river management programmes in Italy. Recently, however, there has been increasing consideration of regional sediment issues, sediment continuity and channel morphodynamic as an integral part of flood protection schemes and future river management. An example is represented by the research project presented in this paper, regarding the mountain basin of the Vara River, in the north-western Italy. The main purpose of the present work is to study the influence of a composite storage tank, designed for flood control, on the sediment transport processes, so that it would be possible to evaluate the impact of the latter on the surrounding lands. We perform a physical model characterized by erodible bed, considering different configurations of the storage tank and different hydrological scenarios with return period ranging from 5 to 200 years. The physical model has been designed to reproduce a fluvial reach 1 km long in geometric scale equal to 1:62.5 and Froude similitude, so to conserve inertial and gravitational effects. We have fed the model with unsteady water and sediment discharge and the bottom topography of the river has been left free to evolve. We also use a mixture of sediments characterized by various diameters to evaluate the selective transport of the different granulometric fraction.

1 INTRODUCTION

Until quite recently, river management programmes in Italy rarely address sediment transport processes in the design of flood control structures. Some recent works (Siviglia et al, 2004; Federici et al, 2004), indicate as there is an increasing awareness of regional sediment issues and channel morphodynamics as an integral part of river management. In this context a research project has been funded by the Agenzia Interregionale per il Fiume Magra, which is the Agency in charge of the management of the basins of the Vara River and the Magra river in the North-west Italy. The general aim of the project was to obtain precise knowledge on the hydro-morphodynamic behaviour of some flood defense systems to be constructed in the Vara River. This knowledge should be a useful tool for engineers in the designing of such a structures in order to reduce the peak discharges.

Due to the complexity of such phenomena and particularly to the irregular geometry of such natural systems this phenomenon is not amenable to analytical treatment. Predicting the consequences of this human intervention on sediment transport could be achievable using either scale models or numerical models. In this paper we present preliminary results obtained employing a physical model of a reach of the Vara River. Different flood control solutions have been tested under controlled laboratory conditions as opposed to costly field programs.

The rest of the paper will proceed as follows. In the next section we give a short general description of the reach of the Vara River analyzed in the paper. In section 3 we describe the main features of the physical model and the measurements technique. Next we present the experimental observations and, then, concluding remarks follow.

2 STUDY AREA

The river Vara is an Apenninic gravel bed-river in northwest Italy. It is the main tributary of the Magra River with a basin which drains an area of about 572 Km^2. It presents an elongated shape in direction NW-SE. Reaches morphologically controlled by the massive presence of man-made structures alternates with reaches with either single threated morphologies with alternate bars or confined meanders. In the final reach, 6 Km upstream the confluence, the river flows in a large valley where a braided morphology is developed. The morphological changes that occurred during the two last centuries along the lower reach shows similarities with the recent channel evolution of many gravel-bed rivers with an initial braided morphology draining from the Apennines and Alps

(Rinalid et al, 2005). In the present study, a river reach which extends for about 1 Km in the upper Apennine part is considered. A single thread morphology with no man-made structures characterised this site. In order to quantitatively described the site under investigations, morphological and topographical surveys has been provided.

2.1 *Topographic survey*

A topographic survey has been carried out in the region of interest, with the aim to gather precise information about the geometry of the main channel and the floodplain on the right hydrographic bank, which is the region where the floodwater is stored. In fact, the construction of the physical model described in the next section has been based on the above survey. The measurements of this survey have been collected in 15 cross-sections with a spatial interval of about 60 m. Furthermore, contours lines with an interval of 1.25 m, has been provided.

2.2 *Sediment survey*

Both subarmor and armor samples have been collected. Subarmor layer characterization is of interest to asses the composition of bed-load transport while surface layer measurements are needed to estimate surface roughness (Parker, 1990). For characterizing the armor layer many sampling methods are available in the literature (Bunte & Abt, 2001; Wolman, 1954; Leopold, 1970). The statistical sampling called pebble counts have been employed. It suggests to pick up a certain number of surface particles, depending on the dimensions of the largest pebble, along parallel transepts, for details see (Bunte & Abt, 2001). The latter method provides precise and reproducible measurements. Characterisation of the sub-armor layer has been done collecting a sample of sediment by means of a mechanical excavator and then sieving the collected material. The minimum sampling mass (in Kg), according to the American standard ASTM *C*136–71 is

$$m_s = 2596D_{n,max}^{1.5}$$

where $D_{n,max}$ (in m) is the nominal diameter of the particles retained on the largest sieve size. The nominal diameter is the diameter of the sphere with the same volume of the particle. In Figure 1 the grain size distribution of both the subarmor and the armor is shown.

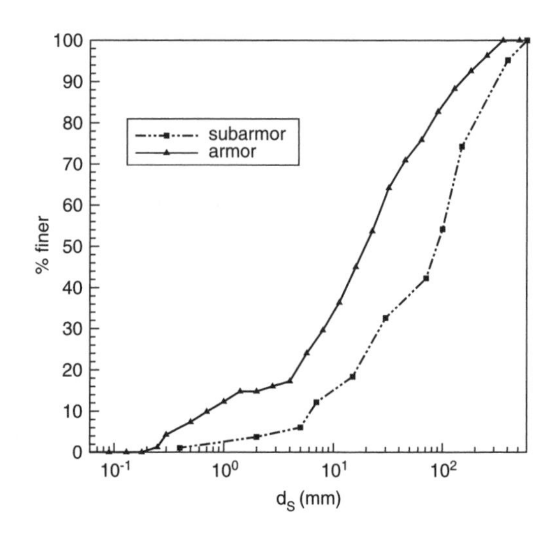

Figure 1. Grain size distribution.

3 PHYSICAL MODEL AND EXPERIMENTAL MEASUREMENTS

In order to account for scale effects, the model has been designed preserving the relevant dimensionless number when free surface flows are studied, i.e. the Froude number defined as:

$$F = \frac{Q}{\Omega\sqrt{g\Omega/b}} = \frac{U}{\sqrt{g\Omega/b}}.$$

where Q is the volumetric discharge, U is the depth-averaged velocity, Ω is the cross-sectional area, b is the width of the flow and g is the gravitational acceleration. Therefore, we impose the Froude number of the model to be equal to the Froude number of the prototype, indicated in the following with the subscript $_P$, i.e.:

$$F_P = \frac{Q_P}{\Omega_P\sqrt{g\Omega_P/b_P}} = \frac{Q}{\Omega\sqrt{g\Omega/b}} = F \tag{1}$$

Once, the geometric reduction for the length scale $\lambda = L_P/L$ is set, the scaling for the other relevant quantities is derived from the relationship (1), we have then obtained the scaling for the volumetric discharge and the time, which reads:

$$\frac{Q_P}{Q} = \lambda^{5/2} \qquad \frac{T_P}{T} = \lambda^{1/2}. \tag{2}$$

In the present study the geometrical scale λ has been set equal to 62.5. The river reach under investigation is about 1 Km long and the area of the river-basin bounded by the contour line corresponding to a value of 325 m asl is about 0.2 Km^2. Therefore, the physical model is about 16 m long and 6 m wide. The floodplain on the right hydrographic bank and the orography

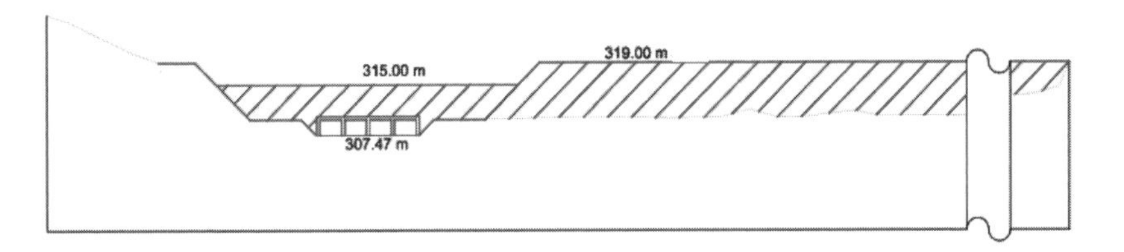

Figure 2. Sketch of the transverse dam, where it is visible the bottom gate and the top spillway. The height reported in the picture is to be intended as asl.

have been built starting from the measured contour surfaces, which are spaced 2 cm with the adopted scaling. To reproduce the latter, 1 × 2 m sheets of polystyrene −2 cm thick – have been properly shaped and superimposed and, finally, they have been joined with light concrete in order to smooth out any discontinuity. A different building process has been adopted for the river reach, due to its more complex geometry. In particular, the reach has been reproduced starting from the 15 measured cross-sections and, eventually, obtaining the reconstruction of the river bed composed by the proper sediment. The bed has been prepared in two different ways depending on the kind of experiment, i.e. homogeneous or heterogeneous sediment. A long narrow area next to the river reach is sparsely wooded, the effect of the latter has been accounted for not allowing the river bed to evolve and reproducing a proper flow resistance for the case of vegetated channels, gluing sand of an appropriate diameter. In particular, starting from the vegetation density, the mean diameter of the trees and following Righetti et al (2004), we were able to obtain an equivalent coefficient of resistance typical of the area under investigation. We have then obtained a corresponding value for the Gauckler-Strickler coefficient k_S and, eventually, the diameter of sand able to produce the required resistance.

Two different configuration of the flood defense system are studied: the first configuration, defined "L1" in the following, is composed by a transverse dam with a bottom gate and a top spillway; the second configuration, defined "M1", is composed by the same transverse dam together with a longitudinal embankment, which has a lateral spillway near the dam. The dimensions of the gate are 16 × 2.8 m, the top spillway of the dam is placed at a height of about 7.5 m over the river bed. In Figure 2 is reported a sketch of the transverse dam. The lateral spillway of the configuration M1 is about 150 long and has an height of 314.5 m asl.

In Figure 3 a sketch of the physical model is shown, where it is visible the flood-plain on the right hydrographic bank, the transverse dam and the longitudinal embankment.

Different hydrological scenarios have been simulated during the experiments, reproducing hydrographs with return period ranging from 5 to 200 years. Starting from the bed topography measured during the

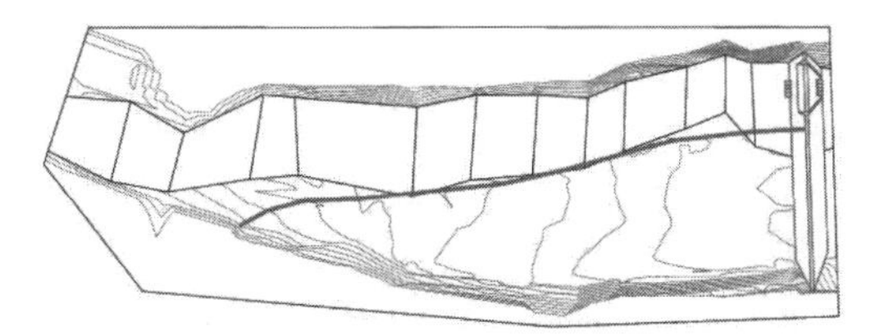

Figure 3. Plan view of the physical model, where it is clearly visible the main channel of the river, the flood-plain, the transverse dam and the longitudinal embankment.

specific survey, we have concentrated our attention on the evolution of the river bed in response to the specific hydrographs. In order to seek possible correlations between the characteristics of the hydrograph and the bed evolution, sequence of experiments with the same hydrograph without restoring the initial bed have been performed, in particular for return period of 15 and 30 years.

The river bed at the end of each run has been measured using a laser sensor mounted on a two dimensional rail able to cover the entire area of the model. The precise planimetric position of the laser has been measured by means of a total station. The raw output was a sparse matrix of bed elevation, which was postprocessed in order to interpolate the scattered data onto a regular grid using a linear interpolation method. In this way, it was easier to produce contour map of the bed elevation and to extract transverse and longitudinal sections.

4 EXPERIMENTAL OBSERVATIONS

A first set of experiments has been carried out not allowing the bed to evolve. The main goal of these experiments was mainly to evaluate the rating curve of the transversal dam and to quantify the hydraulic efficiency of the two configurations in terms of the decrease of the flood peak corresponding to different hydrological scenarios. These sequence of experiments are not reported here for the sake of brevity, but in the conclusion few results will be discussed in comparison with the measurements obtained with erodible bottom.

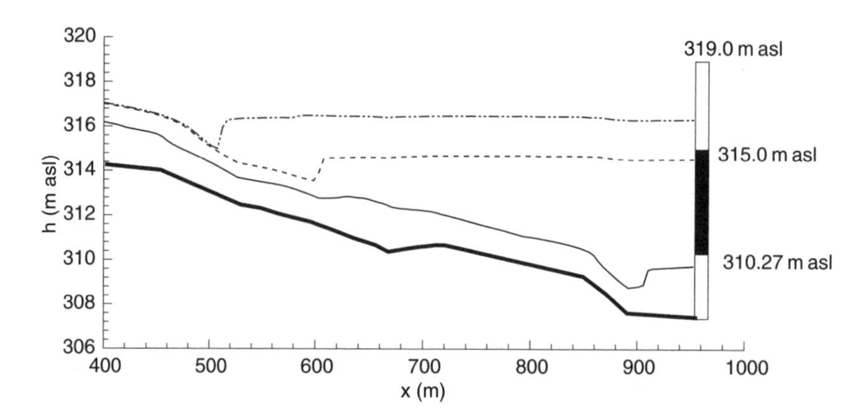

Figure 4. Example of free surface profile at different times during a flood event. The plot represents a 1D numerical simulation obtained with a fixed bed.

Regarding the latter, two different sets of experiments have been performed: the first set has been carried out employing uniform sediment with diameter equal to the d_{50} of the prototype, which correspond to a sand with diameter of about 1.5 mm in the model; for the second set a mixture of three grain sizes has been used, namely three equal parts in weight representing the d_{90}, d_{50} and d_{16} of the prototype. The main goal of the second kind of experiment is to investigate possible processes of grain sorting induced by the differential sediment transport process that occurs during the simulation of a single hydrograph.

4.1 *General observations*

In Figure 4 is shown a numerical simulation of the flow during an intense flood event, in particular with a return period equal to 200 years, which is reported only to describe briefly the flow conditions, especially in terms of the profile of the free surface, which occur during a single event. While the free surface level is below the height of the bottom gate (310.27 m asl) the flow behaves as in the case of a localized contraction due to the presence of bridge piers. In the latter case, the flood-plain is not interested by the flow. When the free surface level exceeds the height of the bottom gate, the dam starts to act in such a way to necessarily slow down the flow, increasing the free surface level, in this case the flood water begins to fill the storage volume. The profile of the free surface is quite horizontal moving from the dam upstream, until it reaches the fast current flowing downstream forming an hydraulic jump, as shown in Figure 4.

The dynamics of the sediment transport processes is strongly affected by the presence of a flood defense of the kind studied in the present work, as it should be easy to guess. In fact, the artificial decreasing of the flow velocity due to the backwater effect, causes the sediment transport capacity of the flow to decrease considerably. It is quite straightforward to imagine that the presence of a dam is able to induce a significant deposit upstream. In the following we briefly show the preliminary experimental observation of the evolution of the bed configuration due to a sequence of events with the same return period in the two configurations (L1 and M1).

4.2 *Uniform sediment*

In Figure 5 and 6 are reported the erosion-deposition maps in the case of a sequence of events with return period $T_R = 30$ years for the configuration L1 and M1 respectively, obtained at the end of runs with uniform sediment. Only after the first run small areas where bed erosion occurred are present, shown in the figure with dashed contour lines. From the second run, the entire area is in deposition. Sensible difference appears between the two configurations due to the presence of a longitudinal embankment in the M1 configuration. In fact, the main role of the embankments is to restrain the sediment transport within the main river reach, not allowing the sediment to occupy part of the flood plain as occurs in the configuration L1. In the latter, successive events cause the entire flood-plain to be interested by deposition. From the contour maps it is quite clear the appearing of three important deposits, whose intensity reaches height of 3–5 m. However, the latter seems to tend to an equilibrium value, in fact, from the second run, the deposit still increases but at a slower rate.

The patterns in terms of macroscale bedforms is reproduced almost identically during a sequence of events with return period greater or smaller than 30 years. No results are shown of the latter experiments for the sake of brevity. However, they do not add any significant information, the only difference regards the intensity of the deposits that occur.

To investigate the possibility of the existence of an equilibrium condition for the river bed, we have

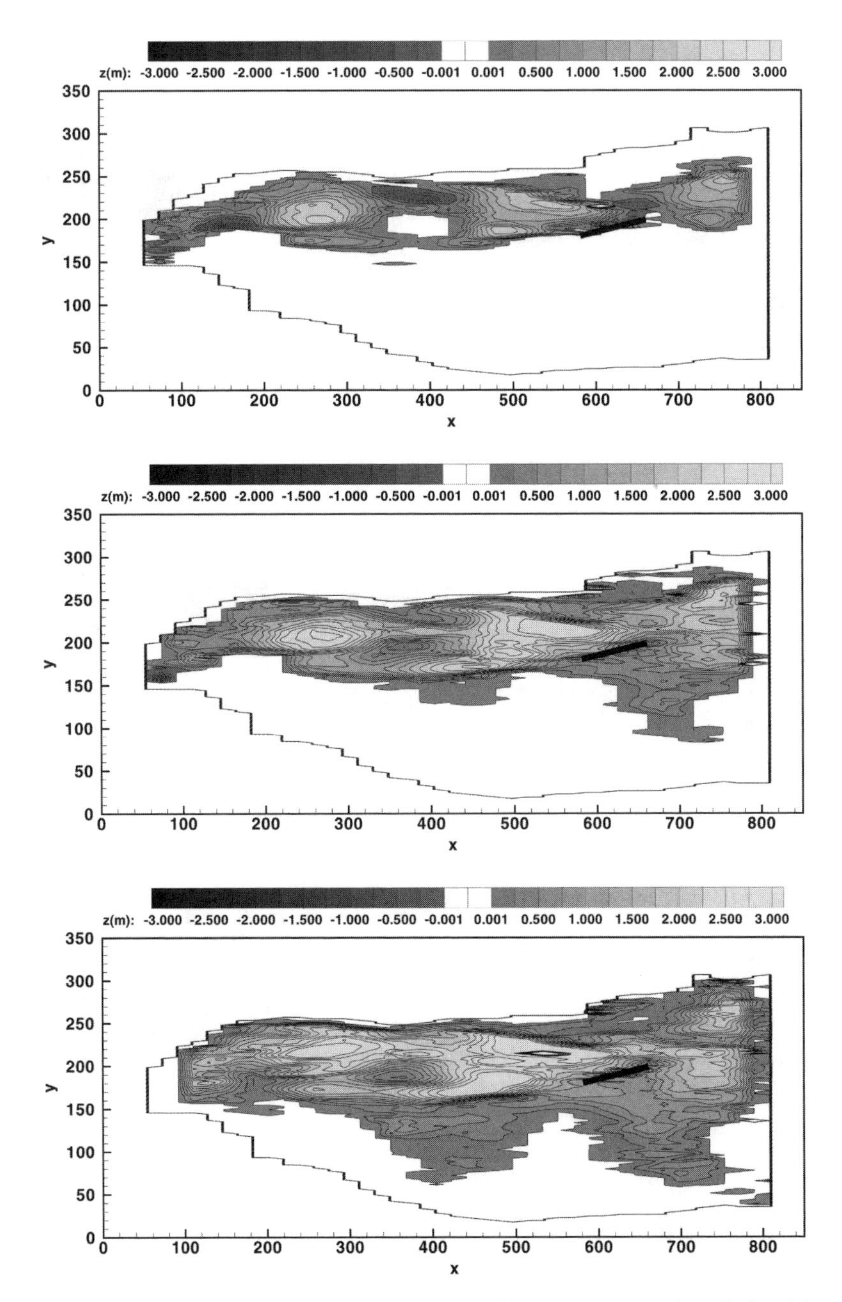

Figure 5. Configuration L1, hydrograph $T_R = 30$ years: erosion-deposition contour map at the end of run1 (top panel), run3 (mid panel) and run5 (bottom panel).

evaluated for each run the mean slope of the entire reach, regardless the transverse slope. The results of this analysis are presented in Figure 7, where the lines representing the mean river bed are plotted for all the runs performed in the two configurations, again for a sequence of events with return period of 30 years. In both cases the mean slope tends to decrease showing a general process of aggradation of the entire reach, aggradation that is due to the influence of the dam. Moreover, the rate of decreasing of the mean slope seems to reduce with the successive events. This can be interpreted as a tendency of the river bed to adjust itself

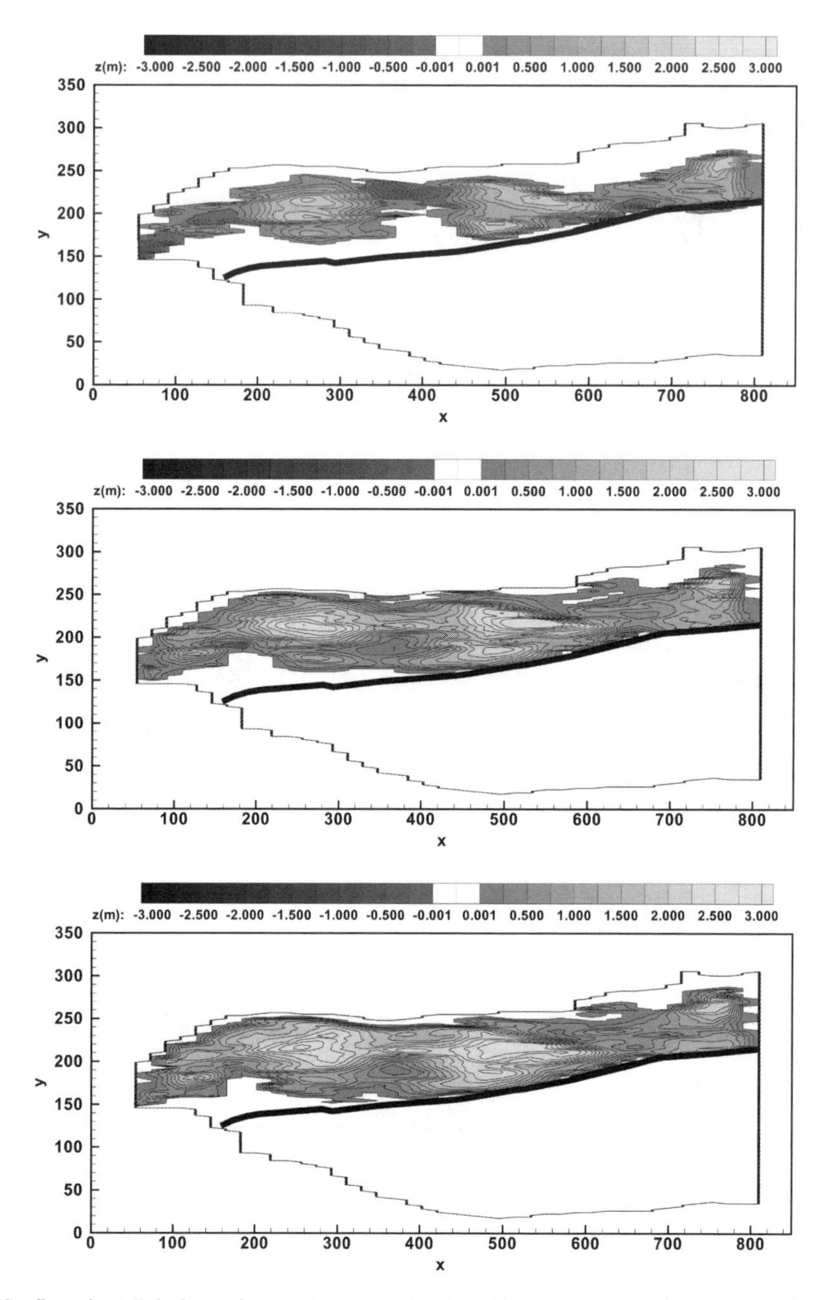

Figure 6. Configuration M1, hydrograph $T_R = 30$ years: erosion-deposition contour map at the end of run1 (top panel), run3 (mid panel) and run5 (bottom panel).

onto an equilibrium slope. At the end of the present section we will comment further on the latter result. We now intend to deepen the description of the bed patterns observed. If the mean slope is removed from the corresponding erosion-deposition contour map, we are left with a sequence of quite clear macroscale bedforms that resemble the characteristic pattern of a series of central bars. Figure 8 shows the longitudinal distribution of the difference between right side and left side measured bed elevation Δz filtered with the mean slope. In all the runs performed three central bars were observed with maximum amplitude ranging

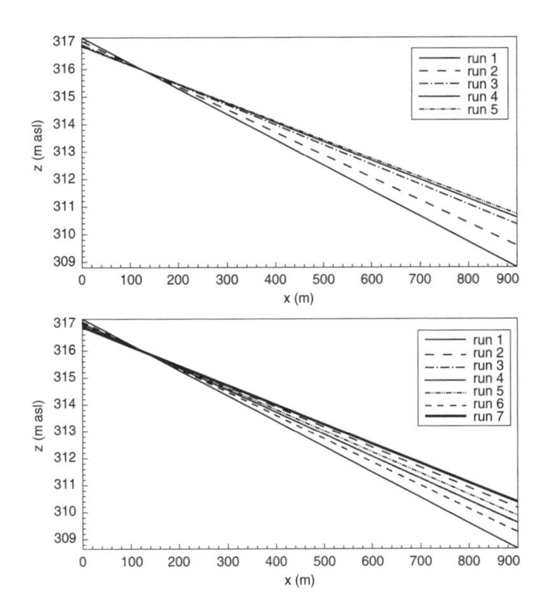

Figure 7. Evolution of the mean slope of the river reach during a sequence of runs with hydrographs with $T_R = 30$ years: configuration L1 (top panel) and configuration M1 (bottom panel).

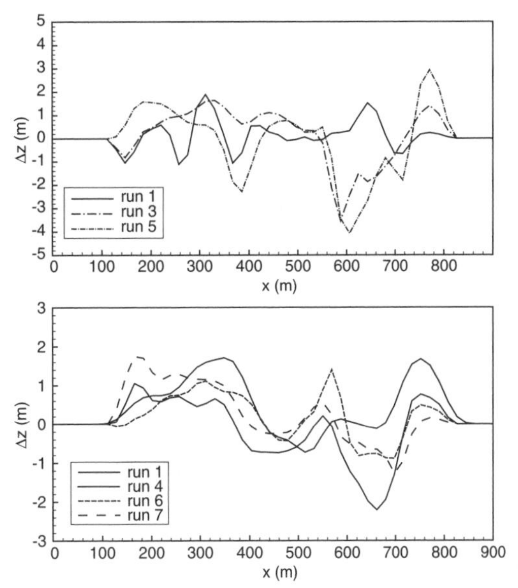

Figure 8. Examples of longitudinal profiles of the difference between the right side and left side measured bed elevation Δz: configuration L1 (top panel) and configuration M1 (bottom panel).

from 2 to 4 m, depending on the flood event and on the different runs during a sequence of events. The central bars seems to migrate upstream with successive events. However, the bed was measured only at the end of each run, thus, we could not measured the changes that occurred during the single event. We could only observe qualitatively a transient behaviour while the experiment was running; in particular, the maximum deposit corresponded to the peak of the hydrograph, while its tail tended to lower the intensity of each deposit. Further analysis of the present data are required to compared the wavelength and the amplitude of the observed bars with the existing theories.

A final comment is dedicated to the existence of an equilibrium of the river bed after a series of events with the same return period. This limiting condition should imply an equilibrium in the sediment transport in terms of sediment discharge. The analysis of the global balance of volume of sediment transported through the model is an important parameter to understand if an equilibrium is reached. We have fed the physical model with a time dependent sediment discharge in equilibrium with the liquid discharge during the entire flood event. We have monitored this feeding condition providing the very first part of the physical model to maintain the same bed elevation, thus, implying a local condition of equilibrium. Then, we have collected the sediment that went through the bottom gate of the dam. By a direct comparison of the volume of sediment fed into the model and went out of the end of the model, we were able to understand how far was the river reach from an equilibrium condition. A first observation that it is worthy to mention regarding this analysis is that for the case L1, which is the configuration without any longitudinal embankment, no sediment is able to pass through the gate, not even after five successive events. This is due to the fact that the entire flood-plain may be considered as a storage volume not only for the liquid discharge, but also for the sediment discharge. As it is shown in Figure 5 the deposit covers more and more area of the flood-plain. In the case of configuration M1, we have found a different behaviour; in fact, starting from the third run, a sensible amount of sediment started to pass through the dam. After 7 successive events, the volume collected after the dam was about the 70% of the volume fed into the model. The latter result suggests that the presence of longitudinal embankment play a fundamental role for the sediment balance, helping the river to find a condition of equilibrium.

4.3 *Heterogeneous sediment*

As mentioned above, a second set of experiments has been performed using a sediment mixture. The aim of these experiments was to investigate possible grain size sorting along the river reach. For this purpose the mixture was prepared with three different diameter reproducing three characteristic size of the

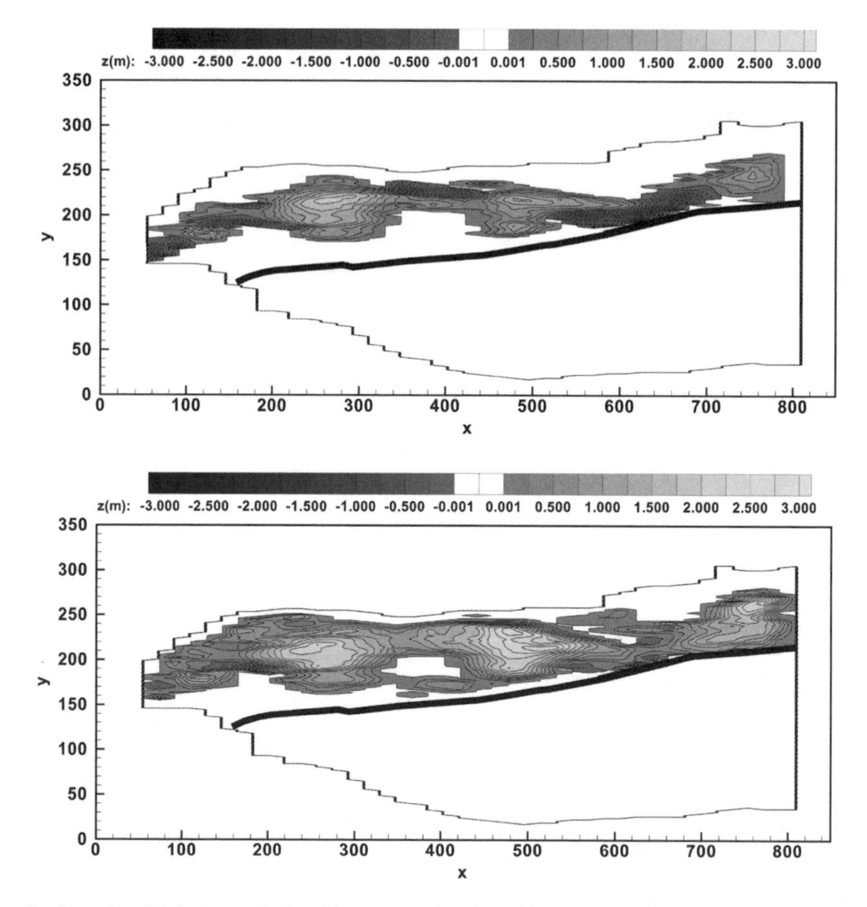

Figure 9. Configuration M1, hydrograph $T_R = 30$ years: erosion-deposition contour map at the end of run1 (top panel) and run4 (bottom panel) in the case of graded sediment.

measured distribution shown in Figure 1. The erosion-deposition map for a sequence of events with return period of 30 years with the configuration M1 is shown in Figure 9. If on one hand, no significant difference are appreciable in the planimetric distribution of the river bedforms, on the other hand, the intensity of the deposit is noticeably different. Despite of the almost identical feeding of the model in terms of weight of sediment, the volume of sediment involved in each run is quite different in the case of graded sediment, due to the smaller porosity of the mixture compared to the uniform sediment. In fact the volume of a sediment mixture might decrease of one quarter for the same weight of uniform sediment. This explain the minor intensity of the measured deposit and scour. Regarding the sorting process that the river reach could undergo, two different analysis were performed on the measured data. A photographic survey has been carried at the end of each run, which together the erosion-deposition map can help us in the understanding of possible surface sorting. However, due to the highly three dimensional character of both longitudinal and vertical sorting, a piece of information could be obtained from the sieving analysis of the bed samples collected at different locations along the model. Starting from the surface distribution of the particle size, obtained by analyzing the recorded images, we have observed the following features: (i) the coarser sediments tend to accrete on the bar crest, while the finer tends to cover the bar trough; (ii) a longitudinal sorting acts in such a way to select the medium and the finer sediment, ultimately, resulting in the disappearance of the coarser particles close to the dam. This observations agree with the experimental measurements reported by Lanzoni (2000), with the difference that in the cited experiments a mixture of only two particles diameters was employed. Moreover, the sieving analysis of the subarmor despite a certain scatter in the vertical and longitudinal distribution of sediment particles, seems to confirm the observations reported above on the armor grain size distribution. The analysis of the sample has been carefully carried out trying to

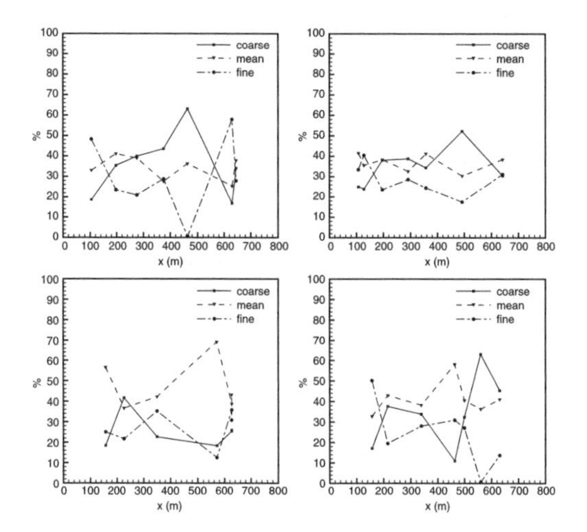

Figure 10. Configuration M1, hydrograph $T_R = 30$ years: grain size distribution of the subarmor for runs 1–4 (from top to bottom).

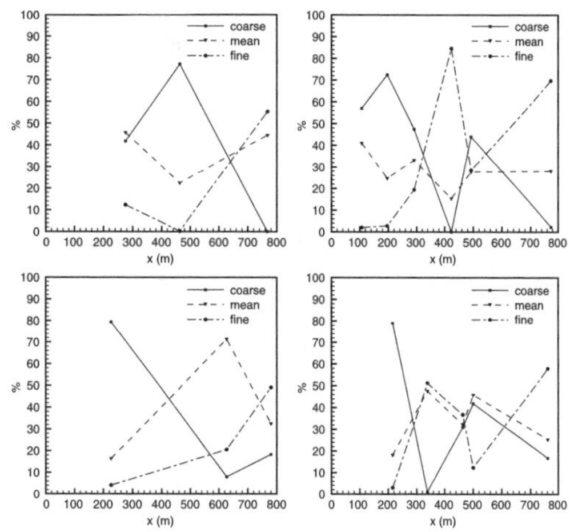

Figure 11. Configuration M1, hydrograph $T_R = 30$ years: grain size distribution in the deposition layer for runs 1–4 (from top to bottom).

isolate the deposition layer from the lower layer, which is supposed to be mainly the bulk mixture. The sediment divided between deposition layer and subarmor bulk layer has been separately analysed. In Figure 10 and 11 the longitudinal grain size distributions for the subarmor and deposition layer, respectively, are shown for a sequence of events with $T_R = 30$ years, in the configuration M1. Two main conclusions can be drawn by the inspection of Figures 10 and 11: (i) the size distribution of the subarmor in some cases is quite far from the initial composition of the bulk mixture; this suggests that during a single event the bed undergoes either erosion or deposition processes that lead to a slightly different sediment composition at the end of the flood event; (ii) on the contrary, the particle size distribution of the deposition layer shows a clear longitudinal sorting which tends to select the mean and the finer particles moving downstream.

As a final comment, it is worthy to note how the efficiency of the flood water storage volume under investigation, defined as the lowering of the flood peak, has been changed noticeably when the bed was allowed to evolve. In fact, we have observed efficiency up to 20%, which means a considerable decreasing in the flood peak, when we had performed experiments with fixed bed. However, the influence of the formation of important deposits cause the efficiency to decrease down by half. This suggests that flood defense systems should be designed considering the possible role of the sediment transport.

5 CONCLUSIONS

In the present contribution we have reported preliminary experimental observation of the influence of the presence of a flood defense system (flood water storage volume) on the sediment transport dynamics of a natural river. Different hydrological scenarios with different designs of the flood control system have been simulated employing a physical model, which reproduces a reach of about 1 Km of the river Vara (North-western Italy), allowing the river bed to evolve. The latter has been prepared both with uniform sediment and graded sediment typical of the river Vara in the reach under study. The measured bed evolution together with volume measures of the transported sediment and the longitudinal and vertical grain size distributions, allowed a physical insight of the sediment transport phenomena and how they are influenced by an artificial structure as the one analysed. The main results can be summarized as follows: (i) the presence of a transverse dam impose an abrupt change of the flow resulting in an important increasing of the free surface elevation, especially close to the dam; the latter, is responsible for a decreased transport capacity of the liquid current, ultimately, causing wide area of deposition in both the configuration of the structures investigated; (ii) there is a strong tendency of the river reach to decrease its mean bed slope, showing, at least in the configuration M1, a trend toward a condition of equilibrium, helped by the presence of a lateral embankment; (iii) macroscale bedforms form during the flood events, whose features resemble the formations of central bars; (iv) longitudinal and vertical sorting process acts such that the coarser sediment tend to disappear in the armor layer moving downstream.

It is worthy to note, as a general comment, that the presence of a transverse dam interrupts almost completely the sediment continuity of a natural river, possibly causing sediment impoverishment downstream.

The presence of a longitudinal embankment may help in restoring, at least partially, the natural sediment transport.

REFERENCES

BILLI, P. 1992 Variazione areale delle granulometrie e dinamica degli alvei ghiaiosi: metodologie di campionamento ed analisi dei primi risultati.*Proceedings of Convegno Fenomeni di Erosione e Alluvionamenti degli Alvei Fluviali*, Ancona 14–15 October 1991, 91–106.

BUNTE, K. & ABT, S.R. 2001 Sampling surface and subsurface particle-size distributions in wadable gravel- and cobble-bed streams for analyses in sediment transport, hydraulics, and streambed monitoring. U.S. Department of Agriculture, Forest Service, Rocky Mountain Research Station, General Technical Report RMRS-GTR-74, 428.

FEDERICI B., SIVIGLIA A., COLOMBINI M., TELESCA P. & SEMINARA G. 2004 Morphodynamics and flood control. A case study: the Cittadella bridge in Alessandria. *Proceedings of XXIX Convegno nazionale di Idraulica e Costruzioni idrauliche*, Trento, Italy, 7–10 September 2004, 549–554.

LANZONI, S. 2000 Experiments on bar formation in a straight flume. *Water Resources Res.*, **36(11)**, 3351–3363.

LEOPOLD, L.B. 1970 An improved method for size distribution of stream bed gravel. *Water Resources Res.*, **6(5)**, 1357–1366.

PARKER, G. 1990 Surface-based bedload transport relation for gravel rivers. *J. Hydr. Res.*, **20(4)**, 417–436.

RIGHETTI, M., ARMANINI, A. & RAMPANELLI, L. 1992 Effetto della scabrezza di parete in alvei vegetati. *Proceedings of XXIX Convegno nazionale di Idraulica e Costruzioni idrauliche*, Trento, Italy, 7–10 September 2004, 549–554.

RINALDI, M., SIMONCINI, C. & SOGNI D. 2005 Variazioni morfologiche recenti di due alvei ghiaiosi appenninici: il F.Trebbia ed il F.Vara. *Geogr.Fis.Dinam.Quat.*, **Suppl.VII**, 313–319.

SIVIGLIA, A., FEDERICI, B., BECCHI, I. & RINALDI, M. 2004 Sediment transport and morphodynamics of the Tanaro River, northwestern Italy. *Proceedings of the Symposium on Sediment Transfer through the Fluvial System*, Moscow 2004, 308–315.

WOLMAN, M.G. 1954 A method of sampling coarse riverbed material. *Eos Trans. AGU*, **35**, 951–956.

Coastal morphodynamics

River, Coastal and Estuarine Morphodynamics: RCEM 2005 – Parker & García (eds)

Influence of cross-shore sediment movement on long-term shoreline change simulation

H. Kang & H. Tanaka
Dept. of Civil Engineering, Tohoku University, Sendai, Miyagi, Japan

ABSTRACT: In a numerical prediction using shoreline change model, the sediment transport coefficient, K, is calibrated through a comparison with measured data, although measured data contain both beach evolution caused by longshore and cross-shore sediment transports. Therefore, there is a high possibility that the value of K thus calibrated include some error due to cross-shore sediment movement. In this study, the sediment transport coefficient is examined by using measured raw data and separated shoreline representing shoreline evolution caused by longshore sediment transport.

1 INTRODUCTION

Incident waves propagate from the southeast in Sendai Coast, located in the northeast area of Japan. These generate northward sediment movement along the coast. And sediment movement that comes from south is intercepted by coastal structures on the beach such as jetties and breakwaters. Also, sediment support is rapidly reduced from river, recently. On these accounts, beach erosion has been severely occurring in Sendai Coast (Uda et al. 1990). Therefore, it is necessary to clarify long-term trend of beach evolution in this area to make a plan for beach protection. Hence, to understand the complex characteristic of topography change, survey of beach topography has been being carried out twice in a month in the study area.

In general, topography change data obtained by surveying include both effects due to longshore sediment movement and due to cross-shore sediment movement. For this reason, it is very difficult to understand topography change by measured raw data. If the complex characteristic of topography change can be separated into beach evolution due to longshore sediment transport and that due to cross-shore sediment transport, the complex characteristic of topography change can be more clarified and easily understood. For this reason, study on analysis of time series data from the survey of water depth has widely carried out using EOF (Empirical Orthogonal Function) to separate complex filed data into simply data by numerous researchers such as Winant et al. (1975), Aubrey (1979) and Aubrey et al. (1980). Tanaka and Mori (2000) subtracted shoreline change caused by longshore sediment movement and cross-shore sediment movement from measured data.

In this study, shoreline evolution caused by longshore sediment transport is extracted from complex measured raw data by EOF method. Moreover, calibration of sediment transport coefficient is carried out for three data sets to examine the effect on shoreline evolution caused by cross-shore sediment transport: (1) extracted data by EOF method representing shoreline evolution cased by longshore sediment transport, (2) measured raw data surveyed twice a month at relatively short interval and (3) measured data surveyed at the interval of one year, considering conventional surveying interval of coastal morphology.

2 STUDY AREA AND MEASURED DATA

2.1 *Study area*

Figure 1 shows the study area located on the northeast coast of Japan, spanning from the southern part of Sendai Port to Natori River with total length about

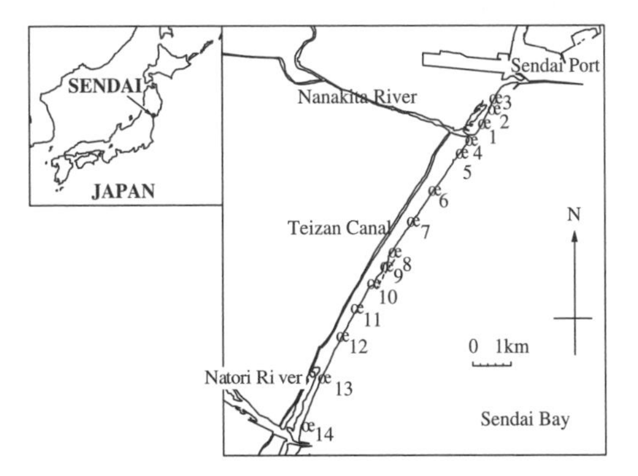

Figure 1. Study area and measured stations.

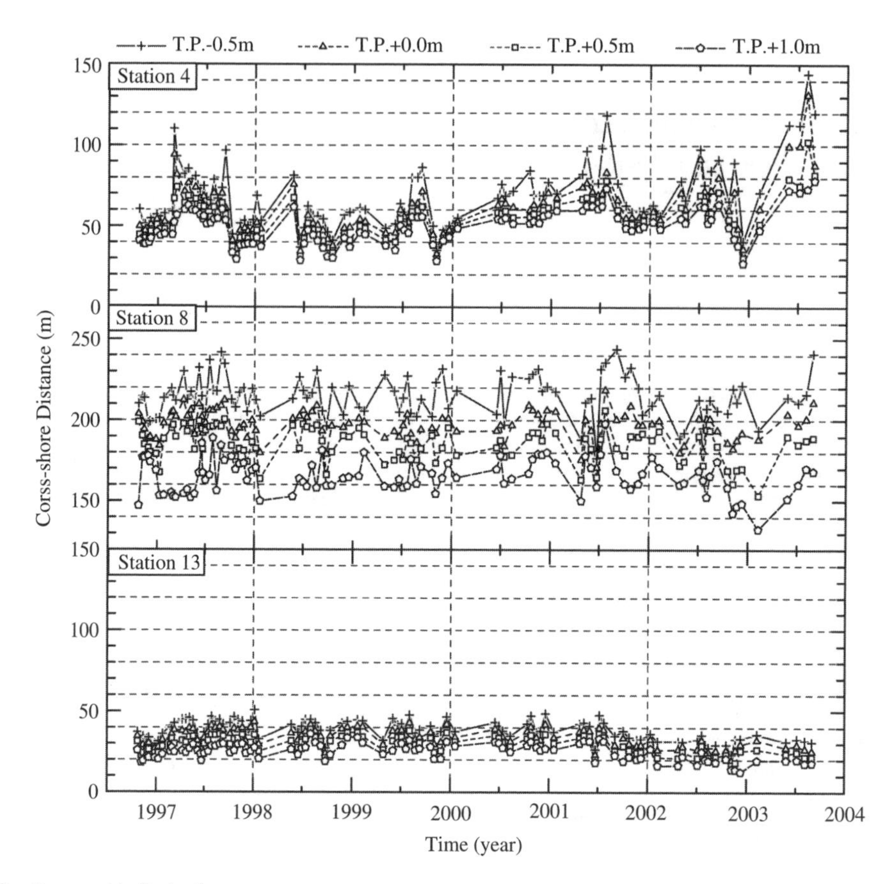

Figure 2. Topographic fluctuation.

12 km. Incident waves propagate from southeast, because Sendai Coast faces southeast and the Oshika Peninsula is located on northern part of Sendai bay, so that it obstructs waves that come in from north-east. Therefore, the sediment movement moves from south to north as predominant longshore sediment movement along the coast (Tanaka, 1995).

The breakwater at Sendai Port bound the northern end of the study area and the jetties at the Natori River mouth and the breakwater at Yuriage Port bound the southern end of the study area.

Longshore sediment transport that comes in from south is completely interrupted by breakwater of Yuriage Port where water depth is 12 m, because effective topography change due to longshore sediment transport occurs within 10 m in this area (Tanaka, et al. 1995). Therefore, Study area can be considered as an independent sediment transport area.

There are Nanakita River and offshore breakwaters about 2 km and 6 km from Sendai Port in the study area, respectively. These also have an effect on longshore sediment movement in this area. The numbers in Figure 1 denote the survey stations. Survey has been being carried out twice a month since 1996.

2.2 *Topographic fluctuation*

Figure 2 shows the fluctuation of shoreline defined in terms of T.P. −0.5 m, T.P. +0.0 m, T.P. +0.5 m, and T.P. +1.0 m at St. 4, St. 8, and St. 13, respectively. Where T.P. (Tokyo Peil) denotes the Japanese measuring system of elevation based on mean sea level in Tokyo Bay. St. 4 and St. 8 in Figure 2 show that short-term shoreline change remarkably fluctuates. It fluctuates maximum up to 70 m. Also, advances or erosion occurs in long-term shoreline change. Hence, if it is possible that shoreline evolution due to cross-shore sediment transport and due to longshore sediment transport are separated from measured raw data, beach evolution is definitely understood. Also, when the empirical sediment transport coefficient, K, in shoreline change model is calibrated based on measured raw data such as data at St. 4 and St. 8, there is high possibility that

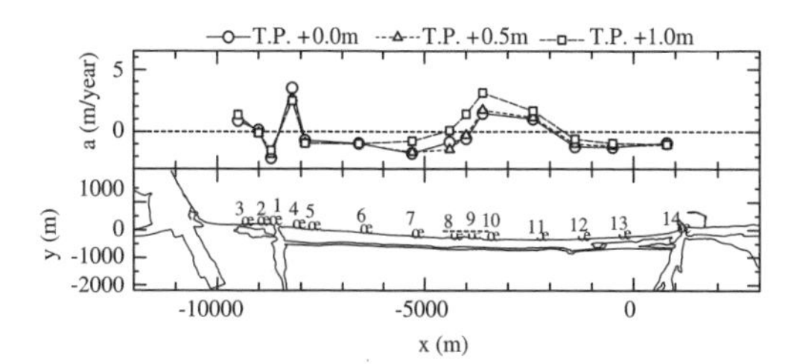

Figure 3. Rate of shoreline change.

the K value includes some error due to cross-shore sediment movement.

The shoreline at St. 4 advances seaward about 2.9 m/year, and fluctuation of shoreline is widely varied due to Nanakita River. St. 8 also shows big fluctuation due to gentle slope as shown in the middle of Figure 2. However, it is comparatively stable in comparison with the other stations in long-term change as rate of shoreline change is about −0.4 m/year, because of the influence of detached breakwaters. Fluctuation of shoreline at St. 13 is less than that of the others as shown in bottom Figure 2, because beach slope is steep. And shoreline continuously retreats with 1.0 m/year in long-term change.

2.3 *Long-term shoreline change*

Figure 3 shows that the rate of long-term shoreline change, a, is calculated using least square method from measured data at every station in order to investigate the long-term shoreline change. This result, it has positive value at St. 4, St. 9, St. 10, and St. 11, so that deposition occurs on these stations. On the contrary, it has negative value at the other stations, so that erosion occurs on the other stations as shown in Figure 3.

These can be expressed that shoreline evolutes due to the northward sediment movement that is predominant longshore sediment movement. Right hand side of breakwaters is advance seaward, because breakwaters interrupt the northward longshore sediment transport. The other hand retreat of shoreline occurs on the left hand side of breakwater. Also, the northward longshore sediment transport is interrupted by terrace in front of Nanakita River mouth, so that advance of shoreline occurs on the right hand of Nanakita River mouth and retreat of shoreline occurs on the left hand of it. And sediment deposition can be seen at St. 3 caused by the effect of reflection wave from breakwater at Sendai Port.

3 EMPIRICAL ORTHOGONAL FUNCTION ANALYSIS

EOF analysis has widely been used in various fields of study. Winant et al. (1975) investigated the relation between eigenfunctions and characteristics of topography change, such as bar, berm and terrace. And the usefulness of the eigenfunctional representation was confirmed as a concise method of representation beach profile changes (Aubrey, 1978). Hashimoto and Uda (1982) proposed a new model to analyze beach profile changes due to longshore and on-offshore sand transports using EOF. Up to now, EOF analysis has been applied to separate data in the direction of cross-shore (e.g., Winant et al., 1975, Hashimoto and Uda, 1982). In this study, EOF analysis is applied to separate shoreline position data in the direction of along the coast.

EOF analysis can be applied to analyze shoreline evolution data to determine their fluctuation characteristics through spatial and temporal eigenfunctions. The shoreline position can be expressed in terms of superposition of eigenfunction as follows:

$$y(x,t) = \sum_{n=1}^{n_x} C_n(t) e_n(x) \tag{1}$$

where $y(x,t) = y_s(x,t) - \bar{y}(x)$, $y_s(x,t)$: the distance from origin point to shoreline, $\bar{y}(x)$: the mean shoreline position), n_x denotes the number of survey station, and $C_n(t)$ and $e_n(x)$ are the temporal and spatial eigenfunctions, respectively. To generate these functions, a correlation matrix A is obtained as follows:

$$A = [a_{ij}] = \frac{1}{n_x n_t} \sum_{t=1}^{n_t} y_{it} y_{jt} \tag{2}$$

where n_t denotes the number of surveyed times. Matrix A possesses a set of eigenvalues, λ_n, and a set of corresponding eigenfuncton which are defined by the matrix equation.

$$Ae_n = \lambda_n e_n, \quad C_n(t) = \sum_{n=1}^{n_t} y(x,t) e_n(x) \tag{3}$$

Contribution, R_n, determines the importance of the data and is defined as follows:

$$R_n = \frac{\lambda_n}{\sum_{i=1}^{n_x} \lambda_i} \tag{4}$$

3.1 *The rate of change of second EOF component*

In this study, the second EOF component that represents shoreline evolution cased by longshore sediment transport is focused on (Kang et al. 2004). The contributions of second EOF components are shown in Table 1. The second temporal eigenfunction, $C_2(t)$, is shown in Figure 4. The rate of second temporal eigenfunction is calculated using least square method

Table 1. Contribution of second EOF component.

T.P. (m)	−0.5	+0.0	+0.5	+1.0
Contribution (%)	15.2	17.8	14.9	16.2

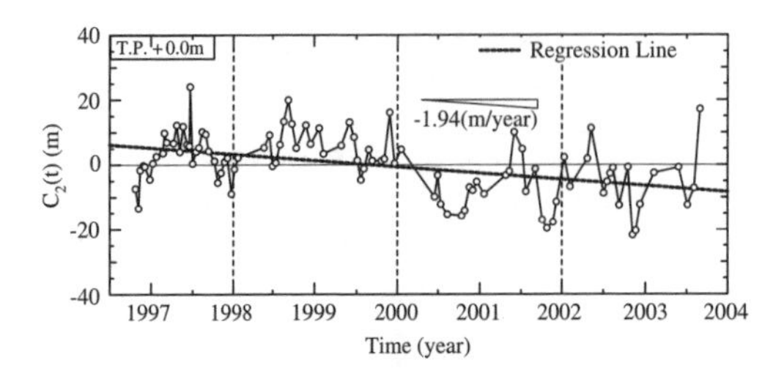

Figure 4. Second temporal eigenfunction.

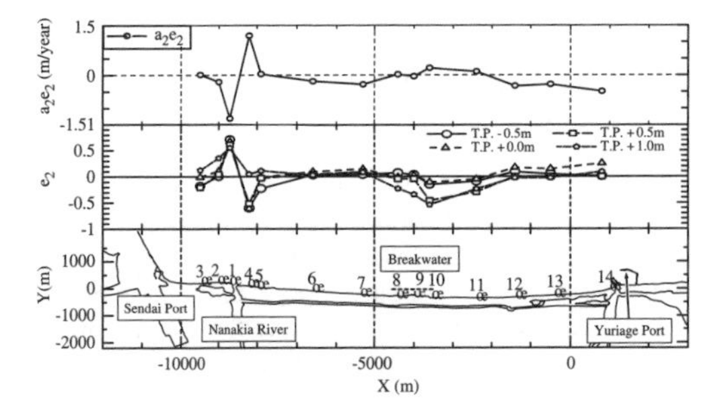

Figure 5. Second spatial eigenfunction and the rate of second EOF component.

as $C_2(t) = a_2t + b$. And then, the rate of second EOF component in long-term change is obtained as follows:

$$\frac{d\{C_2(t)e_2(x)\}}{dt} = a_2e_2(x) \tag{5}$$

where a_2 is the rate of second temporal eigenfunction in long-term change. It is obtained as −1.9 m/year. And $e_2(x)$ denotes second spatial eigenfunction as shown in the middle of Figure 5. The rate of second EOF component in long-term change, $a_2e_2(x)$, is obtained by both the second spatial eigenfunction, $e_2(x)$, and the rate of change of the second temporal eigenfunction, a_2.

The rate of second EOF component is shown in upper Figure 5. It corresponds to shoreline evolution caused by effect of coastal structures and river mouth morphology considering the northward sediment movement that is predominant longshore sediment transport in this area.

Moreover, the rate change of second EOF eigenfunction, $a_2e_2(x)$, shows the similar shape of the rate of long-term shoreline change, a, in Figure 3. Therefore, second EOF component can express long-term shoreline evolution.

4 SHORELINE CHANGE MODEL

Shoreline change model is that beach evolution is represented by the change of the shoreline, which is estimated from the longshore sediment transport rate. And the shoreline change model is a numerical prediction model based on numerical solution of the sediment or sand continuity equation and the sediment transport equation. The governing equation for the shoreline position y_s is given by (Horikawa, 1988):

$$\frac{\partial y_s}{\partial t} + \frac{1}{D}\left(\frac{\partial Q}{\partial x} \mp q\right) = 0 \tag{6}$$

where x is the longshore coordinate, D denotes depth of closure (m), t denotes time, Q denotes the volume rate of longshore sediment transport (m^3/s) and q denotes cross-shore sediment transport rate. The volume rate of longshore sediment transport used the most widely used predictive expression that relates the incident breaking wave energy flux to the volume rate of longshore sediment transport in the following (CERC, 1984):

$$Q = K(Ec_g)_b \sin\alpha_{bs} \cos\alpha_{bs} \tag{7}$$

where, K denotes the empirical coefficient of sediment transport as treated calibration, $(Ec_g)_b$ is the incident wave engergy flux evaluated at the breaker line and α_{bs} is the breaking wave angle to the shoreline.

4.1 *Calibration empirical coefficient of sediment transport*

Empirical coefficient of sediment transport, K, is calibrated using measured raw data that include shoreline evolution caused by longshore sediment transport and cross-shore sediment transport, generally. However, shoreline evolution remarkably fluctuates up to maximum 70 m, as shown in Figure 2. It seems that shoreline evolution due to cross-shore sediment transport have influence on the calibration of sediment transport coefficient based on measured raw data. Therefore, calibration of K is carried out for three data sets in order to examine influence of cross-shore sediment movement on calibration of K: Data (1) are extracted data by EOF method representing shoreline evolution caused by longshore sediment transport (— in Figure 6), data (2) are measured raw data surveyed twice a month at relatively short interval (+ in Figure 6) and data (3) are measured data surveyed at the interval of one year, considering conventional surveying interval of coastal morphology (• in Figure 6).

Wave conditions are given average value during a measured period and initial shoreline is obtained from aerial photo in October 1996. And wave transformation is calculated by the wave ray method. Under these

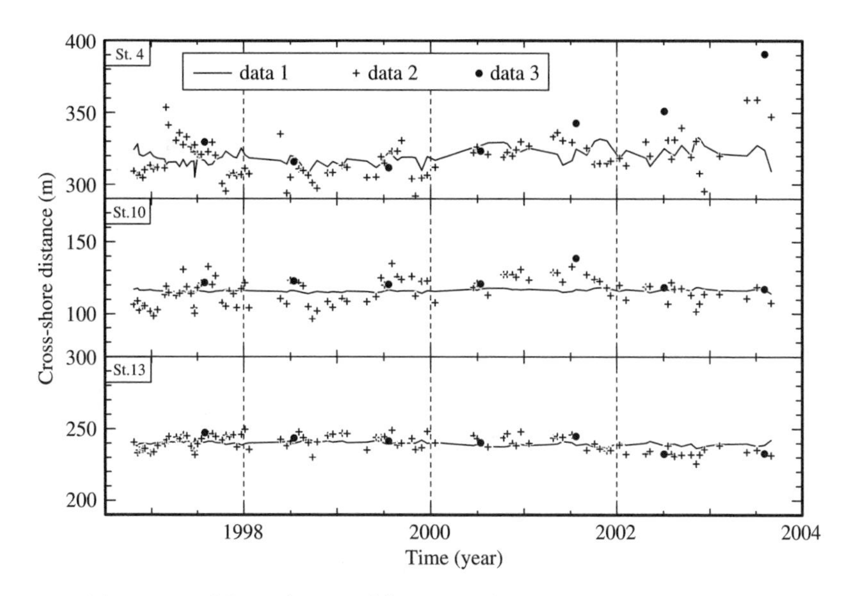

Figure 6. Extracted data, measured data and measured data surveyed once a year.

conditions, shoreline positions are calculated after 6–7 year from October 1996. Shoreline positions are calculated with K varied from 0.01 to 0.09 for calibration of K. And to find the optimum value of K, error is calculated between calculated shoreline position and three data sets.

Case 1: an error, E_{case1}, between each calculated shoreline change and data (1) is calculated as following:

$$E_{case\,1} = \sqrt{\frac{\sum_{i=1}^{N_t}\sum_{j=1}^{N_x}(y_{s(cal)(i,j)} - y_{s(sep)(i,j)})^2}{N_t N_x}} \tag{8}$$

where $y_{s(cal)}$ is calculated shoreline position, $y_{s(sep)}$ is shoreline position based on extracted data by EOF method representing shoreline evolution caused by longshore sediment transport, data (1), and N_t is the number of survey times during about 7 years from 1996 to 2003, and N_x is total station number.

Case 2: an error, E_{case2}, is calculated between each calculated shoreline and data (2).

$$E_{case\,2} = \sqrt{\frac{\sum_{i=1}^{N_t}\sum_{j=1}^{N_x}(y_{s(cal)(i,j)} - y_{s(data\,2)(i,j)})^2}{N_t N_x}} \tag{9}$$

where $y_{s(data2)}$ is measured raw data surveyed twice a month at relatively short interval data.

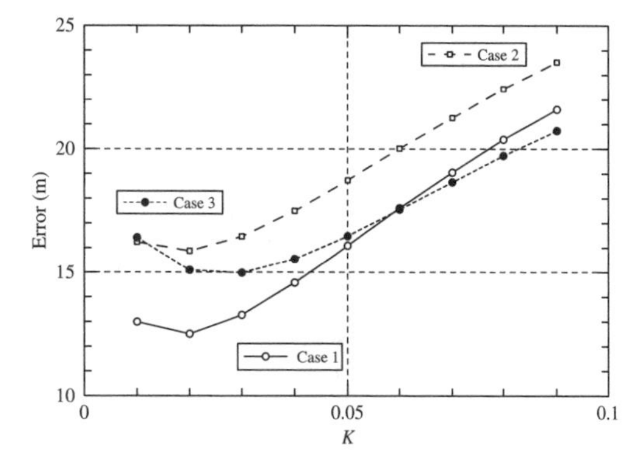

Figure 7. Relationship between error and sediment coefficient.

Case 3: an error, E_{case3}, is calculated between each calculated shoreline and data (3).

$$E_{case\,3} = \sqrt{\frac{\sum_{i=1}^{N_t}\sum_{j=1}^{N_x}(y_{s(cal)(i,j)} - y_{s(data\,3)(i,j)})^2}{N_t N_x}} \tag{10}$$

where $y_{s(data3)}$ is measured data surveyed at the interval of one year, considering conventional surveying interval of coastal morphology.

Figure 7 shows relationship between the empirical sediment transport coefficient and the error in each case. Both case 1 and case 2 show same optimum value of $K(=0.02)$. However, the error is bigger in case 2 than that in case 1 on whole area. Data (2) in case 2 include beach evolution cased by cross-shore sediment transport, so that the error shows bigger than that in case 1. However, data (2) is measured twice a month at relatively short interval, so that the optimum value of K is same as that in case 1.

Optimum value of K in Case 3 is 0.03. It is bigger than that in case 1 and case 2 because data (3) include beach evolution caused by cross-shore sediment transport and is measured once a year, considering conventional surveying interval of coastal morphology.

According to these results, when empirical coefficient of sediment transport, K, is calibrated in shoreline change model based on measured raw data including shoreline evolution caused by both cross-shore sediment transport and longshore sediment transport, empirical coefficient of sediment transport include some error due to cross-shore sediment movement.

5 CONCLUSIONS

In this study, when sediment transport coefficient, K, is calibrated in shoreline change model, the effect of shoreline evolution due to cross-shore sediment transport is examined using three data sets. Results are summarized as follows:

(1) When calibration of K is carried out using extracted data by EOF method, case 1, the error is smaller in whole area than that of the other cases. Because separated data is shoreline evolution caused by longshore sediment transport.

(2) When sediment transport coefficient, K, is calibrated using measured raw data surveyed twice a month at relatively short interval data, case 2, the optimum value of K is same value as that obtained by case 1, because survey is carried out in a relatively short interval. However, the error is bigger than that based on extracted data by EOF method because data (2) include influence of shoreline change due to cross-shore sediment transport.

(3) According to case 3, calibration of K is carried out using measured raw data surveyed once a year, considering conventional surveying interval of coastal morphology, the optimum value of K is bigger than that in case 1 and case 2, because it includes some error caused by shoreline change due to cross-shore change.

It is concluded based on these results that shoreline evolution due to cross-shore sediment transport has effect on calibration of K value. Therefore, it is important that raw survey data are separated into a part of data caused by longshore sediment transport and cross-shore sediment transport, when K value is calibrated in shoreline change model.

REFERENCES

Aubrey, D.G. 1979. Seasonal patterns of onshore/offshore sediment movement. *J. Geophysical Research*, Vol. 84, No. C10: 6347–6354.

Aubrey, D.G., Inman, D.L. & Winant, C.D. 1980. The statistical prediction of beach changes in Southern California. *J. Geophysical Research*, Vol. 85, No. C6: 6264–6276.

Coastal Engineering Research Center 1977. *Shore Protection Manual*. U.S. Army Corps of Eng., U. S. Govt. Printing Office, 3 Vols.

Hashimoto, H. & Uda, T. 1982. Description of beach changes using an empirical predictive model of beach profile changes. *Proc. 18th costal Eng. Conf., ASCE*: 1405–1418

Horikawa, K. 1988. *Nearshore Dynamics and Coastal Processes*. Tokyo Univ. Press: 522.

Kang, H., Tanaka, H. & Sakaue, T. 2004. Investigation of shoreline change and sediment budget on Sendai Coast. *Annual Journal of Coastal Engineering, JSCE*, Vol. 51, 536–540 (In Japanese).

Tanaka, H. & Shoto, N. 1991. Field measurement of the complete closure at the Nanakita River mouth in Japan. *International Symposium on Natural Disaster Reduction and Civil Engineering*: 67–76.

Tanaka, H. & Mori, T. 2001. Separation of shoreline change caused by cross-shore and longshore sediment transports. *Proc. Coastal Dynamics*: 192–201.

Tohoku Regional Bureau, Ministry of Land, Infrastructure and Transport & Miyagi Prefecture Civil Engineering Department 2000. *Technical Report for Sendai Coast*: 59–69 (In Japanese).

Uda, T., Omata, A. & Minematsu, M. 1990. The crisis of sandy beaches along Sendai Bay area. *proceedings of coastal engineering, JSCE*, Vol. 37, 479–483 (In Japanese).

Winant, D.C., Inman, D.L. & Nordstrom, C.E. 1975. Description of seasonal beach changes using empirical eigenfunction. *J. Geophysical Research*, Vol. 80, No. 15: 1979–1986.

River, Coastal and Estuarine Morphodynamics: RCEM 2005 – Parker & García (eds)
© 2006 Taylor & Francis Group, London, ISBN 0 415 39270 5

The presence of a coastal harbor structure in the village of Pecem – Northeast of Brazil – and its influence in the coastal line changes of that region

Chagas, Patrícia & Souza, Raimundo
Department of Hydraulics and Environmental Engineering – Center of Technology – UFC – Campus do Pici, Fortaleza, Ceará, Brazil

ABSTRACT: The construction of a harbor in coastal areas, when compromises the continuity of the transport of sediments, constitute a complex hydraulic engineering problem. This research evaluated, through mathematical models and through environmental monitoring, the possible changes in the dynamic balance of the sandy beaches of Pecém – Ceará, in function of the presence of a port structure in that area. The results showed that the effect of shelter of the port benefits, relatively to the sediment transport, to the Village of Pecém. In that coastal region, already in intense erosive process, the presence of port helps the deposition of sediments in the place. The effect of that construction can also be observed more to West of the Village, where the area of protection of the breakwater off-shore tends to disappear, allowing the gradual increasing of the coastal sediment transport rates between Pecém and Taíba.

1 INTRODUCTION

One of the basic conditions for the existence of a port, in a coastal area, is that it must be at a sheltered place. When bays or great estuaries do not exist and it is needed a port, it enters the engineering, creating the shelter conditions required, as in the case of the State of Ceará. However, the construction of port in backs of sand constitutes one of the more compounds hydraulic engineering problems, fact due to the multiplicity of present variables in the process of transport of sediments.

The presence of a great port terminal brings changes in the dynamics wave energy, generating, in such way, a possible place of accumulation of sand. As the balance of a coast is made, by the passage of sediments along the beach, in the case of occurrence of the imprisonment of these materials, in a special point, one has, consequently, a silting up in the windward direction of the structure, and erosions at the leeward of it.

In those conditions, Porto of Pecém comes as a construction capable to promote alterations in that coastal area. In such away, if preventive actions are not taken, in that area, with an intense coastal transport, it can suffer damages, of the environmental point of view. Thus, to minimize the impacts on the coastal morphology, it was chosen, then, for building a breakwater around 16 meters depth and located 2 km of the coast line, interlinked to the beach by a bridge (Pitombeira, 1995).

The present study evaluates the possible changes in the dynamic balance of the sandy beaches of Pecém and, starting from mathematical models, it analyses the annual taxes of transport of sediments along the coast of Pecém, in function of the presence of a port structure in that area.

That environmental monitoring was realized using sophisticated tools of high precision, such as directional wave rider and the electromagnetic current meters S4, besides the numeric models, MIKE 21 and LITPACK developed by Danish Hydraulic Institute – DHI, of Denmark. It was also made an attendance through dynamic beach profiles, for two years, of the evolution of the coastline in Pecém.

The results found allowed to conclude that the effect of the shelter of the port will be beneficial to the Town of Pecém, because it will allow the formation of a point of sand, connected the earth, through the sediment deposition in that area. The studies showed, yet, that there is necessity of an attendance of long period, so that, it can avoid possible damages to the environment and to the society.

2 COASTAL DYNAMIC

The resulting from the performance of the waves, of the tides, of currents and of the winds on the coastal area is called of coastal dynamics. In the whole Northeastern Coast of Brazil and, in matter in the coast of

Ceará, the performance of the waves and winds on these beaches, in natural conditions, she accomplish promoting formation of beach profiles and movement of sediments, longitudinally, in the coastal area. The strip line of the beach is the area where element of the dynamic coastal acts in a more intense way. The beaches are deposits of sediments, usually sandy, and they play an important part in the protection of the coast against the sea erosion.

When the man intervenes, somehow, in that coastal dynamics, a process of environmental transformation begins, where the active factors suffer mutations, trying to reestablish the balance, that, in spite of its dynamics, it presents limited and vulnerable conditions.

2.1 *Waves*

One of the main processes present in the coastal dynamics is the incidence of the wave movement regimes on the lines of sandy backs. That complex phenomenon, with very defined energy, plays important part in the coastal transport of sediments, considering that it is very sensitive to the height, the period and the direction of waves. As the wave moves, they can suffer deformations due to decrease of depth of the water, in the area of propagation of the wave, or due to obstacles in their course.

The influence of promontories, points, islands, among other natural accidents, interferes, significantly, in the regime of active waves. Among the artificial obstacles, the construction of disconnected breakwaters from the beach, with great depths (structures off-shore), could be mentioned that, usually, provokes changes in the hydraulic conditions of the area, among the port structure and the beach.

2.2 *Tides*

The oscillations in the height of the level of water in the sea constitutes an important mechanisms of modification of the coastline. Wave of tides is defined as being the crest and depression of a wave whose length is of hundreds of kilometers. The crest of the wave is called commonly of high tide, while the depression of the wave is called of low tide.

The waves of tides are generated by the originating from action of the gravitational force of the system earth-moon and of the system earth-sun. Due to proximity of the moon in relation to earth, the influence it, on the earth, is larger than the one of the sun, in spite of his size.

2.3 *Currents*

Three types of currents exist: the currents of tides that only influence the transport of sediments close to the sea arms; the currents generated by the waves, main responsible for the transport of sediments; and the oceanic currents that rarely affect the transport of sediments.

2.4 *Winds*

The winds represent important physical parameter that composes fundamental part of the coastal dynamics. Their presence are decisive in the formation and in the intensity certain marine waves, in the composition of the spectra of energy that you/they compose the turbulent movement, as well as in the feeding and balance processes of the beaches. Along the coast line, the wind plays a fundamental part, mainly, in the sense of, through a drag process, to act, in a decisive way, in the free particles, transporting this material for other coastal areas, maintaining, thus, a continuous balance process in those areas.

3 AREA OF STUDY

Porto of Pecém was built in the Tip of Pecém, in the Municipal district of São Gonçalo of Amarante, in the coast of Ceará, at the distance of approximately 50 km to west of Fortaleza (measured by the coast line), with coordinates 3° 33′ S and 38° 41′ W.

The Beach of Pecém, has a coast line with two main alignments:

- In the previous passage (east side) the tip of Pecém has an approximate direction of 326° with the true north;
- In the subsequent passage (west side) the tip of Pecém has an approximate direction of 270° with the true north.

3.1 *Port facilities*

Porto of Pecém is of the type "off-shore", in other words, out away from the beach, in way to minimize the damages on the line of the coast. In that context, the coastal current, formed between the breaking waves area and the coast line, it will continue to pass underneath of the bridge, feeding the beaches of the coast west. The port facilities are located in the tip of Pecém at a distance of 2000 m, in the northeast direction, and its general arrangement consists of an access bridge, 2 (two) mooring piers and a "L" formatted breakwater.

3.2 *Temporary harbor*

A small terminal of Temporary Embankment (TEP) was built with length of approximately 400 m to This of the Tip of Pecém, taking advantage a rocky promontory present over there. That harbor has the protection of a breakwater, which extends until 200 m of the line of current coast.

Some simulations were made to verify the viability of the harbor, and it was verified that it blocked the coastal transport of the Tip of Pecém, being, soon, necessary to be removed, after the construction of the port facilities.

4 METHODOLOGY

The field works in the area of Pecém were accomplished, through measurements of waves, winds, currents, tides and risings of beach profiles, to characterize the hydrodynamic processes and to monitor the evolution of the coast line in the coastal area of Pecém.

In 1995 occurred the beginning of the development of a campaign of measurements of currents and the following station collectors of environmental parameters were implanted:

- Anemograph Station;
- Tide Gage Station.

The objective of that environmental monitoring went to make available the data of the hydrodynamic, morphologic and sediment conditions of the area of interest, to do a characterization of the atmosphere, and to define the mark of the modeling study to be executed.

The studies, by the mathematical models, to evaluate the coastal impact, caused by the construction of Porto of Pecém, were done using a computational system of last generation. Those software, LITPACK and MIKE 21, were developed by Danish Hydraulic Institute, of Denmark.

The software LITPACK is an integrated system of modeling of coastal processes and dynamics of the coastline, capable to manage interventions in the coast line such as:

- Impact assessment caused by coastal construction;
- Optimization of projects of creation of beaches;
- Optimization of works of coastal protection;
- Project and optimization of recovery of beaches for artificial feeding.

The module MIKE 21 is a professional package of software in hydraulic engineering, applied in the coastal hydraulics and oceanography, environmental hydraulics, processes sediment and waves. Of this program the computational modules, PMS (Refro-diffraction) and HD (Hydrodynamic), were used.

5 ENVIRONMENTAL MONITORING

Environmental Monitoring: For this study the field data measured, from 1995 to 1999, were used.

5.1 *Wind*

In this period (4 years) of continuous measurements of winds in the place of Pecém 31.599 data were registered schedules, with efficiency of 88%. Those data allow characterizing the winds dominant in the place. It was verified that in the project area the main direction of the wind is E-ESE. The performance of the dominant winds of short period is responsible for the generation of waves (sea) close to the coast. As the alignment of the coast obeys the line are SE-NO, the winds have outstanding performance in the sense earth-sea and longitudinally in the body of the beach.

The original winds of the quadrants SE-S, they are responsible for the transport of sediments along the beaches and they also contribute in the release of the sediments of the continent to the sea, feeding like this the coastal transport.

5.2 *Tides*

It was observed that, in 99% of the time, the levels of water were above the zero, corresponding to the own level of reduction of DHN. It was also verified that the main effect of the tides is the change of the pattern of transport of sediments, which are established in agreement with the change in the level of water. During the low tide the breaking waves happens far away from the backs, in the located in front of the city of Pecém.

5.3 *Currents*

It was observed after measurements of speeds and directions of the currents to the square of Pecém, that the tide was not so significant in the currents of the area, what means that the changes of tides (inundation or ebb tide) did not cause expressive variations in the behavior of the currents. The currents of tides are very low, of the order of 0,1 m/s, and it does not have significant effect on the conditions of coastal transport of sediments.

5.4 *Waves*

The present of two types of waves was verified happening on the coast of the place. In the months of March to October the arrival of short waves prevails (sea), with periods in the strip of 2–8 seconds and directions of waves of 80–120°N, while in the months of November to February it is noticed the arrival of waves swell, with periods in the strip of 10–20 seconds and directions of waves 20–45°N, those last loaded of energy capable of erode beaches and to provoke damages in ships and wharf.

That season condition of waves, during the year, echoes directly in the coastal morphological dynamics of the area of Pecém. During the prevalence of sea waves, that happen oblique the coast line, occurs the process of accumulation of sediments of the side West of the Tip of Pecém. While the swell waves, arrivals usually of Northern Hemisphere and happening frontally to the city of Pecém, it favors the retreat of that material. However, if the duration of swell waves goes larger than the amount of available material will begin an intense erosive process in the place.

5.5 *Transport of sediments*

As it is known, the balance of the beaches is dynamic. If we break the sequence, of the transport of sediments, there will be an answer of the nature the such aggression.

The city of Pecém has already been suffering, for the last fifteen years, an unbalance process among the amount of material transported in the beach and the amount of material deposited in the dunes. That erosive process was resulted of a growth quite disordered, where several houses and divisions into lots were installed on movable dunes in the section West of the area. That inadequate occupation harmed the wind transport of sediments, and it consumed a strip of more than 150 meters of beach, moving forward on houses, restaurants, trades and destroying the streets.

The coastal transport along the beaches around Pecém is a complex process due to the combination of great angles of incidence of the waves and of the irregular bathymetry. The presence of swell waves type, with incidence angles very different from the sea waves type, brings more complex scenery.

The main aspect of the field of transport of sediments along the place of Pecém is the supply of sand of the part southeast through the Tip of Pecém. Due to the orientation of the coast in this area, the beaches are highly exposed to the sea waves type coming of more oriental directions.

Those waves getting promontory, lose speed and, consequently, lose the transport capacity. Then, the coastal currents, generated by the breaking wave and adjusted by the wind, come off of the coast line and they, only, return, approximately, 1 km at the west direction.

The orientation of the coastline in the area of Pecém is of approximately 355° N in front of city and 55° N along the coastal strip Tip of Pecém. The angles of approach of the swell waves type vary between 20 and 45° N. It is due to that configuration of the coastline that the coastal currents, generated by the hydrodynamic forces of breaking waves, transport the sediments to far away from the Tip of Pecém, in other words, East on the oriental side and West on the western side.

The performance of the waves type swell feels mainly in the transport of the sediments deposited by the waves type sea. During the periods of prevalence of the waves type sea he/she grows a great point of sand on the western side of the Tip of Pecém and the formation of a maceió, that it is eroded for the swell waves type.

When the period of performance of the waves type swell is too much long, it happens the combined performance of sea and swell, hindering the transfer of sediments for the promontory, and favoring the erosion, in front of beach of Pecém. Being like this, the importance of the waves type swell is verified for the coastal balance of sediments.

For the case of the coastal area of Pecém, the conditions of winds are stable. This implicates in an accentuated sediment transport along the beach. In relation to the direction of the winds it was observed that the winds prevail E-ESE. The transport on the East side of the Tip of Pecém, where the orientation of the coastline is of the order of 60° N, is done in the direction of the earth. Of the side West, where the beach is almost parallel to the direction of the wind, that transport happens along the coastline.

The presence of TEP naturally represents a blockade to the passage of sediments through the Tip of Pecém, and as the transport of sediments in the place is quite significant, a deposition of sediments the wind direction side of the construct is verified. When this accumulation arrives to great landings it will be processed the outline of the tip of the breakwater and the entrance of the sediments in the interior basin of TEP.

6 RESULTS AND DISCUSSIONS

The results of the studies, through the mathematical models, served as base to evaluate the annual taxes of transport of sediments along the coast of Pecém, besides evaluating the impacts in the coastal morphology, coming of the implantation of Porto of Pecém.

Already the risings of beach profiles were made to monitor the evolution of the coastline in the adjacencies of the Port Terminal of Pecém.

6.1 *Studies through the mathematical models*

The module MIKE 21 PMS was used to simulate the propagation of waves in coastal areas and to determine the modifications that the field of waves suffers when they find some obstacle type, such as ridges, breakwaters, wet, etc.

The annual climate of waves representative of Pecém was used as given of entrance for the calculation of the transport of sediments, being taken into account the two orientations of the beach.

It was verified after the simulations with MIKE 21 PMS that the effect of protection of the port can be seen clearly, in the proximities of the breakwater, for that situation the height of the wave is strongly reduced. That shelter area collects the area between the Tip of

Pecém and the city of Pecém. The presence of the port doesn't affect the propagation of swell waves a lot, the area of maximal protection of the port for that wave type is to 1,5 km to the Tip of Pecém.

For the study of the compounds drainage fields in the neighborhoods of the port and around the Tip of Pecém MIKE was used 21 HD, 2D hydrodynamic modeling system, for simulation of levels of water and drainages. The model solves the equations no permanent of drainage, integrated along the depth. Each simulation was accomplished being admitted the level of constant water along the time.

After the simulations, with MIKE 21 HD, it was observed that the breakwater off-shore reduces the speeds strongly of the coastal current to West of the Tip of Pecém. In that point the currents come off of the coastline, returning the join again at a distance of approximately 1 km to west of the same.

LITPACK STP was used to calculate the annual taxes of transport of sediments due to combined action of the waves and of the currents. This model is linear, what means that the coastal transport of sediments is calculated along isolated beach profiles (1D), being used for such the coastal climates of waves determined by MIKE 21 PMS.

The results showed that the presence of waves sea and swell contribute to the balance of the coastal equilibrium and that the directions of sea incidence, more oriental, contribute to the accumulation of sediments in front of the city of Pecém. The swell waves counterattack these depositions, favoring the erosion of the point of sand formed. However, if the swell prevalence lasts for a time sufficiently long, the coastline, in front of city of Pecém, erodes, because the coastal drift from the occident exceeds the sediments around the Tip of Pecém.

A transport of sediment rate, of the order of 350.000 m^3/ano, is transported in the sense are SE-NO, along the Tip of Pecém. The effect of shelter of the port will do with part of that sediment (in the order of 90.000 m^3/ano) comes off of the coastline, forming a linked point to the earth.

After the construction of the breakwater off-shore it should be waited a moving back from the coastline to West of the Tip of Pecém 2–3 m/ano. In the last decades, the coastline of Pecém has been suffering an erosive process, in the order of 4 to 6 meters/year.

7 CONCLUSIONS

In the present article the studies were described accomplished to evaluate the possible changes in the dynamic balance of the sandy beaches of Pecém caused by the presence of the harbor structure.

The results getting from the mathematical modeling of coastal impact on the coastline of Pecém, allowed to conclude that the new port has a strong shelter effect on waves sea, with periods, in the strip, from 2 to 8 seconds and directions of waves, from 80 to 120°N, arrivals most of time from the oriental directions. In that shadow area that is in front of city of Pecém, the speeds of the coastal current decrease significantly, favoring the deposition of sediments.

Along the whole coastline an annual sediment transport rate is waited – due to combinations of sea and swell waves – in the order of 350.000 m^3/ano. Due to that effect of shelter of the port, 250.000 m^3/ano are transported, approximately, to the side western of the Tip of Pecém. A volume of the order of 90.000 m^3/ano of sand is accumulated in front of city of Pecém.

To West of the city of Pecém there is an estimate of erosion rate, in the order of 2 to 3 m/ano, which is a little smaller than the current taxes of erosion. When the effect of protection of the port disappears, between Pecém and Taíba, the taxes of coastal transport will increase gradually, returning the initial situation.

The results of the environmental monitoring of the beach profiles showed that after the Terminal of Temporary Embankment had been removed, the late accumulation of sediments along the oriental side, was decreasing and the sediment going down in the direction of the side sheltered by the harbor. During the two years of operation of the temporary wharf, the erosion taxes increased significantly, having been necessary the transport of sand, artificially, to the side West of the Tip of Pecém.

Therefore, one can conclude that in a close future the influence of that harbor on the coast line should be minima.

REFERENCES

CHAGAS, P. F. Influência da Estrutura Portuária sobre os Processos Hidrodinâmicos na Região Costeira do Pecém – Ce. Fortaleza, 2000. *Dissertação (Mestrado em Recursos Hídricos)* – Curso de Pós-Graduação em Engenharia Civil, Universidade Federal do Ceará.

INPH. Monitoramento Ambiental. *Relatório de Medições de Ondas realizadas na Ponta do Pecém-CE*. Rio de Janeiro, 1999.

INPH. Monitoramento Ambiental. *Relatório de Medições de Ventos realizadas na Ponta do Pecém-CE*. Rio de Janeiro, 2000.

PITOMBEIRA, E. S. Litoral de Fortaleza – Ceará – Brasil: um exemplo de degradação. In: Simpósio Sobre Processos Sedimentares e Problemas Ambientais na Zona Costeira Nordeste do Brasil, 1, 1995. *Anais*... Universidade Federal de Pernambuco. Centro de Tecnologia e Geociências, 1995. 176 p.p. 59–62.

PITOMBEIRA, E. S., VIEIRA, L. A de A. Modificaciones en la línea de costa por implantacion de espigones en la playa de Pirambú – Fortaleza. In: CONGRESSO CHILENO DE INGENIERÍA HIDRÁULICA, 12., 1995, Santiago. *Anais*... Santiago: Sociedad Chilena de Ingeniería Hidráulica, 1995.

PITOMBEIRA, E. S. Sistema de "By Pass" de Areia em Decorrência da Construção do Quebra-mar Protetor do Embarcadouro Provisório do Porto do Pecém, Ceará-Brasil. In: SIMPÓSIO BRASILEIRO DE RECURSOS HÍDRICOS, 12, 1997, Vitória. *Anais*... Vitória: ABRH, 1997. V.4. p. 463–469.

VALENTINI, E. M. Avaliação de Processos Litorâneos e Conseqüências para o Gerenciamento Costeiro do Ceará. Rio de Janeiro, 1994. *Tese (Doutorado em Ciências em Engenharia Costeira)* – Universidade Federal do Rio de Janeiro.

River, Coastal and Estuarine Morphodynamics: RCEM 2005 – Parker & García (eds)
© 2006 Taylor & Francis Group, London, ISBN 0 415 39270 5

The formation of beach cusps by waves with their crests parallel to shorelines

N. Izumi
Tohoku University, Sendai, Japan

A. Tanikawa
Fuji Film Software Co. Ltd., Kanagawa, Japan

H. Tanaka
Tohoku University, Sendai, Japan

ABSTRACT: Beach cusps are a series of arcuate scallops roughly uniformly spaced on beaches. They are inseparably related with nearshore current systems. Nearshore current systems form a variety of circulation patterns, which have been attracting many researchers' interest since a long time ago. In several theories proposed to explain the formation of nearshore current systems, Hino attributed their formation to the dynamic instability, and explained it in terms of linear stability analysis. In this study, his theory is revisited and improved in several aspects. The analysis proposed in this study successfully shows that beach cusps can be formed on beaches even when the wave crests are parallel to the shoreline.

1 INTRODUCTION

Beach cusps are repeated structures of arcuate shorelines. They are classified into several types depending on the scale of wavelength in the longshore direction. Beach cusps with wavelengths of few meters to few tens of meters are so-called "beach cusps" (see Fig. 1) while cusps with larger scales are called "mega cusps", "large cusps" or "giant cusps." A large number of studies have been performed in terms of geomorphology as well as coastal engineering. It is known that, as shown in Figure 2, they strongly pertain to nearshore current systems.

A nearshore current system consists of currents in the vicinity of beaches such as longshore currents and rip currents. It is thought that longshore and rip currents are generated by the radiation stress of waves. They are commonly observed to be regularly spaced in the longshore direction depending on wave conditions. Due to the sediment transport by nearshore current systems, periodic patterns of a shoreline configuration appear on beaches. Those are considered to be beach cusps.

Theories of the generation of nearshore current systems proposed so far are divided largely into two types; one is based on compulsive external causes, and the other is based on self-excited internal causes. The former includes theories in which a wave field loses its uniformity due to an external cause such as edge waves

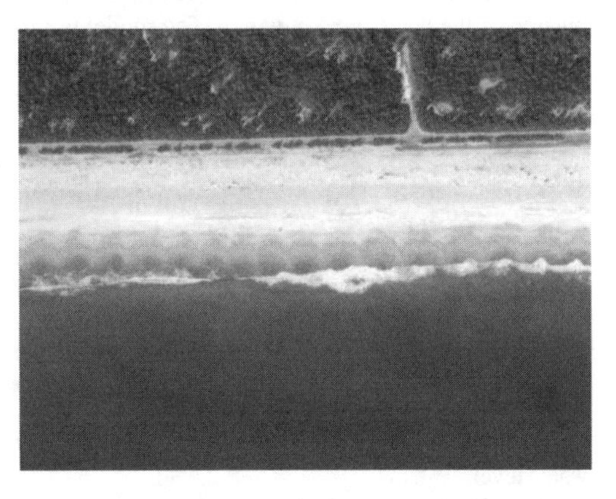

Figure 1. Beach cusps observed on the Sendai Coast. 30 Dec., 2000.

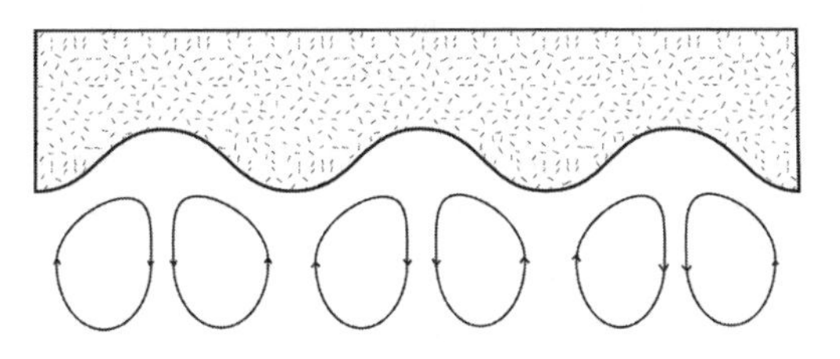

Figure 2. Conceptual diagram of beach cusps and a near shore current system.

and cross waves, resulting in the spatial variation of the radiation stress. Meanwhile, the latter includes theories in which the fluid dynamic instability causes a break of the uniformity of radiation stress even without an external cause, resulting in the generation of rip currents. A famous example of the latter is Hino's study (1973, 1974, 1975). He provided a physical explanation of the generation of beach cusps in terms of linear stability analysis which studies the growth of lateral perturbations imposed on a shoreline uniform in the longshore direction. According to his analysis, beach cusps with wavelengths of one to several times the width of the wave breaking zone are formed when waves come into the wave breaking zone with some angle. Though Hino's theory is innovative in that a rational explanation of the generation of nearshore current systems and beach cusps are provided in terms of the dynamic instability for the first time, the theory still has a room to be improved. For instance, Hino's theory cannot explain the fact that beach cusps are observed to be formed even though waves come into the wave breaking zone with their crests parallel to the shoreline in experiments performed under controlled conditions.

This study shows a possibility that beach cusps are generated even if the incident direction of waves is perpendicular to the shoreline by an appropriate consideration of the matching condition between inside and outside of the wave breaking zone. The model is further improved by including an accurate estimation of the bed shear stress outside of the wave breaking zone.

2 THE CONCEPTUAL MODEL

We assume that waves come into a beach of a constant slope as shown in Figure 3. The incident direction of waves is assumed to be perpendicular to the shoreline. Due to the radiation stress, the wave setup occurs inside of the wave breaking zone while the wave setdown appears outside of the wave breaking zone. As a result, the mean water level is slightly deviated from the still water level.

We take the $\tilde{x}$ and the $\tilde{y}$ axis in the cross-shore (onshore-offshore) and the longshore direction, respectively. We stipulate that the shoreline shifted by the wave setup and the wave breaking point are located at $\tilde{x}=0$ and $\tilde{x}=\tilde{L}_B$, respectively. If the still water depth and the mean water level are denoted by $\tilde{h}$ and $\tilde{\zeta}$ respectively, the condition that the total depth $(\tilde{h}+\tilde{\zeta})$ vanishes at the shoreline can be written in the form

$$\tilde{h} = -\tilde{\zeta} \equiv -\tilde{Z}_M \quad \text{at} \quad \tilde{x}=0 \tag{1}$$

where $\tilde{Z}_M$ is the mean water level at the shoreline.

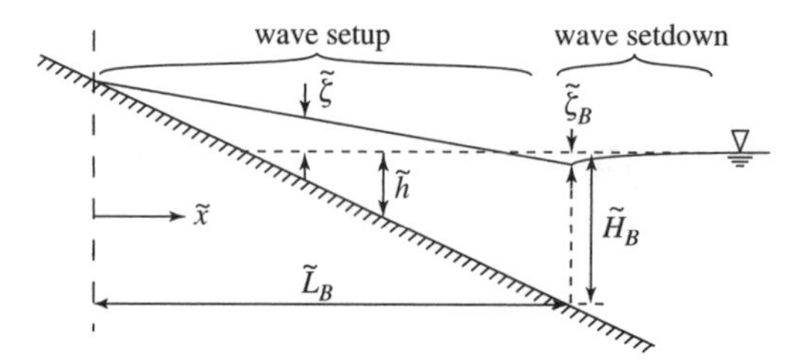

Figure 3. The mean water level near a shoreline.

At the wave breaking point, the ratio between the depth and the wave height is assumed to be constant, such that

$$\frac{2\tilde{a}}{\tilde{h}+\tilde{\zeta}} = \beta \quad \text{at} \quad \tilde{x} = \tilde{L}_B \tag{2}$$

where $\tilde{a}$ is the wave amplitude (a half of the wave height) just before the wave breaks, β is a parameter of the order of unity, the value of which depends on the conditions of waves offshore and the bottom slope. Though the bottom slope changes by the perturbation imposed on the bottom topography as described later, β is assumed not to change due to the perturbation for simplicity.

Denoting the still water depth at the wave breaking point by $\tilde{H}_B$, we have

$$\tilde{h} = \tilde{H}_B \quad \text{at} \quad \tilde{x} = \tilde{L}_B \tag{3}$$

As described earlier, the cross-shore profile of the beach is assumed to be linear before beach cusps are formed. With the use of the above equation and equation (1), $\tilde{h}$ is written in the form

$$\tilde{h} = \frac{\tilde{H}_B + \tilde{Z}_M}{\tilde{L}_B}\tilde{x} - \tilde{Z}_M \tag{4}$$

As mentioned earlier, the wave setup and wave setdown occur inside and outside of the wave breaking zone by the radiation stress. As a result, the mean water level once descends toward the wave breaking point, and ascends toward the shoreline. It is suspected that this state is so unstable that the uniformity of the water surface and bed elevations in the longshore direction is easily broken. In this study, we impose sinusoidal perturbations on the base state unperturbed shoreline configuration, and study the growth of the perturbations in terms of linear stability analysis. The growth rate is expected to be a function of wavelength. If the growth rate is maximized for a certain wavelength (dominant wavelength), a perturbation with the dominant wavelength grows faster than that with any other wavelength, determining the shoreline configuration. The fully-developed perturbation with the dominant wavelength is considered to be beach cusps.

3 FORMULATION

3.1 *Flow equations*

Flow in a wave field can be described by

$$\tilde{u}\frac{\partial \tilde{u}}{\partial \tilde{x}}+\tilde{v}\frac{\partial \tilde{u}}{\partial \tilde{y}} = -g\frac{\partial \tilde{\zeta}}{\partial \tilde{x}}-\frac{\tilde{\tau}_x}{\rho(\tilde{h}+\tilde{\zeta})} -\frac{1}{\rho(\tilde{h}+\tilde{\zeta})}\left(\frac{\partial \tilde{S}_{xx}}{\partial \tilde{x}}+\frac{\partial \tilde{S}_{xy}}{\partial \tilde{y}}\right) \quad (5)$$

$$\tilde{u}\frac{\partial \tilde{v}}{\partial \tilde{x}}+\tilde{v}\frac{\partial \tilde{v}}{\partial \tilde{y}} = -g\frac{\partial \tilde{\zeta}}{\partial \tilde{y}}-\frac{\tilde{\tau}_y}{\rho(\tilde{h}+\tilde{\zeta})} -\frac{1}{\rho(\tilde{h}+\tilde{\zeta})}\left(\frac{\partial \tilde{S}_{xy}}{\partial \tilde{x}}+\frac{\partial \tilde{S}_{yy}}{\partial \tilde{y}}\right) \quad (6)$$

$$\frac{\partial \tilde{u}(\tilde{h}+\tilde{\zeta})}{\partial \tilde{x}}+\frac{\partial \tilde{v}(\tilde{h}+\tilde{\zeta})}{\partial \tilde{y}}=0 \quad (7)$$

where $\tilde{u}$ and $\tilde{v}$ are the depth-averaged velocity in the $\tilde{x}$ and $\tilde{y}$ directions respectively, $\tilde{S}_{ij}$ $(i,j=x,y)$ is the radiation stress tensor, $\tilde{\tau}_i$ $(i=x,y)$ is the bed shear stress vector, ρ is the density of water, and g is the gravity acceleration ($=9.8\,\mathrm{m/s^2}$).

3.2 *The time variation of bed topography*

The time variation of bed topography is assumed to be expressed as

$$\frac{\partial \tilde{h}}{\partial \tilde{t}}=\frac{\partial \tilde{q}_{bx}}{\partial \tilde{x}}+\frac{\partial \tilde{q}_{by}}{\partial \tilde{y}} \quad (8)$$

where $\tilde{t}$ is time, $\tilde{q}_{bi}$ $(i=x,y)$ is the bed material transport vector. Among a variety of studies performed on the bed material transport, we employ a simple model in which the bed material flux is proportional to the velocity vector, such that

$$(\tilde{q}_{bx},\tilde{q}_{by})=C_s\,(\tilde{u},\tilde{v}) \quad (9)$$

where C_s is a parameter with a dimension of length, which is assumed to be a constant.

3.3 *Radiation stress*

When waves come into beaches with their crests parallel to the shoreline, the radiation stress is expressed as

$$\tilde{S}_{xx} = \tilde{E}\left(\frac{2\tilde{c}_g}{\tilde{c}}-\frac{1}{2}\right) \quad (10)$$

$$\tilde{S}_{xy} = 0 \quad (11)$$

$$\tilde{S}_{yy} = \tilde{E}\left(\frac{\tilde{c}_g}{\tilde{c}}-\frac{1}{2}\right) \quad (12)$$

where $\tilde{E}$ is the energy per unit width, $\tilde{c}$ and $\tilde{c}_g$ are the wave velocity and the group velocity, respectively. The energy per unit width $\tilde{E}$ is

$$\tilde{E}=\frac{1}{2}\rho g\tilde{a}^2 \quad (13)$$

and the wave velocity $\tilde{c}$ can be obtained from the dispersion relation described by

$$\tilde{\sigma}^2=g\tilde{k}\tanh\tilde{k}\tilde{h} \quad (14)$$

where $\tilde{\sigma}$ is the angular frequency which is not a function of space, and therefore, from equation (14), the wavenumber $\tilde{k}$ satisfies the relation of the form

$$\tilde{k}\tanh\tilde{k}\tilde{h}=\tilde{k}_\infty \quad (15)$$

where the subscript ∞ denotes variables far offshore. From the above equation, it is found that, once h is given, the wavenumber $\tilde{k}$ is uniquely determined. The relation among the angular frequency $\tilde{\sigma}$, the wavenumber $\tilde{k}$ and the wave velocity $\tilde{c}$ is

$$\tilde{\sigma}=\tilde{c}\tilde{k} \quad (16)$$

With the use of (15) and (16), $\tilde{c}$ is written in the form

$$\tilde{c}=\tilde{c}_\infty\tanh\tilde{k}\tilde{h} \quad (17)$$

where

$$\tilde{c}_\infty=\frac{\tilde{\sigma}}{\tilde{k}_\infty}=\left(\frac{g}{\tilde{k}_\infty}\right)^{1/2} \quad (18)$$

From (17) and (18), the wave velocity $\tilde{c}$ is found to be a function only of the depth.

The group velocity $\tilde{c}_g$ and the wave velocity $\tilde{c}$ satisfy the following relation:

$$\frac{\tilde{c}_g}{\tilde{c}}=\frac{1}{2}\left[1+\frac{2\tilde{k}(\tilde{h}+\tilde{\zeta})}{\sinh 2\tilde{k}(\tilde{h}+\tilde{\zeta})}\right] \quad (19)$$

If the total depth in the wave breaking zone is assumed to be sufficiently small, the above equation can be reduced to $\tilde{c}_g\approx\tilde{c}$, and the radiation stress in the wave breaking zone is expressed as

$$\tilde{S}_{xx}=\frac{3}{2}\tilde{E} \quad (20)$$

$$\tilde{S}_{yy}=\frac{1}{2}\tilde{E} \quad (21)$$

where $\tilde{E}$ is estimated as follows. Dimensional analysis and experiments show that the wave amplitude $\tilde{a}$ in the wave breaking zone is expressed as

$$2\tilde{a} = \gamma(\tilde{h} + \tilde{\zeta}) \tag{22}$$

where γ is a constant of the order of unity. From (13), E takes the form

$$\tilde{E} = \frac{1}{8}\rho g \gamma^2 \left(\tilde{h} + \tilde{\zeta}\right)^2 \tag{23}$$

The radiation stress is then reduced to

$$\tilde{S}_{xx} = \frac{3}{16}\rho g \gamma^2 (\tilde{h} + \tilde{\zeta})^2 \tag{24}$$

$$\tilde{S}_{xy} = 0 \tag{25}$$

$$\tilde{S}_{yy} = \frac{1}{16}\rho g \gamma^2 (\tilde{h} + \tilde{\zeta})^2 \tag{26}$$

Hino (1973, 1974, 1975) assumed that the effect of the radiation stress outside of the wave breaking zone is negligibly small. We also employ this approximation in this study.

3.4 *Bed shear stress*

Liu and Dalrymple (1978) have proposed a bed shear stress formula applicable to the case that unidirectional flow is sufficiently weak compared with wave motion. Their formula is written in the form

$$\tilde{\tau}_x = \rho C_f \langle \tilde{w} \rangle \left[\left(1 + \cos^2\theta\right)\tilde{u} + \sin\theta\cos\theta\tilde{v}\right] \tag{27}$$

$$\tilde{\tau}_y = \rho C_f \langle \tilde{w} \rangle \left[\sin\theta\cos\theta\tilde{u} + \left(1 + \sin^2\theta\right)\tilde{v}\right] \tag{28}$$

where θ is the incident angle of waves, which vanishes when the wave crest is parallel to the shoreline. The orbital velocity due to waves averaged over one wave period $\langle \tilde{w} \rangle$ takes the form

$$\langle \tilde{w} \rangle = \frac{2}{\pi}\tilde{w} \tag{29}$$

where $\tilde{w}$ is the maximum orbital velocity of waves. From the small amplitude wave theory, $\tilde{w}$ can be written as

$$\tilde{w} = \frac{\tilde{a}\tilde{\sigma}}{\sinh \tilde{k}(\tilde{h} + \tilde{\zeta})} = \frac{\tilde{a}g}{\tilde{c}\cosh \tilde{k}(\tilde{h} + \tilde{\zeta})} \tag{30}$$

Outside of the wave breaking zone, the assumption $\tilde{\zeta} << \tilde{h}$ is employed. Equation (30) is then reduced to

$$\tilde{w} = \frac{\tilde{a}\tilde{\sigma}}{\sinh \tilde{k}\tilde{h}} \tag{31}$$

The wave amplitude $\tilde{a}$ is derived from the fact that the energy flux $\tilde{F}$ is constant as follows. The energy flux is expressed as

$$\tilde{F} = \tilde{E}\tilde{c}_g = \frac{1}{2}\rho g \tilde{a}^2 \frac{\tilde{c}}{2}\left(\frac{2\tilde{k}\tilde{h}}{\sinh 2\tilde{k}\tilde{h}} + 1\right) \tag{32}$$

The energy flux far offshore is

$$\tilde{F} = \frac{1}{4}\rho g \tilde{a}_\infty^2 \tilde{c}_\infty \tag{33}$$

Assuming the energy flux $\tilde{F}$ is constant in every cross-section, (17), (32) and (33) are reduced to

$$\begin{aligned} \tilde{a}^2 &= \tilde{a}_\infty^2 \frac{\tilde{k}}{\tilde{k}_\infty}\left[\frac{2\tilde{k}(\tilde{h} + \tilde{\zeta})}{\sinh 2\tilde{k}(\tilde{h} + \tilde{\zeta})} + 1\right]^{-1} \\ &= \tilde{a}_\infty^2 \frac{\tilde{k}}{\tilde{k}_\infty}\left(\frac{2\tilde{k}\tilde{h}}{\sinh 2\tilde{k}\tilde{h}} + 1\right)^{-1} \end{aligned} \tag{34}$$

The variation of $\tilde{k}$ depending on $\tilde{h}$ can be described by (15), and the variation of $\tilde{a}$ depending on $\tilde{h}$ are obtained from the above equation. Substituting (34) into (30), we obtain

$$\tilde{w} = \frac{\tilde{a}_\infty \tilde{\sigma}}{\sinh \tilde{k}\tilde{h}}\left(\frac{\tilde{k}}{\tilde{k}_\infty}\right)^{1/2}\left(1 + \frac{2\tilde{k}\tilde{h}}{\sinh 2\tilde{k}\tilde{h}}\right)^{-1/2} \tag{35}$$

Inside of the wave breaking zone, the assumption of shallow water waves can be applied, so that cosh $\tilde{k}(\tilde{h} + \tilde{\zeta}) \approx 1$. The following relation is then obtained:

$$\tilde{w} = \frac{\tilde{a}g}{\tilde{c}} \tag{36}$$

The wave velocity $\tilde{c}$ can be written in the form

$$\tilde{c} = \left[g\left(\tilde{h} + \tilde{\zeta}\right)\right]^{1/2} \tag{37}$$

If the amplitude is approximated by (22), $\tilde{w}$ is expressed by

$$\tilde{w} = \frac{\gamma}{2}\left[g\left(\tilde{h} + \tilde{\zeta}\right)\right]^{1/2} \tag{38}$$

4 NON-DIMENSIONALIZATION

We introduce the non-dimensionalization described by

$$(\tilde{u}, \tilde{v}, \tilde{c}, \tilde{c}_g, \tilde{w}) = \left(g\tilde{H}_B\right)^{1/2}(u, v, c, c_g, w) \tag{39a}$$

$$(\tilde{x},\tilde{y}) = \tilde{L}_B(x,y) \tag{39b}$$

$$\left(\tilde{h},\tilde{\zeta},\tilde{a}\right) = \tilde{H}_B(h,\zeta,a), \quad \tilde{k} = \tilde{H}_B^{-1}k \tag{39c,d}$$

$$\tilde{\sigma} = \frac{g}{\tilde{H}_B}^{1/2}\sigma, \quad \tilde{t} = \frac{\tilde{H}_B\tilde{L}_B}{C_s\left(g\tilde{H}_B\right)^{1/2}}t \tag{39e,f}$$

In addition, $\tilde{S}_{ij}$, $\tilde{E}$ and $(\tilde{\tau}_x, \tilde{\tau}_y)$ are normalized as

$$\left(\tilde{S}_{ij},\tilde{E}\right) = \rho g\tilde{H}_B^2(S_{ij},E) \tag{40}$$

$$(\tilde{\tau}_x,\tilde{\tau}_y) = \rho g\frac{\tilde{H}_B^2}{\tilde{L}_B}(\tau_x,\tau_y) \tag{41}$$

With the above equations, (5)–(8) are normalized as follows.

4.1 *Outside of the wave breaking zone*

The normalized governing equations outside of the wave breaking zone are

$$u\frac{\partial u}{\partial x} + v\frac{\partial u}{\partial y} = -\frac{\partial \zeta}{\partial x} - \frac{\tau_{bx}}{h} \tag{42}$$

$$u\frac{\partial v}{\partial x} + v\frac{\partial v}{\partial y} = -\frac{\partial \zeta}{\partial y} - \frac{\tau_{by}}{h} \tag{43}$$

$$\frac{\partial uh}{\partial x} + \frac{\partial vh}{\partial y} = 0 \tag{44}$$

$$\frac{\partial h}{\partial t} = \frac{\partial u}{\partial x} + \frac{\partial v}{\partial y} \tag{45}$$

where the approximation $\zeta << h$ has been used to obtain (42)–(44). The bed shear stress vector is

$$(\tau_{bx},\tau_{by}) = Cw(2u,v) \tag{46}$$

where

$$C = \frac{2C_f\tilde{L}_B}{\pi\tilde{H}_B} \tag{47}$$

$$w = \frac{a_\infty k_\infty^{1/2}}{\sinh kh}\left(\frac{2kh}{\sinh 2kh}+1\right)^{-1/2} \tag{48}$$

4.2 *Inside of the wave breaking zone*

The normalized governing equations inside of the wave breaking zone are written in the forms

$$u\frac{\partial u}{\partial x} + v\frac{\partial u}{\partial y} = -\frac{\partial \zeta}{\partial x} - \frac{\tau_x}{h+\zeta} - \frac{1}{h+\zeta}\frac{\partial S_{xx}}{\partial x} \tag{49}$$

$$u\frac{\partial v}{\partial x} + v\frac{\partial v}{\partial y} = -\frac{\partial \zeta}{\partial y} - \frac{\tau_y}{h+\zeta} - \frac{1}{h+\zeta}\frac{\partial S_{yy}}{\partial y} \tag{50}$$

$$\frac{\partial u(h+\zeta)}{\partial x} + \frac{\partial v(h+\zeta)}{\partial y} = 0 \tag{51}$$

$$\frac{\partial h}{\partial t} = \frac{\partial u}{\partial x} + \frac{\partial v}{\partial y} \tag{52}$$

where the radiation stresses S_{xx} and S_{yy} are

$$S_{xx} = \frac{3\gamma^2}{16}(h+\zeta)^2 \tag{53}$$

$$S_{yy} = \frac{\gamma^2}{16}(h+\zeta)^2 \tag{54}$$

The bed shear stress vector is expressed as

$$(\tau_x,\tau_y) = \frac{1}{2}C\gamma(h+\zeta)^{1/2}(2u,v) \tag{55}$$

5 ASYMPTOTIC EXPANSIONS

We assume that, in the base state, the bed profile in the cross-shore direction is assumed to be linear and uniform in the longshore direction as shown in Figure 3. A small sinusoidal perturbation is imposed on the base state. Thus, u, v, ζ and h are expanded as

$$\begin{aligned}(u,\zeta,h) &= (0,\zeta_0,h_0) \\ &\quad + A\,(u_1(x),\zeta_1(x),h_1(x))\,\mathrm{e}^{pt}\cos\lambda y\end{aligned} \tag{56}$$

$$v = Av_1(x)\,\mathrm{e}^{pt}\sin\lambda y \tag{57}$$

where A is the amplitude of the perturbation which is assumed to be small, p and λ are the growth rate and the wavenumber of the perturbation. When the growth rate of perturbation p is positive, the base state is unstable for the perturbation, while the base state is stable when p is negative. When p is a positive function of λ, and there is a dominant wavenumber which maximizes p, the perturbation associated with the dominant wavenumber grows faster than others. The dominant wavelength corresponds to the wavelength of beach cusps formed on beaches consequently.

5.1 *One-dimensional base state*

Substituting (56) and (57) into (42)–(45), and dropping terms of $O(A)$ and higher, we obtain the following equations at $O(1)$:

$$\frac{\partial \zeta_0}{\partial x} = 0 \tag{58}$$

$$\frac{\partial \zeta_0}{\partial y} = 0 \tag{59}$$

Because the effect of radiation stress is neglected outside of the wave breaking zone, the wave setdown does not occur in the base state. Thus, the mean water level in the base state ζ_0 is constant in space. That is

$$\zeta_0 = 0 \tag{60}$$

The mean water level at the wave breaking point Z_B is then expressed as

$$Z_B = 0 \tag{61}$$

Equations inside of the wave breaking zone are reduced to

$$\frac{\mathrm{d}\zeta_0}{\mathrm{d}x} = -\frac{1}{h_0 + \zeta_0}\frac{\mathrm{d}S_{xx0}}{\mathrm{d}x} \tag{62}$$

The radiation stress inside of the wave breaking zone S_{xx0} is

$$S_{xx0} = \frac{3\gamma}{16}(h_0 + \zeta_0)^2 \tag{63}$$

Substituting the above equation into (62), we obtain

$$\frac{\mathrm{d}\zeta_0}{\mathrm{d}x} = -\frac{3\gamma^2}{8}\frac{\mathrm{d}(h_0 + \zeta_0)}{\mathrm{d}x} \tag{64}$$

Integrating the above equation with the condition $h = 1$ and $\zeta = 0$ at the wave breaking point, we obtain ζ_0 in the form

$$\zeta_0 = -K(h_0 - 1), \tag{65}$$

$$K = \frac{3\gamma}{3\gamma + 8} \tag{66}$$

where h_0 is written in the form

$$h_0 = (1 + Z_M)x - Z_M \tag{67}$$

Substituting the above equation into (65), we obtain

$$\zeta_0 = -K(1 + Z_M)(x - 1) \tag{68}$$

Substituting $x = 0$ into the above equation, we found Z_M described by

$$Z_M = \frac{K}{1 - K} \tag{69}$$

Figure 4 shows the base state solutions ζ_0 and h_0 when $\gamma = 1$. Because the radiation stress is ignored outside of the wave breaking zone, only wave setup can be seen in the figure.

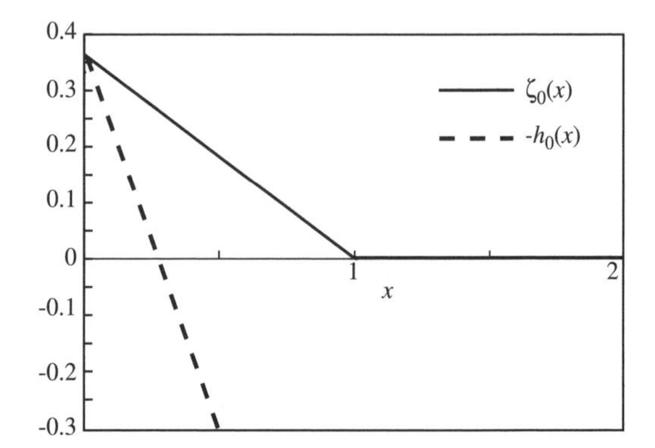

Figure 4. The base state solutions ζ_0 and h_0.

5.2 *The two-dimensional perturbation problem*

5.2.1 *Perturbation equations*

The perturbation equations outside of the wave breaking zone at $O(A)$ are

$$\frac{\mathrm{d}\zeta_1}{\mathrm{d}x} + \frac{2Cw_0}{h_0}u_1 = 0 \tag{70}$$

$$-\lambda\zeta_1 + \frac{Cw_0}{h_0}v_1 = 0 \tag{71}$$

$$h_0' u_1^{\mathrm{o}} + h_0\frac{\mathrm{d}u_1}{\mathrm{d}x} + \lambda h_0 v_1 = 0 \tag{72}$$

$$ph_1 - \frac{\mathrm{d}u_1}{\mathrm{d}x} - \lambda v_1 = 0 \tag{73}$$

where $'$ denotes the total derivative with respect to x.

Meanwhile, the perturbation equations inside of the wave breaking zone are

$$\frac{C}{(h_0 + \zeta_0)^{1/2}}u_1 + \frac{3\gamma}{8}\frac{\mathrm{d}h_1}{\mathrm{d}x} + \left(1 + \frac{3\gamma}{8}\right)\frac{\mathrm{d}\zeta_1}{\mathrm{d}x} = 0 \tag{74}$$

$$\frac{C}{(h_0 + \zeta_0)^{1/2}}v_1 + \frac{\gamma}{8}\lambda h_1 + \left(1 + \frac{\gamma}{8}\right)\lambda\zeta_1 = 0 \tag{75}$$

$$(h_0' + \zeta_0')u_1 + (h_0 + \zeta_0)\frac{\mathrm{d}u_1}{\mathrm{d}x} + \lambda(h_0 + \zeta_0)v_1 = 0 \tag{76}$$

$$ph_1 - \frac{\mathrm{d}u_1}{\mathrm{d}x} - \lambda v_1 = 0 \tag{77}$$

5.2.2 *Boundary conditions at the shoreline*

The location of the shoreline is slightly shifted due to the perturbation imposed on the base state. The perturbed location of the shoreline is expressed as

$$x = A\xi_1 \mathrm{e}^{pt}\cos\lambda y \tag{78}$$

The total depth should vanish at the shoreline, such that

$$h + \zeta = 0 \quad \text{at} \quad x = A\xi_1 \mathrm{e}^{pt} \cos \lambda y \tag{79}$$

Expanding the above equation with respect to A, we obtain the following equation at $O(1)$:

$$h_0(0) + \zeta_0(0) = 0 \tag{80}$$

At $O(A)$, we find

$$\xi_1 = -\frac{h_1(0) + \zeta_1(0)}{h_0'(0) + \zeta_0'(0)} \tag{81}$$

The velocity u vanishes at the shoreline; thus

$$u = 0 \quad \text{at} \quad x = A\xi_1 \mathrm{e}^{pt} \cos \lambda y \tag{82}$$

Expanding the above equation around $x = 0$ with respect to A, we obtain the following equations at $O(1)$ and $O(A)$, respectively:

$$u_0(0) = 0 \tag{83}$$

$$u_1(0) + u_0'(0)\xi_1 = 0 \tag{84}$$

In the case that the wave crests are parallel to the shoreline as assumed in this study, because $u_0(0) = 0$ and $u_0'(0) = 0$, (84) becomes identical to the boundary conditions introduced by Hino (1973, 1974, 1975), which is written as

$$u_1(0) = 0 \tag{85}$$

It should be noted, however, that (85) cannot be used because $u_0(0) \neq 0$ and $u_0'(0) \neq 0$ when the wave crests are not parallel to the shoreline.

5.2.3 *Matching conditions at the wave breaking point*

As perturbations are imposed on the still water depth, the mean water level, and the wave amplitude, the location of the wave breaking point is also perturbed. In this study, we derive the displacement of the wave breaking point due to perturbations by expanding the wave breaking condition (4), and introduce the effect of the displacement of the wave breaking point by matching solutions inside and outside of the wave breaking zone.

We assume that the location of the wave breaking point are shifted from $x = 1$ to

$$x = 1 + A\chi_1 \cos \lambda y \tag{86}$$

Equation (4) is then reduced to

$$2a^{\mathrm{o}} = \beta(h^{\mathrm{o}} + \zeta^{\mathrm{o}}) \quad \text{at} \quad x = 1 + A\chi_1 \cos \lambda y \tag{87}$$

where the superscript o denotes variables outside of the wave breaking zone. Expanding (87) around the unperturbed wave breaking point ($x = 1$) with respect to A, we obtain the following equation at $O(1)$:

$$2a_0^{\mathrm{o}}(1) = \beta[h_0^{\mathrm{o}}(1) + \zeta_0^{\mathrm{o}}(1)] \tag{88}$$

At $O(A)$, we find

$$\chi_1 = -\frac{2a_1^{\mathrm{o}}(1) - \beta\left[h_1^{\mathrm{o}}(1) + \zeta_1^{\mathrm{o}}(1)\right]}{2a_0^{\mathrm{o}\prime}(1) - \beta\left[h_0^{\mathrm{o}\prime}(1) + \zeta_0^{\mathrm{o}\prime}(1)\right]} \tag{89}$$

where a_0^{o} and $a_0^{\mathrm{o}\prime}$ are obtained as follows. Equation (34) is reduced to

$$a = \frac{\sqrt{2}\cosh kh}{(\sinh 2kh + 2kh)^{1/2}} a_\infty \tag{90}$$

From the above equation, we find

$$a_1^{\mathrm{o}} = \frac{\partial a_0}{\partial h_0} h_1^{\mathrm{o}} \tag{91}$$

$$a_0^{\mathrm{o}\prime} = \frac{\partial a_0}{\partial h_0} h_0^{\mathrm{o}\prime} \tag{92}$$

$$\frac{\partial a_0}{\partial h_0} = 2\sqrt{2}k_0 \sinh 2k_0 h_0 \times \frac{k_0 h_0 \sinh k_0 h_0 - \cosh k_0 h_0}{(\sinh 2k_0 h_0 + 2k_0 h_0)^{5/2}} a_\infty \tag{93}$$

Equation (89) is the displacement of the location of the wave breaking point due to perturbations.

The still water depth h has to be continuous inside and outside of the wave breaking zone. Therefore, at $O(1)$ and $O(A)$ respectively, we obtain

$$h_0^{\mathrm{o}}(1) = h_0^{\mathrm{i}}(1) \tag{94}$$

$$h_1^{\mathrm{o}}(1) + h_0^{\mathrm{o}\prime}(1)\chi_1 = h_1^{\mathrm{i}}(1) + h_0^{\mathrm{i}\prime}(1)\chi_1 \tag{95}$$

where the superscript i denotes variables inside of the wave breaking zone. The beach slope in the base state is assumed to be a constant in the x direction; thus

$$h_0^{\mathrm{o}\prime}(1) = h_0^{\mathrm{i}\prime}(1) \tag{96}$$

Therefore, (95) is reduced to

$$h_1^{\mathrm{o}}(1) = h_1^{\mathrm{i}}(1) \tag{97}$$

The mean water level ζ is also continuous at the wave breaking point. The following equations are obtained at $O(1)$ and $O(A)$, respectively:

$$\zeta_0^{\mathrm{o}}(1) = \zeta_0^{\mathrm{i}}(1) \tag{98}$$

$$\zeta_1^{\mathrm{o}}(1) + \zeta_0^{\mathrm{o}\prime}(1)\chi_1 = \zeta_1^{\mathrm{i}}(1) + \zeta_0^{\mathrm{i}\prime}(1)\chi_1 \tag{99}$$

For the velocities u and v, we obtain the following relations:

$$u_1^{\mathrm{o}}(1) = u_1^{\mathrm{i}}(1) \tag{100}$$

$$v_1^{\mathrm{o}}(1) = v_1^{\mathrm{i}}(1) \tag{101}$$

Equations (97), (100) and (101) imply that h_1, u_1 and v_1 are continuous at $x = 1$. Therefore, no correction is needed in the existing theory (Hino 1973; Hino 1974; Hino 1975). However, (99) shows that ζ_1 is discontinuous at $x = 1$, so that a careful matching of the solutions is necessary. This condition has not been considered in the existing theory. It should be noted that (100) and (101) have to be corrected when the incident direction of waves is not perpendicular to the shoreline.

5.2.4 *Boundary conditions far offshore*

Perturbed quantities vanish far offshore. Thus, the following conditions must be satisfied:

$$u_1 = v_1 = h_1 = \zeta_1 = 0 \quad \text{as} \quad x \to \infty \tag{102}$$

5.2.5 *Solution*

We can obtain the growth rate p by solving the perturbation equations (70)–(77), which are four ordinary differential equations with respect to four variables, with the boundary conditions (85) and (102), and the matching conditions (97), (99), (100) and (101). They are found to form a Strum-Liouville type eigenvalue problem with the eigenvalue p. We solve the eigenvalue problem by the use of the spectral method (Boyd 2001) with the Chebyshev polynomial. The perturbed variables u_1, v_1, h_1 and ζ_1 are expressed by the Chebyshev polynomials in the forms

$$u_1 = \sum_{n=0}^{N} c_n^{(1)} T_n(\eta), \quad v_1 = \sum_{n=0}^{N} c_n^{(2)} T_n(\eta) \tag{103a, b}$$

$$h_1 = \sum_{n=0}^{N} c_n^{(3)} T_n(\eta), \quad \zeta_1^{\mathrm{i}} = \sum_{n=0}^{N} c_n^{(4)} T_n(\eta), \tag{103c, d}$$

$$\zeta_1^{\mathrm{o}} = \sum_{n=0}^{N} c_n^{(5)} T_n(\eta), \tag{103e}$$

where

$$\eta = \frac{x+1}{x-1} \tag{103f}$$

Substituting the above equations into the governing equations and the boundary and matching conditions, and evaluating those at the Gauss-Lobatto points ($\eta_j = \cos(j\pi/N), j = 0, 1, 2, \ldots, N$), we obtain a generalized eigenvalue problem in the form

$$\mathcal{L}\mathbf{u} = p\mathcal{R}\mathbf{u} \tag{104}$$

where $\mathcal{L}$ and $\mathcal{R}$ are $5(N+1) \times 5(N+1)$ matrices, and $\mathbf{u}$ is a $N+1$ vector written in the form

$$\mathbf{u} = \Big[c_0^{(1)}, \cdots, c_N^{(1)}, c_0^{(2)}, \cdots, c_N^{(2)}, c_0^{(3)}, \cdots, c_N^{(3)}, c_0^{(4)}, \cdots, c_N^{(4)}, c_0^{(5)}, \cdots, c_N^{(5)}\Big]^T \tag{105}$$

Solving (104) by the QZ method, we obtain the eigenvalue p. We have found that sufficient accuracy was achieved when $N = 20$.

6 RESULTS AND DISCUSSION

The growth rate p as a function of the wavenumber λ and the parameter C is shown in Figure 5. In the figure, we assumed $\tilde{a}_\infty = 0.34$, $\tilde{k}_\infty = 0.18$.

The growth rate p is found to be positive everywhere, and is maximized for a finite value of λ. While,

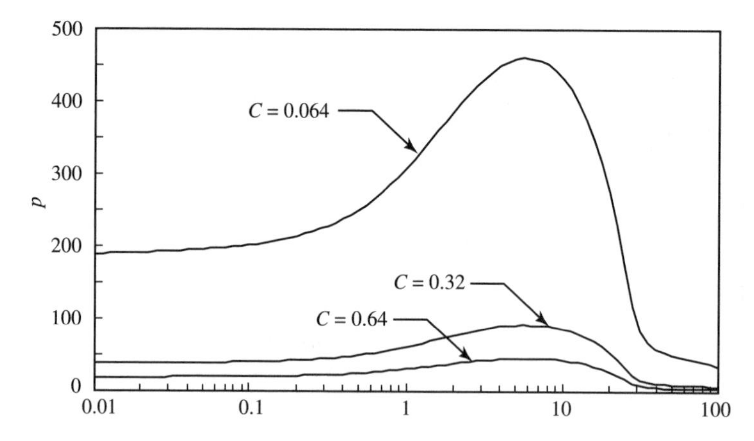

Figure 5. The growth rate of perturbation p as a function of λ and C. $a_\infty = 0.34$, $k_\infty = 0.18$.

in Hino's theory (1975), the dominant wavelength associated with the maximum growth rate is infinitely small when the incident wave angle vanishes, the dominant wavenumber is found to be finite in this analysis. In addition, the dominant wavenumber is around 6.3 regardless of the value of C. The wavelength of the cusp $\tilde{L}_C$ can be written in the form

$$\tilde{L}_C = \frac{2\pi \tilde{L}_B}{\lambda} \tag{106}$$

The implication of the above equation is that, when λ is around 6, the wavelength of cusps is equivalent to the width of the wave breaking zone.

In addition, the growth rate itself decreases with increasing C. Because C is expressed as

$$C = \frac{2C_f \tilde{L}_B}{\pi \tilde{H}_B} \tag{107}$$

$C = 0.064$, 0.32 and 0.64 correspond to $C_f = 0.01$, 0.05 and 0.1 respectively if $\tilde{H}_B/\tilde{L}_B = 0.1$, and $\tilde{H}_B/\tilde{L}_B = 0.1$, 0.02 and 0.01 respectively if $C_f = 0.01$. It is found that, as the friction coefficient C_f increases, and the beach slope $\tilde{H}_B/\tilde{L}_B$ decreases, the growth rate of perturbation decreases and the beach becomes more stable.

According to Tamai's experiments (1975), it is found that cusps are formed even when the incident wave angle vanishes, and that the wavelength of cusps is equivalent to the width of the wave breaking zone. These experimental results are consistent to the results obtained in this analysis. Meanwhile, the wavelength of cusps and the spacing of rip currents in the field is found to be 1.5–8 times the width of the wave breaking zone (Horikawa et al. 1975). The wavelength obtained in the present analysis is rather small compared with field observation.

According to Hino's analysis (1975), though the growth rate does not have a peak when the incident angle of waves vanishes, a peak of the growth rate comes to appear when the incident angle becomes larger than a few degrees. In addition, as the incident angle increases, the dominant wavenumber decreases. This implies that the wavelength of cusps increases with increasing incident angle. This suggests that an increase in the incident angle of waves amplifies the wavelength of cusps. In experiments, it is possible to control the incident angle. However, it is most unlikely that the incident angle is exactly zero; rather, it is more natural that the incident angle has some value. Therefore, it is suggested that the wavelength of cusps observed in the field is large because the incident angle is not zero. Though this analysis treats of the case that the incident angle vanishes, it is expected that the wavelength of cusps increases and agrees well with observations if the effect of the incident angle is included. This should be studied in future.

7 CONCLUSION

The major results of this study are as follows:

- Considering the displacement of the location of the wave breaking point due to perturbation by including matching conditions, we improved the results of analysis and obtained a finite dominant wavenumber even if the incident angle of waves is perpendicular to shorelines.
- The theory is further improved by introducing a formula of the bed shear stress both inside and outside of the wave breaking zone.
- It is found that the wavelength of cusps is equivalent to the width of the wave breaking zone when the incident angle of waves vanishes. This is consistent with experimental results.
- A beach becomes more stable, as the bed friction coefficient increases and the beach slope decreases.

REFERENCES

Boyd, J. P. (2001). Chebyshev and fourier spectral methods, 2nd ed. pp. 668.

Hino, M. (1973). Theory of the nearshore current system, part 3, – a simplified theory –. *Proceedings of 20th Conference of Coastal Engineering*, 339–343.

Hino, M. (1974). The theory of the generation of the nearshore current system. *Proceedings of JSCE 255*, 17–29.

Hino, M. (1975). Theory of the nearshore current system and the beach cusps considering the response. *Proceedings of JSCE 237*, 87–98.

Horikawa, K., T. Sasaki, S. Hotta, and H. Sakuramoto (1975). Study on the nearshore current system, part 2, – field observation –. *Proceedings of 22nd Conference of Coastal Engineering*, 135–139.

Liu, P. L. F. and R. A. Dalrymple (1978). Bottom frictional stress and longshore currents due to waves with large angles of incidence. *J. Marine Res. 366*, 357–375.

Tamai, S. (1975). Study on the formation of beach cusps. *Proceedings of 22nd Conference of Coastal Engineering*, 135–139.

River, Coastal and Estuarine Morphodynamics: RCEM 2005 – Parker & García (eds)
© 2006 Taylor & Francis Group, London, ISBN 0 415 39270 5

Watertable monitoring on a beach equipped with a dewatering system: relationship between watertable elevation and beach morphology, preliminary results

A. Lambert
Université de Provence, CEREGE UMR CNRS

V. Rey
Université de Toulon La Valette, ISITV

O. Samat & M. Provansal
Université de Provence, CEREGE UMR CNRS

ABSTRACT: In February 2004, two Ecoplage systems were installed for the first time one the Mediterranean French coast, one the microtidal beaches of bay of Agay and the Garronette (Saint-Raphaël, Sainte-Maxime, Var, France). Those sites are monitored by the University of Provence and the University of Toulon La Vallette.

The subject of this study is to highlight the relations between watertable movements and weather-marine dynamics, in order to understand the action of a beach dewatering system (engineered by the Danish Geotechnical Institute for the Ecoplage French Society) on the sedimentary behaviour of the studied site.

The experimental protocol comprises three types of measurements:

1. The position of the watertable in the beach is evaluated by a line of six pressures sensors on a cross-shore section, located under the MWL, from the middle of the beach to the offshore limit of the swash.
2. The morphometry of the site is evaluated by measurements on fixed stakes, as well as high resolution daily Digital Elevation Model.
3. An ADCP gives the spectral characteristics of swell at the entry of the bay (25 m depth), a S4ADW provides the same data at 3.5 m depth on the same azimuth as the pressure sensors. Moreover, we can start and stop the pumps with the request.

Results: When the system is off, the comparison between the time data of watertable levels and the weather-sea data shows that the movements of the watertable are related to the conditions of real tide (barometric set-up and astronomical tide) and swell. Without real tide influence, the parameter swell height parameter is the most influential. When the system is on, the watertable is lowered from 7 cm to 15 cm upstream downstream compared to its positions without pumping, in equal weather-sea conditions. Our measurement shows that the influence of the wave height parameter becomes less influent than the period parameter.

The dewatering impact on morphology is a laminar fixing of the material gone up from the lower swash zone to the mid and upper swash zone, which progrades horizontally at speeds of 5 to 15 cm/day.

1 INTRODUCTION

Seventy percent of the sandy littorals are in retreat since 1950. The erosion of these low coasts – which often concentrate demographic and economic stakes, is a real problem for their managers. Until 1990, the solution was to build rocky structures like groins, breakwater, seawall, seadykes ("Hard engineering"). Those equipments, effective in the short term period, appeared detrimental for their environment at medium and long term: phenomena of undermining of the structure's toes which cause their collapse; phenomena of "overerosion" generated by non-dissipation and offshore reflection of wave energy which drag sediment offshore; phenomena of washing away and skirting of groins by the sea. Moreover, those equipments are extremely expensive to build and maintain, and their landscape impact is very significant. They are thus not possible within the framework of policies of sustainable development of coasts with strong environmental and/or economic stakes (tourism).

In order to fight against the erosion of sandy beaches, some new methods, known as "alternative" are elaborate, in particular systems known as "beach dewatering systems". Two systems of this type (called "Ecoplage" in France) have been installed on an experimental basis for the first time in the French Mediterranean Sea, on the test-sites of Garonnette (Sainte-Maxime, Var) and of bay of Agay (Saint-Raphaël, Var).

The operation and the real impacts of the watertable dewatering on beach morphology are still badly known. We will thus be interested in the existing relations between the evolutions of the beach watertable and the weather-marine characteristics, with and without operation of the system, and then we will evaluate the role of the system on the sedimentary behaviour of a portion of the beach of Agay.

2 SITE PRESENTATION

The bay of Agay (fig. 1) is located on the French Mediterranean coast, in the volcanic shield of Estérel chain. The beach material consists of sands resulting from ophiolites deterioration, of a characteristic red colour. This material comes from the bed rock, located between 3 and 5 m under the beach surface, as well as contributions of the Agay River, which drains the soil on a catchments area of 54 km^2.

The hydrodynamic environment is characterized by the low amplitude of the tides (0.4 m of maximum marling, semi-diurnal tide with day anomaly), typical of the French microtidal Mediterranean coasts. The beach is relatively sheltered and only east and south-east swells can affect his morphology. However, the autumn and spring storms are dangerous because they are accompanied by a 1 m MWL rise, phenomenon known under the name of "blow of Labé", which can be at the origin of significant damages on this low coast.

The beach (fig. 2), located in bay bottom, has a crescent morphology. It is 600 m wide from east to west.

The emerged beach is relatively narrow, between 10 and 35 m in the west of bay, near the mouth of Agay, and of 15 with 7 m in the centre and the west. Bathymetry has a "basin" morphology, with a stronger slope of the bottom in the centre than in the west and the east. The small bottoms are characterized by relatively strong slopes, without bar, pertaining to the "reflexive" type of the classification of Shorts and Wright. The swell breaking is generally of surging or collapsing type, with a very short surf zone (2 with 5 m according to the swell). The sedimentary material consists of fine sand (mean: 0.18 mm) and of ballast (mean: 0.4 to 2 cm).

Figure 1. Site location.

Figure 2. Vertical view from the beach.

3 EXPERIMENTAL PROTOCOL

Because of the heaviness of the device, we cannot instrument the site permanently. We thus carry out two campaigns from 19 to January 21 2005, then from the 9 to March 18 2005. Only data of the first campaign are presented here.

In order to follow the movement evolution of the watertable levels, we established (fig. 3) a cross-shore line of six piezo-resistive, high-sensitivity, low range (0–250 mbar) vented pressure sensors. These sensors

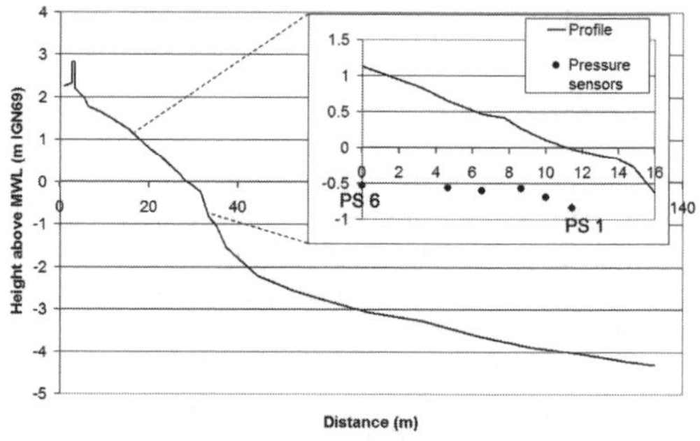

Figure 3. Profile morphology and pressure sensors locations within the beach.

are fixed inside a stainless cylindrical component, equipped with a sand screw, which makes it possible to bury the sensors within the beach until a depth of −0.8 m under the watertable level. They were calibrated for salt water density, in order to give full scale response from 0 to 258.25 cm. These sensors are connected to an Almemeo data logger which record the six ways continuously at a 1.66 Hz per way frequency. At the beginning of campaigns, these instruments are distributed from the lower swash zone to the middle of the beach.

The head of the sensors is protected by a plastic cap which is bored of kind to avoid the formation of air bubble against the membrane at the burying time. In order to maintain the membrane free of sediment (Turner and Nielsen, 1997), we fix sediment net of 0.045 mm around the cap. During the tank tests, the sensors whose caps were filled with silicone oil are proven impossible to gauge. We thus leave the membrane (H1 stainless steel, rust free) to the direct contact of sea water.

The weather-marine data are gathered by a RDI Workhorse Sentinel 600 kHz ADCP at the entry of bay (−25 m) and by a S4ADW located in the axis of the profile defined by the line of pressure sensors, with −3.5 m. These sensors provide us the characteristics of swell and water height at the entry of bay and before the surf zone. The data of S4 are unfortunately not exploitable for the first campaign.

The site morphology is surveyed by micro-grid DEMs on very small time scale (four times per day), on a zone extending from 25 m eastward westward of the pressure sensor line. Sand heights measurements from fixed stakes are also taken in the axis of the profile given by the sensors, with one hour interval.

We also have the ability to start and stop the dewatering system, which is stopped of the 19/01/05,13h40 to the 20/01/05, 12h15, standard time.

4 WATERTABLE HYDRODYNAMICS

4.1 *Weather-marine dynamics characterization*

The spectral data of swell (fig. 4) are calculated from the ADCP recordings located at the entry of bay. We regard these values as the deep water conditions.

The period of study (19/01/05 09h00 to the 21/01/05 16h00) is affected by a climate of short swells (averaged T_p: 6.2 S) and low height (averaged $H_o = 0.48$ m), coming from south-east (146°).

The calculation of ξ_0 (Battjes, 1974, Okazaki and Sunamura, 1991), and of ε (Carrier and Greenspan, 1958; Wright and Al., 1979) make it possible to confirm the observations of the type of breaking: with $1.91 < \xi_0 < 7.66$, and $0.11 < \varepsilon < 0.99$, 95% of the breaker are of surging or collapsing type.

- Surf similarity parameter:

$$\xi_0 = \frac{\tan\beta}{\sqrt{H_0 / L_0}} \quad (1)$$

$$\text{With} \quad Lo = \frac{gT^2}{2\pi} \approx 1.56T^2 \quad (2)$$

- Surf scaling factor:

$$\varepsilon = \frac{a_0\omega^2}{g\tan^2\beta} \quad (3)$$

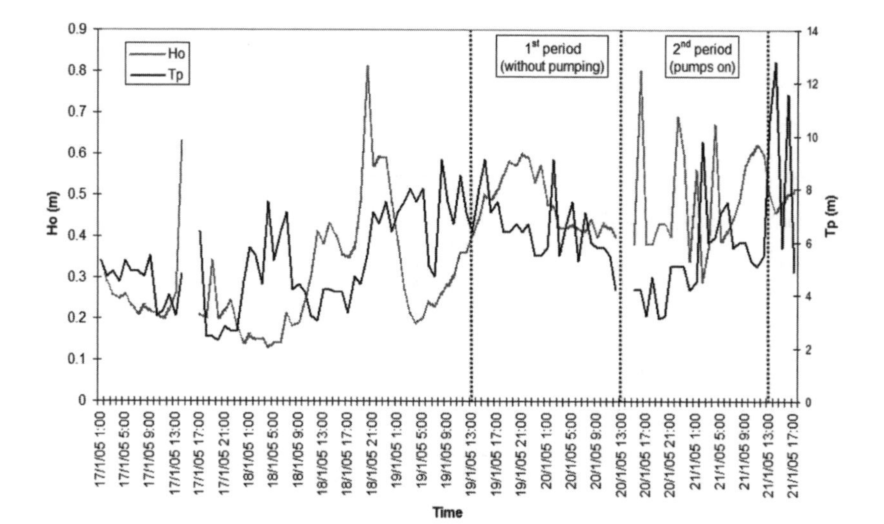

Figure 4. Swell time series.

With

$$\omega = \frac{2\pi}{T} \quad (4)$$

and β, the nearshore slope (here, 3.4°).

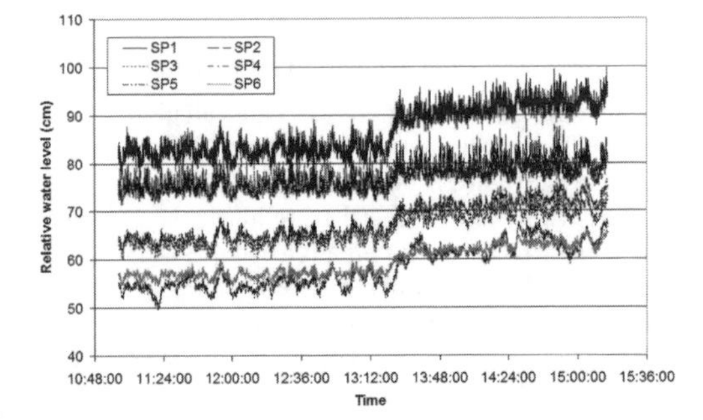

Figure 5. Pressure sensor response to dewatering system stop.

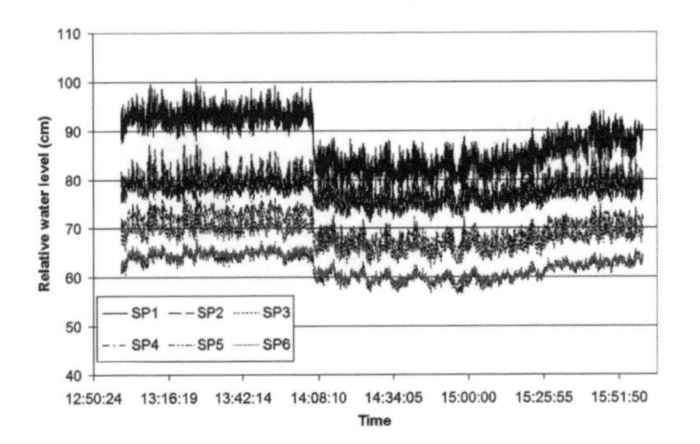

Figure 6. Pressure sensor response to dewatering system start.

These indices classify both the bay of Agay in the reflective beaches category (Wright et al, 1984).

4.2 *Influence of Ecoplage system on the watertable kinematics*

The observation of the pressure sensors recordings, that has been converted to water depths, makes it possible to highlight the instantaneous effect of the dewatering on the watertable of beach: at the time of the stop of pumping (fig. 5) after one week of uninterrupted operation, we observes a very fast rise (2 cm/minute) in the level of the watertable. This recording is synchronous for each sensor. In 5 minutes, the level of the tablecloth rose of 10 centimetres for the whole of the sensors.

The pumps are restarted 24 hours after their stop (fig. 6). The influence on the tablecloth is instantaneous: it is folded back in a synchronous way between all the sensors in 2.5 minutes. The folding back is not homogeneous on the whole of the sensors, its values decrease upstream: the PS 1 (offshore) present an instantaneous lowering of 10 cm, this value not being any more but of 5 cm on the level of the PS 6 (onshore).

In order to study the phenomena at the origin of the watertable beats, the time series of the watertable levels are averaged with a hourly time step, then cut out in two periods corresponding to the stop then the restart of the drainage, P1 and P2.

4.3 *Relationship between watertable position and weather-marine data*

The goal of this section is to carry out a comparison between the two periods P1 and P2, in order to:

1. To try to connect the observations of watertable levels to simple hydrodynamic parameters, without

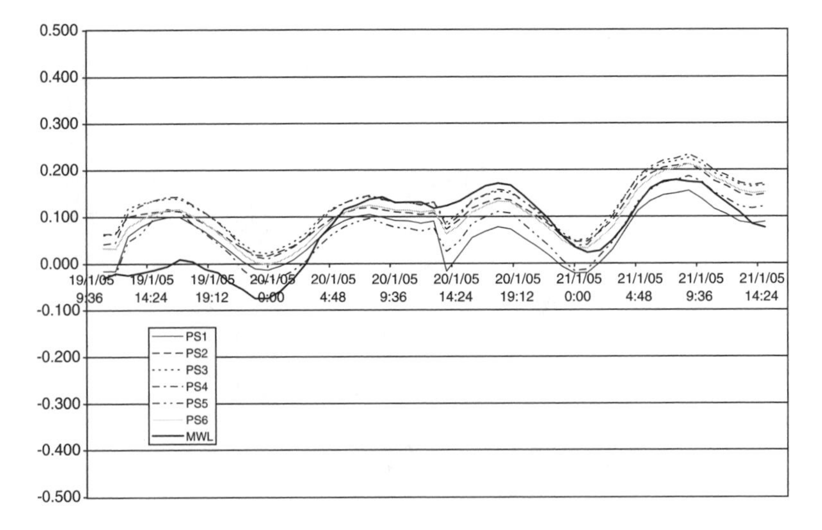

Figure 7. Hourly averaged data of pressure sensors and MWS from ADCP.

consideration for bed fluidization and capillarity parameters, which are probably modified by the drainage.

2. To compare the correlations obtained between the two periods in order to highlight the role of the dewatering in term of modification of the effects of the marine dynamics on the watertable levels.

The pressure sensors recordings realised over one hour present the evolution of the watertable level for the duration of the campaign (fig. 7).

The watertable movements are naturally dependent (PS1, lower swash zone to PS6, middle of the beach) on the hourly time water level (ADCP).

The coefficients of determination (Table 1) indeed translate the influence of the MWS which is the principal component (70%) of watertable time movements, with or without dewatering.

In order to highlight the other components of the signal, we deduce the influence of the MWS from the recorded watertable levels recorded such as $Lw^n r_t = PS_t^n - MWS_t$ (5).

Table 1. Determination coefficients between pressure sensors data and MWS position.

	MWS	
R^2	Without Ecoplage (P1)	With Ecoplage (P2)
PS1	0.76	0.77
PS2	0.77	0.78
PS3	0.74	0.78
PS4	0.73	0.78
PS5	0.65	0.80
PS6	0.78	0.77

The vertical shift of the watertable at the time of the stop of pumping is definitely visible on figure 8.

The time series obtained are correlated:

- with the time characteristics of the swell provided by the ADCP (Ho, Tpic, Dpic)
- to the wave length L_0 computed according to linear airy wave theory for deep water waves
- to the breaking height H_b computed after Komar and Gaughan (1972) formula
- like with two morphodynamic indices computed from these data: the surfing similarity parameter ξ and the surfing scaling Factor ε computed for breaker (ξ_b, ε_b) and deep water conditions (ξ_0, ε_0).

The wave set-up at the shoreline is computed from Gourlay's (1992) formula.

With

- Wave set-up: (Gourlay, 1992)

$$\overline{\eta}_s = 0.35 H_0 \xi_0^{0.4} \tag{5}$$

- For breaker conditions, we use H_b (Komar and Gaughan, 1972) as:

$$Hb = \frac{0.563Ho}{(Ho/Lo)^{1/5}} \tag{6}$$

For the first period (without dewatering), the obtained coefficients of determination (Table 2) highlight the existing correlation between the breaking height of swell and the watertable levels variations: the computed breaker heights explain to 90% these movements. One obtains the same values for the set-up. That

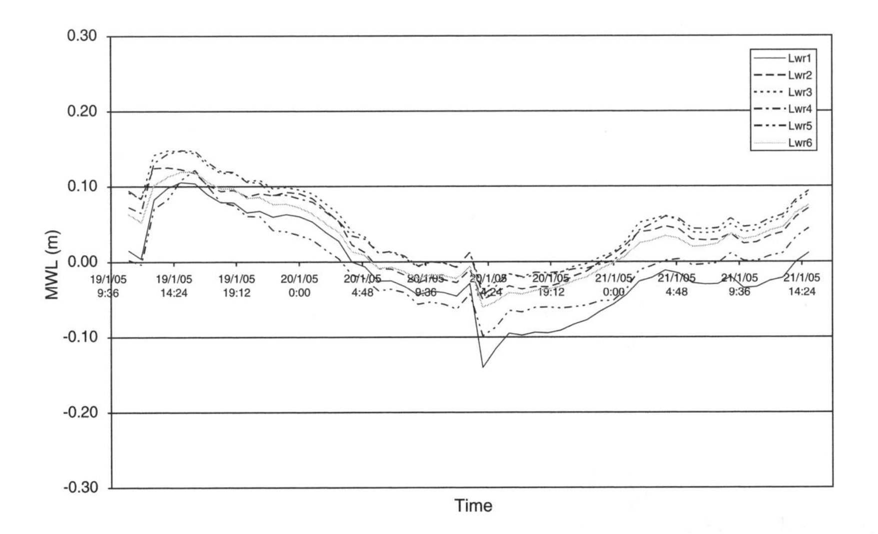

Figure 8. Pressure sensor recording rectified from MWS.

Table 2. Determination coefficients between watertable levels and wave data during first period.

P1	Lw_1r	Lw_2r	Lw_3r	Lw_4r	Lw_5r	Lw_6r
H_o	0.85	0.86	0.85	0.83	0.82	0.86
H_b	0.9	0.9	0.91	0.9	0.9	0.91
T_p	0.47	0.45	0.48	0.48	0.49	0.45
D_p	0.07	0.04	0.09	0.1	0.09	0.06
L_o	0.45	0.43	0.46	0.46	0.47	0.43
H_o/L_o	−0.13	−0.1	−0.14	−0.16	−0.18	−0.12
ξ_0	0.15	0.13	0.16	0.17	0.18	0.13
H_b/L_o	−0.13	−0.1	−0.14	−0.16	−0.18	−0.12
ξ_b	0.15	0.13	0.16	0.17	0.18	0.13
ε_0	−0.13	−0.1	−0.14	−0.6	−0.18	−0.12
set up	0.9	0.9	0.91	0.9	0.9	0.91

Table 3. Determination coefficients between watertable levels and wave data during second period.

P2	Lw_1r	Lw_2r	Lw_3r	Lw_4r	Lw_5r	Lw_6r
H_o	−0.14	−0.16	−0.12	−0.09	−0.8	−0.08
H_b	0.38	0.38	0.42	0.43	0.44	0.44
T_p	0.74	0.73	0.77	0.74	0.73	0.74
D_p	−0.18	−0.17	−0.16	−0.17	−0.16	−0.14
L_o	0.62	0.63	0.67	0.65	0.64	0.66
H_o/L_o	−0.9	−0.9	−0.88	−0.84	−0.82	−0.87
ξ_0	0.68	0.69	0.71	0.67	0.65	0.66
H_b/L_o	−0.89	−0.89	−0.88	−0.83	−0.81	−0.85
ξ_b	0.7	0.71	0.73	0.68	0.66	0.69
ε_0	−0.88	−0.89	−0.88	−0.83	−0.81	−0.85
set up	0.38	0.37	0.42	0.43	0.44	0.44

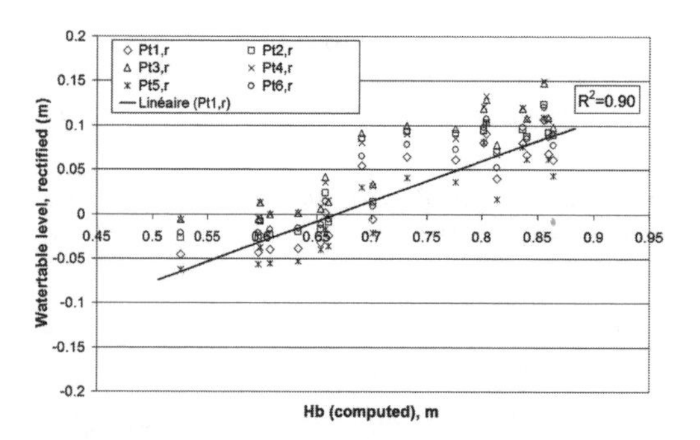

Figure 9. Correlation between H_b and watertable levels during first period.

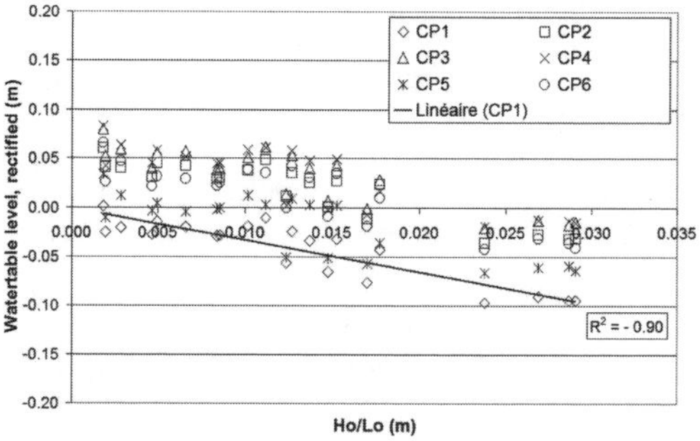

Figure 10. Correlation between H_o/L_o and watertable levels during second period.

is explained by the existing linear relation between these two parameters, such as:

$$\overline{\eta}_s = 0.3965 Hb \qquad (7)$$

The others coefficients of determination does not allow the description of significant correlations, except for H_o, which is related to the calculation of H_b.

During this first period, the computed values of set-up and swell breaking height thus explains all the two 90% of the hourly movements of the watertable deduced from the oscillations of the MWS. For the studied data, the movements of the watertable thus seem correlated with the height parameter and – in a few proportions – with the period of swell (fig. 9).

For the second period (with dewatering), the highest coefficients of determination are obtained for the ratios H_o/L_o, H_b/L_o, like for ε_0, with values ranging between −0.8 and −0.9 (Table 3).

These numbers illustrate the role of the swell period parameter in the watertable oscillations during this second period: indeed, all these indices are functions of the ratio H_o/T_p^2, and T_p present positive coefficients of determinations (ranging between 0.72 and 0.77) with the watertable movements.

According to (1), (3), for H_o constant, the values of H_o/L_o (fig. 10), H_b/L_o and ε_0 decrease for T_p increasing. The variations of the watertable level thus seem correlated contrary to the ratio H_o/T_p^2.

The weight of T_p in this ratio is proportionally more significant than that of H_o because the values of T_p are squared. It thus seems that the variations of swell period are the principal component of the watertable oscillations measured during this second period.

Those differences between the two period's components of the watertable levels could be ascribable with heterogeneity inside the wave conditions.

The study of simple statistical parameters (Table 4) from the wave spectral data shows that swell dynamics of the first period are relatively weak and invariants: (averages: $H_o = 0.48$ m, wideness of the time series: 0.2 m, $T_p = 6.5$ s, *wideness* = 4.9 m, $L_o = 68.11$ m, *wideness*, 101.67).

These observations also apply to the second period, to which the averages (Table 5) are extremely close to the first. (averages: $H_o = 0.49$ m, wideness of the

Table 4. Statistical parameters of wave's data for the first period.

P1	Average	RMS	Extent	Min	Max	Sum	Samples
H_o	0.48	0.07	0.2	0.4	0.6	11.07	23
H_b	0.72	0.1	0.34	0.52	0.86	16.61	23
T_p	6.5	1.19	4.9	4.2	9.1	149.6	23
D_p	137.74	46.6	194	89	283	3168	23
L_o	68.11	25.57	101.67	27.52	129.18	1566.46	23
H_o/L_o	0.01	0	0.01	0	0.01	0.18	23
ξ_0	3.82	0.71	2.69	2.69	5.39	87.93	23
H_b/L_o	0.01	0	0.01	0.01	0.02	0.27	23
ξ_b	3.1	0.46	1.74	2.35	4.09	71.39	23
ε_0	0.35	0.09	0.38	0.19	0.57	7.94	23
set up	0.29	0.04	0.13	0.21	0.34	6.59	23

Table 5. Statistical parameters of wave's data for the second period.

P2	Average	RMS	Extent	Min	Max	Sum	Samples
H_o	0.49	0.05	0.51	0.29	0.8	12.27	25
H_b	0.69	0.13	0.47	0.45	0.93	17.24	25
T_p	5.84	2.3	6.7	3.1	9.8	146.1	25
D_p	156.18	58.93	289	16	305	3436	22
L_o	61.17	55.27	131.83	14.99	146.82	1529.35	25
H_o/L_o	0.01	0.01	0.03	0	0.03	0.32	25
ξ_0	3.48	1.54	5.75	1.91	6.52	87.11	25
H_b/L_o	0.02	0.01	0.03	0	0.03	0.41	25
ξ_b	2.85	0.98	3.64	1.78	5.43	71.31	25
ε_0	0.49	0.25	0.88	0.11	0.99	12.33	25
set up	0.27	0.06	0.19	0.18	0.37	6.83	25

time series: 0.51 m, $T_p = 5.84$ s, *wideness* = 6.70 m, $L_o = 61.17$ m, *extent* = 131.83). Only the extents of the series, a quarter to a third more significant, distinguish them.

In the absence of fundamental difference, the wave parameters of the two periods thus seem relatively homogeneous. Their lights variations don't explain differences in origin of the watertable beats from one period to another.

If we regard the two periods as comparable from their dynamics point of view, only the beach dewatering (except influence of the MWS) seems likely to influence the watertable hydrodynamics.

In period 1, without dewatering, the watertable evolutions are strictly driven by the short wave heights, without considerations of the wave period parameter.

In second period, the watertable evolutions are connected to the wave steepness, and in particular to the increase of wavelength for a given height, which varies like the squared wave period. That suggests that watertable movements are due to the application period of a given water height on the beachface, either only with the wave height, which does not seem any more to influence the watertable beats in the same proportions.

5 RELATIONSHIP WITH MORPHOLOGY

5.1 *Method and site description*

Ten high accuracy DEMs were carried out from 19/01/19/05 to 01/21/05 (Table 6, figs 11 to 20), on a rectangular zone (area: 1708 m^2) extended from the back beach to bottoms of −1.8 m. The dots density is of 0.5 dots/m^2 in the swash and surf zone, 0.25 dots/m^2 on the emerged beach.

The DEMs are computed with the SURFER 9 software. The interpolation function that fits the best the data distributions are deduced from their semi-variogram. The chosen griding method is the kriging, with a griding step of 10 × 10 cm. This very fine resolution makes it possible in theory to seize the evolutions of micro-sized morphological features (such as the micro-cusp, berm scarp), thus to allow very short time scale comparisons.

We suppose that the strand and small bottoms morphology is the result of the previous day's wave climate, and that it is influenced by the operation of Ecoplage, uninterrupted since two months.

The comparison of the DEMs carried out one hour before the system stop, 24 hours after the stop, and 24 hours after the restarting of the pumps, should enable us to seize the impact of dewatering on the morphology of the site, at equal wave climate.

The first DEM presents a site whose back beach is not very wide (22 m) with the significant slope (3.43°).

Table 6. Date and time of the DEM's data collect.

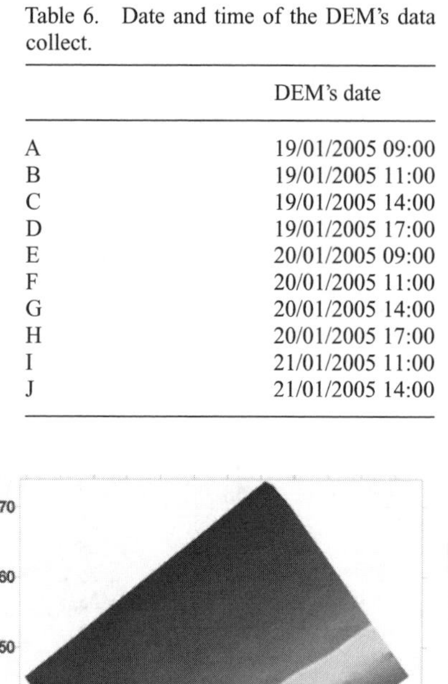

	DEM's date
A	19/01/2005 09:00
B	19/01/2005 11:00
C	19/01/2005 14:00
D	19/01/2005 17:00
E	20/01/2005 09:00
F	20/01/2005 11:00
G	20/01/2005 14:00
H	20/01/2005 17:00
I	21/01/2005 11:00
J	21/01/2005 14:00

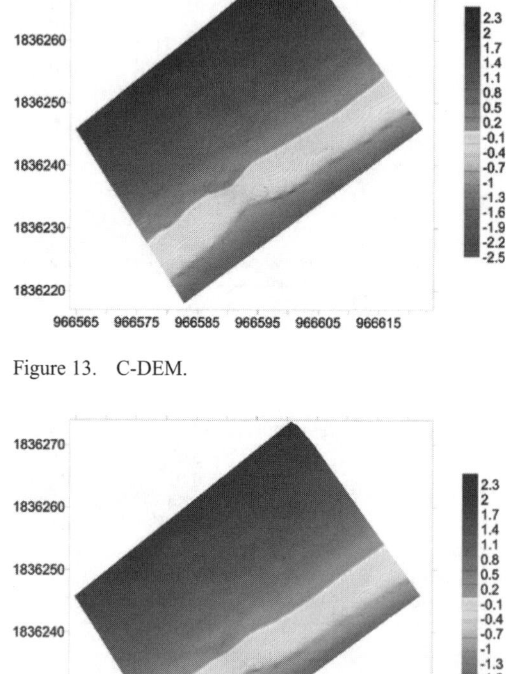

Figure 13. C-DEM.

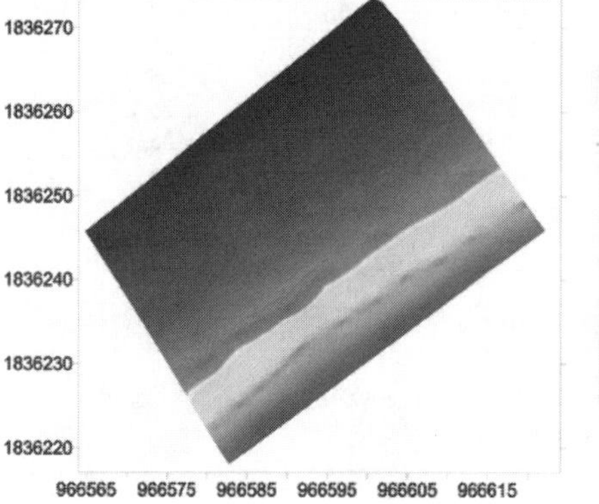

Figure 11. A-DEM.

Figure 14. D-DEM.

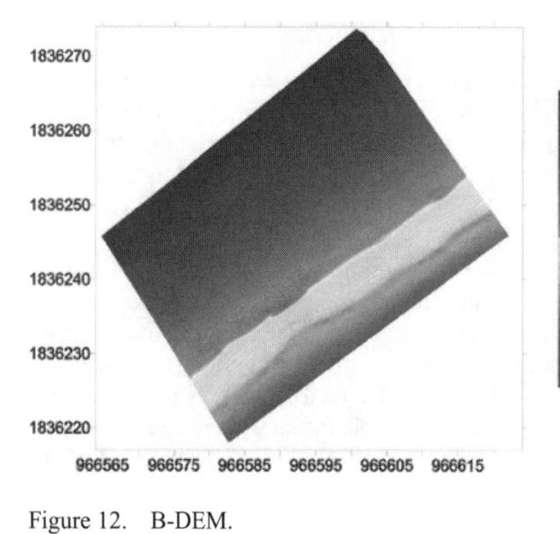

Figure 12. B-DEM.

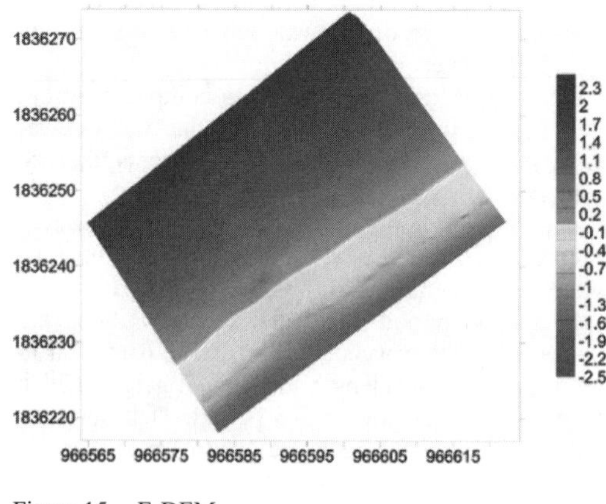

Figure 15. E-DEM.

It does not present a transition with the swash zone, which is relatively short (8 m) with the same slope (3.41°). It finishes at sea by a brutal change of incline and a steeper slope (14°) from −0.2 to −0.8 m. During the campaign, this zone constitutes the breaker zone. From −0.8 m to −2.2 m, the slope becomes milder (5.7°), to lead at the end of the surf zone. The surf zone total extent is of 12 m, for a slope of 7.1°. The nearshore zone (−2.2 with −7 m) presents a mild slope (3.3°). The morphology of the immersed beach is thus overall concave, with a logarithmic profile, without neither bar nor longshore feature, except a micro-cusp on the south-west of the DEM.

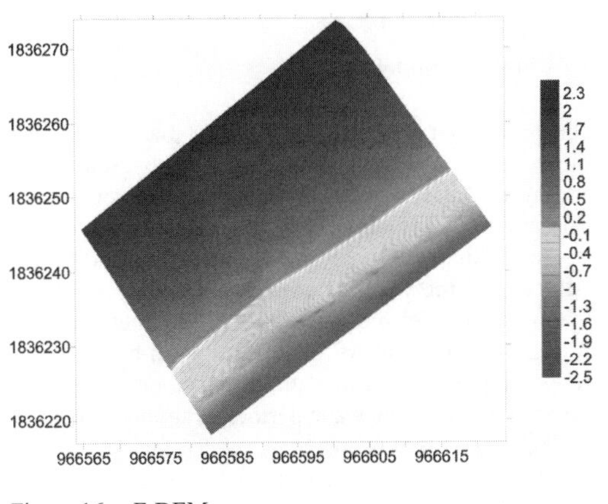

Figure 16. F-DEM.

The immersed beach and the swash zone have a similar surface granulometry, with an average grain of 0.17 mm. The step consists of coarser materials (sands, average grain 0.17 mm, 40% of the total dry weight, ballast, average diameter 1 cm, 60% of the total dry weight). Granulometry becomes again brutally finer (average grain 0.15 mm) after the step and that the slope becomes milder again. Cross-shore trenches showed the existence of a positive vertical grano-classification within the swash zone, up to 25 cm of the surface where five centimetres thick coarse and fine material rolls superimpose themselves until the surface.

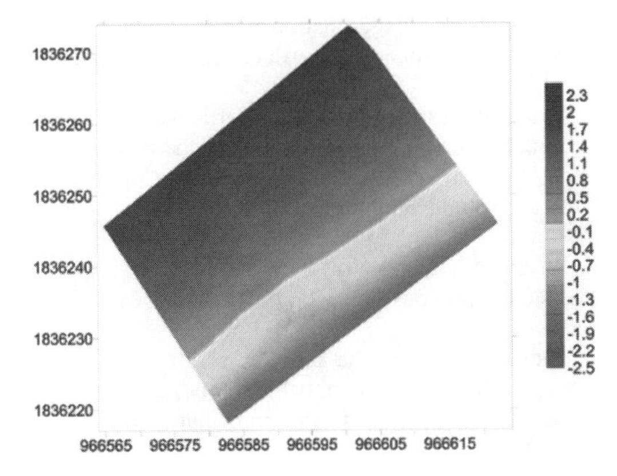

Figure 17. G-DEM.

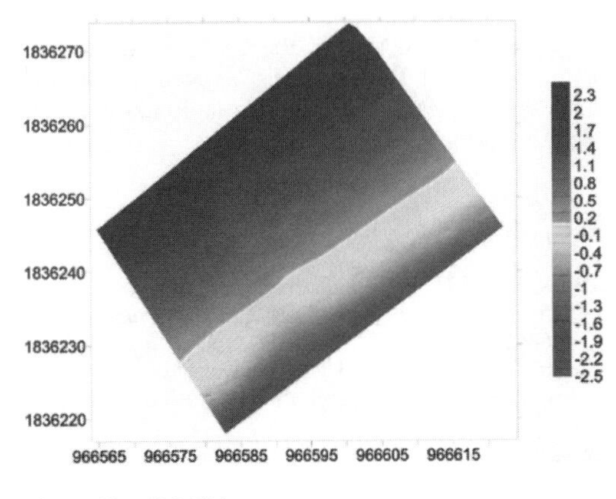

Figure 18. H-DEM.

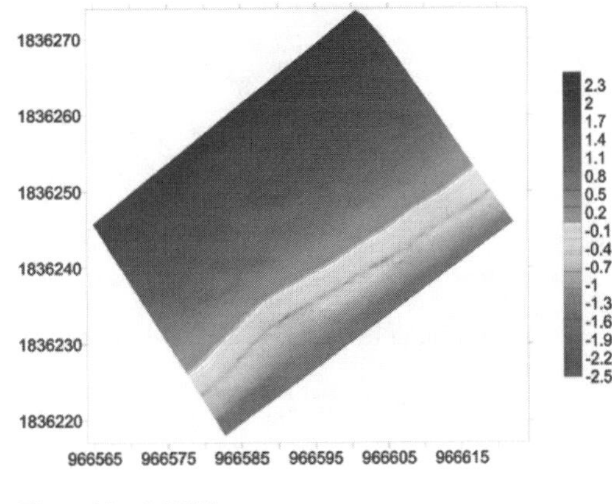

Figure 19. I-DEM.

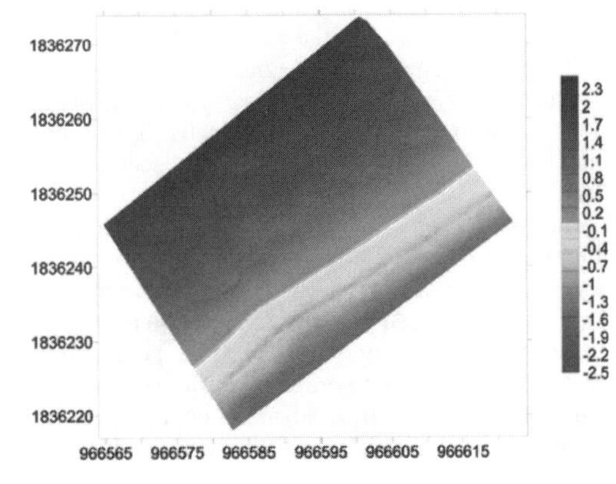

Figure 20. J-DEM.

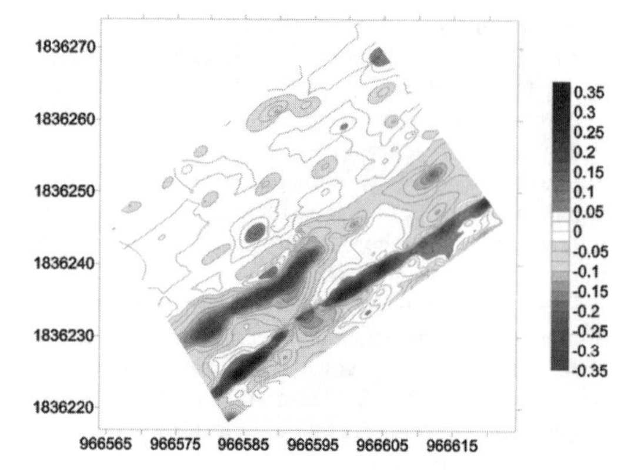

Figure 21. Altitude differential between A and E-DEMs.

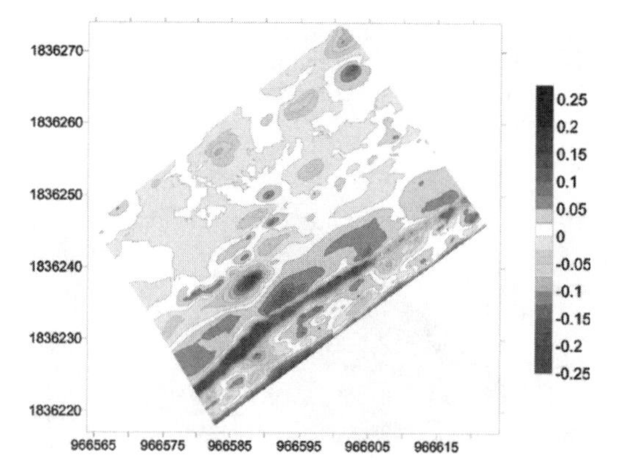

Figure 22. Altitude differential between E and J-DEMs.

5.2 *Results*

The comparison of the DEMs carried out one hour before and 24 hours after the stopping of the pumps (fig. 21) highlights a very clear erosion ($Z = -10$ cm) on the whole swash zone, with a maximum on the micro-cusp in the west of the zone (-0.05 to -0.3 m at the crescent point). This degradation of high and median swash zone is accompanied by deposits in the lower swash zone and on the step, which prograde and whose slope change with the surf zone becomes less clear.

The comparison of the DEMs carried out 24 hours after the stop and 24 hours after the dewatering resumption (fig. 22) lets appear an opposite behaviour: the zone of the swash presents an elevation of 0.05 at 0.15 m, whereas the step and the lower swash zone are eroded during this period. The observed accretion located at the DEM offshore boundary is an interpolation artifact and was not noted in reality.

During the dewatering stop, we thus notes a directed sedimentary transit from intermediate and higher swash zone, which loses material, to the lower swash zone and the step, which prograde.

With the dewatering resumption, the step and the lower swash zone lose material whereas intermediate and higher swash zone shows a progradation.

6 CONCLUSION

The preliminary study of the watertable levels data shows that the beach dewatering system established in Bay of Agay is at the origin of an instantaneous ten-centimetres folding back of the watertable.

This folding back has also for effect to modify the influence of the wave parameters on the levels of the beach watertable: the wave height, parameter dominating the variations of the watertable in the absence dewatering, seems to lose its influence on watertable with the profit of the wave period parameter when the dewatering system is on.

These two combined effects (folding back of 10 cm of the mean watertable level and modification of the influence of the wave parameters) seems generate, over 24 hours, a laminar aggradation of the higher and intermediate swash zones, with values ranging between 5 and 15 cm.

These preliminary results, in particular the swash aggradation and the modifications of the wave influence, are to be confirmed by processing the data *r* from the second campaign carried out over a ten days period.

With a modelling aim, they must also be supplemented by drain position data compared to the pressure sensors.

The influence of the long-term Ecoplage effect was the subject of a tri-annual monthly survey whose results will be published in an immediate future.

ACKNOWLEDGEMENTS

We thank the PACA Area, the CETE Méditerranée in the person of Frederic Pons, as well as the students of The Master "Géomorphologie des Milieux Méditerranéns".

REFERENCES

Battjes, J A, 1974. Surf Similarity. *Proceedings 14th International Conference on Coastal Engineering*. ASCE, 466–480.

Carrier, G F, Greenspan, H P, 1958. Water waves of finite amplitude on a sloping beach. *Journal of Fluid Mechanics*, 4, 97–109.

Komar, P D, Gaughan, M K, 1972. Airy wave theory and breaker height prediction. *Proceedings 13th International Conference on Coastal Engineering*. ASCE, 405–418.

Gourlay, M R, 1992. Wave set-up, wave run-up and beach ground water table: Implications between surf zone hydraulics and ground water hydraulics. *Coastal Engineering*, 17, 93–144.

Okazaki, S, Sunamura, T, 1991. Re-examination of breaker type classification on uniformly inclined laboratory beach. *Journal of Coastal Research*, 7, 559–564.

Turner, Ian L, Nielsen, Peter, 1997. Rapid water table fluctuations within the beach face: implications for swash zone sediment mobility? *Coastal Engineering*, 32 (1997) 45–59.

Wright, L D, Short, A D, Green, M O, 1984. Short-term changes in the morphodynamic states of beaches and surf zone: An empirical predictive model. *Marine Geology*, 62 (1985) 339–364.

Wright, L D, Chappell, J, Thom, B G, Bradshaw, M P, and Cowell, P J, 1979. Morphodynamics of reflective and dissipative beach and inshore systems, Southeastern Australia. *Marine Geology*, 32, 105–140.

River confluences and distributaries

River, Coastal and Estuarine Morphodynamics: RCEM 2005 – Parker & García (eds)
© 2006 Taylor & Francis Group, London, ISBN 0 415 39270 5

Scaling of confluence dynamics in river systems: some general considerations

B.L. Rhoads
Department of Geography, University of Illinois, Urbana, IL, USA

ABSTRACT: The arrangement of streams in networks is a fundamental spatial property of fluvial systems. As water moves through a drainage network, it is forced to converge at confluences. Over the past twenty years a salient body of theoretical, experimental, and field research has emerged on the fluvial dynamics of stream confluences. This work has begun to reveal the complex three-dimensionality of flow at confluences and to connect flow structure to the morphodynamics dynamics of confluences. It has also shown that coherent flow structures of various scales can exist at confluences and that the presence or absence of particular structures can vary both from confluence to confluence and from time to time at a specific confluence.

Despite these advances, to date most studies have focused either on laboratory confluences or on small natural confluences. "Small" as used here implies that field measurements in the confluence can be obtained by wading into the flow (depth < 1.0–1.5 m) or by spanning the confluence with a small bridge (width < 12–18 meters). Few, if any, process-based field investigations of the fluvial dynamics of confluences of large rivers have been conducted and data do not exist to evaluate the relevance of conclusions derived from small-scale confluence studies for understanding the dynamics of large-river confluences. The extent to which findings for small confluences provide insight into the fluvial dynamics of large-river confluences is a scaling issue. Past work on the scaling of channel form and fluvial processes suggests that scaling relations for confluences are likely to be fundamentally nonlinear and therefore complex.

This paper discusses potential differences between the dynamics of large-river confluences and the confluences of laboratory channels or small streams. The discussion draws on research results for small confluences and on general principles regarding scaling relations in river systems. This synthesis of extant knowledge provides the basis for identifying hypotheses regarding possible similarities and differences between the dynamics of small stream confluences and large river confluences. Issues that are addressed include three-dimensional flow structure, turbulence, mixing, and patterns of erosion and deposition within confluences. A research project on large-river confluence dynamics that seeks to evaluate the hypotheses is briefly described.

1 INTRODUCTION

The arrangement of streams in networks is a fundamental spatial property of fluvial systems. As water moves through a drainage network, it is forced to converge at confluences. This forced convergence of flow produces a complex hydrodynamic environment, or confluence hydrodynamic zone (CHZ) (Kenworthy & Rhoads 1995) within the immediate vicinity of the junction. Until recently, relatively little was known about the dynamics of flow through confluences or the relationship of local hydrodynamic conditions to confluence geomorphology. Over the past twenty years a salient body of theoretical (Bradbrook et al. 1998, Bradbrook et al. 2000a,b, Lane et al. 1999; Bradbrook et al. 2001, Weerakoon et al. 1991, Weerakoon & Tamai 1989), experimental (Best 1987, Best 1988, Best & Reid 1984, Best & Roy 1991, Biron et al. 1996a,b, McLelland et al. 1996), and field (Ashmore et al. 1992, Biron et al. 1993a,b, Biron et al. 2002, Bristow et al. 1993, De Serres et al. 1999, Gaudet & Roy 1995, Kenworthy & Rhoads 1995, Rhoads 1996, Rhoads & Kenworthy 1995, 1998, 1999, Rhoads & Sukhodolov 2001, 2004, Roy & Bergeron 1990, Roy et al. 1988, Sukhodolov & Rhoads 2001) research has emerged on the fluvial dynamics of stream confluences. This work has begun to reveal the complex three-dimensionality of flow at confluences and to connect flow structure to the fluvial dynamics of confluences. It has also shown that coherent flow structures of various scales can exist at confluences and that the presence or absence of particular structures can vary both from confluence to confluence and from time to time at a specific confluence.

Despite these advances, to date most studies have focused either on laboratory confluences or on small natural confluences. "Small" as used here implies that field measurements in the confluence can be obtained by wading into the flow (depth < 1.0–1.5 m) or by spanning the confluence with a small bridge (width < 12–18 meters). Few, if any, process-based field investigations of the fluvial dynamics of confluences of large rivers have been conducted and data do not exist to evaluate the relevance of conclusions derived from small-scale confluence studies for understanding the dynamics of large-river confluences. The extent to which findings for small confluences provide insight into the fluvial dynamics of large-river confluences is a scaling issue. Past work on the scaling of channel form and fluvial processes suggests that scaling relations for confluences are likely to be fundamentally nonlinear and therefore complex.

This paper reviews past work on confluence dynamics to provide a context for the development of hypotheses about the fluvial dynamics of large-river confluences. It also describes a research project to address these hypotheses.

2 CONFLUENCE DYNAMICS, PLANFORM GEOMETRY AND MOMENTUM FLUX RATIO

Experimental work in small laboratory confluences (Best 1987; Mosley 1976) has demonstrated the important influence of confluence planform geometry and momentum flux ratio of the incoming flows on fluvial dynamics within the CHZ. Confluence planform geometry includes the angle between the confluent channels (junction angle, α) and the orientation of the these channels relative to the receiving channel (planform symmetry) (Fig. 1). Symmetrical confluences have Y-shaped planforms ($\theta_1 \approx \theta_2$), whereas asymmetrical confluences have a receiving channel that is nearly collinear with one of the upstream channels (θ_1 or $\theta_2 = 0$). Momentum flux rato (M_r) is defined as $\rho \mathbf{Q}_2 \mathbf{V}_2 / \rho \mathbf{Q}_1 \mathbf{V}_1$ where $\mathbf{Q}$ is discharge ($m^3 s^{-1}$), $\mathbf{V}$ is mean velocity ($m\ s^{-1}$), ρ is density of water ($kg\ m^3$), and 1 and 2 refer to the main stem and tributary, respectively.

Confluence flow structure is characterized by numerous phenomena including flow stagnation at the junction apex, flow deflection in the center of the confluence, a mixing interface or shear layer between the confluent flows, lateral separation of flow from the channel banks at the downstream junction corner or corners, shear layers bounding the regions of flow stagnation and lateral separation, flow acceleration through the confluence, separation of flow from the channel bed on the lee-side of step-like avalanche faces bounding a region of scour within the center of the confluence, spiral flow or helical motion associated with curvature of flow streamlines, and progressive recovery of the flow downstream from the confluence

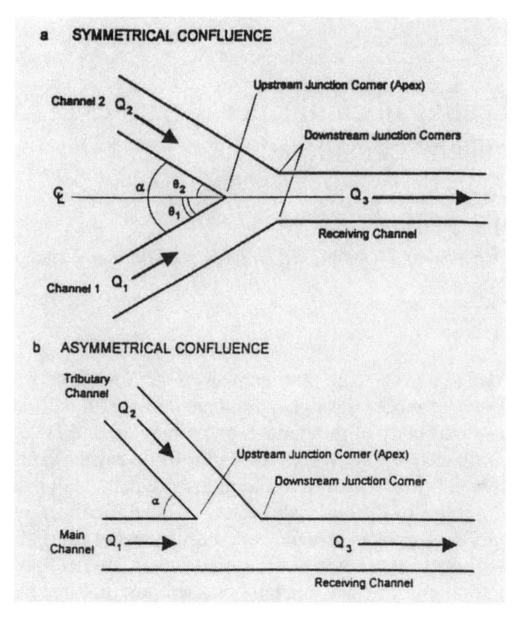

Figure 1. Confluence planform.

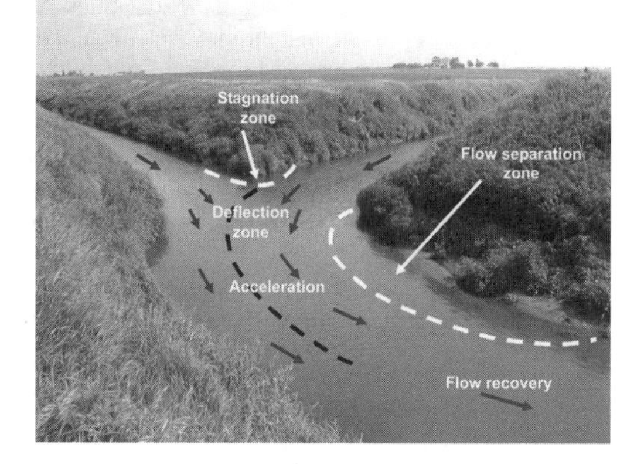

Figure 2. Spatial zonation of flow structure at a typical asymmetrical confluence, Kaskaskia River and Copper Slough (KRCS), East Central Illinois, USA.

(Fig. 2). As momentum ratio and junction angle vary, the spatial position and extent of these various hydrodynamic features also varies (Best & Reid 1984). At asymmetrical confluences, increases in momentum ratio or junction angle lead to increased penetration of flow from the lateral tributary into the confluence, enhanced flow separation at the downstream junction corner, and strong acceleration of flow through the confluence. Increases in junction symmetry tend to promote symmetry of flow on each side of the confluence. Field studies have confirmed general aspects of this conceptual model for small confluences (Ashmore et al. 1992, Bridge 1993, DeSerres et al. 1999, Rhoads 1996, Rhoads & Kenworthy 1995, 1998, Rhoads & Sukhodolov 2001, Roy & Bergeron 1990, Roy et al. 1988), but the extent to which it applies to large-river confluences is uncertain.

An important, yet poorly understood aspect of confluence dynamics is the effect of scale-related changes in channel geometry on flow structure. Downstream hydraulic geometry relations (Leopold & Maddock 1953) predict that bank full channel width (W), channel depth (D) and mean velocity increase nonlinearly, but at different rates, as power functions of discharge (Q): $W = aQ^b$, $D = cQ^f$, $V = kQ^m$ where $b > f > m$. The result is a progressive geometric distortion of channel shape (W/D) with increasing channel size. Thus, large rivers tend to be much wider relative to their depths than small streams. Most experimental and field studies have focused on conditions typical of headwater confluences, where disparities in width-depth relations among incoming tributary channels and the receiving channel downstream are not pronounced. In particular, channel widths generally are only one order of magnitude greater than channel depths in these systems. In large river systems, channel widths exceed channel depth by two or more orders of magnitude, thereby changing fundamentally the geometric constraints of the channel bed and banks on the flow.

The influence of the planform geometry of converging streams on flow structure has been a major focus of past work. Flow structure at stream confluences has been viewed as analogous to flow through two meander bends placed back to back (Ashmore & Parker 1983, Mosley 1976, Bridge 1993). As flows entering a confluence merge and mutually deflect one another, streamlines within each flow will curve, producing a centrifugal force oriented orthogonally to the path of curvature. Superelevation of the water surface produced by flow curvature generates a pressure gradient force that balances the centrifugal acceleration, but only in a depth-averaged sense. Local imbalance between the two forces over depth results in spiral motion of the flow.

At an asymmetrical confluence, the centrifugal effect produces superelevation of the water surface along the mixing interface near the junction apex and along the wall of the main channel opposite the downstream junction corner, and a region of low-pressure develops near the downstream junction corner (Bradbrook et al. 2000a, Hagar 1989, Weerakoon et al. 1991). This pattern of water surface topography generates twin helical cells within the central region of small asymmetrical confluences (Rhoads 1996, Rhoads & Kenworthy 1998, Rhoads & Sukhodolov 2001). Downstream, curvature of flow from the lateral tributary produces strong helicity over the inner portion of the receiving channel (Rhoads & Sukhodolov 2001). At symmetrical confluences, both flows curve to a similar extent (assuming the momentum ratio equals one), and spiral flow is well-developed on both sides of the mixing interface – the locus of water-surface superelevation (Ashmore et al. 1992; Bridge & Gable 1992). At small confluences, complicating factors, such as protruding bank vegetation, can disrupt helical motion even when the confluent flows curve substantially (Rhoads & Sukhodolov 2001).

The extent to which helical motion is important at large river confluences has yet to be explored. Simple considerations suggest that helical motion should be equally important at large and small confluences. The force-balance for helical motion in a curved open-channel flow is $\overline{U}^2/r \propto \Delta E/W$ where $\overline{U}$ is mean velocity of the flow, r is radius of curvature, W is width of flow, and ΔE is the amount of superelevation of the water surface. The ratio W/r is a scaling parameter that provides a scale-independent measure of bend sharpness. Thus, for a confluence flow with constant relative curvature (W/r), ΔE should be proportional to $\overline{U}^2$, thereby maintaining a constant intensity of helicity over scale. However, in contrast to bend flow, superelevation at the upstream junction corner occurs via mounding of the flows against one another, rather than against a solid boundary such as a channel bank (Weerakoon & Tamai 1989), and superelevation may not conform to the linear force-balance relation for bend flow.

The consideration that the width-depth ratio (W/D) at large confluences should greatly exceed W/D at small confluences leads to two alternative hypotheses regarding helical motion: **Flow Hypothesis 1 (FH1)** – the overall strength of helical motion is weaker at large than at small confluences for the same degree of relative flow curvature because the relative proportion of the flow width over which mounding of the convergent flows occurs is greater at large than at small confluences; **Flow Hypothesis 2 (FH2)** – helical motion has the same intensity as at small confluences, but is confined to a small proportion of the total flow width, reflecting pronounced, but localized mounding of the flow near the upstream junction apex.

3 CONFLUENCE BED MORPHOLOGY

The morphology of small field and mobile-bed laboratory confluences has been documented in detail and commonly is characterized by a well-developed region of scour within the confluence (Ashmore & Parker 1983, Best 1986; Best 1988, Biron et al. 1993b, Bristow et al. 1993, Mosley 1976, Rhoads 1996, Rhoads & Kenworthy 1995, 1998, Rhoads & Sukhodolov 2001, Roy & Bergeron 1990, Roy et al. 1988). The development of scour is a combination of flow acceleration through the confluence, high turbulence within the shear layer, and high bed shear stress associated with downward transport of high-momentum fluid toward the bed by helical motion of the flow, but the individual contributions of these factors to scour have yet to be unraveled. All of the factors are dependent on geometric relations among the channels forming the confluence, the planform geometry of the confluence, and the momentum-flux ratio.

At a concordant confluence, the beds of the two tributaries slope into the scour hole uniformly, whereas at a discordant confluence the bed of one of the tributaries is elevated relative to the bed of the other tributary and flow moves over a step into the main channel. If either the side slopes of the scour hole or the discordant step are steep enough flow separation can occur in the lee of these features (Best & Roy 1991, McLelland et al. 1996). The main effect of this separation is to create a cross-stream pressure gradient near the bed that produces distortion of the shear layer, or mixing interface, between the two flows (Biron et al. 1996a, b, Biron et al. 1993a, Bradbrook et al. 1998, Bradbrook et al. 2001). The magnitude of this distortion relative to flow width is such that it apparently disrupts helical motion that otherwise would be induced by flow curvature (Biron et al. 1993a). On the other hand, flow separation from the bed in the lee of the step might also generate wake vortices with helical motion (McLelland et al. 1996).

Patterns of scour at small asymmetrical confluences are highly sensitive to variations in momentum flux ratio. Events with $M_r > 1.0$ result in protrusion of a wedge of sediment into the confluence, confinement of the scour hole to the far bank of the confluence, and decrease in the depth of scour (Fig. 3). Conversely, events with $M_r < 1.0$ produce deep scour holes aligned with the junction-angle bisector that curve downstream into the receiving channel (Fig. 4).

The development of bars at the downstream junction corner at asymmetrical confluences is another prominent morphological feature at many small confluences. Laboratory experiments suggest that such bars are the result of sedimentation within the separation zone (Best 1986, 1988). This type of bar growth requires entrainment of suspended sediment into the separation zone from the adjacent freestream. In contrast, field studies indicate that bar development at the downstream junction corner involves deposition of bedload (Rhoads & Kenworthy 1995). Moreover, such bars often have a coarse surface layer and the particle size of the surface layer decreases from the bar head to the bar tail (Mayer 1995) (Fig. 5). This sedimentological structure is similar to that of point bars, which again suggests an analogy between the dynamics of confluences and the dynamics of meander bends. According to this model, sediment flux convergence resulting from spatial decreases in bed shear

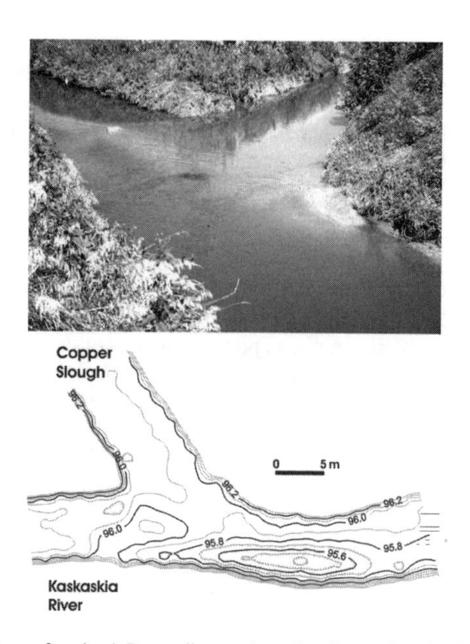

Figure 3. (top) Protruding wedge of sediment into KRCS confluence in response to high M_r and dominance of flow in Copper Slough on right. (bottom) bed morphology for high M_r showing confinement of scour to bank opposite Copper Slough (contour in meters, arbitrary datum, November 1989).

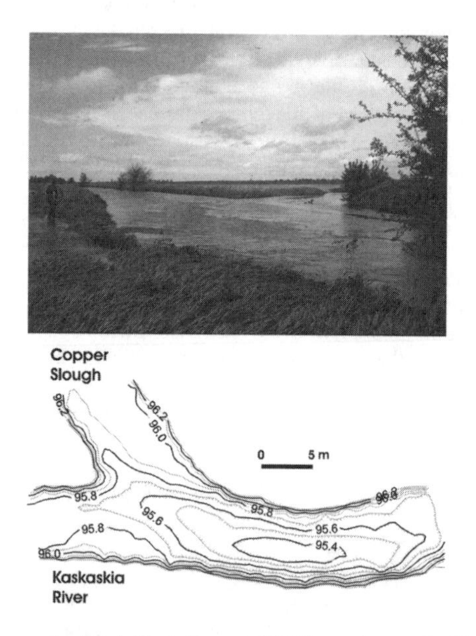

Figure 4. (top) High flow at KRCS confluence in May 1990 with $M_r < 1.0$. (bottom) bed morphology for low M_r showing scour hole orientation changing from parallel to junction-angle bisector within the confluence to parallel to receiving channel downstream. (contour in meters, arbitrary datum, May 1990).

Figure 5. Junction-corner bar with coarse gravel on surface and downstream fining of surficial material.

Figure 6. Bar-unit complex at confluence of Kaskaskia River and Copper Slough, IL showing subaerial deposits and water from Copper Slough (right) cascading over bar surface in the form of a riffle within the confluence.

stress around the downstream junction corner leads to progressive deposition of bedload and a decrease in particle size of deposited material along the inner bank of the receiving channel. Field observations suggest that junction-corner bars are the subaerial component of large bar units that extend into the confluence and wrap around the downstream junction corner (Fig. 6), similar to the bar units in meandering rivers that comprise the riffle and point bar components of bed morphology (Dietrich 1987).

Few studies have examined bed morphology at large confluences, the extent to which characteristic patterns that develop at small confluences are typical of large-river junctions, and the influence of bed morphology on flow dynamics within these confluences. Geometric scaling is not self-similar in rivers; therefore, linear scaling of bed morphology (i.e. deep scour holes with steep side slopes) seems unlikely. Large rivers may be able to accommodate additions of flow from tributaries without substantial accelerative

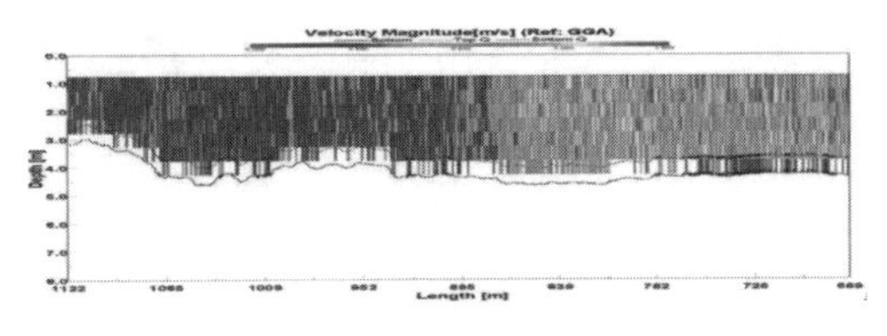

Figure 7. Streamwise velocity transect at confluence of Vermilion River and Wabash River, Indiana showing abrupt change in velocity magnitude at shear layer between slow Vermilion flow (left) and fast Wabash flow (right).

effects that produce deep scour and the large width-depth ratio of these rivers relative to shear-layer widths may limit the relative spatial extent of any shear-layer related scour (**Bed-Morphology Hypothesis 1 – BH1**). Bed morphology at small confluences often is highly responsive to changes in momentum flux ratio, changing configuration dramatically in response to individual hydrological events (Rhoads 1996). Such changes, if also common at large river confluences, could be important in regulating the downstream flux of bed material within drainage networks. However, the dynamics of large-river confluences have yet to be ascertained; inertial considerations suggest that given the large amounts of sediment mobilization or storage required to effect major changes in bed configuration, large-river confluences may not be as responsive to individual hydrological events as small confluences (**Bed-Morphology Hypothesis 2 – BH2**).

4 TURBULENCE, FLOW STRUCTURE AND MIXING

The collision of converging flows at confluences makes these locations one of the most highly turbulent hydrodynamic environments within fluvial systems. A characteristic feature of river confluences is the development of a distinct shear layer between the two flows. At experimental or natural confluences the shear layer is readily identified by abrupt lateral changes in velocity (Fig. 7) or by high levels of turbulence kinetic energy (TKE) within this layer compared to levels of TKE in the ambient flow (Biron et al. 1996b; De Serres et al. 1999; Sukhodolov & Rhoads 2001). In depth-limited environments, the dynamics of the mixing layer are strongly influenced by one dimension of the flow being much less than the other (Barbarutsi & Chu 1998; Chu & Barbarutsi 1988; Uijttewaal & Booij 2000; Uijttewaal & Tukker 1998). The effects of friction constrain mixing layer growth, limiting the width of the mixing layer downstream of the point of initiation. Findings from field investigations suggest that results of experimental work on depth-limited shear layers are generally applicable to confluence shear layers (Rhoads & Sukhodolov 2004), but that

Figure 8. Shear layer between Vermilion River (left) and Wabash River (right) illustrating shear-layer vortices.

Figure 9. Mixing interface at confluence of Wabash River and White River, Illinois-Indiana.

the angled nature of flow convergence at stream confluences compared to parallel flow in experimental studies of shear layers may have an important influence on low-frequency shear-layer dynamics. In particular, mutual backwater effects of the flows on one another may lead to low-frequency oscillations of the layer that are dependent upon the geometry of the confluence and momentum flux ratio (Biron et al. 2002; Rhoads & Sukhodolov 2004). Given that shear-layer vortices scale with depth and channel width-depth ratio increases with river size (Fig. 8), the proportion of the flow width occupied by the shear layer at a given distance downstream of the confluence, as well as the proportional rate of lateral expansion (rate of expansion per unit channel width) of the shear layer downstream of the confluence, should be less for large rivers than for small rivers (**Mixing Hypothesis 1 – MH1**).

Transverse mixing in rivers, including confluences, is governed mainly by turbulent diffusion and advective transport (Rutherford 1994). The development of helical motion associated with flow curvature can produce large-scale transverse advection, greatly increasing rates of mixing at confluences (Kenworthy & Rhoads 1995; Rhoads & Kenworthy 1995; Rhoads & Sukhodolov 2001). If bed discordance exists, distortion of the shear layer by lateral pressure gradients near the bed also can enhance advective mixing (Gaudet & Roy 1995; Biron et al. 2004). These advective mechanisms can produce substantial mixing of the two flows within several channel widths of the confluence. In contrast, distinct mixing interfaces often are visible or detectable in large rivers many kilometers downstream from a confluence (MacKay 1970).

Recent work implies that lateral mixing immediately downstream of large-river confluences is dominated by turbulent diffusion with advective mixing occurring farther downstream in meandering sections of the river far downstream from the confluence (Rathbun & Ronstad 2004). In the absence of strong advective mixing within the confluence, shear-layer turbulence can dissipate before the two confluent flows are mixed (Rhoads & Sukhodolov 2004). Thus, a distinction exists between the *shear layer*, defined on the basis of turbulence characteristics, and the *mixing interface*, defined on the basis of a contrast in constitutive properties of the fluid (temperature, conductivity, chemical solutes) (Sukhodolov & Rhoads 2001) (Fig. 9). It is hypothesized that dissipation of shear-layer turbulence before complete mixing is more common in large rivers than in small rivers due to the absence in large river confluences of strong advective effects, such as helical motion or lateral pressure gradients associated with bed discordance, that enhance rates of mixing relative to turbulence diffusion within the shear layer (**Mixing Hypothesis 2 – MH2**).

5 LARGE-RIVER CONFLUENCE RESEARCH

The preceding discussion has argued that a critical need exists for process-based field investigations of large-river confluences to explore how confluence dynamics change with scale. Such an investigation has been initiated at several confluences in the midwestern United States. The general objective of the project is to examine how time-averaged three-dimensional flow structure, bed morphology, shear-layer dynamics and turbulence, and mixing of the two flows vary with confluence planform geometry and junction angle, and with the momentum flux ratio of the incoming flows. The research represents a companion project to a study examining the fluvial dynamics of the Parana River-Paraguay River confluence in Argentina (http://earth.leeds.ac.uk/research/seddies/largeconf/index.htm).

Specific objectives of the research are to: (1) examine the influence of variations in confluence planform geometry (junction symmetry, junction angle,

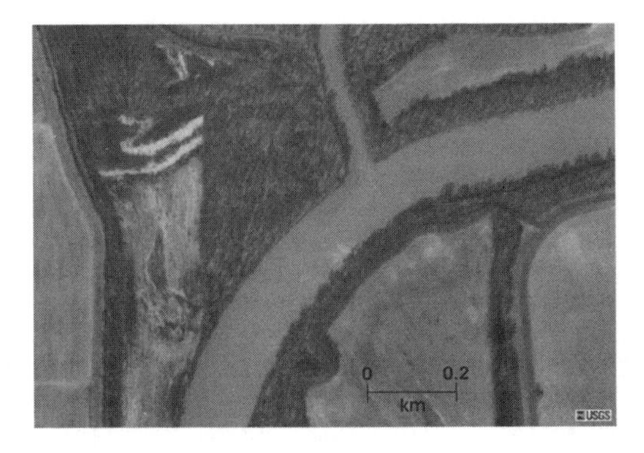

Figure 10. Confluence of Wabash River and Embarras River (top).

Figure 11. Confluence of the Wabash River and Vermilion River (left).

channel curvature) on patterns of time-averaged three-dimensional flow through large river confluences, (2) determine the morphology of the channel bed at large river confluences, changes in bed morphology in response to variations in momentum flux ratio and the interaction between bed morphology and three-dimensional flow structure, (3) examine the spatial and temporal structure of turbulence within the shear layer at large river confluences, (4) document patterns of mixing at large river confluences and the relation of mixing to time-averaged three-dimensional flow structure and to shear-layer turbulence, and (5) use the field data as inputs to an advanced CFD model of three-dimensional flow to test the capability of the model to simulate flow through natural large-river confluences. These objectives will be evaluated through comparison of research results with findings for small field or experimental confluences to determine similarities and differences in fluvial dynamics at different scales. Fulfillment of these six objectives will contribute to basic knowledge in fluvial geomorphology by providing a rigorous evaluation of the validity of conceptual models of confluence dynamics derived from experimental work for describing the fluvial dynamics of large-river confluences.

Three confluences along the Wabash River in Indiana-Illinois will serve as study sites for the project: the confluence of the Wabash River and White River (Fig. 9), the confluence of the Wabash River and Embarras River (Fig. 10) and the confluence of the Wabash River and Vermilion River (Fig. 11). This set of sites encompasses differences in junction planform, junction angle, and momentum flux ratio – the key factors affecting the dynamics of small confluences. Data collection at each site will involve: (1) characterization of confluence hydrological relations based on nearby gaging-station records, (2) surveying of confluence bed morphology and morphologic change, (3) measurements of time-averaged three-dimensional velocities and water temperatures upstream, downstream and throughout each confluence for hydrological events with different momentum flux ratios, and (4) detailed measurements of turbulence and mixing in the shear layer/mixing interface within and immediately downstream of each confluence. Data analysis will focus on three-dimensional characterization of the spatial structure of bed morphology, flow structure and mixing at each confluence for different momentum-ratio events. Results of the analysis will provide the basis for addressing specific hypotheses related to flow structure (**FH1**, **FH2**), bed morphology (**BH1**, **BH2**) and mixing (**MH1**, **MH2**) at large river confluences. Numerical modeling will complement the field investigation by allowing findings for site-specific locations to be understood more completely within a theoretical context and by illustrating how generalizable processes manifest themselves differently in site-specific contexts.

6 CONCLUSION

Over the past 30 years, studies of laboratory junctions and field investigations at small confluences have contributed greatly to knowledge of confluence dynamics. However, basic considerations of changes in channel geometry with scale suggest that scaling relations for confluence dynamics are likely to be non-linear. Hypotheses about potential differences between large and small confluences can be derived from these considerations and from knowledge derived from studies of small confluences. Testing of these hypotheses requires field investigations focusing on collection and analysis of data from large-river confluences. Such studies, including the one described in this paper, are now underway and should lead to an improved understanding of how confluence dynamics do or do not vary with scale.

ACKNOWLEDGMENTS

Material presented in this paper was developed as part of several projects supported by the National Science Foundation through grants BCS-9024225, BCS-9710068, OISE-0097059 and BCS-0453316.

REFERENCES

Ashmore, P.E. & Parker, G. 1983. Confluence scour in coarse braided streams. *Water Resources Research* 19: 392–402.

Ashmore, P.E., Ferguson, R.I., Prestegaard, K.L., Ashworth, P.J. & Paola, C. 1992. Secondary flow in anabranch confluences of a braided, gravel-bed stream. *Earth Surface Processes and Landforms* 17: 299–311.

Barbarutsi, S. & Chu, V.H. 1998. Modeling transverse mixing layer in shallow open-channel flows. *Journal of Hydraulic Engineering* 124: 718–727.

Best, J.L. 1986. The morphology of river channel confluences. *Progress in Physical Geography* 10: 157–174.

Best, J.L. 1987. Flow dynamics at river channel confluences: implications for sediment transport and bed morphology. In F.G. Ethridge, R.M. Flores & M.D. Harvey (eds), *Recent Developments in Fluvial Sedimentology:* 27–35. Special Publication 39. Tulsa, OK: The Society of Economic Paleontologists and Mineralologists.

Best, J.L. 1988. Sediment transport and bed morphology at river channel confluences. *Sedimentology* 35: 481–498.

Best, J.L. & Reid, I. 1984. Separation zone at open-channel junctions. *Journal of Hydraulic Engineering* 110: 1588–1594.

Best, J.L. & Roy, A.G. 1991. Mixing-layer distortion at the confluence of channels of different depth. *Nature* 350: 411–413.

Biron, P., De Serres, B. & Best, J.L 1993a. Shear layer turbulence at unequal depth channel confluence. In N. J. Clifford, J. R. French and J. Hardisty (eds), *Turbulence: Perspectives on Flow and Sediment Transport*: 197–213. Chichester: John Wiley & Sons Ltd.

Biron, P., Roy, A.G., Best, J.L. & Boyer, C.J. 1993b. Bed morphology and sedimentology at the confluence of unequal depth channels. *Geomorphology* 8: 115–129.

Biron, P., Best, J.L. & Roy, A.G. 1996a. Effects of bed discordance on flow dynamics at open channel confluences. *Journal of Hydraulic Engineering* 122: 676–682.

Biron, P., Roy, A.G. & Best, J.L. 1996b. Turbulent flow structure at concordant and discordant open-channel confluences. *Experiments in Fluids* 21: 437–446.

Biron, P.M., Richer, A., Kirkbride, A.D., Roy, A.G. & Han, S. 2002. Spatial patterns of water surface topography at a river confluence. *Earth Surface Processes and Landforms* 27: 913–928.

Bradbrook, K.F., Biron, P., Lane, S.N., Richards, K.S. & Roy, A.G. 1998. Investigation of controls on secondary circulation and mixing processes in a simple confluence geometry using a three-dimensional numerical model. *Hydrological Processes* 12: 1371–1396.

Bradbrook, K.F., Lane, S.N. & Richards, K.S. 2000a. Numerical simulation of three-dimensional, time-averaged flow structure at river channel confluences. *Water Resources Research* 36: 2731–2746.

Bradbrook, K.F., Lane, S.N., Richards, K.S., Biron, P.M. & Roy, A.G. 2000b. Large Eddy Simulation of periodic flow characteristics at river channel confluences. *Journal of Hydraulic Research* 38: 207–215.

Bradbrook, K.F., Lane, S.N., Richards, K.S., Biron, P.M. & Roy, A. G. 2001. Role of bed discordance at asymmetrical river confluences. *Journal of Hydraulic Engineering-Asce* 127: 351–368.

Bridge, J.S. 1993. The interaction between channel geometry, water flow, sediment transport and deposition in braided rivers. In J.L. Best and C.S. Bristow (eds), *Braided Rivers*: 13–71. London: Geological Society Special Publication 75.

Bridge, J.S. & Gable, S.L. 1992. Flow and sediment dynamics in a low sinuosity, braided river - Calamus River, Nebraska Sandhills. *Sedimentology* 39: 125–142.

Bristow, C.S., Best, J.L. & Roy, A.G. 1993. Morphology and facies models of channel confluences. In M. Marzo and C. Puidefabregras *Alluvial Sedimentation*: 91–100. Vol. 17, Special Publications of the International Association of Sedimentology. Malden, MA: Blackwell.

Chu, V.H. & Barbarutsi, S. 1988. Confinement and bed-friction effects in shallow turbulent shear layers. *Journal of Hydraulic Engineering* 114: 1257–1274.

De Serres, B., Roy, A.G., Biron, P.M. & Best, J.L. 1999. Three-dimensional structure of flow at a confluence of river channels with discordant beds. *Geomorphology* 26: 313–335.

Dietrich, W.E. 1987. Mechanics of flow and sediment transport in river bends. In K.S. Richards (ed), River Channels Form and Process: 179–224. Oxford: Basi Blackwell.

Gaudet, J.M. & Roy, A.G. 1995. Effect of bed morphology on flow mixing length at river confluences. *Nature* 373: 138–139.

Hagar, W.H. 1989. Transitional flow in channel junctions. *Journal of Hydraulic Engineering* 115: 243–259.

Kenworthy, S.T. & Rhoads, B.L. 1995. Hydrologic control of spatial patterns of suspended sediment concentration at a stream confluence. *Journal of Hydrology* 168: 251–263.

Lane, S.N, Bradbrook, K.F., Richards, K.S., Biron, P.A. & Roy, A.G. 1999. The application of computational fluid dynamics to natural river channels: three-dimensional versus two-dimensional approaches. *Geomorphology* 29: 1–20.

Leopold, L.B. & Maddock, Jr., T. 1953. *The Hydraulic Geometry of Stream Channels and Some Physiographic Implications.* U.S. Geological Survey Professional Paper 252. Washington, D.C.: U.S. Government Printing Office.

MacKay, J.R. 1970. Lateral mixing of the Liard and Mackenzie rivers downstream from their confluence. *Canadian Journal of Earth Sciences* 7: 111–124.

Mayer, D.R. 1995. Hydrological Control of Spatial Patterns of Surficial Bed-material at a Stream Confluence. M.S. Thesis, Urbana, IL: University of Illinois.

McLelland, S.J, Ashworth, P.J. & Best, J.L. 1996. The origin and downstream development of coherent flow structures at channel junctions. In P. J. Ashworth, S. J. Bennett, J. L. Best and S. J. McLelland (eds), *Coherent Flow Structures in Open Channels*: 459–490. Chichester: John Wiley and Sons.

Mosley, M.P. 1976. An experimental study of channel confluences. *Journal of Geology* 84: 535–562.

Rathbun, R.E. & Ronstad, C.E. 2004. Lateral mixing in the Mississippi River below the confluence with the Ohio River. *Water Resources Research* 40: W05207, doi: 10.1029/2003WR002381.

Rhoads, B.L. 1996. Mean structure of transport-effective flows at an asymmetrical confluence when the main stream is dominant. In P. J. Ashworth, S. J. Bennett, J. L. Best & S. J. McLelland (eds), *Coherent Flow Structures in Open Channels*: 491–517. Chichester: John Wiley.

Rhoads, B.L. & Kenworthy, S.T. 1995. Flow structure at an asymmetrical stream confluence. *Geomorphology* 11: 273–293.

Rhoads, B.L. & Kenworthy, S.T. 1998. Time-averaged flow structure in the central region of a stream confluence. *Earth Surface Processes and Landforms* 23: 171–191.

Rhoads, B.L. & Kenworthy, S.T. 1999. On secondary circulation, helical motion and Rozovskii-based analysis of time-averaged two-dimensional velocity fields at confluences. *Earth Surface Processes and Landforms* 24: 369–375.

Rhoads, B.L. & Sukhodolov, A.N. 2001. Field investigation of three-dimensional flow structure at stream confluences: 1. Thermal mixing and time-averaged velocities. *Water Resources Research* 37: 2393–2410.

Rhoads, B.L. & Sukhodolov, A.N. 2004. Spatial and temporal structure of shear-layer turbulence at a stream confluence. *Water Resources Research* 40: W06304, doi:10.1029/2003WR002811.

Roy, A.G. & Bergeron, N. 1990. Flow and particle paths in a natural river confluence with coarse bed material. *Geomorphology* 3: 99–112.

Roy, A.G., Roy, R. & Bergeron, N. 1988. Hydraulic geometry and changes in flow velocity at a river confluence with coarse bed material. *Earth Surface Processes and Landforms* 13: 583–598.

Rutherford, J.C. 1994. *River Mixing*. Chichester: John Wiley.

Sukhodolov, A.N. & Rhoads, B.L. 2001. Field investigation of three-dimensional flow structure at stream confluences 2. Turbulence. *Water Resources Research* 37: 2411–2424.

Uijttewaal, W.S.J. & Booij, R. 2000. Effects of shallowness on the development of free-surface mixing layers. *Physics of Fluids* 12: 392–402.

Uijttewaal, W.S.J. & Tukker, J. 1998. Development of quasi two-dimensional structures in a shallow free-surface mixing layer. *Experiments in Fluids* 24: 192–200.

Weerakoon, S.B., Kawahara, Y. & Tamai, N. 1991. Three-dimensional flow structure in channel confluences of rectangular section. *Proceedings of the XXIV Congress, IAHR*, A373–380, Madrid, Spain.

Weerakoon, S.B. & Tamai, N. 1989. Three-dimensional calculation of flow in river confluences using boundary-fitted coordinates. *Journal of Hydroscience and Hydraulic Engineering* 7: 51–62.

River, Coastal and Estuarine Morphodynamics: RCEM 2005 – Parker & García (eds)
© 2006 Taylor & Francis Group, London, ISBN 0 415 39270 5

Hydraulic properties of a bifurcated channel system with different bed slopes under sub-critical flow

K. Kobayashi, K. Hasegawa & K. Moriya
Hokkaido university, Sapporo, Japan

ABSTRACT: This study investigates mainstream alternation by examining the hydraulic properties of a bifurcated channel on a flat bed and on a movable bed, under sub-critical flow. We consider a bifurcation model and apply momentum equations to it. The results show that the system has two solutions for discharge ratio r: one for sub-critical flow, and the other for super-critical flow. We performed flat-bed and movable-bed experiments. The flat-bed experiments showed that the discharge ratio r remains constant during each run, a result that agrees fairly well with the theoretical solution under sub-critical flow. The theoretical solution means that discharge into the steeper channel branch is greater than that into the other channel. The movable-bed experiments showed that r oscillates with time for each run. And mainstream alternation occurred only in one Run which bed form is dune.

1 INTRODUCTION

Rivers have various watercourse forms, and one is bifurcated channel. One cause of such bifurcation is the presence of a bar or island. Channel bifurcations or islands in alluvial rivers are unstable. Mainstream alternation is a phenomenon in which the greater flow shifts back and forth between the channel branches. This alternation often occurs also in braided mountain rivers, where the flow is supercritical. In such cases the alternation arises as a result of hydraulic conditions immediately upstream of the bifurcation and as a result of sediment transport into the bifurcation.

The mechanisms of alternation in mountain rivers have been investigated by Hasegawa et al. They noted that one of the branches the bifurcated channel closes, they experimented on the relationship between hydraulic conditions and flow alternation, and they analyzed such alternation using momentum equations under super-critical flow. Hasegawa et al. gave four solutions for water discharge ratio and water depth under supercritical plural flow: (1) both channels have supercritical flow, (2) hydraulic jump occurs in only one channel, and then the flow becomes subcritical, (3) hydraulic jump occurs in only the other channel, and then the flow becomes subcritical, and (4) hydraulic jump occurs in both channels, and then the flow becomes subcritical. However, mainstream alternation in rivers on alluvial plains can result from hydraulic conditions and channel configuration downstream of the bifurcation, e.g., channel branching.

The hydraulic properties at a channel bifurcation on a movable bed with bifurcated channels of different bed slope under subcritical flow has been investigated by several researchers, most recently Bolla, Repetto and Tubino. They found very interesting features of bifurcation systems, such as the existence of plural solutions to discharge ratio and instability of watercourse. They propose a physical nodal point condition to describe channel bifurcation on a movable bed, using a one-dimensional approach. At high values of Shields stress, there is only one solution for discharge ratio when both of the bifurcated channels are open, and the solution is stable. But as the Shields stress decreases, two new stable solutions appear, for which water and sediment discharge in the two branched channels are unbalanced. In this case, the previous solution is unstable.

This study investigates mainstream alternation in terms of the hydraulic properties of a bifurcated channel on a flat bed and on a movable bed under subcritical flow. We considered a bifurcation model in which the main channel is divided into two-branched channels of equal width but different bed slopes, then we employ momentum equations for two control-volume sections to obtain the discharge ratio r and the upstream water depth h_0. And we discuss flume experiments of flat bed and movable bed performed under sub-critical flow.

2 MOMENTUM ANALYSIS MODEL

We consider a bifurcation model in which the main channel of width B is divided into two branched channels of equal width and but different bed slopes

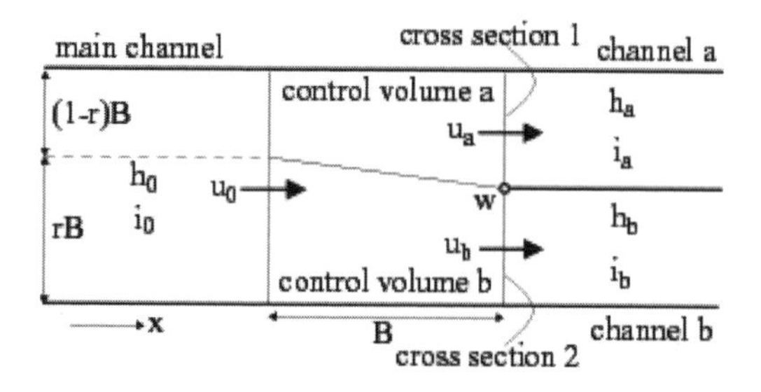

Figure 1. Bifurcation model.

(Figure 1). The channel width B is equal to the distance from the bifurcation to the cross-section upstream of the bifurcation where the flow is unaffected by the bifurcation. The flow velocity at and upstream of that unaffected cross-section is uniform (u_0). Further, we assumed that at the cross-section where the flow is unaffected by the bifurcation, the streamline can be divided into two flows whose widths are in accordance with the ratio of discharge in the two channels after the bifurcation.

We apply momentum equations to two control volumes: a and b. The water depth h_0 is an unknown variable under sub-critical flow.

$$\rho u_a^2 h_a \frac{B}{2} - \rho u_0^2 h_0 (1-r)B = \frac{1}{2}\rho g h_0^2 (1-r)B + \frac{1}{2}\left(\frac{1}{2}\rho g h_0^2 + \frac{1}{2}\rho g h_w^2\right)\left(r - \frac{1}{2}\right)B - \frac{1}{2}\rho g h_a^2 \frac{B}{2} + \rho g V_a i - T_{ax} \tag{1}$$

$$\rho u_b^2 h_b \frac{B}{2} - \rho u_0^2 h_0 r B = \frac{1}{2}\rho g h_0^2 r B - \frac{1}{2}\left(\frac{1}{2}\rho g h_0^2 + \frac{1}{2}\rho g h_w^2\right)\left(r - \frac{1}{2}\right)B - \frac{1}{2}\rho g h_b^2 \frac{B}{2} + \rho g V_b i - T_{bx} \tag{2}$$

where ρ is water density, g is the gravitational acceleration; u_a and u_b are the respective flow velocities at cross sections 1 and 2; i_0, i_a, i_b are the respective bed slopes of the main channel, branched channel a and branched channel b; h_w is the water depth at point w; V_a and V_b are the respective control volumes a and b; and T_{ax} and T_{bx} are the respective totals of shear stress acting in the x direction for control volumes a and b.

The continuity equation is in this form:

$$(1-r)Bh_0u_0 = \frac{B}{2}h_a u_a \tag{3}$$

$$rBh_0u_0 = \frac{B}{2}h_b u_b \tag{4}$$

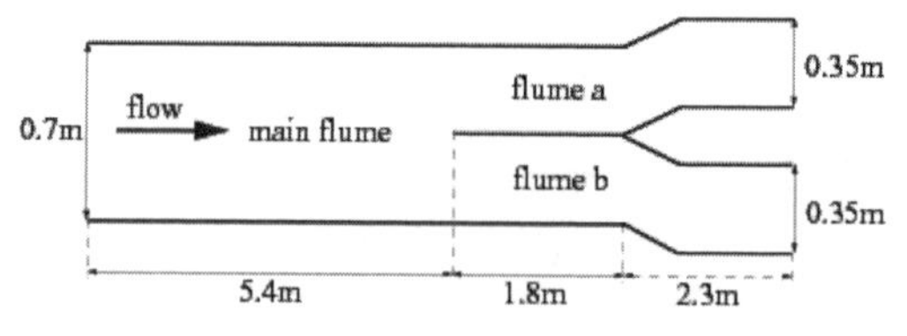

Figure 2. Experimental flume: The main flume is divided into two branches; the bed slope of Flume (b) is steeper than that of Flume (a).

We assume that h_a and h_b are approximated uniform flow depths; hence, they are in the form:

$$h_a = \left\{\frac{2n(1-r)Q_0}{B\sqrt{i_a}}\right\}^{\frac{3}{5}} \tag{5}$$

$$h_b = \left\{\frac{2nrQ_0}{B\sqrt{i_b}}\right\}^{\frac{3}{5}} \tag{6}$$

Furthermore we consider that the water depth h_w takes an average of h_a and h_b, and the sum of the gravitational term and the friction term is zero. Hence, Equations (1) and (2) can be rewritten:

$$\left(\frac{r}{2} - \frac{3}{4}\right)h_a h_0^3 + \left\{\frac{4q_0^2(1-r)^2}{g} + \left(\frac{1}{8} - \frac{r}{4}\right)h_a^2 h_b + \left(\frac{1}{16} - \frac{r}{8}\right)h_a h_b^2 + \left(\frac{9}{16} - \frac{r}{8}\right)h_a^3\right\}h_0 - \frac{2q_0^2(1-r)h_a}{g} = 0 \tag{7}$$

$$\left(-\frac{r}{2} - \frac{1}{4}\right)h_b h_0^3 + \left\{\frac{4q_0^2 r^2}{g} + \left(\frac{r}{8} - \frac{1}{16}\right)h_a^2 h_b + \left(\frac{r}{4} - \frac{1}{8}\right)h_a h_b^2 + \left(\frac{7}{16} + \frac{r}{8}\right)h_b^3\right\}h_0 - \frac{2rq_0^2 h_b}{g} = 0 \tag{8}$$

The solutions of r and h_0 can be solved by Equations (7) and (8), using the Newton-Raphson method.

3 FLAT-BED EXPERIMENTS

3.1 *Experimental summary*

Experiments were performed under sub-critical flow in a flume (Figure 2) with an upstream channel 5.4 m

Table 1. Flat-bed experimental condition.

	Water discharge Q (liter/sec)	*Bed slope of flume a* i_a	Frude number F_r	Shields stress τ_*
Run.1-1	1.87	1/724	0.453	0.035
Run.1-2	2.84	1/724	0.416	0.049
Run.2-1	1.68	1/1110	0.453	0.032
Run.2-2	2.57	1/1110	0.524	0.039
Run.3-1	1.63	1/2448	0.378	0.036
Run.3-2	0.96	1/2448	0.360	0.026

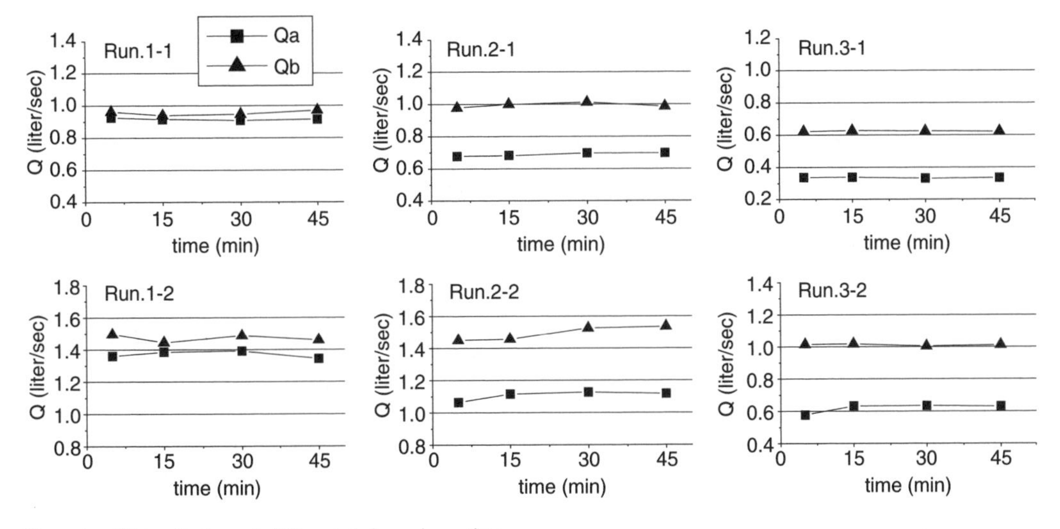

Figure 3. Water discharge in bifurcated channels vs. time.

long and 0.7 m wide, and a downstream divided into two channels each 4.1 m long and 0.35 m wide but with different bed slopes. Sand of uniform grain size (0.5 mm) was laid flat throughout the channel. We set the Shields stress at below 0.05 to make the bed condition flat. The experimental conditions are shown in Table 1. We measured the water discharge of each branch at the outlet using a container, and the water depth 1 m upstream of the branch by point gauge.

3.2 *Experimental result*

We plotted the water discharges in Flume (a) and Flume (b) against time (Figure 3). The figure shows that discharge in Flume (b), the steeper of the two flumes, is greater than that in Flume (a). Also, the discharge is roughly constant for each, despite small fluctuations resulting from measurement error.

4 COMPARISON BETWEEN EXPERIMENTAL AND THEORETICAL RESULTS

4.1 *Solutions for momentum equations*

Figure 4 shows that solutions of water discharge ratio r and upstream water depth h_0 for Run 1-1 obtained using momentum Equations (7) and (8) plot in the range where the h_0-value exceeds zero. The solution for Equation (7) is shown as the solid line, and the solution for Equation (8) is the broken line. The intersections of the two lines are solutions r and h_0 obtained using the momentum analysis model, the symbol ▲ is the solution under sub-critical flow, and the symbol • is the solution under super-critical flow. The results show that the system has two solutions for r and h_0: one for sub-critical flow, and the other for super-critical flow.

In Figure 5-a, r is plotted against the ratio of bed slopes i_b/i_a at $h_b = 1/531$ and $Q = 2.0$ (liter/sec). The figure shows that r increases as i_b/i_a increases.

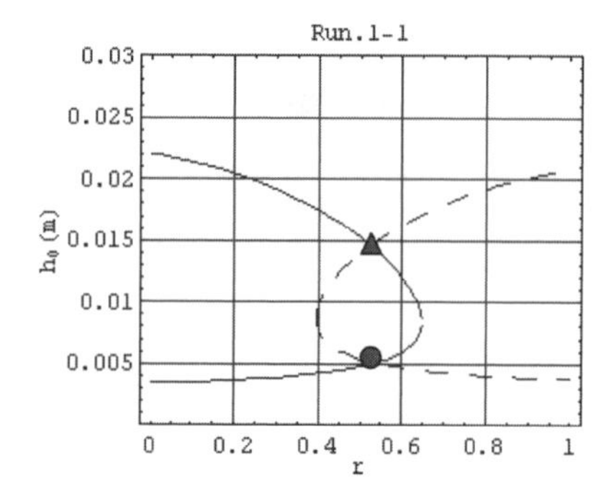

Figure 4. Solutions for momentum equations in Run 1-1.

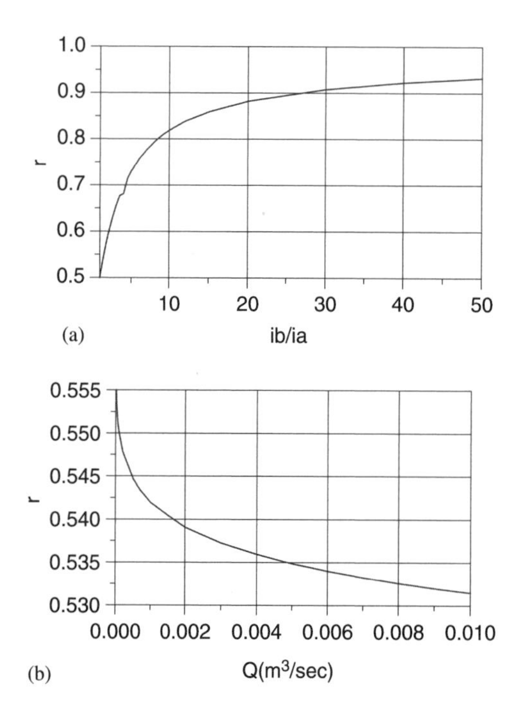

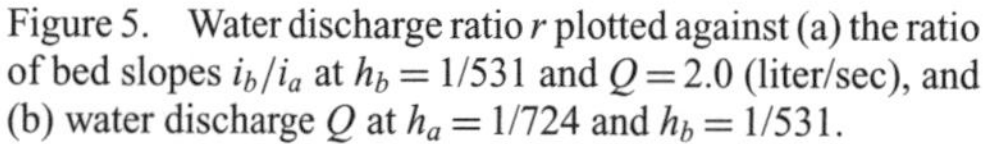

Figure 5. Water discharge ratio r plotted against (a) the ratio of bed slopes i_b/i_a at $h_b = 1/531$ and $Q = 2.0$ (liter/sec), and (b) water discharge Q at $h_a = 1/724$ and $h_b = 1/531$.

In Figure 5-b, the water discharge ratio r is plotted against water discharge Q at $h_a = 1/724$ and $h_b = 1/531$. The figure shows that r decreases as Q increases.

4.2 *Comparison between experimental and theoretical results*

Figure 6-a compares theoretical r values from momentum equations to experimental r values, and Figure 6-b compares h_0 values from momentum equations to experimental h_0 values. According to Figure 6, the theoretical values agree with the experimental values, which shows that the momentum analysis model reproduced the observed trends. The theoretical h_0 value is roughly equal to uniform flow depth; therefore, it is assumed that h_0 has little effect on bifurcation.

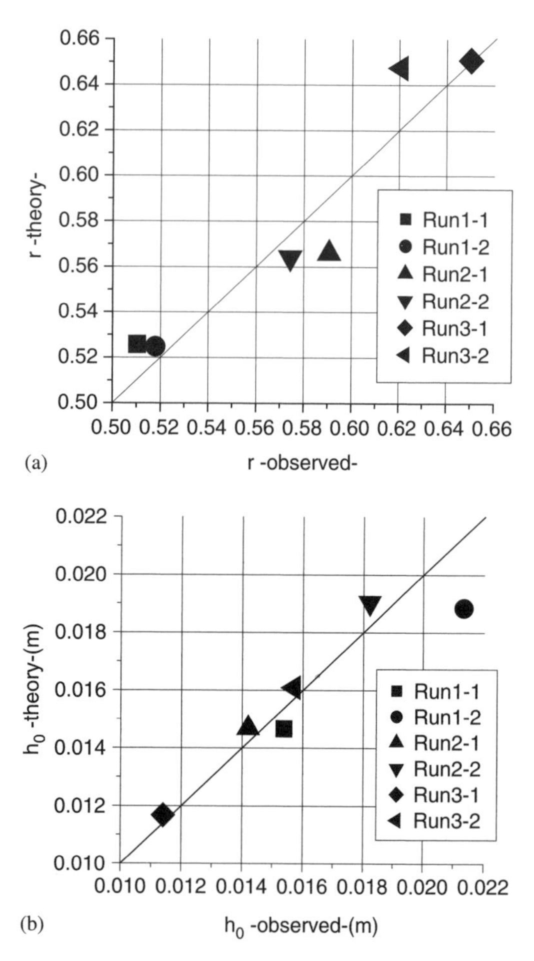

Figure 6. Comparison between experimental and theoretical results.

5 MOVABLE-BED EXPERIMENTS

We performed movable-bed experiments in 3 series, each with 11 runs. In Series 1, 2 and 3, the bed slope of Flume (a) was 1/724, 1/1110, and 1/2448, respectively. In Series 1, 2 and 3 for Flume (b), the bed slope was 1/531. Experimental conditions are shown in Table 2. The time was long only for Run 2-3 (600 min); for the other runs it was 90–120 min.

Table 2. Movable-bed experimental conditions.

	Bed forms	Q (liter/sec)	i_a	F_r	τ_*
Run1-3	Alternating bars	6.72	1/724	0.512	0.057
Run1-4	Double-row bars	9.74	1/724	0.642	0.083
Run1-5	Dune	17.56	1/724	0.729	0.112
Run1-6	Dune	18.31	1/724	0.765	0.113
Run2-3	Double-row bars	8.43	1/1110	0.689	0.072
Run2-4	Double-row bars	12.33	1/1110	0.677	0.093
Run2-5	Dune	14.86	1/1110	0.677	0.106
Run3-3	Double-row bars	10.40	1/2448	0.674	0.091
Run3-4	Double-row bars	12.88	1/2448	0.667	0.097
Run3-5	Dune	16.00	1/2448	0.670	0.118
Run3-6	Dune	18.16	1/2448	0.790	0.109

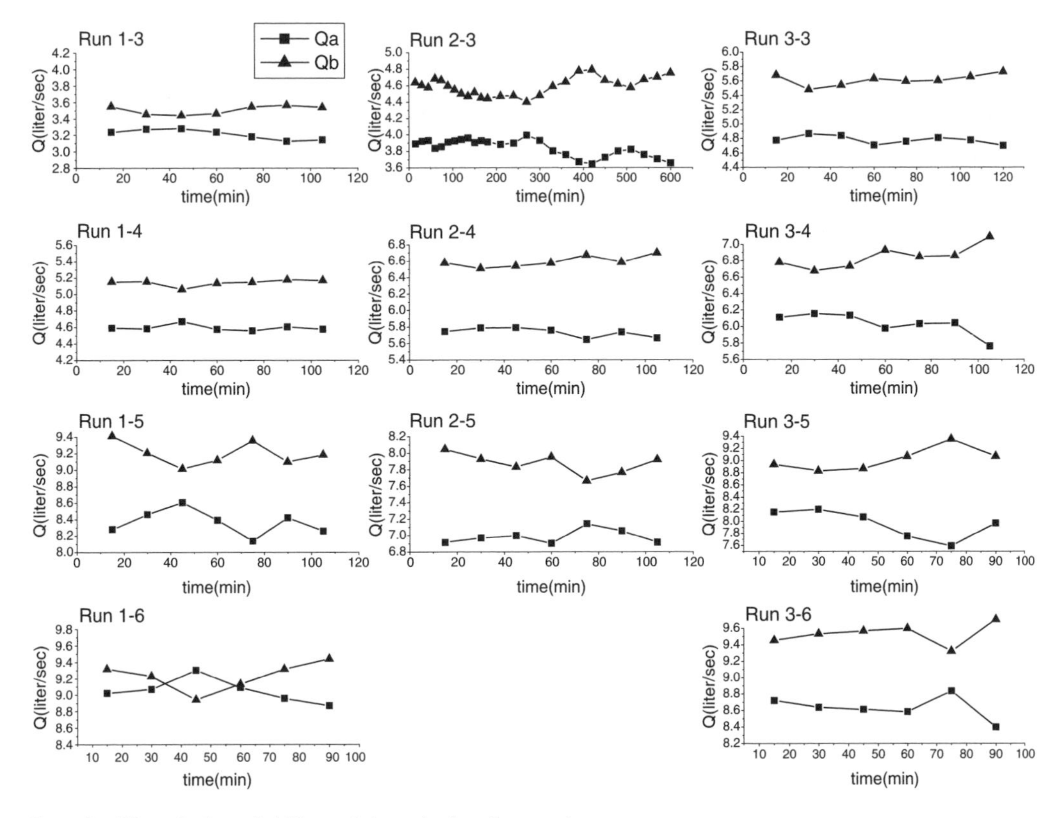

Figure 7. Water discharge in bifurcated channels plotted versus time.

In the movable-bed experiments, results showed that r oscillates with time. In Run 1-6, the discharge ratio changed so much that mainstream alternation occurred. It is noteworthy that the bed configuration was dune in this case. Therefore, we can say that the alternation arises from the peculiar instability of the bifurcation system, rather then from flow deviation resulting from the presence of alternating bars. Mainstream alternation occurred only in Run 1-6, for which the bed slopes ratio i_b/i_a is small, so mainstream alternation may occur when the difference between bed slopes is small.

6 CONCLUSIONS

(1) We applied momentum equations to a bifurcation model in order to investigate the effect of downstream channel condition on the discharge ratio r and on the upstream water depth h_0. The results show that the system has two solutions for r and h_0: one for sub-critical flow, and the other for super-critical flow.

(2) Results of flat-bed experiments show that the discharge ratio r remains constant for each run, and that it agrees fairly well with the theoretical solution for sub-critical flow; this means discharge into the steeper branched channel is greater than that into the less steep branched channel.

(3) The upstream water depth h_0 is roughly equal to uniform flow depth; therefore, it is assumed that bifurcation has little effect upstream of the bifurcation.

(4) Results of movable-bed experiments show that r oscillates with time for each run.

(5) Mainstream alternation was observed when the bed form was dune. Therefore, the alternation arises from the peculiar instability of the bifurcation system, rather than from flow deviation resulting from the presence of alternating bars.

REFERENCES

Bolla Pittaluga, M., Repetto, R. & Tubino, M. 2001. Channel Bifurcation In One-Dimensional Models: A Physically Based Nodal Point Condition. 2nd IAHAR-RCEM Symposium 2001. pp 305–314.

Bolla Pittaluga, M., Repetto, R. & Tubino, M. 2003. Channel bifurcation in braided rivers: Equilibrium configurations and stability. WATER RESOURCES RESEARCH, VOL.39, NO.3, 1046, doi:10.1029/2001WR001112.

Hasegawa, K. & Mizugaki, T. 1993. Changes of Bifurcated in Mountainous Rivers. Proc. of Workshop on Hydraulics and Hydrology in Cold Region Denver USA 1993: 205–216.

Hasegawa, K. & Mizugaki, T. 1994. Analysis of Bloking Phenomena in Bifurcated Channels Found in Mountainous River. Proc. of 1994 National Conf. On Hydraulic Eng.

Hasegawa, K., Hirose, K. & Meguro, H. 2003. Experimental and analysis on alternating mainstream change in the bifurcated channel in mountain rivers. Annual Journal of Hydraulic Engineering, JSCE. VOL47: 755–760.

Meguro, H., Hasegawa, K. & Otuka, T. 2001. Experimental reproduce of the braided channels and bedforms in a mountain river. 2nd IAHR Symposium on River, Coastal and Estuarine Morphodynamics: 733–760.

Meguro, H., Hasegawa, K. & Nakamura, K. 2002. Experimental study on bifurcated channel changes and sediment transport in a mountain river. Annual Journal of Hydraulic Engineering, JSCE. VOL46: 755–760.

Wang, Z. B., Fokkink, R. J., DeVries, M. & Langerak, A. 1995. Stability of river bifurcations in 1 D morphodynamics models. J. Hydraulic Research. 33: 739–750.

River, Coastal and Estuarine Morphodynamics: RCEM 2005 – Parker & García (eds)
© 2006 Taylor & Francis Group, London, ISBN 0 415 39270 5

Study on flow at a river confluence

Kosuke Masujin & Yasuyuki Shimizu
Hokkaido University of Technology, Sapporo, Japan

Kiyoshi Hoshi
Foundation of Hokkaido River Disaster Prevention Research Center, Hokkaido, Japan

ABSTRACT: The confluence of two rivers is prone to flooding, because the flow there is complex. This is particularly true at times of high discharge. This study investigates the flow at such locations. Field investigation was conducted at the confluence of the Atsubetsu and Biu rivers, which was the most severely damaged of any location in the Hidaka district during Typhoon Etau in 2003. Flood flow direction was estimated from the orientation of fallen trees, and water level was estimated from evidence of water marks on the walls of houses. A flume experiment was conducted on the confluence of two rivers. The flow velocity and riverbed elevation in the flume were measured precisely. Numerical analysis using 0-equation and k-ε turbulent models was performed, and the experimental results were compared with calculated results. The numerical analysis using the k-ε model agreed more closely with the experimental results than did the numerical analysis using the 0-equation model. Numerical models were applied to the confluence of the Atsubetsu and Biu rivers. The k-ε model was better than the 0-equation model at reproducing the results of onsite observation. It was shown that the model can be employed as a sophisticated tool to analyze the flow at the confluence of two rivers.

1 INTRODUCTION

River confluences are prone to flooding, because the flow direction is complex. For example, the Nakdong River in Korea flooded at its confluence with the Shinbang River as a result of high discharge caused by typhoon rainfall in 2002. In that case, large amounts of sediment deposited at the confluence obstructed the flow of the Shinbang River, causing flooding there. In the case of the Atsubetsu River in Hokkaido, Japan, flooding resulted from the high discharge caused by rainfall from Typhoon Eatu in August 2003. The confluence of the Atsubetsu and Biu rivers suffered particularly heavy damage. Flooding occurred because the main stream is narrow there and the branch joins the main stream at a right angle.

Study of flow and sediment transport at a river confluence is an important issue in river engineering. This study consists of three items: field investigation after flood, a flume experiment, and numerical analysis. Field investigation was conducted at the confluence of the Atsubetsu and Biu rivers after flooding from Typhoon Etau in 2003. Field investigation estimated flood flow direction from the orientation of fallen trees, and water level from evidence of water marks on the walls of houses. The flume experiment was conducted to study flow at the confluence of the two rivers. The branch of this flume joined the mainstream at a right angle, as is the case for the confluence of the Atsubetsu and the Biu Rivers. Flow direction, flow velocity and water depth were measured precisely. The numerical analysis used a shallow-water flow equation of two-dimensions. The analysis was able to calculate hydraulic phenomenon using discharge data of a main stream and a branch. Calculated results were compared with experimental results. This numerical model was applied to the river confluence of the Atsubetsu and Biu rivers, and flood flow was calculated.

Flow near the river confluence was examined and clarified.

2 FIELD INVESTIGATION

2.1 *Typhoon Etau on 9 August 2003 and Atsubetsu River in Hokkaido*

Typhoon Etau reached the Hidaka area of Hokkaido on 9 August 2003. In Hidaka, 400 mm of rain was observed. Because the banks of the Atsubetsu River there are low and incompletely protected, flood damage was severe. The Atsubetsu River is a mid-size river. The low-water channel is 34.7 km long, and the basin is 266 km^2 in area. The area of field investigation on the Atsubetsu River was 15 km from the mouth. This area includes the confluence of the Atsubetsu and Biu rivers.

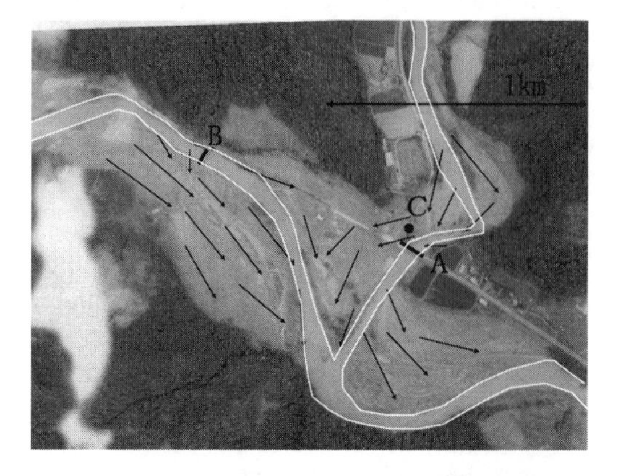

Figure 1. The confluence of the Atsubetsu and Biu rivers.

Figure 2. A damaged house at point C.

2.2 *Field investigation*

The following were observed: riverbed deposition and scouring, inundation, damage to houses and bridges, flow direction, and water depth. Figure 1 is a post-flood aerial photo of the confluence. The main stream, the Atsubetsu River, flows from left to right; the branch, the Biu River, flows from top to bottom. The Atsubetsu River narrows after the confluence. The branch joins the mainstream at a right angle. The topography makes the location prone to flooding. In Figure 1, arrows show the flow direction from field investigation. There are bridges at points A and B. These obstructed the flow of driftwood and sediment. Figure 2 shows a flood-damaged house at point C in Figure 1. The house was inundated by flooding that resulted when the flow became obstructed by the bridge at point A.

3 FLUME EXPERIMENT

To apply a numerical model to a river confluence, a flume experiment was conducted. Figure 3 shows an

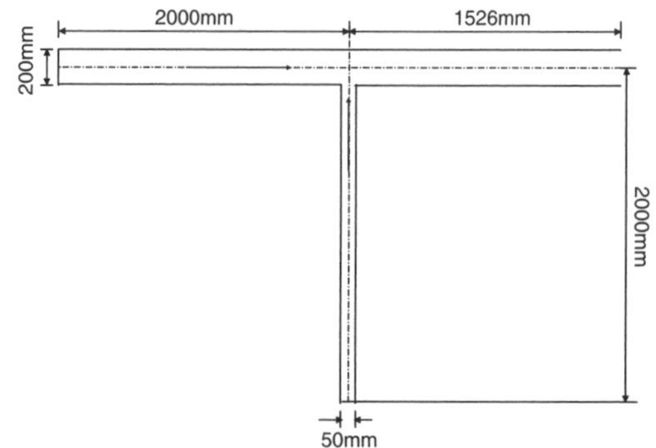

Figure 3. Experimental flume.

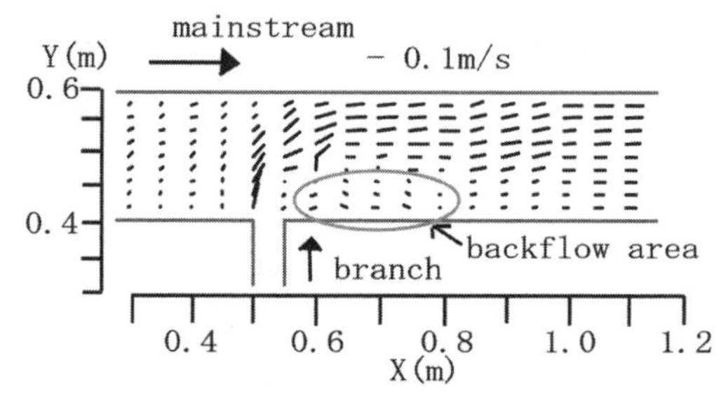

Figure 4. Velocity vector observed in a flume experiment.

experimental flume. The main stream width is 20 cm. The branch width is 5 cm. The branch joins the main stream at a right angle, just as the Biu River joins the Atsubetsu River. The flow velocity was examined at 60% of water depth by electromagnetic velocimeter. Examination points numbered 160, and they were spaced at 5 cm intervals in the longitudinal direction and 2 cm intervals in the transverse direction. The water depth was examined by point gauge. The main stream slope was 1/100; and the branch slope was 1/50. The main stream discharge was 1.14 l/s; the branch discharge was 0.71 l/s. A weir was made at the farthest downstream point. This means that the main stream and the branch had supercritical flow.

Figure 4 shows a velocity vector observed by a flume experiment. An area of backflow formed on the right side of the main stream after the confluence. The results of experiments are described in more detail in the next chapter.

4 NUMERICAL MODEL

4.1 *Numerical calculation method and conditions*

Flow was calculated using a two-dimensional shallow-water flow momentum equation and a continuity equation. The coefficient of eddy viscosity, which is

Table 1. Coefficient used with k-ε model.

C_μ	$C_{1\varepsilon}$	$C_{2\varepsilon}$	σ_k	σ_ε
0.09	1.44	1.92	1.0	1.3

included in the continuity equation, was calculated using the 0-equation and k-ε turbulent models. In using the 0-equation turbulent model, the equation was solved for the coefficient of eddy viscosity (Equation (1)). In using the k-ε turbulent model, the equation was solved for the coefficient of eddy viscosity (Equation (2)).

$$\nu_t = \frac{\kappa}{6} U^* h \tag{1}$$

$$\nu_t = C_\mu \frac{k^2}{\varepsilon} \tag{2}$$

k and ε are solved for using equations (3) and (4).

$$\frac{\partial k}{\partial t} + U\frac{\partial k}{\partial x} + V\frac{\partial k}{\partial y} = \frac{\partial}{\partial x}\left(\frac{\nu_t}{\sigma_k}\frac{\partial k}{\partial x}\right) + \frac{\partial}{\partial y}\left(\frac{\nu_t}{\sigma_k}\frac{\partial k}{\partial y}\right) + P_h + P_{kv} - \varepsilon \tag{3}$$

$$\frac{\partial \varepsilon}{\partial t} + U\frac{\partial \varepsilon}{\partial x} + V\frac{\partial \varepsilon}{\partial y} = \frac{\partial}{\partial x}\left(\frac{\nu_t}{\sigma_\varepsilon}\frac{\partial \varepsilon}{\partial x}\right) + \frac{\partial}{\partial y}\left(\frac{\nu_t}{\sigma_\varepsilon}\frac{\partial \varepsilon}{\partial y}\right) + C_{1\varepsilon}\frac{\varepsilon}{k}P_h + P_{\varepsilon v} - C_{2\varepsilon}\frac{\varepsilon^2}{k} \tag{4}$$

P_{kv} and $P_{\varepsilon v}$ are solved for using equations (5) and (6).

$$P_{kV} = C_k \frac{U^{*3}}{h} \tag{5}$$

$$P_{\varepsilon V} = C_\varepsilon \frac{U^{*4}}{h^2} \tag{6}$$

ν_t is coefficient of eddy viscosity. κ is the Kalman coefficient. U* is velocity of friction. h is water depth. x and y are coordinates. U is flow velocity in the x direction. V is flow velocity in the y direction. C_μ, $C_{1\varepsilon}$, $C_{2\varepsilon}$, σ_κ, σ_ε are coefficients shown in Table 1.

Discharge was given as the boundary condition of the upper stream. The water level was given as the boundary condition of the lower stream. Discharge was given as 1.14 l/s in the main stream and 0.71 l/s in the branch. These discharges were the same as those in the experiment. The Manning coefficient of roughness was set as 0.011. Figure 5 shows the calculation grid. The calculation interval was 0.001 sec. The results calculated between the 60th sec and the 90th sec were averaged.

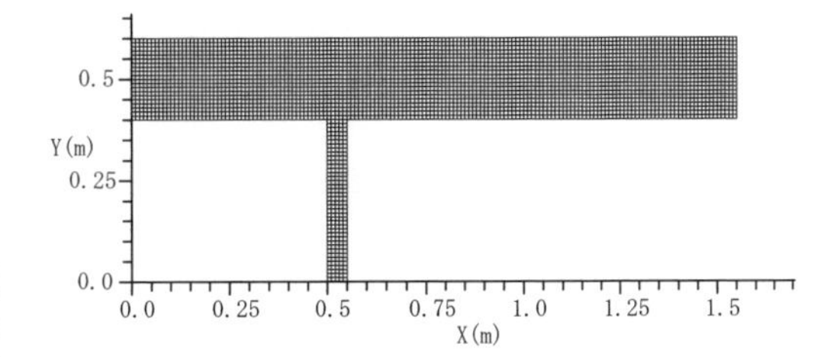

Figure 5. Calculation grid of flume experiment.

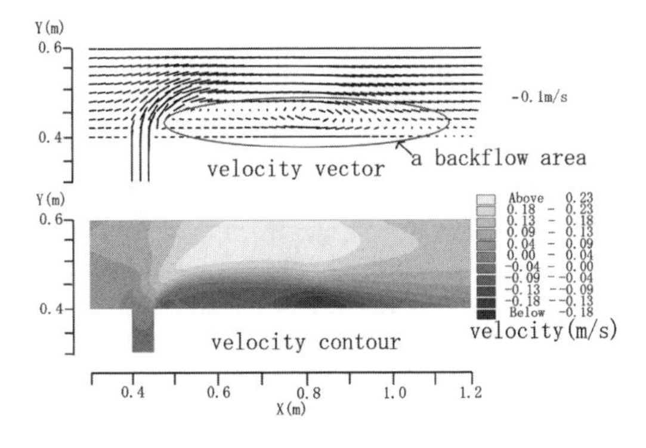

Figure 6. Velocity vector and contour of calculation results using 0-equation model.

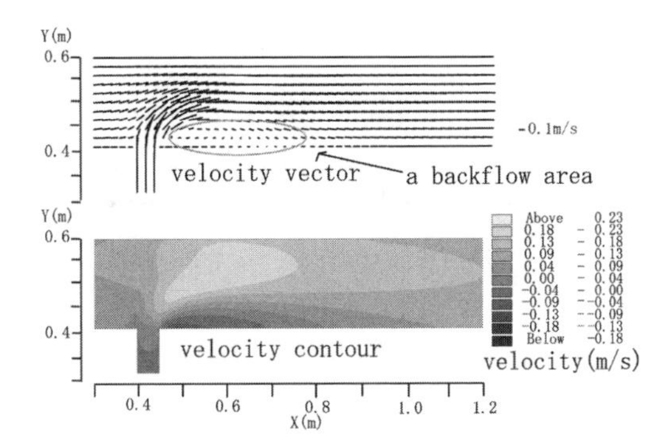

Figure 7. Velocity vector and contour of calculation results by using k-ε model.

4.2 *Calculation results*

Figures 6 and 7 show velocity vectors and contours for the calculated results. Figure 6 shows the results calculated using the 0-equation model. Figure 7 shows the results calculated using the k-ε model. Velocity vectors show flow velocity in the x direction, in order to clarify the backflow area. Although both models showed a backflow area after the confluence of two rivers, its width and length differ between the models. Figure 8 shows the calculated and experimental water level at the center of the main stream. Calculated results were those obtained using the 0-equation model, because

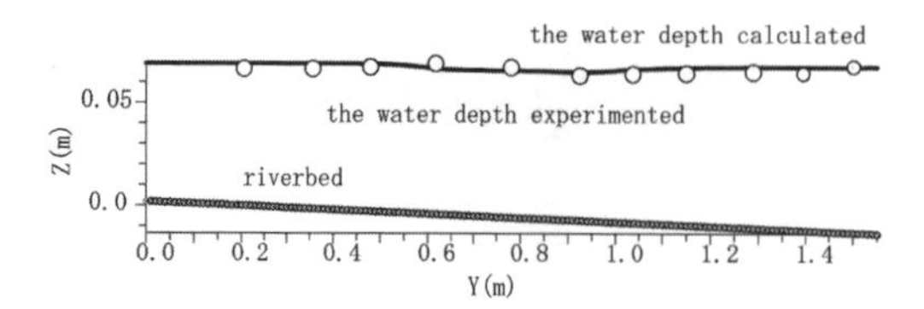

Figure 8. Water level: calculated and experimental.

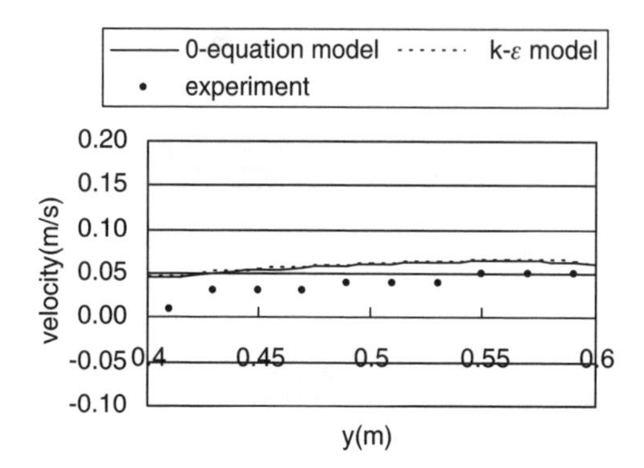

Figure 9. Flow velocity at $x = 0.4$ m: longitudinal component.

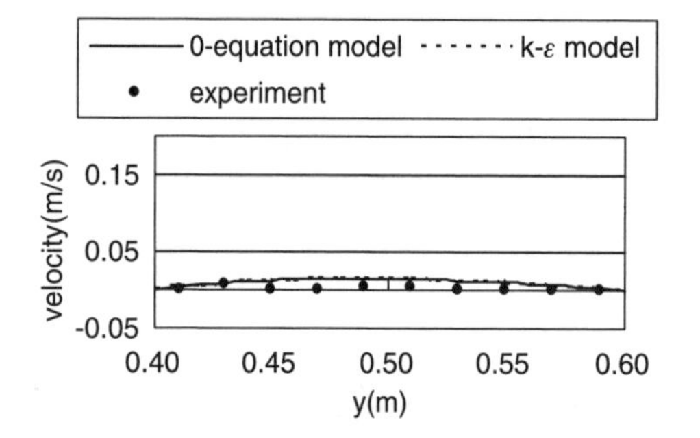

Figure 10. Flow velocity at $x = 0.4$ m: transverse component.

the water level calculated using the k-ε model did not differ from that calculated using the 0-equation model. Calculated velocity is compared with experimental velocity in Figures 9 to 14. The x and y coordinates corresponded to those in Figure 5. The solid line shows flow velocity calculated using the 0-equation model. The broken lines show flow velocity calculated using the k-ε model. The dotted lines show flow velocity obtained experimentally. Experimental results agreed with results calculated using the 0-equation model and the k-ε model before the confluence of two rivers (at the $x = 0.4$ coordinate). After the confluence of the two rivers (at $x = 0.6$ and $x = 0.8$ coordinates) the results calculated using both turbulence models reproduced a backflow area after the confluence. Results calculated using the k-ε model agreed more closely with

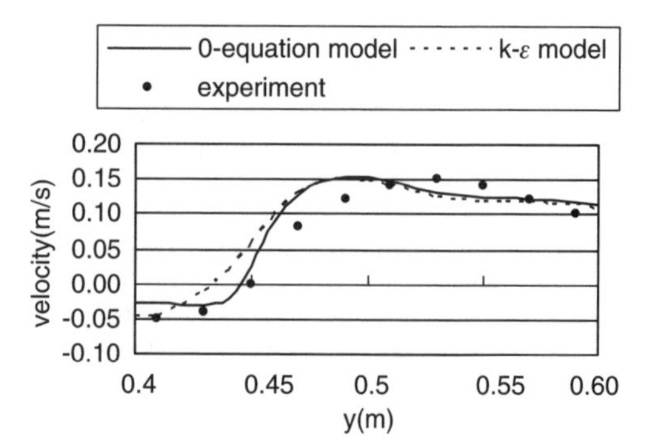

Figure 11. Flow velocity at $x = 0.6$ m: longitudinal component.

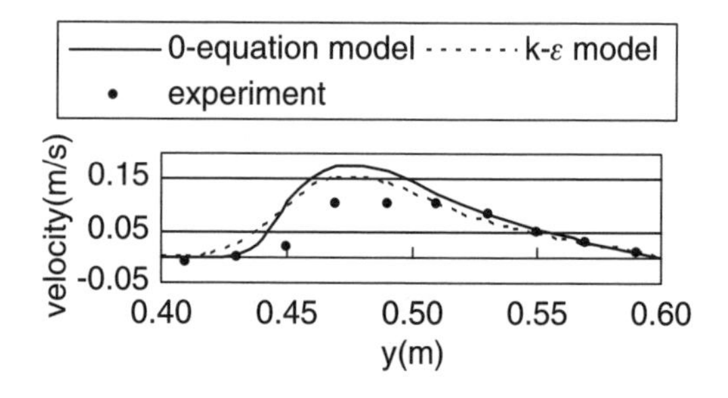

Figure 12. Flow velocity at $x = 0.6$ m: transverse component.

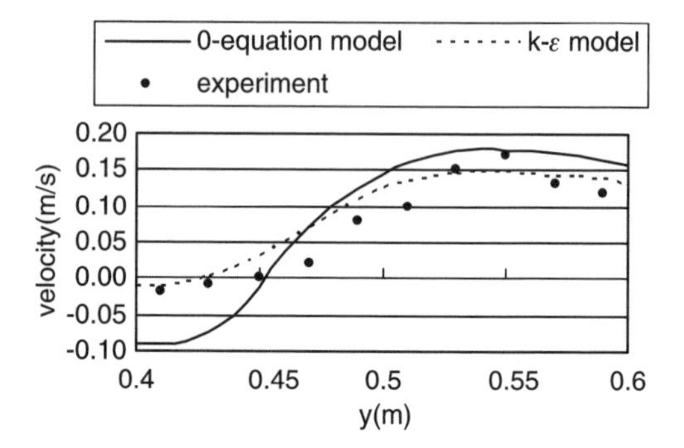

Figure 13. Flow velocity at $x = 0.8$ m: longitudinal component.

the experimental results than did results calculated using the 0-equation model. Results calculated using 0-equation model overestimate the backflow area.

Results calculated using the two turbulence models were similar to the experimental results. The k-ε model is better than the 0-equation model at reproducing the backflow area. In the next section, calculation using k-ε turbulent model is mainly used. But calculation using 0-equation model is also used because it involves fewer variables.

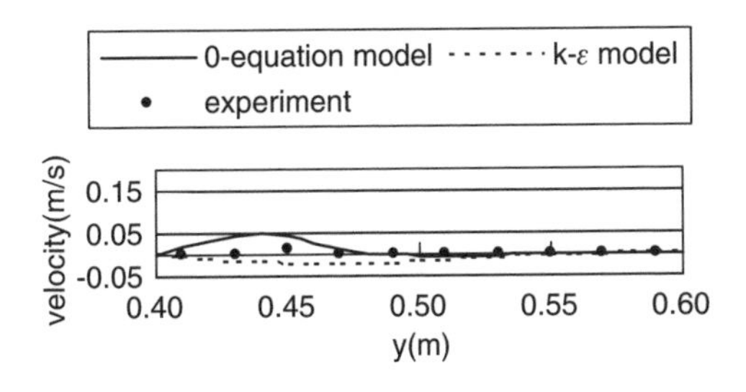

Figure 14. Flow velocity at x = 0.8 m: transverse component.

5 NUMERICAL MODEL OF RIVER CONFLUENCE APPLIED TO A REAL RIVER

5.1 *Numerical conditions*

A numerical model of river confluence was applied to the confluence of the Atsubetsu and Biu rivers. Flood discharge was the maximum discharge calculated by Wongsa Sanit. The main stream discharge was 863 m^3/s, and the branch discharge was 732 m^3/s. The water level at the downstream end of the calculation grid was 49.5 m, which was obtained by field investigation. The Manning coefficient of roughness was 0.03 in the low-water channel and 0.065 in the high-water channel. The time interval of calculation was 0.01 sec, and the duration for calculation was 5000 sec. For this duration of calculation, results calculated between the 1000th and 3000th sec were averaged. The calculation grid included the high-water channel (Figure 15).

5.2 *Calculation results and observed results*

Figure 16 shows the calculated and observed water level at center of the main stream. In this figure, S is the longitudinal length of the riverbed of the main stream. Solid lines show riverbed elevation and calculated water surface elevation. Small circles show observed water surface elevation. Calculated results agreed closely with observed results. Figure 17 shows calculated velocity contours and vectors. The main stream flows straight and floods because it is not able to be diverted. Because this flow hits the flow of the branch, the point after the confluence has a backflow area, as in the flume experiment. Therefore, the width of the main stream was narrow along the entire length. The narrow width at the confluence hindered flow. Flooding was attributed to these factors. The reason that a house at point C was damaged was because the bridge at point A (Figure 1) obstructed the flow, causing the flow to be diverted. Calculation was conducted under the assumption that flow was obstructed by the bridge at point A and that the river bed height there had risen to the height of the bridge. The other conditions

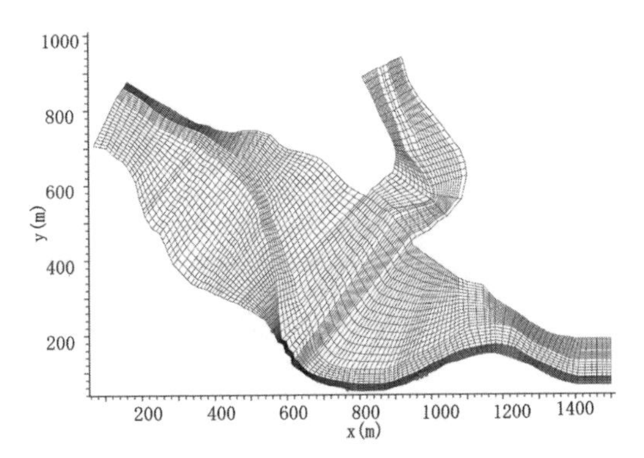

Figure 15. Calculation grid of the confluence of the Atsubetsu and Biu rivers.

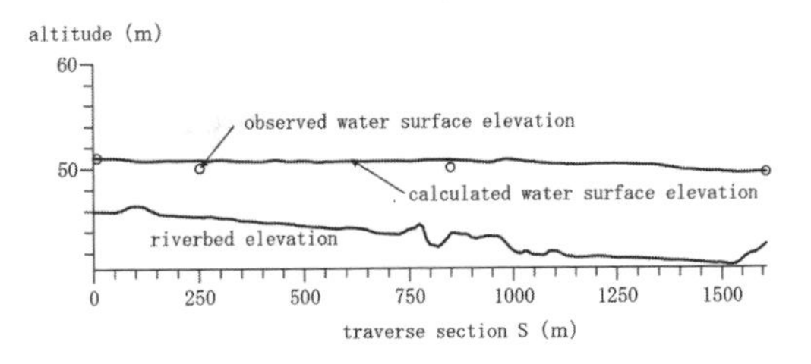

Figure 16. Comparison of calculated water surface elevation and experimental water surface elevation.

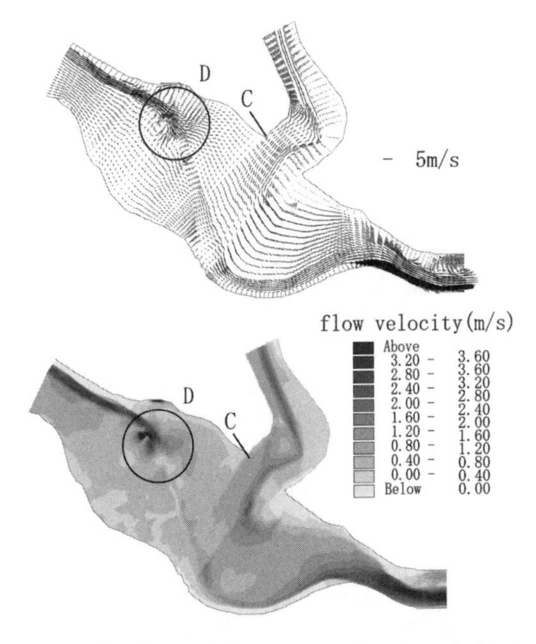

Figure 17. Flow velocity contours and vectors calculated using the k-ε model.

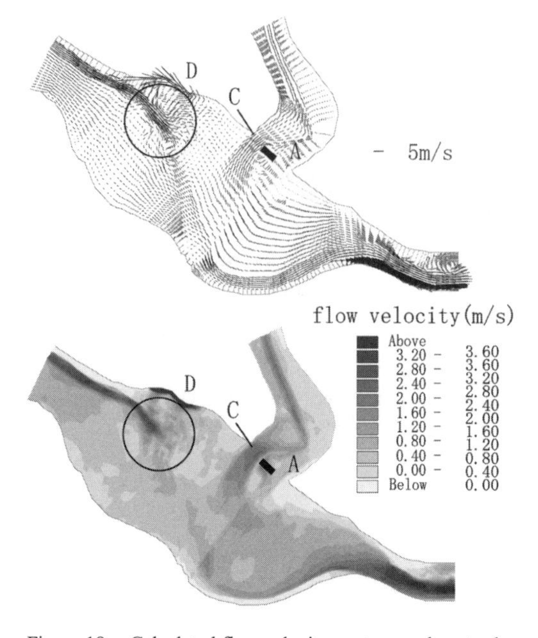

Figure 18. Calculated flow velocity contour and vector by using k-ε model (dammed by a bridge).

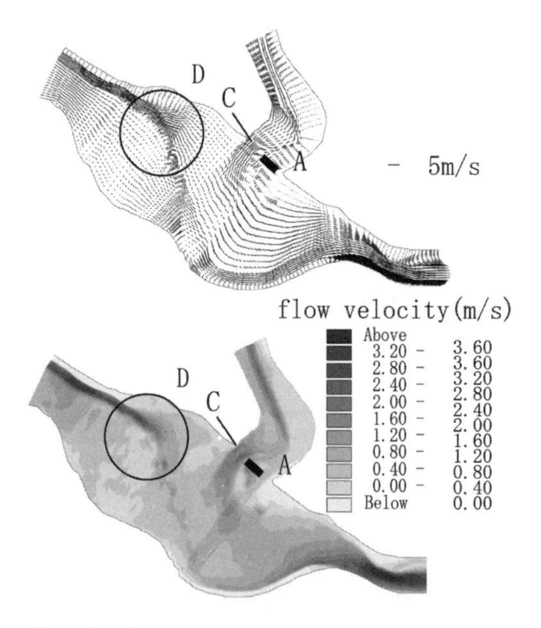

Figure 19. Calculated flow velocity contour and vector by using 0-equation model (dammed by a bridge).

were the same as in previous calculations. Figure 18 shows the calculation results under the assumption that flow was obstructed by the bridge at point A. The flow of the branch was diverted into the house at point A, because the bridge at point A was dammed up.

Figure 19 shows results calculated using the 0-equation model. Results calculated using the k-ε model were compared with those calculated using the 0-equation model. The k-ε model results show greater turbulence than the 0-equation model results at two locations. One is area D, which flooded at a bend in the main stream. The other one is area C, which flooded as a result of the flow obstruction at point A. Although flow velocity and flood direction were not investigated in detail, we did find that the flow velocity and direction at both locations was complex, as evidenced by the serious damage to houses at points C and D. Calculations to reproduce flooding at the confluence of the Atsubetsu and Biu rivers were conducted. The flow direction, water depth and flood that was obstructed by the bridge were reproduced well.

6 CONCLUSION

This study examined a river confluence by field investigation, flume experiment and numerical calculation. A numerical calculation model can serve as a powerful tool for analysing the flow at a river confluence. River improvement requires that hydraulic phenomena be measured at a real river confluence. Examination of river improvement plans and expected flood flow can be conducted easily using this numerical calculation.

REFERENCES

Fujita, I. 1990. Study of mechanism of open channel flow in the vicinity of the river confluence. *Kobe*: Doctoral Thesis for Kobe Univ.

Itakura, T. 1984. Investigations of some turbulent diffusion phenomena in rivers. *Sapporo*: Hokkaido Univ.

Kimura, I., Hosoda, T., Muramoto, Y. and Sakurai, T. 1997. Vortex formation processes in open channel flow with a side discharge using turbulence models, *Annual J. of Hydraulic Eng., Japan Society of Civil Engineers*, 41: 717–722.

Rodi, W. 1993. Turbulence models and their application in hydraulics. IAHR AIRH Monograph.

Wongsa, S., Shimizu, Y. and Murakami, Y. 2004. Runoff and sediment transport characteristics in a river system in the Hidaka region of Hokkaido during Typhoon Etau in 2003. *Annual J. of Hydraulic Eng., Japan Society of Civil Engineers*, 48(2): 1099–1104.

River, Coastal and Estuarine Morphodynamics: RCEM 2005 – Parker & García (eds)
© 2006 Taylor & Francis Group, London, ISBN 0 415 39270 5

Secondary flow at a scour hole downstream a bar-confluence (Paraná River, Argentina)

R.N. Szupiany & M.L. Amsler
Consejo Nacional de Investigaciones Científicas y Técnicas (CONICET), Argentina
Facultad de Ingeniería y Ciencias Hídricas, UNL, Argentina

J.J. Fedele
Saint Anthony Falls Laboratory, University of Minnesota, Minneapolis, USA

ABSTRACT: Most field studies of the flow at confluences were performed in small streams with limited depths and widths essentially because of the lack of proper instrumentation to make them at larger streams. In this paper a preliminary analysis and results of secondary current measurements at a section located across the scour hole downstream of an asymmetrical bar-confluence in a large river (Paraná River, Argentina), are reported. A methodology based on moving-vessel measurements with an Acoustic Doppler Profiler enabled to obtain representative time-averaged velocities at a sufficient number of verticals. The Rozovskii procedure was used to treat the ADP point velocity data. According to the results well defined secondary currents were captured with a pattern similar to those observed in laboratory and smaller natural streams. Some morphologic implications and the extent of results are discussed.

1 INTRODUCTION

Relatively recent contributions have shown that nodal units or channel sectors where flow converges and then expands in braided rivers, would be "key" points which govern largely the hydrosedimentological behavior in this type of streams. This statement is based principally on investigations at river confluences where were identified a diversity of coherent flow structures interacting one another and generating diverse patterns of sediment transport and deposition. Thus the hydrodynamic conditions within the confluence region would control the transfer and distribution of sediment loads downstream and hence the channel morphology variations in that direction (Ashmore & Parker 1983, Best 1988, Mosley 1976, Rhoads 1996).

Most reported studies experimental and numerical, aimed at analyzing the nature of the flow at river confluences, has been performed in small streams with limited depths and widths (Best 1988, Best & Roy 1991, Mclelland et al. 1996, Rhoads 1996, Rhoads & Sukhodolov 2001, Bradbrook et al. 2000, Bradbrook et al. 2001, Huang et al. 2002). These investigations generally recognize the existence of several typical flow features at confluences, namely: (a) a zone of flow stagnation near the upstream junction corner; (b) shear and mixing layers where the two flows combine; (c) a zone of separated flow below the downstream junction corner and immediately after the avalanche faces of both tributaries; (d) secondary currents where the flow converges. The confluence bed morphology relates with its hydrodynamics through: (i) a scour hole normally along the strip of maximum velocities where both flows begin to mix; (ii) avalanche faces at the mouth of each tributary which dip into the scour hole; (iii) sediment deposition in the stagnation zone; (iv) a bar formed within the flow separation zone. The governing gross variables of flow structure and channel morphology would be the confluence angle, its planform asymmetry, the flow and sediment discharge ratio of each tributary (Mosley 1976, Best 1988, Rhoads 1996) and the bed concordance degree of the two confluent streams (Biron et al. 1996).

The study of confluences in large rivers unlike the case in small streams, are extremely scarce essentially because of a lack of proper instrumentation. Few papers dealing with coherent flow structures at large scale currents like those of Richardson et al. (1996), Richardson (1997), Richardson & Thorne (1998, 2001) and Parsons et al. (2004), may be mentioned. These studies are largely based on the results obtained with devices like the Acoustic Doppler Profiler (ADP). Although the Doppler technology proved to be useful for accurate determinations of flow discharges in large streams, the reliability of its signals for other calculations such as turbulent characteristics, secondary currents or concentrations of suspended sediment, still remain highly untested.

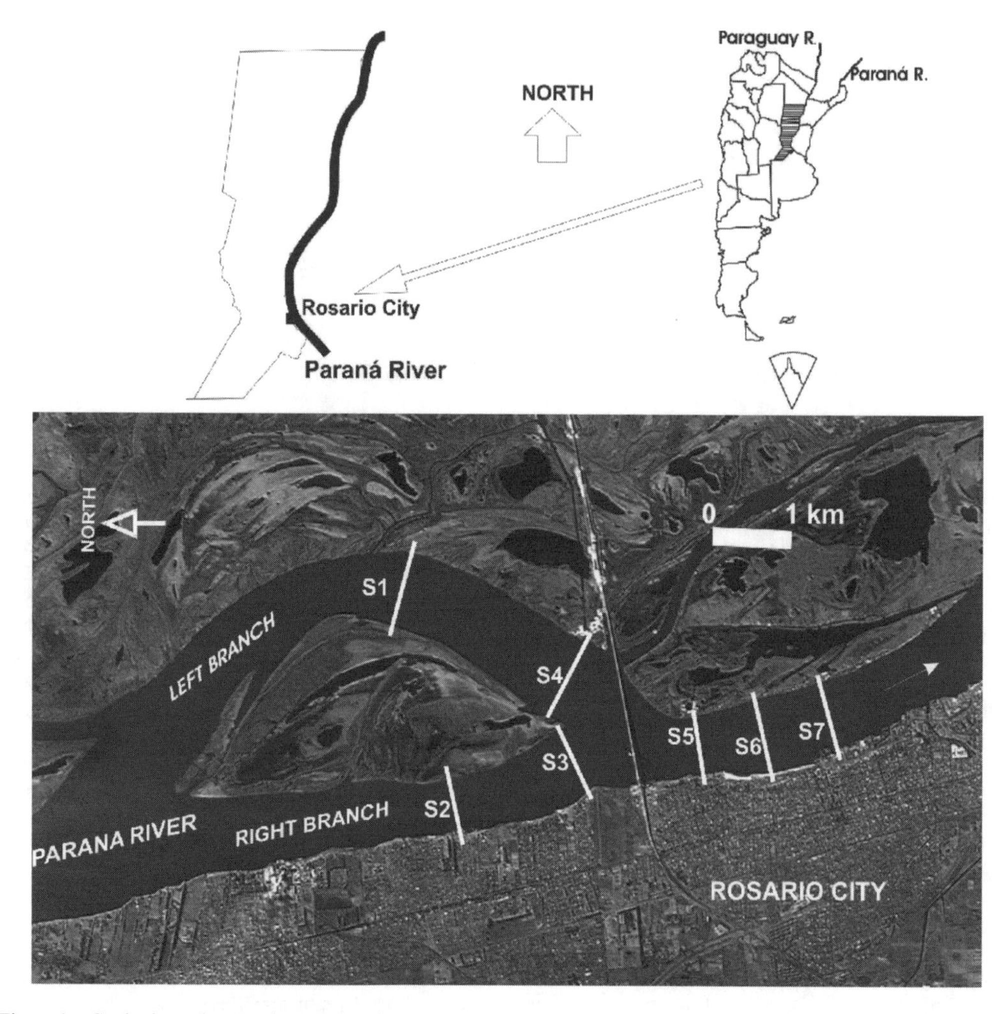

Figure 1. Study site and measuring cross sections.

In this paper, a preliminary analysis and results of secondary currents measured with an ADP at a central scour hole downstream of an asymmetrical bar-confluence in the Parana River, Argentina, are presented. A methodology that allows for obtaining representative values of the time-averaged velocities with moving-vessel measurements was used. The results reported herein, are part of a study to address the interactions between the main and secondary currents, sediment transport distribution and morphologic features at nodal units in large braided rivers.

2 THE PARANÁ RIVER

The Paraná River is one of the largest rivers in the world (Schumm & Winkley 1994) with a drainage basin of $2.3 \times 10^6\,km^2$ which includes part of four countries: Brasil, Bolivia, Paraguay and Argentina. Downstream the confluence with the Paraguay River till its mouth (Fig. 1), the mean discharge is of $19500\,m^3\,s^{-1}$, the water surface slope is in the order of 10^{-5} and the channel bed is composed largely by fine and medium sands (Drago & Amsler 1998). The main channel pattern in that sector was classified as braided with a meandering thalweg (Ramonell et al. 2002). From a plain view it looks like a succession of wide and narrow sections with mean widths and depths ranging between 600–2500 m and 5 to more than 12 m, respectively.

3 DESCRIPTION OF FIELD MEASUREMENTS

Although the Acoustic Doppler Profilers (ADP), widely spread during the last decade, enable to obtain accurate discharge measurements in large streams, further testing and methodologies are still needed if representative values of flow parameters other than the discharge, have to be computed.

In this study, the 3D field velocity at seven cross sections located along a braid-bar confluence in the Paraná River near Rosario city, Argentina (Fig. 1), was obtained with an ADP of 1000 kHz. Moving-vessel measurements were performed. Only the results obtained at section S5 placed across the scour hole downstream of the confluence, are presented herein. The equipment was coupled with a DGPS and an echo-sounder. A specific methodology designed in order to obtain representative values of the time-averaged velocities was used (Szupiany & Amsler 2005). Essentially, the procedure requires that each cross section be run at least five times to obtain horizontal velocity data with errors between 5–10%, compared with fixed-vessel measurements of mean velocities during a relatively long time interval. The methodology requires the ADP to be set with a cell size of 0.75 m and an averaging interval of 10 seconds. Discharges values in both, left and right branches, and a bathymetric map of the whole confluence zone were also obtained.

4 SECONDARY CURRENTS AND ANALYSIS OF VELOCITY DATA

Secondary currents are defined as those occurring all around a plane perpendicular to the main stream axes (Prandtl 1952). Generally they are classified into two types: secondary currents of Prandtl's first kind which received considerable attention in curved channels and meandering rivers (the interaction between centrifugal and pressure gradient forces drive the motion in this case) and secondary currents of Prandtl's second kind observed in straight open-channel flows and generated by the non-homogeneity and anisotropy of turbulence (Nezu & Nakagawa 1993).

In spite of its simple definition, the accurate computation of secondary currents in natural streams has been a problem in river engineering since long time ago. The main flow direction in this type of streams does not normally follow a parallel track to the central channel line. Particularly in the typical confluences and/or expansions of braided rivers, the flow divides into two or more dominant directions with diversity of maximum velocities and discharges. These features make it difficult to define the main and secondary planes, aside from the problem of a possible interaction between the two kinds of secondary circulation produced by the particular morphology of confluences described abode (Lane et al. 2000).

The recent acoustic devices to measure the flow velocity allow for the determination of its components respecting different coordinate systems. So users select the reference system in the case of the Acoustic Velocity Velocimeters (ADV), while ADPs supply the velocity components respecting a North-East-Vertical coordinate system when used with the moving-vessel measuring procedure. The horizontal components rarely agree with the main and secondary flow planes in the last situation.

Several criteria have been advanced in order to define the secondary plane in natural channel. Discussing this topic, Lane et al. (2000) adopted the following classification for the different existing definitions:

- The centerline definition. The secondary circulation is defined as any persistent component of velocity that is not parallel or tangential to the centerline of the channel and also not parallel to and in the same downstream direction of the overall slope (Bhowmik 1982).
- The Rozovskii definition. The secondary circulation is obtained on the basics of individual vertical profiles by defining the primary velocity direction such that the secondary currents in one direction, are balanced by those in the opposite direction to produce zero secondary discharge for the profile. Implicitly, the direction of primary velocity at each vertical is equivalent to the direction of the depth-average flow velocity. With this procedure different planes result at each vertical across a section (Rozovskii 1954, Hey & Thorne 1975, Bathurst et al. 1977, Bathurst et al. 1979, Rhoads & Kenworthy 1995, 1998, Rhoads 1996).
- The zero net cross-stream discharge definition. The primary velocity in this case have such a direction that there is no secondary discharge for the whole section (Paice 1990, Markham & Thorne 1992a,b).
- The discharge continuity definition. On the basics of continuity, the cross-stream velocity field is determined by orienting each cross-section such that the net secondary discharge between adjacent cross-sections is zero (Dietrich & Smith 1983, 1984, Thorne & Rais 1984).

Richardson et al. (1996) suggest also another method based on a secondary current function that enables the identification of different planes in multi-threaded cross sections.

Much discussion still remains regarding which of the above methods supply the most realistic secondary velocity field. In this study, a large alluvial river with its constraints for data collection, the discharge continuity procedure was discarded since it requires rather closely spaced cross-sections. Respecting the zero net cross-stream discharge criteria it was not tested yet at this step of the investigation. Thus the application of the Rozovskii procedure is only present here, computing the primary (v_P) and secondary (v_S) components of each point velocity in each vertical as follows:

$$v_p = v\cos(\theta - \alpha) \quad (1)$$

$$v_s = v\sin(\theta - \alpha) \quad (2)$$

where, v ($\sqrt{v_N^2 + v_E^2}$), point velocity vector; θ, orientation of v respecting the north; v_N, v_E, north and eastward components of the point velocity vector; α, orientation respecting the north of the depth average velocity vector. Following Rhoads & Kenworthy (1998), α was determined from the direction of the depth-average velocity vector defined by averaging v_N and v_E separately over the flow depth for each vertical.

Table 1. Main hydraulics characteristics at the confluence.

Characteristic parameters	S4	S3	S5
Mean flow velocity ($m\,s^{-1}$)	0.90	1.12	1.25
Maximum flow velocity ($m\,s^{-1}$)	1.40	1.50	1.55
Width (m)	680	750	900
Discharge ($m^3\,s^{-1}$)	7036	10610	17646
Mean depth (m)	10.7	12.8	18.8
Maximum depth (m)	18	16	31
Junction angle (degrees)	65	–	
Flow momentum ratio, M*	0.53	–	
Discharge ratio	0.67		

*$M = (\rho Q_L V_L)/(\rho Q_R V_R)$, where ρ, water density ($kg\,m^{-3}$), Q, total discharge ($m^3\,s^{-1}$), V, cross-section average velocity ($m\,s^{-1}$) and the subscripts L and R denote left and right branch.

5 RESULTS

5.1 *Hydraulic and morphologic conditions at the confluence.*

The hydraulic conditions of the Paraná River at the confluence during the measurements are summarized in Table 1.

The channel bed morphology as obtained from echo sounder records (recording frequency: 2 seg.) at cross sections each spaced 100 m, is shown in Figure 2).

The scour hole at section S5 with a maximum depth of 31 m is the most outstanding morphologic feature of the confluence. Note that it is displaced towards the left bank a fact discussed below related with the flow momentum and discharge ratios values.

The avalanche faces at the mouth of each tributary are hardly noticeable. On the right branch, bed downslope maximum angles of 6°–7° were measured for this feature. A zone of shallow depths is also notable at the upstream junction corner.

5.2 *Secondary currents*

Figure 3a shows two counter-rotating surface-convergent secondary cells clearly depicted within the cross hole where the two flows converges.

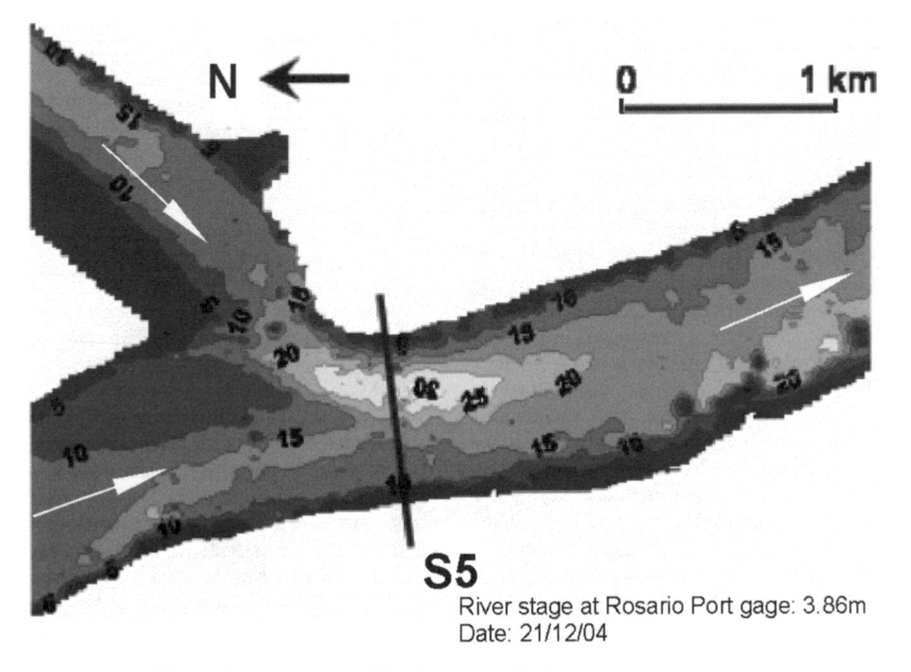

Figure 2. Bathymetry of the confluence (interval of depth contours: 5 m).

In the inner cell lying between ~50–200 m from the left bank, maximum secondary velocities in the order of 0.19 m s^{-1} were measured. This maximum reduced to 0.15 m s^{-1} in the outer cell spreading between ~300–410 m from the left bank.

A distinct downwelling zone defining the mixing interface of the two cells ~50–60 m width, is seen in Figure 3b. Maximum vertical velocities up to 0.08 m s^{-1} were measured in this region. Outside the scour hole towards the right bank, secondary circulation was not well defined.

Since velocities in Figure 3a,b are averages of five ADP's transects, results of a single transect are shown in Figure 4 for comparison purposes. These data were also obtained applying the Rozovskii procedure as shown by Equations (1) and (2). Though a fairly well

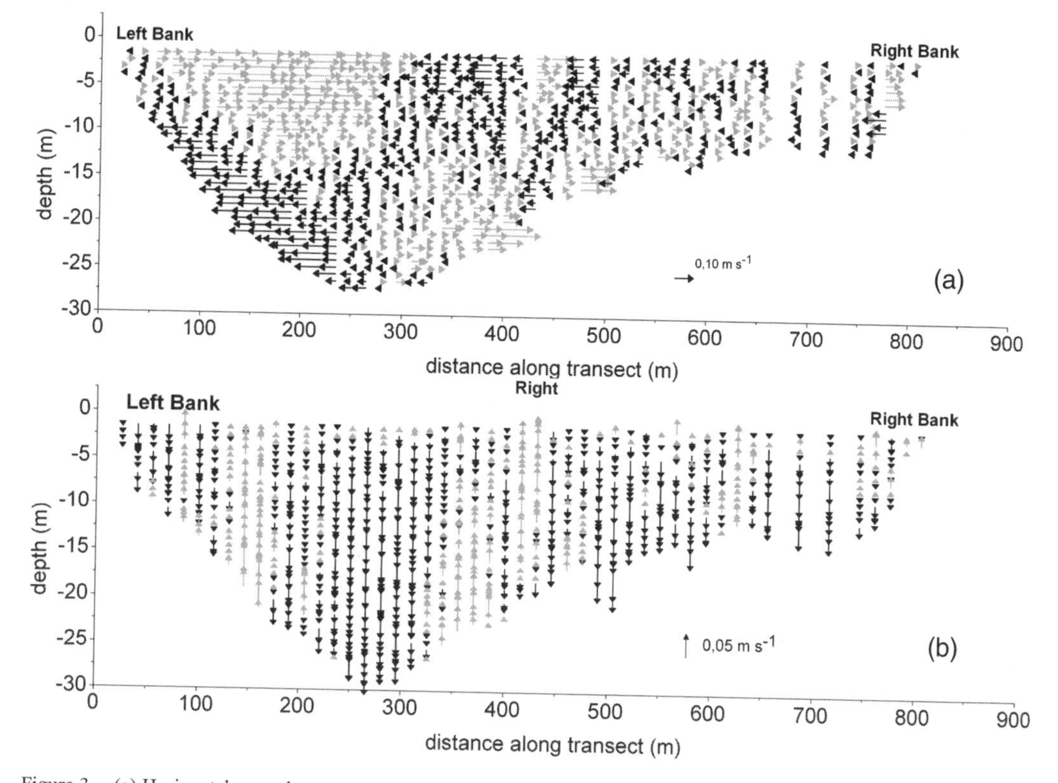

Figure 3. (a) Horizontal secondary currents in section S5. (b) Vertical secondary currents in section S5. (Average values of 5 transects).

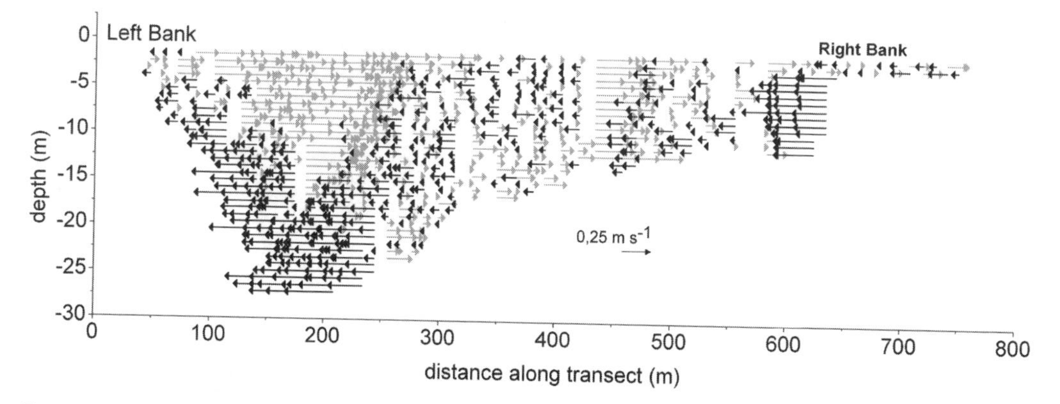

Figure 4. Horizontal secondary currents in section S5 (values of 1 transect).

defined inner cell appears, its secondary velocity values are higher than those in Figure 3a with maximums up to $0.47\,m\,s^{-1}$. On the other hand the outer cell is hardly noticeable.

6 CONCLUDING REMARKS

Though it is recognized that a debate still exists concerning the most proper procedure to represent realistic secondary flow fields from data obtained at meandering rivers and confluence zones (Lane et al. 2000), the secondary circulation in the scour hole downstream of a bar-confluences in a large river was obtained applying the Rozovskii method. The secondary velocities pattern and intensities also involve an averaging method specifically designed to record representative mean flow velocity values with moving-vessel ADP measurements (Szupiany & Amsler 2005). In this manner a large number of average point primary velocities at a given cross-section of a large river is possible rapidly to obtain.

Within this context, the results presented herein must be viewed as the first step of an investigation that includes intercomparison of methods to check which one better represents the secondary structures at confluences and nodes of large rivers like the Paraná. Making allowance for the statement above, two counter-rotating surface convergent secondary currents clearly characterize the mean flow structure at the scour hole downstream the bar confluence studied in the Paraná River. Similar patterns have been observed in the laboratory and in small natural streams (McLelland et al. 1996, Rhoads 1996, between others). The two secondary cells locate on either side of the mixing interface at ~260–300 m from left bank.

Apparently the left secondary cell strength is larger than the right one, something that could be related with the flow momentum value, M (<1). This condition would have important morphologic consequences since the scour hole appears displaced towards the left bank thus preventing the existence of the separation zone and bar region observed downstream the junction corner in small streams (Best 1987, McLelland et al. 1996). Notwithstanding, there is a tendency of the separation zone to reduce when the confluence discharge ratio is lower than 1 in the Best (1987) laboratory experiments. Moreover, the data of his figure 5 predicts a 0 dimensionless width of this zone with diminishing discharge ratios for the different tested junction angles. Bathymetric measurements (FICH 2004) complemented with visual inspections by the authors, revealed an intense erosive process on the left bank at S5 (Fig. 2).

This observation would be also an indirect evidence in the Paraná River of the statement by Rhoads (1996), referring to the two secondary cells as a mechanics to transport high-velocity, near-surface fluid downward toward the bed over the zone of maximum channel depth and inward along the bed toward each bank. Note in Figure 3a that the secondary velocities near the bed of the left cell appear to be larger than those near the water surface.

Finally, it was shown that measurements of a single transect with ADP at a given cross section are not sufficient to obtain a clear pattern of the secondary flow structure. The previous remarks were based on results of a secondary current field established with the averaging procedure outlined above.

REFERENCES

Ashmore, P.E. & Parker, G. 1983. Confluence scour in coarse braided streams. *Water Resources Research*, 19, 342–402.

Bathurst, J.C., Thorne, C.R. & Hey, R.D. 1977. Direct measurement of secondary currents. *Nature*, 269: 504–506.

Bathurst, J.C., Thorne, C.R. & Hey, R.D. 1979. Secondary flow and shear stress at river bends. *Journal of the Hydraulics Division.* American Society of Civil Engineers, 105: 1277–1295.

Best, J.L. 1987. Flow dynamics at river channel confluences: implications for sediment transport and bed morphology. *In:* Recent Developments in Fluvial Sedimentology, SEPM Special Publication 39 (eds: Etheridge, F.G., Flores, R.M. and Harvey, M.D., 27–35.

Best, J.L. 1988. Sediment transport and bed morphology at river channel confluences, *Sedimentology*, 35, 481–498.

Best, J.L. & Roy, A.G. 1991. Mixing-layer distortion at the confluence of channels of different depth. *Nature*, 350, 6317, 411–413.

Bhowmik, N.G. 1982. Shear stress distribution and secondary currents in straight open channel. In Gravel-bed Rivers, Hey R.D., Bathurst, J.C., Thorne, C.R. (eds). Wiley: Chichester, 31–55.

Biron, P., Roy, A.G. & Best, J.L. 1996. Turbulent flow structure at concordant and discordant open channel confluence. *Exp. Fluids*, 21, 437–416.

Bradbrook, K.F, Lane, S.N. & Richards, K.S. 2000. Numerical simulation of three-dimensional, time-averaged flow structure at river channel confluences. *Water Resources Research*, vol. 36, 9, pp 2731–2746.

Bradbrook, K.F., Lane, S.N., Richards, K.S., Biron, P.M. & Roy, A.G. 2001. Role of bed discordance at asymetrical river confluences. *Journal of Hydraulic Engineering*, May, 351–368.

Dietrich, W.E. & Smith, J.D. 1983. Influence of the point bar on flow through curved channel. *Water Resources Research* 19: 1173–1192.

Dietrich, W.E. & Smith, J.D. 1984. Processes controlling the equilibrium bed morphology in river meanders. In River Meanders, Elliott C.M. (eds). American Society of Civil Engineers: New York; 759–769.

Drago, E. & Amsler, M.L. 1998. Bed sediment characteristics in the Parana and Paraguay rivers. *Water Internat.* 23, 174–183.

FICH (Facultad de Ingeniería y Ciencias Hídricas) 2004. Estudios Hidráulicos y Morfológicos Zona Isla de la Invernada – Río Paraná. Subsecretaría de Puertos y Vías

navegables. Universidad Nacional del Litoral. Informe Final. Santa Fe, Argentina.
Hey, R.D. & Thorne, C.R. 1975. Secondary flow in rivers, channels. *Area*, 7: 191–195.
Huang, J., Weber, L.J. & Lai, Y.G. 2002. Three-dimensional numerical study of flows in open-channel junctions. *Journal of Hydraulic Engineering*, March, 268–280.
Lane, S.N., Bradbrook, K.F., Richards, K.F., Biron, P.M. & Roy, A.G. 2000. Secondary circulation cells in river channel confluence: measurement artifacts or coherent flow structures?. *Hydrological Processes*, 14, 2047–2071.
Markham, A.J. & Thorne, C.R. 1992a. Geomorphology of gravel-bed river bends. In Dynamics of Gravel-bed Rivers, Billi, P., Hey, R.D., Thorne, C.R., Tacconi, P. (eds). Wiley: Chischester; 433–450.
Markham, A.J. & Thorne, C.R. 1992b. Reply to discussion of geomorphology of gravel-bed river bends. In Dynamics of Gravel-bed Rivers, Billi, P., Hey, R.D., Thorne, C.R., Tacconi, P. (eds). Wiley: Chischester; 452–456.
Mclelland, S.J., Ashworth, P.J. & Best, J.L. 1996. The Origin and Downstream Development of Coherent Flow Structures at Channel Junction. In Coherent flow structures in open channels, Ashworth, Bennett, Best and McLelland (eds), Wiley and Sons, 491–519.
Mosley, M.P. 1976. An experimental study of channel confluences, *Journal of Geology*, 84, 535–562.
Nezu, I. & Nakagawa H. 1993. *Turbulence in Open-Channel Flows*. Balkema, A.A. Rotterdam, Netherlands.
Paice, C. 1990. Hydraulic control of river bank erosion: an environmental approach. PhD dissertation, School of Enviroment Sciences, University of East Anglia, Norwich.
Parsons, D.R., Best, J.L., Lane, S.N., Hardy, R.J., Orfeo, O. & Kostaschuk, R. 2004. The morphology, 3D flow structure and sediment dynamics of a large river confluence: the Paraná and Río Paraguay, NE Argentina (p. 43–48). *River Flow 2004*. Proc. of the Sec. Int. Conf. on Fluvial Hydraulics (Napoly, Italia) Eds. Greco, M., Carraveta A. & Della Morte, R. Vol. 1. A.A. Balkema Publ. London.
Prandtl, L. 1952. *Essentials of fluid dynamics*, Blackie, London, 452 pp.
Ramonell, C.G., Amsler, M.L. & Toniolo, H. 2002. Shifting modes of the Paraná River thalweg in its middle/lower reach. *Zeitschrift fur Geomorphologie*. Suppl.-Bd. 129, 129–142.
Rhoads, B.L. 1996. Mean structure of transport-effective flows at an asymmetrical confluence when the main stream is dominant. In Coherent flows structure in Open Channel, Ashworth, P.J., Bennett, S.J., Best, J.L., McLelland, S.J. (eds). Wiley and Sons; 491–517.
Rhoads, B.L. & Kenworthy, S.T. 1995. Flow structure in asymmetrical stream confluence. *Geomorphology*, 11: 273–293.
Rhoads, B.L. & Kenworthy, S.T. 1998. Time-average flow structure in the central region of a stream confluence. *Earth Surface Processes and Landforms*, 23: 171–191.
Rhoads, B.L. & Sukhodolov, A.N. 2001. Field investigation of three-dimensional flow structure at stream confluences: 1. Thermal mixing and time-averaged velocities. *Water Resouces Research*, vol. 37, 9, 2393–2410.
Richardson, W.R. 1997. Secondary flow and channel change in braided rivers. Thesis submitted to the University of Nottingham for the degree of Doctor of Philosophy.
Richardson, W.R. & Thorne, C.R. 1998. Secondary Currents around a Braid Bar in the Brahmaputra River, Bangladesh. *Journal of Hydraulic Engineering*, 124: 3, 325–328.
Richardson, W.R. & Thorne, C.R. 2001. Multiple thread flow and channel bifurcation in a braided river: Brahmaputra-Jamuna River, Bangladesh. *Geomorphology*, 38,185–196.
Richardson, W.R., Thorne, C.R. & Mahmood, S. 1996. Secondary flow and channel changes around a bar in the Brahmaputra River, Bangladesh. In Coherent flow structures in open channels, Ashworth, P.J., Bennett, S.J., Best, J.L., and McLelland, S.J. (eds), Wiley and Sons, 519–545.
Rozovskii, I.L. 1954. Concerning the Question of velocity distribution in stream bends. DAN URSR (Report of the Academy of Sciences of the Ukraine SSR), 1.
Schumm, S.A. & Winkley, B.R. 1994. The character of large alluvial rivers. p. 1–9. In: Schumm, S.A. & Winkley, B.R. (eds.): The variability of large alluvial rivers, ASCE, 467 pp.
Szupiany, R.N. & Amsler, M.L. 2005. Estrategia de medición del campo de velocidades en un gran río con la tecnología acústica Doppler. XX° Congreso Nacional del Agua, Mendoza, Argentina.
Thorne, C.R. & Rais, S. 1984. Secondary current measurements in a meandering river. In River Meandering. Elliott C.M. (eds). American Society of Civil Engineers. New York, 675–686.

River, Coastal and Estuarine Morphodynamics: RCEM 2005 – Parker & García (eds)
© 2006 Taylor & Francis Group, London, ISBN 0 415 39270 5

Experimental observations on channel bifurcations evolving to an equilibrium state

Walter Bertoldi, Alessio Pasetto, Luca Zanoni & Marco Tubino
Department of Civil and Environmental Engineering, University of Trento, Italy

ABSTRACT: Bifurcation is the process that generates and maintains a braided pattern and determines the distribution of flow and sediment transport along the downstream branches. According to the results of recent theoretical investigations, under bedload dominated conditions bifurcations are likely to display an unbalanced configuration, which results in an uneven partition of flow discharge in downstream channels (Bolla Pittaluga et al. 2003). In this work we present the results of two series of experimental runs which have been planned with the aim of investigating how the hydraulic parameters and the morphological response of the upstream channel affect the geometry, the water partition and the stability of the node. We have devoted specific attention to examine the role of free bars developing in the incoming channel on the stability of the node. Moreover, we have quantified the characteristics time scales of the evolutionary process of a bifurcation. We have found that two different effects control water and sediment distribution in the case of a self-formed Y-shaped bifurcation. Initially the migration of free bars induces relatively rapid fluctuations of the discharge partition in the downstream branches. Once a strongly unbalanced configuration establishes at the node, the upstream morphodynamic influence induced by the bifurcation slows down the speed of bars and modifies their structure. As a result, further evolution of the bifurcation occurs on a longer timescale.

1 INTRODUCTION

The Y-shaped and X-shaped configurations (i.e. subsequent channel bifurcations and confluences) are relevant, peculiar features of braiding, that are often adopted as single units for detailed analysis (Ashmore 1991; Ferguson et al. 1992; Bristow and Best 1993; Ferguson 1993). In a different but complementary perspective Stojic et al. (1998) and Ashmore (2001) point out how braiding can be seen as a rapid succession of single-thread, weakly meandering branches shifting processes, which display a high degree of complexity due to the possibility of the channel to bifurcate and consequently to distribute water and sediment in the whole plain.

The bifurcation process has been recently the subject of various investigations, which have been devoted to examine both the conditions which lead to the onset of flow divergence (e.g. Richardson and Thorne 2001; Bertoldi and Tubino 2005a) and the configurations achieved by a bifurcation at equilibrium, along with their stability, through theoretical (Wang et al. 1995; Bolla Pittaluga et al. 2003), experimental (e.g. Federici and Paola 2003; Hirose et al. 2003) and field works (Zolezzi et al. 2005). The above works provide a consistent picture of several distinctive features characterizing channel bifurcations in gravel-bed rivers: they are almost invariably asymmetrical, which implies an uneven partition of flow and sediment discharge in downstream distributaries, and highly unstable, in particular at low values of the Shields stress of the incoming flow. Furthermore, the bed topography just upstream of the bifurcation is invariably subject to aggradation, leading to the establishment of a transverse slope which drives flow and sediment transport towards one branch, preferentially.

A recent attempt to reproduce such asymmetries is due to Bolla Pittaluga et al. (2003) within the framework of a simple one-dimensional theory of a Y-shaped configuration. The theory is restricted to the case of fixed-banks. A suitable schematization of the node is introduced, whereby a quasi two-dimensional approach is used to model the partition of sediment transport, which accounts for the transverse exchange of flow and sediment due to topographical effects just upstream the flow divergence. The theory is able to predict the existence of unbalanced equilibrium configurations in which one of the downstream tributaries carries a major part of water discharge and is invariably characterized by lower mean bed elevation. According

Figure 1. An evolving bifurcation in a braided river laboratory model: the main channel shifted from the left to the right side of the floodplain in a time span of about 1 hour. Flow is from bottom to top.

to theoretical results the occurrence of this uneven flow partition is mainly controlled by the hydraulic and morphological parameters of the upstream flow and, in particular, it is fostered for lower values of the sediment mobility (ϑ) and relatively high values of the width to depth ratio (β).

Theoretical findings are summarized in Figure 2 in terms of the Shields parameter, ϑ_a, and width ratio, β_a, of the upstream channel, for different values of the upstream longitudinal slope, S. A threshold line can be determined which separates the region where the only stable solution implies the equal partition of flow in downstream branches from the region where two more unbalanced solutions appear, which are found to be invariably stable, while the former solution becomes unstable.

The temporal evolution of a braided network is strongly conditioned by the evolution of the bifurcations and is characterized by a relatively rapid shift of channels and migration of nodes: both channel morphology and the spatial and temporal patterns of the sediment transport are controlled by bar/pool features on a short scale and by confluence/bar features on a larger scale (Ashmore 2001). Ferguson et al. (1992) have documented through field observations on a chute and lobe configuration how a sediment wave can affect flow distribution in downstream branches, leading to the reversal of the lateral asymmetry. On the other hand, the creation of new channels and the modification of the local structure of a bifurcation cause temporal variations of the sediment transport rate over macro and mega-scales (e.g., Hoey and Sutherland 1991).

Given the above picture, two main questions arise. Does a typical timescale exist which can be inherently associated to the evolutionary process of a bifurcation? How the above scale is conditioned by the influence of bed and channel processes, like the occurrence of steady and migrating bed forms?

In order to check the suitability of the theoretical model of Bolla Pittaluga et al. (2003) we have performed various experiments under controlled laboratory conditions (details are given in Bertoldi and Tubino (2005b), in preparation). The experiments have been also designed to ascertain up to what extent the equilibrium configurations may be influenced by further ingredients which have been neglected so far, namely the effect of width adjustment to flow discharge and the role of bedforms and channel processes. Furthermore, experimental observations provide an estimate of the timescale required to achieve the equilibrium configuration and allow for a comparison between the evolutionary timescale of the node and those typical of other processes occurring in a braided network, like bed forms migration, width changes and the upstream channel development (curvature, orientation of the main flow, channel shift).

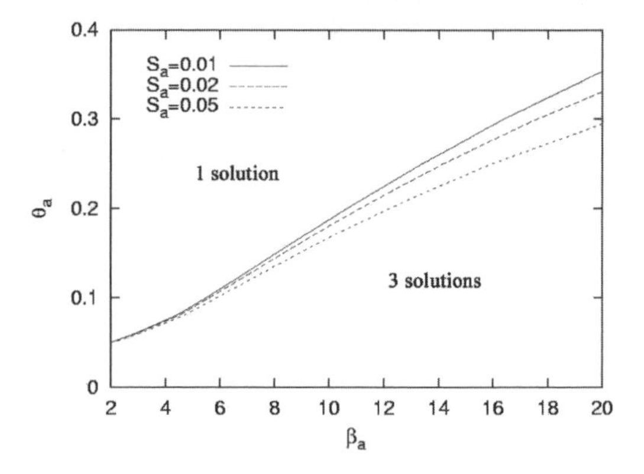

Figure 2. Equilibrium configurations of a symmetrical Y-shaped bifurcation with fixed banks, as determined by the model of Bolla Pittaluga et al. (2003): below the threshold lines two unbalanced solutions are possible along with the solution corresponding to equal flow partition in downstream branches.

In the present paper we analyze in detail the interaction process between bars developing in the upstream channel and the bifurcation, focusing the attention on the mutual effects exerted on the migration speed of bars and on the discharge distribution at the node. To fulfill this aim three series of experimental runs have been performed, in which we have investigated both

(a) *F* runs (b) *M* runs (c) *L* runs

Figure 3. Images of the three experimental configurations built inside the flume.

a configuration with fixed banks and a configuration in which the branches were free to evolve in planform. The rest of paper is organized as follows: in the next chapter the experimental setup and procedure are described; experimental results are presented in sections 3 and 4, where we first describe the equilibrium configurations and then we analyze the timescale of the evolutionary process of the node and the interaction with migrating bars.

2 EXPERIMENTAL SET UP AND PROCEDURE

The experiments have been carried out in a large laboratory flume, 25 m long and 2.90 m wide, which has been constructed in the Hydraulic Laboratory of the University of Trento in order to reproduce gravel-bed braided rivers in a more controlled and easy to monitor environment. The hydraulic model has been designed in Froude similitude, paying particular attention to reproduce some relevant ingredients as the hydraulically rough character of the boundary, the dominance of bed load transport and the typology of bedforms.

The water discharge has been supplied by a pump, regulated by an inverter, that could set discharge values from 0.5 to 20 liters per second. At the upstream end of the flume, the first meter has been devoted to dissipate the kinetic energy of the incoming flow. Furthermore, the sediment input has been provided by an open circuit, made up by a volumetric sand feeder with three screws that convey the sand into the flume through a diffuser. To ensure a constant and well defined input, particularly for relatively low sediment rates, only dry sand has been fed. At the downstream end a chute has been used to convey the flow and sediment transport in a submerged tank where we have measured the weight of the transported sediment through a strain gauge. In all the runs the bed was constituted of a well-sorted quartz sand, characterized by a mean diameter d_{50} of 0.63 mm.

A high accuracy rail has been mounted to support a carriage driven by a motor, whereby the measuring instruments can be moved along the longitudinal, transverse and vertical directions. In particular, the bed topography has been surveyed by a laser profiler driven by a software in order to automatically measure bed elevation on a regular grid (in the present case the adopted spacing was 1 cm in the transverse direction and 10 cm in the longitudinal direction).

In order to get a measure of the discharge flowing in each downstream channel, the flow has been conveyed in two different tanks equipped with a triangular mill weir, where the flow depth has been measured by a pressure sensor device. Furthermore, the evolution of the channel planform has been surveyed with a photographic monitoring by a digital camera located 4 m above the sand bed and supported by another movable carriage.

In the present paper we report on the results of experiments performed with three different Y-shaped configurations which have been built inside the flume in order to study the evolution of a bifurcation both in fixed-banks and in self-formed channels.

In the first series we have built a bifurcation with fixed walls, rectangular cross section and mobile sand bed (this set of runs will be denoted in the following with the initial F). This series was specifically aimed at providing an experimental validation of the one-dimensional theoretical model of Bolla Pittaluga et al. (2003). The upstream channel was 5 m long and 0.36 m wide and the two downstream branches were 0.24 m wide. The bifurcation angle was set to 30 degrees. We observe that the width of the channels has been chosen in order to reproduce typical configuration observed on natural bifurcations: in fact, rational regime theories predict a non-linear relationship between the flow discharge and the channel width, implying a ratio of 1.3 between the total width of the downstream channels and the upstream width, in the case of a symmetrical configuration (Ashmore 2001).

The above configuration has been subsequently modified to investigate the behaviour of a bifurcation which is free to evolve also in planform (these experiments will be denoted by M). We have first removed the fixed banks in the downstream branches and traced the channels on the sand-plain, but we have kept the same fixed upstream channel of F-experiments in order to maintain the control on the morphological and hydraulic parameters of the incoming flow. In this case the node can modify its shape, orientation and position while an additional degree of freedom is related to the adjustment of downstream branches width to the flowing discharge.

In the third examined configuration we have also removed the banks of the upstream channel and traced all the channels on the sand-plain, in order to fully explore the interaction between the bifurcation and the bed and channel processes occurring within the incoming branch. In particular, in these runs (which will be denoted by L) the mutual interplay between the bifurcation and the alternate bars developing in the upstream channel has been identified more clearly. We note that in this case, since the banks are not cohesive, the timescale of planform development is comparable with that of bed deformation.

For each configuration different experimental runs have been performed, changing the water discharge Q and the longitudinal bed slope S such as to reproduce values of the relevant dimensionless parameters typical of gravel-bed braided rivers, namely the Shields parameter ϑ and the aspect ratio β which are defined as follows:

$$\vartheta = \frac{\rho u_*^2}{(\rho_s - \rho) g d_s}, \qquad \beta = \frac{b}{D}, \tag{1}$$

where b is the half channel width, D is the reach averaged value of the water depth, d_s is the diameter of the sediment, ρ_s and ρ are the sediment and the water density, respectively, g is gravity and u_* is the average friction velocity.

In all the experiments the water discharge has been kept constant during the run; furthermore, the sediment supply has been imposed according to the transport capacity of the incoming flow, which has been estimated using Parker formula (Parker 1990). The same symmetrical configuration has been used as initial condition for all runs; the experiments have been performed until the establishment of a fairly stable distribution of discharge in the downstream branches. At the end of each run the final bed topography has been surveyed in detail using the laser profiler.

In Table 1 the values of the experimental conditions for the whole set of experiments are reported. In the same table we also summarize the main characteristics of the equilibrium configuration achieved by the bifurcation in each run. They are given in terms of the ratio rQ of the water discharges Q_b and Q_c flowing in the downstream channels and of the maximum value $\Delta\eta$ of the transverse difference in bed elevation measured at the inlet of downstream branches, normalized with the flow depth of the upstream branch. In particular, the latter parameter provides a measure of the transverse bed deformation which has been almost invariably observed to characterize the region close to the bifurcation. The inlet step $\Delta\eta$ has been evaluated both considering the difference in the elevation of the thalweg line within the first few cross sections of the downstream branches and computing the difference between the values of the average bed elevation at the node obtained through linear interpolation of the measured longitudinal bed profiles of the two distributaries. In all cases the estimates resulting from these two methods have been found to be comparable. In the following, as a convenient measure of discharge asymmetry, we will also employ the parameter

$$A_Q = \frac{Q_b - Q_c}{Q_b + Q_c}, \tag{2}$$

that has the advantage of being bounded within -1 and $+1$.

3 EQUILIBRIUM CONFIGURATION OF A BIFURCATION

The in continuum monitoring of the water discharge in the two downstream branches has been used to obtain a detailed characterization of the bifurcation evolution and to check the achievement of a stable configuration. We first discuss the results obtained in the F runs, namely those with fixed banks. Two distinct behaviours have been clearly identified: in some runs the bifurcation remained stable, keeping a balanced partition of water and sediment in downstream branches (these runs are denoted with closed symbols in Figure 4); in others runs the bifurcation evolved toward an unbalanced configuration (open symbols in Figure 4), which was characterized by an asymptotic value of the discharge ratio lower than 1 and by a distinctive asymmetry in the bed configuration, the bed elevation at the inlet of the main downstream channel being invariably located at a lower level. An

Table 1. Experimental conditions and relevant dimensionless parameters of the three series of experiments.

Run	S	Q [l/s]	D [m]	β	ϑ	ds	rQ	$\Delta\eta$
F3_18	0.0031	1.8	0.0167	10.8	0.0459	0.0377	0.29	0.6943
F3_20	0.0027	2	0.0189	9.5	0.0425	0.0334	0.47	0.5940
F3_21	0.0029	2.1	0.0191	9.4	0.0453	0.0331	0.56	0.5144
F3_23	0.0027	2.3	0.0194	9.3	0.0524	0.0324	0.73	0.3652
F3_25	0.0027	2.5	0.0215	8.4	0.0487	0.0293	0.65	0.3117
F3_29	0.0032	2.9	0.0223	8.1	0.0599	0.0283	0.80	0.1526
F3_37	0.0034	3.7	0.0254	7.1	0.0721	0.0248	0.91	0.0433
F3_45	0.0038	4.5	0.0277	6.5	0.0873	0.0227	0.99	0.0217
F3_61	0.0029	6.1	0.0363	5.0	0.0855	0.0173	0.97	0.0193
F7_06	0.0061	0.6	0.0068	26.3	0.0420	0.0920	0.00	1.3075
F7_07	0.0063	0.7	0.0075	23.9	0.0462	0.0837	0.00	1.3698
F7_08	0.0073	0.8	0.0077	23.3	0.0559	0.0816	0.00	1.7256
F7_09	0.0077	0.9	0.0083	21.7	0.0606	0.0758	0.25	1.3177
F7_10	0.0071	1	0.0093	19.4	0.0574	0.0678	0.05	1.2276
F7_12	0.0072	1.2	0.0102	17.7	0.0655	0.0620	0.45	0.8212
F7_13	0.0077	1.3	0.0105	17.2	0.0734	0.0602	0.50	0.7726
F7_15	0.0075	1.5	0.0113	15.9	0.0787	0.0556	0.50	0.3969
F7_17	0.0073	1.7	0.0126	14.3	0.0784	0.0500	0.45	0.7219
F7_20	0.0077	2	0.0134	13.4	0.0943	0.0471	1.00	0.2350
F7_24	0.0072	2.4	0.0153	11.7	0.0985	0.0411	1.00	0.0950
M3_19	0.0027	1.9	0.0179	10.1	0.0431	0.0352	0.14	0.5461
M3_21	0.0032	2.1	0.0182	9.9	0.0511	0.0347	0.22	0.4640
M3_23	0.0030	2.3	0.0195	9.2	0.0516	0.0322	0.02	0.8186
M3_25	0.0029	2.5	0.0209	8.6	0.0522	0.0301	0.20	0.5287
M3_28	0.0028	2.8	0.0225	8.0	0.0549	0.0280	0.04	0.7249
M3_33	0.0033	3.3	0.0236	7.6	0.0678	0.0266	0.14	0.5538
L7_08	0.0063	0.8	0.0155	11.1	0.0771	0.0405	0.70	0.2959
L7_10	0.0062	1	0.0141	14.4	0.0707	0.0447	0.60	0.1277
L7_12	0.0066	1.2	0.0151	15.9	0.0803	0.0417	0.40	0.3777
L7_15	0.0060	1.5	0.0162	16.8	0.0802	0.0389	0.55	0.1638

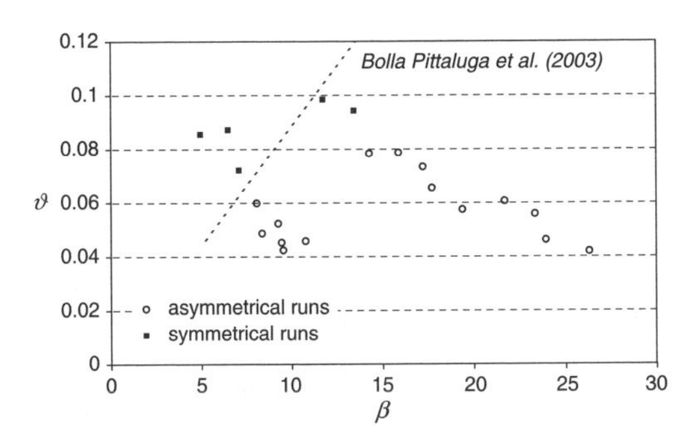

Figure 4. The experimental conditions of runs F are plotted in terms of the aspect ratio and the Shields stress of the incoming flow: open symbols denote those runs in which an unbalanced equilibrium configuration has been reached, while closed symbols correspond to balanced equilibrium configurations. The dashed curve represents the critical line below which the balanced solution is no longer stable as predicted by Bolla Pittaluga et al. (2003).

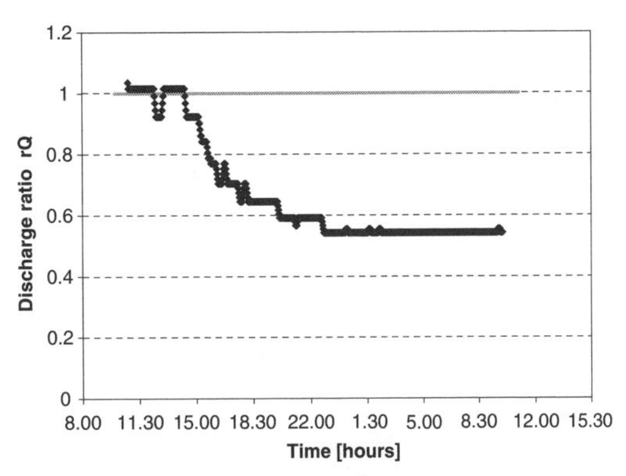

Figure 5. An unstable bifurcation evolving to an unbalanced configuration: the discharge ratio rQ measured in run $(F3-21)$ is plotted as a function of time.

example of an unstable bifurcation is reported in Figure 5, where the time evolution of the discharge ratio rQ for the unbalanced run $F3-21$ is reported. The corresponding asymmetric bed topography is plotted in Figure 6: we may note that a sharp difference in the the average bed elevation of downstream branches occurs at the node.

It is worth noting that the experimental results confirm the theoretical predictions of the model of Bolla

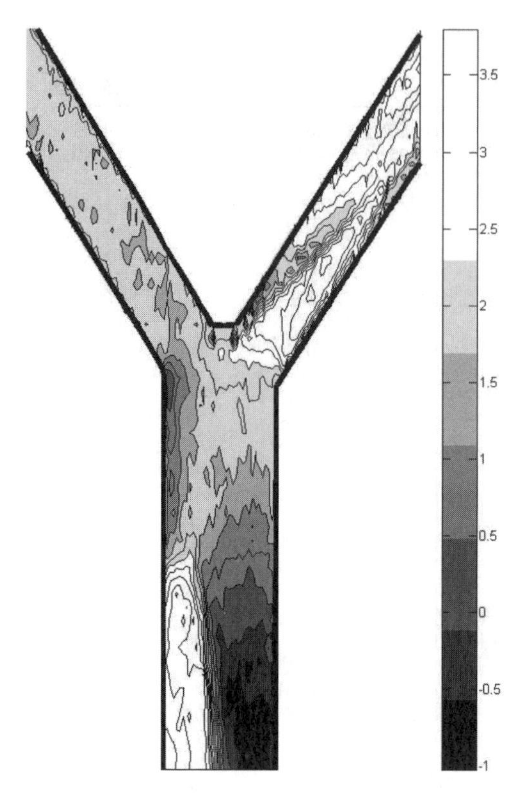

Figure 6. The final bed topography in the unbalanced run ($F3-21$).

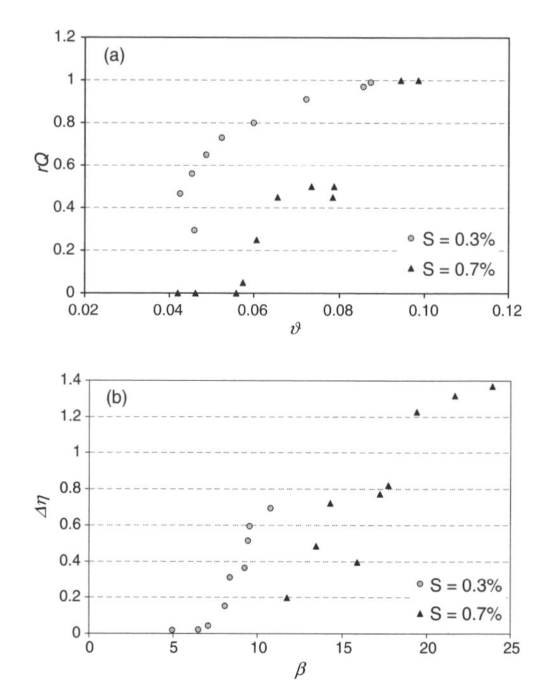

Figure 7. The equilibrium configurations achieved in runs F: (a) the discharge ratio rQ is plotted versus the Shields stress of the upstream channel; (b) the dimensionless inlet step $\Delta\eta$ is plotted versus the aspect ratio of the upstream channel.

Pittaluga et al. (2003). As shown in Figure 4 the theoretical threshold line below which the balanced solution is no longer stable replicates with fairly good accuracy the distinction between balanced and unbalanced runs resulting from the experiments. We also note that the agreement is better for those runs $F3$ in which migrating bars were not observed to develop in the upstream channel. As described in more detail in the next section, the formation of free bars (which invariably occurred in the runs $F7$ and $L7$) can strongly affect the bifurcation evolution. More in general, both the theoretical results of Bolla Pittaluga et al. (2003) and present experimental findings suggest that, whenever the node is not affected by backwater effects, the system composed by an upstream channel dividing into two branches is mainly controlled by the morphological and hydraulic parameters of the upstream stream. This is shown in Figure 7 where the values of the discharge ratio rQ and of the inlet step $\Delta\eta$ corresponding to the final equilibrium configuration are plotted as functions of the Shields parameter and of the aspect ratio of the incoming flow, respectively. They both display a fairly regular increasing trend. We note that for a given channel slope, increasing values of ϑ correspond to decreasing values of β. Furthermore, the stronger is the degree of asymmetry of the equilibrium configuration, the smaller is the discharge ratio rQ. Hence, strongly unbalanced bifurcations invariably display larger values of the inlet step.

We now try to characterize the timescale of the evolutionary process of a bifurcation. To this end we analyze the observed temporal behaviour of the discharge ratio rQ and we also include in the analysis the results of the experimental runs M. In order to identify a suitable timescale which may be inherently associated with the bifurcation we first need to filter out the obvious dependence of the process on Shields parameter which is embodied in the dependence of bed load intensity on the above parameter. In fact, through Exner equation, any morphological process involving bed development proceeds at a speed which is mainly controlled by bed load intensity. Hence, it is appropriate to rescale the time variable with the timescale of bed evolution, as determined through Exner equation:

$$(1-p)\frac{\partial \eta}{\partial t}+\frac{\partial q_s}{\partial x}=0\,, \tag{3}$$

where p is sediment porosity, η is bed elevation, q_s is the sediment transport rate, per unit width, t is time and

x the longitudinal coordinate. In dimensionless form the latter equation reads:

$$\frac{\partial \eta^*}{\partial t^*} + \frac{\partial q_s^*}{\partial x^*} = 0 \,, \tag{4}$$

where the following scaling has been adopted:

$$\eta^* = \frac{\eta}{D} \,, \qquad x^* = \frac{x}{b} \,, \qquad q_s^* = \frac{q_s}{\Phi \sqrt{g \Delta d_s^3}} \,, \tag{5}$$

where Φ is bedload intensity and Δ the relative density of the sediment, b is the half channel width and D is the reach averaged value of the water depth. Consequently the dimensionless time t^* is defined as:

$$t^* = \frac{t}{\tau} \,, \tag{6}$$

where

$$\tau = \frac{bD}{(1-p)\Phi \sqrt{g \Delta d_s^3}} \,. \tag{7}$$

Present experimental findings suggest that, in terms of the above scaled variable, the time behaviour of the discharge ratio toward the equilibrium can be approximated with a decreasing exponential function of the form:

$$rQ = (rQ_0 - rQ_e) \exp\left(-\frac{t^*}{\mathrm{T}}\right) + rQ_e \,, \tag{8}$$

where rQ_0 and rQ_e denote the initial and the equilibrium values of the discharge ratio, respectively, and T is the timescale of the bifurcation. The latter is determined through a best fit procedure, using a least square approximation. Figure 9 shows an example of the time evolution of an unbalanced run along with the corresponding exponential interpolating curve. The resulting values of T display a clear dependence on the relevant parameters of the incoming flow, such that larger values of the timescale correspond to smaller (larger) values of β (ϑ). This means that, in terms of the scaled time variable, the evolution of the bifurcation is faster when the equilibrium configuration is more unbalanced.

The above result can be given a physical explanation in terms of the morphodynamic influence exerted by the bifurcation. According to the results of Zolezzi and Seminara (2001) the above influence is mainly felt upstream or downstream with respect to a planimetric discontinuity, like that associated with an abrupt change of channel curvature, depending upon the width ratio β falling above or below the resonant value β_r defined in Blondeaux and Seminara (1985). The presence of a bifurcation is another example of a planimetric discontinuity which implies a sudden deviation of the main flow orientation. Hence, under supercritical conditions, namely large values of β,

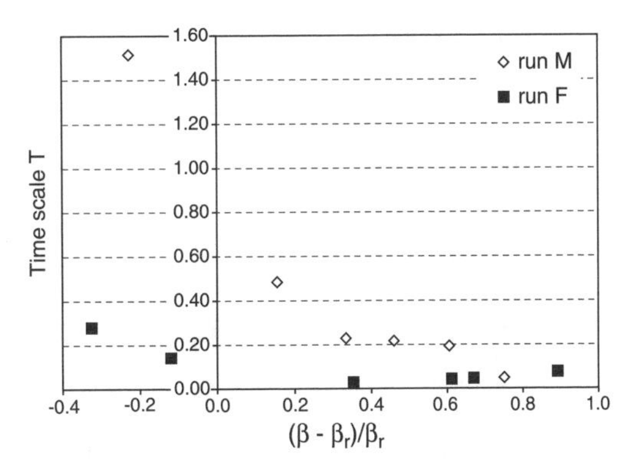

Figure 8. The timescale of the bifurcation as a function of the distance from the resonant conditions.

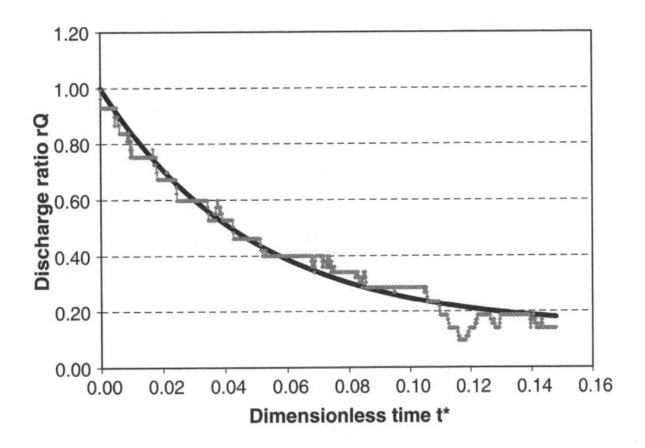

Figure 9. The time evolution of an unbalanced bifurcation ($M3-19$) along with the corresponding interpolating exponential curve.

the bifurcation is more likely to produce a stronger upstream effect which may lead to the establishment of a relatively large transverse bed slope just upstream the node. This in turn affects the bifurcation, leading to a more unbalanced configuration, and may foster a more rapid evolution.

The above speculations are confirmed by the results reported in Figure 8 where the timescale T is plotted as a function of the parameter $(\beta - \beta_r)/\beta_r$ which measures the distance from the resonant conditions. We note that the experimental runs falling in the neighbour of the resonant conditions or below the threshold are invariably characterized by relatively large timescales. We also note that the points corresponding to the lowest values of the aspect ratio represent balanced runs, in which the timescale has been determined analyzing the response of the system to a perturbation, like a concentrated sediment input in one of the two downstream branches. The experimental runs M are invariably characterized by larger values of T, which may be a consequence of the process of width adjustment of downstream branches which is inhibited in runs F.

4 INTERACTION BETWEEN BIFURCATION EVOLUTION AND CHANNEL PROCESSES

In the experimental runs L all the channels were free to change their width and alignment. As pointed out before we have mainly concentrated the attention on the interaction between the bed forms (mainly alternate bars) developing in the channels and the hydraulic and topographic configuration of the bifurcation. All the experiments within this set have been performed with a longitudinal slope equal to 0.7% and have been suitably designed in order to achieve, after the initial widening process, the threshold value of the aspect ratio for the formation of free alternate bars (see for example Colombini et al. 1987).

In the whole set of runs the Y-shaped unit has been subject to a similar evolution. At first, the node shifts downstream, due to the initially high erosion rate of the incoming flow. Consequently, the angle between the downstream branches, which was initially set at 30 degrees, increases its amplitude, reaching a maximum value ranging about 60–70 degrees. As discussed below this phase corresponds to the situation in which we have detected the most unbalanced distribution of water discharge in downstream branches. The subsequent evolution determines a decrease of the amplitude of the bifurcation angle, that eventually reaches a value between 45 and 60 degrees. These values agree with the values of the bifurcation angles observed in other laboratory experiments (Federici and Paola 2003, Bertoldi and Tubino 2005a) and with field data from braided networks. Observed values of the bifurcation angle (maximum and final configuration) are reported in Figure 10, where the angles have been defined in terms of the measured profile of the inner banks.

As the bifurcation evolves, the single branches modifies their planimetric configuration and gradually assume a weakly meandering pattern. As predicted by stability theories (e.g. Blondeaux and Seminara (1985)), the longitudinal wavelength of curvature variations has been found to be mainly controlled by the width of the channel. Such a dependence was clearly recognizable in those runs characterized by a strong unbalanced discharges distribution and therefore by different widths of the distributaries.

In each run within this set of fully-mobile experiments the measured time variation of the discharge partition within the downstream branches has displayed a quite similar behaviour, characterized by a distinctive sequence of phases in which the effect of migrating bars and the morphodynamic influence exerted by the bifurcation have been clearly identifiable. An example of such behaviour is reported in Figure 11 where we plot the observed variations of discharge asymmetry A_Q, defined in Equation 2, in run $L7-08$.

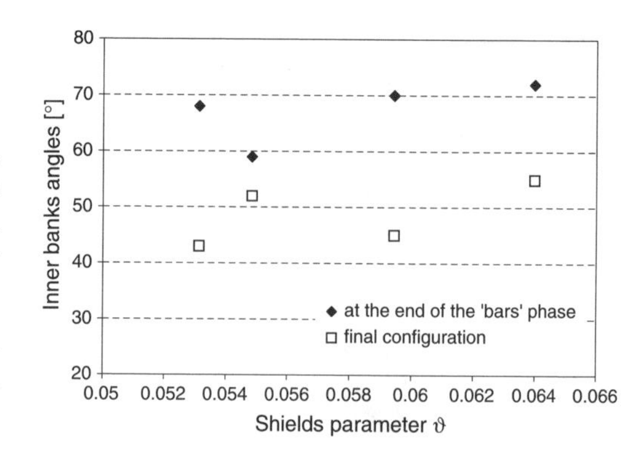

Figure 10. The bifurcation angle at two subsequent stages of runs L.

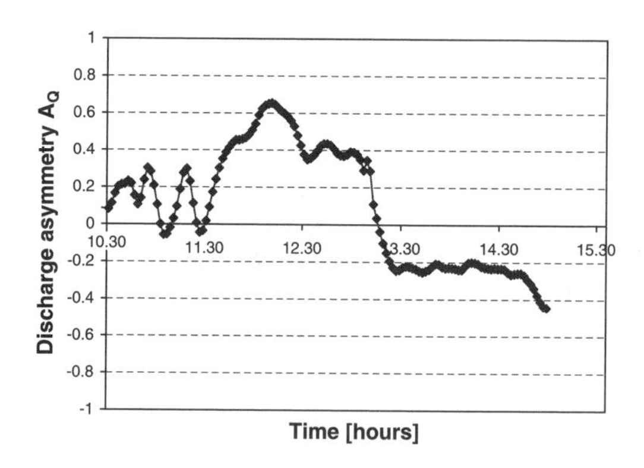

Figure 11. The time evolution of the discharge asymmetry A_Q observed in run $(L7-10)$.

In the first phase the discharge asymmetry fluctuates around the value of 0, that represents the balanced configuration, due to the relatively fast migration of alternate bars which eventually reach the inlet of downstream branches and switch the conveyance direction. However, the above configuration is unstable such that the uneven partition of the discharge leads to the establishment a dominant channel, whose discharge and sediment transport capacity gradually increase. As a consequence, the incoming bars are directed preferentially in the main downstream branch, until two consecutive bars enter in the channel and a sudden switch of flow conveyance occurs, causing the peak of the discharge asymmetry which is clearly recognizable in Figure 11.

At the same time, the bars are in turn influenced by the bifurcation. The above influence can be quantitatively investigated in terms of the observed variations of the wavelength and of the migration speed. In Figure 12(a) we give an example of the decreasing trend of the migration speed measured by direct observations during the run $L7-10$. According to theoretical

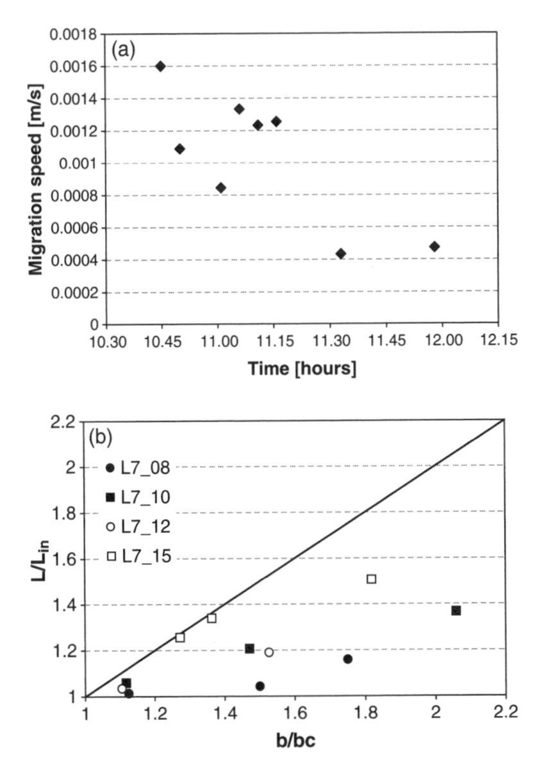

Figure 12. Main features of alternate bars developing in the upstream channel: a) migration speed as a function of time; b) measured bar wavelength, scaled with the initial length, as a function of the channel width, scaled with the threshold value for bar formation.

results of Colombini et al. (1987) a slight decrease of bar speed could be caused by channel widening; however, in this case we observe a marked and rapid decrease, due to the influence of the bifurcation. In Figure 12(b) the measured variation of bar wavelength with respect to its initial value is plotted versus the channel width (which is normalized with the threshold value bc for the formation of free bars). We note that linear theories would predict a linear dependence of bar wavelength on channel width, the critical wave number being almost constant. On the contrary, the longitudinal wavelength keeps almost constant, although the width of the channel increases, reaching a value greater than twice the initial width. Consequently the difference between observed values and theoretical estimates increases during the run.

The transverse structure of bars is also modified by the influence of the bifurcation that induces a quasi-symmetrical forcing. For this reason bars are likely to develop a more central pattern, which furthers the migration of two consecutive bars in the same downstream channel. Furthermore, the bars slow down

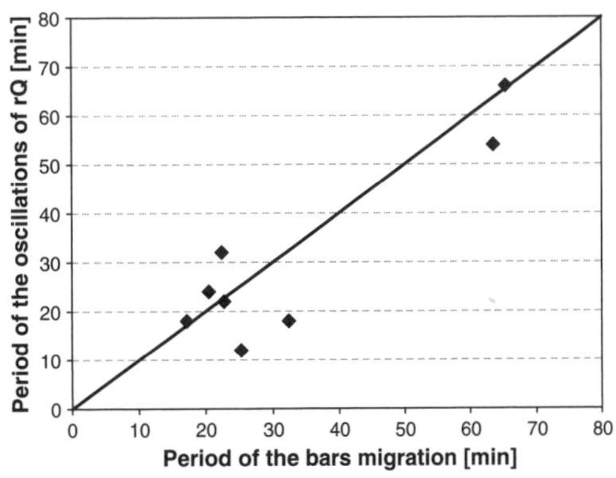

Figure 13. Comparison between the period of oscillation of the discharge ratio and the period of alternate bars measured in run ($L7-10$).

and induce the development of a weakly meandering pattern in the upstream branch, which also displays regular width variations. This configuration in turn influences the bed form pattern, leading to a further inhibition of their migration. An example of the above evolutionary process is given in Figure 14, where the geometry of channels and the bar fronts have been derived from the images acquired through the digital camera located above the flume.

The influence of bars is distinguishable more clearly in the initial stage of the process, when the bifurcation keeps on the average a balanced configuration and the upstream channel is almost straight. The strong link between the train of bars migrating in the channel and the fluctuations of the discharge ratio is highlighted in Figure 13, where the observed period of the oscillations of rQ is compared with the measured period of bars. In the first stage the free bars are characterized by a migration period of about 20–30 minutes, which corresponds to the measured period of the discharge ratio oscillations. This link keeps valid also when bar dynamics is influenced by the bifurcation, as shown by the two points reported in Figure 13 which refer to a subsequent stage where the period reaches a value of about one hour.

In the last phase the bifurcation evolves on a longer timescale and the discharge ratio follows a more regular decreasing trend which may be related both to an inherent dynamic response of the bifurcation and to the slow planimetric evolution of the weakly meandering upstream branch. In this context the behaviour of flow partition at the node can be interpreted through the same interpolation procedure adopted in the analysis of the experimental runs F and M (see Figure 9). However, in this case the process is still influenced by the presence of bed forms in the upstream tributary,

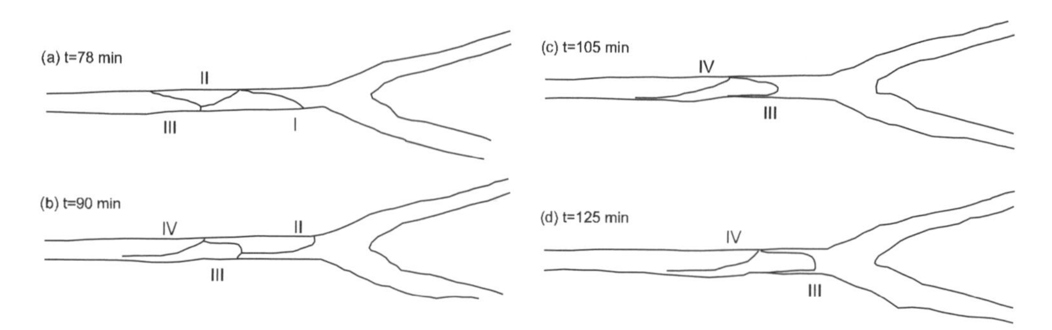

Figure 14. Sketch of bar development and migration observed in the run ($L7-10$).

which imply the occurrence of small scale fluctuations of the discharge ratio. The measured values of the intrinsic timescale T of the bifurcation in the L runs are comparable with those observed in the fixed walls experiments. On the average, a slower evolution occurs in this case, with values of T ranging between 0.3 and 0.6, while in the M runs we measured values ranging between 0.2 and 0.4. The difference can be due to the effect exerted by the upstream channel. We note that due to the continuous evolution of the planimetric configuration, the values of the reference dimensionless parameters have been calculated in this case with reference to the width of the upstream channel measured at the end of the run.

5 CONCLUSIONS

The detailed experimental observations performed on a Y-shaped configuration both with fixed banks and with erodible bank channels provide a consistent picture of the evolutionary process of a gravel-bed river bifurcation and allow for an assessment of the possible equilibrium configurations toward which the system can be driven and of the role played by bed and channel unit processes. In particular, the main results can be summarized as follows:

1. A Y-shaped bifurcation typically displays an unbalanced equilibrium configuration, particularly at relatively low values of the Shields stress, in agreement with the theoretical predictions of the one-dimensional model of Bolla Pittaluga et al. (2003). In the absence of backwater effects such equilibrium is mainly controlled by the morphological and hydraulic parameters of the incoming flow. Beside an unbalanced discharge distribution in downstream branches, we have invariably observed the occurrence an asymmetric bed configuration close to the bifurcation, characterized by the establishment of a transverse step at the inlet of the downstream distributaries.
2. In terms of a scaled time variable, which filters out the dependence on Shields stress inherent to the sediment transport process, a Y-shaped bifurcation has an intrinsic evolution timescale which depends on the aspect ratio of the upstream channel. In particular the above scale depends on the distance of the aspect ratio from the resonant conditions. Supercritical runs are invariably characterized by stronger bed deformation just upstream the node, which induces a faster development, whereas near resonant flow conditions determine larger values of the timescale.
3. The dynamic behaviour of the bifurcation may be strongly affected by the development and migration of bars within the channels. Bar migration initially affects the discharge distribution at the node and determines a fluctuation of the discharge ratio rQ whose period is controlled by the migration of the bars. In turn, the bifurcation acts as a planimetric discontinuity which affects bar development, leading to a sharp decrease of their speed and to a modification of their structure; bars become more centered while their longitudinal wavelength keeps almost constant, in spite of the widening of the channel.
4. A Y-shaped bifurcation with erodible banks invariably attains unbalanced configurations. In this case various stages can be identified in the evolutionary process of the bifurcation, each characterized by a distinctive timescale resulting from the effect of various external forcings (bar migration, channel axis development, width adjustment).

ACKNOWLEDGMENTS

This work has been developed within the framework of the 'Centro di Eccellenza Universitario per la Difesa Idrogeologica dell'Ambiente Montano – CUDAM' and of the projects 'Morfodinamica delle reti fluviali – COFIN2001' and 'La risposta morfodinamica di sistemi fluviali a variazioni di parametri ambientali – COFIN 2003', co-funded by the Italian Ministry of University and Scientific Research (MIUR) and the University of Trento and of the project 'Rischio

Idraulico e Morfodinamica Fluviale' financed by the Fondazione Cassa di Risparmio di Verona, Vicenza, Belluno e Ancona. The authors gratefully acknowledge the fundamental support of Guido Zolezzi and of Stefano Miori, Stefania Baldo, David Marchiori and the staff of the Hydraulic Laboratory of the University of Trento who helped in the execution of the experimental runs.

REFERENCES

Ashmore, P. (1991). How do gravel-bed rivers braid? *Canadian Journal of Earth Sciences*, *28*, 326–341.

Ashmore, P. (2001). *Braiding phenomena: statics and kinetics*, pp. 95–121. Gravel-bed Rivers V, Mosley M.P. (eds), New Zealand Hydrological Society, Wellington.

Bertoldi, W. and M. Tubino (2005a). Bed and bank evolution of bifurcating channels. *Water Resources Research 41 (7)*, W07001 doi:10.1029/2004WR003333.

Bertoldi, W. and M. Tubino (2005b). River bifurcations: experimental observations on equilibrium configurations, *in preparation*.

Blondeaux, P. and G. Seminara (1985). A unified bar-bend theory of river meanders, *Journal of Fluid Mechanic 112*, 363–377.

Bolla Pittaluga, M., R. Repetto, and M. Tubino (2003). Channel bifurcation in braided rivers: equilibrium configurations and stability. *Water Resources Research 39 (3)*, 1046, doi:10.1029/2001WR001112.

Bristow, C. and J. Best (1993). *Braided rivers: perspectives and problems*, pp. 1–9. Braided Rivers: form, process and economic applications, J.L. Best and C.S. Bristow (Eds.), Geological Society Special Publication, 75.

Colombini, M., G. Seminara, and M. Tubino (1987). Finite-amplitude alternate bars. *Journal of Fluid Mechanic 181*, 213–232.

Federici, P. and C. Paola (2003). Dynamics of bifurcations in noncohesive sediments. *Water Resources Research 39 (6)*, 1162, doi:10.1029/2002WR001434.

Ferguson, R. (1993). *Understanding braiding processes in gravel-bed rivers: progress and unresolved problems*, pp. 73-87. Braided Rivers: form, process and economic applications, J.L. Best and C.S. Bristow (Eds.), Geological Society Special Publication, 75.

Ferguson, R., P. Ashmore, P. Ashworth, C. Paola, and K. Prestegaard (1992) Measurements in a braided river chute and lobe. 1. Flow pattern, sediment transport and channel change. *Water Resources Research 28 (7)*, 1877–1886.

Hirose, K., K. Hasegawa, and H. Meguro (2003, 1–5 September). Experiments and analysis on mainstream alternation in a bifurcated channel in mountain rivers. In *Proceedings 3 International Conference on River, Coastal and Estuarine Morphodynamics*, pp. 571–583. Barcelona (Spain).

Hoey, T. and A. Sutherland (1991). Channel morphology and bedload pulses in braided rivers: a laboratory study. *Earth Surface Processes and Landforms 16*, 447–462.

Parker, G. (1990). Surface-based bedload transport relation for gravel rivers. *Journal of Hydraulic Research 28*, 417–436.

Richardson, W. and C. Thorne (2001). Multiple thread flow and channel bifurcation in a braided river: Brahmaputra-Jamuna River, Bangladesh. *Geomorphology 38*, 185–196.

Stojic, M., J. Chandler, P. Ashmore, and J. Luce (1998). The assessment of sediment transport rates by automated digital photogrammetry. *Photogrammetric Engineering and Remote Sensing 64 (5)*, 387–395.

Wang, B., R. Fokking, M. De Vries, and A. Langerak (1995). Stability of river bifurcations in 1D morphodynamics models. *Journal Hydraulic Research 33 (6)*, 739–750.

Zolezzi, G., W. Bertoldi, and M. Tubino (2005). *Morphological analysis and prediction of channel bifurcations*. Submitted for publication in "Braided Rivers 2003", Blackwell.

Zolezzi, G. and G. Seminara (2001). Downstream and upstream influence in river meandering. PART 1. General theory and application of over deepening. *Journal of Fluid Mechanic 438*, 183–211.

River, Coastal and Estuarine Morphodynamics: RCEM 2005 – Parker & García (eds)
© 2006 Taylor & Francis Group, London, ISBN 0 415 39270 5

Morphological evolution of bifurcations in gravel-bed rivers with erodible banks

Stefano Miori
University of Trento, Italy

Rodolfo Repetto
University of L'Aquila, Italy

Marco Tubino
University of Trento, Italy

ABSTRACT: In a recent paper Bolla Pittaluga et al. (2003a) proposed a physically based formulation for the nodal point conditions to be adopted, in the context of a one-dimensional approach, to describe channel bifurcations in gravel-bed rivers. The authors employed the nodal conditions to study the equilibrium configurations of a simple bifurcation and their stability, assuming erodible bed and fixed banks. With the present contribution we extend the model proposed by Bolla Pittaluga et al. (2003a) to the case of channels with erodible banks, i.e. channels which may adapt their width to the actual flow conditions. Such an extension is of a particular interest for river bifurcations in gravel-bed braided streams, in which bed evolution and bank erosion processes may occur over comparable time scales. Moreover, we investigate how the mechanism of generation of the bifurcation may affect its equilibrium configurations, showing that significant differences may arise if the bifurcation is formed through the incision of a new channel or as a consequence of the formation of a central deposit into a channel. In both circumstances, however, the common feature of the equilibrium configurations reached by the system is a strongly unbalanced partition of water and sediment discharges into the two branches. Such a finding agrees with experimental and field observations according to which stable symmetrical bifurcations are seldom observed in natural gravel-bed rivers.

1 INTRODUCTION

In recent years much attention has been devoted to the study of river bifurcations. In particular, field and experimental works have brought some insight on the basic features that characterize bifurcations in gravelbed rivers, which the present work is concerned with. Such observations, some of which will be briefly reviewed below, are of great importance as they may serve as the basis for theoretical models aimed at describing the morphological evolution and characteristics of river bifurcations.

It is first worth recalling the classic experimental work by Bulle (1926), recently reconsidered by de Heer & Mosselman (2004) who attempted to reproduce his results numerically. Bulle (1926) performed experiments in a straight flume from which a minor branch diverted. Banks were kept fixed and the author measured water discharge and bed-load distribution into the branches. The experiments showed the existence of a strongly three-dimensional flow at the bifurcation which markedly affected bed load characteristics, typically inducing a disproportionate sediment supply to the diverting channel. Notice that the above effect is probably enhanced by the sharp deviation of the wall at the bifurcation and by the quite large angle that the bifurcating branch formed with the main channel in the author's experimental configuration; one might therefore expect the above effect to be less intense in natural environments.

Recently Federici & Paola (2003) have performed a series of experiments aimed at studying the behavior of bifurcations in gravel-bed braided rivers. They worked first on a laboratory model of a braided river and then performed a second series of experiments focusing on the behavior of single bifurcations, generated in a diverging channel as consequence of the formation of a central deposit. Most of the well defined bifurcations that formed in the braiding model had the following common features: the average flow velocities and average water depths at the mouth of the two downstream channels were similar but the widths of the two branches were generally markedly unequal, as were water and sediment discharges. Moreover, experiments performed on single bifurcations showed that, for large values of the Shields parameter in the upstream straight channel, the bifurcation was stable and, typically, fairly symmetrical. For small values of the Shields parameter, on the other hand, the system was invariably unstable and one of the two branches was often closed, to be eventually reopened later. This process occurred repeatedly and randomly. The authors also observed that, in correspondence of the bifurcation, water level was approximatively constant, while bed elevation often displayed a sudden step between the mouths of the two downstream branches.

Very recent field observations performed by Zolezzi et al. (2005) on channel bifurcations in braided networks confirm most of the above experimental results. The field campaigns were carried out in the Sunwapta river (Alberta, Canada) and the Ridanna creek (Italy). The bed of both rivers is mainly constituted of gravel and sediment transport essentially occurs as bedload. Seven bifurcations were analyzed in detail. The authors report that the main recurring feature of the observed bifurcations is a strong asymmetry of their morphological configuration: one of the two branches is typically larger than the other and it carries most of water and sediment discharge. The authors also point out the existence of a so called "inlet step", i.e. a difference in bed elevation between the inlets of the two downstream branches. The inlet step amplitude is found to depend on the asymmetry of discharge distribution and to scale with the mean upstream water depth. The presence of such a difference in bed elevation between the inlets of the two downstream branches significantly affects the bed load transport close to the bifurcation, thus playing a significant role in the sediment distribution within the branches.

In order to apply one-dimensional models to the study of braided rivers, suitable nodal point conditions must be imposed at the bifurcations. Obviously, the problem of catching the main features of real bifurcations by means of highly simplified onedimensional nodal point conditions is not a trivial one. The first attempt in this direction was made by Wang et al. (1995), who studied the possible equilibrium configurations of a channel bifurcation along with their stability, mainly focusing their attention on fine sediment rivers. Wang et al. (1995) observed that five conditions are required at the bifurcation point, in order for the problem to be well posed. They proposed a set of conditions among which a so called "nodal point condition" played a particularly important role on the bifurcation evolution. In the authors' approach such a condition has an empirical foundation. More recently, Bolla Pittaluga et al. (2003a)[1] have reconsidered the problem, proposing an alternative formulation for the nodal point conditions, which is based on physical considerations and is aimed at describing the behavior of gravel bed bifurcations. In analogy to the work by Wang et al. (1995) the nodal conditions are employed to study the equilibrium configurations of a simple bifurcation and their stability. The approach proposed by Bolla Pittaluga et al. (2003a) is briefly recalled in the next section.

Both Wang et al. (1995) and Bolla Pittaluga et al. (2003a)'s models are based on the assumption of fixed banks, i.e. the planimetrical geometry of the bifurcation is prescribed. This actually implies that the time scale of planimetrical changes is assumed to be much larger than that of bed evolution. This is a well accepted assumption in the cases of cohesive or vegetated banks as both cohesion and vegetation contribute to make bank erosion a much slower process than bed deformation. In gravel-bed braided rivers, however, the above processes have comparable time scales and it is thus reasonable to expect the width of each channel in a braiding network to be, if not in equilibrium with, at least strongly affected by the hydraulic conditions. This also implies that, during the morphological evolution of a bifurcation, one might expect both bed elevation and channel widths to change. Experimental and field observations suggest, however, that a gravel-bed channel significantly widens when the discharge increases, but the counterpart process, i.e. channel narrowing as a consequence of a discharge fall, is by far less intense. This asymmetry of channel behavior as a response to increasing or decreasing discharge suggests that the time history of a bifurcation may influence its possible equilibrium configurations. In other words the mechanism of generation of a bifurcation, either due for instance to the incision of a new channel or to the deposition of a central bar, may control the final configuration to which the system evolves.

In the present contribution we develop the above ideas and extend the model proposed by Bolla Pittaluga et al. (2003a) to the case of channels with erodible banks, i.e. to channels adapting their width to flow conditions. This is done making use of empirical relationships as it will be described in the following. Moreover, we study the time evolution of a bifurcation from a given initial configuration allowing channel widening when the discharge increases and inhibiting the possibility of channel narrowing. Accounting for the latter effect it is shown that bifurcations generated from the incision of an initially narrow branch diverting from the main channel evolve toward significantly different equilibrium configurations with respect to bifurcations formed as a consequence of a central deposition in a wide channel. Whatever the initial configuration of the system, however, the equilibrium configurations finally reached by the system

[1] See also Bolla Pittaluga et al. (2003b).

display common features. In particular it is invariably found that the system evolves toward a non symmetrical solution in which one of the two branches carries much more water and sediment than the other. This is in agreement with the field and laboratory observations cited above. The presence of the "inlet step" is also reproduced by the model, whatever the initial condition of the system.

It is also worth to stress another novelty of the present model with respect to the work by Bolla Pittaluga et al. (2003) concerning the approach adopted to describe the time evolution of the system. The above authors, following Wang et al. (1995), described the evolution of a river bifurcation as a sequence of uniform flow conditions, i.e. they neglected variations of flow characteristics along each branch and assumed scour or deposition to be uniformly spread out along each channel. An alternative approach has been proposed by Hirose et al. (2003): the authors study the evolution of a bifurcation in terms of local variables, thus neglecting the role of eventual downstream boundary conditions at the outlets of the diverting branches. We follow herein a similar approach which seems more suitable to describe gravelbed braided rivers, where backwater effects play a minor role. The main difference between the present approach and that proposed by Bolla Pittaluga et al. (2003) concerns the evaluation of the time scale of the system evolution: in the authors' approach such a time scale crucially depends on the length of the downstream branches while, in the present case, it is set by the nodal point conditions at the bifurcation.

2 THE MODEL PROPOSED BY BOLLA PITTALUGA ET AL. (2003)

In the following we briefly recall the model proposed by Bolla Pittaluga et al. (2003a) that will be extended to account for channel width adjustments to local flow conditions in the next section. Let us consider a channel bifurcation and denote with the subscript letter a the upstream channel and with the letters b and c the downstream branches. As pointed out in the previous section, in correspondence of a bifurcation, the flow field and bed topography are three-dimensional and rapidly varying. Bolla Pittaluga et al. (2003a) attempt to account for the complexity of such a flow employing a "quasi two-dimensional approach" close to the bifurcation. They consider the last reach of the main channel, with a length αb_a, with b_a being the width of channel a, as divided into two longitudinal adjacent strips as shown in Figure 1. The length αb_a should equal the upstream distance from the bifurcation at which the bed topography is unaffected by the presence of the channel division. Experimental investigations by Bolla Pittaluga et al. (2003a) along

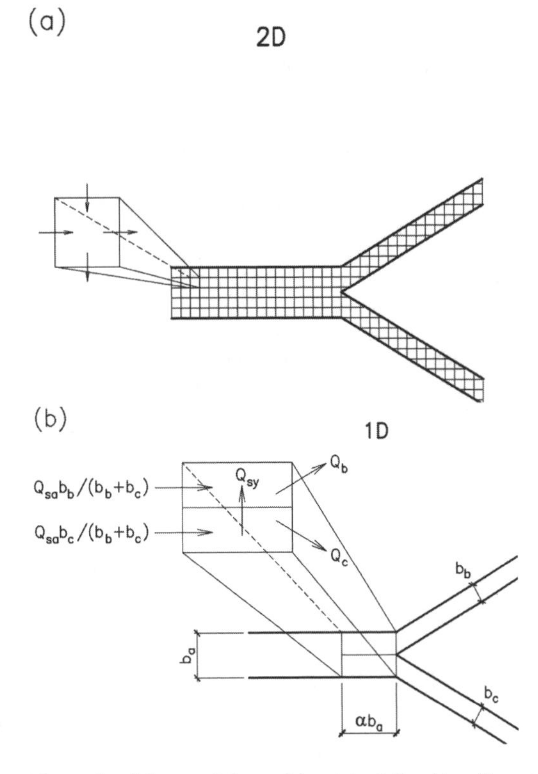

Figure 1. Scheme of the nodal point relationships (from Bolla Pittaluga et al. [2003a]).

with the observations reviewed in the previous section suggest that α should range between 1 and 3 (in the following we will assume $\alpha = 1$). The length of the two longitudinal cells is therefore very small if compared with the longitudinal variations of channel morphology and flow characteristics. The flow throughout the network can thus be reasonably described through a one-dimensional model and the three-dimensionality of the flow field close to the bifurcation is accounted for, in the simplified manner described in the following, only to formulate suitable nodal point conditions at the bifurcation.

Water and sediment discharges in channel a, upstream of the two longitudinal cells, are assumed as uniformly distributed in the cross-section and a possible transverse flux of water and sediment between the two cells is accounted for. In particular, the transverse exchange of sediments is evaluated on the basis the classical approach employed to describe two-dimensional bedload transport over an inclined bed, originally proposed by Ikeda et al. (1981), according to which one may write

$$q_y = q_x \left[V \left(U^2 + V^2 \right)^{-1/2} - \frac{r}{\sqrt{\vartheta}} \frac{\partial \eta}{\partial y} \right], \tag{1}$$

where q_x and q_y are longitudinal and transverse sediment discharges per unit width, U and V are the corresponding velocity components, ϑ is the Shields parameter, η is bed elevation and r is an empirical constant. The second term between square brackets accounts for the effect of gravity related to transverse bed slope $\partial\eta/\partial y$, and is responsible for the deviation of particle trajectories with respect to bottom stress direction. The constant r in equation (1) has been experimentally determined and it ranges between 0.3 and 1 (e. g. Ikeda et al., 1981; Talmon et al., 1995). Bolla Pittaluga et al. (2003a) compute the transverse velocity V as the ratio between the average transverse water discharge Q_y and the cross-section A_{abc} of the surface which divides longitudinally the two cells. Q_y is evaluated by the authors through a liquid mass balance applied to one of the two cells, say cell b, as follows

$$Q_y = Q_b - Q_a \frac{b_b}{b_b + b_c}, \tag{2}$$

and $A_{abc} = \alpha b_a D_{abc}$, where D_{abc} is the average depth computed as

$$D_{abc} = \frac{1}{2}\left(\frac{D_b + D_c}{2} + D_a\right). \tag{3}$$

Moreover, the transverse bed gradient is computed through the bulk relationship

$$\frac{\partial\eta}{\partial y} = \frac{\eta_b - \eta_c}{(b_b + b_c)/2}, \tag{4}$$

with η_b and η_c being bed elevations at the inlets of channels b and c, respectively. Using the above expressions it is readily obtained the following form for equation (1)

$$q_y = q_a \left[\frac{Q_y D_a}{Q_a \alpha D_{abc}} - \frac{2r(\eta_b - \eta_c)}{\sqrt{\vartheta_a}(b_b + b_c)}\right], \tag{5}$$

which is used to evaluate the transverse sediment exchange per unit length between the two adjacent cells of the final reach of channel a (see figure 1).

Bolla Pittaluga et al. (2003a) then propose the following five nodal point conditions at the bifurcation: i) water discharge balance, ii) iii) water level constancy, iv) v) Exner equation applied to both cells of the final reach of channel a, which are written as follows:

$$\frac{1}{2}(1-p)\left(1 + \frac{b_a}{b_b + b_c}\right)\frac{d\eta_i}{dt} + \frac{q_i - q_a\left(\frac{b_a}{b_b+b_c}\right)}{\alpha b_a} + \mp\frac{q_y}{b_i} = 0 \qquad (i = a, b). \tag{6}$$

It is worth noting that the above relationships are highly simplified versions of the original twodimensional ones and slightly different equations could have been written following exactly the same approach. For instance in many places the term $b_b + b_c$ could be replaced with b_a, without any conceptual difference. Therefore, the above equations should not be expected to invariably provide quantitatively correct results on the behavior of a bifurcation. However, the main goal of the above approach is to enlighten the basic mechanisms that govern the behavior of a bifurcation. It will be shown later that many of the features that characterize real gravel-bed channel bifurcations are actually well described by the model if width adaptation to flow conditions is suitably accounted for.

Bolla Pittaluga et al. (2003a) have employed the above nodal point conditions to determine the equilibrium configurations of a bifurcation, under the assumptions of movable bed, fixed banks and uniform flow conditions throughout the network. The stability of such equilibrium configurations was also investigated. The authors found that, for a given value of the Shields stress ϑ_a and of the slope S_a in the upstream reach, a symmetrical solution always exists, in correspondence of which the two downstream branches are fed with the same water and sediment discharges[2]. Such a solution is stable for relatively large values of the Shields parameter ϑ_a of the upstream channel. As ϑ_a is progressively decreased two further solutions appear, in addition to the symmetrical one, which are characterized by a strong unbalance of discharges into the two branches. Bolla Pittaluga et al. (2003a) show that, in this case, the symmetrical solution becomes unstable while the two unbalanced solutions are invariably stable.

3 FORMULATION OF THE PROBLEM

In the present contribution we extend the work by Bolla Pittaluga et al. (2003a) to the case of channels with erodible banks, i.e. channels which may adapt their width to the actual flow conditions. To this end we first need a further ingredient: a suitable relationship between channel width and hydraulic and morphological characteristics of the flow.

For the case of single thread channels many so called "regime relationships" have been proposed in the literature, which allow to compute an equilibrium value of channel width. Very few contributions exist, however, concerning channels in gravel-bed braided

[2] In the paper Bolla Pittaluga et al. (2003a) the Chezy coefficient C_a was prescribed instead of channel slope S_a. However it can be shown that results are qualitatively similar if the channel slope is prescribed, a choice which seem more appropriate to describe gravel-bed channel networks.

rivers. Regime relationships may be distinguished in two different classes: rational and empirical. The former are based on somewhat simplified solutions of the governing equations for water and sediment motion while the latter are based on regressions on field data. As an example, the rational regime relationships proposed by Henderson (1966), Parker (1979) and Griffiths (1981), respectively, are reported below.

$$b = 2.06\, QS^{7/6} d_s^{-3/2}, \tag{7}$$

$$b = 6.8\, QS^{1.22} d_s^{-3/2}, \tag{8}$$

$$b = 5.28\, QS^{1.26} d_s^{-3/2}. \tag{9}$$

Notice that the above formulas have an identical structure: the exponents of dimensional variables are indeed the same. Griffiths (1981) points out that the differences in the coefficients between the above equations are due to the use of different physical constants, resistance relations and sediment entrainment criteria. All the coefficients in the above relationships have been chosen to correctly reproduce the width of single thread channels.

Chew & Ashmore (2001) have tested various regime formulas on field data collected on the braided Sunwapta river. According to the authors empirical formulas based of single thread channels data fail in describing longitudinal river width changes, while rational regime formulas perform better to reproduce field data.

A relationship specifically designed for single channels in gravel-bed braided rivers has been proposed by Ashmore (2001) on the basis of a statistical regression on field data from braided rivers. Such a regression yielded

$$b = 0.087\, \Omega^{0.559} d_s^{-0.445}, \tag{10}$$

where Ω is the stream power, defined as γSQ with γ being water specific weight. The above formula is probably the most accurate relationship available at the moment to evaluate channel width in braided rivers and it will be adopted in the following, though the use of a different relationship would lead to qualitatively similar results.

As pointed out in the introduction experimental and field observations suggest that the process of channel widening due to increasing values of water discharge is by far more intense than channel narrowing when the discharge decreases. When we model the time evolution of the system we account for such an effect by allowing channel widening, according to the relationship (10), and inhibiting channel narrowing. Therefore, in our model, channel widths can not decrease in time.

We wish to employ the nodal point conditions described above to study the evolution of a channel bifurcation with erodible banks and its possible equilibrium solutions. We formulate the mathematical problem of time evolution of the system adopting a different approach with respect to that originally proposed by Wang et al. (1995) and later reconsidered by Bolla Pittaluga et al. (2003a). In order to investigate the stability of the equilibrium solutions of a channel loop the above authors assumed that the time evolution of the whole network could be described as a sequence of uniform flow conditions. In other words, variations of water depth along each channel were neglected and the authors described scour and deposition in the downstream branches as uniformly distributed along the whole channel lengths. Therefore, they applied the sediment continuity equation for the two branches in global form, only accounting for the sediment discharge entering each channel from its mouth end and leaving it from the outlet section. On the other hand, in the present work, we follow the line of reasoning proposed by Hirose et al. (2003), and assume that the time evolution of the bifurcation may be described in terms of local variables, i.e. without concern of the behavior of the downstream channels. This approach seems more suitable to describe bifurcations in gravel-bed braided rivers, where backwater effects play fairly a minor role with respect to the local conditions in controlling the behavior of the system. The main difference between the present approach and that followed by Bolla Pittaluga et al. (2003a) concerns the time scale of evolution of the system. In fact, in the authors' approach, such a time scale crucially depends on the lengths of the downstream branches while, in the present case, it is set by the nodal point conditions.

We study the evolution of the bifurcation assuming that the upstream channel is in steady uniform flow conditions and its width is in equilibrium with the hydraulic conditions, according relationship (10). We then employ the nodal point conditions proposed by Bolla Pittaluga et al. (2003a), relationship (10) applied to both branches and uniform flow conditions at the inlet of the two downstream channels. The latter assumption may seem to be in contradiction with the aim of studying the local behavior of the bifurcation. In fact, however, it simply implies that a rating curve is adopted to describe the relationship between water depth and discharge at the inlet of each branch, an assumption which is sensible to model real bifurcations. Uniform flow equations could, therefore, be replaced by any other rating curve relationships, without any conceptual change to what follows. Summarizing, we need to solve the following system of equations

$$Q_a = Q_b + Q_c, \tag{11}$$

$$\frac{1}{2}(1-p)\left(1+\frac{b_a}{b_b+b_c}\right)\frac{dD_i}{dt}= \quad (12)$$

$$=\frac{q_i-q_a\left(\frac{b_a}{b_b+b_c}\right)}{\alpha b_a}\pm\frac{q_y}{b_i} \qquad (i=b,c),$$

$$Q_i=b_iC_iD_i\sqrt{gR_iS_i} \qquad (i=b,c), \quad (13)$$

and equation (10) applied to channels b and c. In order to write equations (12) in the present form (substituting $d\eta_i/dt$ with $dh_i/dt - dD_i/dt$) we have used the nodal point condition $h_b = h_c$ and have assumed that the time variation of h_a (the surface elevation in the final section of channel a) is negligible with respect to changes of the bed levels at the inlet of the downstream channels. In fact this implies that we neglect the overall raising or lowering of the bifurcation eventually induced by downstream boundary conditions. The sediment discharges per unit width q_b and q_c in equation (12) are assumed equal to the transport capacities at the inlet of each branch and are evaluated employing the transport formula proposed by Parker (1990). The above equations must be solved for the unknowns Q_b, Q_c, D_b, D_c, b_b, b_c, S_b and S_c. It appears that we have 8 unknowns and 7 equations, therefore the problem is under-determined and we need to impose a further condition. This reflects the physical requirement of a downstream boundary condition to completely determine the system configuration. To close the mathematical problem it is reasonable to prescribe a value of the ratio S_b/S_c between the slopes of the branches. Physically, such a ratio is likely to be controlled by the mechanism of generation of the bifurcation and by the topographical characteristics of the network rather than by the evolution of the bifurcation itself. All plots presented in the paper have been computed adopting $S_b = S_c = 1$.

4 DISCUSSION OF THE RESULTS

In the case of a bifurcation with fixed banks, considered by Bolla Pittaluga et al. (2003a), three dimensionless parameters need to be prescribed in order to completely define the system in dimensionless form. For instance, one may set the values of the Shields stress ϑ_a, of the width parameter β_a and of the slope S_a in the upstream channel. With the inclusion of a further relationship, relating channel width to actual flow conditions, the system loses a degree of freedom and two of the above parameters suffice to completely characterize the system. In Figures 2(a),(b) the relationship between the Shields parameter ϑ and the aspect ratio β (for a given value of the slope S) is plotted, according to the formulas proposed by Ashmore (2001) and Griffiths (1981), respectively. The lines represent the possible states of the system.

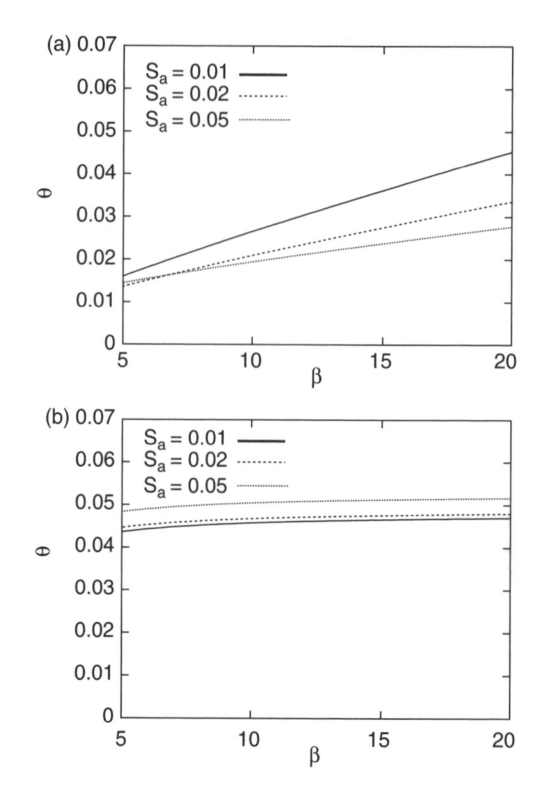

Figure 2. Shields parameter versus channel aspect ratio for different values of channel slope. (a) relationship proposed by Ashmore [2001], (b) relationship proposed by Griffiths [1981].

Figure 2(a) shows that, according to the formula proposed by Ashmore (2001), the Shields parameter slightly increases with the width ratio. This is also true for any rational formula, though the variability of the Shields parameter is, in this case, extremely limited as shown in figure 2(b). It is worth noting that both in figures 2(a) and 2(b) very small values of the Shields parameter (close to the incipient condition for sediment motion) are predicted, whatever the aspect ratio of the channel. Recalling the fact that Bolla Pittaluga et al. (2003a) found that the symmetrical configuration of a bifurcation is unstable for low values of the Shields stress ϑ_a, the above observations suggest that bifurcations with erodible banks in gravel-bed braided rivers should, almost invariably, admit only asymmetrical stable solutions.

Let us now investigate such a problem. We first consider a symmetrical bifurcation and assume the width of each channel to be initially in equilibrium, according to relationship (10). Physically we may think of a bifurcation of this kind to form as a consequence of the generation of a central deposit within the main channel which induces a fairly symmetrical division of the

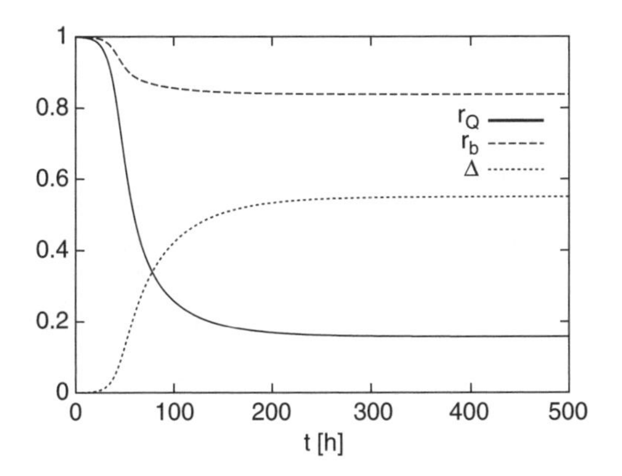

Figure 3. Time evolution of the parameters r_Q, r_b and Δ, starting from a symmetrical initial condition ($S_a = 0{:}01$, $\beta_a = 20$).

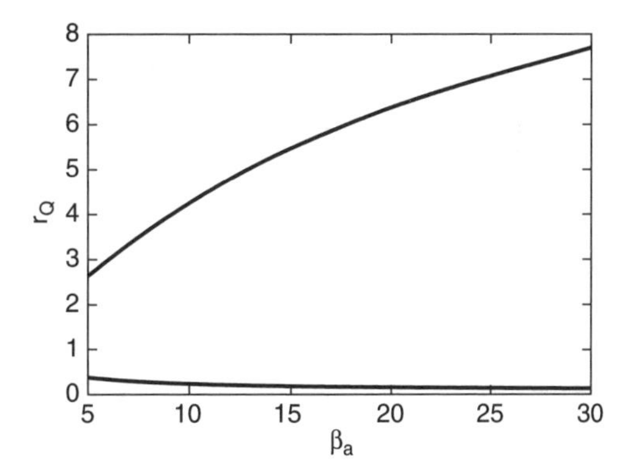

Figure 4. Equilibrium value of r_Q reached by the system starting from a symmetrical initial condition as a function of β_a ($S_a = 0.01$).

flow into two streams. It is easily shown that a symmetrical equilibrium solution of the system always exists according to our model. However, such a solution is found to be invariably unstable and the system always evolves towards other asymmetrical solutions. This confirms the intuitive argument reported above. In Figure 3 the time evolution of the bifurcation is shown, for a given value of β_a and S_a, in terms of the parameters $r_Q = Q_b/Q_c$ and $r_b = b_b/b_c$, which represent the ratios between water discharges and channel widths in the two branches, respectively. The model predicts the existence of two stable and asymmetrical solutions for any value of β_a, which are reciprocal to each other (for graphical reasons only one of such solutions is reported in the figure). In the figure the evolution in time of the dimensionless difference $\Delta = |\eta_b - \eta_c|/D_a$ between bed elevations at the inlets of the two branches is also reported. It appears that, starting from an initial condition in which bed elevation at the mouths of the two bifurcating channels is the same, the model predicts the progressive generation of an inlet step, in agreement with field observation by Zolezzi et al. (2005).

In Figure 4 the parameter r_Q, relative to the two reciprocal equilibrium solutions which are reached by the system starting from an initially symmetrical configuration, is plotted versus the aspect ratio of channel a. Notice that the Shields parameter of the upstream channel ϑ_a is not kept fixed but it assumes the values reported in Figure 2(a) for $S_a = 0.01$. The figure shows that the solutions become more unbalanced as β_a increases.

The same equilibrium configurations are shown in Figure 5 in terms of the parameter r_b. Notice that the ratio between channel widths keeps relatively close to unity. This is due to the fact that we inhibit the narrowing of the branch fed with the lower discharge and, thus, changes of r_b are only induced by the, relatively limited, widening of the other channel.

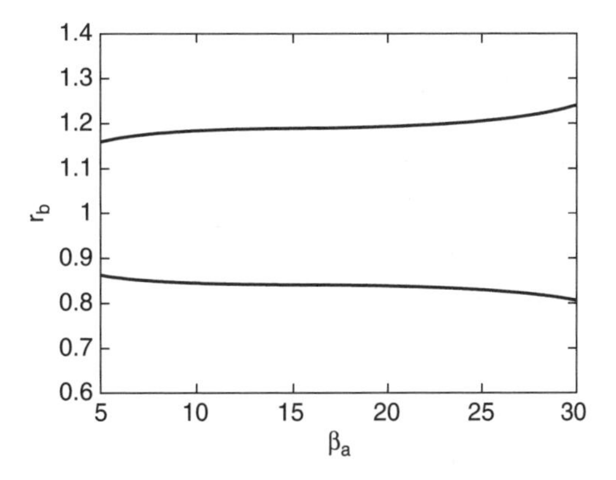

Figure 5. Equilibrium value of r_b reached by the system starting from a symmetrical initial condition as a function of β_a ($S_a = 0.01$).

In Figure 6 the variation of the inlet step height Δ, in equilibrium conditions, is plotted versus β_a. In agreement with field observations, the magnitude of the inlet step increases as the unbalancement of discharges into the downstream branches grows.

As it has been anticipated in the Introduction the different response of channel width to increasing or decreasing water discharge, which is assumed herein, may make the final configuration reached by the system dependent on the initial conditions. In Figures 7 and 8 the time evolution of a bifurcation, starting from different initial values of the channel widths ratio $(r_b)_{t=0}$, is shown in terms of r_b and r_Q, respectively. As inferred by intuitive speculations, it appears that the final equilibrium configuration which is reached

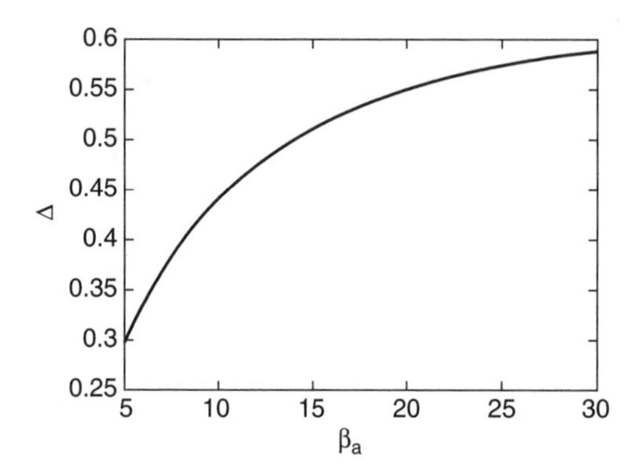

Figure 6. Equilibrium value of Δ reached by the system starting from a symmetrical initial condition as a function of β_a ($S_a = 0.01$).

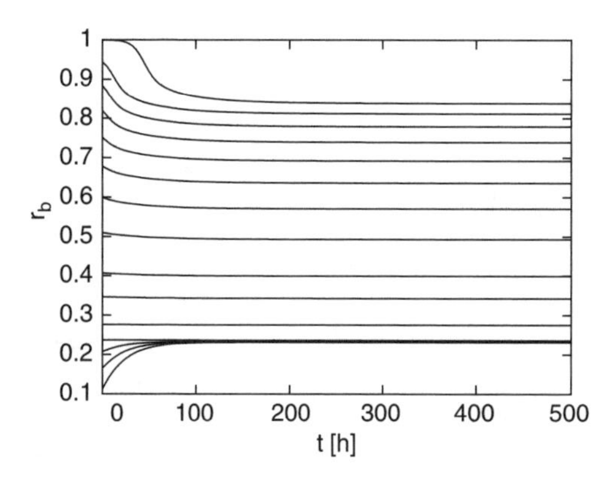

Figure 7. Time evolution of the ratio r_b starting from different initial conditions ($S_a = 0.01$).

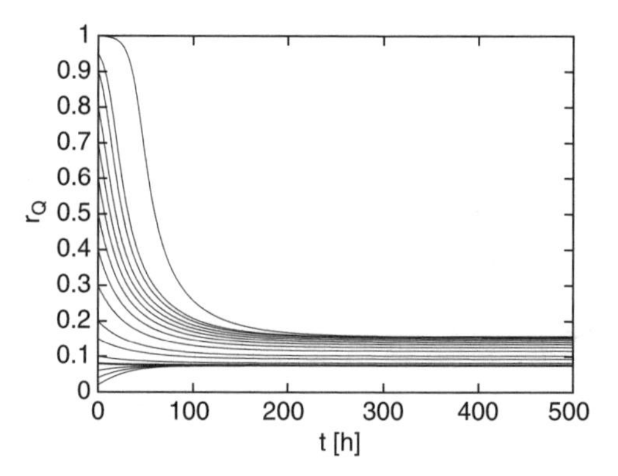

Figure 8. Time evolution of the ratio r_Q starting from different initial conditions ($S_a = 0.01$).

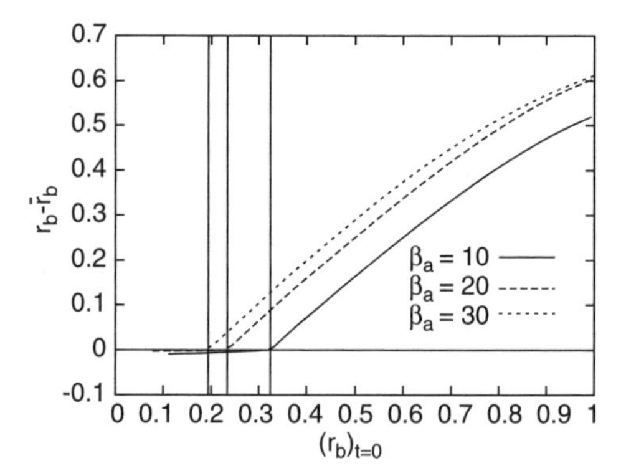

Figure 9. Final equilibrium configurations in terms of $\bar{r}_b - \bar{r}_b$ as a function of the initial value of the ratio between channel widths $(rb)_{t=0}$, for different values of β_a ($S_a = 0.01$).

by the bifurcation actually depends on initial conditions. Note that, in both figures, the flex point which is displayed by the curves starting close to the points $(r_b)_{t=0} = 1$ and $(r_Q)_{t=0} = 1$ is due to the fact the symmetrical configuration of the bifurcation is an unstable solution of the system. In the plots we also report, with a thick line, the equilibrium configuration which is reached by the system if channel width is evaluated, at each time step, through the relationship (10), regardless to the fact that the channel is widening or narrowing. The corresponding values of r_b and r_Q are denoted with $\bar{r}_b$ and $\bar{r}_Q$, respectively.

Figure 7 shows that if $\bar{r}_b < (r_b)_{t=0} < 1$, the range of values of r_b finally reached by the system is quite wide, i.e. initial conditions play, in this case, a significant role on the final equilibrium configuration of the bifurcation. On the other hand if $(r_b)_{t=0} < \bar{r}_b$ the system evolves toward a solution characterized by an equilibrium value of r_b which is very close to $\bar{r}_b$. This reflects the fact that, in the latter case, we prescribe a very small initial width of the narrow branch which, thus, freely widens until the equilibrium solution is reached. Physically this may represent the evolution of a bifurcation formed by incision of a new narrow channel through a chute cutoff or an avulsion process.

The above considerations may be conducted in analogy for r_Q, referring to Figure 8. Quite interestingly, though, even in the case of $\bar{r}_Q < (r_Q)_{t=0} < 1$, the range of values of r_Q finally reached by the system is fairly narrow. Therefore, the equilibrium value of the ratio between discharges into the two branches is not strongly dependent on initial conditions.

The above results are summarized in Figures 9 and 10 where the equilibrium value of the differences $r_Q - \bar{r}_Q$ and $r_b - \bar{r}_b$ are plotted versus the initial value of r_b, for different values of β_a. The vertical lines reported in the figures indicate the values of $\bar{r}_b$ for each value of β_a. It is confirmed that the influence of initial conditions on the final configuration reached by the system is extremely weak when $(r_b)_{t=0} < \bar{r}_b$, i.e. when, at the initial stage, one of the two branches is much narrower than the other. The influence of initial conditions

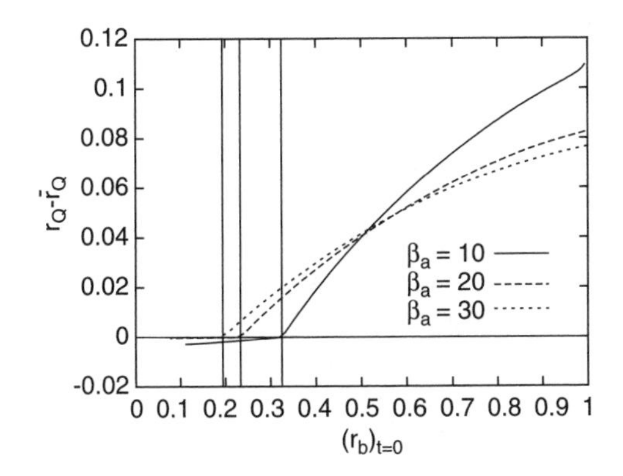

Figure 10. Final equilibrium configurations in terms of $r_Q - \bar{r}_Q$ as a function of the initial value of the ratio between channel widths $(r_b)_{t=0}$, for different values of β_a ($S_a = 0.01$).

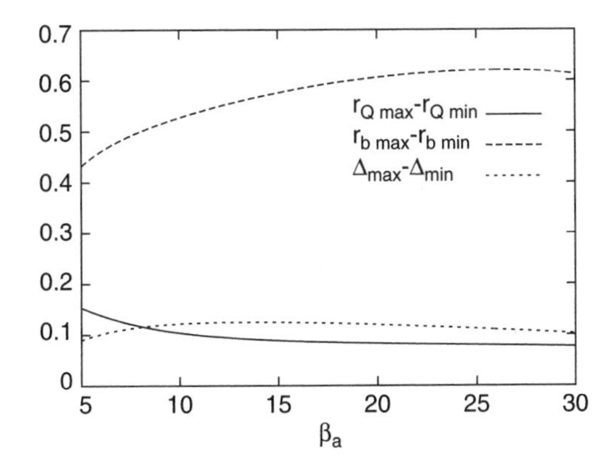

Figure 11. Range of equilibrium values of r_b, r_Q and Δ versus the aspect ratio of the upstream channel β_a ($S_a = 0.01$).

grows the more the initial configuration of the system is close to symmetry. It is also apparent that such an effect is more marked as far as channel widths are concerned, while, whatever the initial conditions, the final value of the ratio r_Q between water discharges into the two branches keeps confined within a relatively narrow range.

Finally, in Figure 11, we plot the range of equilibrium values of r_b, r_Q and Δ, versus the aspect ratio of the upstream channel. Referring, for instance, to the variable r_b, for a given value of β_a, we plot the difference $r_{b\max} - r_{b\min}$, with $r_{b\max}$ being the equilibrium value of r_b reached by the system when the initial condition is symmetrical ($(r_b)_{t=0} = 1$) and $r_{b\min}$ the corresponding equilibrium value obtained when, initially, one of the two branches is much narrower than the other ($(r_b)_{t=0} \sim 0$). The figure shows that the dependence of the above variables on the aspect ratio of channel a is fairly weak. Moreover, while $r_{Q\max} - r_{Q\min}$ monotonically decreases with β_a, both $r_{b\max} - r_{b\min}$ and $\Delta_{\max} - \Delta_{\min}$ have a maximum in the considered range of values of β_a. It is important to stress that the model shows that, even in the fairly idealized case of constant discharge considered herein, it is inappropriate to think of an "equilibrium configuration of a bifurcation". Rather, it is found that a range of possible equilibrium solutions are possible, depending on the initial conditions of the system.

5 CONCLUSIONS

In the present work we have employed a onedimensional model to analyze the equilibrium configurations of a simple channel bifurcation in a gravel-bed river. We have adopted the nodal point conditions proposed by Bolla Pittaluga et al. (2003) and have extended their model in order to account for width adjustments to the actual flow conditions. The time evolution of the system has been computed employing a local analysis, following the approach proposed by Hirose (2003), whereby system development is governed by the nodal point conditions. Therefore, downstream conditions are disregarded in the model. This seems a suitable approach to describe bifurcations in gravel-bed mountainous rivers whose dynamics is essentially controlled by local parameters and backwater effects play a fairly minor role. Channel width has been computed according to regime relationships proposed in literature. Moreover, following field and experimental observation, when computing the time evolution of the bifurcation we have allowed channel widening but have inhibited channel narrowing.

The model shows than, when the hypothesis of fixed banks is removed, the system only admits of stable configurations which are strongly asymmetrical, with one of the two branches being larger than the other and carrying most of water and sediment discharges. Moreover, the model invariably predicts the generation, at the bifurcation, of an inlet step, i.e. a difference in bed elevation between the mouths of the two downstream channels. Such an inlet step is of utmost importance in governing the partition of sediment discharge. Both findings are in good qualitative agreement with field and laboratory observations.

As a consequence of the different response of channel width to increasing or decreasing water discharge it is also found that the initial morphology of the bifurcation influences the final configuration reached by the system. In particular a bifurcation generated from the incision of a new, initially narrow, channel evolves toward a different configuration with respect to a bifurcation which, at its initial stage, is fairly symmetrical. In particular, differences mainly arise in the equilibrium value of the ratio between channel widths while, whatever the initial configuration of the system, the final partition of water discharge assumes values comprised in a fairly narrow range.

It is worth noting that the model is unable to reproduce the disturbance induced on the bifurcation by the presence of bars migrating in the upstream channel, whose importance has been documented experimentally by Bolla Pittaluga et al. (2003). Indeed, the formation of alternating bars in the upstream channel may eventually lead to the complete closure on one of the two branches when the equilibrium configuration of the system is strongly unbalanced.

ACKNOWLEDGMENTS

This work has been developed within the framework of the "Centro di Eccellenza Universitario per la Difesa Idrogeologica dell'Ambiente Montano- CUDAM", of the project "La risposta morfodinamica di sistemi fluviali a variazioni di parametri ambientali COFIN 2003" co-funded by the Italian Ministry of University and Scientific Research and the University of Trento and of the project "Rischio idrogeologico e morfodinamica fluviale RIMOF" co-funded by Fondazione Cassa di Risparmio di Verona Vicenza Belluno e Ancona and the University of Trento.

REFERENCES

Ashmore, P. E. (2001). Braiding phenomena: statics and kinetics. In M. P. Mosley (Ed.), Gravel-bed rivers V, pp. 95–121. New Zealand Hydrologic Society.

Bolla Pittaluga, M., R. Repetto, and M. Tubino (2003a). Channel bifurcation in braided rivers: equilibrium configurations and stability. Water Resour. Res. 39(3), 1046–1059.

Bolla Pittaluga, M., R. Repetto, and M. Tubino (2003b). Correction to "Channel bifurcation in braided rivers: equilibrium configurations and stability". Water Resour. Res. 39(11), 1323.

Bulle, H. (1926). Untersuchungen über die geschiebeableitung bei der spaltung von wasserläufen. Technical report, VDI Verlag, Berlin. in German.

Chew, L. C. and P. E. Ashmore (2001). Channel adjustment and a test of rational regime theory in a proglacial braided stream. Geomorphology 37, 43–63.

de Heer, A. and E. Mosselman (2004). Flow structure and bedload distribution at alluvial diversions. Proceedings of "River Flow 2004", Napoli (Italy).

Federici, B. and C. Paola (2003). Dynamics of bifurcations in noncohesive sediments.Water Resour. Res. 39, No. 6(1162), 3–15.

Griffiths, G. A. (1981). Stable-channel design in gravel-bed rivers. J. Hydrol. 52, 291–305.

Henderson, F. M. (1966). Open channel flow. McMillan, New York.

Hirose, K., K. Hasegawa, and H. Meguro (2003). Experiment and analysis of mainstream alternation in a bifurcated channel in mountain rivers. In Proc. 3rd IAHR Symposium RCEM, Volume 1, Barcelona, pp. 571–583.

Ikeda, S., G. Parker, and K. Sawai (1981). Bend theory of river meanders. part i – linear development. J. Fluid Mech. 112, 363–377.

Parker, G. (1979). Hydraulic geometry of active gravel rivers. J. Hydraul. Div. (ASCE) 105, 1185–1201.

Parker, G. (1990). Surface-based bedload transport relation for gravel rivers, J. Hydraul. Res. 20, 417–436.

Talmon, A., M. C. L. M. van Mierlo, and N. Struiskma (1995). Laboratory measurements of direction of sediment transport on transverse alluvial-bed slopes. J. Hydraul. Res. 33(4), 519–534.

Wang, Z. B., R. J. Fokkink, M. D. Vries, and A. Langerak (1995). Stability of river bifurcations in 1d morphodynamics models. J. Hydraul. Res. 33(6), 739–750.

Zolezzi, G., W. Bertoldi, and M. Tubino (2005). Morphological analysis and prediction of river bifurcations. In "Braided Rivers 2003".

Bars and braiding

River, Coastal and Estuarine Morphodynamics: RCEM 2005 – Parker & García (eds)

Effects of size heterogeneity of bed materials on mechanism to determine bar mode

Atsuko Teramoto
Matsue National College of Technology, Shimane, Japan

Tetsuro Tsujimoto
Nagoya University, Aichi, Japan

ABSTRACT: In fluvial fan rivers, various types of sandbars are formed, and depending on their patterns, we have to consider management of the river system from the viewpoint of flood protection as well as conservation of river ecosystem. In spite of many researches into sand bars, it has not been well understood how the characteristics of sand bars are sensitive to various boundary conditions. This paper focuses on the bar mode as the characteristic of a sandbar, while the size heterogeneity of bed materials is focused on as a surrounding condition. We employed three approaches to clarify the effect of size heterogeneity on the mechanism determining the bar mode: flume experiment, numerical simulation and linear instability analysis. These approaches show the effects of bed material components on bar mode, with the higher number of bar mode appearing in the condition where the bed is composed of materials with wider size of distribution.

1 INTRODUCTION

These days, river environments have changed as a result of human activity. For example, bed degradation, increase of vegetation on bars, and change of bar morphology can occur in a river with a large dam, as shown in Fig. 1. Therefore, many researchers have focused on the mechanisms causing changes to river environments, including morphology, in order to restore or to recover the original environment. In any discussion on changes to river environments alternate and multiple bars are very important because they are characteristic of most alluvial rivers in Japan.

So far, there has been much research focused on bar formation. Most studies are conducted on the basis of a bed-load composed of uniform sediment. However, riverbeds are actually composed of a mixture of materials and the change in components can become significant when the effects of dams are taken into account. So we need to investigate and understand the differences between bars formed on beds composed of uniform sediments and those developed on beds of mixed grain size. Also, we need to indicate the condition of bed materials that we should consider to be able to predict bar formation.

There is less current research that has focused on the bar formation on beds composed of mixed grain size and most researchers have compared the difference in bar height and length, and migration speed between beds composed of uniform and mixed materials (e.g., Lanzoni & Tubino (1999), Lanzoni(2000)). Therefore, the focus of this paper is to determine the effects of size heterogeneity of bed materials on mechanisms that determine bar mode (Number of bar mode (n) is explained in Fig. 2). First, we explain a numerical simulation method for describing of bar formation. This numerical simulation is applied to both uniform and heterogeneous sediments. By numerical simulation, we indicate the effect of size heterogeneity of bed materials on the mechanism to determine the bar mode. Linear instability analysis, considering

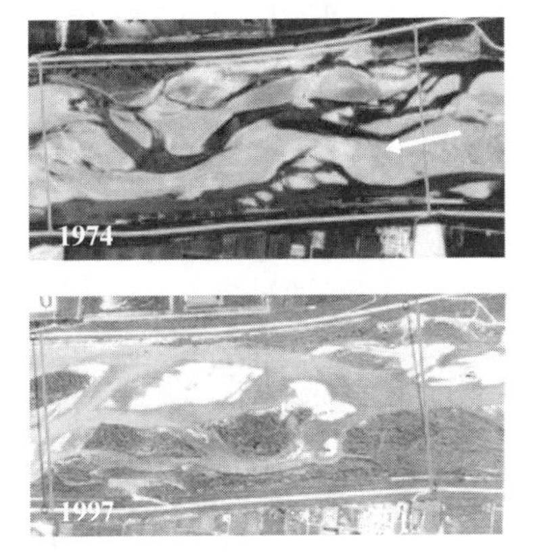

Figure 1. Example of change of river environment in Japan.

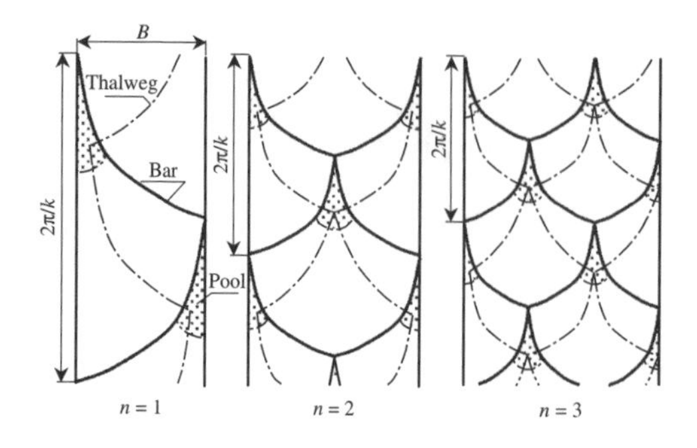

Figure 2. Alternate (left) and multiple bars (middle and right).

size heterogeneity of bed materials, promotes more understanding.

2 USING NUMERICAL SIMULATION TO DESCRIBE BAR FORMATION PROCESS

Numerical simulation is useful for understanding the behaviors of bars and there are several methods in use. However, it was clarified that some methods fail to describe the actual phenomena appearing in real rivers (Teramoto & Tsujimoto, 2003). In this section we explain the method of describing bar formation and the numerical simulation model we recommend to use. We also demonstrate the validity of this method and the simulation model by comparing experimental results.

2.1 *Method of numerical simulation on bar formation process*

To simulate bar formation, some type of disturbance is necessary to be presumed. This procedure in our method is to insert small disturbance following normal distribution to initial bed elevation and inlet flow. Disturbance applied to initial bed elevation is enough to describe initial bar formation in numerical simulation, but continuity of bar migration in any hydraulic condition cannot be described without disturbance to inlet flow (Teramoto & Tsujimoto, 2003). Thus in this numerical simulation, continuous disturbance is also applied to inlet flow from the start of the simulation.

2.2 *NHSED2D model formulation*

The NHSED2D model comprises two main parts: the flow model and the sediment transport and bed variation model. Firstly, the flow is computed over an initial bed configuration until it converges. Then, the sediment transport and the resultant bed variation are calculated using the pre-calculated flow field. After the bed is renewed, the flow is re-calculated on the modified bed. These procedures are repeated to obtain successive bed deformation.

2.2.1 *Flow model of NHSED2D model*

The 2D depth-averaged flow model has the following significant features:

- The finite volume method is employed to describe the governing equation.
- The fractional step method is employed in order to obtain the stable and accurate flow field (Ferziger, *et al.* (1997)).
- To prevent numerical oscillation due to collocated grid arrangement, the approach proposed by Rhie and Chow (1983) to interpolate the flux of mass at the cell surface is used.
- The QUICK scheme is employed to interpolate the convection of momentum at the cell surface.

The governing equations of 2D depth-averaged surface flow can be described by the following equations:

$$\frac{\partial q_x}{\partial t}+\mathrm{div}\left(q_x\frac{\mathbf{q}}{h}-\frac{\mathbf{T}_x}{\rho}\right)=-gh\frac{\partial \zeta}{\partial x}-\frac{C_f}{h^2}q_x|\mathbf{q}| \tag{1}$$

$$\frac{\partial q_y}{\partial t}+\mathrm{div}\left(q_y\frac{\mathbf{q}}{h}-\frac{\mathbf{T}_y}{\rho}\right)=-gh\frac{\partial \zeta}{\partial y}-\frac{C_f}{h^2}q_y|\mathbf{q}| \tag{2}$$

$$\frac{\partial \zeta}{\partial t}+\mathrm{div}\mathbf{q}=0 \tag{3}$$

where t = time; (x, y) = the streamwise and lateral coordinates respectively; (q_x, q_y) = the x and y components of line discharge; $\boldsymbol{q}$ = the line discharge vector, z is the water surface elevation; h = the flow depth; (T_x, T_y) = the x and y components of the Reynolds stress tensor; g = the gravity acceleration; and C_f = the resistance coefficient of bed surface.

2.2.2 *Uniform sediment and bed variation models of NHSED2D model*

The time variation of bed elevation can be described by the following sediment continuity equation:

$$(1-n_e)\frac{\partial z}{\partial t}=-div\mathbf{q}_b \tag{4}$$

where n_e = the porosity; z = the bed elevation; and q_b = the sediment flux vector, which is estimated using the formula from Ashida & Michiue (1971). The x and y components of the sediment flux (q_{Bx}, q_{By}) are estimated by the following equation:

$$q_{Bx}=q_B\cos\varphi\,,\qquad q_{By}=q_B\sin\varphi \tag{5a,b}$$

where $\boldsymbol{q}_B$ = the total bed load transport rate per unit width; and j = the angle of bed-load movement.

The effect of transverse bed slope on the sediment transport is taken into account following Nakagawa *et al.* (1986). The model also includes the effect of secondary flow caused by the curvature of streamlines using Engelund's equation (1974). The angle of bed load movement is expressed by the following equation:

$$\varphi = \tan^{-1}\left(\frac{V}{U} - N_* \frac{h}{r}\right) - \tan^{-1}\left(\sqrt{\frac{\tau_{*c}}{\mu_d \mu_f \tau_*}} \frac{\partial z}{\partial n}\right) \quad (6)$$

where (U, V) = the velocity components in the x and y directions respectively; N_* = the coefficient of the strength of secondary flow (=7.0 as given by Engelund (1974)); r = the curvature radius of the streamline; μ_f and μ_d = respectively the static and kinetic friction coefficients of sand grains respectively, τ_* = the Shields number; τ_{*c} = the critical Shields number; and n = the coordinate normal to the stream line.

After calculating bed variation, the bed slope angles $\tan^{-1}(\Delta z/\Delta x)$ and $\tan^{-1}(\Delta z/\Delta y)$ between adjacent grids on the bed are compared with the angle of repose ϕ (about 45 degrees). In the case where the bed slope angle is greater than the angle of repose ϕ, the bed is assumed to collapse to an angle corresponding to ϕ.

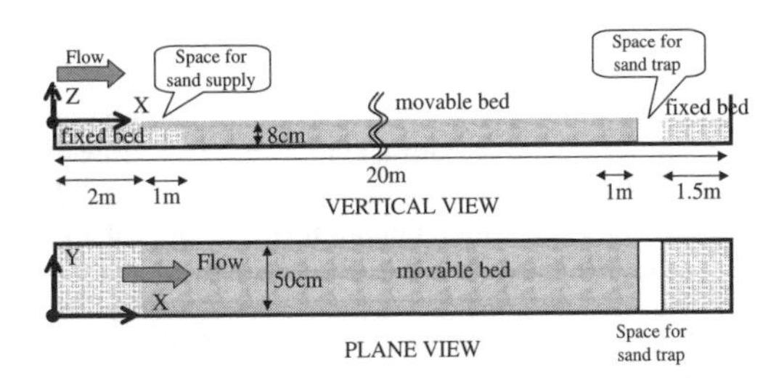

Figure 3. Experimental set up.

Table 1. Experimental conditions.

Case no.	Q m^3/s	h_0 cm	B/h_0	u_0 m/s	u_* m/s	Mode no.
A	0.0008	0.62	81.27	0.26	0.025	2–3
B	0.001	0.70	71.09	0.28	0.026	2
C	0.0015	0.90	55.74	0.33	0.030	1
D	0.002	0.11	46.90	0.38	0.032	1

2.3 *Comparison between experimental results and simulation*

2.3.1 *Hydraulics condition*

The experiment was conducted in the 20 m long and 50 cm wide laboratory flume, shown in Figure 3. Numerical simulation carried out over a domain more than 70 m long and 50 cm wide. Bed material is uniform sand of diameter 0.88 mm. The channel slope was kept at 1/100. Experimental hydraulic conditions are shown in Table 1. "Mode Number" in the table 1 indicates the initial number of bar mode (n) estimated from the linear instability analysis employed by Kuroki & Kishi(1984).

In the flume experiments, the measurement area in the middle of channel is 8 m long. A laser plotter was used to measure the bed elevation, and the pictures were taken every 10 minutes from 8 m to 12 m to get an understanding of the bar formation process.

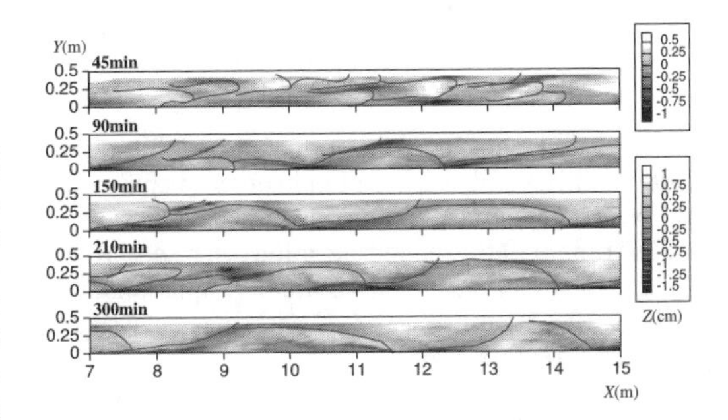

Figure 4. Contour of bed elevation (Case B).

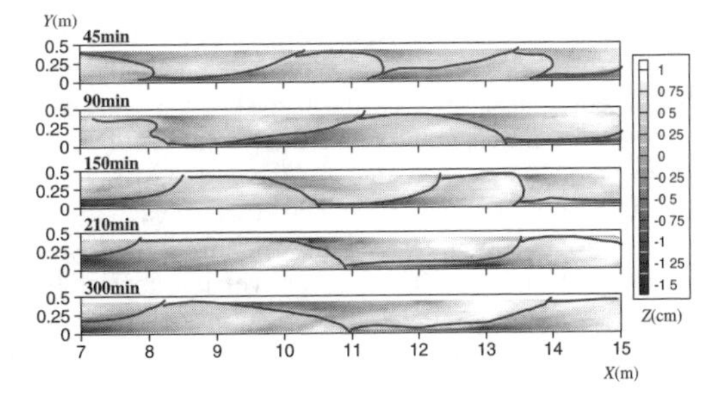

Figure 5. Contour of bed elevation (Case D).

2.3.2 *Bar formation process in experiment*

The experimental results for bed topography are shown in Figures 4 & 5.

In Cases A & B, the bar formation process was as follows:

(1) Poorly developed multiple (double-row) bars appeared after more than 10 minutes.
(2) Multiple bars migrated and developed.
(3) Multiple bars changed to alternate bars.
(4) Alternate bars developed and migrated. Emerged area appeared on some bars.

This phenomenon that the multiple bars that appear initially develop and change into alternate bars has been reported in recent papers (e.g., Watanabe & Kuwamura (2004)). As these results indicate, under these conditions, the final bar mode number is different from the initial one which reflects the multiple bars.

In Cases C & D, the following bar formation process was observed.

(1) Alternate bars formed initially.
(2) They migrated and developed.

2.3.3 *Bar formation process in numerical simulations*

Numerical simulation was carried out for the same set of conditions as the experimental work. The results of numerical simulation are shown in Figures 6 and 7. The phenomenon where double-row bars appear initially and disappear with bar development as in Case B is shown in Figure 6. In the both cases alternate bars finally developed. A comparison between the growing process of bar height and length for experimental and numerical simulation results is given in Figures 8–11. These results show that bar formation predicted by numerical simulation is almost same as that observed in experiment except for the lag-time before onset of bar formation. This indicates that the numerical simulation model can describe the bar formation process well.

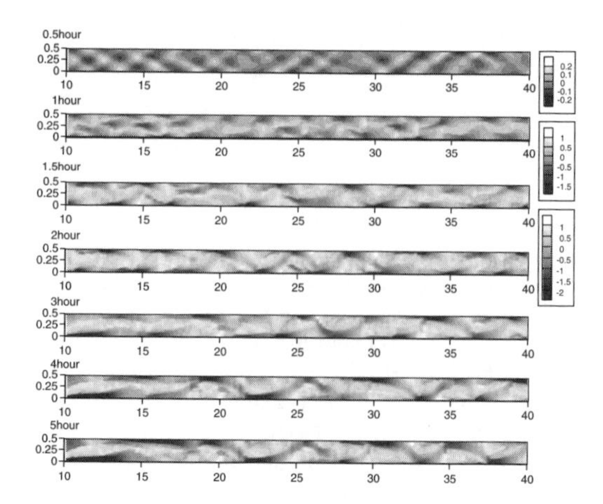

Figure 6. Bed topography in Case B (Numerical simulation).

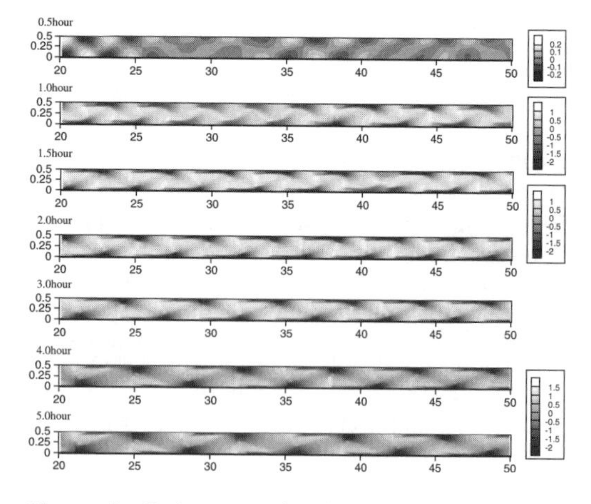

Figure 7. Bed topography in Case D (Numerical simulation).

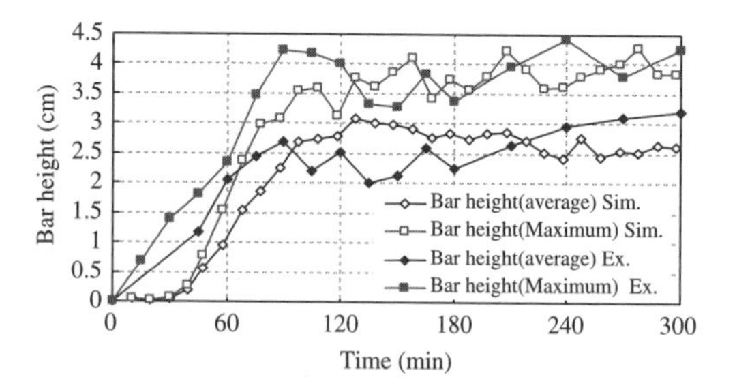

Figure 8. Growth of bar height in Case B.

3 NUMERICAL SIMULATION OF BAR FORMATION CONSIDERING BED MATERIALS

3.1 *NHSED2D model for mixed sediment and bed variation models*

We modified the NHSED2D model designed for uniform sediment and bed variation in order to apply it to various bed material components.

The time variation of bed elevation can be described by the following sediment continuity equation considering sediment discharge of each class classified according to grain size:

$$(1-n_e)\frac{\partial z}{\partial t} = -\sum_{i=1}^{N} div\mathbf{q}_{bi} \tag{7}$$

where q_{bi} = the sediment discharge of sediment class i; (q_{bxi}, q_{byi}) = the sediment discharge components in the x and y directions respectively; and N = the total number of sediment class classified according to grain size.

The non-dimensional sediment discharge of each grain size q_{bi*} is estimated by the Ashida-Michiue equation (8) and critical bed shear stress of each grain size τ_{*ci} calculated by Ashida-Michiue's modified equation of Egiazarroff's equation which is applied to Iwagaki's formula. q_{bi*} is distributed in the direction of sediment movement.

$$q_{bi*} \equiv \frac{q_{bi}}{\sqrt{(\sigma/\rho-1)gd_i^{\,3}}} = p_i 17\tau_{*i}^{\frac{3}{2}}\{1-\tau_{*ci}/\tau_{*i}\}\{1-\sqrt{\tau_{*ci}/\tau_{*i}}\} \tag{8}$$

where the subscripts i implies the quantities concerned with the i-th fraction in the graded materials, σ = relative density of sand; ρ = fluid density; d_i = sediment diameter of class I; and p_i = possessory rate of sediment class i.

In the condition that the thickness of the exchange layer is presumed to be equal to the maximum sediment diameter D, the sediment-sorting process with bed change can be modeled as shown in Figure 4. The possessory rate of sediment will be changed with a

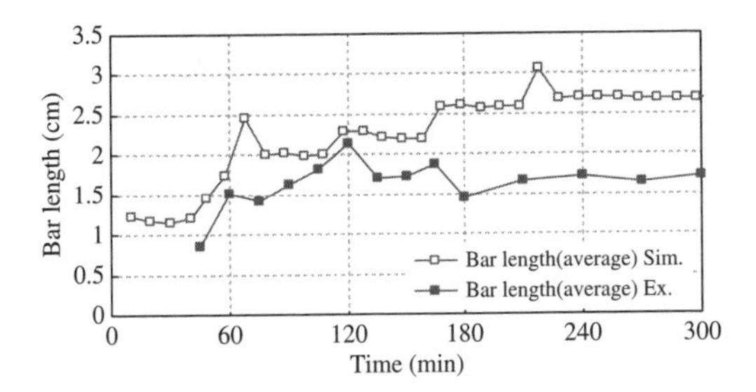

Figure 9. Change of bar length in Case B.

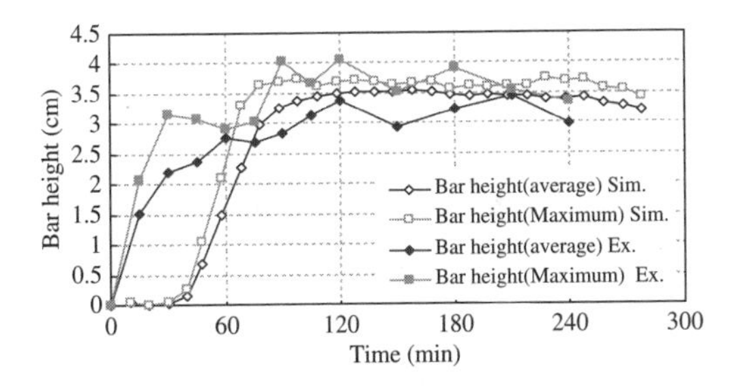

Figure 10. Growth of bar height in Case D.

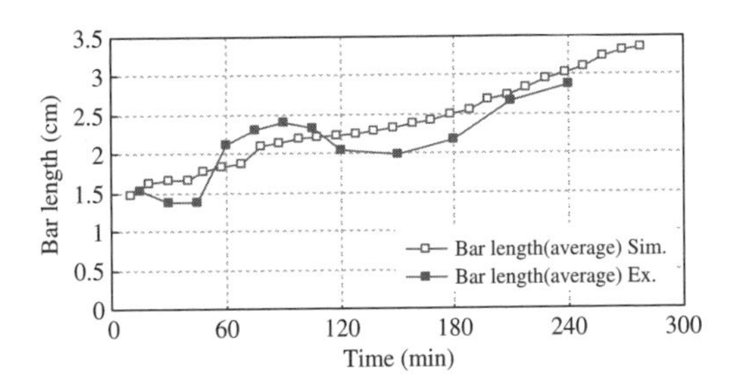

Figure 11. Growth of bar length in Case D.

different pattern between the cases of bed aggradation and degradation as the following equations show.

$$p_i(t+\Delta t)=\frac{p_i(t)D\Delta x\Delta y+\Delta q_i(t)-\Delta R_i}{D\Delta x\Delta y} \tag{9}$$

$$\Delta R_i=\begin{cases} p_i(t)\sum \Delta q_{bi}(t) & (\partial z/\partial t>0) \\ p_{i0}\sum \Delta q_{bi}(t) & (\partial z/\partial t<0) \end{cases} \tag{10}$$

$$\Delta q_{bi}(t)\equiv -\frac{1}{1-n_e}\left(\frac{\partial q_{bxi}}{\partial x}+\frac{\partial q_{byi}}{\partial y}\right)\Delta x\Delta y\Delta t \tag{11}$$

where $p_i(t)=$ the possessory rate in the exchange layer; $p_{i0}=$ the possessory rate in the sedimentation layer; and $q_{bi}(t)=$ sediment discharge of sediment class i. p_{i0} should be determined by the condition of sedimentation layer. Δt should restrict as following

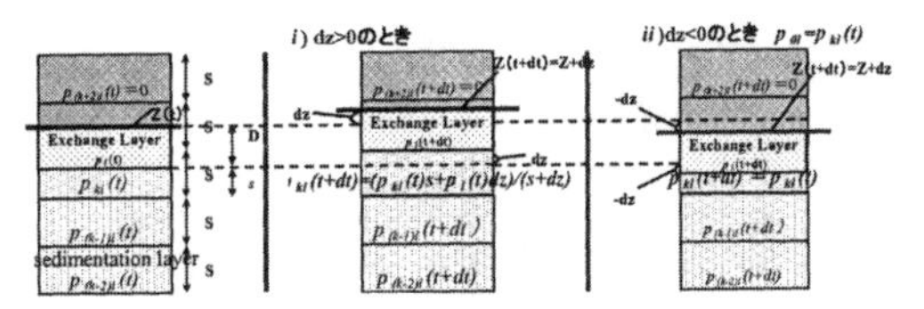

Figure 12. Sedimentation layer.

equation (12). It means that Δt will be conditioned such that the bed displacement of bed should be less than the thickness of the exchange layer and that outflow of sediment is less than content of each class sediment in exchange layer.

$$\begin{cases} \Delta t\le D\Big/\left(\dfrac{\partial z}{\partial t}\right) \\ \left|\dfrac{\partial z_i}{\partial t}\cdot\Delta t\right|\le p_i(t)\cdot D \end{cases} \tag{12}$$

We initially set the grid in vertical direction to record the sedimentation layer. p_{i0} in equation (10) is calculated from the record of distribution of sediment diameter in sedimentation layer. In the case that the bottom of the exchange layer is in the sedimentation layer k at $t=t$, p_{i0} is calculated in Figure 12. Only in the case that bed aggradation occurs, exchange layer replace and sedimentation layer renew and record. On the other hand, under conditions where bed degradation occurs, p_{i0} is calculated from the record of sedimentation layer and exchange layer replacement. The thickness of the sedimentation layer s, set up initially, is defined as more than the maximum sediment diameter and the thickness of exchange layer.

3.2 *Validity of numerical simulation of bar formation process on the bed composed of various materials*

3.2.1 *Experimental condition*

The flume experiments were done a channel set up as shown in Figure 3. Erodible bed was filled with the sediment composed of two classes of sand (diameters 2.0 mm and 0.6 mm) proportions such that the mean diameter was 0.88 mm. Hydraulic conditions are shown in Table 2.

3.2.2 *Comparison between results of experiment and numerical simulation*

Bed topology resulting from experiment and numerical simulation are shown in Figures 13 and 14. The bar formation process in both experiment and numerical simulation is as follows.

(1) Appearance of weakly developed multiple bars.
(2) Bars migrated and developed.

Table 2. Experimental conditions.

Case no.	Q m^3/s	h_0 cm	B/h_0	u_0 m/s	u_* m/s	Mode no.
C-2	0.0014	0.86	58.09	0.33	0.029	1

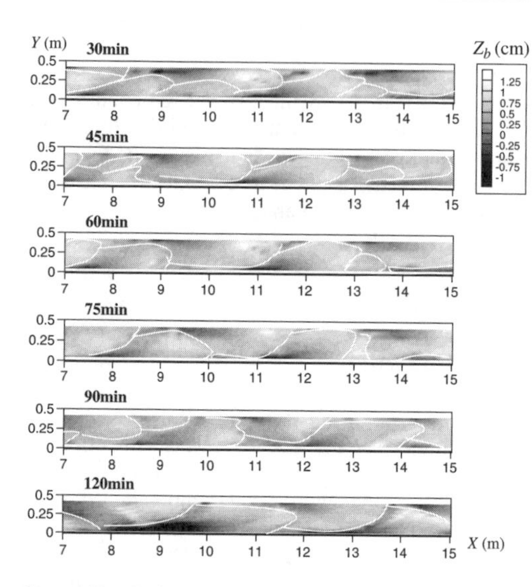

Figure 13. Bed topography Case C-2 (Experiment).

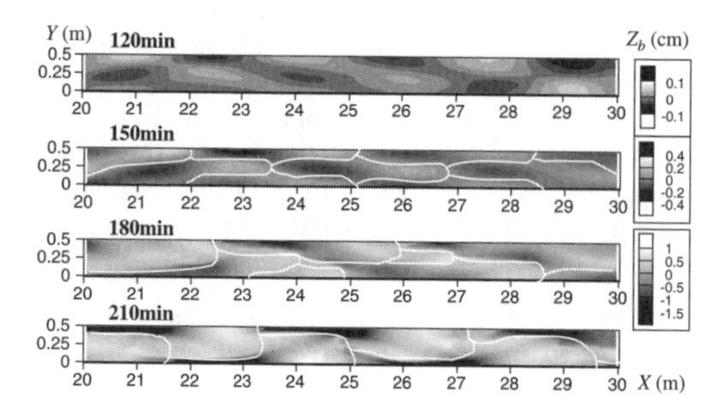

Figure 14. Bed topography Case C-2 (Numerical simulation).

(3) Multiple bars changed to alternate bars.
(4) Alternate bars developed and migrated.

Figure 14 is a picture taken after the water supply was stopped. Darker color points in Figure 14 are coarser materials. Spatial distributions of fine and coarse sand are very clearly demonstrated. It is clear that the bar front is at right angles to the belt of coarser sand in experimental results as shown in Figure 15. Spatial distributions of fine and coarse sand in experiment are similar to those in numerical simulation.

Our model for sediment mixture is useful to describe the bar formation process on a bed composed of mixed fine and coarse sand.

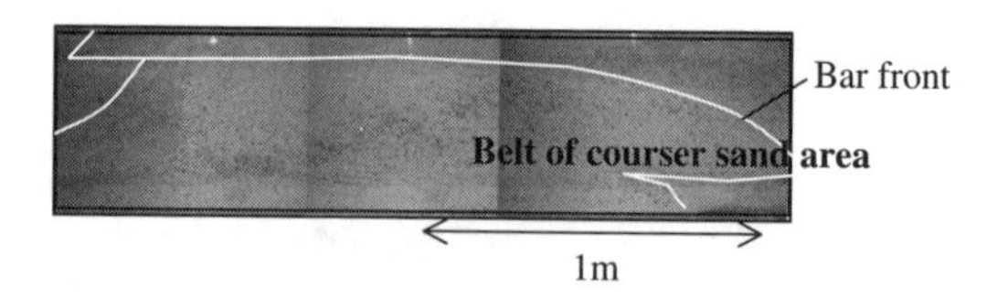

Figure 15. Special distribution of coarse sand on the bar (Experiment).

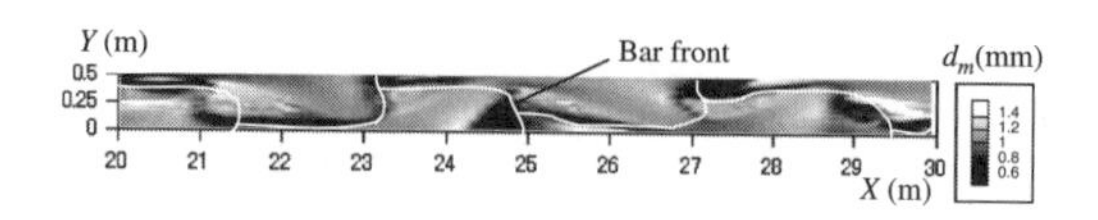

Figure 16. Relationship between sediment mean diameter and bar front (Numerical simulation).

4 EFFECT OF SIZE HETEROGENEITY OF BED MATERIALS TO BAR MODE-NUMERICAL SIMULATION

4.1 *Multiple bar formation*

In this section, we discuss our investigation into the effects of size heterogeneity of bed material on bar formation, especially bar mode. In considering the difference of bar mode according to bed material, we utilized results from experiments conducted by Fujita (1989). Fujita's experimental conditions were the following.

- Channel width is 3 m.
- Water discharge is 30.75 l/s.
- Sediment diameter is 0.88 l/s.
- Bed slope is 1/200.

In Fujita's experiment, multiple bars (double-row bars $n = 2$) appeared and were kept for long time as shown in Figure 17.

The results of numerical simulation carried out in the same conditions as Fujita's experiment are shown in Figure 18. The results of numerical simulation are almost the same as the experimental results. Multiple bars (n = 2) appear and develop with bar migration.

4.2 *Investigation of the effect of size heterogeneity of bed materials*

We repeated the numerical simulation, again using Fujita's experimental conditions to investigate the effects of bed load size heterogeneity on bar formation. We considered representative classes of bed material in the condition that mean diameter is same as Fujita's experiment. The investigated bed material composition is shown in Figure 19. d_i in Figure 19 is sand diameter of sediment class i and p_i is possessory rate of sediment class i. We investigated bar formation by using numerical simulation as described in Section 3.

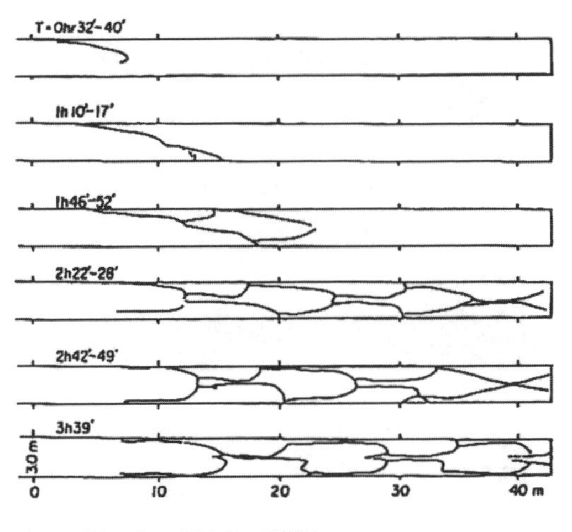

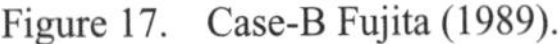

Figure 17. Case-B Fujita (1989).

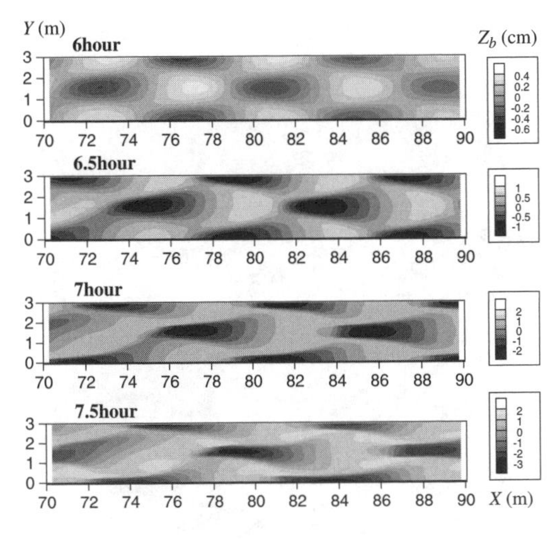

Figure 18. Bed topology in Fujita experimental case (Numerical simulation).

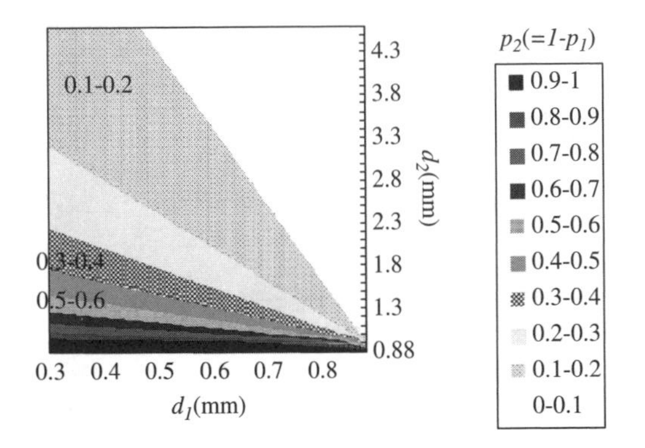

Figure 19. The relationship between possessory rate and sediment diameter composed of bed.

Table 3. Bed material conditions.

Case no.	d_1 (mm)	d_2 (mm)	$\beta = d_2/d_1$	p_1	p_2
NS	0.6	1.5	2.5	0.689	0.311
NN	0.6	2.0	10/3	0.800	0.200
NG	0.6	3.0	5	0.983	0.117
SN	0.4	2.0	5	0.700	0.300
GN	0.8	2.0	5/2	0.333	0.667

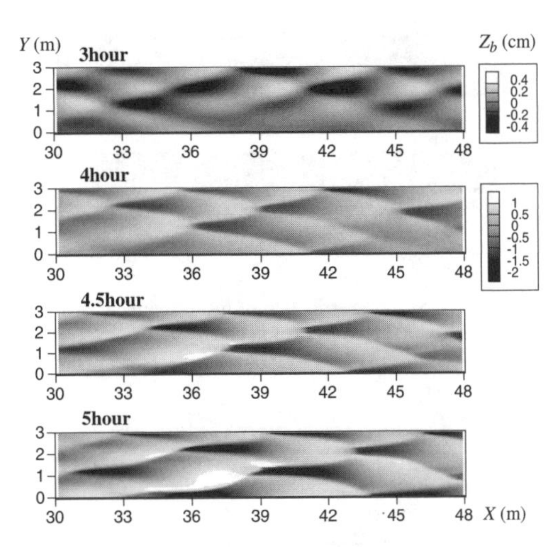

Figure 20. Spatial distribution of mean sediment diameter in the condition that $d_1 = 0.6$ mm and $d_2 = 1.5$ mm.

Figures 20, 21 shows the bar mode number appearing in Case NS is almost 3. It is almost 4 in Case NN (Figures 22, 23) and about 5 in Case NG (Figures 24, 25). Considering the results of Cases NS, NN and NG from numerical simulations, we found the following. Higher mode number f bar appears and develops with bar migration, in the case that bed materials composed of the courser sand.

On the other hand, considering the results of Cases SN, NN, GN, we understand that higher mode number of bars appear develop in the case that bed materials composed of finer sand(Figures 22, 26).

5 EFFECT OF SIZE HETEROGENEITY OF BED MATERIALS TO BAR MODE – LINEAR INSTABILITY ANALYSIS

5.1 *Linear instability analysis considering sediment mixture*

To better understand the phenomena described above, we tried the linear instability analysis considering sediment mixture in order to make the above phenomena more understood. We applied the method for linear instability analysis of bar formation on a bed composed

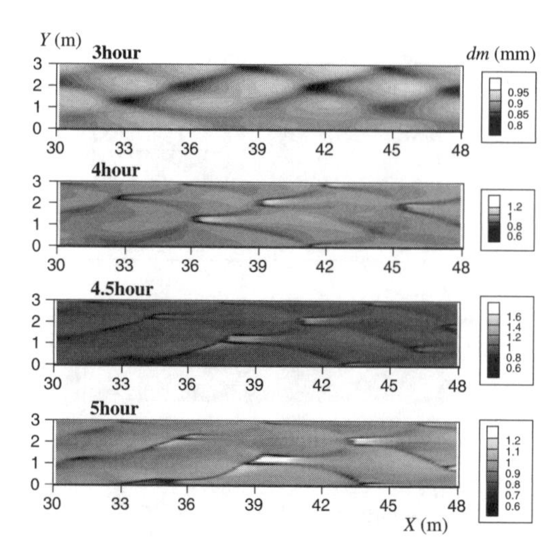

Figure 21. Spatial distribution of mean sediment diameter in the condition that $d_1 = 0.6$ mm and $d_2 = 2.0$ mm.

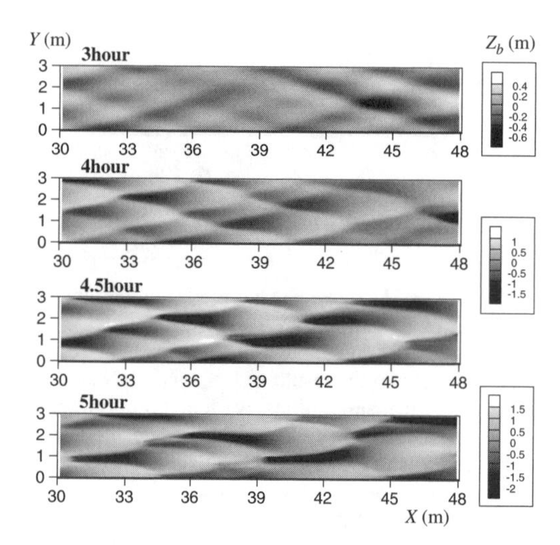

Figure 22. Bed topography in the condition that $d_1 = 0.6$ mm and $d_2 = 2.0$ mm.

of uniform sediment, as carried out by Kuroki & Kishi (1980), to bar formation on a bed composed of two sediment sizes.

5.1.1 *Governing equation*

If we consider flow in a straight channel with constant width B and nonerodible banks, the basic equations are the following, (13) to (18) for shallow-water flow and continuity equation of sediment transport considering 2 classes of sediment diameter.

$$u\frac{\partial u}{\partial x}+v\frac{\partial u}{\partial y}=gI-\frac{\tau_x}{\rho h}-g\frac{\partial}{\partial x}(h+\eta) \quad (13)$$

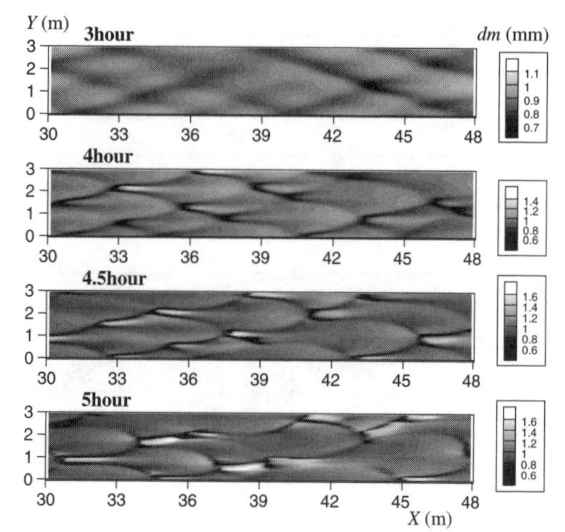

Figure 23. Spatial distribution of mean sediment diameter in the condition that $d_1 = 0.6$ mm and $d_2 = 2.0$ mm.

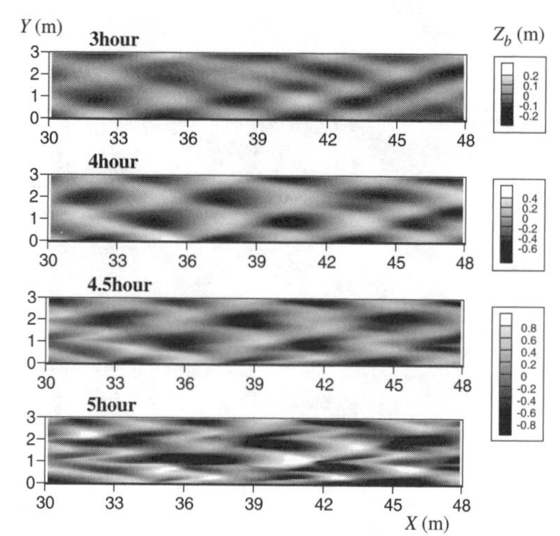

Figure 24. Bed topography in the condition that $d_1 = 0.6$ mm and $d_2 = 3.0$ mm.

$$u\frac{\partial v}{\partial x}+v\frac{\partial v}{\partial y}=-\frac{\tau_y}{\rho h}-g\frac{\partial}{\partial y}(h+\eta) \quad (14)$$

$$\frac{\partial uh}{\partial x}+\frac{\partial vh}{\partial y}=0 \quad (15)$$

$$\frac{\partial \eta}{\partial t}+\frac{1}{1-n_a}\left\{\left(\frac{\partial q_{bx1}}{\partial x}+\frac{\partial q_{by1}}{\partial y}\right)+\left(\frac{\partial q_{bx2}}{\partial x}+\frac{\partial q_{by2}}{\partial y}\right)\right\}=0 \quad (16)$$

$$D\frac{\partial p_i}{\partial t}=\left(\frac{\partial \eta_i}{\partial t}-p_i\frac{\partial \eta}{\partial t}\right)$$

$$\left(\because p_i+\frac{\partial p_i}{\partial t}\Delta t=\left(\frac{\partial \eta_i}{\partial t}\Delta t+p_i D\right)\Big/\left(D+\frac{\partial \eta}{\partial t}\Delta t\right)\right) \quad (17)$$

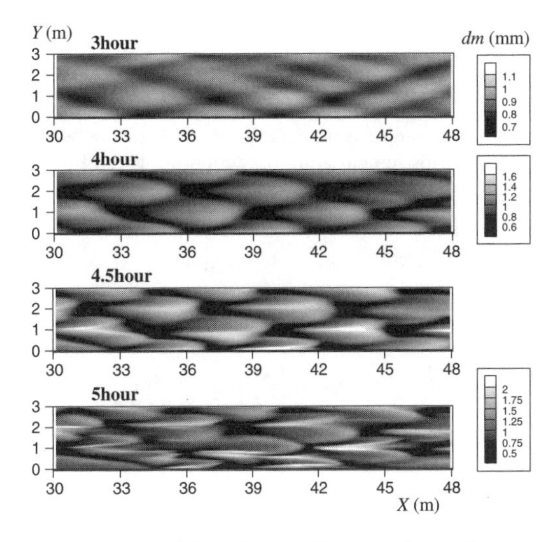

Figure 25. Spatial distribution of mean sediment diameter in the condition that $d_1 = 0.6$ mm and $d_2 = 3.0$ mm.

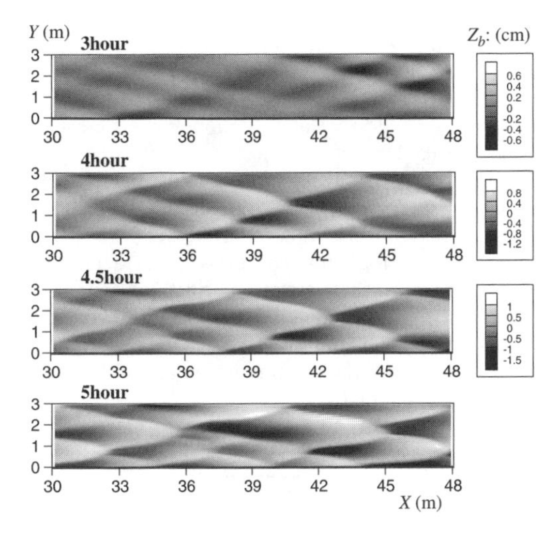

Figure 26. Bed topography in the condition that $d_1 = 0.4$ mm and $d_2 = 2.0$ mm.

$$\sum p_i = 1 \tag{18}$$

where (x, y) = the streamwise and lateral coordinates respectively; (u, v) = depth-averaged velocity components in the axial and radial directions respectively; h = water depth; η = local depth; (τ_x, τ_y) = bottom shear stress in the axial and radial directions respectively; (q_{bxi}, q_{byi}) = sediment discharge of class i, I = average bed slope; n_a = sediment porosity; g = gravitational acceleration; D = exchange layer; and p_i = possessory rate of sediment class i.

The following normalization and imposing perturbation to normal flow has been used to derive the

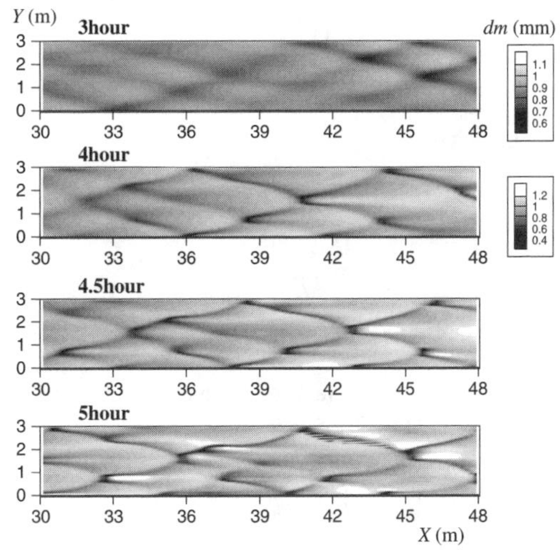

Figure 27. Spatial distribution of mean sediment diameter in the condition that $d_1 = 0.4$ mm and $d_2 = 3.0$ mm.

above equation. The subscript 0 denotes in the uniform flow condition and the tilde denotes a n-dimensional perturbation term.

$$u = U_0(1+\tilde{u}), \quad v = U_0 \cdot \tilde{v} \tag{19}$$

$$\tau_x = \tau_{b0}(1+\tilde{\tau}_x), \quad \tau_y = \tau_{b0} \cdot \tilde{\tau}_y \tag{20}$$

$$h = H_0(1+\tilde{\varepsilon}-\tilde{\eta}), \quad \eta = H_0 \cdot \tilde{\eta}, \quad \varepsilon = H_0 \cdot \tilde{\varepsilon} \tag{21}$$

$$q_{Bxi} = q_{B0i}(1+\tilde{q}_{Bxi}), \quad q_{Byi} = q_{B0i} \cdot \tilde{q}_{Byi} \tag{22}$$

$$p_i = p_{bi}(1+\tilde{p}_i) \tag{23}$$

On substituting the above equations from (19) to (23) into equations from (13) to (18), the following equations were derived

$$Fr^2 \frac{\partial \tilde{u}}{\partial \tilde{x}} + v\frac{\partial \tilde{\varepsilon}}{\partial \tilde{x}} + I(\tilde{\eta} - \tilde{\varepsilon} + \tilde{\tau}_x) = 0 \tag{24}$$

$$Fr^2 \frac{\partial \tilde{v}}{\partial \tilde{x}} + v\frac{\partial \tilde{\varepsilon}}{\partial \tilde{y}} + I \cdot \tilde{\tau}_y = 0 \tag{25}$$

$$\frac{\partial \tilde{u}}{\partial \tilde{x}} + \frac{\partial \tilde{v}}{\partial \tilde{y}} + \frac{\partial \tilde{\varepsilon}}{\partial \tilde{x}} - \frac{\partial \tilde{\eta}}{\partial \tilde{x}} = 0 \tag{26}$$

$$\frac{\partial \tilde{\eta}}{\partial \tilde{t}} + A_{s1}\left(\frac{\partial \tilde{q}_{bx1}}{\partial \tilde{x}} + \frac{\partial \tilde{q}_{by1}}{\partial \tilde{y}}\right) + A_{s2}\left(\frac{\partial \tilde{q}_{bx2}}{\partial \tilde{x}} + \frac{\partial \tilde{q}_{by2}}{\partial \tilde{y}}\right) = 0 \tag{27}$$

$$\frac{\partial \tilde{p}_1}{\partial \tilde{t}} + \frac{A_{s1}}{\tilde{D}}\left(\frac{1}{p_{10}} - 1\right)\left(\frac{\partial \tilde{q}_{bx1}}{\partial \tilde{x}} + \frac{\partial \tilde{q}_{by1}}{\partial \tilde{y}}\right) - \frac{A_{s2}}{\tilde{D}}\left(\frac{\partial \tilde{q}_{bx2}}{\partial \tilde{x}} + \frac{\partial \tilde{q}_{by2}}{\partial \tilde{y}}\right) = 0 \tag{28}$$

$$\frac{\partial \tilde{p}_2}{\partial \tilde{t}} - \frac{A_{s1}}{\tilde{D}}\left(\frac{\partial \tilde{q}_{bx1}}{\partial \tilde{x}} + \frac{\partial \tilde{q}_{by1}}{\partial \tilde{y}}\right) + \frac{A_{s2}}{\tilde{D}}\left(\frac{1}{p_{20}} - 1\right)\left(\frac{\partial \tilde{q}_{bx2}}{\partial \tilde{x}} + \frac{\partial \tilde{q}_{by2}}{\partial \tilde{y}}\right) = 0 \quad (29)$$

$$\sum p_{i0}\tilde{p}_i = 0 \quad (30)$$

where the following normalization has been used:

$$\tilde{x} = x/H_0\,,\ \ \tilde{y} = y/H_0\,,\ \ \tilde{t} = t \cdot U_0/H_0$$
$$\tilde{D} = D/H_0\,,\ \ Fr^2 = U_0^{\,2}/(gH_0),$$
$$A_{si} = q_{Bi0}/(1-n)U_0H_0 \quad (31)$$

The unknown parameters in above equations are following 12 parameters

$$\tilde{u}\,,\ \tilde{v}\,,\ \tilde{\varepsilon}\,,\ \tilde{\eta}\,,\ \tilde{\tau}_x\,,\ \tilde{\tau}_y\,,$$
$$\tilde{q}_{bx1}\,,\ \tilde{q}_{by1}\,,\ \tilde{q}_{bx2}\,,\ \tilde{q}_{byi}\,,\ \tilde{p}_1\,,\ \tilde{p}_2 \quad (32)$$

There are already 6 equations (from (24) to (29)) for 12 unknown parameters. We need 6 more equations, so we consider the following contents.

5.1.2 *Flow resistance*
Firstly, the expression for $\tilde{\tau}_x$ has been obtained from resistance law.

$$\frac{\tau_{b0}}{\rho}\sqrt{(1+\tilde{\tau}_x)^2 + \tilde{\tau}_y} = \left\{U_0\sqrt{(1+\tilde{u}_x)^2 + \tilde{v}_y} \Bigg/ \left(\varphi\frac{H_0(1+\tilde{\varepsilon}-\tilde{\eta})}{d_{m0}\left(1+\frac{p_{10}\tilde{p}_1 + p_{20}\beta\tilde{p}_2}{p_{10}+p_{20}\beta}\right)}\right)\right\}^2 \quad (33)$$

where

$$\varphi(h/k_s) \equiv U/u_* \quad (34)$$

The above equation (13) is linearized to the following equation:

$$\tilde{\tau}_x = \alpha_1\tilde{u} - \alpha_2(\varepsilon - \eta - \alpha_1 \cdot \tilde{p} - \alpha_2 \cdot \tilde{p}_2) \quad (35)$$

where,

$$\alpha_1 = 2\,,\ \ \alpha_2 = 2\frac{H_0/d_{m0}}{\varphi(H_0/d_{m0})} \cdot \left.\frac{d\varphi}{d(h/d)}\right|_{h/d==H_0/d_{m0}}$$

$$\alpha_{61} = \frac{p_{10}}{p_{10}+p_{20}\beta}\,,\quad \alpha_{62} = \frac{p_{20}\beta}{p_{10}+p_{20}\beta} \quad (36)$$

Considering Manning-Strickler's equation as resistance law, we can get $\alpha_2 = 1/3$.

5.1.3 *The direction of sediment transport*
Secondly, let us consider the direction of sediment movement. The direction of water flow and sediment movement and that of resistance to the sediment particle are generally different from each other. Assuming that the resistance of the sediment particle does not vary with grain size, we consider the direction of sediment movement for each grain size in the same way as Kuroki & Kishi. R = the angle between resistance force and x-direction; and ψ_i = the angle between sediment movement and x-direction are given by the following equation:

$$\tan R = \frac{\tau_y}{\sqrt{\tau_x^{\,2} + \tau_y^{\,2}}} \approx \tilde{\tau}_y \quad (37)$$

$$\tan\psi_i = \frac{q_{Byi}}{\sqrt{q_{Bxi}^{\,2} + q_{Byi}^{\,2}}} \approx \tilde{q}_{Byi} \quad (38)$$

Considering momentum equation of sediment transport on bed slope, R and ψ_i can be given as follows:

$$\tilde{\tau}_y = \tilde{v} + \alpha_3\frac{\partial\tilde{\eta}}{\partial\tilde{y}}\,,\quad \alpha_3 = \frac{3}{2}\frac{\phi(\tau_{*0})}{\mu\varphi(H_0/d_{m0})\sqrt{\tau_{*0}}}. \quad (39)$$

$$\tilde{q}_{Byi} = \tilde{v} - \alpha_{4i}\frac{\partial\tilde{\eta}}{\partial\tilde{y}}\,,\quad \alpha_{4i} = \frac{1}{\mu}\sqrt{\frac{\tau_{*ci}}{\tau_{*0i}}\frac{\mu}{\mu_c}} \quad (40)$$

where μ_c (=0.8) = static frictional coefficient of sand grains respectively; and μ (=0.5) = kinetic frictional coefficient of sand grains respectively. And $\tau_{*i} = \tau_{b0}/(\rho sgd_i)$; and τ_{*ci} = non-dimensional bed shear stress and non-dimensional critical bed shear stress values for sediment class i. Critical non-dimensional bed shear stress is defined for each grain size by the following equation (41). s = specific weight of a grain in fluid; and $\phi(\tau_{*i}, \tau_{*ci})$ = sediment rate as given by the Ashida-Michiue formula (1972) as the following equation (42).

$$d_i/d_m \geq 0.4 \quad \frac{\tau_{*ci}}{\tau_{*cm}} = \left\{\frac{\ln 19}{\ln 19(d_i/d_m)}\right\}^2$$
$$d_i/d_m < 0.4 \quad \frac{\tau_{*ci}}{\tau_{*cm}} = 0.85(d_i/d_m) \quad (41)$$

$$\phi(\tau_{*i}, \tau_{*ci}) = 17\tau_{*i}^{\frac{3}{2}}\left(1 - \frac{\tau_{*ci}}{\tau_*}\right)\left(1 - \sqrt{\frac{\tau_{*ci}}{\tau_*}}\right) \quad (42)$$

where τ_{*cm} is non-dimensional critical bed shear stress for mean diameter and is presumed not to change from initial condition. This value is given by Iwagaki's formula (1956).

5.1.4 *Non-aquarium sediment transport*
The last 2 equations are given by non-dimensional sediment transport formula. In general, sand particle motion along the bed surface adapts to hydraulic conditions from starting movement point to deposit point.

However, it is not so easy to consider it. Therefore, local sediment discharge is estimated by using concept of lag-distance δ. Assuming that lag-distance for sediment class i equals to the mean step length, it is given by the following Einstein's formula (1974).

$$\delta_i = \lambda_i = 100 d_i \{1 - \exp(-0.391/\tau_{*i0})\}^{-1}; \; \tau_{*i0} > \tau_{*ci} \tag{43}$$

By adopting the lag-distance above, the sediment transport rate is represented by the following equation.

$$\frac{q_{bi}(x)}{\sqrt{sgd_i^3}} = p_i(x - \delta_i) \cdot \phi\{\tau_{*i}(x - \delta_i)\} \tag{44}$$

Under the uniform-flow conditions, sediment transport rate for sediment class i is written as below:

$$q_{bi0} / \sqrt{sgd_i^3} = p_{i0} \cdot \phi\{\tau_{*i0}\} \tag{45}$$

Substituting equation (42) into (44),

$$\tilde{q}_{bxi}(\tilde{x}) = \tilde{p}_i(\tilde{x} - \tilde{\delta}_i) + \alpha_{5i} \cdot \tilde{\tau}_{*i}(\tilde{x} - \tilde{\delta}_i) \tag{46}$$

Where non-dimensional lag-distance: $\tilde{\delta}_i = \delta_i / H_0$, coefficient α_{5i} is the following expression:

$$\alpha_{5i} = \frac{\tau_{*i0}}{\phi(\tau_{*i0})} \frac{d\phi}{d\tau_{*i}} \bigg|_{\tau_{*i} = \tau_{*i0}} \tag{47}$$

5.1.5 *Comparison between results of experiment and numerical simulation*

Apply a small perturbation to the bed surface as expressed by the following equation (48).

$$\tilde{\eta} = \hat{\eta} \cdot \cos(l\tilde{y}) \cdot \exp\{ik(\tilde{x} - c\tilde{t})\} \tag{48}$$

where $\hat{\eta}$ is amplitude of small perturbation (real number) and c $(= C_r + iC_i)$ is non-dimensional complex moving velocity, C_r is non-dimensional moving velocity (real number) and kC_i is growth rate (real number). k and l are the wave numbers of – and direction, and are defined by the following equations.

$$k = \frac{2\pi H_0}{2L}, \quad l = \frac{m\pi H_0}{B} \tag{49}$$

where L is wavelength and m is number of division for $\tilde{y}$-direction.

Given perturbation to the bed surface elevation as expressed by equation (48), it contributes to the fluctuations of unknown parameters given by the following equations:

$$\begin{pmatrix} \tilde{\varepsilon} \\ \tilde{u} \\ \tilde{\tau}_x \\ \tilde{q}_{bxi} \\ \tilde{p}_2 \end{pmatrix} = \begin{pmatrix} \hat{\varepsilon} \\ \hat{u} \\ \hat{\tau}_x \\ \hat{q}_{bxi} \\ \hat{p}_2 \end{pmatrix} \cdot \cos(l\tilde{y}) \cdot \exp\{ik(\tilde{x} - c\tilde{t})\} \tag{50}$$

$$\begin{pmatrix} \tilde{v} \\ \tilde{\tau} \\ \tilde{q}_{byi} \end{pmatrix} = \begin{pmatrix} \hat{v} \\ \hat{\tau}_y \\ \hat{q}_{byi} \end{pmatrix} \cdot \sin(l\tilde{y}) \cdot \exp\{ik(\tilde{x} - c\tilde{t})\} \tag{51}$$

The amplitude of fluctuation attach to "^" over the character and is a real number. Substituting the equations (48), (50) and (51) into the equations from (24) to (30) and the equations (34), (38), (39), (46), we arrive at the simultaneous equation.

$$A \cdot \begin{pmatrix} \hat{u} \\ \hat{v} \\ \hat{\varepsilon} \\ \hat{\eta} \\ \hat{\tau}_x \\ \hat{\tau}_y \\ \hat{q}_{bx1} \\ \hat{q}_{by1} \\ \hat{q}_{bx2} \\ \hat{q}_{by2} \\ \hat{p}_2 \end{pmatrix} = 0 \tag{52}$$

Solvability for amplitude of each parameter is ensured provided the determinant of the coefficient matrix $\boldsymbol{A}$ of the left side equals 0. The determinant of coefficient matrix $\boldsymbol{A}$ leads to the following equation (53).

$$c^2(i\,A_1 + A_2) + c(i\,B_1 + B_2) + (i\,C_1 + C_2) = 0 \tag{53}$$

where coefficients A_1, A_2, B_1, B_2, C_1, C_2 of equation (53) were calculated with Mathematical(*Ver*.4.2), so matrix $\boldsymbol{A}$ and these coefficients are omitted.

Substituting $c = C_r + iC_i$ in equation (53), value of C_r and can be solved. When kC_i is more than 0, the amplitude of a small perturbation is fluctuating and growing with time. Therefore bars will develop in this condition.

5.2 *Results of linear instability analysis*

Using this linear instability analysis, we try to explain the phenomena, indicated by results of numerical simulation, that a combination of finer and coarser bed material components promotes a higher bar mode number.

Figure 28 shows the number of superior bar mode estimated by linear instability analysis to appear initially when the bed material is composed of sediment with diameters d_1 and d_2. The white lines in this figure are the same ratio of the two bed material components. This graph shows that various combinations of bed

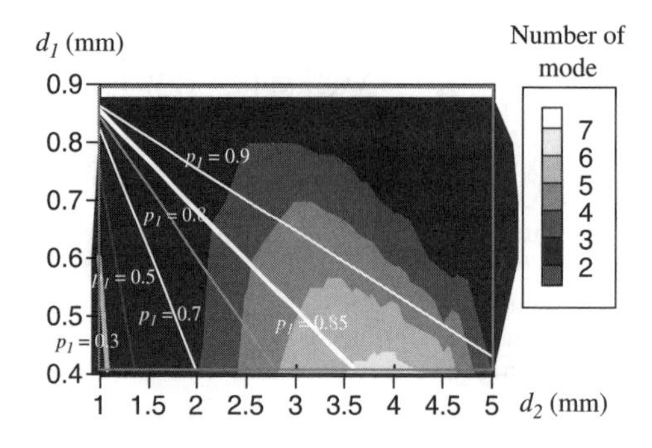

Figure 28. Number of superior bar mode.

material components promote different number of bar mode. It also shows that existence of finer and coarser sand as bed material component promote higher number of bar mode in the condition that the ratio of two classes of bed materials and mean diameter is same. This result is the same as that derived from numerical simulation.

The only difference between our linear instability analysis and the analysis conducted by Kuroki & Kishi (1984) is that of considering sediment discharge estimated according to sand diameter. Therefore, we found out that the different motion of sand particles resulting from the different diameters promoted this phenomenon.

6 CONCLUSIONS

We employed the numerical simulation method of describing bar formation process. When compared to experimental results, the method of describing bar formation process by numerical simulation is shown to have high validity. This method was also applied to the sediment-transport model for a sediment mixture, and we described bar formation on a bed composed of two different sand diameters.

Using the method of describing the bar formation process and the model for a sediment mixture, we investigated bar formation on the bed composed of various combinations of 2 sediment diameters in the condition that double-row bars will develop. Results of numerical simulation show that the existence of courser material promotes a higher bar mode number in the condition that mean diameter and sand diameter of the fine material are constant. On the other hand, it shows that the existence of finer material also promotes a higher bar mode number in the condition that sand diameter of the course material is constant. As results, we found that higher number of bar mode develops on the bed composed of a combination of coarser and finer sediment with numerical simulations. This phenomenon can be explained by linear instability analysis considering the different motion characteristics of each sand diameter. Therefore, we can conclude that different motions of each particle according to sediment diameter can promote higher number of bar mode in the condition where the bed is composed of materials with wider size of distribution.

REFERENCES

Ashida, K. & M. Michiue, 1971. Studies on bed load transportation for non-uniform sediment and riverbed variation. Annual, Disas. Prev. Res. Inst., Kyoto Univ., 14B: 259–273 (in Japanese).

Colunvini, M, G. Seminara & M. Tubino, 1987. Finite-amplitude alternate bars. Jour. Fluid Mech., Vol. 181: 213–232.

Einstein, H. A. 1942. Formulas for the transportation of bed loas, Trans. ASCE, Vol. 107, paper no. 2140: 567.

Engelund, F., 1974. Flow and bed topography in channel beds. Jour. Hydraul. Div., ASCE: 1631–1648.

Ferziger, J. H. & M. Peric, 1997. Computational method for fluid dynamics, Springer.

Fujita, Y. 1989. Bar and Channel Formation in Braided Stream, *the American Geophysical Union*: 417–462.

Kuroki, M. & T. Kishi, 1984. Theoretical study on regime classification of meso-scale bed forms, *Proc. JSCE*, Vol. 342: 87–94 (in Japanese).

Lanzoni, S. & M. Tubino 1999. Grain sorting and bar instability, *J. Fluid Mech.*, Vol. 393: 149–174.

Lanzoni, S. 2000. Experiments on bar formation in s straight flume. Water Resourses Res., Vol. 36, No. 11: 3351–3363.

Nakagawa, H., T. Tsujimoto & S. Murakami, 1986. Non-equilibrium bed load along side bank, Proc. 3rd Int. Sym. River Sedimentation, Jackson, Mississippi, USA: 1029–2065.

Rhie, C. M. & W.L. Chow, 1983. A numerical study of the turbulent flow past an isolated airfoil with trailing edge separation. AIAA J. 21: 1525–1532.

Teramoto, A. & Tsujimoto, T. 2003. Simulation methodologies to describe bar initiation and development. *Journal of Applied Mechanics* Vol. 6 : 975–982 (in Japanese).

Watanabe, Y. & Kuwamura, T. 2004, Experimental study on mode reduction process of double-row bars, *Annual Journal of Hydraulic Engineering*, JSCE: 997–1002 (in Japanese).

River, Coastal and Estuarine Morphodynamics: RCEM 2005 – Parker & García (eds)
© 2006 Taylor & Francis Group, London, ISBN 0 415 39270 5

Mode-decrease process of double-row bars

Y. Watanabe & K. Kuwamura
Civil Engineering Research Institute, Sapporo Japan

ABSTRACT: The hydraulic experiment data were subjected to weakly nonlinear analysis in order to clarify the reproducibility of the process of bar mode decrease. The estimated equilibrium bar height and bar migration rate agree closely with the observed ones. The following facts became clear from this analysis. The initial bar growth rate must be calculated precisely if the process of sandbar mode decrease is to be investigated. The changes in bar wavelengths during that process can be calculated by linear stability analysis.

Weakly nonlinear analysis can predict the process of bar mode decrease when the perturbation parameter is less than 1. The weakly nonlinear analysis accurately reproduces the ratio of each mode of bar component to that of the total of all modes of bar components.

1 INTRODUCTION

The locations and scales of local scouring are largely influenced by the shapes of sand bars. To understand those locations and scales, it is necessary to understand the shapes of sand bars, which will also clarify the movements of scouring. Sand bar morphology has been studied for flood control and water utilization, such as for determination of the revetment embedded depth and of proper locations of river structures. In addition, study of riverbed configuration has become indispensable for examination of the river environment in recent years, because the riverbed is the foundation of the river eco-system.

Regarding the transformation and deformation of meso-scale riverbed configuration, past study addressed the sand bar forward migration. However, except for studies on large-scale river-channel modifications such as channel widening, little research has focused on the changes in configuration of sand bars themselves. Even where such research has been done, the results have rarely been applied. River-channel planning and maintenance today call for a certain tolerance of river-channel migration toward taking advantage of a river's natural environment and flood control ability.

Given such background, many studies have been made in recent years on sandbar behavior during flooding, and on sandbar evolution. Conventionally, the initial growth rate during sandbar development is used to classify the regions of meso-scale riverbed configuration. Various hydraulic test (Fujita 1989, 1992) and numerical simulations (Takebayashi 2001) have reported that under the condition in which sandbars of various modes initially form, the mode is reduced over time (hereinafter: mode reduction). This suggests that in an actual river even under hydraulic conditions in which double-row bars are expected to form, mode reduction has occurred or is in progress. This means that there may be a situation in which the sandbar mode differs from the theoretical classification of the sandbar mode based on the initial growth rate of riverbed configuration. This requires much attention for revetment installation planning and for river environment preservation.

Watanabe et al. (2004) experimented on the sandbar formation process for meso-scale riverbed configuration by setting several hydraulic conditions under which alternate bars or double-row bars formed in certain regions. This study expands on our previous research by examining sandbar mode reduction in greater detail. To this end, we examined the applicability of weakly nonlinear analysis in describing sandbar development. Weakly nonlinear analysis has conventionally been used to reproduce sandbar shape (e.g., bar height) at equilibrium. Many studies have attempted weakly nonlinear analysis of sandbars. Izumi et al. (2002) applied the amplification expansion method for weakly nonlinear analysis of double-row bars at equilibrium. Another aim of this study is to examine how well the growth rate expansion method describes sandbar mode reduction process. To this end, the process is analyzed using the method proposed by Colombini et al. (1987)

Table 1. Hydraulic conditions and bed forms of experiments.

Run	$\tilde{Q}$ cm^3/s	$\tilde{D}_0$ cm	$\tilde{T}$ min.	I_w	I_b	C_f	β	d_s	ϑ	Bed forms	$\tilde{L}_b$ m	$\tilde{Z}_b$ cm	λ	Z_b
S-10-20	1660	0.59	20	1/86	1/80	0.0069	76	0.13	0.055	D	2.42	2.1	1.17	3.56
S-10-40	1660	0.62	40	1/82	1/80	0.0084	73	0.12	0.060	D	2.66	2.1	1.06	3.39
S-10-60	1660	0.46	60	1/83	1/80	0.0034	98	0.17	0.044	D	2.56	2.2	1.10	4.78
S-10-80	1660	0.36	80	1/84	1/80	0.0016	125	0.21	0.034	D+A	3.76	2.5	0.75	6.94
S-10-120	1660	0.61	120	1/83	1/80	0.0079	74	0.12	0.059	D+A	5.83	2.6	0.48	4.26
S-10-240	1660	0.49	240	1/84	1/79	0.0040	92	0.16	0.047	A	3.30	4.8	0.86	9.80
S-10-480	1660	0.50	480	1/84	1/79	0.0043	90	0.15	0.047	A	6.60	5.2	0.43	10.40
S-10-960	1660	0.60	960	1/80	1/78	0.0078	75	0.13	0.060	A	4.88	4.3	0.58	7.17
S-20-20	3250	1.15	20	1/84	1/80	0.0136	39	0.07	0.109	D	2.70	2.1	1.05	1.83
S-20-40	3250	0.96	40	1/85	1/80	0.0078	47	0.08	0.090	D+A	2.68	2.4	1.06	2.50
S-20-60	3250	0.86	60	1/84	1/79	0.0057	52	0.09	0.082	D+A	3.53	3.7	0.80	4.30
S-20-80	3250	1.10	80	1/85	1/80	0.0118	41	0.07	0.103	D+A	5.89	3.4	0.48	3.09
S-20-120	3250	0.96	120	1/84	1/80	0.0079	47	0.08	0.091	A	5.41	4.3	0.52	4.48
S-20-240	3250	0.91	240	1/84	1/79	0.0067	49	0.08	0.086	A	7.50	5.6	0.38	6.15
S-20-360	3250	0.96	360	1/83	1/79	0.0080	47	0.08	0.092	A	6.83	4.5	0.41	4.69
S-20-840	3250	1.00	840	1/82	1/80	0.0092	45	0.08	0.097	A	8.18	5.5	0.35	5.50
S-30-20	5270	1.20	20	1/83	1/80	0.0060	38	0.06	0.115	D	4.65	2.8	0.61	2.33
S-30-40	5270	1.37	40	1/85	1/79	0.0086	33	0.06	0.129	D+A	5.40	6.1	0.52	4.45
S-30-60	5270	1.27	60	1/84	1/78	0.0070	35	0.06	0.121	A	6.00	5.2	0.47	4.09
S-30-80	5270	1.39	80	1/85	1/81	0.0090	32	0.05	0.130	A	6.30	5.2	0.45	3.74
S-30-120	5270	1.20	120	1/83	1/79	0.0060	38	0.06	0.115	A	5.48	4.7	0.52	3.92
S-30-240	5270	1.15	240	1/86	1/81	0.0051	39	0.07	0.107	A	5.78	5.3	0.49	4.61
S-30-600	5270	1.20	600	1/85	1/76	0.0058	38	0.06	0.113	A	8.40	6.1	0.34	5.08
S-40-20	7600	1.64	20	1/84	1/80	0.0072	27	0.05	0.156	D	2.87	2.1	0.99	1.28
S-40-40	7600	1.78	40	1/85	1/81	0.0091	25	0.04	0.167	A	5.70	6.3	0.50	3.54
S-40-60	7600	1.80	60	1/85	1/80	0.0094	25	0.04	0.169	A	4.39	3.6	0.64	2.00
S-40-80	7600	1.84	80	1/84	1/80	0.0102	24	0.04	0.175	A	5.78	5.1	0.49	2.77
S-40-120	7600	1.63	120	1/86	1/80	0.0069	28	0.05	0.151	A	7.80	5.4	0.36	3.31
S-40-180	7600	1.60	180	1/87	1/81	0.0065	28	0.05	0.147	A	9.45	5.2	0.30	3.25
S-40-330	7600	1.50	330	1/83	1/83	0.0056	30	0.05	0.144	A	6.80	5.8	0.42	3.87
S-50-20	10350	2.02	20	1/85	1/80	0.0072	22	0.04	0.190	A	4.80	3.1	0.59	1.53
S-50-40	10350	2.03	40	1/86	1/79	0.0072	22	0.04	0.188	A	4.20	2.6	0.67	1.28
S-50-60	10350	2.10	60	1/83	1/80	0.0083	21	0.04	0.202	A	5.03	4.0	0.56	1.90
S-50-80	10350	1.91	80	1/86	1/82	0.0060	24	0.04	0.177	A	6.38	5.1	0.44	2.67
S-50-120	10350	1.82	120	1/84	1/81	0.0053	25	0.04	0.173	A	5.93	5.5	0.48	3.02
S-50-240	10350	1.80	240	1/79	1/82	0.0055	25	0.04	0.182	A	6.68	5.6	0.42	3.11

D: Double-row bars, A: Alternate bars.

2 OUTLINE OF EXPERIMENT ON SANDBAR MODE REDUCTION

Table 1 shows the hydraulic conditions and the sandbar morphologies in the hydraulic experiment on sandbar mode reduction from double-row bar to alternate bar conducted by Watanabe et al. (2004). In Table 1, $\tilde{Q}$ is discharge, $\tilde{D}_0$ is water depth, $\tilde{T}$ is duration, I_w is water gradient, I_b is bed slope, C_f is drag coefficient of riverbed, $\beta = \tilde{B}/\tilde{D}_0$, $\tilde{B}$ is half the channel width (=45 cm), $d_s = \tilde{d}_s/\tilde{D}_0$, $\tilde{d}_s$ is diameter of bed material (0.76 mm), ϑ is dimensionless shear stress, $\lambda = 2\pi\tilde{B}/\tilde{L}_b$ and $Z_b = \tilde{Z}_b/\tilde{D}_0$. $\tilde{L}_b$ and $\tilde{Z}_b$ are the average wavelength and wave height of sandbars in the observed section of channel. After reaching equilibrium as double-row bars, the bars transform into alternate bars during a transition period. Both double-row bars and alternate bars formed in tests S-10-80, S-10-120, S-20-40, S-20-60, S-20-80 and S-30-40. Therefore, the average wavelengths and wave heights in these tests are the average for double-row bars and alternate bars. The last number of each test name is the duration of water flow, in minutes. Figures 1 and 2 show the observed bed topographies of Run S-20 and Run S-40. The flow velocity vectors of final stage are overwritten.

During mode decreasing period, one of the two scoured areas that form on each side of the watercourse is buried by aggregation. This results in the mode becoming that of alternate bars. The mode decrease

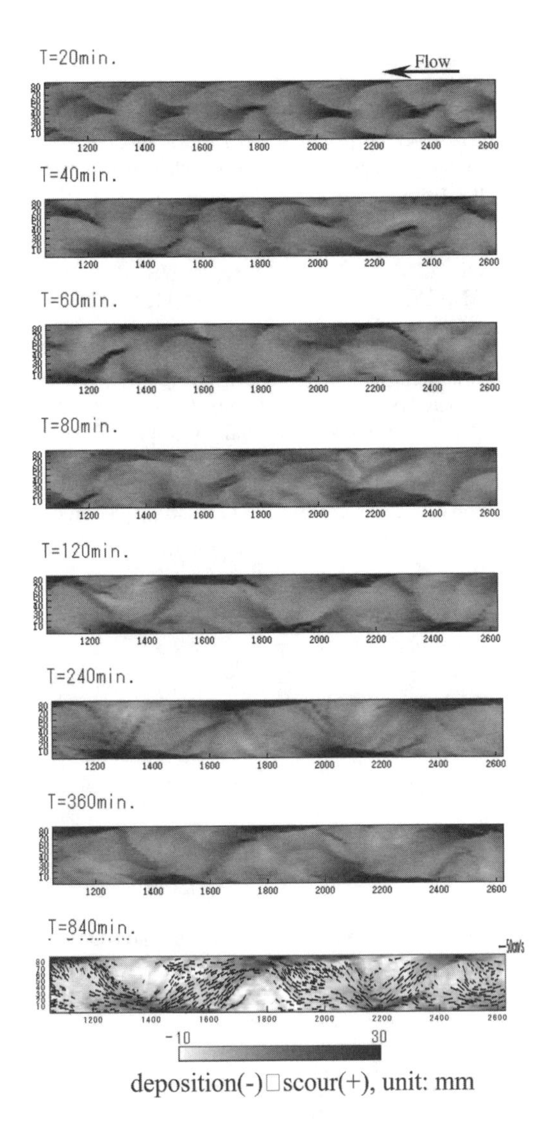

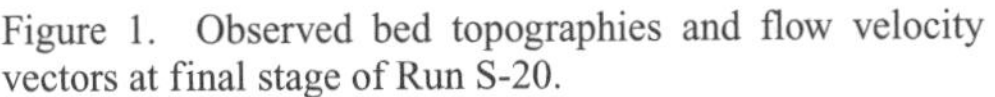

Figure 1. Observed bed topographies and flow velocity vectors at final stage of Run S-20.

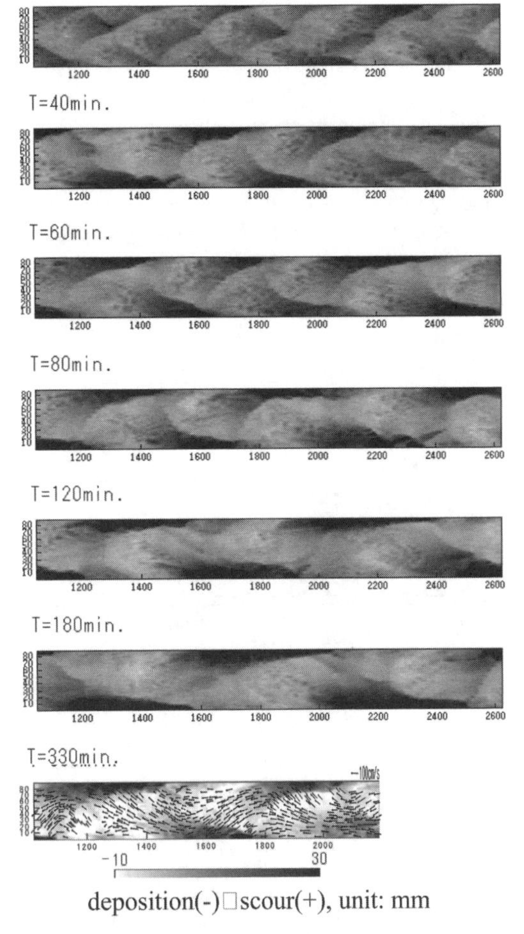

Figure 2. Observed bed topographies and flow velocity vectors at final stage of Run S-40.

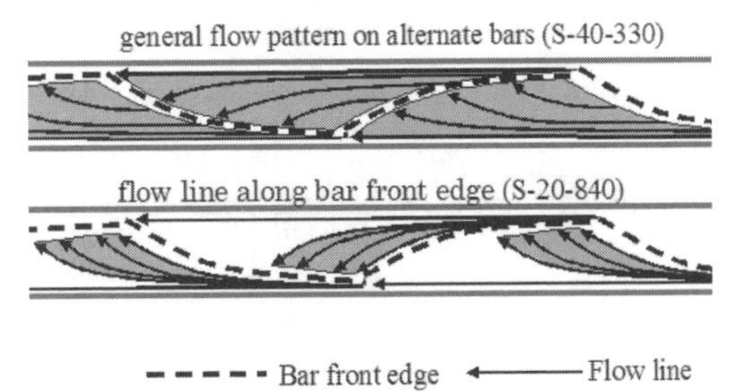

Figure 3. General flow pattern on alternate bars and flow pattern on bars which are generated by mode-decreases.

tends to start later and take longer at larger ratios of width to depth. In general, the flow line on the alternate bars across the bar front edges. This type of flow line is shown in Run S-40-330. However Run S-20-840 which caused mode decrease has two flow lines. One is along the bar front edges and another is along the side walls. The flow line along the bar front edges forms a meandering flow which is like a meandering channel. Figure 3 expresses the difference in the shape of a flow line exemplarily.

We calculated the dimensionless bedload transport, ϕ, for a unit width and the drag coefficient of riverbed, C_f, and compared them with ϕ and C_f from Equations 1 and 2, which are given by Meyer-Peter and Müller, and Engelund and Hansen, respectively.

$$\phi = 8(\vartheta - \vartheta_{cr})^{3/2} \tag{1}$$

$$C_f = \frac{1}{\left[6 + 2.5\ln\left(\frac{1}{2.5d_s}\right)\right]^2} \tag{2}$$

where, ϑ_{cr} is dimensionless critical shear stress. We use the experimental value of 0.038.

Figure 4 plots changes in bedload transport versus time elapsed after water flow started. Except for S-40 and S-50, the bedload variation is small and is around average values. S-40 and S-50 have large variations of bedload transport. Figure 5 compares the experimental and analytical values of bedload transport.

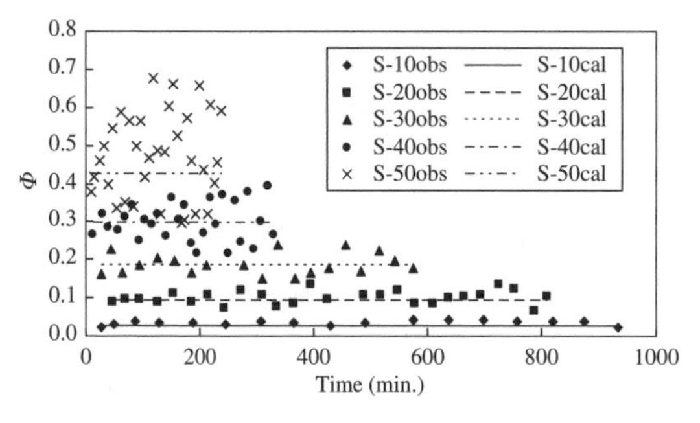

Figure 4. Temporal changes in bedload transport.

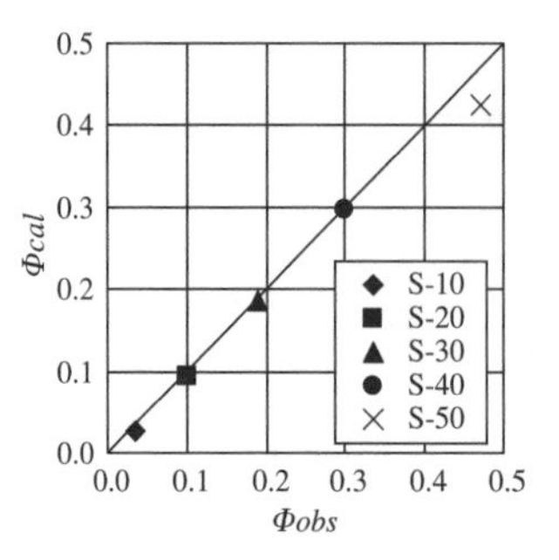

Figure 5. Comparison of experimental and calculated bedload transport.

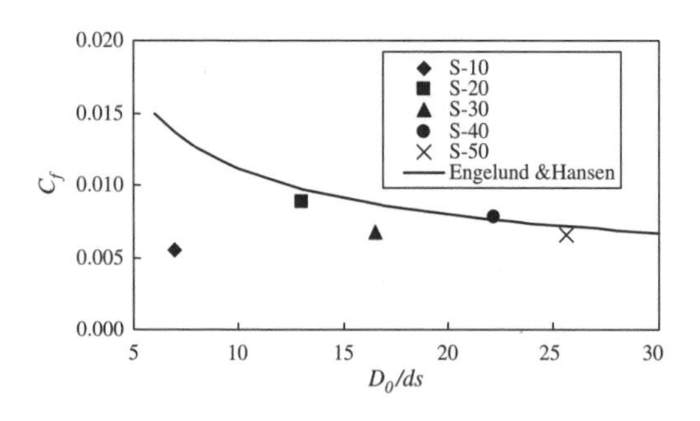

Figure 6. Comparison of experimental and calculated bed resistance.

They closely correspond. Figure 6 compares the experimental and analytical values of drag coefficients of riverbed. Except for S-10, the experimental and analytical values correspond. S-10 has a shallow water depth and emerged bars form, which seem to affect the test results.

3 OUTLINE OF WEAKLY NONLINEAR ANALYSIS USING THE GROWTH RATE EXPANSION METHOD

For a straight channel whose width is $2\tilde{B}$, when the coordinate system shown in Figure 7 is used, we obtain the steady two-dimensional shallow-water flow equations whose diffusion terms are omitted (Equations 3 and 4), the continuity equation of flow (Equation 5), and the continuity equation of bedload (Equation 6).

$$\tilde{U}\frac{\partial \tilde{U}}{\partial \tilde{x}} + \tilde{V}\frac{\partial \tilde{U}}{\partial \tilde{y}} + \tilde{g}\frac{\partial \tilde{H}}{\partial \tilde{x}} + \frac{\tilde{\tau}_x}{\tilde{\rho}\tilde{D}} = 0 \tag{3}$$

$$\tilde{U}\frac{\partial \tilde{V}}{\partial \tilde{x}} + \tilde{V}\frac{\partial \tilde{V}}{\partial \tilde{y}} + \tilde{g}\frac{\partial \tilde{H}}{\partial \tilde{y}} + \frac{\tilde{\tau}_y}{\tilde{\rho}\tilde{D}} = 0 \tag{4}$$

$$\frac{\partial(\tilde{U}\tilde{D})}{\partial \tilde{x}} + \frac{\partial(\tilde{V}\tilde{D})}{\partial \tilde{y}} = 0 \tag{5}$$

$$\frac{\partial \tilde{\eta}}{\partial \tilde{t}} + \frac{\partial \tilde{Q}_{bx}}{\partial \tilde{x}} + \frac{\partial \tilde{Q}_{by}}{\partial \tilde{y}} = 0 \tag{6}$$

where, $\tilde{t}$ is time, $\tilde{x}$ and $\tilde{y}$ are the longitudinal and transverse axes of coordinates, $\tilde{U}$ and $\tilde{V}$ are flow velocity coordinates in the $\tilde{x}$ and $\tilde{y}$ direction, $\tilde{H}$ is water level, $\tilde{D}$ is water depth, $\tilde{\eta}$ is riverbed height ($= \tilde{H} - \tilde{D}$), $\tilde{\tau}_x$ is shear stress in the $\tilde{x}$ direction, $\tilde{\tau}_y$ is shear stress in the $\tilde{y}$ direction, $\tilde{Q}_{bx}$ and $\tilde{Q}_{by}$ are bedload transport rate in the $\tilde{x}$ and $\tilde{y}$ direction, $\tilde{\rho}$ is density of water, and $\tilde{g}$ is acceleration of gravity. Parameters capped with "$\sim$" have dimensions.

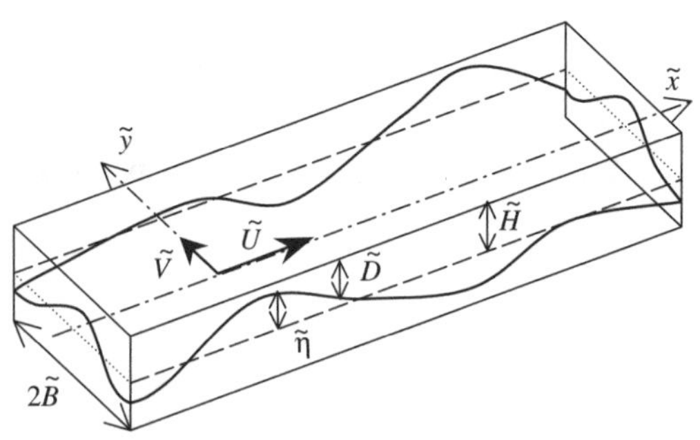

Figure 7. Definitions of coordinate system of the flow model.

Equations 3, 4, 5 and 6 are made dimensionless using the parameters of uniform flow on a flat riverbed: $(U,V)=(\tilde{U}_0,\tilde{V})/\tilde{U}$, $D=\tilde{D}/\tilde{D}_0$, $H=\tilde{H}/(F_0^2\tilde{D}_0)$, $(Q_{sx}, Q_{sy})=(\tilde{Q}_{sx},\tilde{Q}_{sy})/(\Delta\tilde{g}\tilde{d}_s^3)^{1/2}$, $(\tau_x,\tau_y)=(\tilde{\tau}_x,\tilde{\tau}_y)/\tilde{\rho}\tilde{U}_0^2$, $(x,y)=(\tilde{x},\tilde{y})/\tilde{B}$, $t=\tilde{t}/(\tilde{B}/\tilde{U}_0)$. Here, the suffix 0 indicates a value for steady flow on flat bed. Δ is the submerged specific gravity of the riverbed material. U, V, H and D in these equations are substituted for by their expansions whose perturbation parameters are $\varepsilon^{1/2}$, as shown by Equations 7, 8, 9, 10 and 11, to derive a Landau-Stuart-type differential equation (Equation 12) on the order of $\varepsilon^{3/2}$

$$(U,V,H,D)=(1,0,H_0,1)+\varepsilon^{1/2}(U_1,V_1,H_1,D_1)+\varepsilon^{2/2}(U_2,V_2,H_2,D_2)+\varepsilon^{3/2}(U_3,V_3,H_3,D_3) \quad (7)$$

$$(U_1,V_1,H_1,D_1)=\begin{cases} A_{(T)}(S_1,C_1,S_1,S_1)E_1+\text{c.c.} & ;m=1 \\ A_{(T)}(C_2,C_2,C_2,C_2)E_2+\text{c.c.} & ;m=2 \end{cases} \quad (8)$$

$$(U_2,V_2,H_2,D_2)=\begin{cases} \{A_{(T)}{}^2E_2[(C_2u_{22},S_2v_{22},C_2h_{22},C_2d_{22})+(u_{02},v_{02},h_{02},d_{02})]+\text{c.c.} \\ \quad +A_{(T)}\overline{A}_{(T)}[(C_2u_{20},S_2v_{20},C_2h_{20},C_2d_{20})+(u_{00},v_{00},h_{00},d_{00})] \\ \quad +(0,0,0,H_{00})\} & ;m=1 \\ \{A_{(T)}{}^2E_2[(S_4u_{22},S_4v_{22},S_4h_{22},S_4d_{22})+(u_{02},v_{02},h_{02},d_{02})]+\text{c.c.} \\ \quad +A_{(T)}\overline{A}_{(T)}[(S_4u_{20},S_4v_{20},S_4h_{20},S_4d_{20})+(u_{00},v_{00},h_{00},d_{00})] \\ \quad +(0,0,0,H_{00})\} & ;m=2 \end{cases} \quad (9)$$

$$(U_3,V_3,H_3,D_3)=\begin{cases} \left[A_{(T)}{}^2\overline{A}_{(T)}(S_1,C_1,S_1,S_1)E_1+\text{c.c.}\right]+\text{h.h.} & ;m=1 \\ \left[A_{(T)}{}^2\overline{A}_{(T)}(C_2,C_2,C_2,C_2)E_2+\text{c.c.}\right]+\text{h.h.} & ;m=2 \end{cases} \quad (10)$$

$$(S_m,C_m,E_m)=\left(\sin\left(\frac{1}{2}\pi my\right),\cos\left(\frac{1}{2}\pi my\right),\exp[mi(\lambda x-\omega t)]\right) \quad (11)$$

$$\frac{dA_{(T)}}{dT}+\alpha_1A_{(T)}+\alpha_2A_{(T)}{}^2\overline{A}_{(T)}=0 \quad (12)$$

where, $c.c.$ is the complex conjugate of the term that immediately precedes it, m is the sand bar mode number, and $A_{(T)}$ is the amplitude of small perturbation at time T. T is a time scale introduced to express the variation of small perturbation. Its relation to the time scale of flow, t, is given by Equation 13.

$$t=\varepsilon T \quad (13)$$

It is assumed that the scale of time required for a sand bar to develop is much greater than the time for flow. Equation 14 is the solution of Equation 12.

$$|A|=\sqrt{\frac{-\text{Re}(\alpha_1)}{\text{Re}(\alpha_2)-a_0\text{Re}(\alpha_1)\exp[-2\text{Re}(\alpha_1)T]}} \quad (14)$$

where, $|A|_0$ is assumed to be the initial value of small perturbation, and a_0 is given by Equation 15.

$$a_0=\frac{1}{|A|_0{}^2}+\frac{\text{Re}(\alpha_2)}{\text{Re}(\alpha_1)} \quad (15)$$

Linear analysis cannot give the third-order term of $A_{(T)}$. It gives only initial growth rate of small perturbation. Consequently, the solution is given in a form that omits α_2. This means that the unstable region is the only region that is described by linear analysis. As shown by Equation 16, the perturbation parameter in nonlinear analysis, $\varepsilon^{1/2}$, is the square route of the ratio ε obtained by diving the difference between β_c (the minimum river width/depth ratio for which the initial perturbation is neutral, i.e., whose initial growth rate Ω is 0 in linear analysis) and β (the river width/depth ratio under the test conditions) by β_c.

$$\beta=\beta_c(1+\varepsilon) \quad (16)$$

For application of weakly nonlinear analysis, the river width/depth ratio β needs to be close to β_c. In light of this, examinations hereinafter include the range of ε for which the hydraulic conditions can be reproduced by weakly nonlinear analysis. The wave number of bars λ is given by Equation 17. However, here we assume $\lambda_1=0$.

$$\lambda=\lambda_c+\varepsilon\lambda_1 \quad (17)$$

As has been pointed out by Pornprommin et al. (2004), irregularity in sandbar development is shown to result from the interference of sandbars of different mode that occupy the same channel. To examine double-row bars, ideally the interaction between small perturbations of mode-1 sandbars and mode-2 sandbars should be considered. For simplicity, this study omitted such consideration.

4 APPLICATION OF WEAKLY NONLINEAR ANALYSIS

To apply weakly nonlinear analysis to our test, we assumed that Equations 1 and 2 give bedload transport and riverbed resistance, respectively. These equations give analytical values that are nearly equal to values obtained experimentally. S-10 was excluded from analysis because emerged bars formed during the water flow test and the riverbed resistance differs from the value given by Equation 2. We set $|A|_0$ as 0.0001.

4.1 *Linear analysis and experimental results*

First, examination was made by focusing on the differences in magnitude of initial growth rate for mode-1 and mode-2 wave components given by linear analysis in Section 3. Figure 8 shows neutral curves where the initial growth rates of mode-1 and mode-2 wave components are 0 for S-30, S-40 and S-50. The broken lines are experimental values of β. Figure 9 shows initial growth rates of mode-1 and mode-2 wave components for S-30, S-40 and S-50. Figures 8 and 9 indicate that the initial growth rates of mode-1 and mode-2 wave components are nearly equal under S-40 conditions. Under the S-50 condition, the initial growth rate of mode-1 wave components exceeds that of mode-2 wave components, whereas under the S-30 and S-20 conditions, the initial growth rate of mode-2 wave components exceeds that of mode-1 wave components. For S-30, double-row bars form in the early stages of water flow and transform into alternate bars; for S-50, alternate bars form in the early stages of water flow. This corresponds with the initial growth rates of mode-1 and mode-2 wave components shown in Figure 9. We confirmed that for sand bars formed in the early stages of water flow, the mode can obtained by linear analysis.

Figure 10 compares changes in observed wave number with time and wave number $\lambda_{c1}, \lambda_{c2}, \lambda_{\beta 1}, \lambda_{\beta 2}$ calculated by linear analysis. Here, λ_{c1} and λ_{c2} are wave numbers whose initial growth rates of mode-1 and mode-2 wave components are zero for the minimum river width/depth ratio β_c. $\lambda_{\beta 1}$ and $\lambda_{\beta 2}$ are

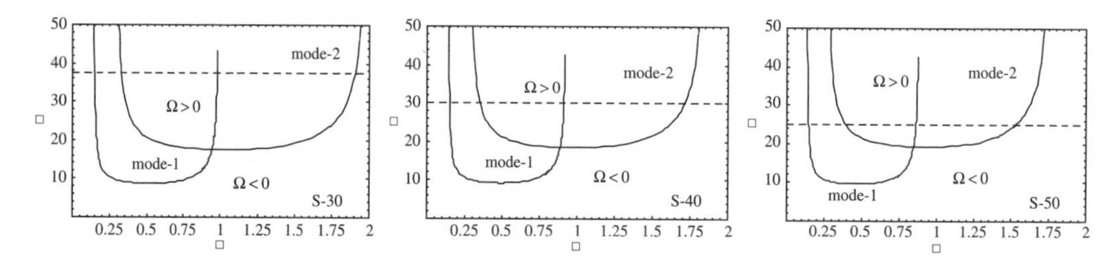

Figure 8. Neutral curves of initial growth rates calculated by the linear analysis.

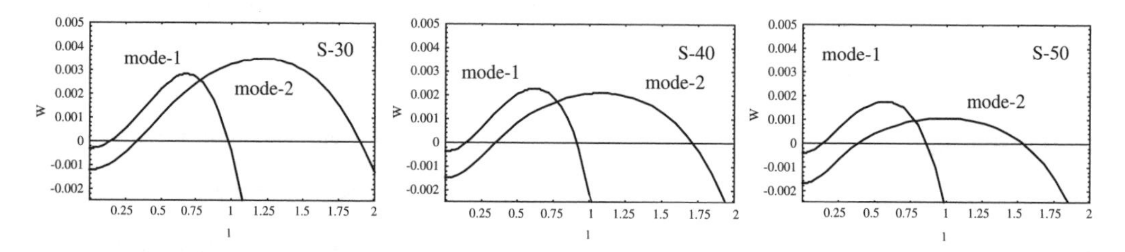

Figure 9. Initial growth rates by experimental condition.

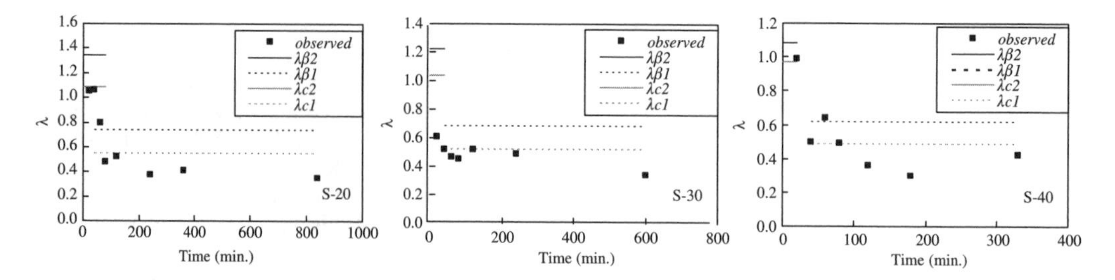

Figure 10. Bar wave numbers from experiments and linear stability analyses.

wave numbers whose initial growth rates of mode-1 and mode-2 wave components are at the maximum for the experimental condition β. Figures 8 indicates that the wave numbers observed during the experiment can be reproduced better using λ_{c1} and λ_{c2} than using $\lambda_{\beta 1}$ and $\lambda_{\beta 2}$. This supports the assumption that $\lambda_1 = 0$, made for Equation 14 when performing the weakly nonlinear analysis. Table 2 gives the results of linear analysis.

4.2 *Weakly nonlinear analysis application to sandbar development, and experimental results*

The tests to which weakly nonlinear analysis was applied in this study have a theoretical problem, because the perturbation parameter $\varepsilon^{1/2}$ is 1 or more for mode-1 in all tests, and is 1 or more for mode-2 in S-20 and S-30. However, we applied the weakly nonlinear analysis to determine how large the perturbation parameter $\varepsilon^{1/2}$ to which these equations apply can be, as shown in Table 2.

Figure 11 compares the amplitude of bed waves obtained by weakly nonlinear analysis to the results of the Fourier spectrum analysis of bed forms (Hasegawa 1982) in the experiments. The predominant transverse and longitudinal wave numbers from the Fourier spectrum analysis of the experimental riverbed are (1, 1), (2, 0), (2, 2), (4, 0). Figure 11 compares these predominant wave numbers. Under the S-50 conditions, weakly nonlinear analysis seems able to reproduce most sandbar changes with time; however, under the S-20 conditions, whose $\varepsilon_2^{1/2}$ exceeds 2, it cannot reproduce the amplitude of each wave and the timing of mode change from double-row bar to alternate bar.

Table 2. Results of linear analysis.

		mode-1				mode-2			
Run	β	β_{c1}	λ_{c1}	$\lambda_{1\beta}$	$\varepsilon_1^{1/2}$	β_{c2}	λ_{c2}	$\lambda_{2\beta}$	$\varepsilon_2^{1/2}$
S-20	45.0	8.17	0.54	0.73	2.12	16.33	1.09	1.34	1.33
S-30	37.5	8.74	0.52	0.68	1.81	17.48	1.04	1.22	1.07
S-40	30.0	9.25	0.49	0.62	1.50	18.50	0.97	1.08	0.79
S-50	25.0	9.60	0.46	0.57	1.27	19.20	0.93	0.99	0.55

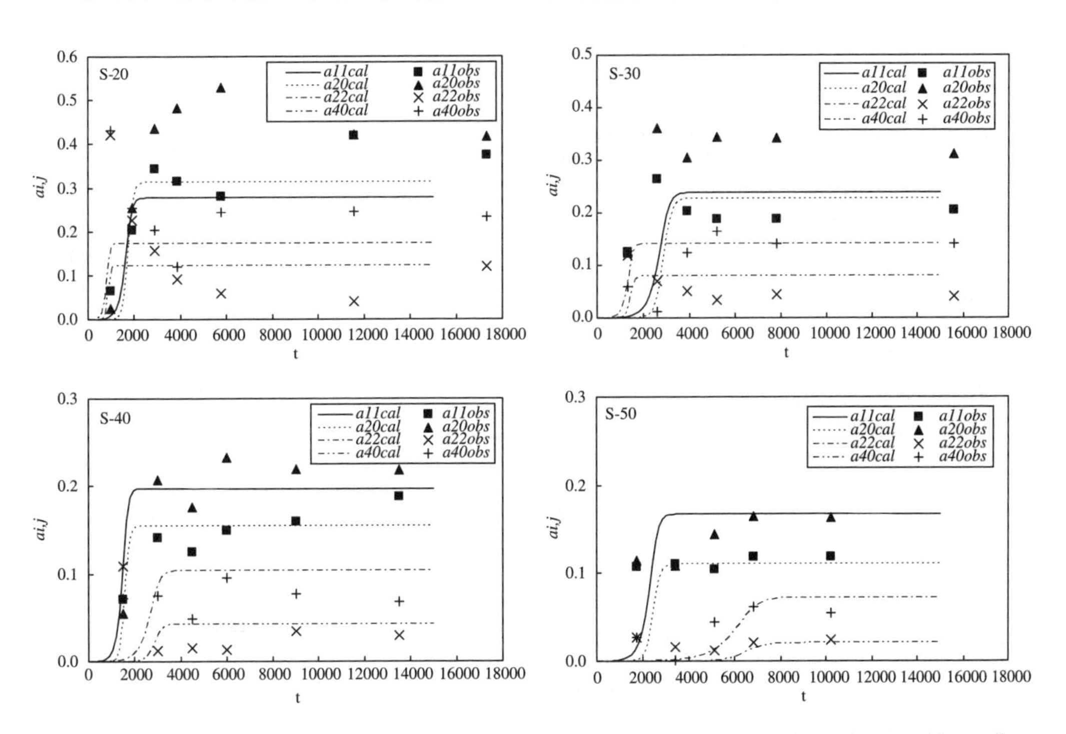

Figure 11. Changes in major components of waves in the mode-reduction experiments, and the results of weakly non-linear analysis.

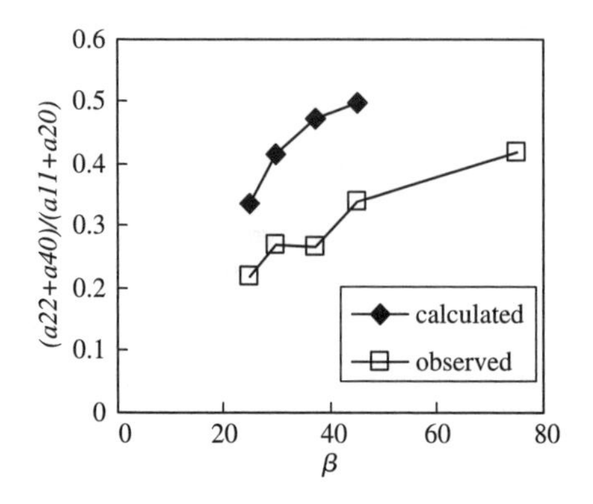

Figure 12. Change with β of major wave component of river-bed form at equilibrium.

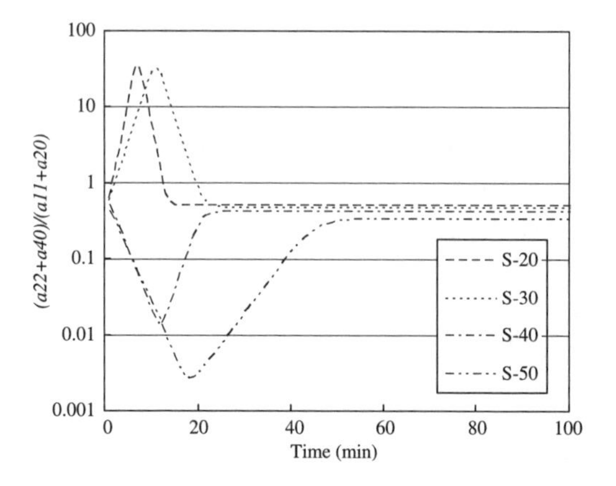

Figure 13. Temporal changes in contribution rate of mode to riverbed form.

Under the S-30 conditions, whose $\varepsilon_2^{1/2}$ is nearly 1, the analysis reproduces the timing of sandbar mode change and other experimental values, to a certain degree. Under the S-40 conditions, whose $\varepsilon_2^{1/2}$ is smaller than that of S-30, the occurrence of the mode-2 waves is reproduced later in the analysis results than in the experimental results and this is the cause for the disagreement between the experimental and analytical results. This can be seen in the comparison of initial growth rates (Figure 9): Mode-1 wave components are slightly larger than mode-2 wave components under the S-40 conditions. This shows that to reproduce sandbar mode change, the initial growth rate calculation must be highly accurate.

Because comparison of the predominant components of different wave modes enables sandbar mode classification of a riverbed (Watanabe and Kuwamura 2004), we examined the contribution of each mode of sand bar to riverbed form at equilibrium. Figure 12 compares analytical and experimental results for changes in predominant components of sand bars with β. The weakly nonlinear analysis gives slightly larger values for mode-2 wave components than the experimental values; however, that analysis does reproduce the trend of mode changes with β. To clarify the changes in sandbar mode in greater detail, temporal changes in the contribution ratio of wave components for sand bars of each mode, which are given by weakly nonlinear analysis, are shown in Figure 13. When sandbar mode changes, the contribution ratio of mode-2 wave components increases upon the start of water flow; then, the contribution ratio of mode-2 wave components gradually is reduced, and the mode of the sand bar is also reduced. When the contribution ratio of the mode-1 wave components is predominant at the beginning of water flow, the contribution ratio of the mode-1 wave components increases with time; then, mode-2 wave components starts to contribute to the sand bars, bringing the sandbar to equilibrium. In both cases, as shown in Figure 12, contribution ratios converge to between 0.1 and 0.6 at which ratios the contribution ratio of mode-1 wave components slightly predominates over that of mode-2 wave components. This means that under the hydraulic condition whose initial growth rate indicates that mode-2 wave components are expected to form, the wave components of mode 2 remain even after the sandbar mode has been reduced from mode 2 to mode 1. The greater such tendency, the larger the β.

5 CONCLUSION

In this study we conduct stability analysis for hydraulic experiments on sandbar mode reduction.

To examine the sandbar mode reduction, calculation of the initial growth rate of sand bar components in each mode must be highly accurate. We confirmed that the wavelength of sand bar immediately before and after mode reduction is not the wavelength that has the maximum growth rate under the given hydraulic condition; rather, it is the minimum wavelength , where perturbation of each mode is able to grow. Judging from the results, it is possible to apply weakly nonlinear analysis to double-row bars by means of growth rate expansion, and such analysis reproduces the experimental results until the perturbation parameter of mode-2 sand bars reaches about 1. When the riverbed is at equilibrium, the analytical contribution ratio of mode-2 wave components tends to be larger than the experimental results, but the trend of change with reproduces the trend of experimental values.

ACKNOWLEDGEMENTS

We would like to thank the Hokkaido Regional Development Bureau of the Ministry of Land, Infrastructure and Transport, for commissioning research to CERI; the Foundation of River & Watershed Environment Management, for a grant; and the Japan Society for the Promotion of Science, for research aid (Basic research (B) (1) 16360242). We would like to extend our appreciation to Prof. Norihiro Izumi for his invaluable assistance on the analysis.

REFERENCES

Colombini, M., G. Seminara and M. Tubino: Finite amplitude alternate bars. *Journal of Fluid Mechanics*. Vol.181, pp.213–232, 1987.

Fujita, Y., N. Nagata and Y. Muramoto: Formation process of braided streams, *Proc. Annual Journal of Hydraulic Engineering*, JSCE, Vol.36, pp.23–28, 1992. (in Japanese with English abstract)

Fujita, Y. Bar and channel formation in braided streams, *River meandering*, S. Ikeda and G. Parker, eds., Water Resource Monograph, Vol.12, AGU, pp.417–462, 1989.

Hasegawa, K. and I. Yamaoka: Experiments and analysis on the characteristic of developed alternate bars, *Proc. Annual Journal of Hydraulic Engineering*, JSCE, Vol.26, pp.31–38, 1982. (in Japanese with English abstract)

Izumi, N. and A. Pornprommin: Weakly nonlinear analysis of bars with the use of the amplitude expansion method, *Journal. Hyd., Coastal and Env. Eng.*, No.712/II-60, JSCE, pp.73–86, 2002. (in Japanese with English abstract)

Pornprommin, A., N. Izumi and T. Tsujimoto: Weakly nonlinear analysis of multimodal fluvial bars, *Annual Journal of Hydraulic Engineering*, JSCE, Vol.48, pp.1009–1014, 2004.

Takebayashi, H., S. Egashira and T. Okabe: Stream formation process between confining banks of straight wide channels, *2nd Proc. RCEM*, IAHR, pp.575–584, 2001.

Watanabe, Y. and T. Kuwamura: Experimental study on mode reduction process of double-row bars, *Annual Journal of Hydraulic Engineering*, JSCE, Vol.48, pp.997–1002, 2004. (in Japanese with English abstract)

River, Coastal and Estuarine Morphodynamics: RCEM 2005 – Parker & García (eds)

Effect of bar topography on hyporheic flow in gravel-bed rivers

A. Marzadri, G. Vignoli, A. Bellin & M. Tubino
Dipartimento di Ingegneria Civile ed Ambientale, Universitá di Trento

ABSTRACT: Solute mass exchange between river stream and the underneath hyphorheic zone has a profound impact on the river ecosystem. In this study we analyse, by means of a Lagrangian approach, the effect of the hyporheic flow on solute transport along a river reach. Hyporheic flow is generated by an uneven distribution of the water pressure at the riverbed, which is tied dynamically with the bed topography. We focus here on a gravel-bed river developing free bars, assuming for simplicity that water discharge is constant and that the typical time scale of subsurface flow is small with respect to the morphological time scale controlling bar development and migration. In such a situation the bed topography can be assumed as fixed in time. Furthermore, since in most practical situations hyporheic flow is a small fraction of the total streamflow, surface and subsurface flows can be uncoupled, with the latter reducing to a Darcyan flow controlled by the water head at the riverbed. First the flow equation is solved analytically in a three dimensional domain which is vertically unbounded and with the horizontal dimensions equal to the bar wavelength and the river width. Then the velocity field is obtained through the Darcy's equation. Solute transport in the hyporheic zone is then modeled numerically by particle tracking. We show that the residence time of the solute is Log-Normally distributed, with the mean depending on the same set of parameters describing the flow field and bed topography, namely the aspect ratio of the channel, the relative roughness and the parameter of Shields. This result is valid locally on a spatial scale of the order of the distance between two consecutive bars.

1 INTRODUCTION

The hyporheic zone is a saturated area surrounding the stream and providing the linkage between the river and the aquifer. It is a rich ecotone exposed to nutrients and contaminants dissolved in the stream water. Furthermore, hyporheic and riparian zones exert control over the export rate of nutrients and the attenuation of contaminant concentration in the stream (Alexander, Smith, and Schwarz 2000). As a consequence, modeling transport of nutrients and contaminants along the river calls for the inclusion of the effect of the hyporheic zone.

In gravel-bed rivers the pressure at the bed surface is non-uniform with zones of high and low pressure at the upstream and downstream ends of the bedform, respectively. This creates a complex flow system within the hyporheic zone which is connected to the steamflow through an alternate sequence of downwelling and upwelling zones reflecting the bed morphology.

Focusing a single wavelength of the bedform one should note that the mean residence time of water within the hyporheic zone is by orders of magnitude larger than the renewing time of the water between successive downwelling and upwelling zones. As a consequence solutes are temporarily stored within the hyporheic zone, thus creating long tailing in the breakthrough curve of solute concentration in streamwater (Haggerty, Wondzell, and Johnson 2002).

Traditionally transport of a non-reactive solute along the stream has been modeled by means of the following advection-dispersion equation, resulting from applying mass balance to the stream (Bencala and Walters 1983):

$$\frac{\partial C}{\partial x}+\frac{\partial uC}{\partial x}+\frac{\partial vC}{\partial y}+\frac{\partial wC}{\partial z}-\frac{\partial}{\partial x}\left(D_x\cdot\frac{\partial C}{\partial x}\right)$$

$$-\frac{\partial}{\partial y}\left(D_y\cdot\frac{\partial C}{\partial y}\right)-\frac{\partial}{\partial z}\left(D_z\cdot\frac{\partial C}{\partial z}\right)=J_S \tag{1}$$

where C is the solute concentration, $\vec{u}=(u,v,w)$ is the seepage velocity, $D_i^T, i=x,y,z$ are the three components of the macrodispersion tensor, and finally J_S is a term resulting from idealizing the mass exchange in terms of a mean exchange rate with a well-mixed hyporheic zone of constant volume. However, the hyporheic zone is far from being well mixed, as shown experimentally by (Tonina and Buffington 2005).

An alternative approach consists in modeling solute transport by particle tracking, a Lagrangian approach in which the solute mass is decomposed into a large number of non-interacting particles which are tracked from the entry point within the downwelling area to the exit point in the following upwelling area (for a comprehensive discussion of the Lagrangian approach in subsurface hydrology see e.g. (Dagan 1989; Rubin 2003)). In this context mass exchange between the stream and the hyporheic zone is characterized by the probability density function of the particle residence time within the hyporheic zone. In this work we analyze, by means of a simplified – yet realistic – model, the residence time pdf of the solute particle, and discuss its implication on solute transport along mountain streams with bed topography made of alternate bars.

2 GROUNDWATER FLOW MODEL

2.1 *Flow equation*

A three-dimensional stationary flow in a homogeneous aquifer with constant hydraulic conductivity K is used to illustrate flow-induced exchange in the hyphorheic zones of a stream with alternating bars. Similar analysis have been performed by (Elliott and Brooks 1997) for a river developing dunes. The stationarity hypothesis implies that the stream water discharge is constant or slowly varying over times of the order of the particle residence time. The governing equation and the associated boundary conditions are:

$$\frac{\partial^2 h}{\partial x^2} + \frac{\partial^2 h}{\partial y^2} + \frac{\partial^2 h}{\partial z^2} = 0 \tag{2}$$

subject to (see Figure 1)

$$\frac{\partial h}{\partial y} = 0 \quad y = \pm B \tag{3}$$

$$h(-L/2, y, z) = h(L/2, y, z) \tag{4}$$

where $h(x, y, z)$ is the hydraulic head, $2B$ is the channel width, and L is bar wavelength. The boundary condition at the bed surface $z = \eta(x, y)$, which coincides with the upper boundary of the porous media, is of given pressure $h(x, y, z = \eta(x, y))$. However, the complexity of the bed topography precludes obtaining closed form solutions of the equation (2). To overcome this difficulty and obtain closed form solutions for the flow field within the hyporheic zone we approximate the river bed as a flat surface at $z = 0$. The resulting computational domain is then bounded in the vertical direction at $z = 0$ and unbounded in the opposite direction ($z < 0$). The hydraulic head at the bottom of the stream, which is given by the weakly nonlinear theory of (Colombini et al. 1987), is then applied to the horizontal plane $z = 0$. Note that the plane $z = 0$ coincides

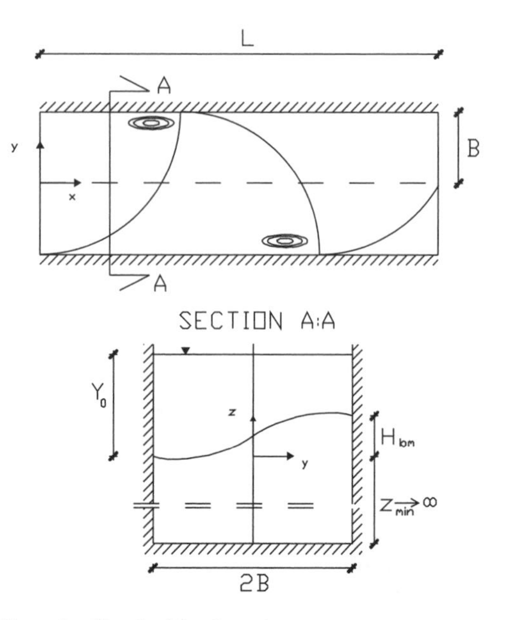

Figure 1. Sketch of the channel structure.

with the average of the bed elevation η in the reference system introduced by (Colombini, Seminara, and Tubino 1987) and adopted here for convenience. Thus, the computational domain differs slightly from the physical domain. However, the impact of this approximation on the travel distance is expected to be small to negligible such that it can be employed safely to obtain closed for solutions of the equation (2). In a manuscript in preparation we investigate the effects of this approximation on the flow field and on the residence time pdf of the solute particle within the hyporheic zone.

Once the distribution of h within the hyporheic zone as been obtained the velocity field can be computed through the Darcy's equation:

$$\vec{u} = -\frac{K}{\vartheta} \cdot \nabla h \tag{5}$$

where ϑ is the porosity.

2.2 *Dirichlet boundary conditions*

The formation of alternate bars is controlled by the aspect ratio $\beta = B/Y_0$, where Y_0 is the mean flow depth, and alternate bars are formed when β exceeds a threshold value β_c which value depends on two quantities: the dimensionless Shields stress

$$\theta = \frac{\tau_0}{(\rho_S - \rho)\, g D_S} \tag{6}$$

and the relative submergence

$$d_S = \frac{D_S}{Y_0} \tag{7}$$

where D_s is the mean grain size of the sediments, τ_0 is the shear stress at the bed, ρ and ρ_s are the density of clean water and sediments, respectively, and g is the gravitational acceleration. An analytical expression of the head at the streambed has been provided by (Colombini, Seminara, and Tubino 1987) as superimposition of Fourier harmonics in the following form:

$$h(x,y,0) = \sum_{n=0,2}^{N_y} \sum_{m=0,2}^{N_x} a_{nm} \cos(m\lambda x - \varphi^{nm})$$

$$\cdot \cos\left(\frac{n\pi y}{2}\right) + a_{11}\cos(\lambda x - \varphi^{11}) \sin\left(\frac{\pi y}{2B}\right) \tag{8}$$

where N_x and N_y are the number of harmonics considered in the expansion along the two planar coordinates, $a_{nm}(a_{11}, a_{22}, a_{02}, a_{20})$ are the dimensionless amplitude function of the bars in the transverse (n) and longitudinal modes (m), λ is the wave number of the bars, and $\varphi^{nm}(\varphi^{11}, \varphi^{22}, \varphi^{02}, \varphi^{20})$ are the dimensionless phase-difference of the perturbation. We retain in the expansion only the terms of order up to the second in both directions, since according to (Ikeda 1982) and (Colombini et al. 1987) higher-order terms exert a negligible impact on h.

2.3 *Solution of the flow field*

The solution of the flow equation (2) with the boundary conditions discussed above in the following:

$$h^*(x^*,y^*,z^*) = a_{11}^* \sin\left(\frac{\pi}{2}y^*\right)\cos(\lambda^* x^* - \varphi^{11})$$

$$\cdot \exp\left(\sqrt{(\lambda^*)^2 + \left(\frac{\pi}{2}\right)^2} z^*\right)$$

$$+ \sum_{n=0,2}^{N_y} \sum_{m=0,2}^{N_x} a_{nm}^* \cos\left(n\frac{\pi y^*}{2}\right)\cos(m\lambda^* x^* - \varphi^{nm})$$

$$\cdot \exp\left(\sqrt{(m\lambda^*)^2 + \left(\frac{n\pi}{2}\right)^2} z^*\right) \tag{9}$$

where the following dimensionless quantities have been introduced: $(x^*, y^*, z^*) = (x/B, y/B, z/B)$, $h^* = h/B$, $a_{nm}^* = a_{nm}/Y_0$, and $\lambda^* = \lambda B$.

Note that the solution is composed of the superimposition of four components with periodicity in both longitudinal and transverse directions, and that h decays exponentially with the depth.

The dimensionless velocity field can be computed by differentiating the equation (9) with respect to the dimensionless spatial coordinates:

$$\overrightarrow{u^*} = (u^*, v^*, w^*) = \left(-\frac{\partial h*}{\partial x*}, -\frac{\partial h*}{\partial y*}, -\frac{\partial h*}{\partial z*}\right) \tag{10}$$

$$u^*(x^*,y^*,z^*) =$$

$$= \lambda^* a_{11}^* \sin\left(\frac{\pi}{2}y^*\right)\sin(\lambda^* x^* - \varphi^{11})$$

$$\cdot \exp\left(\sqrt{(\lambda^*)^2 + \left(\frac{\pi}{2}\right)^2} z^*\right)$$

$$+ \sum_{n=0,2}^{N_x=2} \sum_{m=0,2}^{N_y=2} m\lambda^* a_{nm}^* \cos\left(n\frac{\pi y^*}{2}\right)\sin(m\lambda^* x^*$$

$$- \varphi^{nm}) \exp\left(\sqrt{(m\lambda^*)^2 + \left(\frac{n\pi}{2}\right)^2} z^*\right) \tag{11}$$

$$v^*(x^*,y^*,z^*) =$$

$$= -\frac{\pi}{2} a_{11}^* \cos\left(\frac{\pi}{2}y^*\right)\cos(\lambda^* x^* - \varphi^{11})$$

$$\cdot \exp\left(\sqrt{(\lambda^*)^2 + \left(\frac{\pi}{2}\right)^2} z^*\right)$$

$$+ \sum_{n=0,2}^{N_x=2} \sum_{m=0,2}^{N_y=2} n\frac{\pi}{2} a_{nm}^* \sin\left(n\frac{\pi y^*}{2}\right)\cos(m\lambda^* x^*$$

$$- \varphi^{nm}) \exp\left(\sqrt{(m\lambda^*)^2 + \left(\frac{n\pi}{2}\right)^2} z^*\right) \tag{12}$$

$$w^*(x^*,y^*,z^*) =$$

$$= -\sqrt{(\lambda^*)^2 + \left(\frac{\pi}{2}\right)^2}\, a_{11}^* \sin\left(\frac{\pi}{2}y^*\right)\cos(\lambda^* x^* - \varphi^{11})$$

$$\cdot \exp\left(\sqrt{(\lambda^*)^2 + \left(\frac{\pi}{2}\right)^2} z^*\right)$$

$$- \sum_{n=0,2}^{N_x=2} \sum_{m=0,2}^{N_y=2} \sqrt{(m\lambda^*)^2 + \left(\frac{n\pi}{2}\right)^2}\, a_{nm}^*$$

$$\cdot \cos\left(n\frac{\pi y^*}{2}\right)\cos(m\lambda^* x^* - \varphi^{nm})$$

$$\cdot \exp\left(\sqrt{(m\lambda^*)^2 + \left(\frac{n\pi}{2}\right)^2} z^*\right) \tag{13}$$

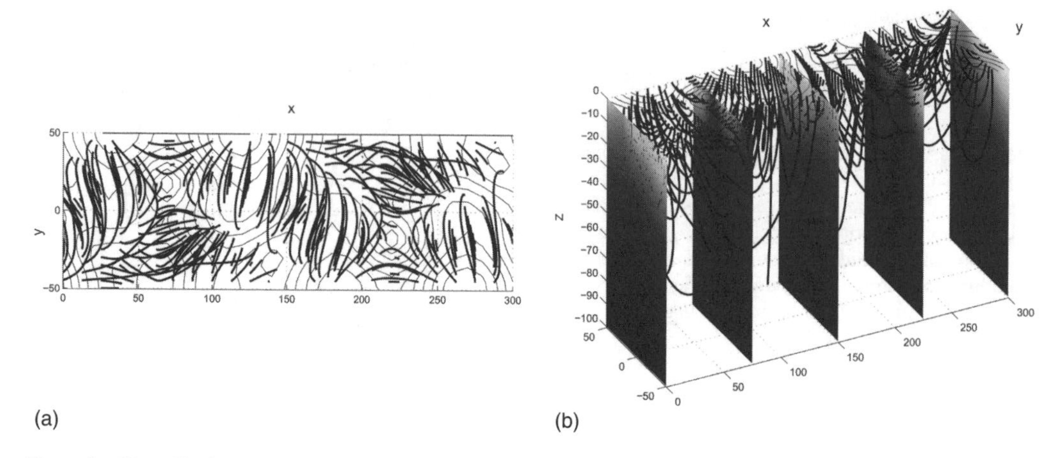

Figure 2. Normalized streamlines distribution over the bed-river surface (a) and in the hyporheic zone (b) for an indefinite sediment layer.

where

$$(u^*, v^*, w^*) = \frac{1}{u_m}(u, v, w) = \frac{B\,\vartheta}{K \cdot Y_0}(u, v, w) \quad (14)$$

is the dimensionless velocity field.

Figure 2a shows the planar view the streamlines resulting from the velocity field described by the equations (11), (12) and (13), while a three dimensional view is shown in Figure 2b. The resulting flow field is symmetric because of the symmetry of the forcing term and with a complex structure dictated by the pressure distribution at the bed surface. The magnitude of the velocity decays exponentially with depth as for the hydraulic head (see Figure 2b).

3 RESIDENCE TIME DISTRIBUTION

We consider the transport of a nonreactive solute within the hyporheic zone by adopting a Lagrangian approach. The mass flux of solute entering the hyporheic zone through the downwelling area is decomposed in a large number of particles which move according to the velocity field. We take advantage here of a result from groundwater hydrology: the mass flux of a nonreactive solute at the exit surface of a control volume, resulting from a pulse injection of an unitary mass of solute within the entry surface, is given by the probability density function (pdf) of the residence (travel) time of the particles within the volume (Jury and White 1986; Dagan, Cvetkovic, and Shapiro 1992). In our case the control volume is the hyporheic zone, the exit surface is the upwelling area, while the solute enters the control volume through the downwelling area. In other words, the concentration breakthrough curve at the upwelling area can be written as a function of the residence time pdf of the solute particles which entered the control volume through the downwelling area as follows:

$$C_F(\tau) = \int_0^{\tau} C_0(\tau')\, f(\tau - \tau')\, d\tau' \quad (15)$$

where f is the pdf of the residence time τ and C_0 is the concentration flux entering through the downwelling area. Under the hypothesis that the solute is well mixed in the stream $C_0 = C = const$, such that equation (15) simplifies to:

$$\frac{C_F(\tau)}{C} = \int_0^{\tau} f(\tau')\, d\tau' = F(\tau) \quad (16)$$

which is the Cumulative Distribution Function (CDF) of the particle residence time τ. Thus the residence time pdf embeds the dynamics controlling transport of nonreactive solutes in the hyporheic zone. Note that F in equation (16) represents the fraction of the particles released at time $t = 0$ with the pulse injection that at time t are within the control volume. In order to delimitate downwelling and upwelling areas the horizontal plane at $z = 0$ is subdivided in a fine grid with spacing $\Delta x = L/400$, $\Delta y = 2B/200$ and the vertical component of the velocity field is computed at each node of the grid (See Figure 3). According to the orientation of the spatial coordinates w is negative in the downwelling area and positive in the upwelling area. As a consequence, the water flux q into the bed assumes the following expression:

$$\begin{cases} q(x, y, z) = -w & \text{if} - w \geq 0 \\ q(x, y, z) = 0 & \text{if} - w < 0 \end{cases} \quad (17)$$

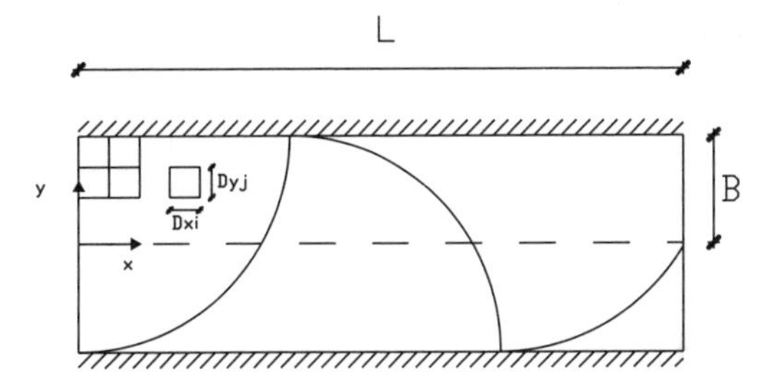

Figure 3. Planar view of the computational domain.

Hence the dimensionless water flux is defined as follows:

$$q^*(x^*, y^*, 0) = \frac{q(x, y, 0)}{u_m} = -w^*(x^*, y^*, 0) \qquad (18)$$

where for the convenience we assumed positive the water flux entering into the hyporheic zone.

3.1 *Solute transport within the hyporheic zone*

We approximate the residence time CDF as follows:

$$F(t) \simeq \frac{\int_A q^*(x_0^*, y_0^*, 0) R(x_0^*, y_0^*, t)\, dx^*\, dy^*}{\int_A q^*(x_0^*, y_0^*, 0)\, dx^*\, dy^*} \qquad (19)$$

where A is the area of the downwelling surface. In equation (19) R is a binary variable which is equal to 1 if the particle released at the location (x_0^*, y_0^*) is within the control volume at the time t since solute injection, and 0, if the particle exited the control volume through the upwelling surface at an early time. The expression (19) converges to the CDF of the residence time as the number of particles grows large and the mass carried by a single particle tends to zero. We utilized for the simulations a computational grid small enough to stabilize the CDF computed by using the equation (19). The resulting number of particles released in the downwelling area ranged from $NP = 16840$ to $NP = 0$ depending on the values assumed by the channel parameters.

Each particle was tracked from the upstream downwelling area to the following upwelling area by solving numerically the following set of integrodifferential equations:

$$X(t; \mathbf{a}) = \int_0^t u\,[X(\tau; \mathbf{a}), Y(\tau; \mathbf{a}), Z(\tau; \mathbf{a})]\, d\tau,$$

$$Y(t; \mathbf{a}) = \int_0^t v(X(\tau; \mathbf{a}), Y(\tau; \mathbf{a}), Z(\tau; \mathbf{a}))\, d\tau, \qquad (20)$$

$$Z(t; \mathbf{a}) = \int_0^t w(X(\tau; \mathbf{a}), Y(\tau; \mathbf{a}), Z(\tau; \mathbf{a}))\, d\tau,$$

where $(X(t;\mathbf{a}), Y(t;\mathbf{a}, Z(t;\mathbf{a}))$ is the trajectory of the particle released at time $t = 0$ at the location $\mathbf{a}$ within the downwelling area. We solved numerically the system of equations (20) by using the following explicit scheme:

$$X(t; \mathbf{a}) = X(t - \Delta t; \mathbf{a}) +$$

$$u\,[X(t - \Delta t; \mathbf{a}), Y(t - \Delta t; \mathbf{a}), Z(t - \Delta t; \mathbf{a})]\ \Delta t,$$

$$Y(t; \mathbf{a}) = Y(t - \Delta t; \mathbf{a}) +$$

$$v\,[X(t - \Delta t; \mathbf{a}), Y(t - \Delta t; \mathbf{a}), Z(t - \Delta t; \mathbf{a})]\ \Delta t,$$

$$Z(t; \mathbf{a}) = Z(t - \Delta t; \mathbf{a}) +$$

$$w\,[X(t - \Delta t; \mathbf{a}), Y(t - \Delta t; \mathbf{a}), Z(t - \Delta t; \mathbf{a})]\ \Delta t$$

(21)

Thus, the trajectories of the particles are obtained by applying recursively the set of equations (21) with a variable time step, as required by the highly variable curvature of the streamlines (see Figure 2). In fact, short streamlines with high curvature require a small time step in order to reproduce accurately the particle trajectory, on the other hand particles following long streamlines with small curvature can be tracked by using larger time steps, without compromising the accuracy, such that to reduce the computational effort. The computations are performed in dimensionless form with the dimensionless time given by: $t^* = tK/(B\beta\vartheta)$. After a number of preliminary simulations we ended up with the following rule of thumb for selecting dynamically the time step during the particle tracking:

$$\Delta t^* = \frac{0.07}{D} \left(\frac{0.01}{|\kappa(s)|^2 + 10^{-6}} \right) \qquad (22)$$

where $k(s)$ is the curvature of the streamline. All the computations are performed in dimensionless form with the dimensionless time step.

4 RESIDENCE TIME PDF

Figure 4 shows the experimental Cumulate Frequency Distribution (CFD) of the residence time τ obtained by releasing $NP = 16480$ particles uniformly distributed over the downwelling area. Visual inspection of the figure suggests the Log-Normal distribution as a possible probability model for τ, which pdf is given by:

$$f(\tau) = \frac{1}{\sqrt{2\pi\sigma_z^2}\,\tau} \cdot \exp\left(-\frac{(ln\tau - \mu_z)^2}{2\sigma_z^2} \right) \qquad (23)$$

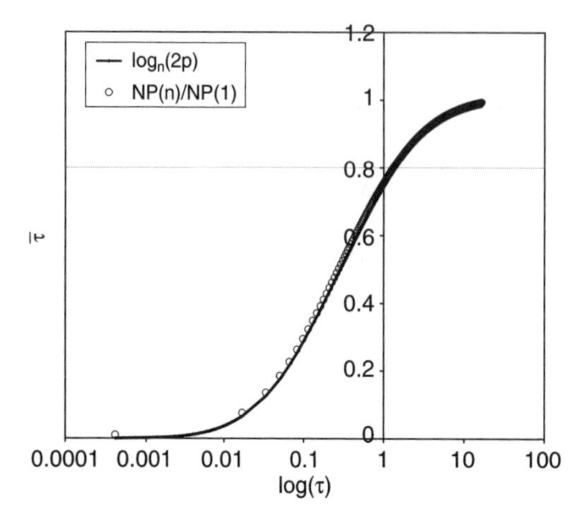

Figure 4. Cumulative probability function of the particle residence time.

where μ_z and σ_z^2 are the mean and the variance of the RV $z = ln\tau$, respectively. In addition, Figure 4 shows the CDF of τ obtained by matching μ_z and σ_z^2 with the sample mean and variance, respectively:

$$\mu_z = ln\left[\frac{\overline{\tau}}{\sqrt{1+\sigma_\tau^2/\overline{\tau}^2}}\right] \tag{24}$$

$$\sigma_z^2 = ln\left[1+\frac{\sigma_\tau^2}{\overline{\tau}^2}\right] \tag{25}$$

where $\bar{\tau}$ and σ_τ^2 are the sample mean and variance of the residence time:

$$\overline{\tau} = \frac{1}{NP}\sum_{j=1}^{NP}\tau_j,\quad \sigma_\tau^2 = \frac{1}{NP-1}\sum_{j=1}^{NP}(\tau_j-\overline{\tau})^2 \tag{26}$$

In equation (26) τ_j is the residence time of the particle number j in a collection of NP particles. The resulting pdf satisfied both the chi-square and the Kolmogorov–Smirnov goodness-of-fit tests with a 5% level of significance (Conover 1999).

5 DISCUSSION AND CONCLUSIONS

We analyze now how the parameters characterizing the bar dynamics influence the residence time moments. Figure 5 shows $\bar{\tau}$ as a function of the relative submergence for several values of θ and $\beta = 15$. Simulations are performed by using the following three values of d_s: 0.01, 0.05 and 0.10 which are typical of mountain streams. We observe that the influence of θ on $\bar{\tau}$ is more pronounced for small relative submergence

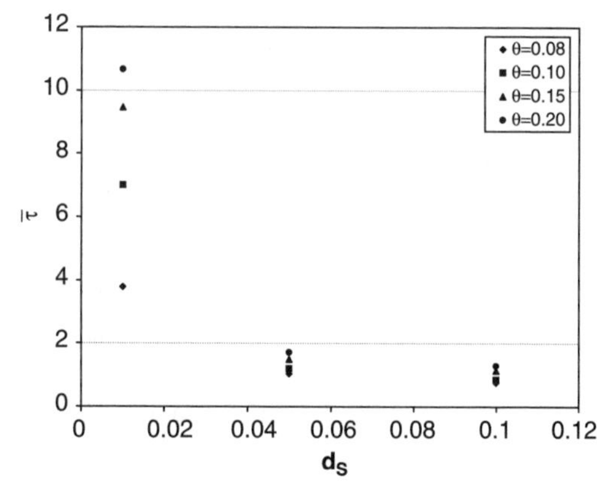

Figure 5. Mean residence time $\bar{\tau}$ versus the relative submergence for several values of θ and $\beta = 15$.

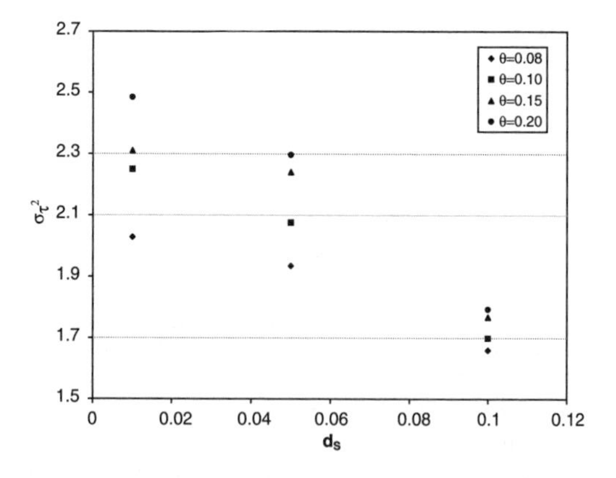

Figure 6. Variance of the particle residence time σ^2 versus the relative submergence for several values of θ and $\beta = 15$.

where $\bar{\tau}$ increases significantly with θ. On the other hand, for large values of d_s the mean residence time is almost insensitive to the value assumed by θ. Similar conclusions can be drawn for σ_τ^2, shown in Figure 6, even though the effect of θ persists over larger values of d_s with respect to $\bar{\tau}$.

Figures 7a and 7b shows $\bar{\tau}$ and σ_τ^2 as a function of θ for several values of β and $d_s = 0.01$. We observe that $\bar{\tau}$ reduces as β increases such that for a constant water depth wider channels result in smaller mean residence times in the hyporheic zone. However, the influence of β on $\bar{\tau}$ is much more pronounced when the shields stress is large. A similar behavior is observed for σ_τ^2 as shown in Figure 7b. Increasing d_s reduces the influence of β on both $\bar{\tau}$ and σ_τ^2, as shown in Figures 8a, 8b and 9a, 9b for $d_s = 0.05$ and 0.10, respectively. From this preliminary investigation we conclude that the residence time of solutes within the hyporheic zone depends in a complex manner form the parameters controlling bed

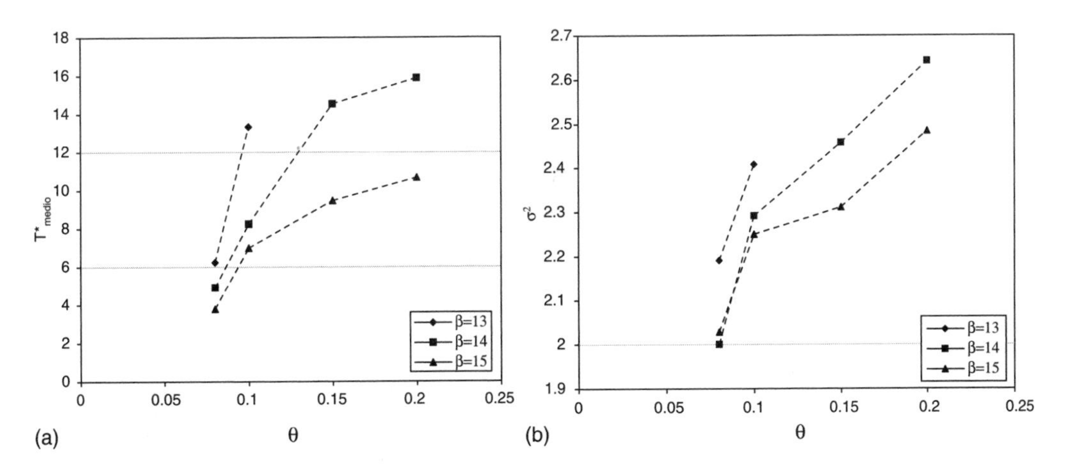

Figure 7. Mean (a) and variance (b) of the residence time versus the Shields stress ($d_s = 0.01$).

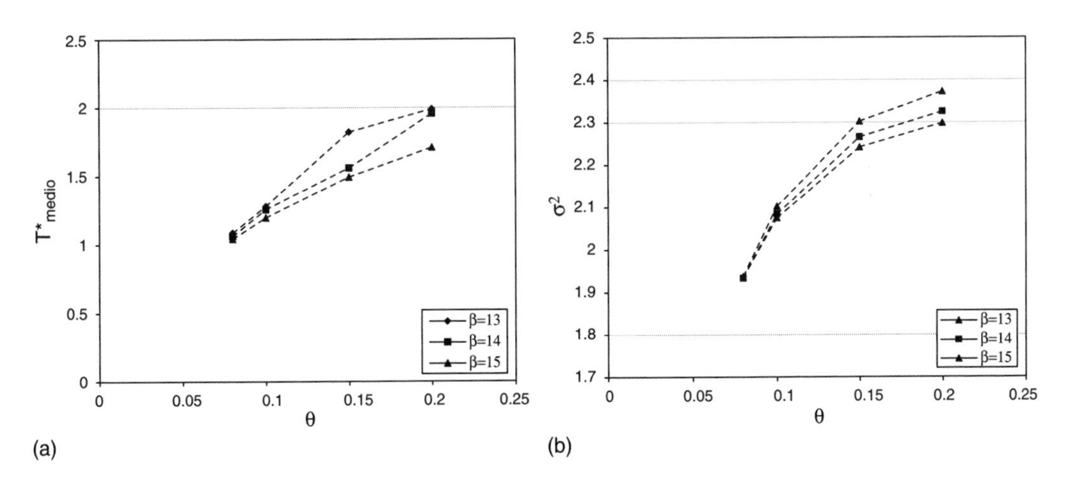

Figure 8. Mean (a) and variance (b) of the residence time versus the Shields stress ($d_s = 0.05$).

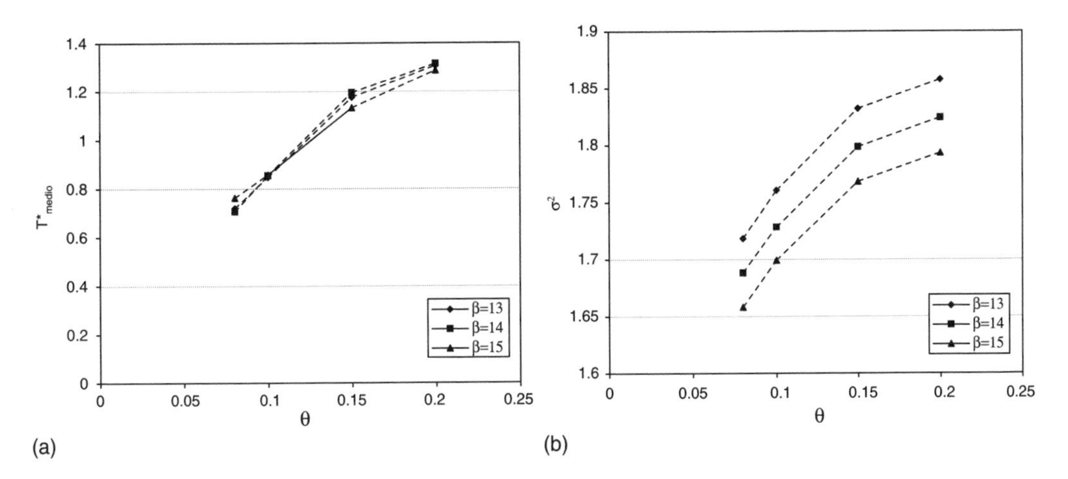

Figure 9. Mean (a) and variance (b) of the residence time versus the Shields stress ($d_s = 0.10$).

morphology. In particular $\bar{\tau}$, depends strongly on θ while the effect of β is significant for small values of d_s and become small to negligible at larger d_s values.

REFERENCES

Alexander, R. B., R. A. Smith, and G. E. Schwarz (2000). Effect of stream channel on the delivery of nitrogen to the gulf of mexico. *Nature 403*, 758–761.

Bencala, K. E. and R. A. Walters (1983). Simulation of solute transport in a mountain pool-and-riffle stream: a transient storage model. *Water Resour. Res. 19*(3), 718–724.

Colombini, M., G. Seminara, and M. Tubino (1987). Finite-amplitude alternate bars. *Journal of Fluid Mechanics 181*, 213–232.

Conover, W. J. (1999). *Practical nonparametric statistics, Third edition*. John Wiley and Sons, New York.

Dagan, G. (1989). *Flow and Transport in Porous Formations*. Springer–Verlag, New York.

Dagan, G., V. Cvetkovic, and A. M. Shapiro (1992). A solute flux approach to transport in heterogeneous formations: 1. The general framework. *Water Resour. Res. 28*(5), 1369–1376.

Elliott, A. H. and N. H. Brooks (1997). Transfer of nonsorbing solutes to a streambed with bedforms: theory. *Water Resour. Res. 33*, 123–136.

Haggerty, R., S. M. Wondzell, and M. A. Johnson (2002). Power-law residence time distribution in the hyporheic zone of a 2nd-order mountain stream. *Geophysical Research Letters 29*(13), 1640, 10.1029/2002GL014743.

Ikeda, S. (1982). Prediction of alternate bar wavelength and height. *Rep. Dept. Found. Engng. and Const. Engng, Saitama Univ. 12*, 23–45.

Jury, W. A. N. G. S. and R. E. White (1986). A transfer function model of solute transport through soil: 1. Fundamental concepts. *Water Resour. Res. 22*(2), 243–247.

Rubin, Y. (2003). *Applied Stochastic Hydrogeology*. Oxford University Press, New York.

Tonina, D. and J. M. Buffington (2005). Hyporheic exchange in gravel-bed rivers with pool-riffle morphology: A three-dimensional model. *Water Resour. Res. 41*(–), –. submitted.

River, Coastal and Estuarine Morphodynamics: RCEM 2005 – Parker & García (eds)
© 2006 Taylor & Francis Group, London, ISBN 0 415 39270 5

Numerical simulation of alternating bars in straight channels

R.M.J. Schielen
Ministry of Transport, Public Works and Water Management, Institute of Inland Water Management and Waste Water Treatment, Arnhem, The Netherlands

T. van Leeuwen & P.A. Zegeling
Mathematical Institute, Utrecht University, Utrecht, The Netherlands

ABSTRACT: The formation of alternating bar patterns as they exist in natural rivers and channels is by now quit well understood. It is the result of an instability mechanism between the flow and the sediment, and analytically, predictions can be made about the critical width-to-depth ratio, the critical wavelength, and even a modulation behaviour, both in time and space. A disadvantage of the analytical method is that it is, due to the nonlinearities in the model, the computations become immense, and therefore the analysis is often restricted to straight or mildly curved channels. Therefore, we apply a numerical approach on the most simple model that still predict alternating bars. The equations of flow and bed-evolutions decouple, and we use the most simple numerical methods available: Lax-Friedrichs to solve the flow field, and forward in time, central in space for the bed-evolution. By applying a stability analysis on the numerical scheme, we are able to choose a suitable combination of temporal and spatial steps such that from any initial perturbation of the bed, the alternating bar pattern arises. Also the modulated behaviour is reproduced, provided that the computational domain is long enough, just as the analytical theory predicts. It turns out that high-frequency 1D patterns tend to grow as well, although the theory predicts that they are stable. It is shown that this is a numerical artefact which can be suppressed by the right combination of spatial and temporal step size. These results are the starting point for a series of numerical experiments in which we want to study the behaviour of bed forms in the neighbourhood of bifurcation points. As a simple first step, a numerical bifurcation point is examined, which is just a set of grid points which behave as a impermeable wall. In a next step, an actual bifurcation in two branches is studied, where the width of the branches relative to the width of the main channel, and the angle of bifurcation are important parameters. Results of these experiments are shown and discussed.

1 INTRODUCTION

There is a large number of papers explaining the existence and evolution of alternating bars in straight channels and rivers. It is well known that these bars arise as a result of an instability mechanism in which there is an interaction between the flow and the erodible bed. In Schielen et al. 1993, it is shown that there is a critical width-to-depth ratio R_c below which a straight bed and uniform flow is stable with respect to arbitrary, small perturbations in the bed and the flow field. For $R > R_c$ the bed is unstable and arbitrary perturbations will grow up to a certain amplitude forming an alternating bar pattern. If z describes the bed, then we have the following formula for the evolution:

$$z = A(\xi, \tau)e^{ik_c x + w_c t} \cos(\pi y) \tag{1}$$

where k_c is a critical wavenumber, ω_c the critical frequency and A satisfies the so called Ginzburg-Landau equation. The analytical analysis for a straight channel, uniform flow and a uniform bed is straightforward but tedious.

From the analytical point of view, one can only state that arbitrary perturbations will evolve on the morphological timescale to the pattern of alternating bars. How this proces really takes place is something that the analytical approach does not tell. In order to study this process, a numerical approach is necessary. To get insight in the numerical proces is the main subject of this paper.

As is indicated in Schielen et al. 1993, simple models which describe the flow and bed-interactions in a straight channel, and which only contain the essential elements that are responsible for the main features and patterns have the advantage that analytical approaches are still possible. In a numerical analysis, there is actually not really the need to keep the models as simple as possible. However, very often it is seen that the models describing the physical processes are made unnecessary complex, and they are subsequently analysed by most complex numerical schemes. We strongly believe that without proper insight in the numerical method, it happens that results are misinterpreted for it may not be clear at forehand whether the results are due to the properties of complex numerical schemes or really physical artefacts.

The inspiration of this work comes from two sources. One is the analytical work in Schielen et al. 1993, which is just available as check for any numerical analysis of alternating bars. The other one is (Federici 2002), who actually performs a thorough (numerical) study of alternating bars. The numerical analysis however, was based on a staggered grid approach which originated from a numerical model developed by Delft Hydraulics in the Netherlands. In studying this, we experienced that it is difficult to really get to the bottom of the model, and in some basic experiments with general available numerical packages (like for instance Delft3D, a product of Delft Hydraulics), we didn't succeed in creating alternating bars from arbitrary conditions. Moreover, if bars where forced, they eventually decayed due to (possible numerical) diffusion. It was exactly this phenomenon that stimulated us to build a numerical model from scratch, to integrate the equations in such a way that any diffusion (if present) could be traced back to the numerical scheme or the original equations. At forehand, the idea was that without artificial numerical diffusion finite amplitude alternating bars should develop. Hence, the goal of this paper is to develop a simple numerical method to solve the model proposed in Schielen et al. 1993. By keeping the numerical method as simple as possible we hope to be able to gain some more insight in the validity of the solution given by the numerical method. Only through analysis of the equations and the numerical method it is possible to discern these numerical artefacts from real solutions. It is not the goal of this paper to propose an optimal method to investigate all aspects of these alternating bars. It is a first step, in which we hope to be able to point out possible improvements.

In the next section, we will describe the physical model that we use to describe the evolution of alternating bars, and the simplifications that go along with the analysis. In section 3, we will very briefly summarize the analytical results of Schielen et al. 1993. In section 4, we discuss the numerical method and the (numerical) stability properties of the method. In section 5, some typical results are discussed and a comparison with the analytical results is made. Finally, in section 6, conclusions and some remarks for further study are made.

2 THE MODEL

As this paper focusses on numerical aspects, the derivation of the model is not described in all aspects. We refer to Schielen et al. 1993 for more details. The

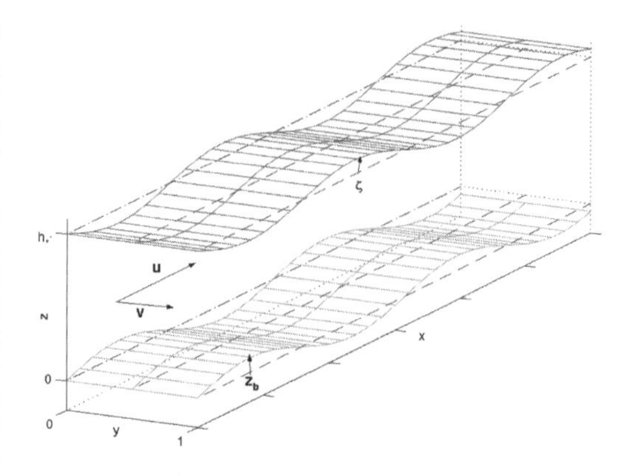

Figure 1. Systematic representation of the situation. For brevity we use $h = \zeta + h_* - z_b$ in the rest of the chapter.

channel geometry is given in figure 1. The width-to-depth ratio is such, that there is little variation in the vertical z-direction, which justifies averaging over z. We will denote the depth-averaged velocities in the x and y direction by u and v respectively, ζ denotes the water surface above the reference level h_* and z_b the bottom elevation. For brevity we also use $h = \zeta + h_* - z_b$ to denote the waterdepth. The liquid is assumed to have unit mass. The width of the channel is scaled to 1 throughout the paper.

An analysis of conservation of mass and momentum leads to three coupled equations for u, v and ζ. These are generally known as the St. Venant or shallow water equations. Conservation of sediment leads to the fourth equation, which depends on the many possible ways of modelling the sediment transport. In essence, it reads

$$\frac{\partial z_b}{\partial t} + \frac{\partial q_x}{\partial x} + \frac{\partial q_y}{\partial y} \tag{2}$$

with

$$\mathbf{Q} = \begin{pmatrix} q_x \\ q_y \end{pmatrix} = \sigma |\mathbf{U}|^b \left(\frac{\mathbf{U}}{|\mathbf{U}|} - \begin{pmatrix} \frac{\partial z_b}{\partial x} \\ \frac{\partial z_b}{\partial y} \end{pmatrix} \right). \tag{3}$$

where $\mathbf{U} = \begin{pmatrix} u \\ v \end{pmatrix}$, σ and b are sediment parameters. This is one of the many possible flux formulae, for more details we refer to (Rijn 1993). Here the flux is proportional to the flow velocity. A term to incorporate effects from the bed profile is also added; a steep slope in the downstream direction will reduce out flux of sediment.

2.1 *Scaling and boundary conditions*

The variables are made dimensionless and are scaled such that the phenomena that we want to study (i.e.

the alternating bars) will be solutions of the scaled model. Hence, we scale z_b with the unperturbed water depth and introduce a morphological timescale. After some straightforward elaborations, we end up with the model that we use as a starting point for the numerical analysis:

$$\kappa\frac{\partial u}{\partial t}+u\frac{\partial u}{\partial x}+v\frac{\partial u}{\partial y}+\frac{\partial \zeta}{\partial x} = -CR(\frac{u|\mathbf{U}|}{h}-1),$$

$$\kappa\frac{\partial v}{\partial t}+u\frac{\partial v}{\partial x}+v\frac{\partial v}{\partial y}+\frac{\partial \zeta}{\partial y} = -CR\frac{v|\mathbf{U}|}{h},$$

$$\kappa\frac{\partial h}{\partial t}+\frac{\partial uh}{\partial x}+\frac{\partial vh}{\partial y} = 0,$$

$$\frac{\partial z_b}{\partial t}+\frac{\partial q_x}{\partial x}+\frac{\partial q_y}{\partial y} = 0. \tag{4}$$

where:

$$h = F^2\zeta + 1 - z_b, \tag{5}$$

$$\mathbf{Q} = |\mathbf{U}|^b(\frac{\mathbf{U}}{|\mathbf{U}|}-\frac{1}{R}\begin{pmatrix}\frac{\partial z_b}{\partial x}\\ \frac{\partial z_b}{\partial y}\end{pmatrix}) \tag{6}$$

and C is the drag coefficient, $F=u_*/\sqrt{gh_*}$, the Froude number, $k=\sigma u_*^b/h_*$, the ratio of the timescale of flowadaption and the morphological timescale (which is small) and $R=1/h_*$ the width-to-depth ratio. We will refer to (4) as 'the model' in the rest of this paper. For convenience we will use the following notation:

$$\kappa\frac{\partial \mathbf{w}}{\partial t}+\mathbf{L}_w\mathbf{w} = \mathbf{F}_w, \tag{7}$$

$$\frac{\partial z_b}{\partial t}+L_z z_b = F_z. \tag{8}$$

where:

$$\mathbf{w} = (u,v,\zeta)^T, \quad \mathbf{L}_w = \begin{pmatrix} L_{uu} & 0 & L_{u\zeta}\\ 0 & L_{vv} & L_{v\zeta}\\ L_{\zeta u} & L_{\zeta v} & L_{\zeta\zeta}\end{pmatrix}, \tag{9}$$

$$\mathbf{F}_w = (CR,0,0)^T \tag{10}$$

with obvious expressions for L_{uu}, $L_{u\zeta}$ etcetera and

$$L_z = \frac{1}{R}(a_1(\frac{\partial^2}{\partial x^2}+\frac{\partial^2}{\partial y^2})+a_2\frac{\partial}{\partial x}+a_3\frac{\partial}{\partial y}), \tag{11}$$

$$a_1 = -|\mathbf{U}|^b, a_2 = -\frac{\partial|\mathbf{U}|^b}{\partial x}, a_3 = -\frac{\partial|\mathbf{U}|^b}{\partial y}, \tag{12}$$

$$F_z = -\frac{\partial u|\mathbf{U}|^{b-1}}{\partial x}-\frac{\partial v|\mathbf{U}|^{b-1}}{\partial y}. \tag{13}$$

The boundary conditions are found by realising that the walls of the channel should be impermeable for water and sediment. Hence:

$$\left.\begin{matrix} v=0\\ \frac{\partial z_b}{\partial y}=0\end{matrix}\right\}\text{if}\quad y=0,\quad y=1. \tag{14}$$

where $y=0$ and $y=1$ are the boundaries of the rescaled width of the channel. Note that the system of four equations decouples for $\kappa\to 0$. This is due to the fact that the morphological timescale is much longer than the characteristic timescale of the flow. We will use this fact in both the analytical approach as well as in the numerical analysis. It means that the flow field adapts relatively fast to a fixed bed, after which a timestep in the numerical integration of the bed equation is done, with fixed values of the flow and waterlevel. Then the flow adapts again to the changed bed-configuration, etcetera.

3 ANALYTICAL RESULTS

Let us briefly summarise the analytical results. We perform a stability analysis around the basic solution of a uniform flow and a flat bed ($u=1, v=0, \zeta=0, z=0$). Hence, we substitute:

$$\phi = \phi_0+\epsilon\phi', \phi = (u,v,\zeta,z) \tag{15}$$

and linearise. Then we substitute

$$\phi' = f(y)e^{ikx+\omega t} \tag{16}$$

with $f(y)=(f_u(y), f_v(y), f_\zeta(y), f_{z_b}(y))$ and with k real and ω complex. Reducing the four equations to one equation for z leaves us with an eigenvalue problem, which is readily solved. Eventually, we end up with a solution for z_{b0}:

$$z_{b0} = A\cos(p\pi y)e^{ikx+\omega t}, \quad p=0,1,2,\ldots. \tag{17}$$

The eigenvalue problem also leads to a dispersion relation between ω and k. The $p=1$ configuration resembles a alternating bar pattern, higher values of p lead to central bars and more complex configurations. The $p=0$ mode is of no physical value and is always stable (i.e. decaying in time). Plotting this relation gives the so called neutral stability curve, which divides the k, R-plane in a part in which the uniform flow and flat bed is linearly stable, and in a part in which the basic pattern is unstable (see figure 2). Weakly nonlinear theory then describes the evolution of the unstable pattern slightly above critical conditions. 'Slightly' is in this perspective a relative term. It

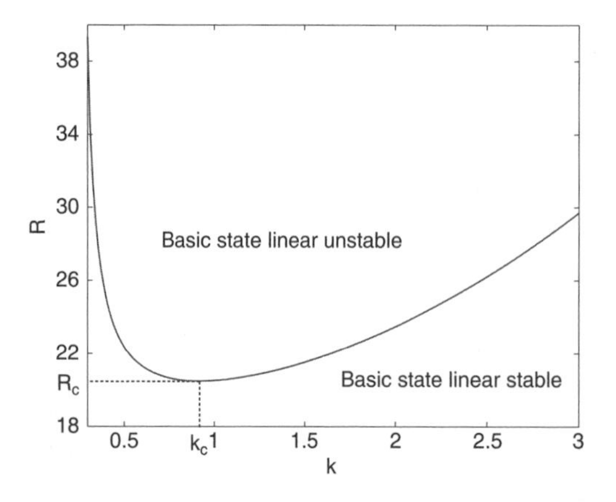

Figure 2. The neutral curve for $p=1$ ($b=5$, $C=0.007$) separates the linearly stable ($Re(\omega)<0$) and unstable ($Re(\omega)>0$) modes. The first mode to become unstable is $k_c \approx 0.9\ldots$ at $R=R_c \approx 20.5$..

is argued in Schielen et al. 1993 that with $R_c \approx 20.5$, weakly nonlinear theory can still be applied for values of R up to 40 or 50. This is important, because it means that in a relative large part of the k, R-space, the numerical results can be compared with the weakly nonlinear theory. The essence of the weakly nonlinear theory states that the temporal and spatial evolution of the linearly unstable pattern is described by a amplitude-function $A(\xi, \tau)$, with ξ and τ slow spatial and temporal coordinates (related to the distance $r=R-R_c$). The amplitude A satisfies the Ginzburg-Landau equation:

$$\frac{\partial A}{\partial \tau} = rA + \alpha \frac{\partial^2 A}{\partial \xi^2} + \beta |A|^2 A \tag{18}$$

The coefficients α and β are complex and can be expressed in terms of the model-coefficients.

4 DISCRETISATION AND NUMERICAL STABILITY

In this section we will perform a stability analysis similar to that of section 3 on the numerical method. This way we get some insight in the behaviour of the numerical method and we can see under what circumstances the numerical method is consistent with the analytical model.

As already explained, the flow and bed-equation (7)–(8) decouple, and the flow field can be solved separately from the bed equation. Thus, we first solve the flow field, and then we let the bed evolve. Numerically the equations are solved by approximating the functions $\mathbf{w}$ and z on a rectangular grid. This grid has $N=N_x \times N_y$ gridpoints, where N_x and N_y are the number of gridpoints in the x and y direction respectively. We call the distance between the gridpoints Δx and Δy. We denote functionvalues on the grid as: $f(i\Delta x, j\Delta y, t) \equiv f_{i,j}(t)$. For every gridpoint an equation can be derived by replacing the spatial derivatives with finite central differences:

$$\frac{\partial f}{\partial x}(i\Delta x, j\Delta y, t) \equiv \partial_x f_{i,j}(t) = \frac{f_{i+1,j}(t) - f_{i-1,j}(t)}{2\Delta x} \tag{19}$$

$$\frac{\partial^2 f}{\partial x^2}(i\Delta x, j\Delta y, t) \equiv \partial_{xx} f_{i,j}(t) = \frac{f_{i+1,j}(t) - -2f_{i,j}(t) + f_{i-1,j}(t)}{\Delta x^2} \tag{20}$$

Thus we get a system of N equations for every variable (u, v, ζ, z). The equations are integrated over time step by step by replacing the temporal derivative by a finite difference. We call the stepsize in time Δt and the functionvalues on the grid as $f(i\Delta x, j\Delta y, n\Delta t) \equiv f^n_{i,j}$.

4.1 *The flow*

We solve the flow equation (7) on the characteristic timescale (i.e. $k=1$). Note that it would be possible to solve the stationary part of (7) directly but this would involve solving a large nonlinear algebraic system. To avoid this we use a simple numerical scheme to converge (in time) to the equilibrium. The method we use is Lax-Friedrichs. This is indeed the simplest method available for this type of equations. There are two main disadvantages for this method. The first one is that there is a limit on the size of the timestep Δt. This means that if we choose Δt too large, small errors will grow exponential. The timestep Δt is determined by the so-called CFL condition which depends on the speed of the travelling surface waves. The second disadvantage is that the method exhibits diffusion. This means that the solution is slightly damped. For more details on this method see, for example Strikwerda 1989.

4.2 *The bed*

To solve the bed equation we also use a very simple numerical method, namely the FCTS (Forward in Time, Central in Space) method. Using this simple method enables us to analyse the numerical growth factor of different modes, and to compare these with the analysis of section 3.

To analyze the numerical growth factor we discretize the bed equation (8). For easier analysis we drop

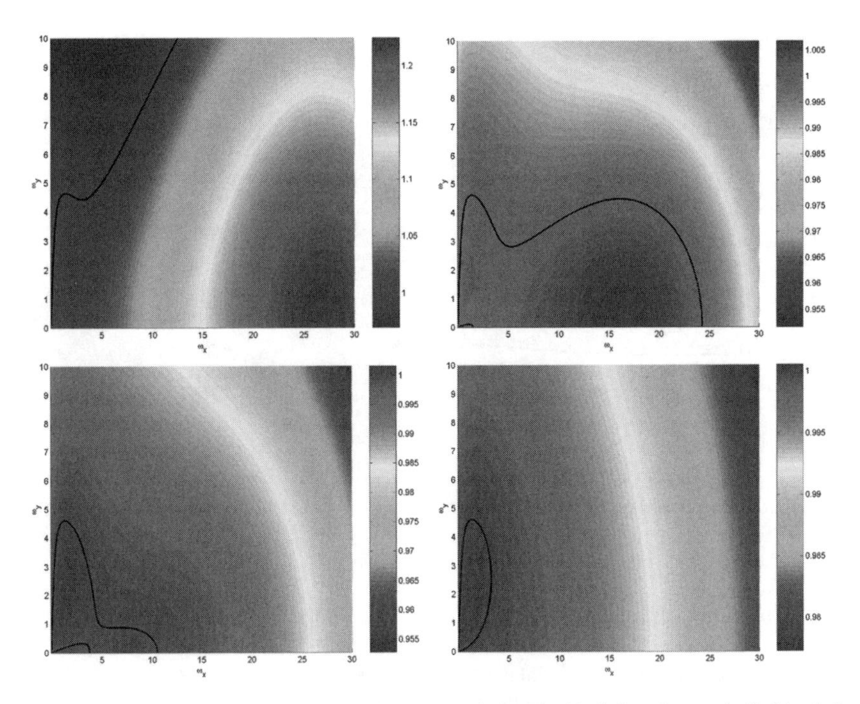

Figure 3. $|g(\theta,\phi)|$ for $R=30$ and $\nu=0.5$, 0.2, 0.15, 0.05 respectively. The black line denotes $|g(\theta,\phi)|=1$, hence it is a separatrix between growing and decaying modes. We see that for $\nu=0.5$, $\nu=0.2$ and $\nu=0.15$ $|g(\theta,\phi)|>1$ for modes other than the critical. For $\nu=0.05$ only the critical modes grow.

the nonlinear terms. The equation for every gridpoint now reads:

$$z_{i,j}^{n+1} = (1+\frac{2\Delta t}{R}(\partial_{xx}+\partial_{yy}))z_{i,j}^{n} + \Delta t(b\partial_x u_{i,j}^{n} + \partial_y v_{i,j}^{n}). \quad (21)$$

To study the growth of the different modes we substitute:

$$z_{i,j}^{n} = g^n \hat{z} e^{i(\omega_x i\Delta x + \omega_y j \Delta y)} \quad (22)$$

and likewise for u, v, ζ. Elaborating this yields:

$$g(\theta,\phi) = 1+\tfrac{\Delta t}{R}(\tfrac{\cos(\theta)-1}{\Delta x^2}+\tfrac{\cos(\phi)-1}{\Delta y^2}) - i\Delta t(b\tfrac{\hat{u}}{\hat{z}}\tfrac{\sin(\theta)}{\Delta x}+\tfrac{\hat{v}}{\hat{z}}\tfrac{\sin(\phi)}{\Delta y}) = 0. \quad (23)$$

with $\theta=\omega_x\Delta x$, $\phi=\omega_y\Delta y$.

We get expressions for $\hat{u}/\hat{z}$, $\hat{v}/\hat{z}$ by solving the discretised, linearised flow equations $M\Phi=I$ with the first and second column of M as follows:

$$\begin{pmatrix} i\frac{\sin(\omega_x\Delta x)}{\Delta x}+2CR \\ 0 \\ i\frac{\sin(\omega_x\Delta x)}{\Delta x} \end{pmatrix}, \begin{pmatrix} 0 \\ i\frac{\sin(\omega_x\Delta x)}{\Delta x}+CR \\ i\frac{\sin(\omega_y\Delta y)}{\Delta y} \end{pmatrix},$$

the third column of M as

$$\begin{pmatrix} i\frac{\sin(\omega_x\Delta x)}{\Delta x} \\ i\frac{\sin(\omega_y\Delta y)}{\Delta y} \\ iF^2\frac{\sin(\omega_x\Delta x)}{\Delta x} \end{pmatrix},$$

and with

$$\Phi = \begin{pmatrix} \hat{u} \\ \hat{v} \\ \hat{\zeta} \end{pmatrix}, I = \begin{pmatrix} -CR\hat{z} \\ 0 \\ i\frac{\hat{z}\sin(\theta)}{\Delta x} \end{pmatrix}.$$

Thus we get an expression for $g(\theta,\phi)$. Please note that we have assumed here that the linearised flow equations are solved exactly, which is not the case in practice.

Modes (θ,ϕ) with $|g(\theta,\phi)|<1$ will decay with time, and are unstable modes. Modes with $|g(\theta,\phi)|>1$ will grow and are stable modes.

It is clear from figure 3 that if Δt is too large modes with relatively large θ and $\phi=0$ are stable. According to the analysis of section 3 these $p=0$ modes should be always unstable. We should therefore choose Δt small enough so that the numerical and analytical growth factors are consistent. To study this we can plot a numerical curve for the different modes $p=0$,

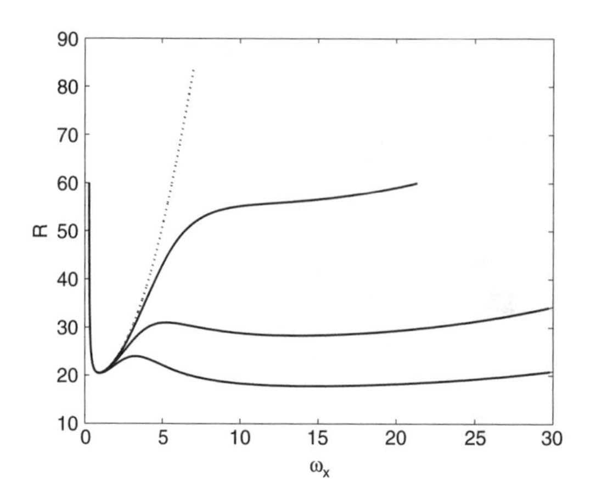

Figure 4. The numerical neutral curves (solid lines) converge to the analytical neutral curve (dotted line) as Δt decreases.

$1, 2, \ldots$ by plotting $|g(\theta, p\pi)| = 1$ for various R, see figure 4.

With this growth factor g and the numerical neutral curve we have the tool we need to justify our numerical results.

5 RESULTS AND COMPARISON WITH ANALYTICAL OUTCOME

In this chapter we will present the experiments done with the entire system. This means that we alternately let the flow adapt to the bed, and evolve the bed for one step. We have assumed that the flow adapts instantly to changes in the bed, this means that we should evolve the bed over a very short time interval only. Note that due to the fact that the domain has a finite length and we impose periodic boundary conditions, not all the modes in longitudinal direction fit. Hence, we cannot expect that bars with exactly the critical wavelength will develop, but a wavenumber close to the critical wavelength will be selected. Surprisingly enough, this is not always the closest mode to k_c. Due to the fact that the analytical k_c cannot be selected, also the critical width-to-depth ratio R_c (above which the bars start to grow) is somewhat higher than might be expected from the theoretical analysis. In figures 5, the evolution of an arbitrary bed configuration towards finite amplitude bars is depicted. In some experiments, an interesting aspect happens. If one starts with a pattern with a wavenumber that is larger than the one that is selected if one would start with an arbitrary bed, then the amplitude of the configuration decreases. After some time, the bars slow down and the surplus on bars is 'kicked out'

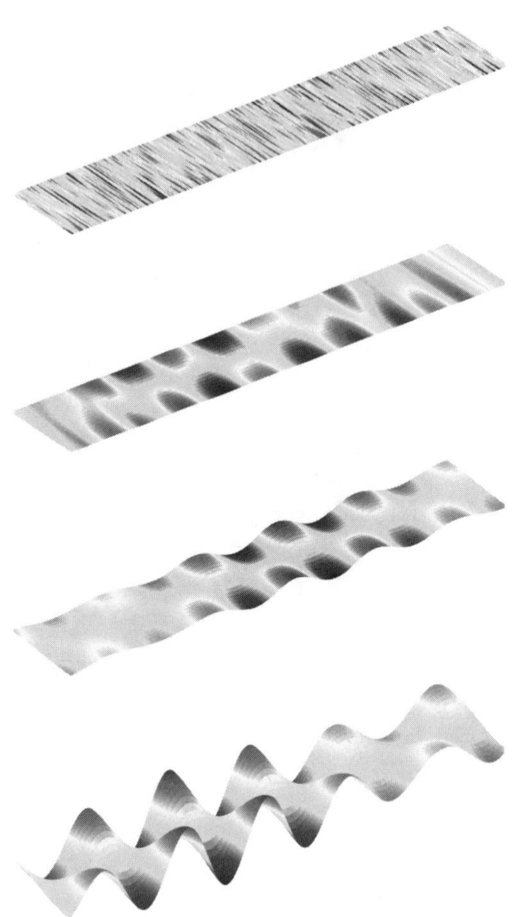

Figure 5. A typical example of the evolution of bars starting from an arbitrary initial bed.

after which the amplitudes increases again. This phenomenon was already observed in Federici 2002. From the theory in Schielen et al. 1993 it is also clear that if the domain is sufficiently long, a pattern of bars will evolve which is not uniform in amplitude. We have seen this in experiments with for instance $R = 50$. An important remark is, that in all the experiments, it seems that for very long simulation time, the 0-mode in the longitudinal direction will grow. This means that mass is not conserved, a phenomenon which is of course not contained in the original set of equations. Hence this is also a numerical artefact, which could be dealt with if one adapts the numerical approach with special mass-conserving schemes. Another interesting point is whether this approach allows for central-bar configurations. In principle, for sufficiently height width-to-depth ratios (i.e. sufficiently shallow configurations), is should be possible from theoretical point

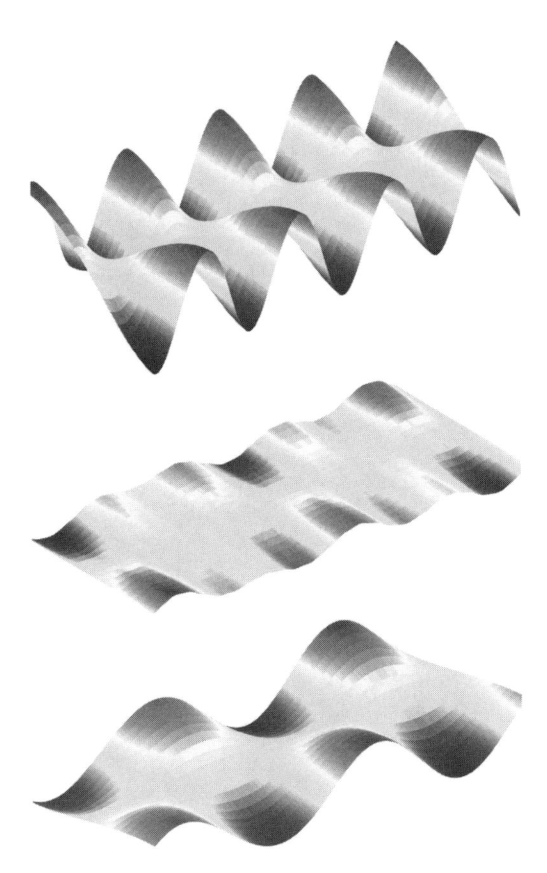

Figure 6. Typical adaptation of a bar pattern in a finite domain.

of view. However, from the *numerical* stability analysis is seems that the alternating bar-modes always have a higher amplification mode than the central bar modes. Hence, even if we start with a central bar pattern close to the expected value, it turns out that after some simulation time, the alternating bar pattern prevails.

6 CONCLUSIONS AND FURTHER RESEARCH

We have seen that even with a basic numerical scheme it is possible to reproduce some of the analytical results from Schielen et al. 1993. Alternating bars did emerge by themselves, without having to force them, as in Rijn 1998. We are also able to more or less determine when the bars will emerge by analyzing the numerical stability of the scheme. However, the wavenumber of the bars was not exactly the same and emerged at a higher value of R as the theory predicted. The bars did behave as the theory predicted, having a uniform amplitude for a domain smaller that the critical domain length, and a non-uniform amplitude for a domain that was larger. The numerical schemes that we studies where among the most simple ones that could still do the job. They enabled us however, to pinpoint almost every anomalie that we saw in the experiments to a numerical artefact (mostly a wrong timestep in either flow or bed). Some questions however remain, such as the selected wavenumber of the bars with respect to the critical wavenumber as described by the analytical analysis and the apparent non-existence of the central modes. Extension of this study is in progress. We are also studying the possibility to allow for open boundary conditions, which will presumably give more freedom to the selected wavenumber of the bars, since it is no longer forced by the periodic conditions. This requires however solving some Rieman-integrals on the boundary which is not completely straightforward. Also from numerical point of view some questions remain. Is an explicit method indeed the best choice, and can we fasten up the simulation? After having completely understood the straight configurations, an extension to slightly curved channels lies at hand, such that the results described in (Schielen 1995) can be tested. But also an extension to bifurcation (or confluention) points could be interesting, especially from more practical points of view. The main river in the Netherlands for instance the river Rhine bifurcates two times in relatively short stretch in three branches, while the part of the Rhine in Germany is relatively straight. The diversion of sediment around these points is important for shipping and industry. The analysis of the behaviour of bars in the neighbourhood of these bifurcation points may shed some light on this issue. Finally, one might think about introducing an erosion equation (and hence a third timescale!) to simulate meander formation.

REFERENCES

Federici, B. (2002). Topics on fluvial morphodynamics. *Ph.D Thesis, University of Genua.*

Rijn, L. (1993). Principles of sediment transport in rivers, estuaries and coastal areas. *Aqua Publications, Amsterdam.*

Rijn, L. (1998). Test computations with delft2drivers. *Report 9.6600, WL | Delft Hydraulics.*

Schielen, R. (1995). Non-linear stability analysis and pattern formation in morphological models. *Ph.D thesis, Utrecht University.*

Schielen, R., Doelman, A. and Swart, H. d. (1993). On the nonlinear dynamics of free bars in straight channels. *J. Fluid Mech. 252.*

Strikwerda, J. (1989). Finite difference schemes and partial differential equations. *Chapman and Hall.*

River, Coastal and Estuarine Morphodynamics: RCEM 2005 – Parker & García (eds)
© 2006 Taylor & Francis Group, London, ISBN 0 415 39270 5

Effect of river bed degradation and aggradation on transformation of alternate bar morphology

H. Miwa
Department of Civil Engineering, Maizuru National College of Technology, Kyoto, Japan

A. Daido
East Asian Office of Technology, Kyoto, Japan

T. Katayama
Advanced Faculty of Civil and Industrial System Engineering, Maizuru National College of Technology, Kyoto, Japan

ABSTRACT: Many studies on alternate bars have been conducted under an equilibrium sediment supply condition which corresponds to the water discharge, and discussed phenomena in an equilibrium state. However, bar morphology in an actual river is not always formed under such condition. Therefore, it is important to investigate how the bar morphology formed by some water and sediment discharge condition varies by the other conditions. In this study, effects of river bed degradation or aggradation and water discharge on alternate bar transformation are investigated by means of flume tests using uniform and non-uniform sediment. Results are summarized as follows: (1) In the experiments, the river bed degradation rate of the non-uniform sediment bed is larger than that of the uniform one, and the aggradation rates of both sediment beds are almost the same. (2) The river bed degradation under the low-water discharge causes the low-watercourse formation, and that under the high-water discharge causes the restriction of alternate bar development. And, the aggradation under the high-water discharge causes the increase of the alternate bar wavelength. (3) The increase and the decrease of wavelength of alternate bars in consequence of the river bed degradation are remarkable in the non-uniform sediment bed due to grain sorting.

1 INTRODUCTION

Investigation into effects of change of water and sediment discharge from upper reach of a river on river bed variation at lower reach is important for an appropriate sediment management at a river basin. For example, the river bed downstream of a dam varies dynamically by discharge and sediment flushing from the dam, and influences river environment. In particular, alternate bars cause some problems in a flood, whereas they also play an important role as fish and plant habitat in a low-water discharge. Therefore, studies on alternate bars have been conducted from various aspects, which are formation condition, development process, characteristics of bar shape and migration, etc. (e.g. Muramoto & Fujita 1978, Fujita et al. 1981, Ikeda 1984, Miwa & Daido 1995). Many of these works were conducted under an equilibrium sediment supply condition which corresponds to the water discharge and river condition, and discussed phenomena in an equilibrium state. However, bar morphology in an actual river is not always formed under such condition. Bar transformation in a flood and river bed degradation downstream of a dam are its examples. Therefore, it is important to investigate how the bar morphology formed by some water and sediment discharge condition varies by the other conditions.

Some investigations about a variation of river bed with alternate bars due to change of water discharge and sediment supply rate have been conducted. Watanabe et al. (2002) and Miwa et al. (2002) performed hydraulic experiments for formation and transformation of alternate bars under unsteady flow conditions, and clarified response characteristics of wavelength and height due to change of water discharge. In these studies, since the sediment supply rate corresponded to the sediment transport capacity, the effect of aggradation and/or degradation on alternate bar transformation was not considered. Uchijima & Hayakawa (1987) investigated the low-watercourse formation and the local scour development caused by the alternate bar deformation under the low-water discharge. Yuki et al. (1992) examined the phenomenon similar to Uchijima and Hayakawa's work, and discussed the variation process of the low-watercourse and the local scour depth. In recent years, Shimizu et al. (2004) discussed the formation process of the low-watercourse in the bed with alternate bars under the dynamic equilibrium condition through a numerical simulation. These studies mainly investigated the effect of water discharge, and did not mention the effect of sediment supply. On the other hand, Teramoto & Tsujimoto (2004) pointed out the importance of a response of river bed on the condition different from the condition which formed the river bed, and discussed the effect of the water discharge and sediment supply condition on alternate bar transformation. Miwa et al. (2004) investigated the effect of a sediment supply on the alternate bar deformation and the meandering streambed formation under low-water discharge form a similar viewpoint, and discussed the emerged bar formation and the migration property of pools. In addition, they discussed the effect of grain sorting on that process.

From previous studies, it is found that investigation into effect of a water discharge and sediment supply condition is important for considering stability and instability of river bed. Alternate bars in a flood might not be equilibrium, and the imbalance between a water discharge and a sediment discharge influences bar behavior through a degradation or an aggradation. Therefore, it is necessary to investigate the effects of the degradation and aggradation on the transformation of bar morphology. In this study, we investigate the transformation process of alternate bars in consequence of river bed degradation and aggradation which are caused by an imbalance between sediment supply rate and sediment transport capacity. Experiments are conducted by using uniform and non-uniform sediment with same mean grain diameter in order to examine effects of grain sorting on that process. The degradation and aggradation are caused by supplying the non-equilibrium sediment discharge to the upstream of the channel with equilibrium alternate bars. By investigating the river bed variation process in detail first, we discuss the transformation processes of alternate bars. Effects of the degradation and aggradation on the geometric scale and migration velocity of alternate bars are examined next. Effect of the grain sorting on these phenomena is also discussed.

2 EXPERIMENTAL SET-UP AND PROCEDURE

Experiments were conducted in a straight rectangular open channel with a length of 12 m, a width of 0.2 m and a depth of 0.3 m. Two kinds of sediment with almost same mean grain diameter were used in the experiments. Their grain size distributions are

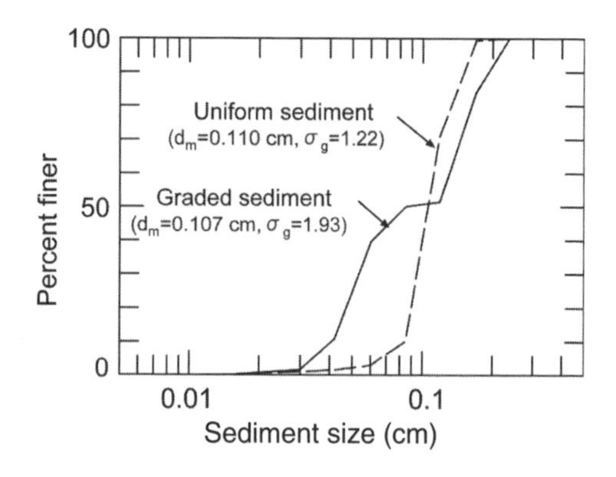

Figure 1. Grain size distributions of the sediment used.

shown in Figure 1. The uniform sediment has the mean grain diameter d_m of 0.110 cm and the geometric standard deviation σ_g (= $(d_{84}/d_{16})^{1/2}$, in which d_{84} and d_{16} are the grain sizes for which 84% and 16% of the sediment is finer respectively) of 1.22. The non-uniform sediment has $d_m = 0.107$ cm and $\sigma_g = 1.93$. Both sediments have the same specific gravity σ_s of 2.65.

The sand bed was flattened with a scraper, and was set to the slope of 0.0167 (1/60) before the experiment. At first, alternate bars were made to develop under the prescribed water discharge. The sediment was externally fed into the flow at the upstream end of the channel in order to keep the average bed level constant. After the formation of stable alternate bars has been confirmed, the longitudinal profile of the water surface was measured with a supersonic displacement sensor mounted on the self-propelled carriage. Longitudinal profiles of the bed surface were measured with a laser displacement sensor mounted on that carriage at intervals of 1 cm in the transverse direction of the channel after stopping the water feeding. The origin of the measurements ($x = 0$) was set at a point of 3.65 m down from the upstream end of the channel, and the measuring reach was from $x = 0$ m to 7.95 m. A mean flow depth, reach averaged values of bed slope, water surface slope and energy slope were calculated form the measured profiles of the water surface and the bed surface.

We investigated the alternate bar transformation process with river bed degradation and aggradation by changing the sediment supply rate, next. The degradation was caused by no sediment supply and by reducing the sediment supply rate to half of the equilibrium sediment supply rate in the above experiment for alternate bar formation, and the aggradation was caused by increasing the sediment supply rate to three times of that. The water discharge was the same as it in the alternate bar formation experiment. The shape of alternate bars and the direction of sediment transport were

Table 1. Experimental conditions on alternate bar development process.

Case	Sediment	Q_w cm^3/s	Q_{Bin} g/min	h_m cm	Fr	I_s $\times 10^{-2}$	τ_*	T min
UA40-A	U	400	70	0.71	1.07	1.62	0.063	62
UA40-B	U	400	70	0.71	1.07	1.60	0.063	61
UA80-B	U	800	200	1.13	1.06	1.60	0.099	88
UA80-D	U	800	200	1.14	1.05	1.64	0.103	63
MA40-A	NU	400	85	0.69	1.11	1.59	0.062	71
MA40-B	NU	400	85	0.65	1.22	1.63	0.061	222
MA80-B	NU	800	250	1.12	1.08	1.57	0.091	52
MA80-D	NU	800	250	1.06	1.17	1.59	0.096	80

Table 2. Experimental conditions on alternate bar transformation process.

Case	Sediment	Q_w cm^3/s	Q_{Bin} g/min	h_m cm	Fr	I_s $\times 10^{-2}$	τ_*	T min
UA40-A	U	400	35	0.79	0.91	1.44	0.063	646
UA40-B	U	400	0	0.83	0.84	1.39	0.066	230
UA80-B	U	800	100	1.25	0.91	1.41	0.097	265
UA80-D	U	800	600	1.08	1.14	2.61	0.155	789
MA40-A	NU	400	43	0.62	1.31	1.46	0.049	224
MA40-B	NU	400	0	0.67	1.16	1.30	0.048	437
MA80-B	NU	800	125	0.97	1.34	1.29	0.069	307
MA80-D	NU	800	750	1.14	1.05	2.39	0.151	648

sketched at any time during the experiment. The shape and location of emerged bars were also sketched, if any. The sediment discharge was specified by caching the sediment at the downstream end of the channel every 5 minutes. The measurements of longitudinal profiles of the water surface and the bed surface were conducted by the above mentioned methods, and were repeated during the experiment. The origin of the temporal (t = 0) is the starting time of the experiment for alternate bar transformation.

The experimental conditions at the time of the end of the experiments for alternate bar development and transformation are listed in Table 1 and Table 2. Case UA80-D and Case MA80-D are the experiments for aggradation process, and the other cases are the experiments for degradation process. U and NU in the "Sediment" column mean uniform sediment and non-uniform sediment respectively. The hydraulic values in Table 2 are the averaged value in degradation or aggradation reach. In the tables, Q_w = water discharge, Q_{Bin} = sediment supply rate, h_m = mean flow depth, F_r = Froude number, I_s = bed slope, τ_* = dimensionless bed shear stress (= $(h_m I_e)/((\sigma/\rho - 1)d_m)$, in which I_e = energy slope, σ = sediment density, ρ = fluid density, g = acceleration of gravity); and T = duration of experiment.

3 EFFECT OF SEDIMENT SUPPLY ON RIVER BED VARIATION

An imbalance between the water discharge and the sediment transport capacity causes the longitudinal bed variation. Situations of this variation in the experiment are shown first. Figure 2a shows the temporal variations in transversely averaged longitudinal bed profile in Case UA80-B and Case MA80-B. The datum of the bed level is the transversely averaged bed at x = 795 cm. The solid line denoted by "initial" is the bed profile of the stable alternate bars in the experiment for alternate bar formation, and that is the initial bed of the experiment for alternate bar transformation. The bed degradation progresses gradually from upstream in both cases. A similar tendency was also found in other cases except Case UA80-D and Case MA80-D. During the experiment, the degradation in Case UA80-B reaches approximately x = 500 cm and that in Case MA80-B reaches approximately x = 600 cm. The comparison of temporal variations in degradation rate at some places from the initial bed is shown in Figure 2b. It is found that the degradation velocity of the non-uniform sediment bed is larger than that of the uniform one. This means that the erosion in the non-uniform sediment bed is

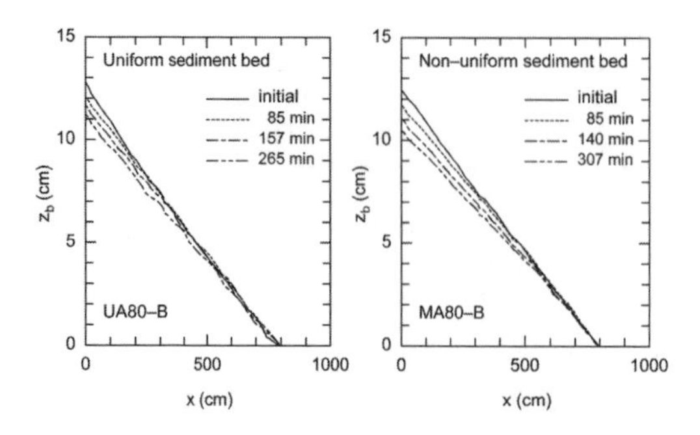

Figure 2a. Temporal variations in longitudinal bed profile in degradation process.

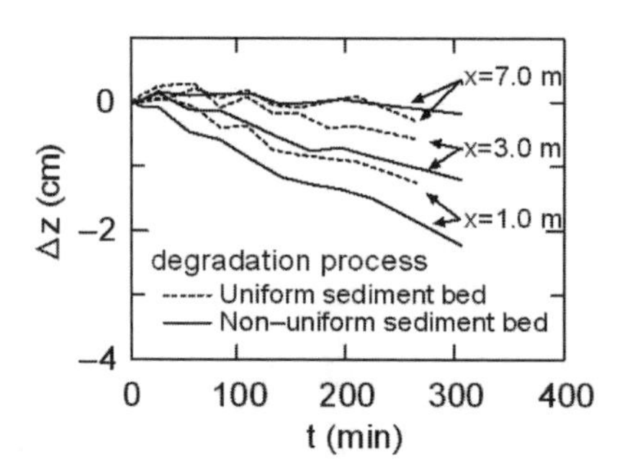

Figure 2b. Temporal variations in degradation rate from initial bed (Case UA80-B & Case MA80-B).

more active than that in the uniform one. Degradation in general is restricted by bed armoring. However, in this experiment, the mobility of coarse grains was greater than that of fine grains because of the "Exposure effect" and the "Smoothing effect" defined by Ikeda & Iseya (1988). These effects were also confirmed in the other experiments. As a result, the bed armoring was weak and partial even if it was formed. In particular, in case of low-water discharge, many of coarse grains tend to be deposited on high-elevation parts of the bed, and contribute to formations of an area without sediment transport and a low-watercourse. As a result of these, the ratio of fine grains in the bed surface sediment at the bed with sediment transport is relatively large, and the bed is eroded easily in comparison with the uniform sediment bed. It is guessed that this is a cause of larger degradation rate than the uniform sediment bed.

Figure 3a shows the transversely averaged longitudinal bed profile in Case UA80-D and Case MA80-D. The bed aggradation progresses gradually from upstream in both cases. The bed slope of the upper reach in each case is larger than that of the lower reach until about 300 minutes, whereas the bed slope is almost the same in all reach after that. The final bed profiles are in equilibrium, and the equilibrium bed slope of Case UA80-D is larger than that of Case

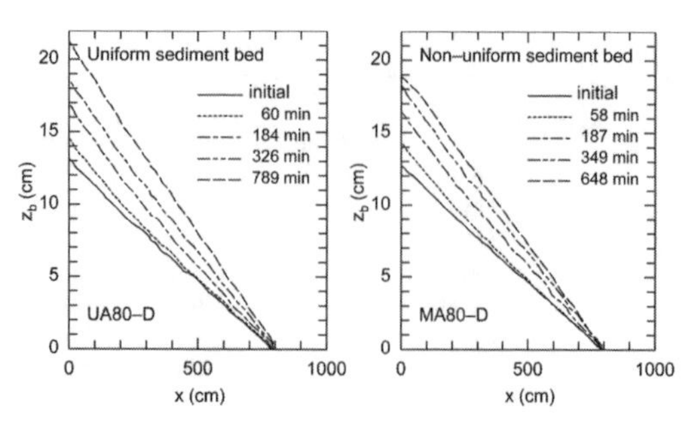

Figure 3a. Temporal variations in longitudinal bed profile in aggradation process.

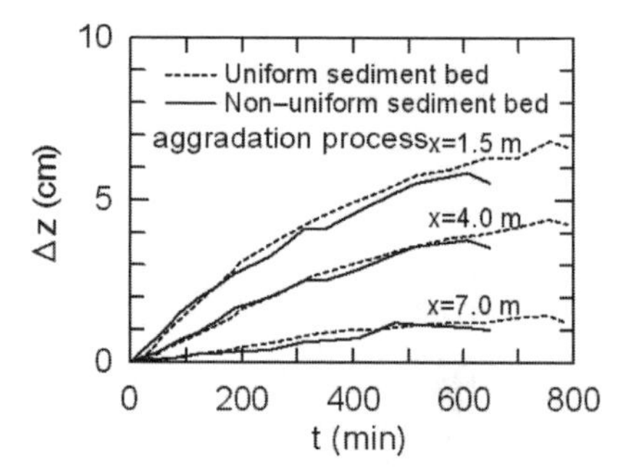

Figure 3b. Temporal variations in aggradation rate from initial bed (Case UA80-D & Case MA80-D).

MA80-D. The comparison of temporal variations in aggradation rate at some places from the initial bed for both cases is shown in Figure 3b. It is found that the aggradation velocity decreases with the passage of time and presents almost the same tendency with both cases.

4 TRANSFORMATION OF ALTERNATE BAR MORPHOLOGY WITH RIVER BED DEGRADATION AND AGGRADATION

The transformation process of alternate bars with the degradation and with aggradation of the river bed is described next. The difference in that process due to water discharge scale is also described.

Figure 4 shows the temporal changes in bed topography and sediment path in the degradation process under the low-water discharge. The datum level of the bed ($z = 0$) is the average bed level of the initial alternate bars, and the positive of z indicates a higher bed level than the datum level. The part enclosed by the solid line indicates an emerged bar, and the arrow line shows a direction of sediment transport obtained from experimental observation. Case UA40-A and Case UA40-B are the experiments in the uniform sediment

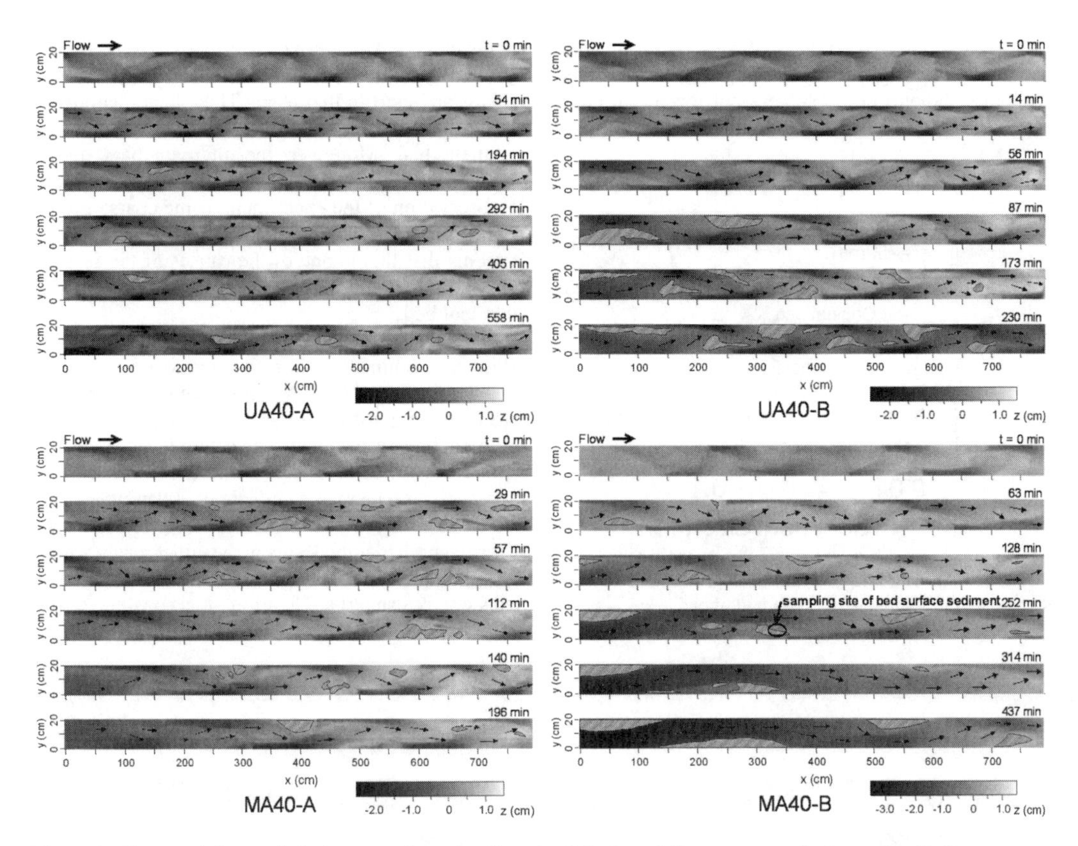

Figure 4. Temporal changes in bed topography and sediment path in degradation process under low-water discharge.

bed, and Case MA40-A and Case MA40-B are the experiments in the non-uniform sediment bed.

The cases of the uniform sediment bed are described first. It can be found that the wavelength gradually increases with degradation in both cases. In particular, this tendency is remarkable in the upper reach where the degradation progresses. In Case UA40-A with the low sediment supply rate, the bed was mainly under the water surface during the experiment, and it disappeared after a while even if the emerged bar appeared. There were little sediment transport on high-elevation parts of the bed, and the sediments were mainly transported along the low streambed. This sediment transport promoted the alternate bar transformation, and the length of the high-elevation part without sediment transport increased. Therefore, wavelength of the alternate bars corresponds to that length. In Case UA40-B without sediment supply, the low-watercourse had developed from upstream together with the emerged bar formation. Therefore, the wavelength of alternate bars corresponds to the meandering wavelength of the low-watercourse. The migration velocity of alternate bars gradually decreased with the degradation, and the alternate bars hardly migrated due to the low-watercourse formation. The wavelength of the alternate bars in this case tends to be longer than that of Case UA40-A.

The cases of the non-uniform sediment bed are described next. It is found that the wavelength gradually increases with degradation in both cases as in the cases of the uniform sediment bed. In particular, since the deep and stable low-watercourse had developed from upstream together with the emerged bar formation in Case MA40-B without sediment supply, the wavelength is considerably long. And, the wavelength of alternate bars in Case MA40-A with the low sediment supply rate is a little longer than that in Case UA40-A. In the non-uniform sediment bed presented here, high-elevation parts without sediment transport and emerged bars were formed easily and were relatively stable in comparison with the uniform sediment bed because coarse grains tend to be deposited on the high elevation-parts of the bed. Figure 5 is an example of grain size distribution of the bed surface sediment in such part. It is considered that these contribute to the alternate bar transformation in the non-uniform sediment bed.

Figure 6 shows the temporal changes in bed topography and sediment path in the degradation process under the high-water discharge. Since the sediment was transported on all places of the bed during the experiment, emerged bars and high elevation parts without sediment transport were not formed in both cases. In Case UA80-B of the uniform sediment bed,

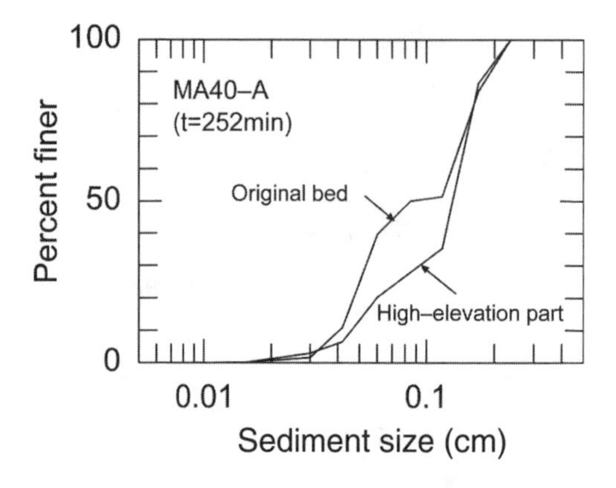

Figure 5. An example of grain size distribution of bed surface sediment in high-elevation part of bed.

the development of alternate bars tends to be restricted by the degradation in the upper reach, and the wavelengths are short in that reach. This phenomenon can also be seen in Case MA80-B of the non-uniform sediment bed. Moreover, the alternate bars disappeared occasionally in the non-uniform sediment bed. In observation of the experiment, some coarse grains deposition areas were formed with the degradation. It seems that this is one of the causes of the restriction of alternate bar development in the non-uniform sediment bed.

Figure 7 shows the temporal changes in bed topography and sediment path in the aggradation process. Case UA80-D and Case MA80-D are the experiment in the uniform sediment bed and in the non-uniform sediment bed, respectively. The datum level of the bed is its average bed level at each time. It is found that the wavelength increases until about 300 minutes, and decreases a little after that in both cases. The increase of wavelength mainly occurs in the upper reach, where the aggradation rate is large and the increasing rate of bed slope is also large. Therefore, it is considered that

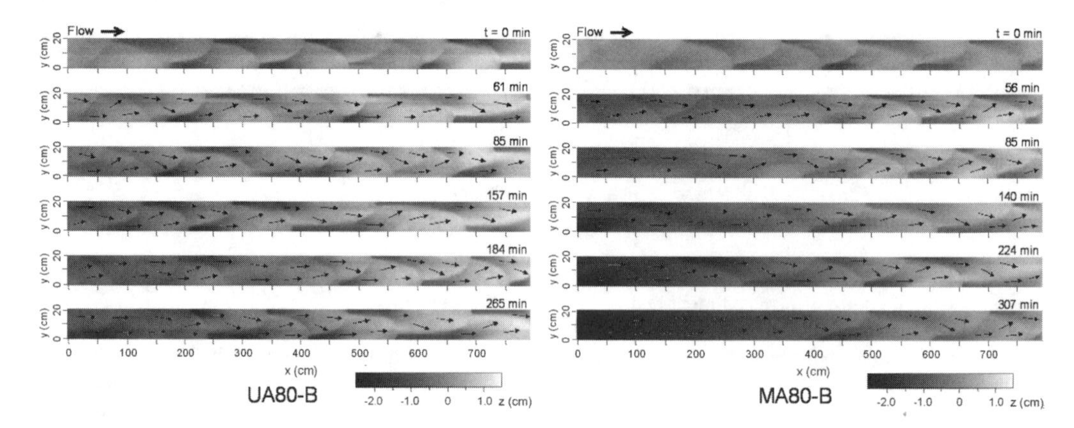

Figure 6. Temporal changes in bed topography and sediment path in degradation process under high-water discharge.

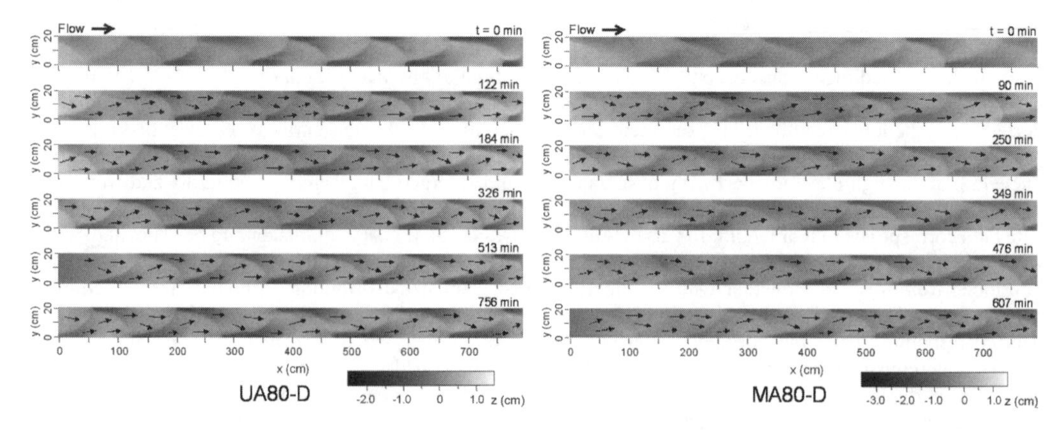

Figure 7. Temporal changes in bed topography and sediment path in aggradation process under high-water discharge.

aggradation rate, bed slope (or bed shear stress) and bar development rate influence this phenomenon. The increase of migration velocity of alternate bars with the aggradation was also observed in the experiments.

As mentioned above, the degradation and aggradation influence the alternate bar transformation through changes of the bed slope and water discharge. We discuss the effect of these changes on the alternate bar transformation. The bar regime criteria diagram presented by Kuroki & Kishi (1984) and change of bar regime variables due to changes of bed slope and water discharge are shown in Figure 8. The symbol denotes the initial hydraulic condition. The arrow line on bed slope denotes the shifting direction of hydraulic condition with changing of bed slope. The arrow lines are calculated from the following resistance equation on bed with alternate bars presented by Miwa & Daido (1995);

$$f = 0.216\left(\frac{R}{d_m}\right)^{-1/3} \quad (1)$$

where f = Darcy-Weisbach friction factor (= $8(u_m/u_*)^{-2}$), u_m = mean flow velocity, u_* = shear velocity (= $(gh_mI_e)^{1/2}$); and R denotes a hydraulic radius.

As for the experiments in this study, the decrease of bed slope is found to causes a decrease of mobility of sediment in both cases of the low-water discharge and the high-water discharge. The decrease of the mobility takes place easily in the low-water discharge,

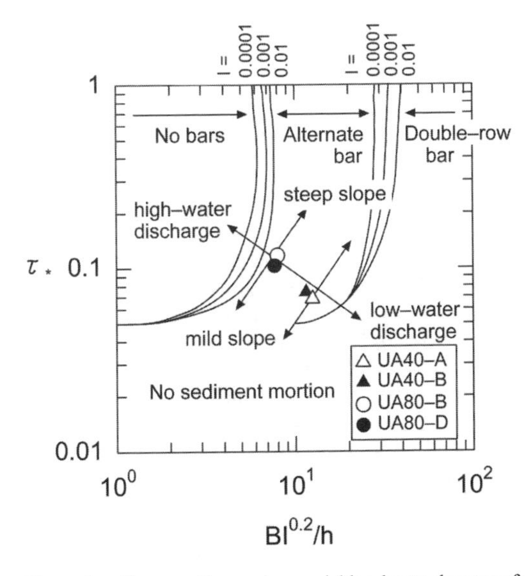

Figure 8. Change of bar regime variables due to changes of bed slope and water discharge on diagram for regime criteria on bars. Adapted form Kuroki and Kishi (1984).

and causes the formation of bed area without sediment transport, the immobilization of bars and the low-watercourse formation. Cases UA40-A, UA40-B, MA40-A and MA40-B are correspond to this case. The dimensionless critical shear stress τ_{*c} used in the regime criteria diagram is 0.05, but it is 0.034 for the mean grain size of the sediment in this study. Then, the regime criteria ranges extend to downward, and the bar regime in a high-water discharge shifts to the "No-bars" range together with degradation. It is considered that Cases UA80-B and MA80-B correspond to this case. The bed morphology in both cases is classified into "Semi-bar" in the other regime criteria diagram presented by Muramoto and Fujita (1978). In the case of increase of bed slope, alternate bars develop according to change of hydraulic condition. Cases UA80-D and MA80-D correspond to this case. In addition, it is possible that the bar regime in low-water discharge shifts from the alternate bar range to the double-row bar range together with increase of bed slope due to aggradation.

5 VARIATIONS IN WAVELENGTH AND HEIGHT OF ALTERNATE BARS

Geometrical scale of alternate bars is discussed next. The temporal variations in mean wavelength L and mean wave height H in the degradation process and the aggradation process are shown in Figures 9, 10 and 11. In each figure, the symbol of white circle denotes the wavelength and height in all of the measurement reach until the degradation or aggradation reaches to the measurement reach, and the symbol of black circle denotes the wavelength and height in the reach with degradation or aggradation. A wavelength was measured as a bar length, but it was measured as a longitudinal distance between pools in both banks when a low-watercourse was developed. A wave height was measured as the difference between the highest bed level and the lowest bed level in the transverse section which has the lowest bed level in a reach of one wavelength long.

Figure 9 shows the results in the degradation process under the low-water discharge condition. As shown in Figure 4, the low-watercourse was formed in Cases UA40-B and MA40-B, but it did not formed in Cases UA40-A and MA40-A. In Case UA40-A and Case MA40-A, the wavelength gradually increase with degradation, but the wave height does not vary so much. It is found that the formation of the non-sediment transport area formed in these cases contributes to the increase of wavelength, but does not contribute to the wave height. The similar tendency can also be seen in Case UA40-B and Case MA40-B before the low-watercourse is formed. However, not

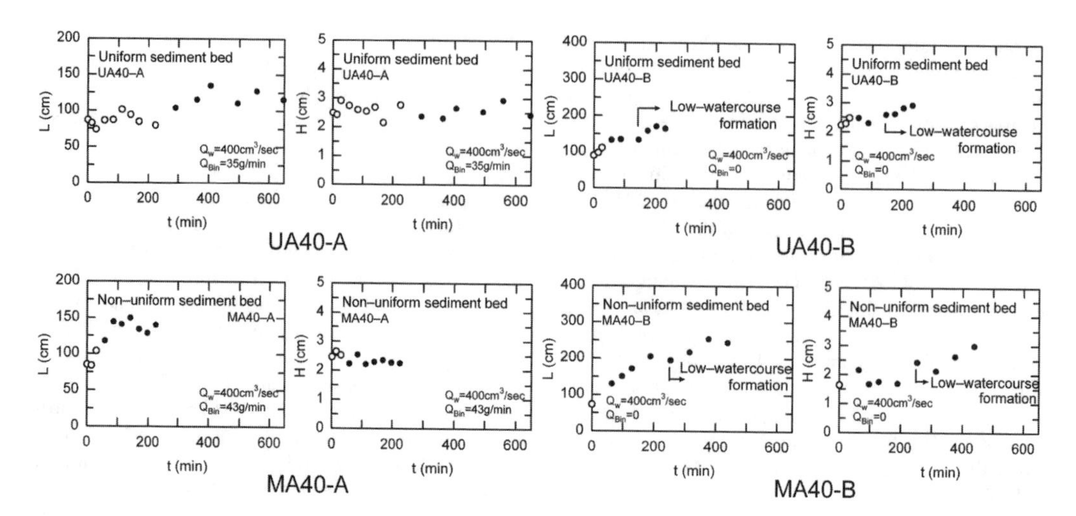

Figure 9. Temporal variations in wavelength and height of alternate bars in degradation process under low-water discharge.

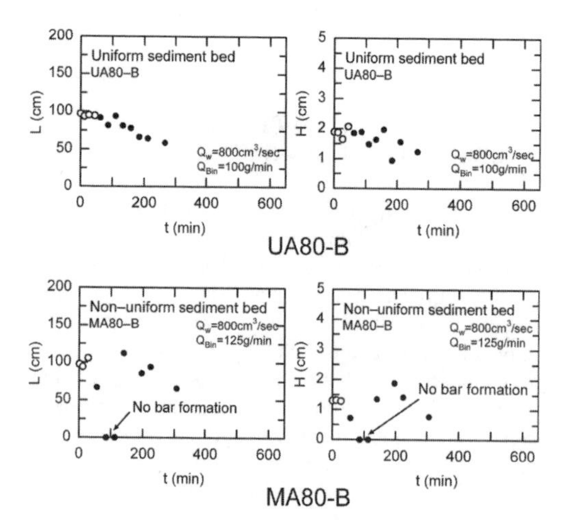

Figure 10. Temporal variations in wavelength and height in degradation process under high-water discharge.

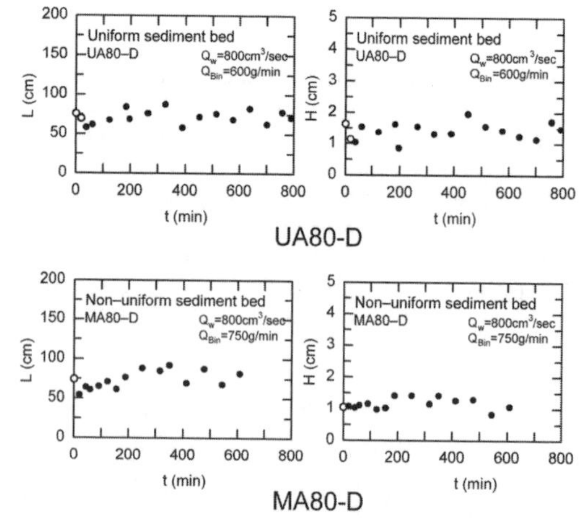

Figure 11. Temporal variations in wavelength and height in aggradation process under high-water discharge.

only wavelength but also wave height increases when the low-watercourse is formed. In particular, the increase of the wavelength and height, which are correspond to meandering wavelength and depth, of the low-watercourse is remarkable in the non-uniform sediment bed because emerged bars in the non-uniform sediment bed were relatively stable. As it was mentioned above, bed armoring did not take place because of high mobility of coarse grains. It is considered that the increase of the wavelength and height is restricted if bed armoring take place due to reduce of the water discharge.

Figure 10 shows the results in the degradation process under the high-water discharge condition. The wavelength and height gradually decrease with degradation in Case UA80-B. And they vary considerably in Case MA80-B. These phenomena were caused by the restriction of alternate bar development. In particular, alternate bars in the non-uniform sediment bed did not develop easily, and were partial even if they developed. And, it is considered that the formation of coarse grains deposition area also influenced the restriction of alternate bar development. Therefore, alternate bar formation was considerably unstable in the decreasing reach. These results suggest that the bar regime of both cases shifted from the "Alternate bar" range to the "No bars" range in Figure 8 by the decrease of bed slope with the degradation.

Figure 11 shows results in the aggradation process under the high-water discharge. We can see the similar variation trend for the wavelength and height in the uniform sediment bed and the non-uniform sediment bed. That is, the wavelength gradually increases after starting the experiment in both cases. This tendency is found till approximately 300 minutes when the aggradation speed is relatively large. The wavelength does not change so much thereafter. On the other hand, the increase of wave height is not always distinct in both cases. From these results, it can be considered that the large increasing rate of bed slope affects the increase of wavelength of alternate bars.

6 CONCLUSIONS

The results obtained in this study are summarized as follows:

1. In the experiments, the river bed degradation rate of the non-uniform sediment bed is larger than that of the uniform one. This is caused by relatively large ratio of fine grains in the bed surface sediment as a result of grain sorting.
2. The wavelength of alternate bars increases gradually with the river bed degradation under the low-water discharge because the streambed like a low-watercourse where little sediment transport exists in the high-elevation parts of the bed is formed, but not the wave height. The wavelength and height increase further when the low-watercourse has developed.
3. The wavelength and height decrease with the degradation under the high-water discharge because the development of alternate bars is restricted. In particular, the alternate bar development is unstable in the non-uniform sediment bed.
4. The wavelength increases gradually with the river bed aggradation under the high-water discharge because the development of alternate bars is promoted. This occurs at relatively large increasing rate of bed slope.
5. The increase and the decrease of the wavelength of alternate bars in consequence of the river bed degradation are remarkable in the non-uniform sediment bed. It is considered that these are influenced by grain sorting.
6. The migration velocity of the alternate bars decreases with the river bed degradation, and that increases with the aggradation. The decrease of migration velocity relates to the immobilization of alternate bars and the formation of low-watercourse.

ACKNOWLEDGEMENT

This study was founded by the Grant-in-Aid for Scientific Research (C) (No.14550518) from Japan Society for the Promotion of Science (JSPS). The writers are thankful for this support.

REFERENCES

Ikeda, S. 1984. Prediction of Alternate Bar Wavelength and Height. Journal of Hydraulic Engineering, ASCE, 110(4), 371–386.

Ikeda, H. & Iseya, F. 1988. Experimental study of heterogeneous sediment transport. *Environmental Research Center Papers*, University of Tsukuba, 12: 1–50.

Fujita, Y., Muramoto, Y. & Horiike, S. 1981. Study on the process of alternating bar development. *Annuals of DPRI*, Kyoto University, 24(B-2): 411–731 (in Japanese).

Kuroki, M. & Kishi, T. 1984. Regime criteria on bars and braids in alluvial straight channels. *Proceedings of JSCE*, 342: 87–96 (in Japanese).

Miwa, H. & Daido, A. 1995. Fluctuations in bed-load transport rate in channel with alternate bars. *Management of Sediment-Philosophy, Aims, and Techniques*, Oxford & IBH publishing: 641–648.

Miwa, H., Yokogawa, J. & Daido, A. 2002. Alternate bar transformation due to change of water discharge in graded and uniform sediment beds. *Proceedings of the 5th International Conference on Hydro-Science and -Engineering*, V: CD-ROM.

Miwa, H., Daido, A. & Yokogawa, J. 2004. Effect of sediment supply on alternate bar deformation and meandering streambed formation. *Proceedings of the 9th international Symposium on River Sedimentation*, III: 1345–1352.

Muramoto, Y. & Fujita, Y. 1978. The classification of mesoscale river bed configuration and the criterion of its formulation. *Proceedings of the 22nd Japanese Conference on Hydraulics*, JSCE, 22: 275–282 (in Japanese).

Shimizu, Y., Osada, K. & Takanashi, T. 2004. Numerical study on the formation of low-watercourse in a straight channel with alternate bars. *Annual Journal of Hydraulic Engineering*, JSCE, 48: 1027–1332 (in Japanese).

Teramoto, A. & Tsujimoto, T. 2004. A study on the transformation of bar morphology affected by flow and sediment discharge. *Advances in River Engineering*, JSCE, 10: 273–278 (in Japanese).

Uchijima, K. & Hayakawa, H. 1987. Characteristics deformation of alternate bars due to low flow. *Proceedings of the 31st Japanese Conference on Hydraulics*, JSCE, 31: 683–688 (in Japanese).

Watanabe, Y., Sato, K., Hoshi, K. & Oyama, F. 2002. Experimental study on bar formation under unsteady flow conditions. *Proceedings of the International Conference on Fluvial Hydraulics, River Flow 2002*, 2: 813–823.

Yuki, T., Ashida, K., Egashira, S. & Okabe, K. 1992. Formation and transformation mechanism of low-waterway. *Annual Journal of Hydraulic Engineering*, JSCE, 36: 75–80 (in Japanese).

Reservoirs and dams

River, Coastal and Estuarine Morphodynamics: RCEM 2005 – Parker & García (eds)
© 2006 Taylor & Francis Group, London, ISBN 0 415 39270 5

Enhanced flushing of sediment deposits in the backwater reach of reservoirs

C.J. Sloff
WL | Delft Hydraulics and Delft University of Technology, Delft, The Netherlands

H.R.A. Jagers
WL | Delft Hydraulics, Delft, The Netherlands

E. Kobayashi
JP Business Service Corporation, Fukagawa, Tokyo, Japan

Y. Kitamura
J-POWER/EPDC, Chigaski Research Institute, Chigasaki, Japan

ABSTRACT: In hydropower and water-supply reservoirs sedimentation processes do not only occur in the storage area, but can extend far upstream along the backwater reach. This leads to undesired impacts such as a rise of flood levels. Using the erosive force of floods part of these deposits can be flushed downstream to the reservoir. Efficiency of flushing operations can be increased by mechanically pushing the deposits from bars and banks into the flushing channel. During flushing these displaced sediments propagate downstream as free bars in a combined one-dimensional (1D) and two-dimensional (2D) way. Analytical analysis shows that the 1D celerity exceeds that of 2D propagation significantly. Simulations with a 2D computational model, including sediment sorting effects and unsteadiness of flow, are required for more profound analysis. Application of a 2D computational model to flushing operations in Sakuma Reservoir in Japan shows that these effects contribute significantly to the efficacy of flushing.

1 INTRODUCTION

In most reservoirs that are constructed by damming a river (such as water-supply or hydropower reservoirs), sedimentation processes lead to an undesired capacity loss. A major part of the sediment is deposited at the head of the reservoir where it directly affects the active storage. These deltaic sediment deposits in reservoirs develops usually progressively in both downstream and upstream direction, e.g. see figure 1. Because of its direct threat to the capacity and lifetime of a reservoir, most attention is paid to the downstream progress of the deltaic deposits into the active storage. Nevertheless, the deposition extending in upstream direction, i.e. the tail reach, becomes increasingly important when it causes water levels over significant reaches. Due to sedimentation in the backwater reach local water-surface elevations are increased, creating additional backwater and deposition even further upstream. This feedback mechanism allows the depositional environment to propagate much further upstream than the initial backwater curve might suggest.

For example, in the Sanmenxia Dam in the Yellow River in China rapid reservoir sedimentation occurred during the first year of operation (around 1960). After changing reservoir operation rules such that incoming sediment could be flushed, these processes could be stopped. Nevertheless, it is shown by Wang and Hu (2004) that a bed-level rise between 3 to 5 m occurred at the city of Tongguan about 110 km upstream of the dam. As a consequence the lower Weihe River, which flows in the Yellow River at Tongguan, also suffers from severe backwater sedimentation leading to an increased flooding risk of the city of Xi'an which is an ancient capital of China. An exemplary cross-section from the paper of Wang and Hu is shown in figure 2. Alternating sedimentation and erosion periods were observed in this reach (following the fall and rise of bed-levels at Tongguan) which were characterized by sedimentation and erosion waves that traveled retrogressively toward the upstream of the Weihe River at a speed of about 10 km per year.

The backwater-induced sedimentation processes in the Weihe River did not only cause a rise of flood

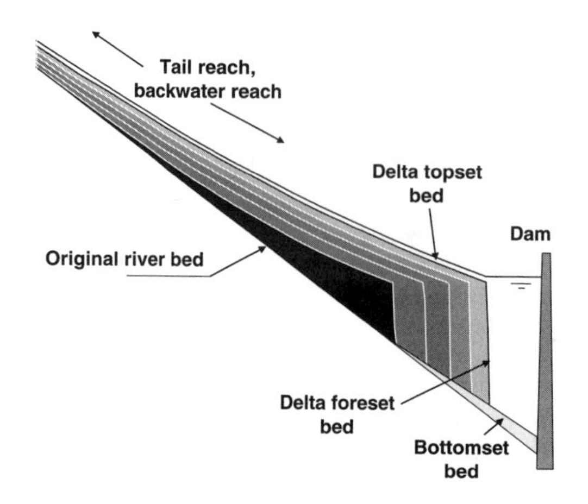

Figure 1. Definition sketch of development of sediment deposits in a reservoir, extending in upstream direction.

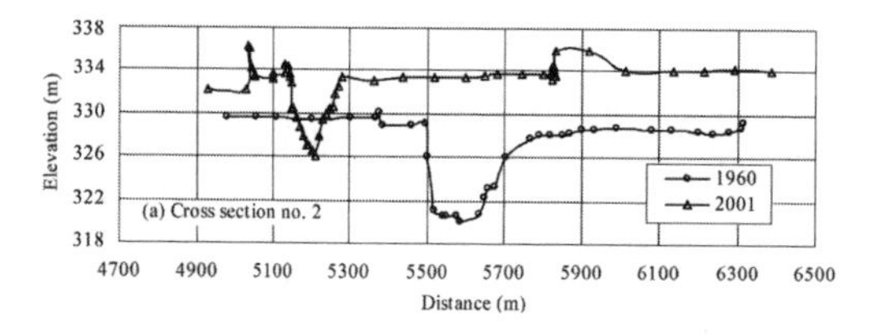

Figure 2. Aggradation in the Lower Weihe River at a cross-section 21 km upstream of Tongguan (Wang and Hu, 2004).

levels, but also changes in the channel pattern. Between 1960 and 1969 the sinuosity of the meandering river channel decreased from 1.65 to 1.06. The meanders were buried and channel became more straight and unstable. Other impacts that can be expected in general are for instance listed by Fan and Morris (1992) and Hotchkiss (2004). Among those are environmental effects (rise of water table, wetland habitat creation, etc.), (socio-) economical effects (decreased navigational clearance, problems with water intakes, relocation of communities, etc.).

For the purpose of preventing an unacceptable increase of flood levels in the backwater reach of reservoirs it is necessary to take appropriate measures, for instance by reduction of the sediment supply, by locally increasing the transport capacity, or by mechanical removal. An effective measure is often the use of erosive force of flood waters to flush part of the sediment in the backwater reach, and in the reservoir. The effect is brought about by reversal of the flow deceleration by backwater which causes the sedimentation, i.e. by lowering the base level (downstream water levels) and/or by increasing the transport capacity locally.

However, permitting only limited water losses and accounting for environmental constraints, the efficiency of these operations require careful planning and preferably have to be combined with additional measures. Tools to support the planning and efficacy of these operations are mainly mathematical models and physical models that are able to reproduce (to some extent) the relevant processes. However, due to the complexity of the processes the predictive capabilities of these tools have proven to be limited. It has to be judged whether more advanced tools and new modelling techniques will lead to improvement.

This paper focuses on the result of a study that aims at a better understanding and prediction of flushing processes in the backwater reach, combined with mechanical support to enhance the efficacy of the sediment removal. Both mathematical simulation using the Delft3D modelling system, and prototype experiments in the Tenryuu River in Japan have been used to carry out this study.

2 PROCESSES AND RELEVANT FACTORS AFFECTING FLUSHING SEDIMENT FROM BACKWATER REACH

By raised water levels in the reservoir the backwater effect in the upper reach causes decelerating flow, and consequently deposition of supplied sediment at a long reach. The primary effect is of a one-dimensional (1D) nature. That means that the aggradation can be considered at first in terms of cross-sectionally averaged values, causing very roughly an equivalent rise of water levels. Note that in the long run, by a decrease of bed-slope, changes in channel pattern and roughness characteristics (e.g. vegetation and change in grain size), the rise of water levels can sometimes even more than the rise of bed levels.

Considering that the primary source of aggradation is the 1D backwater-effect, it is a most obvious measure to operate the reservoir in such a way that the backwater effects are reduced or even reversed during reasonable long periods. This means that the flow-deceleration is temporarily removed to sluice sediments, or, even better, that flow acceleration is created to flush the deposited sediment. Alternative measures to reduce the supply of sediment (blocking the sediment upstream), or to remove sediments by dredging, are often much less efficient.

When looking into more detail to the flushing process, it is often seen that by lowering the base level (reservoir level) a flushing channel is incised into the river bed by retrogressive erosion (e.g., Sloff, 1991). Particularly in relatively wide river reaches the scouring effect of flushing is limited to this narrow channel, and deposits stored in banks and flood plains are not affected. These higher deposits are usually fine sediments that are supplied during floods. Furthermore, by flushing operations more sediment is brought

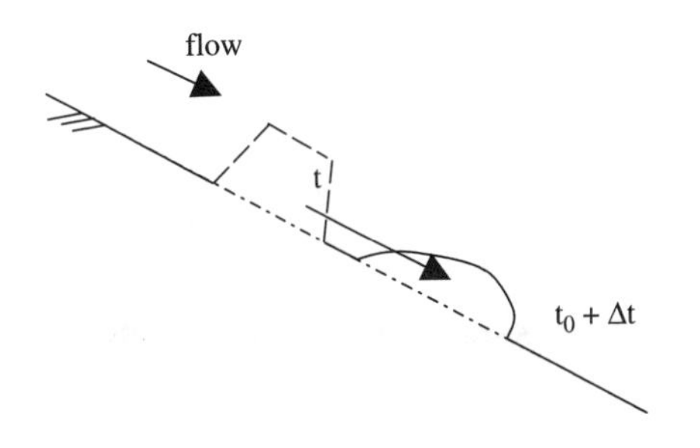

Figure 3. 1D propagation of a dumped shoal.

down into the reservoir, increasing the propagation of the delta into the active storage and just displacing the problem to the reservoir. Although it is often appropriate for sediment management to bring the sediments to the reservoir, sometimes other constraints limit the available time window, acceptable water-level draw down and allowed sediment concentrations. For instance economical losses due to reduction of hydropower or environmental pressure on downstream reaches may restrict the feasibility of flushing.

In such situations flushing efficiency can be enhanced if deposits that are not reached by the scouring flow are pushed into the flushing channel mechanically. For example such a supporting action was mentioned by Scheuerlein (1987) in respect to flushing the deposits in the reservoir itself. The morphological response to such actions depends on the amount and position of the displaced sediments. It is a combination of the following two effects:

(1) Sediments are placed evenly distributed over the flow width (for example the width of the flushing channel) forming a shoal. The shoal will propagate in downstream direction as a deforming 1D disturbance, illustrated schematically in figure 3. The propagation celerity or wave speed of this disturbance can be expressed by:

$$c_b = \frac{ns_0}{h_0\left(1 - Fr_0^2\right)} \tag{1}$$

where $c_b =$ 1D celerity; Fr_0 is Froude number; h_0 is average depth; $s_0 =$ transport rate in undisturbed flow; and n is the power of the flow velocity in the sediment-transport formula, which follows if the transport relation is approximated by a power law of the flow velocity, i.e. $s_0 = m \cdot u_0^n$ with u_0 is average flow velocity.

(2) If sediments are moved from shallow parts to the deep parts of a cross-section as shown schematically in figure 4, also a downstream propagating disturbance is created. However, the response in this case is of two-dimensional (2D) nature, leading to a much lower celerity than that of the 1D disturbance. The celerity of this disturbance traveling downstream as a type of free bar follows from the non-stationary solution

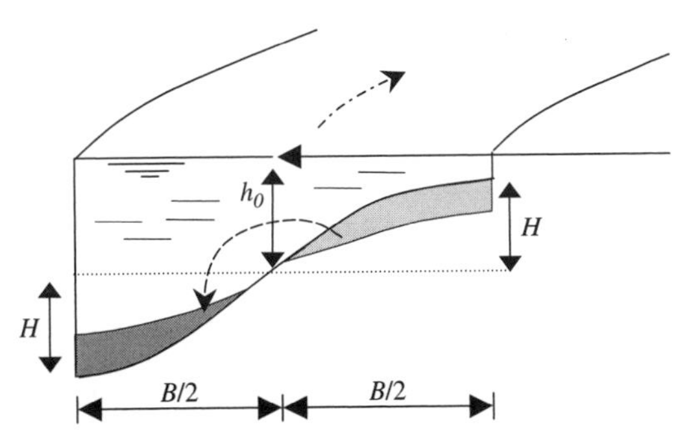

Figure 4. Displacement of sediment within a cross-section in a river bend.

of the linear analysis of the 2D differential equations for flow, transport and morphology (e.g., Struiksma and Crosato, 1989). Consider the homogeneous part of the linearized time-dependent equation for bed development, not considering the forcing terms, following Crosato (1990):

$$\lambda_s \frac{\partial^2 H}{\partial x^2} + \left(1 - \frac{n-3}{2}\right)\frac{\partial H}{\partial x} + \frac{1}{\lambda_w} H + \left(\frac{h_0 \lambda_s}{s_0}\right)\frac{\partial^2 H}{\partial x \partial t} + \left(\frac{h_0 \lambda_s}{s_0 \lambda_w}\right)\frac{\partial H}{\partial t} = 0 \tag{2}$$

in which $H =$ near-bank deformation of water-depth or amplitude of cross-section; $t =$ time; $x =$ downstream coordinate; and λ_w, λ_s are the characteristic adaptation lengths of tangential velocity and bed-topography deformations respectively, written as:

$$\lambda_w = \frac{C^2 h_0}{2g} \tag{3}$$

$$\lambda_s = \frac{f(\theta)}{\pi^2}\left(\frac{B}{h_0}\right)B \tag{4}$$

in which $C =$ Chézy coefficient; $g =$ gravity; $B =$ width; and function $f(\theta)$ expresses the effect of transverse bed-slope on the direction of sediment transport, and is based on the Shields parameter $\theta = u_0^2/(C^2 \Delta D_m)$ with D_m is the mean grain size, and Δ is the relative density of sediment defined by $\Delta = (\rho_s - \rho_w)/\rho_w$ where ρ_s, ρ_w are the densities of sediment and water respectively. The solution of this equation in the linear analysis is written as:

$$H(x,t) = \hat{H}(t) e^{+\frac{t}{T}} \sin k'(x - c_r t) \tag{5}$$

in which T is the time scale, k' is the wave number of a bed disturbance, and c_r the celerity of a 2D bed disturbance. The wave number k' is related to the length of bars or wave length of typical bends, L_p, and expressed as $2\pi\lambda_w/L_p$. The time scale T is negative for damped and positive for growing bars, and can be expressed by

$$T = -T_0\left(1 - \frac{1}{2}\frac{(n-1)k'^2}{1+k'^2}\frac{\lambda_s}{\lambda_w}\right)^{-1} \tag{6}$$

in which $T_0 = \lambda_s h_o/s_o$ the ratio λ_s/λ_w is also called the interaction parameter, and increases for instance for large width to depth ratios (eventually resulting in positive values of T). For free bars the 2D celerity can now be written as:

$$c_r = \frac{s_0}{h_o}\frac{k'^2 - \left(\frac{n-3}{2}\right)}{k'^2 + 1} \tag{7}$$

This celerity is much lower than the 1D celerity expressed by equation (1). This can be explained by the room that is created on the other bank, allowing the flow to evade.

In general the type of response can be a mixture of the two effects described above, but it is also possible that one of the effects dominates. The actual celerity of the disturbances generated by the displacement of sediment is not always well presented by the simple relations from linear analysis shown above. The effect of unsteadiness of the flow and of graded sediment may lead to a more complex behavior.

Lanzoni and Tubino (1999) show that progressive coarsening which takes place along the upstream face of the bars, by selective transport of different grain fractions, induces a damping of the celerity, a reduction of the growth rate and a shortening of bar wave lengths. This has been confirmed by their theory and laboratory experiments for graded sediment.

The effects of grain sorting and flow unsteadiness can best be studied using mathematical models. In the following sections is shown how the analytical approach compares to a numerical modelling approach applied to a prototype flushing experiment.

3 FLUSHING EXPERIMENT IN SAKUMA RESERVOIR IN JAPAN

3.1 *Introduction*

Sakuma dam is located in the Shizuoka region in Central Japan, and is the largest of the 5 hydropower dams that are constructed in the main course of the Tenryuu River. The Sakuma dam is 155.5 m high and 294 m long concrete gravity dam, constructed in 1956 near the end of a narrow gorge directly downstream from the Inada Valley at about 70 km from the river mouth (see figure 5). The maximum power output is 350 MW. The initial storage capacity of the reservoir amounted 327 Mm^3, but in 2004 only about 213 Mm^3 remained as about 114 Mm^3 is lost due to sedimentation. In Figure 6 is shown how the delta and backwater deposits developed in time. The reservoir upstream of Sakuma is of a gorge type, with a narrow elongated shape. It is assumed that the backwater-effect of the dam is effective in the reach between about 20 km to 30 km upstream of the dam.

Figure 5. Sakuma dam.

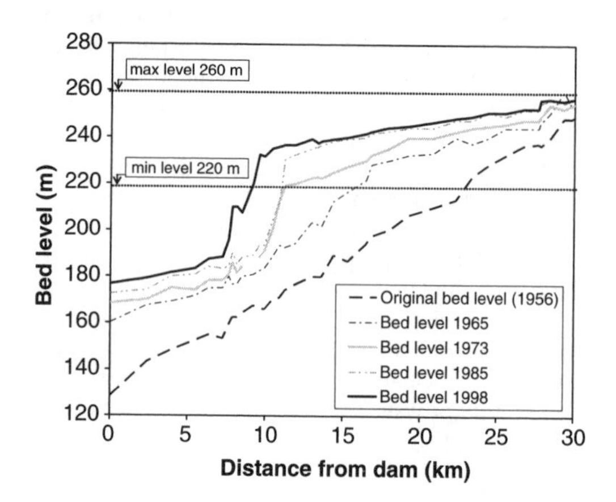

Figure 6. Delta growth observed in Sakuma reservoir, Japan.

The average annual inflowing water volume is about 5 Gm^3/year, carrying large amounts of sediment to an amount of more than 1.8 Mm^3/year presently. Since 1979 a sediment management program has been in force, aiming by sediment rejection measures a river-bed equivalent to that in 1970, established against a design flow of 7,700 m^3/s in the backwater reach. The

Figure 7. Sediment displacement from point bar towards outer bend using caterpillars.

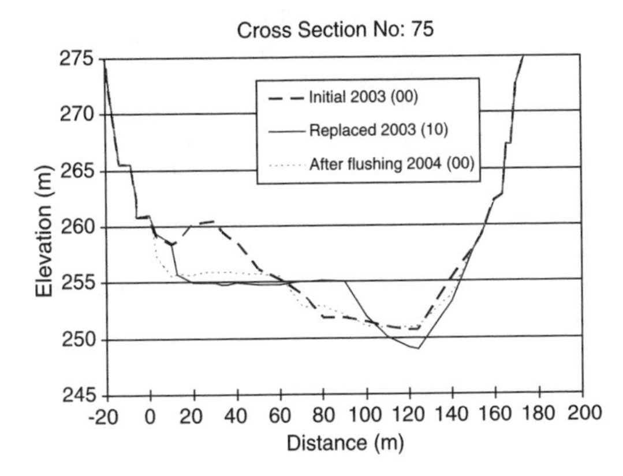

Figure 8. Sediment displacement at cross-section 75, at 27,845 m from the dam.

applied measures to remove accumulated sediments from the backwater reach:

- Flushing or sand-sweeping by natural down flow of accumulated sediment, using the sweeping force of the water discharge in spring, with reservoir lowered during winter period (middle of February to end of March). Target is 1 to 2 Mm^3/year.
- Transferring about 0.4 Mm^3/year of sediment from the top-set bed to the reservoir by 2 pump dredgers.
- Mining about 0.4 Mm^3/year of gravel from the top set bed by pump dredgers.

During the low-flow period of 2004 sediment deposits in three inner-bend areas in the backwater reach of Sakuma reservoir have been moved towards the outer bend by means of shovels. These areas are located about 24, 27 and 28 km upstream of the dam (cross-sections 61, 70 and 75 to 78). The shifted sediment dikes at each area amounted between 10,000 and 20,000 m^3, with a total of about 85,000 m^3, and thickness of 2 m to more than 6 m.

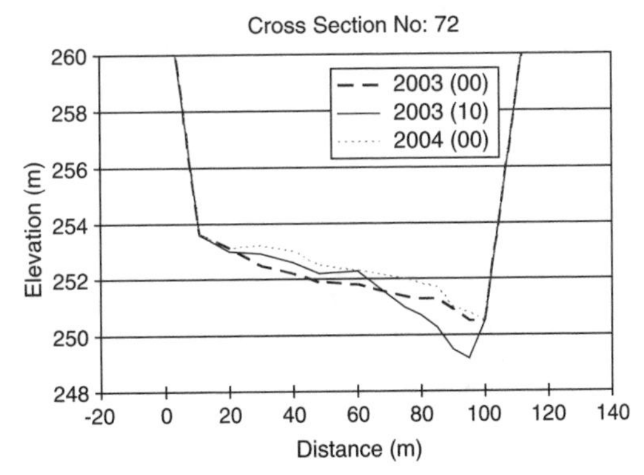

Figure 9. Sediment displacement at cross-section 72, at 27,286 m from the dam.

In figures 8 and 9 is shown how two cross-section directly at the excavation site and 600 m downstream have developed in time. Due to flushing part of cross-section 75 is cleared, but the point bar has not been restored yet. Downstream cross-sections do not show significant changes that can be attributed solely to these excavations, since natural variations have a similar magnitude. In Table 1 an estimation has been given on some characteristic parameters for 1D and 2D development of the bed disturbances during typical flood conditions.

Parameter	Value
Discharge	1200 m^3/s
Flow velocity	1.7 m/s
Depth	5.8 m
Chézy value	30 $m^{1/2}/s$
Froude number	0.23
Shields value ($D_m = 5$ mm)	0.4
Transport rate	0.001 m^2/s (0.13 m^3/s)
Celerity 1D disturbance c_b	80 m/day
λ_w	258 m
λ_s	163 m
k'	1.63
T'	−135 days
Celerity 2D disturbance c_r	2.6 m/day

The values in this table indicate that under these conditions (that may occur a few days a year) the impacts of the sediment displacement propagate only a few tenths to hundreds of meters downstream. Furthermore the negative time scale T' suggests that these effects are damped out. This might explain the limited effects in prototype. Furthermore the sediment in this reach is graded, and sediments from the point bar consist mainly of fine size fractions. The actual behavior of the created sediment dikes may therefore be different than the simple analytical approach suggests. Therefore more detailed computational model simulations have been carried out to reveal this. The results of this model are discussed in the following sections.

3.2 *2D morphological modelling with graded sediment*

Simulations for the Sakuma reservoir have been carried out using a 2D depth-averaged morphological schematization in the Delft3D modelling system. The approach and theoretical background are for instanced presented in Sloff et al. (2001). For simulation of sorting effects this model applies the basic bed-layer concept similar to that of Hirano (1971) as shown in figure 10, in which the bed is subdivided in transport layer with thickness δ_a, and a non-moving substratum that is either schematized with a homogeneous composition, or it is schematized by a number of sublayers for which a bookkeeping system the substrate composition, taking into account the history of its deposits.

In this figure p_i is the probability of a size fraction, z_b is the average bed level, and ϕ_i is the vertical sediment flux through the interface. The active layer represents the upper layer containing the material which is taking part in the actual sediment-transport process, and its thickness is generally related to the average height of bed forms (dunes, ripples). Changes in substrate during a simulation occur for instance when due to sedimentation processes material transported from upstream is deposited and added to the substratum.

The local bed-load transport rate per fraction is described using a standard transport formula. For the two-dimensional approach the Ashida and Michiue formula, prepared for graded sediment simulations in two dimensions following Egashira et al. (1997), are applied. It should be remarked that extension of Ashida and Michiue formula to two-dimensions is not unambiguous, and permits alternative formulations. Particularly variations with respect to the slope effects

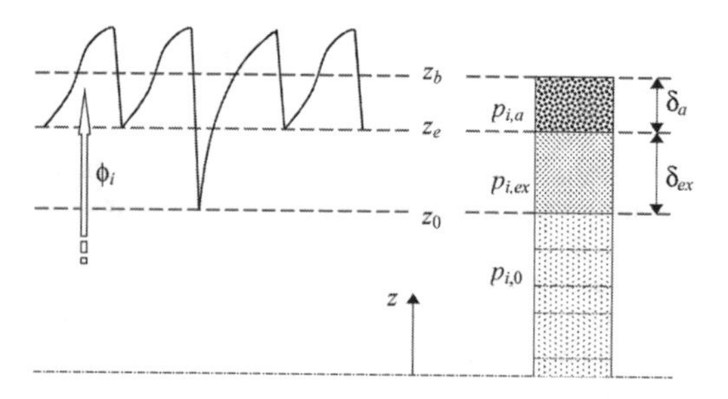

Figure 10. The bed-layer schematization for graded-sediment modelling, where sub-script i is associated to sediment size fraction i.

on transport direction have been applied to improve the results. This relates to the downhill gravitational transport component, which causes the transport direction and transport magnitude in a 2D model to not necessarily coincide with the direction of the bed shear stress of the flow, and the transport capacity on a horizontal bed. In the Ashida and Michiue formula these bed-inclination effects influence the transport rate through a correction factor K_c, and the direction of sediment transport β_i through an expression β_i. The direction of sediment transport is found to be of major importance for the development of typical 2D morphological features, as has been shown in section 2. In graded-sediment simulations the complex interaction between lateral sorting and spatial variations in bed topography do create an additional degree of freedom that is difficult to manage.

The computational model is using a finite-difference approach for which the river and reservoir are schematized on a curvi-linear grid. For helical flow or spiral flow in bends, a typical 3D flow feature that affects the direction of transport through the direction formulas mention above, a parameterization is applied that simulates the production and advection/diffusion process of spiral flow intensity in curved flow.

The simulations must include the effect of the discharge hydrograph because the main erosion of the displaced sediments occurs during flood events (when the deposits are submerged).

3.3 *2D Simulation for flushing operations in Sakuma reservoir*

The flushing of displaced sediments occurs particularly during flood periods. During these events the full cross-section (both the shallow bars as the deep outer bends) are submerged, and the actual propagation of the features is of a mixed 1D and 2D nature (see section 2). Simulations have been carried out for the flood periods following the sediment-displacement operations.

The model schematization has been set up for the reach where the sediment-displacement activities have been carried out, i.e. including the cross-sections 84 (29.5 km upstream of the dam) to 61 (24 km from dam). To allow the simulation of downstream migration of the sediment dike near cross-section 61, the model reach had to be extended several kilometers downstream of cross-section 61. Also some upstream extension is needed for absorbing the boundary effects. The initial bathymetry of the model was taken from digital cross-section data of year 2004, and interpolated to the curvi-linear grid. Cross sections were measured at a varying mutual distance between about 100 and 600 m. For the bed composition in Sakuma reservoir the data from the granulometric survey of 1999 were used. Samples were taken from the top-set bed of the delta to a depth of several meters below the bed, and numbered according to the representative cross-section number where the samples were taken. Figure 11 shows the averaged grain size distributions of some samples of the top-set bed that can be considered representative for various locations along the delta. The fore-set of the delta is located between 6.5 and 10 km upstream of the dam. The grain-size distribution of the bed is schematized in the model using 11 size fractions.

At the inflow boundary the observed time-series of discharges can be introduced; this is useful for reproducing actual events. When the focus is more on the general behavior of the system, it is often better to use a slightly adapted version of the hydrograph, namely such that it can be expressed by a step-wise discharge variation. This reduces the computational time but does not significantly affect the large-scale morphological results. A further sub-division is to separate low-flow periods from flood periods: the low flow periods may not have to be simulated because sediment transport does not occur during that period. For the relevant simulation period the maximum discharge peak (October 2004) was about 3000 m^3/s. However, for the purpose of analysis it has been chosen to represent the flood with an average characteristic discharge of 1200 m^3/s. It should further be noted that during the period of highest flood peaks the reservoir level has not been drawn down, and the river bed (including the point bars) is fully submerged.

Figures 12 and 13 shows the impact of the displaced sediment on the local morphology after about 48 hours of morphological activity at a constant discharge of 1200 m^3/s. The figures show the difference in bed level as computed with and without the sediment displacement (the computation without sediment displacement is the reference simulation). Flow direction is from the top of the figure to the bottom of the figures. The sediment deposits have spread out over a larger area, having a downstream migration

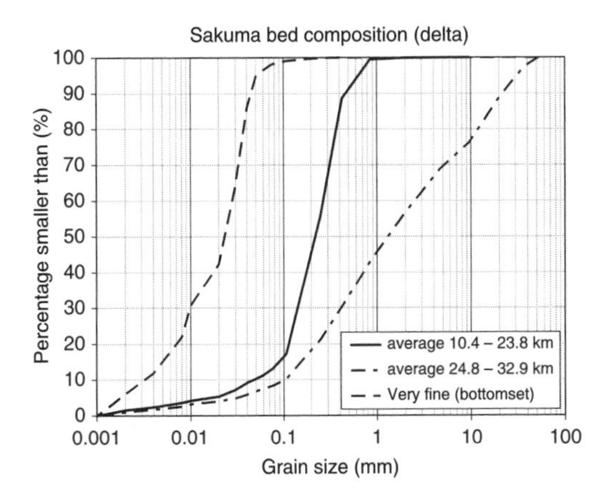

Figure 11. Granulometric survey of the Sakuma-Reservoir top-set bed: grain size distributions average for some sections (distance relates to distance from dam).

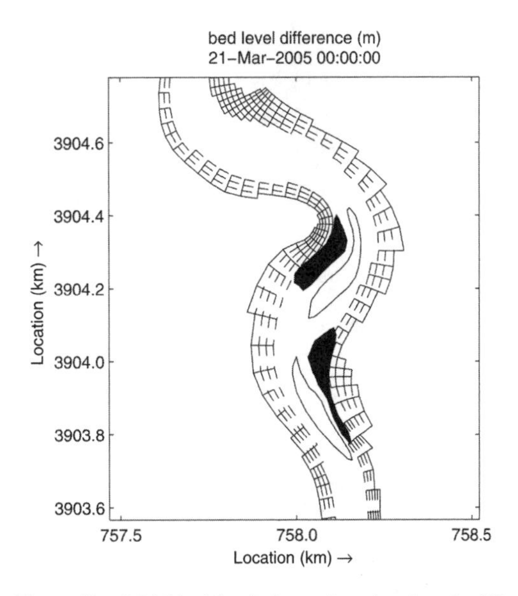

Figure 12. Initial bed level change introduced to the 2D model: i.e., difference in bed level compared to reference situation (contours are +1 m and −1 m: black areas: lowered bed levels; white areas: raised bed levels).

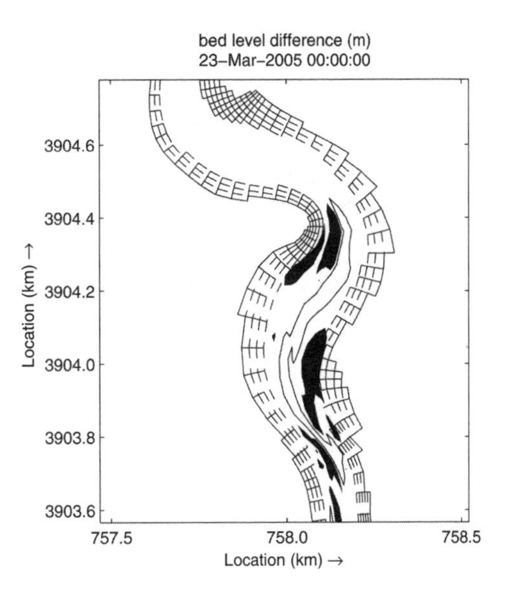

Figure 13. Difference in bed level after 2 days of flood, computed with displaced sediments compared to the reference simulation (contours are +1 m and −1 m: black areas: lowered bed levels; white areas: raised bed levels).

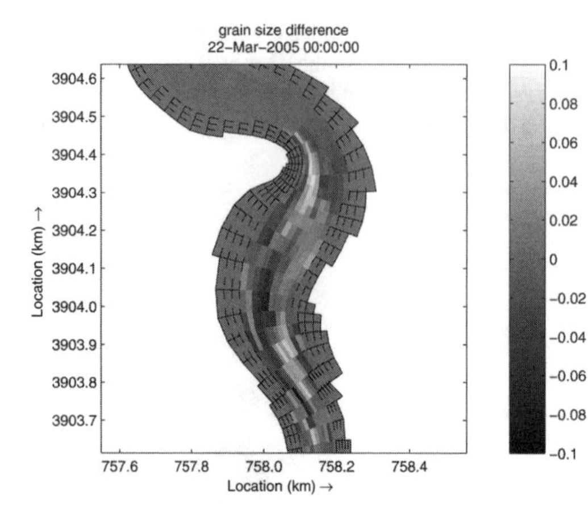

Figure 14. Difference in grain size after 2 days of flood, computed with displaced sediments compared to the reference simulation (light areas: lower mean grain size; dark areas: higher mean grain sizes, with grain sizes in meter).

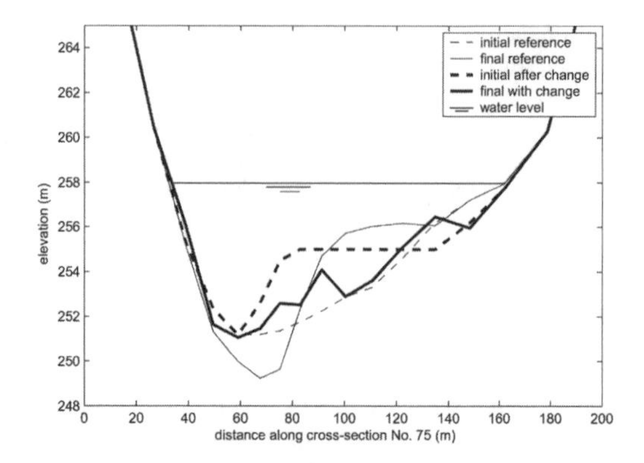

Figure 15. Simulated results after 2 days for situation without sediment displacement (reference) and with displacement (after change) at cross-section 75, at 27,845 m from the dam.

speed of approximately 100 to 200 m per day. Sedimentation occurs at inner bends (where sediment was removed) and erosion occurs in the deeper part of the bend (where sediment was placed).

The results in figure 13 show that the effect of displaced sediments has propagated downstream, but that the bed disturbances still remain strongly present. The observed celerity was found to be in the order of less than 20 m/day, indicating a dominance of 2D propagation behavior for which previously analytical celerities of about 3 m/day were computed, versus 80 m/day for 1D propagation. A rapid adaptation occurs at the transition from the upstream bend to the downstream bend, where a front travels down over a distance of 50 m in about 4 hours, after which it merges with the downstream effects and stagnates. At this crossing the effects are expected to behave more in a 1D way which explains the higher celerities.

The 2D redistribution that characterizes the 2D behavior of these locations is also revealed by the resulting grain-size distribution. In figure 14 the difference is shown in mean grain size for a flushing simulation with and without the displaced sediments. Although the figure shows a more or less scattered result, it can be seen that after displacing sediments from the inner to the outer bend, generally the (deepened) inner bends show a coarsening of the bed whereas the (filled) outer bends show a fining. The reason for this behavior can be explained from the redistribution of flow and transport, where more water flows through the lowered inner bend and less flow through the shallower outer bend. Also slope effects contribute to this behavior. Furthermore, the redistribution of grain sizes and the resulting spatial variation of sediment sizes are believed to contribute to a further stabilization of the cross-section as shown in section 2.

The development of cross-sections in these simulations appeared to agree quite well with the observed cross-sections as shown in figures 8 and 9. For instance in figure 15 the simulated results for cross-section 75 have been presented.

Nevertheless, by comparing figure 15 to figure 8 it can be observed that in the prototype the sediments that were pushed into the outer bend have been eroded after the flood, whereas in the model these sediments are still present. Possible reasons for this deviation can be a finer composition of the displaced material in prototype (that is more easily flushed), and 1D-type of erosion that occurs during the rising stage of the flood before the cross-section is fully submerged (and flow is concentrated in the outer bend). The model allows for further analysis with respect to these propositions.

4 CONCLUSIONS

By studying the processes and phenomena that occur when sediment flushing is enhanced by mechanically displacing sediments from high areas to the flushing channel, several opportunities have been uncovered that may increase the efficacy of these operations. The analysis based on analytical and numerical models has shown that the displaced sediment leads to a combined 1-D and 2-D response of the riverbed. In situations with reasonably uniform sediment, where transport processes are not affected by sorting processes, the 1-D propagation is faster and more effective than the 2D behavior. With respect to enhancement of flushing operations, it can be concluded that the efficacy may therefore be improved if sediments are distributed evenly over the cross-section of the flushing channel, and water levels are lowered to such an extent that the flow is concentrated within this channel.

In practice the river bed often consists of graded sediments, and by transverse sorting processes finer sediments are found on the higher parts of point bars, whereas the coarse sediment are usually found at the deep outer bends and the front-edge of the bars. As fine sediments are more easily eroded during flushing operations, it is therefore more effective to particularly push the fine sediments from the high bars into the deeper part of the cross-section.

In more uniform sediment situations the mechanically modified cross-section is expected to gradually return to its original shape soon. In graded sediment situation the time scale of this adaptation process is larger because of a redistribution of sediment fractions which stabilizes the bed.

To simulate the relevant processes a 2D or 3D mathematical model is the appropriate tool. For practical applications the effect of sediment sorting and unsteadiness of flow need to be accounted for, as these effects appear to have an important impact on the morphological response to the cross-section modifications introduced to enhance the efficacy of flushing.

REFERENCES

Crosato, A. (1990) Simulation of meandering river process. Comm. on Hydraulic and Geotechnical Engrg., Report No. 90-3, Delft University of Technology, Faculty of Civil Engrg., ISSN 0169–6548.

Egashira, S.H., Jin and K. Ashida (1997) Numerical model for river mouth sand-bar flushing and its application to river mouth regulation. Proc. 27th congress of IAHR, Theme B, Vol. 1, pp. 719–724, San Francisco, aug. 10–15, 1997.

Fan, J. and G.L. Morris (1992) Reservoir sedimentation. I: Delta and Density Current Deposits. J. Hydr. Engrg., ASCE, Vol. 118, No. 3, pp. 354–369.

Hirano, M. (1971) River bed degradation with armouring. Trans. of JSCE, Vol. 3, Part 2.

Hotchkiss, R.H. (2004) Challenges of managing sediment deposition upstream from large hydroelectric projects. Proc. of the Ninth Intern. Symposium on River Sedimentations, Oct. 18–21, Yichang, China, Vol. I, pp. 243–249.

Lanzoni, S. and M. Tubino (1999) Grain sorting and bar instability. J. Fluid Mech., Vol. 393, pp. 149–174.

Scheuerlein, H. (1987) Sedimentation of reservoirs- methods of prevention, techniques of rehabilitation. First Iranian Symp on Dam Engrg., Teheran, Iran, June 1987.

Shinjo, T. and Y. Fujita (2004) Studies of sediment in large-scale reservoir for power generation. In: Greco, Carravetta & Della Morte (eds.) River Flow 2004. Proc. Vol. I, pp. 731–738.

Sloff, C.J., Reservoir sedimentation: a literature survey. Comm. on Hydr. and Geotech. Engrg., Report No. 91–2, ISSN 0169-6548, DUT, Delft, The Netherlands.

Sloff, C.J., H.R.A. Jagers, Y. Kitamura and P. Kitamura (2001) 2D morphodynamic modelling with graded sediment. Proc. of 2nd IAHR Symp. on River, Coastal and Estuarine Morphodynamics 10–14 sept. 2001, Obihiro, Japan.

Struiksma, N. and A. Crosato (1989) Analysis of 2-D Bed Topography Model for Rivers. In: River Meandering, AGU, Water Resources Monograph 12, Eds. S. Ikeda and G. Parker, pp. 153–180.

Wang, Z. and C. Hu (2004) Interactions between fluvial systems and large scale hydro-projects. Proc. of the Ninth Intern. Symposium on River Sedimentations, Oct. 18–21, Yichang, China, Vol. I, pp. 46–64.

River, Coastal and Estuarine Morphodynamics: RCEM 2005 – Parker & García (eds)
© 2006 Taylor & Francis Group, London, ISBN 0 415 39270 5

Fine-grained sand behavior in a dam reservoir at the middle reaches of a river

T. Shimada, Y. Watanabe & H. Yasuda
Civil Engineering Research Institute of Hokkaido, Sappro, Japan

ABSTRACT: Sedimentation in dam reservoirs is a matter of great concern in river management, because sediment deposits in reservoirs tend to be less prone to transport than riverbed materials, even for grains of the same diameter. Hydraulic experiments were conducted on the fall velocity of sediment and the entrainment rate of suspended sediment using material deposited in a dam reservoir. It was found that the fall velocity of sediment and the entrainment rate of suspended sediment are smaller than conventional values that are currently used. These results are used to calculate the bed evolution at the Nibutani Dam reservoir that resulted from a flood in 2003. The calculated results largely agree with field-survey-based results.

1 INTRODUCTION

In recent years, changes in flood flow capacity caused by riverbed degradation and aggregation, and sediment budget imbalances between the upper and lower reaches have been identified as issues requiring attention in river administration. Sediment transport in rivers is most active during floods, at which time sediment particles are supplied downstream in great quantity. Field surveys on sediment transport during floods are being performed at various locations. The Civil Engineering Research Institute of Hokkaido (CERI) has been performing surveys during floods on the Saru River, in the Hidaka region of Hokkaido, Japan. The river has a dam reservoir at the middle reaches (Figure 1). Through the survey, data such as observation data during floods and from cross-sectional surveying have been accumulated in an effort to understand the movement of suspended sediments (hereinafter: SS) and nutrients. Figure 2 shows the Saru River basin. At Nibutani Dam (Figure 3), which is at the middle reaches of this river, annual sedimentation is 1 million m^3, and this is an issue for river maintenance. The flood caused by Typhoon Etau (Typhoon No. 10) in August 2003, which had the highest discharge on record, dumped 2.4 million m^3 of sediment into the dam reservoir. This research is a first step toward estimating the sediment budget throughout the river basin, which has a dam reservoir at the middle reaches. Such estimate would be based on flood observation records for the Saru River. The research intends to reach an understanding of the transport of fine, cohesive soil particles that accumulate in the dam reservoir. Such transport has not been well understood. Undisturbed samples of sedimentary deposits from the bottom of August 2003 flood were carefully taken from the Nibutani Dam Reservoir. Sedimentation and

Figure 1. Location of the Saru river.

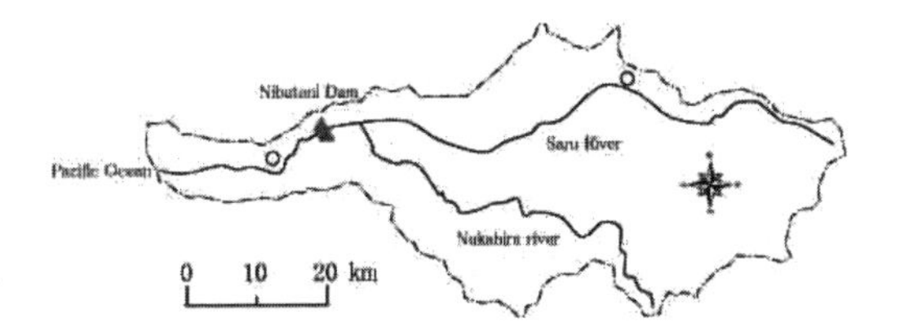

Figure 2. Saru river basin.

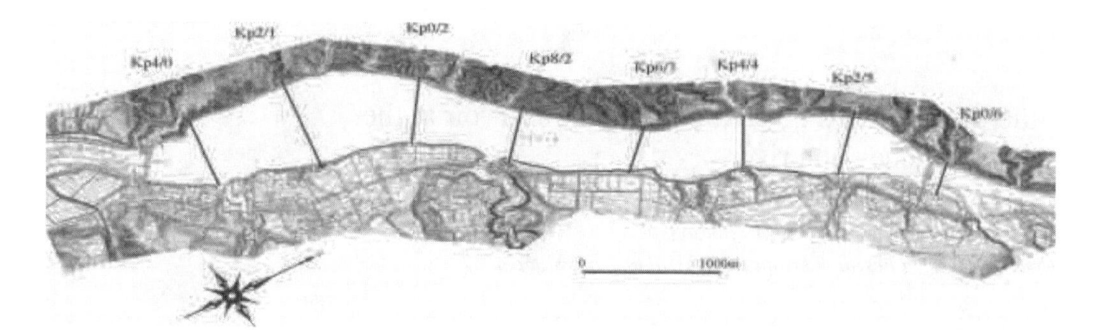

Figure 3. Nibutani dam reservoir.

entrainment experiments were performed on the sediment particles, and estimation equations for sedimentation and entrainment in the Nibutani Dam Reservoir were derived to clarify whether the current equation for calculating the bed variation is applicable. Furthermore, using the experiment results, attempts were made to reproduce the sedimentation and entrainment that occurred in the dam reservoir during the flood in August 2003.

2 EXPERIMENT ON SOIL PARTICLE SEDIMENTATION

2.1 *Outline of the experiment on sedimentation*

The water tank used in the experiment is shown in Figure 4. Bed material sampled from the Nibutani Dam Reservoir was divided according to grain size. We used clay and silt components that passed through a 0.0075-cm sieve mesh and fine sediment component that passed through a 0.0250-cm sieve mesh but was retained by a 0.0075-cm sieve mesh. The sieved samples were made into instilment samples with deionized water to produce a suspension of 10 g/L. For each experiment, 1 to 2 mL of sample was instilled into a water tank filled with deionized water. The sedimentation of soil particles was recorded by video camera, and the diameter and fall velocity of each soil particle was measured by image processing. Fall velocity of soil particle is greatly influenced by water temperature. Therefore, the experiment was performed in a constant-temperature room at a constant room temperature of 20°C. The classification of sediment by grain size is shown in Table 1.

2.2 *Results of the experiment on sedimentation*

The results of the experiment are shown in Figure 5.

Results calculated using Rubeyfs equation (1), the generally used fall velocity equation, are plotted in the figure for comparison.

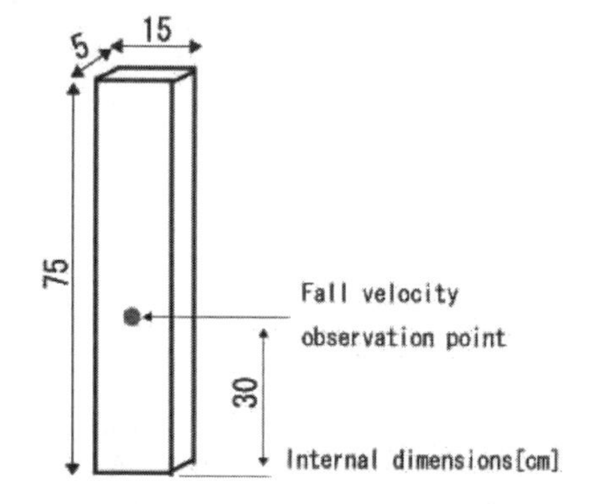

Figure 4. Water tank used in the sedimentation experiment.

Table 1. Classification of sediment by grain size.

	Grain size range (cm)
Clay	0.0000 ‘ 0.0005
Silt	0.0005 ‘ 0.0075
Fine sand	0.0075 ‘ 0.0250
Medium-grain sand	0.0250 ‘ 0.0850
Coarse sand	0.0850 ‘ 0.2000

$$\begin{cases} w_{fi} = \left(\sqrt{\frac{2}{3} + \frac{36\nu^2}{sgd_i^3}} - \sqrt{\frac{36\nu^2}{sgd_i^3}}\right)\sqrt{sgd_i} & : d_i \leq 0.1cm \\ w_{fi} = \sqrt{\frac{2}{3}}\sqrt{sgd_i} & : d_i > 0.1cm \end{cases} \quad (1)$$

Where, w_{fi} is the fall velocity, ν is the kinematic viscosity coefficient of the water, s is the specific gravity of the soil particle, d_i is the soil grain-size (hereinafter, the subscript i indicates a physical quantity for grains whose size is d_i), and g is gravitational acceleration. Figure 5 shows that the fall velocities of sediment particles whose gain-size classifications are gclayh and gsilth are smaller than those obtained by Rubeyfs equation, while the fall velocity of sediment particles whose grain-size classification is gfine sandh approaches that calculated using Rubeyfs equation.

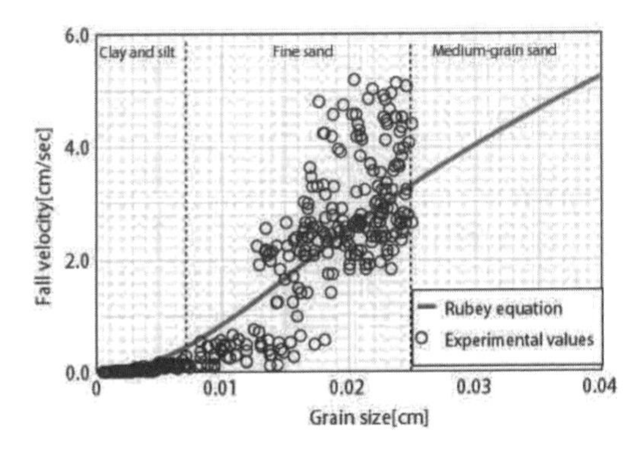

Figure 5. Results of the fall velocity experiment.

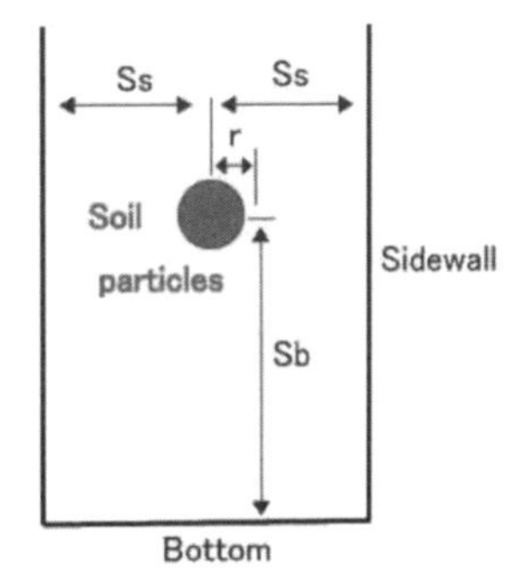

Figure 6. Influence from the bottom and side surfaces of the water tank.

2.3 *Estimation of fall velocity*

Based on the above experiment, estimation of an equation for fall velocity of soil particles in the Nibutani Dam Reservoir was attempted. Rubey performed an experiment and obtained Equation (2), which shows the relation between the coefficient of friction C_D of the soil particle and the Reynolds number R_e.

$$C_D = \frac{24}{R_e} + 2 \tag{2}$$

Where, $R_e = w_{fi}d_i/\nu$. The relation between the drag coefficient C_D and the Reynolds number R_e was estimated by using the soil grain-size and fall velocity obtained in this experiment. However, there is the possibility that the fall velocities were influenced to some degree by the experimental setup. Therefore, estimation of the drag coefficient needs to incorporate such influences. Factors that can influence values measured in such an experiment include the sediment concentration in the water tank and influences from the bottom and sidewall surfaces of the water tank. First we address sediment concentration. Figure 4 shows the water tank used in the sedimentation experiment. It is assumed that after the sample is instilled, by the time the sample reaches the point of observation, which is 30 cm from the bottom, the sample will have spread evenly within the water tank. The sediment concentration is about 0.0003% when it passes the observation point, and since the focus is on samples that are finer than fine sand, it is not regarded as influencing the fall velocity [1]. Furthermore, the use of deionized water is thought to prevent the occurrence of electrochemical forces. Next we address the influence of the bottom and side surfaces of the water tank on the measured fall velocities (Figure 6). In correcting the influence from the bottom surface, the use of Brennerfs equation (3) [2] is proposed.

$$K_{fb} = 1 + \frac{9}{8}S_{rb} \tag{3}$$

Where K_{fb} is the correction factor for the fall velocity, $S_{rb} = r/s_b$, r is the radius of the soil particle, and s_b is the distance between the bottom surface of the water tank and the center of the soil particle (30 cm in this experiment). Correction of fall velocity is needed when $S_{rb} > 15$; however, under the experimental conditions, only sediment gains greater than 300 cm in diameter would be influenced by the bottom surface. Therefore, it is not necessary to consider the influence of the bottom surface of the water tank. In correcting the influence from the side surfaces of the water tank, the use of McNownfs [3] equation (4) is proposed.

$$K_{fs} = 1 + 1.006S_{rs} \tag{4}$$

Where $S_{rs} = r/s_s$, s_s is the distance between the side surfaces of the water tank and center of the soil particle (2.5 cm in this experiment). The average value of the individual K_{fs} obtained from the experiment results was approximately 1.003. Using this correction factor, the observed fall velocity of the individual soil particle is corrected in Equation (5).

$$w_{fk} = K_{fs}w_f \tag{5}$$

Where, w_{fk} is the corrected fall velocity and w_f is the observed fall velocity. From this result, the relation between the calculated drag coefficient of resistance and the Reynolds number R_e is shown in Figure 7. The coefficients of resistance calculated from the experimental results tend to be higher than those calculated using Rubeyfs equation. By power approximation of the experimental values through the least-squares method, Equation (6) is obtained.

$$C_D = \frac{43.18}{R_e^{1.05}} \tag{6}$$

Under the assumption that fluid resistance is proportional to the square of fall velocity,the momentam equation is shown as Equation (7)[4].

$$M\frac{dw_1}{dt} = (M - m)g - \frac{1}{2}m\frac{dw_1}{dt} - \frac{1}{8}\pi d^2 \rho C_D w_1^2 \tag{7}$$

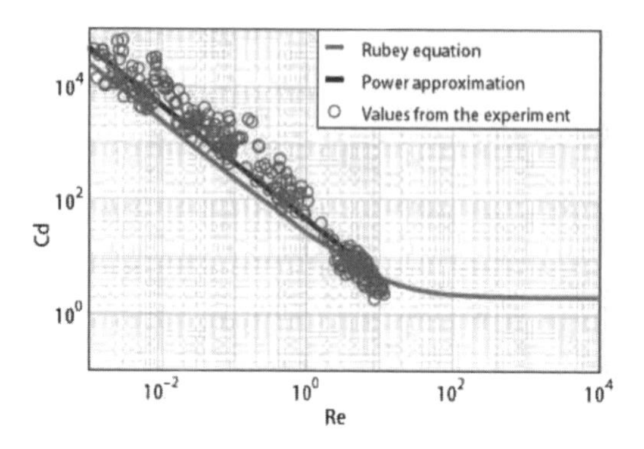

Figure 7. Resistance C_D of soil grains.

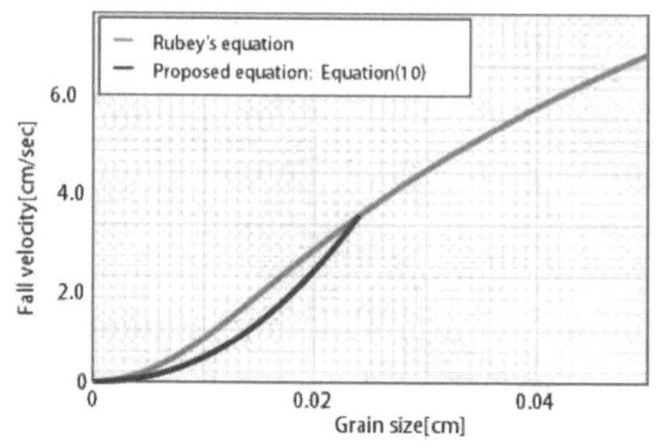

Figure 8. Relation between Rubeyfs equation and the equation proposed by the authors.

Where, M is the mass of soil particle, m is the mass of water, w_1 is the fall velocity of soil particle , d is the soil grain-size, and t is time. By rearranging Equation (7), dw_1/dt is derived as in Equation (8).

$$\frac{dw_1}{dt} = -\frac{\frac{1}{8}\pi d^2 \rho C_D}{M + \frac{1}{2}m} w_1^2 + \frac{(M-m)g}{M + \frac{1}{2}m} \tag{8}$$

For the fall velocity w_1 to become the final fall velocity w_0, the acceleration in the fall direction needs to be 0 (i.e. $dw_1/dt = 0$). Therefore, to obtain the equilibrium fall velocity, the right-hand side of Equation (8) becomes 0, and Equation (9) is obtained

$$w_0 = \left(\frac{4}{3}\frac{gd}{C_D}\left(\frac{\sigma}{\rho} - 1\right)\right)^{1/2} \tag{9}$$

By substituting Equation (6) into the above equation, Equation (10) is derived.

$$w_{fi} = \left(\frac{1}{32.38}\right)^{1.05} \left(\frac{\sqrt{sgd_i^3}}{\nu}\right)^{1.11} \sqrt{sgd_i} \tag{10}$$

Where w_{fi} is the fall velocity, ν is the water kinematic viscosity, s is the specific gravity of the soil particle ($=(\rho_s - \rho)/\rho$), d_i is the soil grain-size, and g is the gravitational acceleration. The fall velocity represented in the form of Equation (10) is shown in Figure 8. The values according to Rubeyfs equation are shown for comparison.

3 EXPERIMENT ON SOIL PARTICLE ENTRAINMENT

3.1 *Outline of the entrainment experiment*

An entrainment experiment on grains from the Nibutani Dam Reservoir was performed at an inclined flume owned by CERI. The flume is 25.0 m long, 1.0 m wide, and 1.0 m deep (Figure 9). Bed material from the

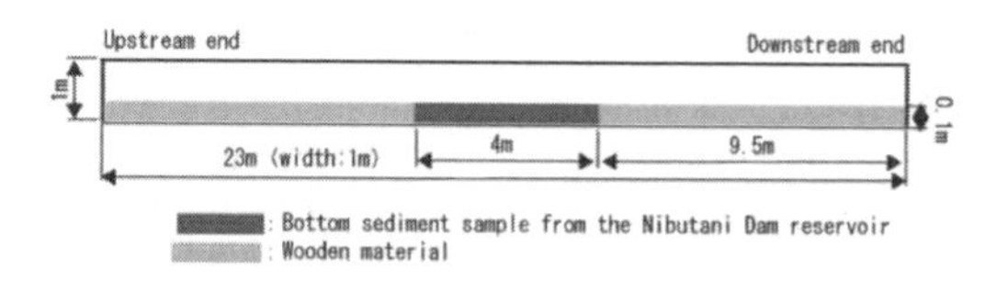

Figure 9. Testing waterway used in the entrainment experiment (longitudinal profile).

Table 2. Parameters in the entrainment experiment.

	Case 1	Case 2	Case 3
Discharge (m^3/s)	0.10	0.20	0.22
Bed slope	1/5000	1/5000	1/400
Water depth (cm)	75	55	21
Duration of water flow (min)	30	30	30

bottom of dam reservoir was spread at a section (1.0 m wide, 4.0 m long) that extends from 9.5 m to 13.5 m from the upstream end. Water was made to flow for 30 minutes under the conditions shown in Table 2. The experiment was repeated twice under each condition. Riverbed elevation at the section with bottom sediment is measured longitudinally at 9 points on the left and right bank. A total of 7 measurements are made from the beginning of the water flow and every 5 minutes thereafter. To obtain the displaced amount of bed material, a detailed measurement of the riverbed elevation is performed before and after the water flow. Measurements are taken at 5 points in the transverse direction and 9 points in the longitudinal direction. From the results, the volume of bed material is calculated for the section with the bottom sediment before and after the water flow. This difference is regarded as the transport volume. Grain-size analysis of the bed material was also performed. To measure the concentration of suspended sediment, water was sampled at one point in the waterway during water flow. It

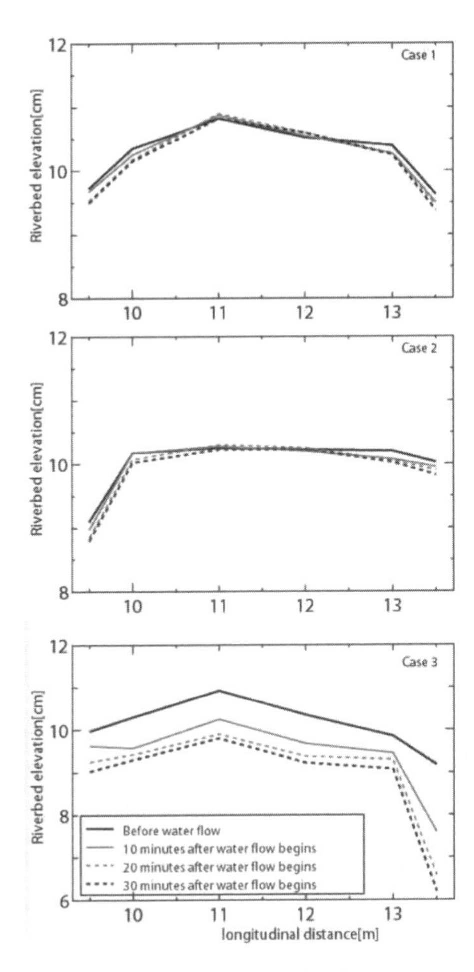

Figure 10. Changes in the riverbed elevation.

was sampled 4 times: once before water flow, and every 10 minutes after the start of flow. Grain-size analysis of the suspended sediment was also performed.

3.2 *Results of the entrainment experiment*

The average values of bed elevation at the left and right bank during water flow are shown in Figure 10. It can be observed that there are large changes in riverbed elevation at the initial stages of the water flow. The transport volume of the bottom sediment according to grain size and the average SS concentration in the water are shown in Figure 11. The figure shows that the SS concentration in the water increases with time, and that the entrainment decreases with time. To estimate an entrainment rate, it is necessary to account for the greater sediment entrainment at the initial stage of the water flow, which declines with time.

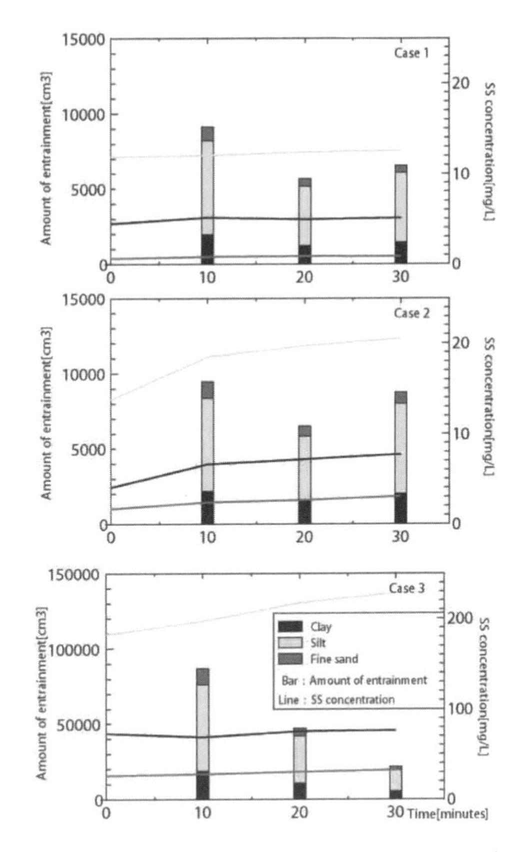

Figure 11. Movement of bottom sediment and changes in SS concentration during water flow.

3.3 *Estimation of the entrainment rate*

An entrainment rate for the Nibutani Dam Reservoir was estimated from the results of the experiment. The transport volume in the riverbed (including suspended sediment) can be expressed by Equation (11).

$$\frac{\partial z}{\partial t} = -\frac{1}{1-\lambda}\left(q_{sui} - w_{fi}c_i\right) \tag{11}$$

Where, q_{sui} is the entrainment from the riverbed per unit area and time, according to grain size, z is the riverbed elevation, λ is the void content, and c_i is the concentration of suspended sediment for each grain size. By substituting the experimental results into Equation (11), the entrainment rate q_{sui} can be obtained. Equation (12) is Itakurafs equation [5]'[6] for the entrainment rate. Although this equation does not account for cohesive soil, when it was used in the calculation for riverbed evolution at the Saru River [7], the equation accurately reproduced the observed phenomenon. To confirm that this equation can be applied to very-fine-grained soil, which is the subject of this

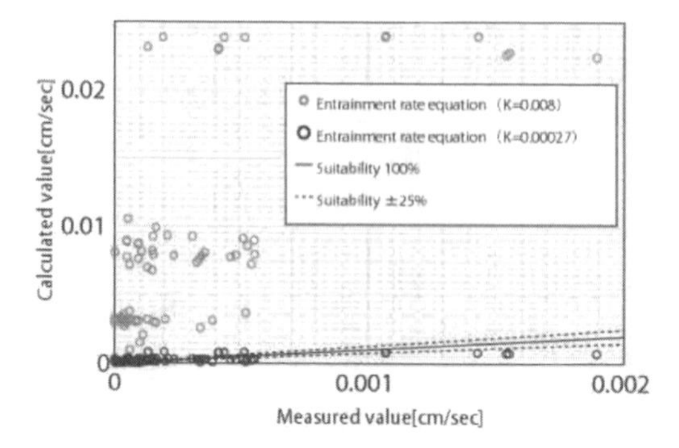

Figure 12. Comparison of entrainment rate.

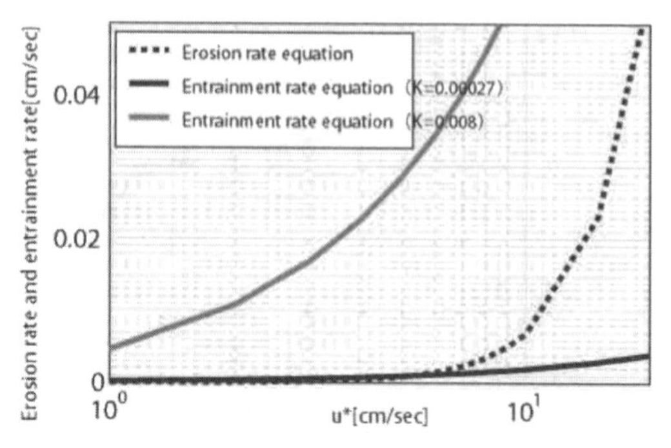

Figure 13. Comparison of entrainment rate and erosion rate.

research, comparisons were made between entrainment rates calculated from the experiment results and those calculated using Itakura's equation.

$$q_{sui} = p_i K \left(\alpha_* \frac{\rho_s - \rho}{\rho_s} \frac{g d_i}{u'_*} \Omega_i - w_{fi} \right) \tag{12}$$

Where p_i is the rate at which grains of diameter di exit, ρ_s is the density of suspended sediment, ρ is the density of water, u'_* is the effective friction velocity, $K = 0.008$, and $\alpha_* = 0.14$. Furthermore, Ω_i is represented in the form of Equation (13).

$$\Omega_i = \frac{\tau'_{*i}}{B_{*i}} \frac{\int_{a'}^{\infty} \xi \frac{1}{\pi} (-\xi^2)\, d\xi}{\int_{a'}^{\infty} \frac{1}{\pi} (-\xi^2)\, d\xi} + \frac{\tau'_{*i}}{B_{*i}\eta_0} - 1 \tag{13}$$

Where, τ'_{*i} is the critical dimensionless effective shear stress for each grain size, $\eta_0 = 0.5$, B_{*i} is the conversion factor in using the friction velocity as the velocity for calculating the lift force, $\xi = z/h$, $a' = B_{*i}/\tau_{*i} - 1/\eta_0$, and τ_{*i} is the dimensionless shear stress for each grain size. u'_* and τ'_{*i} were expected to be the total shear stress, minus the form drag components such as sand waves. However, in this experiment, no sand waves occurred, so the total resistance was treated as being equivalent to the total shear stress. By substituting Equation (11) into Equation (12), the theoretical entrainment rate is obtained. A comparison with the actual entrainment rate is shown in Figure 12 ($K = 0.008$). The entrainment rate calculated using Itakurafs equation overestimates that rate relative to the result calculated using the experimental results. From these results, the entrainment equation for sedimentary deposits in the Nibutani Dam Reservoir is drawn. Using the entrainment rate calculated from the experimental results shown in Figure 12, the factor K in Equation (12) is calculated using the least-squares method. The result is $K = 0.00027$. When the entrainment calculated with this factor is compared with that calculated from the experiment shown in Figure 12 ($K = 0.00027$), it can be observed that the values are more appropriate when $K = 0.00027$. Itakurafs equation is not intended for cohesive soil, but Sekine et al. [8] are continuing research on the erosion of cohesive soil. Equation (14) is the erosion velocity derived from the results of Sekinefs experiment. The erosion velocities calculated using that equation are compared to those calculated using Equation (12).

$$E_s = \alpha R_{wc}^{2.5} u_*^3 \tag{14}$$

Where E_s is erosion rate, α is a factor that reflects the sample (grain size, material, etc.) and water temperature, R_{wc} is water content, and u_* is friction velocity. The erosion rate equation obtained for S.A. clay ($d_{60} = 0.0016$ cm) is compared to that obtained using the estimated entrainment rate equation (Figure 13). This was calculated under the condition of 20°C water temperature, factor $\alpha = 3.89 \times 10^{-5}$, and water content of 50%. Figure 13 includes the entrainment velocities calculated using Itakurafs K. Under this condition, the estimated entrainment rate and erosion rate are compared. The results diverge at values of u_* greater than 10 cm/sec. This needs to be examined in terms of the equationfs range of application. However, for this dam reservoir, it is assumed that the friction velocity is less than 10 cm/sec even during large-scale flooding. As a result, it is thought that reproduction of behavior for very fine sediment is possible by changing the factor K in Itakurafs equation, where originally it was thought that cohesive soil had to be considered.

4 APPLICATION IN CALCULATING SEDIMENT TRANSPORT IN THE DAM RESERVOIR

4.1 *Outline of the model*

To confirm that the fall velocity and entrainment rate estimated in the previous sections agree with observed phenomenon, the calculation to reproduce sediment transport in the Nibutani Dam Reservoir caused by the Saru River flood of August 2003 was attempted. This flood occurred from late at night on August 9, 2003, to early the following morning. Heavy rainfall from Typhoon Etau (Typhoon No. 10) caused the most severe flood on record, which exceeded the design high-water level. Basically, at least Vertical 2-dimensional calculation should be performed when flow-related movement of sediment particles is modeled. However, the suitability of the estimated equations has to be confirmed, so 1-dimensional unsteady flow calculation was performed. The basic equations of open channel unsteady flow are the continuous equation (15), and momentom equation (16).

$$\frac{\partial A}{\partial t} + \frac{\partial Q}{\partial x} = 0 \tag{15}$$

$$\frac{1}{g}\frac{\partial}{\partial t}\left(\frac{Q}{A}\right) + \frac{\partial}{\partial x}\left(\frac{1}{2g}\left(\frac{Q}{A}\right)^2\right) = -\frac{\partial (h+z)}{\partial x} - \frac{n^2Q^2}{R^{\frac{4}{3}}A^2} \tag{16}$$

Where Q is discharge, h is depth, n is coefficient of roughness of manning, R is hydrulic radius, A is cross-sectional area. Since the calculation of riverbed evolution needs to incorporate the behavior of suspended materials, the bed load and suspended sediment were considered. For the amount of bed load, Ashida and Michiue [9] equations (17) were used for each grain size.

$$\frac{q_{bi}}{\sqrt{sgd_i^3}} = p_i 17 \tau_{*i}'^{\frac{3}{2}} \left(1 - \frac{\tau_{*ci}}{\tau_{*i}}\right)\left(1 - \frac{u_{*ci}}{u_*}\right) \tag{17}$$

Where q_{bi} is the bed load per unit width by grain size, τ_{*ci} is the critical dimensionless shear stress for each grain size, and u_{*ci} is the critical friction velocity for each grain size. Furthermore, the previously mentioned Equation (12) was used in calculating the amount of suspended sediment. We set $K = 0.00027$ and the fall velocity equation was used as Equation (10). With the above, the amount of bed variation is calculated from the continuity equation for the total sediment transport (Equation (18)).

$$\frac{\partial z}{\partial t} + \frac{\Psi}{1-\lambda} = 0 \tag{18}$$

$$\Psi = \frac{1}{B}\frac{\partial \sum_i (q_{bi}B)}{\partial x} + \sum_i (q_{sui} - w_{fi}c_{bi})$$

Where, c_{bi} is the concentration of suspended sediment for each grain size near the riverbed. The continuity equation for the sediment transportation for each grain size is Equation (19).

$$\delta \frac{\partial p_i}{\partial t} + p_i^* \frac{\partial z}{\partial t} + \frac{\Psi}{1-\lambda} = 0 \tag{19}$$

Where δ is the thickness of the exchange layer.

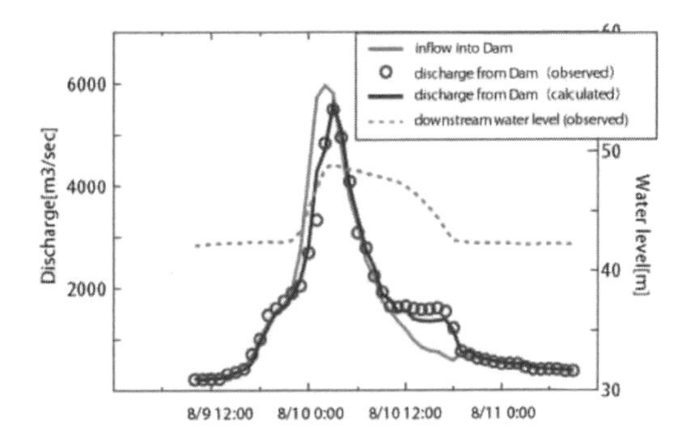

Figure 14. Calculated discharge.

4.2 *Calculation conditions and results*

The data used in calculation are those observed for the flood on the Saru River in August 2003, where the peak flow into the dam was approximately 6,000 m^3/sec. The range of calculation was from the dam to immediately below the check dam, which is at the upstream end of the reservoir. The cross section of the dam used in calculation is the bed elevation and width of the reservoir, which was decided according to cross-sectional survey performed before the June 2003 flood. The conditions of the calculations for the upstream is inflow at the dam and the downstream end is the water lebel during the flood. Furthermore, for the boundary condition used in the calculation of bed variation, the observed SS data for each grain size were used for suspended sediment at the upstream end, and the data observed in July 2003 were used for the distribution of grain size of the initial bed material. It is thought that the SS data includes wash load. However, the sediment budget obtained using the SS values observed by Ogawa et al. [10] shows roughly the same values as those obtained by survey. In light of this result, the total SS data for each grain size was used as the concentration of the suspended sediment. Figure 14 compares the observed and calculated discharge and water level.The calculated resulets adequately reproduce the observed flowfield. Figure 15 compares the observed and calculated SS load discharged from the dam orifice. However, the observed discharge does not include values from around the peak period up to the recession period. In Figure 15 no observation data were available between the peak and the initial period of recession, so we values the

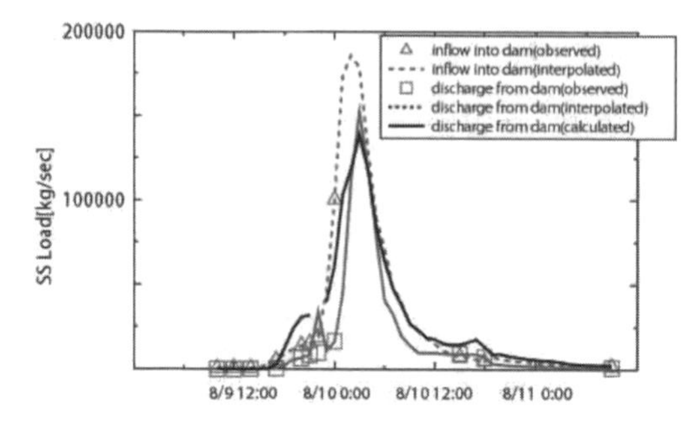

Figure 15. Comparison of calculated and observed SS discharge.

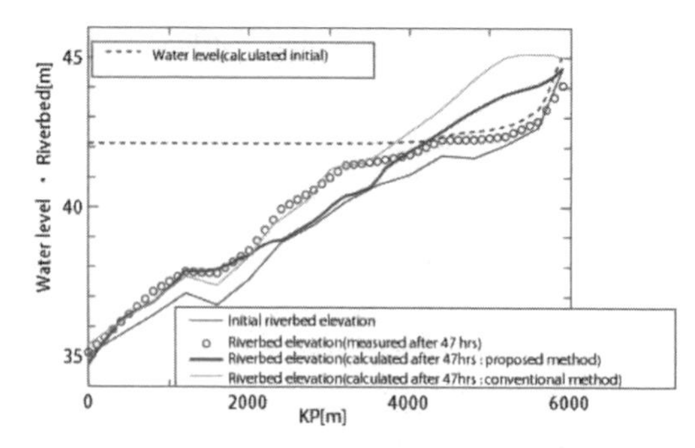

Figure 16. Calculated bed evolutions.

Table 3. Comparison of sediment.

	Sediment [thousand m^3]	Comparison w/surveyed results [%]
Cross-sectional surveying	2,355	–
Budget from SS observation	2,285	97.0
Calculated (proposed method)	2,519	96.1
Calculated (conventional method)	3,645	155.0

calculated results using estimated resulets by Ogawa et al. The behavior of the observed values (which include values estimated by Ogawa et al. and the calculated values seems to be similar, except for the SS load from the initial rise to the recession period. At this point, it is not possible to clarify which values are more suitable for comparison with the inflow values: the calculated values or the observed (including estimated) discharge values. This is an issue for future study. Bed variations calculated according to the method mentioned above are shown in Figure 16, and the amount of sedimentation is shown in Table 3. The results obtained using Itakurafs formula ($K = 0.008$), which is the conventional method of calculating bed

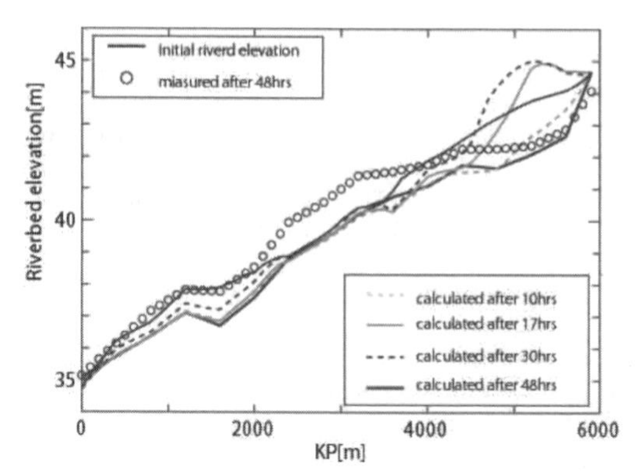

Figure 17. Temporal changes in amount of sediment.

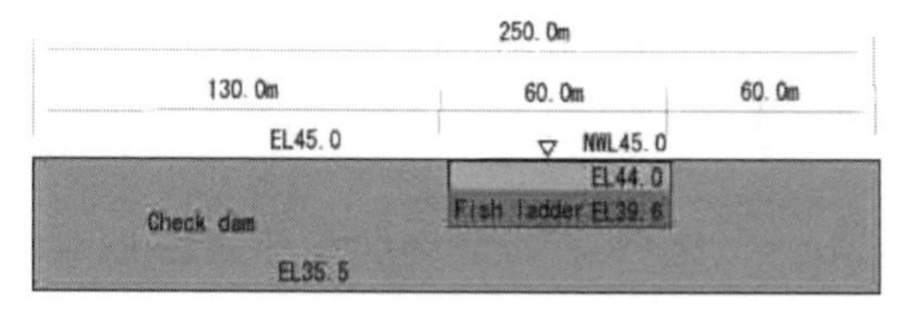

Figure 18. Cross-section of the check dam looking upstream.

variation, are shown for comparison. There are large discrepancies from measured values in the positions and amounts of sedimentation calculated using the conventional formula. In contrast, for our proposed calculation method, which uses the factor $K = 0.00027$ in the equation for overall sediment entrainment, the calculated sedimentation adequately reproduces the observed sedimentation. For the sedimentation trends, the calculation reproduces the observed trends for the downstream end of the dam reservoir, but the calculation results are too low for the middle reaches and too high for the upstream end. To clarify the reason of this difference, the temporal changes in sedimentation are shown in Figure 17. We see that sedimentation at the upstream end of the dam starts around the time of the flood peak. Sediment was transferred downstream during the flood recession period, but the calculated reservoir degradation was not as great as that which was observed. The reason for this considered follows the cross-sectional width at the upstream end is fixed at 250 m in the conditions for calculation. In fact, there is a check dam (Figure 18) at this location, and up until the normal water level of 45 m, the flow width is normally about 60 m. As shown in Figure 14, the water level at the upstream end during the peak flood is higher than 45 m; therefore, the cross-sectional width will be 250 m. However, when the water level is lower than 45 m during the initial rise and recession period of the flood, the flow width is 60 m. Therefore, it is thought that the actual shear stress was greater than

Figure 19. The river immediately downstream of the check dam.

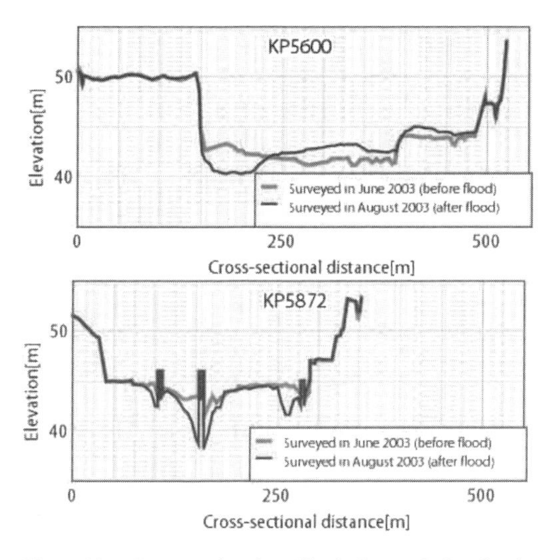

Figure 20. Cross-sectional profiles before and after flood.

that in the calculation and that sediments accumulated at the upstream end were transported to the middle reaches of the reservoir during the flood peak. However, this was not modeled in the calculation. Figure 19 shows a photo taken in August 13, 2003, of the flow into the dam reservoir through the spill way. Figure 20 shows the results of cross-sectional survey performed in June and August 2003, before and after the flood. From the fact that severe riverbed degradation was observed near the check dam after the flood. This phenomenon shows the validity of the above-mentioned consideration.

Lastly, the reproducibility of grain-size composition for the reservoir bed was studied. Figure 21 shows the distribution of average grain sizes at the bottom of the dam reservoir. In observations after the flood, the grain sizes are greater at the upstream end of the reservoir, but there is no significant difference

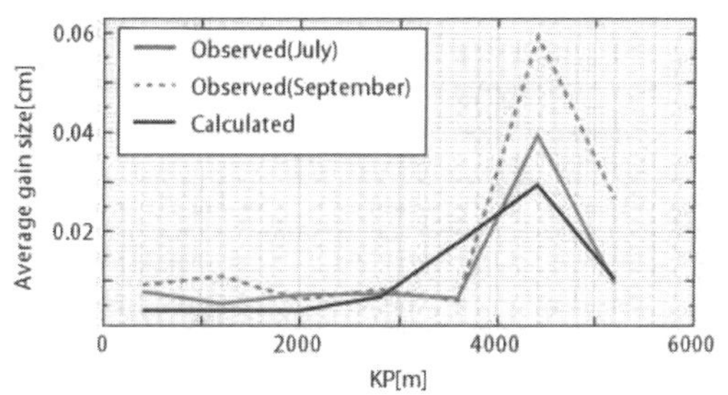

Figure 21. Distribution of average grain size at the dam reservoir bed.

in distribution at the downstream end. The calculated grain sizes are finer than the observed grain sizes at the upstream end of the reservoir and become coarse at the middle reaches, and tend to be similar to observed results at the lower reaches. This is because of the influx of fine-grained sediments and excessive sedimentation at the upper reaches, as mentioned previously, which made the grain size small. Furthermore, it is thought that these sediments did not flow downstream, and that they therefore made the grain size large at the middle reaches.

5 CONCLUSION

To clarify the behavior of sediment particles in this dam reservoir, entrainment and sedimentation experiments were performed on fine-grained soil, using samples taken from the reservoir. This research used the results of those experiments to reproduce phenomena observed during floods. The calculations largely reproduced the observed phenomena in terms of sedimentation amount, but there were some discrepancies, particularly in the location of sedimentation at the upstream end of the reservoir. There were discrepancies also in SS discharged from the dam. This is because missing observation data around the flood peak were supplemented, making the comparison less straightforward. It was possible to reproduce the phenomenon, to a certain degree, using 1-dimensional calculation for unsteady flow. However, the great depth of the dam reservoir means that there might be a vertical distribution in variations of flow velocity and concentration. It is thought that a more accurate calculation of the behavior of sediments within the dam reservoir would be possible by vertical 2-dimensional calculation.

REFERENCES

1. Fidleris, AVD and Whitmore. CRDLDF. The Physical Interaction of Spherical Grain of Suspensions A RheolD, ActaA, Vol.1, CNo.4/6, C1961D.

2. Brenner, CH.F. The Slow Motion of a Sphere Through a Viscous Fluid Towards a Plane Surface CChem. Eng.Sci. CVol.16, 1961.
3. McNown, CJ.S.F. Grain in Slow Motion CLa Houille Blanche CVol.6, CNo.5, C1951.
4. Ishihara, T. and Honma, H. (ed.) (1958): gApplied Hydraulicsh Second Volume, Maruzen, pp. 7–8.
5. Itakura, T. and Kishi, TF. Open Channel Flow with Suspended Sediments, CProc.ASCE, CVol.106, CNo.HY8, Cpp.1325–1343, C1980.
6. Itakura, T. (1984): gResearch on the Diffusion of Turbulent Current in Riversh, Reports of the Civil Engineering Research Institute of Hokkaido, No. 83.
7. Machino, S., Shimizu, Y., Kuroki, M., Fujita, M., Yoshida, Y. (2000): gResearch on Sediment Transport and Bed Variation in Rivers with a Damh, Research Papers for the Japan Society of Civil Engineers, No.656/-52, pp. 61–72.
8. Sekine, M. (2004): gResearch on the Erosion Mechanism in Naturally Accumulated Cohesive Soil in Actual Rivers and a Method for Site-Testing of Erosionh, Scientific Research Grant for Fiscal Year 2002 to 2003 [Infrastructure Research (C) (2)], Report on the Results of the Research.
9. Ashida, K. and Michiue, M. (1972): gBasic Research on the Flow Resistance of Moving Bed and Amount of Bed Loadh, Research Papers and Reports for the Japan Society of Civil Engineers, No. 206, Japan Society of Civil Engineers, pp. 59–69.
10. Ogawa, N. and Watanabe, Y. (2004): gThe Behavior of SS at Nibutani Dam during Typhoon Etau of 2003h, Research Papers on River Engineering, Vol. 10, (accepted on September 9, 2004).

River, Coastal and Estuarine Morphodynamics: RCEM 2005 – Parker & García (eds)
 ISBN 0 415 39270 5

Field observations and modeling of mountain stream response to control dams in Venezuela

J.L. López, Z. Muñoz & M. Falcon
Institute of Fluid Mechanics, Faculty of Engineering, Universidad Central de Venezuela, Caracas, Venezuela

ABSTRACT: In February 2005, intensive rainfall took place along the Avila Mountain in the Vargas State, Venezuela, over the same area affected by the 1999 catastrophic debris flows. Flash floods and small scale debris flows were generated again in some of the streams, causing flooding of water and sediment in the urban areas. The damage was minimized due to sediment control dams that were built after the 1999 event. This paper describes the main characteristics of the February 2005 floods and the effects on the sediment control dams built in Vargas State. A mathematical model, developed for simulation of flow and sediment transport in mountain streams, is applied to reproduce the deposition patterns and grain size distribution changes upstream of sediment control dams. A hypothetical case is presented to show the applicability of the model. The paper also illustrates the process of sedimentation at some of the dams and the attempts to use the numerical model to reproduce the observed data.

1 INTRODUCTION

The north coastal range of Venezuela is located in the northern extreme of South America, and runs parallel to the Caribbean Sea (Figure 1). The Avila Mountain attains elevation of up to 2800 m above sea level and is very steep, with channel average slopes of the order of 30%. In December 15 and 16 of 1999, thousands of landslides were triggered by heavy rainfalls along this coastal range. These landslides generated flash floods and debris flows in 24 streams along fifty kilometers of a very narrow coastal strip in the State of Vargas causing destruction of many towns and killing an estimated of 15,000 people. Large quantity of sediments, woody debris and fractured rocks were eroded and transported by the flows down to the valleys and fans, and to the sea. The volume of sediment deposition in the alluvial fans has been estimated in the order of 20 million cubic meters (Lopez, et al. 2003). As a consequence of this event, the government of Venezuela started a program for the construction of sediment control dams, in order to protect the widely populated urban areas located on the alluvial fans. At present, 24 sediment control dams have been built, most of them gabion dams.

Figure 1. Aerial view from Ikonos satellite image of the north coastal range of Venezuela (longitudinal extension is approximately 30 km).

An experimental basin was established in the San Jose de Galipan basin of the Avila Mountain, in order to collect water and sediment data to study and investigate the mechanism of debris flow formation and sediment transport in mountain areas. The basin has an area of 14 km^2 and the observation system, completed in 2002, consists of a network of seven (7) rain gauges distributed throughout the basin and two (2) water level gages in the streams. Figure 2 shows the approximate location of the rain gages in the basin. A sediment control dam, built downstream in the canyon of the Galipan stream, has been used for sediment volume accumulation surveys and sampling of bed material, since completion of the dam in 2003.

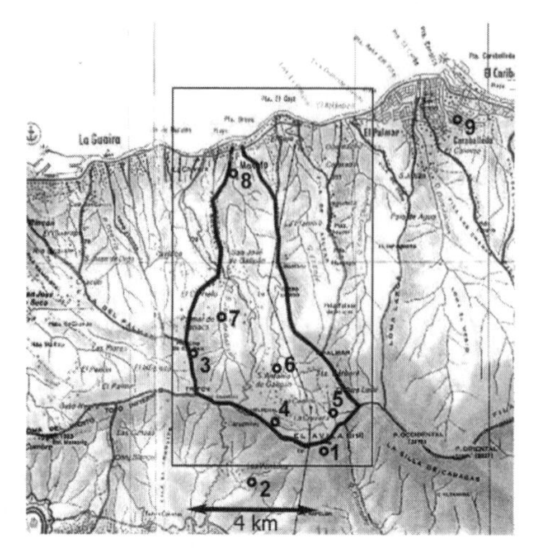

Figure 2. Approximate location of rain gages in the San Jose de Galipan experimental basin.

Table 1. Daily rainfall measurements during February 2005 at stations located in or near San Jose de Galipan basin (number between parenthesis indicate location in the map of Fig.

	Daily Precipitation in mm				
Station	Feb. 7	Feb. 8	Feb. 9	Feb.10	Cumulative
Humboldt (1)	39.2	51.4	26.1	42.6	159.4
Picacho (3)	60.0	38.3	15.7	37.0	152.8
Los Venados (2)	37.6	45.0	6.9	29.0	118.5
San Francisco (5)	56.0	51.5	52.2	60.9	220.7
San Isidro (4)	48.3	44.3	23.8	54.0	171.3
San Jose (7)	80.3	88.6	149.9	106.2	425.0
Macuto (8)	44.0	102.0	175.0	110.0	432.0
Caraballeda (9)	38.0	75.0	159.0	110.0	382.0

Figure 3. View of Macuto dam in the San Jose de Galipan stream. Left picture (a) is taking from downstream; right picture (b) shows level of sediment accumulation by January 2005.

2 THE STORM OF 2005

The extraordinary precipitations that occurred in February 7–10th, 2005, extended over the north-central coastline of Venezuela, and the mountain region in the west, causing great damages in the states of Vargas, Miranda, Merida, Tachira, Carabobo, Falcon, Zulia and Yaracuy, and also in the Capital District of Caracas. Only in the state of Tachira, about 20 bridges collapsed or were severely damaged. The major effects occurred in the coastal area of Vargas State and in the town of Santa Cruz de Mora, located in the Andes region of Venezuela. The origin of the storm is associated to the presence of cold fronts originated in the North Atlantic Ocean.

No significant precipitation occurred before or after those days. It was basically a 4-day storm concentrated between the 7th and 10th day of February. Table 1 shows daily precipitation records measured at the stations located in the San Jose de Galipan basin or near it. The station of Macuto, located in the lower part of the basin, reported the largest precipitation, amounting to 431 mm in the 4-day storm. The mean annual rainfall in the coastal range, near sea level, is about 520 mm. A frequency analysis for the daily annual maximum values indicated that the return period was in the order of 100 years. The measured data indicates that the rainfall was larger below 1000 m elevation. Due to the orographic effect of the Avila Mountain, the amount of rainfall at higher elevations is usually greater than at sea level, but it seems not to be the case for the 2005 storm.

The 1999 event left thousands of scars on the slopes of the Avila Mountain. However, a few new scars were observed after the 2005 storm. Partial reactivations of old scars are noted but in small magnitude. Most of the landslides were generated in the lower part of the basin, where the soil and weathered rock with a lower vegetation cover is easily eroded. Many of small landslides and rock falls were observed along the coastal road that runs parallel to the sea. However, a large volume of sediment material was transported to the urban areas in the alluvial fans, blocking bridges and filling the channels downstream. The predominant process seems to have been the remobilization of sediment deposits left by the 1999 storm in the upstream reaches of the Avila Mountain basins.

3 SEDIMENT CONTROL DAMS

At present, 24 sediment control dams have been built in the basins of Vargas State, most of them of the gabions type. The height of the structures varies between 3 and 7 m. Field observations made after the February storm indicated that approximately 50% of the dams were completely full of sediment, which is an indication of the large sediment yield capacity of the basins. Figure 3 shows the Macuto Dam, a 7 m high

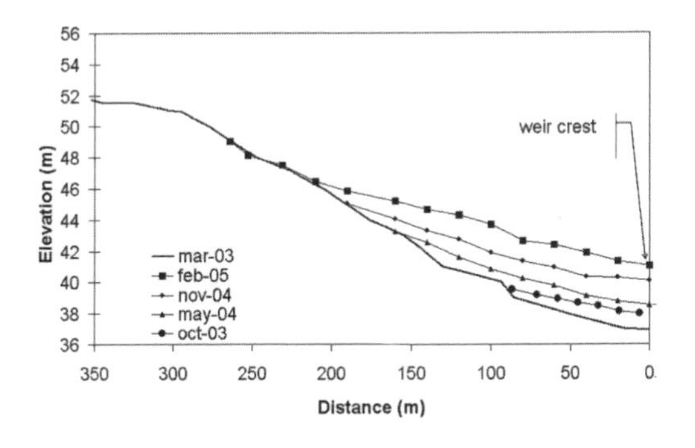

Figure 4. Temporal changes in bed profiles upstream of Macuto dam.

gabion dam, built in the canyon of San Jose de Galipan Basin. The construction of the dam was completed in March 2003 (Figure 3a), and by January 2005 it was almost full of sediment (Figure 3b). The bed level was at 0.75 m from reaching the top of the weir. No extraordinary floods occurred in the basin during the almost 2-year period after dam construction. The rapid process of sedimentation is associated to the lack of windows or openings in the body of the dam to allow passing of normal sediment-laden flows.

Bed level changes have been obtained by field survey upstream of the Macuto dam. Figure 4 shows the longitudinal bed profiles obtained at different dates. The February flood finished filling the dam with sediment and decreased the bed slope. The original slope of 4.5%, for the immediate reach upstream of Macuto dam, was reduced to 2.9%, which is approximately equal to 2/3 of the original one.

4 THE MATHEMATICAL MODEL

A one-dimensional mathematical model for the simulation of flow and sediment transport in mountain streams (Lopez & Falcon, 1999) is applied to analyze the impact of sediment control dams in a river channel. The model assumes quasi-steady flow and that supercritical regime is seldom maintained in high-gradient natural channels, allowing for a simple method of computing the water surface profiles. Resistance to flow is determined as a function of flow depth and grain diameter. Thus, the sediment model calculates the temporal variations of the mean bed levels and of the grain size distribution. Two sediment layers are defined in the bed to account for the change in grain composition (Figure 5). The upper layer of the bed is called the mixing layer where the exchange of sediment between the flow and the bed occurs, and whose initial thickness is assumed to be equal to $2D_{max}$. The second layer lies below the upper one and is called the subsurface layer, which provides material for the replacement of

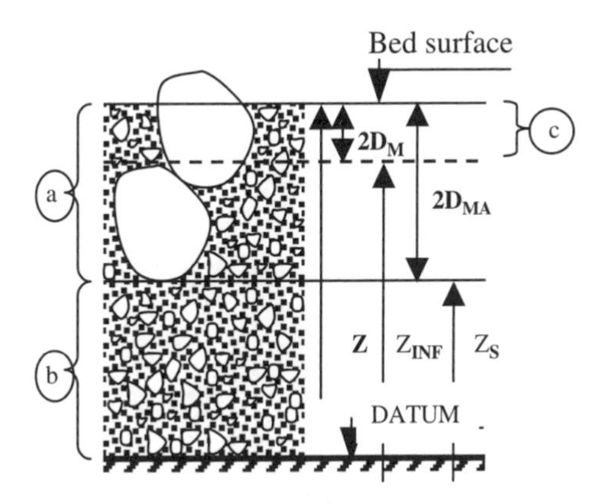

Figure 5. Definition diagram for bed layers: (a) is the mixing layer; (b) is the subsurface layer; (c) is the bed surface layer.

the mixing layer. The bed particles are supposed to be eroded from a surface bed layer whose thickness is assumed to be twice the mean diameter D_m. Track of the different bed layers at each time step is kept with the variables Z, Z_{inf}, and Z_s (see Figure 5).

The criterion presented by Aguirre-Pe & Fuentes (1993) for determining a representative critical diameter, Dcr, is used. Then, the grain size distribution of the moving part of a variable-depth bed mixing layer, for $D_{min} < D < D_{cr}$, is determined, and a Shoklitsch type bed material load is calculated at each cross section of the reach under study. The computational procedure keeps track of the temporal evolution of the grain size distribution within each subreach. The model considers two layers, because generally a gravel sublayer exists below the surface, and track is kept of the position of the interface. If the mixing layer becomes very small, what is left of it is mixed with the gravel sublayer to a given thickness and the temporal process continues. Numerical experiments showed that the armoring phenomenon is well reproduced.

5 CASE STUDY

A preliminary study is conducted to investigate the potential use of the mathematical model to reproduce bed and grain size changes due to construction of sediment control dam.

A hypothetical channel reach 2000 m long, 25 m wide, and mean bed slope of 4% is used to test the model capability. The bed material, ranging from boulders to sand, has a $D_{min} = 0.074$ mm, $D_{max} = 0.85$ m, $D_{50} = 0.117$ m, $D_{16} = 0.012$ m, corresponding to an initial circular distribution (Lopez and Falcon, 1999). The geometric and sediment characteristics of the

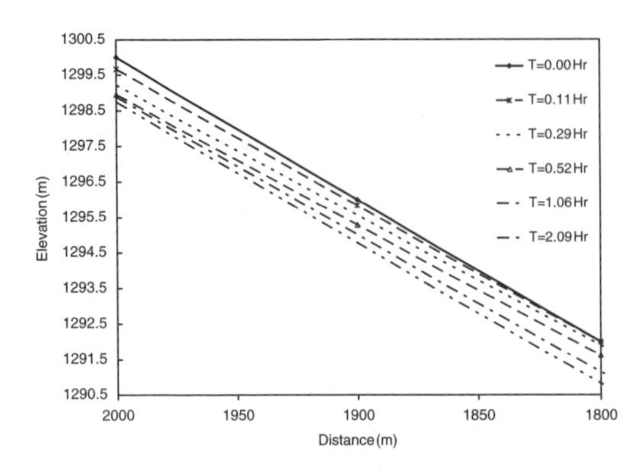

Figure 6. Variation in time of the computed bed profiles downstream of the dam.

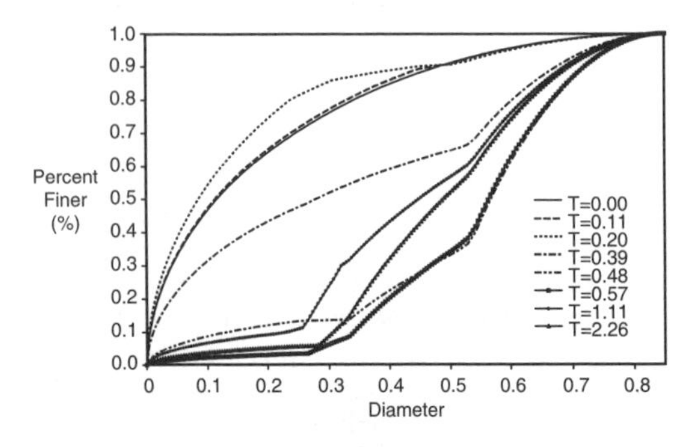

Figure 7. Computed grain size distribution curves at x = 1800 m downstream of the dam.

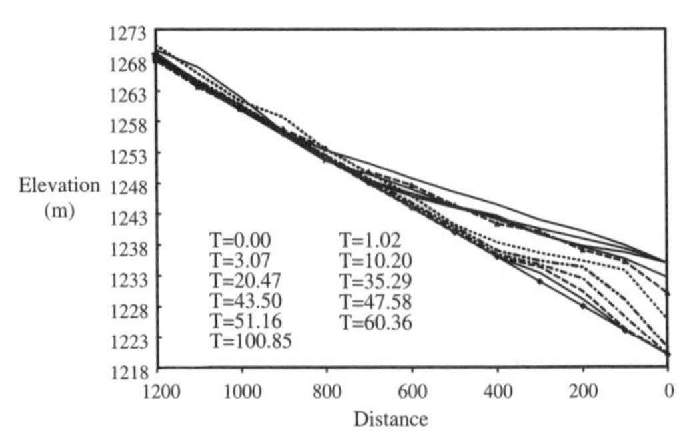

Figure 8. Variation in time of the calculated bed profiles upstream from a dam.

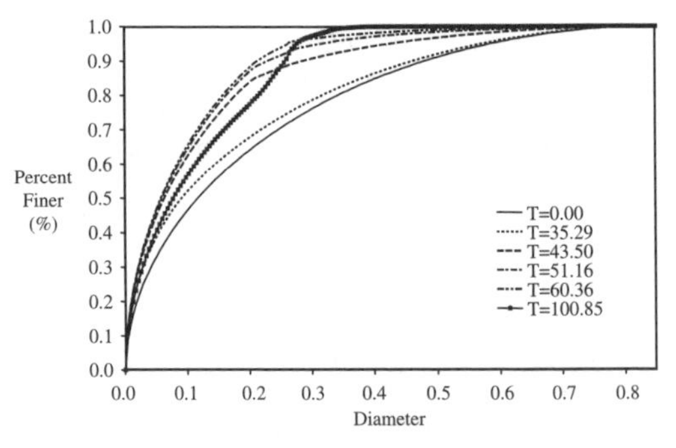

Figure 9. Computed grain size distribution curves at a section just upstream from the dam.

channel reach are similar to the reach upstream of Macuto dam.

A first case of bed response downstream of a dam is considered, by assuming no sediment discharge as the upstream boundary condition and a constant water flow of 200 m^3/s. The downstream boundary condition is given by a constant water depth corresponding to the uniform flow regime. The resulting bed changes are depicted in Figure 6, which shows the computed bed profiles at different times. A maximum degradation of 1.4 m is obtained in the most upstream section, and after six hours the bed remains stable. The downstream degradation extends for a distance of 1300 m approximately from the dam. The time variation of the grain size distribution is shown in Figure 7 for a section located 200 m downstream of the dam (at x = 1800 m). The model indicates a tendency to armouring. The finer part of the bed material is removed and the D_{50} has been increased from 0.12 m to 0.57 m at time equals 0.57 hours. Then, a process of refining is induced by the incorporation of new material from the subsurface layer, after the mixing layer is completely depleted. At time equals 1.11 hours, the D_{50} has been decreased to 0.45 m and then the process of coarsening starts again increasing the size up to a final stationary value of 0.49 at 2.26 hours.

A second case is considered in which the model is applied to predict the aggradation process upstream of a dam, for the same channel reach as the previous case. A constant input of water and sediment is defined as the upstream boundary condition, corresponding to 200 m^3/s and 2000 kg/s, respectively. A constant water level is imposed as the downstream boundary condition. Figure 8 shows the calculated bed profiles for different time steps. The bed evolution shows the motion of the sediment wave up to the foot of the dam. At t = 51.16 hours, the dam is completely full of sediment. The temporal variation of the grain size distribution at a section just upstream from the dam is depicted in Figure 9, showing the refinement of the bed material. After the sediment accumulation reaches the top of the dam, the finer fractions of the bed material starts to be eroded, and this is represented by the coarsening of the bed.

6 APPLICATION TO MACUTO DAM

A preliminary application to the San Jose de Galipan stream, upstream of Macuto dam is presented. The

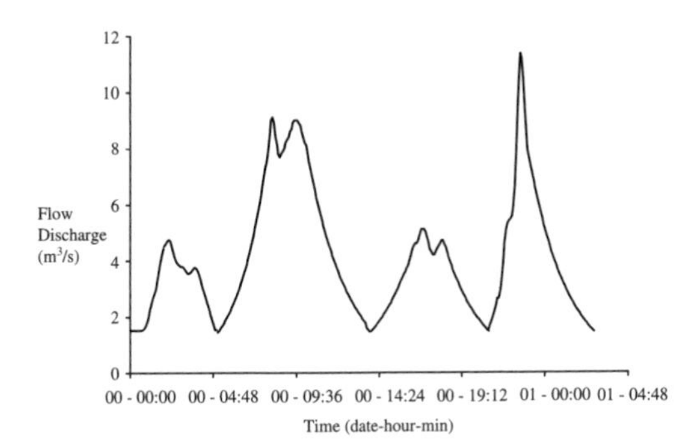

Figure 10. Flood hydrograph for the significant storms in the San Jose de Galipan basin between March 2003 and January 2005.

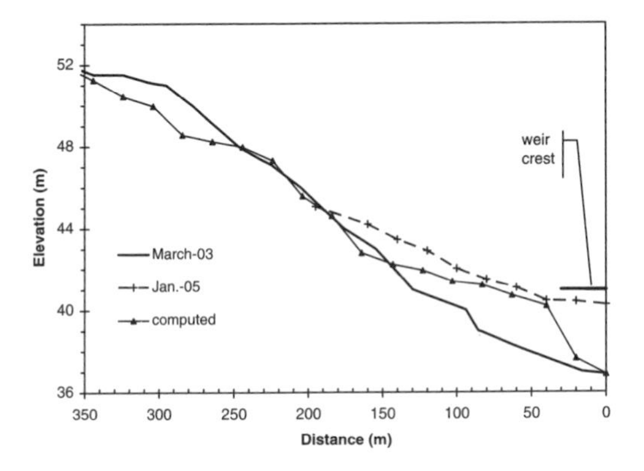

Figure 11. Comparison between observed and computed bed profiles upstream of Macuto dam.

channel reach used for the simulation is represented by 34 sections, for a total length of 1340 m. Space intervals for the cross sections are between 20 and 50 m. Bed width ranges from 20 to 60 m. The original mean bed slope is about 4.5%. The bed material is composed of sand, gravel, cobbles and boulders up to 0.65 m. Construction of a small gabion dam, 7 m high, was finished in March 2003. The precipitation occurring between March 2003 and January 2005 was measured by the rainfall stations installed in the basin. A rainfall-runoff model was used to generate the respective flow hydrographs. Flow discharges less than 1.5 m^3/s were truncated from the input hydrograph. Figure 10 shows the most significant floods that occurred in that period, which were used as the upstream boundary condition for the simulation model. Critical depth is used as the downstream boundary condition over the broad crested dam. A sediment input given by the Shoklitsch equation is specified at the upstream end of the river reach.

The computed and observed bed profiles upstream of Macuto dam are shown in Figure 11. The computer model using a Shoklitsch equation did not predict total sedimentation of the dam after passage of the floods. However, López and Falcón (1999), in accordance with Bathurst et al. (1987, Fig. 15.8) found that a reasonable adjustment could correct the level of sedimentation at the upstream side of the dam. Further analysis is underway to improve the results.

7 CONCLUSIONS

As a consequence of the 1999 debris-flow disaster in Venezuela, government agencies initiated a program of building sediment control dams in the State of Vargas, to protect large urban areas established on the alluvial fans. At present, 24 sediment control dams have been built. Due to the large sediment yield capacity of the basins, the dams have been subjected to a very rapid process of sedimentation. Field observations are made to describe the extraordinary storm of February 2005 and the effect on the sediment control dams. A mathematical model is applied to simulate the stream response to dam construction. A hypothetical case shows the capacity of the model to reproduce an alternate process of coarsening and refining of the bed material, when the river bed is subjected to degradation. A preliminary application to a river reach, in the San Jose de Galipan basin, upstream of Macuto dam, has been presented. Results show that the Shoklitsch equation underestimates the sediment transport capacity of the stream. Further analyses are under way to improve the results of the simulation model.

REFERENCES

Aguirre-Pe, J. & Fuentes R. 1993. Stability and weak motion of riprap at a channel bed. In C.Thorne, S. Abt, F. Barends, S. Maynord, & K. Pilarczyk (eds), *River, coastal and shoreline protection. Erosion control using riprap and armourstone*. Wiley, New York, 77–92.

Bathurst, J.C., Graf, W.H. & Cao, H.H. 1987. Bed load discharge equations for steep mountain rivers. In C.R. Thorne, J.C. Bathurst, & R.D. Hey (eds), *Sediment Transport in Gravel Bed Rivers*, Wiley, New York, 453–491.

López, J.L., Pérez, D. & García, R. 2003. Hydrology and geomorphologic evaluation of the 1999 debris flow event in Venezuela. In Rickenmann & Chen (eds), *Debris-Flow Hazards Mitigation: Mechanics, Prediction and Assessment; Proc. Third Intern. Conf., Davos, Switzerland, Sept. 13–15.*

López, J.L. & Falcón, M. 1999. Calculation of bed changes in mountain streams. *Journal of Hydraulic Engineering*, Vol. 125, No.3, March. pp. 263–270.

River, Coastal and Estuarine Morphodynamics: RCEM 2005 – Parker & García (eds)
© 2006 Taylor & Francis Group, London, ISBN 0 415 39270 5

Evaluating the effectiveness of engineering measures against water level drop downstream of the Three Gorges Dam using a 2-D numerical model

Hong Wang, Fangzhen Huang & Xuejun Shao
Department of Hydraulic Engineering, Tsinghua University, Beijing, China

ABSTRACT: This paper presents results of a numerical study on the effectiveness of proposed engineering measures, such as rock pavements on the river bed, to maintain the water stage in the alluvial channels downstream of the Three Gorges Dam. The numerical model was validated using field observations, and predictions of water level changes under various engineering schemes are presented to illustrate the most effective layout for maximum water level rise at the Yichang Station near the Gezhouba Dam, where navigation depth at the shiplocks will be a major concern in the near future.

1 INTRODUCTION

This study simulates water flows in the Yangtze River channel downstream of the Three Gorges Project,

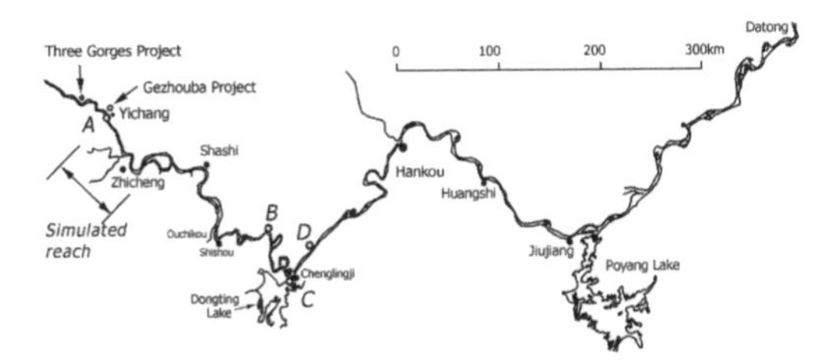

Figure 1. The 1100 km long section of the middle and lower Yangtze River reaches to be affected by bed scour after the impoundment of the Three Gorges Reservoir.

where the river runs through a transitional zone from the mountainous area to the great Wuhan plain (Fig. 1). The operation of the Three Gorges Reservoir will dominate the seasonal variation of water stages in this studied reach between the City of Yichang and the downstream of Zhicheng station. The purpose of this study is to predict the consequent long-term impact on the navigation conditions, by using a 2-D numerical simulation method.

The lower Jingjiang River section is located at about 300 km downstream of the Three Gorges Reservoir. Observed and predicted runoff and sediment transport at stations A, B, C, D are listed in Tables 1 and 2, respectively.

After the Three Gorges Reservoir started its impoundment in June 2003, significant reduction in sediment concentration in the downstream reaches has

Table 1. Observations at 4 gauge stations on the Middle Yangtze River before the construction of the Three Gorges Reservoir (Pan 2001).

	(a) Runoff				(b) Sediment				(c) Average sediment concentration (kg/m^3)			
	Averaged annual runoff (km^3) at each Station				Averaged annual sediment transport (million tons)				Averaged annual sediment transport (million tons)			
Period	A	B	C	D	A	B	C	D	A	B	C	D
1956–1966	439	322	313	629	548	333	59.6	414	1.25	1.04	0.19	0.66
1967–1972	416	336	298	631	493	355	52.5	431	1.18	1.06	0.18	0.68
1973–1980	430	360	279	634	499	394	38.4	463	1.16	1.09	0.14	0.73
1981–1988	439	382	258	633	555	448	32.7	482	1.27	1.17	0.13	0.76
1989–1995	428	387	270	650	411	356	27.6	367	0.96	0.92	0.10	0.56

* For the locations of stations A, B, C, D please see Figure 1.
** Data for Station C is the confluent flow from Dongting Lake into the Yangtze River.

Table 2. Predicted water and sediment discharge after impoundment of the Three Gorges Reservoir and comparison with observations before construction of the reservoir.

Years after impoundment	(a) Annually averaged Discharge ($\times 10^3 m^3/s$) Before	After	(3)/(2) (%)	(b) Total sediment transport per year (million tons) Before	After	(c) Sediment concentration (kg/m^3) Before	After	(8)/(7) (%)
(1)	(2)	(3)	(4)	(5)	(6)	(7)	(8)	(9)
1–10	14.4	14.0	97.2	556	165	1.22	0.37	30.3
11–20	14.4	13.9	96.5	556	156	1.22	0.36	29.5
21–30	14.4	13.9	96.5	556	178	1.22	0.4	32.8
31–40	14.4	13.9	96.5	556	207	1.22	0.47	38.5
41–50	14.4	13.9	96.5	556	256	1.22	0.58	47.5
51–60	14.4	13.9	96.5	556	323	1.22	0.74	60.7
61–70	14.4	13.9	96.5	556	372	1.22	0.85	69.7
71–80	14.4	13.9	96.5	556	408	1.22	0.93	76.2
81–90	14.4	13.9	96.5	556	420	1.22	0.96	78.7
91–100	14.4	13.9	96.5	556	427	1.22	0.97	79.5

been observed. The river bed erosion in the downstream of Gezhouba Project will become more and more serious and the water stage downstream of Gezhouba Project will get even lower than the that at present, leading to a more difficult navigation condition.

The water level drops downstream of the Gezhouba Project will be crucial for the navigation situation through the Gezhouba shiplock, as the water depth there has already been reduced by 1m since the Gezhouba Dam was completed in 1983 (Lu et al, 2002). If the Three Gorges Project starts a new round of bed scour and water stage reduction, the navigation operation through the Gezhouba Dam will be frequently interrupted during the dry season. The purpose of this study is to find out proper countermeasures for the problem of water stage drop.

In this numerical simulation, the water flow is calculated for a 63 km long river reach from the Yichang station to the Zhicheng station (Fig. 1). The channel pattern in this section is mostly straight with several river bends and major bars, where the Yangtze River runs out of the mountainous area into the Great Jianghan Plain.

2 MATHEMATICAL MODEL

The governing equations used in the mathematical model express the conservation of horizontal momentum and mass. For numerical convenience, these equations have been vertically integrated to give equations in terms of the depth mean horizontal velocity components of the total water column. Assuming that the water is of constant density and applying the hydrostatic approximation, the momentum equations can be vertically integrated to give (Lin & Falconer 2001).

$$\frac{\partial UH}{\partial t}+\beta\left(\frac{\partial U^2H}{\partial x}+\frac{\partial UVH}{\partial y}\right)-fVH+gH\frac{\partial \eta}{\partial x} + \frac{gn^2U\sqrt{U^2+V^2}}{H^{1/3}}+\varepsilon\left(\frac{\partial^2 UH}{\partial x^2}+\frac{\partial^2 UH}{\partial y^2}+\frac{\partial^2 VH}{\partial x\partial y}\right)=0 \quad (1)$$

$$\frac{\partial VH}{\partial t}+\beta\left(\frac{\partial UVH}{\partial x}+\frac{\partial V^2H}{\partial y}\right)-fUH+gH\frac{\partial \eta}{\partial y} + \frac{gn^2U\sqrt{U^2+V^2}}{H^{1/3}}+\varepsilon\left(\frac{\partial^2 VH}{\partial x^2}+\frac{\partial^2 VH}{\partial y^2}+\frac{\partial^2 UH}{\partial x\partial y}\right)=0 \quad (2)$$

where: U, V = depth mean velocity components in the x, y horizontal coordinate directions, respectively; H = total depth of flow; β = correction factor for nonuniformity of the vertical velocity profile, is given as 1.016; f = Coriolis parameter, a constant; g = gravitational acceleration; η = water surface elevation above chart datum; ρ = water density; n = Manning's bed roughness coefficient; ε = depth mean eddy viscosity (in m^2/s).

Similarly, the depth integrated conservation equations of mass can be written as:

$$\frac{\partial UH}{\partial x}+\frac{\partial VH}{\partial y}+\frac{\partial \eta}{\partial t}=0 \quad (3)$$

The numerical model employs a finite difference method with a regular grid system. The finite difference equations corresponding to the differential

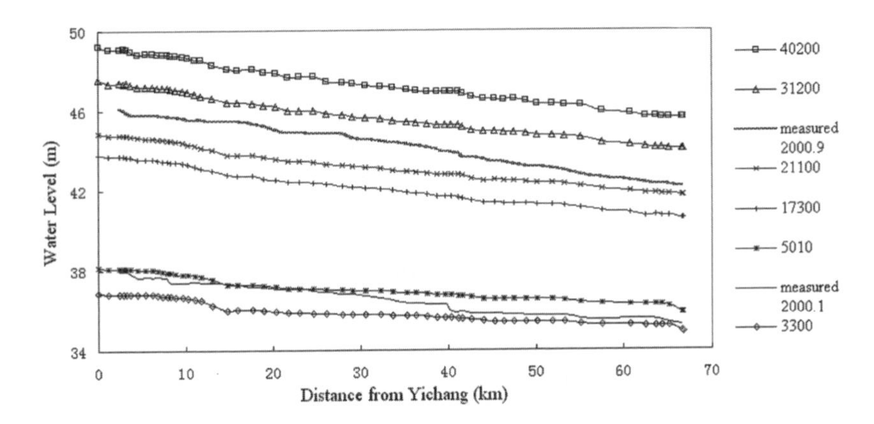

Figure 2. Calculated and measured water surface profiles.

equations were expressed in an alternating direction implicit form.

3 VALIDATION

In this paper, a contour map was used as the initial topography (digitized using a grid system with a size of 50 m × 50 m) between Yichang station and Zhicheng station, based on measurements of submerged topography in October, 1998 (scale = 1:10000). The daily observed hydrological data from 1998 to 1999 at both stations were used as the boundary condition. Water levels at Yichang station during the same period was used as the upstream boundary conditions and Zhicheng station's as the downstream ones. Calculated values of depth-averaged velocity, cross-sectional discharge and water stage were verified against measurements.

The roughness and eddy viscosity values adopted for the river channel simulations between Yichang and Zhicheng cannot be determined by any existing empirical method, since the flow resistance of the mobile river bed is related to flow discharge in a very complex way, and it is necessary to calibrate the roughness coefficients in the numerical model against various discharges on a case by case basis. In this study a trail-and-error method was used to calibrate the parameters. The roughness and eddy viscosity were determined for six discharges, $Q = 3300\,m^3/s$, $5010\,m^3/s$, $17300\,m^3/s$, $21100\,m^3/s$, $31200\,m^3/s$ and $40200\,m^3/s$, with the corresponding water stages obtained from field observations (Table 1). In the numerical simulation the coefficient K_s is used, which is related to the Chezy roughness coefficient C and Manning's n as follows (Graf, 1998):

$$C = \frac{1}{n} R^{1/6} \tag{4}$$

$$C = \sqrt{8g}\frac{1}{\sqrt{f}} \qquad \frac{1}{\sqrt{f}} = 2.0\log\left(\frac{R}{Ks}\right) + 2.2 \tag{5}$$

$$\frac{1}{n}R^{1/6} = \sqrt{8g}\left(2.0\log\left(\frac{R}{Ks}\right) + 2.2\right) \tag{6}$$

in which R = hydraulic radius, which can be replaced by the average depth of water. For different discharges, the hydraulic radius is different. In this simulation the value of n is a function of both K_s and R. Therefore the calibration procedures include adjustments of the K_s values for each discharge. The calculated and measured water surface profiles from Yichang to Zhicheng for the six discharges are shown in Figure 2.

Water stage at the Zhicheng station will be much lower when proposed river engineering works are carried out downstream, which includes the removal of several major gravel bars that block the navigation channels (Navigation Office for Changjiang River, 2002). Such water level drop will propagate upstream to the Gezhouba shiplock and cause navigation difficulties because of the water depth reduction there.

4 EFFECTS OF ROCK PAVEMENTS ON THE STAGE-DISCHARGE RELATIONSHIPS

The proposed countermeasures for water level drop due to channel training work include paving several channel reaches with a layer of cobble-sized rocks on the bed (Fig. 3). In this numerical simulation, the water level rises due to such measures are predicted for the proposed five reaches between Yichang and Zhicheng, whose locations are shown in Figure 4.

The calculations are based on the assumption that the pavement will be applied only to those parts of the river bed which are submerged when the discharge is 5010 m^3/s. After the pavements are completed, the surface of paved areas will be elevated by 0.5 m to 1.0 m with increased roughness.

The effects of various pavement schemes are estimated as shown in Table 3. Assuming that K_s equals to the median grain size D_{50} of rocks used for pavements, and the roughness height K_s is increased from 5.7 mm to 78 mm for pavements using medium-sized rocks, the Chezy coefficient n are found to have increased by 35%. If larger rocks are used for the pavements, e.g., K_s becomes 150 mm, the Chezy coefficient n will be increased by 48%.

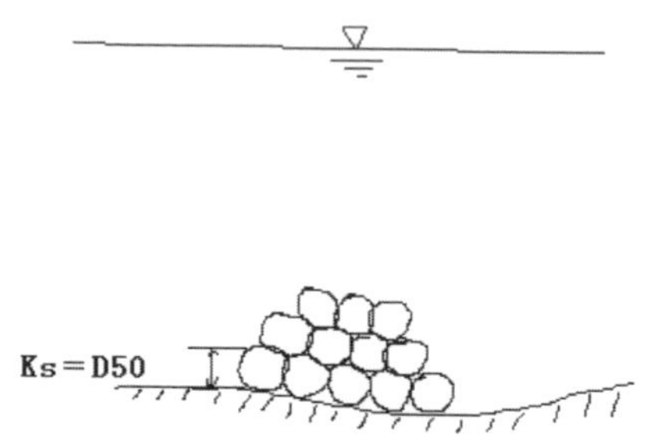

Figure 3. Proposed counter measures: bed pavement.

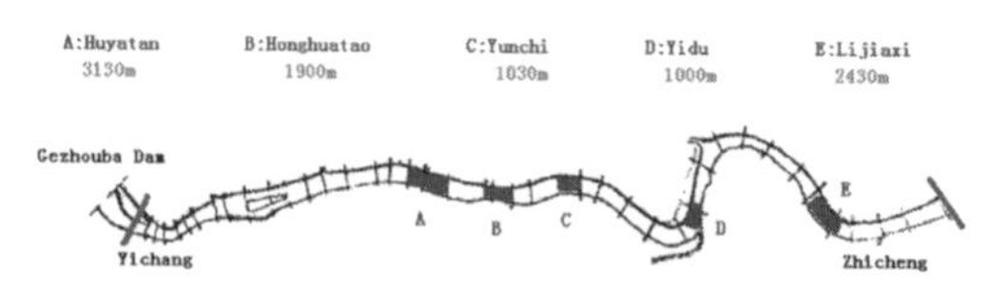

Figure 4. Proposed locations of channel sections to be paved.

Table 3. Combinations of river bed pavements: single and double location(s).

Series number	Paveded sections	Thickness (m)	□ WL of Yichang (m)	
			1.35n	1.48n
s1	A	1.0	0.128	0.15
s2	B	1.0	0.031	0.039
s3	C	1.0	0.037	0.037
s4	D	1.0	0.043	0.052
s5-1	A + B	0.5 + 0.5	0.11	0.14
s5-2	A + B	1.0 + 1.0	0.14	0.18
s6	A + D	0.5 + 0.5	0.077	0.119
s7	B + C	1.0 + 1.0	0.042	0.059

5 IMPACTS OF DOWNSTREAM CHANNEL TRAINING WORKS

The sand-and-gravel bars in the Lujiahe reach (downstream of the Zhicheng station) will be removed to improve navigation conditions, and when such projects are completed, the water level at the Zhicheng station will be lowered by about 0.40 m for the discharge of 5010 m^3/s. Water level at the upstream Yichang station will fall by 0.14 m in response. It is easy to see from Table 4 and Table 5 that, amount the proposed countermeasure schemes, s10-1 and s10-2 are the most effective ones. The corresponding water surface profiles and hydraulic slope in the entire 63 km section of the Yangtze River channel are shown in Figure 5 and Figure 6.

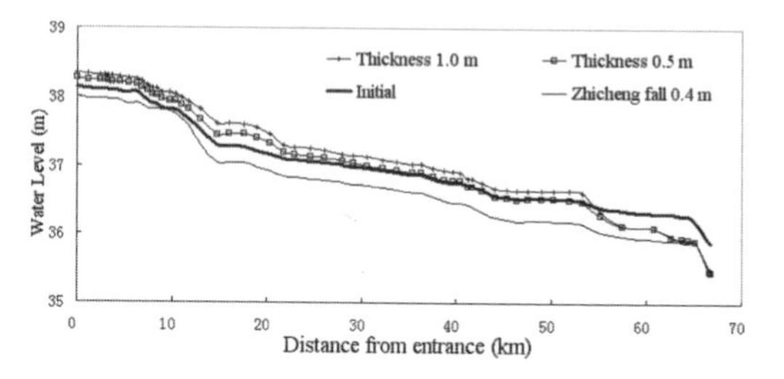

Figure 5. Calculated water surface profiles with pavement measures.

Table 5. Water levels with and without pavement measures.

Location	Initial Water Level Z (m)	No Scheme		Thickness = 0.5 m		Thickness = 0.1 m	
		ΔZ (m)	Z (m)	ΔZ (m)	Z (m)	ΔZ (m)	Z (m)
Yichang	38.13	−0.14	37.99	0.13	38.26	0.21	38.34
Linjiangxi	37.29	−0.26	37.03	0.15	37.44	0.31	37.6
Huyatan	37.15	−0.23	36.92	0.17	37.32	0.3	37.45
Honghuatao	37.01	−0.26	36.75	0.05	37.06	0.14	37.15
Yunchi	36.91	−0.26	36.65	0.03	36.94	0.14	37.05
Yidu	36.72	−0.30	36.42	−0.02	36.70	0.09	36.81
Baiyang	36.53	−0.33	36.20	−0.01	36.52	0.11	36.64
Zhicheng	35.88	−0.40	35.48	−0.40	35.48	−0.40	35.48

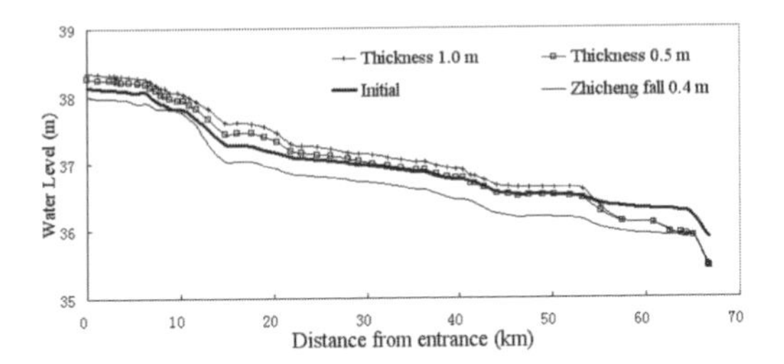

Figure 6. Calculated surface profile for various pavement roughnesses.

Simulation results indicate that the water level at Yichang will be maintained, or even increased, if the bed pavements are applied in the chosen sections of the river channel. At the same time, the velocity and hydraulic slope in the river channel will not increase significantly, therefore it will have little effects on the local navigation conditions.

6 CONCLUSIONS

The fluvial processes downstream the Gezhouba project during the last 20 years have already significantly affected the navigation conditions downstream because of the bed scour after the completion of the project. Engineering measures must be taken to prevent the same problem happening again in the next 20 years after the Three Gorges Project are completed. This paper presents numerical simulations of fluvial process in the Yangtze River reach downstream of the Three Gorges Dam, between Yichang and Zhicheng, with a two-dimensional equation system. The effect of roughness increase due to rock pavement are simulated by using the median diameter of the rocks as the roughness height, K_S, in the equation system.

ACKNOWLEDGEMENTS

This research was supported by the National Natural Science Foundation of China, under Project No. 50479004 and its Science Fund for Creative Research Groups (Grant No. 50221903), and their support is gratefully acknowledged.

REFERENCES

Lin, B. & Falconer, R.A. 2001. Numerical modelling of 3-d tidal currents and water quality indicators in the Bristol Channel. *Proceedings of the Institution of Civil Engineers-Water and Maritime Engineering*, 148(3):155–166.

Graf, W.H. 1998. *Fluvial hydraulics*. Chichester: John Wiley&Sons Ltd.

Lu Yongjun, Chen Zhicong, Zhao Lianbai, Shao Xuejun, Yang Meiqing, Li Yunzhong. 2002. Impact of the Three Gorges Project on the Water Level and Navigation Channel in the Near-dam Reach Downstream The Gezhouba Project. *Engineering Sciences*, 4(10):67–72.(in Chinese with English abstracts).

Pan, Qing-shen. 2001. Study on evolution of middle and lower reaches of Yangtze River in recent fifty years. *Changjiang Kexueyuan Yuanbao* (*Journal of the Yangtze River Scientific Research Institute*). 18(5):18–22. (In Chinese with English Abstracts).

River, Coastal and Estuarine Morphodynamics: RCEM 2005 – Parker & García (eds)
 ISBN 0 415 39270 5

Sedimentation in a series of dam reservoirs and their trap efficiency

K. Ohashi
Graduate School of Engineering, Gifu University, Gifu, Japan

Y. Fujita
River Basin Research Center, Gifu University, Gifu, Japan

ABSTRACT: On the Kiso river, one of major rivers in Japan, many dams had been constructed since 1924, and their reservoirs have stored more than 100 Mm^3 sediment as a whole. Sediment yield and transport ability in a drainage area are tried to be estimated from past surveying data of reservoir sedimentation and hydrological one, focusing on sedimentation rate at early stage of sedimentation. Sedimentation rate in the early stage shows good correlation with sediment runoff potential as a parameter of sediment yield and transport ability, and leads to define a tentative trap efficiency of dam, which is clarified to be related with the dimension of the dam structure as well as behaviors of suspended sediment particles in dam reservoir during floods.

1 INTRODUCTION

The Kiso river is located in Chubu District of Japan whose trunk length is 229 km, and the total area of 9200 km^2 of its drainage basin is divided into four almost same magnitude of sub basins, corresponding to the mainstream and three major tributaries, namely the Hida river, the Nagara river and the Ibi river from east. The Kiso main river has its headwater in Mt. Hachimori (2246 m) and drainage area of 2452 km^2, flowing through the Kiso valley between Kiso mountain range on the east side and Mt. Ontake (3063 m) on the west side. On the Kiso mainstream and on the Ohtaki river, its largest tributary, a series of dams were constructed from 1924 to 1963 in order to use plenty of water flow mainly for power generation, as eight dams on the mainstream and five on the Ohtaki. Sedimentation and hydrological data had been analyzed since 1930s to clarify sediment yield and transportation of the drainage basin and the data is recorded for more than 50 years. Correlations are examined between time series of annual sedimentation and those of hydrological quantities, total rainfall per year, annual maximum of daily rainfall, annual maximum of discharge and annually total of inflow volume. Unfortunately, results cannot show significant correlations because of lack of these hydrological data of individual floods, which are necessary to estimate trap efficiency. There are many other sedimentation data of old dam without neither flood discharges nor the durations. It is important to find significance from these data, and some methods are proposed to clarify sedimentation.

2 OUTLINE OF A SERIES OF DAM RISERVOIRS AND THEIR SEDIMENTATION PROCESS

2.1 *Outline of dams and reservoirs*

On the Ohtaki river, which is one of major tributaries in the Kiso mainstream basin, there are five dams, Miure, Ohtakigawa, Makio, Tokiwa and Kiso, while six dams on the Kiso mainstream, Yomikaki, Ochiai, Ohi, Kasagi and Maruyama, are located at the downstream reach from the confluence with the Ohtaki river to Maruyama dam site. In addition to them, there are quite new dams, Misogawa dam near the headwater and Agigawa dam on the Agi river, a tributary influent at just downstream of Ohi dam. There also is Inagawa dam on the Ina river, having a small reservoir capacity. These three dams are excepted from present investigation because of small amount of sedimentation and of scarcity of data. Locations of these dams are shown in Figure 1.

The total storage capacities of Miure, Makio and Maruyama dam are more than 60.0 Mm^3, that of Ohi dam is 30.0 Mm^3, and Kasagi 14.0 Mm^3. Those of the other dams are about 4.0 Mm^3. Table 1 lists for each dam the constructed year, dimensions of dam structure, catchment area and total storage capacity of reservoir. The highest dam is Makio, the height of which is 104 m, while those of the second and the third highest Maruyama and Miure are 98.2 m and 83.2 m, respectively, and crest heights of these three dams are beyond 70 m.

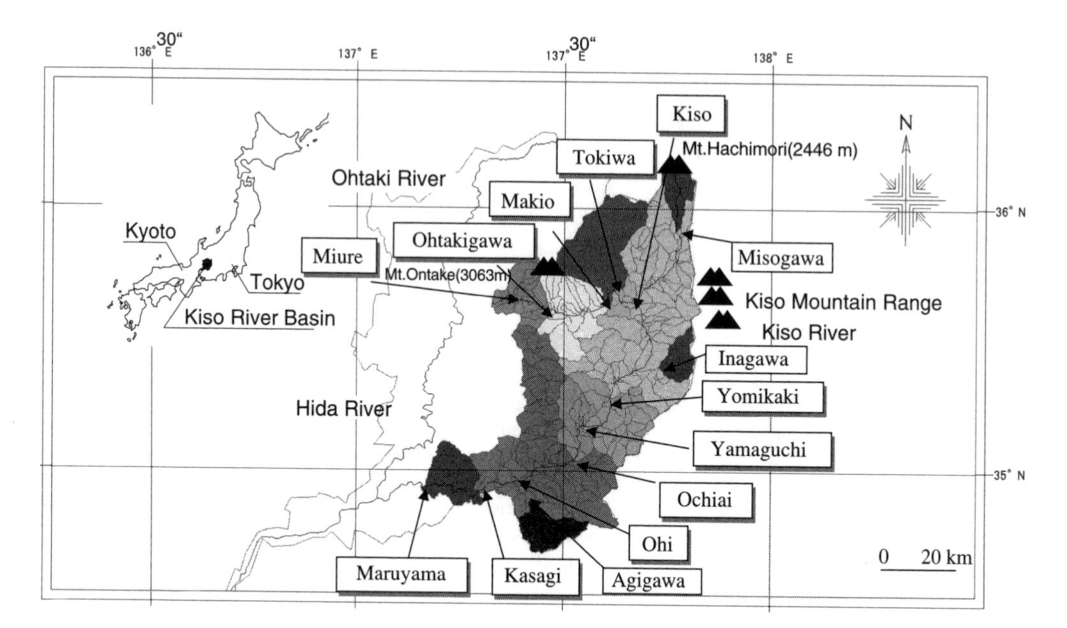

Figure 1. Location of the Kiso river basin and dam sites in its basin.

Table 1. Outline of dams and their reservoirs in the Kiso main river basin.

Dam	River	Build Year	Height (m)	Length (m)	Crest Height (m)	Drainage Area (km^2)	Storage Capacity (Mm^3)
Miure	Outaki	1945	83.2	290	74.6	69.4	62.216
Ohtakigawa	Outaki	1948	18.2	80	9.7	114.2	0.589
Makio	Outaki	1961	104.5	260	89.5	304	75.000
Tokiwa	Outaki	1941	24.1	111.9	15	553.7	1.288
Kiso	Outaki	1968	35.2	132.5	20.5	578.9	4.367
Yomikaki	Kiso	1960	32.1	293.8	17.3	1341.8	4.358
Yamaguchi	Kiso	1957	38.6	181.4	26.1	1534.5	3.484
Ochiai	Kiso	1926	33.334	215.1	23.972	1747	3.872
Ohi	Kiso	1924	53.384	275.8	38.007	2055.3	29.400
Kasagi	Kiso	1936	40.804	154.9	28.727	2301.2	14.121
Maruyama	Kiso	1955	98.2	260	74.5	2409	79.520

2.2 *Volumetric changes in annual sedimentation*

Sedimentation in the above eleven dam reservoirs has last 1925, when Ohi dam was constructed. The annual and accumulate sediment data are shown Figure 2. Unfortunately, there are no sedimentation data in Ohi and Ochiai dam reservoir in early stage before 1930s because almost of all people did not aware of this problem, and annual sedimentation volumes are lead from accumulate sediment volume as mean value. It shows that when a dam is built newly in the upstream reach from the dam, annual sedimentation volume of the downstream dam tends to decrease. After the reservoir of upstream dam is filled with sediment, annual sedimentation of downstream dam increases again. As Sedimentation progresses, accumulated sediment volume converges and fluctuates at a certain value, and annual sediment volumes do not always show positive values. This condition is called an equilibrium state of sedimentation. A period approaching to the equilibrium state is determined by the ratio of total storage capacity of dam reservoir to sediment input. Therefore, a small dam like Ohtakigawa dam becomes the equilibrium condition only for several years.

2.3 *Correlation between sedimentation and hydrological data*

First of all, correlations are examined to clarify sedimentation phenomena between time series of annual sedimentation and those of hydrological quantities,

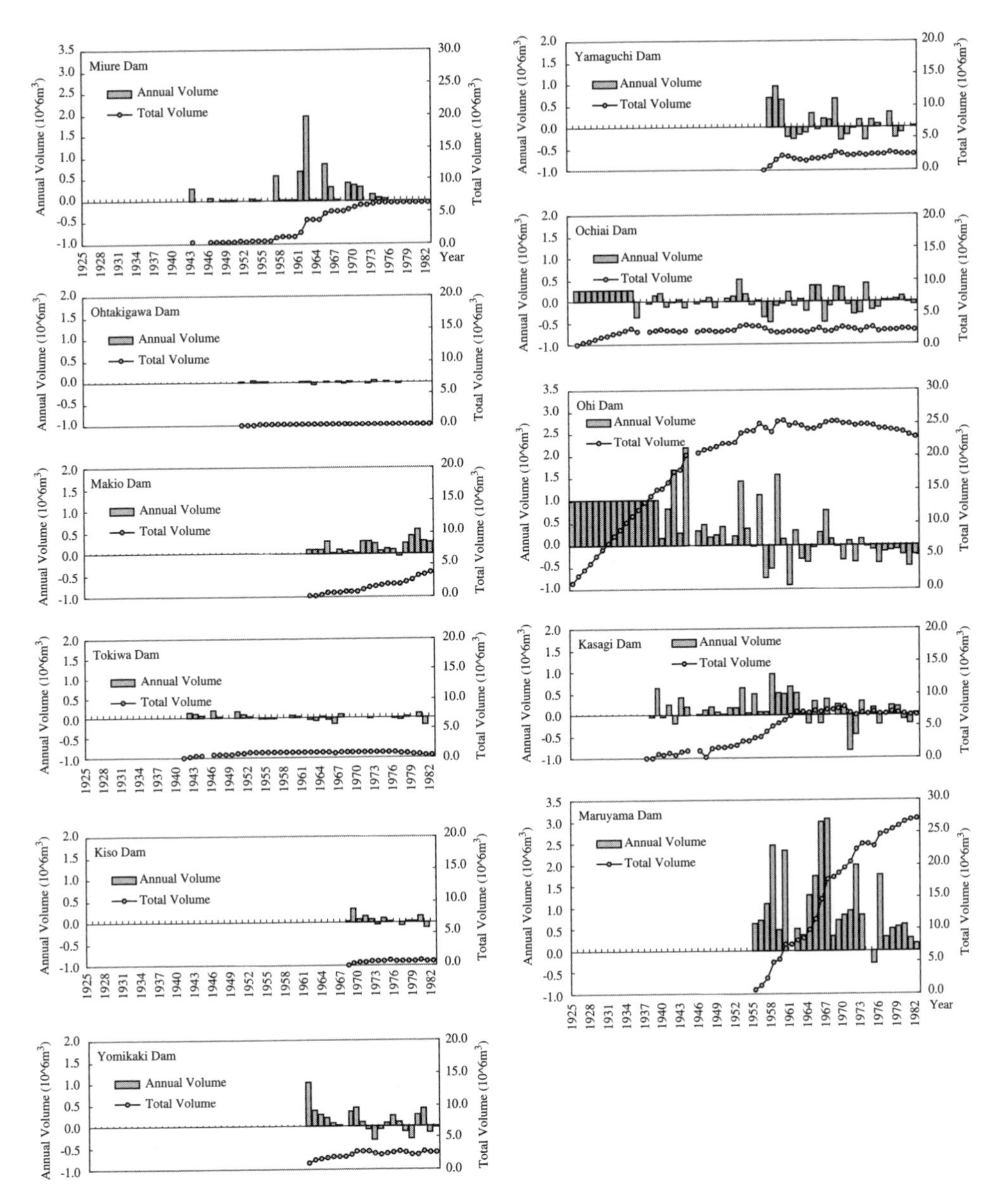

Figure 2. Annual and Total Sedimentation Volume.
There is no Sedimentation data of Ochiai dam for 8 years and Ohi dam for 14 years. The data in the period is given by mean sedimentation volume.

annual total rainfall, annual maximum of diurnal rainfall, annual maximum of discharge, annual total inflow volume and sedimentation rate. Figures 3 and 4 show correlation between annual sediment volume and annual maximum discharge, and between annual sediment volume and annual total inflow respectively. Circle marks are data at early stages of sedimentation and cross marks indicate others. This classification is explained later. It is expected before these that annual maximum discharge correlate with sediment volume more or less, because most of sediment volume is transported in flood, and/or annual maximum of discharge indicate its magnitude. However, signification cannot be found on any dam sites as shown Figure 2. Sediment volume is not estimated by the only annual maximum discharge, because duration and frequency are important factors of it. Another correlation is not also significant because sediment volume

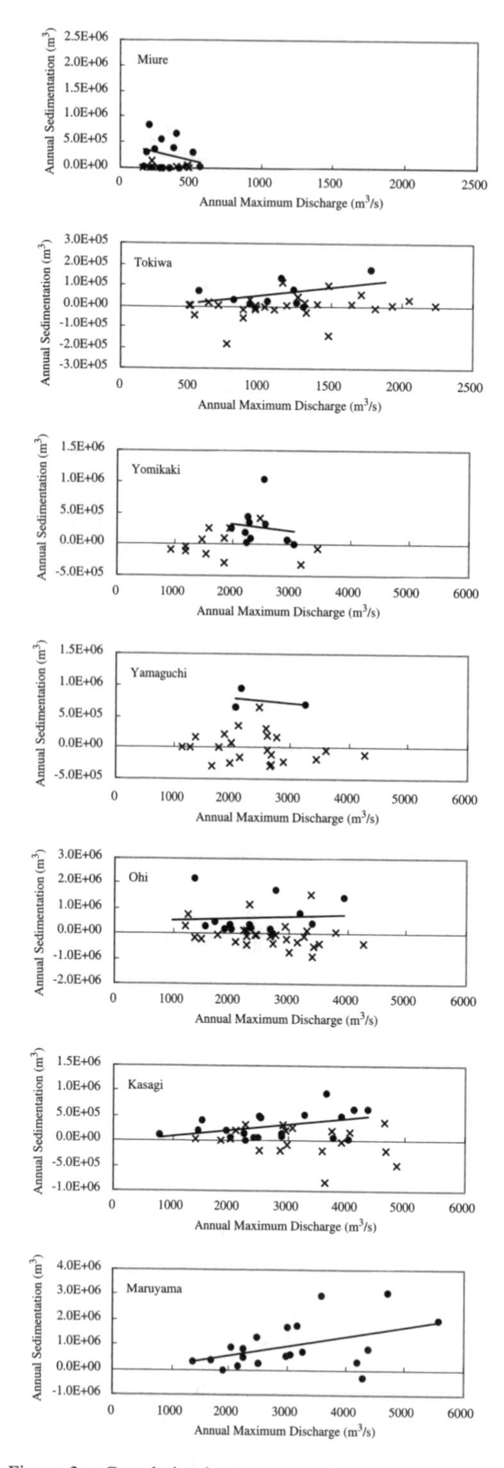

Figure 3. Correlation between annual sedimentation volume and annual maximum discharge in major 6 dam sites especially. Attention should be paid to a difference of longitudinal and transverse scale.

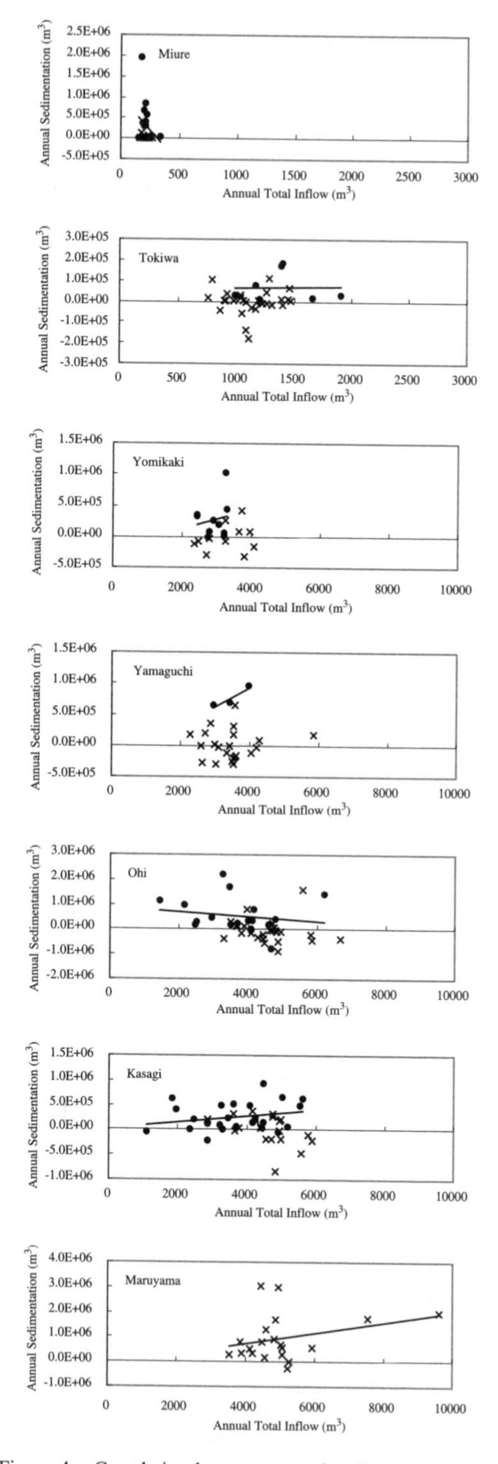

Figure 4. Correlation between annual sedimentation volume and annual total inflow in major 6 dam sites especially. Attention should be paid to a difference of longitudinal and transverse scale.

increases exponentially as flood discharge increase although annual total inflow includes implication of that duration and frequency.

3 ASPECTS OF SEDIMENT TRANSPORT FROM A SERIES OF DAM RESERVOIR SEDIMENTATION

3.1 *Sedimentation rate in the early stage*

It is difficult to analyze sedimentation process quantitatively in case that a lot of dams are constructed in a basin serially, because annual changes in sedimentation volume do not always correspond to fluctuations of hydrological quantities, being affected by changes in the trap efficiency of dams in the upstream reach. Therefore, in order to estimate strictly sediment yield from a drainage basin and transported volume to a dam site, or sediment input to the dam reservoir, sedimentation process must be focused in a period in which a trap effect of the dam is large enough to catch almost of all sediment transported into the reservoir and the influence of dams in upper reach is little on sediment transport conditions. Generally, the trap efficiency of dam is the highest immediately after the construction when the reservoir depth is kept deep to make flow gentle enough to deposit rather fine sediment. As sedimentation progresses to raise riverbed, water depth decreases to increase flow velocity and bed shear stress in the reservoir, and volume of sediment trapped there is lowered. Accordingly, for the purpose of evaluation of sediment transport rate at the dam site, it is necessary to pay attention to changes in total sedimentation volume during a period from the time dam has constructed when sediment trap effect is the greatest to that the sedimentation has approached to the equilibrium state when the trap efficiency has been extinguished.

Here, we consider annual sedimentation rate in early stage to estimate sediment transportation at a dam site under natural condition without it. The early stage is defined as a period from the year of dam construction to that when annual sedimentation volume showed a minus value for the first time, and the period when effects of dams in the upstream reach were large is excluded from it. Table 2 lists results periods of the early stage obtained from the mentioned above and annual sedimentation rates in the early stage as the mean value in this period. Table 2 also shows a tentative estimate of trap efficiency in this period provided that Miure and Maruyama dam reservoirs could catch all of sediment due to their large initial capacities compared with inflow rates.

According to this table, Miure dam, located in the uppermost reach of the Ohtaki river, has a high value of sedimentation rate in the early stage, implying that it entrapped directly a large amount of sediment produced in steep slope area around Mt. Ontake. However, in the downstream reach from it, sedimentation rates in the early stage show an increasing tendency proportional to basin scale. As aforementioned, dam reservoirs have high ability of sediment capture just after their construction, sedimentation rate in the early stage can be considered to indicate mean annual sediment yield in the basin of the reservoir. In the following section we discuss a parameter controlling sediment production expressed by this value.

3.2 *Sediment runoff potential*

Sediment yield is a phenomenon that earth, sand and rock are put into a condition to be movable, or to be sediment or debris, by various external forces such as physical and chemical weathering, eluviation, gravity action and crustal movement. Sediment particles are carried mainly by water, and because it is difficult to distinguish sediment yield in a basin from sediment transport their definitely, they are regarded to be a part of a phenomenon and included as "sediment runoff in

Table 2. Period of the early stage, annual sedimentation rate in it and tentative trap efficiency.

Dams	Period	Sedimentation Volume (m^3)	Sedimentation rate (m^3/year)	Trap Efficiency Et (%)
Miure	1943–1971	6,039,300	208,252	100.0
Ohtakigawa	1945–1955	105,300	9,573	4.6
Makio	1951–1982	3,749,000	170,409	81.8
Tokiwa	1941–1948	566,500	94,417	45.3
Kiso	1968–1972	644,500	161,125	77.4
Yomikaki	1960–1970	2,889,500	321,056	34.0
Yamaguchi	1957–1960	2,283,000	761,000	80.7
Ochiai	1926–1935	2,388,000	265,333	28.1
Ohi	1929–1956	25,021,000	781,906	82.9
Kasagi	1957–1962	3,100,000	620,000	65.8
Maruyama	1962–1972	18,858,100	942,905	100.0

a basin" for discussion here. When the sediment runoff phenomenon is considered in a basin scale overall, the most important factor is mechanical energy including both kinetic energy and potential energy of water existing in the basin which can move and transport sediment particles. This potential energy is originated from that of precipitation on the basin and it changes into kinetic energy at a rate of descending the basin as a current to convey sediment particles. The total potential energy in a basin depends on its area and distribution of both altitude and precipitation, mainly rainfall here. Therefore, the total potential energy of rainfall in a basin can be defined as sediment runoff potential. Relation between the sedimentation rate in the early stage and the sediment runoff potential is examined below.

As a method to express altitude distribution of a basin, Hypsometric curve, an area – altitude distribution curve, is often used. Then, in order to obtain the hypsometric curve for each dam basin, the Kiso mainstream basin is divided into 285 unit basins and the altitude date are embedded into each unit basin with the aid of GIS software. These data used here are offered from the Geographical Survey Institute, the Ministry of Land, Infrastructure and Transport, Japan, such as divides of the unit basins decided from 1/2500 topographical maps and information of altitude also gathered from 50 m mesh data of elevation above the sea level. The unit basins are classified into individual dam basins and hypsometric curve of each dam basin is completed as shown in Figure 5 to compare characteristic of the curve with that of sedimentation.

Since the sediment runoff potential is assumed as the total potential energy of the rainfall in the basin, it is approximated by a product of the average year rainfall in the basin and integral value of the hypsometric curve from the basin minimum altitude equivalent to a dam reservoir level to the basin highest altitude. It is expressed as follows:

$$\text{Sediment Runoff Potential} = \rho R g \int_{h_{min}}^{h_{max}} A(h)dh$$

where ρ is density of water; R = mean annual rainfall; A = basin area; g = acceleration of gravity.

The calculated values of the runoff potential are listed in Table 3. In Figure 6 the sediment runoff potential is correlated with the sedimentation rate in the early

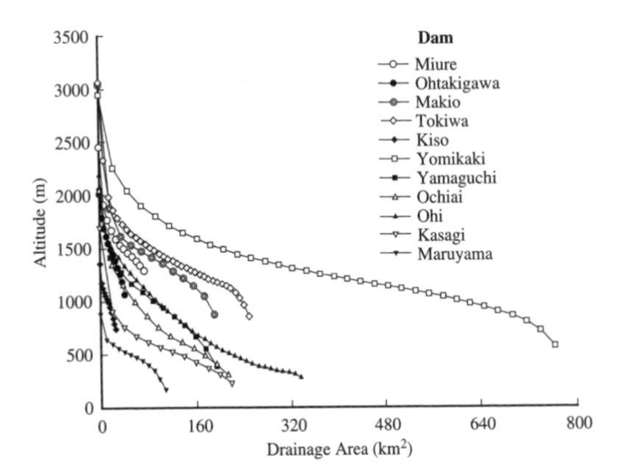

Figure 5. Hypsometric curves of individual dam basins.

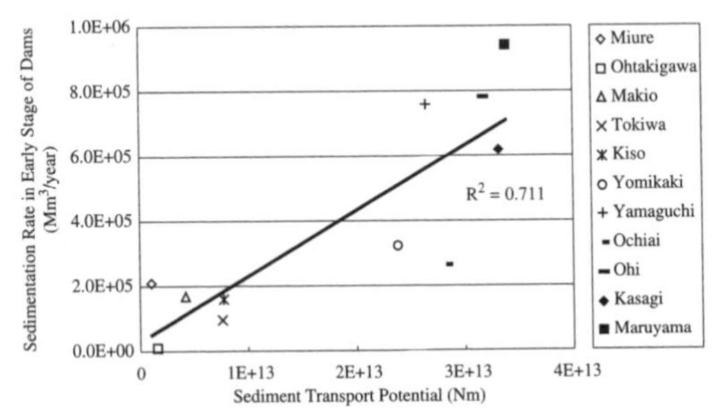

Figure 6. Correlation between transportation the sediment runoff potential and the sedimentation rate in the Early Stage.

Table 3. Runoff potential of each drainage basin.

Dam	Integral of Hypometric Curve (km^3)	Mean Annual Rainfall (mm)	Sediment Transport Potential of Individual dam (Nm)	Accumulate Sediment Transport Potential (Nm)
Miure	31.695	3255.6	1.011.E+12	1.011.E+12
Ohtakigawa	17.307	3546.7	6.016.E+11	1.613.E+12
Makio	111.995	2385.5	2.618.E+12	4.231.E+12
Tokiwo	160.679	2152.2	3.389.E+12	7.620.E+12
Kiso	6.403	1876.2	1.177.E+11	7.738.E+12
Yomikaki	630.308	2604.4	1.609.E+13	2.383.E+13
Yamaguchi	127.080	2113.0	2.631.E+12	2.646.E+13
Ochiai	110.900	1766.2	1.920.E+12	2.838.E+13
Ohi	176.918	1924.1	3.336.E+12	3.171.E+13
Kasagi	82.085	1891.2	1.521.E+12	3.323.E+13
Maruyama	40.650	1644.6	6.552.E+11	3.389.E+13

stage. It shows a clear correlation between them around a regression line except for a very low plot of Ochiai dam which was constructed very early and does not have sedimentation data for about twenty years just after the dam construction, leading to a small value of the sedimentation rate in the early stage. It also demonstrates an interesting feature that plots of dams with large capacity of reservoirs, such as Miure, Maruyama, appear in an upper region than the regression line while data of those of small ones are plotted below the line. This difference implies that though we focused on the sedimentation rate in the early stage to estimate quantity of sediment inflow from sedimentation volume, after all the rates are influenced also by absolute capture abilities relating to dimensions of dam body as a hydraulic structure even just after the construction when dam reservoirs have a high capture effect. In other words, the result indicates that the trap efficiency is not a function of only a basin area and a storage capacity, as concluded in the previous studies, but that including dimensions and types of dam structure.

4 TRAP EFFICIENCY AND ITS CONTROLLING FACTORS

4.1 *Trap efficiency of dam reservoirs*

In general, it is difficult to evaluate the trap efficiency of dam reservoirs before their construction because sediment transport was rarely measured correctly not only at the dam site but also near suitable sites and because exact estimation of sediment yield in the catchment area also is very hard work, almost impossible. Measurement and estimation of sediment yield and transport become possible practically by the use of sedimentation data and by some sophisticated observations at the dam site after their construction. In case of the present study, the trap efficiency can be estimate postulated that dam reservoirs with large capacity such as Maruyama reservoir catch almost all sediment input. However, that in the early stage of old dams cannot be estimated because, for example, in case of the oldest Ohi dam constructed in 1925, as 80 years have already passed, sediment inflow from the upstream and outflow from the dam were not recorded or lost. Therefore, tentative trap efficiency in this early period is estimated, as mentioned in 3.1, provided that initial capacities Miure and Maruyama dam reservoirs of 79.5 Mm^3 and 62.2 Mm^3, respectively, are large enough to catch all of sediment, being compared with water inflow, and it seems from the sedimentation data that these situations have not so changed even in the present stages.

Then, provided also that sedimentation rate in the early stage of Maruyama and Miure dam express total sediment volume transported through the Kiso mainstream for dam reservoirs in the upstream reach had been filled before its construction and the Ohtaki river for it is located in the uppermost reach and its catchment area is large, a estimate of the tentative trap efficiency in the early stage is given as ratio of the sedimentation rate in the early stage of each dam to that of Maruyama or Miure dam. Obtained values are listed in Table 2, as aforementioned.

4.2 *Relation between trap efficiency and dimensions of dam structure*

As the possibility that trap efficiency elates to dam structure is made clear in the preceding chapter, actual transportation phenomena should be examined to derive factors having dominant influences on trap efficiency. Sediment particle passing a dam site must be carried either through intake for water use as generation, irrigation, etc or over the spillway. As a large quantity of sediment is transported to the down stream usually not by ordinary discharge but by flood discharge, most of sediment particles passing a dam site are conveyed through spillway overflowing the dam crest. Therefore, trap efficiency is presumed to depend on a height that the sediment particle can be suspended and that of overflow crest of dam primarily. Unfortunately, about suspension heights of sediment particle cannot be discussed for lack of detailed surveying data, grain size composition of riverbed and hydrological data floods at the time of dam construction. Consequently, height of overflow crest of a dam is assumed to be an index representing the dimension of dam structure.

In Figure 7, the tentative trap efficiency is compared with the height of overflow crest and a certain tendency is discernible in their correlation. The tentative trap efficiency shows sudden raises in the vicinity of 20 m of the height of overflow crest and approach to 100% when the height beyond 40 m in both cases of the Kiso mainstream and the Ohtaki Thus, it is confirmed that the heights of overflow crest influence the tentative trap efficiency greatly.

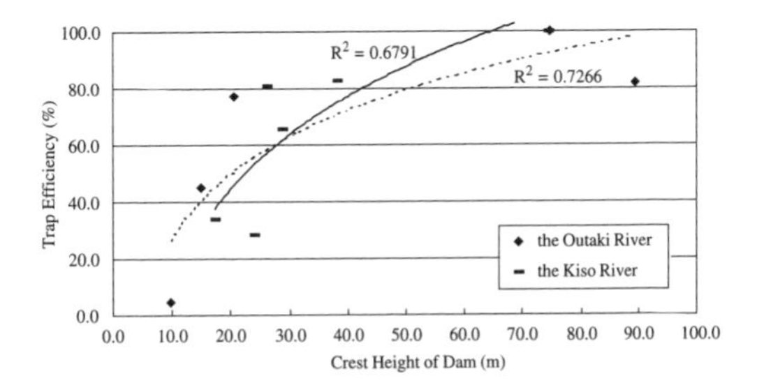

Figure 7. Correlation between tentative trap efficiency in the early stage and height of overflow crest.

5 CONDITION OF SEDIMENT SUSPENSION IN MARUYAMA DAM RESERVOIR

5.1 *Condition of maximum suspension height*

Suspension heights of sediment particles of individual grain size are considered to be one of dominant factors affecting the trap efficiency and can be evaluate from numerical analysis based on turbulent diffusion equation. However, since such a simulation require a lot of effort, a very simple estimation of maximum suspension height is introduced here to judge whether a certain size of sediment particle is able to pass over the dam crest from reservoir bed under a given flood condition. As condition for a sediment particle to be suspended load is, in general, that vertical turbulence intensity approximated by shear velocity u_* become greater than its terminal fall velocity w_0, and a half of suspended particles are raised by turbulence while the other half lowered, so the maximum height is thought to be attained by a particle raised always at a vertical velocity of $u_* - w_0$ during longitudinal convection by mean flow velocity before arrival at the dam site or deposit on the bed downstream in the reservoir. Thus, the finest particle sizes wholly entrapped in reservoir can be predicted under various flood conditions.

In the following section, using the detailed records in Maruyama dam data, loci of particles which attain

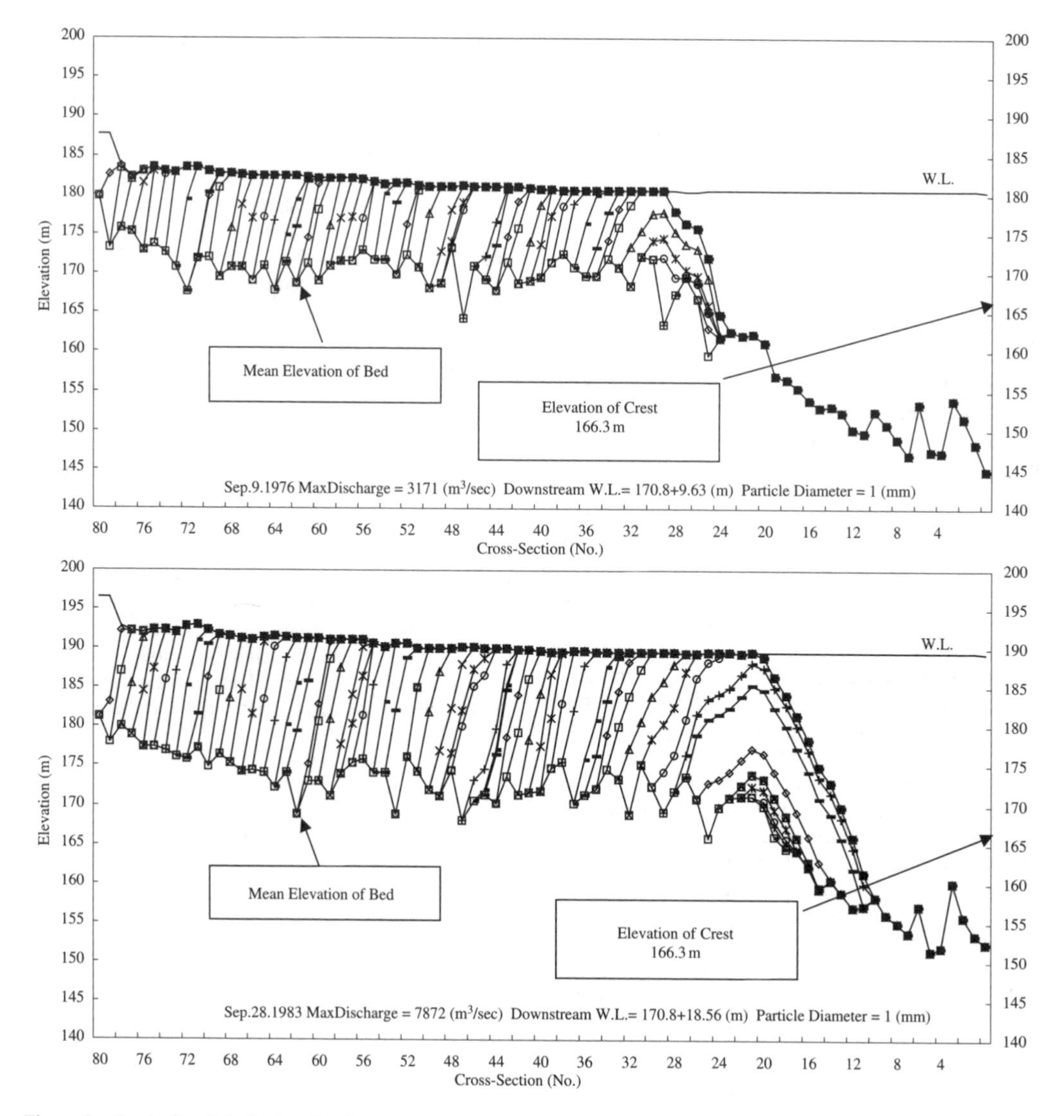

Figure 8. Loci of particle for two floods.

the maximum heights are traced and compared with the overflow crest height postulating several floods discharge corresponding to experienced floods.

5.2 *Analysis of sediment suspension for actual floods in Maruyama dam reservoir*

The particle loci are traced for floods in duration from 1976 to 1999 when a detailed flood record remains. Here, loci of sediment particles of 1 mm grain size calculated for two floods are presented as examples, which are for a flood in 1976 with peak discharge of 3170 m^3/s as rather high frequent floods and that in 1983 recording the greatest peak discharge of 7870 m^3/s after the dam construction. Figure 8 demonstrates the loci of sediment particles starting from the riverbed at every surveying cross section. In the figure, when the shear velocity decreases to be below the fall velocity according to increase in cross sectional area, particles begin to descend toward the riverbed, even being once raised up to the water surface level. In Maruyama reservoir, under the hydraulic conditions of 1976 of comparatively frequent flood, the fall velocity exceeds the shear velocity from No. 30 section to the downstream to stop entrainment of particle on the bed and from No. 23 section no particle exists in the flow. It means that particles of grain size coarser than 1.0 mm were not transport in the downstream reach from No. 23. This feature agrees with characteristics of grain size distribution of bed materials shown in Figure 9, where there are almost no particles coarser than 1.0 mm in No. 25 section though a fraction of that particle size occupies more than 30% in No. 35 section. In addition, even in the case of the 1983 greatest flood not only there was no outflow of the particle of 1.0 mm grain size from dam reservoir overflowing the crest, but also all of the particles begin to fall in No. 20 section and cease from suspension around No. 10 section. As mentioned above, it is concluded that such an analysis for each particle size can predict overall feature of sediment transportation in dam reservoir and grain size which can flow out into downstream from dam reservoir. This simple analysis applying to various floods and reservoirs can help discussion on trap efficiency with high precision.

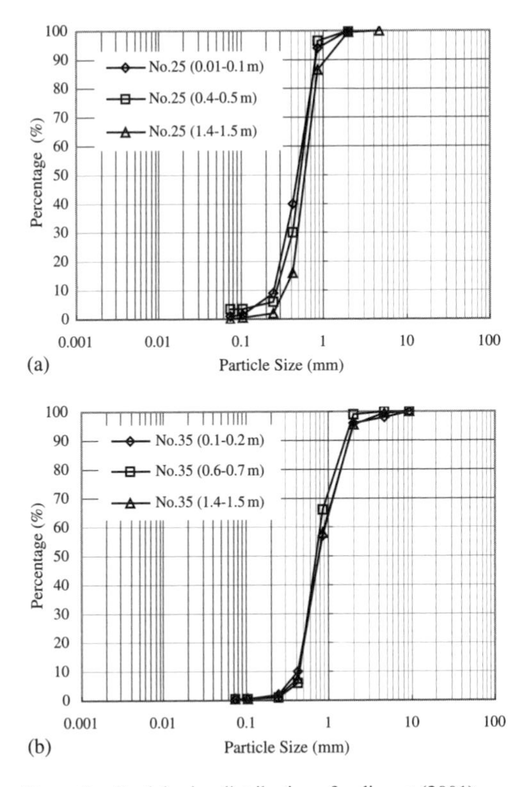

Figure 9. Particle size distribution of sediment (2001).

6 CONCLUSIONS

Sedimentation and trap efficiency in a series of dams on the Kiso main river are studied being based on the past surveying and hydrological data, especially, focusing on those at early stage of dam reservoir sedimentation. The results clarified in the present study are the followings:

1. The sedimentation volume, both total and annual, in a series of dam reservoirs shows almost no correlation with hydrological data.
2. A sedimentation rate in the early stage was defined to discuss sediment transportation in a basin from specific feature of time series of annual sedimentation as that during a period the rate approaches to an equilibrium state.
3. The sediment rate in the early stage is highly correlated with sediment runoff potential, namely the total potential energy of rainfall in the basin evaluated with hypsometric curves, and the correlation suggests an influence of scale and structure of dam.
4. Tentative trap efficiency is calculated from the sedimentation rate in the early stage and it has an interesting relation with height of the overflow crest from the dam base.
5. Behaviors of suspended sediment particles in dam reservoir are analyzed simply under flood conditions to clarify transport limits of certain grain size sediment, and predicted limits agree well to characteristics of particle size distributions of bed materials in the reservoir.

In the near future, vertical two-dimensional flow analysis will be performed and the results will be compared those obtained in the present study to clarify sediment movement mechanisms in dam reservoir, and the results will be applied to predict the sedimentation in new Maruyama dam, which is planed to be constructed 24.3 m higher than present one.

ACKNOWLEDGEMENTS

Maruyama dam and Reservoir Management Office provided organized documents and permitted to use them to prepare this paper. Graduated students belong to our River Engineering Laboratory at Gifu University helped with the data analyses. The authors wish to express deep gratitude to them for their valuable assistance.

REFERENCES

Japan Dam Foundation. 1999 The Dam Almanac: 134–136.

Ministry of Construction Chubu Regional Headquater & Maruyama Dam and Reservoir Management Office & Central Consult Inc. 1983 Report of sediment transportation in Maruyama Dam in 1983: 19–62.

Chubu Fukken Inc. 2002 Report of examination of sedimentation in 2001: reference materials, cross-sectional surveying data, 1–25.

Chubu Fukken Inc. 2003 Report of examination of sedimentation in 2002: reference materials No.1, 1–132: reference materials No.2, 1–13.

Kira, H. 1982 Dam Sedimentation and Prevention. Morikita Publication, 104–10.

Turbidity currents and submarine morphodynamics

River, Coastal and Estuarine Morphodynamics: RCEM 2005 – Parker & García (eds)
© 2006 Taylor & Francis Group, London, ISBN 0 415 39270 5

Incision dynamics and shear stress measurements in submarine channels experiments

P. Lancien, F. Métivier & E. Lajeunesse
Institut de Physique du Globe de Paris, France

M.C. Cacas
Institut Français du Pétrole, France

ABSTRACT: We report observations on the incision dynamics of subaqueous channels in small-scale laboratory experiments, showing the initial incision phase, followed by a widening phase associated with an erosion-deposition wave progradation. In order to better characterize the inception phase, we make measurements on the gravity current prior to incision. Thanks to the particle-with-shadow tracking, a new technique using settling particles to gather information on the velocity field, we obtain the lowest part of the downstream velocity vertical profile with a good accuracy. The velocity gradient in the viscous sublayer gave us access to the first direct shear stress measurements in a gravity current.

1 INTRODUCTION

Submarine channels systems are the conduits that allow turbidity currents to transport and deposit material eroded from the continents in the deep sea. However, little is known about the physical processes of their formation, and consequential questions remain unsolved: what are the physical conditions and the dynamics of submarine erosional channels inception? Are they produced by catastrophic flood events or could a steady-state sediment-charged current be sufficient? What are the velocities and densities required to erode? In other terms, what is the necessary shear stress distribution on the bed? How does the channel select his width and his depth?

The lack of answers is mainly due to the difficulty in making measurements on active or abandoned channels, and the probably long time scale needed to develop these structures compared to a human time scale. Because of these drawbacks, researchers have focused on experimental and numerical studies of turbidity currents (Kneller et al. 1999, Kneller & Buckee 2000).

Métivier et al. (2005) managed to reproduce subaqueous channels in small-scale laboratory experiments. This paper complements this previous work, focusing on the description of some incision dynamics aspects and introducing velocity profiles and shear stress measurements.

2 EXPERIMENTAL SETUP

The experimental setup consists of a 100×50 cm incline, immersed in a $200 \times 50 \times 50$ cm flume filled with fresh water mixed with fine particles. Once the powder has settled, a layer is slowly raked upward on the incline to drape it by a uniform sediment blanket. Then, a sustained gravity current is injected at the top of the ramp and flows over this erodible bed. We control the three main parameters of the experiment: the slope of the plane, the input flow rate, and the input flow specific gravity.

The visualization is made from above, using two different devices depending on the time scale of the event type we want to study. For the capture of long time scale events, like the channel formation, a digital camera is used to take one picture each minute. For short time scale events, like the particles transport, close movies of the plane are taken, using a video camera instead.

This setup is rather simple, but the experimental difficulties lie in the choices made for the nature of the sediments and the density current.

2.1 *The sediments*

The particles composing the sediment blanket are made of a mixture of plastic polymer and titanium oxide, achieving a density of 1080 kg/m^3. Most of the grain diameters are comprised between 15 and 40

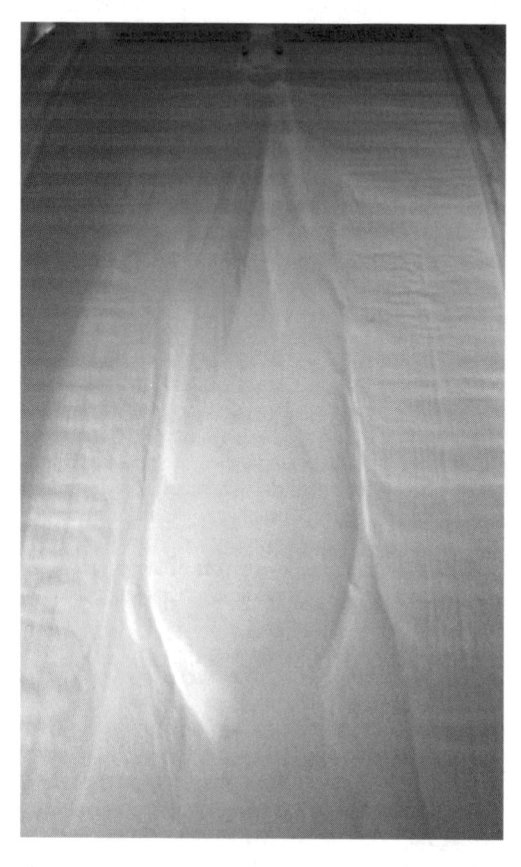

Figure 1. Typical channel and frontal lobe obtained with our experimental setup. The current inlet is visible at the top.

microns. Their low specific gravity in the water, combined with their small size, make these particles very easily erodible.

2.2 *The gravity current*

The turbidity current is simulated using salt water, without particles. Indeed it is the density contrast which is responsible for the erosional power of the turbidites. In nature, sediments play an indirect role for the erosion, in the sense that they induce this density contrast.

Using the settling velocity V_s, the typical velocity of the density current U, and its typical height H, it is possible to construct a characteristic length:

$$X = \frac{UH}{V_s}$$

X is a settling length; it's the typical horizontal distance a suspended particle of the density current can achieve

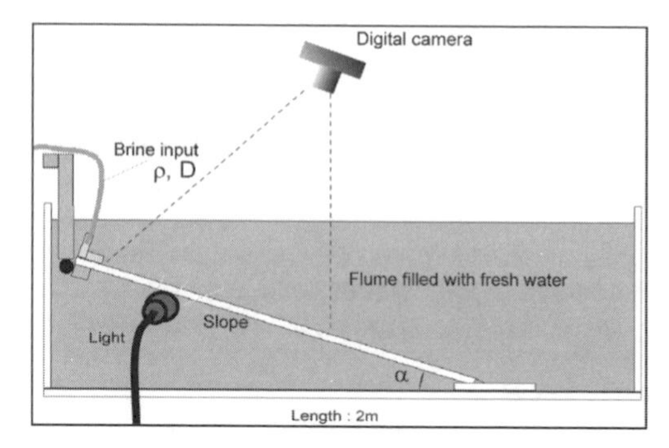

Figure 2. Experimental setup, not to scale.

before reaching the bottom. For small scale experiments, X is less than a few centimetres, even when attempting to lower V_s with very small and light particles. When particles are mixed with the water at the input of our experimental setup, they settle down just a few centimetres after the entrance, and no channel can be formed.

The employ of brine to produce the density contrast figures out this scale problem. Because the brine is transparent, the second benefit of this method is to enable visualization of the bed during the whole experiment, and then to allow us to study the incision dynamics.

3 INCISION DYNAMICS

We have shown in a previous work (Métivier et al. 2005) that a steady-state density current can induce self-channelization. Indeed, in these experiments, the sediment blanket is initially flat. When the brine is introduced, there is a development phase in which the density current spreads over the bed and then the channel inception phase begins. In the meantime, the sediment eroded during the channel formation is deposited at the base of the ramp, and constructs a depositional lobe. Here we present a few aspects of this channel incision dynamics.

3.1 *Measurement method*

In order to perform dynamical acquisitions of the topography, we use an overhead projector to project parallel sinusoidal fringes on the incline. A digital camera takes a picture from above every minute. On these pictures, the height variations induce phase modulations of the sinusoidal signal. At a point A, the phase difference $\Delta\varphi$ due to a sediment layer height h is given by (see figure 3 for notations):

$$\Delta\varphi = \varphi_A - \varphi_B = 2\pi \frac{h}{\lambda . \cos\alpha}$$

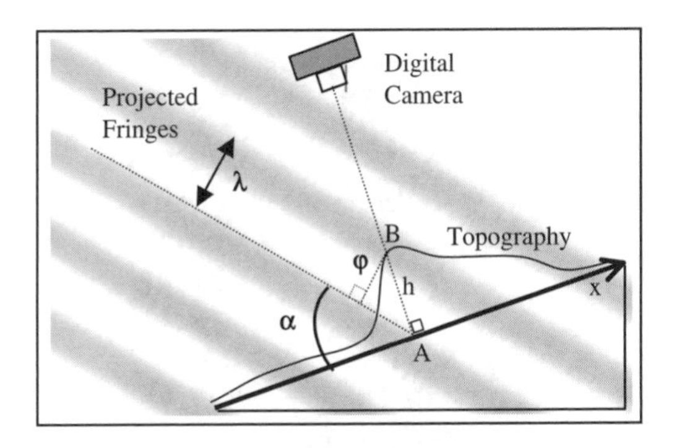

Figure 3. Principle of the optical acquisition technique.

Then with an algorithm based on the Fourier transform we compute a phase demodulation and deduce the contours and their evolution during the experiment. Furthermore, with differences between successive maps, we can elaborate time-varying maps of sedimentation and erosion rates in the system.

3.2 *Results*

Figure 4 shows a typical channel and the time evolution of the sediment layer height for three cross sections. Here is the description of the successive phases observed during such experiments:

The channel inception phase begins rather suddenly, about 15 minutes after the experiment has been started (Fig. 4b, c). Before this moment, the density current was spreading on the bed. Erosion and particle transport were present, but remained weak and relatively well distributed on a large array. Nevertheless we observed that during this development phase the erosion is greater in the middle of the density current than on its sides. Slowly then, the active erosion array is reduced and the density current becomes more focused. This process leads to the beginning of the incision phase.

The first significant incision is located at the top of the ramp. It is followed by a positive feedback mechanism facilitating further erosion. The cross section on figure 4b shows a linear widening phase of 40 minutes. The cross section on figure 4d is taken downstream in comparison with the 4b, and the incision phase is much more abrupt. While, for the 4b cross section, the lateral erosion is responsible for the widening of the sides, on the other hand there is no widening phase for the 4d cross section: the channel is already incised upstream, and this abrupt incision phase is the result of its progradation. This progradation is manifest on the down-slope cross section displayed on figure 4c. An erosion-deposition wave is propagating downstream: both the frontal lobe and the channel incision seem to move forward at the same velocity, which is about 2.5 mm/minute.

In the last phase, the channel width reaches a steady state: the erosion takes place on the bed instead of on the sides. It's a deepening phase. This steady state has been observed with most of the experiments we have made, but not with all of them. It is likely that those experiments were not run for a sufficiently long time.

3.3 *Aspect ratio*

We compared the aspect ratio (i.e. the width over depth ratio) of our experimental channels to natural ones, using the figure of Clark & Pickering (1996), and we found a good agreement (fig. 5). Our values are between 5 and 10, which means that our small-scale channels have morphology similar to natural ones.

3.4 *Meandering*

In some cases, the channel first shows a straight morphology, and then its sinuosity increases through time. We are not currently able to extract the parameter which determines the meandering, but we suspect the initial topography imperfections to play a consequent role in the triggering of this instability.

4 INCISION CONDITIONS

Varying two main parameters, the incline slope and the input flow rate, we ran about twenty experiments in order to see whether or not a channel could be incised. This roughly delineates a sort of phase diagram, plotted on figure 6.

In order to obtain channels, we have to provide high slope and low input flow rate. These two conditions allow the density current to focus instead of spread over the bed. The high slope force the current to flow downward, and the low input flow rate minimizes the jet effects. Both together provide the best conditions for channel incision.

Thus, this diagram is meaningful. However, the slope is not physically relevant here, because this parameter strongly depends on our experimental setup. We wanted to plot this phase diagram with non-dimensional parameters instead, and introduce somehow the erosion power. This could be done taking the shear stress exerted on the bed into account to calculate the Shields parameter.

In order to do this, we have started to study the velocity structure of the current, prior to incision.

5 VELOCITY PROFILE

To better characterize the brine flow and the shear stress on the bed just before the incision phase, we used the same experimental setup as before, without sediment layer on the incline. This way we can have a better

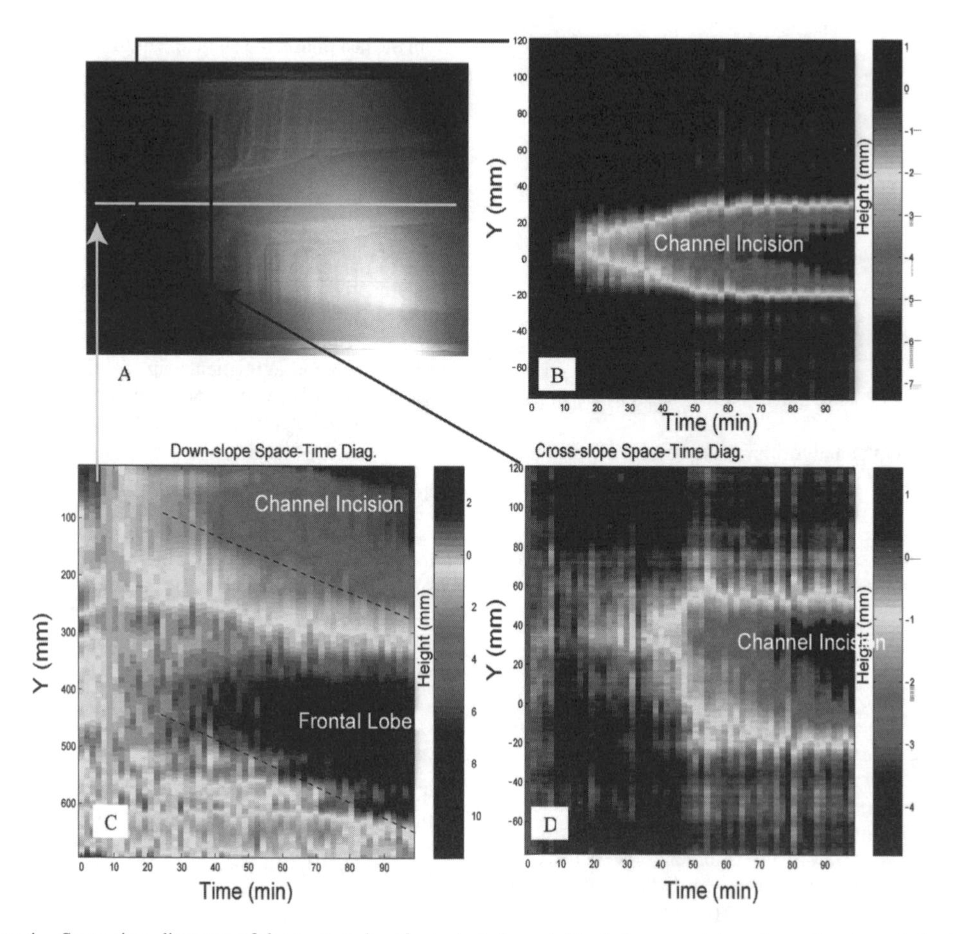

Figure 4. Space-time diagrams of the topography of a typical channel. The left-upper picture shows the channel obtained 100 minutes after the experiment start up, viewed from above (brine entrance is on the left side). Three cross sections are located on this picture. One is down-slope and two are cross-slope. The three other graphs are the space-time diagrams showing the evolution of these cross sections through time. Frontal lobe and channel incision phase are located on the three diagrams. (Parameters: slope = 17°, flow rate = 1.7 g/s, brine density = 1.025).

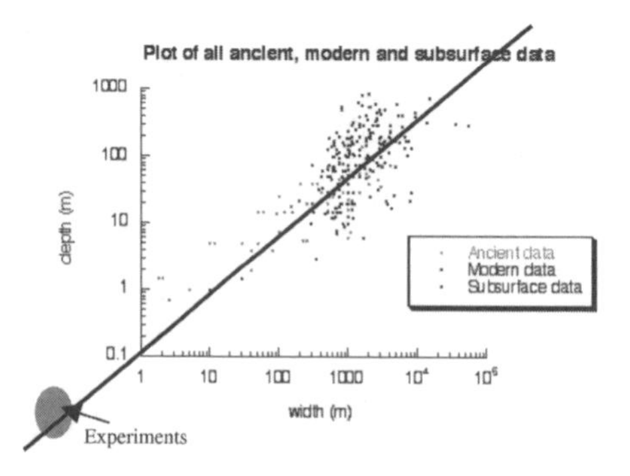

Figure 5. Subaqueous channel aspect ratio, after Clark & Pickering (1996). The dots represent real submarine channels, while the gray area represents the range of our experiments. The line indicates an aspect ratio (width/depth) of 10.

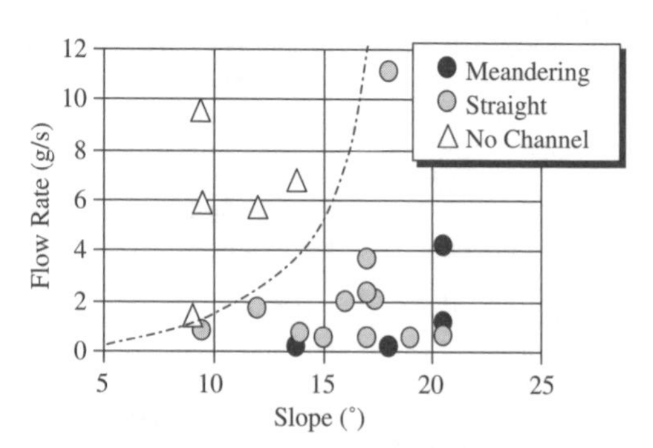

Figure 6. Phase diagram of the channel formation. The solid line has been added as a guide to visualization.

control on the topography, which remains flat to simulate the initial conditions of the with-sediment-blanket experiments.

5.1 *Measurement method*

Here we present the particle-with-shadow tracking technique, a new simple experimental method for measuring the velocity vertical profile, applied on the gravity current with a good accuracy.

We drop some particles in the water over the part of the incline we want to study. The particles sink very slowly because of both their low specific gravity and their small size. Before reaching the bottom, they pass through the brine current and thereby give information on the horizontal velocity field.

The incline is lit by a precise light source, with a small incidence angle. Making close movies of a small part of the incline from above, we track a falling particle and its shadow, on each picture (fig. 7a). The particle height above the plane is deduced directly from the particle-shadow distance. Knowing the particle position in 3D through time, and because the particle traverses the brine flow, we can reconstruct the velocity vertical profile. Only one particle is sufficient to obtain the profile, but the use of several particles for the treatment provides a better accuracy.

The choice for the incidence angle of the light source is consequential: with a very small angle, the shadow is far from the particle, and then a very good accuracy can be achieved when measuring the particle height above the incline. However it becomes more difficult to track the shadow when the particle is high.

The particle-with-shadow tracking is much easier and accurate with the use of space-time diagrams (fig. 7b). These diagrams allow visualizing the evolution of a pixel line with time. Then it is convenient to select and follow falling particles from picture to picture.

5.2 *Results*

Figure 8a shows the vertical profile measured with the tracking method. We managed to obtain it for the lowest part of the density current, where the velocity decreases: all data on this graph are under 0.5 mm high from the bed.

6 SHEAR STRESS MEASUREMENT

We use the downstream velocity profile in order to determine the Shields stresses exerted on the bed during the initial incision phase. We obtain the decreasing

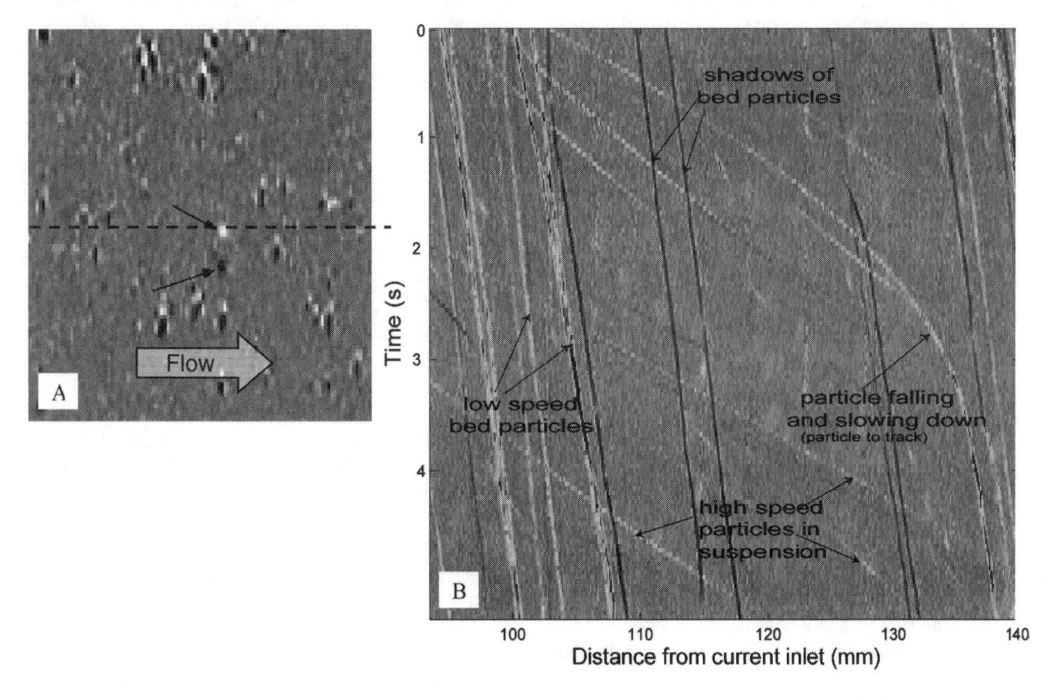

Figure 7. a) Picture extracted from a close movie of the incline taken from above. The arrows point a falling particle with its shadow. The dashed line locates the pixel line used for the space-time diagram. b) Space-time diagram of a down-slope pixel line, showing particles in white and shadows in black.

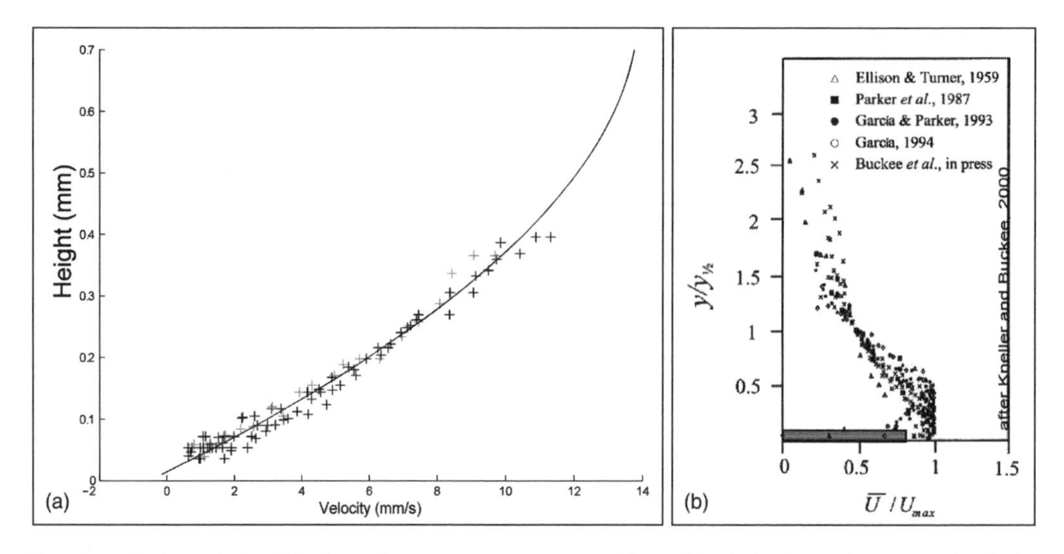

Figure 8. Velocity vertical profiles for gravity currents. a) Lowest part of the profile, obtained using the particle-with-shadow tracking technique on five different particles. Solid line is a 2nd order polynomial fit. (Parameters: slope = 13.7°, flow rate = 0.6 g/s, brine density = 1.025). b) Normalized profiles, after Kneller & Buckee (2000). The gray box locates the area of the profile.

velocity in the lowest part of the flow with a good accuracy. In this part, the Reynolds number is about 1. Thus, the velocity gradient we have located in the viscous sublayer. Then we have direct access to the shear stress τ, using the definition:

$$\tau = \mu \frac{\partial U}{\partial z}\bigg|_{z=0}$$

Where U is the velocity, z is the height, and μ is the dynamic viscosity.

Figure 9 represents the range of values obtained in our experiments on a Shields diagram. This diagram uses non-dimensional parameters τ^* and D^*:

$$\tau^* = \frac{\tau}{\Delta\rho . g . D . \cos\alpha}$$

$$D^* = D \sqrt[3]{\frac{\Delta\rho . g}{\rho . \nu^2}}$$

Where D is the particle diameter, α is the incline slope, g is the gravity, ν is the cinematic viscosity, ρ is the brine density and $\Delta\rho$ is the particle-brine density contrast. τ^* is the non-dimensional shear stress, and D^* is the non-dimensional particle diameter.

On the same diagram we plotted the Shields curve, which gives, for a given particle diameter, the critical shear stress to provide in order to erode, and we plotted an estimation made after the data of Khripounoff et al. 2003, who reported measurements on natural channels.

In both experimental and natural cases, the shear stresses observed are well above the critical shear stress.

7 FORESIGHTS

We showed that a steady-state density current can produce self-channelization, for a sufficiently high slope and a low input flow rate. After a slow development phase, the incision and widening phase suddenly begins, while a frontal lobe progrades. In some cases, a steady state can finally be obtained.

Our channels have aspect ratios similar to natural ones, and from velocity profiles we measured shear stress values about ten times above the critical shear stress.

Further experiments need to be made in order to better characterize the dense current prior to incision, measuring height, width, velocities and shear stress variations when changing the slope and the flow rate. The velocity profiles obtained have to be extended higher to measure the maximum velocity and the typical height of the current. The effect of dilution by fresh water entrainment at the current inlet has to be investigated in more details to provide better control on the density.

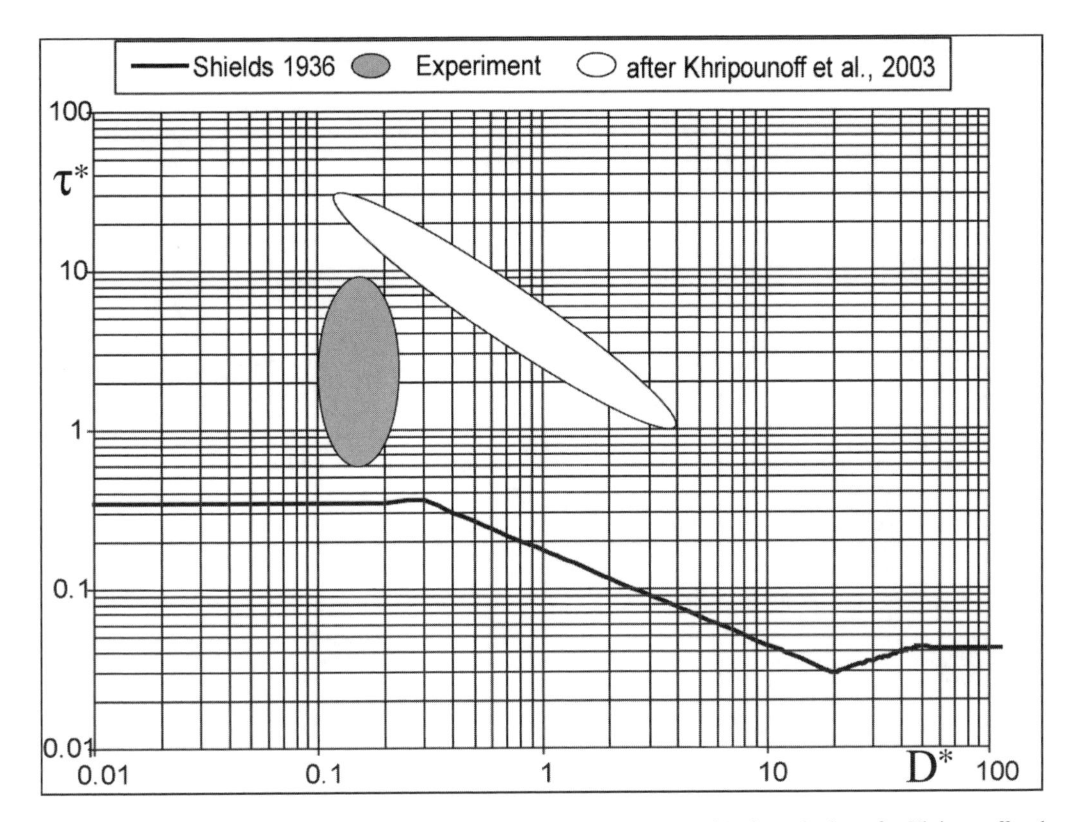

Figure 9. Shields diagram for our experiments and for the Zaïre submarine channel (estimated values after Khripounoff et al. (2003)).

At last, we would like to establish a phase diagram specifying the incision conditions in terms of non-dimensional shear stresses.

REFERENCES

Bonnecaze, R.T. & Lister, J.R. 1999, Particle-driven gravity currents down planar slopes. *Journal of Fluid Mechanics*. 19: 445–467.

Clark, J.D. & Pickering, K.T. 1996, *Submarine Channels: Processes and Architecture*. London: Vallis Press.

Ellison, T.H. & Turner, J.S. 1959, Turbulent entrainment in stratified flows. *Journal of Fluid Mechanics*. 6: 423–448.

Garcia, M.H. & Parker, G. 1993, Experiments on the entrainment of sediment into suspension by a dense bottom current. *Journal of Geophysical Research*. 98: 4793–4807.

Garcia, M.H. 1994, Depositional turbidity currents laden with poorly sorted sediment. *Journal of Hydraulic Engineering*. 120: 1240–1263.

Khripounoff, A. et al. 2003, Direct observation of intense turbidity current activity in the Zaire submarine valley at 4000 m depth. *Marine Geology*. 194: 151–158.

Kneller, B. et al. 1999, Velocity structure, turbulence and fluid stresses in experimental gravity currents. *Journal of Geophysical Research*. 104(C3): 5381–5391.

Kneller, B. & Buckee, C. 2000, The structure and fluid mechanics of turbidity currents: a review of some recent studies and their geological implications. *Sedimentology*. 47(1s): 62–94.

Métivier, F. et al. 2005, Submarine canyons in the bathtub. *Journal of Sedimentary Research*. 75(1): 6–11.

Parker, G. et al. 1987, Experiments on turbidity currents over an erodible bed. *Journal of Hydraulic Research*. 25: 123–147.

River, Coastal and Estuarine Morphodynamics: RCEM 2005 – Parker & García (eds)
© 2006 Taylor & Francis Group, London, ISBN 0 415 39270 5

Numerical experiments on subaqueous cyclic steps due to turbidity currents

S. Kostic
National Center for Earth-Surface Dynamics, University of Minnesota, Minneapolis, USA

G. Parker
Dept. of Civil & Env. Engineering & Dept. of Geology, University of Illinois, Urbana, Illinois, USA

ABSTRACT: A Froude-supercritical flow over an erodible bed may be subject to an instability which gives rise to the formation of cyclic steps, i.e. coherent, quasi-permanent trains of upstream-migrating steps bounded by internal hydraulic jumps in the flow above them. Cyclic steps have been observed and explained in bedrock streams and alluvium, but the possibility of cyclic steps due to turbidity currents in the subaqueous setting has not been explored. Yet turbidity currents are intrinsically more biased toward supercritical flows than rivers, suggesting that cyclic steps are at least as likely to be found in the subaqueous setting as in steep bedrock channels and alluvium. Here a numerical model is employed to explore these rhythmic bedforms generated by 20, 45 and 110-micron turbidity currents in an experimental setting. The results of numerical simulations offer an insight into what can be expected from laboratory experiments on cyclic steps.

1 INTRODUCTION

Cyclic steps are rhythmic bedforms which manifest the fact that supercritical flow over an erodible bed is inherently unstable. Each cyclic step is bounded by an internal hydraulic jump, or a zone over which the flow makes a rather sharp transition from supercritical upstream of it ($Fr > 1$) to subcritical downstream of it ($Fr < 1$). Consequently the flow over each step can be divided by a point where $\mathbf{Fr} = 1$ into an upstream subcritical and a downstream supercritical zone. The slower subcritical zone induces net sediment deposition through enhanced deposition or suppressed incision. The faster supercritical zone induces net sediment erosion through suppressed deposition or enhanced incision. This interplay between deposition and incision results in a coherent, quasi-permanent train of steps migrating upstream as illustrated in Fig. 1.

Cyclic steps have been observed and explained in a variety of subaerial settings. For example, trains of purely incisional rhythmic steps in the beds of bedrock streams have been observed in the field (Fig. 2a; Wohl, 2000), explained theoretically (Parker & Izumi, 2000) and modeled in the laboratory (Fig. 2b, Koyama & Ikeda, 1998; Brooks, 2001). Cyclic steps bounded by hydraulic jumps in alluvium have been also observed in the field (Fig. 3a; Winterwerp et al. 1992). The setting was that of sheet open channel flow over a sandy bed. These features have been modeled in the

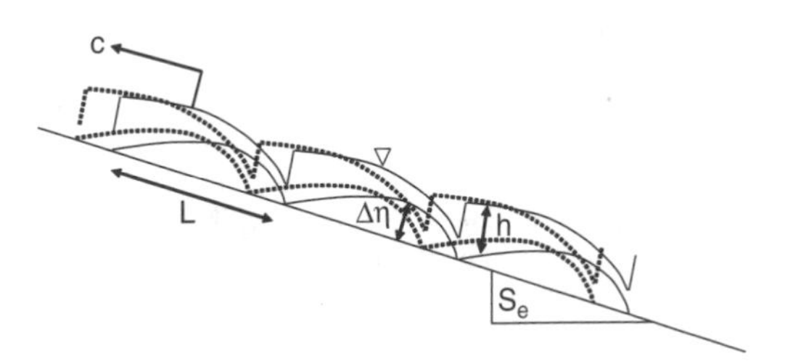

Figure 1. Sketch of cyclic steps locked in place by internal hydraulic jumps.

Figure 2. Incisional cyclic steps in bedrock: a) Big box canyon, Wohl, 2000. b) Experiment, Koyama & Ikeda, 1998.

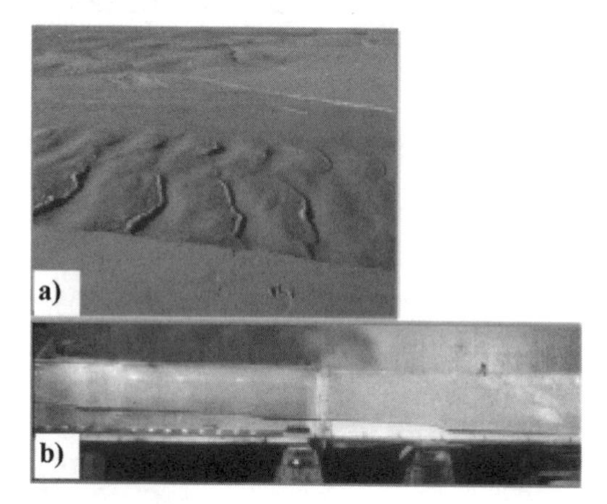

Figure 3. Cyclic steps in alluvium: a) Calais, France (image courtesy H. Capart). b) Experiment, Taki & Parker, in press.

laboratory as well (Fig. 3b, Taki & Parker, in press), and explained theoretically by Sun & Parker (in press). Cyclic steps in alluvium can form under conditions of bed aggradation or degradation, or they may simply migrate updip in the absence of net bed level change.

The possibility of cyclic steps in the subaqueous setting has not been investigated. Submarine sediment waves, which are commonly observed at the base of continental slopes and on the back side of levees of channels on submarine fans, have been often characterized as antidunes or antidune-like features associated with turbidity currents (e.g. Kubo & Nakajima, 2002, Fedele & Garcia, 2001, Lee et al., 2002; Normark et al., 2002). This is in part because turbidity currents are the submarine analog of rivers, and most fluvial antidunes, just like cyclic steps in the subaerial setting, can be characterized as rhythmic bedforms that i) are associated with Froude-supercritical flow and ii) migrate upstream. However, the sediment waves present in many submarine canyon-fan systems differ from antidunes in two important ways; i) they appear to be long-wave phenomena, with wavelengths that are one or two orders of magnitude larger than the depth of flow that created them, and ii) they invariably migrate upstream in orderly self-preserving trains. Fluvial antidunes on the other hand tend to be short-length, ephemeral features that rarely leave a depositional record indicating a clear and coherent train of upstream migrating waveforms.

2 DEPOSITIONAL RESPONSE OF A TURBIDITY CURRENT TO A SLOPE BREAK

Kostic & Parker (in press) have recently completed a parametric study on the response of a turbidity current to a slope break. The assumed configuration

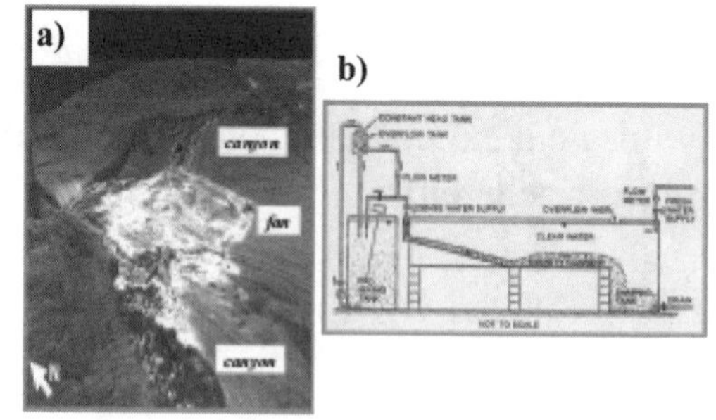

Figure 4. a) Seismic image of a submarine canyon/fan/canyon system on the continental slope off the Niger River, Africa (image courtesy C. Pirmez). b) Experimental configuration of Garcia, 1993.

is summarized in Fig. 4a, and is common to many submarine canyon-fan systems. Net erosive turbidity currents often carve submarine canyon into the continental slope, while net depositional turbidity currents emplace submarine fans on lower slopes farther downstream. It is assumed that the fan is channelized, and channels act to limit lateral spread of turbidity currents. The canyon-fan configuration of Fig. 4a with a channelized fan can be reproduced in an experimental flume, with a relatively high-slope upstream reach followed by a low-slope downstream reach (Fig. 4b).

The analysis of Kostic & Parker (in press) has revealed that a slope decline in the downstream direction can under certain conditions cause a turbidity current to i) undergo an internal hydraulic jump, and ii) leave a depositional signal of that transition. Depending on the flow conditions the jump may form upstream of the canyon-fan transition, be coincident with it or occur downstream of the slope transition. A steady-state analysis of purely depositional turbidity currents has shown that sediment deposition suppresses the ability of a supercritical flow to decelerate and subcritical flow to accelerate. Therefore these currents are less prone to hydraulic jumps than the conservative density currents to which they are related (Kostic & Parker, submitted). As opposed to deposition, erosion enhances the ability of a supercritical flow to decelerate and subcritical flow to accelerate. Thus, an interplay between deposition and erosion dictates whether an underflow undergoes an internal hydraulic jump or not. In general, turbidity currents driven by sediment fine enough to satisfy the condition $v_s/U_o < 2.67 \cdot 10^{-3}$ regularly display an internal hydraulic jump induced by the slope transition. Here v_s denotes sediment fall velocity, and U_o is the depth-averaged velocity at the inflow boundary. Coarser-grained currents, for which the ratio $v_s/U_o > 4.9 \cdot 10^{-3}$, do not undergo a transition to subcritical flow due to the rapid rate of sediment deposition on the bed. A class of turbidity currents with ratios

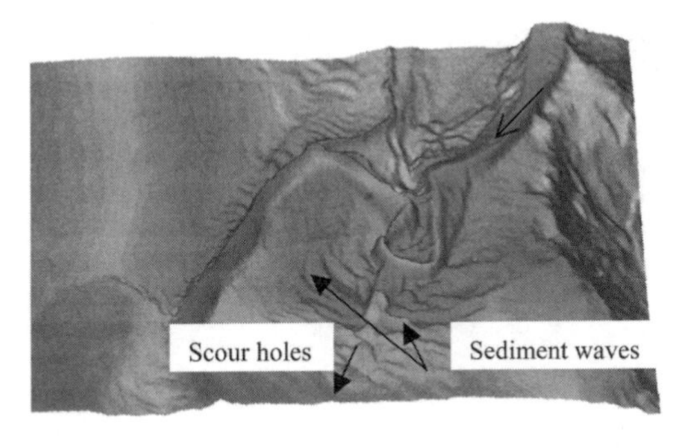

Figure 5. Monterey submarine canyon in the vicinity of the Shepard meander, (image courtesy W. Normark).

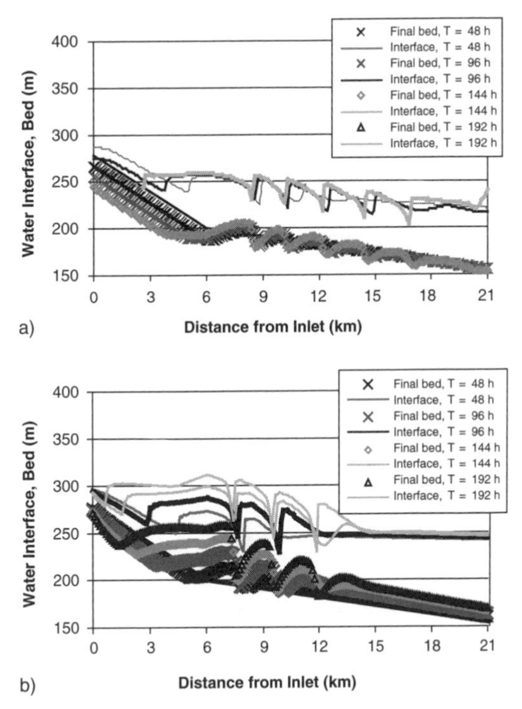

Figure 6. a) Net-erosional series of scour holes in the Monterey East Channel. b) Net-depositional sediment waves on the outside levee of the Shepard bend.

$2.67 \cdot 10^{-3} \leq v_s/U_o \leq 4.9 \cdot 10^{-3}$ can go either way, depending on other flow parameters. When a jump does occur, it leaves a depositional signal in terms of an upstream-facing step. The depositional response is a result of a step increase in net depositional rate across the jump due to step decrease in shear stress, and thus bed erosion rate, across that zone. The signal can be either discernible or undetectable. Turbidity currents that satisfy the crude condition $\Delta\tau^* > 0.3$ leave a discernible signature of the canyon-fan transition. Here $\Delta\tau^* = (\Delta u_* R_f / v_s)^2$ denotes the step drop in Shields number across an internal hydraulic jump, where $R_f = v_s / \sqrt{RgD}$ is a dimensionless particle fall velocity. Furthermore, R is the submerged specific gravity of sediment, g is gravitational acceleration, and D is a characteristic grain size of the sediment.

In addition to numerous experimental-scale simulations, Kostic and Parker (in press) have performed trial field-scale simulations in the subaqueous setting with both the "three equation" and the "four equation" model of turbidity current dynamics of Parker et al. (1986). These runs at highly depositive conditions led to very similar results: a hydraulic jump and a backward-facing step in response to the slope break. Under conditions of increasingly erosive underflows, however, the "four equation" model generated a series of upstream-migrating steps punctuated by hydraulic jumps. This result provided the first hint that many upstream-migrating submarine bedforms, including those commonly found on the back side of levees of submarine channels, may actually be subaqueous cyclic steps rather than antidunes.

The numerical model of Kostic & Parker (in press) has been recently applied to the Monterey Submarine Channel off Monterey, California, in an attempt to characterize i) sediment waves formed on the outside levee of the Shepard meander, as well as ii) a linear series of scour-shaped depressions in the Monterey East Channel, which is excavated into the levee (Fig. 5). The goal of the simulations was not to reproduce the precise sequence of events by which these features formed, but rather to provide evidence that both features can be associated with erosion and deposition by supercritical flows stripped off large and thick turbidity currents moving through the Shepard meander (Fildani et al., submitted). The results of numerical simulations suggest the following. The train of scours in the Monterey East Channel appear to be net-erosional cyclic steps, and the sediment waves on the outside levee of the Shepard bend appear to be net-depositional cyclic steps (Fig. 6). By implication, the net-depositional sediment waves observed on the levees of many other submarine channels are also likely net-depositional cyclic steps.

3 NUMERICAL MODEL

The bed configuration of interest here is shown in Fig. 4b. The sloping upstream portion represents a loose surrogate for a submarine canyon, and the horizontal downstream portion is a loose surrogate for a submarine fan. The abrupt decrease in slope increases the likelihood of a hydraulic jump occurring in the computational domain. The width of the turbidity current is constant to depict a channelized fan.

The dynamics of the turbidity current emanating from a submarine canyon and debouching onto a submarine fan is described here in terms of the layer-averaged "four-equation" model (e.g. Fukushima et al., 1985, Parker et al., 1986), which is here fully coupled with the Exner equation of conservation of bed sediment. The "four-equation" model encompasses conservation relations for water mass, streamwise momentum, suspended sediment mass and turbulent kinetic energy. The turbidity current is assumed to be sufficiently dilute, i.e. the volume concentration c satisfies the condition $c \ll 1$, to allow the use of Boussinesq approximation in the equation of motion. The governing equations are

$$\frac{\partial h}{\partial t}+\frac{\partial Uh}{\partial x}=e_w U \tag{1}$$

$$\frac{\partial Uh}{\partial t}+\frac{\partial U^2 h}{\partial x}=-\frac{1}{2}Rg\frac{\partial Ch^2}{\partial x}-RgCh\frac{\partial \eta}{\partial x}-u_*^2 \tag{2}$$

$$\frac{\partial Ch}{\partial t}+\frac{\partial UCh}{\partial x}=v_s\left(e_s-r_o C\right) \tag{3}$$

$$\frac{\partial Kh}{\partial t}+\frac{\partial UKh}{\partial x}=u_*^2 U+\frac{1}{2}U^3 e_w-\varepsilon_o h+ \\ -Rgv_s Ch-\frac{1}{2}RgChUe_w-\frac{1}{2}Rghv_s\left(e_s-r_o C\right) \tag{4}$$

$$\left(1-\lambda_p\right)\frac{\partial \eta}{\partial t}=v_s\left(r_o C-e_s\right) \tag{5}$$

In the above relations the dependent variables are the current depth h, the depth-averaged velocity U, the depth-averaged volumetric concentration of suspended sediment C, the depth-averaged turbulent kinetic energy per unit mass K and the bed elevation η. Furthermore t is time, x is a bed-attached streamwise coordinate, λ_p is sediment porosity and r_o is an order-one multiplicative constant.

The dimensionless parameter e_w characterizes the rate of entrainment into the turbidity current of ambient water from above. The following form is used here for e_w (e.g. Fukushima et al., 1985);

$$e_w=\frac{0.00153}{0.0204+\frac{1}{Fr_d^2}} \tag{6}$$

where Fr_d is the densimetric Froude number defined as

$$Fr_d=\frac{U}{\sqrt{RgCh}} \tag{7}$$

Shear velocity u_* is dynamically linked to the mean kinetic energy per unit mass of the flow turbulence by a prescribed dimensionless parameter α.

$$u_*^2=\alpha K \tag{8}$$

Here α is set equal to 0.1, a value suggested by Parker et al. (1986).

Equation (4) of the conservation of turbulent kinetic energy requires one more closure relation for the viscous dissipation rate ε_o (e.g. Launder & Spalding, 1972):

$$\varepsilon_o=\beta\frac{K^{1.5}}{h} \tag{9}$$

where according to Fukushima et al. (1985) and Parker et al. (1986).

$$\beta=\left[\frac{1}{2}e_w\left(1-\frac{1}{Fr_d^2}-2\frac{c_f^*}{\alpha}\right)+c_f^*\right]\Big/\left(\frac{c_f^*}{\alpha}\right)^{1.5} \tag{10}$$

In the above relation c_f^* is an "equivalent" value of the coefficient of bed friction as defined by Fukushima et al. (1985).

In (3), (4) and (5) e_s describes the rate of sediment entrainment into suspension by a turbidity current, and is calculated using the following formulation by Garcia & Parker (1991, 1993):

$$e_s=\frac{aZ^5}{1+\frac{a}{0.3}Z^5} \tag{11}$$

where

$$Z=\alpha_1\frac{u_*}{v_s}\mathrm{Re}_p^{\alpha_2} \tag{12}$$

In the above relations a is a constant equal to 1.3×10^{-7}, and α_1 and α_2 are constants given as

$$\left(\alpha_1,\alpha_2\right)=\begin{cases}(0.586, 1.23), & \mathrm{Re}_p\le 2.36\\ (1.0\quad, 0.6\ ), & \mathrm{Re}_p>2.36\end{cases} \tag{13}$$

Re_p denotes a particle Reynolds number defined as

$$\mathrm{Re}_p=\sqrt{RgD}\,D/\nu \tag{14}$$

where ν is the kinematic viscosity of water.

The numerical model presented here involves a grid that deforms in space and time. The outflow boundary is located at, and moves with the underflow front $(x=s)$ until it propagates beyond a computational domain of specified length L. Subsequent to this the outflow boundary is simply the downstream end of the specified domain. Initial and boundary conditions for the numerical model are discussed in more detail in Kostic & Parker (in press). An initial front position is specified, and the initial values of the dependent variables h, U, C and K at all nodal points up to the initial position of the front are set to their values h_o, U_o, C_o and K_o at the inflow boundary located at $x = 0$. The initial canyon-fan configuration is specified in terms of a proximal zone with a length of 0.8 m and a slope of 0.010, and a distal zone with a length of 5.2 m and a horizontal slope. The initial bed profile is then interpolated to the grid $0 \leq x \leq s$. For a supercritical inflow boundary three physical conditions must be specified. They are defined as follows:

$$h\left(x=0,t\right)=h_o,\; U\left(x=0,t\right)=U_o,\; C\left(x=0,t\right)=C_o \tag{15}$$

At the outflow boundary two physical conditions are required as long as the turbidity current has not propagated out of the specified domain length L, which is 6 m. These conditions are given by:

$$U\left(x=L,t\right)=\dot{s},\; \eta\left(x=L,t\right)=\eta_o(L) \tag{16}$$

where $\dot{s}$ denotes the front velocity, and η_o is the antecedent bed elevation as yet unmodified by the turbidity current. The flume of Fig. 4b ends in a free outfall, and once the current covers the entire length of the domain the underflow depth at the last grid point must be equated to the critical depth, such that

$$h\left(x=L,t\right)=\frac{U(x=L,t)^2}{Rg\,C(x=L,t)} \tag{17}$$

The remaining numerical boundary conditions are obtained from the flow field by means of first order extrapolation (Hirsch, 1990).

The governing equations (1)–(5) together with closure relations (6)–(14) and the previously discussed initial and boundary conditions, are solved numerically by means of the ULTIMATE QUICKEST scheme (Leonard, 1979, 1991). This explicit finite-volume upwind algorithm is particularly suitable for highly advective unsteady flows problems. A more comprehensive discussion of the numerical model can be found in Kostic & Parker (2003).

4 NUMERICAL EXPERIMENTS

The numerical experiments performed below can be considered to be a precursor to a more thorough study of subaqueous cyclic steps at laboratory scale. They were performed in order to define i) plausible experimental flow conditions and an initial channel geometry that would generate at least one discernible cyclic step in the experimental flume, and b) the maximal run time so that the resulting deposit does not exceed available

Table 1. Input parameters for the numerical model.

Run #	ho (cm)	Uo (cm/s)	Co	Frdo	Do (μm)	Run time (h)
1	3	25	0.04	1.749	20	5
2	3	20	0.04	1.435	45	3
3	6	30	0.04	1.522	110	2

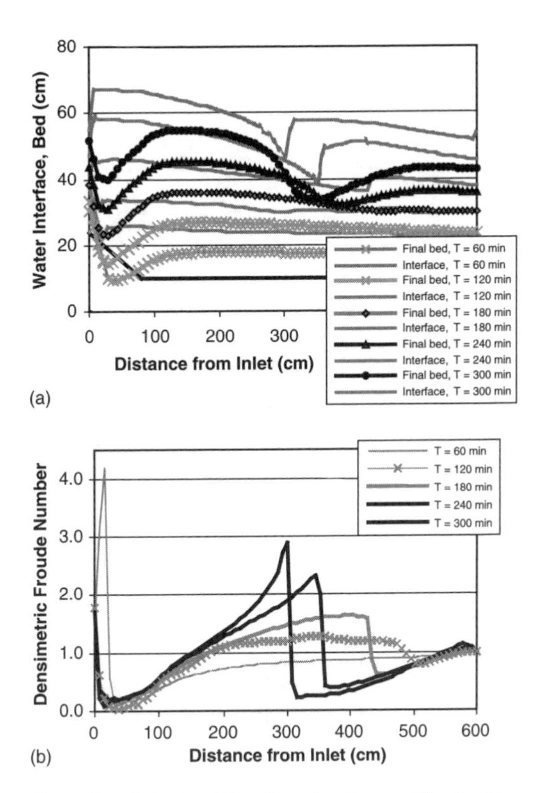

Figure 7. a) Bed and interface elevation profiles for Run 1 (20 μm). b) Downstream variation in densimetric Froude number for Run 1.

channel depth. The input parameters for the numerical model are given in the Table above.

The numerical experiments presented here reveal that a turbidity current carrying uniform sediment can under the right conditions trigger the formation of cyclic steps. They furthermore illustrate that turbidity currents driven by sediment varying in size from silt to sand can generate these rhythmic bedforms.

In addition to the input parameters of the Table, the following assumptions were made: $R = 1.65$, $\lambda_p = 0.5$ and $T = 20°C$. Fig. 7a and Fig. 8a, b summarize profiles of bed elevation η and interface elevation $\eta + h$ for all three numerical experiments. The effective grain size are equal to 20 μm ($v_s = 0.35$ mm/s), 45 μm ($v_s = 1.79$ mm/s), and

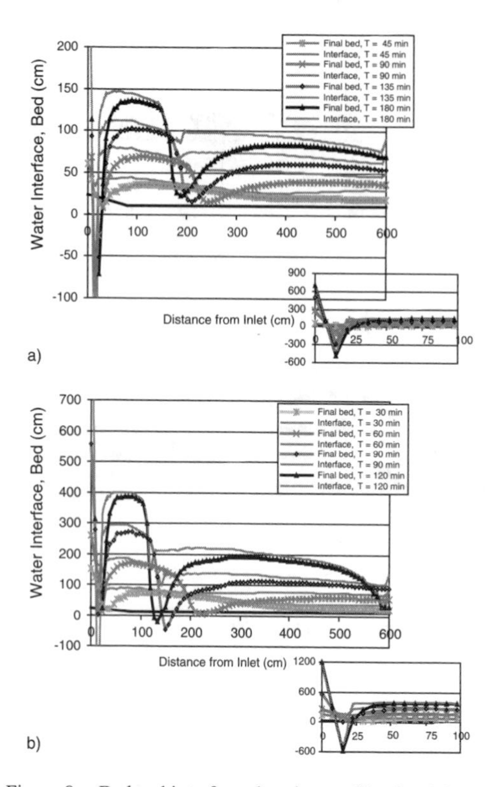

Figure 8. Bed and interface elevation profiles for a) Run 2 (45 μm) and b) Run 3 (110 μm).

110 μm ($v_s = 8.72$ mm/s) respectively. The break between proximal and distal slope triggers the formation of sediment waves. In all three cases the waves are net-depositional, with the exception of a short net-erosional zone upstream where the flow tends to increase the proximal initial slope. This trend is subtle for currents driven by 20 μm sediment, and becomes more pronounced for coarser-grained turbidity currents. Two complete sediment waves are apparent by the end of the 4th hour in Run 1, 90 min in Run 2, and 60 min in Run 3. In the last experiment a third step is also seen to be forming downstream after 2nd hour. The plot of the downstream variation of the densimetric Froude number for Run 1 (Fig. 7b.) illustrates that each wave is indeed a cyclic step bounded both upstream and downstream by a hydraulic jump. The limited channel depth of 78 cm allows for a total experimental time of 5 hours, 30 min and 15 min for Run 1, 2 and 3 respectively.

5 CONCLUSIONS

The results of three numerical experiments on subaqueous cyclic steps created by turbidity currents at laboratory scale are summarized here. They suggest that turbidity currents driven by a single grade of sediment ranging from silt to sand can trigger the formation of multiple cyclic steps. In a sufficiently shallow and short flume, however, it may not be possible for more than one cyclic step to be present at any given time. Taki and Parker (in press) found that in sufficiently short reaches trains of fluvial steps also devolved into a single step that repeatedly formed and migrated upstream.

ACKNOWLEDGEMENTS

This research was funded in part by the National Center for Earth-surface Dynamics, a Science and Technology Center funded by the National Science Foundation of the USA.

REFERENCES

Brooks, P. C. 2001. Experimental study of erosional cyclic steps. M.S. thesis, University of Minnesota.

Fedele, J. J. & Garcia, M. H. 2001. Bedforms and density underflows in the marine environment, Proceedings 2nd IAHR Symp. on River, Coastal and Estuarine Morphodynamics, Obihiro, Japan, September 10–14.

Fildani, A., Normark, W. R., Kostic, S. & Parker, G. Submitted Channel formation by flow stripping: Large scale scour features along the Monterey East Channel and their relation to sediment waves. Sedimentology, preprint available at http://www.ce.umn.edu/~parker/preprints.htm.

Fukushima, Y., Parker, G. & Pantin, H. M. 1985. Prediction of ignitive turbidity currents in Scripps Submarine Canyon, Marine Geology 67, 55–81.

Garcia, M. H. & Parker, G. 1991. Entrainment of bed sediment into suspension. Journal of Hydraulic Engineering, 117(4), 414–435.

Garcia, M. & Parker, G. 1993. Experiments on the entrainment of the sediment into suspension by a dense bottom current. Journal of Geophysical Research, 98, 4793–4807.

Hirsch, 1990. Numerical Computation of Internal and External Flows. Vol. 2, Computational Methods for Inviscid and Viscous Flows. John Wiley, New York.

Kostic, S. & Parker, G. 2003. Progradational sand-mud deltas in lakes and reservoirs: Part 1. Theory and numerical modeling. Journal of Hydraulic Research, 41 (2), 127–140.

Kostic, S. & Parker, G. in press. The response of turbidity currents to a canyon-fan transition: Internal hydraulic jumps and depositional signatures. Journal of Hydraulic Research; http://www.ce.umn.edu/~parker/preprints.htm.

Kostic, S. & Parker, G. submitted. Conditions under which a supercritical turbidity current traverses an abrupt transition to vanishing bed slope without a hydraulic jump. Journal of Fluid Mechanics, preprint available at http://www.ce.umn.edu/~parker/preprints.htm .

Koyama, T. & Ikeda, H. 1998. Effect of riverbed gradient on bedrock channel configuration: A flume experiment. Proc. Environmental Research Center, Tsukuba University, Japan, 23, 25–34. (In Japanese).

Kubo, Y. & Nakajima, T. 2002. Laboratory experiments and numerical simulations of sediment-wave formation by turbidity currents. Marine Geology, 192 (2002), 105–121.

Launder, B. E. & Spalding, D.B. 1972. Lectures in Mathematical Models of Turbulence. Academic Press, London.

Lee, H. J., Syvitski, J. P. M., Parker, G., Orange, D., Locat, J., Hutton, E.W.H. & Imran, J. 2002. Distinguishing sediment waves from slope failure deposits: field examples, including the 'Humboldt slide, and modelling results, Marine Geology 192 (2002) 79–104.

Leonard, B. P. 1979. A stable and accurate convection modeling procedure based on quadratic upstream interpolation. Comp. Methods in Applied Mechanics and Engineering, 19, 59–98.

Leonard, B. P. 1991. The ULTIMATE conservative difference scheme applied to unsteady one-dimensional advection. Comp. Methods in Applied Mechanics and Engineering, 88, 17–77.

Normark, W. R., Piper, D. J. W., Posamentier, H., Pirmez, C. & Migeon, S. 2002. Variability in form and growth of sediment waves on turbidite channel levees. Marine Geology 192, 23–58.

Parker, G., Fukushima, Y. & Pantin, H. M. 1986. Self-accelerating turbidity currents. Journal of Fluid Mechanics 171, 145–181.

Parker, G. & Izumi, N. 2000. Purely erosional cyclic and solitary steps created by flow over a cohesive bed. Journal of Fluid Mechanics, 419, 203–238.

Sun, T. & Parker, G. in press. Transportational cyclic steps created by flow over an erodible bed. Part 2. Theory and numerical simulation. Journal of Hydraulic Research, preprint available at http://www.ce.umn.edu/~parker/preprints.htm.

Taki, K. & Parker, G. in press. Transportational cyclic steps created by flow over an erodible bed. Part 1. Experiments. Journal of Hydraulic Research, preprint available at http://www.ce.umn.edu/~parker/preprints.htm.

Winterwerp, J. C., Bakker, W. T., Mastbergen, D. R. & van Rossum, H. 1992. Hyperconcentrated sand-water mixture flows over erodible bed. J. Hydr. Engrg., 119(11), 1508–1525.

Wohl, E. E. 2000. Substrate Influences on Step-Pool Sequences in the Christopher Creek Drainage, Arizona. Geology, 108, 121–129.

River, Coastal and Estuarine Morphodynamics: RCEM 2005 – Parker & García (eds)
© 2006 Taylor & Francis Group, London, ISBN 0 415 39270 5

The role of erosion rate formulation on the ignition and subsidence of turbidity current

A. Yi & J. Imran
Dept. of Civil and Environmental Engineering, University of South Carolina, Columbia, South Carolina, USA

ABSTRACT: An empirical relationship describing the entrainment of suspended sediment is required even if a vertical structure model is used for modeling turbidity currents. Several relationships can be found in the literature for estimating the entrainment rate of suspended sediment. Turbidity currents at laboratory scale are primarily depositional and the model prediction is not sensitive to the specified entrainment relationship. At the field scale however, turbidity currents are more likely to be erosional and the predicted results can be quite sensitive to the entrainment relations used in the model. Here, the four-equation model of turbidity current is utilized to revisit the ignition and subsiding conditions of turbidity currents originally described in the seminal work of Parker et al. (1986). Three different sediment entrainment models are considered. The ignition conditions are studied for a variety of parameters such as sediment size, slope and inflow current thickness. It has been found that steeper slopes and finer sediment gives lower values of ignition velocity and sediment concentration. The ignition Richardson number is found to be strongly dependent on the slope and insensitive to the sediment size. The ignition condition has been found to be quite sensitive to the entrainment relation used in the computation. The phase diagrams describing ignition conditions, and igniting and subsiding regions obtained from different erosion models are generally similar. The auto-suspension generation line dividing the zones of ignition and subsidence are found to be different for different entrainment relationship. As a result, a turbidity current predicted to be subsiding by one entrainment relation could turn out to be igniting when a different entrainment model is used. This has important implication in the modeling of turbidity current as the same numerical model can give completely different prediction depending on which entrainment relation is used.

1 INTRODUCTION

Turbidity currents are sediment-laden bottom flows which derive their driving force from the force of gravity acting on the suspended sediment. These currents play an important role in transporting sediments in lakes and the ocean. Unlike temperature or salinity driven density currents, turbidity currents are not conservative because the suspended sediment can freely exchange with bed sediments. If the turbidity current is sufficiently swift, it may entrain more sediment from the bed than it deposits, and the current may accelerate further. As a result, the current will enter into a self-acceleration cycle. On the other hand, a weak turbidity current may lose much of its suspended sediment to the bed, thus reducing the gravitational force. As a result, the current will decelerate and eventually vanish after all of the suspended sediments have settled out.

Both Pantin (1979) and Parker (1982) described the mechanism by which a longitudinally uniform current self-accelerates in time. Pantin (1979) carried out a phase-plane analysis to delineate the subsiding and igniting turbidity current fields. Parker (1982) identified a critical "ignition" condition for the onset of an erosive flow. Fukushima et al. (1985) developed numerical models of steady, spatially developing turbidity currents and applied those to the Scripps Submarine Canyon to predict the onset of ignitive acceleration and the downstream development of turbidity currents. Parker et al. (1986) derived the depth averaged three and four equation models of steady spatially developing turbidity current in the downstream direction and presented a detailed analysis of the ignition state of turbidity currents over a wide range of conditions. The depth-averaged model of turbidity current requires closure relationships for water entrainment rate, as well as the deposition rate and the entrainment rate of suspended sediment. A non-depth-averaged model (e.g. Huang et al. 2005) does not need the explicit specification of water entrainment rate or the deposition rate of suspended sediment. However, a relationship for the entrainment of suspended sediment must be specified in all turbidity current models. Whether a turbidity current reaches an igniting or subsiding state may depend strongly on the choice of the entrainment relationship used in the numerical model. Parker et al. (1986) used the sediment entrainment formula of Akiyama & Fukushima (1985) in their analyses.

In this study, we revisited the work of Parker et al. (1986) by considering three different sediment entrainment formulae. The purpose of this study is to demonstrate the sensitivity of model results to the sediment entrainment relationship. In addition to the Akiyama-Fukushima (1985) formula, we considered the Garcia-Parker (1993) model and the Smith-McLean (1977) model for sediment entrainment. Other closure relationships remained same as the ones used by Parker et al. (1986).

2 FOUR-EQUATION MODEL OF TURBIDITY CURRENTS

The layer-averaged equations of Parker et al. (1986) describing the balance of fluid mass, momentum, sediment mass and turbulent kinetic energy can be written as follows.

$$\frac{\partial h}{\partial t}+\frac{\partial hU}{\partial x}=e_w U \tag{1a}$$

$$\frac{\partial hU}{\partial t}+\frac{\partial hU^2}{\partial x}=-\frac{1}{2}Rg\frac{\partial h^2 C}{\partial x}+RghCS-u_*^2 \tag{1b}$$

$$\frac{\partial hC}{\partial t}+\frac{\partial hUC}{\partial x}=v_s(E_s-r_0 C) \tag{1c}$$

$$\frac{\partial hK}{\partial t}+\frac{\partial hKU}{\partial x}=u_*^2 U+\frac{1}{2}U^3 e_w-\varepsilon_0 h- Rgv_s hC-\frac{1}{2}RghCUe_w-\frac{1}{2}Rghv_s(E_s-r_0 C) \tag{1d}$$

where h is the thickness of the current measured normal to the bed, u is the depth-averaged velocity in the x direction, C denotes the depth-averaged volume concentration of suspended sediment in the current, K is the depth-averaged value for the turbulent kinetic energy per unit mass, S is the bed slope, and v_s is the fall velocity of the sediment particle.

The water entrainment rate e_w is a dimensionless variable calculated as

$$e_w=\frac{0.00153}{0.0204+Ri} \tag{2}$$

where Ri is the bulk Richardson number defined as:

$$Ri=\frac{RghC}{U^2} \tag{3}$$

For a fully turbulent current, the bed shear velocity u_* can be related to the mean turbulent kinetic energy K by the simple closure:

$$u_*=C_f U^2=\alpha K \tag{4}$$

where α is approximated to be constant for a given current with values between 0.05 and 0.5. Following Parker et al. (1986), it is set equal to 0.1.

The deposition rate for particles of uniform size is equal to the product of their fall velocity and the near-bed concentration C_b as $D=v_s C_b$. The near-bed concentration is related to the depth or layer-averaged suspended sediment concentration by the following expression

$$C_b=r_0 C \tag{5}$$

For open channel flow, the equation derived by Rouse (1937) is widely used to estimate the vertical distribution of suspended sediment in the water column. Parker et al. (1986) obtained the following approximate form for r_0 by integrating the Rouse equation from 0.05 h to h:

$$r_0=1+31.5(u_*/v_s)^{-1.46} \tag{6}$$

To estimate the fall velocity, the following empirical relation developed by Dietrich (1984) is used.

$$v_s=(gR\nu W_*)^{1/3} \tag{7}$$

with

$$D_*=gRD_s^3/\nu^2 \tag{8}$$

$$Log(W_*)=-3.7617+1.92944\log(D_*)-0.09815(\log(D_*))^2-0.00575(\log(D_*))^3+0.00056(\log(D_*))^4 \tag{9}$$

The turbulent dissipation rate ε_0 is calculated using the following equations of Parker et al. (1986) as:

$$\varepsilon_0=\beta\frac{K^2}{h} \tag{10}$$

where the dimensionless parameter β is given as

$$\beta=\frac{\frac{1}{2}e_w\left(1-Ri-2\frac{C_{f*}}{\alpha}\right)+C_{f*}}{\left(\frac{C_{f*}}{\alpha}\right)^{\frac{3}{2}}} \tag{11}$$

where C_{f*} is the equilibrium coefficient of bed friction.

Researchers have developed relationships to parameterize the entrainment rate of sediment from the bed (e.g. Smith and McLean, 1977; van Rijn, 1984, Akiyama and Fukushima, 1985; Celik and Rodi, 1988; Garcia, 1990, Garcia and Parker, 1993) primarily based on the measurement of near-bed equilibrium concentration.

As mentioned earlier, Parker et al. (1986) performed their analysis using the Akiyama and Fukushima (1985) relationship. Here, we consider three different entrainment models developed by Akiyama and Fukushima (1985), Garcia and Parker (1993) and Smith and McLean (1977) to study the sensitivity of the model results to the use of a specific entrainment relationship. Akiyama and Fukushima (1985) used data from equilibrium flume experiments and river suspensions to develop the following relationship:

$$E_s = \begin{cases} 0.3 & Z > Z_m \\ 3\times10^{-12}\times Z^{10}\left(1-\dfrac{Z_c}{Z}\right) & Z_c < Z < Z_m \\ 0 & Z < Z_c \end{cases} \quad (12)$$

with

$$Z = R_p^{\ 0.5}\left(\frac{u_*}{v_s}\right) \quad (13)$$

where R_p is the particle Reynolds number given by

$$R_p = \frac{\sqrt{RgD_s}\,D_s}{\nu} \quad (14)$$

Z_c is a critical value for the onset of significant suspension, approximately equal to five and Z_m denotes a maximum value of Z beyond which E_s becomes constant, approximately equal to 13.2.

The relationship proposed by Garcia (1990) and Garcia and Parker (1993) has been developed based on data from both open channel and turbidity current experiments. The non-dimensional entrainment rate for uniform sediment proposed by Garcia and Parker (1993) can be expressed as

$$E_s = \frac{1.3\times10^{-7}Z^5}{1+4.33\times10^{-7}Z^5} \quad (15)$$

$$Z = \alpha_1\frac{u_*}{v_s}\left(R_p\right)^{\alpha_2} \quad (16)$$

The parameters α_1 and α_2 in the above equation depend on the particle Reynolds number and are given as

$$\begin{cases}\alpha_1\\ \alpha_2\end{cases} = \begin{cases}1.0\\ 0.6\end{cases} \quad \text{for} \quad R_p > 2.36$$
$$\begin{cases}\alpha_1\\ \alpha_2\end{cases} = \begin{cases}0.586\\ 1.23\end{cases} \quad \text{for} \quad R_p \le 2.36 \quad (17)$$

The Smith and McLean (1977) model of sediment entrainment is based on open channel data and is an estimate of equilibrium concentration very close to the bed. The entrainment relation ship can be expressed as

$$E_s = 0.65\frac{\gamma_0\left(\dfrac{\tau_s^*}{\tau_c^*}-1\right)}{1+\gamma_0\left(\dfrac{\tau_s^*}{\tau_c^*}-1\right)} \quad (18)$$

where:

$$\gamma_0 = 0.0024; \quad \tau_s^* = \frac{u_*}{RgD_s} \quad (19)$$

The critical Shields stress τ_c^* can be obtained from a fit to the shields curve developed by Brownlie (1981). This fitted equation of the Shields diagram can be expressed as

$$\tau_c^* = 0.22R_p^{\ -0.6} + 0.06\exp\left(-17.77R_p^{\ -0.6}\right) \quad (20)$$

For convenience, from now on, the Akiyama-Fukushima, Garcia-Parker and the Smith-McLean relationship will be denoted respectively as E-I, E-II and E-III.

3 IGNITION CONDITIONS AND THE ENTRAINMENT RELATIONSHIP

Steady non-uniform turbidity currents developing in the downstream direction are considered herein. Following Parker et al. (1986), the governing equations under the steady non-uniform flow condition can be expressed in the following forms

$$\frac{dh}{dx} = \frac{-RiS+e_w(2-\frac{1}{2}Ri)+\dfrac{u_*^{\ 2}}{U^2}+\dfrac{1}{2}\dfrac{v_s}{U}r_0Ri\left(\dfrac{\psi_e}{\psi}-1\right)}{1-Ri} \quad (21a)$$

$$\frac{d\psi}{dx} = \frac{\psi v_s}{hU}\left(\frac{\psi_e}{\psi}-1\right) \quad (21b)$$

$$\frac{dU}{dx} = \frac{RiS-e_w(1+\frac{1}{2}Ri)-\dfrac{u_*^{\ 2}}{U^2}-\dfrac{1}{2}\dfrac{v_s}{U}r_0Ri\left(\dfrac{\psi_e}{\psi}-1\right)}{1-Ri}\cdot\frac{U}{h} \quad (21c)$$

$$\frac{dK}{dx} = [\frac{1}{2}e_w(1-Ri)+\frac{u_*^{\ 2}}{U^2}-e_w\frac{K}{U^2}-\frac{\varepsilon_0 h}{U^3}-$$
$$Ri\frac{v_s}{U}-\frac{1}{2}\frac{v_s}{U}r_0Ri\left(\frac{\psi_e}{\psi}-1\right)]\cdot\frac{U^2}{h} \quad (21d)$$

$$\psi_e = \frac{E_s hU}{r_0} \quad (22)$$

Here, $\psi = hUC$ is the volumetric suspended sediment transport rate per unit width and ψ_e denotes the equilibrium value of ψ at which neither erosion nor deposition would occur.

With specified upstream inflow conditions U_0, ψ_0 and h_0, the above equations could be solved by integrating along the downstream direct for supercritical turbidity currents. When the settling velocity v_s is set equal to zero, these equations describe a simple conservative flow which has the equilibrium solutions characterized by a constant Richardson number Ri, constant values of U and ψ and a current thickness that increases linearly in the downstream direction (e.g. Ellison and Turner 1959). However, for turbidity currents, such equilibrium solution would exist only if the deposition and the entrainment rate of suspended sediment are equal which is difficult to achieve.

Parker et al. (1986) considered a pseudo-equilibrium condition by setting the right hand sides of the momentum, sediment mass conservation and the turbulent kinetic energy equations equal to zero at $x = 0$. For a given upstream current thickness h_0, a group roots of U, ψ and K are found. The lowest finite values of U, ψ and K denoted by U_I, ψ_I and K_I are considered to be the ignition values. The ignitive state does not represent a true equilibrium, but it has been proved to be useful in identifying igniting versus subsiding of turbidity currents and could be used as an indicator to determine whether a current with given upstream conditions eventually ignites or subsides.

Given specified dimensionless number S, C_{f*}, α and h_0/D_s, ignition values of U_I, ψ_I and K_I can be found as described above and the ignition concentration C_I can be found from the relationship $\psi_I = h_0 U_I C_I$. The ignition values have been calculated by using three different erosion rate formulas are explored and compared for a variety of input parameters.

Typical values of U_I, ψ_I, C_I, Ri_I, and K_I computed from the four equation model with different erosion rate formulations for a different combinations of D_s, S, and h_0 are shown in Table 1. It is seen that the three sets of ignition values are close for some cases while different for others. Consider the case $D_s = 0.03$ mm, $C_{f*} = 0.004$, Slope $= 0.05$, and $h_0 = 1$ m. It is seen that ignition velocity yielded by E-II and E-III are about double the value yielded by E-I. While for the case $D_s = 0.1$ mm, $C_{f*} = 0.004$, Slope $= 0.1$ and $h_0 = 5$ m, the ignition velocity obtained with E-II and E-III are respectively 24% and 35% less than that obtained with E-I. Clearly, there are large differences between the ignition values yielded by different erosion rate formulas. Under a given set of inflow conditions, a turbidity current predicted to be an igniting flow could quite well turn out to be a subsiding flow if different sediment entrainment models are used.

Fig. 1(a) and (b) show U_I versus h_0/D_s for the sediment size $D_s = 0.1$ and 0.03 mm respectively with S set

Table 1. Typical values at ignition for E-I,E-II and E-III.

Ds	h_0			EI			EII			EIII		
(mm)	(m)	C_{f*}	S	U_I(m/s)	ψ_I(m^2/s)	K_I(m^2/s^2)	U_I(m/s)	ψ_I(m^2/s)	K_I(m^2/s^2)	U_I(m/s)	ψ_I(m^2/s)	K_I(m^2/s^2)
0.1	5.0	0.004	0.1	0.581	1.64×10^{-3}	1.03×10^{-2}	0.440	7.01×10^{-4}	5.31×10^{-3}	0.380	4.48×10^{-4}	3.66×10^{-3}
0.1	1.0	0.004	0.1	0.664	2.46×10^{-3}	1.40×10^{-2}	0.599	1.80×10^{-3}	1.11×10^{-2}	0.747	3.52×10^{-3}	1.83×10^{-2}
0.1	0.5	0.004	0.1	0.705	2.96×10^{-3}	1.61×10^{-2}	0.699	2.88×10^{-3}	1.57×10^{-2}	1.456	2.65×10^{-2}	7.69×10^{-2}
0.06	0.6	0.004	0.05	0.410	8.78×10^{-4}	4.71×10^{-3}	0.439	1.08×10^{-3}	5.56×10^{-3}	0.882	9.06×10^{-3}	2.68×10^{-2}
0.06	1.5	0.004	0.025	0.510	2.42×10^{-3}	4.62×10^{-3}	0.530	2.72×10^{-3}	5.22×10^{-3}	0.748	8.09×10^{-3}	1.41×10^{-2}
0.06	10	0.004	0.1	0.280	1.85×10^{-4}	2.51×10^{-3}	0.198	6.41×10^{-5}	1.12×10^{-3}	0.235	1.08×10^{-4}	1.68×10^{-3}
0.03	1.0	0.004	0.05	0.137	3.28×10^{-5}	5.67×10^{-4}	0.245	1.94×10^{-4}	2.08×10^{-3}	0.270	2.62×10^{-4}	2.57×10^{-3}
0.03	0.5	0.004	0.05	0.144	3.89×10^{-5}	6.45×10^{-4}	0.295	3.40×10^{-4}	3.09×10^{-3}	0.412	9.41×10^{-4}	6.27×10^{-3}
0.03	1.0	0.01	0.025	0.121	3.90×10^{-5}	5.94×10^{-4}	0.189	1.72×10^{-4}	2.19×10^{-3}	0.205	2.28×10^{-4}	2.73×10^{-3}
0.03	0.4	0.01	0.025	0.127	4.61×10^{-5}	7.01×10^{-4}	0.231	3.33×10^{-4}	3.65×10^{-3}	0.419	2.18×10^{-3}	1.44×10^{-2}

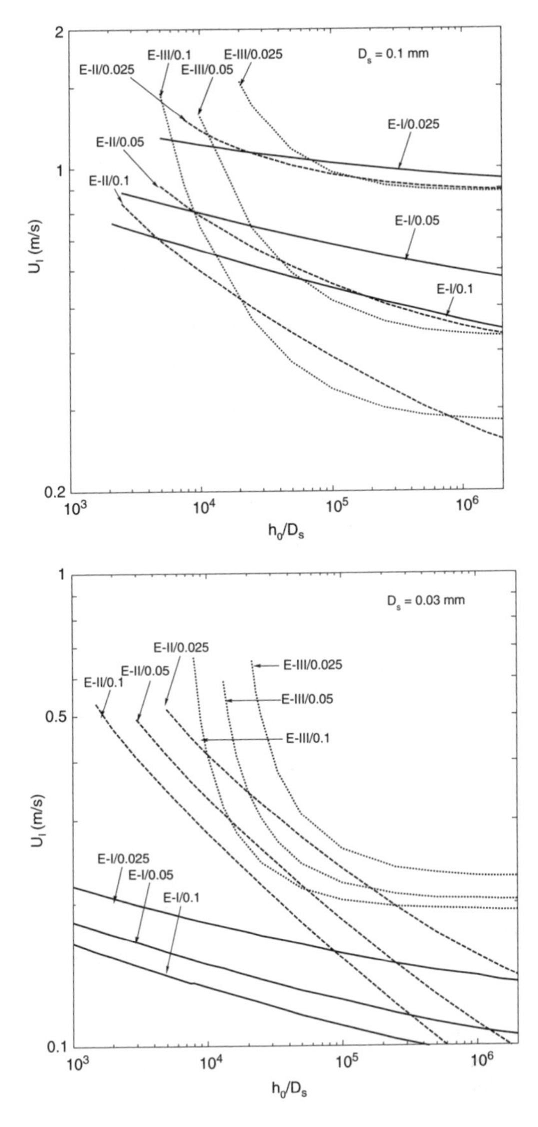

Figure 1. (a) Plot of U_I *vs* h_0/D_s for $D_s = 0.1$ mm; (b) Plot of U_I *vs* h_0/D_s for $D_s = 0.03$ mm. S serves as auxiliary parameter and C_f* is set equal to 0.004.

as an auxiliary parameter. C_{f*} is set equal to a typical value 0.004 in both cases.

Figs. 2(a) and (b) show C_I versus h_0/D_s for the corresponding cases. As expected, higher slope and finer sediment size lead to lower values of U_I and C_I in computation performed using all three sediment entrainment model. On the other hand, achieving the ignition state on low slopes with coarse sediment requires large values of ignition velocity and concentration. When the inflow current thickness is small, use of E-III leads to very large ignition values. In other

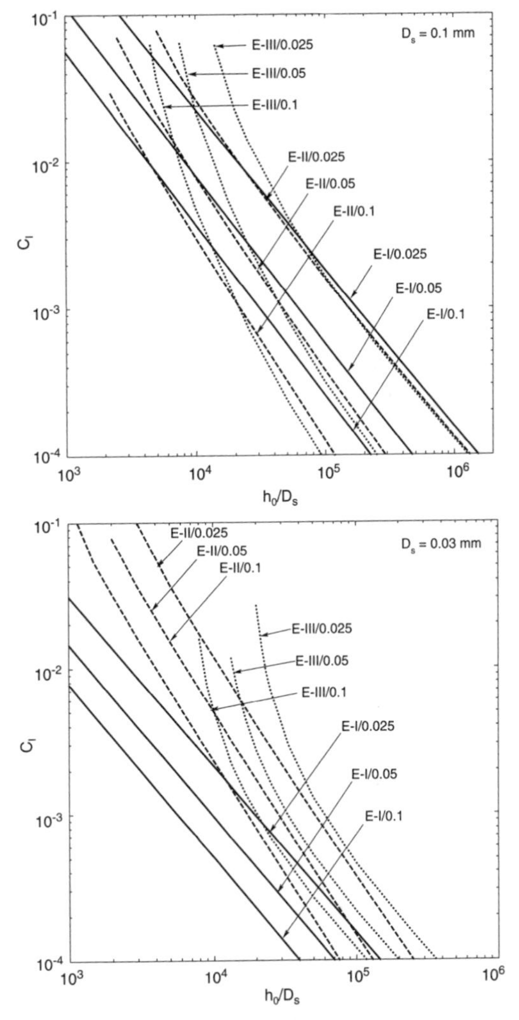

Figure 2. (a) Plot of C_I *vs* h_0/D_s for $D_s = 0.1$ mm; (b) Plot of C_I *vs* h_0/D_s for $D_s = 0.03$ mm. S serves as auxiliary parameter and C_{f*} is set equal to 0.004.

words, it is difficult achieve an ignitive state with E-III if the inflow current thickness is small.

As seen in Figs. 1(a) and 2(a), if the inflow current thickness is relatively large, E-I leads to higher ignition values compared to E-II and E-III in case of the larger sediment size of 0.1 mm. The constant slope lines from an individual entrainment formulation are generally parallel. It is evident that under the condition of large inflow current thickness and coarse sediment, E-III will easily lead to ignition condition compared to E-I and E-II. In case of the finer sediment size $D_s = 0.03$ mm, it is observed from Figs. 1(b) and 2(b) that ignition values predicted with E-III are larger than those predicted with E-II for a given slope.

Similarly, larger ignition values are predicted with E-II compared to E-I. In other words, it is difficult to erode fine sediment with E-III and E-II compared to E-I. This is not unexpected. In E-II, the coefficients α_1, and α_2 are adjusted for smaller values of particle

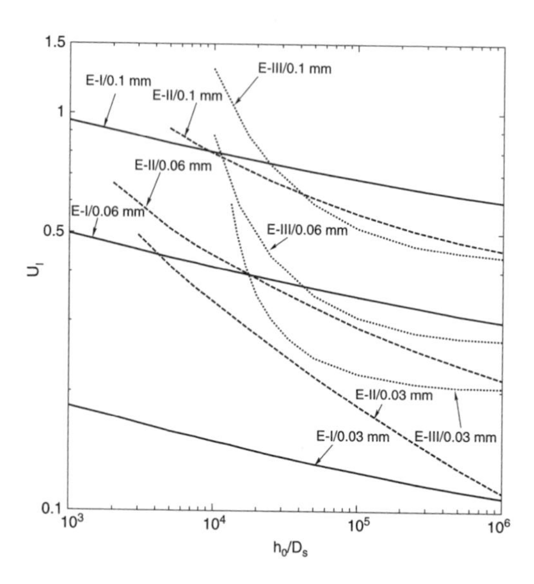

Figure 3. Plot of U_I *vs* h_0/D_s for $D_s = 0.03$ mm, 0.06 mm, and 0.1 mm. S and C_{f*} are respectively set equal to 0.05 and 0.004.

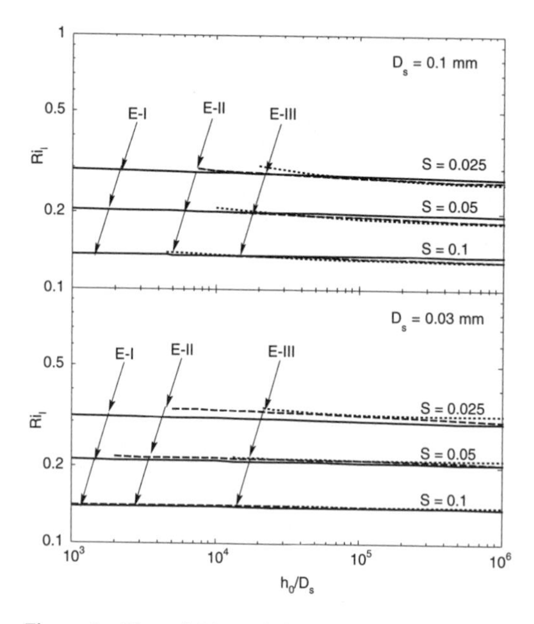

Figure 4. Plot of Ri_I *vs* h_0/D_s, S serves as auxiliary parameter and C_{f*} is set equal to 0.004.

Reynolds number (Garcia and Parker 1993) and in E-III, the entrainment threshold is set by the critical Shields stress which becomes large for fine particles.

The ignition velocity U_I versus h_0/D_s is plotted in Fig. 3 for three different grain sizes of 0.03 mm, 0.06 mm, and 0.1 mm. The slope has been fixed at 0.05 and the ignition values have been obtained using E-I, E-II, and E-III. It is seen here that the ignition velocity changes over a much wider range in case of E-I compared to E-II and E-III for the range of grain size variation considered here. The distance between the lines of constant grain size is smallest in case of E-III.

The ignition Richardson number Ri_I is plotted versus h_0/D_s in Fig. 4 for two different grain sizes of 0.03 mm and 0.1 mm. Three different slopes are considered with C_{f*} set as a constant equal to 0.004. The ignition Richardson number Ri_I is found to be almost independent of h_0, and D_s. The selection of entrainment rate formula does not make much difference either. Rather, the ignition Richardson number is found to be strongly dependent on the incipient slope.

4 PHASE PLANE ANALYSIS

The phase plane analysis carried out by Parker et al. (1986) and Pantin (1979) provided a better understanding of the behavior of a turbidity current. Here, phase diagrams have been plotted with results obtained using E-I, E-II, and E-III. Two different sets of input conditions have been considered. The well-known fourth-order Runge-Kutta method has been used to solve (21) for the downstream development of currents with prescribed upstream values of h_0, U_0, ψ_0, and K_0. All the solutions are projected onto the phase plane (U, ψ), which respectively are normalized by ignitions values U_I, ψ_I. All the currents with specified inflow conditions are found to either ignite or subside in the phase plane. In order to allow for the projection of the solutions onto a (U/U_I, ψ/ψ_I) phase plane and reduce the number of cases to be computed, K_0 is not selected arbitrarily but estimated by the following rule (Parker et al. 1986)

$$\frac{K_0}{U_0^{\,2}} = \frac{K_I}{U_I^{\,2}} \tag{23}$$

The phase diagram of Fig. 5(a) was obtained with the input conditions: $D_s = 0.1$ mm, $h_0 = 5$ m, $C_{f*} = 0.004$ and $S = 0.1$ and that of Fig. 5(b) was obtained with: $D_s = 0.03$ mm, $h_0 = 1$ m, $C_{f*} = 0.004$ and $S = 0.05$. All diagrams are plotted in the normalized form (U_I/U versus ψ_I/ψ). AGL represents the autosuspension generation line while CL represents the convergence line. The convergence line actually is a narrow band representing the trends of downstream current development rather than a single line. The convergence line beyond the ignition point is computed from ignition values. Currents with ignition values as the specified upstream conditions are igniting flows. The circle at (1, 1) represents the ignition point. Other circles show the locations where the autosuspension generation lines cross with convergence line.

The phase plane can be divided into two regions of subsiding and igniting fields. Flows starting at points above the AGL line eventually accelerate and ignite while flows starting at points below the line will decelerate and subside. The position of AGL is very sensitive to the selection of erosion rate formula. In Fig. 5(a), the AGLs for E-II and E-III almost overlap while the AGL for E-I is positioned above parallel to the others. In Fig. 5(b), AGL for E-III and E-II are parallel with the former positioned above the latter. The AGL for E-I crosses that of E-III at approximately $\psi/\psi_I = 0.4$. The positions of the convergence lines (CL) and the line corresponding to Richardson number equal to unity are insensitive to erosion rate formulas. Arrow lines A-I, A-II, A-III and C-I, C-II, C-III correspond to points A and C of the following figure. In Fig. 5(a), A-I, computed using E-I subsides and merges with the CL while A-II computed using E-II ignites and merges with the CL and A-III computed with E-III stays above the convergence line. In Fig. 5(b), C-I remains in the ignitive field while C-II and C-III quickly subside. It is clear that while keeping other conditions same, the choice of different sediment entrainment relationships can lead to completely different downstream development of a turbidity current.

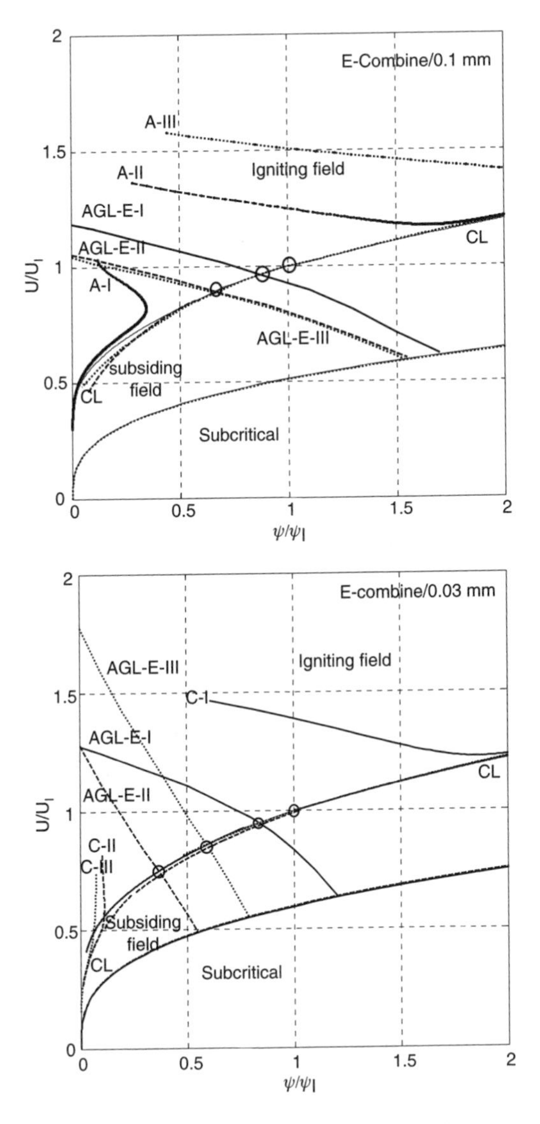

Figure 5. (a) Phase diagram computed for the case: $D_s = 0.1$ mm, $h_0 = 5$ m, $C_{f*} = 0.004$ and $S = 0.1$. (b) Phase diagram computed for the case: $D_s = 0.03$ mm, $h_0 = 1$ m, $C_{f*} = 0.004$ and $S = 0.05$.

5 CROSSFIELD OF IGNITION AND SUBSIDENCE

The AGL-s computed with three entrainment relations are plotted in the dimensional form in Figs. 6(a) and (b), in order to determine the cross field of the igniting and subsiding regions. The terminology cross field denotes the area between the igniting and the subsiding zones bounded by the AGL-s. The input conditions

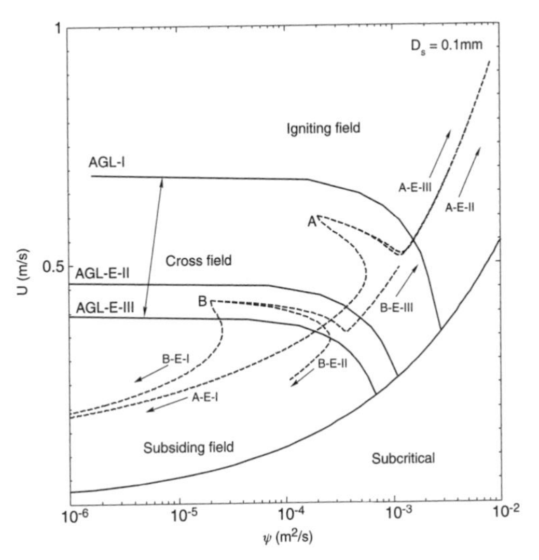

Figure 6. (a) Cross field for case $D_s = 0.1$ mm, $h_0 = 5$ m, $C_{f*} = 0.004$, and $S = 0.1$.

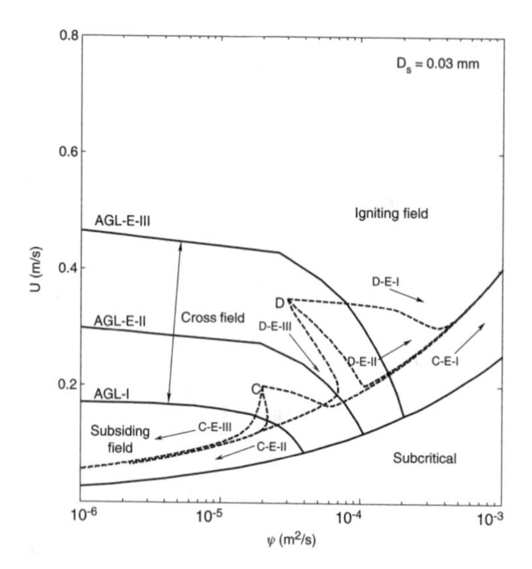

Figure 6. (b). Cross field for case $D_s = 0.03$ mm, $h_0 = 1$ m, $C_{f*} = 0.004$, and $S = 0.05$.

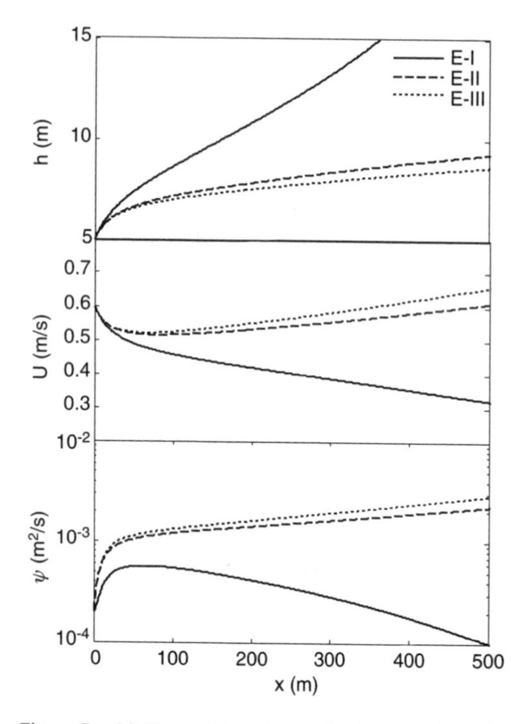

Figure 7. (a) Current downstream development from the point A of Fig. 6(a).

are same as those used in plotting Fig. 5. For the case with coarser sediment, the upper and the lower bound of the cross field is marked respectively by AGL-I and AGL-II as seen in Fig. 6(a). For the case with finer; sediment, the opposite is observed as seen in Fig. 6(b). Different arrow lines in these figures indicate that depending on the location of the starting point with respect to different AGL-s, inside the cross field, a turbidity current can become either ignitive or subsiding.

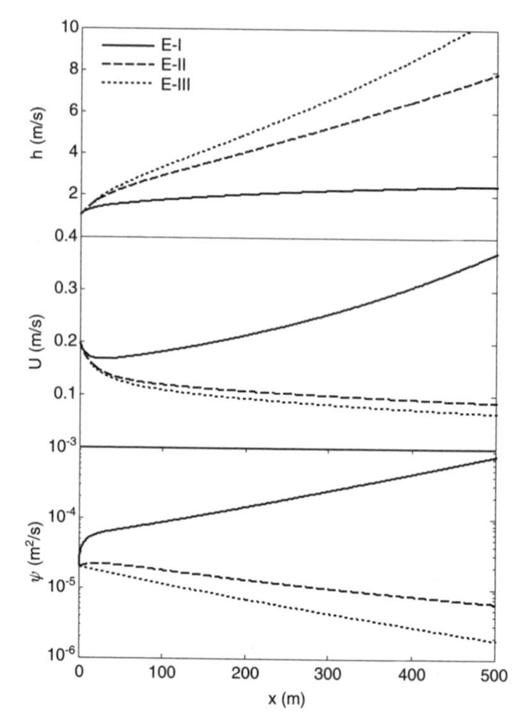

Figure 7. (b) Current downstream development from the point C of Fig. 6(b).

Fig. 7(a) shows the downstream development of a current commencing from the points A of Fig. 6(a). The turbidity current with the prescribed upstream condition: $U_0 = 0.6$ m/s, $\psi_0 = 2.0 \times 10^{-4}$ m^2/s, and $K_0 = 8.0 \times 10^{-3}$ m^2/s^2 entrains sediments and accelerates if E-II or E-III is selected as the sediment entrainment model but loses sediment and decelerates if E-I is used. The current thickness increases at a much faster rate in case of E-I compared to the other two due to dilution caused by the loss of sediment. Fig. 7(b) shows the evolution of a current commencing from point C of Fig. 6(b) with $U_0 = 0.2$ m/s, $\psi_0 = 2.0 \times 10^{-5}$ m^2/s, and $K_0 = 1.0 \times 10^{-3}$ m^2/s^2. The current becomes ignitive in the case when E-I is used and becomes depositional in the cases when E-II or E-III is used.

6 SUMMARY AND CONCLUSIONS

There has been a proliferation in the development and application of numerical models for predicting the seabed morphology. Irrespective of the level of sophistication, a turbidity current model requires empirical relationships to describe the sediment entrainment from the bed. In the present analysis, the classical work of Parker et al. (1986) has been revisited with three different sediment entrainment formulas. The four equation model has been considered. It has been found that for the same input conditions, the ignition values obtained with different models can vary widely. The phase plane analysis has shown that different entrainment relations give different autosuspension generation lines or AGL-s but lead to the same convergence line. The relative positions of the AGL-s also change with the change in sediment size and other inflow conditions. It is therefore, important that researchers take a cautious approach in interpreting numerical model results on turbidity current profiles deposits they create especially at the field scale. The entrainment models perform similarly when compared to the measured data at the laboratory scale (e.g. Garcia 1990) yet they lead to very different behavior of simulated turbidity current when applied at the field scale.

The work demonstrates that there is clear need for further research on the topic of sediment entrainment. The authors of this paper hope that there will be renewed interest in identifying and or developing appropriate entrainment formulas that can correctly predict the behavior of a turbidity current.

ACKNOWLEDGMENT

Funding from the National Science Foundation (OCE-0134167) is gratefully acknowledged.

REFERENCES

Akiyama, J. & Fukushima, Y. 1985. Entrainment of noncohesive bed sediment into suspension. External Memo. No. 175, St. Anthony Falls Hydraulic Laboratory, University of Minnesota, Minneapolis, USA.

Celik, I. & Rodi, W. 1988. Modeling suspended sediment transport in nonequilibrium situation. J. Hydr. Engr. ASCE 114(10), 1157–1191.

Ellison, T. H. & Turner, J. S. 1959. Turbulent entrainment in stratified flows. J. Fluid Mech. 6, 423–448.

Fukushima, Y., Parker, G. & Pantin, H. M. 1985. Prediction of ignitive turbidity currents in Scripps Submarine Canyon. Mar. Geol. 67, 55–81.

Garcia, M. 1990. Depositing and eroding turbidity sediment driven flows: Turbidity currents, Proj. Rep. 306, St. Anthony Falls Hydraulic Lab., Univer. Of Minn., Minneapolis.

Garcia, M. & Parker, G. 1993. Experiments on the entrainment of the sediment into suspension by a dense bottom current. J. of Geophysical Res., 98(C3), 4793–4807.

Garcia, M. 1994. Depositional turbidity currents laden with poorly sorted sediment. J. Hydr. Eng. ASCE, Vol. 120, 1240–1263.

Huang, H., Imran, J. & Pirmez, C. Numerical model of turbidity currents with a deforming bottom boundary. J. Hydr. Eng. ASCE, Vol. 131, 2005.

Pantin, H. M. 1979. Interaction between velocity and effective density in turbidity flow: phase-plane analysis, with criteria for autosuspension. Mar. Geol. 31, 59–99.

Parker, G. 1978. Self-formed straight rivers with equilibrium banks and mobile bed. Part 1. The sand-silt river. J. Fluid Mech. 89, 109–125.

Parker, G. 1982. Conditions for the ignition of catastrophically erosive turbidity currents. Mar. Geol. 46, 307–327.

Parker, G., Fukushima, Y. & Pantin, H. M. 1986. Self-accelerating turbidity currents. J. Fluid Mech. 171, 145–181.

Smith, J. D. & McLean, S. R. 1977. Spatially averaged flow over a wavy surface, Journal of Geophysical Research, 83, 1735–1746.

Turner, J. S. 1973. Buoyancy effects in fluid. Cambridge University Press, Cambridge, 367 pp.

van Rijn, L. C. 1984. Sediment Transport, Part II: Suspended load transport, J. Hydr. Eng ASCE, Vol. 110(11), 1613–1641.

River, Coastal and Estuarine Morphodynamics: RCEM 2005 – Parker & García (eds)
© 2006 Taylor & Francis Group, London, ISBN 0 415 39270 5

Theory for a clinoform of permanent form on a continental margin emplaced by weak, dilute muddy turbidity currents

G. Parker
University of Illinois, Dept. of Civil & Environmental Engineering & Dept. of Geology, Urbana, Illinois, USA

ABSTRACT: A clinoform is a prograding sedimentary deposit. Of interest here are the clinoforms that constitute the building blocks of continental margins. A continental margin consists of a continental shelf, i.e. a shallow region near the shore of a continent that gently slopes seaward, followed by a continental slope, i.e. a steeper region that drops off into deep water. In many cases the continental slope itself constitutes a clinoform that progrades into deeper water. In other cases, and especially near river mouths, a given margin may have several coexisting clinoforms. Many margin clinoforms tend to be muddy. Here a theory of clinoform progradation is developed. The clinoform is assumed to be continuous in the along-shelf direction. A wave-current boundary layer is assumed to stir up a dilute suspension of mud, which then moves slowly as a sheet down the shelf slope in the direction of deep water. Near the shelf-slope rollover the turbidity current loses wave agitation, and forms a net-depositional flow that results in clinoform progradation. When simplified to the case of a horizontal deep-water basement, constant sea level and negligible tectonism, the theory allows a solution for a clinoform of permanent form, i.e. one that progrades without changing shape.

1 INTRODUCTION

A clinoform is a prograding sedimentary deposit that has a sigmoidal (i.e. S-like) shape (Driscoll et al., 2004). The focus of this paper is on clinoforms on continental margins. An example is shown in Figure 1. The image shows a seismic-reflection profile in the vicinity of the Ganges-Brahmaputra Delta (Kuehl et al., 1997). It shows not only the present-day clinoform, but also how it has prograded seaward in recent geologic history.

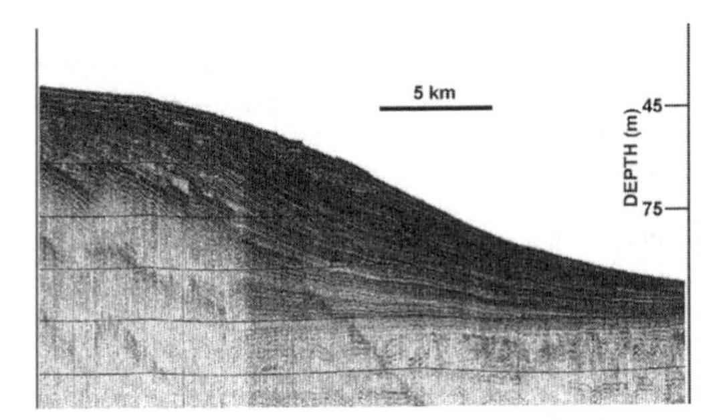

Figure 1. Seismic-reflection profile of the prograding clinoform in the vicinity of the Ganges-Brahmaputra delta. From Kuehl et al. (1997).

Every continent is surrounded by a margin consisting of a relatively shallow continental shelf, which slopes gently seaward, followed by a continental slope, which slopes more steeply seaward. This geometry is illustrated in Figures 2a and 2b. The precise morphology of the continental margin in any region depends

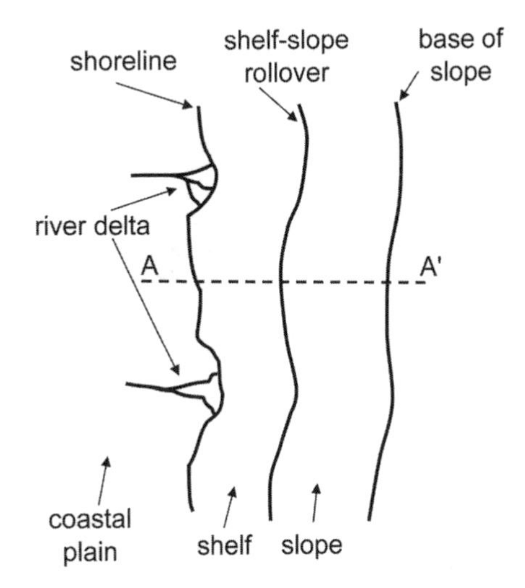

Figure 2a. Plan view of a continental margin with a continuous shelf and slope.

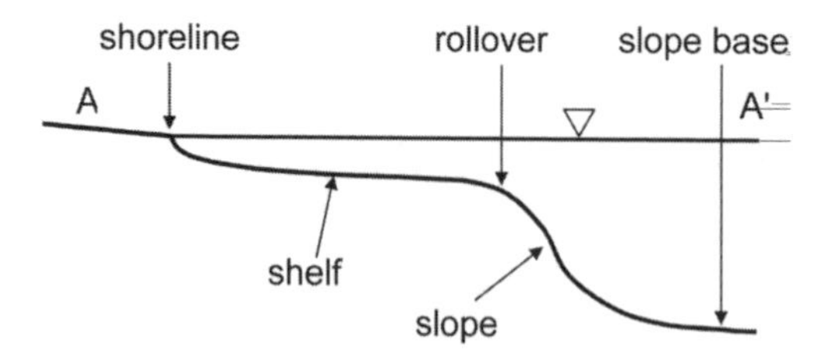

Figure 2b. Cross-sectional view of the continental margin of Figure 2a from point A to point A'.

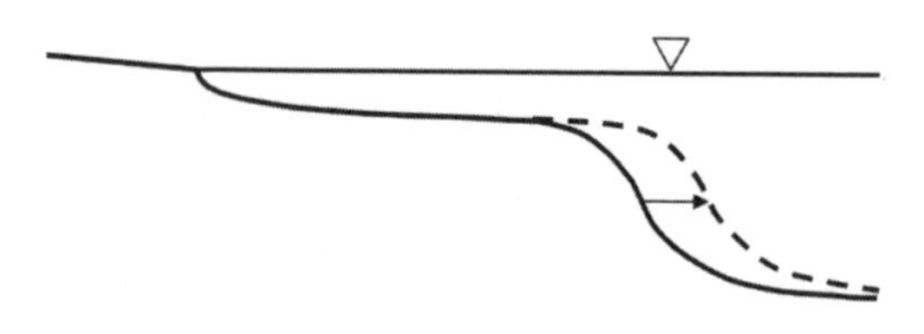

Figure 3a. Case for which the margin contains a single prograding clinoform which is synonymous with the slope.

upon such factors as sediment supply and tectonic regime, with shelves tending to be much narrower along active, uplifting margins (e.g. west coast of North America) than along passive, slowly subsiding margins (e.g. east coast of North America: Pratson and Haxby, 1997; O'Grady et al., 2000). While a margin may show bulges in the vicinities of modern or recent deltas, as illustrated in Figure 2a, the margin itself forms a continuous band around a continent. This is in contrast to most freshwater lakes, where local deltas and shelves tend to form distinct entities that do not merge to form a continuous margin.

Present-day conditions correspond to an interglacial period, with sea level about 120 m higher (high stand) than at the last glacial maximum some 18,000 years ago (low stand; e.g. Bard et al., 1996). As a result, large portions of continental shelves along modern passive margins consist of a drowned coastal plain that was constructed by fluvial processes during low stand. Other large portions of continental shelves are, however, being constructed by subaqueous processes under modern high-stand conditions by means of clinoform progradation. Examples include continental shelves near the Amazon and Ganges-Brahmaputra deltas (Kuehl et al., 1997, Kuehl, 2003; Figure 1), along the Italian margin facing the Adriatic Sea (Cattaneo et al., 2004) and in the Gulf of Papua (Aller and Aller, 2004).

The shelf-slope complex can be thought of as an agglomeration of prograding clinoforms. In the simplest configuration, the continental slope itself corresponds to a single clinoform (Figure 3a). In other cases multiple clinoforms are found on a single shelf-slope complex, as shown in Figure 3b (e.g. Nittrouer et al.,1996).

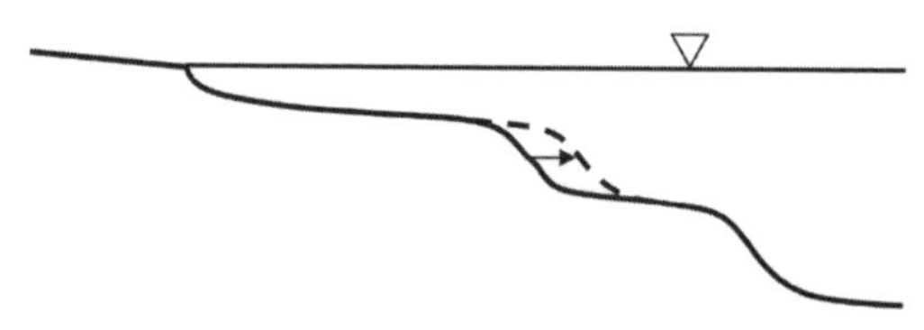

Figure 3b. Case for which the margin contains multiple clinoforms, with only one actively prograding.

2 OCEANS, LAKES AND CONTINENTAL MARGINS

Sediment-laden river water is heavier than fresh, sediment-free water at the same temperature. Consider the case of sediment-laden river water during a flood. As the river enters a lake or reservoir, the coarser material (typically sand and gravel) deposits to form the topset and foreset of a localized delta. Under conditions which are easily realized in freshwater lakes and reservoirs, however, the muddy (and thus heavy) water can plunge to form a turbidity current, which then immediately heads for deep water. This process is schematized in Figure 4.

An example of a reservoir in which plunging turbidity currents emplace a muddy bottomset is Lake Mead, USA. The depositional structure of Lake Mead is shown in Figure 5 (Smith et al., 1954). The plunging turbidity currents have been documented by e.g. Grover and Howard (1938) and Bell (1942); the structure of the resulting turbidites has been recently studied by Twichell et al. (2005).

Neither Twichell et al. (2005) nor other authors note a tendency for the sediment entering Lake Mead from the Colorado River and other sources (e.g. the Las Vegas Wash) to form a well-developed, continuous bench, or shelf ringing the lake. One might argue that shelves are constructed mostly by fluvial processes at low stand and then drowned at high stand to create continuous subaqueous margins. One might further argue that lakes do not undergo the cyclic fluctuations in level characteristic of the ocean. Lake Mead has, however, undergone significant water level variation since filling (e.g. Twichell et al., 2005). A likely reason that river deltas do not merge into continuous shelves in freshwater lakes is the relative ease (compared to rivers entering salty ocean water) with which a river laden with suspended mud plunges and heads toward deep water. It is not implied here that all river flows entering lakes plunge to form muddy turbidity currents. When the suspended sediment concentration in a river is sufficiently low and lake water is sufficiently colder than river water, the resulting density barrier can prevent plunging. Rather, it is argued that plunging is likely to be far more common in the case of rivers entering freshwater lakes than in the case of rivers entering the ocean.

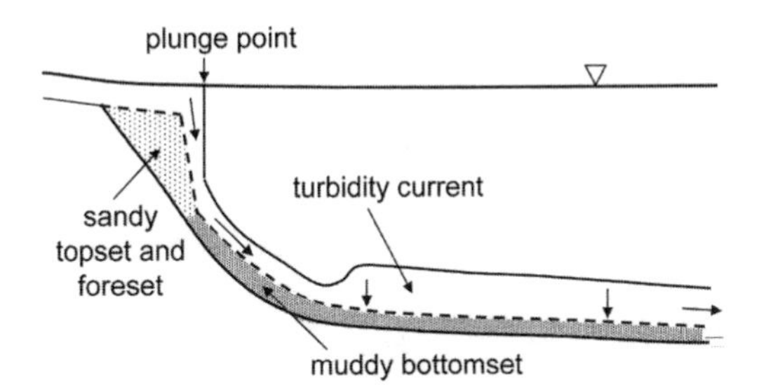

Figure 4. Illustration of a turbidity current formed as a muddy river plunges into a freshwater lake.

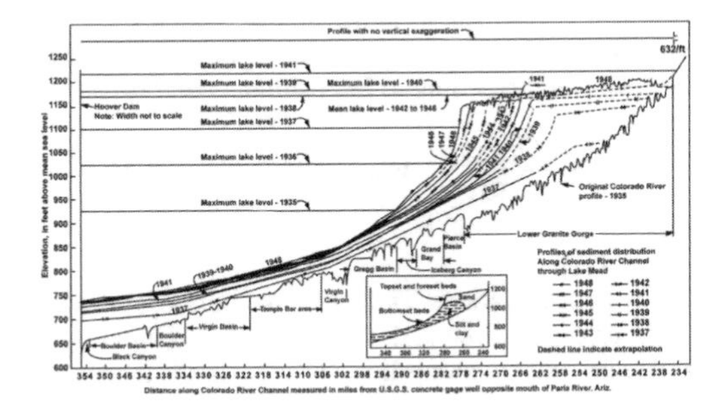

Figure 5. Evolution of sedimentary deposits in Lake Mead from the closing of Hoover Dam until 1948. Adapted from Smith et al. (1954).

Seawater is heavier than the water in most rivers as they enter the sea, even when they are heavily laden with suspended sediment. As a result, seawater creates a formidable barrier to plunging. An overall average for the density ρ_{sw} of seawater is 1.027 tons/m^3. Let C denote the mean volume concentration of suspended sediment in the river water, X denote the concentration in mg/liter, ρ_s denote the material density of the sediment and ρ denote the density of fresh water. The density ρ_r of river water is given as

$$\rho_r = \rho(1 + RC) \tag{1a}$$

where

$$R = \rho_s/\rho - 1 \tag{1b}$$

For natural sediments R is usually close to 1.65. The relation between C and X is

$$X = 1x10^6(R+1)C \tag{2}$$

Based on the above values, the minimum concentrations C and X necessary for a river flow to plunge in seawater (hyperpycnal flow) is $C = 0.0163$, or $X = 43{,}360$ mg/liter. The criterion for hyperpycnal flow can be lowered somewhat by considering seawater

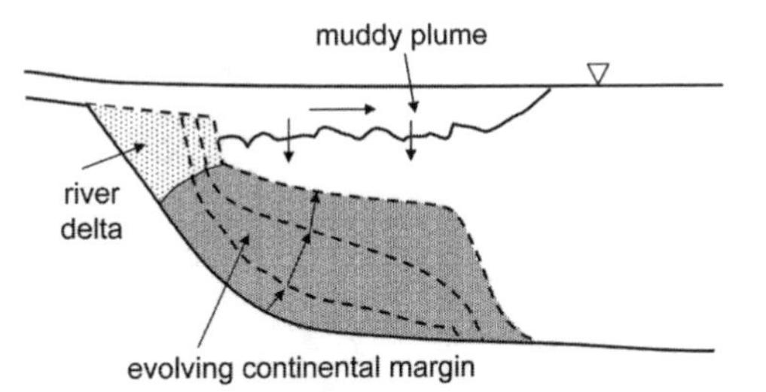

Figure 6. Diagram illustrating the evolution of a continental margin as a hypopycnal river flows into the sea.

of lower salinity and a vertical gradient in the suspended sediment concentration profile of the river. The inescapable conclusion is, however, that river flows that plunge directly into the ocean are rare, and with the exception of the Yellow River, China are nearly all restricted to active, and often rapidly uplifting margins (Mulder and Syvitski, 1995).

In most case of interest, then, and particularly along passive margins, the river water is hypopycnal, and thus forms a surface plume as it enters the sea. As a result, the sediment it carries eventually settles out within a few km of the shore. Initially the sediment may deposit at a level below wave base. In time, however, the deposit builds up to a level at which the sediment can be resuspended by waves, as schematized in Figure 6. Once this state is reached, along-shore currents can form a local bench into a continuous shelf that joins delta to delta, as schematized in Figure 2a.

3 THE HYPOTHESIS: WEAK, DEPOSITIONAL SHEET TURBIDITY CURRENTS AS THE AGENTS OF CLINOFORM PROGRADATION

Figure 7 shows cross-margin profiles of bed slope angle versus distance for five characteristic profiles of passive continental margins (O'Grady et al., 2000). In each case a shelf with a low seaward slope is followed by a rollover to a slope with a steeper slope. Having said this, it is clear that in every case the shelf is not horizontal, but tends to slope seaward. This characteristic proves essential for the arguments below.

It is assumed here that the shelf, or at least that part of the shelf that is above the active clinoform, is above wave base. The shelf sediment is then subject to suspension by a sustained wave-current boundary layer. This suspension can be expected to form a weak turbidity current forced by wave agitation that is extensive (sheet-like) in the along-shelf direction and that gradually flows downslope toward the clinoform. Once the turbidity current slips below wave base at the shelf-slope rollover, however, it loses agitation. The weak, muddy turbidity current may then rapidly dissipate and

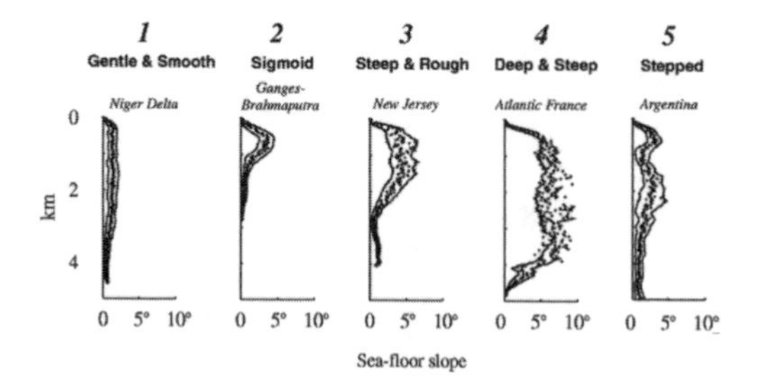

Figure 7. Plots of seafloor slope angle versus offshore distance for five characteristic types of passive continental margins. From O'Grady et al., (2000).

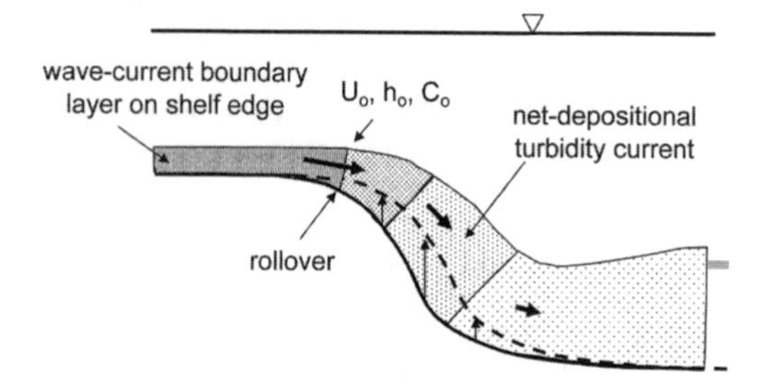

Figure 8. Definition diagram for a net-depositional muddy sheet turbidity current rolling off the shelf.

deposit its sediment on the clinoform. This picture is schematized in Figure 8.

4 MODEL FOR SUSTAINED, DILUTE, NET-DEPOSITIONAL MUDDY TURBIDITY CURRENTS

As noted above, it is assumed that a sustained wave-current boundary layer resuspends mud on the shelf near the shelf-slope rollover. This muddy bottom layer then flows slowly under the weight of its own suspended mud toward the shelf-slope rollover illustrated in Figure 8. The upstream end of the domain of interest is located at this rollover, beyond which the effect of wave-current agitation is assumed to vanish.

Let x denote cross-margin (seaward) distance from the rollover point. Because the resuspension event is assumed to be sustained, the resulting turbidity current is also assumed to be sustained, and the flow is treated as quasi-steady. Let h denote the layer thickness of the turbidity current, U denote layer-averaged flow velocity and C denote layer-flux-averaged volume concentration of mud. The turbidity current is assumed to be dilute in the sense that $C << 1$ and $RC << 1$. The mud is assumed to have flocculated to a single effective fall velocity v_s; Hill (1998) suggests a value of 1 mm/s for marine muds.

The governing equations for flow mass balance, momentum balance and balance of suspended sediment mass take the following respective forms (e.g. Parker et al., 1986);

$$\frac{dUh}{dx} = e_w U \tag{3}$$

$$\frac{dU^2h}{dx} = -\frac{1}{2} Rg \frac{dCh^2}{dx} + RgChS - C_f U^2 \tag{4}$$

$$\frac{dUhC}{dx} = v_s(E - r_o C) \tag{5}$$

In (3) e_w denotes a dimensionless coefficient of entrainment of ambient water from above, which can be specified according to Fukushima et al. (1985) as

$$e_w = \frac{0.00153}{0.0204 + \mathbf{Ri}} \quad , \quad \mathbf{Ri} = \frac{Rgq_s}{U^3} \quad , \quad q_s = UhC \tag{6a,b,c}$$

In the above relations **Ri** denotes the bulk Richardson number of the flow and q_s denotes the volume mud transport rate per alongshelf distance. In (4) g denotes the acceleration of gravity, S denotes bed slope, given as

$$S = -\frac{d\eta}{dx} \tag{7}$$

where η denotes bed elevation and C_f denotes a dimensionless coefficient of bed friction. In (5) r_o is a dimensionless, order-one number defined as

$$r_o = c_b / C \tag{8}$$

where c_b is a near-bed volume concentration of mud. In addition, the parameter E denotes a dimensionless rate of resuspension of mud, here parameterized as

$$E = (U/U_r)^n \tag{9}$$

where U_r denotes a reference velocity (at which the volume resuspension rate per unit bed area per unit time $v_s E$ becomes equal to the fall velocity v_s of the mud), and n is an exponent.

A solution of (3), (4) and (5) to determine the flow over any given bed profile $\eta(x)$ requires specification of the following boundary conditions at the rollover point;

$$h\big|_{x=0} = h_o \quad , \quad U\big|_{x=0} = U_o \quad , \quad C\big|_{x=0} = C_o \tag{10a,b,c}$$

Note that according to (6c), the upstream value of q_s is given as

$$q_s\big|_{x=0} = q_{so} = U_o h_o C_o \tag{10d}$$

Once the flow over the bed is determined, the evolution of the bed can be calculated from the Exner equation of conservation of bed sediment;

$$(1-\lambda_p)\frac{\partial \eta}{\partial t} = v_s(r_o C - E) \tag{11}$$

where λ_p denotes the porosity of the deposited bed sediment.

The time t in (11) refers to the time during which a sustained event is taking place. It can be assumed that most of the time the shelf is quiescent. The difference between event time t and effective total time t_{ef} (which includes the time when the system is quiescent) is treated with an intermittency factor $I_e \leq 1$ corresponding to the fraction of time an event is actually taking place (Paola et al., 1992); thus

$$t_{ef} = t / I_e \tag{12}$$

Two aspects of the above relations deserve some discussion. The resuspension relation defined by (9) should pertain to freshly-deposited sediment. A key feature of the present analysis is a turbidity current that is net depositional, but not purely depositional. The porosity λ_p in (11), on the other hand, refers to the long-term value that evolves after the deposited mud has had a chance to consolidate.

The weakest part of the above model is the relation (9) for the resuspension of marine mud. At present there is no generally accepted set of relations for the resuspension of marine muds (e.g. Winterwerp and Kranenburg, 2002). The form of (9) allows, however, for parametric adjustment to obtain net-depositional turbidity currents on clinoforms.

5 PROGRADING CLINOFORM OF PERMANENT FORM

The essence of any phenomenon can be elucidated by studying it under the simplest conditions under which it forms. To this end it is assumed that:

- the seafloor basement is horizontal;
- tectonism is neglected;
- sea level is taken to be constant; and
- climate remains constant.

Any of the above assumptions can be relaxed with relative ease at a later time, as long as the necessary input parameters are known.

Under such conditions a clinoform can be reasonably assumed to evolve toward a permanent form with constant progradation rate. Let c denote the progradation rate during resuspension events, and let c_{ef} be the long-term effective average progradation rate, such that

$$c_{ef} = I_e c \tag{13}$$

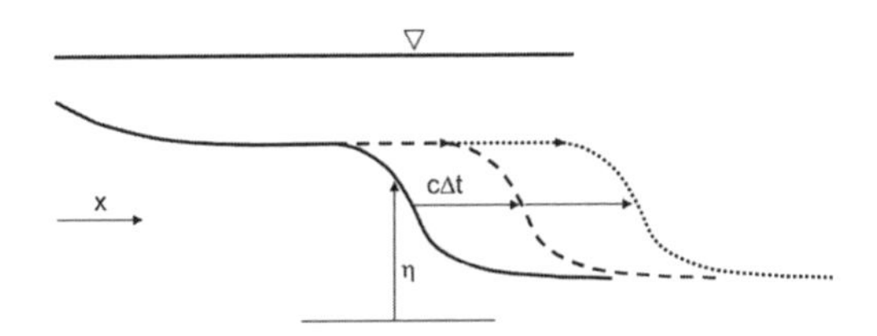

Figure 9. Definition diagram for a clinoform of permanent form prograding with constant speed c.

During resuspension events, the time variation of clinoform shape is assumed to simplify to

$$\eta(x,t) = \eta_p(x - ct) \tag{14}$$

where the subscript "p" denotes a solution of permanent form. That is, the clinoform progrades at constant speed c without changing shape, as shown in Figure 9.

In order to implement the solution for a clinoform of permanent form the following coordinate transformation is introduced;

$$\bar{x} = x - c\bar{t} \quad , \quad \bar{t} = t \tag{15a,b}$$

Equations (3), (4), (5) and (11) are transformed accordingly, and the dependence on time $\bar{t}$ is dropped. Upon reduction with (15), this results in the relations

$$\frac{dUh}{d\bar{x}} = e_w U \tag{16}$$

$$\frac{dU^2h}{d\bar{x}} = -\frac{1}{2} Rg \frac{dCh^2}{d\bar{x}} - RgCh\frac{d\eta_p}{d\bar{x}} - C_f U^2 \tag{17}$$

$$\frac{dUhC}{d\bar{x}} = v_s(E - r_o C) \tag{18}$$

$$-(1-\lambda_p)c\frac{d\eta_p}{d\bar{x}} = v_s(r_o C - E) \tag{19}$$

Equations (18) and (19) can be reduced with (6c) to obtain the form

$$(1-\lambda_p)c\frac{d\eta_p}{d\bar{x}} = \frac{dq_s}{d\bar{x}} \tag{20}$$

As illustrated in Figure 10, the above equation can be integrated from the rollover point $\bar{x} = 0$, where $\eta_p = \eta_r$, to $\bar{x} = \infty$, by which point all of the sediment from the turbidity current has settled out and η_p is equal to constant basement elevation η_b. This results in the relation

$$q_{so} = (1-\lambda_p)c(\eta_r - \eta_b) \tag{21}$$

Equation (21) thus implies a simple relation between the supply rate of mud q_{so}, the elevation drop across the clinoform $\eta_r - \eta_b$ and the progradation speed c.

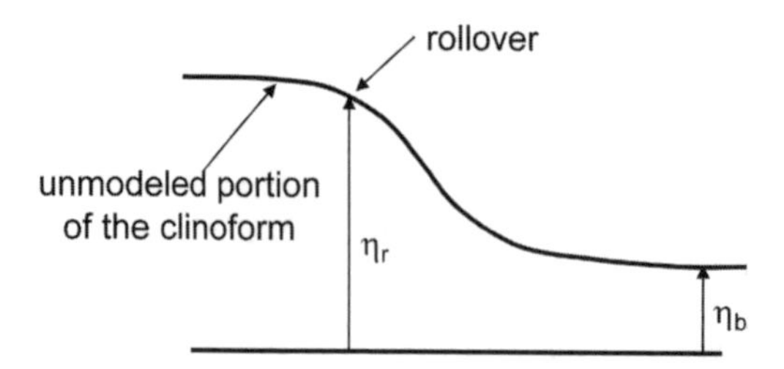

Figure 10. Definition diagram illustrating clinoform bed elevation at the rollover point η_r and basement elevation η_b.

It should be noted that the difference $\eta_r - \eta_b$ is not precisely equal to the elevation difference across the clinoform, because η_r refers to the elevation at the rollover point just below wave base, and not where the clinoform actually joins the shelf. As a result the upper limb of the sigmoidal shape of the clinoforms modeled below is somewhat truncated.

Substituting (19) into (17) and reducing (16) ~ (18) with (6c), the following three relations governing the shape of a clinoform of permanent form are obtained:

$$\frac{dh}{d\bar{x}} = \frac{-P_1 + e_w\left(2 - \frac{1}{2}\mathbf{Ri}\right) + C_f - P_2}{1 - \mathbf{Ri}} \tag{22}$$

$$\frac{h}{U}\frac{dU}{d\bar{x}} = \frac{-P_1 + e_w\left(1 + \frac{1}{2}\mathbf{Ri}\right) - C_f + P_2}{1 - \mathbf{Ri}} \tag{23}$$

$$\frac{dq_s}{d\bar{x}} = -r_o\frac{v_s}{U}\frac{q_s}{h}\left(1 - \frac{UEh}{r_o q_s}\right) \tag{24}$$

where

$$P_1 = \mathbf{Ri}\frac{v_s}{(1-\lambda_p)c}\frac{q_s}{Uh}\left(1 - \frac{UEh}{r_o q_s}\right) \tag{25a}$$

$$P_2 = \frac{1}{2}\mathbf{Ri}\, r_o\frac{v_s}{U}\left(1 - \frac{UEh}{r_o q_s}\right) \tag{25b}$$

The boundary conditions on the above equations are

$$h|_{\bar{x}=0} = h_o \quad , \quad U|_{\bar{x}=0} = U_o \quad , \quad q_s|_{\bar{x}=0} = q_{so} = U_o h_o C_o \tag{26a,b,c}$$

The solution of permanent form is obtained by a) specifying g, R, v_s, C_f, r_o, U_r and n, b) specifying the rollover values h_o, U_o and C_o, c) specifying a progradation rate c, d) solving (22), (23) and (24) for h, U and q_s stepwise in the downstream direction until U and C become negligibly small (the turbidity current has died), e) computing bed slope S from (7) and (19), i.e.

$$S = -\frac{d\eta_p}{d\bar{x}} = \frac{v_s(r_o C - E)}{(1-\lambda_p)c} \tag{27}$$

and f) integrating (27) for bed elevation η_p subject to the condition

$$\eta_p|_{\bar{x}=0} = \eta_r \tag{28}$$

The analysis specifies a value for basement elevation η_b for each specified value of progradation rate c and mud delivery rate q_{so} in accordance with (21).

Numerical experiments reveal that solutions for clinoforms of permanent form can be obtained from (22) ~ (26) only when the turbidity current is subcritical, i.e. $\mathbf{Ri} > 1$ everywhere.

6 SAMPLE IMPLEMENTATION

The above formulation was implemented numerically using a simple Euler step method in a spreadsheet. Two cases, Case 1 and Case 2 were chosen to illustrate the results of the calculations. The following values were used in both cases: $R = 1.65$, $v_s = 1$ mm/s, $C_f = 0.002$, $r_o = 1.5$, $\lambda_p = 0.4$, $U_r = 6$ m/s, $n = 2$ and $\eta_r = -30$ m (rollover point is at a depth of 30 m below msl).

The values of U_o, h_o, C_o and c used for Case 1 were 0.2 m/s, 6 m, 0.001 and 5×10^{-5} m/s, respectively; these yielded values for q_{so}, rollover Richardson number $\mathbf{Ri}_o = Rgq_{so}/U_o^3$ and long-term average progradation rate c_{ef} based on an assumed value of I_e of 0.00274 (corresponding to events lasting a total of one day per year) of 0.0012 m^2/s, 2.43 and 4.32 m/year respectively.

In Case 2 the values of h_o and C_o remained the same as Case 1, but U_o and c were both halved to 0.1 m/s and 2.5×10^{-5} m/s, respectively, resulting in the respective values of q_{so}, $\mathbf{Ri}_o$ and c_{ef} (again computed with I_e taking a value of 0.00274) of 0.0006 m^2/s, 9.72 and 2.16 m/year. In both cases the parameters were selected to yield an elevation drop $\eta_r - \eta_b$ of 40 m from the rollover to the basement.

The results for Case 1 are given in Figures 11a–d and 12. The profiles for elevation η and bed slope S are given in Figure 11a. The elevation profile takes the form of a truncated sigmoid, the rest of which is realized in the unmodeled zone above the rollover point. A hypothetical curve for the unmodeled zone has been added to Figure 11a. For this case 95% of the elevation drop from rollover to basement is realized 2410 m from the rollover point. The bed slope and slope angle at the rollover are 0.0130 and 0.743°, respectively. The maximum values for bed slope and slope angle are 0.0372 and 2.13°, respectively.

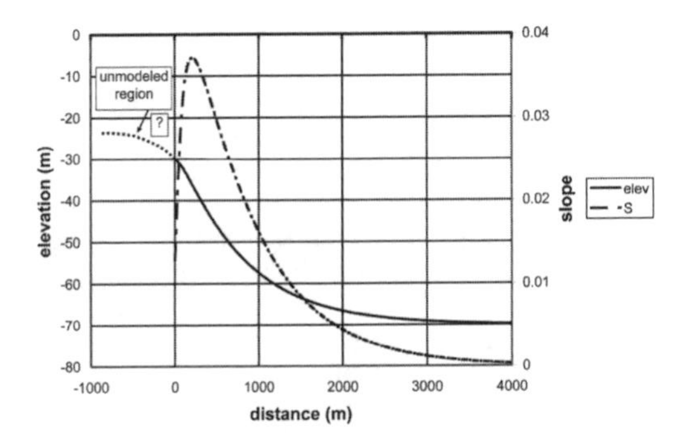

Figure 11a. Plot of bed elevation η and slope S versus x for Case 1.

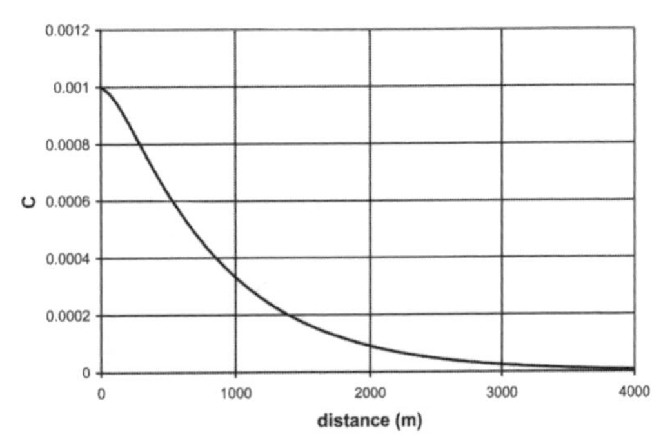

Figure 11c. Plot of volume mud concentration C versus x for Case 1.

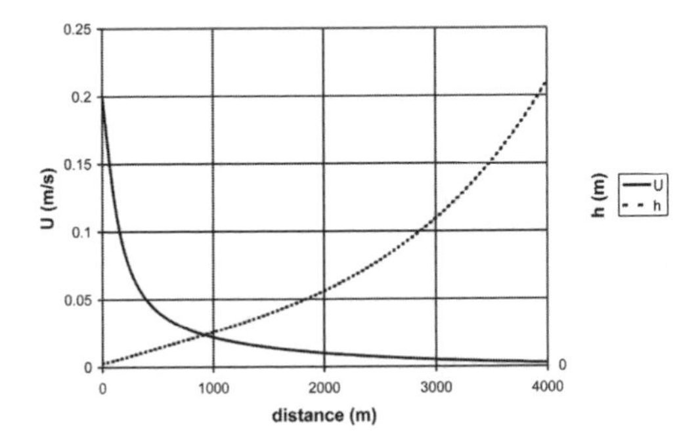

Figure 11b. Plot of flow velocity U and thickness h versus x for Case 1.

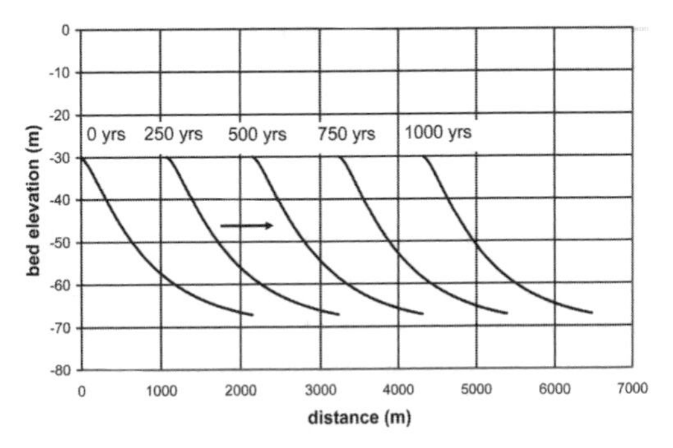

Figure 11d. Plot of clinoform progradation over 1000 years for Case 1.

Figure 11b shows profiles of U and h versus distance x; Figure 11c shows the corresponding profile for C. It is seen that flow velocity U and mud volume concentration C decline sharply away from the rollover; as the current dies its thickness h sharply increases until the current is completely dissipated. The values of h predicted for the outer region of the clinoform are unrealistically large, but in point of fact pertain to nearly-vanishing current.

Figure 11d shows the predicted evolution of the clinoform over 1000 years, during which time it progrades outward by 4320 m (based on events that are active for one day per year on average).

Resuspension of sediment plays a crucial role in the sigmoidal shape of the clinoform. Figure 12 shows a plot of the volume deposition rate of sediment per unit bed area per unit time $D_s = v_s r_o C$, the volume resuspension rate of sediment per unit bed area per unit time $E_s = v_s E$ and the bed slope S. The point of maximum bed slope S is denoted with a vertical line. Upstream of this point E_s is of the same order of magnitude as D_s, and bed slope increases downslope. Not far downstream of this point E_s becomes negligible compared to D_s, and bed slope decreases (approximately exponentially) downstream.

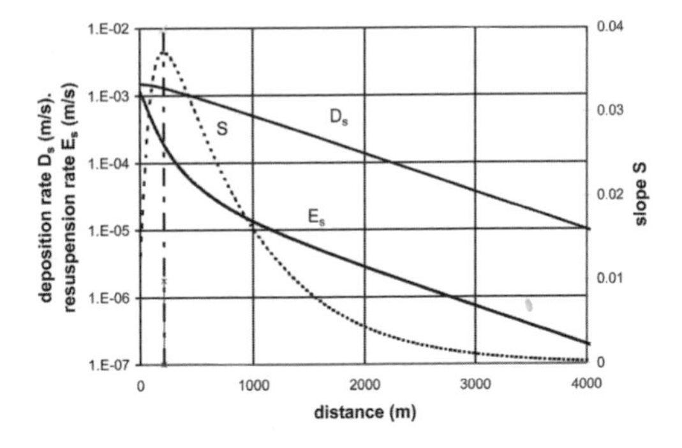

Figure 12. Plot of volume deposition rate D_s and volume resuspension rate E_s, both per unit bed area per unit time, along with bed slope S. The vertical line denotes the point of maximum bed slope. The calculations are for Case 1.

Figure 13 shows the profiles for elevation η and bed slope S for Case 2, for which both c and U_o have been halved compared to Case 1. The resulting clinoform has the same elevation drop from rollover to basement as Case 1, but is shorter and steeper. The bed slope and slope angle at the rollover are 0.0815 and 4.66°, respectively; the corresponding maximum values are 0.0931 and 5.33°. For this case, 95% of the elevation

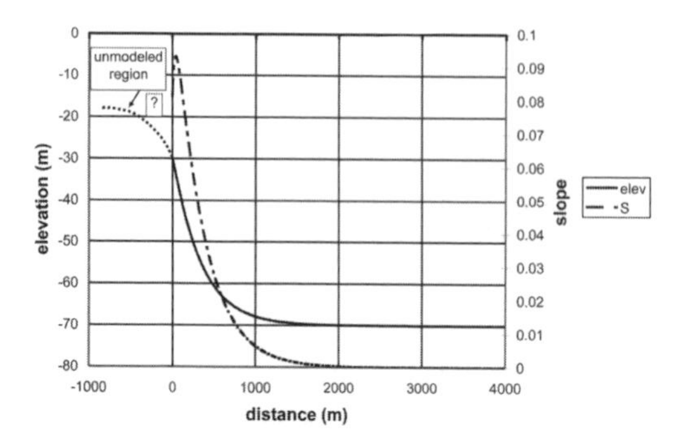

Figure 13. Plot of bed elevation η and slope S versus x for Case 2.

drop from rollover to basement is realized just 1015 m from the rollover point.

7 DISCUSSION

The theoretical/numerical model presented here is not the first to describe the evolution of clinoforms on continental margins. Pirmez et al. (1998) describe a model of clinoform evolution that relies on a combination of advection, diffusion and settling of sediment. The SedFlux model of O'Grady and Syvitski (2003) combines such diverse factors as sedimentation from plumes, turbidity currents and debris flows to describe clinoform evolution; they report that hemipelagic sedimentation from buoyant river plumes exerts the strongest control on over clinoform shape. Swenson and Pratson (2003) use an advection-diffusion model to study clinoform response to sea level change. Friedrichs and Wright (2004) and Friedrichs and Scully (in press) model clinoforms associated with the deposition of fluid muds. The present model is the first to specifically isolate the role of dilute, muddy turbidity currents (as opposed to more highly concentrated fluid muds) as an agent for muddy clinoform formation. It is likely that both dilute muddy turbidity currents and fluid mud flows both play important roles in the generation and progradation of muddy clinoforms, and thus muddy shelf-slope complexes along continental margins.

Evidence has been accumulating for the role of net-depositional turbidity currents in the construction and progradation of clinoforms. Nittrouer et al. (1996) and Kuehl et al. (1996) have emphasized the role of gravity-driven flows of both fluid muds and more dilute muddy suspensions in the modern progradation of the Amazon clinoform. Kuehl (2003) notes the following: "...although the physical mechanism may be distinct for the Amazon and Ganges-Brahmaputra, in both cases gravity-driven flows contribute directly to clinoform growth." Mohrig (2003) similarly argues for the role of turbidity currents in building sandy clinoforms.

It should be noted that while the present model is net-depositional, it is not purely depositional. The suppression of deposition by resuspension of sediment plays an important role in the upper part of the clinoform described here. The clinoform of Figure 11a shows an upper region where slope increases with seaward distance. This region vanishes, and the sigmoidal shape is lost in the limit of vanishing mud resuspension.

Sheet turbidity currents need not be net depositional, nor need they be of mud only. Izumi (2004) has described the process of incision of an active margin by net-erosional sheet turbidity currents. Kostic et al. (2002) have shown how an overriding, predominantly muddy turbidity current can reduce the slope of a sandy clinoform.

8 CONCLUSIONS

A theoretical/numerical model describing a clinoform of permanent form emplaced by weak, net-depositional turbidity currents is developed and implemented. It is shown that even in a net-depositional model, sediment resuspension must be included if the upper limb of the sigmoidal shape characteristic of clinoforms is to be obtained. The model is not intended to describe the only mechanism for clinoform progradation. It allows, however, the detailed study of a mechanism that is likely to play an important role in the formation and evolution of muddy continental slopes.

ACKNOWLEDGEMENTS

This paper is a contribution to the research effort of the National Center for Earth-surface Dynamics (NCED), a National Science Foundation (NSF) Science and Technology Center (EAR-0120914). The research was supported by both NCED and a Margins Source-to-Sink grant from NSF (EAR-0203296). Comments from R. Aller, J. Crockett, R. Geyer, C. Friedrichs, D. Mohrig, C. Nittrouer, R. Slingerland, D. Swift and P. Wiberg proved valuable in formulating the above ideas.

REFERENCES

Aller, J.Y. & Aller, R.C. 2004. Physical disturbance creates bacterial dominance of benthic biological communities in tropical deltaic environments of the Gulf of Papua. *Continental Shelf Research*, 24(19), 2395–2416.

Bard, E., Hamelin, B., Arnold, M., Montaggioni, L., Cabioch, G., Faure, G. & Rougerie, F. 1996. Deglacial sea-level record from Tahiti corals and the timing of global meltwater discharge. *Nature*, 382, 241–244.

Bell, H.S. 1942. Stratified flow in reservoirs and its use in preventing of silting. *Misc. Publ.* 491, USDA, USGPO, Washington, D.C.

Cattaneo, A., Trincardi, F., Langone, L., Asioli, A. & Puig, P. 2004. Clinoform generation on Mediterranean margins. *Oceanography*, 17(4), 105–117.

Driscoll, N., Milliman, J., Slingerland, R., Walsh, J.P. & Babcock, J. 2004. Preliminary shipboard results MV220402: clinoform sequence, stratigraphy in a modern foreland. Report, available at http://sio.ucsd.edu/png/science/results.cfm.

Friedrichs, C.T. & Wright, L.D. 2004. Gravity-driven sediment transport on the continental shelf: implications for equilibrium profiles near river mouths. *Coastal Engineering*, 51, 795–811.

Friedrichs, C.T. & Scully, M.E. in press. Modeling deposition by wave-supported gravity flows on the Po River subaqueous delta: from seasonal floods to prograding clinoforms. Continental Shelf Research.

Fukushima, Y., Parker, G. & Pantin, H. M. 1985. Prediction of ignitive turbidity currents in Scripps Submarine Canyon. *Marine Geology*, 67, 55–81.

Grover, N. & Howard, C. 1938. The Passage of Turbid Water Through Lake Mead. *Transactions* American Society of Civil Engineers, 103, 720–790.

Hill, P.S. 1998. Controls on floc size in the coastal ocean. *Oceanography*, 11, 13–18.

Izumi, N. 2004. The formation of submarine gullies by turbidity currents. *Journal of Geophysical Research*, 109, C03048, doi:10.1029/2003JC001898.

Kostic, S., Parker, G. & Marr, J. G. 2002. Role of turbidity currents in setting the foreset slope of clinoforms prograding into standing fresh water. *Journal of Sedimentary Research*, 72(3), 353–362.

Kuehl, S.A., Nittrouer, C.A, Allison, M.A. Faria, L.E.C., Dukat, D.A., Jaeger, J.M., Pacioni, T.D., Figueiredo, A.G. & Underkoffler, E.C. 1996. Sediment deposition, accumulation, and seabed dynamics in an energetic fine-grained coastal environment. *Continental Shelf Research*, 16(5/6), 787–815.

Kuehl, S.A., Levy, B.M., Moore, W.S. & Allison, M.A. 1997. Subaqueous delta of the Ganges-Brahmaputra river system. *Marine Geology*, 144: 81–96.

Kuehl, S. 2003. Role of mass movement in shelf clinoform growth; the Amazon and Ganges-Brahmaputra examples. *Abstracts with Programs*, Geological Society of America, 35(6), 624.

Mohrig, D. 2003. Reconstructing processes associated with turbidity currents building sandy clinoforms in the Cretaceous Ferron Sandstone, Utah. *Abstracts with Programs*, Geological Society of America, 35(6), 625.

Mulder, T. & Syvitski, J.P.M. 1995. Turbidity currents generated at river mouths generated during exceptional discharge to the world oceans. *Journal of Geology*, 103, 285–299.

Nittrouer, C.A., Kuehl, S.A., Figueiredo, A.G., Allison, M.A., Sommerfield, C.K., Rine, J.M., Faria, L.E.C & Silveira, O.M. 1996. The geological record preserved by Amazon shelf sedimentation. *Continental Shelf Research*, 16(5/6), 817–841.

O'Grady, D.B, Syvitski, J.P.M., Pratson, L.F. & Sarg, J.F. 2000. Categorizing the morphologic variability of siliciclastic passive continental margins. *Geology*, 28(3), 207–210.

O'Grady, D.B. & Syvitski, J.P.M. 2003. Sensitivity of clinoform geometry to sedimentary forcing mechanisms; insights from numerical models. *Abstracts with Programs*, Geological Society of America, 35(6), 626.

Paola, C., Heller, P. L. & Angevine, C. L. 1992. The large-scale dynamics of grain-size variation in alluvial basins. I: Theory. *Basin Research*, 4, 73–90.

Parker, G., Fukushima, Y. & Pantin, H.M. 1986. Self-accelerating turbidity currents. *Journal of Fluid Mechanics*, 171, 145–181.

Pirmez, C., Pratson, L.F. & Steckler, M.S. 1998. Clinoform development by advection-diffusion of suspended sediment: modeling and comparison to natural systems. *Continental Shelf Research.*

Pratson, L.F. & Haxby, W.F. 1997. Panoramas of the seafloor. *Scientific American*, 276(6), 82–87.

Smith, W.O., Vetter, C.P., Cummings, G.B. & others. 1954. Comprehensive survey of Lake Mead: 1948-1949. *Professional Paper* 295, U.S. Geological Survey, 253 p.

Swenson, J. & Pratson, L.F. 2003. Clinoform response to sea level; phase relations between shoreline and rollover, development of compound clinoforms, and the timing of margin progradation. *Abstracts with Programs*, Geological Society of America, 35(6), 626.

Twichell, D.C., Cross, V.A., Hanson, A.D., Buck, B.J., Zybala, J.G. & Rudin, M.J. 2005. Seismic generation and lithofacies of turbidites in Lake Mead (Arizona and Nevada, U.S.A.), an analogue for topographically complex basins. *Journal of Sedimentary Research*, 75(1), 134–148.

Winterwerp, J.C. & Kranenburg, C. 2002. Fine Sediment Dynamics in the Marine Environment. *Proceedings in Marine Science* 5, Elsevier Science, Amsterdam, 713 p.

River, Coastal and Estuarine Morphodynamics: RCEM 2005 – Parker & García (eds)
© 2006 Taylor & Francis Group, London, ISBN 0 415 39270 5

Internal structure of massive division in sediment gravity flow deposits visualized by grain fabric mapping

H. Naruse & F. Masuda
Department of Geology & Mineralogy, Graduate school of Science, Kyoto University
KitashirakawaOiwakecho, Sakyo-ku, Japan

ABSTRACT: This study investigated microscopic feature of the massive structure in experimental and natural turbidites and debris-flow deposits. For analysis of the hidden internal structure of massive sedimentary units, we used a grain fabric mapping method. The method was applied to the analysis of the experimental and natural gravity flow deposit, revealing a number of characteristic features of this type of massive sedimentation. Although all the samples appear structureless by macroscopic observation, the grain fabric map of the debris-flow deposits reveals a range of sedimentary features. On the other hand, grain-fabric maps of turbidites are quite monotonous. All turbidites show uniformly gentle imbrication and no internal layers or thrusts can be observed. This analysis reveals that massive gravity flow deposits actually contain a range of distinctive structures, which are important for solving the problems of distinguishing the deposits of sandy debris flows and high-concentration turbidity currents.

1 INTRODUCTION

Sediment gravity flow is a critical factor for understanding submarine sediment transport processes and evolution of the depositional system in the deep-sea environment, and is a major pathway to transport sediments of various grain-sizes from the subareal – shallow marine environment to the deep-sea depositional system. Because the sediment gravity flow encompasses many different flow types (Mulder & Alexander, 2001), various types of deposits are observable in the deep-sea environment.

Among gravity flow processes, high-density sediment gravity flows, which often results macroscopically featureless (massive) deposits, represent a major sediment transport process in the deep sea environment. Despite the common occurrence of massive sandy units (Stow & Johansson, 2001), there remains some controversy as to the sedimentary processes that result in their formation. Rapid grain-by-grain deposition of high-density turbidity current (Lowe 1988; Arnott & Hand 1989; Allen 1991) and "freezing" deposition of debris flow (Postma 1986; Shanmugam 2000) have been proposed as processes forming massive structures, the true depositional mechanism is impossible to determine by conventional observations because massive structures by definition do not exhibit characteristic features and thus appear similar regardless of the depositional process (Sohn, 1997). Because competences of sediment transportation are largely different between turbidity currents and debris-flows, their recognition is important problem in deep sea sedimentology.

The macroscopically featureless nature of massive structures is, of course, the reason why such structures are so difficult to analyze. Massive structures are formed by depositional processes that do not impart any grain sorting, and as such do not produce visible sedimentary features that may aid interpretation. Therefore, to find traces of such "invisible" processes, depositional properties other than grain size should be examined.

A new visualization method is applied in the present study to deposits produced under controlled depositional conditions and deposits collected from natural strata resulted from submarine sediment gravity flow deposits for comparison. The technique, called grain fabric mapping, elucidates the "invisible" sedimentary structure by measuring local oscillations of the grain fabric rather than grain size or color (Naruse & Masuda, in press). Grain fabric is defined here as the tendency of orientation of grain elongation axes. The method is implemented by automated digital image analysis of thin sections for grain fabric measurement. Automated analysis has been made possible by recent progress in image analysis, allowing numerous grains to be analyzed over a large area (e.g. van den Berg et al. 2003).

2 MATERIAL AND METHOD

Four samples, (1) the experimental debris-flow deposit, (2) experimental turbidite, (3) massive sandstone and (4) high-density turbidite were examined in this study. Samples 1 and 2 were produced by the flume experiments, and samples 3 and 4 were collected from strata, the Cretaceous Nemuro Group, distributed in the eastern Hokkaido, northern Japan (Naruse, 2003).

2.1 *Sample descriptions*

2.1.1 *Sample 1 – Experimental debris-flow deposit*

Debris-flow was generated in an experimental channel, which was 2 m long, 0.25 m wide and 0.35 m deep (Fig. 1). Channel sloped 30°, and its opened mouth was on the board sloped 10°, which is 1.8 m long, 0.9 m wide. A lock gate was set at the middle part of the channel. Tap water was mixed with about 70 w.% sediments in the upstream side of the lock gate. Sediment used was artificial sand with small amount of natural beach sand consisting mainly of quartz particles. The sediment and water mixture was well dispersed by hand until the sediment was completely liquidized. Then, the gate was unlocked and the debris-flow was released from the upstream end of the channel. Debris-flow moves downslope in the channel, and spread over the board after it reached mouth of the channel.

Debris-flow moved on the board for 9 seconds. Velocity of the flow head was measured from digital video equipments. Initial velocity was about 1.47 m/s and it decayed with following an exponential function. Mean velocity was 0.09 m/s. After forming 1.05 m long and 0.55 m wide lobe of the sediment, the flow stopped.

Thickness of the lobe was about 2 cm in average. Many wrinkles, which formed during flow movement, were observed on the surface of the residual sediment lobe. Wrinkles align transverse direction to the down-current direction in the central part of the lobe, and they were arranged radially in marginal area. Internal structure in the lobe was completely massive. No lamination was observed in the vertical section of the lobe, and both normal and inverse grading were not obvious.

2.1.2 *Sample 2 – Experimental turbidite*

This sample was collected from the experimental turbidite. Turbidity currents were generated in an experimental flume of Osaka University; it is 10 m long, 0.2 m wide and 0.5 m deep. A clear small tank with a lock gate was set at the upstream end of the flume. Sediment used was artificial sand consisting mainly of quartz particles (almost 100%). The sediment and water mixture was well dispersed by hand until the sediment was completely suspended. Tap water was mixed with about 10 vol.% sediments in the closed tank. However, average density of the released flows was

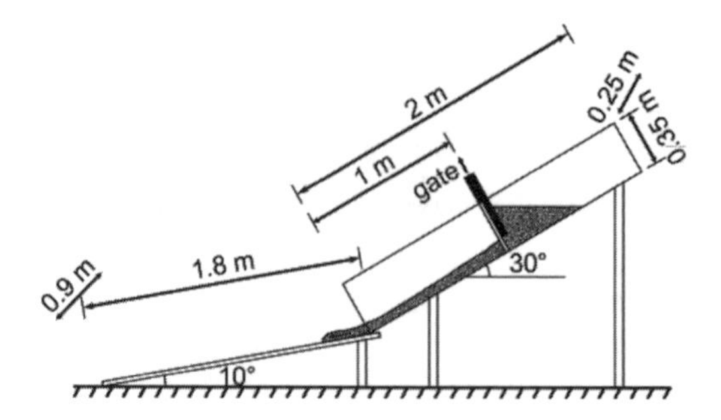

Figure 1. Channel and board for simulation of debris flow.

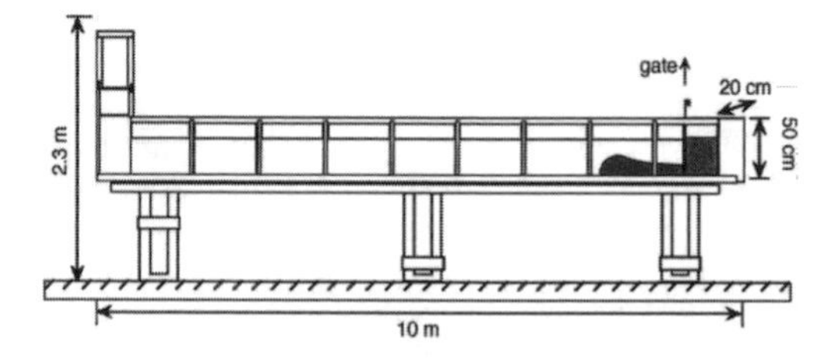

Figure 2. Experimental flume for generating turbidity currents.

less than 10% because complete mixing of sediment with water is technically impossible. Accumulation of the rest sediments inside the tank suggests that higher density of sediment and water mixture might be created in the basal part of the tank because of sediment settling during a short-time lag between stopping of the mixing and unlocking the gate; density of the head was probably higher than 10% and might attain that of high-density flows as reported by Middleton (1967).

When the gate was unlocked, the turbidity current was released from the tank. The flow reached near the sampling points (200 cm downstream points) about 5 seconds after the gate was unlocked and head velocity around the site was about 0.3 m/s. The flow had a denser layer appeared at about 20 cm behind the head in the basal part of the flow. The thickness of the basal density layer was about 5–10 mm.

Accumulated turbidites most thicken at the ca. 60 cm downstream point from the gate, and then thin toward down-current direction. Each turbidite bed has a distinct graded massive division. It is followed by faint parallel lamination near the top of the bed in a few beds. An inversely graded unit less than 1 mm thick was recognized at the bed base.

2.1.3 *Sample 3 – massive sandstone*

This sample was collected from poorly sorted massive coarse sandstone bed that includes granules and large deformed mud clasts (maximum 50 cm in thickness, 3 m in length), which deposited in the proximal basin plain facies association (Naruse, 2003). It represents the environment close to foot of the slope, where channel-levee system did not develop. The strata show flat geometry there.

The beds range from 0.3 to 3 m in thickness. They usually extend laterally with uniform thickness. For example, one sandstone bed belonging to this facies is correlated by a key tuff bed, and is traceable between two areas separated by 6 km distance. Thickness of the sandstone bed changes only 12% within 6 km distance. Many groove casts are observed on the bottom surface of this facies.

The facies of the sampled bed is interpreted as a sediment gravity flow deposit (Lowe, 1982; Mulder & Alexander, 2001). The sediment gravity flow is inferred to have had a high particle concentration (Shanmugam, 1996; Kneller and Buckee, 2000; Mulder & Alexander, 2001) because abundant granules and coarse sands in this facies must have grain-supporting mechanisms related to the particle concentration besides turbulence (dispersive pressure, buoyancy and matrix cohesion) (Lowe, 1982). Although it is difficult to estimate paleo-flow state from macroscopic features of this facies, it is tentatively suggested that this facies represents "sandy debris-flow" deposits (Shanmugam, 1996) because this facies often includes large intra-basinal clasts, which seems to be too fragile in the turbulent state flow to preserve.

2.1.4 *Sample 4 – High-density turbidite*

This sample was collected from a poorly sorted pebbly coarse sandstone bed. Sampling was conducted from completely massive division. The facies ranges from 30 cm to 10 m in thickness, and occurs only in channel-levee complex facies association (Naruse, 2003). The channel-levee complex facies association is developed in the foot of slope – lower slope environment, which is composed of channel-fill deposits showing upward fining succession and spilt-over deposits (levee). Most of beds in this facies are deposited in the channel environment, and appear to be massive. Many mud clasts of less than 10 cm in diameter are included in this facies. The facies overlies the underlying mudstone with an irregular erosional surface, and its bottom surface shows flute casts.

The facies of the sampled bed is interpreted as a turbidite (Middleton & Hampton, 1976; Lowe, 1982). Graded bedding, a characteristic feature of turbidites, is common in this facies. Flute marks on the bottom surface and other erosional features substantiate that the sediment gravity flow was turbulent state (Dzylynski & Sanders, 1962). The turbidity current must have a high particle concentration, because transportation of abundant granules and coarse sands in this facies needs grain-supporting mechanisms related to high particle concentration (i.e. dispersive pressure, buoyancy and matrix cohesion) (Middleton & Hampton, 1976; Lowe, 1982). Therefore, this facies can be interpreted as a highly concentrated (high-density) turbidite (Lowe, 1982).

2.2 *Fabric measurements*

Although digital image analysis for grain identification and measurement is not new (e.g. Russ 2002), recent advances in digital technology have lead to remarkable progress in the sophistication of image analysis, and it is now possible to determine a range of grain properties such as size, shape, orientation and spatial arrangement with a high degree of automation (Francus 1998). In the visualization method presented in this study, the grain fabric is characterized by measuring the direction of grain elongation in images. As grain analysis by the present method employs existing imaging techniques, it is described here only briefly for completeness. The analysis process involves sample preparation, image acquisition, image processing (preprocessing, classification and postprocessing), and image analysis.

2.2.1 *Sample preparation and image acquisition*

The unconsolidated samples were prepared as thin sections for microprobe observation. Sample surfaces were cemented using rapidly solidifying low-viscosity glue. The samples were then carefully removed from the channel or mouth slope, dried completely, and impregnated slowly and carefully with epoxy resin. The completely cemented samples were then split into specimens for measurement of grain imbrication (vertical flow, parallel section).

Thin sections were observed and photographed by scanning electron microscopy (SEM; JSM-6100, JEOL). Back-scattered electron (BSE) images were used for analysis. SEM observations were made at a beam accelerating voltage of 25 keV, with a tilt angle of and working distance of 34 mm. Images were obtained in 256-level grayscale from the SEM video output channel at a magnification of 50× and resolution of 1365 × 1024 pixels (594 pixels = 1 mm). The observations were repeated over the entire area of the sample, and a final mosaic image of the entire thin section was produced by joining images geometrically.

2.2.2 *Image processing*

A binary image was produced from the 256-level grayscale BSE image mosaics, in which black pixels represent the matrix and white pixels represent clastic grains. The image processing procedures involved preprocessing, classification, and postprocessing steps (described below), and all processing was performed on a computer using the public-domain ImageJ program (http://rsb.info.nih.gov/ij/).

As a preprocessing step, image filtering (median filter) was applied to the acquired mosaic images to reduce random electronic noise introduced during BSE image acquisition. Image classification was then performed by dividing the image into regions corresponding to grain sections and the matrix. With regard to sample 1 and 2, the resin appeared distinctively darker than the clastic grains in the BSE images and the boundaries between the matrix and grains were clearly apparent due to the low atomic weight of the epoxy resin fixing the sediment matrix. On the other hand, matrices of sample 3 and 4 are not enough distinctive to analyze automatically (Fig. 3), so that their BSE images were manually traced to separate grains from matrices. The classified mosaics were then converted to binary images.

Touching grains in the classified binary images were split by postprocessing. Grain contact, resulting mainly from overlapping projections of grains, noise and limited image resolution, is a serious problem in automatic grain analysis, and several algorithms for separating touching grains have been proposed (e.g. van den Berg et al. 2002). Opening process (Russ 2002) was found to be sufficient to separate merged grains in the BSE images obtained in the present study. The opening process combines erosion (removing grain pixels touching matrix pixels) followed by dilation (adding grain pixels adjacent to matrix pixels), and is effective for opening gaps between features that are just in contact. After automatic postprocessing, the images were checked visually and manual modifications applied where necessary to correct grain shapes.

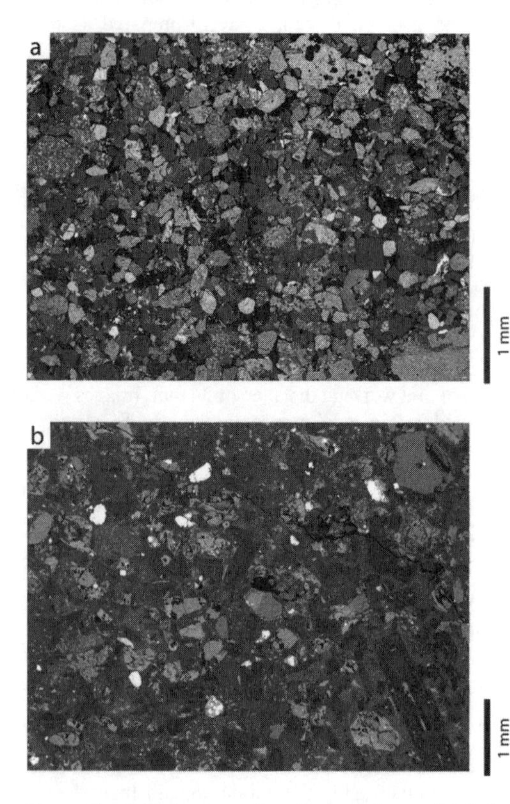

Figure 3. BSE images of vertical sections of (a) the high-density turbidite (sample 4) and (b) massive sandstone (sample 3).

2.2.3 *Image analysis*

The location and elongation direction of grains were obtained as quantitative image analysis data. The location was taken as the average of the x and y coordinates of all white pixels (grain pixels), and the elongation direction was obtained as the angle between the primary axis of an ellipse fitted to the grain by Hough transform and the line parallel to the x axis of the image. All measured data are considered to be apparent two-dimensional characteristics of three-dimensional features. The present study therefore only examines the apparent features of the debris flow samples assuming equivalency to the three-dimensional structure. All analyses were performed automatically using the image analysis software ImageJ.

2.2.4 *Grain fabric mapping*

The grain fabric mapping method visualizes the "invisible" internal structure of massive sedimentary units as numerous small lines indicating the orientation of the grain fabric (mean direction and vector concentration of grain elongation directions) in small local regions in the thin section (Naruse & Masuda, in press). These lines are obtained automatically by image analysis, allowing a map of the grain fabric of the entire deposit to be constructed.

The grain fabric indicators are generated by statistical treatment of grain elongation directions in a circular sampling window. The mean direction $\bar{\theta}$ and orientation strength R are calculated as follows.

$$\bar{\theta} = \arctan\left(\Sigma \sin\theta_i / \Sigma \cos\theta_i\right) \quad (1)$$

$$R = \sqrt{\left(\Sigma \sin\theta_i\right)^2 + \left(\Sigma \cos\theta_i\right)^2} \quad (2)$$

where θ_i is the measured elongation direction. The Watson-Stephens U^* test (Watson 1961; Stephens 1970) is applied to this subset of data, and if the test is passed, the obtained directional is accepted as the preferred orientation. The 95% confidence interval of the mean direction is then calculated using the equation of Fisher & Lewis (1983):

$$\bar{\theta} \pm \arccos\left(1 - U_1(5)/2\kappa R\right) \quad (3)$$

where $U_1(5)$ is the upper 5% point of a χ_1^2 distribution and κ is the concentration parameter of the von Mises distribution estimated from the measured data set (Upton & Fingleton 1989). After calculation of statistic values, the map is constructed. One line is drawn in the center of each sampling window if the data exhibit a statistically significant preferred orientation, with the orientation of the line representing the mean direction and length representing the vector concentration. The sampling window is then moved slightly, and the statistical treatment and line drawing is repeated. By this process, a map of the grain fabric of the section is constructed.

The fabric mapping method was fully applied into the Sample 1 and 2. Because of incompleteness of image analysis, especially insufficient number of measured grains, the Sample 3 and 4 were impossible to be examined by this method using small circular window, so that they were examined by horizontally elongated windows to reveal vertical variation of their grain fabric.

3 RESULTS AND DISCUSSION

3.1 *Grain fabric characteristics of samples and inferences of their formational processes*

3.1.1 *Sample 1 – Experimental debris-flow deposit*

The grain fabric mapping method employed in this study revealed grain fabric of the debris flow deposit varies remarkably even within the same thin section. Sample 1 displayed a steepening-upward tendency of grain imbrication angle (Fig. 4), where grains are oriented nearly parallel to the bedding plane at the based of the section (where the vertical velocity gradient during debris flow would have been highest), and imbricated with a high upcurrent angle (about 45°) in the middle and upper horizons.

As a whole, grains in this sample exhibit an obvious imbrication (Fig. 4). The imbrication angle tends to steepen upward, and the section can be divided vertically into three parts based on the mode of imbrication. In the lowermost part (0–3 mm), grains imbricate with a gentle angle (0–30°) that tends to steepen upward. In the intermediate part (3–14 mm), grains are uniformly imbricated with a steep angle (about 45°). In the uppermost part (14–18 mm), the grains form various fabric patterns, with most imbricated with steep angles (45–90°) but some regions exhibiting no preferred imbrication (not statistically significant). These three horizons are separated by gradual transitions.

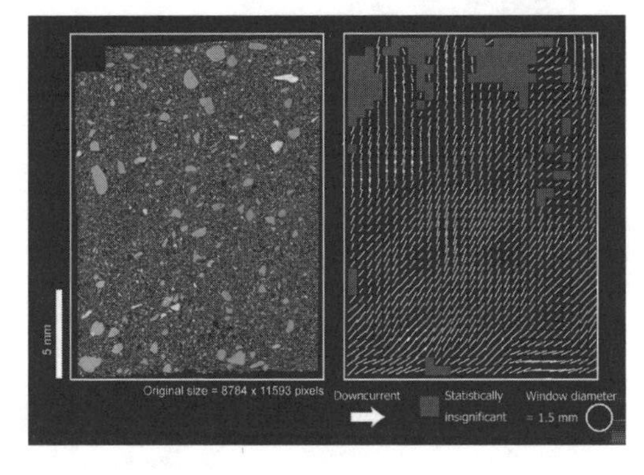

Figure 4. (a) BSE image and (b) grain fabric map of the experimental debris-flow deposit (Sample 1). Detailed explanation is in the text.

Two simplified theoretical models have been proposed to explain the formation of preferred grain fabric in debris flow deposits: a grain/grain interaction model (Rees 1968) and a grain rotation model (Lindsay 1968). It should be noted that it is difficult to predict the behavior of granular materials in fluids precisely due to the large number of factors involved, including fluid movement, grain/fluid interaction, grain collision and grain friction (Middleton & Hampton 1976; Lowe 1982; Mulder & Alexander 2001).

In the grain/grain interaction model proposed by Rees (1968), grain fabric formation in debris flow (grain flow) is governed by grain collision. As grains in the upper region of laminar flow are transported faster than those in the lower region, grains tend to collide with grains lower in the flow, resulting in the orientation of the mean collision axis in an upcurrent-upper direction. According to the experimental results of Bagnold (1954), the mean collision axis in sheared grain flow is inclined 30–40° upcurrent (Taira 1989), irrespective of the flow shear rate or position in the flow. As a result, grains in debris flow are theoretically expected to show relatively high-angle (30–40°) upcurrent imbrication. Indeed, Taira (1989) reported that experimental debris flow deposits exhibit higher imbrication angles (15–35°) than traction sedimentation deposits (10–20°). This high upcurrent imbrication angle (~45°) was also observed in the middle and upper horizons of Sample 1 in the present study, supporting the grain/grain interaction model.

However, it is also known that grains in laminar flow can acquire preferred orientation without grain collision. Lindsay (1968) calculated the rotation of ellipsoidal grains in laminar flow using the theory of Jeffrey (1922) in an analysis of the grain fabric of cohesive debris flow deposits. In this case, grains rotate vertically due to the vertical velocity gradient of laminar flow, and the rotation velocity is minimum when the long axes are oriented parallel to the flow direction. Therefore, the long axes of grains in adequately sustained flow tend to become oriented flow parallel with slight oscillations. Such flow-parallel orientation fabrics have been reported for several debris flow deposits (Lindsay 1968; Enos 1977), and the fabric maps in the present paper also show that grains in the lower horizon of Sample 1 are oriented parallel to the flow direction. The high particle density is considered to induce nearly continuous grain contact and therefore hinder grain collision.

Both grain collision and rotation can be expected to occur in the debris flow, and the properties of the debris flow may therefore vary locally even in single-layered flow due to variations in the particle density and velocity gradient. Most debris flows are considered to exhibit intermediate properties between these two end-member flow types, and thus fabric formation can be expected to occur by a combination of these mechanisms. Indeed, the experimental debris flow analyzed in the present paper exhibits a heterogeneous grain-fabric even though it should exhibit primarily inertia debris flow characteristics due to the scarcity of mud as a matrix. It therefore appears that the presence of such a range of structures in the grain fabric map reflects the internal heterogeneity of flow properties, such as velocity gradient and particle concentration.

3.1.2 *Sample 2 – Experimental turbidite*

The grain fabric mapping method illustrated that grain fabric of the experimental turbidite are quite homogeneous (Fig. 5). Sample 2 showed a monotonous tendency of grain imbrication angle, where grains are monotonously imbricated with a gentle upcurrent angle in the whole area of the thin section.

Sampled sand bed was 8.4 mm thick. In the vertical section, 13,358 grains were measured. Total result of fabric analysis showed that this deposit had obvious imbrication. The grain fabric map shows that a tendency of the imbrication overall in the bed is "massive". Measurements of all intervals show moderate value (8–15°) of the imbrication angle (Fig. 5).

Turbidite is the deposit from turbulent sediment gravity flow, in which the bottom surface rises continuously by independent settling of particles (Kneller & Branney, 1998). Grain and/or fluid interactions are negligible there because turbidite is not freezed deposits but deposited grain-by-grain. Thus, imbrication of grains in turbidite is acquired mainly by different process from those of the debris-flows.

As a result of grain-by-grain deposition, grains are not packed perfectly horizontal, so that settled grains dip either up-current or down-current. The down-current dip position is less stable because the lift and drag forces will tend to flip over the particle (Middleton, 1977). Thus, tangential shear stress from overlying flow imbricates the grains up-current (Middleton & Southard, 1977). In this process, acquired imbrication angle depends on relief of packed grains, and is probably restricted in relatively low angle. Arnott & Hand (1989) experimentally produced massive sand layer which deposited from grain-by-grain sedimentation. They reported 10–20° grain imbrication from their experimental deposits.

Grain fabric analysis of this study is concordant with inference of fabric formational process described above. Continuous rise of the imbrication forming layer (bed-flow boundary) results the bed that is homogeneously composed of imbricated particles from the bottom to the top.

3.1.3 *Sample 3 – Massive sandstone*

Sampled sandstone bed was 150 mm in thickness. Number of measured grains was 16,643 in the vertical section. Total result of fabric analysis showed that grain fabric of this deposit was dominated by up-current imbrication. Vector mean of the imbrication showed relatively high angle (12.84°), and vector concentration percentage was relatively low (9.044%).

Collected data were analyzed to test upward changes in the vector mean of imbrication angle by every vertical interval of 3.6 mm thick, and were diagrammed (Fig. 6). The diagram suggested that a bed

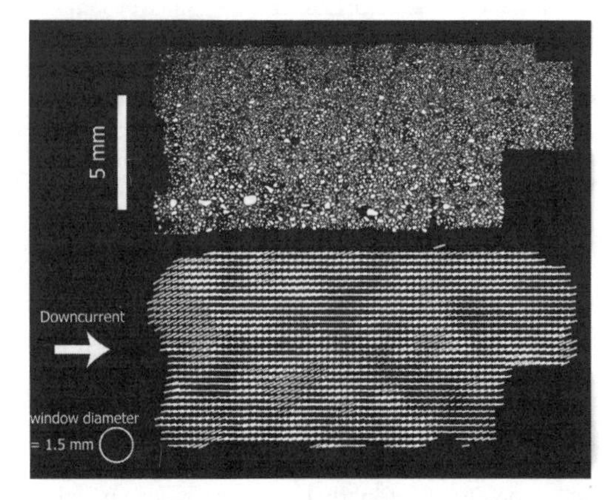

Figure 5. (a) BSE image and (b) grain fabric map of the experimental turbidite (Sample 2). Detailed explanation is in the text.

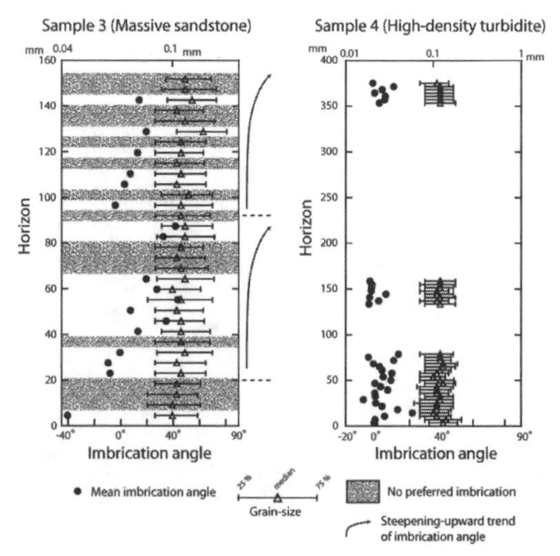

Figure 6. Upward changes of grain-size and vector mean of imbrication angle in the massive structure.

can be divided into two different segments: (1) lower, (2) upper parts. (1) The lower part (0–20 mm) showed nearly chaotic fabric. There were no vertical trends in imbrication angle both by visually and statistically. Vector concentration was generally low (5–14%). (2) In the upper part, most of intervals showed statistically significant preferred imbrication. There are two subsets (20–90, 90–150 mm), which of steepening-upward trend in vector mean angle, and vector concentration keeps relatively high value (12–25%).

Although the origin of lower chaotic part is unknown, steepening-upward trend of grain imbrication is quite resembles to the result of Sample 1, the experimental debris-flow. Heterogeneity and relatively steep angle of grain imbrication may suggests that this sample was deposited from laminar-state flow by "freezing" sedimentation.

3.1.4 *Sample 4 – High density turbidite*

Sampled sandstone bed was 375 mm thick. 10,249 grains were measured in the horizontal and vertical section respectively. Total result of fabric analysis shows that high density turbidite shows up-current imbrication. Vector mean of the imbrication angle shows very low value (0.471°), and vector concentration percentage is moderate (25.836%).

Collected data were analyzed to test upward changes in the vector mean of imbrication angle by every vertical interval of 3.6 mm thick, and were diagrammed (Fig. 6). The diagram suggests that modes of the imbrication are similar overall in the bed. Measurements of almost all intervals show low value (−2–5°) of the imbrication angle, and those of vector concentration percentage are always very high (20–40%).

Monotonous variation and low-angle grain imbrication are completely same characteristics of the experimental turbidite, thus suggesting grain-by-grain deposition. Lower angle of imbrication of this sample may be produced by compaction after deposition.

3.2 *Debris-flow and turbidite types of grain fabric*

Present study revealed various structures in macroscopically structureless deposits, suggesting efficiency of the grain fabric mapping method to study massive sediment gravity flow deposits.

Although results of present study are preliminary because of limited numbers of samples, there seems to be 2 types of grain fabric in sediment gravity flow deposits: the debris-flow deposit and turbidite types. Debris-flow deposit type is characterized by heterogeneous grain fabric and relatively high-angle imbrication. Grain fabrics of Sample 1 and 3 are classified into this type. Heterogeneous grain fabric is probably reflects heterogeneity in the property of laminar-state particulate flow such as particle concentration and/or velocity gradient. High-angle imbrication is a feature of fabric formational processes in the debris-flow. On the other hand, the turbidite type of grain fabric is characterized by monotonous and relatively low-angle imbrication angle. Grain fabrics of Sample 2 and 4 are classified into this turbidite type. Their characteristics are indicative of imbrication forming processes effective in grain-by-grain deposition (Arnott & Hand, 1989).

4 CONCLUSION

Four samples of macroscopically structureless sediment gravity flow deposits, experimental debris-flow deposit, turbidite, massive sandstone and high-density turbidite, were microscopically analyzed to reveal their "invisible" internal structure. Transverse thin sections were prepared and imaged by SEM. Grain fabric mapping was then performed automatically using the processed back-scattered electron images. Although the massive sedimentary samples appeared structureless by macroscopic observation, the grain fabric map revealed various sedimentary structures, including a steepening-upward trend of grain imbrication angle in debris-flow deposit and massive sandstone samples with very low angle imbrication in the basal horizon indicative of high shear rate flow. In contrast, the experimental turbidite and high-density turbidite, which collected from strata, show quite monotonous fabric maps, indicating low-angle upcurrent grain imbrication. Thus, the grain fabric mapping technique has the ability to visualize a range of sedimentary structures that are not observable with the naked eye or by analyses of grain size or color. As such, grain fabric mapping appears to be particularly useful for the analysis of massive deposits, and the method proved to be effective for the discrimination of sediment gravity flow deposits produced by different processes.

ACKNOWLEDGEMENT

Gratitude is extended to the staff and researchers of the Department of Geology & Mineralogy, Kyoto University, and in particular to Dr. H. Maeda for guidance throughout this study. Dr. T. Sakai and M. Yokokawa are also gratefully acknowledged for helpful discussion and advice. This research was supported in part by a Grant-in-Aid for Young Scientists (B) (16740287, 2004) from the Ministry of Education, Science, Sports, Culture and Technology of Japan.

REFERENCES

Allen, J.R.L., 1991. The Bouma division A and the possible duration of turbidity currents. *Journal of Sedimentary Petrology* 61: 291–295.

Arnott, R.W.C. & Hand, B.M., 1989. Bedforms, primary structures and grain fabric in the presence of suspended sediment rain. *Journal of Sedimentary Petrology* 59: 1062–1069.

Bagnold, R.A., 1954. Experiments on a gravity-free dispersion of large solid sheres in a Newtonian fluid under shear. *Proceed. Roy. Soc. London Ser. A. Math. Phys. Sci.* 225: 49–63.

Dzulynski, S. & Sanders, J.E., 1962. Current marks on firm mud bottoms. *Trans. Connecticut Acad. Arts and Sci.* 42: 57–96.

Enos, P., 1977. Flow regimes in debris flow. *Sedimentology* 24: 133–142.

Francus, P., 1998. An image-analysis technique to measure grain-size variation in thin sections of soft clastic sediments. *Sedimentary Geology* 121: 289–298.

Jeffery, G.B., 1922. The motion of ellipsoidal particles immersed in a viscous fluid. *Proceed. Roy. Soc. London Ser. A. Math. Phys. Sci.* 102: 161–179.

Kneller, B.C. & Buckee, C., 2000. The structure and fluid mechanics of turbidity currents. A review of some recent studies and their geological implications. *Sedimentology* 47: 62–94.

Kneller, B.C. & Branney, M.J., 1995. Sustained high-density turbidity currents and the deposition of thick massive sands. *Sedimentology* 42: 607–616.

Lindsay, J.F., 1968. The development of clast fabric in mudflows. *Journal of Sedimentary Petrology* 38: 1242–1253.

Lowe, D.R., 1982. Sediment gravity flows; II, Depositional models with special reference to the deposits of high-density turbidity currents. *Journal of Sedimentary Petrology* 52: 279–297.

Lowe, D.R., 1988. Suspended-load fallout rate as an independent variable in the analysis of current structures. *Sedimentology* 35: 765–776.

Middleton G.V. & Hampton M.A., 1976. Subaqueous sediment transport and deposition by sediment gravity flows. In Stanley D.J. & Swift D.J.P. (ed.), *Marine sediment transport and environmental management*: 197–218. New York: John Wiley & Sons.

Middleton, G. V. & Southard, J. B., 1977. Mechanics of sediment movement. In Middleton & Southard (ed.), *SEPM Short Course* 3. Tulsa: Soc. Sed. Geol.

Mulder, T. and Alexander, J., 2001. The physical character of subaqueous sedimentary density flows and their deposits. *Sedimentology* 48: 269–299.

Naruse, H., 2003. Cretaceous to Paleocene depositional history of North-Pacific subduction zone; reconstruction from the Nemuro Group, eastern Hokkaido, northern Japan. *Cretaceous Research* 24: 55–71.

Postma, G., 1986. Classification for sediment gravity-flow deposits based on flow conditions during sedimentation. *Geology (Boulder)* 14: 291–294.

Rees, A.I., 1968. The production of preferred orientation in a concentrated dispersion of elongated and flattened grains. *J Geol* 76: 457–465.

Russ, J.C., 2002. *The image processing handbook – 4th ed.* Florida: CRC Press.

Shanmugam, G., 2000. 50 years of the turbidite paradigm (1950s–1990s); deep-water processes and facies models; a critical perspective. In Stov D.A.V. & Mayall M. (ed.), *Deep-water sedimentary systems; new models for the 21st century*: 285–342. Oxford: Pergamon.

Sohn, Y.K., 1997. On traction-carpet sedimentation. *Journal of Sedimentary Research* 67: 502–509.

Stow, D.A.V. & Johansson, M., 2000. Deep-water massive sands; nature, origin and hydrocarbon implications. In Stow D.A.V. & Mayall M. (ed.), *Deep-water sedimentary systems; new models for the 21st century*: 145–174. Oxford: Pergamon.

Taira, A., 1989. Magnetic fabrics and depositional processes. In Taira A. & Masuda F. (ed.), *Sedimentary facies in the active plate margin*: p. 43–77. Tokyo: Terra Sci. Publ. Co.

Upton, G.J.G. & Fingleton, B., 1989. *Spatial data analysis by example – volume 2 categorical and directional data.* New York: John Wiley & Sons.

van den Berg, E.H., Bense, V.F. & Schlager, W., 2003. Assessing textural variation in laminated sands using digital image analysis of thin sections. *Journal of Sedimentary Research.* 73: 133–143.

River, Coastal and Estuarine Morphodynamics: RCEM 2005 – Parker & García (eds)
© 2006 Taylor & Francis Group, London, ISBN 0 415 39270 5

Dynamics of sand sedimentation resulting from turbidity currents caused by explosive submarine volcanic eruptions

A. Cantelli & G. Parker
Dept. of Civil & Environmental Engineering & Dept. of Geology, University of Illinois, Urbana, Illinois, USA

S. Johnson
Dept. of Civil Engineering, University of Minnesota, Minneapolis, Minnesota, USA

J.D.L. White
Geology Department, University of Otago, Dunedin, NZ

B. Yu
Chengdu Inst. of Mountain Disaster and Envir., Chengdu, China

ABSTRACT: Deposits of turbidity currents induced by volcanic eruptions under water are increasingly recognized in settings ranging from lakes to the deep sea, and have been referred to as varieties of "eruption-fed density currents." The deposits result from explosive subaqueous eruptions that through condensentation and entrainment produce aqueous eruption plumes and then collapse to form turbidity currents or other density flows. These flows emplace layers of typically glassy volcanic ash. Features of deposits from these turbidity currents reflect their specific origin from volcanic eruptions that inject heat and porous volcanic particles into the water column, with both the difference in current temperature, and in particle characteristics, affecting current dynamics and deposit features. Laboratory studies on density currents performed at St. Anthony Falls Laboratory (University of Minnesota) were staged in order to address the effects of variations in grain density and water temperature on flow properties and depositional processes. Laboratory runs utilized natural grain populations from previously studied volcaniclastic deposits for which bedding features, bulk grain size and particle settling-velocity distributions are known. Corresponding runs were performed under the same flow conditions, but utilizing silica sand that was sieved to match the grain size distribution of the volcaniclastic material. Each grain type was used in a series of runs in which a hot density current enters a cold environment, and a series in which a cold current enters a cold environment. A distinction among deposits of the four sets of currents is displayed on the basis of density of the grains driven by the current and the density contrast between the current and ambient water. Experiments for which the inflow contained hot water and low-density sediment tended to place the center of mass of the deposit proximally, whereas experiments for which the inflow contained cold water and high-density sediment tended to deposit mass distally. Results of the study have special relevance to work on submarine volcaniclastic deposits, but fundamental aspects of how grain properties and water temperature affect the driving force of density currents are also illustrated.

1 INTRODUCTION

Turbidity currents are a specific subset of density currents differentiated by the fact that the density of the current is determined by the presence of suspended sediment. Many aspects of turbidity currents have previously been studied because turbidity currents are considered responsible for various morphologic features in marine and lake environments, and because of their role in forming large proportions of earth's sedimentary rock. The general concepts of density currents have been studied by many authors (e.g. Schmidt 1911; von Karman 1940; Keulegan 1957; Ellison & Turner 1959; Hinze 1960; Plapp & Mitchell 1960; Chu, Pilkey & Pilkey 1979; Lüthi 1981). Turbidity currents are driven by the downslope component of gravity as it pulls the suspended sediment, and thus the surrounding water, downstream. This suspended sediment is free to exchange with bed sediment. Depending upon conditions, turbidity currents can produce both net bed erosion and net bed deposition (Pantin 1979, Parker 1982, Garcia 1985).

In addition to the concentration of suspended sediment in the water column, the dynamics of a turbidity current generated by an underwater volcanic eruption are controlled by other parameters. The water carrying the sediment tends to be hot, and thus lighter than the ambient water. This tends to reduce the driving force of the turbidity current. In addition, particles produced by subaqueous eruptions tend to be lighter than most natural sediments, have highly angular shapes and be riddled with cavities (vesicles). Thus particle density and shape, along with water temperature, are important variables for turbidity current flow dynamics. The extreme values for both current temperatures and particle shape and density produced by explosive subaqueous volcanic eruptions make eruption-fed volcaniclastic deposits a particularly promising subject for research.

In the subaerial environment, explosive volcanic eruptions produce primary pyroclastic deposits formed from eruptive fragmentation followed by single-stage transport through the ambient atmosphere. In contrast, subaqueous explosive eruptions produce turbidity currents that are responsible for a wide range of deposit geometries of volcanic fragments. Various types of density currents generated by submarine volcanic eruptions have been described (White, 2000), with each type of current responsible for a unique set of depositional features. In particular, three main classes of density current exist, each associated with a different kind of eruption. The first comprises subaqueous pyroclastic flows, which are high-temperature and necessarily high-particle-concentration gas-solid flows driven by the excess density of the current relative to the water. They are generated by sustained explosive and gas-thrust column eruptions. The second, and most common, consists of eruption-fed turbidity currents comprising dilute to high-concentration particulate gravity flows with water as the continuous intergranular phase. These form directly both from sustained subaqueous eruptions and from intermittent tephra jets. A third set involves lava-fed density currents characterized by a continuous water phase and low to high particle concentrations generated by fragmentation of lavas during rapid effusive eruption.

The purpose of this paper is to investigate the dynamics and the resultant deposits generated by turbidity currents induced by volcanic eruptions. This is accomplished by analyzing data from a series of laboratory experiments using both regular sand and pyroclastic sediment and involving varied temperature difference between inflowing and ambient water.

2 EXPERIMENTAL SET UP

Turbidity current experiments were performed at the St. Anthony Falls Laboratory (University of Minnesota) in a 0.2 m wide and 9.5 m long channel suspended inside a larger glass-walled tank (main tank; Figure 1). The channel slope was fixed at 6 degrees. Dimensions of the exterior tank are 10 m long, 3 m high and 0.6 m wide. Experiments were performed by releasing turbidity currents from a head tank with a capacity of 200 liters by way of an automated gate with a width of 0.15 m and a height of 0.11 m. The head tank was filled with a slurry of water and sediment which was released to form the turbidity current. Before release it was constantly agitated in order to maintain the sediment suspension.

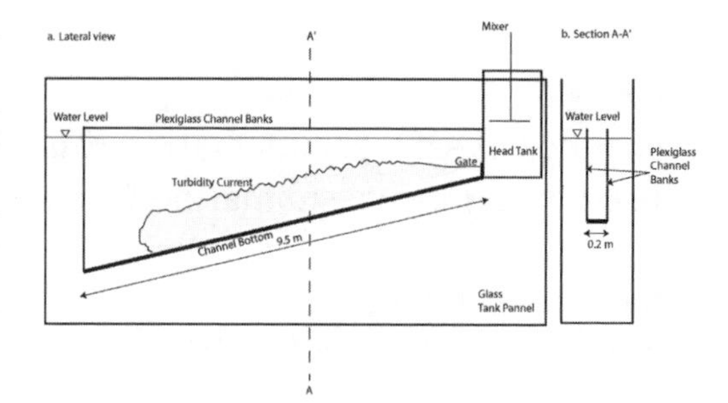

Figure 1. Facilities used at the St. Anthony Falls Laboratory.

Table 1. Summary of the experiments performed.

Series	Runs	Sediment	ΔT*	R_t
1	5	Silica Sand	0°C	0.02045
2	5	Silica Sand	62°C	0.0016
3	3	Pyroclastic	0°C	0.00763
4	3	Pyroclastic	62°C	−0.01103

* Temperature difference between the turbidity current slurry in the head tank and the ambient water temperature in the main tank.

Four series of runs were performed; each series consisted of a number of turbidity currents with successive deposits stacked one on top of another. Each turbidity current in a series had the same input conditions. As shown in Table 1, two types of sediment (silica sand and pyroclastic volcanic ash) and two temperature conditions (no temperature difference between head tank water and main tank water, and head tank water that was 62°C hotter than the water in the main tank) yielded four series.

Of these four series, Series 1 corresponds most closely to a pure turbidity current carrying siliciclastic material and Series 4 corresponds most closely to the result of an explosive volcanic eruption.

Pyroclastic sediment: This sediment is unique because of its low specific density and porous shape. It was collected from Pahvant Butte, Utah, where it was originally emplaced by a subaqueous eruption (White, 1996). Common silica sand has a specific density

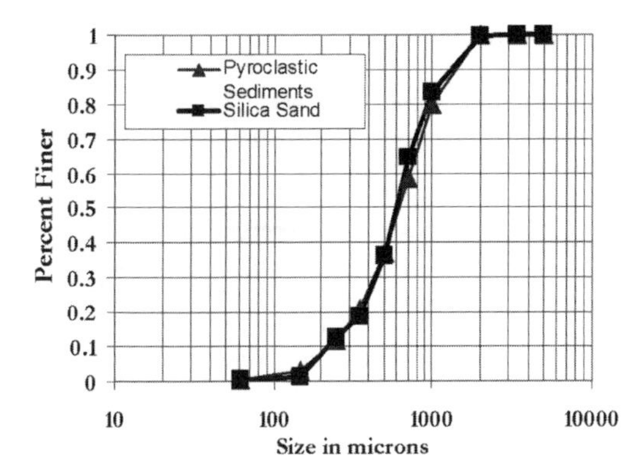

Figure 2. Grain size distributions of the silica sand the pyroclastic sediment.

of 2650 kg/m^3, but the pyroclastic sediment averages 2230 kg/m^3. This difference can be explained by the vesicularity (tendency for air pockets to be trapped in a grain of sediment) of pyroclastic sediment which is typical of material of volcanic origin. Though not addressed by the experiments reported here, it is of value to note that during a subaqueous eruption particles would *initially* all have vesicles filled with gas. Particle cooling and vapor condensation would rapidly draw water into open vesicles, but internal vesicles would remain as open cavities contributing to reduced particle density (Whitham and Sparks, 1986; Manville et al., 1998).

In order to compare the effect of the two very different kinds of sediment, the silica sand was sieved to match the grain size distribution of the pyroclastic material. Figure 2 illustrates the comparison between the grain size distributions of the two sediments.

Hot water: Temperature difference between sediment-laden water and the ambient water around it is an important variable in turbidity currents induced by a volcanic eruption (White, 2000). Let ΔT denote the temperature difference between the water in the head tank and that in the main tank. The water was heated inside the head tank until the desired temperature was obtained. In two of the series ΔT was zero; in the other two ΔT was equal to 62°C. That is the water in the head tank was maintained 62°C hotter than the ambient water in the main tank. Though the range of initial temperatures in the field is an order of magnitude higher than that possible in the laboratory, the difference used here is sufficient to capture important aspects of the dynamics of field scale turbidity currents. This is because heat is transferred very rapidly from particles initially at c. 1200°C to water (e.g. Gudmundsson, 2003), yielding water at sub-boiling temperatures for turbidity currents arising from the eruption.

Data acquisition consisted of measuring water temperature in the exterior main tank/channel and in the head tank before each run, capturing video of the advancing turbidity currents using a combination of four synchronized video cameras, measuring deposit profiles using a point gage, acquiring a series of digital images from the side of the channel, and capturing underwater digital photographs from above the channel in order to ascertain the presence of bedforms. All of the images were corrected for optical distortion using appropriate software.

3 CALIBRATION OF THE INITIAL CONDITIONS

A hot current injected into a colder environment is subject to an upwardly buoyant force due to the density difference between hot water and cold water. For hot turbidity currents, this force acts to reduce the current's excess density, impeding propagation of the current along the bottom of the channel, and ultimately converting the density current into a buoyant plume that rises up toward the water surface (e.g. Sparks et al., 1993). The fluid buoyancy effect is initially overcome by the density of the sediment fraction of the turbidity current. By determining the correct balance between sediment concentration and water temperature, it is possible to create a turbidity current with the following behavior: it exits the head tank with a sediment concentration that is sufficient to maintain downslope flow and attachment to the bottom. Further downslope, as sediment is lost through deposition, the density difference between the current and the ambient water decreases, and eventually reverses near a point where the suspension detaches from the bottom of the channel and rises upward as a hot, buoyant but sediment laden plume.

Because this effect is inevitable during the full runout of a natural volcanic current, reproducing this behavior in the laboratory setting is essential in order to investigate the effect of the temperature difference on the resultant deposit.

In Figure 3a, the effect of temperature on water density is plotted. The density of a turbidity current ρ_{tc} is a function of the water density ρ_w, sediment density, ρ_s, and the sediment volume concentration, C, as follows:

$$\rho_{tc} = \rho_w(1-C) + \rho_s C = \rho_w\left(1+\frac{\rho_s-\rho_w}{\rho_w}C\right) \quad (1)$$

Using ρ_{wh} and ρ_{wa} to denote hot and cold (ambient) water density, respectively, the turbidity current density in the head tank assumes a different value in each of our series.

Whether or not a turbidity current transporting water that is hotter than the ambient water continues

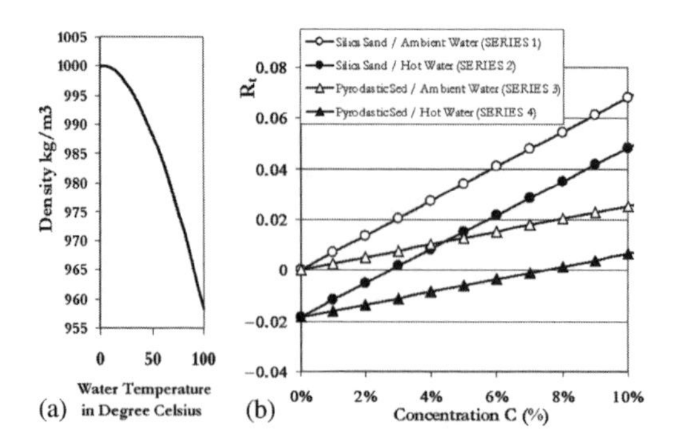

Figure 3. (a) Relation between density of fresh water and water temperature and (b) fractional excess density R_t of a turbidity current as a function of the volume sediment concentration C. The four cases plotted correspond to a) silica sand with $\Delta T = 0°C$ (ambient and head tank at 12°C); b) silica sand with $\Delta T = 62°C$ (ambient water at 12°C and head tank at 74°C); c) pyroclastic material with $\Delta T = 0°C$ (ambient and head tank at 12°C) and d) pyroclastic material with $\Delta T = 62°C$ (ambient water at 12°C and head tank at 74°C).

downslope along the bed or tends to lift off to become a buoyant plume is governed by its the fractional excess density R_t, defined as:

$$R_t = \frac{\rho_{tc} - \rho_{wa}}{\rho_{wa}} \tag{2}$$

In Figure 3b, the value of R_t for the water-sediment mixture in the head tank is plotted as a function of the sediment volume concentration C for the four series considered here.

The plot shows the competing effect of sediment and temperature in determining the fate of a current released from the headbox. When $C = 0$ and $\Delta T = 0$ the fluid in the headbox is neither dense nor buoyant ($R_t = 0$), and no density-driven current results upon release from the headbox. In Series 1 (silica sand) and 3 (pyroclastic sediment), $\Delta T = 0°C$; as long as the volume sediment concentration $C > 0$ the fractional density difference R_t is also greater than zero, so that release from the headbox results in a turbidity current hugging the bottom of the channel. Such turbidity currents should continue to propagate downslope until all or nearly all the sediment has deposited.

In Series 2 (silica sand) and 4 (pyroclastic sediment), $\Delta T = 62°C$. As a result, R_t is less than zero in the absence of sediment, and release from the headbox results in a buoyant plume. As sediment concentration C in the headbox increases, however, a point is reached beyond which $R_t > 0$, so that release from the headbox results in a bottom-hugging turbidity current proximal to the headbox. As sediment settles out farther downstream, however, C declines, so that R_t eventually reverses sign from positive to negative and the turbidity current is converted into a buoyant plume.

In Figure 3b, the difference in specific density between the silica and the pyroclastic sediment is reflected in the different slopes of the two lines for each sediment type. The simple calculation of Equation 2 was used to set the experimental parameters in the headbox. More specifically, a headbox value for volume concentration C of suspended sediment of 0.03 was selected so as to ensure that the currents or Series 1 and 3 were well-defined, pure turbidity currents with the ability to run out a significant distance before depositing, and the current of Series 2 commenced as turbidity currents but was converted to a buoyant plume farther downstream as its sediment settled out.

In the case of Series 4, consideration needs to be given to the runs since, as shown in Table 1, the fractional excess density R_t is negative upon initial release of the 3% current. This value predicts an initial buoyant plume; however, due the following controls, the flow first exits the head box as a bottom flow partly driven by sediment. First, the head difference between head tank and the main tank gives momentum to the flow in the downstream direction. Second, the grains of the flow are sufficiently coarse such that, upon release of the head gate, the sediment concentration assumes a value higher than 3% toward the bottom of the channel, while it is less than 3% toward the top of the current.

Therefore, this value of concentration maximized the use of the facilities and the need to compare all the series using the same input parameters.

4 EXPERIMENTAL RESULTS

Figure 4a shows a turbidity current from Series 3, for which $\Delta T = 0$. The current flowed downslope, always attached to the bottom of the channel. This current, while strongly depositional in nature, ran the entire 10 m of the tank. Figure 4b shows a turbidity current from Series 2, for which $\Delta T = 62°C$; the effect of the temperature difference is visible from the interface that divides the hot turbidity current from the clear cold water. The turbidity current of Figure 4b detached from the bottom of the channel approximately 4.5 m downstream from the inlet gate, feeding a plume that contained a low concentration of suspended sediment and rose to the water surface.

The hot turbidity currents (Series 2 and 4) showed patterns of downstream variation in head velocity that differ significantly from the cold currents (Series 1 and 3). Let X denote a down-channel streamwise coordinate, Z denote a coordinate upward-normal to the bed, as defined in Figure 4, and V_X and V_Z denote the values of head velocity in the corresponding directions. Figure 5 shows the downstream variation in V_X

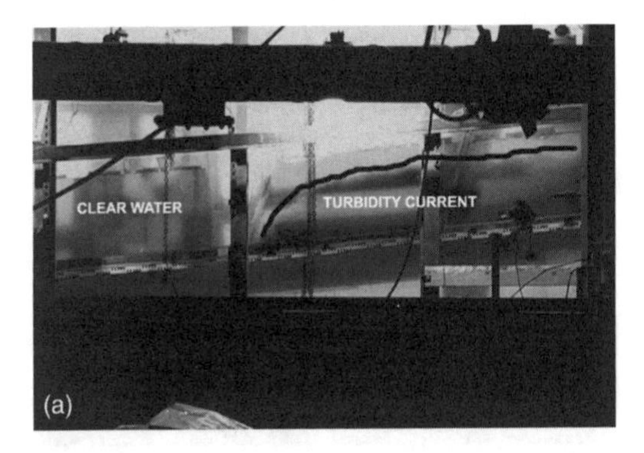

Figure 4. (a) A turbidity current with no temperature difference (Series 3 run a, $\Delta T = 0°C$) remains attached to the bottom of the channel. The flow is from right to left and (b) a turbidity current with an upstream temperature difference ΔT of 62°C tends to billow upward as it propagates downslope (Series 2 run a). Note that although the bed slopes to the left, the interface between the sediment-laden and clear water slopes to the right. The flow direction is from right to left.

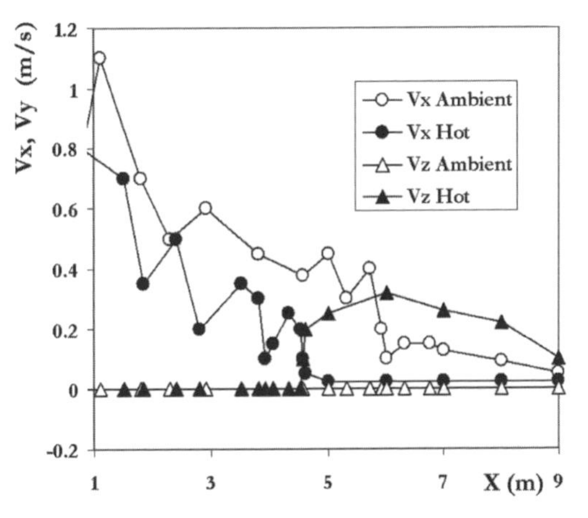

Figure 5. Comparison of head propagation velocity in the X (circles) and Z (triangles) directions for hot (thick line) and ambient temperature (thin line) runs. Data from Run 1 of Series 1 and 2.

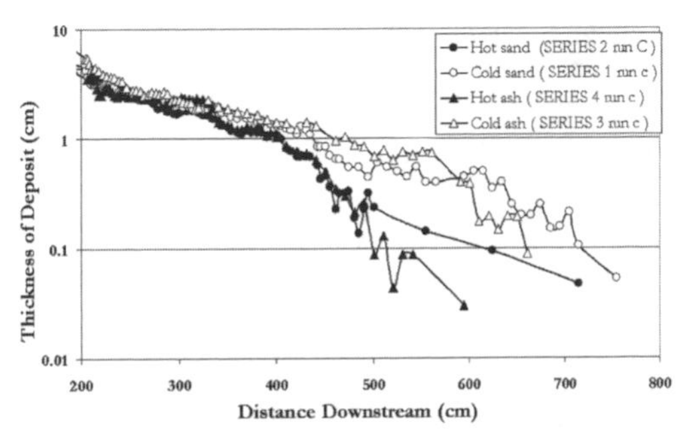

Figure 6. Deposit profiles for the third run (c) of each series.

and V_z for the first run of Series 1 ($\Delta T = 0°C$; cold current) and the first run of Series 2 ($\Delta T = 62°C$; hot current). In the case of the cold current, $V_x > 0$ and drops monotonically to the end of the tank, whereas in the case of the hot current, V_x is everywhere lower then the corresponding value for the cold current, and drops to nearly zero near $X = 4.55$ m. In addition, V_Z is essentially negligible for the cold current, but becomes substantial downstream of the point $X = 4.55$ m for the hot current as a buoyant plume forms. For simplicity, this location is referred to as the "detachment point". Head propagation velocities were consistent in trend as well as order of magnitude for all runs in the series.

The turbidity currents were net-depositional in all runs. Analysis of the resultant deposits illustrates different features corresponding to temperature difference and sediment type.

Figure 6 shows downstream profiles of deposit thickness for the third run (c) of each series.

It is apparent from Figure 6 that for the same sediment type, deposit thickness declined much more rapidly downstream in the case of the hot flows ($\Delta T = 62°C$). This illustrates the tendency of the temperature difference to arrest the current. The difference in deposit structure between hot and cold flows was stronger for the case of the volcanic ash than in the case of the silica sand. A weaker, but still discernible difference in deposit structure is evident in comparing different sediment types at the same temperature difference. The very distal deposit thickness for the flow with volcanic ash is less than that for the corresponding flow with silica sand. Both trends are expected; hot water tends to arrest a turbidity current, and volcanic ash is lighter than silica sand, yielding a weaker downslope gravitational force acting on the current.

Although not shown here, the deposits also showed a tendency for characteristic grain size to decrease downstream.

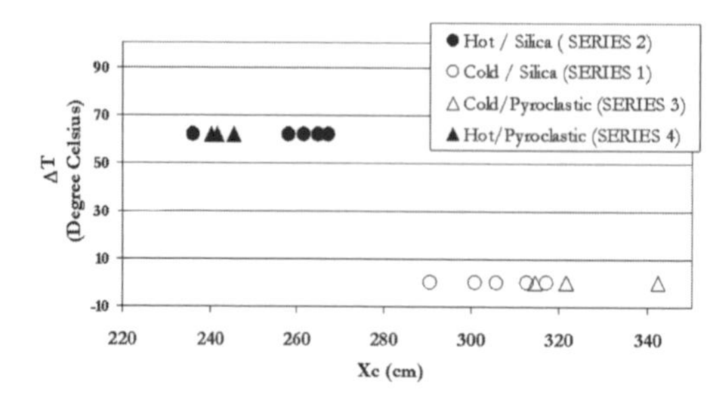

Figure 7. Center of mass X_c of the deposit as a function of temperature difference ΔT. Here circles represent silica sand runs, triangles represent pyroclastic sediment runs, open symbols denote cold runs ($\Delta T = 0°C$) and black symbols denote hot runs ($\Delta T = 62°C$).

The long profile of the thickness of the deposit of each run was used to compute a streamwise position X_c of the center of mass of the deposit. A small value of X_c implies that the sediment deposited more proximally (weaker turbidity current) and a large value of X_c implies that the sediment deposited more distally (stronger turbidity current). Figure 7 demonstrates that this center of mass was consistently displaced upstream in the hot runs (Series 2 and 4) as compared to the cold runs (Series 1 and 3).

The deposits of cold turbidity currents thinned gradually until the end of the channel. In the runs performed with hot turbidity currents, the detachment point (where the hot turbidity currents transformed to buoyant plumes and detached from the channel) marked the threshold between two distinct depositional zones; the upstream zone, where deposit thickness and sediment grain size decreased rapidly, and the downstream zone, where the deposit consisted of a thin, uniform layer of fine sediment.

These two zones of the deposits of the hot currents were formed by quite distinct depositional processes: the proximal deposit upstream of the detachment point was formed by direct deposition from the turbidity current, whereas the distal deposit was formed by settling of suspended sediment through the water column from the buoyant plume. In the laboratory runs, this predominantly fine-grained sediment ultimately settled more or less uniformly across the entire length of the channel, capping the proximal deposit and forming the entirely of the distal deposit. Figure 8 shows the deposit downstream of the detachment zone for a hot run (Series 4, Run c), as well as the deposit of a cold run of the corresponding series at the same distance downstream of the headbox (Series 3, Run c). Note that no detachment occurred in the case of the cold run. It is clear from the figure that the deposit of the hot turbidity current (left) is finer and less sorted than the deposit of the cold current (right). The arrows show the direction of the flow.

Figure 8. Deposit downstream of the detachment zone for the hot and cold turbidity current runs.

The patterns in Figure 8 have field analogs. In particular, field observations at Pahvant Butte and Black Point volcanoes (White, 1996; White, 2000) show relatively steep-sided accumulations of beds that nevertheless have shallow dips; the inference is that abruptly terminating beds are stacked up near the underwater vent, with an abrupt transition outward to much lower depositional rates producing thin and fine-grained beds of wide dispersal. At Black Point this transition is reflected in a change, over less than 1 km laterally, from a near-source deposit comprising ~200 m of thick beds of gravelly sand-grade ash to a <5 m thick one of ripple-laminated silty sand-grade ash (White, 2000).

5 CONCLUSIONS

In order to address the effects of variations in grain density and water temperature on flow properties and depositional processes, laboratory studies on density currents were performed at St. Anthony Falls Laboratory (University of Minnesota). Experiments were performed under a fixed set of flow conditions utilizing both pyroclastic material and silica sand that was sieved to a grain size distribution equivalent to the pyroclastic material. A series of runs in which a hot density current entered a cold environment, and a series in which a cold current entered a similarly cold environment were performed using each grain type. The deposits of the four sets of currents displayed a distinction based on density of the grains driving the current and the density contrast between the inflowing current and the ambient water.

The dynamics of the current and properties of the consequential deposit displayed the importance of the effect of temperature difference. In particular, a temperature gradient associated with a hot, sediment-laden current flowing into cold ambient water had the effects of a) reducing the streamwise head propagation velocity of the turbidity current, b) inhibiting the runout distance of the current and c) creating a buoyant plume with an upward velocity that drove the turbidity current from the channel bottom upward near the point of maximum runout.

The temperature difference had a direct influence on the deposit. For example, the center of mass of the deposit was located more proximally when a hot current entered a colder environment, while the center of mass was located distally when the temperature difference was zero.

In the case of a temperature difference, the resultant deposit showed two distinguishable zones. Upstream of the point of maximum runout the deposit was formed directly by the net-depositional turbidity current. The deposit downstream of this point consisted of grains that settled out of the buoyant plume.

ACKNOWLEDGEMENTS

The authors are grateful to Benjamin Erickson for help with the experiments. This paper is a publication of the National Center for Earth-Surface Dynamics (NCED), which is funded by the National Science Foundation (NSF). Preparation of this paper was entirely supported by NSF. JDL White also received support from the NZ Public Good Science Fund contract C05X0402 for part of this work.

REFERENCES

Chu, F.H., Pilkey, W.D. & Pilkey, O.H. 1979. An analytical study of turbidity current steady flow. Mar. Geol. 33, 205–220.

Ellison, T.H. & Turner, J.S. 1959. Turbulent entrainment in stratified flows. J. Fluid Mech. 6, 423–448.

Garcia, M.H. 1985. Experimental study of turbidity currents. MS Thesis, Dept. of Civil and Mineral Engineering, University of Minnesota, USA.

Gudmundsson, M.T. 2003. Melting of ice by magma-ice-water interactions during subglacial eruptions as an indicator of heat transfer in subaqueous eruptions. In: J.D.L. White, J.L. Smellie and D.A. Clague (Editors), Explosive Subaqueous Volcanism. American Geophysical Union Monograph. American Geophysical Union, Washington D.C.

Hinze, J.O. 1960. On the hydrodynamics of turbidity currents. Geol. Mijnb. 39e, 18–25.

Keulegan, G.H. 1957. An experimental study of the motion of salinewater from locks into fresh water channels. US National Bureau of Standards Report, 5168.

Lüthi, S. 1981. Some new aspects of two-dimensional turbidity currents. Sedimentology 28, 97–105.

Manville, V., White, J.D.L., Houghton, B.F. & Wilson, C.J.N. 1998. The saturation behaviour of pumice and some sedimentological implications Sed. Geol. 119, 5–16.

Pantin, H.M. 1979. Interaction between velocity and effective density in turbidity flow: phase-plane analysis, with criteria for autosuspension. Mar. Geol. 31, 59–99.

Parker, G. 1982. Conditions for the ignition of catastrophically erosive turbidity currents. Mar. Geol. 46, 307–327.

Plapp, J.E. & Mitchell, J.P. 1960. A hydrodynamic theory of turbidity currents. J. Geophys. Res. 65, 983–992.

Schmidt, W. 1911. Zer Mechanik der Boen. Z. Meteorol. 28, 355–362.

von Karman, T. 1940. The engineer grapples with nonlinear problems. Bulletin of the American Mathematical Society, 46, 615–683.

White, J.D.L. 1996. Pre-emergent construction of a lacustrine basaltic volcano, pahvant butte, utah (USA) Bull. Volcanol. 58, 249–262.

White, J.D.L. 2000. Subaqueous eruption-fed density currents and their deposits. In: Processes in physical volcanology and volcaniclastic sedimentation: modern and ancient Precambrian Res. 101, 87–109.

Whitham, A.G. & Sparks, R.S.J. 1986. Pumice Bull. Volcanol. 48, 209–223.

River, Coastal and Estuarine Morphodynamics: RCEM 2005 – Parker & García (eds)
© 2006 Taylor & Francis Group, London, ISBN 0 415 39270 5

Controls on geometry and composition of a levee built by turbidity currents in a straight laboratory channel

D. Mohrig, K.M. Straub & J. Buttles
Department of Earth, Atmospheric and Planetary Sciences, MIT, Cambridge, USA

C. Pirmez
Shell International Exploration and Production, Inc., Houston, USA

ABSTRACT: Experimental results are presented that quantify properties of a channel levee built by depositional turbidity currents. Nine currents of constant initial thickness and composition were released into a pre-existing channel and levee growth on the channel bank was observed. Rates of sediment deposition on the proximal levee, grain size, and taper of levee deposits were all found to be inversely related to local channel depth. These relationships were controlled by 1) the thickness of the depositing current relative to the local channel depth and 2) the vertical profiles of concentration and size for grains suspended within the current. Change in local depth confined more or less current within the channel and determined the fraction of current present at and above the levee-crest elevation where it could act as a sediment source for bank construction. The results are used to develop a simple framework for interpreting submarine levees.

1 INTRODUCTION

Many submarine channels have banks defined by prominent levees. It is generally agreed that these topographic features are built by turbidity currents that spill out of channels and deposit sediment while moving across the overbank surfaces. Unfortunately, direct observations of currents constructing these channel margins are not yet available and specific processes responsible for building submarine levees must be inferred through analyses of their geometry and composition. The laboratory experiment described here resolves, at a reduced scale, many of the interactions between turbidity currents and bottom topography that lead to construction of channel-bounding levees. Analysis of experimental data focuses on establishing the connections between levee thickness, levee taper and composition (grain size). These levee characteristics are in turn related to the following physical properties of the depositing currents: 1) current thickness relative to the local depth of the channel; and 2) vertical profiles of sediment concentration and grain size for the solid particles suspended within the interiors of currents. The experimental results are intended to compliment recent quantitative studies of natural submarine levees including those by Skene et al. (2002) and Pirmez and Imran (2003) in order to improve the accuracy with which processes associated with the levee construction can be estimated from the morphology and composition of the levees themselves.

2 EXPERIMENTAL SETUP

The experiment was conducted in a tank 5 m long, 5 m wide and 1.2 m deep (Fig. 1). The initial channel form consisted of 6 concrete segments that were laid end to end, producing a 3 m reach. This channel was trapezoidal is cross-section (Fig. 1). Average width and depth were 0.685 m and 0.050 m, respectively. A map of the initial channel topography is presented as Figure 2A. Each current passed through a momentum reduction box before entering the channel (Fig. 1) so that each flow was a sediment-laden plume driven by buoyancy alone. A 0.4 m deep moat prevented current reflections off of tank sidewalls. The tank remained filled with water throughout the experiment.

The experiment consisted of 9 depositional turbidity currents composed of fresh water plus suspended sediment with an initial volume concentration of 1.5×10^{-2}. The sediment was crushed silica flour with D1, D5, D16, D50, D84, D95, and D99 equal to 1.4 μm, 2.4 μm, 6 μm, 29 μm, 59 μm, 89 μm, and 133 μm, respectively. Each current was pre-mixed in a holding reservoir and introduced to the system via a constant-head tank that ensured a steady entrance discharge throughout each 1 m^3 release of sediment + water. The bulk density for each current as it entered the water-filled tank was 1024 kg/m^3 and current thickness, H, at the channel entrance ($x = 0$ m) was set at 9.0×10^{-2} m. Additional initial conditions for the nine currents are summarized in Table 1.

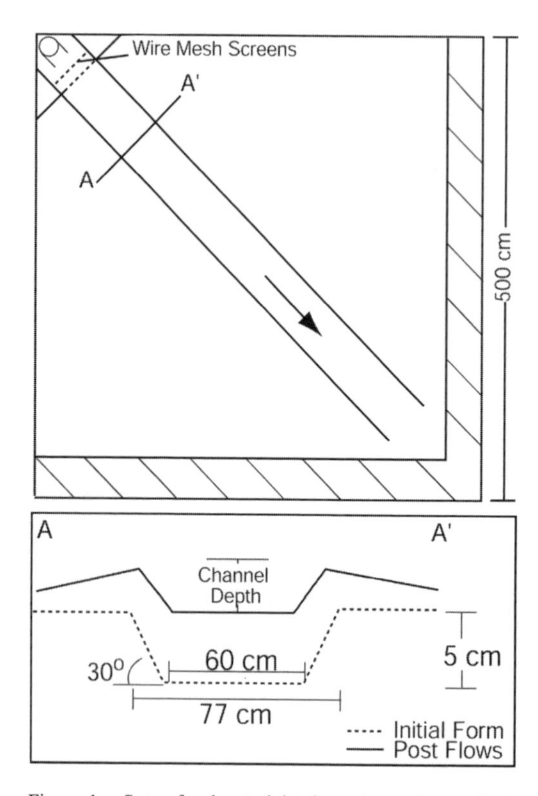

Figure 1. Setup for the straight channel experiment. (top) Plan view of straight channel configuration. Currents enter in top left corner of tank and pass through an excess-momentum diffuser before entering the channel. The moat running along far sidewalls of the tank is marked by a hatched pattern. Piping in the moat removes current from the tank, inhibiting reflections. (bottom) Initial form and an approximate final geometry of a channel cross-section.

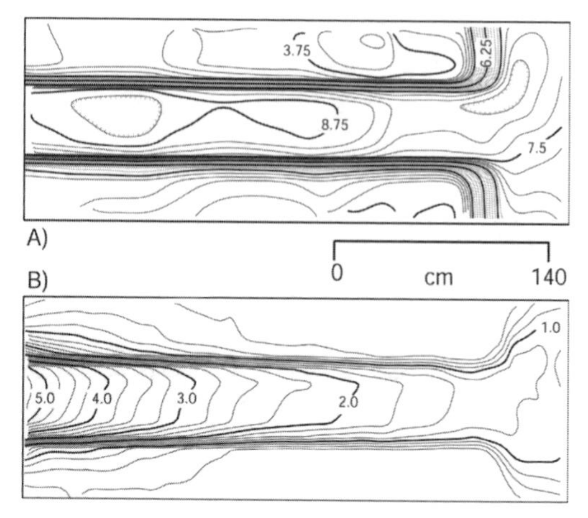

Figure 2. Maps of the 1.27 m × 3.59 m study area. Flow was from left to right. (A) Initial channel topography. Contour interval is 0.5 cm. (B) Thickness of sediment deposited from the first 8 turbidity currents. Contour interval is 0.5 cm.

Table 1. Characteristics of experimental currents.

Current	u (m/s)	Flow duration (hr)	*Fr*	*Re*
1	5.0×10^{-2}	16×10^{-2}	3.4×10^{-1}	4.5×10^{3}
2	6.0×10^{-2}	16×10^{-2}	4.1×10^{-1}	5.4×10^{3}
3	5.5×10^{-2}	14×10^{-2}	3.7×10^{-1}	5.0×10^{3}
4	5.5×10^{-2}	14×10^{-2}	3.7×10^{-1}	5.0×10^{3}
5	5.0×10^{-2}	14×10^{-2}	3.4×10^{-1}	4.5×10^{3}
6	7.0×10^{-2}	8.9×10^{-2}	4.7×10^{-1}	6.3×10^{3}
7	11×10^{-2}	6.8×10^{-2}	7.4×10^{-1}	9.9×10^{3}
8	7.0×10^{-2}	8.9×10^{-2}	4.7×10^{-1}	6.3×10^{3}
9	12×10^{-2}	6.9×10^{-2}	8.1×10^{-1}	11×10^{3}

The geometric scaling for our experimental system was set at 1/1000. Maximum width, depth and length for the laboratory channel correspond to natural scales of 770 m, 50 m, and 3 km. Additional model-current properties can be compared to natural or prototype systems using three dimensionless parameters; the densimetric Froude number, *Fr*, the ratio of particle fall velocity to the shear velocity, w_s/u_*, and the Reynolds number, *Re*. An approximate dynamic similarity between the currents of different scale is ensured by setting $Fr_{(\text{model})} = Fr_{(\text{prototype})}$ (Graf, 1971). This equality yields prototype values for mean streamwise velocity, *u*, current thickness, *H*, and current duration of 2.2 m/s, 50 m, and 2.8 hr for current #8. Sediment transporting conditions were mapped between scales by setting $w_s/u_{*(\text{model})} = w_s/u_{*(\text{prototype})}$, where $u_*^2 = C_D u^2$, C_D is a bed friction coefficient and w_s was calculated using Dietrich (1982). Values of $C_{D(\text{prototype})} = 2 \times 10^{-3}$ and $C_{D(\text{model})} = 2 \times 10^{-2}$ were used to account for the weak dependence of bed friction coefficient with turbidity-current scale (Parker et al., 1987; Garcia, 1994). Resulting values for D5, D50, and D95 in the natural system are 7 μm, 101 μm, and 434 μm. $Re_{(\text{model})}$ (Table 1) was always large enough to satisfy the approximate Reynolds similarity for fully turbulent gravity currents proposed by Parsons and Garcia (1998).

An acoustical system was used to produce 5 maps of the channel form; at the beginning of the experiment, after current #1, current #2, current #3, and current #8. The bathymetric measurements were collected with a 1 MHz ultrasonic transducer connected to a pulse/receiver box. Each map was built from 23205 points collected on a 14 × 14 mm grid and had a vertical resolution of about 100 μm (Fig. 2). Maps of deposit thickness were produced by differencing successive bathymetric measurements (Fig. 2B).

Average streamwise velocity was measured at the channel entrance using a Sontek ADV (Table 1). A

Sontek PC-ADP was used to measure the streamwise velocity profile at multiple centerline points down the channel. Profile data for current #8 showed that total current thickness remained $9.0 \pm 0.8 \times 10^{-2}$ m halfway down the channel ($x = 1.5$ m) and the velocity maximum was located 2.5×10^{-2} m above the bed.

Following current #9 the tank was drained and samples of the resulting sedimentary deposit were collected for grain-size analysis. The samples were taken from the uppermost deposit and represent deposition associated with final currents. All of the sediment samples were processed using a Horiba LA-300 laser particle-size analyzer.

3 EXPERIMENTAL RESULTS

A primary objective of this experiment was to connect levee development on the channel banks to spatial and temporal variation in the sediment-transport system. One property that varied consistently throughout the experiment was the thickness of currents compared to the local channel depth, h. Currents became increasingly thick relative to the channel topography because initial current thickness was held constant while channel depth systematically decreased via sedimentation. Channel depth decreased through time because sedimentation rates were always greatest inside of the channel, on the channel bed. The cross-section in Figure 3 documents this decrease in channel depth as a function of current number. Figure 3 also documents levee growth with the passage of successive turbidity currents.

All levee data presented here were collected from the bank marked by arrows in Figure 3. Cross-sectional profiles of the levee built by the first 8 currents are shown in Figure 4. These profiles record a consistent change in thickness and taper with streamwise distance. Only the first 0.2 m of the levee was measured, a lateral distance equal to 29% of the channel width. Trends in levee properties are therefore restricted to the most proximal section of the constructional feature.

Streamwise or longitudinal profiles of levee thickness are shown in Figure 5B. These two profiles are associated with the endpoints for the cross-sections in Figure 4. Both profiles (Fig. 5B) show a systematic reduction in levee thickness with distance down the channel. Levee taper was measured perpendicular to the channel direction and calculated by first subtracting the points on one profile from points on the other and then dividing these thickness differences by the lateral distance separating the two profiles. Values for levee taper are shown in Figure 5C and systematically decreased with distance down the channel. It is worth noting that the measured taper would be equal to levee surface slope if the bank of the original channel form had been perfectly flat and horizontal.

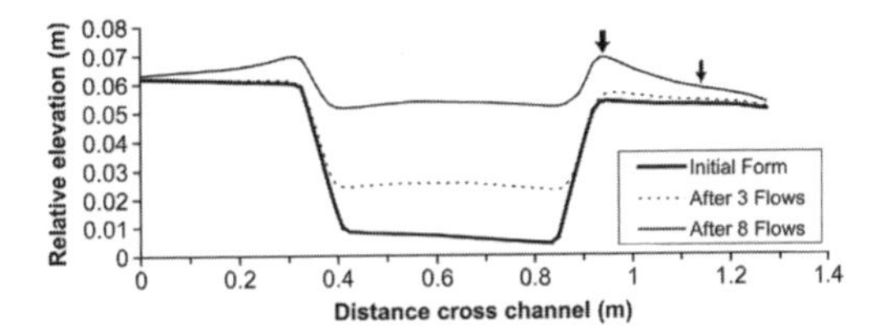

Figure 3. Channel cross section at beginning of experiment, after deposition from 3 currents and after deposition from 8 currents (looking upstream). Section was located 0.35 m from channel entrance. The thick arrow marks the crest of the depositional levee. Grain-size data reported here is from this crest-line profile. The thin arrow marks the position of a second longitudinal profile located 0.2 m from the levee crest-line.

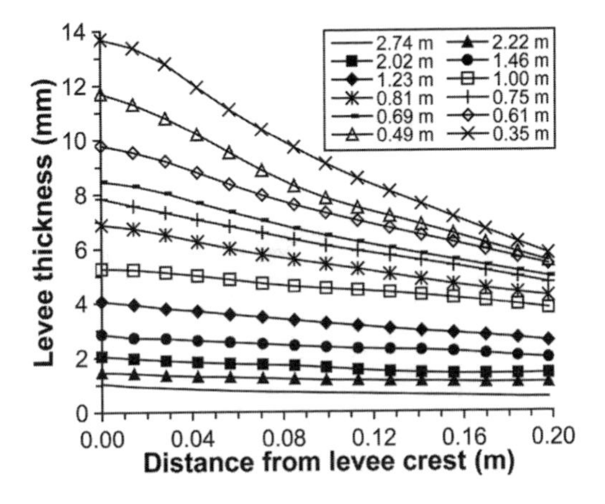

Figure 4. Cross-sectional profiles of the depositional levee at 12 different locations down the channel. Levee thickness and taper decrease with increasing distance from the channel entrance. The longitudinal sections defining the beginning and ending points for all of these levee profiles are marked by the two arrows in Figure 3.

Sediment samples were collected from the uppermost portion of the levee-crest deposit following the release of all nine currents. These samples defined a consistent reduction in the particle size with distance from the channel entrance (Fig. 5D). A smaller number of sediment samples were also collected from the levee along transects running perpendicular to the channel centerline. Size analysis of these samples showed a fining of the levee deposit with distance from its crest line. Unfortunately the number of samples collected was insufficient to accurately resolve this lateral trend in the grain size of the levee.

Sedimentation on the channel bed by the first 8 currents produced a channel form with a local depth that varied with downstream distance (Fig. 5A). This variation in local depth affected levee growth. Data collected from along the channel can be group together

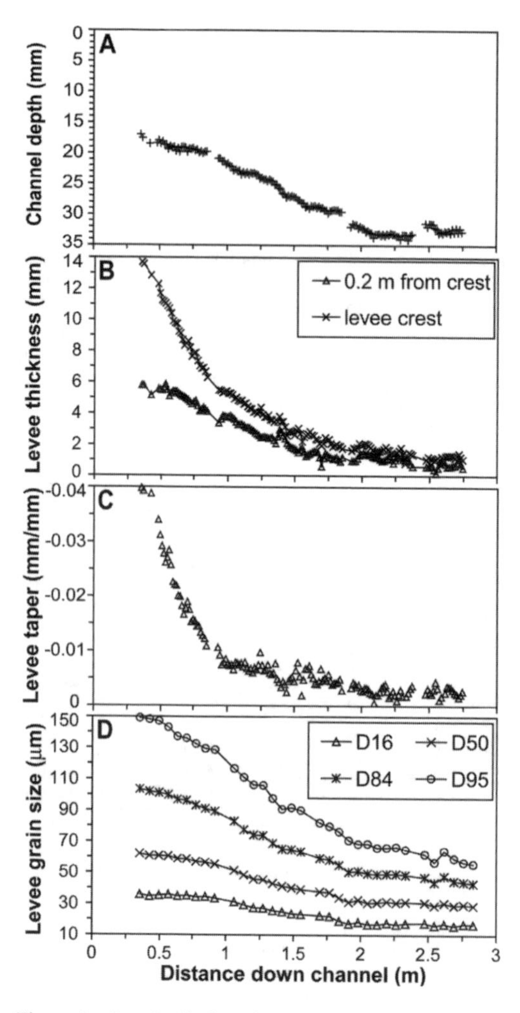

Figure 5. Longitudinal profiles from the channel bed and the levee following sedimentation by 8 currents. (A) Channel depth. Depth equals the elevation difference between the levee crest and the channel bed (see Figure 1). (B) Thickness of levee deposit along the two sections marked by arrows in Figure 3. (C) Lateral taper (thinning) of the proximal levee. Levee taper was calculated by taking the elevation difference between associated points on the two profiles in (B) and dividing this elevation drop by the lateral distance separating the two profiles, 20 mm. (D) Grain size of sediment on the levee crest. D50 is the diameter of the median particle size. D16, D84 and D95 are the representative diameters of the size fractions larger than 16%, 84%, and 95% of the deposited grains, respectively.

to show how sedimentation rate at the levee crest varied as a function of local channel depth. Figure 6 shows that the sedimentation rate at the levee crest was inversely related to local channel depth. Points on the levee crest-line elevated farthest above the bed of the channel had measurably less deposition than

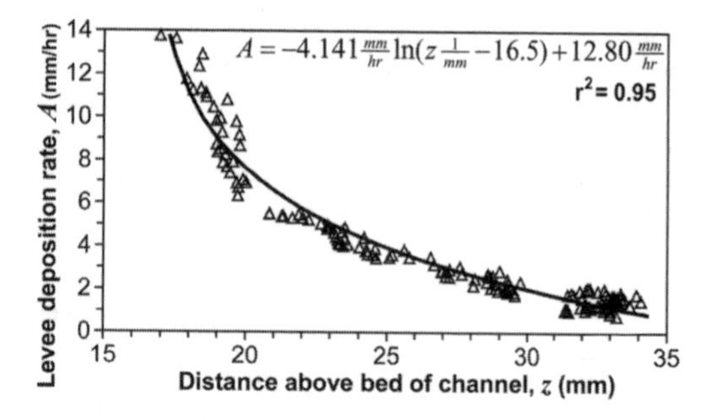

Figure 6. Average sedimentation rate on the studied levee crest as a function of the local channel depth. Graph assembles data from all measured points down the edge of the channel. A best-fit line defining the curve is also shown.

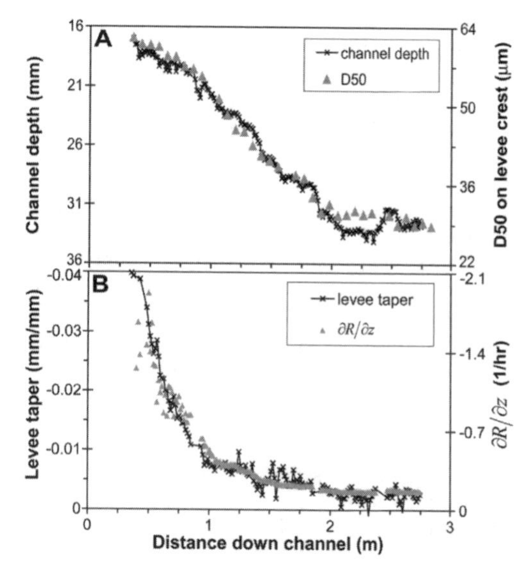

Figure 7. Comparative trends from the experimental channel. (A) Longitudinal variation in channel depth (Fig. 5A) versus median grain size on the levee crest (Fig. 5D). (B) Longitudinal variation in levee taper (Fig. 5C) versus the gradient in sedimentation rate (2) at $z = h$ (Fig. 5A) for all points down the channel.

points where the vertical separation between crest and bed was smaller.

4 INTERPRETATION OF RESULTS

One remarkable property of the experimental system was the correlation found between grain-size on the levee crest-line and downstream change in local channel depth. From inspection of Figure 7A it is clear that the trend for median particle size is a very close match to the longitudinal profile of depth. Particularly compelling is the break in slope present in both trends at about $x = 2$ m. We envision channel depth controlling levee composition in the following way. The overall size of particles within a turbidity current decreases with distance above its base and a changing local depth acts to confine more or less current within the channel, leaving a finer or coarser-grained portion of current at and higher than the levee crest elevation where it can act as a sediment source for bank construction. This interpretation is supported by the measurements of current thickness collected with the Sontek PC-ADP. These profiles of current velocity record an approximately constant current thickness down the entire length of the channel centerline, suggesting that change in local values of relative turbidity-current thickness, H/h, is primarily related to change in local channel depth.

Deposition rate along the levee crest has already been shown to vary as a function of vertical distance above the local bed of the channel (Fig. 6) so we now turn our attention to processes controlling the magnitude of the levee taper. This taper represents a reduction in deposition rate with distance from the levee crest. Can this spatial change in rate be related simply to a vertical structure within the depositing current? Levee deposition rate, A, in Figure 6 is transformed from a property of the bed into a property of the associated turbidity current through a simple correction for bed porosity. The resulting sedimentation rate, R, for the depositing current is

$$R = (1-p)A = -2.692\tfrac{mm}{hr}\ln(z\tfrac{1}{mm} - 16.5) + 8.322\tfrac{mm}{hr} \quad (1)$$

where p is bed porosity $= 0.35$ and z is vertical distance above the base of the current measured in mm. Equation (1) defines a vertical structure for the representative current and taking the derivative of (1) with respect to z, defines a gradient in sedimentation rate equal to

$$\frac{\partial R}{\partial z} = \frac{-2.692\frac{1}{hr}}{z\frac{1}{mm} - 16.5} \quad (2)$$

Values for (2) are calculated at the levee-crest position, $z = h$, for every point down the channel and plotted in Figure 7B. The measured values of levee taper are also plotted in Figure 7B and it can be seen that taper and $\partial R/\partial z$ are strongly correlated. This correlation is consistent with the lateral taper of the proximal levee being a consequence of the local gradient in sedimentation rate (2) for that fraction of the current moving out onto the bank of the channel. In other words, the vertical structure of the current at the position of the levee crest is translated into a laterally varying deposition rate that builds the levee taper via the lateral advection and settling of suspended particles onto the levee surface. The local sedimentation rate for a strongly depositional turbidity current can be approximated as $R = w_s\varepsilon_s$, where w_s the characteristic settling velocity for the assemblage of depositing particles and ε_s is

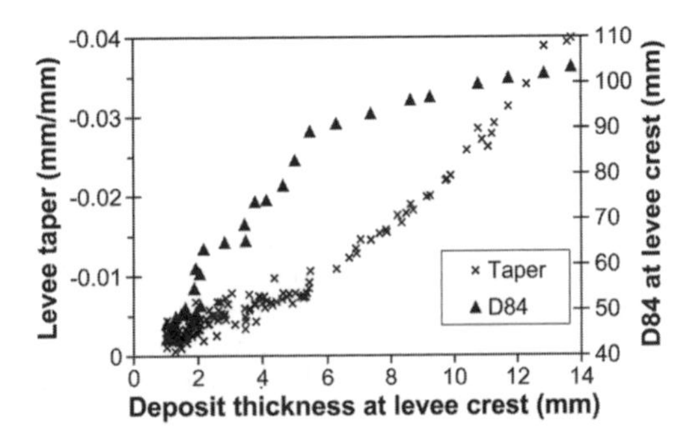

Figure 8. Correlations between levee taper (Fig. 5c), grain size (Fig. 5d) and thickness of the levee-crest deposit.

their volume concentration in the flow directly above the aggrading surface. Levee taper is therefore a product of both the vertical distribution of particle sizes within the supra-channel current and the vertical structure of suspended-sediment concentration within an upper fraction of the flow.

The reduced spatial and temporal scales of the laboratory system allowed for sampling that connected the depositing currents to the developing levee form. Submarine channels are always relatively under-sampled with many components of the system unresolved. Direct measurements of or quantitative estimates for the properties of channel-building turbidity currents are particularly hard to come by. The most abundant data from natural levees are measurements of their geometry. These data may or may not be complimented by boreholes that can provide grain size information at some small number of geographic locations. With these limitations in mind, data from the experimental levee is summarized in Figure 8. This figure highlights the connections between levee thickness, levee taper and levee grain size. The correlations are very clear; 1) variation in levee taper and unit levee thickness is directly related, 2) variation in levee grain-size and unit levee thickness is directly related, and 3) variation in levee taper and levee grain-size is directly related. These results are not surprising, but they form the core of a simple framework for interpreting submarine levees. A measurable change in the thickness of beds composing levee strata should have associated with it a measurable change in the taper of the levee-building beds. If this is observed, a significant change in the grain size of the deposit is predicted, even if borehole confirmation is not available. Additionally, sandier levees should exhibit significantly larger tapers than levees built from muddier deposits.

5 CONCLUSIONS

Analyses of levee deposits associated with submarine channels have been used to estimate the vertical structure of sediment concentration and grain size for the solid particles suspended within the turbidity currents that built them (Hiscott et al., 1997; Skene et al., 2002; Pirmez and Imran, 2003). The laboratory experiment presented here supports the use of levee deposits in reconstructing properties of depositing flows by documenting the ties between levee thickness, levee taper and levee grain size and the physical properties of the currents. A particularly important flow parameter was the relative thickness of the current. A relatively thick, steep and coarse-grained levee was constructed at locations along the channel where the currents were significantly thicker than the channel was deep and a relatively thin, weakly tapered and fine-grained levee formed where the channel was deeper (Fig. 5). Levee geometry and composition were related to both the absolute concentration and size of the particles suspended in the depositing current at the elevation of the levee crest and to the vertical gradients in concentration and size. The relationships established here need to be tested against geometric and compositional data from natural levees and should be used to refine existing models of channel development that are based on properties of the confining levees.

The levee trends reported here were measured in a straight channel and not affected by any cross-channel variation associated with channel bends. Variability in levee form and composition induced by irregularity in channel plan-form is bound to complicate the connection between properties of the depositing flows and the levees they construct. At the very least, plan-form irregularity increases the number of measurements necessary to resolve any systematic change in levee properties through space or time. The experiment here indicates that geometric and compositional trends for a levee are best defined close to the levee crest. We suggest that future studies of levees on submarine channels should focus on that portion of the form within one channel width of the channel sidewalls. This proximal section of levees has been under sampled (Pirmez and Imran, 2003) and an emphasis here should provide information that can be used to significantly improve our understanding of submarine channel evolution.

ACKNOWLEDGEMENTS

Support for our research was provided by Shell International Exploration and Production, Inc. and the STC Program of the National Science Foundation under Agreement Number EAR-0120914.

REFERENCES

Dietrich, W.E. 1982. Settling velocity of natural particles. *Water Resources Research* 18(6): 1615–1626.

Garcia, M.H. 2004. Depositional turbidity currents laden with poorly sorted sediment. *Journal of Hydraulic Engineering* 120: 1240–1263.

Graf, W.H. 1971. *Hydraulics of sediment transport*. New York: McGraw-Hill.

Hiscott, R.N., Hall, F.R. & Pirmez, C. 1997. Turbidity current overspill from the Amazon Channel: texture of the silt/sand load, paleoflow from anisotrophy of magnetic susceptibility, and implications for flow processes. In R.D. Flood, D.J.W. Piper, A. Klaus and L.C. Peterson (eds.), *Proceedings of the ODP Sci. Results 155*. College Station: Ocean Drilling Program. 53–78.

Parker, G., Garcia, M., Fukushima, Y. & Yu, W. 1987. Experiments on turbidity currents over an erodible bed. *Journal of Hydraulic Research* 25: 123–147.

Pirmez, C. & Imran, J. 2003. Reconstruction of turbidity currents in Amazon Channel. *Marine and Petroleum Geology* 20: 823–849.

Skene, K., Piper, D. & Hill, P. 2002. Quantitative analysis of variations in depositional sequence thickness from submarine channel levees. *Sedimentology* 49: 1411–1430.

River, Coastal and Estuarine Morphodynamics: RCEM 2005 – Parker & García (eds)

Propagation and final deposition of granular flow: dam-break experiments with water and gas-fluidized grains

O. Roche
Institut de Recherche pour le Développement, Laboratoire Magmas et Volcans, Université Blaise Pascal, France, and Departments of Geology and of Civil Engineering, Universidad de Chile, Chile

S. Montserrat, Y. Niño & A. Tamburrino
Department of Civil Engineering, Universidad de Chile, Chile

ABSTRACT: Experimental results on propagation and final deposition of initially fluidized granular flows are presented and discussed. In the experiments, an initially fluidized bed of nearly spherical glass beads with sizes of about 45 to 90 μm is suddenly released into a horizontal channel where particles are finally deposited. During initial stages of granular flows, depth and front velocity behave similarly to those of an equivalent flow of a Newtonian fluid. This is proved by comparison with experiments, conducted in the same facility, in which water was released under the same conditions. A constant Froude number of about 2.58 is observed for the fluidized granular flows, which is somewhat lower than the value of about 3.2 observed for the water flows. This is explained in terms of reduced resistance in the latter case relative to the former. Fluidized granular flows generate deposits with a nearly horizontal bulge and a front wedge with a slope of about 2 to 4%. It is argued that fines-rich, dense pyroclastic flows may emplace as fluid gravity currents, except at late stages just before final deposition. Such fluid-like behavior, with negligible inter-particle friction, is likely to promote high flow mobility even on nearly horizontal slopes. Several characteristics of pyroclastic deposits are consistent with the present experimental results.

1 INTRODUCTION

Pyroclastic density currents consist of hot mixtures of volcanic gas and particles and represent an important natural hazard. They encompass a range of phenomena, from dense gas-fluidized pyroclastic flows to dilute turbulent flows dominated by the gaseous phase (Branney & Kokelaar 2002). This study focuses on dense pyroclastic flows rich in fine particles (ash) that generate deposits called ignimbrites, but may also have implications for a wider range of flows. Examples of pyroclastic deposits are found, for instance, at the Lascar Volcano in northern Chile, as a consequence of an eruption occurred in 1993 (Denniss et al. 1998, Calder 2000). This type of flow is generated during explosive eruptions from collapsing fountains. The fluidizing gas can be generated at the vent as well as being ingested at the flow front or released by the particles (Wilson 1984).

Recently, the first author has reported results of experiments on the behavior of a granular flow over a horizontal surface in a confined channel of constant width that is suddenly released from an initial state of fluidized bed (Roche et al. 2004). It has been shown that under certain conditions, specially for small particle size, initially gas-fluidized granular flows can behave in a similar way to buoyancy-driven gravity currents of a Newtonian fluid. It is inferred that these conditions may represent the behavior of fines-rich pyroclastic flows for most of their duration, implying a negligible effect of particle internal friction in them. The rebuilding of particles contact chains, as porous pressure decreases due to defluidization, may be manifested in significant basal sedimentation, which reduces the flowing ability and promotes flow deceleration. The flows eventually enter a final stopping phase when they decline in energy due a rapid increase in internal friction.

In this paper results of a study that extend and complement the previous experimental study on initially fluidized granular flows are reported. In particular, the outcome of a number of dam-break type experiments, conducted by suddenly releasing water in the same experimental apparatus and under the same conditions at which the granular flows are generated, are presented and analyzed in order to facilitate a direct comparison of characteristics of both of these flows. Finally, characteristics of the deposits generated by the granular flows are also presented and discussed.

2 EXPERIMENTAL STUDY

The experimental apparatus (Fig. 1) consists of a 10 cm wide and 3 m long rectangular channel of variable slope, and a reservoir of the same width and variable length where the granular material is fluidized by introducing an air flux through a 10 mm thick porous plate. The fluidized granular material is released by means of the sudden opening of a gate, thus creating a flow that propagates along the channel. No air flux is provided from the base of the channel and the flow defluidizes as it propagates.

The initial bed, prior release, can be fluidized to various degrees with U_g/U_{mf} ranging from 0 to 25, where U_g denotes the gas velocity (defined as the ratio between the air flow rate and the reservoir cross sectional area) and U_{mf} denotes the minimum fluidization velocity, corresponding to the value of U_g for which the total weight of the bed is supported by the gas flux and the internal friction of bed particles vanishes. The experiments reported in this paper correspond to the condition of incipiently non-expanded fluidized beds ($U_g = U_{mf}$). Other experiments have shown that the initial bed expansion has very little influence on the subsequent flow emplacement. For $U_g/U_{mf} = 0$ to 25 at a given initial bed height, the flow run-out is nearly constant although the bed is highly expanded and agitated at the highest degrees of fluidization. A run-out increase of only about 5% is observed when the initial bed is slightly expanded at $U_g/U_{mf} = 1$ to 2 (Roche et al. 2004).

The particles used are nearly spherical glass beads with a density, ρ_s, of 2500 kg/m^3 and grain size, d_s, in the range from 45 to 90 μm.

The experiments were recorded using both regular and high-speed digital video cameras, with time resolutions of 30 (or 25) and 90 Hz, respectively. Video images of the flows were digitized and analyzed using software specially developed for this purpose, which automatically tracks the front position and determines the main kinematic properties of the flow, such as depth and front velocity. Longitudinal profiles of the deposits were taken at the end of the experiments.

A number of experiments were conducted, in which similar volumes of water were suddenly released into the same channel and the resulting flows were registered using the same visualization technique.

The experimental conditions reported here correspond to a channel of horizontal slope, a reservoir length $x_0 = 10$ cm, and different initial reservoir heights, h_0, in the range between 10 and 25 cm for the granular flow experiments and 10 and 40 cm for the water flow ones.

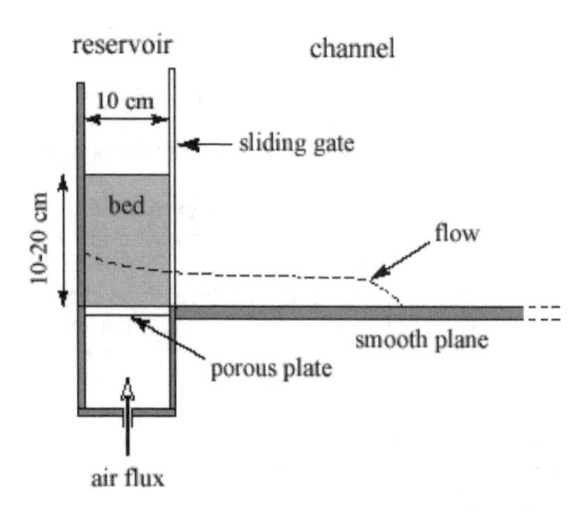

Figure 1. Experimental apparatus.

3 RESULTS

3.1 *Granular flow experiments*

As the gate of the reservoir is suddenly opened and the fluidized bed is released from the reservoir, a granular flow front develops that advances into the channel (Fig. 2). The granular flows thus generated occur in

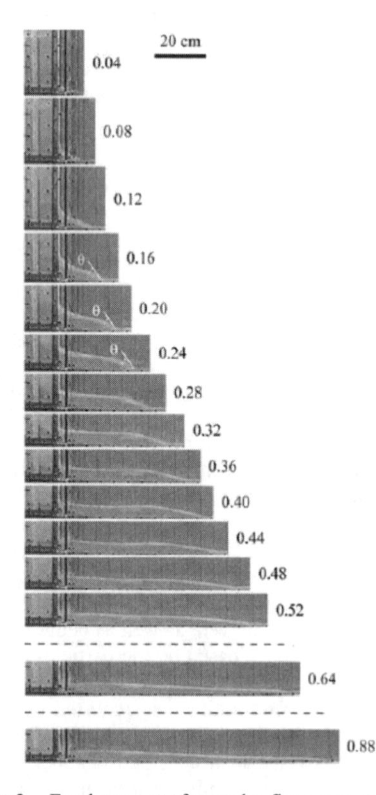

Figure 2. Emplacement of granular flows generated from incipiently non-expanded fluidized bed of $h_0 = 20$ cm. Numbers indicate time in seconds.

three distinct phases (Fig. 3). There is an initial phase during which the flow is generated from the lower part of the fluidized bed. The thickness of the flow front rapidly decreases downstream and longitudinal acceleration is close to $1g$. Then, a second phase begins, in which the flows propagate at nearly constant height and front velocity and the fluidized bed in the reservoir collapses. The flows have a well defined front with an initial slope of about 50–60°, which decreases with time.

In this second phase, current height, h_f, was estimated from the analysis of video images, by measuring an average flow depth in the nearly horizontal flow reach that is observed upstream of the front (Fig. 2). Front velocity was estimated from the slope of the curve relating front position and time (Fig. 3). From these data, the front Froude number, $Fr = u_f/(gh_f)^{1/2}$, was determined, as well as the dimensionless current height, h_f/h_0. During the second phase, which lasts for only a few tenths of a second, the h_f/h_0 ratio has a value of about 0.19–0.22, independent of the initial height h_0, and the front velocity, u_f, is about 0.9 and 1.6 m s^{-1}

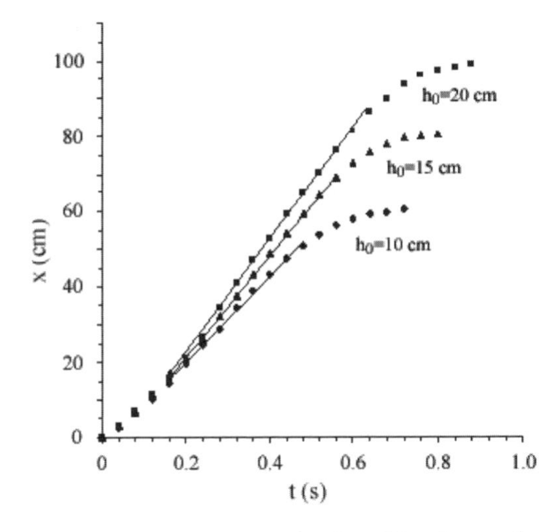

Figure 3. Front position as a function of time for granular flows generated from fluidized beds with different heights. Straight lines indicate the constant velocity phase.

for values of h_0 of 10 and 20 cm, respectively. Corresponding values of Fr are about 2.58, independent of the initial height h_0.

Once the bed height in the reservoir has decreased to the same height as the flow, a third phase occurs, during which the flow decelerates and progressively acquires a wedge shape (Fig. 2). This corresponds to the stopping phase of the flow, during which deposition occurs very rapidly and the flow stops.

The deposits in the reservoir are horizontal due to the original fluidized state (Fig. 4). The maximum bed height is located further downstream in the channel, resulting in a bulge with a commonly flat and nearly horizontal upper surface. An upstream slope of 1° to 2° is observed in some cases. The relative length of this proximal bulge decreases as the initial bed height increases. Video analysis reveals that this bulge forms at late stages of emplacement. The deposits also exhibit a front wedge with a slope of about 2 to 4% (Fig. 4). Deposits commonly have a flat front perpendicular to the channel side walls when viewed from above. The front is slightly curved or slanted in some cases, so that a mean value of the front position is taken to determine the final extent of the deposit. In contrast, non-fluidized granular flows generate wedge shaped deposits with a maximum height at the rear of the reservoir. The surface slope, in this case, is close to that given by the angle of repose of the granular material (~28.5°) in the reservoir and it decreases downstream to be of only a few degrees at the front. The longitudinal extent, x_f, of the deposits generated by the fluidized granular currents are about twice as large as those observed for the non-fluidized ones. The ratio x_f/h_0 of the deposits of fluidized granular flows decreases almost linearly from about 7 to 5.5 as the aspect ratio h_0/x_0 increases from 1 to 2.

3.2 *Water flow experiments*

As the gate of the reservoir is suddenly opened and water is released from the reservoir, a front develops that advances into the channel, similarly to what is observed in the granular flows (Fig. 5). Contrary

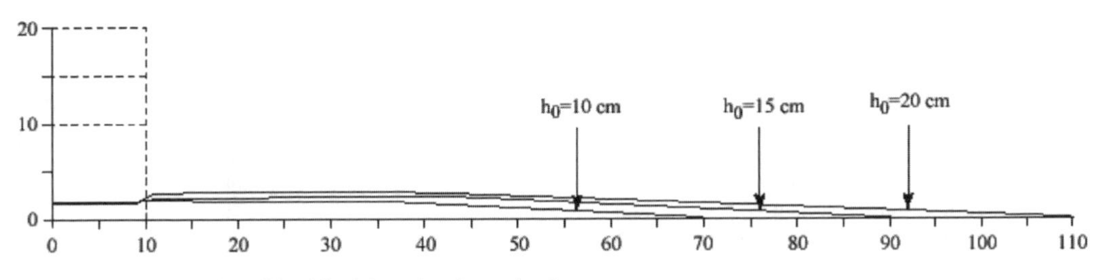

Figure 4. Longitudinal profile of final deposits of granular flows.

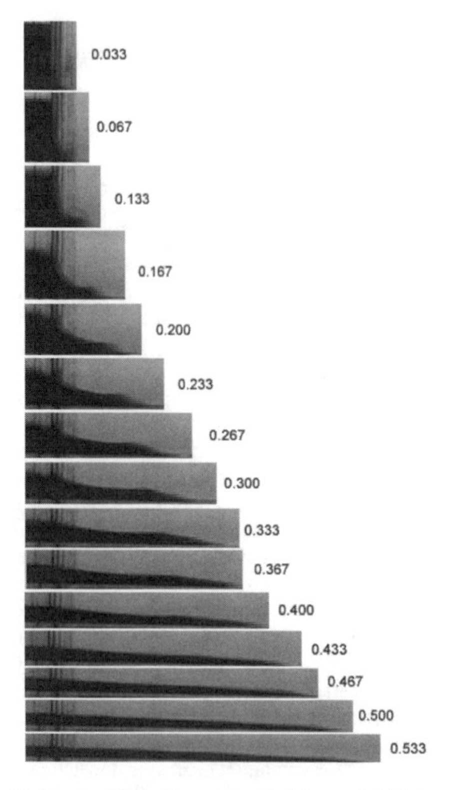

Figure 5. Water flows generated from an initial reservoir height of $h_0 = 20$ cm. Numbers indicate time in seconds.

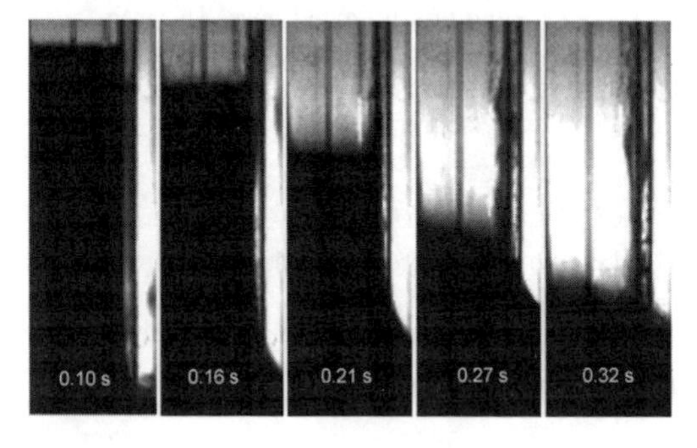

Figure 6. Vertical descent of water level within the reservoir for water flow with $h_0 = 0.4$ m. Numbers indicate time in seconds.

to what is predicted by depth-averaged hydrostatic models of the dam-break problem, water level does not appear to rotate around a fixed point located at a height equal to 4/9 h_0 at the gate position, with a negative wave propagating upstream within the reservoir (e.g., Ritter's model, see Stocker 1957). Instead, the water level seems to fall vertically within the reservoir (Fig. 6).

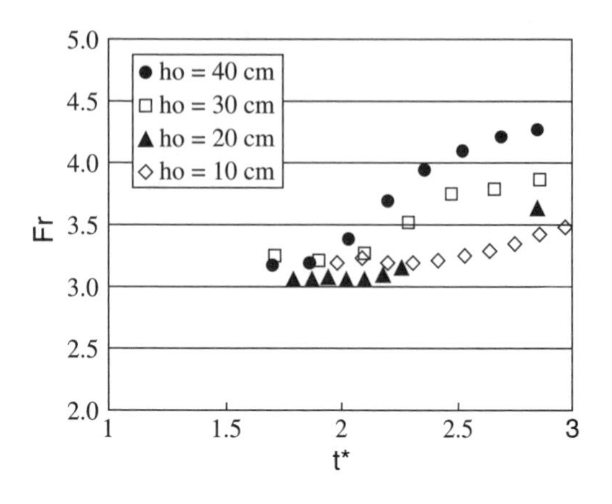

Figure 7. Water flows Froude number variation as a function of dimensionless time $t^* = t/t_2$ for different initial reservoir heights.

Current height, h_f, was estimated, similarly as in the granular flow experiments, by averaging flow depths measured in the middle portion of the reach extending from the flow front to a point located 10 cm downstream from the gate position. Again, front velocity was estimated from the slope of the curve relating front position and time, and from these data, front Froude number and dimensionless current height were estimated. Three time scales can be introduced in this problem. The first one is $t_0 = x_0/(g\ h_0)^{1/2}$ defined by Roche et al. (2004) (see also Rottman & Simpson 1983). This scale, however, does not provide a good collapse of the time evolution curves for the ratio h_f/h_0 associated to different values of the aspect ratio of the initial water volume in the reservoir, h_0/x_0. The second time scale defined is $t_1 = (x_0/g)^{1/2}$. This scale does provide a good general collapse of the time evolution curves of the dimensionless current height for different h_0/x_0 values. In these curves an initial stage of nearly constant current height is identified, once the flow is established in the channel, which also corresponds to a stage of nearly constant velocity and Froude number. This stage is equivalent to that identified as of constant velocity for the granular flows. It is observed that a better scaling of the initial constant velocity stage is obtained by using a third time scale, defined as $t_2 = (h_0/g)^{1/2}$ (Figs. 7, 8). In the present experiments, this initial stage of constant height and velocity extends in the range of dimensionless time $t^* = t/t_2$ of about 1.5 to 2.3. The equivalent stage in the granular flows seems to extend for slightly longer dimensionless times, up to a value of t^* of about 2.5.

During the initial constant velocity stage, the dimensionless current height in the water flows has a value in the range 0.16 to 0.19, which seems to be independent of the ratio h_0/x_0, and is slightly less than the value of about 0.20 observed in the granular

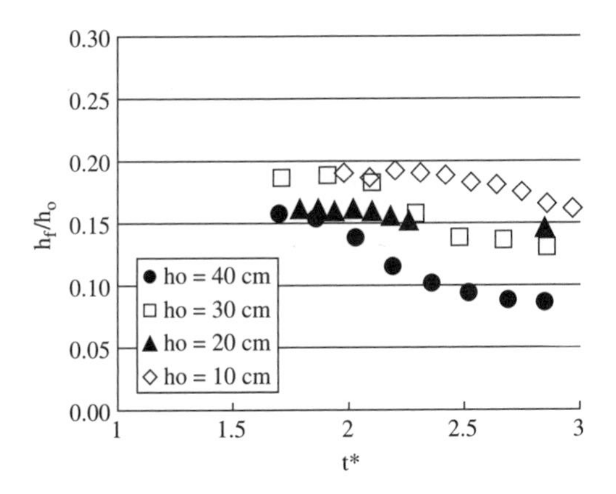

Figure 8. Water flows dimensionless height variation as a function of dimensionless time $t^* = t/t_2$ for different initial reservoir heights.

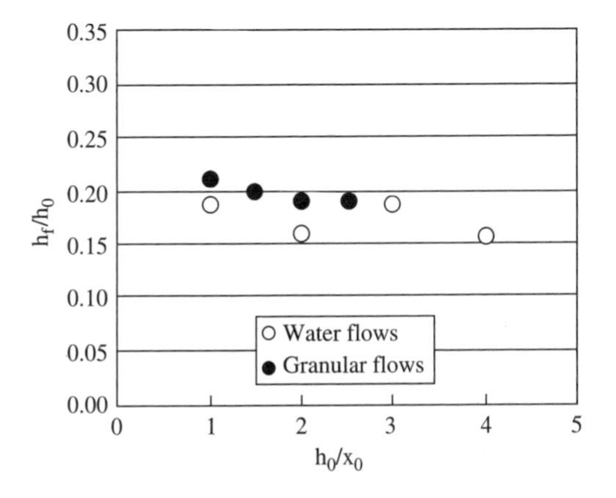

Figure 9. Dimensionless flow height as a function of the aspect ratio of the initial flow volume within the reservoir.

flow experiments (Figs. 8, 9). The corresponding front velocity in this stage scales with $(gh_0)^{1/2}$. Indeed, the values of the Froude number defined as $Fr_0 = u_f/(g\, h_0)^{1/2}$ for the water flows are in the range 1.3 to 1.4, independent of the ratio h_0/x_0, and only slightly less than the value of $\sqrt{2}$ expected for orifice flow (Fig. 10). In the granular flows the values of Fr_0 are lower than in the water flows, with a value of about 1.15, also independent of h_0/x_0 (Fig. 10).

As a consequence of the different values of h_f/h_0 and Fr_0 observed in the water and granular flow experiments, the corresponding values of the current Froude number, Fr, differ in both cases. Recall that $Fr = Fr_0/(h_f/h_0)^{1/2}$. Values of Fr in the water flows are in the range 3.1 to 3.2, independent of the ratio h_0/x_0, while in the granular flows they are about 2.58, also independent of this ratio (Fig. 10).

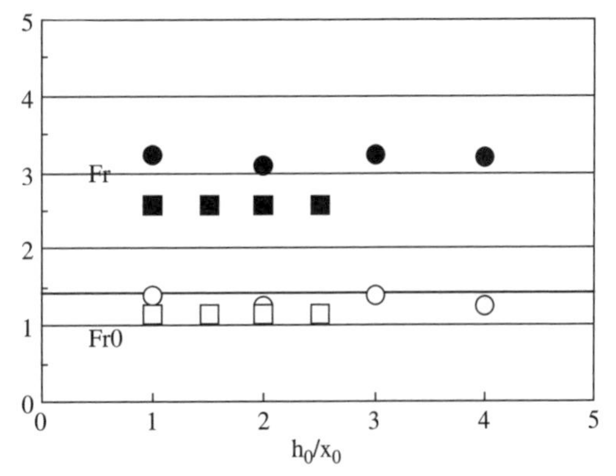

Figure 10. Values of the front Froude number, Fr (black symbols) and Fr_0 (white symbols) as a function of the aspect ratio of the initial flow volume within the reservoir. Squares correspond to granular flows, circles to water flows. The horizontal black line corresponds to $\sqrt{2}$.

4 DISCUSSION

4.1 *Granular flow behavior*

Results of water flow experiments confirm that the observed fluidized granular flow behavior is similar to that of a Newtonian fluid, with some differences probably related to the energy dissipation mechanisms operating in each case.

According to Iverson & Denlinger (2001) in gravity driven flows of dense grain-fluid mixtures, internal stresses are governed by viscous interstitial fluid stresses, collisions between grains, and Coulomb friction derived from gravitational contact between grains. It is argued that in fluidized mixtures, pore fluid pressure can eliminate the latter effect, as the grain weight is fully supported by this pressure. In such case, the Bagnold number is a dimensionless parameter that can be used to estimate the relative effect of fluid viscosity and interparticle collisions: $N_B = (c_v^{1/3}/(c_{vmax}^{1/3} - c_v^{1/3}))^{1/2} \rho_s \gamma^Y d_s^2/\mu$, where $\underline{c}_v$ and c_{vmax} are the volume solid concentration and maximum value of c_v, respectively, γ^Y denotes the shear rate, which in the present flows can be estimated as: u_f/h_f, and μ denotes air dynamic viscosity. Even though it is difficult to have a good estimate of the ratio c_v/c_{vmax} in the present experiments, it is argued that it should be close to 1, as the initial bed is only marginally expanded at the moment of release. Assuming a 1% percent expansion of the granular flows (i.e., $c_v/c_{vmax} = 0.99$) yields values of N_B in the range 420 to 840. As these values are higher than the limit value of 45 proposed by Bagnold for the macroviscous regime, it is clear that the present granular flows are not dominated by interstitial fluid viscosity. Likewise, the N_B values are close to or higher than 450, which implies that the flows would be dominated mainly by interparticle collision (Iverson & Denlinger 2001).

Despite the latter result, it seems unlikely that the present granular flows are sustained by interparticle collisions during their fluid-like behavior phase. According to Eames & Gilbertson (2000), an adequate description of fluidized granular flow stresses requires to take into account, not only internal friction and interparticle collision effects, but also momentum exchange between the particles and the aerating gas.

In fluidized granular flows the so-called granular temperature, or particle kinetic energy, may be generated by interparticle collisions, wall interactions, and the effect of the fluidizing gas flow, as long as it provides bubbles to disturb the particles (Eames & Gilbertson 2000). In the present flows, the later effect can be neglected as the initial bed expansion was only marginal and no bubble formation was observed at low fluidization velocities. Eames & Gilbertson (2000) show that for conditions similar to those of the present granular flows, collisional stress derived from granular temperature production is negligible compared to momentum exchange between the fluidizing gas and particles. This means that, most probably, the present flows are sustained by such momentum exchange during their fluid-like behavior phase, which would be maintained as the flow deaereates through pore fluid pressure diffusion. This momentum exchange would correspond basically to drag forces exerted by the air flowing through the grains, similar to those occurring in buoyancy-driven flows in porous media (Eames & Gilbertson 2000).

In any case, the present granular flows appear to exhibit a higher resistance than the equivalent water flows, their associated h_f/h_0 values being higher and their Fr_0 values lower than those of the water flows. The energy dissipation mechanisms in the fluidized granular flows would be related mainly to interparticle collision and air-grain viscous drag. As pore fluid pressure is diffused out of the current, Coulomb friction between grains becomes the dominant energy dissipation mechanism and grain deposition leads to the rapid stopping phase of the granular flows.

4.2 *Implications for pyroclastic flows*

Pyroclastic deposits commonly contain large amounts of fine particles as many of them are matrix-supported (Sparks 1976). It is important to emphasize that the amount of fine particles in the parent flows is up to twice that in the deposits and loss of fines is caused by elutriation during transport (Sparks & Walker 1977). Recent investigations have revealed that the fine, poorly sorted ash matrix of an ignimbrite displays a behavior similar to that of the present granular flows when fluidized at high temperature up to $\sim$550°C and under shear, with typical bubble free expansion and constant deaeration rate (Druitt et al. 2004).

Based on the present evidence, it is argued that fines-rich (i.e., matrix-supported), dense pyroclastic flows may emplace as fluid gravity currents, except at late stages just before final deposition. Such fluid-like behavior, with negligible inter-particle friction, is then likely to promote high flow mobility even on nearly horizontal slopes. This picture is consistent with the concept of Sparks (1976) that pyroclastic flows can be considered as a mixture of coarse clasts suspended in fluidized fines.

Several characteristics of pyroclastic deposits are consistent with the present experimental results. Fines-rich pumice flows commonly have longer run-outs and smaller upper surface slopes than that of fines-depleted, block-and-ash flows of similar volume (Calder et al. 1999, Calder et al. 2000). The fluid-like emplacement of large flows, suggested by the flat-topped morphology of ignimbrites deposited on initial surface slopes of only a few degrees (Branney & Kokelaar 2002), is similar to that observed in the present experiments.

The interparticle and gas-particle interactions discussed above are independent of the length scale. However, some other parameters operating at larger scale may also influence the emplacement mechanisms of pyroclastic flows. Deaeration and pore pressure diffusion have minor effects in experiments. However, time scales of deaeration and pore pressure diffusion both scale with the length scale. Large, thick flows in nature will then have disproportionally longer deaeration and pore pressure diffusion timescales and may have longer run-outs than the small flows in experiments. Deaeration will be particularly important in flows having an expansion of the order of tens of percent (Druitt et al. 2004). In contrast, high pore pressure will be sustained in flows having a very low permeability, e.g. highly concentrated, highly polydisperse, and/or with a large fine particle mass fraction (Iverson & Denlinger 2001). At large scale, high flow speed will promote significant particle collisions, which will generate high stresses and may cause important particle segregation, which, in turn, could influence the flow motion. There probably also exists a maximum size of the blocks that can be transported in a matrix of fines, and this may also influence the flow dynamics.

5 CONCLUSIONS

Water gravity currents and initially fluidized granular flows generated by the sudden opening of a gate show strong similarities during the initial stages of flow development. Front velocity and flow height are nearly constant, after the current is established, during a dimensionless time t^* that extends up to a value of about 2.3 for the water flows, or 2.5 for the granular flows. During this constant velocity phase, flow depth to initial bed height ratio, h_f/h_0, is about 0.17 for the water flows and about 0.20 for the granular flows. Corresponding front Froude numbers, Fr, are about 3.2 for the water flows and about 2.6 for the granular flows. The present experimental evidence indicates that granular flows exhibit a higher resistance than the equivalent water flows. An analysis of the characteristics of the present granular flows shows that, during the constant velocity phase, they would correspond to flows with negligible internal Coulomb friction, as the particle weight is supported by some air flow caused by the release of the initial pore fluid pressure. Likewise, viscous effects in the interstitial fluid would be negligible compared with interparticle collision effects, however such effects would be, in turn, negligible compared to those associated with momentum exchange between the fluidizing gas and particles.

In the granular flows, a final stopping phase occurs. As pore fluid pressure is diffused out of the current, Coulomb friction between grains becomes the dominant energy dissipation mechanism and grain deposition leads to a rapid deceleration of the flow, which progressively acquires a wedge shape until the motion ceases. The granular flow deposits exhibit a bulge with a commonly flat and nearly horizontal upper surface and a front wedge with a slope of about 2 to 4%. The longitudinal extent of the deposits generated by the fluidized granular currents are about twice as large as those observed for non-fluidized ones. The ratio x_f/h_0 of the deposits of fluidized granular flows decreases almost linearly from about 7 to 5.5 as the aspect ratio h_0/x_0 increases from 1 to 2.

It is argued that fines-rich, dense pyroclastic flows may emplace as fluid gravity currents, except at late stages just before final deposition. Such fluid-like behavior, with negligible inter-particle friction, is likely to promote high flow mobility even on nearly horizontal slopes. Pyroclastic flows could then be considered as a mixture of coarse clasts suspended in fluidized fines. Besides, several characteristics of pyroclastic deposits are consistent with the present experimental results. The fluid-like emplacement of large pyroclastic flows, suggested by the flat-topped morphology of ignimbrites deposited on slopes of only a few degrees, is similar to that observed in the present experiments.

ACKNOWLEDGMENTS

The authors gratefully acknowledge support from MECESUP in the form of a Ph.D. fellowship for the second author, FONDECYT Project 1040494, IRD, and the Departments of Geology and of Civil Engineering of the University of Chile.

REFERENCES

Branney, M.J. & Kokelaar, P. 2002. Pyroclastic Density Currents and the Sedimentation of Ignimbrites. *Mem. Geol. Soc. London*, 27, 152 pp.

Calder, E.S., Cole, P.D., Dade, W.B., Druitt, T.H., Hoblitt, R.P., Huppert, H.E., Ritchie, L., Sparks, R.S.J. & Young, S.R. 1999. Mobility of pyroclastic flows and surges at the Soufrière Hills Volcano, Montserrat. *Geophys. Res. Lett.*, 26, 537–540.

Calder, E.S., Sparks, R.S.J. & Gardeweg, M.C. 2000. Erosion, transport and segregation of pumice and lithic clasts in pyroclastic flows inferred from ignimbrite at Lascar Volcano, Chile. *J. Volcanol. Geotherm. Res.*, 104, 201–235.

Denniss, A.M., Harris, A.J.L., Rothery, D.A., Francis, P.W. & Carlton, R.W. 1998. Satellite observations of the April 1993 eruption of Lascar volcano. *Int. J. Remote Sensing*, 19(5), 801–821.

Druitt, T.H., Bruni, G., Lettieri, P. & Yates, J.G. 2004. The fluidization behavior of ignimbrite at high temperature and with mechanical agitation. *Geophys. Res. Lett.*, 31, doi: 10.1029/2003GL018593.

Eames, I. & Gilbertson, M.A. 2000. Aerated granular flow over a horizontal rigid surface. *J. Fluid. Mech.*, 424, 169–195.

Iverson, R.M. & Denlinger, R.P. 2001. Flow of variably fluidized granular masses across three-dimensional terrain 1. Coulomb mixture theory. *J. Geoph. Res.*, 106(B1), 537–552.

Roche, O., Gilbertson, M.A., Phillips, J.C. & Sparks, R.S.J. 2004. Experimental study of gas-fluidized granular flows with implications for pyroclastic flow emplacement. *J. Geoph. Res.*, 109, doi: 10.1029/2003JB002916.

Rottman, J.W. & Simpson, J.E. 1983. Gravity currents produced by instantaneous releases of a heavy fluid in a rectangular channel. *J. Fluid Mech.*, 135, 95–110.

Sparks, R.S.J. 1976. Grain size variations in ignimbrites and implications for the transport of pyroclastic flows, Sedimentology, 23, 147–188.

Sparks, R.S.J. & Walker, G.P.L. 1977. The significance of vitric-enriched air-fall ashes associated with crystal-enriched ignimbrites. *J. Volcanol. Geotherm. Res.*, 2, 329–341.

Stocker, J.J. 1957. Water vaves. New-York, Interscience.

Wilson, C.J.N. 1984. The role of fluidization in the emplacement of pyroclastic flows, 2, Experimental results and their interpretation. *J. Volcanol. Geotherm. Res.*, 20, 55–84.

River, Coastal and Estuarine Morphodynamics: RCEM 2005 – Parker & García (eds)
© 2006 Taylor & Francis Group, London, ISBN 0 415 39270 5

Author index